Micromechanics of Heterogeneous Materials

Micromechanics of Heterogeneous Materials

Valeriy A. Buryachenko
Unviersity of Dayton Research Institute

Springer

Valeriy A. Buryachenko
University of Dayton Research Institute
300 College Park
Dayton, OH 45469-0168
USA

Library of Congress Control Number: 2007922743

ISBN 978-0-387-36827-6 e-ISBN 978-0-387-68485-7

Printed on acid-free paper.

9 8 7 6 5 4 3 2 1

springer.com

Elena, Andrey, Alexander, Nina, Irina
for their love and support over the years

Preface

Materials that are either manufactured or occur in nature and used both in industry and in our daily lives (metals, rocks, wood, soil, suspensions, and biological tissue) are very seldom homogeneous, and have complicated internal structures. Although the combination of two or more constituents to produce materials with controlled distinguish properties has been exploited since at least ancient civilization, modern composite materials were developed only a few decades ago and have found intensive application in contemporary life and in all branches of industry. Establishment of a link between the structure and properties in order to understand which kind of structure provides the necessary properties is an objective of "micromechanics," which exploits information about microtopology and properties of constituents of the heterogeneous medium for development of mathematical models predicting the macroproperties. The problem of micromechanical modeling of the mechanical properties of engineering materials is today a crucial part of the design process, and sample testing is usually performed only during the final stage for validation of the "virtual" design. An accuracy of the classical "trial and error" testing method of the new materials and constructions is no longer be affordable in modern industry and science.

Owing to wide applications of composite materials, their modeling has been developed very intensively over recent decades, as reflected in the numerous papers and books only partially presented in the reference section of this book. A variety of materials and approaches appearing in apparently different contexts and among different scientific disciplines (solid mechanics, geophysics, solid-state physics, hydromechanics, biomechanics, chemical technology, etc.) do not allow the opportunity to investigate adequately the whole field in a single book. In light of this, I was challenged with a natural question as to why it was necessary to write another book and what is the difference between this book and the ones published earlier. In parallel with this book, there are a few fundamental books combining readily applicable results useful for material scientists with a significant contribution to progress in theoretical research itself. However, a fundamental difference of this book is a systematic analysis of statistical distributions of local microfields rather than only effective properties based on the average field concentrator factors inside the phases.

The uniqueness of this book consists of the development and expressive representation of statistical methods quantitatively describing random structures which are most adopted for the subsequent evaluation of a wide variety of macroscopic transport, electromagnetic, and elastic properties of heterogeneous media. The popular methods in micromechanics are essentially one-particle ones that are invariant with respect to statistical second and higher order quantities examining the association of one particle relative to other particles. This book expressively reflects the explosive progress of modern micromechanics resulting from the development of image analyses and computer-simulation methods on one hand and improved materials processing on the other hand, since processing controls the prescribed microstructure. With the appearance of new experimental techniques, it is now possible to study the microtopology of disordered materials much more deeply to understand their properties. Modern techniques are also available to design materials with morphological properties that are suitable for planned applications. This progress in micromechanics is based on methods allowing for statistical mechanics of a multiparticle system considering n-point correlation functions and direct multiparticle interaction of inclusions, and the book presents a universally rigorous scheme for both analyses of microstructures and prediction of macroscopic properties which leaves room for correction of their individual elements if improved methods are utilized for the analysis of these individual elements. The book successfully combines advanced numerical methods for the analysis of a finite number of interacting inhomogeneities in either the bounded or unbounded domain with analytical methods.

It should be mentioned that there are two coupled classes of micromechanical problems for which averaging is critical. Averaging is usually suitable for predicting effective elastic properties. However, failure and elastoplastic deformations will depend on specific details of the local stress fields when fluctuation is important. In the framework of computational micromechanics in such a case, one must check the specific observable stress fields for many large system realizations of the microstructure with the use of an extremal statistical technique. A more effective approach used in analytical mechanics is the estimation of the statistical moments of the stress field at the interface between the matrix and inhomogeneities. The inherent characteristics of this interface are critical to understanding the failure mechanisms usually localized at the interface. In the framework of a unique scheme of the proposed multiparticle effective field method (MEFM), we attempted to analyze the wide class of statical and dynamical, local and nonlocal, linear and nonlinear micromechanical problems of composite materials with deterministic (periodic and nonperiodic), random (statistically homogeneous and inhomogeneous, so-called graded) and mixed (periodic structures with random imperfections) structures in bounded and unbounded domains, containing coated or uncoated inclusions of any shape and orientation and subjected to coupled or uncoupled, homogeneous or inhomogeneous external fields of different physical natures.

I do not pretend to cover the whole field of micromechanics in this book (restricted by my interests); there are many other topics in micromechanics, of much industrial and scientific importance, that are either treated schematically or only mentioned. In particular, the homogenization theory of periodic structures, the geometrically nonlinear problems, flow in porous media, viscoelasticity problems, and cross-property relations are not considered at all in this book while stochastic geometry, variation methods, propagation of waves in composites, and multiscale discrete modeling are not treated in depth they deserve and sometimes were only mentioned. Interested readers are referred to the references cited in the appropriate sections to achieve a deeper understanding of these topics. This book finalizes my research in micromechanics that began with the papers published in 1986. It was written piece by piece at different times and in different countries facing new challenging problems in micromechanics. The book is suitable as a reference for researchers in different disciplines (applied mathematicians; physicists; geophysicists; material scientists; and electrical, chemical, civil, and mechanical engineers) working in micromechanics of heterogeneous media and providing a rigorous interdisciplinary treatment through experimental investigation. The book is also appropriate as a textbook for an advanced graduate course.

I gratefully acknowledge the financial support for 30 years of research in micromechanics of heterogeneous materials by the National Research Council, and the Air Force Office of Scientific Research (AFOSR) of the United States, Austrian Society for the Promotion of Scientific Research, Max-Planck-Society of Germany, as well as by the Ministry of Higher Education, National Academy of Science, and the Ministry of Machine Construction of USSR and Russia. This book would not have been possible without this support. The excellent cooperation with the team at Springer is gratefully acknowledged; I owe special thanks to E. Tham, C. Simpson, and C. Womersly for their careful and patient editing of this book. Also, thank you Elsevier for allowing permission to use Figs. 3.1–3.3, 4.12, 4.13, 5.1–5.4, 8.1, 9.5, 9.6, 11.1–11.8, 12.2, 13.5–13.11, 14.1–14.4, 15.1–15.5, 15.8–15.11, Tables 11.1–11.5 originally published in [133], [134], [136], [137], [146], [168], [179], [182], [184], [190], [191], [633]; Birkhäuser Verlag AG for permission to use Figs. 4.10, 4.11, 10.1–10.6, 15.7, originally published in [145], [152]; SAGE Publication for permission to use Figs. 4.4, 4.6–4.10, 12.4–12.6, originally published in [166], [167]; ASME for allowing permission to use Figs. 12.11, 15.7, 16.10–16.14, originally published in [138], [187]; Begell House Inc. for permission to use Figs. 4.1–4.3, 12.3, 12.12, originally published in [144], [153], [195]; Springer Science and Business Media for allowing permission to use Figs. 12.1, 12.7–12.10, 15.6, 16.6, 16.7, 18.7–18.19 originally published in [131], [155], [185].

My scientific interests go back to my student days in Moscow State University, where I had the good fortune of being nurtured by brilliant teachers and scientists such as, e.g., V. I. Arnold, G. I. Barenblat, A. A. Ilyushin, A. N. Kolmogorov, S. P. Novikov, A. N. Shiryaev, and my first research adviser O. A. Oleinik. In later years I met most of the key players in the development of the subject, many of whom have had a profound influence on my own work. I would like to acknowledge G. J. Dvorak, J. Fish, W. Kreher, E. Kröner, V. I. Kushch, A. M. Lipanov, N. J. Pagano, V. Z. Parton, F. G. Rammerstorfer, T. D. Shermergor, S. Torquato, and G. J. Weng for the helpful and inspiring exchanges of ideas, fantastic collaboration, stimulating discussions, encouraging remarks, and thoughtful criticism I have received over the years. At the risk of forgetting to include some names, I thank the staff of Air Force Research

Laboratory, Ohio (and especially MLBC) and University of Dayton Research Institute: K. L. Andersen, J. Baur, V. T. Bechel, J. Braun, S. Chellapilla, A. Crasto, S. Donaldson, T. Gibson, R. Hall, R. Y. Kim, R. J. Kerans, K. Lafdy, S. Lindsay, B. Maruyama, D. B. Miracle, O. O. Ochoa, V. Y. Perel, B. Regland, B. P. Rice, E. Ripberger, R. Roberts, A. Roy, G. Schoeppner, J. E. Spowart, K. Strong, G. P. Tandon, T. Benson Tolle, K. Thorp, M. Tudela, R. A. Vaia, and T. Whitney for the productive collaboration and cordial hospitality.

Last, but certainly not least, my deepest gratitude goes to my wife Elena and our son Andrey for their love, devotion, care, tolerance, and understanding that the most beneficial scientific insights often come at time that were destined to be spent with the family. Finally, I extend my greatest thanks to my father Alexander, mother Nina, and sister Irina for all the encouragement and support they have provided me throughout my life. I am pleased to dedicate this book to them.

I am aware of the fact that the book may contain controversial statements, too personal or one-sided arguments, inaccuracies, and typographical errors. Any the remarks and comments regarding the book will be fully appreciated (buryach@woh.rr.com).

Valeriy Buryachenko
Dayton, Ohio
July 2006

Contents

1

Introduction

1.1 Classification of Composites and Nanocomposites

To introduce readers to the subject of composite (or heterogeneous) materials it is necessary to begin with the main notions and definitions which are presented with the understanding that any definition and classification scheme is neither comprehensive nor complete.

The engineering level of composite material definition is given in ASTM D 3878-95c: "Composite material. A substance consisting of two or more materials, insoluble in one another, which are combined to form a useful engineering material possessing certain properties not possessed by the constituents." This definition is usually added by the notion of length scale when the characteristic sizes of structural elements (the number of which is statistically large) are much smaller than characteristic sizes of composite medium but much larger than the molecular dimensions with the result that the mechanical state of structural elements is described by the equations of continuous mechanics.

The materials used for structural applications are very seldom homogeneous, and because of this, the problem of modeling the mechanical properties of engineering materials has to be investigated in the framework of heterogeneous materials. The chief purpose of introduction of fillers into composite materials (CM) is to create materials with a prescribed set of physico-mechanical properties, which can be controlled, for example, by varying the type of base matrix and fillers, and its particle size distribution, shape, and arrangement (see for details, e.g., [310], [384], [1135], [1141]). Thanks to considerable advances in CM research, their use in a broad range of industries – machin building, construction, aerospace technology, etc. – has become extensive. Development of an assortment of CMs, and decisions concerning their possible scope of applicability and service conditions involve research into a very wide spectrum of problems: (1) choice of the matrix, fillers, functional adjuvants, design of the optimal microtopology of CM for creating materials with prescribed properties; (2) processing of a particular CM into constructions, with optimization of the process parameters to provide the best tradeoff between the properties of each of the constituents of the CM; and (3) prediction of the serviceability of a construction under prescribed service conditions.

By now an almost incomprehensible amount of data on the properties and processing technology of the individual constituents and CM has been amassed. However, the diversity of different materials with different geometrical and mechanical properties of constituents is such that it is useful to have a classification of these materials to choose the appropriate modeling tool. A number of classification schemes are known that are distinguished by the criteria used for these classifications: geometry of structural elements, their mechanical properties, and their manufacture. The description of CM encountered in materials science can be useful to mechanical engineers if one avoids excessive intractable details and considers at first the geometrical characteristics and then the mechanical properties of constituent phases. Such a classified description should be favorable to a qualitative understanding of the relation between microstructure and properties on one hand and of the relation between process parameters and microstructure on the other.

This section outlines a general classification of CM based on the geometry of structural elements, their mechanical properties, and their manufacture.

1.1.1 Geometrical Classification of Composite Materials (CM)

Geometrical microstructure of composites can be classified according to the following features: (1) nature of structural element connectivity; (2) the characteristic relative sizes and shapes of structural elements; and (3) arrangement of structural elements.

The nature of structural element connectivity defines three groups of composite materials depending on the existence of a phase "containing" the others. To the first group, one can assign the composite with one continuous over the volume constituent called the matrix reinforced by the isolated inhomogeneities of the different shapes. Such composites are called the matrix ones. The second group of composites forms the skeletal (or mutually penetrated or multipercolated) composites where at least two phases make a monolithic frame so they can be seen as containing the other. The laminate composites as well as open cellular structures such as some metallic foam or porous silicon [383] are particular cases of percolated structures. A more complicated structure is intrinsic to entangled materials such as cellulose-reinforced polymers [384], metallic wools, and glass wool. These materials have both a network connecting them to the penetrated structure, and also an incomplete network with only one connection to the network tree. It should be mentioned that variation of the constituent concentration can lead to changeover from one connectivity group to another. For example, closed cell foams [383] can be considered as porous metal matrix composites, while the open cell foams have a percolated type of structure. The third group of composites is characterized by a structure in which there is no contact among the elements of each constituent and the different phases are contiguous but none can be seen as containing the other. For example, polycrystalline single-phase metal containing the structural elements of the same anisotropic crystals with different orientations of crystallographic axes pertain to the third group of composite materials, which in this case are called the heterogeneous materials.

The grains fill the whole space and no element with one orientation contains the other element with another orientation.

The lack or existence of a continuous matrix in CM is an essential geometrical feature, but the aspect ratio that may characterize the phases is also important. The matrix composites can be classified by the characteristic relative sizes and shapes of structural elements, as continuously reinforced (typically plies and continuous tapes and fibers), short fiber (or chopped whiskers) reinforced (aspect ratio noticeably larger than 1), particle reinforced (aspect ratio around 1), flake or platelet reinforced (aspect ratio smaller than 1), and hybrid reinforcement. If the ratio of characteristic sizes of the reinforcement and the composite materials is taken as a dimension of the composite structure, then the particle, fiber, and laminated composites are called zero-dimensional, one-dimensional, and three-dimensional ones, respectively. Polyreinforced (simple and hybrid) CMs contain reinforcements from different materials. In the simple polyreinforced CM, the reinforcements from different materials have the same dimensions. In the hybrid CM the reinforcements from different materials have different dimensions (for example, fibers and particles). The ultimate objective of hybridization is to create and combine an engineered proportion of material components for a range of applications, including the following hybrid systems: ternary alloys, hybrid particulate composites, glass fiber-containing toughened plastics, liquid crystalline polymer hybrid composites, and nanocomposites (for references see [1194]).

The composites can be further structured for identification by noting the arrangement of structural elements. A distinction is made between the periodic (or deterministic) and random structure composites. For example, for woven fiber-reinforced composites, one can indicate the following structures: periodic, cylindrical, braided, orthogonal interlock, and angle interlock structures. Such a perfectly regular structure, of course, does not exist in actual cases, although the periodic modeling can be quite useful, as it provides rigorous estimations with *a priori* prescribed accuracy for various material properties. The random structure composites, in turn, should be subdivided into statistically homogeneous structures that are translationally invariant under a shift of the space and into statistically inhomogeneous (or functionally graded, FG) composites in which the arrangement (e.g., the concentration and orientation) of elements is position-dependent. Statistically homogeneous composites are divided on statistically isotropic one which structure is invariant with respect to the rotation and statistically anisotropic with anisotropy assessing from directional measurements, such as, e.g., variation of the chord size distribution with their orientation. A particular sort of FG composites is the clustered (or aggregated) composites in which the areas with high and small concentration of reinforcements can be separated. In a polymeric matrix the filler (even if present in minimum quantities) is always more or less agglomerated. The ability of the filler to form the structural network is largely responsible for its reinforcing action [688]. On the other hand, the structuralization in melt increases the viscosity of the material and hampers its processability. Microscopic studies have confirmed the existence of two types of primary structures in CMs, i.e., filler aggregates with particles bound together firmly enough, and agglomerates—systems of weakly interrelated aggregates. For a fixed filler content, the viscosity and moduli of solid CM based on the same matrix with agglomerates is always higher than that of the "well-dispersed" sample (for references see, e.g., [310], [688]). However, the mentioned behavior of increasing the effective stiffness with the agglomeration of the spherical particles changes drastically with variation of the inclusion shape. So, well known experimental data point to decrease of effective elastic properties with the growth of clustering (when the aligned clusters contain the parallel nanoplates with a small aspect ratio $\alpha \sim 0.001$), with the maximum effective stiffness of silicate nanocomposites corresponding to the completely exfoliated structures [992]. At the same time, the strength of CM is usually lower in materials with agglomerates of spherical particles than in CM with uniformly distributed filler. Clustered composites in the simplest case have so-called ideal cluster structure that contains either finite or infinite, deterministic or random ellipsoidal domains called particle clouds distributed in the composite matrix. In so doing the concentration of particles is a piecewise and homogeneous one within the areas of ellipsoidal clouds and composite matrix. An alternative case of clustered structures is a dual scale one existing, e.g., in Al7%Si alloy [264], [384], where the elementary unit of the microstructure is a core of almost pure aluminum, surrounding by a thick layer that can be considered as CM consisting of aluminum matrix reinforced by silicon particles. The ratio of linear sizes of the core and the layer as well as the ratio of the volume fraction of Si particles to the Al matrix in the layer are defined by both the composition of the alloy and technological features.

As an example of composites combining random and periodic structure (also called periodic structure with random imperfections), we can indicate the initially periodic structure with randomly distributed damage as well as composites with small random perturbation of inclusion location with respect to the periodic grid due to manufacturing errors. From another aspect, the laminated composites usually classified at a mesoscale level as either periodic or deterministic structures can be considered at the microscale level as the FG composites with the statistically inhomogeneous packing of fibers in each ply. Nanocomposites are randomly reinforced by the nanofibers which in turn have a periodic or deterministic structure.

This leads to the notion of coupling of different scales. The growing recognition that the properties of materials involve different scales expressively reflects the explosive progress of modern nano-and micromechanics resulting from the development of image analysis and computer-simulation methods on one hand and both the advanced experimental technique (e.g., computed X-ray tomography and electron microscopy) and improved materials processing on the other hand, since processing controls the prescribed structure. One recalls that at the upper level, the collective results of events or processes taking place at the lower level are registered and are measured on a finer scale. Although in material modeling, one can distinguish four levels of simulation: electronic structure, atomistic, microscale, and macroscale, micromechanics of CMs is usually concentrated just on the micro-macro connection of the multiscale concept when the continuum mechanics technique is appropriate.

1.1.2 Classification of Mechanical Properties of CM Constituents

The composites also can be categorized with respect to the mechanical properties of constituents according to the following features: polymer matrix composites (PMCs), metal matrix composites (MMCs), ceramic matrix composites (CMCs), carbon-carbon (CC) and hybrids. The polymer resins fall into two groups: thermosets and thermoplastics.

The cured thermoset matrix has polymer chains that have become irreversibly crosslinked in a three-dimensional frame like a space truss. A cured thermoset matrix has the following mechanical properties: elasticity modulus 2.5–10.0 GPa, tensile strength 30–120 MPa, compressive strength 80–250 MPa, and density 1200–1400 kg/m^3. Among the most often used thermoset resins are polyester, vinylester, epoxy, phenolic, polyurethane, and polybutadiene. Thermoplastic resins are processed at high temperatures and solidify on cooling. That is, polymer crosslinking does not occur, and thermoplastic resins remain plastic, and can be reheated and reshaped. The solidified thermoplastic resins have high tensile strength (50–100 MPa), high stiffness (1.4–4.2 GPa), and high density 1100–1700 kg/m^3. The most widely used thermoset resins are polyethylene, polystyrene, polypropylene, acrilonitride-butadiene styrene, acetate, polycarbonate, polyvinyl chloride, polysulfone, polyphenylene sulfide, and nylon.

MMCs have several advantages [1060], [1079]: high strength, elastic moduli, fracture toughness, and impact properties; high electrical and thermal conductivity; low sensitivity to temperature change or thermal shock; high surface durability, and low sensitivity to surface flaws.

Reinforcement of the matrix composites is accomplished by the fillers carrying the majority of the loading. Conventional fillers are separated into three categories: reinforcing, active, and inert. The reinforcing class includes mainly fiber composites. Disperse fillers may also perform the reinforcing function, and then they are called active. The basic properties of commonly used fibers are as follows: continuous glass fibers have a diameter 3–20·10^{-6} m, tensile strength 2–6 GPa, stiffness 50–100 GPa, and 2400–2600 kg/m^3. They are used in fabrication of composite elements in the form of elementary yarns, rovings, and cloth. The advantage of glass fibers is in their high tensile strength and compressive strength, low cost of the materials and manufacturing, and good compatibility with polymeric matrix. The disadvantages are associated with their low stiffness and heat stability. The carbon fibers have the following characteristics: 1.5–7.0 GPa tensile strength, 150–800 GPa elasticity modulus, 1500–2000 kg/m^3 density. They have higher stiffness in comparison with glass fibers, and are characterized by a negative expansion coefficient. Their mechanical properties do not deteriorate with temperature increases up to 450°C, a fact that allows the use of carbon fibers in composites with both polymeric and metal matrixes. Boron fibers have 2–4 GPa tensile strength, 370–430 GPa modulus of elasticity, and 2500–2700 kg/m^3 density. The main advantage of boron fibers is their high stiffness and compression strength but they are rather expensive, more brittle and less processable than glass fibers. Boron fibers are used with both polymeric and metal matrixes in the form of elementary or hybrid threads which are bundles of parallel continuous fibers wound with an auxiliary glass thread. Organic fibers (e.g., Kevlar) with different polymer compositions and method of molding can be obtained with 2–4 GPa tensile strength, 70–150 GPa modulus of elasticity, 1410–1450 kg/m^3 density. Organic fibers preserve their initial characteristics up to 180°C. They are characterized by high tensile strength and stiffness in comparison with glass fibers but under compression, composites with organic fibers have substantially lower properties than fiberglass composites.

It should be emphasized that the interfaces are not always as perfect as sometimes assumed in micromechanical society, they may have different tension and shear strength as well as different constitutive equations describing their mechanical properties. The strength must be higher in CMs with fillers featuring the absolute adhesion to the matrix than that in systems with little or no adhesion. The relative elongation and specific impact strength must, on the contrary, go up with increasing adhesion. In so doing, the modulus of CMs is significantly less sensitive to the adhesion properties of the interphase. The most promising method of modification of CM is by pretreating the filler or adding to the matrix specific depressants or modifiers with the aim of creating chemical bonds at the interphase. It has been convincingly demonstrated that on the filler-polymer boundary there exist adsorbed layers of polymer characterized by a density greater than that of the unfilled matrix. Macromolecular fixation on the filler surface, their orientation in the direction coplanar or normal to the surface, and condensation result in "hard" interphases. Alternatively, selective sorption by the filler sorption of one of the matrix low molecular components may lead to plasticization of the boundary layers and appearance of "soft" interphases [310]. The role of interfaces increases with reduction of a reinforcement size.

Fiber nanotube composites. Carbon fibers are currently considered to be most promising. The recent discovery of carbon nanotubes (CNTs) by Iijima [490] has gained ever-broaded interest due to providing unique properties generated by their structural perfection, small size, low density, high strength, and excellent electronic properties (see, e.g., comprehensive reviews in [907], [1083], [1093]), depending on the diameter and chirality [208], [881], [1219]. Indeed, the longitudinal Youngs modulus of CNTs falls between 0.4 and 4.15 TPa, and a tensile strength approaching 100 GPa. Carbon nanotubes occur in two distinct forms, single-walled nanotubes (SWNT), which are composed of a graphite sheet rolled into a cylinder and multi-walled nanotubes (MWNT), which consist of multiple concentric graphite cylinders. Compared with multi-walled nanotubes, single-walled nanotubes are expensive and difficult to obtain and clean, but they have been of great interest owing to their expected novel electronic, mechanical, and gas adsorption properties [306]. Nanofibers can have a number of different internal structures, wherein graphene layers are arranged as concentric cylinders, nested truncated cones, segmented structures, or stacked coins (see, e.g., [738]). External morphologies include kinked and branched structures and diameter variation. A high aspect ratio of CNT and extraordinary mechanical properties (strength and flexibility) make them ideal reinforcing fibers in nanocomposites, which can produce advanced nanocomposites with improved stiffness and strength [1092], [1099]. Tensile test on composites show that 1 wt% nanotube additions result in 40 and 30% increase in elastic modulus and strength, respectively, indicating significant load transfer across the nanotube matrix interface. An additional significant impact of CNT on the effective properties of nanocomposites is based on the fact that CNTs provide very high interfacial area if embedded in a matrix, and assemblies with unique architectures might be constructed by interconnected CNTs. It is emphasized in [864] that the planar graphite surface structure of these nanotubes provides relatively few sites suitable for facile chemical modification. It is believed that graphitic CNTs are perhaps more attractive additives for fabricating polymer/carbon nanofiber composite materials of enhanced mechanical and physical properties due to their unusual atomic structure. Depending on the metal catalyst and thermal syntheses conditions, CNTs can be grown with widths ranging from 5 to 1000 nm and lengths ranging from 5 to 10 μm. The design and fabrication of these materials are performed on the nanometer scale with the ultimate goal to obtain highly desirable macroscopic properties.

Silicate clay nanocomposites. Polymeric composites reinforced with clay crystals of nanometer scale recently attracted tremendous attention in material engineering. A particular class of nanocomposites is clay nanocomposites (see Chapter 18 and [351], [721], [992], [1099], [1159], [1160] where additional references can be found). In general, layered silicate-reinforced presenting many challenges for mechanics are conditionally divided into three types (see [198], [375], [486]): conventional composites where the layered silicate is presented as the original aggregate state of the clay particles (tactoids) with no intercalation of the matrix material into the layered silicate; intercalated nanocomposites (modeled in Chapter 18) where the matrix material is inserted in the form of thin (a few nm) layers into the space between the parallel silicate layers of extremely small thickness ($\sim$ 1 nm); completely exfoliated or delaminated nanocomposites where the individual silicate platelets of 1 nm thickness are randomly dispersed in a continuous polymer matrix. Single clay layers were proposed to be an ideal reinforcing agent due to their extremely small aspect ratio (α =0.001–0.01), the nanometer filler thickness being comparable to the scale of the polymer chain structure, as well as to their high stiffness (200–400 GPa). A doubling of the tensile modulus and strength is achieved, and the heat distortion temperature increases by 100°C for nylon-layered silicate nanocomposites containing as little as 2% vol. nanoplates [651], [1217]. However, exfoliation of layered minerals is seriously hampered by the fact that sheet-like materials exhibit a strong tendency to agglomeration due to their large contact surfaces. Generally, exfoliated nanocomposites demonstrate better properties than intercalated ones at the same nanoplate concentration. Vaia and Giannelis [1125] emphasized that the presence of many chains at interfaces means that much of the polymer is really "interphase-like" instead of having bulk-like properties that modify the thermodynamics of polymer chain confirmations and kinetics of chain motion. Since an interface limits the confirmations that polymer molecules can adopt, the properties of polymer in this interface are fundamentally different from that of polymer far removed from the interface. Consequently, almost all matrixes in CMs with a few volume percent of nanofiller can be considered as nanoscopically confined interfacial polymer. Thus, many mechanical, physical, and chemical factors could potentially formulate the properties of nanocomposites, and a better understanding of their relative contribution is needed.

1.1.3 Classification of CM Manufacturing

Classification of composites with respect to the manufacturing route divides them into "artificial" (i.e., man-made) and "natural" (i.e., in situ) composites. "Artificial" composites include long and short fiber composites, particulate and plate composites, interpenetrating multiphase composites, cellular and foam solids, and concrete. The manufacturing features can lead to significant variation of the microtopology and mechanical properties of CM. For example, depending on the thermomechanical treatment, the grains of a polycrystalline material can be isotropic spheroid (after full recrystallization and grain growth) or flake (after rolling) or elongated "cigars" (after extrusion). Among the "natural" composites, one can mention polycrystals, soil, sandstone, ice, wood, bone, cell aggregates, and the Earth's crust. Because the mechanical properties and the microtopology of composites are assumed to be known, their classification based on the manufacturing features will not be considered (see for details [1135], [1141]).

1.2 Effective Material and Field Characteristics

A wide class of phenomena for microinhomogeneous media are described by linear partial differential equations with random rapidly oscillating coefficients that can be reduced to linear integral equations by the use of a Green's function for a selected problem for a homogeneous comparison medium in bounded or infinite domains. The replacement of a governing integral equation with rapidly oscillating coefficients by an equation with either constant or smoothly varying "overall" coefficients is a main problem of micromechanics.

In classical elasticity, under assumption of geometrical and physical linearity, the relations between the local stress $\boldsymbol{\sigma}(\mathbf{x})$ and strain $\boldsymbol{\varepsilon}(\mathbf{x})$ tensors of microinhomogeneous media are established by Hooke's law: $\boldsymbol{\sigma}(\mathbf{x}) = \mathbf{L}(\mathbf{x})\boldsymbol{\varepsilon}(\mathbf{x})$, $\boldsymbol{\varepsilon}(\mathbf{x}) = \mathbf{M}(\mathbf{x})\boldsymbol{\sigma}(\mathbf{x})$, where $\mathbf{L}(\mathbf{x})$ and $\mathbf{M}(\mathbf{x}) = \mathbf{L}^{-1}$ are the known local stiffness and compliance fourth-order tensors. Effective tensors of elastic properties $\mathbf{L}^*$ and $\mathbf{M}^*$ of statistically homogeneous medium are determined as the proportionality factors $\langle\boldsymbol{\sigma}(\mathbf{x})\rangle = \mathbf{L}^*\langle\boldsymbol{\varepsilon}(\mathbf{x})\rangle$, $\langle\boldsymbol{\varepsilon}(\mathbf{x})\rangle = \mathbf{M}^*\langle\boldsymbol{\sigma}(\mathbf{x})\rangle$ between the macroscopic stresses $\langle\boldsymbol{\sigma}(\mathbf{x})\rangle$and strains $\langle\boldsymbol{\varepsilon}(\mathbf{x})\rangle$averaged over the volume $\bar{w}$ of a large sample w containing statistically large numbers of inhomogeneities. Thus, at the macroscale significantly exceeding the length scale of the material inhomogeneity (microscale), the behavior of a microinhomogeneous medium can be described by that of a suitably chosen equivalent homogeneous medium with the macroscopic stresses and strains slowly varying according to the macroscopic geometry and boundary conditions. In so doing, the local stress and strains, and other values, can be split into slow and fast (marked by primes) constituents that lead to the average constitutive equations presented in the following form: $\langle\boldsymbol{\sigma}\rangle = \langle\mathbf{L}\rangle\langle\boldsymbol{\varepsilon}\rangle + \langle\mathbf{L}'\boldsymbol{\varepsilon}'\rangle$, $\langle\boldsymbol{\varepsilon}\rangle = \langle\mathbf{M}\rangle\langle\boldsymbol{\varepsilon}\rangle + \langle\mathbf{M}'\boldsymbol{\sigma}'\rangle$, where $\mathbf{f}' = \mathbf{f} - \langle\mathbf{f}\rangle$ $[\mathbf{f} = \boldsymbol{\varepsilon}(\mathbf{x}), \boldsymbol{\sigma}(\mathbf{x}), \mathbf{L}(\mathbf{x}), \mathbf{M}(\mathbf{x})]$. Obviously, fluctuating parts of stresses $\boldsymbol{\sigma}'$ and strains $\boldsymbol{\varepsilon}'$ can be expressed through the random influence functions:

$$\boldsymbol{\varepsilon}'(\mathbf{x}) = \mathbf{A}^*(\mathbf{x})\langle\boldsymbol{\varepsilon}\rangle, \quad \boldsymbol{\sigma}'(\mathbf{x}) = \mathbf{B}^*(\mathbf{x})\langle\boldsymbol{\sigma}\rangle, \tag{1.1}$$

which exist (but are generally unknown) due to linearity of the problem. Substitution of the micro-macro linkages (1.1) into the right-hand sides of the averaged constitutive equations leads to representations for the effective stiffness and compliance:

$$\mathbf{L}^* = \langle \mathbf{L} \rangle + \langle \mathbf{L}'\mathbf{A}^* \rangle, \quad \mathbf{M}^* = \langle \mathbf{M} \rangle + \langle \mathbf{M}'\mathbf{B}^* \rangle. \tag{1.2}$$

Thus, the problem of effective properties estimation is reduced to finding $\mathbf{A}^*$ (or $\mathbf{B}^*$), which is turns out some functional involving all points $\mathbf{y}$ surrounding the considered point $\mathbf{x}$. The mentioned problem of many bodies appearing in many sections of astronomy, mechanics, and physics can be generally solved only approximately.

The problem formulated in the preceding, which is central to our understanding of micromechanics, looks very trivial. However, this illusion of triviality disappears if we recall the first steps of research in this field. Voight [1142] proposed in 1889 to consider the average of the elastic moduli of the crystallites in a polycrystal as effective elastic moduli of the aggregate $\mathbf{L}^* = \langle \mathbf{L} \rangle$. Much later in 1929, Reuss [934] proposed an analogous operation for obtaining the effective compliance $\mathbf{M}^* = \langle \mathbf{M} \rangle$. Much later still, Hill [459] proved in 1952 that Voight and Reuss estimations are upper and lower bounds of the true effective elastic moduli of the polycrystals: $\langle \mathbf{M} \rangle^{-1} \leq \mathbf{L}^* \leq \langle \mathbf{L} \rangle$.

The random values involved in the estimation of effective properties (1.2) can be classified as two sets: the values consisting of the material tensors (e.g., $\mathbf{L}, \mathbf{M}$) and some functions of these tensors, and the values formed by the field tensors (e.g., $\boldsymbol{\varepsilon}, \boldsymbol{\sigma}$). The distribution parameters of random material tensors can be evaluated through the known properties of constitutive phases and their random topological structure. The distributions of the random field tensors are expressed via the random material tensors by the use of the appropriate equations of elasticity theory. The simplest characteristic of a random variable is a one-point probability density whose first statistical moments defined the mean and dispersion. For example, the Voight and Reuss estimations are based on the first one-point distribution functions such as the phase volume fractions. Significantly more detailed information is contained in the multipoint probability densities incorporating all points of the medium. So, a knowledge of at least two-point distributions of the material tensors is necessary for estimation of the corrections introduced the terms $\langle \mathbf{L}'\mathbf{A}^* \rangle$ and $\langle \mathbf{M}'\mathbf{B}^* \rangle$ into the effective properties representations (1.2). The progress in micromechanics of random structure composites is based on the methods of statistical mechanics of a multiparticle system considering n-point correlation functions and direct multiparticle interactions of inhomogeneities.

For the sake of brevity, we intentionally sacrificed rigorousness for simplification of a presentation of basic concepts of micromechanics. For example, the average can be introduced as either the volume or ensemble averages $\{\mathbf{f}\} = (\bar{w})^{-1} \int_w \mathbf{f}(\mathbf{x})\, d\mathbf{x})$, $\langle \mathbf{f} \rangle = n^{-1} \sum_{i=1}^n \mathbf{f}_i(\mathbf{x})$, respectively, where $\mathbf{f}_i$ is an i's realization of a random process in the point $\mathbf{x}$. However, we will demonstrate that the size of RVE depends on the problem being considered and the method used for solution of this problem, and, because of this, RVE is simply a notation convenience reflecting a nature of nonlocal interaction between microelements that interact without touching of elements as in the case of local interaction.

Probably the first demonstration of the nonlocality principle was the discovery of the Newtonian law of gravitation, which implies the existence of interaction between bodies remote from one another. A similar law for electric forces was demonstrated by Coulomb a century later.

The nonlocal nature of the effective constitutive law is manifested most conspicuously in the case of the eventual abandonment of the hypothesis of statistically homogeneous fields leading to a nonlocal coupling between statistical averages of the stress and strain tensors when the statistical average stress is given by an integral of the field quantity weighted by some tensorial kernel, i.e., the *nonlocal* effective elastic operator $\boldsymbol{\mathcal{L}}^*$: $\langle\boldsymbol{\sigma}\rangle(\mathbf{x}) = \int \boldsymbol{\mathcal{L}}^*(\mathbf{x},\mathbf{y})\langle\boldsymbol{\varepsilon}\rangle(\mathbf{y})\, d\mathbf{y}$. Perhaps the most challenging issue is how micromechanics can contribute to the understanding of the bridging mechanism between the coupled scales, which is described by the nonlocal constitutive equations involving the parameters of a relevant effective nonlocal operator with the microstructure. Considering dispersed media, this approach makes intuitive sense since the stress at any point will depend on the arrangement of the surrounding inclusions. Therefore the value of the statistical average field will locally depend on its value at the other points in its vicinity. This approach is especially appropriate if the inclusion number density varies over distances that are comparable to the particle size.

It should be mentioned that a multiscale hierarchy of composite materials is typically investigated via the use of the RVE concept when parameters of larger scale models are measured or calculated on a smaller scale. The RVE concept serving as a bridge between different scales is rigorously justified if a field scale (field inhomogeneity) infinitely exceeds a material scale (material inhomogeneity). However, even in this case in a large class of problems, the length scales cannot be separated in this way; the coupling between them is strongly "bidirectional." It takes place, for example, when the microscopic structure (such as, e.g., the plastic strain and damage accumulation) develops due to some macroscopic forces. The coupling phenomenon is manifested even more visually in the case of comparable scales of the material scale of analyzed structures and the inhomogeneity scale of internal stresses, which of necessity yields a nonlocal character of the constitutive law in the area of coupling of the mentioned scales, and, therefore, the popular basic assumption acting as a bridge between adjusted scales can be considered as an approximation of the real nonlocal constitutive law.

We need to emphasize the emphasized an importance of analysis of statistical distributions of local microfields rather than only effective properties, which in actuality is a secondary problem deriving from estimation of the average field concentrator factors inside the phases. Averaging is usually suitable for predicting effective elastic properties. However, failure and elastoplastic deformations will depend on specific details of the local stress fields when fluctuation is important. A determination of failure mechanism usually localized at the interfaces is essentially important for the ceramic and metal matrix composites that have relatively weak interfaces and multiple interphase regions. Plastic deformations usually emanating in the vicinity of the interphases lead to loss of the homogeneity of the mechanical properties of the constituents when the local properties of the phases become position dependent. In the framework of computational micromechanics in such a case, one must check the specific observable stress fields for many large system realizations of the microstructure with the use of an extremal statistical technique. A more effective approach used in analytical mechanics is the estimation of the statistical moments of the stress field at the interface between the matrix and inhomogeneities. The inherent characteristics of this interface are critical to understanding the failure mechanisms usually localized at the interface.

The widespread accessibility of a unique level of computing power provides promotes the argument that modeling and simulation, properly formulated and performed, can be a credible partner to the traditional approaches of theory and experiments. Micromechanical modeling and simulation of random structures are becoming more and more ambitious as a result of advances in modern computer software and hardware. On one hand, some models have been developed with the goal of minimizing empirical elements and assumptions. Specifically, several examples are available in the area of nonlocal and nonlinear problems such as localization of the plastic strains and failure based on the estimations of local stress concentration factors. In many cases, the resolution of microscopic phenomena has led to improved accuracy, and offers the possibility of solving previously intractable problems. On the other hand, there are ambitions to attack increasingly large systems. Such methods, usually referred to as computational micromechanics, are based on the wide exploitation of Monte Carlo simulation with forthcoming numerical analysis for each random realization of multiparticle interactions of microinhomogeneities. However, at the present capacities of computer hardware and software, they are practical only for realizations containing no more than a few thousand inhomogeneities. In the case of periodic structures, the asymptotic unit cell homogenization method coupled with the finite element method (FEM) has been justified to be the most successful tool especially for the establishment of a link between the unit cell microstructure with periodic boundary conditions and overall properties of the heterogeneous media. Such a perfectly periodic problem, of course, does not exist in actual cases, although periodic modeling can be quite useful, since it provides rigorous estimations with *a priori* prescribed accuracy for various material properties. However, if the microscopic crack is considered, it should lead to the unrealistic assumption that the unit cell model with a crack is repeated. The case of a single microcrack (imperfection) in a composite material of periodic structure was investigated in the framework of the scheme of "matching rules" coupling the stresses in the outer and inner regions of the cell containing the imperfection. Implementation of the mentioned scheme accompanied by the Monte Carlo simulation has prohibitive computation costs for the periodic structures with random field of imperfections. Because of this, combining opportunities of computational micromechanics with basic assumptions of analytical micromechanics is very promising and considered in this book. The starting point of any average scheme is the choice of the elastic properties of the comparison medium that usually coincides with the homogeneous matrix. This allows one to restrict the estimation of stress fields to the fields just inside the inclusions. The choice of the comparison medium in the form of the perfect periodic structure has several benefits, which are considered in this book. These corrections incorporate specially adopted versions of the finite element analysis (FEA) method of analysis of multiple interacting inclusion problems into the modified versions of the MEFM combining a well-established theory of homogenization of periodic media with the methodology of averaging of random structures in the infinite domains.

Another barrier limiting Monte Carlo simulation appears in the analysis of statistically inhomogeneous random structures. Statistically homogeneous media are typically reduced to the averaging over a set of pseudorandom generations based on the determination of the "exact" values of the effective moduli of a periodic unit cell containing a few tens of particles. Trivialness of this problem is explained by the *a priori* known fact that the effective elastic moduli is a constant relating the average stresses and average strains. The case of the statistically inhomogeneous random structure is essentially more complicated because the effective elastic properties are now described by a nonlocal integral operator, which is a two-point one with *a priori* unknown structure. However, combining the MEFM with the numerical solution for a finite number of inclusions (considered in this book) presents no difficulties in the analysis of the indicated problem. In parallel with computational micromechanics mentioned above, classical or analytical micromechanics are also widely used and schematically considered in the next section.

1.3 Homogenization of Random Structure CM

For random structure composites, in general, the problem of estimation of effective properties cannot be solved exactly and is usually treated in an approximate manner. According to Willis [1187], the numerous methods in micromechanics can be classified into four broad categories: perturbation methods, self-consistent methods of truncation of a hierarchy, variational methods, and the model methods among which there are no rigorous boundaries. Perturbation methods are effectively utilized only for the analysis of weakly inhomogeneous media and are based on the theory of a small parameter having controlled accuracy ([69], [70], [699]). Variational methods generate the bounds of effective properties via substitution of approximate fields into the strict energy bounds obtained by the volume average of the energy density with the help of lower-order correlation functions (see the fundamental works [775], [897], [1103], [1184], [1186] and references therein). The variational methods represent the most rigorous trend of micromechanics of statistically homogeneous media; however, in the case of either statistically inhomogeneous structures or strong inhomogeneity, these bounds are not instructive. The model methods based on the replacement of the real random structure of composites by a specific one (see, e.g., [237]) have intuitive physical interpretations; but, majority of model methods are self-contained: usually the modification of either geometrical structure of composites or constitutive equations generates a need for creation of a new model with *a priori* unknown prediction accuracy. The variational and model replacement methods lead to some exact results, however, the analysis of these methods is considered schematically in this book.

We will consider in detail the self-consistent methods based on some approximate and closing assumptions for truncating an infinite system of integral equations involved and its approximate solution. Solution methods considered include a multiparticle effective field method (MEFM), effective medium method, Mori-Tanaka method, differential methods, and some others. The following also tend to concentrate on methods and concepts and their possible generalizations and connections of different methods rather than explicit results. For linear problems, the estimation methods of effective properties (compliance, thermal expansion, stored energy) as well as first and second statistical moments in the components for the general case of nonhomogeneity of the thermoelastic inclusion properties are analyzed. Both statistically homogeneous and statistically inhomogeneous, e.g., functionally graded, composites with a homogeneous matrix are considered, and the dependence of the effective properties (both local and nonlocal) on the concentration of the inclusions is presented. Both the Fourier transform method and iteration method are analyzed. It is likely that all of the methods discussed for linear elastic static problems can be generalized and in many cases are generalized to the dynamic case as well as to coupled phenomena such as thermoelectroelasticity.

Consideration of composites with nonlinear constitutive behavior, including plasticity and creep, requires an acceptance of some additional assumptions. The popular hypothesis of the mean field method takes into account only average stresses in the components. However, the statistical higher moments of stresses are extremely useful for understanding the evaluation of nonlinear phenomena. So, for estimation of nonlinear functions described the nonlinear process (the strength surface in fatigue, the yield function in plasticity, the stress potential in power-law creep, the energy release rate in fracture) one can use the statistical second moment of stresses in the phase. The hypothesis of homogeneity of plastic strains in the components and its generalization for plastic strain concentrations in thin layers around the inclusions are also considered.

It should be mentioned that many micromechanical methods, including self-consistent ones, use the hypothesis (called hypothesis H1) according to which each inclusion is located inside a homogeneous so-called effective field. Exploiting this hypothesis allows for replacement of the real inclusions with, perhaps, an arbitrary shape and structure by the fictitious ellipsoidal homogeneous inclusions with the same average thermoelastic response (see for details Chapter 8). No restrictions are imposed on the microtopology of the microstructure and the shape of inclusions as well as on the inhomogeneity of the stress field inside the inclusions. It will be justified that the most general the MEFM includes as the particular cases the well-known methods of mechanics of strongly heterogeneous media (such as the effective medium and the mean field methods). The MEFM is based on the theory of functions of random variables and Green's functions. Within this method a hierarchy of statistical moment equations for conditional averages of the fields estimated in the inclusions is derived. The hierarchy is established by introducing the notion of an effective field. In this way the interaction of different inclusions is taken directly into account in the framework of the hypothesis of the effective field. The main advantage of the MEFM proposed is a universal calculation scheme of general integral equations, which leaves room for corrections of their individual elements if either the initial integral equations and boundary conditions are changed or more improved methods are utilized for the analysis of these individual elements. In particular, the hypothesis H1 is merely a zero-order approximation of binary interacting inclusions that results in a significant shortcoming of the MEFM. This substantial obstacle will be overcome in the upgraded version of the MEFM in light of the generalized schemes based on the numerical solution of the problem for two inclusions in the infinite media, subjected to the effective field evaluated from forthcoming self-consistent estimations. The mentioned self-consistent methods are usually based on such fundamental notions as the Green's function and Eshelby tensor. However, a combination of the general anisotropy of the matrix and the general shape of randomly located inclusions with continuously variable anisotropic properties poses a significant barrier to the classical approaches based on either analytical or numerical representations for the Eshelby tensor for inclusions. Because of this, the combination of computational micromechanics with analytical micromechanics seems very promising because it allows for exploring the most powerful features of both mentioned groups of methods. The so-called numerical MEFM presents a universally rigorous scheme of both analyses of the microstructures and prediction of macroscopic properties and the statistical distributions of the local stress fields. One of the main advantages of the MEFM is in an efficient calculation of the general integral equations, which allows for hierarchical improvement of the method. The progress of the modern mathematical materials science demands the subsequent development of the MEFM providing a wide implementation of advanced numerical methods. It is possible to get a numerical solution in the form of a table in which each line among a few hundred lines corresponds to an accurate solution for two inclusions in an infinite medium subjected to unit loading. Such a table is be used subsequently for the estimation of local and nonlocal effective properties of statistically homogeneous and inhomogeneous composites in both the bounded and unbounded domains with a wide range of both inclusion concentration and elastic mismatch.

Lastly, we should recognize that the properties of materials involve different scales reflects the explosive character of progress in modern nano- and micromechanics resulting from the development of image analyses and computer-simulation methods on one hand and advanced experimental technique (such as e.g. X-ray tomography and electron microscopy) and improved materials processing (prescribed structure controlled by processing) on the other hand. At the coarse level, the collective results of events or processes taking place at the fine level are accounted for. In material modeling, one can distinguish four levels of simulation: electronic structure, atomistic, microscale, and macroscale. Although this book concentrates on the micro-macro connection of the multiscale concept within the framework of continuum mechanics, we also consider some elements of molecular modeling and bridging nanomechanics to micromechanics in nanocomposites (NCs).

Unlike many other fields in material science, the evolution of NCs to its anticipated level of importance to material engineering significantly depends on the contributions from modeling and simulation. Computational approaches, based on the Newton's equations of motion at the atomistic level [referred to as the molecular dynamics (MD) approach], are currently limited to the nanoscale (where all dimensions are at the nanoscale) and cannot deal with the micro-length scales (which are traditionally analyzed by a continuum mechanics approach). These studies rely on fitting of atomistic simulation results to determine the important elastic parameters such as elastic moduli that implicitly assumes a local nature of constitutive law of continuum mechanics at the nanoscale based in turn on the assumption that a field scale (internal stress inhomogeneity) infinitely exceeds a material scale (molecular inhomogeneity). This popular basic assumption acting as a bridge between nano- and micromechanics is questionable due to comparable scales of the material and field inhomogeneities that with necessity yields nonlocal character of the constitutive law in the area of coupling the mentioned scales. The numerical technique proposed in this book is particularly well suited for a fast and accurate prediction on the effective elastic properties in the random nanocomposites where other existing methods may not be reliable. This goal is accomplished by combining an advanced MD simulation just for one nanoinclusion in the matrix with the methodology of averaging of random functionally graded structures recently developed.

1.4 Overview of the Book

The contents give a good of what this publication is about. Briefly stated, the first chapter discusses the classification of composites and nanocomposites, motivation for its studying and outlines homogenization techniques of random structure composites from various viewpoints. Chapter 2 is a brief version of a solid mechanics course particularly tailored to provide background for the rest of this book. The key tools of random structure composite materials (CMs) such as Green's function, Eshelby and related tensors are considered in Chapter 3 for the problems of static and dynamic elasticity, conductivity, and thermoelectroelasticity. Analytical and numerical solutions for one and a few ellipsoidal and nonellipsoidal heterogeneities in an infinite matrix and in a half space are considered in Chapters 3 and 4.

Chapter 5 describes basic concepts and recent advances in quantitative characterization and numerical simulation of the microstructure of random structure composites. General representations of both the effective properties and effective energy functions are presented through the local stress and strain concentrator factors in Chapter 6. The principal formulae of thermoelastostatics of statistically homogeneous composites satisfying the ergodicity condition adapted to our intended application to composites are discussed; minimum energy principles are derived that lead to variational bounds on all the effective properties in terms of trial fields. General integral equations of micromechanics of CMs with deterministic (periodic and nonperiodic), random (statistically homogeneous and inhomogeneous, so-called graded), and mixed (periodic structures with random imperfections) structures are derived in Chapter 7. Chapter 8 describes the multiparticle effective field method (MEFM) (put forward and developed by the author in the last 20 years) which is based on the theory of functions of random variables and Green's functions. Within this method a hierarchy of statistical moment equations for conditional averages of the stresses in the inclusions is derived. The hierarchy is established by introducing the notion of an effective field.

In this way the interaction of different inclusions is taken directly into account. One observes in Chapters 8 and 9 that the MEFM includes as particular cases the well-known methods of mechanics of strongly heterogeneous media (such as the effective medium and the mean field methods and some others). Generalization of the MEFM presented in Chapter 10 is based on the abandonment of the effective field hypothesis and on a numerical solution of the problem for both one and two inclusions in an infinite medium.

Periodic structures and periodic structures with random imperfections are investigated in Chapter 11 by three different methods; the most general method uses a decomposition of the desired solution on the solution for the perfect periodic structure and on the perturbation produced by the imperfections in the perfect periodic structure. Nonlocal effects in statistically homogeneous and inhomogeneous random structure composites are analyzed in Chapter 12 by the MEFM which also has qualitative benefits following immediately from the consideration of multiparticle interactions. From such considerations it can be concluded that the final relations for effective properties described by nonlocal integral operators depend explicitly not only on the local concentration of the inclusions but also on at least binary correlation functions of the inclusions. Although the effective behavior of the composite is traditionally the main focus of micromechanics, it is also essential to supply insight into the statistical description of the local strains and stresses (considered in Chapter 13 by both the perturbation method and the method of integral equations) which are extremely useful for understanding the evolution of nonlinear phenomena such as plasticity, creep, and damage. The general methods of estimation of both the stress concentrator factors and nonlocal effective operators are presented in Chapter 14 for matrix composites with an arbitrary elastic and thermal mismatch of constituents in a half-space; numerical results were obtained by the proposed method of integral equations in some limiting cases for dilute concentration of inclusions as well as for the elastically homogeneous composites with random residual microstresses in a half-space.

The fundamental role of the statistical averages of the second moments of stress concentration factors in nonlinear analysis is explained by the fact that both the yield surface, fiber/matrix interface failure criterion, and the energy release rate are the quadratic functions of the local stress distributions. The effective limiting surfaces separating the linear and nonlinear behavior arrears of the mentioned nonlinear phenomena are constructed in Chapter 15 with subsequent application in Chapter 16 where evolution of nonlinear phenomena is considered. Chapter 17 presents a summary of similar results in the related problems of conductivity, thermoelectroelasticity, and scattering of elastic waves. An approach of multiscale mechanics of nanocomposites (NC) is presented in Chapter 18; this approach includes the scheme of molecular dynamic simulation accompanied with a bridging mechanism between nano- and micromechanics which are reduced to the continuum mechanics approach incorporated in the MEFM for the analyses of NC with prescribed random orientation of nanofibers as well as clustered clay NCs. A short sketch of limitations and ideas as well as possible generalizations of some basic concepts of micromechanics is presented in the Conclusion.

2

Foundations of Solid Mechanics

A major problem in continuum mechanics consists of predicting the deformations and internal forces arising in a material body subjected to a given set of external fields such as the forces of pure mechanical nature, temperature and electromagnetic fields. This Chapter summaries some of important definitions, relations, and methods commonly employed for the analysis of reformable media although, of course, many aspects of mechanical behavior are left unaccounted for in this approach and referring the unsatisfied reader to monographs in which a level of higher generality is adopted. However, it is hoped that sufficient details are given so that the reader can understand remainder Chapters without constant reference to other books.

As is customary in continuum mechanics studies, material properties and fields are expressed in tensor form in this book. The general technique of tensor analysis is given, e.g. in [99], [1028], [1068] presenting the tensor theory not only as an autonomous mathematical discipline, but also as a preparation for theories of continuum mechanics. The special applications of tensors are described in the books [722], [1114]. The books [21], [364] provide an introduction to the theories of linear elasticity and nonlinear elasticity. More comprehensive treatments of nonlinear and anisotropic elasticity can be found, e.g., in the books [398], [410], [995], [1015], [1082], [832], [722], and [1098]. The coupling effects of mechanical, temperature and electromagnetic fields are considered in [980], [747], [861]. The general reviews of phenomenological plasticity with an extensive list of references are given in [527], [737], [711].

2.1 Elements of Tensor Analysis

This Section provides a remainder of basic notions of tensors of a certain rank which components transforming in a particular way under the transformation of coordinates $(\mathbf{e}_1, \mathbf{e}_2, \mathbf{e}_3) \rightarrow (\mathbf{e}'_1, \mathbf{e}'_2, \mathbf{e}'_3)$.

We concern with the rectangular Cartesian coordinate systems and require that the orthonormalized basis is right-handedness $\mathbf{e}_1 \cdot (\mathbf{e}_2 \times \mathbf{e}_3) = 1$ and $\mathbf{e}_i \cdot \mathbf{e}_i = \delta_{ij}$, where a scalar multiplication $\mathbf{e}_i \cdot \mathbf{e}_j$ denotes the projection of the ith basic vector on the jth one (or vice-versa), and

$$\delta_{ij} = \begin{cases} 1 & \text{if } i = j, \\ 0 & \text{otherwise} \end{cases} \tag{2.1}$$

is the Kronecker delta. The vector (cross) product of two basic vectors $\mathbf{e}_i \times \mathbf{e}_j$ can be defined by the permutation tensor ϵ (summation over repeated Latin indices is implied)

$$\mathbf{e}_i \times \mathbf{e}_j = \epsilon_{ijk}\mathbf{e}_k, \ \epsilon_{ijk} = \begin{cases} +1 & \text{if } ijk \text{ is an even permutation of 123,} \\ -1 & \text{if } ijk \text{ is an odd permutation of 123,} \\ 0 & \text{if any two indices are identical.} \end{cases} \tag{2.2}$$

The vector product of the vectors $\mathbf{a}$, $\mathbf{b}$ and the triple scalar product of the vectors $\mathbf{a}, \mathbf{b}, \mathbf{c}$ can consequently be written as

$$\mathbf{a} \times \mathbf{b} = \epsilon_{ijk} a_i b_j \mathbf{e}_k, \quad (\mathbf{a} \times \mathbf{b}) \cdot \mathbf{c} = \epsilon_{ijk} a_i b_j c_k = \begin{vmatrix} a_1 & b_1 & c_1 \\ a_2 & b_2 & c_2 \\ a_3 & b_3 & c_3 \end{vmatrix}. \tag{2.3}$$

The expansion of another orthonormalized basis $\mathbf{e}'$ ($\mathbf{e}'_i \cdot \mathbf{e}'_j = \delta_{ij}$) with the same origin of coordinates into the old one $\mathbf{e}'_{i'} = g_{i'j}\mathbf{e}_j$ is defined by coefficients $g_{i'j}$ called a *cosine matrix*, because each element $g_{i'j}$ is the cosine of the angle between two corresponding axes

$$g_{i'j} = \mathbf{e}'_i \cdot \mathbf{e}_j = cos\alpha_{i'j}. \tag{2.4}$$

The most frequently used methods of parametric representation of orthogonal transformation of Cartesian coordinates $\mathbf{e}'$ and $\mathbf{e}$ are presented in Appendix A.1. An arbitrary vector $\mathbf{x}$ is completely defined by its magnitude (modulus) and direction in the space

$$\mathbf{x} = \sum x_i \mathbf{e}_i = x_i \mathbf{e}_i, \ x_i = \mathbf{x} \cdot \mathbf{e}_i = |\mathbf{x}| \cos(\mathbf{x}, \mathbf{e}_i). \tag{2.5}$$

The orthogonal matrix $\mathbf{g} = \|g_{ij}\|$, $(i, j = 1, \ldots, 3)$ called the rotation matrix links the coordinates $\mathbf{x}'$ of an arbitrary point in the reference orthonormalized frame $\mathbf{e}'$ ($\mathbf{e}'_i \cdot \mathbf{e}'_j = \delta_{ij}$) with the coordinates $\mathbf{x}$ in the crystal orthonormalized coordinate system $\mathbf{e}$ ($\mathbf{e}_i \cdot \mathbf{e}_j = \delta_{ij}$) with the same origin of coordinates (see for details Appendix A.1)

$$\mathbf{x}' = \mathbf{g}\mathbf{x}. \tag{2.6}$$

The definitions of tensors of order 0 (scalar) ${}^0\mathbf{T}(\mathbf{e}_1, \mathbf{e}_2, \mathbf{e}_3) \equiv^0 \mathbf{T}'(\mathbf{e}'_1, \mathbf{e}'_2, \mathbf{e}'_3)$ and 1 [contravariant vector (2.6)] can be easily generalized to contravariant tensors of order n

$${}^n\mathbf{T} =^n T_{i_1 i_2 \ldots i_n} \mathbf{e}_{i_1} \otimes \ldots \mathbf{e}_{in} =^n T'_{i_1 i_2 \ldots i_n} \mathbf{e}'_{i_1} \otimes \ldots \mathbf{e}'_{in}, \tag{2.7}$$

where the 3^n tensor components in the prime and unprime coordinates are linked by the following transformation

$${}^nT'_{j_1 j_2 \ldots j_n} = g_{j_1 i_1} \cdots g_{j_n i_n} {}^nT_{i_1 i_2 \ldots i_n}. \tag{2.8}$$

A tensor ${}^{n+m}\mathbf{T}$ of rank $n + m$, contravariant of rank n and covariant of rank m, is an object with the induced component transformation

$$^{n+m}T'^{j_1j_2\ldots j_n}_{j'_1j'_2\ldots j'_m} = g_{j_1i_1}\ldots g_{j_ni_n}g_{i'_1j'_1}\ldots g_{i'_mj'_m}\,{}^{n+m}T^{i_1i_2\ldots i_n}_{i'_1i_2\ldots i'_m}. \tag{2.9}$$

It is possible to distinguish four operations with the tensors. The first one joins the summation and multiplication over the number of the tensors $T_{i_1,\ldots,i_n}$ and $S_{i_1,\ldots,i_n}$ of the same order:

$$\lambda T_{i_1,\ldots,i_n} + \beta S_{i_1,\ldots,i_n} = R_{i_1,\ldots,i_n} \;\Leftrightarrow\; \lambda\mathbf{T} + \beta\mathbf{S} = \mathbf{R}, \tag{2.10}$$

where λ and β are scalars. The tensor (or diadic) production of two tensors of orders n and m is a tensor of order $n+m$, defined as

$$\mathbf{T}\otimes\mathbf{S} = (T_{i_1,\ldots,i_n}S_{j_1,\ldots,j_m})\mathbf{e}_{i_1}\ldots\mathbf{e}_{i_n}\mathbf{e}_{j_1}\ldots\mathbf{e}_{j_m}. \tag{2.11}$$

In so doing, in general $\mathbf{T}$ and $\mathbf{S}$ are not commutative, $\mathbf{T}\otimes\mathbf{S} \neq \mathbf{S}\otimes\mathbf{T}$. The third operation called convolution (or contraction) is the action with one tensor reducing his order on two in the framework of the Einstein summation convention, for example

$$T_{i_1,\ldots,i_n,k,k,j_1,\ldots,j_m} = \mathbf{S}_{i_1,\ldots,i_n,j_1,\ldots,j_m}. \tag{2.12}$$

The fourth operation produces the index permutation leads to a new tensor of the same order but with different index order. For example, one can generates only one new tensor (called transposed one) of the second order: $\mathbf{S} = \mathbf{T}^\top \Leftrightarrow S_{ij} = T_{ji}$. In so doing, a second order tensor $\mathbf{T}$ is called symmetric if $\mathbf{T}^\top = \mathbf{T}$, and it is called antisymmetric if $\mathbf{T} = -\mathbf{T}^\top$.

By straightforward check we are able to ascertain that the new objects formed by mentioned operations are really the tensors. These operations can be arranged in the different combinations. For example, the combination of the diadic production and convolution denoting by a dot leads to the following tensors

$$\mathbf{R} = \mathbf{T}\cdot\mathbf{S} \Leftrightarrow R_{ij} = T_{ik}S_{kj}, \tag{2.13}$$
$$\lambda = \mathbf{R}\cdot\cdot\mathbf{T} \Leftrightarrow \lambda = R_{ij}T_{ji}, \tag{2.14}$$
$$\lambda = \mathbf{R}:\mathbf{T} \Leftrightarrow \lambda = R_{ij}T_{ij}, \tag{2.15}$$

where multiple dots are appropriate for complex constructions.

In particular, the simplest tensors $^2\mathbf{T}$ of order 2 are the unit tensor $\boldsymbol{\delta}$ (1.1) and the rotation tensor $\mathbf{g}$ (2.4). They hold the property of *orthogonality*: $^2\mathbf{T}\cdot{}^2\mathbf{T} = \boldsymbol{\delta}$ and *isotropy*: $^2\mathbf{T}' = {}^2\mathbf{T}$. The tensor $\boldsymbol{\delta}$ transform a vector $\mathbf{a}$ into itself while $\mathbf{g}$ transforms the unit base vector $\mathbf{e}_i$ to the unit vector $\mathbf{e}'_i$ of the primed coordinate system: $\mathbf{e}'_i = \mathbf{g}\cdot\mathbf{e}_i$. The simplest tensor of order 4 is the unit tensor $\mathbf{I}$ with the components $I_{ijkl} = \frac{1}{2}(\delta_{ik}\delta_{jl} + \delta_{il}\delta_{jk})$ can be decomposed into the *bulk* and *deviatoric* parts

$$\mathbf{I} = \mathbf{N}_1 + \mathbf{N}_2,\;\; \mathbf{N}_1 \equiv \frac{1}{3}\boldsymbol{\delta}\otimes\boldsymbol{\delta},\;\; \mathbf{N}_2 = \mathbf{I} - \mathbf{N}_1 \tag{2.16}$$

which have the property of orthogonality

$$\mathbf{N}_1:\mathbf{N}_1 = \mathbf{N}_1,\;\; \mathbf{N}_2:\mathbf{N}_2 = \mathbf{N}_2,\;\; \mathbf{N}_1:\mathbf{N}_2 = \mathbf{0}. \tag{2.17}$$

Differentiation denoting by the *del operator* $\partial_i \equiv \partial/\partial x_i$ of a tensor with the suitably smooth components forms another tensors

$$\nabla \cdot \mathbf{T} \equiv \mathrm{Div}\mathbf{T} = (\partial_j T_{ji_1,\ldots,i_{n-1}})(\mathbf{x})\mathbf{e}_{i_1} \otimes \mathbf{e}_{i_2} \ldots \mathbf{e}_{i_{n-1}}, \tag{2.18}$$
$$\nabla \otimes \mathbf{T} \equiv \mathrm{Grad}\mathbf{T} = (\partial_{i_1} T_{i_2,\ldots,i_{n+1}})(\mathbf{x})\mathbf{e}_{i_1} \otimes \mathbf{e}_{i_2} \ldots \mathbf{e}_{i_{n+1}}, \tag{2.19}$$
$$\nabla \times \mathbf{T} \equiv \mathrm{Curl}\mathbf{T} = [\epsilon_{i_1 jk}(\partial_j T_{k,i_2,\ldots,i_n})](\mathbf{x})\mathbf{e}_{i_1} \otimes \mathbf{e}_{i_2} \ldots \mathbf{e}_{i_{n+1}}, \tag{2.20}$$

which are called *divergence, gradient*, and *curl* of the tensor field $\mathbf{T} =^n \mathbf{T}$ and have the orders $n-1, n+1$, and n, respectively. We will also use the other differential operators of the first-and second-order tensors:

$$\mathrm{Def}\mathbf{a} \equiv \frac{1}{2}(\nabla \otimes \mathbf{a} + \mathbf{a} \otimes \nabla) = \frac{1}{2}(a_{i,j} + a_{j,i})\mathbf{e}_i \otimes \mathbf{e}_j, \tag{2.21}$$
$$\mathrm{Inc}\ \mathbf{b} \equiv \epsilon_{ijk}\epsilon_{lmn} b_{jn,km}\mathbf{e}_i \otimes \mathbf{e}_j \tag{2.22}$$

called the operator of *deformation*, and *incompatibility*, respectively.

Analogously to differentiation, the integration of a tensor field $\mathbf{T}$ on both the domain w and their boundary ∂w can be defined in terms of the integration of its components. The *Gauss theorem* relates the action of the *del* operator on the tensor field $\mathbf{T}$ in domain $\mathcal{E}$ to its flux across the boundary $\partial\mathcal{E}$, under suitable smoothness of both the tensor $\mathbf{T}$ and the domain boundary ∂w with the unit vector of an outward normal $\mathbf{n}(\mathbf{s}) \perp \partial\mathcal{E}$ ($\mathbf{s} \in \partial\mathcal{E}$):

$$\int_{\mathcal{E}} \nabla * \mathbf{T}\, d\mathbf{x} = \int_{\partial\mathcal{E}} \mathbf{n} * \mathbf{T}\, ds, \tag{2.23}$$

where $*$ can stand for " $\cdot$ ",$\otimes$, or $\times$. Let $\partial\mathcal{E}'$ be a portion of an oriented surface with unit outward normal $\mathbf{n}$ and with the surface closed edge C. Then for the tensor field $\mathbf{T}$ continuously differentiable in $\partial\mathcal{E}'$ and continuous in C, the generalized Stokes theorem can be presented in the form

$$\int_{\partial\mathcal{E}'} (\mathbf{n} \times \nabla) * \mathbf{T}\, ds = \oint_C \mathbf{T} *\, d\mathbf{l}. \tag{2.24}$$

Equation (2.23) is also generalized for the volume $\mathcal{E}$ and surface $\partial\mathcal{E}$, excluding the points of a discontinuity surface τ with unit normal $\mathbf{n}$ which may be sweeping the body. In a similar manner, the surface $\partial\mathcal{E}'$ with edge C in Eq. (2.24) is accompanied by a discontinuity line γ that may be sweeping the surface. Thus (see e.g., [316]),

$$\int_{\mathcal{E}\setminus\tau} \nabla \cdot \mathbf{T}\, d\mathbf{x} = \int_{\partial\mathcal{E}\setminus\tau} \mathbf{T} \cdot \mathbf{n}\, ds - \int_{\tau} [\mathbf{T}] \cdot \mathbf{n}\, ds, \tag{2.25}$$
$$\int_{\partial\mathcal{E}'\setminus\gamma} (\mathbf{n} \times \nabla) \cdot \mathbf{T}\, ds = \oint_{C\setminus\gamma} \mathbf{T} \cdot\, d\mathbf{l} - \oint_{\gamma} [\mathbf{T}]\, d\mathbf{l}, \tag{2.26}$$

where the brackets $[\cdot]$ denote the jump of its enclosure across the discontinuity surface τ or the discontinuity line γ; $\mathcal{E} \setminus \tau$, $\partial\mathcal{E} \setminus \tau$, and $\partial\mathcal{E}' \setminus \gamma$, $C \setminus \gamma$ exclude points of τ and γ, respectively.

2.2 The Theory of Strains and Stresses

We consider a body occupying the regions $\mathcal{E}_0$ and $\mathcal{E}$, with boundaries $\mathcal{E}_0$ and $\partial\mathcal{E}$, in some fixed reference configuration and in the current one, respectively. Suppose an arbitrary material point $P^0 \in \mathcal{E}_0$ at time $t = 0$ has the spatial coordinates $\mathbf{x}$ in the orthonormalized coordinate system $\mathbf{e}_i$ $(i = 1, 2, 3)$. The point P^0 moves during the time t of deformations to the point $P \in \mathcal{E}$ with coordinates $\boldsymbol{\xi}(t)$ in the same coordinate system. The relations between the coordinates $\boldsymbol{\xi}$ and $\mathbf{x}$ are defined by the smooth one-to-one invertible functions:

$$\boldsymbol{\xi} = \boldsymbol{\xi}(\mathbf{x}, t), \ (\xi_i = \xi_i(x_1, x_2, x_3, t)) \ \text{ and } \ \mathbf{x} = \mathbf{x}(\boldsymbol{\xi}, t), \ (x_i = x_i(\xi_1, \xi_2, \xi_3, t)), \tag{2.27}$$

which are provided by an assumption that the Jacobian

$$J = \det\left\|\frac{\partial(\xi_1, \xi_2, \xi_3)}{\partial(x_1, x_2, x_3)}\right\| = \begin{vmatrix} \partial\xi_1/\partial x_1 & \partial\xi_1/\partial x_2 & \partial\xi_1/\partial x_3 \\ \partial\xi_2/\partial x_1 & \partial\xi_2/\partial x_2 & \partial\xi_2/\partial x_3 \\ \partial\xi_3/\partial x_1 & \partial\xi_3/\partial x_2 & \partial\xi_3/\partial x_3 \end{vmatrix} \tag{2.28}$$

exists at each point of the configuration $\mathcal{E}$ and that $J > 0$ meaning that the material cannot penetrate itself, and that material element of non-zero volume cannot be compressed to a point or expanded to infinite volume during the motion. The independent variables $\mathbf{x}$ and $\boldsymbol{\xi}$ in Eq. (2.27) are referred to as Lagrange and Euler variables, respectively.

We define the strain tensor (specifying the change between the points in the deformable body) by introducing the displacement $\mathbf{u} = \boldsymbol{\xi} - \mathbf{x}$, $u_i = \xi_i - x_i$ as well as the tensors of the deformation gradient $\mathbf{F}$, material displacement gradient $\mathbf{H}$, and spatial displacement gradient ($J \equiv \det\mathbf{F}(\mathbf{x}) > 0$):

$$\frac{\partial\boldsymbol{\xi}}{\partial\mathbf{x}} = \mathbf{F}(\mathbf{x}), \ \frac{\partial\mathbf{u}(\mathbf{x})}{\partial\mathbf{x}} = \mathbf{H}(\mathbf{x}) \equiv \mathbf{F}(\mathbf{x}) - \mathbf{I}, \ \frac{\partial\mathbf{u}(\boldsymbol{\xi})}{\partial\boldsymbol{\xi}} = \mathbf{h}(\boldsymbol{\xi}), \ \mathbf{Fh} = \mathbf{H}, \tag{2.29}$$

respectively. We will reproduce the correspondences between some elements of the regions $\mathcal{E}_0$ and $\mathcal{E}$ in the reference and current configurations, respectively. The equation $d\boldsymbol{\xi} = \mathbf{F}\, d\mathbf{x}$ describes how an infinitesimal line element $d\mathbf{x} \in \mathcal{E}_0$ transforms linearly into the element $d\boldsymbol{\xi} \in \mathcal{E}$. The areas ds and dS of the surface elements of $\partial\mathcal{E}$ and $\partial\mathcal{E}_0$ with the (positive) unit normals $\mathbf{n}$ and $\mathbf{N}$, respectively, are connected by the *Nanson's formula*:

$$\mathbf{n}ds = J\mathbf{BN}dS, \ \mathbf{B} = (\mathbf{F}^{-1})^\top. \tag{2.30}$$

The volume dv in the deformed configuration is related with the volume dV in the reference configuration by the equation

$$div = JdV. \tag{2.31}$$

The ratio of current $|\boldsymbol{\xi}|$ to reference $|d\mathbf{x}|$ lengths of a line element that was in the direction $\mathbf{M}$ in the reference configuration defines the *stretch*:

$$\lambda(\mathbf{M}) \equiv \frac{|d\boldsymbol{\xi}|}{|d\mathbf{x}|} = |\mathbf{FM}| = \ [\mathbf{M} \cdot (\mathbf{F}^\top\mathbf{FM})]^{1/2} \tag{2.32}$$

in the direction $\mathbf{M}$ at $\mathbf{x}$. The quality $\lambda(\mathbf{M}) - 1$ is called the extension ratio in the direction $\mathbf{M}$, while $|d\boldsymbol{\xi}| - |d\mathbf{x}|$ is the extension.

The deformed states of the body in the vicinity of P^0 are defined the strain tensor specified by the displacement $\mathbf{u}$. Working with the material $\mathbf{x}$ and spatial $\boldsymbol{\xi}$ variables we can define the strain tensors in Lagrange's (Green's tensor) and Euler's (Almansi's tensor) representations:

$$\boldsymbol{\varepsilon} = \frac{1}{2}(\mathbf{H}^\top + \mathbf{H} + \mathbf{H}^\top\mathbf{H}),\ \varepsilon_{ij}(\mathbf{x}) = \frac{1}{2}\Big(\frac{\partial u_i}{\partial x_j} + \frac{\partial u_j}{\partial x_i} + \frac{\partial u_k}{\partial x_i}\frac{\partial u_k}{\partial x_j}\Big), \quad (2.33)$$

$$\tilde{\boldsymbol{\varepsilon}} = \frac{1}{2}(\mathbf{h}^\top + \mathbf{h} + \mathbf{h}^\top\mathbf{h}),\ \tilde{\varepsilon}_{ij}(\mathbf{x}) = \frac{1}{2}\Big(\frac{\partial u_i}{\partial \xi_j} + \frac{\partial u_j}{\partial \xi_i} - \frac{\partial u_k}{\partial \xi_i}\frac{\partial u_k}{\partial \xi_j}\Big), \quad (2.34)$$

Hereafter the values referred to the variables $\boldsymbol{\xi}$ are marked by the symbol $\tilde{}$. In many practical applications, it is possible to neglect the products of derivatives in Eqs. (2.33) and (2.34). Then the coincidence of the tensors of Green and Almansi yields the well-known infinitesimal strain expression:

$$\boldsymbol{\varepsilon} = \mathrm{Def}\,\mathbf{u},\ \varepsilon_{ij}(\mathbf{x}) = \frac{1}{2}\Big(\frac{\partial u_i}{\partial x_j} + \frac{\partial u_j}{\partial x_i}\Big). \quad (2.35)$$

If a displacements $\mathbf{u}$ are given, the components of the strain tensors (2.33)-(2.35) are easily calculated by proper differentiation of these quantities. However, if six component of a strain tensor are given, they must be interrelated to serve as an integrability conditions ensuring a compatible set of the three independent displacement components. For the linear infinitesimal strain (2.35), these six interrelations, called *Saint-Venan't compatibility equations*, produce only three independent relations between the six components of the strain:

$$\mathrm{Inc}\boldsymbol{\varepsilon} = \mathbf{0},\quad \epsilon_{ilm}\epsilon_{jpq}\frac{\partial^2\varepsilon_{mn}}{\partial x_l \partial x_q} = 0. \quad (2.36)$$

For an arbitrary symmetric tensor of the second order (including Green and Almansi strain tensors), the characteristic equation $\det\|\varepsilon_{ij} - \varepsilon\delta_{ij}\| = 0$, defining a polynomial of the third degree in ε:

$$\varepsilon^3 - I_\varepsilon\varepsilon^2 + II_\varepsilon\varepsilon - III_\varepsilon = 0 \quad (2.37)$$

has three real roots $\varepsilon_1, \varepsilon_2, \varepsilon_3$ called *principal values.* The coefficients $I_\varepsilon, II_\varepsilon, III_\varepsilon$ of Eq. (2.37) are scalars called *principal basic invariants*, and are given by

$$I_\varepsilon = \mathrm{tr}\boldsymbol{\varepsilon} \equiv \varepsilon_{ii} = \varepsilon_1 + \varepsilon_2 + \varepsilon_3, \quad (2.38)$$

$$II_\varepsilon = \frac{1}{2}\epsilon_{ijk}\epsilon_{ilm}\varepsilon_{jl}\varepsilon_{km} \equiv \frac{1}{2}(I_\varepsilon^2 - \boldsymbol{\varepsilon}:\boldsymbol{\varepsilon}) = \varepsilon_{11}\varepsilon_{22} + \varepsilon_{22}\varepsilon_{33} + \varepsilon_{33}\varepsilon_{11}$$
$$- \varepsilon_{12}^2 - \varepsilon_{13}^2 - \varepsilon_{23}^3 = \varepsilon_1\varepsilon_2 + \varepsilon_2\varepsilon_3 + \varepsilon_1\varepsilon_3, \quad (2.39)$$

$$III_\varepsilon = \det\boldsymbol{\varepsilon} \equiv \frac{1}{6}(2\varepsilon_{ij}\varepsilon_{jk}\varepsilon_{ki} - 3I_\varepsilon\boldsymbol{\varepsilon}:\boldsymbol{\varepsilon} + I_\varepsilon^3) = \varepsilon_1\varepsilon_2\varepsilon_3. \quad (2.40)$$

To each principal value ε_i, the corresponding principal direction $\mathbf{m}_i$ is defined by the equation (i not summed) $\boldsymbol{\varepsilon}\cdot\mathbf{m}_i - \varepsilon_i\mathbf{m}_i = \mathbf{0}$. The tensor $\boldsymbol{\varepsilon}$ can be expressed in diagonal form in the principal triad $(\mathbf{m}_1, \mathbf{m}_2, \mathbf{m}_3)$

$$\boldsymbol{\varepsilon} = \sum_{i=1}^{3} \varepsilon_i \mathbf{m}_i \otimes \mathbf{m}_i, \tag{2.41}$$

which is orthogonal if $\varepsilon_1 \neq \varepsilon_2 \neq \varepsilon_3$: $\mathbf{m}_i \cdot \mathbf{m}_j = \delta_{ij}$. In such a case, the components of the tensor $\boldsymbol{\varepsilon}$ in the principle triad have the form (no sum on i) $\varepsilon_{ij} = \varepsilon_i \delta_{ij}$.

In parallel with the principle basic invariants (2.38)-(2.40) one can introduce other invariants such as, e.g., *basic algebraic invariants*:

$$A_1 = \varepsilon_{ii} = \varepsilon_1 + \varepsilon_2 + \varepsilon_3, \tag{2.42}$$
$$A_2 = \varepsilon_{ij}\varepsilon_{ji} = \varepsilon_1^2 + \varepsilon_2^2 + \varepsilon_3^2, \tag{2.43}$$
$$A_3 = \varepsilon_{ij}\varepsilon_{jk}\varepsilon_{ki} = \varepsilon_1^3 + \varepsilon_2^3 + \varepsilon_3^3 \tag{2.44}$$

as well as the invariant system specified by the intensity e_i of the tensor deviator $\boldsymbol{\varepsilon}'$ and its phase ψ

$$e = (\varepsilon_1 + \varepsilon_2 + \varepsilon_3)/3, \tag{2.45}$$
$$e_i = \frac{1}{3}\sqrt{(\varepsilon_1 - \varepsilon_2)^2 + (\varepsilon_2 - \varepsilon_3)^2 + (\varepsilon_1 - \varepsilon_3)^2}, \tag{2.46}$$
$$\frac{1}{\sqrt{2}} \cos 3\psi = e_i^{-3}(\varepsilon_1 - e)(\varepsilon_2 - e)(\varepsilon_3 - e). \tag{2.47}$$

The following relationships between the invariants hold true:

$$I_\varepsilon = A_1, \; 2II_\varepsilon = a_1^2 - A_2, \; III_\varepsilon = (2A_3 - 3A_2A_1 + A_1^3)/6, \tag{2.48}$$
$$e = \frac{1}{3}A_1, \; e_i = \frac{1}{3}\sqrt{3A_2 - A_1^2},$$
$$\cos 3\psi = \sqrt{2}(9A_3 - 9A_1A_2 + 2A_1^3)(3A_2 - A_1^2)^{3/2} \tag{2.49}$$

Now we will consider some definitions and notations of stress analysis. Let an element of the internal surface with an area ds have a unit normal vector $\mathbf{n}$. The total force $d\mathbf{f}$ acting on this element of surface defines the surface traction $\mathbf{t}^{(n)} = d\mathbf{f}/ds$, where the superscript (n) designates the normal to the surface element. The second-order Cauchy (or Eulerian) stress tensor $\mathbf{T}$ is formally related to the stress traction by the symmetric tensor equation:

$$\mathbf{t}^{(n)} = \mathbf{n} \cdot \mathbf{T}, \quad t_j^{(n)} = n_i T_{ij}. \tag{2.50}$$

The tensor $\mathbf{T}$ is symmetric if and only if balance of moment of momentum holds. Then the stress state in the point $\boldsymbol{\xi}$ can be described either by six components T_{ij} of the tensor $\mathbf{T}$ or by three invariants I_T, II_T, III_T called the *principal stresses* [introduced analogously to Eqs. (2.38)-(2.40), $\mathbf{T} = \boldsymbol{\sigma}$]:

$$I_\sigma = \mathrm{tr}\boldsymbol{\sigma} \equiv \sigma_{ii} = \sigma_1 + \sigma_2 + \sigma_3, \tag{2.51}$$
$$II_\sigma = \frac{1}{2}\epsilon_{ijk}\epsilon_{ilm}\sigma_{jl}\sigma_{km} \equiv \frac{1}{2}(I_\sigma^2 - \boldsymbol{\sigma} : \boldsymbol{\sigma}) = \sigma_{11}\sigma_{22} + \sigma_{22}\sigma_{33} + \sigma_{33}\sigma_{11}$$
$$- \sigma_{12}^2 - \sigma_{13}^2 - \sigma_{23}^3 = \sigma_1\sigma_2 + \sigma_2\sigma_3 + \sigma_1\sigma_3, \tag{2.52}$$
$$III_\sigma = \det\boldsymbol{\sigma} \equiv \frac{1}{6}(2\sigma_{ij}\sigma_{jk}\sigma_{ki} - 3I_\sigma\boldsymbol{\sigma} : \boldsymbol{\sigma} + I_\sigma^3) = \sigma_1\sigma_2\sigma_3 \tag{2.53}$$

and by the Eulerian angles specifying the principle triad.

The Cauchy stress tensor defined by Eq. (2.50) refers to the deformed body in the spatial coordinate system. In a similar manner, we can introduce a *nominal stress tensor* (that is transposed of the first Piola-Kirchhoff stress tensor) associated with an original element of undeformed area:

$$\mathbf{T}^0 = J\mathbf{F}^{-1}\mathbf{T}, \quad T^0_{ij} = J\frac{\partial x_i}{\partial \xi_k}T_{kj}, \tag{2.54}$$

where the Jacobian J is written as a determinant (2.28). The first Piola-Kirchhoff stress tensor is a two-point and, in general, nonsymmetric tensor. For elimination of these two disadvantages, one introduces the symmetric *second Piola-Kirchhoff stress tensor*

$$\boldsymbol{\sigma} = \mathbf{T}^0\mathbf{B} = J\mathbf{B}^\top\mathbf{T}\mathbf{B}, \quad \sigma_{ij} = T^0_{im}\frac{\partial x_j}{\partial \xi_m} = J\frac{\partial x_i}{\partial \xi_m}T_{mn}\frac{\partial x_j}{\partial \xi_n}, \tag{2.55}$$

which may be inverted

$$T_{ij} = J^{-1}\frac{\partial \xi_i}{\partial x_m}\sigma_{mn}\frac{\partial \xi_i}{\partial x_n}, \quad T^0_{im} = \sigma_{ij}\frac{\partial \xi_m}{\partial x_j}. \tag{2.56}$$

2.3 Basic Equations of Solid Mechanics

2.3.1 Conservation Laws, Boundary Conditions, and Constitutive Equation

There are four conservation laws in continuum mechanics: (1) mass conservation, (2) conservation of linear momentum, (3) conservation of angular momentum, and (4) conservation of energy. We will reproduce the first, second, and fourth conservation laws.

The *balance equation of linear momentum* in spatial coordinates in terms of the Cauchy stress tensor $\mathbf{T}$ is

$$\frac{\partial T_{ij}}{\partial \xi_i} + \rho b_i = \rho\Big(\frac{\partial v_i}{\partial t} + v_i\frac{\partial v_j}{\partial \xi_i}\Big), \tag{2.57}$$

where $\mathbf{b}$ is the body force per unit of mass acting upon the volume element; ρ is the mass density at time t related to the initial mass density ρ_0 by the *equation of continuity* (the first conservation law):

$$\rho = J^{-1}\rho_0 \tag{2.58}$$

according to Eq. (2.31). v_i is a velocity of a particle occupying the coordinate ξ_i at some time t found by taking the time derivative and holding the material coordinates constant: $v_i = \partial\xi_i/\partial t$.

Since the boundary conditions in elastic problems are most easily expressed in material coordinates, we shall reproduce an equivalent equation of motion in

Lagrangian coordinates in terms of the first and second Piola-Kirchhiff stress tensors:

$$\frac{\partial T^0_{im}}{\partial x_i} + \rho_0 b_{0m} = \rho_0 \frac{d^2 u_m(\mathbf{x})}{dt^2}, \tag{2.59}$$

$$\frac{\partial}{\partial x_i}\left[\sigma_{ij}\left(\delta_{jm} + \frac{\partial u_m(\mathbf{x})}{\partial x_j}\right)\right] + \rho_0 b_{0m} = \rho_0 \frac{d^2 u_m(\mathbf{x})}{dt^2}, \tag{2.60}$$

respectively, where the time derivative is actually a simple time derivative.

The basic *energy-conservation equation* (or *first principle of thermodynamic*) equates the rate of change of kinetic K and internal energy E to the rate at which surface and body forces do mechanical work P and the rate at which nonmechanical energy Q is transferred (per unit volume):

$$\dot{K} + \dot{E} = \dot{Q} + P, \tag{2.61}$$

where

$$K = \frac{1}{2\bar{\mathcal{E}}}\int_{\mathcal{E}} \rho v_i v_i \, dv, \quad E = \frac{1}{\bar{\mathcal{E}}}\int_{\mathcal{E}} \rho e \, dv, \tag{2.62}$$

$$Q = -\frac{1}{\bar{\mathcal{E}}}\int_{\partial\mathcal{E}} q_k n_k \, ds + \frac{1}{\bar{\mathcal{E}}}\int_{\mathcal{E}} h \, dv = \frac{1}{\bar{\mathcal{E}}}\int_{\mathcal{E}} (-\nabla\cdot\mathbf{q} + h) \, dv, \tag{2.63}$$

$$P = \frac{1}{\bar{\mathcal{E}}}\int_{\mathcal{E}} b_i v_i \, dv + \frac{1}{\bar{\mathcal{E}}}\int_{\partial\mathcal{E}} T_{ij} n_j v_i \, ds, \tag{2.64}$$

where e is a specific internal energy per unit mass, $\mathbf{q}$ is a nonmechanical energy flux tensor defining the rate $q_k n_k ds$ of any nonmechanical energy transmitted outward through the surface element ds with outward directed normal $\mathbf{n}$, and h is heat created per unit volume in the body. In Eq. (2.61) K, E, Q, and P are global quantities represented as integrals over the volume and its surface (2.62)–(2.64), but they also have local analogies (for which we will use the same notations) defined pointwise within the body. Equation (2.61) can be transformed into the differential form by the use of the equation of continuity (2.58) and the equations of motion:

$$\rho\frac{de}{dt} = T_{ij} d_{ij} - \frac{\partial q_j}{\partial \xi_j} + \rho h, \quad d_{ij} = \frac{1}{2}\left(\frac{\partial v_j}{\partial \xi_i} + \frac{\partial v_i}{\partial \xi_j}\right), \tag{2.65}$$

where, $\mathbf{d}$ called the *rate strain tensor*, is the symmetric part of the tensor $v_{i,j}$, and $d/dt = \partial/\partial t + v_i \partial/\partial\xi_i$ is a *material derivative.*

The *second law of thermodynamics* states the existence of the total differential for reversible processes:

$$dS = \frac{dQ}{T}, \tag{2.66}$$

where S is the entropy regarded as a measure of energy dissipation with respect to the absolute temperature T, and

$$\dot{S} + \int_{\partial\mathcal{E}} \left(\frac{\mathbf{q}}{T}\right)\cdot\mathbf{n} \, ds - \int_{\mathcal{E}} \frac{h}{T} \, dv \geq 0. \tag{2.67}$$

The last expression is referred to as the second law of thermodynamics, known as the Clausius-Duhem inequality, which can be rewritten as

$$\int_{\mathcal{E}} \left[\dot{s} + \nabla \cdot \left(\frac{\mathbf{q}}{T} \right) - \frac{h}{T} \right] div \geq 0, \tag{2.68}$$

or

$$T\dot{s} + \nabla \cdot \mathbf{q} - \frac{1}{T}\mathbf{q} \cdot \nabla T - h \geq 0, \tag{2.69}$$

which is the local form of the Clausius-Duhem inequality. Here and henceforth s is the specific entropy density, and the term "specific" will mean "per unit volume". Without loss of generality, it is assumed mes $\mathcal{E} = 1$.

If the right-hand side of motion equations (2.57), (2.59), (2.60) are zero, the resulting equations are called the equilibrium equation. One way to maintain a body in equilibrium is to apply suitable boundary conditions. The typical boundary conditions involves the prescribed displacements $\mathbf{u}$ specifying on part of the boundary $\partial\mathcal{E}^0_{\mathbf{u}} \subset \partial\mathcal{E}^0$, and the stress vector on the remained, $\partial\mathcal{E}^0_{\sigma}$, so that

$$\mathbf{u}(\mathbf{x}) = \mathbf{u}^{\partial}(\mathbf{x}), \quad \text{on} \quad \partial\mathcal{E}^0_{\mathbf{u}}, \tag{2.70}$$

$$\mathbf{T}^{0\top} \cdot \mathbf{N} = \mathbf{t}^{0(\mathbf{N})}(\mathbf{F}, \mathbf{x}) \quad \text{on} \quad \partial\mathcal{E}^0_{\sigma}, \tag{2.71}$$

where $\partial\mathcal{E}^0_{\mathbf{u}} \cap \partial\mathcal{E}^0_{\sigma} = \partial\mathcal{E}^0$, $\partial\mathcal{E}^0_{\mathbf{u}} \cup \partial\mathcal{E}^0_{\sigma} = 0$. A general *configuration dependent loading* (2.70) if reduced to a *dead-load traction* in the surface traction introduced by (2.70) is independent of $\mathbf{F}$.

The aforementioned equations of conservation laws are in general insufficient to determine the body motion produced by given boundary conditions and body forces. They need to be accompanied by a *constitutive equation* characterizing the constitution of the body. It is assumed that there are no stresses and no strains in the initial (virgin) strain of the body. For *Cauchy elastic materials*, it is described by a symmetric single-valued *response function* $\mathbf{G}$: $\mathbf{T} = \mathbf{G}(\mathbf{F})$, which does not depend on the path of deformation. In a special case of Cauchy elasticity called hyperelasticity, there exists a *specific strain-energy function* $w = w(\mathbf{F})$ defined on the space of deformation gradients:

$$\mathbf{T}^0 = \frac{\partial w}{\partial \mathbf{F}}, \quad \mathbf{T} = J^{-1}\mathbf{F}\frac{\partial w}{\partial \mathbf{F}}. \tag{2.72}$$

Equations (2.72) are the constitutive equations for finite-deformation elasticity. The function w defined per unit volume in $\mathcal{E}^0$ represents the work done per unit volume at $\mathbf{x} \in \mathcal{E}^0$ in changing the deformation gradient from $\mathbf{I}$ to $\mathbf{F}$. The function w can be considered as depending on either $\boldsymbol{\varepsilon}$ (2.33) or H_{ij} (2.29). Then we get the representations

$$\sigma_{ij} = \frac{\partial w(\boldsymbol{\varepsilon})}{\varepsilon_{ij}}, \quad T^0_{ij} = \frac{\partial w(\mathbf{H})}{H_{ij}} \tag{2.73}$$

relating the second Piola-Kirchhoff stress tensor $\boldsymbol{\sigma}$ to the Green deformation tensor $\boldsymbol{\varepsilon}$, and the nominal stress tensor $\mathbf{T}^0$ to the material displacement gradient $\mathbf{H}$, respectively.

A material is said to be *isotropic* if there is no orientation effect in the material. In such a case, the strain energy can be presented either as a symmetric function of the principal stretches or as a function of three independent invariants, such as (2.38)–(2.40) or (2.42)–(2.44). For an incompressible material

$J = III_\varepsilon \equiv 1$ holds, and the function w depends on only two independent invariants. In particular, the *Mooney-Rivlind* form of strain energy

$$w = C_1(I_\varepsilon - 3) + C_2(II_\varepsilon - 3), \ (C_1, C_2 = \text{const.}) \tag{2.74}$$

is reduced to the *neo-Hookean* form

$$w = C_1(I_\varepsilon - 3). \tag{2.75}$$

Using Eq. (2.73_1) and the forms of the strain-energy functions (2.74) and (2.75), the stress tensor take the forms $\boldsymbol{\sigma} = -p\mathbf{I} + 2C_1\mathbf{F}\mathbf{F}^\top$ and $\boldsymbol{\sigma} = -p\mathbf{I} + 2C_1\mathbf{F}\mathbf{F}^\top - 2C_2(\mathbf{F}\mathbf{F}^\top)^{-1}$ for Mooney-Rivlin and neo-Hookean materials, respectively; here p is an arbitrary scalar. Many other forms of the strain energy have been analyzed for both incompressible and compressible nonlinear rubberlike solids in [26], [830].

Now we will consider the strain-energy functions of physically nonlinear elasticity defined on the space of infinitesimal strain tensors (2.35). A general quadratic form of the function w composed by the algebraic invariants (2.42) and (2.43) is

$$w = a_1 A_1^2 + a_2 A_2, \tag{2.76}$$

leading to the known constitutive equation (2.73_1) for the isotropic medium:

$$\boldsymbol{\sigma} = \mathbf{L}\boldsymbol{\varepsilon}, \ L = 3\lambda\mathbf{N}_1 + 2\mu\mathbf{I} \tag{2.77}$$

by the use of two, λ and μ, Lame's elastic constants. A more general special form of specific strain energy depending on three basic algebraic invariants (2.42)–(2.44) is the *Murnaghan* form, defined by

$$w - \frac{1}{2}\lambda A_1^2 + \mu A_2 + \frac{a}{3}A_1^3 + bA_1A_2 + \frac{c}{3}A_3, \tag{2.78}$$

where λ and μ are the Lame' elastic constants of the second order, and a, b, c are the elastic constants of the third order. Equations (2.73_1) and (2.78) yield the following tensor constitutive equations:

$$\boldsymbol{\sigma} = \mathbf{L}\boldsymbol{\varepsilon} + \mathbf{L}^{(3)}\boldsymbol{\varepsilon} \otimes \boldsymbol{\varepsilon}, \tag{2.79}$$

$$L_{ijkl} = \lambda\delta_{ij}\delta_{kl} + \mu(\delta_{ik}\delta_{jl} + \delta_{il}\delta_{jk}), \tag{2.80}$$

$$L_{ijklmn} = a\delta_{ij}\delta_{mn}\delta_{kl} + b(\delta_{ij}I_{mnkl} + \delta_{mn}I_{ijkl} + \delta_{kl}I_{mnij}) + cJ_{ijmnkl}, \tag{2.81}$$

$$J_{ijmnkl} = \frac{1}{2}(I_{ipkl}I_{pqmn} + I_{ipmn}I_{pjkl}). \tag{2.82}$$

The Kauderer potential depending on two basic algebraic invariants (2.42)–(2.43) and three constant (λ, μ, γ) is known:

$$w = \frac{1}{2}\lambda A_1^2 + \mu A_2 + \frac{1}{3}\gamma(A_2 - \frac{1}{3}A_1)^2. \tag{2.83}$$

The strain energy (2.83) is associated with constitutive equations

$$\sigma_{ij} = \lambda\varepsilon_{kk}\delta_{ij} + 2\mu\varepsilon_{ij} + 2\gamma\varepsilon_{\rm eq}(\varepsilon_{ij} - \frac{1}{3}\delta_{ij}\varepsilon_{kk}), \tag{2.84}$$

$$\varepsilon_{ij} = \frac{1}{9k}\sigma_{kk}\delta_{ij} + \frac{1}{2\mu}(1 - \frac{\gamma}{9\mu^3}\tau^2)s_{ij}, \tag{2.85}$$

where $k \equiv \lambda + 2\mu/3$ is the bulk modulus, and τ and $\varepsilon_{\rm eq}$ are the stress and strain intensities

$$\tau = \left(\frac{3}{2}\mathbf{s} : \mathbf{s}\right)^{1/2}, \quad \varepsilon_{\rm eq} = \left(\frac{2}{3}\mathbf{e} : \mathbf{e}\right)^{1/2}, \tag{2.86}$$

respectively, where $\mathbf{s} = \mathbf{N}_2\boldsymbol{\sigma}$ and $\mathbf{e} = \mathbf{N}_2\boldsymbol{\varepsilon}$. The constitutive equation (2.84) can be presented in more compact form:

$$\boldsymbol{\sigma} = \mathbf{L}\boldsymbol{\varepsilon} + \boldsymbol{\Psi}(\boldsymbol{\varepsilon}),\ \mathbf{L} = 3k\mathbf{N}_1 + 2\mu\mathbf{N}_2,\ \boldsymbol{\Psi} = 2\gamma\varepsilon^2_{\rm eq}\mathbf{N}_2, \tag{2.87}$$

where the tensor $\boldsymbol{\Psi}$ depending on the effective strains $\varepsilon_{\rm eq}$ describes the nonlinear material properties.

Also used is an asymptotic expansion of the specific strain energy in a Taylor series about the state of zero strain and stress as

$$w = \frac{1}{2}L^{(2)}_{ijkl}\varepsilon_{ij}\varepsilon_{kl} + \frac{1}{6}L^{(3)}_{ijklmn}\varepsilon_{ij}\varepsilon_{kl}\varepsilon_{mn} + \ldots \tag{2.88}$$

assuming $w(\mathbf{0}) = 0$. One can get from Eq. (2.88) that

$$\mathbf{L}^{(n)} = \frac{\partial^{(n)} w}{\partial\boldsymbol{\varepsilon}^{(n)}} \tag{2.89}$$

where $n = 2, 3, \ldots$ and $\boldsymbol{\varepsilon}^{(n)} = \boldsymbol{\varepsilon} \otimes \ldots \otimes \boldsymbol{\varepsilon}$ is n-multiple tensor production of the tensor $\boldsymbol{\varepsilon}$. The equality (2.89) yields the following symmetry properties of the elasticity tensors $\mathbf{L}^{(n)}$ of the $2n$ order with respect to the index pairs

$$L^{(2)}_{ijkl} = L^{(2)}_{jikl} = L^{(2)}_{ijlk} = L^{(2)}_{lkij}, \tag{2.90}$$

$$L^{(3)}_{ijklpq} = L^{(3)}_{jiklpq} = L^{(3)}_{ijlkpq} = L^{(3)}_{ijklqp} = L^{(3)}_{klijpq} = L^{(3)}_{pqklij} = L^{(3)}_{ijpqkl}, \tag{2.91}$$

and so forth. In so doing, the "pair symmetry" described by two first equalities of Eq. (2.90) and corresponding to an interchange of the indices of the first pair and interchange of the indices of the last pair are followed from the symmetry of both the $\boldsymbol{\sigma}$ and $\boldsymbol{\varepsilon}$ tensors. The "diagonal symmetry" corresponding to an interchange of the first pair of indices with the second one is described by the third equality of Eq. (2.90) and defined by Eq. (2.88). Fourth-order tensors with pair and diagonal-symmetry are referred as "full-symmetric" tensors.

2.3.2 The Equations of Linear Elasticity

We will reproduce the reduction of the elasticity theory of finite deformation to a linear theory, which is a special case of small deformations superposed on a finite deformations with the special values $\mathbf{b}_0 = \mathbf{0},\ \mathbf{F}_0 = \mathbf{I},\ \mathbf{T}^0_0 = \mathbf{0}$. Then all stress tensors coincide and can be recognized as a Cauchy stress. All strain tensors likewise reduce to the infinitesimal strain tensor (2.35). Then the equalities

$$\boldsymbol{\sigma} = \mathbf{T} = \mathbf{T}^0, \quad \boldsymbol{\varepsilon} = \tilde{\boldsymbol{\varepsilon}}, \quad J = 1 + \varepsilon_{ii} \tag{2.92}$$

simplify the basic equations of elasticity theory

$$\nabla\boldsymbol{\sigma} + \rho\mathbf{b} = \rho\ddot{\mathbf{u}}, \quad \frac{\partial \sigma_{ij}}{\partial x_j} + \rho b_i = \rho\frac{\partial^2 u_i}{\partial t^2}, \tag{2.93}$$

$$\boldsymbol{\sigma} = \mathbf{L}\boldsymbol{\varepsilon}, \quad \sigma_{ij} = L_{ijkl}\varepsilon_{kl}, \tag{2.94}$$

$$\boldsymbol{\varepsilon} = \text{Def}\,\mathbf{u}, \quad \varepsilon_{ij} = \frac{1}{2}\Big(\frac{\partial u_i}{\partial x_j} + \frac{\partial u_j}{\partial x_i}\Big). \tag{2.95}$$

The mixed boundary conditions on $\partial\mathcal{E}$ with the unit outward normal $\mathbf{n}^{\partial\mathcal{E}}$ will be considered

$$\mathbf{u}(\mathbf{x}) = \mathbf{u}^{\partial\mathcal{E}}(\mathbf{x}), \quad \mathbf{x} \in \partial\mathcal{E}_{\mathbf{u}}, \tag{2.96}$$

$$\boldsymbol{\sigma}(\mathbf{x}) \cdot \mathbf{n}^{\partial\mathcal{E}}(\mathbf{x}) = \mathbf{t}^{\partial\mathcal{E}}(\mathbf{x}), \quad \mathbf{x} \in \partial\mathcal{E}_{\sigma}, \tag{2.97}$$

where $\partial\mathcal{E}_{\mathbf{u}}$ and $\partial\mathcal{E}_{\sigma}$ are prescribed displacement and traction non-intersected boundary conditions such that $\partial\mathcal{E}_{\mathbf{u}} \cup \partial\mathcal{E}_{\sigma} = \partial\mathcal{E}$; $\mathbf{u}^{\partial\mathcal{E}}(\mathbf{x})$ and $\mathbf{t}^{\partial\mathcal{E}}(\mathbf{x})$ are prescribed the displacement on $\partial\mathcal{E}_{\mathbf{u}}$ and traction on $\partial\mathcal{E}_{\sigma}$, respectively; mixed boundary conditions, such as in the case of elastic supports are also possible. When $\partial\mathcal{E}_{\sigma}$ is empty, the mixed boundary conditions (2.96) and (2.97) reduce to the *displacement problem* or the *first* boundary value problem. If $\partial\mathcal{E}_{\mathbf{u}}$ is empty, the boundary conditions (2.96), (2.97) becomes the *traction problem* or the *second* boundary value problem. As usual we shall distinguish the interior from the exterior problem according to whether the body occupies the interior or the exterior domain with respect to $\mathcal{E}$.

Of special practical interest are the homogeneous boundary conditions:

$$\mathbf{u}^{\partial\mathcal{E}}(\mathbf{x}) = \boldsymbol{\varepsilon}^{\partial\mathcal{E}}\mathbf{x}, \ \boldsymbol{\varepsilon}^{\partial\mathcal{E}}(\mathbf{x}) \equiv \text{const.}, \ \mathbf{x} \in \partial\mathcal{E}, \tag{2.98}$$

$$\boldsymbol{\sigma}(\mathbf{x}) \cdot \mathbf{n}^{\partial\mathcal{E}}(\mathbf{x}) = \boldsymbol{\sigma}^{\partial\mathcal{E}}(\mathbf{x}) \cdot \mathbf{n}^{\partial\mathcal{E}}(\mathbf{x}), \quad \boldsymbol{\sigma}^{\partial\mathcal{E}}(\mathbf{x}) = \text{const.}, \ \mathbf{x} \in \partial\mathcal{E}, \tag{2.99}$$

where $\boldsymbol{\varepsilon}^{\partial\mathcal{E}}(\mathbf{x}) = \frac{1}{2}\big[\nabla \otimes \mathbf{u}^{\partial\mathcal{E}}(\mathbf{x}) + (\nabla \otimes \mathbf{u}^{\partial\mathcal{E}}(\mathbf{x}))^{\top}\big]$, $\mathbf{x} \in \partial\mathcal{E}_{\mathbf{u}}$.

2.3.3 Extremum Principles of Elastostatic

The principle of minimum potential energy and the principle of minimum complementary energy completely characterize the solution of the mixed boundary problem (2.93)–(2.97) of elastostatic ($\ddot{\mathbf{u}} \equiv \mathbf{0}$ in Eq. (2.93)). For definition of these energy functions we introduce the notion of admissible fields. A sufficiently smooth (of class C^1) displacement field $\mathbf{u}$ with the displacement boundary conditions:

$$\mathcal{A}^k(\mathbf{u}) = \{\mathbf{u}|\mathbf{u}(\mathbf{x}) \in C^1(\mathcal{E}), \ \mathbf{u}(\mathbf{y}) = \mathbf{u}^{\partial\mathcal{E}}(\mathbf{y}), \ \mathbf{y} \in \partial\mathcal{E}_{\mathbf{u}}\} \tag{2.100}$$

is called a *kinematically admissible displacement field.* A kinematically admissible state is an ordered array $s = \{\mathbf{u}, \boldsymbol{\varepsilon}, \boldsymbol{\sigma}\}$ (forming the set $\mathcal{A}^k(s)$) of kinematically admissible displacement field $\mathbf{u} \in \mathcal{A}^k(\mathbf{u})$ (2.100) generating the symmetric fields

$\boldsymbol{\varepsilon}$ and $\boldsymbol{\sigma}$ according to Eqs. (2.95) and (2.94), respectively. While $\boldsymbol{\varepsilon}(\mathbf{u})$ is compatible, $\boldsymbol{\sigma}(\mathbf{u})$ may not necessarily satisfy the equilibrium equations (2.93).

A sufficiently smooth (of class C^1) stress field satisfying the equilibrium equation (2.93) and the traction boundary condition (2.97):

$$\mathcal{A}^s(\boldsymbol{\sigma}) = \{\boldsymbol{\sigma}|\boldsymbol{\sigma}(\mathbf{x}) \in C^1(\mathcal{E}),\ \nabla \cdot \boldsymbol{\sigma} + \rho\mathbf{b} = \mathbf{0},\ \boldsymbol{\sigma}(\mathbf{y}) \cdot \mathbf{n}^{\partial\mathcal{E}}(\mathbf{y}) = \mathbf{t}^{\partial\mathcal{E}}(\mathbf{y}),\ \mathbf{y} \in \partial\mathcal{E}_\sigma\} \tag{2.101}$$

is referred to as a *statically admissible stress field*. The ordered array $s^s = \{\mathbf{u}, \boldsymbol{\varepsilon}, \boldsymbol{\sigma}\}$ (forming the set $\mathcal{A}^s(s)$) is called a statically admissible state if $\boldsymbol{\sigma} \in \mathcal{A}^s(\boldsymbol{\sigma})$ (2.101) generates the symmetric tensor $\boldsymbol{\varepsilon}$ according to the constitutive equation (2.94) and the displacement field $\mathbf{u}$ is related to $\boldsymbol{\varepsilon}$ by the use of Eq. (2.95). In so doing, $\boldsymbol{\varepsilon}(\boldsymbol{\sigma})$ is not necessarily compatible.

We also define the total strain energy $W(\boldsymbol{\varepsilon})$ and the total stress energy $W^c(\boldsymbol{\sigma})$ by

$$W(\boldsymbol{\varepsilon}) = \frac{1}{2}\int_{\mathcal{E}} \boldsymbol{\varepsilon} : \mathbf{L} : \boldsymbol{\varepsilon}\, d\mathbf{x}, \quad W^c(\boldsymbol{\sigma}) = \frac{1}{2}\int_{\mathcal{E}} \boldsymbol{\sigma} : \mathbf{M} : \boldsymbol{\sigma}\, d\mathbf{x}, \tag{2.102}$$

respectively. Let $\delta\mathbf{u} = \delta\mathbf{u}(\mathbf{x})$ be a virtual, or imaginative, infinitesimal displacement field associated with a virtual strain field from the current state $\delta\boldsymbol{\varepsilon} = \mathrm{Def}(\delta\mathbf{u})$. The components of the virtual displacement vector $\delta\mathbf{u}$ vanish at $\partial\mathcal{E}_{\mathbf{u}}$: $\delta\mathbf{u}(\mathbf{y}) \equiv \mathbf{0}$ at $\mathbf{y} \in \partial\mathcal{E}_{\mathbf{u}}$. It can be shown from the equations of linear and angular momentum, that it holds the *principle of virtual work* (PVW):

$$\int_{\mathcal{E}} \boldsymbol{\sigma} : \delta\boldsymbol{\varepsilon}\, d\mathbf{x} = \int_{\partial\mathcal{E}} \mathbf{t}^{\partial\mathcal{E}_\sigma} \cdot \delta\mathbf{u}\, ds + \int_{\mathcal{E}} \rho\mathbf{b} \cdot \delta\mathbf{u}\, d\mathbf{x}, \tag{2.103}$$

which is a starting point in developing minimum principles.

As a consequence, one can obtain a work–energy relation (theorem of work expended) involving the admissible stress field $\boldsymbol{\sigma}$ and the admissible displacement field $\mathbf{u}$ [$\boldsymbol{\varepsilon}(\mathbf{u})$ is the related strain field] corresponding to the given external forces $\rho\mathbf{b}(\mathbf{x})$ and $\mathbf{t}^{\partial\mathcal{E}_\sigma}(\mathbf{y})$ ($\mathbf{x} \in \mathcal{E},\ \mathbf{y} \in \partial\mathcal{E}_\sigma$):

$$\int_{\mathcal{E}} \boldsymbol{\sigma} : \boldsymbol{\varepsilon}\, d\mathbf{x} = \int_{\partial\mathcal{E}} \mathbf{t}^{\partial\mathcal{E}_\sigma} \cdot \mathbf{u}\, ds + \int_{\mathcal{E}} \rho\mathbf{b} \cdot \mathbf{u}\, d\mathbf{x}, \tag{2.104}$$

which looks similar to the first law of thermodynamics (2.61), although there is no thermodynamic expression in (2.104). We note that $\boldsymbol{\sigma}$ and $\boldsymbol{\varepsilon}$ need not be connected by a specific stress-strain relation.

We will present the classical extremum principles.

Principle of minimum of potential energy. Let Π, called the *potential energy* (corresponding to the given external forces $\rho\mathbf{b}(\mathbf{x})$ and $\mathbf{t}^{\partial\mathcal{E}_\sigma}(\mathbf{y})$, $\mathbf{x} \in \mathcal{E}$, $\mathbf{y} \in \partial\mathcal{E}_\sigma$), be the function defined on the set of kinematically admissible states $\tilde{s}^k = \{\tilde{\mathbf{u}}, \tilde{\boldsymbol{\varepsilon}}, \tilde{\boldsymbol{\sigma}}\} \in \mathcal{A}^k(\tilde{s})$ by the equation

$$\Pi(\tilde{s}^k) = W(\tilde{\boldsymbol{\varepsilon}}) - \int_{\mathcal{E}} \rho\mathbf{b} \cdot \tilde{\mathbf{u}}\, d\mathbf{x} - \int_{\partial\mathcal{E}_\sigma} \mathbf{t}^{\partial\mathcal{E}_\sigma} \cdot \tilde{\mathbf{u}}\, ds. \tag{2.105}$$

Then the actual displacement field $\mathbf{u}$ with corresponding admissible state s^k renders the potential energy Π an absolute minimum

$$\Pi(s^k) \leq \Pi(\tilde{s}^k). \tag{2.106}$$

Principle of minimum of complementary energy. Let Π^c called the *complementary energy* (corresponding to the given boundary displacement $\mathbf{u}^{\partial\mathcal{E}_\mathbf{u}}(\mathbf{y}) \in \partial\mathcal{E}_\mathbf{u}$) be the function defined on the set of statically admissible states $\tilde{s}^s = \{\tilde{\mathbf{u}}, \tilde{\varepsilon}, \tilde{\boldsymbol{\sigma}}\} \in \mathcal{A}^s(\tilde{s})$ by the equation

$$\Pi^c(\tilde{s}^k) = W^c(\tilde{\boldsymbol{\sigma}}) - \int_{\partial\mathcal{E}_\mathbf{u}} \tilde{\mathbf{t}}_\mathbf{n} \cdot \mathbf{u}^{\partial\mathcal{E}_\mathbf{u}} \, ds, \; \tilde{\mathbf{t}}_\mathbf{n} \equiv \tilde{\boldsymbol{\sigma}} \cdot \mathbf{n}^{\partial\mathcal{E}}. \tag{2.107}$$

Then the actual stress field $\boldsymbol{\sigma}$ with corresponding admissible state s^s renders the complementary energy Π^c an absolute minimum

$$\Pi^c(s^s) \leq \Pi^c(\tilde{s}^s). \tag{2.108}$$

2.4 Basic Equations of Thermoelasticity and Electroelasticity

2.4.1 Thermoelasticity Equations

In classical thermodynamics we are concerned with the small neighborhood of thermodynamic equilibrium. We also consider the infinitesimal deformations (2.92). In such a case the first law of thermodynamics (2.65) can be recast in the form (hereafter a unit volume is considered)

$$dE = dQ + \sigma_{ij} d\varepsilon_{ij}, \tag{2.109}$$

To derive the constitutive relations of thermoelasticity theory, we transform (2.109) for the reversible processes (2.66) to the thermodynamic relation:

$$dE = \sigma_{ij} \; d\varepsilon_{ij} + T \; dS \tag{2.110}$$

equating the total differential of the internal energy (2.62_2) and the sum of the increment of the deformation work and the amount of the heat introduced into the considered volume; here T is the absolute temperature and S is the entropy density (2.68). We see that E is a state function of $\boldsymbol{\varepsilon}$ and S for the reversible adiabatic (isentropic, S =const.) processes, and according to the ordinary rules of differentiation, we get

$$\sigma_{ij} = \left.\frac{\partial E}{\partial \varepsilon_{ij}}\right|_S, \quad T = \left.\frac{\partial E}{\partial S}\right|_\varepsilon. \tag{2.111}$$

On the other hand, if the process is isothermal (T =const.), the dependence of the stresses $\boldsymbol{\sigma}$ on the strains $\boldsymbol{\varepsilon}$ is found by introducing the *Helmholtz free-energy function*:

$$F = E - TS, \quad dF = \sigma_{ij} \; d\varepsilon_{ij} - S \; dT \tag{2.112}$$

with the strains and temperature as its independent variables. The value $-TS$ is the irreversible heat energy due to entropy as related to temperature, with the negative sign indicating that the compressive reaction results from thermal expansion in a restrained body. Then, from Eqs. (2.110) and (2.112), we get

$$\sigma_{ij} = \frac{\partial F}{\partial \varepsilon_{ij}}\bigg|_T, \quad S = -\frac{\partial F}{\partial T}\bigg|_\varepsilon. \tag{2.113}$$

Thus, the strain energy density w can be identified with the internal energy E in an isentropic process ($S \equiv$const.) and the free energy F in an isothermal process ($T \equiv$const.). The temperature is determined by Eq. (4.3_2) when using the potential E and the entropy is determined by Eq. (2.113) when using the potential F.

For small deformations and small temperature changes $\theta = T - T^0$ ($\theta/T_0 \ll 1$), $F(\boldsymbol{\varepsilon}, T)$ can be expanded in a power series of its arguments in the neighborhood of the virgin state ($\boldsymbol{\varepsilon} = \mathbf{0}$, $\theta = 0$):

$$F(\boldsymbol{\varepsilon}, T) = \frac{1}{2} L_{ijkl}\varepsilon_{ij}\varepsilon_{kl} + \alpha^T_{ij}\varepsilon_{ij}\theta - \frac{C_\varepsilon}{2T_0}\theta^2, \tag{2.114}$$

where one introduces the notations

$$L_{ijkl} = \frac{\partial^2 F(\mathbf{0}, T_0)}{\partial \varepsilon_{ij}\partial \varepsilon_{kl}}, \quad \alpha^T_{ij} = \frac{\partial^2 F(\mathbf{0}, T_0)}{\partial \varepsilon_{ij}\partial T}, \quad \frac{C_\epsilon}{T_0} = -\frac{\partial^2 F(\mathbf{0}, T_0)}{\partial T^2} \tag{2.115}$$

and C_ε denotes the specific heat at constant strain, which can be also determined as

$$C_\epsilon = \frac{dQ}{dT}\bigg|_\epsilon = \frac{\partial E}{\partial T}\bigg|_\epsilon = T\frac{\partial S}{\partial T}\bigg|_\epsilon. \tag{2.116}$$

The negative sign for the last term on the right-hand side of Eq. (2.114) indicates that a temperature rise leads to a compressive reaction on the restricted body. This term signifies the thermal energy due to temperature while the second term $\boldsymbol{\alpha}^T\boldsymbol{\varepsilon}\theta$ describes the coupling effect between temperature and mechanical deformation. The difference between the isentropic moduli $\tilde{\mathbf{L}}$ and isothermic moduli $\mathbf{L}$:

$$\tilde{\mathbf{L}} - \mathbf{L} = \frac{T_0}{\rho C_\epsilon}\boldsymbol{\alpha}^T \otimes \boldsymbol{\alpha}^T, \quad \tilde{\mathbf{L}} = \frac{\partial^2 E}{\partial \boldsymbol{\varepsilon} \otimes \partial \boldsymbol{\varepsilon}}\bigg|_S, \quad \mathbf{L} = \frac{\partial^2 F}{\partial \boldsymbol{\varepsilon} \otimes \partial \boldsymbol{\varepsilon}}\bigg|_T$$

is of as order of 1% or less for metals and ceramics.

To determine the entropy as a function of $\boldsymbol{\varepsilon}$ and T, consider the total differential of the function $S(\boldsymbol{\varepsilon}, T)$:

$$dS = \frac{\partial S}{\partial \boldsymbol{\varepsilon}}\bigg|_T : d\boldsymbol{\varepsilon} + \frac{\partial S}{\partial T}\bigg|_\epsilon dT. \tag{2.117}$$

Taking Eqs. (2.113) and (2.116) into account and that

$$\frac{\partial \boldsymbol{\sigma}}{\partial T}\bigg|_\varepsilon = \frac{\partial^2 F}{\partial T \partial \boldsymbol{\varepsilon}}\bigg|_\varepsilon = \boldsymbol{\alpha}^T, \tag{2.118}$$

equation (2.117) becomes

$$dS = -\boldsymbol{\alpha}^T : d\boldsymbol{\varepsilon} + \frac{C_\varepsilon}{T}\, dT, \tag{2.119}$$

which after integration under the virgin state conditions $\boldsymbol{\varepsilon} = \mathbf{0}$ and $T = T_0$ gives, for small temperature change ($\theta/T_0 \ll 1$),

$$S = -\boldsymbol{\alpha}^T : \boldsymbol{\varepsilon} + \frac{C_\varepsilon}{T_0}\theta. \tag{2.120}$$

The Gibbs thermodynamic potential per unit volume is defined for the case where the stress and temperature are chosen as the independent state variables

$$G_e = F - TS, \quad dG_e = -\varepsilon_{ij}\, d\sigma_{ij} - S\, dT. \tag{2.121}$$

Since dG_e is a total differential (as dE and dF), it follows that

$$\left.\frac{\partial G_e}{\partial T}\right|_\sigma = -S, \quad \left.\frac{\partial G_e}{\partial \sigma_{ij}}\right|_T = -\varepsilon_{ij}. \tag{2.122}$$

Thus, $-G_e$ can be identified with the complementary energy function W^c, which has the property that $\partial W^c/\partial \sigma_{ij} = \varepsilon_{ij}$. In the neighborhood of the virgin state ($\theta/T_0 \ll 1$, $\mathbf{M} :: (\boldsymbol{\sigma} \otimes \boldsymbol{\sigma}) \ll 1$; $\mathbf{M}$ is the compliance tensor) we have a power series expansion

$$G_e(\boldsymbol{\sigma}, T) = -\frac{1}{2} M_{ijkl}\sigma_{ij}\sigma_{kl} - \beta^T_{ij}\sigma_{ij}\theta + \frac{1}{2} C_\sigma \frac{\theta^2}{T_0} \tag{2.123}$$

depending on the coefficient of thermal expansion $\boldsymbol{\beta}^T$ and the specific heat at constant stress C_σ

$$\boldsymbol{\beta}^T \equiv \left.\frac{\partial \boldsymbol{\varepsilon}}{\partial T}\right|_\sigma = -\mathbf{M}\boldsymbol{\alpha}^T, \quad C_\sigma = \left.\frac{dQ}{dT}\right|_\sigma, \tag{2.124}$$

where C_σ is related to C_ϵ by the equation

$$C_\sigma = C_\varepsilon + L_{ijkl}\boldsymbol{\beta}^T_{ij}\beta^T_{kl}T_0. \tag{2.125}$$

Analogously to the tensor $\mathbf{L}$ (2.90), the complience tensor $\mathbf{M}$ is also full-symmetric.

Substitution of the representation (2.114) into Eq. (2.113_1) yields the general *Duhamel-Newmann form of Hook's law* for an anisotropic body in the form of invertible relations:

$$\boldsymbol{\sigma} = \mathbf{L}\boldsymbol{\varepsilon} + \boldsymbol{\alpha}, \quad \boldsymbol{\varepsilon} = \mathbf{M}\boldsymbol{\sigma} + \boldsymbol{\beta} \tag{2.126}$$

where $\boldsymbol{\alpha} \equiv \boldsymbol{\alpha}^T\theta$, $\boldsymbol{\beta}^\top = \boldsymbol{\beta}/\theta$ is the coefficient of thermal expansion related to $\boldsymbol{\alpha}$ by $\boldsymbol{\alpha} = -\mathbf{L}\boldsymbol{\beta}$, $\alpha_{ij} = -L_{ijkl}\beta_{kl}$; see Eq. (2.124). $\boldsymbol{\beta}(\mathbf{x})$ and $\boldsymbol{\alpha}(\mathbf{x}) \equiv -\mathbf{L}(\mathbf{x})\boldsymbol{\beta}(\mathbf{x})$ are second-order tensors of local eigenstrains and eigenstresses, respectively (frequently called transformation fields) which may arise by thermal expansion, phase transformation, twinning, and other changes of shape or volume of the material. Substitution of the constitutive equation (2.126_1) into the motion equation (2.60) at small deformations (2.92) leads to the Navier equation:

$$\nabla(\mathbf{L}\nabla\mathbf{u})+\rho\mathbf{b}=\rho\ddot{\mathbf{u}}-\boldsymbol{\alpha}^T\nabla\theta,\quad \frac{\partial}{\partial x_j}\left(L_{ijkl}\frac{\partial u_k}{\partial x_l}\right)+\rho b_i=\rho\frac{\partial^2 u_j}{\partial t^2}-\alpha_{ij}^T\frac{\partial\theta}{x_j},\quad (2.127)$$

which should be considered simultaneously with the generalized heat condition equation:

$$\nabla(\boldsymbol{\lambda}\nabla\theta)-C_\varepsilon\dot{\theta}+T_0\boldsymbol{\alpha}^T:\dot{\varepsilon}=-h,\quad \frac{\partial}{\partial x_i}\left(\lambda_{ij}\frac{\partial}{\partial x_j}\theta\right)-C_\varepsilon\frac{\partial\theta}{\partial t}+T_0\alpha_{ij}^T\frac{\partial\varepsilon_{ij}}{\partial t}=-h, \quad (2.128)$$

where h is the heat source function. The coefficient $\boldsymbol{\alpha}^T$ couples the equations of the dynamic thermoelastic problem (2.127) and (2.128). Dropping the term $T_0\boldsymbol{\alpha}^T:\dot{\varepsilon}$ in the left-hand side of Eq. (2.128_1), we obtain the uncoupled equations of dynamic thermoelasticity. Further neglecting the inertial term $\rho\ddot{\mathbf{u}}$ in Eq. (2.127) leads to a system of equations of the quasistatic thermoelastic problem.

The system (2.127) and (2.128) should be accompanied by appropriate boundary and initial conditions. The initial conditions specify the temperature, displacements, and velocities at all points $\mathbf{x}\in\mathcal{E}$ within $\mathcal{E}$: $\theta(\mathbf{x},0)=\theta_0(\mathbf{x})$, $\mathbf{u}(\mathbf{x},0)=\mathbf{u}_0(\mathbf{x})$, $\dot{\mathbf{u}}(\mathbf{x},0)=\mathbf{v}_0(\mathbf{x})$ with the prescribed functions $\theta_0,\mathbf{u}_0,\mathbf{v}_0$ on $\mathcal{E}$. The most widely used boundary condition for the temperature function θ is expressed by

$$\boldsymbol{\lambda}:(\nabla\theta(\mathbf{x},t)\otimes\mathbf{n})+\alpha_{\partial\mathcal{E}}[\theta(\mathbf{x},t)-\theta_{\partial\mathcal{E}}(\mathbf{x},t)]=q_0(\mathbf{x},t) \quad (2.129)$$

with $\theta_{\partial\mathcal{E}}$, q_0 as known functions of the time and point $\mathbf{x}\in\partial\mathcal{E}$. If $q_0\equiv 0$, Eq. (2.129) describes heat exchange of the third kind called Newton's law. The conditions $\alpha_{\partial\mathcal{E}}=\infty$ and $q_0\equiv 0$ reduce Eq. (2.129) to the boundary conditions of the first kind. We obtain the second kind of boundary conditions if $\alpha_{\partial\mathcal{E}}\equiv 0$. The mechanical boundary conditions (2.96) and (2.97) should be added to the thermal boundary conditions (2.129). In so doing, the traction $\mathbf{t}(\mathbf{x})=\boldsymbol{\sigma}(\mathbf{x})\mathbf{n}(\mathbf{x})$ acting on any surface with the normal $\mathbf{n}(\mathbf{x})$ through the point $\mathbf{x}$ can be represented in terms of displacements:

$$\mathbf{t}(\mathbf{x})=\hat{\mathbf{t}}(\mathbf{n},\nabla)\mathbf{u}(\mathbf{x})+\boldsymbol{\alpha}\mathbf{n},\quad \hat{t}_{ik}(\mathbf{n},\nabla)\equiv L_{ijkl}n_j(\mathbf{x})\partial/\partial x_l, \quad (2.130)$$

where $\hat{t}_{ik}(\mathbf{n},\nabla)$ is the so-called "stress operator."

We will consider the uncoupled quasistatic thermoelasticity theory [$\rho\ddot{\mathbf{u}}\equiv\mathbf{0}$ in Eq. (2.127) and $T_0\boldsymbol{\alpha}^T:\dot{\varepsilon}\equiv\mathbf{0}$ in Eq. (2.128)], where the temperature field is determined by Eq. (2.128) with no influence of the latent heat due to the change of strain. It takes place according to so-called *body force analogy* [197] asserting that $\{\mathbf{u},\varepsilon,\boldsymbol{\sigma}\}$ is a solution of the mixed problem of thermoelastostatics corresponding to the external loading conditions $(\mathbf{b},\mathbf{t}^{\partial\mathcal{E}_\sigma},\mathbf{u}^{\partial\mathcal{E}_\mathbf{u}},\theta)$ on $\mathcal{E}\cup\partial\mathcal{E}_\sigma\cup\partial\mathcal{E}_\mathbf{u}$ if and only if $\{\mathbf{u},\varepsilon,\boldsymbol{\sigma}-\boldsymbol{\alpha}\}$ is a solution of the mixed problem of elastostatic corresponding to the external loading $(\mathbf{b}+1/\rho\nabla\boldsymbol{\alpha},\mathbf{t}^{\partial\mathcal{E}_\sigma}-\boldsymbol{\alpha}\cdot\mathbf{n},\mathbf{u}^{\partial\mathcal{E}_\mathbf{u}})$ on $\mathcal{E}\cup\partial\mathcal{E}_\sigma\cup\partial\mathcal{E}_\mathbf{u}$.

To present the extremum principles of linear thermoelastostatic, determine the total strain energy $W(\varepsilon)$ and total stress energy $W^c(\boldsymbol{\sigma})$:

$$W(\varepsilon)=\int_{\mathcal{E}}w(\varepsilon)\,d\mathbf{x},\quad w(\varepsilon)=\frac{1}{2}(\varepsilon-\boldsymbol{\beta}):\mathbf{L}:(\varepsilon-\boldsymbol{\beta}),$$

$$W^c(\boldsymbol{\sigma})=\int_{\mathcal{E}}w^c(\boldsymbol{\sigma})\,d\mathbf{x},\quad w^c(\boldsymbol{\sigma})=\frac{1}{2}(\boldsymbol{\sigma}-\boldsymbol{\alpha}):\mathbf{M}:(\boldsymbol{\sigma}-\boldsymbol{\alpha}) \quad (2.131)$$

yielding Eqs. (2.102) at $\boldsymbol{\alpha} = \boldsymbol{\beta} \equiv \mathbf{0}$. $w(\boldsymbol{\varepsilon})$ and $w^c(\boldsymbol{\sigma})$ are potential functions for determining the stress and strain according to the equations

$$\sigma_{ij} = \frac{\partial w(\boldsymbol{\varepsilon})}{\partial \varepsilon_{ij}}, \quad \varepsilon_{ij} = \frac{\partial w^c(\boldsymbol{\sigma})}{\partial \sigma_{ij}}, \tag{2.132}$$

respectively, which are valid for both elasticity and thermoelasticity; in linear theory $w(\boldsymbol{\varepsilon})$ and $w^c(\boldsymbol{\sigma})$ are equal: $w = w^c$. Positiveness of w and w^c (at $\boldsymbol{\varepsilon}, \boldsymbol{\sigma} \neq \mathbf{0}$) leads to the positive definiteness of $\mathbf{L}$ and $\mathbf{M}$, respectively (i.e. $\boldsymbol{\varepsilon} : \mathbf{L} : \boldsymbol{\varepsilon} > 0$ for $\forall \boldsymbol{\varepsilon} \neq \mathbf{0}$).

The classical extremum principles (2.106) and (2.108) for pure mechanical loading can be generalized for the thermostatic case in the following manner. We determine the potential Π and complementary Π^c energies:

$$\begin{aligned} \Pi(\tilde{s}^k) &= W(\tilde{\varepsilon}) - \int_{\mathcal{E}} \rho \mathbf{b} \cdot \tilde{\mathbf{u}} \, d\mathbf{x} - \int_{\partial\mathcal{E}_\sigma} \mathbf{t}^{\partial\mathcal{E}_\sigma} \cdot \tilde{\mathbf{u}} \, ds, \\ \Pi^c(\tilde{s}^s) &= W^c(\tilde{\boldsymbol{\sigma}}) - \int_{\partial\mathcal{E}_\mathbf{u}} \tilde{\mathbf{t}}_\mathbf{n} \cdot \mathbf{u}^{\partial\mathcal{E}_\mathbf{u}} \, ds, \quad \tilde{\mathbf{t}}_\mathbf{n} \equiv \tilde{\boldsymbol{\sigma}} \cdot \mathbf{n}^{\partial\mathcal{E}}. \end{aligned} \tag{2.133}$$

defined on the sets of kinematically $\tilde{s}^k = \{\tilde{\mathbf{u}}, \tilde{\varepsilon}, \tilde{\boldsymbol{\sigma}}\} \in \mathcal{A}^k(\tilde{s})$ (2.100) and statically $\tilde{s}^s = \{\tilde{\mathbf{u}}, \tilde{\varepsilon}, \tilde{\boldsymbol{\sigma}}\} \in \mathcal{A}^s(\tilde{s})$ (2.101) admissible states, respectively. In so doing one assumes that Eqs. (2.94) is replaced by Eq. (2.126$_1$) to include the $\tilde{\boldsymbol{\sigma}}$ and $\tilde{\varepsilon}$ generated by $\tilde{\mathbf{u}} \in \mathcal{A}^k(\tilde{\mathbf{u}})$ and by $\tilde{\boldsymbol{\sigma}} \in \mathcal{A}^s(\tilde{\boldsymbol{\sigma}})$, respectively. Then the actual state $s = \{\mathbf{u}, \boldsymbol{\varepsilon}, \boldsymbol{\sigma}\}$ renders the potential energy Π and complementary energy Π^c an absolute minimum

$$\Pi(s) \leq \Pi(\tilde{s}^k), \quad \Pi^c(s) \leq \Pi^c(\tilde{s}^s). \tag{2.134}$$

2.4.2 Electroelastic Equations

Another example of the theory of coupled fields is associated with electroelasticity and deals with phenomena caused by interactions between the elastic, electric, and thermal fields. The relations between the mechanical, electrical, and thermal properties can be demonstrated from equilibrium thermodynamics described by the potential *Gibbs function*:

$$G_e = E - \mathbf{E} \cdot \mathbf{D} - ST, \tag{2.135}$$

whose differential form

$$dG_e = \sigma_{ij} d\varepsilon_{ij} - D_i dE_i - S \, dT \tag{2.136}$$

yields the relations

$$\sigma_{ij} = \left. \frac{\partial G_e}{\partial \varepsilon_{ij}} \right|_{\mathbf{E},T}, \quad D_m = -\left. \frac{\partial G_e}{\partial E_m} \right|_{\varepsilon,T}, \quad S = -\left. \frac{\partial G_e}{\partial T} \right|_{\varepsilon,\mathbf{E}}. \tag{2.137}$$

Differentiating Eqs. (2.137) gives

$$\left.\frac{\partial\sigma_{ij}}{\partial E_m}\right|_{\varepsilon,T} = -\left.\frac{\partial D_m}{\partial\varepsilon}\right|_{\mathbf{E},T}, \quad \left.\frac{\partial\sigma_{ij}}{\partial T}\right|_{\varepsilon,\mathbf{E}} = -\left.\frac{\partial S}{\partial\varepsilon_{ij}}\right|_{\mathbf{E},T}, \quad \left.\frac{\partial D_m}{\partial T}\right|_{\varepsilon,\mathbf{E}} = \left.\frac{\partial S}{\partial E_m}\right|_{\varepsilon,T}. \tag{2.138}$$

It was assumed that the strains $\boldsymbol{\varepsilon}$, the electric field intensity $\mathbf{E}$, and the temperature $T = T_0 + \theta$ are considered the independent variables, and the dependent variables will be the stresses $\boldsymbol{\sigma}$, the vector of induction $\mathbf{D}$, and the entropy S. G_e is a state function of $\boldsymbol{\varepsilon}$, $\mathbf{E}$, and T, and the ordinary rules of differentiation yield the explicite representation of $d\boldsymbol{\sigma}$, $d\mathbf{D}$, and dS as the functions of $d\boldsymbol{\varepsilon}$, $d\mathbf{E}$, and dT:

$$\begin{aligned}
d\sigma_{ij} &= \left.\frac{\partial\sigma_{ij}}{\varepsilon_{kl}}\right|_{\mathbf{E},T} d\varepsilon_{kl} + \left.\frac{\partial\sigma_{ij}}{E_m}\right|_{\varepsilon,T} dE_m + \left.\frac{\partial\sigma_{ij}}{\partial T}\right|_{\mathbf{E},\varepsilon} dT,\\
dD_m &= \left.\frac{\partial D_m}{\varepsilon_{kl}}\right|_{\mathbf{E},T} d\varepsilon_{kl} + \left.\frac{\partial D_m}{E_k}\right|_{\varepsilon,T} dE_k + \left.\frac{\partial D_m}{\partial T}\right|_{\varepsilon,\mathbf{E}} dT,\\
dS &= \left.\frac{\partial S}{\varepsilon_{kl}}\right|_{\mathbf{E},T} d\varepsilon_{kl} + \left.\frac{\partial S}{E_m}\right|_{\varepsilon,T} dE_m + \left.\frac{\partial S}{\partial T}\right|_{\varepsilon\mathbf{E}} dT,
\end{aligned} \tag{2.139}$$

For small variations of $\boldsymbol{\varepsilon}$, $\mathbf{E}$, and T in the neighborhood of virgin state $(\boldsymbol{\varepsilon}, \mathbf{E}, \theta = \mathbf{0})$, the partial derivatives can be considered as constants and Eqs. (2.139) becomes integrable. This allows us to obtain the constitutive relations of a deformable piezoelectric medium (see, e.g., [747], [861]):

$$\begin{aligned}
\sigma_{ij} &= L_{ijkl}\varepsilon_{kl} + e_{kij}E_k + \alpha^T_{ij}\theta,\\
D_i &= e_{ijk}\varepsilon_{jk} - k_{ij}E_j + p_i\theta, \quad dS = \alpha_{ij}\varepsilon_{ij} + p_iE_i + c_\varepsilon\theta/T_0,
\end{aligned} \tag{2.140}$$

where the coefficients of the system (2.140) denote the particular derivatives:

$$\begin{aligned}
L_{ijkl} &= \left.\frac{\partial\sigma_{ij}}{\partial\varepsilon_{kl}}\right|_{\mathbf{E},T}, \quad & e_{ijk} &= \left.\frac{\partial\sigma_{ij}}{\partial E_k}\right|_{\varepsilon,T},\\
p_i &= \left.\frac{\partial D_i}{\partial T}\right|_{\varepsilon,\mathbf{E}}, \quad & k_{ij} &= \left.\frac{\partial D_i}{\partial E_j}\right|_{\varepsilon,T},\\
\alpha^T_{ij} &= \left.\frac{\partial\sigma_{ij}}{\partial T}\right|_{\varepsilon,\mathbf{E}}, \quad & \frac{c_\varepsilon}{T_0} &= \left.\frac{\partial S}{\partial T}\right|_{\varepsilon,\mathbf{E}}
\end{aligned} \tag{2.141}$$

The constitutive equations (2.140_1) and (2.140_2) can be inverted

$$\varepsilon_{ij} = M_{ijkl}\sigma_{kl} + d_{kij}D_k + \beta^T_{ij}\theta, \tag{2.142}$$

$$E_i = d_{ijk}\sigma_{jk} - b_{ij}D_j + q_i\theta. \tag{2.143}$$

The coefficients of Eqs. (2.140) and (2.142), (2.143) are denoted as follows: $\mathbf{L}$ and $\mathbf{M}$ are the elastic and compliance tensors, $\mathbf{k}$ and $\mathbf{b}$ are the tensors of dielectric permeability and impermeability, $\mathbf{e}$ and $\mathbf{d}$ are the piezoelectric moduli, $\boldsymbol{\alpha}^T$ and $\boldsymbol{\beta}^T$ are the coefficients of thermoelastic stress and expansion, $\mathbf{p}$ and $\mathbf{q}$ are the pyroelectric coefficients; and c_ε is a specific heat per unit volume at constant strain. To obtain a symmetric matrix of coefficients we replace the electric field $\mathbf{E}$ by $-\mathbf{E}$. Hereinafter in Subsection 2.4.3, the Latin indexes range from 1 to 3, and the Greek ones range from 1 to 4. We will assume that the electric and elastic fields are fully coupled, but temperature enters the problem only as a parameter through the constitutive equation. The tensor coefficients in the above equations are related as $e_{kij} = d_{kmn}L_{mnij}$, $d_{kij} = e_{kmn}M_{mnij}$, $M_{ijmn}L_{mnkl} = I_{ijkl}$. We will use the matrix notation instead of the tensor one, as it is accepted in the theory of elasticity. For notational convenience the elastic and electric variable will be treated on equal footing, and with this in mind we recast the local linear constitutive relations of thermoelectroelasticity for this material in the notation introduced in [38]. For this purpose, we introduce the matrices of generalized stresses $\boldsymbol{\Sigma}$ and strains $\boldsymbol{\mathcal{E}}$ $(i, j = 1, 2, 3)$:

$$\boldsymbol{\Sigma} = \begin{pmatrix} \boldsymbol{\sigma} \\ \mathbf{D} \end{pmatrix}, \quad \Sigma_{ij} = \sigma_{ij}, \quad \Sigma_{i4} = D_i, \tag{2.144}$$

$$\boldsymbol{\mathcal{E}} = \begin{pmatrix} \boldsymbol{\varepsilon} \\ \mathbf{E} \end{pmatrix}, \quad \mathcal{E}_{ij} = \varepsilon_{ij}, \quad \mathcal{E}_{4i} = E_i, \tag{2.145}$$

which are interrelated through the matrices of generalized elastic coefficients $\mathbb{L}_{\alpha\beta\gamma\delta}$, compliance $\mathbb{M}_{\alpha\beta\gamma\delta}$, and generalized coefficients of thermal expansion $\Lambda_{\alpha\beta}$:

$$\boldsymbol{\Sigma} = \mathbb{L}(\boldsymbol{\mathcal{E}} - \boldsymbol{\Lambda}), \quad \boldsymbol{\mathcal{E}} = \mathbb{M}\boldsymbol{\Sigma} + \boldsymbol{\Lambda}, \tag{2.146}$$

where the generalized coefficients can be presented in matrix form:

$$\mathbb{L} = \begin{pmatrix} \mathbf{L} & \mathbf{e}^\top \\ \mathbf{e} & -\mathbf{k} \end{pmatrix}, \quad \mathbb{M} = \begin{pmatrix} \mathbf{M} & \mathbf{d}^\top \\ \mathbf{d} & -\mathbf{b} \end{pmatrix}, \quad \boldsymbol{\Lambda} = \begin{pmatrix} \boldsymbol{\beta}^T\theta \\ \mathbf{q}\theta \end{pmatrix} \tag{2.147}$$

as well as in component form:

$$\mathbb{L}_{ijkl} = L_{ijkl}, \;\; \mathbb{L}_{4ikl} = \mathbb{L}_{i4kl} = e_{ikl}, \tag{2.148}$$

$$\mathbb{L}_{ij4l} = \mathbb{L}_{ijl4} = e_{lij}, \; \mathbb{L}_{44kl} = \mathbb{L}_{kl44} = \; k_{kl}, \tag{2.149}$$

$$\mathbb{M}_{ijkl} = M_{ijkl}, \; \mathbb{M}_{4ikl} = \mathbb{M}_{i4kl} = d_{ikl}, \tag{2.150}$$

$$\mathbb{M}_{ij4l} = \mathbb{M}_{ijl4} = d_{lij}, \; \mathbb{M}_{44kl} = \mathbb{M}_{kl44} = -b_{kl}, \tag{2.151}$$

$$\Lambda_{ij} = \beta_{ij}\theta, \; \Lambda_{4i} = \Lambda_{i4} = q_i\theta \tag{2.152}$$

The generalized static equation and the Cauchy conditions of small generalized deformations are expressed in the form:

$$\nabla_\alpha \Sigma_{\alpha\beta} = f_\beta, \;\; \nabla_4 \equiv 0, \tag{2.153}$$

$$\mathcal{E}_{\alpha\beta} = (\nabla_\alpha U_\beta + \nabla_\beta U_\alpha)/2, \tag{2.154}$$

where $U = (u_1, u_2, u_3, -\phi)^\top$, ϕ is the electrostatic potential, and f_β is a density of generalized body forces. Equations (2.153) and (2.154) should be complemented by the boundary conditions of the first or the second type. Except for notations, these equations coincide with the equations of linear thermoelasticity (2.126). Because of this the theory of piezoelectric composite materials (PCM) retraces at a particular instant the path of development of the theory of microinhomogeneous elastic media, exhibiting substantial progress. In light of the analogy mentioned in our brief survey we will not consider in detail the PCM; one may refer instead to the appropriate scheme of thermoelastic composites.

2.4.3 Matrix Representation of Some Symmetric Tensors

Many material (e.g., elastic moduli (2.90)) and field (e.g., the stresses (2.50)) tensors have internal symmetry with respect to pairs of indices. Although all calculations with these tensors can be done by common methods of tensor analysis (see Section 2.2), it is sometimes convenient to introduce a contracted notation significantly simplifying the calculations. It is possible to form six different pairs ij for permuted indices i, j $(i, j = 1, 2, 3)$. Then the pair of these Latin indices ij can be substituted by a single Greek index α ranging from 1 to 6 according to the Voight notation:

$$\begin{array}{ll} \text{tensor notation } ij = & 11\ \ 22\ \ 33\ \ 23\ 32\ \ 13\ 31\ \ 12\ 21 \\ \text{matrix notation } \alpha = & \ 1\ \ \ \ 2\ \ \ \ 3\ \ \ \ \ \ 4\ \ \ \ \ \ \ \ 5\ \ \ \ \ \ \ \ 6 \end{array}, \tag{2.155}$$

which can be presented in a formula form (no sum on i, j):

$$\alpha = i\delta_{ij} + (9 - i - j)(1 - \delta_{ij}). \tag{2.156}$$

To avoid confusion, it should be mentioned that the prescribed reduction rule $ij \leftrightarrow \alpha$ between the Latin and Greek indices is also used for Latin indices $ij \leftrightarrow k$ $(i, j = 1, 2, 3,\ k = 1, \ldots, 6)$ where the correspondence between the pair indices and one index is provided by the scheme (2.155). We will also use this traditional form of an accordance.

We accept the convention that for the tensors with the prescribed aforementioned symmetry, Greek indices α, β, γ range from 1 to 6 and are connected with the pairs of Latin indices ij, kl and others by relation (2.155). In such a case, the tensor representations of the thermoelastic constitutive equations (2.126) can be presented in matrix notations:

$$\sigma_\gamma = L_{\gamma\delta}\varepsilon_\delta + \alpha_\gamma, \quad \varepsilon_\gamma = M_{\gamma\delta}\sigma_\delta + \beta_\gamma. \tag{2.157}$$

It is also required that the tensor representations of widely used scalar productions $\mathbf{L} : \mathbf{M}$ and $\boldsymbol{\sigma} : \boldsymbol{\varepsilon}$ have an appropriate vector form $L_{\lambda\mu}M_{\mu_\nu} = \delta_{\lambda\nu}$, $\sigma_{ij}\varepsilon_{ij} = \sigma_\lambda\varepsilon_\lambda$ with the standard assumption of summation over the repeated indices. However, it is conceivable to show that if the quantities corresponding to the different tensors are formed by one and the same rule (such as e.g. $\sigma_{ij} = \sigma_\lambda$), the equalities either (2.156) or (2.157) cannot be fulfilled. The mentioned inconsistency can be avoided if the following matrix representations of tensors are used $(i, j, k, l = 1, 2, 3;\ \lambda, \mu = 1, \ldots, 6)$

$$\varepsilon_{ij} = \frac{\varepsilon_\lambda}{2 - \delta_{ij}}, \quad \sigma_{ij} = \sigma_\lambda, \tag{2.158}$$

$$L_{ijkl} = L_{\lambda\mu}, \quad M_{ijkl} = \frac{M_{\lambda\mu}}{(2 - \delta_{ij})(2 - \delta_{kl})}. \tag{2.159}$$

In so doing, the transition to Greek indices for all symmetric second-rank tensors could be done in the unique manner $\alpha_{ij} = (2 - \delta_{ij})^{-1/2}\alpha_\lambda$.

Since the components of the stress and strain tensors are functions of the orientation of the reference system, the elastic coefficients in Eqs. (2.126_1) (and, analogously, in Eq. (2.126_2)) are also tensor functions of this orientation and they can be presented in a new coordinate system:

$$\mathbf{L}' = \mathbf{g}\mathbf{g}\mathbf{L}\mathbf{g}^\top\mathbf{g}^\top, \ L'_{ijkl} = g_{im}g_{jn}g_{ko}g_{lp}L_{mnop} \tag{2.160}$$

with the relevant transformation for the strain tensors:

$$\boldsymbol{\varepsilon}' = \mathbf{g}\mathbf{g}\boldsymbol{\varepsilon}, \ \ \varepsilon'_{ij} = g_{ik}g_{jl}\varepsilon_{kl}. \tag{2.161}$$

These tensors as well as the thermal expansion and thermoelasticity coefficients are transformed using Voight reduced components as

$$\mathbf{L}' = \mathbf{q}^L\mathbf{L}(\mathbf{q}^L)^\top, \quad \mathbf{M}' = \mathbf{q}^R\mathbf{M}(\mathbf{q}^R)^\top, \tag{2.162}$$

$$\boldsymbol{\sigma}' = \mathbf{q}^L\boldsymbol{\sigma}, \quad \boldsymbol{\varepsilon}' = \mathbf{q}^R\boldsymbol{\varepsilon}, \quad \boldsymbol{\alpha}^{T\prime} = \mathbf{q}^L\boldsymbol{\alpha}^T, \quad \boldsymbol{\beta}^{T\prime} = \mathbf{q}^R\boldsymbol{\beta}^T. \tag{2.163}$$

where the fourth-order tensors $\mathbf{q}^L$ and $\mathbf{q}^R = \left[(\mathbf{q}^L)^{-1}\right]^\top$ presented in the form of (6×6) transformation matrices:

$$\mathbf{q}^L = \begin{pmatrix} \mathbf{K}_1 & 2\mathbf{K}_2 \\ \mathbf{K}_3 & \mathbf{K}_4 \end{pmatrix}, \quad \mathbf{q}^R = \begin{pmatrix} \mathbf{K}_1 & \mathbf{K}_2 \\ 2\mathbf{K}_3 & \mathbf{K}_4 \end{pmatrix}, \tag{2.164}$$

are combined by four matrices (3×3):

$$\mathbf{K}_1 = \begin{pmatrix} g_{11}^2 & g_{12}^2 & g_{13}^2 \\ g_{21}^2 & g_{22}^2 & g_{23}^2 \\ g_{31}^2 & g_{32}^2 & g_{33}^2 \end{pmatrix}, \quad \mathbf{K}_2 = \begin{pmatrix} g_{12}g_{13} & g_{13}g_{11} & g_{11}g_{12} \\ g_{22}g_{23} & g_{23}g_{21} & g_{21}g_{22} \\ g_{32}g_{33} & g_{33}g_{31} & g_{31}g_{32} \end{pmatrix},$$

$$\mathbf{K}_3 = \begin{pmatrix} g_{21}g_{31} & g_{22}g_{32} & g_{23}g_{33} \\ g_{31}g_{11} & g_{32}g_{121} & g_{33}g_{13} \\ g_{11}g_{21} & g_{12}g_{22} & g_{13}g_{23} \end{pmatrix},$$

$$\mathbf{K}_4 = \begin{pmatrix} g_{22}g_{33} + g_{23}g_{32} & g_{23}g_{31} + g_{21}g_{33} & g_{21}g_{32} + g_{22}g_{31} \\ g_{32}g_{13} + g_{33}g_{12} & g_{33}g_{11} + g_{31}g_{13} & g_{31}g_{12} + g_{32}g_{11} \\ g_{12}g_{23} + g_{13}g_{22} & g_{13}g_{21} + g_{11}g_{23} & g_{11}g_{22} + g_{12}g_{21} \end{pmatrix}. \tag{2.165}$$

The elements of the matrices $\mathbf{q}^L$ and $\mathbf{q}^R$ can be represented by means of the formulae

$$\frac{1}{2}(g_{i'k}g_{j'l} + g_{i'l}g_{j'k}) = \frac{q^R_{\lambda'\mu}}{2 - \delta_{i'j'}} = \frac{q^L_{\lambda'\mu}}{2 - \delta_{kl}}, \tag{2.166}$$

where the index λ' corresponds to the pair $i'j'$, and μ to the pair kl.

The equations describing the piezoelectric effect (2.142) and (2.143) can be presented in the matrix form:

$$\varepsilon_\gamma = M_{\gamma\delta}\sigma_\delta + d_{\gamma i}E_i + \lambda_\gamma, \quad D_i = e_{i\gamma}\varepsilon_\gamma - k_{ij}E_j + p_i\theta, \tag{2.167}$$

which leads to the following recalculation rules for the coefficients ($jk \leftrightarrow \gamma$): $d_{ijk} = (2 - \delta_{jk})^{-1}d_{i\gamma}$, $\ e_{ijk} = e_{i\gamma}$ with Latin indices ranging from 1 to 3. The transformation formulae for quantities $d_{i\gamma}$ and $e_{i\gamma}$ in the old $\mathbf{e}_k$ and new $\mathbf{e}'_k$ $(k = 1, 2, 3)$ coordinate systems with the cosine matrix g_{ij} (2.4) are

$$d'_{i\gamma} = g_{ij}q^R_{\gamma\delta}d_{j\delta}, \ \ d_{j\delta} = g^\top_{ji}q^L_{\delta\gamma}d'_{i\gamma}, \ \ e'_{i\gamma} = g_{ij}q^L_{\gamma\delta}e_{j\delta}, \ \ e_{j\delta} = g^\top_{ji}q^R_{\delta\gamma}e'_{i\gamma}. \tag{2.168}$$

The recalculation rules and the transformation formulae can be obtained by the same method for other coefficients characterizing piezoelectric properties of crystals (for details see [1015]).

2.5 Symmetry of Elastic Properties

Under an orthogonal transformation (2.6) the elastic stiffnesses L'_{ijkl} referred to the $\mathbf{e}'_i$ coordinate system are described by Eq. (2.160). When $\mathbf{L}' = \mathbf{L}$, i.e.,

$$L_{ijkl} = g_{ip}g_{jq}g_{kr}g_{st}L_{pqrt}, \tag{2.169}$$

the material is said to possesses a *symmetry* with respect to $\mathbf{g}$, which is called a symmetry transformation. If $\mathbf{g}$ is a symmetry transformation, $\mathbf{g}^{-1} = \mathbf{g}^\top$ is also a symmetry transformation, and if $\mathbf{g}$ and $\mathbf{q}$ are two symmetry transformations then $\mathbf{gq}$ is also a symmetry transformation. The set of all symmetry transformations at $\mathbf{x} \in \mathcal{E}$ forms a symmetry group $S_\mathbf{x}$ with the unit transformation $\boldsymbol{\delta}$. The group $S_\mathbf{x}$ always contains the two-element subgroup $\{\boldsymbol{\delta}, -\boldsymbol{\delta}\}$. We describe by $\mathbf{g}(\omega\mathbf{n})$ (A.1.2) a right-handed rotation by the angle ω, $0 < \omega < \pi$, about an axis oriented in the direction of unit vector $\mathbf{n}$. The transformation $\mathcal{R}(\mathbf{n}) = -\mathbf{g}(\pi\mathbf{n})$ is called a reflection in the plane $P(\mathbf{n})$ with the normal unit vector $\mathbf{n}$. A unit vector $\mathbf{n}$ is called an axis of symmetry at $\mathbf{x}$ if $\mathbf{gn} = \mathbf{n}$ for some $\mathbf{g} \in S_\mathbf{x}$, with $\mathbf{g} \neq \boldsymbol{\delta}$. A plane $P(\mathbf{n}_1, \mathbf{n}_2)$, spanned by two mutually orthogonal unit vectors $\mathbf{n}_1$ and $\mathbf{n}_2$, is called a plane of symmetry at $\mathbf{x}$ if $\mathbf{gn}_1 = \mathbf{n}_1$ and $\mathbf{gn}_2 = \mathbf{n}_2$, for some $\mathbf{g} \in S_\mathbf{x}$, with $\mathbf{g} \neq \boldsymbol{\delta}$. The condition symmetry (2.169) introduces a restriction on the elastic moduli $\mathbf{L}$. A *triclinic* material has a minimum symmetry group $S_\mathbf{x}$ formed by the two-element group $\{\boldsymbol{\delta}, -\boldsymbol{\delta}\}$. According to Eq. (2.90), the number of independent elastic coefficients for the general anisotropic linearly elastic triclinic materials is reduced from $3^4 = 81$ to 21, which is the number of terms on the main diagonal of the matrix $\mathbf{L}$ and the five subdiagonals above it: $6+5+4+3+2+1 = 21$. This number is reduced when the symmetry of the crystal class of medium is accounted for. A material is called *monoclinic* if for any $\mathbf{x} \in \mathcal{E}$ the symmetry grope $S_\mathbf{x}$ is formed by the transformations $\pm\boldsymbol{\delta}, \pm\mathbf{g}(\pi, \mathbf{n})$ for any $\mathbf{x} \in \mathcal{E}$. An *orthotropic* has the symmetry group $S_\mathbf{x}$ formed for any $\mathbf{x} \in \mathcal{E}$ by the transformations $\pm\boldsymbol{\delta}, \pm\mathbf{g}(\pi, \mathbf{n}_1), \pm\mathbf{g}(\pi, \mathbf{n}_2), \pm\mathbf{g}(\pi, \mathbf{n}_3)$, where $\mathbf{n}_1, \mathbf{n}_2, \mathbf{n}_3$ are three mutually orthogonal unit vectors. A material is called *transversally isotropic* with respect to the direction $\mathbf{n}$ if its symmetry group $S_\mathbf{x}$ $(\forall \mathbf{x} \in \mathcal{E})$ consists of the transformations $\pm\boldsymbol{\delta}$ and $\pm\mathbf{g}(\omega, \mathbf{n})$ with $0 < \omega < \pi$.

Depending on the number of rotations and/or reflection symmetry a crystal possesses, the 6×6 matrix $\mathbf{L}$ can be represented by one of the eight groups summarized in [821]. If the matrix $\mathbf{L}'$ referred to a different coordinate system $\mathbf{e}'_i$ is desired, the transformation (2.160) is used.

Let us consider polycrystalline aggregates of an orthorhombic system. Then for the classes of symmetry with nine independent elastic constants, the tensors of the elastic moduli and compliance of a crystallite in the crystallographic system of coordinates can be represented in the form (see, e.g., [995]):

$$L_{ijkl} = \sum_n \Big[\lambda_n\delta_{in}\delta_{jn}\delta_{kn}\delta_{ln} + \mu_n(\delta_{in}\delta_{jn}\delta_{kl} + \delta_{ij}\delta_{kn}\delta_{ln}) \\ + \nu_n(\delta_{in}\delta_{jk}\delta_{ln} + \delta_{jn}\delta_{ik}\delta_{ln} + \delta_{in}\delta_{jl}\delta_{kn} + \delta_{jn}\delta_{il}\delta_{kn})\Big], \tag{2.170}$$

$$M_{ijkl} = \sum_n \Big[p_n\delta_{in}\delta_{jn}\delta_{kn}\delta_{ln} + q_n(\delta_{in}\delta_{jn}\delta_{kl} + \delta_{ij}\delta_{kn}\delta_{ln})$$

$$+ r_n(\delta_{in}\delta_{jk}\delta_{ln} + \delta_{jn}\delta_{ik}\delta_{ln} + \delta_{in}\delta_{jl}\delta_{kn} + \delta_{jn}\delta_{il}\delta_{kn})\Big], \qquad (2.171)$$

where the elastic constants λ_n, μ_n, ν_n and p_n, q_n, r_n $(n = 1, 2, 3)$ are defined by the matrix representation of the elastic moduli L_{mn} and stiffness M_{ij} in the crystallographic system:

$$\begin{aligned}
\lambda_1 &= L_{11} + L_{23} + 2L_{44} - (L_{12} + L_{13} + 2L_{55} + 2L_{66}),\\
\lambda_2 &= L_{22} + L_{13} + 2L_{55} - (L_{12} + L_{23} + 2L_{44} + 2L_{66}),\\
\lambda_3 &= L_{33} + L_{12} + 2L_{66} - (L_{13} + L_{23} + 2L_{44} + 2L_{55}),\\
2\mu_1 &= L_{12} + L_{13} - L_{23}, \quad 2\nu_1 = L_{55} + L_{66} - L_{44},\\
2\mu_2 &= L_{12} + L_{23} - L_{13}, \quad 2\nu_2 = L_{44} + L_{66} - L_{55},\\
2\mu_3 &= L_{13} + L_{23} - L_{12}, \quad 2\nu_3 = L_{44} + L_{55} - L_{66},\\
p_1 &= M_{11} + M_{23} + M_{44}/2 - (2M_{12} + 2M_{13} + M_{55} + M_{66})/2,\\
p_2 &= M_{22} + M_{13} + M_{55}/2 - (2M_{12} + 2M_{23} + M_{44} + M_{66})/2,\\
p_3 &= M_{33} + M_{12} + M_{66}/2 - (2M_{13} + 2M_{23} + M_{44} + M_{55})/2,\\
2q_1 &= M_{12} + M_{13} - M_{23}, \quad 8r_1 = M_{55} + M_{66} - M_{44},\\
2q_2 &= M_{12} + M_{23} - M_{13}, \quad 8r_2 = M_{44} + M_{66} - M_{55},\\
2q_3 &= M_{13} + M_{23} - M_{12}, \quad 8r_3 = M_{44} + M_{55} - M_{66},
\end{aligned} \qquad (2.172)$$

In particular, for tetragonal crystals, the relations $L_{11} = L_{22}$, $L_{13} = L_{23}$, $L_{44} = L_{55}$ reduce the number of unknown elastic constants from nine to six:

$$\begin{aligned}
\lambda_1 &= \lambda_2 = L_{11} - (L_{12} + L_{66}),\\
\lambda_3 &= L_{33} + L_{12} + 2L_{66} - 2(L_{13} + 2L_{44}),\\
2\mu_1 &= 2\mu_2 = L_{12}, \quad 2\mu_3 = 2L_{13} - L_{12},\\
2\nu_1 &= 2\nu_2 = L_{66}, \quad 2\nu_3 = 2L_{44} - L_{66}.
\end{aligned} \qquad (2.173)$$

For hexagonal crystals,the addition condition $2L_{66} = L_{11} - L_{12}$ takes place substitution of which into Eq. (2.173) leads to

$$\begin{aligned}
\lambda_1 &= \lambda_2 = 0, \quad \lambda_3 = L_{11} + L_{33} - 2(L_{13} + 2L_{44}),\\
2\mu_1 &= 2\mu_2 = L_{12}, \quad 2\mu_3 = 2L_{13} - L_{12},\\
4\nu_1 &= 4\nu_2 = L_{11} - L_{12}, \quad 4\nu_3 = 4L_{44} + L_{12} - L_{11}.
\end{aligned} \qquad (2.174)$$

For cubic symmetry, substitution of the conditions $L_{22} = L_{33}$, $L_{12} = L_{13}$, $L_{44} = L_{66}$ into Eq. (2.173) gives

$$\begin{aligned}
\lambda_1 &= \lambda_2 = \lambda_3 = L_{11} - L_{12} - 2L_{44},\\
\mu_1 &= \mu_2 = \mu_3 = L_{12}/2, \quad \nu_1 = \nu_2 = \nu_3 = L_{44}/2.
\end{aligned} \qquad (2.175)$$

At last, a material is called elastically isotropic if its elastic properties are independent of direction. In the case of cubic crystal symmetry this is realized when $L_{44} = (L_{11} - L_{12})/2$. The deviation of the cubic lattice from an isotropic one can be quantified by the so-called *Zener anisotropy ratio*

$$Z = \frac{2L_{44}}{L_{11} - L_{12}}. \tag{2.176}$$

Substitution of an additional condition $2L_{44} = L_{11} - L_{12}$ for the isotropic medium into Eq. (2.175) yields

$$\begin{aligned}
&\lambda_1 = \lambda_2 = \lambda_3 = 0, \quad \mu_1 = \mu_2 = \mu_3 = L_{12}/2, \\
&\nu_1 = \nu_2 = \nu_3 = (L_{11} - L_{12})/2.
\end{aligned} \tag{2.177}$$

The representation of elastic moduli in the reference coordinate system (Ox', Oy', Oz') can be obtained by the use of rotation $\mathbf{g} \in O^3$ from the tensors $\mathbf{L}$ (2.170) and $\mathbf{M}$ (2.171) according to Eqs. (2.160) and (2.161), respectively. The final representation is simplified in exploiting of the identities $g_{ij}\delta_{jn} = g_{in}$ and $g_{ij}g_{kj} = g_{ik}$:

$$\begin{aligned}
L'_{ijkl} = \sum_n \Big[&\lambda_n g_{in} g_{jn} g_{kn} g_{ln} + \mu_n (g_{in} g_{jn} \delta_{kl} + \delta_{ij} g_{kn} g_{ln}) \\
&+ \nu_n (g_{in} \delta_{jk} g_{ln} + g_{jn} \delta_{ik} g_{ln} + g_{in} \delta_{jl} g_{kn} + g_{jn} \delta_{il} g_{kn}) \Big],
\end{aligned} \tag{2.178}$$

$$\begin{aligned}
M'_{ijkl} = \sum_n \Big[&p_n g_{in} g_{jn} g_{kn} g_{ln} + q_n (g_{in} g_{jn} \delta_{kl} + \delta_{ij} g_{kn} g_{ln}) \\
&+ r_n (g_{in} \delta_{jk} g_{ln} + g_{jn} \delta_{ik} g_{ln} + g_{in} \delta_{jl} g_{kn} + g_{jn} \delta_{il} g_{kn}) \Big].
\end{aligned} \tag{2.179}$$

where the elastic constants λ_n, μ_n, ν_n and $p_n,\ q_n,\ r_n$ $(n = 1, 2, 3)$ are defined by the matrix representations L_{ij} and M_{ij} in the crystallographic coordinate system (2.172).

In particularly, for the *tetragonal* crystals (2.173),

$$\begin{aligned}
L'_{ijkl} = &L_{12}\delta_{ij}\delta_{kl} + L_{66}(\delta_{ik}\delta_{jl} + \delta_{il}\delta_{jk}) + (L_{11} - L_{12} - 2L_{66}) \textstyle\sum_{ijkl} \\
&+ (L_{33} - L_{11} + 2L_{12} - 2L_{13} + 4L_{66} - 4L_{44}) T_{ijkl} \\
&+ 2(L_{44} - L_{66})(T_{(ik}\delta_{jl)} + T_{(il}\delta_{jk)}) + 2(L_{13} - L_{12}) T_{(ij}\delta_{kl)},
\end{aligned} \tag{2.180}$$

for *hexagonal* crystals (2.174),

$$\begin{aligned}
L'_{ijkl} = &L_{12}\delta_{ij}\delta_{kl} + (L_{11} - L_{12})(\delta_{ik}\delta_{jl} + \delta_{il}\delta_{jk}) \\
&+ (L_{33} + L_{11} - 2L_{13} - 4L_{44}) T_{ijkl} + 2(L_{13} - L_{12}) T_{(ij}\delta_{kl)} \\
&+ (L_{12} - L_{11} + 2L_{44})(T_{(ik}\delta_{jl)} + T_{(il}\delta_{jk)}),
\end{aligned} \tag{2.181}$$

for *cubic* crystals (2.175),

$$L'_{ijkl} = L_{12}\delta_{ij}\delta_{kl} + L_{44}(\delta_{ik}\delta_{jl} + \delta_{il}\delta_{jk}) + (L_{11} - L_{12} - 2L_{44}) \textstyle\sum_{ijkl}, \tag{2.182}$$

and for the *isotropic* material (2.177),

$$L'_{ijkl} = L_{12}\delta_{ij}\delta_{kl} + L_{44}(\delta_{ik}\delta_{jl} + \delta_{il}\delta_{jk}). \tag{2.183}$$

Here one introduces the notations

$$\sum_{ijkl} = \sum_{n=1}^{3} g_{in}g_{jn}g_{kn}g_{ln}, \; T_{ij} = g_{i3}g_{j3}, \; T_{ijkl} = g_{i3}g_{j3}g_{k3}g_{l3}, \tag{2.184}$$

where the tensors T_{ijkl} and $\sum_{ijkl}$ show the properties

$$T_{iikl} = T_{kl}, \; T_{ijkl}T_{klpq} = T_{ijpq}, \; \sum_{iikl} = \delta_{kl}, \; \sum_{ijkl}\sum_{klpq} = \sum_{ijpq}. \tag{2.185}$$

The representation (2.182) for the *cubic* crystals can be also presented in more detailed form:

$$\begin{aligned} L'_{11} &= L_{11} - 2L(g_{11}^2 g_{12}^2 + g_{12}^2 g_{13}^2 + g_{13}^2 g_{11}^2), \\ L'_{12} &= L_{12} + L(g_{11}^2 g_{21}^2 + g_{12}^2 g_{22}^2 + g_{13}^2 g_{23}^2), \\ L'_{44} &= L_{44} + L(g_{12}^2 g_{13}^2 + g_{22}^2 g_{32}^2 + g_{23}^2 g_{33}^2), \; L \equiv L_{11} - L_{12} - 2L_{44}. \end{aligned} \tag{2.186}$$

The inverse transformations from the coefficients $L_{\gamma\nu}$ to $M_{\gamma\nu}$ can be carried out by direct inversion of the matrix of known coefficients because $L_{\gamma\delta}M_{\delta\nu} = \delta_{\gamma\nu}$. However, in some cases the simple relations between the coefficients $L_{\gamma\nu}$ and $M_{\gamma\nu}$ are useful. They are presented below for some symmetry groups described, at most, by seven independent parameters:
Trigonal materials:

$$\begin{aligned} &M_{11} + M_{12} = L_{33}/l, \; M_{11} - M_{12} = L_{44}/l', \; M_{13} = -L_{13}/l, \\ &M_{14} = -L_{14}/l', \;\; M_{33} = (L_{11} + L_{12})/l, \; M_{44} = (L_{11} - L_{12})/l', \\ &l \equiv L_{33}(L_{11} + L_{12}) - 2L_{13}^2, \;\; l' \equiv L_{44}(L_{11} - L_{12}) - 2L_{14}^2. \end{aligned} \tag{2.187}$$

Hexagonal materials:

$$\begin{aligned} &M_{11} + M_{12} = L_{33}/l, \; M_{11} - M_{12} = (L_{11} - L_{12})^{-1}, \; M_{13} = -L_{13}/l, \\ &M_{33} = (L_{11} + L_{12})/l, \; M_{44} = L_{44}^{-1}, \\ &l \equiv L_{33}(L_{11} + L_{12}) - 2L_{13}^2. \end{aligned} \tag{2.188}$$

Cubic materials:

$$\begin{aligned} M_{11} &= (L_{11} + L_{12})(L_{11} - L_{12})^{-1}(L_{11} + 2L_{12})^{-1}, \\ M_{12} &= -L_{12}(L_{11} - L_{12})^{-1}(L_{11} + 2L_{12})^{-1}, \\ M_{44} &= L_{44}^{-1}. \end{aligned} \tag{2.189}$$

In particular, the elastic constants in the isotropic matrix ($\mathbf{A} = \mathbf{L}, \mathbf{M}$; $A_{ij} = L_{ij}, M_{ij}$):

$$\mathbf{A} = \begin{pmatrix} A_{11} & A_{12} & A_{12} & 0 & 0 & 0 \\ A_{12} & A_{11} & A_{12} & 0 & 0 & 0 \\ A_{12} & A_{12} & A_{11} & 0 & 0 & 0 \\ 0 & 0 & 0 & A_{44} & 0 & 0 \\ 0 & 0 & 0 & 0 & A_{44} & 0 \\ 0 & 0 & 0 & 0 & 0 & A_{44} \end{pmatrix}, \tag{2.190}$$

(where $L_{44} = (L_{11} - L_{12})/2, \; M_{44} = 2(M_{11} - M_{12})$) are usually written in the engineering notations:

$$L_{11} = \frac{E}{1+\nu}\frac{1-\nu}{1-2\nu}, \quad L_{12} = \frac{E}{1+\nu}\frac{\nu}{1-2\nu}, \quad L_{44} = \frac{E}{2(1+\nu)}, \tag{2.191}$$

$$M_{11} = \frac{1}{E}, \quad M_{12} = -\frac{\nu}{E}, \quad M_{44} = \frac{2(1+\nu)}{E}, \tag{2.192}$$

where two independent constants *Young's modulus* E and *Poisson's ratio* ν can be also used in the tensor representations ($\mathbf{N}_1 = \boldsymbol{\delta} \otimes \boldsymbol{\delta}/3$)

$$\mathbf{L} = \frac{3\nu E}{(1+\nu)(1-2\nu)}\mathbf{N}_1 + \frac{E}{1+\nu}\mathbf{I}, \quad \mathbf{M} = -\frac{3\nu}{E}\mathbf{N}_1 + \frac{1+\nu}{E}\mathbf{I}. \tag{2.193}$$

Another popular representations of the elastic isotropic properties using *Lame's constants* λ and μ can be formed in both the matrix

$$L_{11} = \lambda + \mu, \ L_{12} = \lambda, \ L_{66} = \mu, \tag{2.194}$$

and tensor

$$\mathbf{L} = 3\lambda\mathbf{N}_1 + 2\mu\mathbf{I}, \quad \mathbf{M} = 3s\mathbf{N}_1 + 2q\mathbf{I}, \tag{2.195}$$

notations where the compliances s and q are expressed in terms of elastic constants:

$$s = \frac{1}{3(3\lambda + 2\mu)}, \quad q = \frac{1}{4\mu}. \tag{2.196}$$

Due to the decomposition of the unit tensor $\mathbf{I}$ into the orthogonal bulk $\mathbf{N}_1$ and deviatoric $\mathbf{N}_2$ tensors (2.16), the isotropic tensors $\mathbf{L}$ and $\mathbf{M}$ can also be decomposed:

$$\mathbf{L} = (3k, 2\mu) \equiv 3k\mathbf{N}_1 + 2\mu\mathbf{N}_2, \quad \mathbf{M} = (3p, 2q). \tag{2.197}$$

where k and μ, called the bulk and shear moduli, are related to the appropriate compliance coefficients p, q by means of the "multiplication table" for the products between the elementary idempotent tensors $\mathbf{N}_1$ and $\mathbf{N}_2$ (2.17). Then the inverse tensor $\mathbf{L}_1^{-1} \equiv \mathbf{M}$, and the product $\mathbf{L}_1 : \mathbf{L}_2$ of such tensors, $\mathbf{L}_1 = (3k_1, 2\mu_1)$ and $\mathbf{L}_2 = (3k_2, 2\mu_2)$, can be estimated by the compact formulae:

$$\mathbf{L}_1^{-1} = (3p, 2q) = \left(\frac{1}{3k}, \frac{1}{2\mu}\right), \quad \mathbf{L}_1 : \mathbf{L}_2 = (9k_1k_2, 4\mu_1\mu_2). \tag{2.198}$$

Table 2.1 shows the relationships between the various elastic constants of isotropic materials.

Table 2.1. Relationships between elastic moduli

Constant	$\lambda,\ \mu$	$K,\ \mu$	$\mu,\ \nu$	$E,\ \nu$	$E,\ \mu$
λ	λ	$K - \frac{2}{3}\mu$	$\frac{2\mu\nu}{1-2\nu}$	$\frac{\nu E}{(1+\nu)(1-3\nu)}$	$\frac{\mu(E-2\mu)}{3\mu-E}$
μ	μ	μ	μ	$\frac{E}{2(1+\nu)}$	μ
K	$\lambda + \frac{2}{3}\mu$	K	$\frac{2\mu(1-\nu)}{3(1-2\nu)}$	$\frac{E}{3(1-2\nu)}$	$\frac{E\mu}{3(3\mu-E)}$
E	$\frac{(3\lambda+2\mu)\mu}{\lambda+\mu}$	$\frac{9K\mu}{3K+\mu}$	$2(1+\nu)\mu$	E	E
ν	$\frac{\lambda}{2(\lambda+\mu)}$	$\frac{3K-2\mu}{6K+2\mu}$	ν	ν	$\frac{E}{2\mu}$-1

The bulk and shear components of the isotropic tensors **L** and **M** are found through a suitable contraction of the indices of **L** and **M** in the following relations

$$3k = \frac{1}{3}L_{iikk}, \quad 2\mu = \frac{1}{5}(L_{ikik} - \frac{1}{3}L_{iikk}), \tag{2.199}$$

$$3p = \frac{1}{3}M_{iikk}, \quad 2q = \frac{1}{5}(M_{ikik} - \frac{1}{3}M_{iikk}), \tag{2.200}$$

which can be used at the estimation of average moduli of isotropic polycrystals.

Transversely isotropic materials have five independent elastic parameters. If the plane of isotropy coincides with the x_1, x_2-plane, the elastic and compliance matrices ($\mathbf{A} = \mathbf{L}, \mathbf{M};\ A_{ij} = L_{ij}, M_{ij}$):

$$\mathbf{A} = \begin{pmatrix} A_{11} & A_{12} & A_{13} & 0 & 0 & 0 \\ A_{12} & A_{11} & A_{13} & 0 & 0 & 0 \\ A_{13} & A_{13} & A_{11} & 0 & 0 & 0 \\ 0 & 0 & 0 & A_{44} & 0 & 0 \\ 0 & 0 & 0 & 0 & A_{44} & 0 \\ 0 & 0 & 0 & 0 & 0 & A_{55} \end{pmatrix}, \tag{2.201}$$

(where $L_{55} = (L_{11} - L_{12})/2,\ M_{55} = 2(M_{11} - M_{12})$) can be presented in engineering notations:

$$L_{11} = l\Big(\frac{1}{EE_3} - \frac{\nu_3^2}{E_3^2}\Big), \quad L_{12} = l\Big(\frac{\nu}{EE_3} + \frac{\nu_3^2}{E_3^2}\Big), \quad L_{13} = l\frac{(1+\nu)\nu_3}{EE_3},$$
$$L_{33} = l\frac{1-\nu^2}{E^2}, \quad L_{44} = \mu_3,\ l = \frac{E^2E_3^2}{(1+\nu)[(1-\nu)E_3 - 2\nu_3^2 E]}, \tag{2.202}$$

$$M_{11} = \frac{1}{E}, \quad M_{12} = -\frac{\nu}{E}, \quad M_{13} = -\frac{\nu_3}{E_3}, \quad M_{33} = \frac{1}{E_3}, \quad M_{44} = \frac{1}{\mu_3}. \tag{2.203}$$

Here the Young modulus and the Poisson ratios in the x_1, x_2-plane are the same, say, $E_1 = E_2 = E$ and $\nu_{12} = \nu_{21} = \nu$, respectively. The corresponding shear modulus $\mu_{12} = \mu_{21} = \mu$ is formed as $\mu = E/2(1+\nu)$. Young modulus, Poisson ratio, and the shear modulus associated with the x_3-direction and a direction in the x_1x_2-plane are denoted by, say, E_3, $\nu_{13} = \nu_{23} = \nu_3$, and $\mu_{13} = \mu_{23} = \mu_3$, respectively.

For the *orthotropic* case, the elastic and compliance matrices are defined by nine independent elastic parameters and can be presented in a coordinate system coincident with the material symmetry directions as ($\mathbf{A} = \mathbf{L}, \mathbf{M};\ A_{ij} = L_{ij}, M_{ij}$):

$$\mathbf{A} = \begin{pmatrix} A_{11} & A_{12} & A_{13} & 0 & 0 & 0 \\ A_{12} & A_{11} & A_{23} & 0 & 0 & 0 \\ A_{13} & A_{23} & A_{11} & 0 & 0 & 0 \\ 0 & 0 & 0 & A_{44} & 0 & 0 \\ 0 & 0 & 0 & 0 & A_{55} & 0 \\ 0 & 0 & 0 & 0 & 0 & A_{66} \end{pmatrix}, \tag{2.204}$$

where, for example, the compliance matrix in engineering terminology becomes

$$\mathbf{M} = \begin{pmatrix} 1/E_1 & -\nu_{21}/E_2 & -\nu_{31}/E_3 & 0 & 0 & 0 \\ -\nu_{12}/E_1 & 1/E_2 & -\nu_{32}/E_3 & 0 & 0 & 0 \\ -\nu_{13}/E_1 & -\nu_{23}/E_2 & 1/E_3 & 0 & 0 & 0 \\ 0 & 0 & 0 & 1/\mu_{23} & 0 & 0 \\ 0 & 0 & 0 & 0 & 1/\mu_{13} & 0 \\ 0 & 0 & 0 & 0 & 0 & 1/\mu_{12} \end{pmatrix}. \tag{2.205}$$

We will now consider a unique representation of the constitutive laws and elastic moduli for a d-dimensional linear isotropic material with bulk modulus $k_{[d]}$, shear modulus $\mu_{[d]}$, Young's modulus $E_{[d]}$, and Poisson's ratio $\nu_{[d]}$ in the space dimensionality d $(i,j,k,l = 1,\ldots,d;\ d = 2,3)$ (see, e.g., [1106]):

$$\sigma_{ij} = (k_{[d]} - \frac{2}{d}\mu_{[d]})\varepsilon_{kk}\delta_{ij} + 2\mu_{[d]}\varepsilon_{ij}, \quad \varepsilon_{ij} = -\frac{\nu_{[d]}}{E_{[d]}}\sigma_{kk}\delta_{ij} + \frac{(1+\nu_{[d]})}{E_{[d]}}\sigma_{ij} \tag{2.206}$$

and

$$\mathbf{L} = dk_{[d]}\mathbf{N}_1 + 2\mu_{[d]}\mathbf{N}_2, \qquad \mathbf{M} = \frac{1+\nu_{[d]}(1-d)}{E_{[d]}}\mathbf{N}_1 + \frac{1+\nu_{[d]}}{E_{[d]}}\mathbf{N}_2, \tag{2.207}$$

where the projection tensors

$$\mathbf{N}_{1|ijkl} = \frac{1}{d}\delta_{ij}\delta_{kl}, \quad \mathbf{N}_{2|ijkl} = \frac{1}{2}(\delta_{ik}\delta_{jl} + \delta_{il}\delta_{jk}) - \frac{1}{d}\delta_{ij}\delta_{kl} \tag{2.208}$$

satisfy the normality conditions (2.17) allowing the inverse of relation (2.207_1)

$$\mathbf{M} = \frac{1}{dk_{[d]}}\mathbf{N}_1 + \frac{1}{2\mu_{[d]}}\mathbf{N}_2. \tag{2.209}$$

Comparison of representations (2.207_2) and (2.209) leads to the interrelations

$$\begin{aligned} k_{[d]} &= \frac{E_{[d]}}{d[1+\nu_{[d]}(1-d)]}, \quad & \mu_{[d]} &= \frac{E_{[d]}}{2(1+\nu_{[d]})}, \\ \frac{1}{E_{[d]}} &= \frac{1}{d^2 k_{[d]}} + \frac{d-1}{2d\mu_{[d]}}, \quad & \nu_{[d]} &= \frac{dk_{[d]} - 2\mu_{[d]}}{d(d-1)k_{[d]} + 2\mu_{[d]}}. \end{aligned} \tag{2.210}$$

Passages to the limits of the positive moduli $k_{[d]}/\mu_{[d]} \to \infty$ and $\mu_{[d]}/k_{[d]} \to \infty$ in the expression (2.210_4) yield the upper and lower limits of $\nu_{[d]}$: $-1 \le \nu_{[d]} \le (d-1)^{-1}$, respectively.

Finally, we will consider either *plane-strain* ($\varepsilon_{11} = \varepsilon_{12} = \varepsilon_{13}$ in Eq. (2.206) with $d = 3$) or *plane-stress* ($\sigma_{11} = \sigma_{12} = \sigma_{13}$ in Eq. (2.206) with $d = 3$) elasticity. If we compare these simplified three-dimensional expressions to relations (2.206) with $d = 2$, we obtained the interrelations:

$$E_{[2]} = \frac{E_{[3]}}{(1+\nu_{[3]})(1+\nu_{[3]})}, \quad \nu_{[2]} = \frac{\nu_{[3]}}{1-\nu_{[3]}}, \quad k_{[2]} = k_{[3]} + \frac{1}{3}\mu_{[3]}, \quad \mu_{[2]} = \mu_{[3]}, \tag{2.211}$$

for plane-strain elasticity and

$$E_{[2]} = E_{[3]}, \quad \nu_{[2]} = \nu_{[3]}, \quad k_{[2]} = \frac{9k_{[3]}\mu_{[3]}}{3k_{[3]} + 4\mu_{[3]}}, \quad \mu_{[2]} = \mu_{[3]} \tag{2.212}$$

for plane-stress elasticity, respectively.

2.6 Basic Equations of Thermoelastoplastic Deformations

2.6.1 Incremental Theory of Plasticity

We assume that the thermomechanical properties of the composite medium with, generally speaking, anisotropic components are described by the theory of small elastic-plastic strains under arbitrarily varying external loading. Additive decomposition of the increments of total strain tensor ε is assumed:

$$d\varepsilon = d\varepsilon^e + d\varepsilon^t + d\varepsilon^p, \tag{2.213}$$

with the increments of the elastic strains $d\varepsilon^e$, the so-called "transformation strains" $d\varepsilon^t$ including thermal strains and plastic strains $d\varepsilon^p$.

The stress increment $d\boldsymbol{\sigma}$ relates to the elastic part of the strain increment $d\varepsilon$ by the elasticity relation given in the form

$$d\boldsymbol{\sigma}(\mathbf{x}) = \mathbf{L}(\mathbf{x})\ d\varepsilon^e(\mathbf{x}), \tag{2.214}$$

where $\mathbf{L}(\mathbf{x})$ is the fourth-order elasticity tensor (2.88).

The transformation strain increment $d\varepsilon^t$ (2.213) may consist of contributions of different physical origins. For example, if only thermal effects are considered, $\varepsilon^t = \boldsymbol{\beta}^T\theta$, where $\boldsymbol{\beta}^T$ is the tensor of linear thermal expansion coefficients and θ is the temperature change from the reference value to the current temperature.

In the six-dimensional stress space, consider a yield surface $f(\mathbf{x},\ldots) = 0$ bonding the region of plastic deformation, in which $f(\boldsymbol{\sigma},\ldots) > 0$; here dots stand for temperature and for internal variables characterizing material hardening. The behavior is elastic if $f < 0$, or if $f = 0$ and $[(\partial f/\partial\boldsymbol{\sigma}) : d\boldsymbol{\sigma} + (\partial f/\partial\theta)d\theta] \le 0$ (for the elastic unloading and neutral loading); elastic-plastic deformations take place under active loading, when $f = 0$ and $[(\partial f/\partial\boldsymbol{\sigma}) : d\boldsymbol{\sigma} + (\partial f/\partial\theta)d\theta] > 0$. In the general case, the yield surface can depend on a variety of tensor $\mathbf{a}^p$ and scalar hardening parameters. The hardening tensor parameter, in particular, may simply coincide with the plastic-strain tensor. Regarding the constitutive equations for the elastic-plastic materials we use the so-called J_2-flow theory with combined isotropic-kinematic hardening. The von Mises form of the yield surface is given by

$$f \equiv \tau - F(\gamma,\theta) = 0, \quad F(0,\theta) = \tau_0(\theta) \tag{2.215}$$

in terms of von Mises effective stress, τ, and the accumulated effective plastic strain increment, γ, respectively, as defined by

$$\tau = \left(\frac{3}{2}s^a_{ij}s^a_{ij}\right)^{1/2}, \quad d\gamma = \left(\frac{2}{3}d\varepsilon^p_{ij}\ d\varepsilon^p_{ij}\right)^{1/2}, \quad \mathbf{s}^a = \mathbf{N}_2(\boldsymbol{\sigma} - \mathbf{a}^p). \tag{2.216}$$

Here τ_0 is the initial yield stress, F is a nonlinear function describing the hardening effect and temperature dependence, for example as in the modified Ludwik equation:

$$F(\gamma,\theta) = \tau_0(\theta) + h(\theta)\gamma^{n(\theta)}, \tag{2.217}$$

where h and n are hardening parameters. $\mathbf{s}^a$ is the active stress deviator; $\mathbf{a}^p$ is a symmetric second order tensor corresponding to the "back-stresses" defining the location of the center of the yield surface in the deviatoric stress space. For evaluation of the back stress tensor $\mathbf{a}^p$ we use Ziegler's assumption:

$$d\mathbf{a}^p = d\gamma A\mathbf{s}^a, \quad A \equiv A(\gamma), \tag{2.218}$$

or Prager's rule

$$d\mathbf{a}^p = Bd\varepsilon^p, \quad B \equiv B(\gamma), \tag{2.219}$$

respectively, with $\mathbf{a}^p = 0$ if $\gamma = 0$. In the case $\mathbf{a}^p \equiv \mathbf{0}$, $h \neq 0$, Eqs. (2.215) and (2.216) are reduced to the isotropic hardening; the case $h \equiv 0$, $\mathbf{a}^p \neq \mathbf{0}$ correspondents to the kinematic hardening. Though the von Mises yield criterion (2.215) is assumed in this study, modifications of the present method for introducing other yield criteria and hardening laws are possible.

Drucker's postulate states that the work of additional stresses is positive over the whole cycle. Then for any stress state $\boldsymbol{\sigma}^0$ within or on the convex yield surface the local maximum principle $(\boldsymbol{\sigma} - \boldsymbol{\sigma}^0) : d\varepsilon^p \geq 0$. Drucker's postulate leads to the necessity of the associated law of plastic flow when the smooth yield function f is taken as plastic potential function from which the incremental plastic strain can be derived as

$$\varepsilon^p = d\lambda \frac{\partial f}{\partial \boldsymbol{\sigma}}, \quad \text{for} \ \ f = 0, \ \ \frac{\partial f}{\partial \boldsymbol{\sigma}} : d\boldsymbol{\sigma} > 0, \tag{2.220}$$

where the plastic flow is fixed in the direction along the normal to the yield surface, while its magnitude $d\lambda$, called a proportionality factor, is undetermined. For consistency during plastic deformation:

$$df \equiv \frac{\partial f}{\partial \boldsymbol{\sigma}} : d\boldsymbol{\sigma} + \frac{\partial f}{\partial \mathbf{a}^p} : d\mathbf{a}^p + \frac{\partial f}{\partial \gamma} d\gamma + \frac{\partial f}{\partial \theta} d\theta = 0. \tag{2.221}$$

Substitution of (2.216), (2.218), (2.219) in (2.221) leads to the determination of $d\lambda$ by

$$d\lambda = G\left(\frac{\partial f}{\partial \boldsymbol{\sigma}} : d\boldsymbol{\sigma} + \frac{\partial f}{\partial \theta} d\theta\right), \tag{2.222}$$

where the proportionality parameter G is defined by the explicit form of hardening rule (2.217)–(2.219). For example, for isotropic-kinematic hardening, (2.217), (2.218), we have

$$G = -\left(\frac{\partial f}{\partial \gamma} + A(\gamma)\frac{\partial f}{\partial \mathbf{a}^p} : \mathbf{s}^a\right)^{-1} \left(\frac{2}{3}\frac{\partial f}{\partial \boldsymbol{\sigma}} : \frac{\partial f}{\partial \boldsymbol{\sigma}}\right)^{-1/2}. \tag{2.223}$$

The associated flow law (2.220) requires the smoothness of the yield surface with a defined normal to the surface. However, the plastic flow corresponding to edges or vertexes of the yield surface $f = 0$ originated by the intersections of smooth surfaces $f_\alpha = 0$ $(\alpha = 1, \ldots)$ can be estimated from superposition principle by Koiter [581]:

$$d\boldsymbol{\varepsilon}^p = \sum_{\alpha=1} d\lambda_\alpha \frac{\partial f_\alpha}{\partial \boldsymbol{\sigma}}, \tag{2.224}$$

$$f_\alpha = 0, \quad \frac{\partial f_\alpha}{\partial \boldsymbol{\sigma}} : d\boldsymbol{\sigma} \geq 0. \tag{2.225}$$

A state of stress on the yield surface is described by a value of zero of one or more yield functions, all other yield functions being negative. In so doing, the nonzero proportionality factors $d\lambda_\alpha$ correspond only to the yield surfaces satisfying the conditions (2.225).

2.6.2 Deformation Theory of Plasticity

It is assumed that the rheological properties of isotropic media are described by the theory of small elastoplastic strains under monotonic, proportional loading when the ratio of the stress components $\sigma_{11} : \sigma_{22} : \sigma_{33} : \sigma_{13} : \sigma_{23} : \sigma_{12}$ is held constant at all time. Specifically, the total strain $\boldsymbol{\varepsilon}$ is written as the sum of elastic $\boldsymbol{\varepsilon}^e$ and plastic $\boldsymbol{\varepsilon}^p$ contributions:

$$\boldsymbol{\varepsilon} = \boldsymbol{\varepsilon}^e + \boldsymbol{\varepsilon}^p, \tag{2.226}$$

where the mean stress and mean strain are linearly related: $\sigma_{ii} = 3k\varepsilon_{ii}$. The local equation for the elastic material state, which relates the stress tensor $\boldsymbol{\sigma}(\mathbf{x})$ and the elastic strain tensor $\boldsymbol{\varepsilon}^e(\mathbf{x})$, is given in the form (2.214) where $\mathbf{L}(\mathbf{x})$ is an isotropic fourth-order elasticity tensor described by the Young's modulus E and the Poisson's ratio ν (2.193).

The relation between flow stress and plastic strains is represented by the expression

$$\tau = \tau_0 + f(\varepsilon^p_{\text{eq}}), \quad f(0) = 0 \tag{2.227}$$

in terms of the von Mises effective stress τ and the effective plastic strain ϵ^p_{eq}, defined as

$$\tau = \left(\frac{3}{2} s_{ij} s_{ij}\right)^{1/2}, \quad \varepsilon^p_{\text{eq}} = \left(\frac{2}{3}\varepsilon^p_{ij}\varepsilon^p_{ij}\right)^{1/2}, \quad \mathbf{s} \equiv \mathbf{N}_2\sigma. \tag{2.228}$$

In Eq. (2.227), τ_0 is the initial yield stress and f is a nonlinear function describing the material's hardening behavior, for example:

$$f(\epsilon^p_{\text{eq}}) = h(\epsilon^p_{\text{eq}})^n, \tag{2.229}$$

where h and n are the strength coefficient and the work-hardening exponent, respectively. The material remains elastic, i.e., $\boldsymbol{\varepsilon}^p \equiv \mathbf{0}$, when $\tau < \tau_0$. In addition, Hencky's flow rule

$$\varepsilon^p_{ij} = \frac{3\varepsilon^p_{\text{eq}}}{2\tau}\sigma_{ij} \tag{2.230}$$

is adopted. A more general relation given in Prager's form is

$$\varepsilon^p_{ij} = \alpha s_{ij} + \beta t_{ij}, \tag{2.231}$$

where $t_{ij} = s_{ik}s_{kj} - 2/3J_2\delta_{ij}$ is the deviation of the square of the stress deviation, and for an isotropic materials α and β are the functions of the invariants $J_2 = s_{ij}s_{ij}/2$ and $J_3 = s_{ij}s_{jk}s_{ki}/3$.

Based on relation (2.226) one may determine the secant modulus as

$$E^s = \left[\frac{1}{E} + \frac{\epsilon^p_{\rm eq}}{\tau_0 + f(\epsilon^p_{\rm eq})}\right]^{-1}. \tag{2.232}$$

Due to the plastic incompressibility the secant bulk modulus k^s is equal to k, and we may accordingly define the secant Poisson's ratio and shear modulus:

$$\nu^s = \frac{1}{2} - \left(\frac{1}{2} - \nu\right)\frac{E^s}{E}, \quad \mu^s = \frac{E^s}{2(1+\nu^s)}. \tag{2.233}$$

Both here and below the superscript ^{s} indicates the calculation of the parameter under consideration with the help of the secant modulus E^s. Therefore, under a monotonic, proportional loading, the plastic state can be described by a single secant modulus, say E^s, and the other two elastic constants. For these loading the incremental flow rule can be integrated to yield the deformation rule.

3

Green's Functions, Eshelby and Related Tensors

3.1 Static Green's Function

Let $f_i(\mathbf{x})$ be the distribution density of volume forces with finite support in an infinite medium with an elastic modulus L_{ijkl}. Then the elastic fields of small displacements are found from the equilibrium equation:

$$\mathcal{L}\mathbf{u} = -\mathbf{f}, \quad (\mathcal{L})_{ik} \equiv \nabla_j L_{ijkl} \nabla_l, \tag{3.1}$$

where $\mathbf{u}(\mathbf{x})|_{|\mathbf{x}|\to\infty} = \mathbf{0}$. The Green's tensor for the displacement $G_{ij}(\mathbf{x})$ in the infinite medium is defined from the relation $(\mathbf{x}, \mathbf{y} \in R^d, \ d = 2, 3)$

$$u_k = G_{ki} * f_i \equiv \int G_{ki}(\mathbf{x} - \mathbf{y}) f_i(\mathbf{y}). \tag{3.2}$$

For the unique definition of the Green's tensor, one specifies the boundary conditions vanishing at infinity. From the equilibrium equation (3.1) and Eq. (3.2), one obtains an equation for the estimation of the Green's function:

$$L_{ijkl} G_{lm,jk}(\mathbf{x}) = -\delta_{im} \delta(\mathbf{x}), \tag{3.3}$$

where $\delta(\mathbf{x})$ is the Dirac delta function equal to zero in the whole space except the point $\mathbf{x} = \mathbf{0}$ and having the property $\int \mathbf{f}(\mathbf{x})\delta(\mathbf{x})\, d\mathbf{x} = \mathbf{f}(\mathbf{0})$ if the integration area contains the point $\mathbf{x} = \mathbf{0}$. Equation (3.3) defines the physical meaning of the Green's tensor component $G_{lm}(\mathbf{x}-\mathbf{y})$ as the component of the displacement vector $u_i(\mathbf{x})$ arising from the unit force directed along the axis x_m and applied at the point $\mathbf{y}$.

Estimation of the Green's tensor $\mathbf{G}$ will be carried out by the use of the Fourier transform

$$\mathbf{g}(\mathbf{k}) = F(\mathbf{g}) = \int \mathbf{g}(\mathbf{x}) e^{-i\mathbf{k}\cdot\mathbf{x}}\, d\mathbf{x},$$

$$\mathbf{g}(\mathbf{x}) = F^{-1}(\widetilde{\mathbf{g}}) = (2\pi)^{-d} \int \mathbf{g}(\mathbf{k}) e^{i\mathbf{k}\cdot\mathbf{x}}\, d\mathbf{k}, \tag{3.4}$$

provided, of course, that the integrals on the right-hand sides of the equations are convergent; here $\xi \cdot \mathbf{x} = \xi_1 x_1 + \ldots + \xi_d x_d$, $i = \sqrt{-1}$, and for the tensorial function $\mathbf{g}(\mathbf{x})$ and its Fourier transform $\mathbf{g}(\mathbf{k})$ one uses the same notations with the replacement of an argument $\mathbf{x}$ over $\mathbf{k}$. Performing the Fourier transform of Eq. (3.3) yielding $L_{ijkl}k_jk_lG_{kn}(\mathbf{k}) = \delta_{ij}$, to say, the problem of estimation of the Fourier transform of the Green's tensor is formally reduced to the solution of a linear algebraic system:

$$K_{ik}G_{kn}(\mathbf{k}) = \delta_{in}, \; K_{ik} = L_{ijkl}k_jk_l. \tag{3.5}$$

Then $G_{kn}(\mathbf{k}) = K^*_{kn}/\det\{\mathbf{K}\}$, where $\mathbf{K}^*$ is a cofactor of an element K_{kn} in a symmetric matrix:

$$\mathbf{K}(\mathbf{k}) = \begin{pmatrix} K_{11}(\mathbf{k}) & K_{12}(\mathbf{k}) & K_{13}(\mathbf{k}) \\ K_{21}(\mathbf{k}) & K_{22}(\mathbf{k}) & K_{23}(\mathbf{k}) \\ K_{31}(\mathbf{k}) & K_{32}(\mathbf{k}) & K_{33}(\mathbf{k}) \end{pmatrix}, \tag{3.6}$$

and $\det\{\mathbf{K}\}$ is a determinant of the Christoffel matrix $\|K(\mathbf{k})\|$ which is positive define. The matrixes $\|\mathbf{K}(\mathbf{k})\|$ and $\|\mathbf{K}^*(\mathbf{k})\|$ are symmetric due to the symmetry of the elastic modulus tensor. The following representations for $\det\{\mathbf{K}\}$ and for $\|\mathbf{K}^*(\mathbf{k})\|$ take place:

$$\begin{aligned} \det\{\mathbf{K}\} &= \epsilon_{mnl}K_{m1}(\mathbf{k})K_{n2}(\mathbf{k})K_{l3}(\mathbf{k}), \\ K^*_{ij}(\mathbf{k}) &= \frac{1}{2}\epsilon_{ijl}\epsilon_{jmn}K_{lm}(\mathbf{k})K_{ln}(\mathbf{k}) = K_{im}K_{mj} - K_{mm}K_{ij} \\ &+ (\epsilon_{mn1}K_{m2}(\mathbf{k})K_{n2}(\mathbf{k}) + \epsilon_{mn3}K_{m1}(\mathbf{k})K_{n2}(\mathbf{k}))\delta_{ij}. \end{aligned} \tag{3.7}$$

Then, passing on from $\mathbf{k}$-representation to the $\mathbf{x}$-representation of the Green's tensor leads to

$$G_{ij}(\mathbf{x}) = (2\pi)^{-3}\int K^*_{ij}(\mathbf{k})(\det\{\mathbf{K}\})^{-1}e^{i\mathbf{kx}}\,d\mathbf{k}. \tag{3.8}$$

As can be seen from (3.5)–(3.7), $K_{ij}(\mathbf{k})$ and $\det\{K(\mathbf{k})\}$ are the homogeneous polynomials of the forth and six order, respectively. Then Eq. (3.8) can be presented in the form of the surface integral over the surface of a unit sphere S_1 in the $\mathbf{k}$-space:

$$G_{ij}(\mathbf{x}) = (8\pi^2)^{-1}\int_{S_1}\delta(\overline{\mathbf{k}}\overline{\mathbf{x}})K^*_{ij}(\overline{\mathbf{k}})(\det\{\mathbf{K}(\overline{\mathbf{k}})\})^{-1}\,d\mathbf{s}(\overline{\mathbf{k}}), \tag{3.9}$$

where $\overline{\mathbf{x}} \equiv \mathbf{x}/|\mathbf{x}|$, $\overline{\mathbf{k}} = \mathbf{k}/|\mathbf{k}|$. The representation (3.9), in turns is reduced to the counter integral over the curve of the intersection $\gamma_1 = \{\boldsymbol{\xi} \in R^3 | \|\boldsymbol{\xi}\| = 1, \; \boldsymbol{\xi}\cdot\mathbf{x} = 0\}$ of the sphere S_1 with the plate orthogonal to the vector $\mathbf{x}$ and passing through the coordinate center:

$$G_{ij}(\mathbf{x}) = \frac{1}{8\pi^2|\mathbf{x}|}\oint_{\gamma_1}K_{ij}(\overline{\mathbf{k}})(\det\{\mathbf{K}(\overline{\mathbf{k}})\})^{-1}\,d\gamma_1(\overline{\mathbf{k}}). \tag{3.10}$$

The integrand in Eq. (3.10) has no singularities and, therefore, the tensorial Green's function can be estimated by a straightforward numerical quadrature, e.g., Romberg integration, that was performed by Barnett [36], who also demonstrated that corresponding contour integral formulae for derivatives of $\mathbf{G}$ can be derived in a similar manner.

Construction of the Green's function in an analytical form is reduced to finding of the roots the algebraic equation of the six orders with the coefficients depending on L_{ijkl}. The roots of this equation are found in the explicit form only for the particular cases of the isotropic materials and the media with a hexagonal symmetry.

Namely, one obtains in an isotropic case from Eqs. (3.6) and (3.7):

$$\begin{aligned} \det\{\mathbf{K}(\mathbf{k})\} &= \mu^2(\lambda+2\mu)|\mathbf{k}|^6, \\ K_{ij}(\mathbf{k}) &= \frac{1}{(2\pi)^3\mu}\Big[\delta_{ij}\int \frac{1}{|\mathbf{k}|^2}e^{i\mathbf{kx}}\,d\mathbf{k} - \frac{\lambda+\mu}{\lambda+2\mu}\int \frac{k_ik_j}{|\mathbf{k}|^4}e^{i\mathbf{kx}}\,d\mathbf{k}\Big]. \end{aligned} \tag{3.11}$$

Estimation of integrals in Eq. (3.11) can be carried out by means of differentiation over x_i and x_j of the identity

$$\pi^{-2}\int \mathbf{k}^{-4}e^{i\mathbf{kx}}\,d\mathbf{k} = -|\mathbf{x}|,$$

leading to the relations

$$|\mathbf{x}|_{,ij} = \frac{1}{\pi^2}\int \frac{k_ik_j}{\mathbf{k}^4}e^{i\mathbf{kx}}\,d\mathbf{k}, \quad |\mathbf{x}|_{,ii} = \frac{1}{\pi^2}\int \frac{1}{\mathbf{k}^2}e^{i\mathbf{kx}}\,d\mathbf{k}. \tag{3.12}$$

This allows us to obtain the final representation for the Green's function:

$$G_{ij}(\mathbf{x}) = \frac{1}{8\pi\mu}\Big[\delta_{ij}|\mathbf{x}|_{,mm} - \varkappa|\mathbf{x}|_{,ij}\Big] = \frac{1}{4\pi\mu|\mathbf{x}|}[(2-\varkappa)\delta_{ij} + \varkappa n_in_j], \quad \varkappa = \frac{\lambda+\mu}{\lambda+2\mu}, \tag{3.13}$$

where $n_i = |\mathbf{x}|_{,i} = x_i/|\mathbf{x}|$.

For completeness we reproduce the compact representation of the Green's function valid for both 2-D and 3-D cases (see, e.g., [28], [214], [1104]):

$$G_{ij}(|\mathbf{x}|) = \frac{1}{4\omega_d(1-\bar\nu)\mu}\Big[(3-4\bar\nu)\gamma(|\mathbf{x}|)\delta_{ij} + \frac{1}{|\mathbf{x}|^{d-2}}n_in_j\Big], \tag{3.14}$$

where

$$\gamma(|\mathbf{x}|) = \begin{cases} -\ln|\mathbf{x}|, & d=2, \\ |\mathbf{x}|^{-1}, & d=3, \end{cases}, \qquad \bar\nu = \begin{cases} \nu, & \text{for 3-D and plane strain,} \\ \frac{\nu}{1+\nu}, & \text{for plane stress problems,} \end{cases} \tag{3.15}$$

and ω_d is the surface area of the unit sphere in R^d.

Construction of the Green's function for the transversally isotropic medium is more complicated. Kröner [605] has obtained a close representation of the Green's function for the case $(L_{1111}L_{3333})^{1/2} - L_{1133} - 2L_{2323} \neq 0$. This problem was analyzed [684], [1178] by the perturbation and Fourier methods, which are generally limited to "small" anisotropy. They found that as $x_1 + x_2 \to 0$ (x_3 is the symmetry axis of a medium), the items in the representation for $G_{ij}(\mathbf{x})$ tend to infinity although the total value of $G_{ij}(\mathbf{x})$ is finite. This drawback is absent in the representation obtained in [854] by the potential method for an arbitrary transversally isotropic media; the compact form of these formulae was proposed in [554]. An alternative to analytic expansions is direct numerical estimation of the contour integral (3.10) which has an obvious drawback for the problems of micromechanics due to the necessity of calculating the integrand at many points, with a nontrivial amount of work at each evaluation. To avoid of this difficulty, one [1191] has evaluated the Green's function and its derivatives by interpolation of precalculated tables. Gray *et al.* [396] proposed a hybrid analytical-numerical perturbation approach based on an expansion of $\mathbf{G}$ in which the "expansion point" (zero-order term) is an isotropic Green's function. The higher order contributions are expressed as contour integrals of matrix products, and evaluated with a symbolic manipulation program. For moderately anisotropic materials, employing the first few terms in the series provides an accurate solution and a fast computational algorithm, which, however, is not competitive with the method [1191] for strongly anisotropic solids. The basic idea of splitting off an isotropic part has been also used by Schclar [972], who treated the anisotropic part of the Green's function as a generalized body force.

3.2 The Second Derivative of Green's Function and Related Problems

3.2.1 The Second Derivative of Green's Function

It should be mentioned that in micromechanics, the second derivative of the Green's tensor $\nabla\nabla\mathbf{G}$ is more likely to be used than the Green's function $\mathbf{G}$. The first gradient of the Green's function $\nabla\mathbf{G}(\mathbf{x}-\mathbf{y})$ found by formal differentiation are the homogeneous function of degree $d-1$ and locally integrable everywhere. At the same time the second gradient $\nabla\nabla\mathbf{G}$ is considered the generalized functions and for the construction of the integral regularized representation of the generalized function of the type of derivatives of a homogeneous regular function one usually considers a scheme proposed in [374] (see also [618]). Then the components of the function representation of an even homogeneous distribution $\mathbf{U}(\mathbf{x}) = \nabla\nabla\mathbf{G}$ of the order $-d$ is defined in the form of a sum of the formal $\mathbf{U}^{(f)}(\mathbf{n})$ and singular $\mathbf{U}^{(s)}$ components depending on the shape of the integration domain in the improper integral involved (which is chosen for definiteness sake as a ball, see, e.g., [123], [995], [1104]) at the regularization of the generalized function $\mathbf{U}(\mathbf{x})$, though itself, $\mathbf{U}(\mathbf{x})$ does not depend on this region shape

$$(G_{ij,kl},\phi) = (G^{(s)}_{ij,kl},\phi) + (G^{(f)}_{ij,kl},\phi), \tag{3.16}$$

$$(G^{(s)}_{ij,kl},\phi) = \int_v G_{ij,kl}(\mathbf{x})\, d\mathbf{x}\phi(\mathbf{0}) = \int_S G_{ij,k}(\mathbf{x})\, d\mathbf{s}_m \cdot \phi(\mathbf{0}),$$

$$(G^{(f)}_{ij,kl},\phi) = \int_v G_{ij,kl}(\mathbf{x})[\phi(\mathbf{x}) - \phi(\mathbf{0})]\, d\mathbf{x} + \int_{R^3\setminus v} G_{ij,kl}(\mathbf{x})\phi(\mathbf{x})\, d\mathbf{x}, \tag{3.17}$$

where $\phi(\mathbf{x})$ is an arbitrary twice differentiable finite function, and V and S are the volume and the surface of integration domain containing an origin of coordinates where the power singularity is located. The integral of the formal constituent of the Green's function converges in the singular point because $\phi(\mathbf{x}) - \phi(\mathbf{0}) \to 0$ as $\mathbf{x} \to \mathbf{0}$. The surface integral of the singular constituent also converges because the singular point is not located at the integration domain.

Let us obtain the decomposition (3.17) for the ellipsoidal domain v described by the equation

$$\mathbf{x}\mathbf{a}^{-2}\mathbf{x} = x_i(\mathbf{a}^{-2})_{ij}x_j = 1 \tag{3.18}$$

where $\mathbf{a}$ is the second-order symmetric tensor with $\det \mathbf{a} > 0$. According to the regularization (3.17), the distribution

$$U_{kijl}(\mathbf{x}) = [\partial_k \partial_l G_{ij}(\mathbf{x})]_{(ki)(jl)} \tag{3.19}$$

can be presented in the form

$$\mathbf{U}(\mathbf{x}) = \mathbf{U}^{(f)}(\mathbf{x}) - \mathbf{P}\delta(\mathbf{x}), \tag{3.20}$$

where $\mathbf{U}^{(f)}(\mathbf{x}) = \nabla\nabla\mathbf{G}^{(f)}$ is a regular function, and $\mathbf{P}$ is a constant tensor

$$\mathbf{P} = -\int \nabla\mathbf{G}(\mathbf{x})\mathbf{n}(\mathbf{x})\, d\mathbf{s}, \tag{3.21}$$

where $\mathbf{n}(\mathbf{x})$ is an outer normal in the point $\mathbf{x} \in S$ to the surface S of the domain v. Transforming Eq. (3.21) by application of the Gauss theorem (2.23) yields

$$\mathbf{P} = \int \nabla\mathbf{G}(\mathbf{x})\nabla V(\mathbf{x})\, d\mathbf{x}, \tag{3.22}$$

where $V(\mathbf{x})$ is a characteristic function of the domain v. The relation (3.22) can be recast in the form

$$\mathbf{P} = (2\pi)^{-3} \int \mathbf{k}\mathbf{G}(\mathbf{k})\mathbf{k}V(\mathbf{k})\, d\mathbf{k} \equiv -(2\pi)^{-3} \int \mathbf{U}(\mathbf{k})V(\mathbf{k})\, d\mathbf{k}, \tag{3.23}$$

where the Parserval equality was exploited. The affine transform $\mathbf{x} = \mathbf{a}\mathbf{y}$ transforms the ellipsoidal domain v into a unit sphere with a characteristic function $V_1(\mathbf{y}) = V(\mathbf{a}\mathbf{y})$ with a Fourier transform $V_1(\mathbf{a}\mathbf{k}) = (\det \mathbf{a})^{-1}V(\mathbf{k})$. Transition of the integration over the unit sphere in the Fourier space transforms Eq. (3.23) to the form

$$\mathbf{P} = -(2\pi)^{-3} \int \mathbf{U}(\mathbf{a}^{-1}\mathbf{k})V_1(\mathbf{k})\, d\mathbf{k}. \tag{3.24}$$

Integration in Eq. (3.24) over the variable $\mathbf{k} = |\mathbf{k}|\bar{\mathbf{k}}$ is carried out over the surface of unit sphere $\bar{\mathbf{k}}^2 = 1$ and over the scalar $|\mathbf{k}|$. Obviously, $\mathbf{U}(\mathbf{a}^{-1}\mathbf{k}) = \mathbf{U}(\mathbf{a}^{-1}\bar{\mathbf{k}})$ and $V_1(\mathbf{k}) = V_1(|\mathbf{k}|)$ as far as $\mathbf{U}(\mathbf{k})$ is a homogeneous function of zero order and $V_1(\mathbf{k})$ is a spherically symmetric, and, therefore,

$$\mathbf{P} = -\frac{1}{4\pi} \int \mathbf{U}(\mathbf{a}^{-1}\bar{\mathbf{k}})\, d\bar{\mathbf{k}}, \tag{3.25}$$

where the integration is performed over the unit sphere.

It might be well to point out that the constant tensor $\mathbf{P}$ (3.21) depends on the shape (but not on the shape) of the ellipsoid v because $\mathbf{U}(\mathbf{k})$ is a homogeneous function of zero order. Then the integral over v in the definition of the formal constituent (3.17) does not depend on the size of v and vanishes at the shrinking of the ellipsoid v into the point $\mathbf{x} = \mathbf{0}$. Thus, one can define

$$(\mathbf{U}^{(f)}, \phi) = \det\{\mathbf{a}\} \fint \mathbf{U}(\mathbf{ax})\phi(\mathbf{ax})\, d\mathbf{x}, \tag{3.26}$$

where the integral in the Cauchy principle sense exists due to the existence of $(\mathbf{U}^{(f)}, \phi)$.

In the particular case of an isotropic medium, we formally note the first and the second derivatives of the Green's function (3.13) in the form

$$\begin{aligned} G_{ij,k}(\mathbf{x}) &= \frac{1}{8\pi\mu}(\delta_{ij}|\mathbf{x}|_{,ppk} - \varkappa|\mathbf{x}|_{,ijk}), \\ G_{ij,kl}(\mathbf{x}) &= \frac{1}{8\pi\mu}(\delta_{ij}|\mathbf{x}|_{,pplm} - \varkappa|\mathbf{x}|_{,ijkl}), \end{aligned} \tag{3.27}$$

where the derivatives from $|\mathbf{x}|$ until the third order can be directly calculated as follows:

$$\begin{aligned} &|\mathbf{x}|_{,i} = n_i \equiv \frac{x_i}{|\mathbf{x}|}, \quad |\mathbf{x}|_{,ij} = n_{i,j} = \frac{1}{|\mathbf{x}|}(\delta_{ij} - n_{ij}), \quad |\mathbf{x}|_{,pp} = \frac{2}{|\mathbf{x}|}, \\ &|\mathbf{x}|_{,ijk} = n_{i,jk} = \frac{1}{|\mathbf{x}|^2}(3n_{ijk} - n_i\delta_{jk} - n_k\delta_{ik} - n_l\delta_{ik}), \quad |\mathbf{x}|_{,ppk} = -\frac{2}{|\mathbf{x}|^2}n_k, \end{aligned} \tag{3.28}$$

The finding of the higher order derivatives is related to the exploiting the regularization (3.16): $|\mathbf{x}|_{,ijkl} = |\mathbf{x}|^{(s)}_{,ijkl} + |\mathbf{x}|^{(f)}_{,ijkl}$. Then for the spherical domain in the regularization (3.16), we get

$$\begin{aligned} |\mathbf{x}|^{(s)}_{,ijkl} &= \frac{-8\pi}{15}\delta(|\mathbf{x}|)\delta_{ijkl}, \quad |\mathbf{x}|^{(s)}_{,ppkl} = \frac{-8\pi}{3}\delta(|\mathbf{x}|)\delta_{kl}, \quad |\mathbf{x}|^{(s)}_{,ppqq} = -8\pi\delta(|\mathbf{x}|), \\ |\mathbf{x}|^{(s)}_{,ijkl} &= n_{i,jkl} = \frac{1}{|\mathbf{x}|^3}\Big[3(n_{ij}\delta_{kl} + n_{ik}\delta_{jl} + n_{il}\delta_{jk} + n_{kl}\delta_{ij} + n_{jl}\delta_{ik} + n_{jk}\delta_{il}) \\ &\quad - \delta_{ijkl} - 15n_{ijkl}\Big], \quad |\mathbf{x}|^{(f)}_{,ppkl} = \frac{2}{|\mathbf{x}|^3}(3n_{kl} - \delta_{kl}), \quad |\mathbf{x}|^{(f)}_{,ppii} = 0. \end{aligned} \tag{3.29}$$

Calculation of the singular constituent $|\mathbf{x}|^{(s)}_{ijkl}$ is carried out exploiting the transfer to a spherical coordinate system

$$(|\mathbf{x}|^{(s)}_{ijkl}, \phi) = \int |\mathbf{x}|^{(s)}_{ijk} n_l |\mathbf{x}|^2 d\omega \phi(\mathbf{0}).$$

The $\phi(\mathbf{0})$ in the singular constituent of derivatives is stipulated by the known relation $(\delta, \phi) = \phi(\mathbf{0})$.

The determined values of derivatives (3.29) allow us to represent the formulae (3.27) in the closed form:

$$
\begin{aligned}
G_{ij,k} &= -\frac{1}{8\pi\mu|\mathbf{x}|^2}[3\varkappa n_{ijk} + (2-\varkappa)n_k\delta_{ij} - \varkappa(n_i\delta_{jk} + n_j\delta_{ik})], \\
G^{(s)}_{ij,kl} &= -\frac{1}{3\mu}\delta(|\mathbf{x}|)\left[\delta_{ij}\delta_{kl} - \frac{\varkappa}{5}\delta_{ijkl}\right], \\
G^{(f)}_{ij,kl} &= \frac{1}{8\pi\mu|\mathbf{x}|^3}\Big\{(2-\varkappa)\delta_{ij}(3n_{kl} - \delta_{kl}) + \varkappa[2I_{ijkl} + 15n_{ijkl} \\
&\quad - 3(n_{ij}\delta_{kl} + n_{ik}\delta_{jl} + n_{il}\delta_{jk} + n_{jl}\delta_{ik} + n_{jk}\delta_{il})]\Big\}
\end{aligned} \tag{3.30}
$$

For completeness we reproduce the compact representation of these tensors valid for both 2-D and 3-D cases of the isotropic media:

$$
\begin{aligned}
G_{ij,k}(\mathbf{x}) &= \frac{1}{4\omega_d(1-\bar{\nu})\mu|\mathbf{x}|^{d-1}}\Big[(4\bar{\nu}-3)\delta_{ij}n_k + \delta_{ik}n_j \\
&= \delta_{jk}n_i + (d-4)n_in_jn_k\Big],
\end{aligned} \tag{3.31}
$$

$$
\mathbf{U}^{(s)}(\mathbf{x}) = \frac{1}{(d+2)[dk^{(0)} + 2(d-1)\mu^{(0)}]}\left[\frac{k^{(0)} + 2\mu^{(0)}}{\mu^{(0)}}d\mathbf{E}^{(1)} - \frac{\alpha}{d}\mathbf{E}^{(2)}\right]\delta(\mathbf{x}), \tag{3.32}
$$

$$
\mathbf{U}^{(f)} = \sum_{p=1}^{6} b^U_p \mathbf{E}^{(p)}, \tag{3.33}
$$

where the values b^U_p $(p = 1, 2, \ldots, 6)$ are defined by the equations:

$$
b^U_1 = \frac{-2d}{2\omega_d[dk^{(0)} + 2(d-1)\mu^{(0)}]|\mathbf{x}|^d}, \quad b^U_2 = \frac{\alpha}{2\omega_d[dk^{(0)} + 2(d-1)\mu^{(0)}]|\mathbf{x}|^d}, \tag{3.34}
$$

$$
b^U_3 = b^U_4 = \frac{-d\alpha}{2\omega_d[dk^{(0)} + 2(d-1)\mu^{(0)}]|\mathbf{x}|^d}, \tag{3.35}
$$

$$
b^U_5 = \frac{-2d(d-2)\alpha}{2\omega_d[dk^{(0)} + 2(d-1)\mu^{(0)}]|\mathbf{x}|^d}, \quad b^U_6 = \frac{d(d+2)\alpha}{2\omega_d[dk^{(0)} + 2(d-1)\mu^{(0)}]|\mathbf{x}|^d}, \tag{3.36}
$$

with the parameters ω_d and $\bar{\nu}$ as defined in Eqs. (3.14), $\alpha = dk^{(0)}/\mu^{(0)} + (d-2)$ is a dimensionless parameter, and the tensors $\mathbf{E}^{(p)}$ are the basic tensors of the special structure that depend on the direction $\mathbf{n}$ (see Appendix A.2).

3.2.2 The Tensors Related to the Green's Function

Analogously for the distribution (called the fundamental stress tensor)

$$
\boldsymbol{\Gamma}(\mathbf{x}) = -\mathbf{L}(\mathbf{I}\delta(\mathbf{x}) + \mathbf{UL}), \quad \Gamma_{ijkl}(\mathbf{x}) = -L_{ijkl}\delta(\mathbf{x}) - L_{ijmn}U_{mnpq}L_{pqkl} \tag{3.37}
$$

we can define the regular and singular constituents

$$
\begin{aligned}
\boldsymbol{\Gamma}(\mathbf{x}) &= \boldsymbol{\Gamma}^{(f)} + \boldsymbol{\Gamma}^{\delta} = \boldsymbol{\Gamma}^{(f)} - \mathbf{Q}\delta(\mathbf{x}), \\
(\boldsymbol{\Gamma}^{(f)}, \phi) &= det\{\mathbf{a}\} \fint \boldsymbol{\Gamma}(\mathbf{ax})\phi(\mathbf{ax})\, d\mathbf{x}, \quad \mathbf{Q} = -\frac{1}{4\pi}\int \boldsymbol{\Gamma}(\mathbf{a}^{-1}\bar{\mathbf{k}})\, d\bar{\mathbf{k}},
\end{aligned} \tag{3.38}
$$

and the integration in Eq. (3.26) is carried out over the unit sphere $\bar{\mathbf{k}}^2 = 1$, and $\mathbf{Q} = \mathbf{L} - \mathbf{LPL}$.

In parallel with the tensor of "fundamental displacement" $\mathbf{G}$, we need the associated tensor of the "fundamental traction" $\mathbf{T}(\mathbf{x},\mathbf{s})$ which is the homogeneous function of degree $d-1$ locally integrable everywhere and given by the relation

$$T_{im}(\mathbf{x},\mathbf{s}) = L^{(0)}_{ijkl} n^S_j(\mathbf{s})\partial_l G_{km}(\mathbf{x}-\mathbf{s}), \quad \mathbf{x},\mathbf{s} \in R^d. \tag{3.39}$$

For isotropic media, the tensor $\mathbf{T}(\mathbf{x},\mathbf{s})$ and the formal constitutive $\mathbf{S}^{(f)} : (\mathbf{x},\mathbf{s})$ of the tensor $\mathbf{S}(\mathbf{x},\mathbf{s}) \equiv \mathbf{L}^{(0)}\nabla\mathbf{T}(\mathbf{x},\mathbf{s})$ have the components $(\mathbf{x}\neq\mathbf{s})$

$$\begin{aligned} T_{ij}(\mathbf{x},\mathbf{s}) &= \frac{1}{2\omega_d(1-\bar\nu)r^{d-1}}\Big[\big((1-2\bar\nu)\delta_{ij} + dn_in_j\big)n_kn^S_k \\ &- (1-2\bar\nu)(n_in^S_j - n_jn^S_i)\Big], \end{aligned} \tag{3.40}$$

$$\begin{aligned} S^{(f)}_{ijk}(\mathbf{x},\mathbf{s}) &= \frac{\mu}{\omega_d(1-\bar\nu)r^d}\Big\{d\big[(1-2\bar\nu)\delta_{ij}n_k + \bar\nu(\delta_{ik}n_j + \delta_{jk}n_i) \\ &- (d+2)n_in_jn_k\big]n_mn^S_m + d\bar\nu n_k[n_in^S_j + n_jn^S_i] \\ &+ (1-2\bar\nu)[dn_in_jn^S_k + \delta_{ik}n^S_j + \delta_{jk}n^S_i] - (1-4\bar\nu)\delta_{ij}n^S_k\Big\}. \end{aligned} \tag{3.41}$$

Here the tensors $\mathbf{T}(\mathbf{x},\mathbf{s})$ and $\mathbf{S}(\mathbf{x},\mathbf{s})$ are defined at the surface $\mathbf{x},\mathbf{s}\in S$ with the unit outward normal $\mathbf{n}^S = (n^S_1,\ldots,n^S_d)^\top$ at $\mathbf{s}\in S$; $r\equiv|\mathbf{x}-\mathbf{s}|$.

3.3 Dynamic Green's and Related Functions

By analogy with the static Green's tensor, (3.3) we introduce its generalization to dynamic problems. For the differential equation of motion

$$\boldsymbol{\mathcal{L}}\mathbf{u} = -\mathbf{f}, \quad (\boldsymbol{\mathcal{L}})_{ik}(\mathbf{x},t) \equiv \nabla_j L_{ijkl}\nabla_l - \delta_{ik}\rho\frac{\partial^2}{\partial t^2}, \tag{3.42}$$

where $\mathbf{u}(\mathbf{x})|_{|\mathbf{x}|\to\infty} = \mathbf{0}$ and the Green's tensor for the displacement $G_{ij}(\mathbf{x},t)$ in the infinite medium with density ρ is defined from the relation ($\mathbf{x},\mathbf{y}\in R^d$, $d=2,3$):

$$u_k = G_{ki} * f_i \equiv \iint\limits_{-\infty}^{\infty} G_{ki}(\mathbf{x}-\mathbf{y},t-t')f_i(\mathbf{y})\,dt' \tag{3.43}$$

assuming vanishing of $\mathbf{G}(\mathbf{x},t)$ at the infinity $|\mathbf{x}|\to\infty$. The differential equation for $\mathbf{G}$ is produced by the action of the operator $(\boldsymbol{\mathcal{L}})_{ik}$ on Eq. (3.43):

$$\boldsymbol{\mathcal{L}}\mathbf{G}(\mathbf{x},t) = -\boldsymbol{\delta}\delta(\mathbf{x})\delta(t), \quad \mathcal{L}_{ik}G_{kj}(\mathbf{x},t) = -\delta_{ij}\delta(\mathbf{x})\delta(t), \quad (t\geq 0), \tag{3.44}$$

and $\mathbf{G}(\mathbf{x},t) = \mathbf{0}$, $G_{kj}(\mathbf{x},t) = 0$, $(t<0)$. Thus, the kj-component of the symmetric Green's tensor $G_{kj}(\mathbf{x},t)$ defines the displacement component in the x_k-direction at point $\mathbf{x}$ and time t, generated by a unit impulsive force applied in the x_j direction at point $\mathbf{x}=\mathbf{0}$ and time $t=0$. Addition of the second equation (3.44_2) is stipulated by a causality principle, according to which the response in the point $\mathbf{x}$ should arise only after applying the unit impulse in the origin.

The construction of the dynamic Green's tensor is carried out by the use of the Fourier transforms over the coordinates and time (for details see [794]):

$$\mathbf{G}(\mathbf{k},\omega) = \iint_{-\infty}^{\infty} \mathbf{G}(\mathbf{x},t)e^{i(\mathbf{k}\cdot\mathbf{x}-\omega t)}\, d\mathbf{x}\, dt,$$

$$\mathbf{G}(\mathbf{x},t) = \frac{1}{(2\pi)^{-d}} \iint_{-\infty}^{\infty} \mathbf{G}(\mathbf{k},\omega)e^{-i(\mathbf{k}\cdot\mathbf{x}-\omega t)}\, d\mathbf{k}\, d\omega.$$

Performing the Fourier transform of Eq. (3.44) yields the algebraic equation

$$L_{ijkl}k_jk_lG_{kn}(\mathbf{k},\omega) - \rho\omega^2 G_{kn}(\mathbf{k},\omega)\delta_{ik} = \delta_{in}, \tag{3.45}$$

whence it follows that

$$G_{ij}(\mathbf{k},\omega) = K^*_{ij}(\mathbf{k},\omega)/|\mathbf{K}(\mathbf{k},\omega)|, \tag{3.46}$$

where $\mathbf{K}^*$ is a cofactor of an element K_{kn} in a symmetric matrix

$$\mathbf{K}(\mathbf{k},\omega) = \begin{pmatrix} K_{11}(\mathbf{k}) - \rho\omega^2 & K_{12}(\mathbf{k}) & K_{13}(\mathbf{k}) \\ K_{21}(\mathbf{k}) & K_{22}(\mathbf{k}) - \rho\omega^2 & K_{23}(\mathbf{k}) \\ K_{31}(\mathbf{k}) & K_{32}(\mathbf{k}) & K_{33}(\mathbf{k}) - \rho\omega^2 \end{pmatrix}, \tag{3.47}$$

with the following representation of the tensor $\mathbf{K}^*$ and the determinant $|\mathbf{K}(\mathbf{k},\omega)|$:

$$\begin{aligned} K^*_{ij} &= K_{im}K_{mj} + (\rho\omega^2)K_{ij} + [(\rho\omega^2)^2 - \rho\omega^2 A + B]\delta_{ij}, \\ |\mathbf{K}(\mathbf{k},\omega)| &= -(\rho\omega^2)^3 + (\rho\omega^2)^2 A - \rho\omega^2 B + C, \end{aligned} \tag{3.48}$$

where

$$\begin{aligned} &A = K_{mm}, \quad C = \epsilon_{mnl}K_{m1}K_{n2}K_{l3}, \\ &B = \epsilon_{mn1}K_{m2}K_{n3} + \epsilon_{mn2}K_{m3}K_{n1} + \epsilon_{mn3}K_{m1}K_{n2}. \end{aligned}$$

By virtue of the fact that the equation $|\mathbf{K}(\mathbf{k},\omega)| = 0$ contains the degrees of $\rho\omega^2$, it has the roots $\rho\omega_1^2$, $\rho\omega_2^2$, $\rho\omega_3^2$. Then removing the parentheses in the representation $|\mathbf{K}(\mathbf{k},\omega)| = \rho^3(\omega_1^2-\omega^2)(\omega_2^2-\omega^2)(\omega_3^2-\omega^2)$ and comparison of the relation obtained with Eq. (3.48_1) leads to $A = \rho(\omega_1^2+\omega_2^2+\omega_3^2)$, $C = \rho^3\omega_1^2\omega_2^2\omega_3^2$, and $B = \rho^2(\omega_1^2\omega_2^2+\omega_2^2\omega_3^2+\omega_1^2\omega_3^2)$. Then Eq. (3.46) can be presented in the form

$$G_{ij}(\mathbf{k},\omega) = \sum_{\nu=1}^{3} \frac{{}^\nu\phi_i{}^\nu\phi_j}{\rho(\omega_\nu^2 - \omega^2)}, \tag{3.49}$$

where

$${}^1\phi_i{}^1\phi_j = \frac{(K_{im} - \rho\omega_2^2\delta_{im})(K_{mj} - \rho\omega_3^2\delta_{mj})}{\rho^2(\omega_1^2-\omega_2^2)(\omega_1^2-\omega_3^2)} \tag{3.50}$$

and ${}^2\phi_i$, ${}^2\phi_j$, ${}^3\phi_i$, ${}^3\phi_j$ are obtained by cyclic permutation of indexes $(1,2,3)$. ${}^\nu\phi_i$ and $\rho\omega_\nu^2$ are the eigenvectors and eigenvalues, respectively, of the matrix $\|\mathbf{K}(\mathbf{k},\omega)\|$ (3.47):

$$(K_{ij}(\mathbf{k},\omega)-\rho\omega_\nu^2\delta_{ij})^\nu\phi_j=0,\quad \sum_{\nu=1}^3 {}^\nu\phi_i{}^\nu\phi_j=\delta_{ij}. \tag{3.51}$$

By use of the inverse Fourier transform, one can find from Eq. (3.49)

$$G_{ij}(\mathbf{x},t)=\frac{1}{(4\pi)^4\rho}\int_{-\infty}^{\infty}\sum_{\nu=1}^3\frac{\exp(-i\omega t)}{\omega_\nu^2-\omega^2}\,d\omega\int_0^\infty|\mathbf{k}|^2d|\mathbf{k}|$$
$$\times\int_{S_1}{}^\nu\phi_i(\bar{\mathbf{k}}){}^\nu\phi_j(\bar{\mathbf{k}})\exp(i\mathbf{k}\cdot\mathbf{x})\;d\mathbf{s}(\bar{\mathbf{k}}), \tag{3.52}$$

where S_1 is the surface of the unit sphere $S_1=\{\mathbf{k}||\bar{\mathbf{k}}|=1\}$, $\bar{\mathbf{k}}\equiv\mathbf{k}/|\mathbf{k}|$, and it is taken into account that ${}^\nu\phi_i{}^\nu\phi_j$ are the homogeneous functions of $\mathbf{k}$ of degree 0.

We present only the final relation for the integral (3.52) obtained by the use of the residue theory:

$$G_{ij}(\mathbf{x},t)=\frac{-H(t)}{8\pi^2\rho}\frac{\partial}{\partial t}\sum_{\nu=1}^3\int_{S_1}\delta(\bar\omega_\nu t-\bar{\mathbf{k}}\cdot\mathbf{x}){}^\nu\phi_i(\bar{\mathbf{k}}){}^\nu\phi_j(\bar{\mathbf{k}})\bar\omega_\nu^{-2}(\bar{\mathbf{k}})\;d\mathbf{s}(\bar{\mathbf{k}}), \tag{3.53}$$

where $\bar\omega_\nu=\omega_\nu/|\mathbf{k}|$, and $H(t)$ is the Heaviside step function.

Let us consider the case of an isotropic medium when

$$\begin{aligned}K_{ij}(\mathbf{k})&=(\lambda+\mu)k_ik_j-\mu\delta_{ij}|\mathbf{k}|^2,\\ |\mathbf{K}|&=(\mu|\mathbf{k}|^2-\rho\omega^2)[(\lambda+2\mu)|\mathbf{k}|^2-\rho\omega^2],\\ \rho\omega_1^2&=(\lambda+2\mu)|\mathbf{k}|^2,\quad \rho\omega_2^2=\rho\omega_3^2=\mu|\mathbf{k}|^2,\end{aligned} \tag{3.54}$$

and we get from (3.50) and (3.51):

$${}^1\phi_i{}^1\phi_j=k_ik_j/|\mathbf{k}|^2,\quad {}^2\phi_i{}^2\phi_j+{}^3\phi_i{}^3\phi_j=\frac{\delta_{ij}|\mathbf{k}|^2-k_ik_j}{|\mathbf{k}|^2}. \tag{3.55}$$

The representation (3.53) can be simplified after substitution of Eqs. (3.55):

$$G_{ij}(\mathbf{x},t)=\frac{-H(t)}{8\pi^2\rho}\frac{\partial}{\partial t}\Big\{\int_{S_1}\delta(c_{\rm L}t-\bar{\mathbf{k}}\cdot\mathbf{x})\bar k_i\bar k_jc_{\rm L}^{-2}\;d\mathbf{s}(\bar{\mathbf{k}})$$
$$+\int_{S_1}\delta(c_{\rm T}t-\bar{\mathbf{k}}\cdot\mathbf{x})(\delta_{ij}-\bar k_i\bar k_j)c_{\rm T}^{-2}\;d\mathbf{s}(\bar{\mathbf{k}})\Big\}, \tag{3.56}$$

where one introduces the notations

$$c_{\rm L}=\sqrt{\frac{\lambda+2\mu}{\rho}},\quad c_{\rm T}=\sqrt{\frac{\mu}{\rho}}, \tag{3.57}$$

and uses the integrals

$$\int_{S_1}\delta(c_{\rm L}t-\bar{\mathbf{k}}\cdot\mathbf{x})\bar k_i\bar k_jds(\bar{\mathbf{k}})=|\mathbf{x}|^{-1}H(|\mathbf{x}|-c_{\rm L}t)\int_0^{2\pi}\bar k_i\bar k_j\;d\varphi.$$

When $c_{\rm L}t-\bar{\mathbf{k}}\cdot\mathbf{x}=0$, we have

$$\bar{k}_i = c_{\rm L} t \bar{x}_i/|\mathbf{x}| + \eta_i, \quad \int_0^{2\pi} \eta_i \eta_j \, d\varphi = \pi(\delta_{ij} - \bar{x}_i \bar{x}_j)(1 - c_{\rm L}^2 t^2 |\mathbf{x}|^{-2}),$$

$$\int_0^{2\pi} \bar{\xi}_i \bar{\xi}_j \, d\varphi = 2\pi (c_{\rm L} t/|\mathbf{x}|)^2 \bar{x}_i \bar{x}_j + \pi(\delta_{ij} - \bar{x}_i \bar{x}_j)(1 - c_{\rm L}^2 t^2 |\mathbf{x}|^{-2}), \tag{3.58}$$

and finally

$$G_{ij}(\mathbf{x}, t) = \frac{H(t)}{4\pi\rho|\mathbf{x}|} \Big[\delta(|\mathbf{x}| - c_{\rm L} t) \bar{x}_i \bar{x}_j c_{\rm L}^{-1} + \delta(|\mathbf{x}| - c_{\rm T} t)(\delta_{ij} - \bar{x}_i \bar{x}_j) c_{\rm T}^{-1}$$
$$+ t|\mathbf{x}|^{-2} (\delta_{ij} - 3\bar{x}_i \bar{x}_j) \{ H(|\mathbf{x}| - c_{\rm L} t) - H(|\mathbf{x}| - c_{\rm T} t) \} \Big]. \tag{3.59}$$

As can be seen, $c_{\rm L}$ and $c_{\rm T}$ introduced by the equalities (3.57) have the meaning of the longitudinal and transverse wave speeds, respectively. The summand with a cofactor $\delta(|\mathbf{x}| - c_{\rm T} t)$ (3.59) describes the impulses where the displacements are perpendicular to the direction of its propagation, and the summand with the cofactor $\delta(\mathbf{x}| - c_{\rm L} t)$ corresponds to the impulses with the transversal direction of displacement. The last summand in Eq. (3.59) describes the perturbation arising between the arrival of impulses produced the longitudinal and transverse displacements.

From Eqs. (3.46) and (3.47) the following relationships between the dynamic and static tensors in the Fourier space $\mathbf{K}(\mathbf{k}, 0) = \mathbf{K}(\mathbf{k})$, $|\mathbf{K}(\mathbf{k}, 0)| = |\mathbf{K}(\mathbf{k})|$ follow and therefore, according to (3.46) and (3.53), $\mathbf{G}(\mathbf{k}, 0) = \mathbf{G}(\mathbf{k})$ leading to the connection between the static and dynamic Green's tensors:

$$\mathbf{G}(\mathbf{x}) = \int_{-\infty}^{\infty} \mathbf{G}(\mathbf{x}, t) \, dt. \tag{3.60}$$

We will demonstrate employing (3.60) that for the isotropic medium Eq. (3.59) immediately yields the representation (3.13) for the static Green's tensor. Actually, integration of Eq. (3.59) over time leads to

$$G_{ij}(\mathbf{x}) = \frac{1}{8\pi\rho|\mathbf{x}|} \Big[\frac{2}{c_{\rm T}^2} \delta_{ij} + (\frac{1}{c_{\rm L}^2} - \frac{1}{c_{\rm T}^2})(\delta_{ij} - \bar{x}_i \bar{x}_j) \Big] \tag{3.61}$$

coinciding with formula (3.13) according to Eqs. (3.57).

Now we will consider the widely occurring case of dynamic process such as the wave or steady-state process when the displacements in the medium are described by the relation $\mathbf{u}(\mathbf{x}, t) = \mathbf{u}(\mathbf{x}) \exp(-i\omega t)$ where $\mathbf{u}(\mathbf{x})$ is the amplitude satisfying to the Helmholtz equation

$$\boldsymbol{\mathcal{L}}^{\omega}(\mathbf{x}) \mathbf{u}(\mathbf{x}) = -\mathbf{f}(\mathbf{x}), \ \mathcal{L}_{ik}^{\omega} = \delta_{ik} \rho \omega^2 + \nabla_j L_{ijkl} \nabla_l. \tag{3.62}$$

If a unit body force applied at the origin of coordinates in the x_m-direction with time dependence $\exp(-i\omega t)$, then the ith component of a displacement at $\mathbf{x}$ can be presented in the form $g_{im}(\mathbf{x}) \exp(i\omega t)$ where $g_{im}(\mathbf{x})$, called the wave Green's tensor, satisfies the motion equation

$$L_{ijkl} g_{km,lj}(\mathbf{x}) + \rho \omega^2 g_{jm}(\mathbf{x}) \delta_{ij} + \delta_{im} \delta(\mathbf{x}) = 0. \tag{3.63}$$

To find the solution of Eq. (3.63) we will apply to the obtained value $\mathbf{G}(\mathbf{k},\omega)$ (3.49) the inverse Fourier transform only over the coordinates

$$g_{ij}(\mathbf{x}) = (2\pi)^{-3} \int_{\infty}^{\infty} G_{ij}(\mathbf{k},\omega) \exp(i\mathbf{k}\cdot\mathbf{x})\, d\mathbf{k}. \tag{3.64}$$

For the isotropic medium, we have from Eq. (3.59):

$$G_{ij}(\mathbf{k},\omega) = \frac{k_i k_j}{|\mathbf{k}|^2(\lambda+2\mu)(|\mathbf{k}|^2 - k_{\rm L})} + \frac{\delta_{ij}|\mathbf{k}|^2 - k_i k_j}{|\mathbf{k}|\mu(|\mathbf{k}|^2 - k_{\rm T})}, \tag{3.65}$$

where $k_{\rm L} = \omega/c_{\rm L}$ and $k_{\rm T} = \omega/c_{\rm T}$ are the wave numbers. Substitution of Eq. (3.65) into (3.44) leads to the final representation for the time-reduced wave Green's tensor for the isotropic medium:

$$g_{ij}(\mathbf{x}) = \frac{1}{4\pi\rho\omega^2|\mathbf{x}|}\left\{k_{\rm T}^2\delta_{ij}e^{ik_{\rm T}|\mathbf{x}|} - |\mathbf{x}|\frac{\partial^2}{\partial x_i\partial x_j}\left[\frac{\exp(ik_{\rm L}|\mathbf{x}|)}{|\mathbf{x}|} - \frac{\exp(ik_{\rm T}|\mathbf{x}|)}{|\mathbf{x}|}\right]\right\}. \tag{3.66}$$

3.4 Inhomogeneity in an Elastic Medium

3.4.1 General Case of Inhomogeneity in an Elastic Medium

Let an inhomogeneity v with an arbitrary shape and characteristic function V be included in an infinite homogeneous matrix. Stresses and strains are related to each other via the constitutive equations $\boldsymbol{\sigma}(\mathbf{x}) = \mathbf{L}(\mathbf{x})\boldsymbol{\varepsilon}(\mathbf{x}) + \boldsymbol{\alpha}(\mathbf{x})$ or $\boldsymbol{\varepsilon}(\mathbf{x}) = \mathbf{M}(\mathbf{x})\boldsymbol{\sigma}(\mathbf{x}) + \boldsymbol{\beta}(\mathbf{x})$ (2.126). All tensors of material properties $\mathbf{g}$ (e.g., $\mathbf{g} = \mathbf{L}, \mathbf{M}, \boldsymbol{\alpha}, \boldsymbol{\beta}$) are decomposed as

$$\mathbf{g} \equiv \mathbf{g}^{(0)} + \mathbf{g}_1(\mathbf{x}) = \mathbf{g}^{(0)} + \mathbf{g}_1(\mathbf{x})V(\mathbf{x}), \tag{3.67}$$

where $\mathbf{g}^{(0)}(\mathbf{x}) \equiv \mathbf{g}^{(0)}$ =const in the matrix ($\mathbf{x} \in v^{(0)}$). It is assumed that $\mathbf{L}(\mathbf{x})$ and $\mathbf{L}^{(0)}$ are positive definite. To hold this property in the cases of a cavity and rigid inclusions, the later are be modeled by the use of limiting passing ($\mathbf{x} \in v$):

$$\begin{aligned} &\mathbf{L}_1(\mathbf{x}) \to -\mathbf{L}^{(0)}, \quad \mathbf{M}_1(\mathbf{x}) \to \mathbf{0}, \ \text{for a cavity},\\ &\mathbf{L}_1(\mathbf{x}) \to \infty, \quad \mathbf{M}_1(\mathbf{x}) \to -\mathbf{M}^{(0)}, \ \text{for a rigid inclusions}. \end{aligned} \tag{3.68}$$

The initial equilibrium equation for the displacement $\mathbf{u}(\mathbf{x})$ has form

$$\begin{aligned} &\boldsymbol{\mathcal{L}}\mathbf{u} + \nabla\boldsymbol{\alpha} \equiv (\boldsymbol{\mathcal{L}}^{(0)} + \boldsymbol{\mathcal{L}}_1)\mathbf{u} + \nabla\boldsymbol{\alpha} = -\mathbf{f},\\ &\nabla_j[\mathbf{L}_{ijkl}(\mathbf{x})\nabla_l u_k(\mathbf{x}) + \boldsymbol{\alpha}_{lk}(\mathbf{x})] = -f_i(\mathbf{x}), \end{aligned} \tag{3.69}$$

where $\boldsymbol{\mathcal{L}}^{(0)} = \nabla\mathbf{L}^{(0)}\nabla,\ \boldsymbol{\mathcal{L}}_1 = \nabla\mathbf{L}_1\nabla$ and the density of external forces is assumed to be a generalized function, decreasing sufficiently rapidly at infinity. The body force tensor $\mathbf{f}$ can be generated, e.g., by either gravitational loads or a centrifugal load. If $\mathbf{f}$ has no singularities of the type of the single or double layer assumed hereafter on the boundary S of the inhomogeneity v, Eq. (3.69) automatically provides conjunction conditions at the boundary $\mathbf{x} \in S$:

$$\mathbf{u}^-(\mathbf{x}^-) = \mathbf{u}^+(\mathbf{x}^+), \quad \mathbf{n}\boldsymbol{\sigma}^-(\mathbf{x}^-) = \mathbf{n}\boldsymbol{\sigma}^+(\mathbf{x}^+), \; \mathbf{x} \in S, \tag{3.70}$$

where $\mathbf{g}^-$ and $\mathbf{g}^+$ ($\mathbf{g} = \mathbf{u}, \boldsymbol{\sigma}$) are the limiting values of the function $\mathbf{g}$ outside and inside, respectively , near the boundary S:

$$\mathbf{g}^-(\mathbf{x}^-) = \lim_{\mathbf{y}\to\mathbf{x}} \mathbf{g}(\mathbf{y}), \quad \mathbf{g}^+(\mathbf{x}^+) = \lim_{\mathbf{z}\to\mathbf{x}} \mathbf{g}(\mathbf{z}), \tag{3.71}$$

where $\mathbf{n}\bot S$ is an outward normal vector on S at the point $\mathbf{x} \in S$, and $\mathbf{y} \in v^{(0)}$, $\mathbf{z} \in v$.

Let $\mathbf{u}^0(\mathbf{x})$ be a displacement field that would be present in a homogeneous medium ($\mathbf{L}_1(\mathbf{x}) = \boldsymbol{\beta}_1(\mathbf{x}) \equiv \mathbf{0}$) under the action of the body force $\mathbf{f}^0$:

$$\boldsymbol{\mathcal{L}}\mathbf{u}^0 = -\mathbf{f}^0. \tag{3.72}$$

This field tends to zero at infinity and can be uniquely presented in the form

$$\mathbf{u}^0 = \mathbf{G} * \mathbf{f}^0, \; u_i^0(\mathbf{x}) = \int G_{ij}(\mathbf{x}-\mathbf{y}) f_j^0(\mathbf{x}) \, d\mathbf{x}, \tag{3.73}$$

where $\mathbf{G}$ (3.2) is the Green's function for the displacement.

For the inhomogeneous displacement field $\mathbf{u}_1(\mathbf{x}) \equiv \mathbf{u}(\mathbf{x}) - \mathbf{u}^0(\mathbf{x})$ produced by the inhomogeneity v, we get from Eqs. (3.69) and (3.72) ($\mathbf{u}_1(\mathbf{x}) \to \mathbf{0}$, as $\mathbf{x} \to \infty$)

$$\begin{aligned} \boldsymbol{\mathcal{L}}^{(0)}\mathbf{u}_1 &= -\boldsymbol{\mathcal{L}}_1\mathbf{u} - \nabla\boldsymbol{\alpha}_1 - \mathbf{f}_1, \\ \nabla_j L^{(0)}_{ijkl} \nabla_l u_{1k}(\mathbf{x}) &= -\nabla_j L_{1ijkl}(\mathbf{x}) \nabla_l u_k(\mathbf{x}) - \nabla_j \alpha_{1ij}(\mathbf{x}) - \mathbf{f}_{1i}. \end{aligned} \tag{3.74}$$

In so doing, according to the potential theory the solutions of Eqs. (3.72) and (3.74) exist and are unique and $\boldsymbol{\varepsilon}^1(\mathbf{x}) = \nabla\mathbf{u}^1(\mathbf{x})$ has at the surface S of the inhomogeneity a jump providing a continuity of the normal constituent of stresses $\boldsymbol{\sigma}(\mathbf{x})$. The right-hand side of Eq. (3.74) is formally equivalent to the presence of a fictitious body force that allows us to reduce Eq. (3.74) to an integral equation by the use of the Green's tensor $\mathbf{G}(\mathbf{x})$ (3.2) for the displacement

$$\begin{aligned} \mathbf{u}_1 &= \mathbf{G} * \nabla(\mathbf{L}_1\boldsymbol{\varepsilon} + \boldsymbol{\alpha}_1) + \mathbf{G} * \mathbf{f}_1, \\ u_{1k} &= \int G_{ki}(\mathbf{x}-\mathbf{y}) \nabla_j [L_{1ijml}\varepsilon_{lm}(\mathbf{y}) + \alpha_{1ij}(\mathbf{x})] d\mathbf{y} + \int G_{ki}(\mathbf{x}-\mathbf{y}) f_{1i}(\mathbf{y}) d\mathbf{y}. \end{aligned} \tag{3.75}$$

Symmetrized derivation of Eq. (3.75) with subsequent integration of the first integral obtained by parts leads to the equations for strains:

$$\begin{aligned} \boldsymbol{\varepsilon}(\mathbf{x}) = \boldsymbol{\varepsilon}^0(\mathbf{x}) &+ \int \mathbf{U}(\mathbf{x}-\mathbf{y})[\mathbf{L}_1(\mathbf{y})\boldsymbol{\varepsilon}(\mathbf{y}) + \boldsymbol{\alpha}_1(\mathbf{y})] \, d\mathbf{y} \\ &+ \int \nabla\mathbf{G}(\mathbf{x}-\mathbf{y})\mathbf{f}_1(\mathbf{y}) \, d\mathbf{y}, \end{aligned} \tag{3.76}$$

where it was used that $\nabla_{\mathbf{y}} = -\nabla_{\mathbf{x}}$. Equation (3.76) can be recast in terms of stresses

$$\sigma(\mathbf{x}) = \sigma^0(\mathbf{x}) + \int \Gamma(\mathbf{x}-\mathbf{y})[\mathbf{M}_1(\mathbf{y})\sigma(\mathbf{y}) + \beta_1(\mathbf{y})]\, d\mathbf{y}$$
$$+ \int \mathbf{L}^{(0)}\nabla\mathbf{G}(\mathbf{x}-\mathbf{y})\mathbf{f}_1(\mathbf{y})\, d\mathbf{y}, \qquad (3.77)$$

by the use of identities $\mathbf{L}_1(\varepsilon-\beta) = -\mathbf{L}^{(0)}\mathbf{M}_1\sigma$ and $\varepsilon = [\mathbf{M}^{(0)}\sigma+\beta^{(0)}]+[\mathbf{M}_1\sigma+\beta_1]$; here the convolution kernels $\mathbf{U}(\mathbf{x}-\mathbf{y})$ and $\Gamma(\mathbf{x}-\mathbf{y})$ are found via Eqs. (3.19) and (3.37), respectively.

Equations (3.76) and (3.77) can be represented in the operator form:

$$\varepsilon = \varepsilon^0 + \mathbf{U}(\mathbf{L}_1\varepsilon + \alpha_1)V, \quad \sigma = \sigma^0 + \Gamma(\mathbf{M}_1\sigma + \beta_1)V, \qquad (3.78)$$

where it was assumed that $\mathbf{f}_1 \equiv \mathbf{0}$ and the operators $\mathbf{U}$ and Γ have the convolution kernels $\mathbf{U}(\mathbf{x}-\mathbf{y})$ (3.19) and $\Gamma(\mathbf{x}-\mathbf{y})$ (3.37).

In particular, for a cavity (3.68_1) and a rigid inclusion (3.68_2), Eqs. (3.78_1) and (3.78_2) are reduced to

$$\varepsilon = \varepsilon^0 - \mathbf{U}(\mathbf{L}^{(0)}\varepsilon + \alpha^{(0)})V, \quad \sigma = \sigma^0 - \Gamma(\mathbf{M}^{(0)}\sigma - \beta_1)V, \qquad (3.79)$$

respectively.

Let us consider a related problem for a gas-saturated pore at the gas pressure p^0 ($\mathbf{L}_1(\mathbf{x}) \to -\mathbf{L}^{(0)}$, $\mathbf{x} \in v$; see (3.68_1)). Substituting the constitutive equation in the pore:

$$\sigma(\mathbf{x}) = (1 - V(\mathbf{x}))\mathbf{L}^{(0)}\varepsilon(\mathbf{x}) - \mathbf{p}(\mathbf{x}), \qquad (3.80)$$

where $p_{ij}(\mathbf{x}) = p^0\delta_{ij}V(\mathbf{x})$, into the equilibrium equation leads to the differential equation for the displacement analogous to Eq. (3.74) ($\mathbf{u}_1(\mathbf{x}) \to \mathbf{0}$, as $\mathbf{x} \to \infty$):

$$\mathcal{L}^{(0)}\mathbf{u}_1 = \mathcal{L}^{(0)}(V\mathbf{u}) + \nabla\mathbf{p} - \mathbf{f}_1,$$
$$\nabla_j L^{(0)}_{ijkl}\nabla_l u_{1k}(\mathbf{x}) = \nabla_j L^{(0)}_{ijkl}[\nabla_l u_k(\mathbf{x})V(\mathbf{x})] + \nabla_j p_{ij}(\mathbf{x}) - f_{1i}, \qquad (3.81)$$

which can be transformed to the integral one similar to obtaining Eq. (3.76):

$$\varepsilon(\mathbf{x}) = \varepsilon^0(\mathbf{x}) - \int \mathbf{U}(\mathbf{x}-\mathbf{y})[\mathbf{L}^{(0)}(\mathbf{y})\varepsilon(\mathbf{y}) + \mathbf{p}(\mathbf{y})]V(\mathbf{y})\, d\mathbf{y}$$
$$+ \int \nabla\mathbf{G}(\mathbf{x}-\mathbf{y})\mathbf{f}_1(\mathbf{y})\, d\mathbf{y}, \qquad (3.82)$$

It should be noted that Eqs. (3.76) and (3.82) coincide with $\beta^{(0)} = \mathbf{0}$, $\mathbf{L}_1(\mathbf{x}) \to -\mathbf{L}^{(0)}$, $\mathbf{L}(\mathbf{x})\beta_1(\mathbf{x}) = \mathbf{p}(\mathbf{x})$ taken at $\mathbf{x} \in v$. Therefore, one needs only to investigate the Eq. (3.76).

It should be mentioned that the integrands in Eqs. (3.76), (3.77), and (3.82) have the bounded support $V(\mathbf{y})$. Therefore, the problem of the estimation of the fields $\sigma(\mathbf{x})$, $\varepsilon(\mathbf{x})$ is solved inside and outside the inclusion v if the fields $\sigma(\mathbf{y})$ and $\varepsilon(\mathbf{y})$ ($\mathbf{y} \in v$) are found inside the domain v.

3.4.2 Interface Boundary Operators

Let us turn to the analysis of the fields $\boldsymbol{\sigma}(\mathbf{x})$ and $\boldsymbol{\varepsilon}(\mathbf{x})$ in the matrix in the vicinity of the sufficiently smooth inclusion surface S. It happened that the relations for the jump of the stresses and strains at the interface between two media with homogeneous elastic moduli $\mathbf{L}^-$ and $\mathbf{L}^+$ and stress-free strains $\boldsymbol{\beta}^-$ and $\boldsymbol{\beta}^+$, respectively, can be explicitly found without a concrete representation of the Green's tensor $\mathbf{G}(\mathbf{x})$ (3.3). Ideal boundary conditions of continuity of displacement and of the normal stress components are assumed

$$u_i^-(\mathbf{x}^-) = u^+(\mathbf{x}^+), \ n_i\sigma_{ij}^-(\mathbf{x}^-) = n_i\sigma_{ij}^+(\mathbf{x}^+), \tag{3.83}$$

at the interface with the unit outward tensor $\mathbf{n}\bot S$ in the point $\mathbf{x} \in S$. Here and below the symbols $^+$ and $^-$ relate to the different boundary sides.

According to [464] (see also [618]) one defines the projective operators $\boldsymbol{\eta}$, $\boldsymbol{\nu}$ and $\mathbf{E}$, $\mathbf{F}$ of the second order and fourth order:

$$\begin{aligned} \eta_{kl} &\equiv n_k n_l, \quad \nu_{kl} \equiv \delta_{kl} - \eta_{kl}, \\ F_{klmn} &\equiv (\nu_{km}\nu_{ln} + \nu_{ln}\nu_{km})/2, \quad E_{klmn} \equiv I_{klmn} - F_{klmn} \end{aligned} \tag{3.84}$$

prescribed at the surface S with the outward normal vector $\mathbf{n}$ at the considered surface point; the tensors F_{ijkl} and E_{ijkl} are symmetric with respect to the replacement $i \leftrightarrow j$, $k \leftrightarrow l$, $ij \leftrightarrow kl$.

By testing we obtain immediately "orthogonal" properties of operators (3.84)

$$\begin{aligned} &\boldsymbol{\eta\eta} = \boldsymbol{\eta}, \quad \boldsymbol{\nu\nu} = \boldsymbol{\nu}, \quad \boldsymbol{\nu\eta} = \mathbf{0}, \ \boldsymbol{\eta} + \boldsymbol{\nu} = \boldsymbol{\delta}, \ \mathbf{E} + \mathbf{F} = \mathbf{I}, \\ &\mathbf{FF} = \mathbf{F}, \ \mathbf{EE} = \mathbf{E}, \quad \mathbf{E}\boldsymbol{\nu} = 0, \quad \mathbf{F}\boldsymbol{\nu} = 0, \quad \mathbf{FE} = 0. \end{aligned} \tag{3.85}$$

Orthogonal tensors $\mathbf{E}$ and $\mathbf{F}$ allows us to decompose an arbitrary tensor of the second order t_{ij} on two constituents:

$$t_{ij} = \tau_{ij} + \theta_{ij}, \quad \tau_{ij} = E_{ijkl}t_{kl}, \quad \theta_{ij} = F_{ijkl}t_{kl}, \tag{3.86}$$

which can be also presented in the form

$$\begin{aligned} \tau_{ij} &= [n_i(\delta_{jk} - n_j n_k) + n_j(\delta_{ik} - n_i n_k) + n_i n_j n_k]t_{kl}n_l, \\ \theta_{ij} &= (\delta_{ik} - n_i n_k)(\delta_{jl} - n_j n_l)t_{kl}. \end{aligned}$$

The decompositions (3.86) are most readily illustrated in the local coordinate system with the origin at the point $\mathbf{x} \in S$ when for the outward normal takes the representation $\mathbf{n} = (0, 0, 1)^\top$. Then the tensorial operator $\mathbf{E}$ conserves the tensor components t_{ij} containing the index 3 and vanishes all others. On the contrary, operator $\mathbf{F}$ conserves components t_{ij}, which do not contain the index 3, and vanishes the components containing index 3. Therefore, the tensors τ_{ij} and θ_{ij} can be named the normal and tangential constituents of the tensor t_{ij}, respectively, with respect to the point $\mathbf{x} \in S$ being considered.

Perfect contact (3.83) between two materials means

$$\mathbf{E}\boldsymbol{\sigma}^+ = \mathbf{E}\boldsymbol{\sigma}^-, \quad \mathbf{F}\boldsymbol{\varepsilon}^+ = \mathbf{F}\boldsymbol{\varepsilon}^-. \tag{3.87}$$

Further, the surface tensors are defined by

$$\mathbf{L}^{\pm}(\mathbf{n}) = \mathbf{L}^{\pm}\boldsymbol{\eta},\ \mathbf{G}^{\pm}(\mathbf{n}) = [\mathbf{L}^{\pm}(n)]^{-1}, \quad \boldsymbol{\Gamma}^{\pm}(\mathbf{n}) = \mathbf{L}^{\pm} - \mathbf{L}^{\pm}\mathbf{U}^{\pm}(\mathbf{n})\mathbf{L}^{\pm},$$
$$U(n)^{\pm}_{klmn} = [n_k G^{\pm}(n)_{lm} n_n]_{(kl)(mn)}. \tag{3.88}$$

The second-order tensor $L_{ij}(\mathbf{n})$ is dimetric and positive definite due to the positiveness of elastic energy, and therefore the tensor $L_{ij}(\mathbf{n})$ has an inverse tensor $G_{ij}(\mathbf{n})$, which is also symmetric and positive definite. By straightforward testing, one can obtain

$$\mathbf{FU} = \mathbf{UF} = \mathbf{0}, \quad \mathbf{E}\boldsymbol{\Gamma} = \boldsymbol{\Gamma}\mathbf{E} = \mathbf{0},$$
$$\mathbf{EU} = \mathbf{UE} = \mathbf{U}, \quad \mathbf{E}\boldsymbol{\Gamma} = \boldsymbol{\Gamma}\mathbf{E} = \mathbf{0}. \tag{3.89}$$

It should be mentioned that the tensors $\boldsymbol{\Gamma}(\mathbf{n})$ and $\mathbf{U}(\mathbf{n})$ are obviously related to the Green's tensors $\boldsymbol{\Gamma}(\mathbf{x})$ (3.37) and $\mathbf{U}(\mathbf{x})$ (3.19). Indeed, let, for example, $\boldsymbol{\Gamma}(\mathbf{k})$ be a Fourier transform of the tensor $\boldsymbol{\Gamma}(\mathbf{x})$. Then $\boldsymbol{\Gamma}(\mathbf{n})$ coincides with the value of $\boldsymbol{\Gamma}(\mathbf{k})$ at the surface of a unit sphere $|\mathbf{k}| = 1$ at $\mathbf{n} = \mathbf{k}/|\mathbf{k}|$. The surface tensors $\boldsymbol{\Gamma}(\mathbf{n})$ and $\mathbf{U}(\mathbf{n})$ are found by the use of the Fourier transforms $\boldsymbol{\Gamma}(\mathbf{k})$ and $\mathbf{U}(\mathbf{k})$ of the involved Green's tensors rather than their original. The tensors $\boldsymbol{\Gamma}(\mathbf{n})$ and $\mathbf{U}(\mathbf{n})$ are always finite and depend only on the normal $\mathbf{n}$ in the point being considered rather than on the size and the shape on the inhomogeneity v.

Estimating the components of the tensors **ELE** and **FMF** in terms of the projective operators leads to

$$(\mathbf{ELE})\mathbf{U} = \mathbf{E} = \mathbf{U}(\mathbf{ELE}), \quad (\mathbf{FMG})\boldsymbol{\Gamma} = \mathbf{F} = \boldsymbol{\Gamma}(\mathbf{FMF}), \tag{3.90}$$

or, in other words:

$$\mathbf{U}(\mathbf{n}) = (\mathbf{ELE})^{-1}\mathbf{E}, \quad \boldsymbol{\Gamma}(\mathbf{n}) = (\mathbf{FMF})^{-1}\mathbf{F}. \tag{3.91}$$

Let us come back to the estimation of the jump of stresses and strains at the interface S. Exploiting the contact conditions (3.87) leads to representation of the constitutive equation in the vicinity of the interface in the form

$$\boldsymbol{\sigma}^{-} = \mathbf{L}^{-}(\mathbf{F}\boldsymbol{\varepsilon}^{+} + \mathbf{E}\boldsymbol{\varepsilon}^{-} - \boldsymbol{\beta}^{-}), \quad \boldsymbol{\varepsilon}^{-} = \mathbf{M}^{-}(\mathbf{E}\boldsymbol{\sigma}^{+} + \mathbf{F}\boldsymbol{\sigma}^{-}) + \boldsymbol{\beta}^{-}. \tag{3.92}$$

Multiplying Eqs. (3.92_1) and (3.92_2) over the tensors **E** and **F**, respectively, yields

$$\mathbf{E}\boldsymbol{\sigma}^{+} = \mathbf{E}\mathbf{L}^{-}(\mathbf{F}\boldsymbol{\varepsilon}^{+} + \mathbf{E}\boldsymbol{\varepsilon}^{-} - \boldsymbol{\beta}^{-}), \quad \mathbf{F}\boldsymbol{\varepsilon}^{+} = \mathbf{F}\mathbf{M}^{-}(\mathbf{E}\boldsymbol{\sigma}^{+} + \mathbf{F}\boldsymbol{\sigma}^{-}) + \mathbf{F}\boldsymbol{\beta}^{-}, \tag{3.93}$$

from which, with inclusion of Eqs. (3.91), it follows

$$\boldsymbol{\varepsilon}^{-} = [\mathbf{I} + \mathbf{U}^{-}(\mathbf{n})(\mathbf{L}^{+} - \mathbf{L}^{-})]\boldsymbol{\varepsilon}^{+} + \mathbf{U}^{-}(\mathbf{n})(\boldsymbol{\alpha}^{+} - \boldsymbol{\alpha}^{-}), \tag{3.94}$$
$$\boldsymbol{\sigma}^{-} = [\mathbf{I} + \boldsymbol{\Gamma}^{-}(\mathbf{n})(\mathbf{M}^{+} - \mathbf{M}^{-})]\boldsymbol{\sigma}^{+} + \boldsymbol{\Gamma}^{-}(\mathbf{n})(\boldsymbol{\beta}^{+} - \boldsymbol{\beta}^{-}). \tag{3.95}$$

The replacement $+ \leftrightarrow -$ in Eqs. (3.94) and (3.99) leads to the representations

$$\boldsymbol{\varepsilon}^{+} = [\mathbf{I} + \mathbf{U}^{+}(\mathbf{n})(\mathbf{L}^{-} - \mathbf{L}^{+})]\boldsymbol{\varepsilon}^{-} + \mathbf{U}^{+}(\mathbf{n})(\boldsymbol{\alpha}^{-} - \boldsymbol{\alpha}^{+}), \tag{3.96}$$
$$\boldsymbol{\sigma}^{+} = [\mathbf{I} + \boldsymbol{\Gamma}^{+}(\mathbf{n})(\mathbf{M}^{-} - \mathbf{M}^{+})]\boldsymbol{\sigma}^{-} + \boldsymbol{\Gamma}^{+}(\mathbf{n})(\boldsymbol{\beta}^{-} - \boldsymbol{\beta}^{+}). \tag{3.97}$$

Substitution of Eqs. (3.94), (3.95) into the right-hand side of Eqs. (3.96), (3.97), respectively, leads to

$$\mathbf{U}^-(\mathbf{n}) - \mathbf{U}^+(\mathbf{n}) = \mathbf{U}^-(\mathbf{n})(\mathbf{L}^+ - \mathbf{L}^-)\mathbf{U}^+(\mathbf{n}), \tag{3.98}$$
$$\mathbf{\Gamma}^-(\mathbf{n}) - \mathbf{\Gamma}^+(\mathbf{n}) = \mathbf{\Gamma}^-(\mathbf{n})(\mathbf{M}^+ - \mathbf{M}^-)\mathbf{\Gamma}^+(\mathbf{n}). \tag{3.99}$$

Let us denote $\mathbf{A}^-(\mathbf{n}) = [\mathbf{I} + \mathbf{U}^-(\mathbf{n})(\mathbf{L}^+ - \mathbf{L}^-)]$, $\mathbf{B}^-(\mathbf{n}) = [\mathbf{I} + \mathbf{\Gamma}^-(\mathbf{n})(\mathbf{M}^+ - \mathbf{M}^-)]$. Then from the expressions (3.94) and (3.95), we get the correlations between the stress and strain fields in the vicinity of the interface:

$$\boldsymbol{\varepsilon}^-(\mathbf{n}) = \mathbf{A}^-(\mathbf{n})\boldsymbol{\sigma}^+(\mathbf{x}) + (\mathbf{A}^-(\mathbf{n}) - \mathbf{I})(\mathbf{L}^+ - \mathbf{L}^-)^{-1}(\boldsymbol{\alpha}^+ - \boldsymbol{\alpha}^-), \tag{3.100}$$
$$\boldsymbol{\sigma}^-(\mathbf{n}) = \mathbf{B}^-(\mathbf{n})\boldsymbol{\sigma}^+(\mathbf{x}) + (\mathbf{B}^-(\mathbf{n}) - \mathbf{I})(\mathbf{M}^+ - \mathbf{M}^-)^{-1}(\boldsymbol{\beta}^+ - \boldsymbol{\beta}^-), \tag{3.101}$$

which are also valid at the replacement $+ \leftrightarrow -$.

In particular for an isotropic medium with the elastic modulus $\mathbf{L} = (3k, 2\mu)$ (2.197) an inversion of the matrix $\mathbf{L}(\mathbf{n})$ may be simplified and we obtain

$$L_{kl}(\mathbf{n}) = \mu\delta_{kl} + \left(k + \frac{\mu}{3}\right) n_k n_l, \quad G_{kl}(\mathbf{n}) = \mu^{-1}\left(\delta_{kl} - \frac{2k+\mu}{3k+4\mu} n_k n_l\right),$$
$$U_{klmn}(\mathbf{n}) = \frac{1}{2\mu}\left(E_{klmn} - \frac{3k-2\mu}{3k+4\mu} n_k n_l n_m n_n\right),$$
$$\Gamma_{klmn}(\mathbf{n}) = 2\mu\left[F_{klmn} + \frac{3k-2\mu}{3k+4\mu}(\delta_{kl} - n_k n_l)(\delta_{mn} - n_m n_n)\right]. \tag{3.102}$$

From Eqs. (3.99) and (3.102) one can obtain the formulae for the strain jump $[\boldsymbol{\varepsilon}] \equiv \boldsymbol{\varepsilon}^+ - \boldsymbol{\varepsilon}^-$ at the interface S between joint isotropic media in the local coordinate system when the normal vector has the coordinate $\mathbf{n} = (0, 0, 1)^\top$:

$$[\varepsilon_{11}] = [\varepsilon_{12}] = [\varepsilon_{22}] = 0, \ [\varepsilon_{13}] = -\frac{\mu^+ - \mu^-}{\mu^-}\varepsilon_{13}^+, \ [\varepsilon_{23}] = -\frac{\mu^+ - \mu^-}{\mu^-}\varepsilon_{23}^+,$$
$$[\varepsilon_{33}] = -\frac{1}{\lambda^- + 2\mu^-}\left\{(\lambda^+ - \lambda^-)(\varepsilon_{11}^+ + \varepsilon_{22}^+ + \varepsilon_{33}^+) + 2(\mu^+ - \mu^-)\varepsilon_{33}^+\right\}, \tag{3.103}$$

where $\lambda^\pm = k^\pm - \frac{2}{3}\mu^\pm$.

3.5 Ellipsoidal Inhomogeneity in the Elastic Medium

Let us consider a particular case of the ellipsoidal homogeneous domain v described by Eq. (3.18) in the Cartesian coordinate system connected to the semi-axes a_1, a_2, a_3 when $\mathbf{a} = \text{diag}(a_1, a_2, a_3)$ (3.18). If the elastic modulus is homogeneous inside the matrix and inclusion: $\mathbf{L}(\mathbf{x}) \equiv \mathbf{L}^{(0)} + \mathbf{L}_1(\mathbf{x})$, $\mathbf{L}_1(\mathbf{x}) = \mathbf{L}_1^+ V(\mathbf{x})$, $\mathbf{L}_1^+ \equiv$ const., then the solutions of some important problems can be found explicitly that is based on the following theorem of polynomial conservation (the p-property) [620] (see also [618]).

Theorem. If the external field $\boldsymbol{\varepsilon}^0(\mathbf{x})$ acting on the inhomogeneity $\mathbf{x} \in v$ is a polynomial of degree m, then the field $\boldsymbol{\varepsilon}(\mathbf{x})$ generated inside v is also polynomial of the same order m.

For the proof of this theorem it is enough to prove that the operator $\mathbf{U}$ with the kernel $\mathbf{U}(\mathbf{x}-\mathbf{y})$ transforms the a polynomial $\mathbf{P}_m(\mathbf{x})$ of degree m into the polynomial of the same degree $\mathbf{P}'_m(\mathbf{x})$ $(\mathbf{x}\in v)$, and the ellipsoid v is a unit ball $v=\Omega_1\equiv\{\mathbf{x}:\ |\mathbf{x}|\le 1\}$. Since $\mathbf{G}(\mathbf{x})$ is an even homogeneous function of the order -1, then for an arbitrary anisotropy of the medium the next representation is valid $\mathbf{U}(\mathbf{x}-\mathbf{y})=\mathbf{g}^0(\mathbf{n})/r^3,\ \ n=(\mathbf{x}-\mathbf{y})/r,\ \ r=|\mathbf{x}-\mathbf{y}|$, where $\mathbf{g}^0$ is an even fourth-order tensor of the unit vector $\mathbf{n}$. For the second derivative of the Green's tensor, the decomposition $\mathbf{U}(\mathbf{x})=-\mathbf{P}\delta(\mathbf{x})+\mathbf{U}^{(f)}(\mathbf{x})$ (3.20) over the singular and regular constituents is used. Since the singular constituent (3.20) holds the p-property, then it is enough to investigate the action of the regular constituent $\mathbf{U}^{(f)}$ on a homogeneous polynomial of degree m, or that is the same on the functions $\mathbf{x}^{(m)}=\mathbf{x}^{\lambda_1}\cdot\ldots\cdot\mathbf{x}^{\lambda_m}$:

$$\mathbf{F}^{(m)}(\mathbf{x})=\int\frac{\mathbf{g}^0(\mathbf{n})}{r^3}(\mathbf{y}^{(m)}-\mathbf{x}^{(m)})V(\mathbf{y})\,d\mathbf{y},\ \ \mathbf{x}\in v. \tag{3.104}$$

Substituting $\mathbf{y}=\mathbf{x}+r\mathbf{n}$ into $\mathbf{y}^{(m)}$ leads to

$$\mathbf{F}^{(m)}(\mathbf{x})=\sum_{k=1}^{m}\mathbf{x}^{(m-k)}\mathbf{J}^{(k)}(\mathbf{x}),\ \ \mathbf{J}^{(k)}=\int\mathbf{g}^0(\mathbf{n})\mathbf{n}^{(k)}r^{k-3}V(\mathbf{y})\,d\mathbf{y}. \tag{3.105}$$

Let us $\mathbf{y}=r^2drd\omega$ and integrate at first over the elementary cone $d\omega(\mathbf{n})$ with the vertex at the point $\mathbf{x}$:

$$\mathbf{J}^{(k)}=\frac{1}{k}\int\mathbf{g}(\mathbf{n})\mathbf{n}^{(k)}\rho^k(\mathbf{x},\mathbf{n})d\omega,\ \ \mathbf{x}\in v, \tag{3.106}$$

where $\rho(\mathbf{x},\mathbf{n})=\mathbf{x}\cdot\mathbf{n}+[1-\mathbf{x}\cdot\mathbf{x}+(\mathbf{x}\cdot\mathbf{n})^2]^{1/2}$ is a distance from the point $\mathbf{x}$ till the unit sphere Ω_1, and

$$\rho^k=\sum_l(-1)^l\frac{(k+l)!}{k!l!}(\mathbf{x}\cdot\mathbf{n})^l[1-\mathbf{x}\cdot\mathbf{x}+(\mathbf{x}\cdot\mathbf{n})^2]^{(k-l)/2}. \tag{3.107}$$

Considering that $\mathbf{g}^0(\mathbf{n})$ is the even function of $\mathbf{n}$, the contribution into $\mathbf{J}^{(k)}(\mathbf{x})$ is provided only by the items of the expansion (3.107) whose production with $\mathbf{n}^k$ is even. These items containing the radical of an even degree have the form $\sum_l A_{(k-2l)}(\mathbf{n})\mathbf{x}^{(k-2l)}$, and, therefore, $\mathbf{J}^{(\mathbf{k})}(\mathbf{x})$ is a polynomial of degree k:

$$\mathbf{J}^{(k)}(\mathbf{x})=\sum_l B^{(k)}_{(k-2l)}\mathbf{x}^{(k-2l)}, \tag{3.108}$$

where the surface integral over the unit sphere

$$B^{(k)}_{(k-2l)}=\frac{1}{k}\int_\omega\mathbf{g}^0(\mathbf{n})\mathbf{n}^{(k)}A_{(k-2l)}(\mathbf{n})\,d\omega \tag{3.109}$$

is a constant tensor. By substituting Eq. (3.108) into Eq. (3.104) we are able to verify that $\mathbf{F}^{(m)}(\mathbf{x})$ is a polynomial of degree m with nonzero coefficient at $\mathbf{x}^{(m)}$ that completes the proof of the theorem. The most important case of the homogeneous external loading $\boldsymbol{\sigma}^0(\mathbf{x})$ =const. was proved by Eshelby [318] for the isotropic matrix and in [574] and [1150] for the anisotropic matrix; the cases of the polynomial fields were considered in [319] and in [20] for the isotropic and anisotropic materials, respectively.

Let us consider the consequence of this theorem when $\boldsymbol{\varepsilon}^0$, $\boldsymbol{\sigma}^0(\mathbf{x})$ =const. and, therefore, the produced fields $\boldsymbol{\varepsilon}^+(\mathbf{x})$ and $\boldsymbol{\sigma}^+(\mathbf{x})$ are also homogeneous. By this means it is necessary to define the action of generalized functions $\mathbf{U}(\mathbf{x})$ and $\boldsymbol{\Gamma}(\mathbf{x})$ in the operator Eqs. (3.78$_1$) and (3.78$_2$) on the finite functions $\boldsymbol{\varepsilon}^+V(\mathbf{x})$ and $\boldsymbol{\sigma}^+V(\mathbf{x})$. According to the regularizations (3.26) and (3.38), the action of the formal constituents $\mathbf{U}^{(f)}$ and $\boldsymbol{\Gamma}^{(f)}$ of tensors $\mathbf{U}$ and $\boldsymbol{\Gamma}$, respectively, on the mentioned functions equal zero, and the effect of the operators $\mathbf{U}$ and $\boldsymbol{\Gamma}$ equivalent to the multiplication on the constant tensors:

$$\mathbf{P} = -\frac{1}{4\pi}\int \mathbf{U}(a^{-1}\bar{\mathbf{k}})\, d\bar{\mathbf{k}}, \quad \mathbf{Q} = -\frac{1}{4\pi}\int \boldsymbol{\Gamma}(a^{-1}\bar{\mathbf{k}})\, d\bar{\mathbf{k}}, \tag{3.110}$$

where integration is performed over the surface of unit sphere $\bar{\mathbf{k}}^2 = 1$ and the regularization domain has a shape of the ellipsoid v. A relationship follows between the tensors $\mathbf{P}$ and $\mathbf{Q}$: $\mathbf{Q} = \mathbf{L}^{(0)} - \mathbf{L}^{(0)}\mathbf{P}\mathbf{L}^{(0)}$, derived from definition (3.37). From Eqs. (3.78$_1$), (3.78$_2$) and (3.110$_1$), (3.110$_2$), it follows the linear algebraic equations relating the generated fields inside the inclusion with the remote fields ($\boldsymbol{\beta}_1 \equiv \mathbf{0}$):

$$(\mathbf{I} + \mathbf{P}\mathbf{L}_1)\boldsymbol{\varepsilon}^+ = \boldsymbol{\varepsilon}^0, \quad (\mathbf{I} + \mathbf{Q}\mathbf{M}_1)\boldsymbol{\sigma}^+ = \boldsymbol{\sigma}^0. \tag{3.111}$$

Let us represent another form of the tensors $\mathbf{P}$ (3.110$_1$) based on the homogeneity of the function $\mathbf{U}(\mathbf{k})$:

$$\int \mathbf{U}(\mathbf{a}^{-1}\bar{\mathbf{k}})d\bar{\mathbf{k}} = \int \mathbf{U}(\mathbf{n})d\bar{\mathbf{k}} = \det \mathbf{a} \int \mathbf{U}(\mathbf{n})\rho^3(\mathbf{n})\, d\mathbf{n}, \tag{3.112}$$

$$\mathbf{n} = \frac{\mathbf{a}^{-1}\bar{\mathbf{k}}}{|\mathbf{a}^{-1}\bar{\mathbf{k}}|}, \quad \rho(\mathbf{n}) = \frac{1}{\sqrt{\mathbf{n}\mathbf{a}^2\mathbf{n}}}. \tag{3.113}$$

The point $\mathbf{x} \in S = \partial v$ belonging to the surface S described by the equation $\mathbf{x}\mathbf{a}^{-2}\mathbf{x} = 1$ (3.18) is related with the normal $\mathbf{n} \perp S$ in this point by the relations

$$\mathbf{x} = \frac{\mathbf{a}^2\mathbf{n}}{\sqrt{\mathbf{n}\mathbf{a}^2\mathbf{n}}}, \quad \mathbf{n} = \frac{\mathbf{a}^{-2}\mathbf{n}}{\sqrt{\mathbf{x}\mathbf{a}^{-4}\mathbf{x}}} \tag{3.114}$$

establishing one-to-one correspondence between the points $\mathbf{x}$ on the surface S and the points $\mathbf{n}$ on the unit sphere $S_1 = \{\mathbf{n} :\ \mathbf{n}^2 = 1\}$. Then the averages of an arbitrary function f over the ellipsoid S and the sphere S_1

$$\langle f(\mathbf{x})\rangle = \frac{1}{4\pi \det \mathbf{a}} \int_v \frac{f(\mathbf{x})}{\rho(\mathbf{x})}\, d\mathbf{x}, \quad \text{and} \quad \langle f(\mathbf{n})\rangle = \frac{\det \mathbf{a}}{4\pi} \int_{S_1} f(\mathbf{n})\rho^3(\mathbf{n})\, d\mathbf{n}, \tag{3.115}$$

respectively, coincide. Then from Eqs. (3.112)–(3.115), we get the representations for the tensors $\mathbf{P}$ and, analogously, for $\mathbf{Q}$:

$$\mathbf{P} = -\langle \mathbf{U}(\mathbf{n})\rangle, \quad \mathbf{Q} = -\langle \boldsymbol{\Gamma}(\mathbf{n})\rangle \tag{3.116}$$

defined by the averages over the unit sphere of the tensors $\mathbf{U}(\mathbf{n})$ and $\mathbf{\Gamma}(\mathbf{n})$ describing the jumps of the strains and stresses on the interface S (3.94)–(3.97).

Let us consider the case of the linear external field $\varepsilon^0_{ij}(\mathbf{x}) = b_{ijk}x_k$ $(\mathbf{x} \in v)$ when, according to the p-property theorem, $\varepsilon_{ij}(\mathbf{x}) = d_{ijk}x_k$, $(\mathbf{x} \in v)$. The tensors b_{ijk} and d_{ijk}, symmetrical over the first pair of indexes, are related by the linear ratio which can be presented in the form:

$$b_{i_1i_2i_3} = 3\Lambda_{i_1i_2i_3j_1j_2j_3}d_{j_1j_2j}(\mathbf{a}^2)_{jj_3}. \tag{3.117}$$

To obtain $\boldsymbol{\Lambda}$, let us multiply Eq. (3.116$_1$) by $\mathbf{x}$ and average the result over v. Then, analogously to the case of the homogeneous field (3.112)–(3.117), we get

$$\boldsymbol{\Lambda} = \langle \rho(\mathbf{n})\mathbf{n}(\mathbf{I} - \mathbf{U}(\mathbf{n})\mathbf{L}_1)\mathbf{n}\rho(\mathbf{n})\rangle. \tag{3.118}$$

In conclusion, we consider one more practical problem for the ellipsoidal inhomogeneity with eigenstrain ($\boldsymbol{\beta}_1(\mathbf{x}) = \text{const.} \neq \mathbf{0}$) subjected to the homogeneous remote fields. From Eqs. (3.78$_1$), (3.78$_2$) and (3.110$_1$), (3.110$_2$), follow the linear algebraic equations related to the generated fields inside the inclusion with the remote fields

$$\boldsymbol{\varepsilon}^+ = \mathbf{A}\boldsymbol{\varepsilon}^0 + \mathbf{C}^\epsilon, \quad \boldsymbol{\sigma}^+ = \mathbf{B}\boldsymbol{\sigma}^0 + \mathbf{C}, \tag{3.119}$$

with obvious correlations between the tensors involved into Eqs. (3.119):

$$\mathbf{A} = (\mathbf{I} + \mathbf{P}\mathbf{L}_1)^{-1} = \mathbf{M}^{(i)}\mathbf{B}(\mathbf{M}^{(0)})^{-1}, \tag{3.120}$$
$$\mathbf{B} = (\mathbf{I} + \mathbf{Q}\mathbf{M}_1)^{-1} = \mathbf{L}^{(i)}\mathbf{A}(\mathbf{L}^{(0)})^{-1}, \tag{3.121}$$
$$\mathbf{C}^\epsilon = -\mathbf{A}\mathbf{P}\boldsymbol{\alpha}_1, \quad \mathbf{C} = -\mathbf{B}\mathbf{Q}\boldsymbol{\beta}_1, \tag{3.122}$$
$$\mathbf{M}_1 = -\mathbf{M}^{(i)}\mathbf{L}_1\mathbf{M}^{(0)} = -\mathbf{M}^{(0)}\mathbf{L}_1\mathbf{M}^{(i)}, \tag{3.123}$$
$$\mathbf{L}_1 = -\mathbf{L}^{(0)}\mathbf{M}_1\mathbf{L}^{(i)} = -\mathbf{L}^{(i)}\mathbf{M}_1\mathbf{L}^{(0)}, \tag{3.124}$$
$$\mathbf{Q} = \mathbf{L}^{(0)} - \mathbf{L}^{(0)}\mathbf{P}\mathbf{L}^{(0)}, \tag{3.125}$$
$$\mathbf{P} = \mathbf{M}^{(0)} - \mathbf{M}^{(0)}\mathbf{Q}\mathbf{M}^{(0)}, \tag{3.126}$$

Let an ellipsoidal inclusion v_i with the homogeneous compliance $\mathbf{M}^+$ be located in an infinite homogeneous matrix with compliance $\mathbf{M}^-$ and loaded by the homogeneous stress $\boldsymbol{\sigma}^0$. Then, according to Eshelby's theorem (with $\boldsymbol{\beta} \equiv 0$), we have the representations for the stresses

$$\boldsymbol{\sigma}^+ = \boldsymbol{\sigma}^0 - \mathbf{Q}(\mathbf{M}^+ - \mathbf{M}^-)\boldsymbol{\sigma}^+, \tag{3.127}$$
$$\boldsymbol{\sigma}(\mathbf{x}) = \boldsymbol{\sigma}^0 + \bar{v}_i\mathbf{T}_i(\mathbf{x}_i - \mathbf{x})(\mathbf{M}^+ - \mathbf{M}^-)\boldsymbol{\sigma}^+, \tag{3.128}$$

and the strains

$$\boldsymbol{\varepsilon}^+ = \boldsymbol{\varepsilon}^0 - \mathbf{P}(\mathbf{L}^+ - \mathbf{L}^-)\boldsymbol{\varepsilon}^+, \tag{3.129}$$
$$\boldsymbol{\varepsilon}(\mathbf{x}) = \boldsymbol{\varepsilon}^0 + \bar{v}_i\mathbf{T}^\epsilon_i(\mathbf{x}_i - \mathbf{x})(\mathbf{L}^+ - \mathbf{L}^-)\boldsymbol{\varepsilon}^+, \tag{3.130}$$

where the tensors $\mathbf{T}^\epsilon_i(\mathbf{x} - \mathbf{x})$ and $\mathbf{T}_i(\mathbf{x}_i - \mathbf{x})$ are defined by the relations

$$\mathbf{T}^\epsilon_i(\mathbf{x}_i - \mathbf{x}) = \frac{1}{\bar{v}_i}\int V_i(\mathbf{y})\mathbf{U}(\mathbf{y} - \mathbf{x})\,d\mathbf{y},$$
$$\mathbf{T}_i(\mathbf{x}_i - \mathbf{x}) = \frac{1}{\bar{v}_i}\int V_i(\mathbf{y})\mathbf{\Gamma}(\mathbf{y} - \mathbf{x})\,d\mathbf{y} \tag{3.131}$$

for the points $\mathbf{x} \notin v_i$.

If $\mathbf{x}$ is a limiting point $\mathbf{x} = \mathbf{x}^-$ in the matrix near the ellipsoidal surface ∂v_i, then, substituting the relations (3.94), (3.95) and (3.130), (3.128) into Eq. (3.129), (3.127), respectively, we obtain

$$\overline{v}_i \mathbf{T}_i^{\epsilon}(\mathbf{x}_i - \mathbf{x}^-) = \mathbf{U}(\mathbf{n})^- - \mathbf{P}, \quad \overline{v}_i \mathbf{T}_i(\mathbf{x}_i - \mathbf{x}^-) = \mathbf{\Gamma}(\mathbf{n})^- - \mathbf{Q}. \tag{3.132}$$

3.6 Eshelby Tensor

3.6.1 Tensor Representation of Eshelby Tensor

The equations (3.119) define the strains and stresses inside ellipsoidal inclusion in terms of the remote fields in the explicit form utilizing the Eshelby' solution in the form obtained in [620] (see also [618]). However, for completeness, we will reproduce another more descriptive method of the *equivalent inclusions* proposed by Eshelby [318], [319], which has played a key role in the micromechanics of composites. Namely, it is assumed that the perturbation of a stress field, introduced by the inhomogeneity v with elastic moduli $\mathbf{L}^{(1)}(\mathbf{x})$ $(\mathbf{x} \in v)$ into the applied remote stress, can be modeled by the fictitious eigenstrains $\boldsymbol{\varepsilon}^*$ localized in the area v with elastic modulus $\mathbf{L}^{(0)}$ called inclusion v. The choice of $\boldsymbol{\varepsilon}^*$ called equivalent eigenstrains is defined from the condition of equivalence of a stress states of the media with the inhomogeneity and inclusion:

$$\mathbf{L}^{(1)}(\mathbf{x})\boldsymbol{\varepsilon}(\mathbf{x}) = \mathbf{L}^{(0)}(\boldsymbol{\varepsilon}(\mathbf{x}) - \boldsymbol{\varepsilon}^*(\mathbf{x})). \tag{3.133}$$

For ellipsoidal inhomogeneity subjected to the homogeneous remote stress $\boldsymbol{\sigma}^0 = \mathbf{L}^{(0)}\boldsymbol{\varepsilon}^0$, the homogeneity of the fields $\boldsymbol{\varepsilon}(\mathbf{x})$ and $\boldsymbol{\varepsilon}^*(\mathbf{x})$ $(\mathbf{x} \in v)$ occurs, and

$$\boldsymbol{\varepsilon} = \boldsymbol{\varepsilon}^0 + \mathbf{S}\boldsymbol{\varepsilon}^*, \tag{3.134}$$

where the well-known Eshelby tensor

$$\mathbf{S} = -\int \mathbf{U}(\mathbf{x} - \mathbf{y})V(\mathbf{y})\, d\mathbf{y}\mathbf{L}^{(0)} \equiv \text{const.}, \quad \mathbf{x} \in v \tag{3.135}$$

is related to the tensors $\mathbf{P}$ (3.110_1) and $\mathbf{Q}$ (3.110_2) by the equations

$$\mathbf{S} = \mathbf{P}\mathbf{L}^{(0)}, \quad \mathbf{S} = \mathbf{I} - \mathbf{M}^{(0)}\mathbf{Q}. \tag{3.136}$$

Substitution of (3.134) into (3.133) yields the equation

$$\mathbf{L}(\mathbf{x})(\boldsymbol{\varepsilon}^0 + \mathbf{S}\boldsymbol{\varepsilon}^*) = \mathbf{L}^{(0)}(\boldsymbol{\varepsilon}^0 + \mathbf{S}\boldsymbol{\varepsilon}^* - \boldsymbol{\varepsilon}^*), \tag{3.137}$$

from which we get the representation for both the equivalent eigenstrain and the strain inside the inhomogeneity

$$\begin{aligned} \boldsymbol{\varepsilon}^*(\mathbf{x}) &= -(\mathbf{L}_1^{(1)}\mathbf{S} + \mathbf{L}^{(0)})^{-1}\mathbf{L}_1^{(1)}\boldsymbol{\varepsilon}^0, \\ \boldsymbol{\varepsilon}(\mathbf{x}) &= (\mathbf{I} + \mathbf{S}\mathbf{M}^{(0)}\mathbf{L}_1^{(1)})^{-1}\boldsymbol{\varepsilon}^0, \end{aligned} \tag{3.138}$$

respectively.

The method of the equivalent inclusions is also appropriate in the case when the inhomogeneity v is subjected to the stress-free strains $\boldsymbol{\beta}_1^{(1)}(\mathbf{x})$ ($\mathbf{x} \in v$, $\boldsymbol{\beta}^{(0)} \equiv \mathbf{0}$). Then the inhomogeneity with the properties $\mathbf{L}^{(1)}$ and $\boldsymbol{\beta}_1^{(1)}$ is modeled as an inclusion with the moduli $\mathbf{L}^{(0)}$ and with the eigenstrain $\boldsymbol{\beta}_1^{(1)}$ subjected to the fictitious eigenstrain $\boldsymbol{\varepsilon}^*(\mathbf{x}) = \boldsymbol{\varepsilon}^* V(\mathbf{x})$. Equivalence of stress states of the inhomogeneity and the inclusions means that

$$\mathbf{L}^{(1)}(\mathbf{x})(\boldsymbol{\varepsilon}(\mathbf{x}) - \boldsymbol{\beta}_1^{(1)}) = \mathbf{L}^{(0)}(\boldsymbol{\varepsilon}(\mathbf{x}) - \boldsymbol{\beta}_1^{(1)} - \boldsymbol{\varepsilon}^*(\mathbf{x})). \tag{3.139}$$

If remote stress $\boldsymbol{\sigma}^0$ and the eigenstrain $\boldsymbol{\beta}^{(1)}$ are homogeneous then Eq. (3.139) is valid if $\boldsymbol{\varepsilon}$ is a solution to the problem of the inclusion with the eigenstrain. According to Eq. (3.134) we have a representation for the strain inside the inclusion

$$\boldsymbol{\varepsilon} = \boldsymbol{\varepsilon}^0 + \mathbf{S}(\boldsymbol{\beta}_1^{(1)} + \boldsymbol{\varepsilon}^*). \tag{3.140}$$

Substitution of (3.140) into (3.139) yields

$$\mathbf{L}^{(1)}(\mathbf{x})\Big[\boldsymbol{\varepsilon}^0 + \mathbf{S}(\boldsymbol{\beta}_1^{(1)} + \boldsymbol{\varepsilon}^*) - \boldsymbol{\beta}_1^{(1)}\Big] = \mathbf{L}^{(0)}\Big[\boldsymbol{\varepsilon}^0 + (\mathbf{S} - \mathbf{I})(\boldsymbol{\beta}_1^{(1)} + \boldsymbol{\varepsilon}^*)\Big] \tag{3.141}$$

from which we can find the equivalent eigenstrain $\boldsymbol{\varepsilon}^*$ and the strains inside the inhomogeneity ($\mathbf{x} \in v$):

$$\boldsymbol{\varepsilon}^*(\mathbf{x}) = -(\mathbf{L}_1^{(1)}\mathbf{S} + \mathbf{L}^{(0)})^{-1}[\mathbf{L}_1^{(1)}\boldsymbol{\varepsilon}^0 - \mathbf{L}^{(0)}\boldsymbol{\beta}_1^{(1)}] - \boldsymbol{\beta}_1^{(1)}, \tag{3.142}$$

$$\boldsymbol{\varepsilon}(\mathbf{x}) = (\mathbf{I} + \mathbf{S}\mathbf{M}^{(0)}\mathbf{L}_1^{(1)})^{-1}[\boldsymbol{\varepsilon}^0 + \mathbf{S}\mathbf{M}^{(0)}\mathbf{L}^{(1)}\boldsymbol{\beta}^{(1)}]. \tag{3.143}$$

The method of equivalent inclusions (3.133)–(3.143) can be easily recast in terms of stresses by replacement of the tensors $\mathbf{L}$, $\boldsymbol{\beta}$, $\boldsymbol{\varepsilon}$, $\boldsymbol{\varepsilon}^0$, $\boldsymbol{\varepsilon}^*$ with the tensors $\mathbf{M}$, $\boldsymbol{\alpha}$, $\boldsymbol{\sigma}$, $\boldsymbol{\sigma}^0$, $\boldsymbol{\sigma}^*$ where $\boldsymbol{\sigma}^*$ is an equivalent eigenstress accompanying the eigenstress $\boldsymbol{\alpha}$ in a similar manner as the equivalent eigenstrass $\boldsymbol{\varepsilon}^*$ accompanying the eigenstrain $\boldsymbol{\beta}$.

The Eshelby tensor depending only on the elastic properties of the matrix is symmetric with respect to the indexes inside the first and the second pairs $S_{ijkl} = S_{jikl} = S_{ijlk}$; however, it is not symmetric, in general, with respect to the exchange of the index pairs $S_{ijkl} \neq S_{klij}$. Moreover, for the isotropic medium, the Eshelby tensor depends only on Poisson ratio which we will see from an explicit representation. To accomplish this, we at first estimate

$$S_{imn} = -\int G_{ij,k}(\mathbf{x} - \mathbf{y})V(\mathbf{y})\, d\mathbf{y} L_{jkmn}^{(0)}, \tag{3.144}$$

where the Green's found $G_{ij}(\mathbf{x})$ (3.13) is known for the isotropic matrix. Substituting of Eq. (3.13) into Eq. (3.144) yields

$$S_{imn} = -\frac{1}{8\pi(1 - \nu^{(0)})}\int g_{imn}(\mathbf{l})V(\mathbf{y})\, d\mathbf{y}, \tag{3.145}$$

where the unit vector $\mathbf{l} = (\mathbf{x} - \mathbf{y})/|\mathbf{x} - \mathbf{y}|$ and one introduces

$$g_{imn}(\mathbf{l}) = (1 - 2\nu^{(0)})(\delta_{im}l_n + \delta_{in}l_m - \delta_{mn}l_i) + 3l_il_ml_n. \tag{3.146}$$

For the internal point $\mathbf{x}$ (3.145) of the ellipsoid v, the volume element $d\mathbf{y}$ can be presented in the form $d\mathbf{y} = r^2drd\omega$, where $r = |\mathbf{x} - \mathbf{y}|$ and $d\omega$ is the differential element of the surface of the unit sphere S_1 with the center in the point $\mathbf{x}$. After integration with respect to r in (3.145), we get

$$S_{imn} = -\frac{1}{8\pi(1 - \nu^{(0)})}\int_{S_1} r(\mathbf{l})g_{imn}(\mathbf{l})d\omega, \tag{3.147}$$

where the positive root of the equation

$$(x_1 + rl_1)^2/a_1^2 + (x_2 + rl_2)/a_2^2 + (x_3 + rl_3)^2/a_3^2 = 1$$

equals $r(\mathbf{l}) = -f/h + (f^2/h^2 + e/h)^{1/2}$, where

$$h = \sum_{i=1}^{3}(l_i/a_i)^2,\ f = \sum_{i=1}^{3} l_ix_i/a_i^2,\ e = 1 - \sum_{i=1}^{3}(x_i/a_i)^2.$$

After substitution of $r(\mathbf{l})$ into Eq. (3.147), $(f^2/h^2 + e/h)^{1/2}$ which is even with respect to l, vanishes because $g_{imn}(\mathbf{l})$ is odd in $\mathbf{l}$. Then

$$S_{imn} = \frac{x_k}{8\pi(1 - \nu^{(0)})}\oint_{S_1}\frac{\lambda_k g_{imn}}{h}d\omega, \tag{3.148}$$

where one introduces the vector $\lambda_k = l_k/a_k^2$ (do not sum over k, $k = 1, 2, 3$). Deriving Eq. (3.148) leads to representation of Eshelby tensor:

$$S_{ijmn} = \frac{1}{16\pi(1 - \nu^{(0)})}\int_{S_1}\frac{\lambda_i g_{jmn} + \lambda_j g_{imn}}{h}d\omega. \tag{3.149}$$

The Eshelby tensor for the 2D case can be obtained by passage to the limit in Eq. (3.149) at $a_1 \to \infty$. It was obtained [525] the representation for 2D Eshelby tensor for the plane stresses by the use of a straightforward following the scheme for the 3D case:

$$S_{ijmn} = \frac{1}{8\pi(1 - \nu^{(0)})}\int_0^{2\pi}\frac{\lambda_i g_{jmn} - \lambda_j g_{imn}}{h}d\omega \tag{3.150}$$

where the parameters g_{imn}, λ_i, h are also determined by the formulae (3.146), and (3.148) where in such a case the indexes i, m, n take the values 1 and 2.

The 3D Eshelby tensor (3.135), (3.149) does not depend on $\mathbf{x}$ as confirmed in the theorem of polynomial conservation. The surface integrals (3.149) are reduced to simple ones:

$$I_1 = \int_{S_1}\frac{l_1^2}{a_1^2h}d\omega = 2\pi a_1a_2a_3\int_0^\infty\frac{1}{(a_1^2 + s)\Delta(s)}ds,$$
$$I_{11} = \int_{S_1}\frac{l_1^4}{a_1^4h}d\omega = 2\pi a_1a_2a_3\int_0^\infty\frac{1}{(a_1^2 + s)^2\Delta(s)}ds,$$
$$I_{12} = 3\int_{S_1}\frac{l_1^2l_2^2}{a_1^2a_2^2h}d\omega = 2\pi a_1a_2a_3\int_0^\infty\frac{1}{(a_1^2 + s)(a_2^2 + s)\Delta(s)}ds, \tag{3.151}$$

where $\Delta(s) = (a_1^2+s)^{1/2}(a_2^2+s)^{1/2}(a_3^2+s)^{1/2}$ and the other integrals I_i and I_{ij} are found by the simultaneous cyclic permutations of indexes $i,j=(1,2,3)$. In turn,the integrals (3.123) can be reduced to the standard elliptic integrals. Indeed, the I_i and I_{ij} integrals are given by the formulae assuming $a_1 > a_2 > a_3$:

$$I_1 = \frac{4\pi a_1 a_2 a_3}{(a_1^2 - a_2^2)(a_1^2 - a_3^2)^{1/2}}[\mathrm{F}(\theta,\mathrm{k}) - \mathrm{E}(\theta,\mathrm{k}],$$
$$I_3 = \frac{4\pi a_1 a_2 a_3}{(a_2^2 - a_3^2)(a_1^2 - a_3^2)^{1/2}}\Big[\frac{a_2(a_1^2-a_3^2)^{1/2}}{a_1 a_3} - \mathrm{E}(\theta,\mathrm{k})\Big], \tag{3.152}$$

and by the equations connecting the tensors I_i and I_{ij}:

$$I_1 + I_2 + I_3 = 4\pi, \quad I_{12} = (I_2 - I_1)/(a_1^2 - a_2^2)$$
$$3I_{11} + I_{12} + I_{13} = 4\pi/a_1^2, \quad 3a_1^2 I_{11} + a_2^2 I_{12} + a_3^2 I_{13} = 3I_1, \tag{3.153}$$

where F and E are the elliptic integrals of the first and the second kind:

$$\mathrm{F}(\theta,\mathrm{k}) = \int_0^\theta \frac{dw}{(1-\mathrm{k}^2\sin^2 w)^{1/2}}, \quad \mathrm{E}(\theta,\mathrm{k}) = \int_0^\theta (1-\mathrm{k}^2\sin^2 w)^{1/2}\, dw,$$
$$\theta = \arcsin\Big(\frac{a_1^2 - a_3^2}{a_1^2}\Big)^{1/2}, \quad \mathrm{k} = \Big(\frac{a_1^2 - a_2^2}{a_1^2 - a_3^2}\Big)^{1/2}. \tag{3.154}$$

The values I_1, I_3 (3.152) and Eqs. (3.153) accompanied by the pertinent equations obtained by the cyclic permutation of indexes $(1,2,3)$ enable the relations for I_i and I_{ij} to be found in terms of I_1 and I_3.

Thus, if the unbounded medium is an isotropically elastic and the semi-axes a_i coincide with the coordinate axes x_i $(i=1,2,3)$ of a rectangular Cartesian coordinate system then the Eshelby tensor $\mathbf{S}$ has the symmetry of an ellipsoid and is defined by the nine nonzero components:

$$S_{1111} = \frac{1}{8\pi(1-\nu^{(0)})}[3a_1^2 I_{11} + (1-2\nu^{(0)})I_1],$$
$$S_{1122} = \frac{1}{8\pi(1-\nu^{(0)})}[a_2^2 I_{12} - (1-2\nu^{(0)})I_1],$$
$$S_{1212} = \frac{1}{16\pi(1-\nu^{(0)})}[(a_1^2+a_2^2)I_{12} + (1-2\nu^{(0)})(I_1+I_2)] \tag{3.155}$$

with the remaining nonzero six components obtained by the cyclic index permutation $(1,2,3)$.

Eshelby tensor for the particular cases has the representations
1) Sphere $(a_1 = a_2 = a_3 = a)$

$$S_{ijkl} = \frac{5\nu^{(0)}-1}{15(1-\nu^{(0)})}\delta_{ij}\delta_{kl} + \frac{4-5\nu^{(0)}}{15(1-\nu^{(0)})}(\delta_{ik}\delta_{jl} + \delta_{il}\delta_{jk}). \tag{3.156}$$

2) Elliptic cylinder $(a_3 \to \infty)$

$$
\begin{aligned}
S_{1111} &= \frac{1}{2(1-\nu^{(0)})}\left[\frac{a_2^2+2a_1a_2}{(a_1+a_2)^2}+(1-2\nu^{(0)})\frac{a_2}{a_1+a_2}\right],\\
S_{2222} &= \frac{1}{2(1-\nu^{(0)})}\left[\frac{a_1^2+2a_1a_2}{(a_1+a_2)^2}+(1-2\nu^{(0)})\frac{a_1}{a_1+a_2}\right],\\
S_{1122} &= \frac{1}{2(1-\nu^{(0)})}\left[\frac{a_2^2}{(a_1+a_2)^2}-(1-2\nu^{(0)})\frac{a_2}{a_1+a_2}\right],\\
S_{2323} &= \frac{1}{2(1-\nu^{(0)})}\frac{2\nu^{(0)}a_1}{a_1+a_2}, \quad S_{3333}=0,\\
S_{2211} &= \frac{1}{2(1-\nu^{(0)})}\left[\frac{a_1^2}{(a_1+a_2)^2}-(1-2\nu^{(0)})\frac{a_1}{a_1+a_2}\right],\\
S_{1133} &= \frac{1}{2(1-\nu^{(0)})}\left[\frac{2\nu^{(0)}a_2}{a_1+a_2}\right], \quad S_{3311}=S_{3322}=0,\\
S_{1212} &= \frac{1}{4(1-\nu^{(0)})}\left[\frac{a_1^2+a_2^2}{(a_1+a_2)^2}+1-2\nu^{(0)}\right],\\
S_{2323} &= \frac{a_1}{2(a_1+a_2)}, \quad S_{3131}=\frac{a_2}{2(a_1+a_2)}.
\end{aligned} \tag{3.157}
$$

3) Penny-shape inclusion ($a_1 = a_2 \gg a_3$):

$$
\begin{aligned}
S_{1111} &= S_{2222} = \frac{\pi(13-8\nu^{(0)})}{32(1-\nu^{(0)})}\frac{a_3}{a_1}, \quad S_{3333}=1-\frac{\pi(1-2\nu^{(0)})}{4(1-\nu^{(0)})}\frac{a_3}{a_1},\\
S_{1122} &= S_{2211} = \frac{\pi(8\nu^{(0)}-1)}{32(1-\nu^{(0)})}\frac{a_3}{a_1}, \quad S_{1133}=S_{2233}=\frac{\pi(2\nu^{(0)}-1)}{8(1-\nu^{(0)})}\frac{a_3}{a_1},\\
S_{3311} &= S_{3322} = \frac{\nu^{(0)}}{1-\nu^{(0)}}\left[1-\frac{\pi(4\nu^{(0)}+1)}{8\nu^{(0)}}\frac{a_3}{a_1}\right],\\
S_{1212} &= \frac{\pi(7-8\nu^{(0)})}{32(1-\nu^{(0)})}\frac{a_3}{a_1}, \quad S_{3131}=S_{2323}=\frac{1}{2}\left[1+\frac{\pi(\nu^{(0)}-2)}{4(1-\nu^{(0)})}\frac{a_3}{a_1}\right].
\end{aligned} \tag{3.158}
$$

In the case of the transversally isotropic media, the scheme of the Eshelby tensor estimation is analogous to Eqs. (3.145)–(3.152) and was performed by [1193].

3.6.2 Eshelby and Related Tensors in a Special Basis

We will consider the representations of the Eshelby and some related tensors in a special **P**-basis (A.2.7) (for details see [618], [545], [986]. Eshelby tensor for spheroidal inclusion (with the semi-axes $a_1 = a_2 = a$, a_3, aspect ratio $\gamma = a_3/a$, and a symmetry axis $\mathbf{n}$ forming the **P**-basis) $\mathbf{S} = (s_1, \ldots, s_6)$ has the components

$$
\begin{aligned}
s_1 &= \frac{1}{2(1-\nu^{(0)})}f_0+f_1, \quad s_2=\frac{3-4\nu^{(0)}}{2(1-\nu^{(0)})}f_0+f_1,\\
s_3 &= \frac{\nu^{(0)}}{1-\nu^{(0)}}f_0-2f_1, \quad s_4=\frac{\nu^{(0)}}{1-\nu^{(0)}}(1-2f_0)-2f_1,\\
s_5 &= 2(1-f_0-4f_1), \quad s_6=1-2f_0+4f_1,
\end{aligned} \tag{3.159}
$$

where

$$f_0=\frac{\gamma^2(1-g)}{2(\gamma^2-1)},\ f_1=\frac{\varkappa^{(0)}\gamma^2}{4(\gamma^2-1)^2}[(2\gamma^2+1)g-3],\ \varkappa^{(0)}=\frac{1}{2(1-\nu^{(0)})},$$

$$g(\gamma)=\begin{cases}\frac{1}{\gamma\sqrt{1-\gamma^2}}\arctan\frac{\sqrt{1-\gamma^2}}{\gamma}, & \text{oblate shape } (\gamma<1),\\ \frac{1}{2\gamma\sqrt{\gamma^2-1}}\ln\frac{\gamma+\sqrt{\gamma^2-1}}{\gamma-\sqrt{\gamma^2-1}}, & \text{prolate shape } (\gamma>1).\end{cases} \tag{3.160}$$

The components of the tensors $\mathbf{P}=(p_1,\ldots,p_6)$ and $\mathbf{Q}=(q_1,\ldots,q_6)$ are

$$\begin{aligned}
&p_1=\frac{1}{2\mu^{(0)}}[(1-\varkappa^{(0)})f_0+f_1],\ p_2=\frac{1}{2\mu^{(0)}}[(2-\varkappa^{(0)})f_0+f_1],\\
&p_3=p_4=-\frac{f_1}{\mu^{(0)}},\ p_5=\frac{1-f_0-4f_1}{\mu^{(0)}},\ p_6=\frac{(1-\varkappa^{(0)})(1-2f_0)+2f_1}{\mu^{(0)}},\\
&q_1=\mu^{(0)}[4\varkappa^{(0)}-1-2(3\varkappa^{(0)}-1)f_0-2f_1],\\
&q_2=2\mu^{(0)}[1-(2-\varkappa^{(0)})f_0-f_1],\quad q_3=g_4=2\mu^{(0)}[(2\varkappa^{(0)}-1)f_0+2f_1],\\
&q_5=4\mu^{(0)}[f_0+4f_1],\ q_6=4\mu^{(0)}[(1+2\varkappa^{(0)})f_0-2f_1].
\end{aligned}\tag{3.161}$$

If $\gamma\ll 1$ (a crack), we can expand the functions f_0 and f_1 in a series over a parameter γ, and retaining terms with the first order of γ reduces the components of the tensor $\mathbf{Q}$ to the following:

$$\begin{aligned}
&q_1=\mu^{(0)}(4\varkappa^{(0)}-1)-\pi\mu^{(0)}\gamma(7\varkappa^{(0)}-2)/4,\\
&q_2=2\mu^{(0)}-\pi\mu^{(0)}\gamma(4-\varkappa^{(0)})/4,\quad q_3=g_4=\pi\mu^{(0)}\gamma(3\varkappa^{(0)}-1)/2,\\
&q_5=\pi\mu^{(0)}\gamma(1+2\varkappa^{(0)}),\quad q_6=\pi\mu^{(0)}\gamma(1+\varkappa^{(0)}).
\end{aligned}\tag{3.162}$$

In another limiting case $\gamma\gg 1$ (a needle), the representations for the tensors $\mathbf{P}$ and $\mathbf{Q}$ also have the form (3.161) where in the functions f_0 and f_1, the main terms of a serial expansion over the small parameter γ^{-1} need to be conserved: $f_0=(\gamma^2+1-\ln 2\gamma)/(2\gamma^2)$, $f_1=-\varkappa^{(0)}(3-2\ln 2\gamma)/(4\gamma^2)$.

The polarization concentration tensors $\mathbf{R}^{v\sigma}=\mathbf{M}_1(\mathbf{I}+\mathbf{Q}\mathbf{M}_1)^{-1}=(r_1^\sigma,\ldots,r_6^\sigma)$ and $\mathbf{R}^{v\epsilon}=\mathbf{L}_1(\mathbf{I}+\mathbf{P}\mathbf{L}_1)^{-1}=(r_1^\epsilon,\ldots,r_6^\epsilon)$ in a special $\mathbf{P}$-basis have the components

$$\begin{aligned}
&r_1^\sigma=\frac{1}{2\Delta_1}\Big[\delta k+\frac{4}{3}\delta\mu+q_6\Big],\ r_2^\sigma=\frac{1}{2\delta\mu+q_2},\ r_5^\sigma=\frac{4}{4\delta\mu+q_5},\\
&r_3^\sigma=r_4^\sigma=-\frac{1}{\Delta_1}\Big[\delta k-\frac{2}{3}\delta\mu+q_3\Big],\ r_6^\sigma=\frac{2}{\Delta_1}\Big[\delta k+\frac{1}{3}\delta\mu+q_1\Big],
\end{aligned}\tag{3.163}$$

$$\begin{aligned}
&r_1^\epsilon=\frac{1}{2\Delta_2}\Big[\frac{\lambda_1+\mu_1}{\mu_1(3\lambda_1+2\mu_1)}+p_6\Big],\ r_2^\epsilon=\frac{2\mu_1}{1+2p_2\mu_1},\\
&r_3^\epsilon=r_4^\epsilon=-\frac{1}{\Delta_2}\Big[-\frac{\lambda_1}{2\mu_1(3\lambda_1+2\mu_1)}+p_1\Big],\ r_5^\epsilon=\frac{4\mu_1}{1+\mu_1p_5},\\
&r_6^\epsilon=\frac{1}{\Delta_2}\Big[\frac{\lambda_1+2\mu_1}{2\mu_1(3\lambda_1+2\mu_1)}+2p_1\Big],
\end{aligned}\tag{3.164}$$

where $\lambda_1=\lambda^{(1)}-\lambda^{(0)}$, $\mu_1=\mu^{(1)}-\mu^{(0)}$, and

$$\Delta_1 = 2\Big[3\delta\mu\delta k + \delta k(q_1 + q_6 - 2q_3) + \frac{\delta\mu}{3}(4q_1 + q_6 + 4q_3) + q_1 q_6 - q_3^2\Big],$$
$$\Delta_2 = \frac{1}{2\mu_1(3\lambda_1 + 2\mu_1)}[1 + (\lambda_1 + 2\mu_1)p_6 + 4(\mu_1 + \lambda_1)p_1 + 4\lambda_1 p_3] + 2p_1 p_6 - 2p_3^2,$$
$$\delta k = k^{(1)}k^{(0)}/(k^{(0)} - k^{(1)}), \quad \delta\mu = \mu^{(1)}\mu^{(0)}/(\mu^{(0)} - \mu^{(1)}), \tag{3.165}$$

For the particular case of a spherical inclusion, all above-mentioned tensors are isotropic $\mathbf{g} = (3g^k, 2g^\mu)$ ($\mathbf{g} = \mathbf{S}, \mathbf{P}, \mathbf{Q}, \mathbf{R}^{v\sigma}, \mathbf{R}^{v\epsilon}$, see Eq. (2.197)) and can be represented in a compact form:

$$\mathbf{S} = \left(\frac{3k^{(0)}}{3k^{(0)} + 4\mu^{(0)}}, \frac{6(k^{(0)} + 2\mu^{(0)})}{5(3k^{(0)} + 4\mu^{(0)})}\right), \tag{3.166}$$
$$\mathbf{P} = \left(\frac{1}{3k^{(0)} + 4\mu^{(0)}}, \frac{3(k^{(0)} + 2\mu^{(0)})}{5(3k^{(0)} + 4\mu^{(0)})}\right), \tag{3.167}$$
$$\mathbf{Q} = \left(\frac{12k^{0)}\mu^{(0)}}{3k^{(0)} + 4\mu^{(0)}}, \frac{2(9k^{(0)} + 8\mu^{(0)})\mu^{(0)}}{5(3k^{(0)} + 4\mu^{(0)})}\right), \tag{3.168}$$
$$\mathbf{A} = \left(\frac{k^{(0)} + k_0^*}{k^{(1)} + k_0^*}, \frac{\mu^{(0)} + \mu_0^*}{\mu^{(1)} + \mu_0^*}\right), \tag{3.169}$$
$$\mathbf{B} = \left(\frac{k^{(0)} + k_0^*}{k^{(1)} + k_0^*}\frac{k^{(1)}}{k^{(0)}}, \frac{\mu^{(0)} + \mu_0^*}{\mu^{(1)} + \mu_0^*}\frac{\mu^{(1)}}{\mu^{(0)}}\right), \tag{3.170}$$
$$\mathbf{R}^{v\epsilon} = \left(3(k^{(1)} - k^{(0)})\frac{k^{(0)} + k_0^*}{k^{(1)} + k_0^*}, 2(\mu^{(1)} - \mu^{(0)})\frac{\mu^{(0)} + \mu_0^*}{\mu^{(1)} + \mu_0^*}\right), \tag{3.171}$$
$$\mathbf{R}^{v\sigma} = -\left(\frac{k^{(1)} - k^{(0)}}{3(k^{(0)})^2}\frac{k^{(0)} + k_0^*}{k^{(1)} + k_0^*}, \frac{\mu^{(1)} \quad \mu^{(0)}}{2(\mu^{(0)})^2}\frac{\mu^{(0)} + \mu_0^*}{\mu^{(1)} + \mu_0^*}\right), \tag{3.172}$$

where

$$k_0^* = \frac{4}{3}\mu^{(0)}, \quad \mu_0^* = \frac{3}{2}\left(\frac{1}{\mu^{(0)}} + \frac{10}{9k^{(0)} + 8\mu^{(0)}}\right)^{-1}. \tag{3.173}$$

It should be mentioned that the representation of Eshelby and related tensors (such as $\mathbf{P}, \mathbf{Q}, \mathbf{A}, \mathbf{B}$) are also known in the Walpole's tensor $\mathbf{B}$-basis (A.2.17) (for details see [908], [1152], [1184]). For example, for an inclusion oriented in direction $\mathbf{e}_1$, Walpoles components of $\mathbf{S}$ are given by

$$\mathbf{S} = (S_{2222} + S_{2233}, S_{1111}, 2S_{2323}, 2S_{1212}, S_{1122}, S_{2211})^W \tag{3.174}$$

where the components S_{ijkl} are presented by Eqs. (3.155)–(3.158). Eshelby tensor for the inclusion oriented in the direction $\mathbf{n}$ can be found using the same components (3.174) and Walpole's basis relative to the direction $\mathbf{n}$ (A.2.17) and (A.2.20). Analogous representations of the Eshelby and related tensors in the basis $\mathbf{E}^{(\alpha)}$ (A.2.1) with an altered numeration were used in [654].

Let both materials of the matrix $v^{(0)}$ and inclusion $v^{(1)}$ be transversally isotropic with the common axis of isotropy directed along the fiber's axis $\mathbf{n}$ and presented in the special $\mathbf{P}$-basis (A.2.7) ($i = 0, 1$):

$$\mathbf{L}^{(i)} = (k^{(i)}, 2m^{(i)}, l^{(i)}, l^{(i)}, 4\mu^{(i)}, n^{(i)}) \tag{3.175}$$

with the parameters related with the tensor components of the tensor $\mathbf{L}^{(i)}$ by Eq. (A.2.9) and (A.2.12). The Eshelby tensor $\mathbf{S}(\mathbf{L}^{(0)})$ for an ellipsoidal inclusion in the transversally isotropic medium was obtained in [1193], while the $\mathbf{P}$-basis representations of the nonzero components of the related tensors $\mathbf{P} = (p_1, \ldots, p_6)$ and $\mathbf{A} = (a_1, \ldots, a_6)$ were proposed in [545] for an infinite cylindrical fiber:

$$p_1 = \frac{1}{4m^{(0)}(k^{(0)} + m^{(0)})}, \quad p_2 = \frac{k^{(0)} + 2m^{(0)}}{4m^{(0)}(k^{(0)} + m^{(0)})}, \quad p_5 = \frac{1}{2\mu^{(0)}}, \tag{3.176}$$

$$a_1 = \frac{1}{2}\Big(1 + \frac{k_1^{(1)}}{k^{(0)} + m^{(0)}}\Big)^{-1}, \quad a_2 = \Big[1 + \frac{m_1^{(1)}(k^{(0)} + 2m^{(0)})}{2m^{(0)}(k^{(0)} + m^{(0)})}\Big]^{-1}, \tag{3.177}$$

$$a_3 = -\frac{l_1^{(1)}}{2(m^{(0)} + k^{(1)})}, \quad a_5 = \frac{2\mu^{(0)}}{\mu^{(0)} + \mu^{(1)}}, \quad a_6 = 1, \tag{3.178}$$

where $f_1^{(1)} = f^{(1)} - f^{(0)}, \quad f = (k, m, l)$.

3.7 Coated Ellipsoidal Inclusion

More detailed consideration of the mechanical behavior of composite materials requires analysis of the interface between the reinforcement and the matrix. These interfaces may represent weak interfacial layer due to imperfect bonding between the two phases and inter-diffusion and/or chemical interaction zones (with properties varying through the thickness and/or along the surface) at the interface between the two phases [558], [1085]. It is well known that the overall effective properties of composite materials are significantly influenced by the properties of the interfaces between the constituents. First, the interface controls the in situ reinforcement's (particles' or fibers') strength and hence the strength of the composite. Second, defects and damage are likely to occur at the interface (e.g., debonding, sliding, and interface cracks, etc.), and these interfacial defects control the degradation of the composite. Therefore, to evaluate more accurately the effective properties of a composite, the behavior and structures of interfaces must be taken into consideration ([230], [914]). In our short survey we consider the problem of a single coated ellipsoidal inclusion inside an infinite matrix.

Classical works dealing with three-phase solids with spherical or cylindrical coated inclusions are discussed, e.g., in [427], [437]. More general cases of mechanical loading, including location dependent transformation were considered in [239], [720] (see references in [62]). The effect of variation of elastic properties with radial distance from the fiber's boundary in continuously reinforced fiber composites ([495], [1085]) was analyzed. The above studies are restricted to isotropic materials and spherical or cylindrical reinforcement shapes. Micata and Taya [763] applied Boussinesq–Sadowsky stress–functions when calculating the stress field for two confocal prolate spheroids embedded in an infinite body.

The thin-layer hypothesis appeared as a principal step in the investigation of coated inclusions, because it allows the use of the well-developed Eshelby [319] theory and Hill [464] interface operators for the general case of anisotropy of the materials being in contact. In his pioneering paper in this direction Walpole

[1151] assumed that the stress and strain components inside the inclusion coincide with those already determined before the coating was introduced. Afterwards this assumption was replaced by the hypothesis of homogeneity of the stress state inside the core (inclusion), and thermoelastic problems were considered in [207], [230], [445], [909]. In most of the papers homogeneous thermoelastic properties were assumed for the coating (a case of the multi-coated inclusion was analyzed in a similar manner, see [690]). In the present work we relax these restrictive assumptions. The case of inhomogeneity of elastic mismatch properties in the coating is a typical situation the production of the coated inclusions and due to thermal and plastic deformations of the matrix near the inclusion. Even in situations in which for a specific reference system (connected to the unit normal and tangential vectors of the surface of the inclusion) homogeneous stress and strain fields may be assumed, the introduction of a global coordinate system requires consideration of inhomogeneous fields in the coating.

Another direction of research in the field of coated inclusion mechanics is dealing with sliding interfaces being intensively treated in [301], [436], [479], [496], [713], [794], [914], [1001]. The elastic field of a single sliding inclusion is solved by distributing Somiliana's dislocation on the interface where the interface shear exceeds the stick limit and the contact condition changes from perfect bonding to perfect sliding with non-continuous elastic fields. Furthermore, Hashin [436] has shown that sliding along a two-dimensional interface is equivalent to the response of some isotropical very thin flexible coating. An arbitrarily anisotropic thin interphase between two anisotropic solids was modeled by Benveniste [58] as a surface between its two neighbouring media by means of appropriately devised interface conditions on it.

We consider in this section (following [184]) thermoelastic solution for a coated heterogeneity subjected to remote homogeneous loading with thin inhomogeneous coatings of the inclusions along the surface. The micromechanical approach is based on the Green's function technique as well as on the interfacial Hill operators.

3.7.1 General Representation for the Concentrator Factors for Coated Heterogeneity

At first, let us consider a single heterogeneity (no restrictions are imposed on the shape and structure of the heterogeneity) v_i in an infinite medium subjected to the homogeneous field $\boldsymbol{\sigma}^0(\mathbf{x}) \equiv \boldsymbol{\sigma}^0 = \text{const.}$ which is assumed to be known. Then in view of the linearity of the problem there exist fourth and second-rank tensors $\mathbf{B}_i(\mathbf{x})$, $\mathbf{R}_i(\mathbf{x})$ and $\mathbf{C}_i(\mathbf{x})$, $\mathbf{F}_i(\mathbf{x})$, such that

$$\boldsymbol{\sigma}(\mathbf{x}) = \mathbf{B}_i(\mathbf{x})\boldsymbol{\sigma}^0 + \mathbf{C}_i(\mathbf{x}), \quad \bar{v}_i\boldsymbol{\eta}(\mathbf{x}) = \mathbf{R}_i(\mathbf{x})\boldsymbol{\sigma}^0 + \mathbf{F}_i(\mathbf{x}), \quad \mathbf{x} \in v_i, \tag{3.179}$$

where (with no sum on i):

$$\mathbf{R}_i(\mathbf{x}) = \bar{v}_i\mathbf{M}_1(\mathbf{x})\mathbf{B}_i(\mathbf{x}), \quad \mathbf{F}_i(\mathbf{x}) = \bar{v}_i[\mathbf{M}_1(\mathbf{x})\mathbf{C}_i(\mathbf{x}) + \boldsymbol{\beta}_1(\mathbf{x})]. \tag{3.180}$$

where we defined the strain polarization tensor $\boldsymbol{\eta}(\mathbf{x}) = \mathbf{M}_1(\mathbf{x})\boldsymbol{\sigma}(\mathbf{x}) + \boldsymbol{\beta}_1(\mathbf{x})$, (which is simply a notation convenience). In the general case the estimation of the tensors $\mathbf{B}_i(\mathbf{x})$, $\mathbf{C}_i(\mathbf{x})$ is a particular problem of the transformation field analysis method [302]. which can be realized by different numerical methods, for example, by finite element analysis (for details see Section 4.1).

We determine an arbitrary imaginary ellipsoid v_i^0 (with the smallest possible volume) with the characteristic function $V_i^0(\mathbf{x})$ inside which the real inclusion $v_i \subset v_i^0$ is to be inscribed. Averaging of Eq. (3.78$_2$) over the volume v_i^0 and exploiting Eshelby theorem [318] leads to the algebraic equation:

$$\langle\boldsymbol{\sigma}\rangle_i^0 = \overline{\boldsymbol{\sigma}} - \frac{\bar{v}_i}{v_i^0}\mathbf{Q}_i^0\langle\boldsymbol{\eta}\rangle_i, \tag{3.181}$$

where $\mathbf{Q}_i^0(\mathbf{x}) = \mathbf{Q}_i^0 \equiv -\bar{v}_i^0\langle\boldsymbol{\Gamma}(\mathbf{x}-\mathbf{y})\rangle_i^0 = \text{const.}$ $(\mathbf{x} \in v_i \subset v_i^0,\ \mathbf{y} \in v_i^0)$ is associated with the well-known Eshelby tensor by $\mathbf{S}_i = \mathbf{I} - \mathbf{M}^{(0)}\mathbf{Q}_i$. Hereafter

$$\mathbf{f}_i \equiv \langle\mathbf{f}(\mathbf{y})\rangle_i = \bar{v}_i^{-1}\int \mathbf{f}(\mathbf{y})V_i(\mathbf{y})\,d\mathbf{y}, \mathbf{f}_i^0 \equiv \langle\mathbf{f}(\mathbf{y})\rangle_i^0 = (\bar{v}_i^0)^{-1}\int \mathbf{f}(\mathbf{y})V_i^0(\mathbf{y})\,d\mathbf{y} \tag{3.182}$$

denotes averaging of some tensor $\mathbf{f}(\mathbf{y})$ over the volume of the regions $\mathbf{y} \in v_i \subset v_i^0$ and $\mathbf{y} \in v_i^0$, respectively; in so doing $\mathbf{x} \in v_i$ is fixed. In deriving Eq. (3.182) it was essentially exploited that the correlation hole v_i^0 is ellipsoidal, and, therefore $\mathbf{Q}_i^0(\mathbf{x}) = \mathbf{Q}_i^0 \equiv$const., and that $\boldsymbol{\eta}(\mathbf{x}) \equiv \mathbf{0}$ at $\mathbf{x} \in v_i^0 \setminus v_i$.

Substitution of Eq. (3.181) into the right-hand side of Eq. (3.179$_1$) averaged over the volume v_i^0 leads to the equation

$$\mathbf{B}_i^0\overline{\boldsymbol{\sigma}} + \mathbf{C}_i^0 = \overline{\boldsymbol{\sigma}}_i - \frac{\bar{v}_i}{\bar{v}_i^0}\mathbf{Q}_i^0\langle\boldsymbol{\eta}\rangle_i, \tag{3.183}$$

yielding the following representation for the average strain polarization tensors:

$$\bar{v}_i\langle\boldsymbol{\eta}\rangle_i = \bar{v}_i^0(\mathbf{Q}_i^0)^{-1}[\mathbf{I} - \mathbf{B}_i^0]\overline{\boldsymbol{\sigma}}_i - \bar{v}_i^0(\mathbf{Q}_i^0)^{-1}\mathbf{C}_i^0, \tag{3.184}$$

and the relations between the averaged tensors (3.179) (no sum on i):

$$\mathbf{R}_i^0 = \bar{v}_i^0(\mathbf{Q}_i^0)^{-1}(\mathbf{I} - \mathbf{B}_i^0), \quad \mathbf{F}_i^0 = -\bar{v}_i^0(\mathbf{Q}_i^0)^{-1}\mathbf{C}_i^0, \tag{3.185}$$

$$\mathbf{B}_i^0 = \mathbf{I} - (\bar{v}_i^0)^{-1}\mathbf{Q}_i^0\mathbf{R}_i^0, \quad \mathbf{C}_i^0 = -(\bar{v}_i^0)^{-1}\mathbf{Q}_i^0\mathbf{F}_i^0. \tag{3.186}$$

For a particular case of the homogeneous ellipsoidal heterogeneity $v_i \equiv v_i^0$:

$$\mathbf{M}_1(\mathbf{x}) = \mathbf{M}_1^{(i)} = \text{const}, \ \boldsymbol{\beta}_1(\mathbf{x}) = \boldsymbol{\beta}_1^{(i)} = \text{const} \quad \text{at } \mathbf{x} \in v_i, \tag{3.187}$$

we get the tensors $\mathbf{B}_i$ (3.121) and $\mathbf{C}_i$ (3.122$_2$) as well as the tensors

$$\mathbf{R}_i = \bar{v}_i\mathbf{M}_1^{(i)}\mathbf{B}_i, \quad \mathbf{F}_i = \bar{v}_i(\mathbf{I} + \mathbf{M}_1^{(i)}\mathbf{Q}_i)^{-1}\boldsymbol{\beta}_1^{(i)}. \tag{3.188}$$

By comparing the relations (3.179) and (3.188), we see that the average thermoelastic response (i.e., the tensors $\mathbf{B}_i$, $\mathbf{C}_i$, $\mathbf{R}_i$, $\mathbf{F}_i$) of any coated inclusion v_i is the same as the response of some fictitious ellipsoidal homogeneous inclusion with thermoelastic parameters:

$$\mathbf{M}_1^{f(i)} = (\mathbf{Q}_i^0)^{-1}((\mathbf{B}_i^0)^{-1} - \mathbf{I}), \quad \boldsymbol{\beta}_1^{f(i)} = -(\mathbf{Q}_i^0)^{-1}(\mathbf{B}_i^0)^{-1}\mathbf{C}_i, \tag{3.189}$$

$$\mathbf{M}_1^{f(i)} = (\mathbf{I}\bar{v}_i - \mathbf{R}_i^0\mathbf{Q}_i^0)^{-1}\mathbf{R}_i^0, \quad \boldsymbol{\beta}_1^{f(i)} = \bar{v}_i^{-1}(\mathbf{M}_1^{f(i)}\mathbf{Q}_i^0 + \mathbf{I})\mathbf{F}_i^0. \tag{3.190}$$

The parameters (3.189) and (3.190) of fictitious ellipsoidal inclusions are simply a notational convenience. No restrictions are imposed on the microtopology of the coated inclusions as well as on the inhomogeneity of the stress state in the coated inclusions. No restrictions are imposed on the microtopology of the inclusions that have any shape and arbitrarily varying (continuously or discontinuously) thermoelastic properties as well as on the inhomogeneity of the stress state in the inclusions. The assumption of an ellipsoidal shape of the region v_i^0, was used only in order to obtain analytical representation (3.125). Moreover, the similar scheme of fictitious ellipsoidal inclusions is realized in [1220] for partially debonded inclusions.

3.7.2 Single Ellipsoidal Inclusion with Thin Coating

In this subsection an analytical method for estimating the tensors $\mathbf{B}(\mathbf{x})$ and $\mathbf{C}(\mathbf{x})$, see (3.179), is developed for the example of a single ellipsoidal inclusion with a thin coating in an infinite matrix loaded by a constant macroscopic stress $\boldsymbol{\sigma}^0$. Let the coated inclusion v_1 consist of an ellipsoidal core $v^i \subset v_1$ with a characteristic function $V^i(\mathbf{x})$ and thermoelastic parameters $\mathbf{M}^i$, $\boldsymbol{\beta}^i \equiv$ const, and a thin coating $v^c \equiv v_1 \setminus v^i$ with a characteristic function $V^c \equiv V_1 - V^i$ and thermoelastic inhomogeneous properties $\mathbf{M}^c(\mathbf{x})$, $\boldsymbol{\beta}^c(\mathbf{x}) \neq$ const (see Fig. 3.1). In

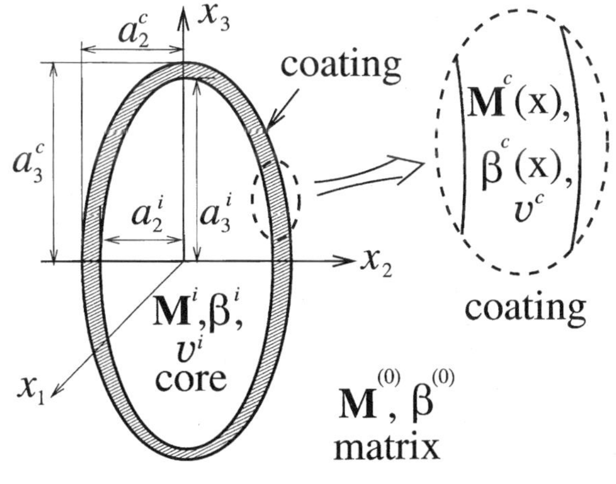

Fig. 3.1. Single coated inclusion.

the considered case of a single inclusion the origin of the coordinate system is chosen to be the center of the inclusion $\mathbf{x}^i = \mathbf{0}$ and the coordinate axes coincide with the axes of the inclusions. In addition to the decomposition (3.67) we define the jump of the material properties $\mathbf{f}$ ($\mathbf{f} = \mathbf{M}, \boldsymbol{\beta}$) across the boundary s^i between the core and the coating as $\mathbf{f}_2(\mathbf{x}) \equiv \mathbf{f}^i - \mathbf{f}^c(\mathbf{x})$. For the single coated inclusion Eq. (3.77) yields

$$\boldsymbol{\sigma}(\mathbf{x}) = \boldsymbol{\sigma}^0 + \int [V^i(\mathbf{y}) + V^c(\mathbf{y})]\, \boldsymbol{\Gamma}(\mathbf{x}-\mathbf{y})\, [\mathbf{M}_1(\mathbf{y})\boldsymbol{\sigma}(\mathbf{y}) + \boldsymbol{\beta}_1(\mathbf{y})]\; d\mathbf{y}. \quad (3.191)$$

In analogy to [207], [230] we find an approximative solution of Eq. (3.191) under the approximative assumption of an homogeneous stress state in the core

$$\boldsymbol{\sigma}(\mathbf{x}) \equiv \boldsymbol{\sigma}^i = \text{const}, \quad \mathbf{x} \in v^i \subset v_1. \tag{3.192}$$

and the thin-layer hypothesis, which means that the characteristic function $V^c(\mathbf{y})$ can be replaced by a surface δ-function (with weighting function ρ) at the outer surface s^i_- of the boundary $s^i = s^i_- \cup s^i_+$ and the volume integral of the continuous function $\mathbf{g}(\mathbf{y}),\ \mathbf{y} \in v^c$ is equal to a surface integral over outer surface s^i_- [374]:

$$\int V^c(\mathbf{y})\mathbf{g}(\mathbf{y})\, d\mathbf{y} = \int S^i_-(\mathbf{s})\mathbf{g}(\mathbf{s})\rho\, d\mathbf{s}, \tag{3.193}$$

where the product of the characteristic function S^i_- of the boundary s^i_- and some continuous function $\mathbf{g}(\mathbf{s}) \equiv \lim \mathbf{g}(\mathbf{y}),\ (\mathbf{y} \to \mathbf{s},\ \mathbf{y} \in v^c,\ \mathbf{s} \in s^i_-)$ is integrated over the surface s^i_-. In the particular case considered hereafter, the weighting function ρ for a domain v^c bounded by two ellipsoidal surfaces with the same center and with identically oriented semi-axes a_j and a^c_j $(j = 1, 2, 3)$, respectively, is estimated in [230] by $(\mathbf{y} \equiv (y_1, y_2, y_3)^\top \in s^i_-)$:

$$\rho(\mathbf{y}) = \left(\frac{y_1^2}{a_1^4} + \frac{y_2^2}{a_2^4} + \frac{y_3^2}{a_3^4}\right)^{-1/2} \sum_{j=1}^{3} \frac{(a^c_j - a_j)}{a_j} \frac{y_j^2}{a_j^2}. \tag{3.194}$$

Under these assumptions the integral equation (3.191) is, after averaging over the domain v^i, reduced to

$$\boldsymbol{\sigma}^i = \boldsymbol{\sigma}^0 - \mathbf{Q}^i \boldsymbol{\eta}^i + \int S^i_-(\mathbf{s})\mathbf{T}^i(\mathbf{x}^i - \mathbf{s})\left[\mathbf{M}^c_1(\mathbf{s})\boldsymbol{\sigma}(\mathbf{s}) + \boldsymbol{\beta}^c_1(\mathbf{s})\right]\rho\, d\mathbf{s}, \tag{3.195}$$

where $\boldsymbol{\eta}^i \equiv \mathbf{M}^i_1\boldsymbol{\sigma}^i + \boldsymbol{\beta}^i_i$. Here and in the following the upper index i for the tensors $\mathbf{Q}^i,\ \mathbf{B}^i,\ \mathbf{T}^i(\mathbf{x}^i - \mathbf{s})$ stands for the calculation of these tensors for the core v^i by the use of the formulae (3.181), (3.121), (3.131_2), respectively. Obviously, discarding the integral term in (3.195) leads to the Eshelby solution.

Taking the properties of the interface operator $\boldsymbol{\Gamma}(\mathbf{n})$ (3.95) and (3.131_2) into account leads to

$$\begin{aligned} \boldsymbol{\sigma}^c(\mathbf{s}) \equiv \boldsymbol{\sigma}(\mathbf{s}) &= \boldsymbol{\sigma}^i + \boldsymbol{\Gamma}(\mathbf{n}, \mathbf{M}^c)\left[\mathbf{M}^i_2(\mathbf{s})\boldsymbol{\sigma}^i + \boldsymbol{\beta}^i_2(\mathbf{s})\right], \\ \bar{v}^i\mathbf{T}^i(\mathbf{x}^i - \mathbf{s}) &= \boldsymbol{\Gamma}(\mathbf{n}, \mathbf{M}^{(0)}) - \mathbf{Q}^i, \\ \boldsymbol{\Gamma}(\mathbf{n}, \mathbf{M}^{(0)})\mathbf{M}^c_1\boldsymbol{\Gamma}(\mathbf{n}, \mathbf{M}^c) &= \boldsymbol{\Gamma}(\mathbf{n}, \mathbf{M}^{(0)}) - \boldsymbol{\Gamma}(\mathbf{n}, \mathbf{M}^c), \end{aligned} \tag{3.196}$$

where $\mathbf{n}$ is a unit outward normal vector on ∂v^i in the point $\mathbf{s}$. Here both interface operators $\boldsymbol{\Gamma}(\mathbf{n}, \mathbf{M}^{(0)})$ and $\boldsymbol{\Gamma}(\mathbf{n}, \mathbf{M}^c)$ are defined by formula (3.131_2) applied with the compliances $\mathbf{M}^{(0)}$ and $\mathbf{M}^c$, respectively; $\mathbf{M}^i_2(\mathbf{s}) \equiv \mathbf{M}^i - \mathbf{M}^c(\mathbf{s}),\ \ \boldsymbol{\beta}^i_2(\mathbf{s}) \equiv \boldsymbol{\beta}^i - \boldsymbol{\beta}^c(\mathbf{s})$.

For the coated inclusion $v_1 = v^i \cup v^c$, according to (3.131_2) the tensor $\boldsymbol{\Gamma}(\mathbf{n})^-$ in (3.131_2) integrated over the coating v^c yields

$$\int V^c(\mathbf{y})\boldsymbol{\Gamma}(\mathbf{n})^-\, d\mathbf{y} = \bar{v}^c\mathbf{Q}^i + \int \left[V_1(\mathbf{y}) - V^i(\mathbf{y})\right] \int V^i(\mathbf{x})\boldsymbol{\Gamma}(\mathbf{x} - \mathbf{y})\, d\mathbf{x}\, d\mathbf{y}. \tag{3.197}$$

Changing the integration sequence and applying Eshelby's theorem, we get from (3.197) $\int V^c(\mathbf{y})\mathbf{\Gamma}(\mathbf{n})^-\, d\mathbf{y} = \overline{v}^c\mathbf{Q}^i + \overline{v}^i(\mathbf{Q}^i - \mathbf{Q}_1)$.

Now Eq. (3.195) reduces to an equation with only one unknown constant tensor $\boldsymbol{\sigma}^i$:

$$\begin{aligned}\boldsymbol{\sigma}^i &= \boldsymbol{\sigma}^0 + \left[\frac{\overline{v}^c}{\overline{v}^i}\mathbf{Q}^i - \mathbf{Q}_1\right]\boldsymbol{\eta}^i - \frac{1}{\overline{v}^i}\mathbf{Q}^i \int S^i_-(\mathbf{s})\left[\mathbf{M}^c_1(\mathbf{s})\boldsymbol{\sigma}^i + \boldsymbol{\beta}^c_1(\mathbf{s})\right]\rho\, d\mathbf{s} \\ &- \frac{1}{\overline{v}^i}\int S^i_-(\mathbf{s})\left[\mathbf{I} + \mathbf{Q}^i\mathbf{M}^c_1(\mathbf{s})\right]\left[\mathbf{I} + \mathbf{\Gamma}(\mathbf{n}, \mathbf{M}^{(0)})\mathbf{M}^c_1(\mathbf{s})\right]^{-1}\mathbf{\Gamma}(\mathbf{n}, \mathbf{M}^{(0)}) \\ &\cdot \left[\mathbf{M}^i_2(\mathbf{s})\boldsymbol{\sigma}^i + \boldsymbol{\beta}^i_2(\mathbf{s})\right]\rho\, d\mathbf{s}. \end{aligned} \tag{3.198}$$

The tensor $\mathbf{Q}_1$ for the ellipsoidal inclusion v_1 is determined by the relation (3.110_2).

In this way we obtain an estimation of the stress distribution inside the coated inclusion $\boldsymbol{\sigma}^i$ and $\boldsymbol{\sigma}^c(\mathbf{s})$, see (3.196_1) and (3.198). Therefore, the stress concentration tensors $\mathbf{B}(\mathbf{x}),\ \mathbf{C}(\mathbf{x})$ in Eq. (3.179) are found to be $\mathbf{B}(\mathbf{x}),\ \mathbf{C}(\mathbf{x}) =$ const at $\mathbf{x} \in v^i$ and $\mathbf{B}(\mathbf{x}),\ \mathbf{C}(\mathbf{x}) \neq$ const at $\mathbf{x} \in v^c$. After that the tensors $\mathbf{R}$ and $\mathbf{F}$ are defined by Eq. (3.179) and the thermoelastic properties of the fictitious homogeneous inclusions, $\mathbf{M}_1^{f(i)}$ and $\boldsymbol{\beta}_1^{f(i)}$, are evaluated by the relations (3.189) and (3.190).

Let us consider a simplification of the solution (3.198) for different particular cases of coated inclusions. According to (3.193) and (3.197) for a homogeneous coating, i.e., $\mathbf{M}^c(\mathbf{x})$ and $\boldsymbol{\beta}^c(\mathbf{x})$ are constant for any $\mathbf{x} \in v^c$, we get from (3.198) and (3.196_1)

$$\begin{aligned}\boldsymbol{\sigma}^i &= \boldsymbol{\sigma}^0 - \mathbf{Q}_1\boldsymbol{\eta}^i - \frac{\overline{v}^c}{\overline{v}^i}\left\{\mathbf{Q}^i\mathbf{M}^c_1\mathbf{Q}^i(\mathbf{M}^c) + \mathbf{Q}^i(\mathbf{M}^c) - \mathbf{Q}^i\right\}\left[\mathbf{M}^i_2\boldsymbol{\sigma}^i + \boldsymbol{\beta}^i_2\right] \\ &+ (\mathbf{I} + \mathbf{Q}^i\mathbf{M}^c_1)\left[\mathbf{Q}_1(\mathbf{M}^c) - \mathbf{Q}^i(\mathbf{M}^c)\right]\left[\mathbf{M}^i_2\boldsymbol{\sigma}^i + \boldsymbol{\beta}^i_2\right], \end{aligned} \tag{3.199}$$

$$\langle\boldsymbol{\sigma}^c\rangle^c = \boldsymbol{\sigma}^i + \left\{\mathbf{Q}^i(\mathbf{M}^c) + \frac{\overline{v}^c}{\overline{v}^i}\left[\mathbf{Q}_1(\mathbf{M}^c) - \mathbf{Q}^i(\mathbf{M}^c)\right]\right\}\left[\mathbf{M}^i_2\boldsymbol{\sigma}^i + \boldsymbol{\beta}^i_2\right]. \tag{3.200}$$

Here the tensors $\mathbf{Q}$ and $\mathbf{Q}(\mathbf{M}^c)$ (with the indices i and 1) are calculated for the compliances $\mathbf{M}^{(0)}$ and $\mathbf{M}^c$, respectively; $\langle(.)\rangle^c$ denotes average over v^c. In the particular case of homothetic surfaces ∂v^i and ∂v^c (when $\mathbf{Q}^i = \mathbf{Q}_1$) the equations (3.199) and (3.200) can be further simplified to become

$$\boldsymbol{\sigma}^i = \boldsymbol{\sigma}^0 - \mathbf{Q}^i\boldsymbol{\eta}^i - \frac{\overline{v}^c}{\overline{v}^i}\left[\mathbf{Q}^i\mathbf{M}^c_1\mathbf{Q}^i(\mathbf{M}^c) + \mathbf{Q}^i(\mathbf{M}^c) - \mathbf{Q}^i\right]\left[\mathbf{M}^i_2\boldsymbol{\sigma}^i + \boldsymbol{\beta}^i_2\right], \tag{3.201}$$

$$\langle\boldsymbol{\sigma}^c\rangle^c = \boldsymbol{\sigma}^i + \mathbf{Q}^i(\mathbf{M}^c)[\mathbf{M}^i_2\boldsymbol{\sigma}^i + \boldsymbol{\beta}^i_2]. \tag{3.202}$$

The relations (3.199)–(3.202) have been proposed previously by a more specific method [230] (see also [177], [183]).

Clearly, the thin-layer hypothesis can be rejected if the elastic properties of the coating and the matrix are the same: $\mathbf{M}^c(\mathbf{x}) \equiv \mathbf{M}^{(0)},\ \mathbf{x} \in v^c$. Then under assumption (3.192) the equation (3.191) can be solved immediately, leading to

$$\boldsymbol{\sigma}^i = \mathbf{B}^i\left[\boldsymbol{\sigma}^0 - \mathbf{Q}^i\boldsymbol{\beta}^i_1 + \int V^c(\mathbf{y})\mathbf{T}^i(\mathbf{x}^i - \mathbf{y})\boldsymbol{\beta}^c_1(\mathbf{y})\, d\mathbf{y}\right], \tag{3.203}$$

$$\langle \boldsymbol{\sigma}^c \rangle^c = \boldsymbol{\sigma}^0 + \frac{\overline{v}^i}{\overline{v}^c}(\mathbf{Q}^i - \mathbf{Q}_1)\boldsymbol{\eta}^i - \frac{\overline{v}^i}{\overline{v}^c}\int V^c(\mathbf{y})\mathbf{T}^i(\mathbf{x}^i - \mathbf{y})\boldsymbol{\beta}_1^c(\mathbf{y})\, d\mathbf{y}$$
$$- \mathbf{Q}_1\langle \boldsymbol{\beta}_1^c \rangle^c, \tag{3.204}$$

where $\mathbf{B}^i = (\mathbf{I} + \mathbf{Q}^i\mathbf{M}_1^i)^{-1}$. For a thin coating we obtain from (3.198) and (3.195):

$$\boldsymbol{\sigma}^i = \mathbf{B}^i\left[\boldsymbol{\sigma}^0 - \mathbf{Q}^i\boldsymbol{\beta}_1^i - \frac{\overline{v}^c}{\overline{v}^i}\mathbf{Q}^i\langle \boldsymbol{\beta}_1^c \rangle^c + \frac{1}{\overline{v}^i}\int S_-^i(\mathbf{s})\boldsymbol{\Gamma}(\mathbf{n}, \mathbf{M}^{(0)})\boldsymbol{\beta}_1^c(\mathbf{s})\rho\, ds\right], \tag{3.205}$$

$$\langle \boldsymbol{\sigma}^c \rangle^c = \boldsymbol{\sigma}^0 + \frac{\overline{v}^i}{\overline{v}^c}(\mathbf{Q}^i - \mathbf{Q}_1)\boldsymbol{\eta}^i - \frac{\overline{v}^c}{\overline{v}^i}\mathbf{Q}^i\langle \boldsymbol{\beta}_1^c \rangle^c$$
$$+ \frac{\overline{v}^c - \overline{v}^i}{\overline{v}^i\overline{v}^c}\int S_-^i(\mathbf{s})\boldsymbol{\Gamma}(\mathbf{n}, \mathbf{M}^{(0)})\boldsymbol{\beta}_1^c(\mathbf{s})\rho\, ds. \tag{3.206}$$

These relations result from (3.203) and (3.204) under the assumption of the thin-layer hypothesis (3.193).

3.7.3 Numerical Assessment of Thin-Layer Hypothesis

Let us consider a single spherical inclusion of the radius a^i with a homothetic spherical coating of the radius a^c with $\mathbf{Q}^i = \mathbf{Q}_1$ and, according to (3.194), $\rho = a^c - a^i$ in an infinite matrix. The elastic properties of the coating coincide with the elastic properties of the isotropic matrix, i.e., $\mathbf{L}^c = \mathbf{L}^{(0)} = (3k^{(0)}, 2\mu^{(0)})$ (2.197). Let $\boldsymbol{\beta}^i = \boldsymbol{\beta}^{(0)} = \mathbf{0}$, and $\boldsymbol{\beta}^c$ has a special form with some physical meaning represented by

$$\boldsymbol{\beta}^c = \gamma\boldsymbol{\delta} + (\chi - \gamma)\mathbf{n}\otimes\mathbf{n}, \tag{3.207}$$

where $\mathbf{n}$ is the unit outward normal vector on ∂v^i.

It can easily be shown that γ and χ are the transformation parameters of the coating in the tangential and normal directions, respectively. If $\gamma = \chi$ this constitutive characterization of the coating corresponds to the particular case of an isotropical thermal expansion, considered by a number of authors (see, e.g., [207], [445]). We will analyze a less trivial case $\gamma \neq \chi$ allowing the existence of a prestress in the coating. This situation is typical for the case of production of coated inclusions separately from the matrix. Moreover, under purely thermal deformations (e.g., $\boldsymbol{\sigma}^0 = \mathbf{0}$) the plastic strains of the matrix near the inclusions also have the form (3.207) (see Chapter 16 and [186]). Clearly, in some local coordinate system connected with the inclusion surface ∂v^i the tensor $\boldsymbol{\beta}^c$ is constant. However, in the global coordinate system $\boldsymbol{\beta}^c$ is a function of the unit normal $\mathbf{n}$ and, therefore, is an inhomogeneous function of the coordinates. Therefore, the system (3.199), (3.200), obtained under the assumption of a homogeneous coating, is not suitable and it is necessary to consider the system for either thin coating, i.e., (3.205) and (3.206), or thick one, i.e., (3.203), (3.204). In the more general case of a thick coating we have

$$\boldsymbol{\sigma}^i = \mathbf{B}^i\left[\boldsymbol{\sigma}^0 + (\chi - \gamma)\int V^c(\mathbf{y})\mathbf{T}^i(\mathbf{x}^i - \mathbf{y})(\mathbf{n}\otimes\mathbf{n})\, d\mathbf{y}\right], \tag{3.208}$$

$$\langle \boldsymbol{\sigma}^c \rangle^c = \boldsymbol{\sigma}^0 - \frac{1}{3}(\chi + 2\gamma)\mathbf{Q}^i\boldsymbol{\delta} - \frac{\overline{v}^i}{\overline{v}^c}(\chi - \gamma)\int V^c(\mathbf{y})\mathbf{T}^i(\mathbf{x}^i - \mathbf{y})(\mathbf{n}\otimes\mathbf{n})\, d\mathbf{y}. \quad (3.209)$$

According to (Eqs. (3.181) and (3.182) we can now find the concentration tensors $\mathbf{B} = \mathbf{I} + \overline{v}^i(\overline{v}_1)^{-1}(\mathbf{B}^i - \mathbf{I})$, and

$$\mathbf{C} = -\frac{1}{3}\frac{\overline{v}^c}{\overline{v}_1}(\chi + 2\gamma)\mathbf{Q}^i\boldsymbol{\delta} + \frac{\overline{v}^i}{\overline{v}_1}(\chi - \gamma)(\mathbf{B}^i - \mathbf{I})\int V^c(\mathbf{y})\mathbf{T}^i(\mathbf{x}^i - \mathbf{y})(\mathbf{n}\otimes\mathbf{n})\, d\mathbf{y}. \quad (3.210)$$

Clearly, the tensor $\mathbf{C}$ (3.210) is an isotropic one: $\mathbf{C} \equiv C\boldsymbol{\delta}$, as well as $\mathbf{C}(\mathbf{x}) \equiv C^i\boldsymbol{\delta},\ \mathbf{x} \in v^i$ and

$$\mathbf{C}(\mathbf{x}) \equiv C^i\boldsymbol{\delta} = (\chi - \gamma)\mathbf{B}^i\int V^c(\mathbf{y})\mathbf{T}^i(\mathbf{x}^i - \mathbf{y})(\mathbf{n}\otimes\mathbf{n})\, d\mathbf{y}, \quad \text{at } \mathbf{x} \in v^i. \quad (3.211)$$

To show the good quality of the estimation resulting from the thin-layer hypothesis let us define a normalized residual average stress ($\boldsymbol{\sigma}^0 \equiv \mathbf{0}$) in the coated inclusion $\sigma^{res} \equiv -\langle \sigma_{11} \rangle^{(1)}/[\gamma 3Q^k(1 - B^{ik})]$, with isotropic tensors $\mathbf{Q}^i = \mathbf{Q}_1 = (3Q^k, 2Q^\mu)$, $\mathbf{B}^i = (3B^{ik}, 2B^{i\mu})$. Average residual stress in the coated inclusion $\langle \boldsymbol{\sigma} \rangle^{(1)}$ can be found by the use of Eqs. (3.181) and (3.182) with $\mathbf{Y} = \mathbf{I}$ (when $\langle \boldsymbol{\sigma} \rangle^{(1)} \equiv \mathbf{C}$). We consider now the particular case of rigid spherical inclusions inside the coating with pure tension prestress ($\chi = 0$) and $\nu^{(0)} = 0.3$. Using the thin-layer hypothesis (3.193) from (3.210) the normalized residual stress is determined to be $\sigma^{res} = \overline{v}^c/\overline{v}_1$ which does not depend on the elastic properties of the matrix. In Fig. 3.2 the parameter σ^{res} is represented as a function of relative thickness of the coating $h = (a^c - a^i)/a^i$. Results obtained by using the thin-layer hypothesis (3.193) are compared with results (3.210) without this approximative assumption. As evident from Fig. 3.2, the thin-layer hypothesis provides an acceptable exactness for a not too thick coating, let's say for $h < 0.2$.

Now the normalized residual stress in the core $\sigma_i^{res} \equiv \langle \sigma_{11} \rangle^i/(3B^{ik}Q^k\gamma)$ is estimated, where the average over the core v^i is considered: $\langle (.) \rangle^i$. Analytically derived results (3.211) under the thin-layer hypothesis (3.193) (for which $v^c/v^i \cong 3h$) are compared in Fig. 3.3 with the one obtained by a more exact approach, Eq. (3.211), and with results from finite element analysis. The numerical results obtained from finite element analysis (presented in [177]) differ from the analytical solution (3.211) by not more than 1%. Increase of the Poisson ratio of the matrix, $\nu^{(0)}$, would move the solid and dashed lines in Fig. 3.3 slightly but not significantly; clearly, the thin-layer approximation $\sigma_i^{res} = 3h$ does not change. In conclusion, the analytical solution (3.210), (3.211) provides a high degree of exactness in the examples considered.

3.8 Related Problems for Ellipsoidal Inhomogeneity in an Infinite Medium

3.8.1 Conductivity Problem

It is well known that the equations describing the steady-state conditions of such processes as heat and mass transfer, electric conductivity and permittivity, and filtration of a Newtonian liquid in undeformable cracked-porous media (see, e.g., [41], [293], [420], [644], [1002], [1007]) are mathematically equivalent. For reasons

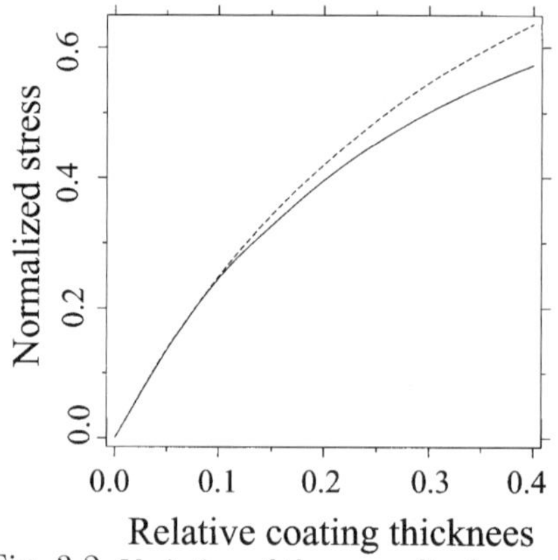

Fig. 3.2. Variation of the normalized average residual stress in the coated inclusion, σ^{res}, vs the relative coating thickness $h = (a^c - a^i)/a^i$. Dotted line: under the assumption of the thin-layer hypothesis, solid line: without this assumption.

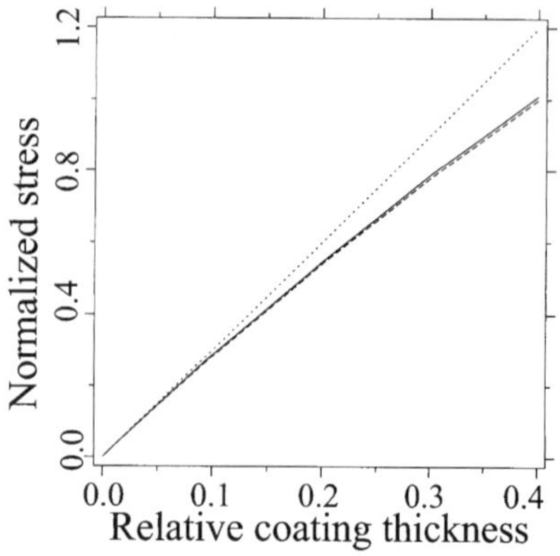

Fig. 3.3. Normalized stress σ_i^{res}, calculated by the use of the thin-layer hypothesis (3.196) (with $v^c/v^i = 3h$, dotted line) and without this assumption (dashed line). Numerical evaluation by the finite element method (solid line).

of this mathematical analogy, we consider the basic equations of steady-state transfer process:

$$\nabla \mathbf{q} = 0, \quad \mathbf{q} = -\boldsymbol{\kappa}\nabla T, \tag{3.212}$$

where T is a potential (e.g., temperature), $\boldsymbol{\kappa}$ is a conductivity tensor, $\mathbf{q}$ is a flux vector, and the first Eq. (3.212) is a transfer equation (Fourier, Ficks, Ohm, Darcy, etc.). Equations (3.212) are simpler field equations than Eq. (3.7) which simplifies its analysis.

For materials admitting an energy density function, the conductivity tensor $\boldsymbol{\kappa}$ must be symmetric: $\varkappa_{ij} = \varkappa_{ji}$. This symmetry arises also from Onsager's (see, e.g., [821]) reciprocity theorem for irreversible processes, in which fluxes are linearly related to intensities. For symmetry with respect to one plane (monoclinic symmetry), say the $x_1 - x_2$ plane, the tensor $\boldsymbol{\kappa}$ has four independent components:

$$\boldsymbol{\kappa} = \begin{pmatrix} \varkappa_{11} & \varkappa_{12} & 0 \\ \varkappa_{12} & \varkappa_{22} & 0 \\ 0 & 0 & \varkappa_{33} \end{pmatrix}. \tag{3.213}$$

For symmetry with respect to three orthogonal plates (orthogonal symmetry), the tensor $\boldsymbol{\kappa}$ has three independent components $\boldsymbol{\kappa} = \mathrm{diag}(\varkappa_{11}, \varkappa_{22}, \varkappa_{33}) \equiv \mathrm{diag}(\varkappa_1, \varkappa_2, \varkappa_3)$ referred to its principal axes. For symmetry with respect to a 90° rotation about one axis (transversally isotropic symmetry), say the x_3-axis, we have two independent components: $\boldsymbol{\kappa} = \mathrm{diag}(\varkappa_1, \varkappa_1, \varkappa_3)$. Finally, for symmetry with respect to any rotations (isotropic symmetry), the conductivity tensor is defined by the scalar $\varkappa_0$: $\varkappa_{ij} = \varkappa_0\delta_{ij}$. All numbers $\varkappa_1, \varkappa_2$, and $\varkappa_3$ are always found to be positive that provides the positive definiteness of the tensor $\boldsymbol{\kappa}$.

Analogous to the elasticity problem (3.3), we define the Green's function as

$$\varkappa_{ij} G_{,ij}(\mathbf{x}) = -\delta(\mathbf{x}), \tag{3.214}$$

where $\delta(\mathbf{x} - \mathbf{y})$ is the delta function. For media with both arbitrary anisotropic and isotropic ($\boldsymbol{\kappa} = \varkappa_0 \boldsymbol{\delta}$) properties, Green's functions G have the representation (see, e.g., [642], [617])

$$G(\mathbf{x}) = \frac{1}{4\pi r^k(\mathbf{x})}, \quad G(\mathbf{x}) = \frac{1}{4\pi \varkappa_0 |\mathbf{x}|}, \tag{3.215}$$

respectively, where $r^k(\mathbf{x}) \equiv (|\boldsymbol{\kappa}| x_i k_{ij} x_j)^{1/2}$, $|\boldsymbol{\kappa}| \equiv \det\boldsymbol{\kappa}$, $\boldsymbol{k} = \boldsymbol{\kappa}^{-1}$. In a case of orthotropic media in coordinates coinciding with the principal axes of $\varkappa_{ij}$, Eq. (3.214) is reduced to $\varkappa_1 G_{,11}(\mathbf{x}) + \varkappa_2 G_{,22}(\mathbf{x}) + \varkappa_3 G_{,33}(\mathbf{x}) = \delta(\mathbf{x})$, which can be solved [224] as $G(\mathbf{x}) = (4\pi)^{-1}(\varkappa_2\varkappa_3 x_1^2 + \varkappa_1\varkappa_3 x_2^2 + \varkappa_1\varkappa_2 x_3^2)^{-1/2}$, by the use of the coordinate transformation $x_i = \sqrt{\varkappa_i}\xi_i$ (no sum over i).

Following [224], we estimate local fields $T(\mathbf{x})$ and $\mathbf{q}(\mathbf{x})$ in an ellipsoidal inclusion v (3.104_1) with conductivity of the matrix $\boldsymbol{\kappa}(\mathbf{x}) \equiv \boldsymbol{\kappa}^{(0)}$ and polynomial eigen-intensity

$$\beta_i^*(\mathbf{x}) = \beta_i + \beta_{ij} x_i x_j + \beta_{ijk} x_i x_j x_k + \ldots, \tag{3.216}$$

($\beta_{i\ldots}$ =const., $\mathbf{x} \in v$) prescribed in the coordinate system coinciding with the principal axes of the ellipsoid v. Using the Green's function technique analogously to the elastic problem (3.2), we can get the representations for the local field in the inclusion and matrix:

$$T(\mathbf{x}) = \frac{1}{4\pi\varkappa_0}\frac{\partial}{\partial x_i}\int_v \frac{\beta_i^*(\mathbf{y})}{|\mathbf{x} - \mathbf{y}|}\, d\mathbf{y}, \quad q_j = \frac{1}{4\pi\varkappa_0}\frac{\partial^2}{\partial x_i \partial x_j}\int_v \frac{\beta_i^*(\mathbf{y})}{|\mathbf{x} - \mathbf{y}|}\, d\mathbf{y}, \tag{3.217}$$

where we consider the isotropic matrix $\boldsymbol{\kappa}(\mathbf{x}) \equiv \varkappa_0 \boldsymbol{\delta}$ ($\mathbf{x} \in R^3$) just for simplicity. Substituting (3.216) into (3.217) yields

$$T(\mathbf{x}) = \frac{1}{4\pi\varkappa_0}\frac{\partial}{\partial x_i}[\beta_i \phi(\mathbf{x}) + \beta_{ir}\phi_r(\mathbf{x}) + \beta_{irs}\phi_{rs}(\mathbf{x}) + \ldots], \tag{3.218}$$

$$q_j(\mathbf{x}) = \frac{1}{4\pi\varkappa_0}\frac{\partial^2}{\partial x_i \partial x_j}[\beta_i \phi(\mathbf{x}) + \beta_{ir}\phi_r(\mathbf{x}) + \beta_{irs}\phi_{rs}(\mathbf{x}) + \ldots], \tag{3.219}$$

where $\phi(\mathbf{x}) = \int_v |\mathbf{x} - \mathbf{y}|^{-1}\, d\mathbf{y}$, $\phi_{rs\ldots t}(\mathbf{x}) = \int_v y_r y_s \ldots y_t |\mathbf{x} - \mathbf{y}|^{-1}\, d\mathbf{y}$,and the Newtonian potential ϕ for the domain v is a solution of the equation $\Delta\phi(\mathbf{x}) = -V(\mathbf{x})$.

It turns out that the theorem of polynomial conservation (the p-property) (see Section 3.5) holds also for the conductivity problem. Moreover, according to Ferrers and Dyson (see, e.g., [794]) ϕ and ϕ_i can be written in terms of elliptic integrals:

$$\phi = \frac{1}{2}[I(\lambda) - x_n x_n I_N(\lambda)], \quad \phi_i = \frac{1}{2} a_I x_i [I_i(\lambda) - x_n x_n I_{Ni}(\lambda)], \ldots \tag{3.220}$$

where

$$I(\lambda) = 2\pi a_1 a_2 a_3 \int_\lambda^\infty \frac{1}{\Delta(s)} ds, \quad I_i(\lambda) = 2\pi a_1 a_2 a_3 \int_\lambda^\infty \frac{1}{(a_i^2 + s)\Delta(s)} ds,$$

$$I_{ij}(\lambda) = 2\pi a_1 a_2 a_3 \int_\lambda^\infty \frac{1}{(a_i^2 + s)(a_j^2 + s)\Delta(s)} ds, \tag{3.221}$$

$\Delta(s) = [(a_1^2+s)(a_2^2+s)(a_3^2+s)]^{1/2}$, and the constant λ is zero for $\mathbf{x} \in v$ and is the largest positive root of $x_1^2(a_1^2+\lambda)^{-1} + x_2^2(a_2^2+\lambda)^{-1} + x_3^2(a_3^2+\lambda)^{-1} = 1$ for $\mathbf{x} \notin v$. One follows Mura's [794] tensorial index notation when uppercase indices always take on the same numbers as the corresponding lowercase ones but are not summed up, e.g. $x_n x_n I_N = x_1^2 I_1 + x_2^2 I_2 + x_3^2 I_3$, $x_i I_i = x_1 I_1$, if $i = 1$.

Turning back to Eqs. (3.218) and (3.219), we can recast the fields $T(\mathbf{x})$ and $\mathbf{q}(\mathbf{x})$ in the form

$$\begin{aligned} T(\mathbf{x}) &= a_i(\mathbf{x})\beta_i + a_{ij}(\mathbf{x})\beta_{ij} + a_{ijk}\beta_{ijk} + \ldots, \\ q_r(\mathbf{x}) &= b_{ri}(\mathbf{x})\beta_i + b_{rij}(\mathbf{x})\beta_{ij} + b_{rijk}\beta_{rijk} + \ldots, \end{aligned} \tag{3.222}$$

where

$$\begin{aligned} a_i(\mathbf{x}) &= \frac{1}{4\pi}\phi_i, \; a_{ij}(\mathbf{x}) = \frac{1}{4\pi}\phi_{j,i}, \; a_{ijk}(\mathbf{x}) = \frac{1}{4\pi}\phi_{jk,i}, \ldots \\ b_{ri}(\mathbf{x}) &= \frac{1}{4\pi}\phi_{,ri}, \; b_{rij}(\mathbf{x}) = \frac{1}{4\pi}\phi_{j,ri}, \; b_{rijk}(\mathbf{x}) = \frac{1}{4\pi}\phi_{jk,ri}, \ldots. \end{aligned} \tag{3.223}$$

The explicit representations of the integrals $I(\lambda), I_{i\ldots}(\lambda)$ for spheroidal inclusions can be found in [224]. It is interesting that for a uniform eigenfield $\beta_i^*(\mathbf{x}) \equiv \beta_i$, the interior intensity $\mathbf{q}(\mathbf{x})$ $(\mathbf{x} \in v)$ (see Eqs. (3.222) and (3.223)) involving the second derivative of the potential $\phi(\mathbf{x})$ is spatially uniform because $\phi(\mathbf{x})$ is simply a quadratic function of $\mathbf{x} \in v$ (see Eq. (3.220$_1$)).

In particular, for the homogeneous ellipsoidal inhomogeneity with the conductivity $\boldsymbol{\kappa}$ inside remote homogeneous field ∇T^0 , the local field ∇T inside the inhomogeneity is also homogeneous:

$$\nabla T = \left[\boldsymbol{\delta} + \mathbf{S}(\boldsymbol{\kappa}^{(0)})^{-1}\boldsymbol{\kappa}_1\right]^{-1} \nabla T^0, \tag{3.224}$$

where $\mathbf{S}$ is a symmetric *depolarization* tensor, which for the isotropic matrix and in the principal axes frame has diagonal structure

$$\mathbf{S} = \mathrm{diag}[M_1, M_2, (1 - M_1 - M_2)] \tag{3.225}$$

with the factors M_i expressed as the classical elliptic integrals [642], [1050]

$$M_i = \frac{1}{2} a_1 a_2 a_3 \int_0^\infty \frac{1}{(a_i^2 + s)\Delta(s)}\, ds. \tag{3.226}$$

For our calculations, we present results for a spheroidal inclusions with the semi-axes $a_1 = a_2 = a$ and a_3 when

$$M_1 = M_2 = M = \frac{(1-g)\gamma^2}{2(\gamma^2 - 1)} \tag{3.227}$$

where $\gamma = a_3/a$ and g is defined by Eq. (3.160). The factors M_i can also be simply evaluated for an elliptical cylinder with $a_3 = \infty$: $M_1 = a_2/(a_1 + a_2)$, $M_2 = a_1/(a_1 + a_2)$. In particular, for a spherical inclusion ($\gamma = 1$) $M = 1/3$, for a cylinder ($\gamma = \infty$) $M = 1/2$, and for a disk-shaped inclusion ($\gamma = 0$) $M = 0$.

We assumed ideal contact between the phases when the temperature and the normal component of the heat flux are continuous at the interphase of the phases. In reality, however, perfect bonding does not exist. The availability of an interphase between the inhomogeneities and the surrounding matrix has also attracted the attention of researchers. In a simplest case the interface is considered as an additional homogeneous isotropic phase. So, one [444] (the case of n-layered isotropic spherical inclusion is considered in [456]) solves the problem for two-layered spheroids consisting of isotropic materials in an isotropic matrix in spherical coordinates using Legendre polynomials. The thickness of the interphase may be comparable to the size of the heterogeneities for some cases, for example, in thick interphases deliberately produced for various purposes. The usual scheme of work on heterogeneous media is based on modeling an inclusion surrounded by an inhomogeneous interphase by mapping it onto an effective homogeneous particle of identical size in order to predict the effective moduli. The problem is simplified in the case of a thin interphase when a boundary value problem for the inclusion-interphase-matrix configuration can be replaced by a simpler problem involving the inclusion and matrix only, plus certain matching conditions simulating the interphase. So, Benveniste and Miloh [64] considered the case of a spherical inclusion with as imperfect thermal contact at a constituent interface, when the boundary conditions on the surface of the inclusion have the form

$$\varkappa_0^{(0)}\frac{\partial T^-}{\partial \mathbf{n}}=\varkappa_0^{(1)}\frac{\partial T^+}{\partial \mathbf{n}}=\beta(T^- - T^+), \tag{3.228}$$

where $\mathbf{n}$ is the outer normal of the inclusion; T^+ and T^- are the limiting values of the potential inside and outside the inclusion, near the surface of the heterogeneity. The case $\beta\to 0$ corresponds to "adiabatic" boundary conditions and $\beta\to\infty$ corresponds to "isothermal" boundary conditions. The boundary condition (3.228) is mathematically equivalent to the surface of a homogeneous inclusion having an infinitesimally thin layer with thickness $\delta \ll a$ and conduction coefficient $\varkappa_0^{\text{int}}=\beta/\delta$. Transforming the results [64], one [165] estimated the average response of such a sphere in the terms of the tensor $\mathbf{R}^t=R_0^t\boldsymbol{\delta}$ defined as $\bar v_1\langle\varkappa_1\nabla T\rangle\equiv\mathbf{R}^t\nabla T^0$ analogously to the tensor $\mathbf{R}^\varepsilon$ (3.164):

$$R_0^t=3\bar v_1\varkappa_0^{(0)}\Big[\beta a\varkappa_0^{(1)}+(\varkappa_0^{(0)}+\varkappa_0^{(1)})\varkappa_0^{(0)}\Big]\Big[\beta a(\varkappa_0^{(1)}+3\varkappa_0^{(0)})+(2\varkappa_0^{(1)}+\varkappa_0^{(0)})\varkappa_0^{(0)}\Big]^{-1}. \tag{3.229}$$

The error that arises when the interphase of finite thickness is replaced by the boundary condition (3.228) was evaluated in [444].

One [1109], [734] treats nonideal contact of a spherical inclusion of the radius a in two limiting case of both the large and small conductivity $\boldsymbol{\kappa}^{\text{int}}=\varkappa_0^{\text{int}}\boldsymbol{\delta}$ of the intercase with a small thickness $\delta\ll a$ [the cases of an ellipsoidal heterogeneity with these nonideal contacts were analyzed in [64], [770]. In the first case with $\varkappa_0^{\text{int}}\to\infty$ being of the order $1/\delta$, the limit

$$C=\frac{1}{ak_0^{(0)}}\lim_{1/\delta,\varkappa_0^{\text{int}}\to\infty}\delta\varkappa_0^{\text{int}} \tag{3.230}$$

remains finite. For the boundary conditions of "superconducting" interface (3.230), the temperature field is continuous at the interface, but the heat flux suffers a jump. In the elastic context Eq. (3.230) corresponds to a rigid "flake" in the matrix when the displacement is continuous, but the normal stress component has a jump on the flake.

In the second case with $\varkappa_0^{\rm int} \to 0$, being of the order δ, as $\delta \to 0$, the limit

$$R = \frac{\varkappa_0^{(1)}}{a} \lim_{\delta, \varkappa_0^{\rm int} \to 0} \frac{\delta}{\varkappa_0^{\rm int}} \tag{3.231}$$

remains finite. For "resisting" boundary conditions (3.231), the heat flux remains continuous, but the temperature suffers a jump on the interface. In the elastic context the resisting condition (3.231) corresponds to a crack where the displacement has a jump, but the normal stress component is continuous on the interface.

The solutions [734], [1109] for the sphere with the boundary conditions (3.230) and (3.231) make it possible to get the representations for the tensor $\mathbf{R}^t = R_0^t \boldsymbol{\delta}$:

$$\begin{aligned} R_0^t &= \frac{3\bar{v}_1 \varkappa_0^{(0)} (\varkappa_0^{(1)} - k^{(0)})}{2k^{(0)} + 3\varkappa_0^{(1)}} + \frac{6\bar{v}_1 (\varkappa_0^{(0)})^2 C}{\varkappa_0^{(1)} + 2(1+C)\varkappa_0^{(0)}}, \\ R_0^t &= \frac{3\bar{v}_1 \varkappa_0^{(0)} (\varkappa_0^{(1)} - k^{(0)})}{2k^{(0)} + 3\varkappa_0^{(1)}} - \frac{3\bar{v}_1 (\varkappa_0^{(0)})^2 R}{\varkappa_0^{(1)} + 2(1+R)\varkappa_0^{(0)}}, \end{aligned} \tag{3.232}$$

for the conditions (3.230) and (3.231), respectively.

3.8.2 Scattering of Elastic Waves by Ellipsoidal Inclusion in a Homogeneous Medium

A fundamental task of nondestructive methods of testing the structure of composites is the establishment of the relation between the elastic-wave propagation microstructural parameters of the material [359], [360]. The structural and material study of the material begins by solving the problem of a single ellipsoidal inclusion in an infinite matrix. For an inclusion of simple shape in an isotropic matrix, the T-matrix approximation was employed [587], [1133], which relates the coefficients in the series expansion, in vector spherical wave functions, of the incident and scattered waves. However, for an anisotropic matrix, [407], [1121], [1181] demonstrated an advantage of an integral equation method, starting from a representation of the scattered field as an integral involving a Green's function for the matrix analogously to the static limit (3.76). In the long wave region (lengths of the propagating waves are more than the characteristic size of inclusions), the first approximation, called the Born (or Rayleigh-Gaus) approximation [408], is obtained by substituting the incident field into the integrand of the appropriate wave analog of Eq. (3.76). In the isotropic case, the method of equivalent inclusions was used in [361] to obtain a generalization of Eshelby's theorem to dynamic problems for moderate wave-number values. Similar results have been obtained for an anisotropic material by means of series expansion, in a small vibration frequency, of the fundamental solution of the wave operator for a homogeneous medium [544], [1181]. For the case in which the mechanical properties of the phases differ slightly, the perturbation method has been used in [1199]. In this section, we will present the problem of the scattering of elastic waves by a homogeneous ellipsoidal inclusion in the longwave region following [169], [542], [544], [1181].

Consider now a homogeneous medium with the elastic moduli $\mathbf{L}^{(0)}$ and density $\rho^{(0)}$ containing a homogeneous ellipsoidal inclusion v with the moduli $\mathbf{L}^{(1)}$ and density $\rho^{(1)}$. The differential equation of motion (3.42) can be presented in the form

$$\boldsymbol{\mathcal{L}}^{(0)}\mathbf{u} = -\boldsymbol{\mathcal{L}}_1\mathbf{u} - \mathbf{f}, \tag{3.233}$$

where $\boldsymbol{\mathcal{L}}^{(0)} = \nabla\mathbf{L}^{(0)}\nabla - \boldsymbol{\delta}\rho^{(0)}\partial^2/\partial t^2$, $\boldsymbol{\mathcal{L}}_1 = \nabla\mathbf{L}_1(\mathbf{x})\nabla - \boldsymbol{\delta}\rho_1(\mathbf{x})\partial^2/\partial t^2$ and the parameters $\mathbf{L}_1(\mathbf{x})$ and $\rho_1(\mathbf{x})$ are nonzero only over the volume v. Equation (3.233) will now be applied to the problem of the scattering of a plane wave by the inclusion v when an incident plane wave with angular frequency ω and amplitude $\mathbf{u}^0$ is expressed by $\mathbf{u}^0(\mathbf{x},t) = \mathbf{u}^0(\mathbf{x})\exp(-i\omega t)$. A steady-state solution $\mathbf{u}(\mathbf{x},t)$ depending upon t through a factor $\exp(-i\omega t)$ satisfies the equation for the wave amplitude $\mathbf{u}(\mathbf{x})$ (see (3.62)) $\boldsymbol{\mathcal{L}}^{\omega(0)}\mathbf{u} = -\boldsymbol{\mathcal{L}}_1^{\omega}\mathbf{u} - \mathbf{f}$, which can be reduced analogously to (3.75) to the following integral equation:

$$\mathbf{u}(\mathbf{x}) = \mathbf{u}^{(0)}(\mathbf{x}) + \int \mathbf{g}(\mathbf{x}-\mathbf{y})\nabla\boldsymbol{\tau}(\mathbf{y})\,d\mathbf{y} + \int \mathbf{g}(\mathbf{x}-\mathbf{y})\boldsymbol{\pi}(\mathbf{y})\,d\mathbf{y}, \tag{3.234}$$

Here we used the time-reduced Green's function (3.64) and an assumption $\mathbf{f}_1(\mathbf{x}) \equiv \mathbf{0}$; the introduced stress polarization tensor $\boldsymbol{\tau}$ and momentum polarization $\boldsymbol{\pi}$

$$\boldsymbol{\tau}(\mathbf{x}) = \mathbf{L}_1(\mathbf{x})\boldsymbol{\varepsilon}(\mathbf{x}), \quad \boldsymbol{\pi}(\mathbf{x}) = \omega^2\rho_1(\mathbf{x})\mathbf{u}(\mathbf{x}) \tag{3.235}$$

vanish in the matrix $R^d \setminus v$. Transforming the first integral by parts and taking the equality $\nabla_{\mathbf{y}} = -\nabla_{\mathbf{x}}$ into account, we obtain

$$\mathbf{u}(\mathbf{x}) = \mathbf{u}^{(0)}(\mathbf{x}) + \int \nabla\mathbf{g}(\mathbf{x}-\mathbf{y})\boldsymbol{\tau}(\mathbf{y})\,d\mathbf{y} + \int \mathbf{g}(\mathbf{x}-\mathbf{y})\boldsymbol{\pi}(\mathbf{y})\,d\mathbf{y},$$
$$u_i(\mathbf{x}) = u_i^{(0)}(\mathbf{x}) + \int \nabla_j g_{ik}(\mathbf{x}-\mathbf{y})\tau_{jk}(\mathbf{y})\,d\mathbf{y} + \int g_{ij}(\mathbf{x}-\mathbf{y})\pi_j(\mathbf{y})\,d\mathbf{y} \tag{3.236}$$

Differentiating both sides of (3.215), we obtain the equation for the strain tensor $\varepsilon_{ij}(\mathbf{x}) = \nabla_{(i}u_{j)}(\mathbf{x})$

$$\boldsymbol{\varepsilon}(\mathbf{x}) = \boldsymbol{\varepsilon}^0(\mathbf{x}) + \int \mathbf{K}^{\omega}(\mathbf{x}-\mathbf{y})\boldsymbol{\tau}(\mathbf{y})\,d\mathbf{y} + \int \nabla\mathbf{g}(\mathbf{x}-\mathbf{y})\boldsymbol{\pi}(\mathbf{y})\,d\mathbf{y},$$
$$\varepsilon_{ij}(\mathbf{x}) = \varepsilon_{ij}^{(0)}(\mathbf{x}) + \int K_{ijkl}^{\omega}(\mathbf{x}-\mathbf{y})\tau_{kl}(\mathbf{y})d\mathbf{y} + \int \nabla_{(i}g_{j)k}(\mathbf{x}-\mathbf{y})\pi_k(\mathbf{y})d\mathbf{y}, \tag{3.237}$$

where $K_{ijkl}^{\omega}(\mathbf{x}) = \nabla_{i)}\nabla_{(j}g_{k)(l}(\mathbf{x})$.

To obtain tractable results, we approximately solve Eqs. (3.236) and (3.237) (following [542], [545]) in the long wave limit when the length of the incident wave significantly exceeds the characteristic size of v. In such a case, keeping only essential terms in the real and imaginary parts of the expansion of $\mathbf{g}$ in ω:

$$\mathbf{g}(\mathbf{x}) = \mathbf{g}_0(\mathbf{x}) + i\omega\mathbf{g}_1(\mathbf{x}) - i\omega^3|\mathbf{x}|^2\mathbf{g}_3(\mathbf{n}), \tag{3.238}$$

where for the arbitrary anisotropic medium we have the following series expansion with respect to a small parameter ω:

$$\mathbf{g}(\mathbf{x}) = \mathbf{g}_0(\mathbf{x}) + \frac{1}{|\mathbf{x}|}\sum_{k=1}^{\infty}\frac{(i\omega|\mathbf{x}|)^k}{(k-1)!}\mathbf{g}_k(\mathbf{n}), \quad \mathbf{g}_0(\mathbf{x}) = \frac{1}{8\pi^2|\mathbf{x}|}\int_{|\xi|=1}\mathbf{L}^{-1}(\boldsymbol{\xi})\delta(\mathbf{n}\boldsymbol{\xi})\,ds_{\xi},$$

$$\mathbf{g}_k(\mathbf{x}) = \frac{(\rho^{(0)})^{k/2}}{16\pi^2}\int_{|\xi|=1}\mathbf{L}^{-(k+2)/2}(\boldsymbol{\xi})|\mathbf{n}\boldsymbol{\xi}|^{k-1}\,ds_{\xi}, \quad L_{ik}(\boldsymbol{\xi}) = L^{(0)}_{ijkl}\xi_j\xi_l, \tag{3.239}$$

where $\mathbf{n} = \mathbf{x}/|\mathbf{x}|$, $\boldsymbol{\xi}$ is a vector on the unit sphere $|\boldsymbol{\xi}| = 1$ and $\mathbf{g}_0$ is the static limit of the operator $\mathbf{g}$ obeying Eq. (3.3): $\mathbf{g}_0(\mathbf{x}) = \mathbf{G}(\mathbf{x})$.

The expansion (3.238) leads to the following expression for the kernel $\mathbf{K}^{\omega}(\mathbf{x})$:

$$\mathbf{K}^{\omega}(\mathbf{x}) = \mathbf{K}_0(\mathbf{x}) + i\omega^3\mathbf{K}_3, \quad K_{3|ijkl} = -\frac{(\rho^{(0)})^{3/2}}{16\pi^2}\int_{|\xi|=1}\xi_{(i}L^{-5/2}_{j)(k}(\boldsymbol{\xi})\xi_{l)}\,ds_{\xi}, \tag{3.240}$$

where $\mathbf{K}_0(\mathbf{x}) = \mathbf{U}(\mathbf{x})$ is a static limit (3.19) of the tensor $\mathbf{K}^{\omega}(\mathbf{x})$ at $\omega = 0$, and $\mathbf{K}_3$ is a constant tensor. In a similar manner, we will find the solution of Eqs. (3.235) and (3.237) in the form of the expansion in ω. We will consider a linear approximation of the incident field

$$\mathbf{u}^0(\mathbf{x}) = \mathbf{u}^0(\mathbf{0}) + \boldsymbol{\varepsilon}^0(\mathbf{0})\mathbf{x}, \quad \mathbf{x} \in v \tag{3.241}$$

with respect to $\mathbf{x}$ in the neighborhood of the center of gravity in the inclusion v. Substituting (3.241) and the series expansion of $\mathbf{u}$ and $\boldsymbol{\varepsilon}$ in ω in Eqs. (3.235) and (3.237) and equating the terms of the same order with respect to ω, we will obtain a linear dependence for the fields $\mathbf{u}(\mathbf{x})$ and $\boldsymbol{\varepsilon}(\mathbf{x})$ inside the ellipsoid v:

$$\mathbf{u}(\mathbf{x}) = \boldsymbol{\lambda}\mathbf{u}^0(\mathbf{0}) + \boldsymbol{\Lambda}(\omega, \mathbf{a})[\mathbf{H}(\mathbf{x}) \otimes \boldsymbol{\varepsilon}^0(\mathbf{0})], \quad \boldsymbol{\varepsilon}(\mathbf{x}) = \boldsymbol{\Lambda}(\omega, \mathbf{a})\boldsymbol{\varepsilon}^0(\mathbf{0}),$$
$$\boldsymbol{\lambda} = \boldsymbol{\delta} + i\omega^3\rho_1\bar{v}\mathbf{g}_1, \quad \boldsymbol{\Lambda}(\omega, \mathbf{a}) = \mathbf{A} + i\omega^3\bar{v}\mathbf{A}\mathbf{K}_3\mathbf{L}_1^{(1)}\mathbf{A}, \tag{3.242}$$

where the terms of order ω^2 in the real parts and of order ω^5 in the imaginary ones are discarded and $\mathbf{H}(\mathbf{x}) = \mathbf{x}$ in the coordinate system with origin at the center of the inclusion v with the volume $\bar{v}$. The tensor $\mathbf{A} = \mathbf{A}(\mathbf{a})$ determined by Eq. (3.120) for the static limit depends on the shape and orientation of the ellipsoid v while the constant tensors $\mathbf{g}_1$ and $\mathbf{K}_3$ do not dpend on the shape and orientation of v.

Substitution of (3.242) into the integrands of Eqs. (3.235) and (3.237) leads to the representation of the wave fields outside the inclusion $\mathbf{x} \notin v$. From (3.242) we obtain, for the ellipsoid v with center $\mathbf{x}_1 = \mathbf{0}$, an analytical representation for the displacement amplitude ($\mathbf{x} \notin v$):

$$\mathbf{u}(\mathbf{x}) = \mathbf{u}^{(0)}(\mathbf{x}) + \mathbf{G}^{\omega}(\mathbf{x})\bar{v}\omega^2\rho_1\boldsymbol{\lambda}\mathbf{u}^0(\mathbf{0}) + \nabla\mathbf{G}^{\omega}(\mathbf{x})\bar{v}\mathbf{L}_1\boldsymbol{\Lambda}\boldsymbol{\varepsilon}^0(\mathbf{0}),$$
$$u_i(\mathbf{x}) = u_i^0 + G^{\omega}_{im}(\mathbf{x})\bar{v}\omega^2\rho_1\lambda_{mk}\mathbf{u}_k^0(\mathbf{0}) + G^{\omega}_{im,k}(\mathbf{x})\bar{v}L_{1|mkpq}\Lambda_{pqln}u^0_{l,n}(\mathbf{0}) \tag{3.243}$$

and, in the general case of anisotropy of the matrix, we can assume a point approximation $\mathbf{G}^{\omega}(\mathbf{x}) = \mathbf{g}(\mathbf{x})$. A more accutrate representation has been obtained for an isotropic matrix in [1030]:

$$\mathbf{G}^{\omega}(\mathbf{x}) = \mathbf{g}^{\mathrm{T}} I^{\mathrm{T}} + \mathbf{g}^{\mathrm{L}} I^{\mathrm{L}},$$
$$\mathbf{g}^{\mathrm{T}} = \frac{1}{4\pi\rho^{(0)}\omega^2}\Big(k_{\mathrm{T}}\frac{\exp(ik_{\mathrm{T}}|\mathbf{x}|)}{|\mathbf{x}|}\boldsymbol{\delta} + \nabla\nabla\frac{\exp(ik_{\mathrm{T}}|\mathbf{x}|)}{|\mathbf{x}|}\Big),$$
$$\mathbf{g}^{\mathrm{L}} = \frac{1}{4\pi\rho^{(0)}\omega^2}\nabla\nabla\frac{\exp(ik_{\mathrm{L}}|\mathbf{x}|)}{|\mathbf{x}|}, \quad I^{\alpha} = \bar{v}^{-1}\int e^{ik_{\alpha}\mathbf{x}\cdot\xi}dv_{\xi}, \qquad (3.244)$$

where $k_\alpha = \omega/c_\alpha$ and c_α (3.43) ($\alpha = L, T$) are the wave numbers and propagation velocities of longitudial and transverse waves in a matrix. As was shown in [361], approximation (3.243) and (3.244) yields an error of not more than 5% in estimating $\mathbf{u}(\mathbf{x})$, $\mathbf{x} \notin v$ for $ka^1 \leq 2$. For the isotropic medium (3.54) with Lamé moduli $\lambda^{(0)}$ and $\mu^{(0)}$, we have

$$(\rho^{(0)})^{k/2} L_{ij}^{-k}(\boldsymbol{\xi}) = \frac{1}{c_{\mathrm{T}}^{2k}}(\delta_{ij} - \xi_i\xi_j) + \frac{1}{c_{\mathrm{L}}^{2k}}\xi_i\xi_j, \qquad (3.245)$$

allowing to obtain the representations

$$\mathbf{g}_k(\mathbf{n}) = \frac{c_{\mathrm{T}}^{-(k+2)}}{4\pi\rho^{(0)}k(k+2)}[(1 + k + \eta^{k+2})\boldsymbol{\delta} + (k-1)(\eta^{k+2} - 1)\mathbf{n}\otimes\mathbf{n}]. \qquad (3.246)$$

In particular, the isotropic tensors $\mathbf{g}_1 = g_1\boldsymbol{\delta}$ and $\mathbf{K}_3 = (3K_3^k, 2K_3^\mu)$ are defined by the expressions

$$g_1 = \frac{2 + \eta^3}{12\pi\rho^{(0)}k_{\mathrm{T}}^3}, \quad 3K_3^k = -\frac{\eta^5}{36\pi\rho^{(0)}k_{\mathrm{T}}^5}, \quad 3K_3^\mu = -\frac{3 + 2\eta^5}{60\pi\rho^{(0)}k_{\mathrm{T}}^5}. \qquad (3.247)$$

For a spherical inclusion, the tensor $\mathbf{A}$ is also isotropic and expressed by Eq. (3.169), and $I^\alpha = [3/(k_\alpha a)^2][\sin(k_\alpha a) - k_\alpha a\cos(k_\alpha a)]$, $(\alpha = \mathrm{L}, \mathrm{T})$.

Further development of the approaches to the scattering of elastic waves by ellipsoidal inclusions deserve attention. A wave generalization of the theorem of polynomial conservation was obtained in [330]; the limiting cases of the cracks, discs, or fibers were considered in [543], [1181]. Binary interaction of inclusions is analyzed in [675] taking the polynomial dependence on coordinates of the displacement field inside the inclusions. For isotropic materials the problems of elastic wave propagation in the cylindrical and spherical inclusions coated by any number of homogeneous layers of constant thickness were considered [102], [575], [819], [1004]. Scattering of anti-plane shear waves by a single piezoelectric cylindrical inhomogeneity partially bonded to an unbounded piezomagnetic matrix are obtained in [280] by means of the wave function expansion method. Recently a comprehensive database (containing more than 700 references) of the T-matrix method since the inception of the technique in 1965 through early 2004 has been published [779].

3.8.3 Piezoelectric Problem

Analogously to the linear elastic composite materials, we will consider the Eshelby tensor for a piezoelectric medium that play a central role in the characterization of piezoelectric composites. A solution of a single inclusion in an infinite piezoelectric solid has been analyzed in [54], [288], [291], [481], [670], [765], and [1157]. Following [670] we will reproduce the relevant schematic representations of the Eshelby and related tensors.

Consider an infinite piezoelectric medium R^d containing an ellipsoidal inclusion v with piecewise constant parameters $\mathbf{g}(\mathbf{x}) = \mathbf{g}^{(0)} + \mathbf{g}_1(\mathbf{x})$ ($\mathbf{g} = \mathbb{L}, \mathbb{M}, \boldsymbol{\Lambda}$) (2.146): $\mathbf{g}^{(0)} \equiv$const., $\mathbf{g}_1(\mathbf{x}) =$const. if $\mathbf{x} \in v$ and $\mathbf{g}_1(\mathbf{x}) = \mathbf{0}$ otherwise; $v = \{\mathbf{x} \,|\, x_i(a^{-2})_{ij}x_j \leq 1\}$, $a_{ij} = a_i\delta_{ij}$ (no sum on i). Analogously to Eq. (3.3) we define the Green's function under isothermal conditions $\nabla\mathbb{L}^{(0)}\nabla\mathbb{G}(\mathbf{x}) = -\delta(\mathbf{x})\text{diag}(\boldsymbol{\delta}, -1)$, which can be calculated through the integral over a unit sphere in a Fourier space:

$$\mathbb{G}(\mathbf{x}) = \frac{1}{8\pi^2}\int_{|\boldsymbol{\zeta}|=1} \mathbb{G}(\boldsymbol{\zeta})\delta(\boldsymbol{\zeta}\cdot\mathbf{x})\, ds(\boldsymbol{\zeta}), \quad \mathbb{G}(\boldsymbol{\zeta}) = \begin{pmatrix} G_{ij}(\boldsymbol{\zeta}) & \gamma_i^\top(\boldsymbol{\zeta}) \\ -\gamma_i(\boldsymbol{\zeta}) & g(\boldsymbol{\zeta}) \end{pmatrix},$$
$$G_{ij} = \left(L_{ij} - \lambda^{-1}H_i h_j\right)^{-1}, \quad \gamma_j = \lambda^{-1}h_i G_{ij}, \quad g = -(\lambda + h_i L_{ij}^{-1}H_j)^{-1},$$
$$L_{ij}(\boldsymbol{\zeta}) = L_{ijkl}^{(0)}\zeta_k\zeta_l, \; H_i(\boldsymbol{\zeta}) = e_{kil}^{(0)}\zeta_k\zeta_l, \; h_j(\boldsymbol{\zeta}) = e_{jkl}^{(0)}\zeta_k\zeta_l, \; \lambda(\boldsymbol{\zeta}) = k_{ij}^{(0)}\zeta_i\zeta_j, \quad (3.248)$$

where $|\boldsymbol{\zeta}| = 1$ is a unit sphere centered at $\boldsymbol{\zeta} = \mathbf{0}$. Then the Eshelby tensor $\mathbb{S} = \mathbb{P}\mathbb{L}$ is expressed through the tensor

$$\mathbb{P} = -\int \boldsymbol{\mathcal{D}}\mathbb{G}(\mathbf{x}-\mathbf{y})\boldsymbol{\mathcal{D}}V(\mathbf{y})\, d\mathbf{y} = \frac{|\det a|}{4\pi}\int_{|\boldsymbol{\zeta}|=1} \rho^{-3}(\boldsymbol{\zeta})\mathbb{U}(\boldsymbol{\zeta})\, ds(\boldsymbol{\zeta}), \quad (3.249)$$

where $\mathbb{U}(\boldsymbol{\zeta}) = \boldsymbol{\zeta}\mathbb{G}(\boldsymbol{\zeta})\boldsymbol{\zeta}$, $\rho(\boldsymbol{\zeta}) = \sqrt{\zeta_i(a^2)_{ij}\zeta_j}$, and $\boldsymbol{\mathcal{D}} = \text{diag}(\text{def}, \text{grad})$. Thus, the generalized homogeneous field inside the homogeneous ellipsoidal inclusion $\boldsymbol{\mathcal{E}}(\mathbf{x})$ ($\mathbf{x} \in v$) (2.145) is expressed via a homogeneous remote filed $\boldsymbol{\mathcal{E}}^0$ by the equation

$$\boldsymbol{\mathcal{E}}(\mathbf{x}) = \mathbb{A}(\boldsymbol{\mathcal{E}}^0 - \mathbb{P}\boldsymbol{\Lambda}_1), \; \mathbb{A} = (\mathbb{I} + \mathbb{P}\mathbb{L}_1)^{-1}, \quad (3.250)$$

where $\mathbb{I} = \text{diag}(\mathbf{I}, \boldsymbol{\delta})$, $\theta \equiv$const.

Dunn [288] has obtained the explicit representations of Eshelby tensor for elliptical cylindrical inclusion and thin-disc inclusion embedded in a transversaly isotropic matrix. Explicit results for the piezoelectric Eshelby tensor are obtained in [765] for the transversely isotropic piezoelectric materials when the elliptic cylindrical inclusion is aligned either along the axis of the anisotropy or perpendicular to this axis. For a spheroidal inclusion in a transversely isotropic piezielectric matrix, the coupled fields inside the inclusion are expressed in [1157] in terms of a system of the linear algebraic equations containing only some single integral. Because of the awkwardness of the obtained Eshelby tensors, interested readers are referred to the cited publications.

4

Multiscale Analysis of the Multiple Interacting Inclusions Problem: Finite Number of Interacting Inclusions

4.1 Description of Numerical Approaches Used for Analyses of Multiple Interacting Heterogeneities

Elastic analysis of local fields in heterogeneous solids containing multiple inclusions subjected to remote loading is of considerable interest in many engineering disciplines. Several techniques have been proposed for computing multi-particle interactions in an infinite medium. In the two-dimensional case the method of Kolosov-Muskhelishvili's complex potentials is highly efficient. Combining the body force method with complex stress function theory, a complex integral equation method was proposed in [284], [468] to study interaction problems between holes and other defects in an infinite or semiinfinite elastic plane. The conformal mapping technique was generalized in [948] for the analysis of elastic fields in two joined half-planes with an inclusion of an arbitrary shape. Sherman [994] constructed an elegant complex variable method based on the Kolosov-Muskhelishvili potentials leading to a singular integral equation for a complex-valued density. An effective numerical algorithm utilizing this method [1214] was developed, and Greengard and Helsing [400] combined Sherman's method with the fast multipole method and adaptive quadrature technique; the methods mentioned can be implemented in two-dimensional piecewise inhomogeneous media with isotropic constituents.

However, in three dimensions there are no analogous powerful methods to analyze the multiparticle interaction problems. Thus, Sternberg and Sadowsky [1041] analyzed the axisymmetric problem for two spherical cavities, whereas the case of spherical inclusions was investigated by Chen and Acrivos [215] by the use of the Boussinesq-Papkovich stress functions and multipole expansion technique in which the solutions are expanded into a series of spherical harmonics with respect to the centers of spheres. In the multipole expansion method [788] the Fredholm integral equation of the second kind was approximated using Taylor's expansion with the simulation of a neighboring inclusion by a singular source of the polarization strain located at the center. Moschovidis and Mura [789] used Taylor's expansion twice: for the transformation field inside the inclusion being considered as well as for the strain fields induced by the neighbors. In a similar integral equation Chen and Young [213] (see also [511], [660]) used a Taylor series expansion of the constrained displacement about the origin of inclusions; the solution of this equation reduced to the determination of the biharmonic potential functions and their derivatives for an isotropic system and for the inclusion morphology under consideration.

In recent years the multiparticle interaction problem was considered via several techniques. McPedran and Movchan [757] developed a method of multipole expansions for an analysis of interacting long cylinders for the two-dimensional case, while the same method for the case of elastic interactions of non-rigid spherical particles with imperfect interfaces was developed by Sangani and Mo [963]. The essence of the method [392] analyzed N interacting spheres is the representation of the matrix displacement vector as a sum of general solutions for a single-particle problem. The latter is expressed by a series of partial vectorial solutions of Navier's equation in a spherical basis. Satisfaction of interfacial boundary conditions leads to an infinite linear algebraic system with a normal determinant. A transversely isotropic solid with a single inclusion subjected to the polynomial external loading was considered Podil'chuk [887], [888] by the use of separation of variables in the properly introduced curvilinear coordinates. The last two methods were essentially developed and generalized by Kushch [628], [629], [630] (see also [631]) to transversely isotropic phases and ellipsoidal inhomogeneities with no restrictions on their number, size, aspect ratio, elastic properties, and arrangement. Application of this approach to a polymer nanocomposite of clustered structure was considered in [155] (see also Chapter 19).

Of course, a realistic model of the problem cannot be solved by analytical methods in the general case of inclusion shape and its coating structure. Powerful numerical methods, such as Finite Element Method (FEM) [371], [379], [412], [1051], [1055], and Boundary Integral Equation (BIE) method [4], [225], [307], [492], [698], [853] are useful for the solution of these problems. However, the FEM (including both the generalized [1051] and extended [1055] versions incorporating special functions which reflect the local character of the solution into the construction of the approximation) requires discretization of the full domain containing the inclusions. The BIE method is ideally suited for macroscale analysis with the well-known advantages of reduction of dimension by one and can easily handle complex geometries along with general loading conditions. Nevertheless, if the inclusion has a complicated coating structure with continuously inhomogeneous layers [217], BIE becomes extremely cumbersome in dealing with the problem. To avoid using the fundamental solution for materials with continuously varying anisotropy, the volume integral equation (VIE) method was proposed in [656], [657], [730], [956] for the analysis of the stress field in an isotropic space containing a finite number of anisotropic inclusions of an arbitrary shape. The hybrid boundary-domain element method [276], [658] (which we will call also the hybrid BIE and VIE method) is applied to the discretization of the integral equations thus obtained, in which the boundary is divided into surface elements while the inner domain is divided into volume elements. The hybrid BIE and VIE methods considered in this chapter enables one to restrict discretization to the inclusions only (in contrast to the FEM), and an inhomogeneous structure of inclusions presents no problem in the framework of the same numerical scheme (compared to the standard BIE method). A version of the hybrid method proposed by Dong and Bonnet [275] involves the fundamental solution for an isotropic half-plane medium. If the number of interacting inclusions n is small enough, the standard numerical method for the solution of the Fredholm integral equation of the second kind can be used. Increasing n to 10^4 leads to the necessity of employing the iterative solution strategies with usually a small number of required iterations (or a combination of the analytical and numerical methods proposed in this article) reducing the operation count from $O(n^3)$ to $O(n^2)$ [872]. In so doing at each iteration the fast multipole method (FMM) [400], [402], for solving the many-body electrostatics problem can be extremely effective which reduces the computing cost from $O(n^2)$ to $O(n)$. Very recently some versions of FMM were generalized for the solution of 2D [400], [866], and 3D (see [363] where additional references can be found) problems of linear elasticity. Since FMM is usually combined with the BIE method, which employs Green's functions for each isotropic homogeneous constituent, it cannot be easily extended to the composite materials containing the inclusions with anisotropic and especially with either spherically or cylindrically anisotropic anisotropic [217], [535] and nonlinear properties, although the VIE method used in this chapter requires no corrections for these cases if the matrix is isotropic.

Additional fundamental difficulties appear in the analysis of micro-macro problems when micro-inclusions and their spacing have a length-scale that is a few orders of magnitude smaller than the lengthscales of the macroscopic problem (geometric, loading, and boundary conditions); related problems are the boundary value problem for a solid containing inclusions leading to the so-called edge effect and inclusions in bimaterials consisting of two semiinfinite solids (see [795] for references). In the consideration of a spherical (or circular) inclusion in an elastic half-space one Jasiuk *et al.* [497], [661] used an infinite series of harmonic functions, which corresponds to an infinite number of eigenstrains; the method is restricted by both regular shape of inclusions and isotropic properties of constituents. A similar problem for a circular tunnel in an isotropic half-space was solved by Verruijt [1138], [1139] via the method of Kolosov-Muskhelishvili's complex potentials. In an alternative approach for the analysis of an inclusion in transversely isotropic bimaterials, Yu *et al.* [1210] uses the fundamental solution for transversely isotropic bimaterials and the equivalent inclusion method by Eshelby [319]; a system of singular integral equations obtained can be solved numerically. Particular cases of spherical and finite cylindrical inclusions in an isotropic bimaterial are investigated in [1153] in detail. A finite number of plies with cylindrical inclusions in a finite body were analyzed by Pagano and Rybicki [850], [928], [953] by the finite element method; they discovered a fundamental role of an edge effect leading to the redistribution of local stresses in a boundary layer region compared to the stress estimations in remote cells. The authors [851], [921] proposed a combination of the effective medium and micromechanical theories leading to reasonably accurate interface stresses. Several authors [922], [924], [974] have proposed a hybrid method connecting a finite element discretization for the near field and a boundary element discretization for a far-field as well as a hierarchical coupling of standard FE meshes on the micro and the macro scales.

In our approach a hybrid method based on the combination of the VIE and BIE methods is proposed for the micro-macro solution of elastostatic 2-D and 3-D *multiscale problems* in bounded or unbounded solids containing interacting multiple inclusions of essentially different scale. The hybrid micro-macro formulation allows decomposition of the complete multiscale problem into two associated subproblems, one residing entirely at the micro-level and the other at the macro-level. The efficiency of the standard iterative scheme of the BIE and VIE methods for the singular integral equation involved is enhanced by the use of a modification in the spirit of a subtraction technique as well as by a judicious choice of the initial analytical approximation for interacting inclusions (micro-level) in an unbounded medium subjected to inhomogeneous loading.

The judicious choice of the initial analytical approximation is based on the decomposition of the original problem for n macro and micro inclusions into n uncoupled problems for a single inclusion v_i $(i = 1, \ldots, n)$ subjected to a complicated so-called effective field $\overline{\boldsymbol{\sigma}}_i(\mathbf{x})$. The approximate solution of the n-particle interaction problem obtained in the framework of the homogeneity hypothesis of effective fields in the vicinity of micro inclusions was chosen as a zero-order approximation for effective fields in the surrounding inclusions v_j $(j = 1, \ldots, n;\ j \neq i)$. This step of our approach coincides with the previous scheme [184], [138], [174] proposed for coated inclusions and generalizes the approaches [79], [515], [525], [572], [941], [1189] for homogeneous ellipsoidal inclusions as well as the approach [618] for the point approximation of interacting inclusions $\mathbf{T}_{ij}(\mathbf{x}_i - \mathbf{x}_j) = \boldsymbol{\Gamma}(\mathbf{x}_i - \mathbf{x}_j)$ (for references see also [349]). To obtain the first-order approximation the inclusion v_i is loaded by inhomogeneous effective stress $\overline{\boldsymbol{\sigma}}_i(\mathbf{x})$ obtained as a superposition of the remote stress and the disturbances caused by the other inclusions loaded by the known homogeneous fields $\overline{\boldsymbol{\sigma}}_j$ $(j = 1, \ldots, n;\ j \neq i)$. This second step is motivated by an assumption made by Kachanov [529] in an analysis of interacting cracks who has used an analytical solution for a single crack loaded by the effective field $\overline{\boldsymbol{\sigma}}_i(\mathbf{x})$. The numerical solution of singular integral equations (SIEs) involved is determined by the BIE and VIE methods with the evaluation of the appropriate integrals with a Gauss quadrature formula. In doing so the singular terms in the Gauss formulae are handled in the spirit of a subtraction technique [227] that eliminates the difficulties connected with the use of singular finite elements in the VIE considered in [656], [657], [658]. The next order approximations are constructed by the iteration method. The accuracy and efficiency of the method are examined through comparison with results obtained from FE methods.

4.2 Basic Equations for Multiple Heterogeneities and Numerical Solution for One Inclusion

4.2.1 Basic Equations for Multiple Heterogeneities

Let a linear elastic body occupy an open bounded domain $w \subset R^d$ with a smooth boundary Γ and with a characteristic function W and space dimensionality d ($d = 2$ and $d = 3$ for 2-D and 3-D problems, respectively). The domain w

contains a homogeneous matrix $v^{(0)}$ and a finite set $X = (v_i)$ of inclusions v_i with characteristic functions V_i and is bounded by the closed smooth surfaces Γ_i $(i = 1, 2, \ldots, n)$; $\Gamma \cap X = \emptyset$. For the sake of definiteness, in the 2-D case we will consider a plane-strain problem. At first no restrictions are imposed on the elastic symmetry[1] of the phases or on the geometry of the inclusions. Stresses and strains are related to each other via the constitutive equations $\boldsymbol{\sigma}(\mathbf{x}) = \mathbf{L}(\mathbf{x})\boldsymbol{\varepsilon}(\mathbf{x}) + \boldsymbol{\alpha}(\mathbf{x})$ or $\boldsymbol{\varepsilon}(\mathbf{x}) = \mathbf{M}(\mathbf{x})\boldsymbol{\sigma}(\mathbf{x}) + \boldsymbol{\beta}(\mathbf{x})$ (2.126). In particular, for isotropic constituents the local stiffness tensor $\mathbf{L}(\mathbf{x}) = (dk(\mathbf{x}), 2\mu(\mathbf{x}))$ (2.207) $(d = 2, 3 = R^d)$ is given in terms of the local bulk modulus $k(\mathbf{x})$ and the local shear modulus $\mu(\mathbf{x})$, and the local eigenstrain $\boldsymbol{\beta}(\mathbf{x}) = \beta_0(\mathbf{x})\boldsymbol{\delta}$ is given in terms of the bulk component $\beta_0(\mathbf{x})$. All tensors $\mathbf{f}$ $(\mathbf{f} = \mathbf{L}, \mathbf{M}, \boldsymbol{\alpha}, \boldsymbol{\beta})$ of material properties are decomposed as $\mathbf{f} \equiv \mathbf{f}^{(0)} + \mathbf{f}_1(\mathbf{x})$; see Eq. (3.67), where $\mathbf{f}_1(\mathbf{x}) = \mathbf{f}_1^{(i)}(\mathbf{x})$ is an inhomogeneous function inside the inclusions v_i $(i = 1, 2, \ldots, n)$. The upper index of the material properties tensor in parentheses shows the number of the respective component. The subscript 1 denotes a jump of the corresponding quantity (e.g. of the material tensor). We assume that the phases are perfectly bonded [see Eq. (3.83)], so that the displacements and the traction components are continuous across the interphase boundaries. The traction $\mathbf{t}(\mathbf{x}) = \boldsymbol{\sigma}(\mathbf{x})\mathbf{n}(\mathbf{x})$ acting on any plane with the normal $\mathbf{n}(\mathbf{x})$ through the point $\mathbf{x}$ can be represented in terms of displacements; see Eq. (2.130). The boundary conditions at the interface boundary will be considered together with the mixed boundary conditions on Γ with the unit outward normal $\mathbf{n}^\Gamma$

$$\mathbf{u}(\mathbf{x}) = \mathbf{u}^\Gamma(\mathbf{x}), \quad \mathbf{x} \in \Gamma_u, \tag{4.1}$$

$$\boldsymbol{\sigma}(\mathbf{x})\mathbf{n}^\Gamma(\mathbf{x}) = \mathbf{t}^\Gamma(\mathbf{x}), \quad \mathbf{x} \in \Gamma_t, \tag{4.2}$$

where Γ_u and Γ_t are prescribed displacement and traction nonintersected boundary conditions such that $\Gamma_u \cup \Gamma_t = \Gamma$, $\Gamma_u \cap \Gamma_t = \emptyset$. $\mathbf{u}^\Gamma(\mathbf{x})$ and $\mathbf{t}^\Gamma(\mathbf{x})$ are prescribed displacement on Γ_u and traction on Γ_t, respectively; mixed boundary conditions, such as in the case of elastic supports are possible. As usual we shall distinguish the interior from the exterior problem according to whether the body occupies the interior or the exterior domain with respect to Γ.

We now summarize the principal formulae of thermostatics in a decomposition form that is appropriate to our intended application [28], [214], [862]:

$$\mathbf{u}(\mathbf{x}) = \mathbf{u}^0(\mathbf{x}) + \mathbf{u}^1(\mathbf{x}), \tag{4.3}$$

$$\mathbf{u}^0(\mathbf{x}) = \int_\Gamma \Big[\mathbf{G}(\mathbf{x}-\mathbf{s})\underline{\mathbf{t}}(\mathbf{s}) - \mathbf{T}(\mathbf{x},\mathbf{s})\mathbf{u}(\mathbf{s})\Big]\, d\mathbf{s}, \tag{4.4}$$

$$\mathbf{u}^1(\mathbf{x}) = \int_w \nabla\mathbf{G}(\mathbf{x}-\mathbf{y})\Big\{\mathbf{L}_1(\mathbf{y})[\boldsymbol{\varepsilon}(\mathbf{y}) - \boldsymbol{\beta}(\mathbf{y})] - \mathbf{L}^{(0)}\boldsymbol{\beta}_1(\mathbf{y})\Big\}\, d\mathbf{y}, \tag{4.5}$$

where $\underline{\mathbf{t}} \equiv \mathbf{t} - \mathbf{n}^{\Gamma\top}\boldsymbol{\alpha}^{(0)}$ is the modified boundary traction, and the tensor of "fundamental displacement" $\mathbf{G}$ (3.3) is the infinite-homogeneous-body Green's function (vanishing at infinity $|\mathbf{x}| \to \infty$) of the Navier equation (3.1) with an

[1] It is known that for 2-D problems the plane-strain state is possible only for material symmetry no lower than orthotropic (see, e.g., [663]) that will be assumed hereafter in 2-D case.

elastic modulus tensor $\mathbf{L}^{(0)}$. The tensor of the "fundamental traction" $\mathbf{T}$ associated with the tensor of "fundamental displacement" $\mathbf{G}$ is given by Eq. (3.40). The field $\mathbf{u}^1(\mathbf{x})$ is a perturbation induced by the inclusions, while the function $\mathbf{u}^0(\mathbf{x})$ is the displacement field which would exist in the domain with homogeneous properties $\mathbf{L}^{(0)}$ (with $\mathbf{L}_1(\mathbf{x}), \boldsymbol{\beta}_1(\mathbf{x}) \equiv \mathbf{0}$) and appropriate boundary conditions (4.1) and (4.2).

Equations (4.3)–(4.5) are the well-known Somigliana identity written for a fictitious "body-force" which can be rewritten as well in terms of stresses:

$$\boldsymbol{\sigma}(\mathbf{x}) = \boldsymbol{\sigma}^0(\mathbf{x}) + \boldsymbol{\sigma}^1(\mathbf{x}), \tag{4.6}$$

$$\boldsymbol{\sigma}^0(\mathbf{x}) = \int_\Gamma \Big[\mathbf{L}^{(0)}\nabla\mathbf{G}(\mathbf{x}-\mathbf{s})\underline{\mathbf{t}}(\mathbf{s}) - \mathbf{L}^{(0)}\nabla\mathbf{T}(\mathbf{x},\mathbf{s})\mathbf{u}(\mathbf{s})\Big]\, ds + \boldsymbol{\alpha}^{(0)}, \tag{4.7}$$

$$\boldsymbol{\sigma}^1(\mathbf{x}) = \int_w \boldsymbol{\Gamma}(\mathbf{x}-\mathbf{y})\boldsymbol{\eta}(\mathbf{y})\, d\mathbf{y}, \tag{4.8}$$

where we defined the strain polarization tensor $\boldsymbol{\eta}(\mathbf{y}) = \mathbf{M}_1(\mathbf{y})\boldsymbol{\sigma}(\mathbf{y}) + \boldsymbol{\beta}_1(\mathbf{y})$, (which is simply a notation convenience). The integral operator kernel, $\boldsymbol{\Gamma}(\mathbf{x}-\mathbf{y}) \equiv -\mathbf{L}^{(0)}\,[\mathbf{I}\delta(\mathbf{x}-\mathbf{y}) + \mathbf{U}(\mathbf{x}-\mathbf{y})\mathbf{L}^{(0)}]$, called the Green's stress tensor, is defined by the Green's tensor, see Eq. (3.37). The stress field $\boldsymbol{\sigma}^0(\mathbf{x})$ (4.7) is derived from the field $\mathbf{u}^0(\mathbf{x})$: $\boldsymbol{\sigma}^0(\mathbf{x}) = \mathbf{L}^{(0)}\boldsymbol{\varepsilon}^0(\mathbf{x}) + \boldsymbol{\alpha}^{(0)}$, $\boldsymbol{\varepsilon}^0 = [\nabla \otimes \mathbf{u}^0 + (\nabla \otimes \mathbf{u}^0)^\top]/2$. The decomposition of the displacement field $\mathbf{u}(\mathbf{x})$ $(\mathbf{x} \in w)$ (4.3) (or stress field $\boldsymbol{\sigma}(\mathbf{x})$ (4.6)) from the fields induced by the boundary conditions and by the fictitious "body-force" was carried out for clarity of the forthcoming presentation in the spirit of the *superposition techniques.*

In particular, for the conditions (4.2) (with $\Gamma_t \equiv \Gamma$) the right-hand side of Eq. (4.7) can be considered as a continuation of $\boldsymbol{\sigma}(\mathbf{x})$, $\mathbf{x} \in \Gamma_t$ (4.2), into w being the stress field that the boundary condition (4.2) would generate in the homogeneous domain w with homogeneous moduli $\mathbf{L}^{(0)}$ (with $\mathbf{L}_1, \boldsymbol{\beta}_1(\mathbf{x}) \equiv \mathbf{0}$). On the other hand, if $\mathbf{u}^0(\mathbf{x})$ (or $\boldsymbol{\sigma}^0(\mathbf{x})$), $\mathbf{x} \in w$, is known then Eqs. (4.3) (or (4.6)) can be considered as the equations for thc field $\mathbf{u}(\mathbf{x})$ (or $\boldsymbol{\sigma}(\mathbf{x})$) induced by the inclusions in an infinite medium loaded by the prescribed field $\mathbf{u}^0(\mathbf{x})$ (or $\boldsymbol{\sigma}^0(\mathbf{x})$). Such decomposition permits one to use separately the known methods of analyses of problems both in an infinite medium and the boundary value problems that will be shown below.

It should be mentioned that the integrand terms $\mathbf{L}_1(\mathbf{y})[\boldsymbol{\varepsilon}(\mathbf{y}) - \boldsymbol{\beta}(\mathbf{y})] - \mathbf{L}^{(0)}\boldsymbol{\beta}_1(\mathbf{y})$ and $\boldsymbol{\eta}(\mathbf{y})$ in Eqs. (4.5) and (4.8), respectively, are nonzero within the inclusions only, since $\mathbf{M}_1$, $\boldsymbol{\beta}_1 \equiv \mathbf{0}$, outside the inclusions. This permits one to restrict consideration of the fields $\boldsymbol{\sigma}^0(\mathbf{x})$, $\boldsymbol{\eta}(\mathbf{x})$ in Eqs. (4.7) and (4.8) to the analysis of the stress state only inside inclusions.

The main advantage of (4.3) and (4.6) is the possibility of making systematic approximations by the use of standard methods of solutions for Fredholm volume integral equations of the second kind in a full space $w = R^d$. The accompanying surface integral equations (4.4) and (4.7) can be treated via the BIE method which is ideally suited for macro-scale analysis and can handle complex geometries along with general boundary conditions. Thus the decomposition of the initial equation into associated equations containing the volume and surface integrals permits the use of a hybrid method based on the combination of well known analytical and numerical methods.

We will consider now a thermoelastic problem for a single inhomogeneity v_i in the infinite homogeneous matrix subjected to the effective field $\overline{\boldsymbol{\sigma}}_i(\mathbf{x})$ coinciding with $\boldsymbol{\sigma}^0(\mathbf{x})$ (4.6) at $w = R^d$. For this purpose we will at first consider the general approach to the solution of this problem and then apply FEA for a concrete solution.

Let the heterogeneity v_i inside the infinite matrix be subjected to a *homogeneous* field $\overline{\boldsymbol{\sigma}}(\mathbf{x}_i)$ (no sum on i)

$$\boldsymbol{\sigma}(\mathbf{x}) = \overline{\boldsymbol{\sigma}}(\mathbf{x}) + \int \boldsymbol{\Gamma}(\mathbf{x}-\mathbf{y})V_i(\mathbf{y})\boldsymbol{\eta}(\mathbf{y})\, d\mathbf{y}. \tag{4.9}$$

Then in view of the linearity of the problem, there exist constant fourth-and second-rank tensors $\mathbf{B}_i(\mathbf{x})$, $\mathbf{R}_i(\mathbf{x})$ and $\mathbf{C}_i(\mathbf{x})$, $\mathbf{F}_i(\mathbf{x})$, such that (see Eqs. (3.179) and (3.180)) (no sum on i; $\mathbf{x} \in v_i$)

$$\boldsymbol{\sigma}(\mathbf{x}) = \mathbf{B}_i(\mathbf{x})\overline{\boldsymbol{\sigma}} + \mathbf{C}_i(\mathbf{x}), \quad \overline{v}_i\boldsymbol{\eta}(\mathbf{x}) = \mathbf{R}_i(\mathbf{x})\overline{\boldsymbol{\sigma}} + \mathbf{F}_i(\mathbf{x}), \tag{4.10}$$
$$\mathbf{R}_i(\mathbf{x}) = \bar{v}_i\mathbf{M}_1(\mathbf{x})\mathbf{B}_i(\mathbf{x}), \quad \mathbf{F}_i(\mathbf{x}) = \bar{v}_i[\mathbf{M}_1(\mathbf{x})\mathbf{C}_i(\mathbf{x}) + \boldsymbol{\beta}_1(\mathbf{x})]. \tag{4.11}$$

The goal of Section 4.2 is a numerical estimation of the tensors $\mathbf{B}_i(\mathbf{x})$, $\mathbf{C}_i(\mathbf{x})$, $\mathbf{R}_i(\mathbf{x})$, $\mathbf{F}_i(\mathbf{x})$.

In a similar manner, the tensors $\mathbf{A}_i(\mathbf{x})$, $\mathbf{R}_i^{\epsilon}(\mathbf{x})$ and $\mathbf{C}_i^{\epsilon}(\mathbf{x})$, $\mathbf{F}_i^{\epsilon}(\mathbf{x})$ can be introduced [compare with Eqs. (3.119)] ($\mathbf{x} \in v_i$)

$$\boldsymbol{\varepsilon}(\mathbf{x}) = \mathbf{A}_i(\mathbf{x})\overline{\boldsymbol{\sigma}} + \mathbf{C}_i^{\epsilon}(\mathbf{x}), \quad \overline{v}_i\boldsymbol{\tau}(\mathbf{x}) = \mathbf{R}_i^{\epsilon}(\mathbf{x})\overline{\boldsymbol{\varepsilon}} + \mathbf{F}_i^{\epsilon}(\mathbf{x}),$$
$$\mathbf{R}_i^{\epsilon}(\mathbf{x}) = \bar{v}_i\mathbf{L}_1(\mathbf{x})\mathbf{A}_i(\mathbf{x}), \quad \mathbf{F}_i^{\epsilon}(\mathbf{x}) = \bar{v}_i[\mathbf{L}_1(\mathbf{x})\mathbf{C}_i^{\epsilon}(\mathbf{x}) + \boldsymbol{\alpha}_1(\mathbf{x})], \tag{4.12}$$

where $\overline{\boldsymbol{\varepsilon}} = \mathbf{M}^{(0)}\overline{\boldsymbol{\sigma}}$, and $\boldsymbol{\tau}(\mathbf{x}) = \mathbf{L}_1(\mathbf{x})\boldsymbol{\varepsilon}(\mathbf{x}) + \boldsymbol{\alpha}_1(\mathbf{x})$ is a stress polarization tensor.

4.2.2 The Heterogeneity v_i Inside an Imaginary Ellipsoid v_i^0

We determine an arbitrary imaginary ellipsoid v_i^0 (the with smallest possible volume) with the characteristic function $V_i^0(\mathbf{x})$ inside which the real inclusion $v_i \subset v_i^0$ is to be inscribed. The results of this section are valid for an arbitrary ellipsoid v_i^0, but in subsequent presentations (e.g., in Chapter 8), v_i^0 is assumed to be connected with statistical distribution of surrounding inclusions in a composite material. Averaging of Eq. (4.9) over the volume v_i^0 and exploiting of the Eshelby theorem [318] leads to the algebraic equation $\langle\boldsymbol{\sigma}\rangle_i^0 = \overline{\boldsymbol{\sigma}}_i - \bar{v}_i(v_i^0)^{-1}\mathbf{Q}_i^0\langle\boldsymbol{\eta}\rangle_i$ (3.181), where $\mathbf{Q}_i^0(\mathbf{x}) = \mathbf{Q}_i^0 = \text{const.}$ (3.125) for the ellipsoidal domain v_i^0 at $\mathbf{x} \in v_i^0$, and the averaging operations $\langle\,\cdot\,\rangle_i$ and $\langle\,\cdot\,\rangle_i^0$ are defined by Eq. (3.182).

Substitution of Eq. (3.181) into the right-hand side of Eq. (4.10$_1$) averaged over the volume v_i^0 leads to the equation $\mathbf{B}_i^0\overline{\boldsymbol{\sigma}}_i + \mathbf{C}_i^0 = \overline{\boldsymbol{\sigma}}_i - \bar{v}_i(\bar{v}_i^0)^{-1}\mathbf{Q}_i^0\langle\boldsymbol{\eta}\rangle_i$ (3.183) yielding the following representation for the average strain polarization tensors:

$$\bar{v}_i\langle\boldsymbol{\eta}\rangle_i = \bar{v}_i^0(\mathbf{Q}_i^0)^{-1}[\mathbf{I} - \mathbf{B}_i^0]\overline{\boldsymbol{\sigma}}_i - \bar{v}_i^0(\mathbf{Q}_i^0)^{-1}\mathbf{C}_i^0,$$
$$\bar{v}_i\langle\boldsymbol{\eta}\rangle_i = \bar{v}_i^0(\mathbf{Q}_i^0)^{-1}[(\mathbf{B}_i^0)^{-1} - \mathbf{I}]\langle\boldsymbol{\sigma}\rangle_i^0 - \bar{v}_i^0(\mathbf{Q}_i^0)^{-1}(\mathbf{B}_i^0)^{-1}\mathbf{C}_i^0, \tag{4.13}$$

and the relations between the averaged tensors $\mathbf{B}_i(\mathbf{x}), \mathbf{C}_i(\mathbf{x}), \mathbf{R}_i(\mathbf{x}), \mathbf{F}_i(\mathbf{x})$ (4.10) presented by Eqs. (3.185) and (3.186).

The estimations $\mathbf{f}_i$ and $\mathbf{f}_i^0$ of the tensors $\mathbf{f}(\mathbf{x})$ averaged over the volumes of the inclusion v_i and over the imaginary ellipsoid v_i^0, respectively [e.g., $\mathbf{f}(\mathbf{x}) = \mathbf{B}_i(\mathbf{x}),\ \mathbf{C}_i(\mathbf{x}),\ \mathbf{R}_i(\mathbf{x}),\ \mathbf{F}_i(\mathbf{x})$] are carried out in a straightforward manner; in so doing $\mathbf{g}_i = \bar v_i^0 \bar v_i^{-1}\mathbf{g}_i^0$ ($\mathbf{g}_i(\mathbf{x}) = \mathbf{R}_i(\mathbf{x}),\ \mathbf{F}_i(\mathbf{x})$) because $\mathbf{M}_1^{(i)}(\mathbf{x}) \equiv \mathbf{0}$ as $\mathbf{x} \in v_i^0 \setminus v_i$. In particular, in the case of nonellipsoidal v_i^0, the tensor $\mathbf{Q}_i^0 \neq$ const. and Eqs. (3.183) and (4.13$_1$) should be recast ($\boldsymbol{\beta}_1 \equiv \mathbf{0}$):

$$\langle\boldsymbol{\sigma}\rangle_i^0 = \overline{\boldsymbol{\sigma}} - \frac{1}{v_i^0}\int \mathbf{Q}_i^0(\mathbf{x})\boldsymbol{\eta}(\mathbf{x})V_i(\mathbf{x})\, d\mathbf{x}, \quad \langle\mathbf{Q}_i^0\boldsymbol{\eta}\rangle_i = \frac{\bar v_i^0}{\bar v_i}[\mathbf{I} - \mathbf{B}_i^0]\overline{\boldsymbol{\sigma}}. \tag{4.14}$$

A homogeneous effective field $\overline{\boldsymbol{\sigma}}_i$ can be produced not only by an applied remote stress field but also by some fictitious loading by a homogeneous eigenstrain concentrated in an arbitrary ellipsoidal domain enclosing v_i. In particular, the ellipsoidal correlation hole v_i^0 can be chosen as such a domain with a fictitious homogeneous eigenstrain $V_i^0(\mathbf{x})\boldsymbol{\beta}^{\rm f}$ ($\boldsymbol{\beta}^{\rm f}$ = const). Then the problem (4.9) is reduced to

$$\boldsymbol{\sigma}(\mathbf{x}) - \int V_i(\mathbf{y})\boldsymbol{\Gamma}(\mathbf{x}-\mathbf{y})\mathbf{M}_1(\mathbf{y})\boldsymbol{\sigma}(\mathbf{y})\, d\mathbf{y} = \int V_i(\mathbf{y})\boldsymbol{\Gamma}(\mathbf{x}-\mathbf{y})\mathbf{M}_1(\mathbf{y})\boldsymbol{\beta}_1(\mathbf{y})\, d\mathbf{y} - \mathbf{Q}_i^0\boldsymbol{\beta}^{\rm f}. \tag{4.15}$$

The solution of Eq. (4.15) can be formally presented in an operator form ($\mathbf{x} \in v_i$):

$$\boldsymbol{\sigma}(\mathbf{x}) = \int\!\!\int V_i(\mathbf{z})\boldsymbol{\mathcal{B}}(\mathbf{x}-\mathbf{y})\boldsymbol{\Gamma}(\mathbf{y}-\mathbf{z})\boldsymbol{\beta}_1(\mathbf{z})\, d\mathbf{z}\, d\mathbf{y} - \mathbf{B}_i(\mathbf{x})\mathbf{Q}_i^0\boldsymbol{\beta}^{\rm f}, \tag{4.16}$$

where it was taken into account that the action of the operator $\boldsymbol{\mathcal{B}} = (\mathbf{I} - \boldsymbol{\Gamma}\mathbf{M}_1)^{-1}V_i$ with the kernel $\boldsymbol{\mathcal{B}}(\mathbf{x})$ at the constant is reduced to the multiplication on a tensor $\mathbf{B}_i(\mathbf{x}) \equiv \int \boldsymbol{\mathcal{B}}(\mathbf{x}-\mathbf{y})\, d\mathbf{y}$. Equation (4.16) allows one to get a representation for the strain polarization tensor:

$$\boldsymbol{\eta}(\mathbf{x}) = \int\!\!\int V_i(\mathbf{z})\mathbf{M}_1(\mathbf{x})\boldsymbol{\mathcal{B}}(\mathbf{x}-\mathbf{y})\boldsymbol{\Gamma}(\mathbf{y}-\mathbf{z})\boldsymbol{\beta}_1(\mathbf{z})\, d\mathbf{z}\, d\mathbf{y} - \mathbf{M}_1(\mathbf{x})\mathbf{B}_i(\mathbf{x})\mathbf{Q}_i^0\boldsymbol{\beta}^{\rm f} + \boldsymbol{\beta}_1(\mathbf{x}). \tag{4.17}$$

Comparing Eqs. (4.10) and (4.16), (4.17) we can see that the tensors $\mathbf{B}_i(\mathbf{x})$, $\mathbf{C}_i(\mathbf{x})$, $\mathbf{R}_i(\mathbf{x})$, $\mathbf{F}_i(\mathbf{x})$ (as well as their averages $\mathbf{B}_i, \mathbf{C}_i, \mathbf{R}_i, \mathbf{F}_i$) are expressed in terms of the operator $\boldsymbol{\mathcal{B}}$. It should be mentioned that our current goal is an establishment of relationships between the tensors $\mathbf{B}_i(\mathbf{x}), \mathbf{C}_i(\mathbf{x}), \mathbf{R}_i(\mathbf{x}), \mathbf{F}_i(\mathbf{x})$ and some another tensors (which will be used in the forthcoming estimations of effective properties of composite materials) with the numerical solutions of different model problems obtained via the popular numerical methods (such as, e.g., FEA). A formal introduction of the integral operator $\boldsymbol{\mathcal{B}}$ is intended only for ascertainment of these relationships rather than for the numerical solution of Eq. (4.9) via the volume integral equation method (see, e.g., Section 4.3). Because of this, a concrete construction of the integral operator $\boldsymbol{\mathcal{B}}$ is not in our current interest.

In the general case the estimation of the tensors $\mathbf{B}(\mathbf{x})$, $\mathbf{C}(\mathbf{x})$, $\mathbf{Q}_i^0$ is a particular problem of the transformation field analysis methods [302] which can be realized by different numerical methods, for example, by finite element analysis.

Indeed, let the inclusion v_i be subjected to homogeneous remote stress $\overline{\boldsymbol{\sigma}}$ = const. with a single nonzero component $\overline{\sigma}_j = 1$; otherwise $\overline{\sigma}_k \equiv 0$

$(j,k=1,\ldots,3d-3; k\neq j)$, and $\boldsymbol{\beta}_1\equiv\mathbf{0}$. We assume that the stress field $\boldsymbol{\sigma}(\mathbf{x})$ $(\mathbf{x}\in v_i)$ inside the inclusion v_i is estimated by some numerical method. Then the tensor $\mathbf{B}_i(\mathbf{x})$ is represented explicitly over the known stress field $\boldsymbol{\sigma}(\mathbf{x})$:

$$B_{i|jm}(\mathbf{x})=\sigma_m(\mathbf{x}) \quad \text{for } \overline{\sigma}_j=1,\ \overline{\sigma}_k\equiv 0\ (j\neq k), \tag{4.18}$$

where $j,k,m=1,\ldots,3d-3$ and $\mathbf{x}\in v_i$, $(i=1,\ldots,n)$.

Furthermore, instead of the real eigenstrain $\boldsymbol{\beta}_1(\mathbf{x})$, let the inclusion be subjected to the fictitious eigenstrain $\boldsymbol{\beta}^{\mathrm{f}}$ such that $\beta^{\mathrm{f}}_j=\beta_{1|j}(\mathbf{x})$ $(\mathbf{x}\in v_i)$ $(j=1,\ldots,d)$; otherwise $\beta^{\mathrm{f}}_k\equiv 0$ $(k=1,\ldots,3d-3; k\neq j)$ and $\overline{\boldsymbol{\sigma}}\equiv\mathbf{0}$. Then we obtain the explicit representation of the tensor $\mathbf{C}_i(\mathbf{x})$ over the known stress field $\boldsymbol{\sigma}(\mathbf{x})$ inside the inclusion v_i:

$$C_{i|jm}(\mathbf{x})=\sigma_m(\mathbf{x}) \quad \text{for } \overline{\boldsymbol{\sigma}}\equiv\mathbf{0},\ \beta^{\mathrm{f}}_j(\mathbf{x})=\beta_{1|j}(\mathbf{x}),\ \ \beta^{\mathrm{f}}_k=0\ (j\neq k), \tag{4.19}$$

where $j=1,\ldots,d$; $k,m=1,\ldots,3d-3$ and $\mathbf{x}\in v_i$, $i=1,\ldots,n$.

At last, for the domain v_i^0 of any shape, the tensor $\mathbf{Q}_i^0(\mathbf{x})\equiv-\bar{v}_i^0\langle\boldsymbol{\Gamma}(\mathbf{x}-\mathbf{y})\rangle_i^0(\mathbf{x})$ $(\mathbf{x}\in v_i\subset v_i^0,\ \mathbf{y}\in v_i^0)$ can be found for $\overline{\boldsymbol{\sigma}}=\boldsymbol{\beta}_1\equiv\mathbf{0}$ from estimations of the stresses inside the inclusion $\mathbf{x}\in v_i^0$ $[\boldsymbol{\sigma}(\mathbf{x})=-\mathbf{Q}^0\boldsymbol{\beta}^{\mathrm{f}}$; see Eqs. (4.10) and (3.119)] subjected to homogeneous fictitious eigenstrain $V_i^0(\mathbf{x})\boldsymbol{\beta}^{\mathrm{f}}$ $(\boldsymbol{\beta}^{\mathrm{f}}=\text{const.})$:

$$\mathbf{Q}^0_{i|mj}(\mathbf{x})=\sigma_m(\mathbf{x}) \quad \text{for } \beta^{\mathrm{f}}_j=1,\ \beta^{\mathrm{f}}_k=0\ (j\neq k), \tag{4.20}$$

where $j=1, k,m=1,\ldots,3d-3$ and $\mathbf{x}\in v_i^0$. In particular, for an ellipsoidal domain v_i^0 usually considered in the subsequent discussion, the tensor $\mathbf{Q}_i^0(\mathbf{x})$ is a constant: $\mathbf{Q}_i^0(\mathbf{x})=\mathbf{Q}_i^0\equiv\text{const.}$

The constant tensor $\mathbf{Q}_i^0$ can also be estimated from the pure elastic problem (4.18) for ellipsoidal domain v_i^0. Indeed, the replacement of the modulus $\mathbf{L}(\mathbf{x})$ $(\mathbf{x}\in v_i^0)$ by the infinite modulus $\mathbf{L}(\mathbf{x})\equiv\infty$ (a rigid inclusion) and estimation of the tensor $\mathbf{B}_i^{\text{rigid}}$ analogously to Eq. (4.18) leads to the evaluation

$$\mathbf{Q}_i^0=[(\mathbf{B}_i^{\text{rigid}})^{-1}-\mathbf{I}]\mathbf{L}^{(0)}. \tag{4.21}$$

For the ellipsoidal domain v_i^0 Eq. (4.21) is more convenient for the estimation of the tensor $\mathbf{Q}_i^0$ than Eq. (4.20) because it already uses what is considered a pure elastic case (4.19), which is necessary anyway. However, for a nonellipsoidal domain v_i^0 the possibility of the estimation of the inhomogeneous tensor $\mathbf{Q}_i^0(\mathbf{x})$ from the pure elastic problem $(\boldsymbol{\beta}_1\equiv\mathbf{0},\ \overline{\boldsymbol{\sigma}}\neq\mathbf{0})$ is questionable.

Thus, the tensor $\mathbf{Q}_i^0$ involved in the eigenstrain problem was estimated from the pure elastic problem (4.21) as well as from the eigenstrain problem (4.20). It turns out that the stress concentrator factor $\mathbf{B}_i(\mathbf{x})$ (4.10_1) can also be found from a fictitious eigenstrain problem rather than only from the pure elastic one (4.18). For this purpose, we will consider a fictitious eigenstrain loading $V_i^0(\mathbf{x})\boldsymbol{\beta}^{\mathrm{f}}$ $(\boldsymbol{\beta}^f=\text{const.})$ accompanied by the conditions $\overline{\boldsymbol{\sigma}}_i=0, \boldsymbol{\beta}_1\equiv\mathbf{0}$. Let the applied eigenstrain $\boldsymbol{\beta}^{\mathrm{f}}$ have a single nonzero component $\beta^{\mathrm{f}}_j=1$ $(j=1,\ldots,d)$; otherwise $\beta^{\mathrm{f}}_k\equiv 0$ $(k=1,\ldots,3d-3; k\neq j)$. Then we form the square matrix (6×6) $\boldsymbol{\Sigma}(\mathbf{x})$ with a j-tuple $(j=1,\ldots,6)$ composed of the six components of the stress vector $\boldsymbol{\sigma}(\mathbf{x})$ generated by the eigenstrain $\boldsymbol{\beta}^{\mathrm{f}}$ with only one nonzero component $\beta^{\mathrm{f}}_j=1$.

Then we obtain the explicit representation of the tensor $\mathbf{B}(\mathbf{x})$ through the known matrices (6×6) of both the stress fields $\boldsymbol{\sigma}^M(\mathbf{x})$ and $\mathbf{Q}_i^0$ $(\mathbf{x} \in v_i^0)$:

$$\mathbf{B}_i(\mathbf{x}) = \boldsymbol{\Sigma}(\mathbf{x})(\mathbf{Q}_i^0)^{-1}. \tag{4.22}$$

The tensor $\mathbf{B}_i(\mathbf{x})$ (and $\mathbf{R}_i(\mathbf{x}) = \bar{v}_i \mathbf{M}_1(\mathbf{x})\mathbf{B}_i(\mathbf{x})$) depends on both the shape and size of the domain v_i^0 of the auxiliary fictitious problem although the tensor $\mathbf{Q}_i^0$ explicitly enters into the right-hand side of Eq. (4.22).

The estimations $\mathbf{f}_i$ and $\mathbf{f}_i^0$ of the tensors $\mathbf{f}(\mathbf{x})$ averaged over the volumes of the inclusion v_i and over the imaginary ellipsoid v_i^0, respectively, ($\mathbf{f}(\mathbf{x}) = \mathbf{B}_i(\mathbf{x}),\ \mathbf{C}_i(\mathbf{x}),\ \mathbf{R}_i(\mathbf{x}),\ \mathbf{F}_i(\mathbf{x})$) are carried out in a straightforward manner; in so doing $\mathbf{g}_i = \bar{v}_i^0 \bar{v}_i^{-1} \mathbf{g}_i^0$ ($\mathbf{g}_i(\mathbf{x}) = \mathbf{R}_i(\mathbf{x}),\ \mathbf{F}_i(\mathbf{x})$) because $\mathbf{M}_1^{(i)}(\mathbf{x}) \equiv \mathbf{0}$ as $\mathbf{x} \in v_i^0 \setminus v_i$.

4.2.3 The Heterogeneity v_i Inside a Nonellipsoidal Imaginary Domain v_i^0

As we will demonstrate later, the inclusions v_i in the composite materials are in general subjected to an inhomogeneous effective field $\overline{\boldsymbol{\sigma}}_i(\mathbf{x})$ (usually assumed to be homogeneous). However, in some popular methods of micromechanics this inhomogeneity of the effective field acting on the considered heterogeneity v_i of a general shape has a special structure that is identical to a stress field generated by some nonellipsoidal fictitious inclusion v_i^0 with a homogeneous eigenstrain. For the modeling of this inhomogeneous field, we will consider an auxiliary problem for a single heterogeneity v_i with vanished applied remote stress and an introduced fictitious homogeneous eigenstrain $V_i^0(\mathbf{x})\boldsymbol{\beta}^{\rm f}$ ($\boldsymbol{\beta}^{\rm f} \equiv$ const.) inside the domain v_i^0. Then the stress field is also described by Eq. (4.9) with the inhomogeneous effective field of a special structure $\overline{\boldsymbol{\sigma}}(\mathbf{x}) = -\mathbf{Q}_i^0(\mathbf{x})\boldsymbol{\beta}^{\rm f}$, where $\mathbf{Q}_i^0(\mathbf{x}) \equiv -\bar{v}_i^0 \langle \boldsymbol{\Gamma}(\mathbf{x}-\mathbf{y})\rangle_i^0$ $(\mathbf{x},\mathbf{y} \in v_i^0)$ is not a constant for the nonellipsoidal domain v_i^0 being considered.

At first we will consider a "pure elastic" problem when the real eigenstrain $\boldsymbol{\beta}_1 \equiv \mathbf{0}$ and $\boldsymbol{\eta}(\mathbf{x}) \equiv \mathbf{M}_1(\mathbf{x})\boldsymbol{\sigma}(\mathbf{x})$. Then

$$\bar{v}_i \langle \boldsymbol{\eta} \rangle_i = \bar{v}_i \langle \mathbf{M}_1 \boldsymbol{\mathcal{B}} * \mathbf{Q}_i^0 \rangle_i \boldsymbol{\beta}^{\rm f}, \tag{4.23}$$

where the integral operator $\boldsymbol{\mathcal{B}} = V_i(\mathbf{I} + \mathbf{M}_1\boldsymbol{\Gamma})^{-1}$ with a kernel $\boldsymbol{\mathcal{B}}(\mathbf{x})$ was defined in Eq. (4.16) and $\boldsymbol{\mathcal{B}} * \mathbf{Q}_i^0 = \int V_i^0(y)\boldsymbol{\mathcal{B}}(\mathbf{x}-\mathbf{y})\mathbf{Q}_i^0(\mathbf{y})\, d\mathbf{y}$, $\mathbf{x} \in v_i$. Thus, the averaging of the tensor $\mathbf{R}_i^{\mathbf{Q}0}(\mathbf{x}) \equiv V_i(\mathbf{x})\boldsymbol{\mathcal{B}} * \mathbf{Q}_i^0(\mathbf{x})$

$$\mathbf{R}_i^{\mathbf{Q}0} = \bar{v}_i \langle \mathbf{M}_1 \boldsymbol{\mathcal{B}} * \mathbf{Q}_i^0 \rangle_i \tag{4.24}$$

can be considered as found if the average strain polarization tensor $\langle \boldsymbol{\eta} \rangle_i$ produced by the fictitious homogeneous eigenstrain $\boldsymbol{\beta}^{\rm f}$ is found via any numerical method (e.g., FEA)

In a similar manner, the stress distribution in the inhomogeneity can be presented in the form $\boldsymbol{\sigma}(\mathbf{x}) = \mathbf{B}_i^{\mathbf{Q}0}(\mathbf{x})\boldsymbol{\beta}^{\rm f}$, $\langle \boldsymbol{\sigma} \rangle_i = \mathbf{B}_i^{\mathbf{Q}0}\boldsymbol{\beta}^{\rm f}$, where $\mathbf{B}_i^{\mathbf{Q}0}(\mathbf{x}) = \boldsymbol{\mathcal{B}} * \mathbf{Q}_i^0(\mathbf{x})$. It should be mentioned that although the estimation of the integral operator $\boldsymbol{\mathcal{B}}$ in a general case of the shape and inhomogeneity of the heterogeneity v_i is not trivial, the evaluation of the tensor $\mathbf{R}_i^{\mathbf{Q}0}$ (4.24) (or that is the same of the tensor $\langle\boldsymbol{\eta}\rangle_i$) should present no problems when exploiting standard numerical methods. An additional feature of the tensor $\mathbf{R}_i^{\mathbf{Q}0}(\mathbf{x})$ is that it is defined at $\mathbf{x} \in v_i \subset v_i^0$, while the operator $\boldsymbol{\mathcal{B}}$ is applied to the domain v_i^0. However, such computational details will not be used in the future because only the tensor $\mathbf{R}_i^{\mathbf{Q}0}$ (rather than operator $\boldsymbol{\mathcal{B}}$) expressed through the numerical solution $\langle\boldsymbol{\eta}\rangle_i$ (4.23) will be used at the estimation of effective properties of composite materials. A tensor $\mathbf{R}_i^{\mathbf{Q}0}$ (4.24) can be expressed through the concentrator factors $\mathbf{R}_i$ and $\mathbf{B}_i$

$$\mathbf{R}_i^{\mathbf{Q}0} = \mathbf{R}_i\mathbf{Q}_i^0, \quad \mathbf{B}_i^{\mathbf{Q}0} = \mathbf{B}_i\mathbf{Q}_i^0, \tag{4.25}$$

respectively, only for ellipsoidal domain v_i^0 when the tensor $\mathbf{Q}_i^0$ is a constant and can be carried out from the sign of averaging in Eq. (4.24), whereas the action of operator $\boldsymbol{\mathcal{B}}$ at the constant is reduced to multiplication over the tensor $\mathbf{B}_i(\mathbf{x})$. Thus, the equality (4.25) enables one to estimate the concentrator factor $\mathbf{R}_i$ through the solution of a fictitious eigenstrain particular problem for the homogeneous domain v_i^0 described by the tensors $\mathbf{R}_i^{\mathbf{Q}0}$ and $\mathbf{Q}_i^0$: $\mathbf{R}_i = \mathbf{R}_1^{\mathbf{Q}0}(\mathbf{Q}_i^0)^{-1}$. In a similar manner the averaging of the tensor $\mathbf{F}_i(\mathbf{x})$ (4.10$_2$) $\mathbf{F}_i = \bar{v}_i\langle\mathbf{M}_1\boldsymbol{\mathcal{B}}*\mathbf{Q}_i\rangle_i\boldsymbol{\beta}_1^{(i)}$ can also be considered as found if the average strain polarization tensor $\langle\boldsymbol{\eta}\rangle_i$ produced by the real homogeneous eigenstrain $\boldsymbol{\beta}_1^{(i)}$ is found. Here the tensor $\mathbf{Q}_i(\mathbf{x}) \not\equiv \text{const.}$ ($\mathbf{x} \in v_i$) for nonellipsoidal domain v_i.

For nonellipsoidal v_i^0, along with the tensors $\mathbf{B}_i(\mathbf{x}), \mathbf{C}_i(\mathbf{x}), \mathbf{R}_i(\mathbf{x}), \mathbf{F}_i(\mathbf{x})$ and their averages estimated in Subsection 4.2.2 by the use, perhaps, of an arbitrary imaginary ellipsoidal domain [as in Eqs. (4.18) and (4.22)], we will need to know the tensor $\mathbf{R}_i^{\mathbf{Q}0}$ (4.24) depending on the nonellipsoidal domain v_i^0. It can be easily performed following the scheme of obtaining the tensor $\mathbf{C}_i(\mathbf{x})$ (4.10$_1$). Indeed, let the domain v_i^0 be subjected to the homogeneous fictitious eigenstrain $V_i^0(\mathbf{x})\boldsymbol{\beta}^{\mathrm{f}}$ such that $\beta_j^{\mathrm{f}} = 1$ ($j = 1, \ldots, d$); otherwise $\beta_k^{\mathrm{f}} \equiv 0$ ($k = 1, \ldots, 3d-3; k \neq j$), and $\overline{\boldsymbol{\sigma}} \equiv \mathbf{0}$. Then we obtain the explicit representation of the tensor $\mathbf{R}_i^{\mathbf{Q}0}$ over the known average strain polarization factor $\langle\boldsymbol{\eta}\rangle_i$:

$$R_{i|j}^{\mathbf{Q}0} = \bar{v}_i^0\langle\eta_j(\mathbf{x})\rangle_i \quad \text{for } \overline{\boldsymbol{\sigma}} \equiv \mathbf{0}, \quad \beta_j^{\mathrm{f}} = 1, \quad \beta_k^{\mathrm{f}} = 0 \ (j \neq k). \tag{4.26}$$

where $j, k, m = 1, \ldots, 3d-3$. It is interesting that for ellipsoidal domain $v_i^0 \supset v_i$, the tensor $\mathbf{R}_i^{\mathbf{Q}0}$ does not depend on the relative size $\bar{v}_i^0/\bar{v}_i$ of the domain v_i^0, while such a dependence is true for the nonellipsoidal domain v_i^0.

Analogously to Eq. (4.26), the tensor $\mathbf{B}_i^{\mathbf{Q}0}(\mathbf{x})$ (4.18) can be evaluated through the known stress distribution inside the heterogeneity $v_i \ni \mathbf{x}$ (i is the inclusion number):

$$B_{i|j}^{\mathbf{Q}0}(\mathbf{x}) = \sigma_j(\mathbf{x}) \quad \text{for } \overline{\boldsymbol{\sigma}} \equiv \mathbf{0}, \quad \beta_j^{\mathrm{f}} = 1, \quad \beta_k^{\mathrm{f}} = 0 \ (j \neq k). \tag{4.27}$$

where six different sets $\sigma_m(\mathbf{x})$ are estimated for six different sets of $\boldsymbol{\beta}^{\mathrm{f}}$ with only one nonzero component $\beta_j^{\mathrm{f}} = 1$, $\beta_k^{\mathrm{f}} = 0$ in the jth set $\boldsymbol{\beta}^{\mathrm{f}}$ ($j, k, m = 1, \ldots, 3d-3$; $k \neq j$).

4.2.4 Estimation of Concentrator Factor Tensors by FEA

It was mentioned in the preceding text that estimation of the tensors $\mathbf{B}(\mathbf{x})$, $\mathbf{C}(\mathbf{x})$, and $\mathbf{Q}_i^0$ can in general be realized by numerical methods only: in the present study, FEA will be utilized. Namely, following the Bubnov-Galerkin approach, we multiply the equilibrium equation with the shape function N_m (associated with nodes with number m of the finite element (FE) mesh, $m = 1, 2, ..., K$, K being a total number of nodes in a given FE model) and then integrate it over the volume V. It gives us a "weak" form of the equilibrium equation:

$$\int_V \nabla \cdot \boldsymbol{\sigma} \; N_m \, dv + \int_V \mathbf{f} N_m \, dv \; = 0, \tag{4.28}$$

where $\mathbf{f}$ is the body force vector. Using the well-known identities $\nabla \cdot (\boldsymbol{\sigma} \; N_m)$ $= \; \nabla N_m \cdot \boldsymbol{\sigma} \; + N_m \nabla \cdot \boldsymbol{\sigma}$, and

$$\int_V \nabla \cdot \boldsymbol{\sigma} \, dv \; = \; \int_S \mathbf{n} \cdot \boldsymbol{\sigma} \, ds, \tag{4.29}$$

one obtains from (4.28) $\int_V \boldsymbol{\sigma} \cdot \nabla N_m \, dv \; = \int_V \mathbf{f} N_m \, dv + \; \int_S N_m \mathbf{n} \cdot \boldsymbol{\sigma} \, ds$. Here, $\mathbf{n}$ is the unit vector, normal to the surface S. After we substitute the constitutive and equilibrium equations into (4.29), we obtain the equation

$$\frac{1}{2} \int_V [\mathbf{L}(\nabla \mathbf{u} + (\nabla \mathbf{u})^\top)] \nabla N_m \, dv = \int_V \mathbf{f} N_m \, dv - \int_V \boldsymbol{\alpha} \cdot \nabla N_m \, dv + \int_S N_m \mathbf{n} \cdot \boldsymbol{\sigma} \, ds. \tag{4.30}$$

The last term in (4.30) is nonzero in cases where the stress condition $\boldsymbol{\sigma} \cdot \mathbf{n} = \mathbf{T}$ is prescribed at the outer boundary.

In accordance with the FEA scheme, the fields $\mathbf{u}$, $\mathbf{p}$, and $\boldsymbol{\alpha}$ on each finite element are approximated as

$$W \; = \; W_p N_p, \quad p = 1, \; 2, \; ..., K^e, \; (W = (\mathbf{u}, \mathbf{f}, \boldsymbol{\alpha})), \tag{4.31}$$

where K^e is the node number of elements numbered e, $e = 1, 2, ..., M$, M is a total number of FEs. By substituting the expressions (4.31) in (4.30), we come to the equations for the nodal values of the displacement vector $\mathbf{u}(\mathbf{r}_k, t) = \mathbf{U}_k(t)$ $\equiv \mathbf{U}_k$:

$$\frac{1}{2} \sum_{e=1}^{M} \int_{V^e} [\mathbf{L}(\nabla(\mathbf{U}_k \mathbf{N}_k) + (\nabla(\mathbf{U}_k \mathbf{N}_k))^\top)] \nabla N_m \, dv$$
$$= \sum_{e=1}^{M} \Big[\int_{V^e} f_k N_k N_m \, dv - \int_{V^e} \alpha_k N_k \cdot \nabla N_m \, dv + \int_{S^e} N_k N_m \mathbf{n} \cdot \boldsymbol{\sigma}_k \, ds \Big]. \tag{4.32}$$

After calculating the volume and surface integrals in the local linear sets (4.32) for each element, we apply the assembling procedure to obtain a global set of linear algebraic equations with the unknown vector $\{\mathbf{U}\}$ of nodal displacements: $[A] \{\mathbf{U}\} = \{\mathbf{R}\}$. Here, the vector $\{\mathbf{R}\}$ is constructed from the right-hand parts of the local equations (4.32), and $[A]$ is $K \times K$ is the symmetric and positively defined matrix, which enables use of the efficient iterative solvers. Specifically, convergence of the Jacobi conjugate gradient solver utilized by us was found to be strongly dependent on the aspect ratio of inclusion. So, for very high values of α^r (say, 1000), the tolerance level of order 10^{-10} must be taken to get the fully convergent solution.

We are coming now to the concrete numerical analysis of the elastic problems for one nonellipsoidal inhomogeneity in an infinite matrix. The 3-D problem for one cylindrical inclusion with the smooth ends will be considered in Chapter 19 as an application to the modeling of nanocomposites. Here we will consider the 2D plane strain problem for the noncircular inclusion. Let the elastic properties of the isotropic components be $E^{(0)} = 1$ GPa, $E^{(1)} = 100$ GPa, $\nu^{(0)} = \nu^{(1)} = 0.45$. Figure 4.1a is a schematic representation of the quarter of the inclusion boundary in the first quadrant described by the curve

$$\begin{cases} (x-l)^2 + (y-l)^2 = (r-l)^2, & \text{for } x > l \text{ and } y > l, \\ |x+y| + |x-y| = 2r, & \text{for } x < l \text{ or } y < l, \end{cases} \tag{4.33}$$

which reduces to a circle and a square in the limiting cases $l = 0$ and $l = r$, respectively. In parallel with the inclusion (4.33), we will consider the inclusion with the same shape but with the different orientation obtained by the rotation of the inclusion (4.33) on the angle $\pi/4$ as shown in Fig. 4.1b, where the quarter of the inclusion in the first quadrant is described by the equation

$$\begin{cases} x^2 + (y - l\sqrt{2})^2 = (r-l)^2, & \text{for } y > (r+l)/\sqrt{2}, \\ (x - l\sqrt{2})^2 + y^2 = (r-l)^2, & \text{for } x > (r+l)/\sqrt{2}, \\ |x| + |y| = \sqrt{2}r, & \text{for } |x| < (r+l)/\sqrt{2} \text{ and } |y| < (r+l)/\sqrt{2}, \end{cases} \tag{4.34}$$

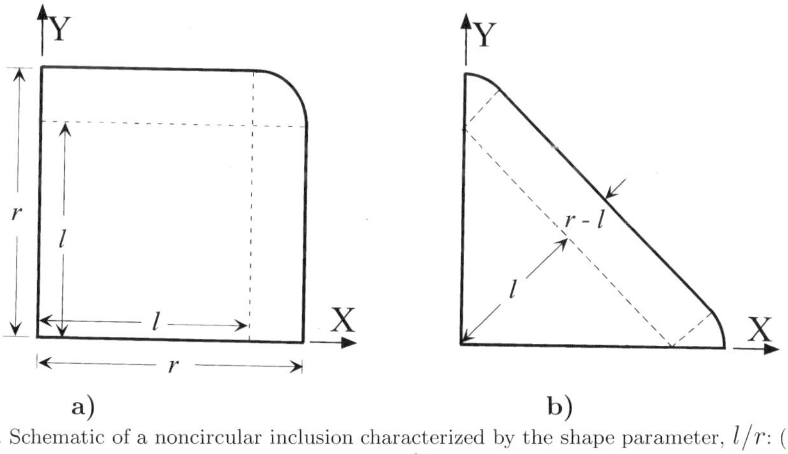

Fig. 4.1. Schematic of a noncircular inclusion characterized by the shape parameter, l/r: (a) initial configuration, (b) configuration rotated by $\pi/4$ about the Z-axis.

The commercial finite element code, ABAQUS [1], was used for the FE modeling with the mesh designed for the noncircular inclusion shape shown in Fig. 4.2b along with an expanded view of the mesh inside and near the inclusion boundary in Fig. 4.2a. The inclusion and matrix materials were assumed to be isotropic, linear elastic, and modeled with plane strain second-order isoparametric (CPE8) elements, while the interface was modeled as perfectly bonded. The infinite dimensions of the matrix were approximated with a length of 40 inclusion diameters: $R = 40r$. Increasing the matrix dimensions further did not significantly change the results. To capture the rapidly changing stresses near the interface, elements with a radial span of 0.0083 inclusion diameters were used near the interface. Symmetry conditions were implemented (i.e., $u_y = 0$ along

$y = 0$ and $u_x = 0$ along $x = 0$, as shown in Fig. 4.2b) so that only one quadrant of the problem required meshing. Typical meshes contained a total of 3200 elements (approx. 20000 d.o.f.s) with 1000 elements in the inclusion and 2200 in the matrix. Normal loading of the matrix boundary in the X- and Y- directions was implemented by applying unit pressure on the corresponding surfaces (e.g. $\sigma_{ij} = \delta_{1i}\delta_{1j}$ and $\sigma_{ij} = \delta_{2i}\delta_{2j}$), such as is illustrated in Fig. 4.2b for $\sigma_{ij} = \delta_{1i}\delta_{j1}$.

Unlike the case of pure normal loading where only a quarter of the problem required meshing, the full problem had to be considered for the shear loading case. The complete view of the FE model used to consider pure shear in the X-direction of the coordinate system (4.33) is shown in Fig. 4.3b along with the

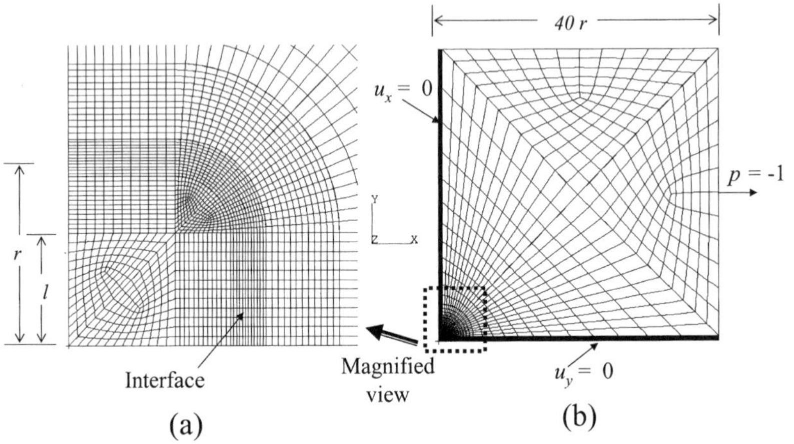

Fig. 4.2 (a) Magnified view and (b) complete view of the FE mesh utilized for solution of a single inclusion in an infinite matrix subjected to normal loading at infinity.

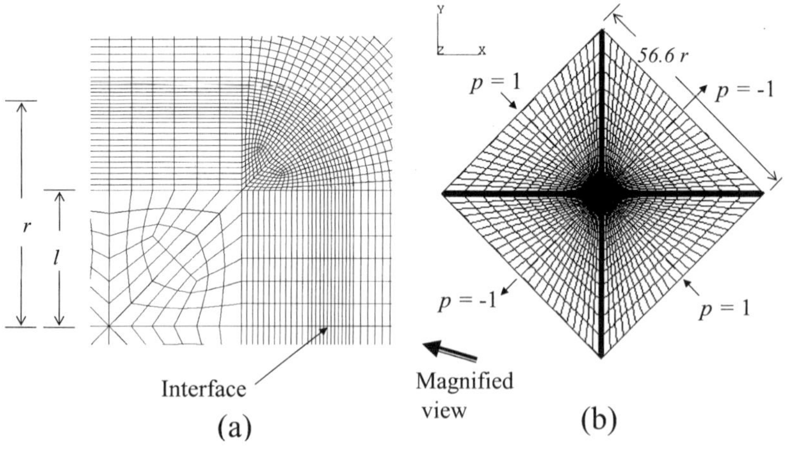

Fig. 4.3 (a) Magnified view and (b) complete view of the FE mesh utilized for solution of a single inclusion in an infinite matrix subjected pure shear in the $X-$direction.

magnified view of the mesh inside the inclusion in Fig. 4.3a. The FE model shown in Fig. 4.3 was obtained from the normal loading case (Fig. 4.2) by keeping the inclusion fixed and simply rotating the outside matrix boundary by 45 degrees about the Z-axis. Typical meshes for the full problem contained a total of 8600 elements (approx. 52000 d.o.f.s) with 2520 elements in the inclusion and 6080 in the matrix. The outside matrix surfaces were subjected to unit pressure loading as shown in Fig. 4.3b to yield a unit shear in the X-direction.

We assume that for all inclusions (4.33) the tensors $\mathbf{Q}_i^0$ are identical and coincide with the tensor $\mathbf{Q}^0$ for the circular inclusion which was estimated both numerically by the use of FEA and analytically by means of the Eshelby tensor. The error of the numerical solution is not more than 0.15%, where the ratio $R/r = 40$ is considered to represent an infinite medium. It was demonstrated that the stress inside the inclusion approaches a constant value for mesh with the inclusion and matrix divisions equal 24 and 72, respectively, along the X-axis.

4.3 Volume Integral Equation Method

4.3.1 Regularized Representation of Integral Equations

The boundary integral representation of displacement field (4.4) contains the weakly singular kernel (logarithmic or r^{-1} for $d = 2, 3$, respectively, $r = |\mathbf{x} - \mathbf{y}|$ is the distance between the source $\mathbf{y}$ and field $\mathbf{x}$ points) and strongly singular kernel (r^{1-d}), while the kernel of volume integral (4.5) is weakly singular (r^{1-d}) for the space dimensionality d. For the stresses (4.6)–(4.8) the order of singularities increases by one, that is, the kernels of the first and second integrand of the boundary integral representation for stresses (4.7) have a strong singularity (r^{1-d}) and hypersingularity r^{-d}, respectively, while the volume integral kernel (4.8) has a strong singularity r^{-d}.

The weak singularities in the surface integrals can be removed by using a transformation of the integration variable, and the integrals being considered are integrable in the ordinary sense even if $\mathbf{x} \to \boldsymbol{\zeta} \in \Gamma$. The strong singularity, however, requires either definition of singular integrals in the Cauchy principal value (CPV) sense or a regularization procedure to eliminate or reduce the order of strong singularities [1017], [1069]. We will use the subtraction technique [882] which is the basis of so-called physical regularization. Then the operation of the Taylor series expansion of the densities about the singular point, subtraction and adding back the singular terms as well as the use of some fundamental solution identities, leads to elimination of CPV integrals that will be shown now for the volume (4.8) and boundary (4.7) integral equations.

At first we will consider Eq. (4.6) assuming that the stress field $\boldsymbol{\sigma}^0(\mathbf{x})$ ($\mathbf{x} \in v_i$, $i = 1, \ldots, n$) is known. A known quadrature method for obtaining an approximate solution of Eq. (4.8) is to evaluate the volume integrals with a Gauss quadrature formula. Then the corresponding equations at the Gauss points will contain a singular term that results when the field point $\mathbf{x}$ and source point $\mathbf{y}$ are coincident: $\mathbf{x} = \mathbf{y}$ (4.8). The difficulties with the troublesome singularities can be avoided if rearrangement of Eq. (4.8) is performed in the spirit of a subtraction technique used in the modified quadrature method [268]:

$$\boldsymbol{\eta}(\mathbf{x}) = \mathbf{M}_1(\mathbf{x})\boldsymbol{\sigma}^0(\mathbf{x}) + \boldsymbol{\beta}_1(\mathbf{x}) + \mathbf{M}_1(\mathbf{x}) \int V_i^0(\mathbf{y})\boldsymbol{\Gamma}(\mathbf{x} - \mathbf{y})\, d\mathbf{y}\boldsymbol{\eta}(\mathbf{x})$$

$$+ \mathbf{M}_1(\mathbf{x}) \int \mathbf{\Gamma}(\mathbf{x}-\mathbf{y})\big[\boldsymbol{\eta}(\mathbf{y}) - V_i^0(\mathbf{y})\boldsymbol{\eta}(\mathbf{x})\big]\, d\mathbf{y}, \quad \mathbf{x} \in v_i, \tag{4.35}$$

where v_i, perhaps, is not an ellipsoid. The equation is valid for any domain $\mathbf{x} \in v_i^0$ with the characteristic function $V_i^0(\mathbf{x})$. We assume that the first integral in (4.35) is easy computable: $\bar{v}_i^0 \langle \mathbf{\Gamma}(\mathbf{x}-\mathbf{y})\rangle_i^0 = -\mathbf{Q}_i^0(\mathbf{x})$ (3.181). In such a case the problem is reduced to computing the second integral in (4.35) which is not a singular as will be shown later. In a general case of the inclusion shape v_i the function $\mathbf{Q}_i^0(\mathbf{x})$ can be found numerically, e.g., by finite element analysis (FEA; see, e.g., Section 4.3). For ellipsoidal domain $v_i^0 \supset v_i$, the first integral on the right-hand side of (4.35) is known and is associated with the well-known Eshelby tensor by Eq. (3.136): $\mathbf{S}_i^0 = \mathbf{I} - \mathbf{M}^{(0)}\mathbf{Q}_i^0$. The assumption of an ellipsoidal shape of the domain v_i^0, was used only in order to obtain an analytical representation of the integral (3.181).

This is because the tensor $\langle \mathbf{\Gamma}(\mathbf{x}-\mathbf{y})^0\rangle_i$ is homogeneous for $\mathbf{x} \in v_i$, $\mathbf{y} \in v_i^0$ for an ellipsoid. For nonellipsoidal inclusions v^{n-e} one could assume that in some parts of the region $v_i^c \subset v_i$ the properties $\mathbf{M}_1(\mathbf{x}) \equiv \mathbf{0}$, $\beta_1(\mathbf{x}) \equiv \mathbf{0}$, i.e., it is sufficient to replace a real nonellipsoidal inclusion $v^{n-e} = v_i \setminus v_i^c$ with an fictitious ellipsoid (with smallest possible volume) and call it the inclusion v_i with a "coating" v_i^c. In so doing, at the estimation of the second integral in (4.35) we keep in mind that $\int \mathbf{\Gamma}(\mathbf{x}-\mathbf{y})(V_i(\mathbf{y})-V_i^0(\mathbf{y}))\, d\mathbf{y} = \mathbf{Q}_i^0 - \mathbf{Q}_i(\mathbf{x})$ ($\mathbf{x} \in v_i$). With the nonessential restriction on the shape of the inclusion v_i mentioned above we can consider without loss of generality an ellipsoidal inclusion $v_i^0 = v_i$; in so doing $\mathbf{M}_1(\mathbf{y}) \equiv \mathbf{0}$, $\beta_1(\mathbf{y}) \equiv \mathbf{0}$ at $\mathbf{y} \in v_i^c \subset v_i$. Then Eq. (4.35) can be rewritten in compact form:

$$\boldsymbol{\eta}(\mathbf{x}) = \boldsymbol{\eta}_i^0(\mathbf{x}) + \int \mathbf{K}(\mathbf{x},\mathbf{y})\big[\boldsymbol{\eta}(\mathbf{y}) - \boldsymbol{\eta}(\mathbf{x})\big]\, d\mathbf{y}, \quad \mathbf{x} \in v_i, \tag{4.36}$$

where $\boldsymbol{\eta}_i^0(\mathbf{x}) = \mathbf{E}_i(\mathbf{x})\boldsymbol{\sigma}^0(\mathbf{x}) + \mathbf{H}_i(\mathbf{x})$, ($\mathbf{x} \in v_i$) is called the external strain polarization tensor in the inclusion v_i, and (no sum on i):

$$\mathbf{K}(\mathbf{x},\mathbf{y}) = \mathbf{E}_i(\mathbf{x})\mathbf{\Gamma}(\mathbf{x}-\mathbf{y})V_i(\mathbf{y}), \tag{4.37}$$

$$\mathbf{E}_i(\mathbf{x}) = \mathbf{M}_1(\mathbf{x})[\mathbf{I} + \mathbf{Q}_i^0\mathbf{M}_1(\mathbf{x})]^{-1}, \quad \mathbf{H}_i(\mathbf{x}) = [\mathbf{I} + \mathbf{M}_1(\mathbf{x})\mathbf{Q}_i^0]^{-1}\boldsymbol{\beta}_1(\mathbf{x}). \tag{4.38}$$

Equation (4.36) is in any case less singular than the original form (4.6) and (4.8), since $\boldsymbol{\eta}(\mathbf{y}) - \boldsymbol{\eta}(\mathbf{x}) = \mathbf{0}$ at the singular point $\mathbf{y} = \mathbf{x}$. Moreover, if $\boldsymbol{\eta}(\mathbf{y}) \in C^{0,\alpha}$ at $\mathbf{y} = \mathbf{x} \in v_i$ (Hölder continuity, i.e., $|\boldsymbol{\eta}(\mathbf{y})-\boldsymbol{\eta}(\mathbf{x})| < K|\mathbf{y}-\mathbf{x}|^\alpha$, $0 < K$, $0 \le \alpha \le 1$) then the subtracted form (4.36) is regular, and the improper volume integral (4.36) over the region $v_i \setminus \omega$ absolutely converges and is independent of the shape of an excluded contracting region ω. We recast the integral equation obtained in [140] in a form (4.36)–(4.38) that is appropriate to our intended application. One has justified, by doing so, the applicability of a standard quadrature rule for the numerical estimation of the integral (4.36). Only for the homogeneous ellipsoidal inclusion (3.187) with $v_i^c \equiv \emptyset$, Eq. (4.36) is reduced to the equation ($\mathbf{x} \in v_i$) $\boldsymbol{\sigma}(\mathbf{x}) = \boldsymbol{\sigma}^0(\mathbf{x}) - \mathbf{Q}_i\mathbf{M}_1^{(i)}\boldsymbol{\sigma}(\mathbf{x}) + \int \mathbf{\Gamma}(\mathbf{x}-\mathbf{y})\mathbf{M}_1(\mathbf{y})\big[\boldsymbol{\sigma}(\mathbf{y}) - \boldsymbol{\sigma}(\mathbf{x})\big]\, d\mathbf{y}$ (when $\mathbf{Q}_i = \mathbf{Q}_i^0$) obtained in [227].

We come now to a similar regularized representation of Eq. (4.4) suggested by Perlin [869] (for references see also [862], [1069], [1214]). One can show that the singularity for the surface integral with the kernel $\mathbf{T}(\mathbf{x},\mathbf{s})$ is removed by the use of the following fundamental identity for an inclusion v_i: $(1-e)\mathbf{I} = -\int_{\Gamma_i} \mathbf{T}(\mathbf{x},\mathbf{s})\, ds$, where $e=0$ and $e=1$ for the internal ($\mathbf{x}\in v_i$) and external ($\mathbf{x}\notin v_i$) problem, respectively. Then Eqs. (4.3) and (4.4) can be rewritten in the form

$$\mathbf{u}(\mathbf{x}) = \mathbf{u}^1(\mathbf{x}) + (1-e)\mathbf{u}(\boldsymbol{\zeta}) + \int_\Gamma \mathbf{G}(\mathbf{x}-\mathbf{s})\underline{\mathbf{t}}(\mathbf{s})\, ds - \int_\Gamma \mathbf{T}(\mathbf{x},\mathbf{s})[\mathbf{u}(\mathbf{s})-\mathbf{u}(\boldsymbol{\zeta})]\, ds, \tag{4.39}$$

where the field $\mathbf{u}^1(\mathbf{x})$ induced by the inclusions is equal to the integral on the right-hand side of Eq. (4.3) and is assumed to be known now for convenience. The integral representation (4.39) is free of CPV integrals and exists in the ordinary sense for $\mathbf{u}(\mathbf{s})$ smooth enough at $\mathbf{x}=\boldsymbol{\zeta}$ even if $\mathbf{x}\to\boldsymbol{\zeta}\in\Gamma$ and we can simply put $\mathbf{x}=\boldsymbol{\zeta}$.

The same subtraction technique can be applied to Eq. (4.7). Assuming $C^{0,\alpha}$ continuity for $\nabla\mathbf{u}(\mathbf{s})$ at $\mathbf{s}=\boldsymbol{\zeta}\in\Gamma$ and subtracting and adding back relevant terms of the Taylor series expansion of $\mathbf{u}(\mathbf{s})$ near $\boldsymbol{\zeta}$ the regularized representation of Eq. (4.7) presented in [1069] can be obtained.

4.3.2 The Iteration Method

Thus we obtain a regular representation of the integral equation (4.3) in (4.36) and (4.39) and can use known methods for their solution. Numerical methods of solving integral equations are based in the first place on the possibility of evaluating the integrals appearing in the equations, no matter which method of solution is used: whether it is the method of successive approximations (when the whole integrand is known at each stage) or the mechanical quadrature method (when the unknown function is assumed to be constant or varying in a certain manner within small regions, which enables one to pass to an integral of a known expression). These approximation methods for integral equations lead to full linear systems. Application of the method of mechanical quadratures is very popular [656], [657], [658] although it requires solving linear systems of very high order even for fairly smooth surfaces of inclusions and a smoothly varying load. Only if the number of unknowns is reasonably small, these equations can be solved by direct methods such as Gaussian elimination. But, in general, a satisfactory accuracy of the approximate solution to the integral equation will require a comparatively large number of unknowns, in particular for integral equations in more than one dimension. Therefore iterative methods for the resulting linear systems will be preferable: first, their implementation requires the retention of only the last iteration and the sum of the preceding ones in the memory of a computer and second, since there is a proof of convergence (in the case of its exact implementation) and the approximate determination of a finite sum of the series reduces to a finite number of quadratures, it may be stated that the numerical algorithm leads to a solution with given accuracy for a sufficiently fine discretization.

Therefore, we solve Eq. (4.36) by the method of successive approximations, which is also called the Neumann series method. With this aim in view we rewrite Eq. (4.36) in symbolic form:

$$\boldsymbol{\eta} = \boldsymbol{\eta}^0 + \boldsymbol{\mathcal{K}}\boldsymbol{\eta}, \tag{4.40}$$

where $(\boldsymbol{\mathcal{K}}\boldsymbol{\eta})(\mathbf{x}) = \int \boldsymbol{\mathcal{K}}(\mathbf{x},\mathbf{y})\boldsymbol{\eta}(\mathbf{y})\, d\mathbf{y}$ defines the integral operator $\boldsymbol{\mathcal{K}}$ with the kernel formally represented as

$$\boldsymbol{\mathcal{K}}(\mathbf{x},\mathbf{y}) = \mathbf{K}(\mathbf{x},\mathbf{y}) - \delta(\mathbf{x}-\mathbf{y})\int V_i(\mathbf{z})\mathbf{K}(\mathbf{x},\mathbf{z})\, d\mathbf{z}. \tag{4.41}$$

We formally write the solution of Eq. (4.40) as

$$\boldsymbol{\eta} = \boldsymbol{\mathcal{L}}\boldsymbol{\eta}^0, \tag{4.42}$$

where the inverse operator $\boldsymbol{\mathcal{L}} = (\mathbf{I} - \boldsymbol{\mathcal{K}})^{-1}$ will be constructed by the iteration method based on the recursion formula

$$\boldsymbol{\eta}_{(k+1)} = \boldsymbol{\eta}^0 + \boldsymbol{\mathcal{K}}\boldsymbol{\eta}_{(k)} \tag{4.43}$$

to construct a sequence of functions $\{\boldsymbol{\eta}_{(k)}\}$ that can be treated as an approximation of the solution of Eq. (4.40). We have used point Jacobi (called also Richardson and point total-step) iterative scheme for ease of calculations. The details of the real iteration method used for the solution of Eq. (4.40) will be presented in Subsection 4.4.4. Usually the driving term of this equation is used as an initial approximation:

$$\boldsymbol{\eta}_{(0)}(\mathbf{x}) = \boldsymbol{\eta}^0(\mathbf{x}), \tag{4.44}$$

which is exact for a homogeneous ellipsoidal inclusion subjected to remote homogeneous stress field $\boldsymbol{\sigma}^0(\mathbf{x}) \equiv \boldsymbol{\sigma}^0 = \text{const}$. It suggests the Neumann series form for the solution $\boldsymbol{\eta}$ of (4.36) $\boldsymbol{\eta} = \boldsymbol{\mathcal{L}}\boldsymbol{\eta}^0$, $\boldsymbol{\mathcal{L}} \equiv \sum_{k=0}^{\infty} \boldsymbol{\mathcal{K}}^k$, where $\boldsymbol{\mathcal{L}}$ is the inverse integral operator with the kernel $\boldsymbol{\mathcal{L}}(\mathbf{x},\mathbf{y})$, and the power $\boldsymbol{\mathcal{K}}^k$ is defined recursively by the condition $\boldsymbol{\mathcal{K}}^1 = \boldsymbol{\mathcal{K}}$ and the kernel of $\boldsymbol{\mathcal{K}}^k$ is $\boldsymbol{\mathcal{K}}_k(\mathbf{x},\mathbf{y}) = \int \boldsymbol{\mathcal{K}}(\mathbf{x},\mathbf{z})\boldsymbol{\mathcal{K}}_{k-1}(\mathbf{z},\mathbf{y})\, d\mathbf{z}$ [882]. The sequence $\{\boldsymbol{\eta}_{(k)}\}$ (4.43) with arbitrary continuous $\boldsymbol{\eta}_{(0)}(\mathbf{x})$ converges to a unique solution $\boldsymbol{\eta}$ if the norm of the integral operator $\boldsymbol{\mathcal{K}}$ turns out to be small "enough" (less than 1) $\|\boldsymbol{\mathcal{K}}\|_{\infty,v_i} \equiv \max_{\mathbf{x}\in v_i} \int |\boldsymbol{\mathcal{K}}(\mathbf{x},\mathbf{y})|\, d\mathbf{y} < 1$ and the problem is reduced to computation of the integrals involved, the density of which is given. In effect the iteration method (4.43) transforms the integral equation problem (4.43) into a linear algebra problem in any case.

The iteration method (4.43) has two known drawbacks. First, the Neumann series ensures the existence of solutions to integral equations of the second kind only for sufficiently small kernels (4.41), and second, in general, it cannot be summed in closed form. Of course Eq. (4.40) can be solved directly by the quadrature method even if this condition of smallness is not valid. However, strongly inhomogeneous problems may lead to much larger numbers of quadrature points, making iteration potentially worthwhile. Moreover, increasing the problem dimensionality (from 2-D to 3-D) raises the number of nodes to the dimensional power and the situation changes radically. As will be shown in the concrete examples, only a few iterations of Eq. (4.43) are necessary; these iterations converge very much faster than a direct inversion of the operator $\mathbf{I} - \boldsymbol{\mathcal{K}}$ by the quadrature method (a comparative analysis of the inversion of the matrices by the iteration and direct methods is presented in [268]).

We now turn to the discussion of boundary integral equations (4.39). This equation can be rewritten in the form (4.40) with obvious subsequent implementation of the method of successive approximations (4.43) and (4.44). This procedure is identical and because of this is not presented. The convergence of successive approximations in elasticity theory has been proved in [650] for the exact evaluation of boundary integrals at each iteration [766], [974]. Perlin [869] investigated the convergence and proposed an effective algorithm to solve the integral equation if the calculations include an error in the accuracy of estimation.

4.3.3 Initial Approximation for Interacting Inclusions in an Infinite Medium

4.3.3.1 A single inclusion subjected to inhomogeneous remote stress. At first, we have considered in Section 4.3 a case of a single ellipsoidal inclusion v_i in an infinite medium subjected to the homogeneous field $\boldsymbol{\sigma}^0(\mathbf{x}) \equiv \boldsymbol{\sigma}^0 = \text{const.}$ which is assumed to be known. The relationships between the averaged tensors $\mathbf{B}_i$, $\mathbf{C}_i$ and $\mathbf{R}_i$, $\mathbf{R}_i$ describing the thermoelastic solution (4.10) and (4.11) were established.

We now turn our attention to the analysis of the inhomogeneous field $\boldsymbol{\sigma}^0(\mathbf{x})$, $\mathbf{x} \in w$. Solving Eq. (4.42) with the initial approximation (4.44) leads to the solution (4.43) by the series of integral operators recurrently coupled. The solution (4.42) obtained inside the region v_i for strain polarization tensor $\boldsymbol{\eta}(\mathbf{x})$ $(\mathbf{x} \in v_i)$ can be rewritten for stress field inside the the inclusion v_i (if $\mathbf{M}_1(\mathbf{x}) \neq \mathbf{0}$) $\boldsymbol{\sigma}(\mathbf{x}) = \mathbf{M}_1^{-1}(\mathbf{x})[\boldsymbol{\eta}(\mathbf{x}) - \boldsymbol{\beta}_1(\mathbf{x})]$, or, taking Eqs. (4.6) and (4.8) into account, the alternative representation for the stress field has the form $\boldsymbol{\sigma}(\mathbf{x}) = \boldsymbol{\sigma}^0(\mathbf{x}) + \int \int \boldsymbol{\Gamma}(\mathbf{x},\mathbf{z})\boldsymbol{\mathcal{L}}_i(\mathbf{z},\mathbf{y})\boldsymbol{\eta}^0(\mathbf{y})\, d\mathbf{y}\, d\mathbf{z}.$ for both inside and outside the inclusion v_i.

For the homogeneous external loading $\boldsymbol{\sigma}^0(\mathbf{x}) = \boldsymbol{\sigma}^0 = \text{const.}$ the integral operator $\boldsymbol{\mathcal{L}}_i$ (4.42) reduces the tensors (4.10) to $(\mathbf{x} \in v_i)$:

$$\mathbf{R}_i(\mathbf{x}) = \bar{v}_i\big(\boldsymbol{\mathcal{L}}_i\mathbf{E}_i\big)(\mathbf{x}), \quad \bar{v}_i\mathbf{F}_i(\mathbf{x}) = \big(\boldsymbol{\mathcal{L}}_i\mathbf{H}_i\big)(\mathbf{x}), \tag{4.45}$$

If $\mathbf{M}_1(\mathbf{x}) \neq \mathbf{0}$ $(\mathbf{x} \in v_i)$ then Eqs. (4.11) and (4.45) lead to the representation of tensors $\mathbf{B}_i(\mathbf{x})$ and $\mathbf{C}_i(\mathbf{x})$ $(\mathbf{x} \in v_i)$:

$$\mathbf{B}_i(\mathbf{x}) = \mathbf{M}_1^{-1}(\mathbf{x})(\boldsymbol{\mathcal{L}}_i\mathbf{E}_i)(\mathbf{x}), \quad \mathbf{C}_i(\mathbf{x}) = \mathbf{M}_1^{-1}(\mathbf{x})\Big[(\boldsymbol{\mathcal{L}}_i\mathbf{H}_i)(\mathbf{x}) - \boldsymbol{\beta}_1(\mathbf{x})\Big]. \tag{4.46}$$

To put this another way, Eqs. (4.45) and (4.46) give the explicit representation of tensors $\mathbf{R}_i(\mathbf{x})$, $\mathbf{F}_i(\mathbf{x})$ and $\mathbf{B}_i(\mathbf{x})$, $\mathbf{C}_i(\mathbf{x})$, respectively, obtained by the use of VIE method for the homogeneous $\boldsymbol{\sigma}^0$.

Moreover, taking Eqs. (4.13) and (4.45) into account we find that for the homogeneous ellipsoidal inclusion v_i the operator $\boldsymbol{\mathcal{L}}_i$ is an identity operator

$$\mathbf{R}_i = \boldsymbol{\mathcal{L}}_i\mathbf{R}_i, \quad \mathbf{F}_i = \boldsymbol{\mathcal{L}}_i\mathbf{F}_i. \tag{4.47}$$

4.3.3.2 Effective fields and the general scheme of the solution for n inclusions in an infinite medium Let us consider a certain set $X = (v_i)$ of inclusions v_i with characteristic functions V_i $(i = 1, \ldots, n)$. To solve the problem through a superposition technique, one replaces the original problem by the superposition of n problems where each involves an infinite space with a single inclusion at the designated location. Then the stress field in the medium is defined from the general integral equations (4.9) and

$$\overline{\boldsymbol{\sigma}}(\mathbf{x}) = \boldsymbol{\sigma}^0(\mathbf{x}) + \sum_{j\neq i} \int \boldsymbol{\Gamma}(\mathbf{x}-\mathbf{y})\boldsymbol{\eta}(\mathbf{y})V_j(\mathbf{y})\, d\mathbf{y}, \tag{4.48}$$

where we defined the effective field $\overline{\boldsymbol{\sigma}}(\mathbf{x})$ ($\mathbf{x} \in v_i, \ldots, v_n$) as a stress field in which the chosen inclusion v_i is embedded. This effective field is a function of all the other positions of the surrounding inhomogeneities, and is a superposition of the remote stress and the disturbances (unknown) caused by the other inclusions of the considered set. No restrictions on the homogeneity of the field $\overline{\boldsymbol{\sigma}}(\mathbf{x})$ were used. The field $\overline{\boldsymbol{\sigma}}(\mathbf{x})$ defined by Eq. (4.48) in the region v_i is nothing more than a notational convenience that may be interpreted as the remote field of the inclusion v_i (analysis of this concept will be considered in Chapter 8). In this field $\overline{\boldsymbol{\sigma}}(\mathbf{x})$, the inclusion v_i behaves as an isolated one, and, therefore, the strain polarization tensor $\boldsymbol{\eta}(\mathbf{x}) = \boldsymbol{\eta}_i(\mathbf{x})$ inside the inclusion v_i can be found in much the same way as a solution of Eq. (4.40):

$$\boldsymbol{\eta}_i = \boldsymbol{\mathcal{L}}_i \overline{\boldsymbol{\eta}}_i, \tag{4.49}$$

where we have introduced the effective strain polarization tensor $\bar{v}_i\overline{\boldsymbol{\eta}}_i = \mathbf{E}_i(\mathbf{x})$ $\overline{\boldsymbol{\sigma}}(\mathbf{x})$ $+\mathbf{H}_i(\mathbf{x})$.

Substituting Eq. (4.49) into (4.48) leads to

$$\overline{\boldsymbol{\eta}}_i = \boldsymbol{\eta}_i^0 + \sum_{j\neq i} \boldsymbol{\mathcal{L}}_{ij}^{\text{out}} \overline{\boldsymbol{\eta}}_j, \tag{4.50}$$

where the kernel of the "outer" integral operator $\boldsymbol{\mathcal{L}}_{ij}^{\text{out}}$ is $\boldsymbol{\mathcal{L}}_{ij}^{\text{out}}(\mathbf{x},\mathbf{y}) = \mathbf{E}_i(\mathbf{x})$ $\int \boldsymbol{\Gamma}(\mathbf{x}-\mathbf{z})\boldsymbol{\mathcal{L}}_j(\mathbf{z},\mathbf{y})V_j(\mathbf{z})\, d\mathbf{z}$ where $\mathbf{y},\ \mathbf{z} \in v_j,\ \mathbf{x} \in v_i;\ i \neq j$.

In principle, the system of Eqs. (4.50) can be analyzed directly, using different numerical methods, but multiparticle problems that involve many inhomogeneities require a great deal of computer time. As mentioned for a general case of a single inhomogeneity the iteration schemes are usually more efficient. We now define a sequence of approximations $\overline{\boldsymbol{\eta}}_{i(k)}$ $(k \to \infty)$ to $\overline{\boldsymbol{\eta}}_i$ as follows:

$$\overline{\boldsymbol{\eta}}_{i(k+1)} = \boldsymbol{\eta}_i^0 + \sum_{j\neq i} \boldsymbol{\mathcal{L}}_{ij}^{\text{out}} \overline{\boldsymbol{\eta}}_{j(k)}, \quad (i = 1, \ldots, n;\ k = 0, 1, \ldots). \tag{4.51}$$

The performance of the iterative scheme proposed can be improved by choice of the initial approximation $\overline{\boldsymbol{\eta}}_{i(0)}$ of the solution of Eq. (4.51) proposed below.

4.3.3.3 Initial approximation for n inclusions in an infinite medium. In the traditional scheme of iteration methods the driving term of the integral equation involved is taken as the initial approximation:

$$\boldsymbol{\eta}_{i(0)}(\mathbf{x}) = \boldsymbol{\eta}_i^0(\mathbf{x}), \quad (i = 1, \ldots, n), \tag{4.52}$$

or, expressing in terms of the interaction problem, the iterative procedure is initiated by assuming that in each inclusion the effective strain polarization tensor is induced by the remote field as if the neighbors are absent. As will be shown in the numerical examples the choice of the initial approximation on the basis of an approximate analytical solution leads to a significant increase of the accuracy in the first steps of an iterative procedure.

With this in mind let us rewrite Eq. (4.48) in the spirit of a subtraction technique ($\mathbf{x} \in v_i$):

$$\overline{\boldsymbol{\sigma}}(\mathbf{x}) = \boldsymbol{\sigma}^0(\mathbf{x}) + \sum_{j\neq i} \mathbf{T}_j(\mathbf{x}-\mathbf{x}_j)\boldsymbol{\xi}_j + \sum_{j\neq i} \int \boldsymbol{\Gamma}(\mathbf{x}-\mathbf{y})\big[\boldsymbol{\eta}(\mathbf{y}) - \boldsymbol{\xi}_j\big]V_j(\mathbf{y})\, d\mathbf{y}. \quad (4.53)$$

Equation (4.53) is valid for any constant tensors $\boldsymbol{\xi}_j$ $(i, j = 1, \dots, n;\ j \neq i)$. For definitiveness, we shall use $\bar{v}_j \boldsymbol{\xi}_j = \mathbf{R}_j \langle \overline{\boldsymbol{\sigma}} \rangle_j + \mathbf{F}_j$ $(j \neq i)$ that can be considered as an approximation of strain polarization tensors in the inclusions v_j in the spirit of the Saint-Venant principle for constant tensors:

$$\boldsymbol{\eta}(\mathbf{y}) \equiv \boldsymbol{\xi}_j, \ (\mathbf{y} \in v_j), \quad (4.54)$$

which can be formulated in an alternative form (see effective field hypothesis (9.21) in Chapter 8):

$$\int \boldsymbol{\Gamma}(\mathbf{x}-\mathbf{y})\boldsymbol{\eta}(\mathbf{y})V_j(\mathbf{y})\, d\mathbf{y} = \mathbf{T}_j(\mathbf{x}-\mathbf{x}_j)(\mathbf{R}_j\langle\overline{\boldsymbol{\sigma}}\rangle_j + \mathbf{F}_j), \ (\mathbf{x} \notin v_j), \quad (4.55)$$

where the tensors $\mathbf{T}_j(\mathbf{x}-\mathbf{x}_j)$ (3.131) and $\mathbf{T}_{ij}(\mathbf{x}_i-\mathbf{x}_j)$, $\mathbf{T}^{\epsilon}_j(\mathbf{x}-\mathbf{x}_j)$, $\mathbf{T}^{\epsilon}_{ij}(\mathbf{x}_i-\mathbf{x}_j)$ introduced here and below ($\mathbf{y} \in v_j$, $\mathbf{x} \notin v_j$):

$$\mathbf{T}_j(\mathbf{x}-\mathbf{x}_j) = \langle \boldsymbol{\Gamma}(\mathbf{x}-\mathbf{y})\rangle_j, \quad \mathbf{T}_{ij}(\mathbf{x}_i-\mathbf{x}_j) = \langle \mathbf{T}_j(\mathbf{x}-\mathbf{x}_j)\rangle_i,$$
$$\mathbf{T}^{\epsilon}_j(\mathbf{x}-\mathbf{x}_j) = \langle \mathbf{U}(\mathbf{x}-\mathbf{y})\rangle_j, \quad \mathbf{T}^{\epsilon}_{ij}(\mathbf{x}_i-\mathbf{x}_j) = \langle \mathbf{T}^{\epsilon}_j(\mathbf{x}-\mathbf{x}_j)\rangle_i \quad (4.56)$$

originated in [1189], and have the analytical representation for d-dimensional ($d = 2, 3$) spherical inclusions in an isotropic matrix presented in Appendix A.3.1. For two identical inclusions the matrix $\mathbf{T}_{ij}$ is symmetric, and $\mathbf{T}_j(\mathbf{x}-\mathbf{x}_j) = \lim \mathbf{T}_{ij}(\mathbf{x}_i - \mathbf{x}_j)$ as $\overline{v}_i \to 0$ and $\mathbf{x} \in v_i$,

In the framework of the hypothesis (4.55) a zero-order approximation of the average effective stresses can be estimated from Eqs. (4.53) and (4.54) $\langle\overline{\boldsymbol{\sigma}}\rangle_{i(0)} = \langle\boldsymbol{\sigma}^0\rangle_i + \sum_{j\neq i} \mathbf{T}_{ij}(\mathbf{x}-\mathbf{x}_j)\big[\mathbf{R}_j\langle\overline{\boldsymbol{\sigma}}\rangle_{j(0)} + \mathbf{F}_j\big]$ leading to explicit representation

$$\bar{v}_i \boldsymbol{\eta}_{i(0)} = \mathbf{R}_i\langle\overline{\boldsymbol{\sigma}}\rangle_{i(0)} + \mathbf{F}_i = \sum_{j=1}^{n} \mathbf{Z}_{ij}\big[\mathbf{R}_j\langle\boldsymbol{\sigma}^0\rangle_j + \mathbf{F}_j\big], \quad (4.57)$$

where the matrix $\mathbf{Z}^{-1}$ has elements $(\mathbf{Z}^{-1})_{ij}$ $(i, j = 1, \dots, n)$ (the elements represent fourth-order tensors; no sum i, j)

$$(\mathbf{Z}^{-1})_{ij} = \mathbf{I}\delta_{ij} - (1-\delta_{ij})\mathbf{R}_j\mathbf{T}_{ij}(\mathbf{x}_i - \mathbf{x}_j). \quad (4.58)$$

The representation for the strain polarization tensor (4.57) can be rewritten as

$$\bar{v}_i \boldsymbol{\eta}_{i(0)} = \bar{v}_i \boldsymbol{\eta}^0_i + \mathbf{R}_i \sum_{j\neq i}^{n} \mathbf{T}_{ij}(\mathbf{x}_i - \mathbf{x}_j) \sum_{k=1}^{n} \mathbf{Z}_{jk}\boldsymbol{\xi}^0_k \bar{v}_k, \quad (4.59)$$

where $\bar{v}_i\boldsymbol{\xi}_i^0 = \mathbf{R}_i\langle\boldsymbol{\sigma}^0\rangle_i + \mathbf{F}_i$.

By this means we have estimated the average strain polarization tensors (4.57) obtained in the framework of the homogeneity hypothesis of effective fields inside the inclusions: $\overline{\boldsymbol{\sigma}}(\mathbf{x}) = \langle\overline{\boldsymbol{\sigma}}\rangle_j$ $(\mathbf{x} \in v_j)$. Substituting (4.57) in (4.10$_2$) leads to the inhomogeneous strain polarization tensors obtained in the framework of the same homogeneity hypothesis of effective fields (4.55):

$$\bar{v}_i\boldsymbol{\eta}_{i(0)}(\mathbf{x}) = \mathbf{R}_i(\mathbf{x})\mathbf{R}_i^{-1}\Big(\sum_{j=1}^n \mathbf{Z}_{ij}\boldsymbol{\xi}_j^0\bar{v}_j - \mathbf{F}_i\Big) + \mathbf{F}_i(\mathbf{x}). \tag{4.60}$$

The initial approximations (4.59) and (4.60) will be used at the analysis of interacting inclusions of commensurable sizes; otherwise at the analysis of a macroinclusion we will use the initial approximation (4.52) (for details see Section 4.5).

If the number of inclusions is rather large (say 10^{2d-1}) the iteration methods can be more powerful than the direct method of the inversion of the $3(d-1)n \times 3(d-1)n$ matrix (4.58). Then

$$\boldsymbol{\eta}_{i(0)} = \mathbf{R}_i\langle\overline{\boldsymbol{\sigma}}\rangle_{i(0)} + \mathbf{F}_i = \sum_{j=1}^n\sum_{k=0}^\infty \hat{\mathbf{Z}}_{ij}^k\boldsymbol{\xi}_j^0\bar{v}_j \tag{4.61}$$

where

$$\hat{\mathbf{Z}}_{ij}^0 = \mathbf{I}\delta_{ij}, \quad \hat{\mathbf{Z}}_{ij}^1 = \bar{\mathbf{Z}}_{ij}, \tag{4.62}$$

$$\hat{\mathbf{Z}}_{ij}^k = \sum_{i_1=1}^n \ldots \sum_{i_k=1}^n \bar{\mathbf{Z}}_{ii_1}\ldots\bar{\mathbf{Z}}_{i_kj}, \qquad \bar{\mathbf{Z}}_{i_pi_q} = \mathbf{R}_{i_q}\mathbf{T}_{i_pi_q}(\mathbf{x}_{i_p} - \mathbf{x}_{i_q}), \tag{4.63}$$

where $k = 2,\ldots;\ i_p, i_q = 1,\ldots,n$. Thus the iteration method (4.61)–(4.63) does not require inversion of the large $3(d-1)n \times 3(d-1)n$ matrix (4.57), instead the many $3(d-1)\times 3(d-1)$ matrices (4.63$_2$) have to be multiplied.

We turn our attention to a discussion of different simplified assumptions for obtaining zero-order approximations (4.57) which are known for the analysis of two interacting homogeneous ellipsoidal inclusions.

The so-called *point approximation* of interacting inclusions [618] is exact for infinitely spaced heterogeneities:

$$\mathbf{T}_{ij}(\mathbf{x} - \mathbf{x}_j) = \mathbf{T}_j(\mathbf{x}_i - \mathbf{x}_j) = \boldsymbol{\Gamma}(\mathbf{x}_i - \mathbf{x}_j), \tag{4.64}$$

and, therefore, $(\mathbf{Z}^{-1})_{ij} = \mathbf{I}\delta_{ij} - (1-\delta_{ij})\mathbf{R}_j\boldsymbol{\Gamma}(\mathbf{x}_i - \mathbf{x}_j)$.

The *approximation of widely space inclusions* (see, e.g., [528], [529]) is defined by a quite restrictive geometrical inequality

$$\varepsilon_r = (\text{inclusion sizes})/(\text{spacing between inclusions}) \ll 1, \tag{4.65}$$

leading to the solution

$$\bar{v}_i\boldsymbol{\eta}_{i(0)} = \bar{v}_i\boldsymbol{\eta}_i^0 + \mathbf{R}_i\sum_{j\neq i}^n \boldsymbol{\Gamma}(\mathbf{x}_i - \mathbf{x}_j)\boldsymbol{\xi}_j^0\bar{v}_j \tag{4.66}$$

Other simplifications can be obtained as a truncation of the expansion (4.61). So, the iteration scheme (4.61)–(4.63) was used in [156] in the analysis of two interacting homogeneous ellipsoidal inclusions, while Willis and Acton [1189] have considered two first iterations of the scheme (4.62) for two spherical homogeneous inclusions; their approximation can be rephrased in our notations as

$$\begin{aligned} \mathbf{Z}_{ij} &= \mathbf{I}\delta_{ij} + (1-\delta_{ij})\mathbf{R}_j\mathbf{T}_{ij}(\mathbf{x}_i-\mathbf{x}_j), \\ \bar{v}_i\boldsymbol{\eta}_{i(0)} &= \bar{v}_i\boldsymbol{\eta}_i^0 + \big[\mathbf{I}+\mathbf{R}_j\mathbf{T}_{ij}(\mathbf{x}_i-\mathbf{x}_j)\big]\mathbf{R}_j\mathbf{T}_{ij}(\mathbf{x}_i-\mathbf{x}_j)\boldsymbol{\xi}_j^0\bar{v}_j. \end{aligned} \tag{4.67}$$

The right-hand side of Eq. (4.67_2) is a sum of terms of order $O(\varepsilon_r^\alpha)$ (if $\varepsilon_r \ll 1$), where the nonnegative integer parameter α varies from 0 to $2d+4$. Then the *asymptotic approximation* can be obtained by truncation of terms having the order less than $O(\varepsilon_r^{d+2})$:

$$\bar{v}_i\boldsymbol{\eta}_{i(0)} = \bar{v}_i\boldsymbol{\eta}_i^0 + \big[\mathbf{I}+\mathbf{R}_j\boldsymbol{\Gamma}(\mathbf{x}_i-\mathbf{x}_j)\big]\mathbf{R}_j\mathbf{T}_{ij}(\mathbf{x}_i-\mathbf{x}_j)\boldsymbol{\xi}_j^0\bar{v}_j. \tag{4.68}$$

For two identical spherical homogeneous inclusions subjected to homogeneous external loading, Eq. (4.68) was obtained by much lengthier means in [515] (3-D case) and in [525] (2-D case).

At last the simplest initial approximation (4.52) in the framework of effective field hypothesis (4.54) follows from Eq. (4.67_2) by neglecting of the second summand in the square brackets in the right-landside of Eq. (4.67_2). All related simplified methods mentioned above yield only a moderate reduction of computer costs and can be derived in the framework of the method proposed (4.57).

4.3.4 First-Order and Subsequent Approximations

To obtain the first-order approximation of strain polarization tensors $\overline{\boldsymbol{\eta}}_{i(1)}(\mathbf{x})$ ($\mathbf{x} \in v_i,\ i = 1,\ldots,n$) we should substitute (4.58) (or (4.52)) into (4.51) which leads to the necessity of evaluation of integrals in the right-landside of Eq. (4.51) if $\overline{\boldsymbol{\eta}}_{i(0)}(\mathbf{x}) \neq$ const. The first iteration (4.51) can be estimated analytically if $\overline{\boldsymbol{\eta}}_{i(0)}(\mathbf{x}) =$ const., ($\mathbf{x} \in v_i$) that occurs exactly for homogeneous ellipsoidal inclusions. Because of this, for the sake of simplicity we will assume that the first-order approximation of a perturbation of the effective field induced by the inhomogeneous neighbors (4.57) (or (4.52)) is the same as the perturbation induced by the homogeneous neighbors with fictitious properties (3.189). Then, taking (4.47) into account, we can obtain the first-order approximation of the effective strain polarization tensor in inclusion $\mathbf{x} \in v_i$ for different initial approximations.

For example, choosing the simplest initial approximation (4.52) leads to the first-order approximation:

$$\overline{\boldsymbol{\eta}}_{i(1)}(\mathbf{x}) = \boldsymbol{\eta}_i^0(\mathbf{x}) + \mathbf{E}_i(\mathbf{x})\sum_{j\neq i}^{n}\mathbf{T}_j(\mathbf{x}-\mathbf{x}_j)\boldsymbol{\xi}_j^0\bar{v}_j. \tag{4.69}$$

Substituting zero-order approximation (4.57) considering interacting effects between inclusions into (4.51) leads to another representation of the first-order approximation:

$$\overline{\boldsymbol{\eta}}_{i(1)}(\mathbf{x}) = \boldsymbol{\eta}_i^0(\mathbf{x}) + \mathbf{E}_i(\mathbf{x})\sum_{j\neq i}^{n}\mathbf{T}_j(\mathbf{x}-\mathbf{x}_j)\sum_{k=1}^{n}\mathbf{Z}_{jk}\boldsymbol{\xi}_k^0\bar{v}_k. \tag{4.70}$$

Substituting (4.70) (or (4.69)) into (4.51) yields the first-order approximation for the strain polarization tensor inside the inclusion v_i

$$\boldsymbol{\eta}_{i(1)}(\mathbf{x}) = \big(\boldsymbol{\mathcal{L}}_i\overline{\boldsymbol{\eta}}_{i(1)}\big)(\mathbf{x}). \tag{4.71}$$

In such a manner, at the estimation of the first-order approximation of the effective strain polarization tensors $\overline{\boldsymbol{\eta}}_{i(1)}(\mathbf{x})$ one first replaces the non-uniform effective fields in the integrand of Eq. (4.53) by their average values. After solving n average effective fields from n linear self-consistent equations (4.70), one is able to evaluate the non-uniform effective field of the first-order approximation (4.71) and thereupon one estimates the strain polarization tensors $\boldsymbol{\eta}_{i(1)}(\mathbf{x})$ (4.71) inside each inclusion being considered.

It should be mentioned that the derivation of the relation (4.70) is based on the hypothesis equivalent to the assumption proposed in [529] for interacting cracks if (and only if) the inclusions are the homogeneous and have an ellipsoidal shape, and subject to a remote homogeneous stress field. This assumption can be formulated in an alternative form. The initial problem for n inhomogeneities can be represented by n problems for an isolated single inclusion inside the infinite medium subject to a remote stress field induced by the remaining inhomogeneities and the real remote stress. The perturbation induced by remaining inclusions to the remote field of a reference inclusion is defined by average equivalent strain polarization tensors. In light of the analogy mentioned, the first-order approximation (4.69) is called *approximation of weak interaction* which can be interpreted as a truncation of "feedbacks": transmission of the $\boldsymbol{\sigma}^0$-induced effective fields from the jth inclusion on the ith inclusion is not followed by feedback.

The iterative solution strategy with the simplest initial approximation (4.69) for the homogeneous ellipsoidal inclusions subjected to mechanical loading ($\boldsymbol{\beta}_1 \equiv \mathbf{0}$) was considered in [148] where this approximation is called as *a simple method of analysis*. The relations (4.70) and (4.71) generalize a similar equation proposed in [941] under the analysis of two spherical cavities subjected to homogeneous remote stress $\boldsymbol{\sigma}^0$. After evaluation of the effective field $\overline{\boldsymbol{\sigma}}_{i(1)}$ (4.70) at the surface of cavities, Rodin and Hwang [941] have solved the problem for a single isolated inclusion by the finite element analysis using the basic idea of Eshelby. However, the field $\overline{\boldsymbol{\eta}}_{i(1)}(\mathbf{x})$ (and, therefore $\overline{\boldsymbol{\sigma}}_{i(1)}(\mathbf{x})$) is defined inside the domain $\mathbf{x}\in v_i$ $(i\neq j,\ j=1,\ldots,n)$, and the use of the finite element analysis becomes too involved as the inclusions themselves become inhomogeneous. At the same time, there is no change in the basic formulation for the method proposed (4.70) and (4.71) for the cavities and the inhomogeneous inclusions.

Obtaining the next approximations is carried out in the framework of a standard schema of the iteration method:

$$\overline{\boldsymbol{\eta}}_{i(k+1)}(\mathbf{x}) = \boldsymbol{\eta}_i^{(0)}(\mathbf{x}) + \sum_{j\neq i}\big(\boldsymbol{\mathcal{L}}_{ij}^{\text{out}}\overline{\boldsymbol{\eta}}_{j(k)}\big)(\mathbf{x}), \quad \boldsymbol{\eta}_{i(k+1)}(\mathbf{x}) = \big(\boldsymbol{\mathcal{L}}_i\overline{\boldsymbol{\eta}}_{i(k+1)}\big)(\mathbf{x}), \tag{4.72}$$

where $\mathbf{x}\in v_i,\ i=1,\ldots,n$.

4.3.5 Numerical Results for Two Cylindrical Inclusions in an Infinite Matrix

The method proposed can be applied to arbitrary packing sequences and shapes of the inclusions. With the non-essential restriction on space dimensionality d and the shape of inclusion we will consider 2-D problems for circular inclusions. The domains of small inclusions v_i $(i = 2, \ldots, n)$ are discretized along the polar angle and the radius in the local polar coordinate system with the centers $\mathbf{x}_i$. Then the points

$$\Big\{(r, \varphi) \mid (p-1)\frac{2\pi}{l} < \varphi < p\frac{2\pi}{l},\ (q-1)\frac{a_i}{m} < r < q\frac{a_i}{m}\Big\} \tag{4.73}$$

$(p = 1, 2 \ldots, l;\ q = 1, 2, \ldots, m)$ represent the elements of Γ_i^{pq} of the meshes Ω_i $(i = 2, \ldots, n)$; this meshes are not optimized, but are efficient. Moreover, the square meshes

$$\Big\{(x_1, x_2)^\top \mid (p-1)\frac{a_i}{l} < x_1 < p\frac{a_i}{l},\ (q-1)\frac{a_i}{l} < x_2 < q\frac{a_i}{l}\Big\}, \tag{4.74}$$

where x_1, x_2 are local coordinates with origins at the fiber centers will be analyzed. We will use piecewise-constant elements of the meshes, which are not very cost efficient but are very easy for computer programming, and the discretization (4.74) permits analysis of nonregular inclusion shapes. For simplicity in estimation of integrals involved we will utilize the basic numerical integrations formulas of Simpson's rule and trapezoidal rule for the uniform (4.73) and (4.74) and nonuniform meshes considered below, respectively.

In order to compare the calculated results with those obtained by other methods, a simple example is considered. In order to check the accuracy of the VIE the influence of the interaction between two circular inclusions (e.g., fibers) with the same radius a in an infinite matrix on the stress field in the inclusions is studied. Let the matrix be titanium Ti ($k^{(0)} = 92.6$GPa, $\mu^{(0)} = 37.9$GPa) which contains two circular SiC ($k^{(1)} = 293.3$GPa, $\mu^{(1)} = 176$GPa) fibers (see, e.g., [248]). Let us consider a plane-strain problem and the pair of inclusions v_1 and v_2 loaded by tension $\boldsymbol{\sigma}_{ij}^0 = \sigma_{11}^0\delta_{1i}\delta_{1j}$ along the line linking the inclusion centers $\mathbf{x}_2 = (Da, 0)^\top$ and $\mathbf{x}_1 = (0, 0)^\top$; D is a normalized distance between the inclusion centers. In the domains v_1 and v_2 we will use the meshes (4.73) and (4.74). Both meshes (4.73) and (4.74) give similar results. In Fig. 4.4 the numerical result obtained by the method proposed, by the VIE method [729] (see also [656], [657], [658]) as well as by FEA are presented. As can be seen in Fig. 4.4 the elastic stress concentration factor $\sigma_{11}(\mathbf{z})/\sigma_{11}^0$ at the point $\mathbf{z} = \arg\min|\mathbf{x} - \mathbf{y}|$ ($\mathbf{x}, \mathbf{z} \in v_2$, $\mathbf{y} \in v_1$) can be quite strong for inclusion spacing $D \approx 2$. As Fig. 4.4 suggests, the first-order approximation (4.49) and (4.52), and especially (4.49) and (4.59), may actually provide a reasonably accurate description of the stress field inside interacting inclusions. The method proposed, the VIE method [656], [657], as well as FEA produce almost identical results.

Up to now we explicitly used the Eshelby solution (3.120)–(3.122) for a single homogeneous ellipsoidal inclusion (3.187) subjected to homogeneous remote stress $\boldsymbol{\sigma}^0 = \text{const}$. Let us change the circular inclusions in the pair being considered by the identical inclusions of another shape described by equation $|x_1|^\alpha + |x_2|^\alpha = 1$ in the Cartesian coordinate system connected to the axes of the symmetry. At the estimation of zero-order approximation (4.60) the average concentrator of the strain polarization tensor $\mathbf{R}$ (4.13) is calculated by the use of the scheme (4.18) and a concept of a fictitious ellipsoidal homogeneous inclusion (3.189). As this takes place, the estimation of the stress concentrator tensor $\mathbf{B}_i(\mathbf{x})$ (4.18) reduced to the evaluation of stress distribution inside the domain v_i. So the stress distributions $\sigma_{11}(\mathbf{x})/\sigma_{11}^0$ in the cross sections $x_2 \equiv 0$ and $x_1 \equiv 0$ of the single isolated inclusion v_1 are represented in Fig. 4.5 for $\alpha = 1.5$, 2 and $\alpha = 6$. Buryachenko [140] calculated normalized stresses (the first (4.70) and (4.71) and the fifth iterations (4.71) and (4.72$_1$)) for $\mathbf{x}_2 - \mathbf{x}_1 = (2.4a, 0)^\top$; $\alpha = 6$. Once again, the first-order approximation provides an acceptable accuracy if $D \geq 2.4$.

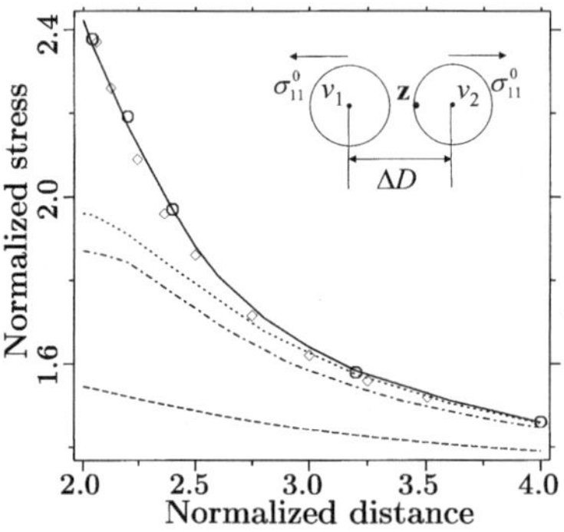

Fig. 4.4. Stress concentration factor $\sigma_{11}(\mathbf{z})/\sigma_{11}^0$ vs normalized distance D: FEA (○), 6th iteration (4.72) (solid curve), VIE method [656] (◇), the first-order approximation (4.49) and (4.59) (dotted curve), the first-order approximation (4.49) and (4.52) (dot-dashed curve) the initial approximation (4.59) (dashed curve).

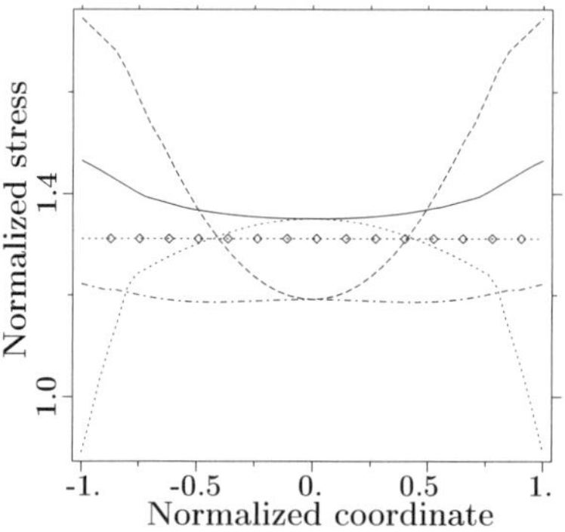

Fig. 4.5. The stress distribution $\sigma_{11}(\mathbf{x})/\sigma_{11}^0$ in the cross sections $x_2 \equiv 0$ (solid and dot-dashed) and $x_1 \equiv 0$ (dot and dashed curves) of a single isolated inclusion v_1 vs x_1/a and x_2/a, respectively. $\alpha = 1.5$ (solid and dotted curves), $\alpha = 6$ (dashed and dot-dashed curves), and $\alpha = 2$ (◇).

Moreover, the consideration of small inclusions with the nonsmooth shape presents no additional difficulties in the realization of the approach proposed. So, Buryachenko [134] (see Chapter 13) considered a particular case of the current approach proposed and applied it to a collinear array of semiinfinite (and because of this, nonperiodic) field of microcracks in an infinite medium.

4.4 Hybrid VEE and BIE Method for Multiscale Analysis of Interacting Inclusions (Macro Problem)

4.4.1 Initial Approximation for the Fields Induced by a Macroinclusion

Up to now the general case of inclusions in the region w bounded by the surface Γ was analyzed without any restrictions on the nature of the region w. If this region is finite, we can consider it as a surface consisting of $n+1$ simple closed surfaces, such that the exterior surface Γ_{n+1}, contains n surfaces Γ_i $(i = 1, \ldots, n)$. If the surface Γ_{n+1} extends to infinity, then the region w occupies the full space R^d. This infinite region, however, is bounded by the interior surfaces $\Gamma_1, \ldots, \Gamma_n$. As an example of our multiscale analysis we will consider the case of an infinitely remote surface Γ_{n+1} assuming that the inclusion v_1 is macroscopically large and the other inclusions $v_2, \ldots, v_n$ are small: $diam\, v_1 \gg diam\, v_i$, $(i = 2, \ldots, n)$. The case of few large inclusions and/or bounded region w can be analyzed in a similar manner. Then the field $\boldsymbol{\sigma}^{00}(\mathbf{x})$ which would exist in the infinite region with homogeneous properties $\mathbf{L}^{(0)}$ $(\mathbf{L}_1(\mathbf{x}), \boldsymbol{\beta}_1(\mathbf{x}) \equiv \mathbf{0})$ and appropriate boundary condition at infinity is assumed to be known.

Let us consider the problem (called the *macro problem*) of estimation of a stress field $\boldsymbol{\sigma}^0(\mathbf{x})$ for the stress field in a full space R^d with the large inclusion v_1. This field can be estimated by different numerical methods. In particular, in the case of the use of the VIE method in Eq. (4.36) we have:

$$\boldsymbol{\eta}^0(\mathbf{x}) = \boldsymbol{\eta}_1^{00}(\mathbf{x}) + \int \mathbf{K}_1(\mathbf{x}, \mathbf{y})\big[\boldsymbol{\eta}^0(\mathbf{y}) - \boldsymbol{\eta}^0(\mathbf{x})\big]\, d\mathbf{y}, \quad \mathbf{x} \in v_1. \tag{4.75}$$

For the homogeneous inclusion v_1 the BIE method is preferable and the displacement field $\mathbf{u}^0(\boldsymbol{\zeta})$ $(\boldsymbol{\zeta} \in \Gamma_1)$ can be found from Eq. (4.39) for the external problem

$$\mathbf{u}^0(\boldsymbol{\zeta}) = \mathbf{u}_1^{00}(\boldsymbol{\zeta}) + \int_{\Gamma_1} \mathbf{G}(\boldsymbol{\zeta} - \mathbf{s})\underline{\mathbf{t}}^0(\mathbf{s})\, ds - \int_{\Gamma_1} \mathbf{T}(\boldsymbol{\zeta}, \mathbf{s})[\mathbf{u}^0(\mathbf{s}) - \mathbf{u}^0(\boldsymbol{\zeta})]\, ds, \ \boldsymbol{\zeta} \in \Gamma_1 \tag{4.76}$$

which should be accompanied by an analogous equation for the internal problem for the inclusion v_1. The problem (4.76) is simplified for both the first (4.1) $(\Gamma_u = \Gamma)$ and second (4.2) $(\Gamma_t = \Gamma)$ boundary value problems when one of the two integrals on the right-landside of Eq. (4.76) is known.

Equations (4.75) and (4.76) can both be solved by the iteration method presented in Subsection 4.3.2. In so doing the mesh used for the numerical solution will be called macro-mesh Ω^0; the size of mesh cells is compatible with the length scale of the inhomogeneity of the field $\boldsymbol{\sigma}^{00}(\mathbf{x})$ $(\mathbf{x} \in v_1)$ as well as with the curvature of the surface Γ_1 and with the length scale of the inhomogeneity of mechanical properties $\mathbf{L}_1(\mathbf{x})$, $\boldsymbol{\beta}_1(\mathbf{x})$ $(\mathbf{x} \in v_1)$. The solution of this macro problem is an inexpensive operation since the fine-scale features induced by small inclusions are eliminated. Moreover, for the homogeneous ellipsoidal inclusion subjected to polynomial field $\boldsymbol{\sigma}^{00}(\mathbf{x})$ the stress field inside the inclusion has the analytical Eshelby's [319] representation; in particular, for the homogeneous field $\boldsymbol{\sigma}^{00}(\mathbf{x}) \equiv \text{const}$, the stress field mentioned is described by Eq. (4.10_1) with the constant stress concentration factor $\mathbf{B}_1$ (3.121) and $\mathbf{C}_1$ (3.122).

Thus, if the stress field $\boldsymbol{\sigma}^0(\mathbf{x})$ $(\mathbf{x} \in v_1)$ inside the macro inclusion v_1 is found, then the stress field in the matrix in the vicinity of the inclusion v_1 is expressed by the formula

$$\boldsymbol{\sigma}^-(\mathbf{n}_1) = \mathbf{B}_1(\mathbf{n}_1)\boldsymbol{\sigma}^+(\boldsymbol{\zeta}) + \mathbf{C}_1(\mathbf{n}_1), \tag{4.77}$$

where $\boldsymbol{\sigma}^-(\mathbf{n}_1)$ and $\boldsymbol{\sigma}^+(\boldsymbol{\zeta})$ are the limiting stress outside and inside, respectively, near the inclusion boundary Γ_1: $\boldsymbol{\sigma}^-(\mathbf{n}_1) = \lim \boldsymbol{\sigma}^0(\mathbf{y})$, $\boldsymbol{\sigma}^+(\boldsymbol{\zeta}) = \lim \boldsymbol{\sigma}^0(\mathbf{z})$, $\mathbf{y} \to \boldsymbol{\zeta}$,

$\mathbf{z} \to \boldsymbol{\zeta}$; $\mathbf{y} \notin v_1$, $\mathbf{z} \in v_1$, $\boldsymbol{\zeta} \in \Gamma_1, \mathbf{n}_1 \bot \Gamma_1$; $\mathbf{n}_1$ is the unit outward normal vector on Γ_1, where the interior region is v_1. The tensors $\mathbf{B}_1(\mathbf{n}_1)$ and $\mathbf{C}_1(\mathbf{n}_1)$ are defined only by the elastic properties of the contacting materials and by the direction of the normal $\mathbf{n}_1$ (see Eq. (3.101)). Eq. (4.77) is also valid for a hole (when $\mathbf{L}^{(1)} \to \mathbf{0}$). To accomplish this it will suffice to carry out the estimations with a reasonably soft inclusion, for example: $\mathbf{L}^{(1)} = 10^{-4}\mathbf{L}^{(0)}$. In the case of the use of the BIE method (4.76) the stress field $\boldsymbol{\sigma}^-(\mathbf{n}_1)$ can be estimated by the use of traction and tangential derivatives of displacement fields over the entire boundary [28].

By this means the stress field in the matrix in the vicinity of the macro inclusion v_1 in the absence of small inclusions (initial approximation) $\boldsymbol{\sigma}^0_{(0)}(\mathbf{x}) \equiv \boldsymbol{\sigma}^0(\mathbf{x})$ is found and can be used to obtain an initial approximation of strain polarization tensors in the small inclusions $\boldsymbol{\eta}_{i(0)}(\mathbf{x})$ (4.60) ($i = 2, \dots, n$). In so doing for the case $diam\, v_1 \gg diam\, v_i$ ($i = 2, \dots, n$) being considered in this paper, the field $\boldsymbol{\sigma}^0_{(0)}(\mathbf{x})$ ($\mathbf{x} \in v_i$) is homogeneous inside the small inclusions v_i, ($i = 2, \dots, n$).

4.4.2 Initial Approximation in the Micro Problem

The estimation of stress fields induced by small inclusions will be called the *micro problem*. In so doing the perturbation field $\mathbf{u}^1_{(0)}(\boldsymbol{\zeta})$ (4.39) can be estimated analytically if $\overline{\boldsymbol{\eta}}_{i(0)}(\mathbf{x}) = \text{const.}$, ($\mathbf{x} \in v_i$, $i = 2, \dots n$) that takes place exactly for a homogeneous ellipsoidal inclusion in an infinite medium. Because of this, for the sake of simplicity we will assume that the zero-order approximation of a perturbation induced by the small inhomogeneities (4.57) (or (4.52)) is the same as the perturbation induced by the fictitious homogeneous ellipsoidal neighbors with the properties (3.189). Hereafter, we will use the approximation (4.57) because its advantage as compared with (4.52) has been determined with the problem of two inclusions in an infinite medium in [140].

Specifically, for the homogeneous ellipsoidal large and small inclusions subjected to a homogeneous stress field $\boldsymbol{\sigma}^{00} \equiv \text{const.}$ at infinity, the initial approximation for the fields induced by the macroinclusion has an analytical representation. So, the displacement field $\mathbf{u}^0_{(0)}(\boldsymbol{\zeta})$ at the boundary, the initial approximation of the stresses in the matrix in the vicinity of the large inclusions $\boldsymbol{\sigma}^-_{(0)}(\mathbf{n}_1)$ and inside the small inclusions $\boldsymbol{\sigma}_{i(0)}(\mathbf{x})$ ($\mathbf{x} \in v_i$, $i > 1$), as well as the effective stress averaged over the volume of the small inclusions $\overline{\boldsymbol{\sigma}}_{j(0)}$ can be found by the use of analytical representations, (3.120), (4.55), (4.57), and (4.77):

$$\mathbf{u}^0_{(0)}(\boldsymbol{\zeta}) = \Big[\mathbf{M}^{(1)}(\mathbf{B}_1\boldsymbol{\sigma}^{00} + \mathbf{C}_1) + \boldsymbol{\beta}^{(1)}_1\Big](\boldsymbol{\zeta} - \mathbf{x}_1), \quad (\boldsymbol{\zeta} \in \Gamma_1), \tag{4.78}$$

$$\boldsymbol{\sigma}^-_{(0)}(\mathbf{n}^{(i)}_1) = \mathbf{B}_1(\mathbf{n}^{(i)}_1)(\mathbf{B}_1\boldsymbol{\sigma}^{00} + \mathbf{C}_1) + \mathbf{C}_1(\mathbf{n}^{(i)}_1), \tag{4.79}$$

$$\boldsymbol{\sigma}_{i(0)}(\mathbf{x}) = \mathbf{M}^{(i)}_1\Big[\sum_{j=2}^{n}\mathbf{Z}_{ij}\big[\mathbf{R}_j\overline{\boldsymbol{\sigma}}_{j(0)} + \mathbf{F}_j\big] - \mathbf{F}_i\Big] + \mathbf{C}_i, \quad (\mathbf{x} \in v_i), \tag{4.80}$$

$$\overline{\boldsymbol{\sigma}}_{j(0)} = \boldsymbol{\sigma}^{00} + \mathbf{T}_{1j}(\mathbf{x}_1 - \mathbf{x}_j)(\mathbf{R}_1\boldsymbol{\sigma}^{00} + \mathbf{F}_1). \tag{4.81}$$

Hereafter $\mathbf{n}_1^{(i)}$ $(i = 2, \ldots, n)$ are the unit outward normal vectors on Γ_1 at the points $\boldsymbol{\zeta}_i \in \Gamma_1$ nearest to the small inclusion v_i: $\boldsymbol{\zeta}_i = \arg\min |\boldsymbol{\zeta} - \mathbf{x}|$, $(\boldsymbol{\zeta}, \boldsymbol{\zeta}_i \in \Gamma_1,\ \mathbf{x} \in v_i,\ i = 2, \ldots, n)$. If $\operatorname{diam} a_i,\ |\boldsymbol{\zeta}_i| \ll \operatorname{diam} a_1$ then the initial stress approximation induced by a large inclusion v_1 is homogeneous near the small inclusions v_i $(i = 2, \ldots, n)$ and Eq. (4.81) is reduced to $\overline{\boldsymbol{\sigma}}_{j(0)} = \boldsymbol{\sigma}^-_{(0)}(\mathbf{n}_1^{(j)})$. The error of such a reduction in the examples considered in Section 4.4 ($a_1 = 10^7 a_2$) is less than $10^{-4}\%$.

4.4.3 The Subsequent Approximations

To obtain the first-order approximation of the displacement field on the boundary of the macroinclusion $\mathbf{u}^0_{(1)}(\boldsymbol{\zeta})$ $(\boldsymbol{\zeta} \in \Gamma)$ we substitute $\boldsymbol{\eta}_{i(0)}(\mathbf{x})$ (4.60) $(i = 2, \ldots, n)$ into (4.5) and then into (4.39) (if we use the BIE method) and solve it by the iteration method. If we use the VIE method then $\boldsymbol{\eta}_{i(0)}(\mathbf{x})$ (4.60) $(i = 2, \ldots, n)$ should be substituted into Eqs. (4.51) and (4.49) with $i = 1$. We shall distinguish the interior from exterior iterations according to whether these iterations are produced for the analysis of a single inclusion (such as (4.43)) or these iterations are fulfilled for estimation of interactions between inclusions (such as (4.51)), respectively.

The first-order approximation of exterior iterations of stresses in the small inclusions can be obtained by the use of stresses $\boldsymbol{\sigma}^0_{(1)}(\mathbf{x})$ $(\mathbf{x} \in v_i,\ i = 2, \ldots, n)$ induced by the macroinclusion found by the BIE (4.7) or by the VIE (4.50) (with $j = 1,\ i = 2, \ldots, n$) methods. As this takes place, the initial interior approximation of an interacting small inclusion is estimated via Eq. (4.59) with the replacement of fields $\langle \boldsymbol{\sigma}^0 \rangle_i$ by $\langle \boldsymbol{\sigma}^0_{(1)} \rangle_i$ $(i = 2, \ldots, n)$. Then the field $\boldsymbol{\sigma}^0_{(1)}(\mathbf{x})$ $(\mathbf{x} \in v_i)$ and the fields produced by surrounding small inclusions v_j $(j = 2, \ldots, n,\ j \neq i)$ are utilized as the initial approximation of the internal problem (4.43) for the inclusion v_i.

Obtaining the next approximations is carried out in the framework of a standard schema of the iteration method (4.72). The iterations (4.78)–(4.81) and (4.72) are realized using the VIE method for the large and each small inclusion. If the analysis of the inclusions is performed via the BIE method then the procedure is identical and so this is not presented. In so doing the integration of Eqs. (4.72) at $k \geq 1,\ i = 1$ are carried out at the graded micro-mesh Ω^{mic}.

We now turn our attention to the estimation of computing costs. In general the case of the quadrature method involved direct inversion of the integral operator (4.40) and (4.50); they are usually solved by a direct Gauss elimination, with the number of operations proportional to $\big(3(d-1)nK\big)^3$, where n is the number of inclusions and K is the number of quadrature points in each inclusion. The obvious approach of reducing this $O(n^3K^3)$ cost is to construct some iteration scheme. For the iteration method described by Eqs (4.43) and (4.51) the operation count per iteration is proportional to $\big(3(d-1)nK\big)^2$. Hence if P iterations are carried out the total cost is proportional to $P\big(3(d-1)nK\big)^2$. In the numerical examples considered $P = 5$ iterations provide the relative error less than 0.2%. Since the norms of integral operators (4.49) and (4.50) are approximately independent of K, the number of iterations is also approximately independent of K and the total cost of the iteration scheme is $O\big(n^2K^2\big)$. Therefore, forms (4.43) and (4.51) would almost always be preferable for computation.

The next step of reducing the $O\big(n^2K^2\big)$ cost is the successful choice of the initial approximation of the iteration scheme (4.51). The cost of obtaining the initial approximation (4.59) is only $O(n^2)$, and the cost of estimation of the strain polarization tensor in reference inclusions (4.71) is $O(K^2)$. Hence, the total cost of producing the first-order approximation (4.70) and (4.71) is $O(nK^2+n^2)$. However, for large n (say 10^{2d-1}) its $O(nK^2+n^2)$ cost dependence can lead to surprisingly long computing times, especially for the 3-D problems. So a large number of interacting inclusions must be taken into account under the analysis of random structure composites. Fortunately, the modern multiparticle methods in micromechanics of random structure composites are based on the evaluation of only two, three, or other small number of interacting inclusions with the following exhaustive search of all possible pairs, triples, etc. interacting inclusions (for references see Chapters 8–10). Let us consider, for example, the two-particle approximation of multiparticle effective field method (MEFM) proposed [156], and the first-order approximation of effective fields induced by surrounding identical inclusions; we recall that this approximation provides an acceptable accuracy of local stresses evaluated. In our current notation the first order approximation of strain polarization tensor $\overline{\boldsymbol{\eta}}_{i(1)}(\mathbf{x})$ (4.70) is estimated as a superposition of effective fields induced by $n-1$ pairs of neighbors obtained in the framework of zero order approximation (4.70); these estimations requiring the inversion of $n-1$ $3(d-1)\times 3(d-1)$ matrices (4.57) are rather cheap in terms of computer cost. Then the total cost of the estimate (4.70) and (4.71) at a single reference inclusion is only $O(nK+K^2)$, which is much lower than the $O\big(nK^2+n^2\big)$ cost of those proposed in this chapter for the direct estimation of the first order approximation (4.51) and (4.53). Recall that the known semi-analytical approach of the MEFM (for references see Chapter 8) is based on the employment of only zero-order approximation of binary interaction of inclusions (4.57), the cost of which is only $O(n)$. A more detailed analysis of different simplifying assumptions for the analysis of random structure composites with an infinite number of inclusions is pursued in the chapters that follow.

4.4.4 Some Details of the Iteration Scheme in Multiscale Analysis

For lack of space, and instead of just presenting the formulas, we shall be satisfied with sketching an outline of the essential steps, in a heuristic rather than rigorous manner.

For a numerical solution of Eq. (4.39) (or (4.51)) we subdivide the macroboundary Γ_1 (or macrodomain v_1) into the micromesh Ω^{mic} and macromesh Ω^0. The boundaries of the elements in the two meshes Ω^0 and Ω^{mic} do not have to coincide and both meshes cover the whole surface Γ_1 (or the region v_1). But, as contrasted to Ω^0, the size of mesh Ω^{mic} cells is compatible with the length scale of the inhomogeneity of the "micro" field $\mathbf{u}^1(\boldsymbol{\zeta})$, $(\boldsymbol{\zeta}\in\Gamma_1)$ (or $\boldsymbol{\sigma}^1(\mathbf{x})$, $\mathbf{x}\in v_1$) produced by small inclusions and localized near the point $\boldsymbol{\zeta}$ nearest the small inclusions. Therefore the mesh Ω^{mic} is strongly inhomogeneous and, as will be shown below, the sizes of its cells can be distinguished from one another by a factor of order of magnitude six and even more. This strongly localized effect of interactions between the macro- and microinclusions permits one to decompose the stress fields in the macroinclusion on the micro- and macrofields associated with the different length scales and plays a fundamental role in multiscale analysis.

In more detail we will consider the solution of Eq. (4.39) for the large hole with the curved boundary arc Γ_1 discretized along the arc in segments Γ_1^j ($j = 0, \pm 1, \pm 2, \ldots, \pm l$) with the arc length

$$l(\Gamma_1^{j+\text{sign}(j)}) = l(\Gamma_1^0)\frac{|\mathbf{s}_j - \mathbf{x}_j^s|^2}{|\mathbf{s}_0 - \mathbf{x}_0^s|^2}, \tag{4.82}$$

where $(0,0) \in \Gamma_1^0$, $\mathbf{s}_j$ are the centers of arc Γ_1^j and $\mathbf{x}_j^s = \arg\min|\mathbf{x} - \mathbf{s}_j|$, $\mathbf{x} \in v_2$. The micromesh Ω_1^{mic} (4.82) takes into account the asymptotic behavior as $O(|\mathbf{s}_j - \mathbf{x}_j^s|^{-2})$ of the integrand in Eq. (4.4) induced by the small inclusion v_2 at $|\mathbf{s}_j - \mathbf{x}_j^s| \to \infty$. The graded mesh Ω_1^{mic} is asymptotically sensitive to the localized nature of the stress distribution in the large inclusion although rigorous analysis of adaptive meshes [1123] is beyond the scope of the current chapter.

For the solution of Eq. (4.40) by the VIE method, without loss of generality, the geometry of the large inclusion is selected as a circular domain with radius a_1 and the center $\mathbf{x}_1 = (0, -a_1)^\top$. We will use a local polar coordinate system with the center $(0,0)^\top$ with a uniform polar angle mesh (as in (4.73)) and an increasing step of the radial mesh along the radius

$$r_1^{j+1} = r_1^j + r_1^1\left[\frac{r_1^1(r_1^j + D)^4}{r_1^j(r_1^1 + D)^4}\right], \quad j = 1, \ldots, l. \tag{4.83}$$

Similar to (4.82), the micro-mesh (4.83) takes into account the asymptotic behavior as $O(|\mathbf{y} - \mathbf{x}|^{-4})$ of the integrand in Eq. (4.40) induced by the small inclusion v_2 at $|\mathbf{y} - \mathbf{x}| \to \infty$, $\mathbf{y} \in v_1$, $\mathbf{x} \in v_2$.

It should be mentioned, however, that the micromesh Ω^{mic} (e.g., defined by Eqs. (4.82) and (4.83)) is refined only in a small part $\mathbf{x} \in v_1' \subset v_1$ of the macroinclusion v_1 nearest to the microinclusion v_2 but on the opposite side $\mathbf{y} \in v_1'' \equiv v_1 \setminus v_1'$ the mesh Ω^{mic} is very coarse, and, perhaps, is more coarse than macromesh Ω^0; so in an example considered in Subsection 4.4.5 we have $l(\Gamma_1^{\pm l}) = 10^3 l(\Gamma_1^0) = 10^2 a_2$. Nevertheless, despite the fact that the kernels of integral equations (4.39) and (4.40) and their densities vanish for sufficiently large $|\mathbf{x} - \mathbf{y}|$, the popular crude approach often used in practice and called truncation of integrals over the region v_1'' is questionable, since in actuality the appropriate densities are presented as either $\boldsymbol{\eta}(\mathbf{x}) - \boldsymbol{\eta}(\mathbf{y})$ or $\mathbf{u}(\mathbf{x}) - \mathbf{u}(\mathbf{y})$, and therefore the value of the "truncated" integrals over the region v_1'' will be finite.

Thus, the meshes in the large and small inclusions are constructed and we come now to the realization of a standard iteration scheme for the solution of the VIE (4.43) and BIE involved. We detected that in the concrete examples of high matrix-inclusion elastic contrast and some others, these standard popular iterative schemes may diverge or converge very slowly (i.e., the iteration scheme (4.39) does not work in general) so that an implementation of the improved algorithm proposed in this chapter becomes more complicated. In such a case, Mikhlin [766] and Perlin [869] proved that the convergence of the scheme (4.43) is provided by their modification $\boldsymbol{\eta}_{(k+1)} = [\boldsymbol{\eta}^0 + \boldsymbol{\eta}_{(k)} + \boldsymbol{\mathcal{K}}\boldsymbol{\eta}_{(k)}]/2$ (for details see [167]). Although the convergence of this method was rigorously proved in [766] for the elastic problems of an arbitrary dimensions, we will demonstrate its effectiveness only for 2-D problems.

4.4.5 Numerical Result for a Small Inclusion Near a Large One

4.4.5.1 Small inclusion near a large hole and the second boundary value problem. Without loss of generality, the geometry of the large hole is selected as a circular domain with radius a_1 and the center $\mathbf{x}_1 = (0, -a_1)^\top$ located in an infinite matrix subjected to the homogeneous stress field $\sigma_{ij}^{00} \equiv \sigma_{11}^{00}\delta_{1i}\delta_{1j}$ at infinity. The small circular inclusion with the center $\mathbf{x}_2 = (0, D)^\top$ has the radius $a_2 < D \ll a_1$. Then the initial approximation for the displacement field $\mathbf{u}^0(\boldsymbol{\zeta})$ at the boundary as well as the initial approximation of the stresses in the matrix in the vicinity of the large inclusion $\boldsymbol{\sigma}^-_{(0)}(\mathbf{n}_1)$ and inside the small inclusion $\boldsymbol{\sigma}_{(0)}(\mathbf{x})$ ($\mathbf{x} \in v_2$) are presented via analytical relations (4.78)–(4.80), respectively, where the compliance $\mathbf{M}^{(1)} \to \infty$ in the hole v_1, $n = 2$, and in Eqs. (4.79) and (4.80) $\mathbf{n}_1 = (0, 1)^\top$. This problem makes it possible to simulate the problem for a circular inclusion in an elastic half-space subjected to a uniform uniaxial stress at infinity parallel to a free edge boundary (*the second boundary value problem*).

For analysis of subproblems with the macro-(hole) and microinclusion (3.120) we will use the BIE and VIE methods, respectively. The domain v_2 is discretized by the mesh either (4.73) or (4.74). For the solution of Eq. (4.39) we discretized the curved boundary arc $\varGamma_1$ along the arc in segments $\varGamma_1^j$ ($j = 0, \pm 1, \ldots, \pm l$) with the arc length (4.82). The first and the subsequent approximations of solutions of Eqs. (4.36) and (4.39) are carried out according to the scheme (4.78) and (4.79). For an illustration we will present the numerical results relating to the stress distribution $\boldsymbol{\sigma}^-_{(0)}(\mathbf{n}_1)$ (4.79). In particular, the case $\sigma^-_{ij(0)}(\mathbf{n}_1) = \delta_{1i}\delta_{1j}$ ($\mathbf{n}_1 = (0, 1)^\top$) being considered corresponds to the uniaxial tension in the absence of the small inclusion in the local coordinate system connected with the hole surface $\varGamma_1$. In Fig. 4.6 the normalized stresses $\sigma_{11}(\mathbf{x})/\sigma^-_{11(0)}(\mathbf{n}_1)$ [at the points with $\mathbf{n}_1 = (0, 1)^\top$, $\mathbf{x} = (0, x_2)^\top \in v_2$, $D = 1.3a_2$] are presented and compared via the method proposed for $a_1 = 10^7 a_2$ [2], for the seventh exterior, third interior, and the first iterations are presented; the convergence of the iteration method proposed can be seen. In the same figure the stress distributions were obtained for $a_1 = 10^2 a_2$ and also for $a_1 = 30a_2$. As can be seen the normalized stresses $\sigma_{11}(\mathbf{x})/\sigma^-_{11(0)}(\mathbf{n}_1)$ for the radius of large hole $10^2 a_2 < a_1 < 10^7 a_2$ are nearly identical and are deemed coinciding with relevant stresses for the inclusion in a half-space (*the second boundary value problem*). Just for comparison with the related results dedicated to the multiscale analysis of interacting inclusions,

---

[2] Just for a spectacular clarification of the value of the ratio $a_1/a_2 = 10^7$, it will be interesting to mention that the ratio of sizes of New York City and the one cent coin is just equal to $0.5 \cdot 10^7$.

ref. [1018] should be mentioned, where the finite element analysis was carried out for two interacting holes differing by the maximum size factor $a_1/a_2 = 50$.

The VIE method requires no corrections for the case of continuously variable anisotropic elastic properties. So the isotropic elastic properties of *SiC* fibers was replaced in [167] by the continuously variable anisotropic properties $L^{(2)anis}_{ijkl}(\mathbf{x}) = L^{(2)}_{ijkl}(1-\delta_{1i}\delta_{2j}\delta_{1k}\delta_{l2}) + \delta_{1i}\delta_{2j}\delta_{1k}\delta_{l2}[L^{(0)}_{1212} + L^{(2)}_{1|1212}(1-\alpha+(2\alpha-1)|\mathbf{x}|)]$, (no sum on i,j,k,l) where $\alpha =$ const., and $\mathbf{x} \in v_2$ is the coordinate of point considered in the local Cartesian coordinate system of the inclusion v_2. It was demonstrated that the anisotropy of the fiber leads to significant redistribution of stress fields and the rate of convergence is equally fast as in Fig. 4.6. The iteration method proposed is ideally adapted to the analysis of the stress state in the inclusions with physically nonlinear elastic properties. So, the approximation of the *secant modulus approach* is a replacement of the tensors $\mathbf{M}^{(i)}(\mathbf{x})$ $(\mathbf{x} \in v_i)$ varying over the constituents v_i with the tensors $\mathbf{M}^{(i)[k+1]}(\mathbf{x}) = \mathbf{M}^{(i)}(\boldsymbol{\sigma}^{[k]}(\mathbf{x}))$ where $[k]$ is the number of an iteration, and the initial approximation $\mathbf{M}^{(i)[0]} = \mathbf{M}^{(i)}(\boldsymbol{\sigma}^{[0]}(\mathbf{x}))$, where $\boldsymbol{\sigma}^{[0]} \equiv \mathbf{0}$. If $||\mathbf{I} - \mathbf{M}^{(i)[k+1]}/\mathbf{M}^{(i)[k]}|| \equiv \xi \ll 1$ then the solution is found; usually $k \le 5$ at $\xi = 10^{-2}$.

Let us replace the small inclusion by a small hole $(\mathbf{M}^{(1)}, \mathbf{M}^{(2)} \to \infty)$ and apply the BIE method for analyses of both the macro-and micro-holes. The limiting case of this problem $(a_1 \to \infty)$ simulates a problem for a circular tunnel in an isotropic-half space which is classical in geomechanics. The normalization of stresses $\sigma^{-}_{11}(\mathbf{s})/\sigma^{-}_{11(0)}(\mathbf{n}_1)$ $(\mathbf{n}_1 = (0,1)^\top,\ \mathbf{s} = (s_1, 0)^\top \in \Gamma_1)$ (see Fig. 4.7)

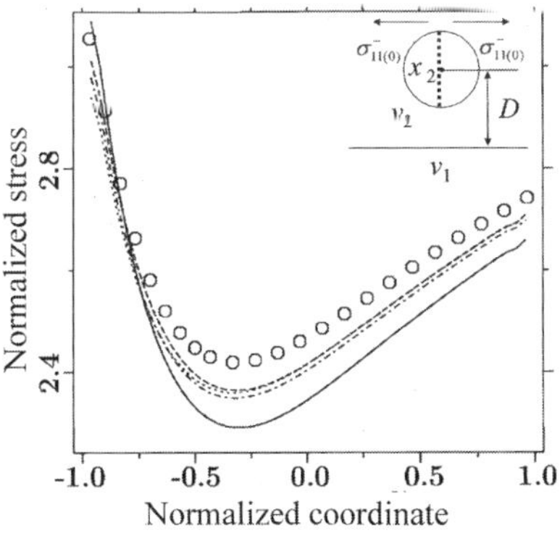

Fig. 4.6 $\sigma_{11}(\mathbf{x})/\sigma^{-}_{11(0)}(\mathbf{n}_1)$ vs x_2/a_2; $D = 1.3a_2$. $a_1 = 30a_2$, 7th iteration (○); $a_1 = 10^2 a_2$, 7th iteration (dotted curve); $a_1 = 10^7 a_2$, 1st iteration (dot-dashed curve); $a_1 = 10^7 a_2$, 3rd iteration (dashed curve); $a_1 = 10^7 a_2$, 7th iteration (solid curve).

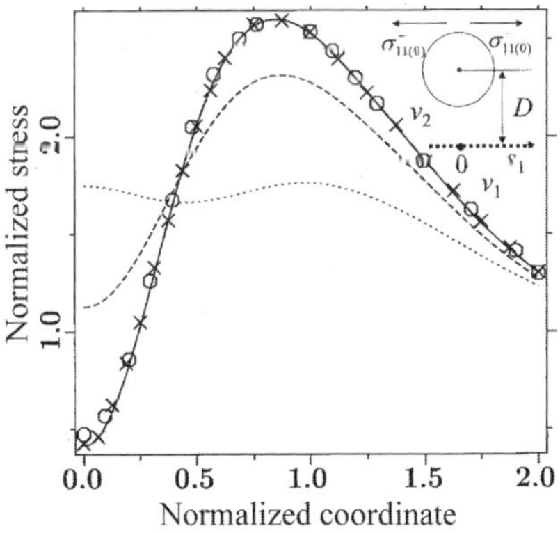

Fig. 4.7. $\sigma_{11}(\mathbf{s})/\sigma^{-}_{11(0)}(\mathbf{n}_1)$ vs s_1/a_2 $(\mathbf{s} = (s_1, s_2)^\top \in \Gamma_1,\ a_1 = 10^7 a_2)$. Mindlin's (×); Telles and Brebbia's (○); method proposed: 1st iteration (dotted curve), 3rd iteration (dashed curve), 7th iteration (solid curve).

makes it possible to compare our results with the analytical result by Mindlin [777] and with numerical estimations [1081] obtained by the BIE used in the fundamental solution for a half-plane. Numerical results were obtained for the essentially inhomogeneous graded micromesh (4.82) with $l(\Gamma_1^{\pm l}) = 10^6 l(\Gamma_1^0)$

for the large hole Γ_1 ($a_1 = 10^7 a_2$, $D = 1.34a_2$) and for the uniform polar angle mesh (4.73) ($r \equiv a_1$) for the small hole. Again, the iteration method proposed converges rapidly and we contend that the first few iterations provide an excellent approximation of the stress field to be solved. This allows one to anticipate that the method proposed is a powerful tool for analysis of multiple interaction inclusion problems in a half-space.

Now we will consider a set X of a finite number of small identical circular holes $v_2, \ldots, v_n$ with the centers $\mathbf{x}_i = (0, D + (i-2)\cdot \Delta D)^\top$. In Fig. 4.8 the normalized stresses $\sigma_{11}(\boldsymbol{\zeta}_i)/\sigma^0_{11(0)}(\mathbf{n}_1)$ are presented at the points with $\mathbf{n}_1 = (0,1)^\top$, $\boldsymbol{\zeta}_i = (0, D - a_2 + (i-2)\cdot \Delta D)^\top \in \Gamma_i$, $i = 2, \ldots, n$ for the parameters $D = 1.66a_2$ and $D = \infty$, $\Delta D = 2.66a_2$, and $n = 102,\ 26$ (the dotted and dot-dashed curves coincide for the values of the number of small holes $i = 2, 3$). Comparison of results obtained for $D = 1.66a_2$ and $D = \infty$ makes it possible to detect two sorts of boundary effects. The first one is known in the theory of functionally graded unbounded materials with either periodic (Chapter 11) or random (Chapter 12) finite clouds X of inclusions where one detected the variation of stress concentration factor over the cross-section of the cloud X that depends also on the size of the cloud X (scale effect). As can be seen in Fig. 4.8, the stress concentration factor in the center of a rather large cloud ($n = 102$) is close to a similar set of inclusions for a uniaxial stressed infinite plate with an infinite row of circular holes [870]. The second sort of boundary effects (called also edge effect) is defined by the presence of the free surface (edge) Γ_1 and was described in [850], [953] (see also Chapter 14). The second effect shows up in conjunction with the first one. The second boundary effect can be more significant than the first one if D is small enough; nevertheless the first effect has a greater long-range action.

It should be mentioned that the effectiveness of the method proposed is defined by the strong localized effect of interactions between the macro-and micro-inclusions. This makes it possible to use essentially inhomogeneous graded meshes Ω_1^{mic}. So the variation of $l(\Gamma_1^{\pm l})$ (4.82) from $10^4 l(\Gamma_1^0)$ to $10^6 l(\Gamma_1^0)$ under the fixed $l(\Gamma_1^0)$ yields close results. As this takes place 90% of all segments $\Gamma_1^{\pm j}$ ($j = 1, \ldots, l$) are localized at distances less than $10a_2$ from the point $(0,0)^\top$.

4.4.5.2 Small inclusion near a large stiff inclusion and the first boundary value problem. Let us replace the large hole considered in the previous subsection by a stiff inclusion with the compliance $\mathbf{M}^{(1)} \ll \mathbf{M}^{(0)}$. For the estimation of stresses inside both the large and small inclusions we will utilize the version of the VIE method proposed. The domain v_2 is discretized by the mesh (4.74) while the domain v_1 with the center $\mathbf{x}_1 = (0, -a_1)^\top$ is discretized in the local polar coordinate system (4.83).

The initial approximations (4.78)–(4.80) are valid for the case being considered as well. We will present our results in the form normalized with respect to $\sigma^-_{ij(0)}(\mathbf{n}_1) \equiv \delta_{2i}\delta_{2j}$, ($\mathbf{n}_1 = (0,1)^\top$) that provides a way to model the problem for a circular inclusion in a half-space with a fixed edge (*the first boundary value problem*). In Fig. 4.9 the normalized stresses $\sigma_{22}(\mathbf{x})/\sigma^-_{22(0)}(\mathbf{n}_1)$ (in the cross section $\mathbf{x} = (0, x_2)^\top \in v_2$, $a_1 = 10^3 a_2$, $D = 1.3a_2$) are presented for high matrix-inclusion elastic contrast, ($\mathbf{L}^{(1)} = \mathbf{L}^{(2)} = 10^6\mathbf{L}^{(0)}$). The first-order approximations provide an acceptable accuracy if $D \geq 1.3a_2$. Numerical results displayed in Fig. 4.9 were obtained by the method proposed for the essentially inhomogeneous graded mesh Ω_1^{mic} when $r_1^l = 10^3 r_1^0$ (4.83).

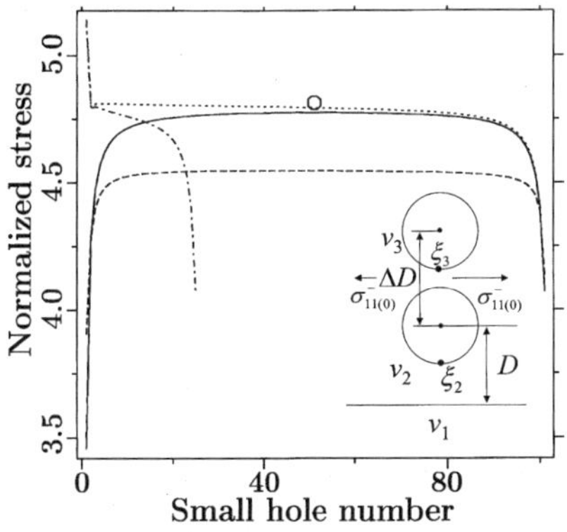

Fig. 4.8. $\sigma_{11}(\zeta_i)/\sigma^0_{11(0)}(\mathbf{x}_i)$ vs the number i of v_i $(i = 2, \ldots, n;\ \Delta D = 2.66a_2)$ estimated at 7th iteration for $n = 102,\ D = \infty$ (solid curve), $n = 102,\ D = 1.66a_2$ (dotted curve), $n = 26,\ D = 1.66a_2$ (dot-dashed curve), and at 1st iteration for $n = 102,\ D = 1.66a_2$ (dashed curve). The result [870] is marked by ∘.

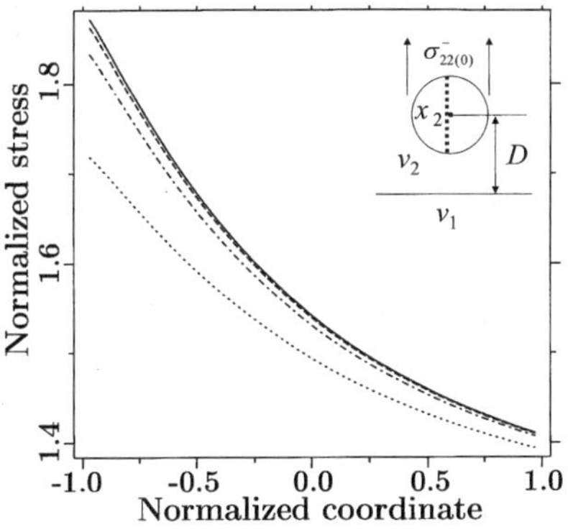

Fig. 4.9. The stress distribution of normalized stresses $\sigma_{22}(\mathbf{x})/\sigma^-_{22(0)}(\mathbf{n}_1)$ in the cross section $\mathbf{x} = (0, x_2)^\top \in v_2$ vs the coordinate x_2 for $D = 1.3a_2,\ a_1 = 10^4 a_2$: first iteration (dotted curve), second (dot-dashed curve), third iteration (dashed curve), seventh iteration (solid curve).

We have considered the case of a stiff macroinclusion ($\mathbf{M}^{(1)} \ll \mathbf{M}^{(0)}$). Without any correction of a computer program we could analyze the general cases of anisotropy of the macro-and micro-inclusions in the isotropic matrix modeling the popular problem of a defect in bimaterials. The case of an anisotropic inclusion embedded in an anisotropic half-space which, in turn, is bonded to an isotropic half-space can be simulated by the anisotropic macroinclusion with an anisotropic microinclusion $v_2 \subset v_1$ which is located, in turn, in an infinite isotropic matrix. Then the micromesh Ω_1^{mic} should be corrected in the neighborhood of the microinclusion $v_2 \subset v_1$.

4.4.6 Discussion

It should be mentioned that the VIE and BIE methods have a series of advantages and disadvantages, and it is crucial for the analyst to be aware of their range of applications. So, the highefficiency of the method proposed is realized only if the accuracy of the first-order approximation (see Fig. 4.4) is sufficient for a researcher and the VIE method is more efficient in the case being considered than other methods. However, for example, for composite materials with isotropic homogeneous inclusions and a multiscale analysis (such as finite number of inclusions in a half space) the BIE method is more effective than the VIE method because it has a well-known advantage of reduction of dimension by one. Because of this the hybrid BIE and VIE method supplied by an iterative solution strategy was presented in this section (see Figs. 4.6–4.9). In the case of homogeneous anisotropic inclusions the BIE can be exploited if the appropriate Green functions are known. On the other hand, the use of a reference isotropic fundamental solution with some residual volume integral [868], [972], eliminates the main well-known advantage of the BIE method of reduction of dimension by one. There are strong grounds for believing that the harnessing of the analytical solution for a single ellipsoidal homogeneous inclusion in either a bimaterial or half-space [661], [1210], will have culminated in a refinement of the first-order approximation considered in Subsection 4.4.2. If both the macro-and microinclusions are isotropic then the use of BIE for both the macro-and microinclusions is more effective. In so doing efficiency of the BIE method is provided by use of an asymptotically adapted micromesh (see, e.g., Eq. (4.82)) supplied by the iteration method. So, the CPU time expended for the estimation of stress distribution in the problem for 102 holes with essentially different sizes with the factor 10^7 (see the dotted curve in Fig. 4.3) equals 1 hour for a PC with a 666 MHz processor; for comparison, Oden and Zohdi [827] exploited a multiprocessor supercomputer thousands of times more powerful for the analysis of the stress field in a 3-D body containing inclusions differing by the factor 20 from a size of the body.

Furthermore, the regular shape of the macroinclusion v_1 in Section 4.7 was considered in order to make it easier to obtain the analytical representation of the initial approximation $\boldsymbol{\sigma}^0_{(0)}(\mathbf{x})$ induced by the macroinclusion in the absence of small inclusions. In the case of nonregular shape of the inclusion v_1 we can use the BIE and/or VIE methods. Then we can analyze a nonsmooth surface Γ_1 in the neighborhood of small inclusions v_i $(i = 2, \ldots, n)$, in particular this part of Γ_1 can have the shape of either a quarter or an eighth space. In addition, the case of a few macroinclusions can be analyzed in the framework of the same scheme as well. So, we can consider two anisotropic macroinclusions with their smooth boundaries close together. In so doing, the small inclusion can be located in a layer between two macroinclusions that have parallel boundaries. In all cases mentioned above the anisotropy of both the micro-and macroinclusions creates no complications and we do not need to know the Green functions for either a half or quarter of the space, or for the layer as used in the standard BIE method.

4.5 Complex Potentials Method for 2-D Problems

As mentioned, the well known Kolosov-Muskhelishvili complex potentials method [472], [798], [966], is now well recognized as a powerful and easy-to-use tool for solving a variety of the plane many-inclusion linear elastic problems with an accurate account for the inclusion-inclusion interaction. For details of this method, we refer to the original Muskhelishvili book [798]; here, due to widespread use of this method, we outline its basic idea and give the necessary formulae following [152].

The modified conditional stress $\boldsymbol{\tau}(\mathbf{x})$ and strain $\boldsymbol{\eta}(\mathbf{x})$ $(\mathbf{x} \in v_i)$ polarization tensors at the fixed second circle inclusions (e.g., fibers) v_j with the center $\mathbf{x}_j$ can be estimated by the use of evaluation of the strains and stresses, respectively, in the fiber v_i if the fiber pair v_i and v_j inside the infinite homogeneous matrix is subjected to homogeneous remote fields ($\boldsymbol{\varepsilon}^{\infty}$ and $\boldsymbol{\sigma}^{\infty}$, respectively) with single nonzero components ε_j^{∞} and σ_j^{∞}, otherwise $\varepsilon_k^{\infty}, \sigma_k^{\infty} = 0$, $(j, k, = 1, 2, 3;\ j \neq k)$.

We keep the basic notations introduced in [798] to write a general solution in the form

$$(u_1 + iu_2) = \bar{\varkappa}\varphi(z) - z\overline{\varphi'(z)} - \overline{\psi(z)}, \tag{4.84}$$

where u_i are the Cartesian components of displacement vector $\mathbf{u} = (u_1,\, u_2)^{\top}$ and $i = \sqrt{-1}$. Also, $z = (x_1 + ix_2)$ is a complex variable representing the point $\mathbf{x} = (x_1,\, x_2)^{\top}$ in the complex plane Ox_1x_2 and $\bar{\varkappa}$ is the constant factor equal to $3 - 4\nu$ or $(3 - \nu)/(1 + \nu)$ depending on whether the plane strain or plane stress problem is considered, respectively. In the polar coordinates (r, ϕ) corresponding to the Cartesian ones $(x_1,\, x_2)$, $z = re^{i\phi}$ and the vector $\mathbf{u}$ components are related by $(u_r + iu_\phi) = (u_1 + iu_2)\, e^{-i\phi}$. In (4.84), the functions $\varphi(z)$ and $\psi(z)$ are the complex potentials; what is important, components of the corresponding to $\mathbf{u}$ stress tensor σ also can be expressed in terms of these potentials:

$$\sigma_{11} + \sigma_{22} = 8\mu \mathrm{Re}\left[\varphi'(z)\right], \quad \sigma_{22} - \sigma_{11} + 2i\sigma_{12} = 4\mu\left[\overline{z}\varphi''(z) + \psi'(z)\right]. \tag{4.85}$$

Here, as well as in (4.84), $\overline{z}$ means the complex conjugate of z : $\overline{z} = (x_1 - ix_2) = re^{-i\phi}$. In what follows, we need also the curvilinear components of the tensor σ given by the formula:

$$\begin{aligned}(\sigma_r - i\sigma_{r\phi}) &= \frac{1}{2}\left[\sigma_{11} + \sigma_{22} - (\sigma_{22} - \sigma_{11} + 2i\sigma_{12})\, e^{2i\phi}\right] \\ &= 2G\left\{2\mathrm{Re}\left[\varphi'(z)\right] - \left[\overline{z}\varphi''(z) + \psi'(z)\right] e^{2i\phi}\right\}.\end{aligned} \tag{4.86}$$

The specific form of φ and ψ is problem-dependent: in our case, we can separate the far field term by taking $\varphi(z) = \Gamma_1 z + \varphi_0(z)$ and $\psi(z) = \Gamma_2 z + \psi_0(z)$, where Γ_1 and Γ_2 are the complex-valued constants and $\varphi_0(z)$ and $\psi_0(z)$ are the functions vanishing at infinity. So, suppose we have the homogeneous remote strain $\langle\varepsilon\rangle$ prescribed. It follows from (4.84) that $(u_1 + iu_2) \to (\bar{\varkappa} - 1)\Gamma_1 z - \overline{\Gamma_2}\overline{z}$ as $|z| \to \infty$. On the other hand, $\mathbf{u} \to \mathbf{u}^{\infty} = \boldsymbol{\varepsilon}^{\infty} \cdot \mathbf{x}$ and we immediately get

$$\Gamma_1 = \frac{\varepsilon_{11}^{\infty} + \varepsilon_{22}^{\infty}}{2(\bar{\varkappa} - 1)}, \quad \Gamma_2 = \frac{1}{2}\left(\varepsilon_{22}^{\infty} - \varepsilon_{11}^{\infty} + 2i\varepsilon_{12}^{\infty}\right). \tag{4.87}$$

Now, we apply the above theory to derive an accurate, asymptotically exact series solution for an infinite solid containing two fibers perfectly bonded with the matrix. First, we introduce the global Cartesian coordinate system Ox_1x_2 arbitrarily. On this basis, locations of the fiber centers are given by Z_1 and Z_2, respectively. To construct a solution in the matrix domain we make use of the generalized superposition principle to write

$$\mathbf{u}^{(0)} = (\bar{\varkappa} - 1)\Gamma_1 z - \overline{\Gamma_2}\overline{z} + \mathbf{u}_1(z - Z_1) + \mathbf{u}_2(z - Z_2) \tag{4.88}$$

where $\mathbf{u}_p$ $(p = 1, 2)$ is the vanishing at infinity disturbance field due to p-th inhomogeneity which can be written in the form (4.84). More specifically, $\mathbf{u}_p(z) = \bar{\varkappa}^{(0)}\varphi_{0p}(z) - z\overline{\varphi'_{0p}(z)} - \overline{\psi_{0p}(z)}$, where the potentials φ_{0p} and ψ_{0p} are taken as a singular part of the Loran's series:

$$\varphi_{0p}\left(z\right)=\sum_{n=1}^{\infty}\frac{A_{n}^{(p)}}{z^{n}},\quad \psi_{0p}\left(z\right)=\sum_{n=1}^{\infty}\frac{B_{n}^{(p)}}{z^{n}},\tag{4.89}$$

with $A_n^{(p)}$ and $B_n^{(p)}$ being the complex-valued series expansion coefficients to be found. It is quite clear that such a choice of φ_{0p} and φ_{0p} provides the proper behavior of $\mathbf{u}^{\infty}$ at infinity.

On the contrary, the displacement field inside the fiber has no singularity and, thus, the regular part should be retained only in the Loran's series expansion of the corresponding complex potentials. Thus, the displacement vector inside the fiber can be represented as

$$\mathbf{u}^{(q)}=\bar{\varkappa}^{(q)}\,\varphi_{q}\left(z_{q}\right)-z_{q}\,\overline{\varphi_{q}'\left(z_{q}\right)}-\overline{\psi_{q}\left(z_{q}\right)},\tag{4.90}$$

where $\bar{\varkappa}_q=\bar{\varkappa}\left(\nu_q\right)$, $z_q=z-Z_q$, $z\in v_q$, $q=1,2,$

$$\varphi_{q}\left(z\right)=\sum_{n=1}^{\infty}C_{-n}^{(p)}\,z^{n},\quad \psi_{q}\left(z\right)=\sum_{n=1}^{\infty}D_{-n}^{(p)}\,z^{n},\tag{4.91}$$

and $C_{-n}^{(p)}$ and $D_{-n}^{(p)}$ are the unknown constants. Similar to $A_n^{(p)}$ and $B_n^{(p)}$, they must be chosen to satisfy the matrix-fiber perfect bonding conditions

$$\left(\mathbf{u}^{(0)}-\mathbf{u}^{(q)}\right)\Big|_{|z_q|=a}=0;\quad \left[\mathbf{t_n}\left(\mathbf{u}^{(0)}\right)-\mathbf{t_n}\left(\mathbf{u}^{(q)}\right)\right]\Big|_{|z_q|=a}=0,\tag{4.92}$$

where $\mathbf{t_n}=\boldsymbol{\sigma}\mathbf{n}$ is the traction vector and a is the radius of fibers. In our case of a circular fiber, the unit normal vector $\mathbf{n}=\mathbf{e}_r$ and $\mathbf{t_n}=\left(\sigma_r,\sigma_{r\phi}\right)^{\top}$.

To derive a resolving set of equations, we substitute the expressions of $\mathbf{u}^{(0)}$ (4.90) and $\mathbf{u}^{(q)}$ (4.90) into the conditions (4.92) and equate the corresponding power series coefficients in the opposite parts of equality. This procedure is, however, not quite straightforward and involves some algebra. To be specific, we consider the first fiber and perform the necessary transformations to obtain the linear equations for $A_n^{(1)}$, $B_n^{(1)}$, $C_{-n}^{(1)}$ and $D_{-n}^{(1)}$; carrying out the boundary conditions on the second fiber–matrix interface follows the same procedure. To accomplish the above plan we obviously need first to represent (4.90) as a function of local variable $z_1=z-Z_1$. Obtaining a local expansion of $\mathbf{u}^{(0)}$ in the vicinity of the point Z_1 uses the formula

$$\frac{1}{\left(z+Z\right)^{n}}=\sum_{k=0}^{\infty}H_{nk}(Z)\,z^{k},\quad H_{nk}(Z)=(-1)^{k}\,\frac{(n+k-1)!}{k!\,(n-1)!}Z^{-(n+k)}.\tag{4.93}$$

to be applied to the last term in (4.90). So, we have

$$\varphi_{02}\left(z-Z_{2}\right)=\sum_{n=1}^{\infty}\frac{A_{n}^{(2)}}{\left(z_{1}+Z_{21}\right)^{n}}=\sum_{k=0}^{\infty}A_{-k}^{(1)}\left(z_{1}\right)^{k},\tag{4.94}$$

where $A_{-k}^{(1)}=\sum_{n=1}^{\infty}A_n^{(2)}\,H_{nk}(Z_{21})$, $Z_{21}=Z_1-Z_2$. The desired representation is

$$\mathbf{u}^{(0)}(z_1) = \mathbf{u}_{01} + \left(\bar{\varkappa}^{(0)} - 1\right)\Gamma_1 z_1 - \overline{\Gamma_2}\overline{z_1} + \bar{\varkappa}_0\varphi_{0\Sigma}(z_1) - z_q\overline{\varphi'_{0\Sigma}(z_1)} - \overline{\psi_{0\Sigma}(z_1)}, \tag{4.95}$$

where

$$\mathbf{u}_{0q} = \left(\bar{\varkappa}^{(0)} - 1\right)\Gamma_1 Z_1 - \overline{\Gamma_2 Z_1}, \quad \varphi_{0\Sigma}(z) = \sum_{n=-\infty}^{\infty}\frac{A_n^{(1)}}{z^n},$$

$$B_{-k}^{(1)} = \sum_{n=1}^{\infty}\left[B_n^{(2)} - \frac{\overline{Z_{21}}}{Z_{21}}(k+n)A_n^{(2)}\right]H_{nk}(Z_{21}), \quad \psi_{0\Sigma}(z) = \sum_{n=-\infty}^{\infty}\frac{B_n^{(1)}}{z^n}. \tag{4.96}$$

Now, we substitute (4.95) and (4.90) for $q = 1$ into the first set of conditions (4.92). Taking into account that at the interface $z_1 = ae^{i\phi}$, $z_1\overline{z_1} = a^2$ and $(z_1)^2 = a^2e^{2i\phi}$, we arrive, after somewhat tedious algebra, at the infinite set of linear equations ($k = 1, 2, \ldots$):

$$\bar{\varkappa}^{(0)}\frac{A_k^{(1)}}{a^{2k}} - a^2(k+2)\overline{A_{-(k+2)}^{(1)}} - \overline{B_{-k}^{(1)}} = \delta_{k1}\overline{\Gamma_2} - a^2(k+2)\overline{C_{-(k+2)}^{(1)}} - \overline{D_{-k}^{(1)}},$$

$$\frac{B_k^{(1)}}{a^{2k}} - \bar{\varkappa}^{(0)}\overline{A_{-k}^{(1)}} - \frac{(k-2)}{a^{2k-2}}A_{k-2}^{(1)} = \delta_{k1}\left[(\bar{\varkappa}^{(0)} - 1)\Gamma_1 + C_{-k}^{(1)}\right] - \bar{\varkappa}^{(1)}\overline{C_{-k}^{(1)}}. \tag{4.97}$$

To satisfy the second conditions (4.92), we recognize first that the stress in the matrix domain can be written as

$$\sigma_{11} + \sigma_{22} = \sigma_{11}^{\infty} + \sigma_{22}^{\infty} + 8\mu_0\mathrm{Re}\left[\varphi'_{01}(z - Z_1) + \varphi'_{02}(z - Z_2)\right],$$

$$\sigma_{22} - \sigma_{11} + 2i\sigma_{12} = \sigma_{22}^{\infty} - \sigma_{11}^{\infty} + 2i\sigma_{12}^{\infty} + 4\mu_0\left[\overline{(z - Z_1)}\varphi''_{01}(z - Z_1)\right.$$
$$\left. + \psi'_{01}(z - Z_1) + \overline{(z - Z_2)}\varphi''_{02}(z - Z_2) + \psi'_{02}(z - Z_2)\right], \tag{4.98}$$

and that the formula for a traction vector $\mathbf{t_n} = (\sigma_r + i\sigma_{r\phi})$ is simply the complex conjugate of (4.86). Substituting (4.98) into (4.86) and carrying out the transformations analogous to those described above, we obtain another set of algebraic equations:

$$\frac{A_k^{(1)}}{a^{2k}} + a^2(k+2)\overline{A_{-(k+2)}^{(1)}} + \overline{B_{-k}^{(1)}} = -\delta_{k1}\overline{\Gamma_2} + \rho\left[a^2(k+2)\overline{C_{-(k+2)}^{(1)}} + \overline{D_{-k}^{(1)}}\right],$$

$$\frac{B_k^{(1)}}{a^{2k}} - \frac{(k-2)}{a^{2k-2}}A_{k-2}^{(1)} + \overline{A_{-k}^{(1)}} = -2\delta_{k1}\Gamma_1 + \rho\left(\delta_{k1}C_{-k}^{(1)} + \overline{C_{-k}^{(1)}}\right), \tag{4.99}$$

where $\rho = \mu_1/\mu_0$. Combining (4.99) with (4.97), we can exclude the unknowns $C_{-k}^{(1)}$ and $D_{-k}^{(1)}$ and obtain the infinite linear system containing the unknowns $A_k^{(1)}$ and $B_k^{(1)}$ only:

$$\frac{\Lambda}{a^{2k}}\left[a^2(k-2)A_{k-2}^{(1)} - B_k^{(1)}\right] + \sum_{n=1}^{\infty}A_n^{(2)}H_{nk}(Z_{21}) = -\delta_{k1}\Gamma_1,$$

$$\frac{\Omega}{a^{2k}}A_k^{(1)} + \sum_{n=1}^{\infty}\left\{\overline{A_n^{(2)}}\left[(k+n)\frac{Z_{21}}{\overline{Z_{21}}}\overline{H_{nk}(Z_{21})}\right.\right.$$
$$\left.\left. - (k+2)a^2\overline{H_{n,k+2}(Z_{21})}\right] + \overline{B_n^{(2)}}H_{nk}(Z_{21})\right\} = -\delta_{k1}\Gamma_1 \tag{4.100}$$

$k = 1, 2, \ldots$; where

$$\Lambda = \frac{\left(2\rho + \bar{\varkappa}^{(1)} - 1\right)}{2\left[\rho\left(\bar{\varkappa}^{(0)} - 1\right) + \left(\bar{\varkappa}^{(1)} - 1\right)\right]} \quad \text{for } k = 1, \quad \Lambda = \frac{\left(\rho + \bar{\varkappa}^{(1)}\right)}{2\left(\rho\bar{\varkappa}^{(0)} - \bar{\varkappa}^{(1)}\right)} \quad \text{for } \ k > 1,$$

and $\Omega = \left(\rho\bar{\varkappa}^{(0)} + 1\right)/[2\left(\rho - 1\right)]$. It is quite obvious that obtaining the second half of the resolving set of equations (corresponding to the second fiber) consists of simply changing the index "1" to "2" and vice versa in (4.100): $Z_{12} = -Z_{21}$. It can be shown that the resulting infinite linear system has normal determinant [550] provided that the nontouching condition $\|Z_{12}\| > 2a$ is satisfied. Thus, its approximate numerical solution can be obtained by the truncation method which assumes retaining in (4.100) and its counterpart the unknowns and equations with $k, n \leqslant n_h$. As n_h increases, these approximate solutions converge to the exact one, providing evaluation of the stress tensor in every point of a composite domain with any desirable accuracy by taking n_h sufficiently large.

In order to check the accuracy of the numerical method the influence of the interaction between two identical circle inclusions (e.g., fibers) on the stress field in the inclusions is studied. Let the elastic properties of the matrix and the inclusions be $\mu^{(0)} = 1$, $\nu^{(0)} = 0.5$ and $\mu^{(1)} = 1000$, $\nu^{(1)} = 0.5$. Let us consider a plane strain problem and the pair of inclusions v_i and v_j loaded by the tension $\boldsymbol{\sigma}^{\infty} = (\sigma_{11}^{\infty}, 0, 0)^{\top}$ along the line linked to the inclusion centers $\mathbf{x}_2 = (Da, 0)^{\top}$ and $\mathbf{x}_1 = (0, 0)^{\top}$, D is a normalized distance between the inclusion centers. In Fig. 4.10 the numerical results of the normal tractional component $\sigma_{rr} = \sigma_{rr}(\phi)$ at the interface of the first fiber $\mathbf{x} = a(\cos(\phi), \sin(\phi))^{\top} \in v_1$ obtained for the numbers of harmonics $n_h = 120,\ 40,\ 20,\ 10$ are presented at $D = 2.01$ that demonstrate an error 5.8%, 35% and 65% of the solution with the $n_h = 40,\ 20$, and $n_h = 10$, respectively, in comparison with the solution corresponding to $n_h = 120$. The CPU times expended for the solution at a PC with a 2.0 GHz processor equal 150 sec and 0.5 sec for $n_h = 120$ and $n_h = 30$, respectively. The error of less than 0.1% is provided by the number of harmonics $n_h = 70$. It can be seen that as the fibers approach each other, the accuracy using the Kolosov-Muskhelishvili complex potentials decreases rapidly, and a small improvement in accuracy up to 0.1% can be provided just by significantly increasing the number of harmonics used to $n_h = 95$ at $D = 2.005$. In a similar manner, the number of approximating harmonics providing an error of less than 0.1%, 1%, and 5% were estimated in Fig. 4.11 as a function of the relative distances between the fibers D. Although the present numerical estimation for the local stress field in the fiber loses accuracy (which can be protected by increasing of the number n_h of harmonics) when the fibers almost touch, this does not affect the accuracy of the estimated effective properties, which is dependent on the smoothing integral operator, and is very insensitive to the exact local stress distribution in the fibers when $D \to 2$ (see Chapter 10).

The effectiveness of the Kolosov-Muskhelishvili method is widely recognized for circle interacting inclusions while the case of the elliptical inclusion is less well understood. The matter here is that the conformal mapping technique working so perfectly for a plane with holes of general shape does not allow straightforward extension on the case of inhomogeneities. Of course, the interaction problem for two elliptic inclusions can be solved by any numerical method such as, e.g., the method of singular integral equations used in [812]. As to an analytical solution, ref. [759] is worthy of notice. For the analysis of two identical, symmetrically placed elliptic inhomogeneities in an infinite plane, they utilized the Papkovich-Neihber representation of general solution in terms of real potentials. Derived by accurate matching of the elastic fields in the matrix and inclusions an infinite set of equations appears, however, to be rather involved and thus difficult to use. Application of complex potentials to solve some particular problems for the interacting elliptic inclusions was considered in [592] in the framework of the point-to-point collocation scheme. Kaloerov [536] has applied least-squares fitting to determine the unknown series expansion coefficients. Subsequent development of the complex potential method was undertaken in [632] in the problem solution of interacting elliptical inclusions with no restrictions on their number, size, aspect ratio, elastic properties, and arrangement. By exact satisfaction of all the interface conditions, a primary boundary-value problem stated on a complicated heterogeneous domain has been reduced to an ordinary well-posed set of linear algebraic equations. A properly chosen form of potentials provides a remarkably simple form of a solution and thus an efficient computational algorithm.

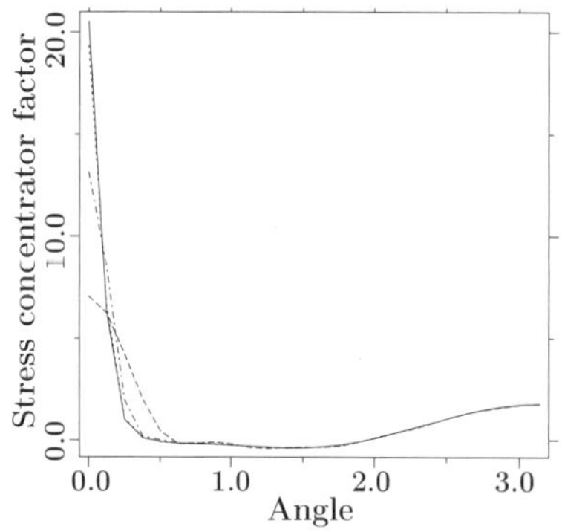

Fig. 4.10. Stress concentration factor $\sigma_{11}(\mathbf{z})/\langle\sigma_{11}\rangle$ at the fiber matrix interface $\mathbf{z} = (a, 0)^\top$ vs φ (radian): $n_h = 120$ (solid curve), $n_h = 40$ (dotted curve), $n_h = 20$ (dot-dashed curve), $n_h = 10$ (dashed curve).

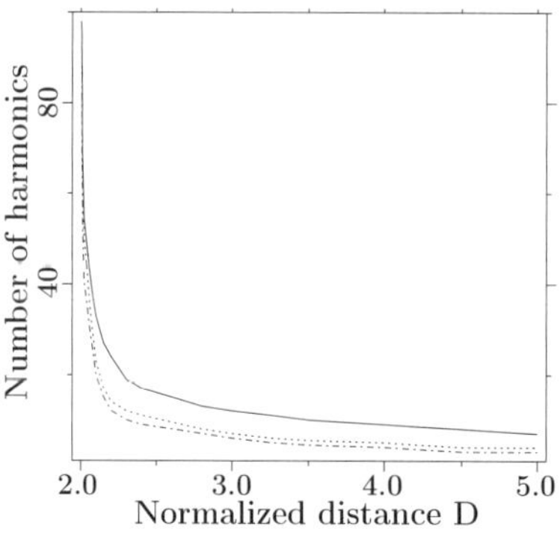

Fig. 4.11. The number n_h of harmonics providing the error less than 0.1% (solid curve), 1.0% (dotted curve), and 5.0% (dot-dashed curve) vs the normalized distance D between the inclusion centers.

Moreover, Kushch *et al.* [633] generalized the complex potential method to a stress analysis in a half plane containing a finite array of elliptic inclusions. The method combines the Muskhelishvili's method of complex potentials with the Fourier integral transform technique. Again, by accurate satisfaction of all the boundary conditions, a primary boundary-value elastostatics problem for a piece-homogeneous domain has been reduced to an ordinary well-posed set of linear algebraic equations. A properly chosen form of potentials provides a remarkably simple form of equations and thus an efficient computational algorithm (for details see [633]).

We reproduce here only one application of the theory developed to the practical problem of a polymer nanocomposite of clustered structure. There, an isolated cluster was idealized as a multilayer stack containing N silicate plates with uniform interlayer spacing (see [992] as well as Chapter 19). The experimentally observed plate thickness was roughly 1 nm, the layer spacing ranged from 2 to over 5 nm, and the number of plates per cluster varied from 1 to 50. In our model, the silicate nanolayers are approximated by the aligned ellipses with aspect ratio $\alpha^r = 0.01$ and the ellipse's major semiaxis $a_1 = 1$, Young's modulus $E^{(1)} = 300$ GPa, and Poisson's ratio $\nu^{(1)} = 0.4$ embedded into *Epox*862 matrix with $E^{(0)} = 3.01$GPa and $\nu^{(0)} = 0.41$. The number of inclusions $N = 15$ and their position is given by $Z_p = 0 + i(\alpha^r + dH + 4(p-1)\alpha^r)$, $p = 1, 2, \ldots, N$. In Figs. 4.12 and 4.13 the average stress concentration factors $\langle\sigma_{11}\rangle_p/\sigma_{11}^{\infty}$ are

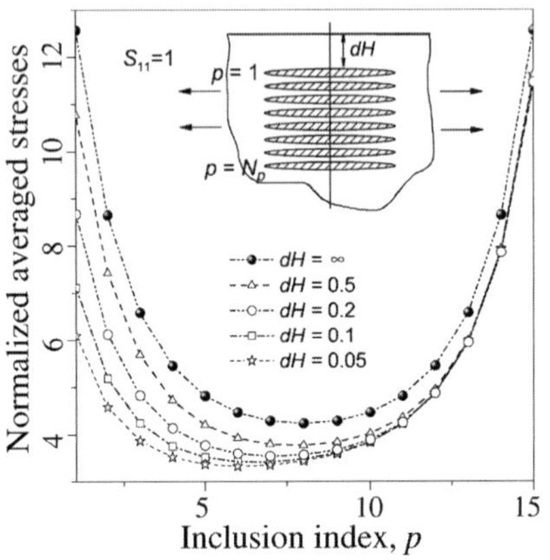

Fig. 4.12. Normalized averaged stress $\langle\sigma_{11}\rangle_p/\sigma_{11}^{\infty}$ in the inclusions of a half plane with a finite cluster of thin hard inclusions.

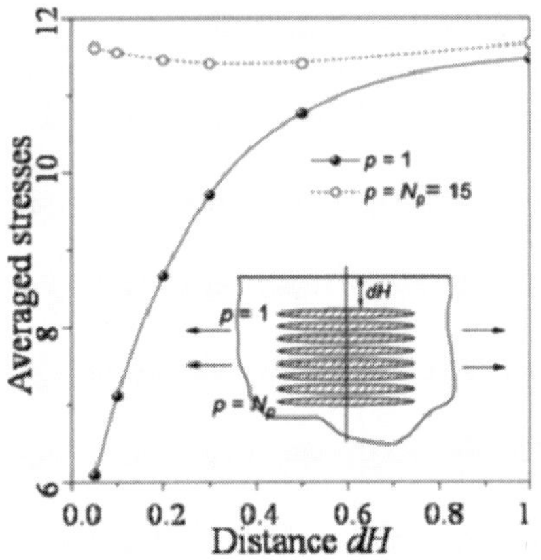

Fig. 4.13. Normalized averaged stress $\langle\sigma_{11}\rangle_p/\sigma_{11}^{\infty}$ in the inclusions with indices $p = 1$ and $p = N$ vs dH.

presented for the different distances dH between the first inclusion surface and flat boundary of a half plane. As the computations show, for a single inclusion in the infinite plane $\sigma_{11}/\sigma_{11}^{\infty} = 34.6$; for $N = 15$, it decreases more than two times even for the outer ellipses. The maximum average stress at $dH = \infty$ is observed in the outer inclusions acting as a shield for the internal inclusions of the cluster: only a small amount of load is transferred to the middle silicate layers. Such a behavior is qualitatively confirmed by 2-D finite element analysis in [992] of three rectangular inclusions in the matrix. The situation is somewhat different when the cluster is placed near the composite's surface. The corresponding data for a finite dH are shown in Fig. 4.12 by the open triangles, circles, squares, and stars. As our numerical analysis shows, the effect of the nearby free boundary manifests itself in considerable (up to two times and more) reduction of average stress in the inclusions closest to the boundary. At the same time, stress in the inclusion on the opposite side of the cluster remains practically unchanged (see Fig. 4.13).

5

Statistical Description of Composite Materials

The quantitative description of the microtopology of heterogeneous media, such as composite materials, porous and cracked solids, suspensions, and amorphous materials, is crucial in the prediction of overall mechanical and physical properties of these materials. After many years of comprehensive study by direct measurements and empirical relations that is extremely laborious, the structures of microinhomogeneous materials are still not completely understood. In composite materials internal structure refers to the size and shape of inhomogeneities, defects, and crystallites (grains); to the distribution of their orientations (textures); and to spatial correlations between these mechanical, geometrical, and crystallographic features [7], [124], [503], [902], [1106], [1107]. Apart from obvious goal of quantitative analysis of the real morphology of random media, this research has been fueled by the fact that these statistical descriptors arise in rigorous bounds of diverse physical quantities [503], [1106]. The observable details of internal structure are entirely dependent upon the length scales that are accessible to the microscopic technique. Physical observation and simulation of internal structures are tractable at many different length scales. A central challenge in material science is to distinguish those aspects of internal structure that form the material properties. Experimentally, the complete description of local stress-strain states is not accessible at the grain level, and the main descriptor is the geometrical one representing statistical quantitative information concerning the orientational and spatial correlation between the structure elements. In the last decades, increasing attention has been paid to the combination of microstructure quantification with stochastic or deterministic numerical simulation of mechanical properties of microinhomogeneous media.

In this chapter the basic concepts and hypotheses associated with quantification of microstructure morphology of CM are introduced in a quite general form applicable to a wide class of both random and deterministic structures. Finally, the methods of stochastic simulation of real random structures of fiber composites are considered.

5.1 Basic Terminology and Properties of Random Variables and Random Point Fields

5.1.1 Random Variables

The basic notion in probability theory is a *probability space* $(\Omega, \mathcal{F}, \mathcal{P})$ comprising three quantities: a sample space Ω, a collection $\boldsymbol{\mathcal{F}}$ of certain subsets of Ω called a σ-algebra of events, and a probability measure $\mathcal{P}$. These quantities are defined as following. In the simplest case, Ω consists of a denumerable collection of elementary events $\omega_1, \omega_2, \ldots$, while nondenumerated *sample spaces* are more common, e.g., $\Omega = \{\omega | 0 \leq \omega \leq 1\}$. The *$\sigma$-algebra* of events $\mathcal{F}$ is a particular collection of events posing the properties: $\mathcal{F}$ is not empty; if $\mathcal{A} \subset \mathcal{F}$ then its compliment is $\mathcal{A}^c = \{\omega | \omega \in \Omega, \omega \notin \mathcal{A}\}$; if $\mathcal{A}_1, \mathcal{A}_2, \ldots \subset \mathcal{F}$ then its union is $\cup_i \mathcal{A}_i = \{\omega | \omega \in \mathcal{A}_i$ for some $i\}$. The *probability measure* $\mathcal{P}$ is a real-valued function mapping each set in $\mathcal{F}$ into the unit interval $[0,1]$: $0 \leq \mathcal{P}(\mathcal{A}) \leq 1$ for every $\mathcal{A} \subset \mathcal{F}$; $\mathcal{P}(\Omega) = 1$; if $\mathcal{A}_1, \mathcal{A}_2, \ldots$ are disjoint sets in $\mathcal{P}$, then $\mathcal{P}(\cup_i \mathcal{A}_i) = \sum_i \mathcal{P}(\mathcal{A}_i)$.

We specify a real *random variable* (r.v.) $X(\omega)$ (called also *structure function*) defined on $(\Omega, \mathcal{F}, \mathcal{P})$ as a mapping from Ω to $\mathcal{R} \equiv \mathcal{R}^1$. This mapping is such that the set $\{\omega | X(\omega) \leq x\}$ is in $\mathcal{F}$ for every choice of $x \in \mathcal{R}$, and probability of this set is well defined for $X(\omega)$ $F(x) = \Pr(\{\omega | X(\omega) \leq x\})$ and called the *probability distribution*, which has the properties: $\Pr(a \leq X < b) = F(b) - F(a)$; $F(x_1) \leq F(x_2)$, if $x_1 \leq x_2$; $\lim_{x\to+\infty} F(x) = 1$; $\lim_{x\to-\infty} F(x) = 0$. If there is a function $f(x) \geq 0$, called the *probability density function* for $X(\omega)$, such that $\Pr(X \in \mathcal{B}) = \int_{\mathcal{B}} f(x)\, dx$ hold for an arbitrary set $\mathcal{B} \subset R$ for which probability $\mathcal{P}(\{\omega | X(\omega) \in \mathcal{B}\})$ is defined, then the random variable $X(\omega)$ is called continuous. The notion of probability density $f(x)$ can also be introduced for a discrete r.v. by setting $f(x) = \sum_{i=1}^n p_i \delta(x - x_i)$, in which x_i denote the possible values of the r.v., p_i are their corresponding probabilities $p_i = \Pr(X = x_i)$, and $\delta(x)$ is the δ-function.

The extension to more than one random variable defined on the same probability space is straightforward. For example, the *joint distribution function* $F(\mathbf{x}) \equiv F_{\mathbf{X}}(\mathbf{x})$ ($\mathbf{x} = [x_1, \ldots, x_n]^\top$) of a r.v. $\mathbf{X}(\omega) = [X_1(\omega), \ldots, X_n(\omega)]^\top$ is defined by

$$F(\mathbf{x}) \equiv F_{X_1,\ldots,X_n}(x_1, \ldots, x_n) = \Pr(\{\omega | \mathbf{X}(\omega) \leq \mathbf{x}\}), \tag{5.1}$$

for all $\mathbf{x} \in R^n$, here the vector inequality $\mathbf{X}(\omega) \leq \mathbf{x}$ means $X_i(\omega) \leq x_i$ for $i = 1, \ldots, n$. For a system of continuous random variables $\mathbf{X}(\omega)$, there can exist a probability density $f(\mathbf{x}) \equiv f_{\mathbf{X}}(\mathbf{x}) \geq 0$ related with $F(\mathbf{x})$ by the formulae

$$f(\mathbf{x}) = \frac{\partial^n F(x_1, \ldots, x_n)}{\partial x_1 \ldots \partial x_n}, \quad F(\mathbf{x}) = \int_{-\infty}^{x_1} \cdots \int_{-\infty}^{x_n} f(y_1, \ldots, y_n)\, dy_1 \ldots dy_n. \tag{5.2}$$

According to the consistency theorem, the functions $F(\mathbf{x})$ and $f(\mathbf{x})$ must satisfy the compatibility conditions $F(x_1, \ldots, x_m) = F(x_1, \ldots, x_m, \infty, \ldots, \infty)$, and $F_{X_{i_1},\ldots,X_{i_n}}(x_{i_1}, \ldots, x_{i_n}) = F_{X_1,\ldots,X_n}(x_1, \ldots, x_n)$ for both any permutation $\{i_1, i_2, \ldots, i_n\}$ of $\{1, 2, \ldots, n\}$ and $\forall m < n$. In particular, the probability density should satisfy the condition

$$f(x_1,\ldots,x_m)=\int_{-\infty}^{\infty}\cdots\int_{-\infty}^{\infty} f(x_1,\ldots,x_n)\,dx_{m+1}\ldots dx_n. \tag{5.3}$$

If $\mathbf{Y}(\omega)=\mathbf{g}(\mathbf{X}(\omega))$ is a random variable defined by $\mathbf{X}(\omega)$ and the function $\mathbf{g}$ mapping R^n into R^m is integrable over Ω with respect to $\mathcal{P}$, we can define the mathematical expectation, which for the continuous r.v. has the form

$$E(\mathbf{g}(\mathbf{X}))=\int_{R^n}\mathbf{g}(\mathbf{x})\,dF_{\mathbf{X}}(\mathbf{x}). \tag{5.4}$$

In particular, let us consider

$$\mathbf{g}(\mathbf{x})=\prod_{i=1}^{n}x_i^{\nu_i}, \tag{5.5}$$

where ν_i, $(i=1,\ldots,n)$ are nonnegative integers. Then $\alpha_{\nu_1,\ldots,\nu_n}=E(X_1^{\nu_1}\ldots X_n^{\nu_n})$, if it exists, is called the moment of order ν_1 in $X_1,\ldots,$ and ν_n in X_n. In the particular case of one r.v. X_k, we can define the mean, variance, and moment of qth order of X_k $(i_l=\delta_{lk},\ j_l=2\delta_{lk},\ p_l=q\delta_{lk},\ l=1,\ldots,n)$:

$$E(X_k)\equiv\bar{x}_k=\alpha_{i_1,\ldots,i_n},\quad \mathrm{var}(X_k)\equiv\sigma_{kk}=\alpha_{j_1,\ldots,j_n}-\alpha^2_{i_1,\ldots,i_n},\quad m_q=\alpha_{p_1,\ldots,p_n}, \tag{5.6}$$

as well as the central qth moment of X_k:

$$\mu_q=E[X_k-E(X_k)]^q. \tag{5.7}$$

Another special case takes place when $\nu_i=\nu_j=1-\delta_{ij}$ and $\nu_k=0$ for $k=1,\ldots,n,\ k\neq i,\ k\neq j$. Then the value

$$\sigma_{ij}=E[(X_i-EX_i)(X_j-EX_j)]\equiv E(X_iX_j)-E(X_i)E(X_j), \tag{5.8}$$

if it exists, is called the *covariance* between X_i and X_j $(i,j=1,\ldots,n,\ i\neq j$. The real symmetric, positive semidefinite matrix (σ_{ij}) $(i,k=1,\ldots,n)$ is called *variance-covariance* (or *dispersion*) matrix of the random vector $\mathbf{X}$. The random variables $X_1,\ldots,X_n$ are said to be uncorrelated if the nondiagonal elements of the covariance matrix are zero.

The probability for the product of n events $\mathcal{A}_1,\ldots,\mathcal{A}_n$ can be estimated through the conditional probabilities $\mathcal{P}(\mathcal{A}_m|\prod_{k=1}^{m-1}\mathcal{A}_k)$:

$$\mathcal{P}(\prod_{k=1}^{n}\mathcal{A}_k)=\mathcal{P}(\mathcal{A}_1)\mathcal{P}(\mathcal{A}_2|\mathcal{A}_1)\mathcal{P}(\mathcal{A}_3|\mathcal{A}_1\mathcal{A}_2)\cdot\ldots\cdot\mathcal{P}(\mathcal{A}_n|\prod_{k=1}^{n-1}\mathcal{A}_k). \tag{5.9}$$

In so doing, an appropriate formula for the probability density of a system can be expressed in terms of the conditional densities according to the chain rule:

$$f(x_1,x_2,\ldots,x_n)=f_1(x_1)f_2(x_2|x_1)f_3(x_3|x_1,x_2)\ldots f_n(x_n|x_1,x_2,\ldots,x_n). \tag{5.10}$$

as well in terms of the conditional probability density of the subsystem of the random variables $(X_1,\ldots,X_k)$, estimated under the assumption that the remaining random variables $X_{k+1},\ldots,X_n$ assume certain values:

$$f(x_1,\ldots,x_n)=f(x_1,\ldots,x_k|x_{k+1},\ldots,x_n)f_{k+1,\ldots,n}(x_{k+1},\ldots,x_n). \qquad (5.11)$$

If for any m, where $m=2,3,\ldots,n$, and any k_j $(j=1,2,\ldots,n)$, $1\le k_1<k_2<\ldots<k_m\le n$, Eq. (5.9) can be simplified $\mathcal{P}(\prod_{k=1}^n\mathcal{A}_k)=\prod_{k=1}^n\mathcal{P}(\mathcal{A}_k)$ then the events $\mathcal{A}_1,\ldots,\mathcal{A}_n$ are said to be independent, and the next properties occur for them: $F(x_1,\ldots,x_n)=F(x_1)\cdot\ldots\cdot F(x_n)$, $f(x_1,\ldots,x_n)=f(x_1)\cdot\ldots\cdot f(x_n)$. If the conditional distribution $f(x_1,x_2,\ldots,x_k|x_{k+1},x_{k+2},\ldots,x_n)$ (5.11) does not coincide with the "unconditional" distribution $f(x_1,x_2,\ldots,x_k)$ (5.3) then there is a statistical dependence between the sets of random variables $X_1,X_2,\ldots,X_k$ and $X_{k+1},\ldots,X_n$. The measure of statistical dependence between random variables vanishing if they are statistically independent is described by multiple correlations of various orders. The covariance $\mathbf{K}(X_1,X_2)=\sigma_{ij}$ of two random variables X_1 and X_2 is defined by Eq. (5.8) vanishing if X_1 and X_2 are independent, when we have $E(X_1X_2)=E(X_1)E(X_2)$. A triple correlation $\mathbf{K}(X_1,X_2,X_3)$ describes statistical dependence on all three random variables X_1,X_2,X_3; vanishes when at least one of the variables is independent of the others, and can be taken in the form of

$$\mathbf{K}(X_1,X_2,X_3)=E(X_1X_2X_3)-\bar{x}_1\sigma_{23}-\bar{x}_2\sigma_{13}-\bar{x}_3\sigma_{12}+2\bar{x}_1\bar{x}_2\bar{x}_3. \qquad (5.12)$$

In a similar manner, the formulae for higher-order correlations can be obtained.

The concrete probability distributions described analytically by distribution function $F(x)$ can take a very different form. We reproduce in Table 5.1 the elementary properties of some distribution functions (discrete and continuous) which have proved to be extremely useful in micromechanics: the Poisson, the binominal, the uniform, the normal (or Gaussian), the lognormal, the gamma distributions. In particular, the probability that an event occurs m times in n independent trial, in which the probability of occurrence of the event is $0\le p\le 1$, is given by the binominal distribution presented in the second line of Table 5.1. The normal r.v. is completely characterized by its mean m and variance σ^2. Since the normal distribution is symmetric with respect to its mean values m, all odd central moments vanish ($\mu_{2k+1}=0$), and, moreover, $\mu_{2k}=(2k-1)!!\sigma^{2k}$ $(k=1,2,\ldots)$. The lognormal r.v. X is defined by the parameters m and ξ, which are the mean and variance of the normal random variable $\ln X$, respectively. The mean $m_X=\exp(m+\xi/2)$ and variance σ_X^2 of a lognormal distribution are given in terms of m and ξ. A probability density $f(x)$ $(0\le x)$ of a gamma distribution $G(a,p)$ is defined by the parameters $a,p>0$ and the gamma function $\Gamma(p)$.

Table 5.1. Some distribution functions

Distr.	Poss. values	Density	Mean	Variance
Poisson	$0<\lambda,\ n=0,.$	$\frac{1}{n!}\lambda^n e^{-\lambda}$	λ	λ
Binomin.	$m=0,1,.,n$	$C_n^m p^m q^{n-m}$	np	npq
Uniform	$a<b-$ real	$(b-a)^{-1}\chi_{[a,b]}(x)$	$(a+b)/2$	$(b-a)^2/12$
Normal	$0<\sigma, m-$real	$(2\pi\sigma)^{-1/2}\exp\left[-\frac{(x-m)^2}{2\sigma^2}\right]$	m	σ^2
Lognorm.	$0<\xi$	$\frac{1}{\xi x\sqrt{2\pi}}\exp\left[-\frac{(\ln x-m)^2}{2\xi^2}\right]$	$\exp(m+\frac{\xi}{2})$	$m_X^2[\exp\xi^2-1]$
Gamma	$0<a,p$	$a^p[\Gamma(p)]^{-1}x^{p-1}\exp(-ax)$	p/a	pa^{-2}

As an example of the system of random variables, we will consider the probability density of a system of n normal random variable $X_1,\ldots,X_n$:

$$f(x_1,\ldots,x_n)=\frac{1}{\sqrt{(2\pi)^n\det(\sigma)}}\exp\Big(-\frac{1}{2}\sum_{i,j=1}^{n}(x_i-\bar{x}_i)\sigma_{ij}^{(-1)}(x_j-\bar{x}_j)\Big), \quad (5.13)$$

where $\det(\sigma)$ is the determinant of the covariance matrix (5.8), and $\sigma_{ij}^{(-1)}$ are the elements of the inverse covariance matrix (5.8): $\sigma_{ij}^{(-1)}=A_{ij}\det(\sigma)^{-1}$, A_{ij} is the cofactor of the element σ_{ij} (5.8). For the normal distribution (5.13), the higher correlations vanish and the moments of third, fourth, ... orders can be expressed by $\bar{x}_i$, σ_{ij}:

$$E(X_kX_lX_m)=\bar{x}_k\bar{x}_l\bar{x}_m+\bar{x}_k\sigma_{lm}+\bar{x}_l\sigma_{km}+\bar{x}_m\sigma_{kl}, \quad (5.14)$$

$$E(X_kX_lX_mX_n)=\bar{x}_k\bar{x}_l\bar{x}_m\bar{x}_n+\sigma_{kl}\sigma_{mn}+\sigma_{km}\sigma_{ln}+\sigma_{kn}\sigma_{lm}+\bar{x}_k\bar{x}_l\sigma_{nm}$$
$$=\bar{x}_k\bar{x}_m\sigma_{ln}+\bar{x}_k\bar{x}_n\sigma_{lm}+\bar{x}_m\bar{x}_n\sigma_{kl}+\bar{x}_l\bar{x}_n\sigma_{km}+\bar{x}_l\bar{x}_m\sigma_{kn}. \quad (5.15)$$

In particular one obtains $M(X_k^3)=(\bar{x}_k)^3+3\bar{x}_k\sigma_{kk}$, and $M(X_k^4)=(\bar{x}_k)^4+3(\sigma_{kk})^2+6(\bar{x}_k)^3\sigma_{kk}$. Many important models are based on the normal distribution. The reason for this popularity originates in the central limit theorem of probability theory, which contends that, under some general conditions, the sum of independent random values is approximately normal distributed if their number tends to infinity.

5.1.2 Random Point Fields

Let each $\omega\in\Omega$ correspond to a realization of random medium that occupies some domain w called the window of observation in d-dimensional Euclidean space, i.e., $w\subset\mathcal{R}^d$. The *random point field* (called also *random process*) studied is denoted by Φ assuming that Φ is a random field with a finite number of points in any bounded area and with the property that different points are not coincide (simple point process). The random set of all points with the coordinates $\mathbf{x}_1,\ldots,\mathbf{x}_n$ is denoted as $\Phi=\{\mathbf{x}_i\}$ and $\mathbf{x}\in\Phi$; Φ_B denotes the restriction of Φ to a given area $B\subset\mathcal{R}^d$. Note that physicists speak of a "field" while statisticians use the term "process" even if no timedependence is considered. In so doing, we will use a parameterization by space ($\mathbf{x}$) rather than (t) coordinate owing to our concern with spatially random microstructure. The goal of statistical researches leading to so-called nonparametric estimators for different statistical parameters is to obtain *unbiased* estimators, which are free of systematic errors. It should be mentioned that the term "random" is sometimes used in a restrictive intuitive sense to mean uniformly distributed points. However, we use the term "random" in its more general sense of "non-deterministic" and not in the sense of specific distribution. Furthermore, since the object with which our processes are concerned are geometrical we are dealing with topics in geometric probability and, in particularly, in point processes [745], [1046].

For a given set $\mathcal{A}\subset\mathcal{R}^d$, $\Phi(\mathcal{A})$ is the random number points of Φ containing in $\mathcal{A}$. An observed sample of Φ_w in the observation window w is denoted by ϕ_w and $\phi(w)$ is the observed number of points in w. $N=\Phi(w)$ denotes the random number of points $\mathbf{x}_i$ falling in w; $\hat{N}(w)$ is the observed number of points in w. A random point field is a random variable in the sense of probability theory.

We will also introduce the notations for some sets and set operations. So, $\chi_{\mathcal{A}}(\mathbf{x})$ stands the *indicator function* of $\mathcal{A}$:

$$\chi_{\mathcal{A}}(\mathbf{x}) = \begin{cases} 1 & (\mathbf{x} \in \mathcal{A}), \\ 0 & (\mathbf{x} \notin \mathcal{A}). \end{cases} \tag{5.16}$$

Translation $\mathcal{A}_{\mathbf{x}}$ is the set $\mathcal{A}$ shifted by the vector $\mathbf{x}$:

$$\mathcal{A}_{\mathbf{x}} = \mathcal{A} + \mathbf{x} = \{\mathbf{y} + \mathbf{x} : \mathbf{y} \in \mathcal{A}\}. \tag{5.17}$$

Let $\mathbf{g}$ be a rotation matrix (see Appendix A.1) rotating the initial coordinate system around o. If $\mathbf{x} \in \mathcal{A}$ is a point of $\mathcal{R}^d$ then $\mathbf{gx}$ is the rotated point, and $\mathbf{g}\mathcal{A}$ is a rotated point field $\mathbf{g}\mathcal{A} = \{\mathbf{y} : \mathbf{y} = \mathbf{gx},\ \mathbf{x} \in \mathcal{A}\}$ The symbols $\oplus$ and $\ominus$ denote the *Mincowski addition* and *subtraction*:

$$\mathcal{A} \oplus \mathcal{B} = \cup_{\mathbf{x}\in\mathcal{A},\mathbf{y}\in\mathcal{B}}\{\mathbf{x} + \mathbf{y}\} = \cup_{\mathbf{y}\in\mathcal{B}}\mathcal{A}_{\mathbf{y}} = \cup_{\mathbf{x}\in\mathcal{A}}\mathcal{B}_{\mathbf{x}}, \tag{5.18}$$

$$\mathcal{A} \ominus \mathcal{B} = \cap_{\mathbf{y}\in\mathcal{B}}\mathcal{A}_{\mathbf{y}} = (\mathcal{A}^c \oplus \mathcal{B})^c. \tag{5.19}$$

For the special important case of a ball $b(o, r)$ of radius r centered at the origin o, the sets $\mathcal{A}_r = \mathcal{A} \oplus b(o, r)$ and $\mathcal{A}^r = \mathcal{A} \ominus b(o, r)$ are known as the *outer parallel* (or *delation*) set of $\mathcal{A}$ and the *inner parallel* (or *erosion*) set of $\mathcal{A}$, respectively.

A quantitative description of the composite microstructure is based on the consideration of the inclusion centers statistically described by the multiparticle probability densities $p_m(\mathbf{x}_1, \ldots, \mathbf{x}_m)$ that give the probability $p_m(\mathbf{x}_1, \ldots, \mathbf{x}_m)$ $d\mathbf{x}_1 \ldots d\mathbf{x}_m$ to find an inclusion center in the vicinities $d\mathbf{x}_1 \ldots d\mathbf{x}_m$ of the point $\mathbf{x}^m = (\mathbf{x}_1, \ldots, \mathbf{x}_m)$. The p_m are the most basic descriptors that characterize the structure of a multiparticle system and have been well studied in the statistical mechanics of liquids [424]. In particular, $p_1 = n$, where n is the number density of inclusions v_i connected with the volume concentration (or fraction) $c = n\bar{v}_i$, $\bar{v}_i \equiv \text{mes}v_i$.

The point field $\varPhi$ defined by the point coordinates can be generalized to the *marked process* denoted by the same letter $\varPhi = \{(\mathbf{x}_i, \mathbf{m}_i)\}$ if each point $\mathbf{x}_i$ of the point process $\{\mathbf{x}_i\}$ has a *mark* $\mathbf{m}_i$ of the mark space $\{\mathbf{m}_i\} = \mathcal{M}$ distinguishing it to a great extent. The most popular marks used in micromechanics of CM are qualitative marks to distinguish different phases; quantitative marks to characterize the object such as, e.g., the orientation, shape, and size. Of course, any marked point process can be interpreted as a nonmarked point process in more general space, namely in the product space $\mathcal{R}^d \times \mathcal{M}$. However, the points and marks are usually treated separately. For example, the translation of the marked point field $\mathcal{A} = \{(\mathbf{x}_i, \mathbf{m}_i)\}$ becomes $\mathcal{A}_{\mathbf{x}} = \{(\mathbf{x}_i + \mathbf{x}, \mathbf{m}_i)\}$.

Although the point pattern analyzed by statisticians are usually finite, the stochastic analysis is simplified by the assumption of an infinity of random field with subsequent accompanying assumptions, namely those of homogeneity and isotropy. A point field $\varPhi = \{\mathbf{x}^m\}$ is called *homogeneous* (or *stationary*) if $\varPhi$ and translated field $\varPhi_{\mathbf{x}}$ (5.17) have the same distribution for $\forall \mathbf{x} \in \mathcal{R}^d$:

$$p_m(\mathbf{x}_1 + \mathbf{x}, \ldots, \mathbf{x}_m + \mathbf{x}) = p_m(\mathbf{x}^m),\ \ m = 1, 2, \ldots;\ \forall \mathbf{x} \in \mathcal{R}^d. \tag{5.20}$$

Sometimes the properties (5.20) are replaced by the hypothesis that the mean point numbers are translation-invariant $E\varPhi(\mathcal{A}) = E\varPhi(\mathcal{A}_{\mathbf{x}})$ for any compact set

$\mathcal{A}$ and all $\mathbf{x} \in \mathcal{R}^d$. For the homogeneous point field Φ, the last property makes it possible to define the *intensity*:

$$\lambda = E\Phi(\mathcal{A})/\bar{\mathcal{A}}, \ (\bar{\mathcal{A}} = \text{mes}\mathcal{A}) \tag{5.21}$$

where the position, size, and shape of $\mathcal{A}$ are irrelevant. The intercity λ noted also n is a mean number of points in the unit cube $[0,1]^d \subset \mathcal{R}^d$. The intensity measure $E\Phi(\mathcal{A})$ plays a similar role in the theory of point fields as the mean does for random variables.

It should be mentioned that mathematical notion of "homogeneous pattern" is wider of the sense of this term used by many physicists and material scientists. For them it means "uniform", which is not the same for, e.g., clustered fields scattered similarly on the whole space (these fields are homogeneous but not uniform). In a general case, the rigorous deviation of the notions homogeneity and uniformity is very difficult and impossible if only one pattern in a bounded observation window w is given. If more than one sample (or a few parts of a sample) is given which can be considered as independent, then at least some aspects of homogeneity can be tested in the framework of the theory of testing of statistical hypotheses.

The isotropy of the point field Φ is defined in a similar manner. The point process Φ is called *isotropic* if

$$\Pr(\Phi \subset \mathcal{F}) = \Pr(\mathbf{g}\Phi \subset \mathcal{F}) \tag{5.22}$$

for any rotation $\mathbf{g}$ of the set Φ around the origin O. The point field which is both homogeneous and isotropic has the same distribution for the rotation around arbitrary points ($\neq O$) and is called motion-invariant. A bonded part Φ_w of Φ is assumed to be derived via a brick from a larger ensemble of random field Φ that is stationary and isotropic.

A stronger property is *ergodicity* ensuring an equality of the spatial average estimated over one a sufficiently large sample w and the mean value over the random fluctuations. Implicitly often supposed ergodic hypothesis enables us the replace ensemble average with volume averaging, assuming that observation window tends to infinity. A necessary condition of ergodicity is $\lim_{w\uparrow R^d} \Phi(w)/\bar{w} = \lambda$, where λ is the intensity and $w \uparrow \mathcal{R}^d$ is means that w contains a sphere of radius r limiting to infinity. A sufficient condition for ergodicity is the some "mixing" property which should be hold for any subsets of a homogeneous point field Φ. In particular, if $\mathcal{A}$ means that the sphere $b(O,r)$ does not contain a point and $\mathcal{B}$ means that the same sphere contains two points then the fulfillment of the mixing condition means

$$\Pr\big(\Phi(b(O,r)) = 0, \ \Phi(b(O,r)) = 2\big) \to \Pr\big(\Phi(b(O,r)) = 0\big)\Pr\big(\Phi(b(O,r)) = 2\big).$$

A popular example of a nonergodic field is the lattice pont fields with the points located at the nodes of a periodic grid, which are randomly positioned in the space (see other examples in [745], [1046], where some philosophical problems of the ergodic hypothesis are discussed).

Thus, we summarized some basic ideas, notations and properties of random point processes [249], [272], [584], [938], [939], [1045], [1047], [1106]. We usually will consider statistically homogeneous (or stationary) and isotropic ergodic random field Φ, keeping in mind that the stationarity, isotropy, and ergodicity can never be tested statistically in their full generality. Stationarity means invariance under arbitrary translation, and isotropy means invariant under arbitrary rotation. The ergodicity ensured that one sample (one point pattern) is sufficient for obtaining statistically secure results, assuming the convergence of results obtained for infinitely expanding observation window w.

5.1.3 Basic Descriptors of Random Point Fields

The packed random structure can be characterized by several parameters, such as packing density, coordination number, radial distribution function, particle cage, inter-particle spacing, and others. The various methods of estimating the effective properties of a composite material use a knowledge of statistical geometrical information about the microstructure. In order to incorporate the spatial arrangement of components in a micromechanical simulation, it is essential to quantitatively characterize the random structure of the composite. The most important factors characterizing the microstructure of composite material are the shape, volume fraction, and arrangement (random or regular) of the components that permit the calculation of bounds of effective moduli.

We will indicate three basic classes of descriptors and their estimators for unmarked point fields: (1) the intensity (mean number of points); (2) two distance distribution functions (for the distance analysis); and (3) several second-order quantities (for the descriptions based on the counting and counting-distance analysis).

The common unbiased estimator of the intensity (5.21) $\hat{n} = \phi(w)/\overline{w}$ ($\overline{w} = \text{mes} w$), which is also consistent (i.e., accuracy increases with increasing window area). The intensity is a simples statistical parameter used in the estimation of effective moduli.

The interaction effects generating the local stresses produced by inclusions are highly sensitive to their locations. In so doing the nonlinear processes such as fracture and fatigue are far more sensitive to local stresses and, therefore, to local variations of microstructure then other mechanical phenomena. Because of this, in the deformation and fracture processes, another important statistical parameter of the location of inclusions is a *nearest neighbor distribution* (NND) function $D(r)$ [901] defined by the density $d^{nd}(r)$ such that $d^{nd}(r)dr$ equals to the probability that there is no other inclusion centroid in a circle of radius r with the center $\mathbf{x}_i$, and there is at least one inclusion centroid in the ring of radius r and $(r + dr)$. In other words, the NND function describes a probability of the shortest of the distances from an arbitrary chosen point $\mathbf{x}_n \in \Phi$ to all other points of Φ:

$$D(r) = \Pr(\text{distance from } \mathbf{x}_n \text{ to nearest point of } \Phi \text{ is less then } r), \ (r \geq 0). \tag{5.23}$$

Many problems in point field theory require the consideration of an arbitrary "typical" point of point field Φ. $D(r)$ is in fact a conditional probability function described by the Palm probability distribution P^o, which can be interpreted as being the probability distribution of point field events given under the condition that a point $\mathbf{x}_n \in \Phi$ called "a typical point" is observed at a specific location in the origin O (or any arbitrary chosen point of $\mathcal{R}^d$) $D(r) = \Pr^o(\Phi(b^0(O,r)) > 1)$, where $b^o(O,r)$ denotes the open ball of radius r centered at the origin O. Palm distribution theory makes precise the notion of a typical point that is heuristically clear. The estimation of $D(r)$ is based on the points $\mathbf{x}_i \in \Phi_w$ $(i = 1, \ldots, N)$ whose distance ρ_i to the nearest neighbor in w is smaller than r, and whose shortest distance ρ_i^∂ to the boundary ∂w is larger that r. Repley [938] proposed the following unbiased estimator for $D(r)$ taking the edge effect into account

$$\hat{D}(r) = \frac{\#\{\rho_i \leq r \wedge \rho_i^\partial > r\}}{\#\{\rho_i^\partial > r\}}, \tag{5.24}$$

where $\#(\cdot)$ is the counting function which tallies the number of points in the specified set. In the estimator (5.24), only the points are analyzed for which the nearest neighbor distance ρ_i is smaller than the smallest distance ρ_i^∂ to the boundary ∂w. The equation (5.24) can be presented in a formalized form:

$$\hat{D}(r) = \sum_{\mathbf{x}_i \in \Phi} \chi_{(O,r)}(\rho_i)\chi_{w \ominus b(O,\rho_i^\partial)}(\mathbf{x}_i)\Big[\sum_{\mathbf{x}_i \in \Phi} \chi_{w \ominus b(O,\rho_i^\partial)}(\mathbf{x}_i)\Big]^{-1}, \tag{5.25}$$

where the eroded window $w \ominus b(O, \rho_i^\partial)$ is defined by the Minkowski subtraction (5.19).

It should be mentioned that, when dealing with applications, it will in general be necessary to use simultaneously several quantities of different types to analyze given point fields. In particular, this mean that basing conclusions upon mean values or distance functions, or second-order quantities only, is not enough in general because none of these quantities can, in general, determine uniquely the distribution of point fields. Therefore, a combined use of intensities, of a distance distribution, and second-order quantity is desirable. Due to the wide utilization of periodic boundary conditions in numerical simulations of random packing, an alternative *toroidal edge correction* is often used (and precisely this method will be exploited hereafter for the elimination of the boundary effect) in which each rectangular region w can be regarded as a torus, so that points on opposite edges are considered to be closed. Then w can be considered to be part of a grid of identical rectangles, forming a border around the pattern inside w. Distances are then measured from the point in the central rectangle w to points in the surrounding periodic rectangles [939].

The average neighbor distance $\langle D \rangle$ is found by the formula $\langle D \rangle = \int_0^\infty rD(r)\,dr$. The nearest neighbor and higher-order neighbor distributions can be estimated experimentally by measuring the frequency of occurrence of different distances of different order neighbors of the inclusions for a large number of inclusions.

The estimation of $D(r)$ (5.24), (5.25) is based on choosing a typical point $\mathbf{x}_i \in \Phi_w$ to measure the distance to the nearest other point of Φ. Instead of the point $\mathbf{x}_i \in \Phi_w$, any point $\mathbf{x} \in w$, which does not necessarily belong to Φ, can be chosen. This leads to the definition of the *spherical contact distribution* function $H_S(r)$, of a homogeneous isotropic point field Φ, which is interpreted as the distribution function of the distance of a random test point independent of the point field to its nearest neighbor $H_S(r) = \Pr(\Phi(b^0(O,r)) > 0)$, where the arbitrary point $\mathbf{x} \in \mathcal{W}$ is chosen in O due to the homogeneity of the field Φ. The function $H_S(r)$ with the boundary conditions $H_S(0) = 0,\ H_S(\infty) = 1$ is monotonically increasing, and the related function $1 - H_S(r)$ characterizes probability of the empty spaces (matrix) within Φ. In the case of a homogeneous isotropic point field Φ, an unbiased estimation for $H_S(r)$ is

$$\hat{H}_S(r) = \big\{\text{mes}[w \ominus b(O,r)] \cap \cup_{\mathbf{x}_i \in \Phi_w} b(\mathbf{x}_i, r)\big\}/\text{mes}[w \ominus b(O,r)].$$

An important step forward to a more refined description and analysis of the point arrangement is the use of *second-order quantities* such as the pair-correlation function, the radial distribution function, and the second-order product density. The widely used informative function which describes the point distribution is the *second-order intensity function* $K(r)$ (also called Ripley's K function) defined as the number of further points expected to be located within a distance r of an arbitrary point divided by the number of points per unit area n. This notion looks like the definition of the nearest-neighbor distance distribution function $D(r)$ (5.23), but we are now interested in numbers of points rather than in distances. Heuristically speaking, $K(r)$ equals the mean number of points that have the distance smaller than $r \geq 0$ from the "typical pont," which itself is excluded:

$$\hat{K}(r) = \frac{1}{N}\sum_{i=1}^{N} \Phi(b(\mathbf{x}_i, r) \setminus \{\mathbf{x}_i\}),\ (N = \Phi(w)). \tag{5.26}$$

Due to the assumed homogeneity of Φ, we may translate the typical point $\mathbf{x}_i$ to the original O, and then again use the Palm probability distribution P^o, introduced in the definition of $D(r)$: $K(r) = n^{-1}E_{P^o}(\Phi(b(O,r)) - 1)$, where the mean is taken with respect to the Palm probability P^o. The unbiased estimator (5.26) has the disadvantage that it cannot be completely obtained from the observation window w because parts of the spheres $b(\mathbf{x}_i, r)$ may be outside w. Since points lying outside the observation window w are unobserved, the latter depends strongly on the shape and the size of w. It is our aim to simulate a typical realization that also includes interaction with the structure element outside of the window while avoiding systematic errors or biases in the estimation procedure. The effect of the edge of the domain w becomes increasingly dominant as the dimension increases. A number of special edge corrections are known. A naive way proposed in *the minus-sampling* method is to consider $w^* = w \ominus b(O,r)$ within domain w and allow measurements from an object in w^* to an object in w. To estimate $K(r)$, points $\mathbf{x}_i$ are used for which the sphere $b(\mathbf{x}_i, r)$ lies completely inside w:

$$\widehat{K}(r) = \sum_{\mathbf{x}_i \in w \ominus b(O,r)} \frac{\Phi(b(\mathbf{x}_i, r) \setminus \{\mathbf{x}_i\})}{\text{mes}(w \ominus b(O,r))},\ (r \geq 0) \tag{5.27}$$

Although the effective sample size is then the number of points in w, the method, of course, leads to a large loss of information because only the inner points in $w \ominus b(O,r)$ are sphere centers. A much better idea of edge correction of the estimator for $K(r)$ was suggested in [938]:

$$\hat{K}(r) = \frac{\bar{w}}{\hat{N}^2} \sum_{i=1} \hat{N} \sum_{j \neq i}^{\hat{N}} w_{ij}^{-1} I_{ij}(r_{ij}), \tag{5.28}$$

where $\hat{N}$ is the number of points $\mathbf{x}_k$ $(k = 1, \ldots \hat{N})$ in the field of observation w with the area $\bar{w}$, $I_{ij}(r_{ij})$ is the indicator function equals 1 if $r_{ij} \leq r$ and zero otherwise where r_{ij} is a distance between the points $\mathbf{x}_i$ and $\mathbf{x}_j$. w_{ij} is the ratio of the circumference contained within w to the whole circumference with radius r_{ij}. For circles intersecting the boundary ∂w, the function w_{ij} compensating for the boundedness of w is less than one and has an explicit formula when the field of observation w is rectangular [272]; the formulae and algorithm for the 3-D case were proposed in [584]. The function $\hat{K}(r)$ (5.28) is obtained by averaging over all inclusions at each value of r. Equation (5.28) is an approximately unbiased estimator, which is free of systematic errors, for sufficiently small r because $N/\bar{w}$ is a slightly biased estimator for n. In 2-D, Diggle [272] recommended an upper limit of r equal to half the length of the diagonal of a square sampling region.

The plots of K-functions may difficult to interprete because their usual power behavior of order r^d for large r. In such a case, there are some clear advantages in using the *L-functions* instead of working directly with K-functions: $L(r) = (K(r)/\omega_d)^{1/d}$, where

$$\omega_d = \frac{\pi^{d/2}}{\Gamma(1 + d/2)} \tag{5.29}$$

is the value of unit sphere in $\mathcal{R}^d$. For example, for $d = 1, 2$ and 3, ω_d equals 2, π, and $4\pi/3$, respectively. The estimators for $L(r)$ (5.12) can be obtained though the estimators for $K(r)$ (5.26)–(5.28).

Some related further second-order quantities, based on the derivatives of the K-function, is known for the analysis of motion-invariant point fields. The *radial distribution function* (RDF) $g(r)$ (called also the pair correlation function) which plays a role similar to the variance in a classical analysis of random variables is defined as the radial distribution of the average number of sphere centers per unit area in a spherical shell. The RDF can be estimated from second-order intensity function as

$$g(r) = \frac{1}{d\omega_d r^{d-1}} \frac{dK(r)}{dr}. \tag{5.30}$$

The RDF is related to the derivative of $K(r)$ (5.14), and is therefore it is more sensitive to changes in the spatial order than is the function $K(r)$. An edge-corrected kernel density estimator for $g(r)$ is govern by $\hat{g}(r) = \hat{\rho}(r)/\widehat{n^2}$, where, for $\Phi(w) = N$:

$$\begin{aligned} \widehat{n^2} &= N(N-1)\bar{w}^{-2}, \\ \hat{\rho}(r) &= \sum_{\mathbf{x}_i, \mathbf{x}_j \in \Phi_w} \frac{k_\epsilon(\|\mathbf{x}_i - \mathbf{x}_j\| - r)}{4\pi \|\mathbf{x}_i - \mathbf{x}_j\|^2 \text{mes}[(w + \mathbf{x}_i) \cap (w + \mathbf{x}_j)]}, \end{aligned} \tag{5.31}$$

with summation over i, j $(i \neq j)$ and the area $w + \mathbf{x}_i$ denoting the region centered at $\mathbf{x}_i$ obtained by translating w, centered at the origin. Here a kernel function $k_\epsilon(s) = [2/(4\epsilon\sqrt{5})][1 - 0.2(s/\epsilon)^2]\chi_{(0,\sqrt{5}\epsilon)}(|s|)$ depends on a suitable ϵ determined empirically such as e.g. $\epsilon = 0.1n^{-1/3}$.

5.2 Some Random Point Field Distributions

5.2.1 Poisson Distribution

Various models have been proposed for the generation of center coordinates of randomly packed spheres. If the sphere radii are small enough then their centroid coordinates will be described by the stationary (or homogeneous) Poisson point process for which: for any bounded region w the number of points of X falling in w follows the Poisson one-dimensional and m-dimensional distributions:

$$P(\Phi(w) = k) = \frac{(n\bar{w})^k}{k!} \exp(-n\bar{w}), \quad k = 0, 1, \ldots \tag{5.32}$$

and for nonoverlapping sets $w_1, \ldots, w_m$, the numbers $\Phi(w_1), \ldots, \Phi(w_m)$ are independent random variables

$$P(\Phi(w_1) = k_1, \ldots, \Phi(w_m) = k_m) = \frac{(n\bar{w}_1)^k}{k_1!} \ldots \frac{(n\bar{w}_m)^k}{k_m!} \exp(-n\sum_{i=1}^{m} \bar{w}_i), \tag{5.33}$$

$(k_1, \ldots, k_m \geq 0)$. These properties imply stationarity and isotropy because of the translation and rotation invariance of volume. A further implication is that the point positions are independent and uniformly distributed within w, when the points are placed in the area where each possible location for a point is equally likely to be chosen, and the location of each point is independent of the location of any other point. The uniform (binomial) point process of a given number $\Phi(w) = N$ of points $\mathbf{x}_1, \ldots, \mathbf{x}_N$, independent and uniformly distributed in w, if for nonoverlapping bounded $w_1, \ldots, w_N \subset w$, $P(\mathbf{x}_1 \in w_1, \ldots, \mathbf{x}_N \in w_N) = (\bar{w}_1 \ldots \bar{w}_N)/(\bar{w})^N$, with the binomial distribution of the number of points in any bounded region $w_j \subset w$:

$$P(\Phi(w_j) = k) = C_N^k (p(w_j))^k (1 - p(w_j))^{N-k}, \; k = 0, 1, \ldots, N, \tag{5.34}$$

where $p(w_j) = \bar{w}_j/\bar{w}$, $E\Phi(w_j) = Np(w_j)$, and $VarN(w_j) = Np(w_j)(1 - p(w_j))$.

A stationary Poisson point process may serve as a reference model for complete spatial randomness and can be easily simulated with a computer. For example, $N(w)$ points are generated with uniform random position in a region $w = [0, L_1] \times \ldots \times [0, L_d]$ as a sequence of $d \cdot N(w)$ independent random numbers $x_1, \ldots, x_{d \cdot N(w)}$ uniformly distributed on $[0, 1]$ and generated by a random number generator. The coordinates of the ith $(i = 1, N(w))$ point are then $\mathbf{x}_i = (L_1 x_{d \cdot i - d + 1}, \ldots, L_d x_{d \cdot i})^\top$. Although the hypothesis of a Poisson set of centers for non-overlapping spheres is not fulfilled for finite sphere radii, it can often be used as a useful approximate description of the observed structures [938]. We recall that for Poisson distribution:

$$H_s(r) = D(r) = 1 - e^{-n\omega_d r^d}, \quad K(r) = \omega_d r^d, \quad L(r) = r, \quad g(r) = 1, \tag{5.35}$$

where the first equality (5.35) means that the point field Φ seen from the position of an arbitrary chosen point $\mathbf{x}$ has the same distribution as the point field with the point $\mathbf{x} \in \Phi$. The expectation mean $\langle D \rangle$ and variation $E(s^2)$ of nearest neighbor distances are estimated as ($d = 2$, see [1032]): $\langle D\rangle = 0.4n^{-1/2}$, $E(s^2) = (4-\pi)/(2\pi n)$. For a Poisson point process, an equivalency $H_s(r) \equiv D(r)$ provides the basis for a test of whether or not a given point pattern is consistent with Poisson process. The Poisson distribution depends only on one parameter n, which has the meaning of the mean number of points per unit volume. A common unbiased estimator $\hat{n} = \phi(w)/\overline{w}$ can be used for calculation of all interesting quantities and distributions mentioned above.

Many important mathematical models of random point fields are constructed from Poisson fields. For infinitely extended stationary field, the following are known: the (i) Boolean model, (ii) non-overlapping sphere model, and (iii) the system of edges or faces of the Poisson Voronoi tessellation [1044], [1047].

The most famous Boolean model is generated from the Poisson field of primary (or *gern*) points $\mathbf{x}_1, \mathbf{x}_2, \dots$ of intensity n. The *grains* form a sequence of independent identically distributed compact sets K_i (such as, e.g., the spheres). The union of all these grains shifted to the gerns is the Boolean model $\Xi = \bigcup_{i=1}^{\infty}(K_i + \mathbf{x}_i)$. Usually the grains K_i are convex, although the non-convex grains are also important (the case where K_i are a finite point set of a Poisson field considered in the next section). The parameters of a Boolean model are the intensity n of the gern and parameters characterizing the grains K_i which can be estimated from the observation of appropriate geometric functionals of Ξ [781]. The generalization of the prescribed Poisson germ-grain model is the case where the primary grains are "tubes" around either the straight lines or planes uniformly distributed in space modeled as an isometric Poisson line field or Poisson plane field. The mentioned Boolean models do not coincide with real microstructures because any overlapping is in conflict with fundamental formation mechanism. However, the "complete randomness" of Boolean models often fits well to low-density approximations of material structures.

Space Dirichlet tessellations subdividing an Euclidian space into n-dimensional bounded convex polytopes (polygons in 2-D case) are widely used to characterize the spatial distribution, size, and shape of a filled phase [380], [379]. One provides a natural and unique approach [380] for defining a particle's neighbors and neighborhood. At first, a Poisson process of gern points is taken and the particle centers are placed at these gern points (the last problem is a nonoverlapping model considered later). The Dirichlet tessellation of the two-dimensional domain w yields a network of convex Voronoi polygons containing one gern point (inclusion) with the center $\mathbf{x}_i$ ($i = 1, \dots, n$) each, at most. The cells of this tessellation consist of those points that are closer to a gern point than to all other gern points. The interior of the Voronoi cell associated with the point $\mathbf{x}_i$ is the region $w_i = \{\mathbf{x} \in w : |\mathbf{x} - \mathbf{x}_i| < |\mathbf{x} - \mathbf{x}_j|, \ \forall j \neq i\}$ that is the neighborhood of $\mathbf{x}_i$. The tessellation is constructed by plotting lines to the centers of all nearby particles and then constructing perpendicular bisecting planes to those lines. Green and Simpson [399] have proposed the algorithm generating Voronoi polygons for n points by computing in $O(n \log n)$ time by tracing boundary adjustment, as a new polygon is fitted into a previously generated set. Ghosh and coworkers (for references see [381]) have developed a material-based Voronoi cell method for directly treating multiple-phase Voronoi polygons as elements in a finite element model for elastic and thermoelastoplastic problems; they suggested a modification to the standard tessellation procedure to preclude the situation where neighboring fibers are substantially different in size and are closely spaced, which may result in polygons that do not completely envelope their corresponding fibers, and may instead "cut" though the fibers. Since each Voronoi cell contains a single particle surrounded by the matrix, the Dirichlet tessellation can be used for a description of a statistical structure of composites in the form of the frequency distribution, for all cells, of the ratio in particle-to-cell volume that is also a measure of particle clustering in microstructure [82]. The Dirichlet tessellation is usually constructed from a set of points. Another partitions called skeletization by influence zones (SKIZs) [354] is built from a continuous and uniform dilatation procedure, starting from the inclusion surface, up to total erosion of the matrix. The SKIZ partitions (in contrast to Voronoi one) ensures that each cell completely envelopes its inclusion, which makes it possible to define a local concentration of inclusions.

In a nonoverlapping spherical model, we will again take a Poisson field Φ of gern point and start a uniform radial growth of grain in each of the points. The grain growth stops when its diameter is the distance to the nearest neighbor in the point field. For this model, called, the Stienen model, the diameter of the typical sphere has the same distribution as the nearest neighbor distance $D(r)$ of the Poisson point field (5.35) that leads to the representations for the mean volume of a typical sphere $\bar{v}_i = 2^{-d}/n$ and the volume concentration $\langle V \rangle = 2^{-d}$. The case of greater volume concentrations of nonoverlapping spheres with non-Poisson field of particle centers will be defined operationally in Section 5.4.

5.2.2 Statistically Homogeneous Clustered Point Fields

Many important mathematical models of random point fields are constructed from Poisson fields. The simplest generalization, called a *mixed Poisson field*, is generated by a Poisson field where the intensity $n(\mathbf{x})$ is a constant for each sample, but $n(\mathbf{x})$ varies from sample to sample. The basis of more usable *Neyman-Scott fields* is generated by the points (called "parent points") of a homogeneous Poisson field of constant intensity n_p. The set of "daughter points" forming the *cluster field* is independently scattered around each parent point according to the same density function (see, e.g., [902]). Because a cluster center is not usually a point in the process, there is no loss of generality in accepting the cluster center as the center of symmetry. For the sake of simplicity assume that the center of representative isotropic cluster is placed at the origin O. Then the coordinates of the ith daughter point are described by a random vector $\mathbf{x}_i$ with the same probability density function $d(\mathbf{x})$. In the simplest case of *Matern cluster field* (called also ideal cluster field), the number k of points has a Poisson distribution (5.32) with parameter μ. Then the positions $\mathbf{x}_i$ are uniformly distributed in the disc $b(O, R)$ of radius R centered at O:

$$d(\mathbf{x}) = (\pi R^2)^{-1}\chi_{b(O,R)}(\mathbf{x}), \quad \text{or} \quad d(r) = 2rR^{-2}\chi_{[0,R]}(r). \tag{5.36}$$

The parent points of a stationary Neyman-Scott field can be easily generated by a Poisson field (see Subsection 5.3.1) of intensity n_p in a delation $w \oplus b(O, R)$

that avoids the edge effect. Then the number of daughter points is determined for each cluster generated in the vicinity of each parent point (in the case of the Matern clusters a Poisson random number of the intensity μ is generated). The coordinates of each daughter point are simulated with a random position in a polar coordinate system $[0, R] \times [0, 2\pi]$ when the radial coordinate has the distribution $d(r)$ (5.36$_2$) and the angle coordinate is uniformly distributed on $[0, 2\pi]$. The pair correlation function $g(r)$ of cluster fields decreases for increasing r with the range characterization less than or equal to $2R$; it is possible that $g(0) = \infty$.

Matern cluster fields are most usable and for them some formulae are derived easily [1047]. So, the intensity n of the field is $n = n_p\mu$. The pair distribution function satisfies $g(r) = 1 + (2\pi n_p r)^{-1} f(r)$, where

$$f(r) = \begin{cases} \frac{4r}{\pi R^2}\left[\cos^{-1}\left(\frac{r}{2R}\right) - \frac{r}{2R}\sqrt{1 - \frac{r^2}{4R^2}}\right], & (0 \le r \le 2R), \\ 0, & (r > 2R). \end{cases} \tag{5.37}$$

In particular, $g(0) = 1 + (\pi n_p R^2)^{-1}$. The K-function has the form

$$K(r) = \pi r^2 + \frac{1}{n_p} \times \begin{cases} 2 + \pi^{-1}[(8z^2 - 4)\cos^{-1} z - 2\sin^{-1} z \\ +4z\sqrt{(1 - z^2)^3} - 6z\sqrt{1 - z^2}] & (r \le 2R), \\ 1, & (r > 2R), \end{cases} \tag{5.38}$$

where $z = r/2R$. We considered the simplest case of Matern cluster fields. The applicability of one or another cluster shape is defined by the physical problem being considered [110]. If the problem of recrystallization or residual stresses, it is reasonable that the the important clusters have a convex shape. If one is concerned by the damage properties of dielectric breakdown, a long chain of broken or conducting particles forming a "percolating network" can be more deleterious than the same number of particles arranged in a convex cluster; for chain clusters not only the local density of particles but also their arrangement are relevant, demonstrated in a limiting case of the "weakest link." Due to the numerous possible cluster configurations and limited prediction models, one usually considers a degenerate case of cluster materials that, nevertheless, has a wide applications. Namely, one analyzes so-called ideal cluster materials [1077] that contain either finite or infinite, deterministic or random ellipsoidal domains called particle clouds distributed in the composed matrix. In so doing the concentration of particles is piecewise and homogeneous within the areas of ellipsoidal clouds and composed matrix. In particular, in Chapter 12 we will consider a single particle cloud with the shape of a thick ply located in an infinite matrix with zero concentration of particles. The ideal cluster configuration may be created by distributing cloud centers deterministically or randomly and placing offspring points randomly and uniformly around parent points within a predetermined area (see, e.g., [981], [1033]). For a description of the cluster arrangement, the notion of "cluster radius" [570] is used for the presentation of a cluster as a particle of the second generation with the radius $R^{\rm clust} = \sum_i^N \sum_{j\neq i}^N |r_{ij}|/N(N-1)$ where i, j are particle numbers; r_{ij} is the interparticle distance; and N is a total number of particles in the cluster. Usually the cluster covers 80–9% of the total volume of the sphere with $R^{\rm clust}$ radius.

5.2.3 Inhomogeneous Poisson Fields

For researchers the statistical description of statistically homogeneous structures is a fascinating subject studied in many papers and books. In contrast, the analogous analysis for statistically inhomogeneous media such as clustered and functionally graded materials pose annoying problems. In such a case, the ergodicity fails, and ensemble and volume averages do not coincide. The degenerate case of this material is a random matrix composite bounded in some directions; in addition, the composite media for the inclusions are located in a region bounded in some directions, although unrestrictedness of the domain of inclusion locations does not preclude statistical inhomogeneity. For example, any laminated composite materials randomly reinforced with aligned fibers in each ply are a statistically inhomogeneous (or functionally graded) material (see Chapter 12 and Fig. 12.2). There are just a few theoretical papers which study structures with a gradient or that clustered objects [333], [416], [452] [916].

An important sort of random point field constructed from a homogeneous Poisson field is an inhomogeneous Poisson field with deterministic differences in the point density. Instead of the intensity n (5.32) an intensity measure Λ on R^d is introduced, so that the measure $\Lambda(\mathcal{B})$ of any measurable set $\mathcal{B}$ has a sense of the number points in $\mathcal{B}$. It is assumed that this field has the same independent property of the random variables $\Phi(w_1),\ldots\Phi(w_n)$ (if $w_i \cap w_j = \emptyset$, for $\forall i \neq j$) as the homogeneous Poisson field. However, the property (5.32) is modified:

$$P(\Phi(w)=k) = \frac{(\Lambda(w))^k}{k!} e^{-\Lambda(w)}, \quad k=0,1,\ldots \tag{5.39}$$

There is often a density $n(\mathbf{x})$ called the *density function* of Λ so that $\Lambda(w) = \int_w n(\mathbf{x})\, d\mathbf{x}$, and $\Lambda(w)$ is the mean number of points in w. In the homogeneous case $n(\mathbf{x}) \equiv n =$ const., an inhomogeneous Poisson field is reduced to a homogeneous Poisson field with intensity n (5.32). However, even if $n(\mathbf{x}) \neq$ const., it has a similar intuitive interpretation as a probability $n(\mathbf{x})\, d\mathbf{x}$ to detect a point of Φ in an infinitesimally small disc of center $\mathbf{x}$ and area $d\mathbf{x}$. In a similar manner, the probability of finding a point in each of two infinitesimally small discs centered at $\mathbf{x}_1$ and $\mathbf{x}_2$ with area $d\mathbf{x}_1$ and $d\mathbf{x}_2$ is $n(\mathbf{x}_1)n(\mathbf{x}_2)\, d\mathbf{x}_1\, d\mathbf{x}_2$. In a similar manner, one can define systems on a finite region $\mathcal{B}$ by choosing the intensity function to be a restriction of n to $\mathcal{B}$: $n(\mathbf{x})\chi_{\mathcal{B}}(\mathbf{x})$. We note that the formulae for the number distribution as in the homogeneous case (see Subsection 5.3.1) holds if $n\bar{w}$ (5.32) is replaced by $\Lambda(w)$. The same is true for $P(\Phi(w)=0) = e^{-\Lambda(w)}$.

The elegant method of simulation of an inhomogeneous Poisson field with intensity function $n(\mathbf{x})$ is based on the following thinning procedure of a homogeneous Poisson field. Let the intensity $n(\mathbf{x})$ be bounded $n(\mathbf{x}) \leq n^*$ and the function $p(\mathbf{x}) = n(\mathbf{x})/n^*$ is the so-called thinning function. This means that at first a homogeneous field of intensity n^* is generated in the observation window w (see Subsection 5.3.1). Then for every point, $0 \leq p(\mathbf{x}) \leq 1$ is estimated and compared with a random number $z \in [0,1]$ independently generated. If $z > p(\mathbf{x})$, the point $\mathbf{x}$ is eliminated; otherwise it is a point of a newly generated inhomogeneous Poisson field. The probability density $n(\mathbf{x})$ as well as all another microstructural functions for the inhomogeneous fields will depend on the absolute positions of their arguments. The statistical descriptions for particulate, statistically inhomogeneous two-phase random media were proposed [416], [916] (see also [333]) by the use of the theory of a general Poisson process [1047] and for some simulated fully penetrable (Poisson distribution) spheres the canonical n-point microstructure function, the nearest-neighbor functions, and the linear-path function were estimated, that, unlike the homogeneous case, will depend on their absolute positions.

In the graded structures there is a naturally given direction (e.g., the z-axis of the Cartesian coordinate system) such that the structure shows a trend-like variability in this direction, while in the planes perpendicular to the z-axis the random field is the homogeneous and isotropic. Many materials consist of layers and have a gradient in their interior or near to their boundaries owing to structural inhomogeneities. For statistical description of the microstructure of ceramic matrix composites containing fiber-rich and fiber-poor regions that have a laminated structure, Yang *et al.* [1205] have proposed so-called the strip model. The key parameters of the strip model are the number densities of fiber in the fiber-poor n^p and fiber-rich n^r regions, the widths of the fiber-poor l^p and fiber-rich l^r regions, and the global number density of the fibers $n \equiv (n^p l^p + n^r l^r)/(l^p + l^r)$. Thus, the number density in a strip model is described by a piecewise constant function. Generalization of this model is based on the introduction of a variable number of densities $n(z)$ defined as the average number of disc centers per unit in a thin ply parallel to the plate $z = 0$.

The most important problem of inhomogeneous Poisson fields is the inverse one of estimation of the intensity function $n(\mathbf{x})$ from the observed data $\mathbf{x}_i$. The determination of the unknown parameters in an analytical representation of an intensity function can be done using the maximum likelihood method. This approach was described in [416], [1047], for graded structures that are inhomogeneous along a particular gradient direction but homogeneous perpendicular to that direction. The choice of the analytical form of $n(\mathbf{x})$ is sometimes unnatural if there is no physical law determining the point density. In such a case, a simple method of kernel estimators for solving this problem can be used, presenting $\hat{n}_h(\mathbf{x})$ and its edge-correction form as

$$\hat{n}_h(\mathbf{x}) = \frac{\phi(b(\mathbf{x}, h))}{\text{mes}(b(\mathbf{x}, h))}, \quad \hat{n}_h(\mathbf{x}) = \frac{\phi(w \cap b(\mathbf{x}, h))}{\text{mes}(w \cap b(\mathbf{x}, h))}. \tag{5.40}$$

The representations (5.40) estimating the mean point density in a sphere centered at $\mathbf{x}$ with radius h depend on the parameter h, and the question arises as to which h leads to reasonable results. A more general case for 2-D problems using a kernel function $k_h(\mathbf{z})$ is $(N = \phi(w))$:

$$\hat{n}_h(\mathbf{x}) = \sum_{i=1}^{N} k_h(\mathbf{x} - \mathbf{x}_i), \tag{5.41}$$

where the Epanecnikov kernel has the "linear" and "quadratic" versions $k_h(\mathbf{y}) = \frac{8}{3\pi h} e_h(\|\mathbf{y}\|)$, $k_h(\mathbf{y}) = e_h(y')e_h(y'')$, $(\mathbf{y} = (y', y''))$, respectively. Here $e_h(t) = \frac{3}{4h}(1 - t^2/h^2)\chi_{(-h,h)}(t)$ and the smoothing parameter h is recommended for a quadratic window w as $h = 0.68N^{-0.2}$. The particular choice $k_h(\mathbf{y}) = \chi_{b(O,h)}(\mathbf{y})(\text{mes}b(O, h))^{-1}$ reduces the estimator (5.41) to Eq. (5.40_1). For graded materials with the graded direction $\mathbf{z}$, the estimator $\hat{n}(z)$ using the Epanecnikov kernel is $\hat{n}_h(z) = \sum_{i=1}^{N} e_h(z - \mathbf{x}_i^z)$, where $\mathbf{x}_i^z$ is the z-coordinate of the ith point.

Usually one analyzes so-called ideal cluster materials that contain either a finite or infinite, deterministic or random ellipsoidal domains called particle clouds distributed in the composite matrix. In so doing the concentration of particles is a piecewise constant and a homogeneous one within the areas of ellipsoidal clouds and composite matrix [652], [989]. The most used descriptor for the clustered and graded materials is a volume concentration of inclusions [256], [780], [932], [1059], which is not enough for the characterization of the micromorphology of fillers, simply because one can present other morpholies with the same descriptor. However, just taking into account binary interacting inclusions effects directly dependent on the radial distribution function allows to detect some fundamentally new nonlocal effects for graded materials (see Chapter 12). In light of this, the further development of a statistical quantitative description for so many prospective clustered and graded materials (see, e.g., [333], [916], [989], [1084], [1205]) is of profound importance in both practically and theoretically sense.

5.2.4 Gibbs Point Fields

For the Poisson fields, the location of a point is independent of the location of any other point. The Gibbs fields studied in both bounded regions and in the whole space are more regular and take the interactions of particles into account (statistically isotropic fields of unmarked spherical particles with the radius a are considered). This approach starts with a potential function $\Phi^{\rm p}(\mathbf{x}^N)$ defining the probability of a configuration $\mathcal{C}^N = \{\mathbf{x}^N\}$, $[\mathbf{x}^N = (\mathbf{x}_1, \ldots, \mathbf{x}_N)]$ with fixed particle number N in a bounded box w [424], [1046], [1106], [1232]. In the absence of an external field, we can assume that the potential function contains only pairwise additive components:

$$\Phi^{\rm p}(\mathbf{x}^N) = \sum_{i<j}^{N} \phi^{\rm p}(|\mathbf{x}_{ij}|) \tag{5.42}$$

because higher-order potentials are negligibly small compared to two-particle terms; $\mathbf{x}_{ij} = \mathbf{x}_i - \mathbf{x}_j$. In the considered case with fixed particle number N and pair potential $\Phi^{\rm p}$ (5.42), the joint distribution of the N particles is the Boltzmann-Gibbs canonical distribution:

$$\varphi(\mathbf{x}^N) = \frac{1}{Z_N} \exp\Big[-\beta \sum_{i=1}^{N} \sum_{j=i+1}^{N} \phi^{\rm p}(|\mathbf{x}_{ij}|) \Big], \tag{5.43}$$

where $\beta = 1/(kT)$ is a reciprocal temperature T, k is Boltzmann's constant, and the normalizing constant Z_N called configurational partitional function ensures that $\varphi(\mathbf{x}^N)$ is a density function

$$Z_N = \int \exp(-\beta \Phi^{\rm p}(\mathbf{x}^N))\, d\mathbf{x}^N,$$

where $\int(\cdot)\,d\mathbf{x}^N$ denotes integration over w^N. Due to the difficulties of computation of Z_N, Ogata and Tanemura [828], [829] proposed an approximate method of the second cluster approximation

$$Z_N = \overline{w}^n(1 - a\overline{w}^{-1})^{n(n-1)/2}, \tag{5.44}$$

where $a = d\omega_d \int_0^\infty \left(1 - e^{-\beta\Phi^{\rm p}(r)}\right) r^{d-1}\,dr$, and $d\omega_d$ is the surface area of the unit ball in d-space (5.29). The positive and negative values of $\phi^{\rm p}(r)$ lead to "repulsion" and "attraction" of particles, so that the interparticle distances r appear rarely and frequently, respectively.

By considering different $\Phi^{\rm p}$ one gets a wide variety of distributions from (5.43). For example, for a system of noninteracting inclusions, we have $\Phi^{\rm p}(\mathbf{x}^N) \equiv 0$ for $\mathbf{x}^N \in w$, and Eq. (5.43) defines a uniform distribution of N particles in the box w. If $\phi^{\rm p}(r) = \infty$ for $r < 2a$ then interpoint distances smaller than $2a$ are impossible. If additionally $\phi^{\rm p}(r) = 0$ at $r > 2a$ then any interaction between the particle is absent and the potential function

$$\phi^{\rm p}(r) = \begin{cases} \infty, & (r \le 2a), \\ 0, & (r > 2a), \end{cases} \tag{5.45}$$

forms a Poisson hard core field (or nonoverlapping sphere model considered in Subsection 5.3.1) that is ergodic and the most popular in micromechanics. The important particular cases of the more complicated potential functions are

$$\phi^{\rm p}(r) = \begin{cases} \infty, & (r \le \sigma), \\ -\epsilon, & (\sigma < r \le 2a), \\ 0, & (r > 2a), \end{cases} \quad \phi^{\rm p}(r) = 4\epsilon\left[\left(\frac{\sigma}{r}\right)^{12} - \left(\frac{\sigma}{r}\right)^{6}\right], \tag{5.46}$$

both of which incorporate both repulsive and attractive interactions of particles. Here the parameter ϵ is the attractive well depth in the minimum in $\phi^{\rm p}(r)$. In Eq. (5.46_1) the parameter σ is a hard-core distance estimated by the minimum interpoint distance in the pattern while $2a$ is a radius of interaction. The parameter ϵ describes a jump at $r = 2a$ of the radial distribution function (5.30) of corresponding Gibbs field $\lim_{r\uparrow 2a} g(r) = e^{\epsilon} \lim_{r\downarrow 2a} g(r)$. Ogata and Tanemura [829] describe maximum-likelihood estimation of parameters for these models in case of sparseness by the use of Eqs. (5.43) and (5.44). The square-well potential (5.46_1) at $\epsilon = 0$ is reduced to the cherry-pit potential in which each sphere has a core of diameter σ that is impenetrable to other cores. The surrounding shells, however, can penetrate other shells and cores that can be appropriately modified to model polymerization-filled composites [310]. A limiting case of the square-well potential (5.46_1) reducing attractive interactions to a delta function at interphase is called the sticky hard-sphere potential, which provides a modeling of aggregation processes. Lennard-Jones potential (5.46_2), widely used in the theories of classical simple liquids and colloidal dispersions [424], [952], vanishes $\phi^{\rm p}(r) = 0$ at r equals the collision diameter σ.

It is rather difficult to estimate the characteristics of the Gibbs fields (such as, e.g., $g(r)$ and $D(r)$) in terms of the intensity n and potential function $\phi^{\rm p}(r)$ determining the character of the point distribution (hard core, cluster, etc.). For the solution of this problem Ornstein and Zernike [839] proposed an integral equation linking dimensionless parameters such as the total correlation function $h(\mathbf{x}) = g_2(\mathbf{x}) - 1$ and the direct correlation function $c(\mathbf{x})$:

$$h(\mathbf{x}_{12}) = c(\mathbf{x}_{12}) + n(c * h)(\mathbf{x}_{12}) \tag{5.47}$$

where $c(\mathbf{x}_{12})$ describes a direct short-range effect of particle at $\mathbf{x}_1$ on one at $\mathbf{x}_2$, and the convolution integral depends on the indirect effects weighted by the intensity and averaged over all possible $\mathbf{x}_3$. This integral can be presented in 2-D case as

$$(c * h)(r) = \int_0^{2\pi} \int_0^{\infty} c(s)h((r^2 + s^2 - 2rs\cos\xi)^{1/2})\, ds\, d\xi.$$

Considering that Eq. (5.47) is a convolution integral equation with constant coefficients, the standard method of its solution is using the Fourier transform method to transform the integral Eq. (5.47) into the division problem of solving the multiplicative algebraic equation

$$\tilde{h}(\mathbf{k}) = \tilde{c}(\mathbf{k}) + n\tilde{c}(\mathbf{k})\tilde{h}(\mathbf{k}) = \frac{\tilde{c}(\mathbf{k})}{1 - n\tilde{c}(\mathbf{k})}, \tag{5.48}$$

where the Fourier transformation $\tilde{\mathbf{f}}(\mathbf{k})$ of a function $\mathbf{f}(\mathbf{x})$ and its inverse are defined by the formulae (3.4).

According to Eq. (5.43), we can put the initial low-density approximation of $c(\mathbf{x}_{12})$ as the Mayer f-function:

$$c^{(0)}(\mathbf{x}_{12}) \approx f(\mathbf{x}_{12}) = \exp(-\beta\Phi^{\rm p}(\mathbf{x}_{12})) - 1 \tag{5.49}$$

closing Eq. (5.47). The next step resides in modification of Eq. (5.49) for better fit of the solution to the results of computer simulation. An appropriate *Percus-Yevick* (PY) approximation [867] contains an additional factor in Eq. (5.48)

$$c^{(0)}(\mathbf{x}_{12}) \approx \Big[\exp(-\beta\Phi^{\rm p}(\mathbf{x}_{12})) - 1\Big] \exp(\beta\Phi^{\rm p}(\mathbf{x}_{12}))g_2(\mathbf{x}_{12}) \tag{5.50}$$

so that $c(\mathbf{x}_{12})$ is zero wherever the potential vanishes.

The numerical method of solution of Eq. (5.47) is based on the solution of Eq. (5.47). The Fourier transform $\tilde{c}(\mathbf{k})$ of an initial approximation $c^{(0)}(\mathbf{x}_{12})$ is calculated and inserted in Eq. (5.48) and an inverse transform leads to an initial approximation $h^{(0)}(\mathbf{x}_{12})$. After that, the appropriate relation between $f(r)$ and $c(r)$ is used to obtain of the first iteration $c^{(1)}(r)$. An obvious iteration scheme continues until convergence is achieved. However, for the hard-core field of 3-D spheres the PY approximation enables one to solve Eq. (5.47) analytically by the Laplace transform method [1174]: $c(\rho) = -a_1 - 6ca_2\rho - \frac{1}{2}ca_1\rho^3$ at $(\rho \le 1)$, and $c(\rho) = 0$ at $(\rho > 1)$, where $a_1 = (1 + 2c)^2(1 - c)^{-4}$, $a_2 = -(1 + c/2)^2(1 - c)^{-4}$, and $\rho = r/(2a)$, $c = n\bar{v}_i$ is a volume concentration of inclusions. The solution $h(r)$ (and $g_2(r)$) can be found by Fourier transformation of $c(r)$, substituting $\tilde{c}(\mathbf{k})$ into (5.48), and inverting the last representation (5.48):

$$h(r) = \frac{1}{2\pi r} \int_0^{\infty} \frac{\tilde{c}(k)k \sin kr}{1 - n\tilde{c}(k)}\, dk, \tag{5.51}$$

where

$$\tilde{c}(k) = -\frac{4\pi}{q^3}\Big\{a_1[\sin q - q\cos q] + \frac{6\pi c a_2}{q}[2q\sin q + (2-q^2)\cos q - 2] + \frac{ca_1}{2q^3}[4q(q^2-6)\sin q - (24 - 12q^2 + q^4)\cos q + 24]\Big\}, \tag{5.52}$$

and $q = 2ak$. Unfortunately, the mentioned problem was not solved analytically for the hard-disk system. A semiempirical modification of the PY estimation of $g_2(r)$ proposed in [1096], [1137] provides a good fit of the theory to the numerical data. The equation of the direct correlation function of hard d-spheres was obtained [44] allowing a unified treatment of both the odd and even space dimensions d. In the particular case $d = 2$, they obtained

$$c(\rho) = -\frac{\partial}{\partial c}[cz((c)]\chi_{[0,1]}(\rho)\Big\{1 - b^2c + \frac{2}{\pi}b^2c\Big[\cos^{-1}\frac{\rho}{b} - \frac{\rho}{b^2}\sqrt{b^2-\rho^2}\Big]\Big\}, \tag{5.53}$$

where $b = b(c)$ is obtained by solving the equation

$$\frac{2}{\pi}\Big[b^2(b^2-4)\sin^{-1}\frac{1}{b} - (b^2+2)\sqrt{b^2-1}\Big] = \frac{1}{c^2}\Big\{1 - 4c - \Big[\frac{\partial}{\partial c}[cz(c)]\Big]^{-1}\Big\} \tag{5.54}$$

with subsequent calculation of the direct and inverse Fourier transform for the functions $\tilde{c}(k)$ and $g_2(r)$, respectively. Baus and Colot [44] analyzed different sophisticated representations for z; the most simple of them is $z(c) = (1 + 0.125c^2)(1-c)^{-2}$.

For completeness, we will reproduce some analytical approximations of numerical simulations and experimental data. For the system of identical spheres (3-D case), the formula

$$g(r) = [1 - \chi_{[0,2a)}(r)]\Big[1 + \Big(\frac{2+c}{2(1-c)^2} - 1\Big)\cdot\cos(\frac{\pi r}{a})e^{2(2-r/a)}\Big], \tag{5.55}$$

was proposed in [541], [545] as a modification of the formula [1180], which takes into account a neighboring order in the distribution of the inclusions (see also [1039]). The representation for the 2-D case of identical circle system [424], [1108] is also known:

$$g(r) = [1 - \chi_{[0,2a)}(r)]\Big\{1 + \frac{4c}{\pi}\Big[\pi - 2\sin^{-1}(\frac{r}{4a}) - \frac{r}{2a}\sqrt{1 - \frac{r^2}{16a^2}}\Big]\chi_{[0,4a]}(r)\Big\}. \tag{5.56}$$

In the limiting case of dilute concentration of inclusions ($c \to 0$) the representations (5.55) and (5.56) are reduced to the well-stirred approximation of noninteracting particles:

$$g(r) = 1 - \chi_{[0,2a)}(r). \tag{5.57}$$

differing from the RDF for a Poisson distribution (5.32) by the availability of "excluded volume" with the center $\mathbf{x}_i$ where $g(\mathbf{x}_i - \mathbf{x}_q) \equiv 0$.

5.3 Ensemble Averaging of Random Structures

The method of ensemble averaging is very popular in micromechanics [16], [40], [41], [66], [120], [358], [453], [465], [610], [612], [900], [1106] and will be summarized below in a form selected for the subsequent presentation. The advantage with respect to volume-averaging is complete generality, which does not require the separation of scalars needed in volume averaging the existence of "an elementary macrovolume dV ... the characteristic linear dimensions of which are many times greater than the nonuniformities a but at the same time much less than the characteristic macrodimensions L of the problem" [810]. For example, Cemlins [200] explicitly demonstrated a highly oscillatory of the volume fraction defined on the basis of volume averaging unless the size of averaging volume is several interparticle distances. However, in the context of ensemble averaging, these spurious oscillations are easily removed by the use of the very natural definition of volume fraction as the probability that the point $\mathbf{x}$ is occupied by the phase of interest. The ensemble averaging produces the results which have a clear meaning in terms of the probability densities, and there is no necessity to require the presence of a great number of particles in physically small volumes of a composite as in the case for volume averaging. Ensemble averaging can be used in a unified way as a basis for the formulation of a variety of approximations of the closure problem in which the unknown variables are expressed in a computable form. Ensemble averaging can be applied both to random and periodic structure composites because an exclusive case of strictly ordered periodic systems can be included in the analysis by allowing for the ensemble distribution function to be expressed in terms of proper delta-functions. The advantage of ensemble averaging is most pronounced in the analysis of statistically inhomogeneous heterogeneous media (bounded or infinite) subjected to essentially inhomogeneous loading by the fields of the traction, temperature, and body forces. Each distinguishing feature of either the composite structure or the parameters of loading mentioned above leads to the loss of the statistical homogeneity of stress fields estimated that makes the use of the volume averaging technique impossible for the purpose of evaluating effective properties. However, a feasible way of generalizing all results and conclusions, obtained for statistically homogeneous composites by the use of ensemble average, to the more complex situation of nonlocal problems is quite straightforward, so that there is no need to devise special methods in order to treat statistically inhomogeneous media.

The development of an adequate mathematical formalism ensures a necessary foundation for averaging local constitutive equations valid in each phase of a composite. It provides a proper statistical smoothing and eliminates excessive details related to exact positions and orientations of all the inhomogeneities as well as to local stress fields in their vicinity. Such a procedure of smoothing in the most general averaging sense has to include the concept of possible states of the system of particles with actually unknown random parameters characterizing the composite structure at the particle level.

The packed random structure can be characterized by several parameters, such as packing density, coordination number, radial distribution function, particle cage, inter-particle spacing, and others. The various methods of estimating the effective properties of a composite material use a knowledge of statistical geometrical information about the microstructure. In order to incorporate the spatial arrangement of components in a micromechanical simulation, it is essential to quantitatively characterize the random structure of the composite. The most important factors characterizing the microstructure of composite material are the shape, volume fraction, and arrangement (random or regular) of the components that permit the calculation of bounds of effective moduli. There are two possible ways of describing the distribution (or configuration $\mathcal{F}$) of constituents in the composite. In the first way, the statistical moments of the probability densities of vectors describing locations, shapes, and orientations of different inhomogeneities in the medium are given. The second way is based on specification of which constituent each point of medium belongs and prescribes the multipoint moments of characteristic function of the region occupied by the inclusions.

5.3.1 Ensemble Distribution Functions

Let a linear elastic infinite body occupying full space R^d contain an open bounded domain $w \subset R^d$ (window of observation) with a boundary Γ and with an indicator function W and space dimensionality d ($d = 2$ and $d = 3$ for 2-D and 3-D problems, respectively). The domain w contains a homogeneous matrix $v^{(0)}$ with indicator function $V^{(0)}(\mathbf{x})$ and a random finite set $\Omega = (v_i)$ $(i = 1, \ldots, N(w))$ of inclusions v_i with centers $\mathbf{x}_i$ and with indicator functions $V_i(\mathbf{y})$ equal to one in $\mathbf{y} \in v_i$ and zero otherwise and bounded by the surfaces Γ_i; $v = \cup v_i$, $v^{(0)} = w \setminus v$ and all domains v_i, $v^{(0)}$ are open $(i = 1, 2, \ldots, N)$. It is assumed that the representative mesodomain w contains a statistically large number of cranes v_i and that the grain can be grouped into components (phases) $v^{(k)}$ $\quad (k = 1, 2, \ldots, N)$ with identical mechanical and geometrical properties (such as the shape, size, orientation, and microstructure of inclusions). Members of the ensemble are symbolized by discernible sets of the vectors $\mathbf{x}_i$ acceptable from a physical point of view. In a more general case, when the inhomogeneities are of the same but arbitrary form and of different size, it is necessary to incorporate into the analysis a set of unit orientation vectors $\mathbf{g}$ locked inside each particle and to introduce likewise a scale of particles $\mathbf{a}$ as the new independent ensemble variables. Thus, the random point field Φ defined by the points $\{\mathbf{x}_i\}$ is determined now by a triplet $\{(\mathbf{x}_i, \mathbf{g}_i, \mathbf{a}_i)\}$. In so doing, the set $\{(\mathbf{g}_i, \mathbf{a}_i)\}$ can be considered as a mark $\{\mathbf{m}_i\} = \{(\mathbf{g}_i, \mathbf{a}_i)\}$ of the marked process $\Phi = \{(\mathbf{x}_i, \mathbf{m}_i)\}$; the mark $\mathbf{m}_i$ can also contain some other information about the particles such as, e.g., their material "properties", particularly elastic moduli and stresses [225], [381]. We will use the term *configuration* $\mathcal{C}^N$ to indicate the set of values of numbers of quantities $\{(\mathbf{x}_i, \mathbf{m}_i)\}$ sufficient to specify uniquely the state of a system. The configuration $\mathcal{C}^N$ is completely described by a detailed distribution function, $\varphi(t, \mathbf{x}^N, \mathbf{m}^N)$, for N particles with the centers $\mathbf{x}^N = (\mathbf{x}_i, \ldots, \mathbf{x}_N)$ and marks $\mathbf{m}^N = (\mathbf{m}_1, \ldots, \mathbf{m}_N)$. Thus, $\varphi(t, \mathcal{C}^N) d\mathcal{C}^N \equiv \varphi(t, \mathbf{x}^N, \mathbf{m}^N)\, d\mathbf{x}^N\, d\mathbf{m}^N$ represent the probability of finding, at time t, the system in a configuration in which the first inclusion is within $d\mathbf{x}_1$, $d\mathbf{m}_1$ of $(\mathbf{x}_1, \mathbf{m}_1)$, etc. The time dependence can describe evolving microstructure that may arise as a result of, e.g., phase transitions and plastic deformations. However, we will usually assume that the microstructures are static or can be considered as static, and the function $\varphi(\mathbf{x}^N, \mathbf{m}^N)$ will be taken to be independent of time. The center locations $\mathbf{x}^N$ are arbitrary within w, except for the restriction $V_i(\mathbf{x})V_j(\mathbf{x}) = 0$ and $V_i(\mathbf{x})V^{(0)}(\mathbf{x}) = 0$ for any numbers $i \neq j$ and $\forall \mathbf{x} \in w$ that result from the mutual conditions of nonoverlapping of inclusions and of impenetrability of the boundaries. The distribution functions $\varphi(\mathcal{C}^N)$ are regarded as the probability density in which to find in the domains $v_1, \ldots, v_N$ the inhomogeneities with the centers $\mathbf{x}^N$ and marks $\mathbf{m}^N$. The normalized condition for $\varphi(\mathcal{C}^N)$ reads in the proper phase space as

$$\int \varphi(\mathcal{C}^N) d\mathcal{C}^N = 1. \tag{5.58}$$

The information contained in the marked random density function $\varphi(\mathcal{C}^N)$ can be too detailed and one has to use a reduced probability density of lower order which can be obtained by integrating $\varphi(\mathcal{C}^N)$ over the positions of $m < n$ inhomogeneous leading to less detailed distribution functions for the remaining $n - m$ particles. In particular, the one- and two-particle distribution functions for one and two fixed particles can be obtained from Eq. (5.3) $(i, j, k = 1, \ldots, N)$

$$\varphi(\mathbf{x}_i, \mathbf{m}_i) = \int \ldots \int \varphi(\mathcal{C}^N) \prod_{k \neq i} d\mathbf{x}_k \; d\mathbf{m}_k, \tag{5.59}$$

$$\varphi(\mathbf{x}_i, \mathbf{x}_j, \mathbf{m}_i, \mathbf{m}_j) = \int \ldots \int \varphi(\mathcal{C}^N) \prod_{k \neq i \neq j} d\mathbf{x}_k \; d\mathbf{m}_k, \tag{5.60}$$

The first equation (5.59) gives the unconditional probability density of the center of a particle with mark $\mathbf{m}_i$ being placed at $\mathbf{x}_i$. The equation (5.60) gives the probability density of an event in which simultaneously the centers of two particles with marks $\mathbf{m}_i$ and $\mathbf{m}_j$ are at $\mathbf{x}_i$ and $\mathbf{x}_j$, respectively. It follows from these definitions and from Eq. (5.59) that all $\varphi(\mathbf{x}^M, \mathbf{m}^M)$ $(M = 1, \ldots, N)$ are normalized.

In addition to probability densities $\varphi(\mathbf{x}^M, \mathbf{m}^M)$ $(M = 1, \ldots, N)$, we will follow to the definition (5.11) and introduce various conditional probability density configurations in which some particles are fixed. For example, the densities $\varphi(\mathbf{x}^{N-1}, \mathbf{m}^{N-1}|\mathbf{x}_i, \mathbf{m}_i)$ and $\varphi(\mathbf{x}^{N-2}, \mathbf{m}^{N-2}|\mathbf{x}_i, \mathbf{x}_j, \mathbf{m}_i, \mathbf{m}_j)$ corresponding to the configurations where the inclusions v_i as well as v_i and v_j are fixed, respectively, $\{\mathbf{x}^{N-1}, \mathbf{m}^{N-1}\}$ and $\{\mathbf{x}^{N-2}, \mathbf{m}^{N-2}\}$ are the hypersurfaces of $\{\mathbf{x}^N, \mathbf{m}^N\}$ which play the role of the phase space for all other $N-1$ and $N-2$ inclusions. For example, $\varphi(\mathbf{x}^{N-1}, \mathbf{m}^{N-1}|\mathbf{x}_i, \mathbf{m}_i)$ is the probability of a configuration $\mathcal{C}^{N-1}$ being found in the range $d\mathcal{C}^{N-1}$ about $\mathcal{C}^{N-1}$ where a particle with the mark $\mathbf{m}_i$ at the point $\mathbf{x}_i$ is fixed. These conditional probability densities are defined analogously to Eq. (5.11):

$$\varphi(\mathbf{x}^{N-1}, \mathbf{m}^{N-1}|\mathbf{x}_i, \mathbf{m}_i) = \varphi(\mathbf{x}^N, \mathbf{m}^m)/\varphi(\mathbf{x}_i, \mathbf{m}_i), \tag{5.61}$$

$$\varphi(\mathbf{x}^{N-2}, \mathbf{m}^{N-2}|\mathbf{x}_i, \mathbf{x}_j, \mathbf{m}_i, \mathbf{m}_j) = \varphi(\mathbf{x}^{N-1}, \mathbf{m}^{N-1}|\mathbf{x}_i, \mathbf{m}_i)/\varphi(\mathbf{x}_i, \mathbf{x}_j, \mathbf{m}_i, \mathbf{m}_j). \tag{5.62}$$

Extension of the chain rule used in Eqs. (5.61) and (5.62) leads to the definition of conditional distributions $\varphi(\mathcal{C}^{N-K}|\mathcal{C}^K)$ $(0 < K < N)$, $\mathcal{C}^K$ and $\mathcal{C}^{N-K}$ are the hypersurfaces of $\mathcal{C}^N$, $\mathcal{C}^K \cap \mathcal{C}^{N-K} = \emptyset$. Let us introduce a conditional probability density ($\mathcal{C}^{N-K} = \mathcal{C}^{N-K-1} \cup \{\mathbf{x}_m, \mathbf{m}_m\}$):

$$\varphi(\mathbf{x}_m, \mathbf{m}_m | \mathcal{C}^K) = \int \varphi(\mathcal{C}^{N-K} | \mathcal{C}^K) d\mathcal{C}^{N-K-1},$$

which is a probability density to find the m-th inclusion with the center $\mathbf{x}_m$ and the mark $\mathbf{m}_m$ in the domain v_m with fixed inclusions $v_1, \ldots, v_K$ with the centers $\mathbf{x}_1, \ldots, \mathbf{x}_K$ and the marks $\mathbf{m}_1, \ldots, \mathbf{m}_K$. The notation $\varphi(\mathbf{x}_m, \mathbf{m}_m|; \mathcal{C}^K)$ denotes the case $\mathbf{x}_m \neq \mathbf{x}_1, \ldots, \mathbf{x}_K$. The condition of nonoverlapping leads to an obvious restriction of the function $\varphi(\mathbf{x}_m, \mathbf{m}_m|; \mathcal{C}^K)$ which would be quite complicated in general, especially in the case of spatial correlation of location $\mathbf{x}_i$ and $\mathbf{m}_i$ of inclusions.

We will consider the limiting behavior and some simplifying assumption for this function. For an "overlapping inclusion" model $\varphi(\mathcal{C}^N) = \prod_{i=1}^N \varphi(\mathbf{x}_i, \mathbf{m}_i)$. A more realistic probability density must account for the mutual exclusiveness of the inclusions. Of course, $\varphi(\mathbf{x}_m, \mathbf{m}_m|; \mathcal{C}^K) = 0$ for values of $\mathbf{x}_m$ lying inside the"excluded volumes" (or "correlation holes") $\cup v_{0i}$ $(i = 1, \ldots, K)$, where $v_{0i} \supset v_i$ with characteristic functions V_{0i} (since inclusions cannot overlap). It will be usually assumed that there is no long-range order in the medium, and that the state probabilities of particles whose separation is large compared with a particle size are independent: $\varphi(\mathbf{x}_m, \mathbf{m}_m|; \mathcal{C}^K) \to \varphi(\mathbf{x}_m, \mathbf{m}_m)$ at $|\mathbf{x}_i - \mathbf{x}_m| \to \infty, \quad i = 1, \ldots, n$. The basic simplifying assumption uses these limiting behaviors as an approximation of the distribution function:

$$\varphi(\mathbf{x}_m, \mathbf{m}_m|; \mathcal{C}^K) = \sum_{i=1}^{K} (1 - \chi_{v_{0i}}(\mathbf{x}_m))\varphi(\mathbf{x}_m, \mathbf{m}_m), \; (l = 1, \ldots, K) \tag{5.63}$$

which is called a well-stirred approximation and is in fact a low-density limit for multiparticle distribution function. It means that the inclusions are forbidden only to overlap; otherwise their locations are statistically independent. The next assumption is the absence of a spatial correlation between location $\mathbf{x}^N$ and mark $\mathbf{m}^N$ of inclusions, i.e., spatial location and mark are statistically independent:

$$\varphi(\mathcal{C}^N) = \varphi_{\mathbf{x}}(\mathbf{x}^N)\varphi_{\mathbf{m}}(\mathbf{m}^N). \tag{5.64}$$

The second mark assumption is that the marks of different inclusions are also statistically independent. This means that the mark distribution functions are also factorized:

$$\varphi_{\mathbf{m}}(\mathbf{m}^N) = \varphi_{\mathbf{m}}(\mathbf{m}_1) \cdot \ldots \cdot \varphi_{\mathbf{m}}(\mathbf{m}_N). \tag{5.65}$$

For statistically homogeneous media, $\varphi(\mathcal{C}^N)$ is translationary invariant $\varphi(\mathcal{C}^N) = \varphi(\mathbf{x}_{12}, \ldots, \mathbf{x}_{1n}, \mathbf{m}^N)$, where $\mathbf{x}_{ij} = \mathbf{x}_i - \mathbf{x}_j$. In particular, for unmarked fields we can define the number density $\varphi(\mathbf{x}_1) \equiv \lim_{\bar{w}\to\infty} N/\bar{w}$ and the n-particle correlation function $g_N(\mathbf{x}^N) = n^{-N}\varphi(\mathcal{C}^N)$, which tends to unity for the system with no long-range order and $|\mathbf{x}_{ij}| \to \infty$ $(1 \le i, j \le N, \; i \neq j)$. The widely used pair correlation function $g_2(\mathbf{x}_{12})$ for statistically isotropic media (5.22) depends on the radial distance $|\mathbf{x}_{12}|$ only:

$$g_2(\mathbf{x}_{12}) = g_2(|\mathbf{x}_{12}|), \tag{5.66}$$

and is called the radial distribution function.

We have considered probability densities treating the particles as distinguishable with states $\mathbf{x}_i, \mathbf{m}_i$ labeled by the index i ($i = 1, \ldots, N$). If N particles in the region w are identical, we should not consider their states as different from each other, and, because of this, we need to introduce suitable corrections to the normalization condition (5.58) to account for the indistinguishability of the states. The number of placement events of distinguishable and identical particles differs by the permutation number and so we have the normalized relation

$$\frac{1}{N!}\int \varphi(\mathcal{C}^N)\, d\mathcal{C}^N = 1 \tag{5.67}$$

where each of the N volume integrals comprising integration with respect to $\mathcal{C}^N$ is taken over the whole of the volume w. The normalization condition (5.67) permits the use of a probability density in which the particles are distinguished. So, the reduced K particle probability distribution with prescribed configuration of K particles can be introduced in much the same way as (5.60):

$$\varphi(\mathcal{C}^K) = \frac{1}{(N-K)!}\int \varphi(\mathcal{C}^N)\, d\mathcal{C}^{N-K}, \tag{5.68}$$

with the integration over all physically permissible configurations $\mathcal{C}^{N-K}$ where the factor $(N-K)!$ accounts for the identicability of states of $N-K$ particles. Taking into account the normalized condition (5.67) in the integration of Eq. (5.68) over the configuration of K particles leads to another normalized relation:

$$\int \varphi(\mathcal{C}^K) d\mathcal{C}^K = \frac{N!}{(N-K)!}, \tag{5.69}$$

which can be used to obtain of the normalized condition for the one-particle probability density $\varphi(\mathbf{x}_1, \mathbf{m}_1)$: $\int \varphi(\mathbf{x}_1, \mathbf{m}_1)\, d\mathbf{x}_1\, d\mathbf{m}_1 = N$ justifying the following definition of particle number density $n(\mathbf{x}_1) = \int \varphi(\mathbf{x}_1, \mathbf{m}_1)\, d\mathbf{m}_1$ as the average number of particles per unit volume.

In parallel with the conditional probability densities $\varphi(\mathbf{x}_m, \mathbf{m}_m|\mathcal{C}^K)$ and $\varphi(\mathbf{x}_m, \mathbf{m}_m|;\mathcal{C}^K)$ of the marked field Φ, other equivalent notations are used. Namely, for the description of the random structures let us introduce at first the multipoint probability densities $\varphi(v_1, \mathbf{x}_1, \ldots, v_n, \mathbf{x}_n)$ that give the probability $\varphi(v_1, \mathbf{x}_1, \ldots, v_n, \mathbf{x}_n)\, d\mathbf{x}_1 \ldots d\mathbf{x}_n$ to find an inclusion center in the vicinity $\mathbf{x}_1 \ldots \mathbf{x}_n$ of the points $\mathbf{x}_1, \ldots, \mathbf{x}_n$, respectively. A conditional probability density $\varphi(v_i, \mathbf{x}_i|v_1, \mathbf{x}_1, \ldots, v_n, \mathbf{x}_n)$ is a probability density to find the ith inclusion with the center $\mathbf{x}_i$ in the domain v_i with fixed inclusions $v_1, \ldots, v_n$ with the centers $\mathbf{x}_1, \ldots, \mathbf{x}_n$. The notation $\varphi(v_i, \mathbf{x}_i|; v_1, \mathbf{x}_1, \ldots, v_n, \mathbf{x}_n)$ denotes the case $\mathbf{x}_i \neq \mathbf{x}_1, \ldots, \mathbf{x}_n$. The notation $\varphi(v_i, \mathbf{x}_i|; v_1, \mathbf{x}_1, \ldots, v_n, \mathbf{x}_n; v^{(0)}, \mathbf{x}_0)$ is considered under the condition that the inclusions $v_1, \ldots, v_n$ are located in the points $\mathbf{x}_1, \ldots, \mathbf{x}_n$, whereas $\mathbf{x}_0 \in v^{(0)}$ is the matrix position vector. In a general case of statistically inhomogeneous media with the homogeneous matrix (for example, for so-called FGM), the conditional probability density is not invariant with respect to translation: $\varphi(v_i, \mathbf{x}_i|v_1, \mathbf{x}_1, \ldots, v_n, \mathbf{x}_n) \neq \varphi(v_i, \mathbf{x}_i|v_1, \mathbf{x}_1 + \mathbf{x}, \ldots, v_n, \mathbf{x}_n + \mathbf{x})$, that is, the microstructure functions depend their absolute positions [916]. In particular, a random medium is called statistically homogeneous in a narrow sense if its multipoint statistical moments of any order are shift-invariant functions of spatial variables. Of course, $\varphi(v_i, \mathbf{x}_i|; v_1, \mathbf{x}_1, \ldots, v_n, \mathbf{x}_n) = 0$ for values of $\mathbf{x}_i$ lying inside the "excluded volumes" $\cup v_{im}^0$ $(m = 1, \ldots, n)$, where $v_{im}^0 \supset v_m$ with characteristic functions V_{0m} (since inclusions cannot overlap), and $\varphi(v_i, \mathbf{x}_i|; v_1, \mathbf{x}_1, \ldots, v_n, \mathbf{x}_n) \to \varphi(v_i, \mathbf{x}_i)$ at $|\mathbf{x}_i - \mathbf{x}_m| \to \infty$, $m = 1, \ldots, n$ (since no long-range order is assumed). If the system is isotropic (5.22), its two-particle distribution function $\varphi(v_m \mathbf{x}_m, v_i, \mathbf{x}_i)$ depends only on $r = |\mathbf{x}_m - \mathbf{x}_i|$; in this case written as $g(r)$ and called the radial distribution function (RDF) (5.66). The well-stirred approximation (5.63) can be presented in these notations as $(v_k \subset v^{(K)})$:

$$\begin{aligned}\varphi(v_k, \mathbf{x}_k|; v_i, \mathbf{x}_i, v_j, \mathbf{x}_j) &= (1 - V_i^0(\mathbf{x}_k))(1 - V_j^0(\mathbf{x}_k))n^{(K)}, \\ \varphi(v_k, \mathbf{x}_k|; v_i, \mathbf{x}_i) &= (1 - V_i^0(\mathbf{x}_k))n^{(K)}. \end{aligned} \tag{5.70}$$

For both spherical and circle inclusions, the RDF $\varphi(v_k, \mathbf{x}_k|; v_i, \mathbf{x}_i)$ is well analyzed theoretically, that is, in agreement with both the results of Monte-Carlo simulations and experimental data. Various approximations have been proposed with varying degrees of success [168], [1106]. Much less investigated is the three-particle distribution function: $g_3(\mathbf{x}_1, \mathbf{x}_2, \mathbf{x}_3) = \varphi(v_1, \mathbf{x}_1, v_2, \mathbf{x}_2, v_3, \mathbf{x}_3)/(n^{(1)})^3$ [84], [393], [1038] because it depends on two ore more variables $g_3(\mathbf{x}_1, \mathbf{x}_2, \mathbf{x}_3) = g_3(r_1, r_2, r_3)$ $(r_1 = |\mathbf{x}_1 - \mathbf{x}_2|,\ r_2 = |\mathbf{x}_1 - \mathbf{x}_3|,\ r_3 = |\mathbf{x}_2 - \mathbf{x}_3|)$, while the two-particle functions depend only on the radial coordinate. The earliest approximation for the three-particle functions of a homogeneous isotropic system was the superposition approximation of Kirkwood [577]:

$$\begin{aligned} g_3(r_1, r_2, r_3) &= g(r_1)g(r_2)g(r_3), \\ \varphi(v_k, \mathbf{x}_k|; v_i, \mathbf{x}_i, v_j, \mathbf{x}_j) &= \varphi(v_k, \mathbf{x}_k|; v_i, \mathbf{x}_i)\varphi(v_k, \mathbf{x}_k|; v_j, \mathbf{x}_j) \end{aligned} \tag{5.71}$$

which overestimated the probability to find the inclusion v_k between the inclusions v_i and v_j for the small distance $|\mathbf{x}_i - \mathbf{x}_j|$. The Kirkwood superposition approximation (5.71) turns out to describe the triplet structure in an inadequate way, and because of this, Bildstein and Kahl [84] have strongly advised not to use it. Approximation (5.71) is rigorous only in the limiting case $|\mathbf{x}_i - \mathbf{x}_j| \to \infty$. Another approximation of triplet-structure

$$\begin{aligned}\varphi(v_k, \mathbf{x}_k|; v_i, \mathbf{x}_i, v_j, \mathbf{x}_j) &= (1 - V_i^0(\mathbf{x}_k))(1 - V_j^0(\mathbf{x}_k)) \\ &\times \max_{R^d \backslash (v_i^0 \cup v_j^0)} [\varphi(v_k, \mathbf{x}_k|; v_i, \mathbf{x}_i)\varphi(v_k, \mathbf{x}_k|; v_j, \mathbf{x}_j)] \end{aligned}$$

is also valid only in the limiting case $|\mathbf{x}_i - \mathbf{x}_j| \to \infty$. We will also use the notation $V(\mathbf{y}|v_1, \mathbf{x}_i; \ldots; v_n, \mathbf{x}_n)$ denoting a random characteristic function of inclusions under the condition that the inclusions $v_1 \ldots \neq v_n$ are located in the domains with the centers $\mathbf{x}_1, \ldots, \mathbf{x}_n$. In particular, the terms with $\mathbf{y} \in v_1$ and $\mathbf{y} \in v_2$ may be isolated in the characteristic function $V(\mathbf{y}|v_1, \mathbf{x}_1; v_2, \mathbf{x}_2)$ with the help of the equality

$$V(\mathbf{y}|v_1,\mathbf{x}_1;v_2,\mathbf{x}_2) = V_1(\mathbf{y}) + V_2(\mathbf{y}) + V(\mathbf{y}|;v_1,\mathbf{x}_1;v_2,\mathbf{x}_2), \tag{5.72}$$

which can be also presented in terms of the conditional probability density:

$$\varphi(v_q,\mathbf{x}_q|v_1,\mathbf{x}_1;v_2,\mathbf{x}_2) = \delta(\mathbf{x}_q-\mathbf{x}_1)+\delta(\mathbf{x}_q-\mathbf{x}_2)+\varphi(v_q,\mathbf{x}_q|;v_1,\mathbf{x}_1;v_2,\mathbf{x}_2). \tag{5.73}$$

5.3.2 Statistical Averages of Functions

Let $\mathbf{g}(\mathbf{x}|\mathcal{C}^N)$ be an arbitrary local physical function of coordinates $\mathbf{x} \in w$ which is continuous inside each constituent, but may have discontinuities on the inclusion surfaces. This function depends in addition on the arrangement of the configuration $\mathcal{C}^N$. The ensemble average conditional average of the function $\mathbf{g}(\mathbf{x}|\mathcal{C}^N)$ over the ensemble can be defined as

$$\langle \mathbf{g}\rangle(\mathbf{x})=\int \mathbf{g}(\mathbf{x}|\mathcal{C}^N)\varphi(\mathcal{C}^N)\, d\mathcal{C}^N, \quad \langle \mathbf{g}|\mathcal{C}^K\rangle(\mathbf{x})=\int \mathbf{g}(\mathbf{x}|\mathcal{C}^N)\varphi(\mathcal{C}^{N-K}|\mathcal{C}^K)\, d\mathcal{C}^{N-K}, \tag{5.74}$$

respectively. Equation (5.74_2) is calculated by integrating over all physically permissible values of $(\mathbf{x}_i,\mathbf{m}_i)$, $i=1,N-K$, compatible with the condition that the centers of the inclusions with marks $\mathbf{m}_1,\ldots,\mathbf{m}_K$ lie at the points $\mathbf{x}_1,\ldots,\mathbf{x}_K$. It follows from Eqs. (5.74)

$$\langle \mathbf{g}\rangle(\mathbf{x}) = \int \mathbf{g}(\mathbf{x}|\mathbf{x}_i,\mathbf{m}_i)\varphi(\mathbf{x}_i,\mathbf{m}_i)\, d\mathbf{x}_i\, d\mathbf{m}_i, \tag{5.75}$$

$$\langle \mathbf{g}|\mathbf{x}_i,\mathbf{m}_i\rangle(\mathbf{x}) = \int \mathbf{g}(\mathbf{x}|\mathbf{x}_i,\mathbf{x}_j,\mathbf{m}_i,\mathbf{m}_j)\varphi(\mathbf{x}_i,\mathbf{x}_j,\mathbf{m}_i,\mathbf{m}_j)\, d\mathbf{x}_j\, d\mathbf{m}_j,\ldots. \tag{5.76}$$

Analogously, the multipoint correlation function of the function $\mathbf{g}(\mathbf{x}|\mathcal{C}^N)$ is defined as

$$\langle \mathbf{g}(\mathbf{x}_1),\ldots,\mathbf{g}(\mathbf{x}_k)\rangle = \int \mathbf{g}(\mathbf{x}_1|\mathcal{C}^N)\cdot\ldots\cdot \mathbf{g}(\mathbf{x}_k|\mathcal{C}^N)\varphi(\mathcal{C}^N)\, d\mathcal{C}^N \tag{5.77}$$

We will also introduce the multipoint "phase" statistical averages over the constituent $\mathbf{x}\in v^{(i)}$ by means of the following formulae:

$$\langle \mathbf{g}\rangle^{(i)}(\mathbf{x}_1,\ldots,\mathbf{x}_k) = \langle \prod_{i=1}^k \chi_{v^{(i)}}(\mathbf{x}_i)\rangle \int \prod_{i=1}^k \chi_{v^{(i)}}(\mathbf{x}_i)\mathbf{g}(\mathbf{x}|\mathcal{C}^N)\varphi(\mathcal{C}^N)\, d\mathcal{C}^N, \tag{5.78}$$

where $\chi_{v^{(i)}}(\mathbf{x})$ is an indicator function of the constituent $v^{(i)}$.

In analysis of statistically homogeneous media (5.20) we shall use an ergodicity hypothesis that will be assumed when the spatial average estimated over one sufficiently large sample (with realization $V^{*(i)}(\mathbf{x})$ $(i=1,\ldots,N)$ and corresponding function $\mathbf{g}^*(\mathbf{x})$) and statistical mean $\langle\mathbf{g}\rangle(\mathbf{x})$ coincide for both the whole volume and the individual constituent:

$$\langle \mathbf{g}\rangle = \lim_{w\uparrow R^d}\frac{1}{\bar w}\int_w \mathbf{g}^*(\mathbf{x})\, d\mathbf{x}, \quad \langle \mathbf{g}\rangle^{(i)} = \lim_{w\uparrow R^d}\frac{1}{\bar v^{(i)}}\int_w \mathbf{g}^*(\mathbf{x})V^{*(i)}\, d\mathbf{x}, \tag{5.79}$$

where $\sum V^{(k)} = \sum V_i \equiv V$, $k = 1,2,\ldots,N$; $i = 1,2,\ldots$. $V^{(k)}$ is the characteristic functions of $v^{(k)}$. The bar appearing above the region represents its measure, e.g. $\overline{v} \equiv$ mes v. $V^{(k)}$ is the characteristic function of $v^{(k)}$. The average over an individual inclusion $v_i \subset v^{(k)}$ $(i = 1,2,\ldots)$ is $\langle(.)\rangle_i = \langle(.)\rangle^{(k)}$. In the case of statistically homogeneous fields, the averages $\langle\mathbf{g}\rangle$ and $\langle\mathbf{g}\rangle^{(i)}$ (5.79) do not depend on $\mathbf{x}$. However, for statistically inhomogeneous fields, the statistical averages $\langle\mathbf{g}\rangle(\mathbf{x})$ (5.74$_1$) and $\langle\mathbf{g}\rangle^{(i)}(\mathbf{x})$ (5.78) depend on $\mathbf{x}$ and do not coincide with the spatial averages (5.79).

In particular, in the case under treatment the structural function is described by the field $\Omega \subset w$ of inhomogeneities with indicator functions $V_i(\mathbf{x})$ $(i = 1,2,\ldots)$ and can be presented by the marked random density function [618], [735], [1049]:

$$\mathcal{C}^N(\mathbf{x},\mathbf{m}) = \sum_{\mathbf{x}_i\in\Omega} \delta(\mathbf{x}-\mathbf{x}_i)\delta(\mathbf{m}-\mathbf{m}_i). \tag{5.80}$$

The standard method of estimation of the statistical average consists of exploiting the ergodic property (5.79) with subsequent estimation of the spatial average. For example, for a simplified unmarked field (5.80), we find that the statistical average

$$\langle\mathcal{C}^N(\mathbf{x})\rangle = \lim_{w\uparrow R^d} \frac{1}{\overline{w}} \int_w \sum_{\mathbf{x}_i\in\Omega} \delta(\mathbf{x}-\mathbf{x}_i)\, d\mathbf{x} = \lim_{w\uparrow R^d} \frac{\hat{N}}{\overline{w}} = n \tag{5.81}$$

defines the intensity of the field Ω; here $\hat{N} = \phi(w)$ is the number of points $\mathbf{x}_i$ in the field of observation w. In a similar manner, by the use of the phase average of the marked field (5.77) with the absence of correlations between location $\mathbf{x}$ and mark $\mathbf{m}$ of inclusions (5.64) and (5.65), we get

$$\langle\mathcal{C}^N(\mathbf{x},\mathbf{m})\rangle = n\varphi_{\mathbf{m}}(\mathbf{m}),$$
$$\langle\mathcal{C}^N(\mathbf{x}_1,\mathbf{m}_1)\mathcal{C}^N(\mathbf{x}_2,\mathbf{m}_2)\rangle = n\varphi_{\mathbf{m}}(\mathbf{m}_1)\delta(\mathbf{x}_{12})\delta(\mathbf{m}_{12}) + \varphi_{\mathbf{x}}(\mathbf{x}_{12})\varphi_{\mathbf{m}}(\mathbf{m}_1)\varphi_{\mathbf{m}}(\mathbf{m}_2),$$

where for brevity we denote $\mathbf{x}_{12} = \mathbf{x}_1 - \mathbf{x}_2$, $\mathbf{m}_{12} = \mathbf{m}_1 - \mathbf{m}_2$, and $\varphi_{\mathbf{x}}(\mathbf{x}_1,\mathbf{x}_2) = \varphi_{\mathbf{x}}(\mathbf{x}_{12})$ for statistically homogeneous fields. We can define $\varphi(v_i) = n\varphi_{\mathbf{m}}(\mathbf{m}_k)$ $(v_i \in v^{(k)})$ as a numberical density $n^{(k)}$ of the inclusions of the kth phase $v^{(k)}$; $c^{(k)}$ is the concentration, i.e., volume fraction, of the component $v^{(k)}$: $c^{(k)} = \langle V^{(k)}\rangle$; $c^{(k)} = \overline{v}_k n^{(k)}$, $c^{(0)} = 1 - c^{(1)}$ $(k = 0,1;\ m = 1,2,\ldots)$.

Below the notation $\langle(.)(\mathbf{x})|v_1,\mathbf{x}_1;\ldots;v_m,\mathbf{x}_m\rangle$ will be used for the conditional average taken for the ensemble of a statistically homogeneous ergodic field $\Phi = (v_i,\mathbf{x}_i)$, on the condition that there are inclusions $v_1,\ldots,v_m$ at the points $\mathbf{x}_1,\ldots,\mathbf{x}_m$ and $\mathbf{x}_1 \neq \ldots \neq \mathbf{x}_m$. The notation $\langle(.)(\mathbf{y})|;v_1,\mathbf{x}_1;\ldots;v_m,\mathbf{x}_m\rangle$ is used for the case $\mathbf{y} \notin v_1,\ldots,v_m$.

5.3.3 Statistical Description of Indicator Functions

Since the overall properties of composite materials are sensitive to the details of the microstructure, the geometrical basis for modeling actual microstructures is needed. We considered the first method of random structure composite description. The second popular statistical description of microinhomogeneous media is based on the specification as to which constituent each point of medium belongs, and on estimations of expectations of products of the indicator function $V^{(i)}(\mathbf{y})$, assuming that the role of the matrix is assigned to phase "0." In a simple case the bounding of effective properties uses the multipoint statistic reducing to only one point probability density (volume fraction) $S_1^{(i)}(\mathbf{y}_1)$ defined as ($i = 1, 2, \ldots, N$): $S_1^{(i)}(\mathbf{y}_1) = \langle V_i(\mathbf{y}_1)\rangle \equiv c^{(i)}$, where the angle brackets $\langle(\cdot)\rangle$ denote an ensemble average, and a substitution of the one-point correlation function by its volume average (i.e., *volume concentration* of the ith phase) is provided by an ergodic assumption. Improved bounds on a variety of different effective properties have been derived in terms of n-point probability densities

$$S_n^{(i)}(\mathbf{y}^n) = \langle V_i(\mathbf{y}_1), \ldots V_i(\mathbf{y}_n)\rangle, \tag{5.82}$$

which are symmetric in their arguments, i.e., the probability of simultaneously finding n points in a specified geometrical arrangement $\mathbf{y}^n \equiv \mathbf{y}_1, \ldots, \mathbf{y}_n$ in one of the phases. For example, the one-point correlation function is the probability that any point lies in any particular. The two-point probability function is the probability that both two points $\mathbf{y}_1$ and $\mathbf{y}_2$ lie in the same phase. These functions provide a method for experimentally determining the low-order multipoint moments for real two-phase media, as considered in [246], where Corson used an unautomated painstaking procedure to compute three-ponit functions from micrographs of cross-sections of composites.

Digital image analysis is now available for estimation of descriptors of the spatial arrangement of microstructural features observed in a transverse cross-section of the material based on the *stereological* technique [76], [1045] allowing us to measure the geometry of such two-dimensional structures more rapidly, more precisely, and more comprehensively than ever before. Classical stereology investigates the spatial structures by planar sections, and statistically analyzes the visible structures from a range of microscopic techniques such as transmission electron microscopy, and scanning tunneling electron microscopy (see for references [1107]). For example, local stereology [500] shows how mean particle volumes can be estimated by length measurement. In general, some important spatial characteristics cannot be estimated stereologically, and statistical methods based on new microscopic techniques (e.g. computed X-ray tomography, magnetic resonance imaging, and confocal microscopy [551], [584], [835], [1107] using three-dimensional measurement were proposed. All these imaging methods are nondestructive, leaving the sample intact and unaltered. For example, by moving the local plane of a confocal scanning laser microscope up or down through the specimen, a stack of serial sections, called a brick, is obtained, from which a 3-D image may be produced without the need to use physical sections.

In a similar manner, an n-point correlation function denoted as $S_n^{(i_1 i_2 \ldots i_n)}(\mathbf{y}^n)$ $= \langle V_{i_1}(\mathbf{y}_1), \ldots V_{i_n}(\mathbf{y}_n)\rangle$, ($i_1, i_2, \ldots, i_n = 1, \ldots, N$) is associated with an event that for n points ($\mathbf{y}_1, \mathbf{y}_2, \ldots, \mathbf{y}_n$) randomly and simultaneously distributed in the medium, the point $\mathbf{y}_1$ belongs to the i_1-th constituent, the point $\mathbf{y}_2$ belongs to the i_2-th constituent, etc. If the medium is assumed to be statistically homogeneous, i.e., $S_n^{(i)}(\mathbf{y}^n)$ is translationary invariant for any $\mathbf{y}$ = const., then $S_n^{(i)}(\mathbf{y}^n) = S_n^{(i)}(\mathbf{y}_1 + \mathbf{y}, \ldots, \mathbf{y}_n + \mathbf{y})$. In addition, the n-point correlation function

$S_n^{(i_1 i_2 \dots i_n)}(\mathbf{y}^n)$ has the limiting properties

$$\begin{aligned} S_n^{(i_1 i_2 \dots i_n)}(\mathbf{y}^n) &= \delta_{i_1 i_n} \cdot \ldots \cdot \delta_{i_{n-1} i_n} S^{(i_n)}(\mathbf{y}_n), \\ S_n^{(i_1 i_2 \dots i_n)}(\mathbf{y}^n) &\to S_1^{(i_n)}(\mathbf{y}_n) \cdot S_{n-1}^{(i_1 i_2 \dots i_{n-1})}(\mathbf{y}^{n-1}), \end{aligned} \tag{5.83}$$

if $\mathbf{y}_{i_j} = \mathbf{y}_{i_n}$ or $|\mathbf{y}_{i_j} - \mathbf{y}_{i_n}| \to \infty$, respectively, for $\forall i_j = (i_1, \dots, i_{n-1})$. Equation (5.83_1) demonstrates that the probability of finding two different phases at by a single point is equal to 0 or is given by one-point probability function if phases are identical. Relation (5.83_2) (denoted as the *no-long order* hypothesis) states that the statistical distributions inside points are statistically independent for large distances between these points i_n and i_j $(i_j = (i_1, \dots, i_{n-1})$.

Since it is possible to determine multipoint probabilities only up to some relatively low order, Kroner [611] proposed classifying materials as uniform "of grade k" if probabilities involving up to k points were known to be translation-invariant. Statistical isotropy of the medium means invariant under arbitrary rotation. It is also known that the conventional bounds on effective properties are given in terms of other types of statistical quantities such as point/q-particle functions, surface-surface correlation functions, nearest-neighbor distribution function, linear-path function, two-point cluster function, chord-length distribution function as well as the generalized n-point distribution function for the system of identical spheres $H_n(\mathbf{y}^m; \mathbf{y}^{p-m}; \mathbf{r}^q)$, which is defined as the correlation associated with finding m points with positions $\mathbf{y}^m$ on certain surfaces within the medium, $p - m$ with positions $\mathbf{y}^{p-m}$ in certain spaces exterior to the spheres, and q sphere centers with positions $\mathbf{r}^q$, $n = p + q$ [1106]. However, although higher-order correlation functions $(N > 3)$ are obtained on theoretical grounds, this is not a very practicable approach.

We will consider some analytical representations of second-order characteristics for two-phase media. For statistically homogeneous media, the functions (also called autocorrelation functions) $S_2^{(i)}(\mathbf{y}_1, \mathbf{y}_2) \equiv S_2^{(i)}(\mathbf{y})$ $(i = 0, 1)$ provide a measure of how the end points of a vector $\mathbf{y} = \mathbf{y}_1 - \mathbf{y}_2$ fall in phase $v^{(i)}$. For statistically isotropic media, one point can be taken at the origin and the functions $S_2^{(i)}(\mathbf{y}) \equiv S_2^{(i)}(|\mathbf{y}|)$ are invariant with respect to the direction of the vector $\mathbf{y}$ and reach their maximum value of $c^{(i)}$ at $|\mathbf{y}| = 0$ with nonmonotonic decay to their asymptotic value of $(c^{(i)})^2$ in the case of *no long-range order* in the medium being assumed. For statistically homogeneous media, it is convenience to define the autocovariance of phase $v^{(i)}$

$$R^{(i)}(\mathbf{y}) = \langle V^{(i)\prime}(\mathbf{x}) V^{(i)\prime}(\mathbf{x}+\mathbf{y}) \rangle = S_2^{(i)}(\mathbf{y}) - (c^{(i)})^2 \tag{5.84}$$

taking the limiting values $R^{(i)}(0) = c^{(0)} c^{(1)}$ and $R^{(i)}(\infty) = 0$; here $V^{(i)\prime}(\mathbf{x}) = V^{(i)}(\mathbf{x}) - c^{(i)}$ is the fluctuating part of the random field $V^{(i)}(\mathbf{x})$, and $\langle (V^{(i)\prime})^2(\mathbf{0}) \rangle = c^{(0)} c^{(1)}$. It can be easily verified that $R^{(0)}(\mathbf{y}) = R^{(1)}(\mathbf{y})$, and, therefore $S_2^{(1)}(\mathbf{y}) = S_2^{(0)}(\mathbf{y}) + (c^{(1)})^2 - (c^{(0)})^2$. The two-point inclusion probability function $S_2^{(1)}(\mathbf{y}_1, \mathbf{y}_2)$ is connected with the two-point correlation even function $\gamma_2(r)$ $(r = |\mathbf{y}_1 - \mathbf{y}_2|)$ for statistically uniform two-phase materials consisting of an isotropic distribution of phases:

$$\gamma_2(r) = \langle (V^{(i)\prime})^2(\mathbf{0})\rangle^{-1}\langle V^{(i)\prime}(\mathbf{0})V^{(i)}(\mathbf{x})\rangle = \frac{1}{c^{(1)}c^{(0)}}[S_2^{(1)}(r) - (c^{(1)})^2], \quad (5.85)$$

which has the limiting values $\gamma_2(0) = 1$ and $\gamma_2(\infty) = 0$. For future use, we will represent some properties of the two-point probabilities $S_2^{(ij)}(\mathbf{y}_1, \mathbf{y}_2)$:

$$\begin{aligned} &S_2^{(0)}(\mathbf{y}_1,\mathbf{y}_2) + S_2^{(01)}(\mathbf{y}_1,\mathbf{y}_2) = S_1^{(0)}(\mathbf{y}_1) = S_2^{(0)}(\mathbf{y}_2,\mathbf{y}_1) + S_2^{(01)}(\mathbf{y}_2,\mathbf{y}_1),\\ &S_2^{(01)}(\mathbf{y}_2,\mathbf{y}_1) + S_2^{(1)}(\mathbf{y}_2,\mathbf{y}_1) = S_1^{(1)}(\mathbf{y}_1) = S_2^{(10)}(\mathbf{y}_1,\mathbf{y}_2) + S_2^{(1)}(\mathbf{y}_1,\mathbf{y}_2), \quad (5.86)\end{aligned}$$

where $S_1^{(0)}(\mathbf{y}) + S_1^{(1)}(\mathbf{y}) = 1$. For statistically homogeneous media

$$\begin{aligned} S_2^{(0)} &= (c^{(0)})^2 + c^{(0)}c^{(1)}g_2(\mathbf{y}),\ S_2^{(01)} = c^{(0)}c^{(1)}(1 - g_2(\mathbf{y})),\\ S_2^{(1)} &= (c^{(1)})^2 + c^{(0)}c^{(1)}g_2(\mathbf{y}), \quad (5.87)\end{aligned}$$

where $g_2(\mathbf{y})$ is a function of $r = |\mathbf{y}|$ only if the distribution of phases is isotropic. It should be mentioned that this statement does not refer to the elastic symmetries of the phases so that the composite could still be elastically anisotropic.

In a similar manner, for N-phase composites, we can analyze a covariance function $S_2^{(ij)}(\mathbf{y}_1, \mathbf{y}_2)$ exhibiting the following properties for statistically homogeneous media:

$$\begin{aligned} &S_2^{(ij)}(\mathbf{y}_1,\mathbf{y}_2) = S_2^{(ji)}(\mathbf{y}_1,\mathbf{y}_2), \quad S_2^{(ij)}(\mathbf{y}_1,\mathbf{y}_2) \equiv S_2^{(ij)}(\mathbf{y}) = c^{(i)}\delta_{ij}, \text{ at } |\mathbf{y}| = 0,\\ &c^{(i)} = \sum_{j=1,N} S_2^{(ij)}(\mathbf{y}_1,\mathbf{y}_2), \quad S_2^{(ij)}(\mathbf{y}) = c^{(i)}c^{(j)}, \text{ at } |\mathbf{y}| = \infty. \quad (5.88)\end{aligned}$$

The covariance function $S_2^{(ij)}(\mathbf{y}_1, \mathbf{y}_2)$ for statistically homogeneous media can be presented in the form $S_2^{(ij)}(\mathbf{y}) = S_2^{(ij)}(r, \omega)$, where $\omega = \mathbf{y}/r$. The autocorrelation function $S_2^{(ii)}(r, \omega)$ generates the isovalue contours at the origin of the covariance space, say $\lim_{r\to 0} S_2^{(ii)}(r, \omega)$, as well as those of the r-derivative, $\lim_{r\to 0} \partial S_2^{(ii)}(r, \omega)/\partial r$, which are related to the shape of a "primary grain" [354]. This primary grain replects the $v^{(i)}$ phase arrangement, even if this phase is not of the inclusion type. It should be mentioned that the shape of the primary grain is not directly given by the shape of the $S_2^{(ii)}(r, \omega)$ covariance isocontour limit, unless these isocontours are of ellipsoidal shape. Analogously, the "cross-covariance" function $S_2^{(ij)}(r, \omega)$ has slopes at the origin related to a contiguity probability between phase $v^{(i)}$ and $v^{(j)}$. In so doing, the isocontours of $S_2^{(i0)}(r, \omega)$ at $r \to 0$ also relate to the inclusion shape of the phase $v^{(i)}$ for the composites with the matrix $v^{(0)}$. A covariance function $S_2^{(ij)}(r, \omega)$ is called r-uniform if their isocontours are homothetic to a same shape, at any length-scale.

In many important cases, the material can be modeled as a dispersion of nonoverlaping and identical spheres that can alternatively be described statistically by means of the multipoint densities $\varphi(\mathbf{x}_1, \ldots, \mathbf{x}_n)$ instead of the multipoint moments (5.82). In partiqular, $\varphi(v_i, \mathbf{x})$ is a number density $n^{(k)} = n^{(k)}(\mathbf{x})$ of component $v^{(k)} \ni v_i$ at the point $\mathbf{x}$, and $c^{(k)} = c^{(k)}(\mathbf{x})$ is the concentration, that is, volume fraction, of the component $v^{(k)}$ at the point $\mathbf{x}$: $c^{(k)}(\mathbf{x}) =$

$\langle V^{(k)}\rangle(\mathbf{x}) = \overline{v}_i n^{(k)}(\mathbf{x})$ $(k = 1,2,\ldots,N;\ i = 1,2,\ldots)$, $c^{(0)}(\mathbf{x}) = 1 - \langle V\rangle(\mathbf{x})$. Hereinafter, we will use the notations $\langle(.)\rangle(\mathbf{x})$ and $\langle(.)|v_1,\mathbf{x}_1;\ldots;v_n,\mathbf{x}_n\rangle(\mathbf{x})$ for the average and for the conditional average taken for the ensemble of a statistically inhomogeneous field $X = (v_i)$ at the point $\mathbf{x}$, on the condition that there are inclusions at the points $\mathbf{x}_1,\ldots,\mathbf{x}_n$ and $\mathbf{x}_1 \neq \ldots \neq \mathbf{x}_n$. Similarly, $V(\mathbf{x}|v_1,\mathbf{x}_1;\ldots;v_n,\mathbf{x}_n)$ is a random characteristic function of inclusions $\mathbf{x} \in v$ under the condition that $v_1 \neq \ldots \neq v_n$. The notations $\langle(.)|;v_1,\mathbf{x}_1;\ldots;v_n,\mathbf{x}_n\rangle(\mathbf{x})$ and $V(\mathbf{x}|;v_1,\mathbf{x}_1;\ldots;v_n,\mathbf{x}_n)$ are used for the case $\mathbf{x} \notin v_1,\ldots,v_n$. The notation for the conditional probability density $\varphi(v_i,\mathbf{x}_i|;v_1,\mathbf{x}_1,\ldots,v_n,\mathbf{x}_n;\mathbf{x}_0)$ is considered under the condition that the inclusions $v_1,\ldots,v_n$ are located at the points $\mathbf{x}_1,\ldots,\mathbf{x}_n$, whereas the matrix position is denoted by $\mathbf{x}_0$.

Sets of expansions are known that yield systematic bounds for material and transport properties [1106] in terms of the $S_n^{(i)}$ and related functions for statistically homogeneous media. The results can then be reexpressed in terms of $\varphi(\mathbf{x}_1,\ldots,\mathbf{x}_n)$ if the equation $S_n^{(i)}(\mathbf{x}^n) \sim \varphi(\mathbf{x}^n)$ is known. In various applications it is necessary to interconnect these two kinds of statistical descriptions. For example, the functions $\varphi(\mathbf{x}_1,\mathbf{x}_2)$ and $\varphi(\mathbf{x}_1,\mathbf{x}_2,\mathbf{x}_3)$ are usually assumed to be known, and the integrals involving the two- and three-point moments $S_2^{(1)}(\mathbf{x}_1,\mathbf{x}_2)$ and $S_3^{(1)}(\mathbf{x}_1,\mathbf{x}_2,\mathbf{x}_3)$ should be estimated in bounds on the effective properties of particular composites. Torquato and Stell [1110] have related the n-point matrix probability function $S_N^{(0)}$ to multidimensional integrals over the infinite set of n-particle probability densities $\varphi(\mathbf{x}_1,\ldots,\mathbf{x}_n)$ of impenetrable spheres of unit radius:

$$S_n^{(0)}(\mathbf{x}_1,\ldots,\mathbf{x}_n) = 1 + \sum_{s=1}^{n} \frac{(-1)^s}{s!} \int \varphi(\mathbf{x}_{n+1},\ldots,\mathbf{x}_{n+s}) \times \prod_{j=n+1}^{n+s} \Big\{1 - \prod_{i=1}^{n}[1 - \chi_{b(0,1)}(\mathbf{x}_{ij})]\Big\}\, d\mathbf{x}_{n+j}. \tag{5.89}$$

The series truncated at the nth term because the particles are not permitted to overlap. The points $\mathbf{x}_{n+1},\ldots,\mathbf{x}_{n+s}$ may be considered as "test" inclusion centers and $\mathbf{x}_{ij} = \mathbf{x}_i - \mathbf{x}_j$. The general series expansion (5.89) was simplified for S_1, S_2, and S_3, in particular:

$$\begin{aligned}
S_1^{(0)}(\mathbf{x}_1) &= 1 - \int \varphi(\mathbf{x}_2)\chi_{b(0,1)}(\mathbf{x}_{12})\, d\mathbf{x}_2 = 1 - n\bar{v}_1,\\
S_2^{(0)}(\mathbf{x}_1,\mathbf{x}_2) &= 1 - n\int \Big\{1 - [1 - \chi_{b(0,1)}(\mathbf{x}_{13})][1 - \chi_{b(0,1)}(\mathbf{x}_{23})]\Big\}\, d\mathbf{x}_3\\
&\quad + \frac{n^2}{2}\int\!\!\int g(r_{34})\Big\{1 - [1 - \chi_{b(0,1)}(\mathbf{x}_{13})][1 - \chi_{b(0,1)}(\mathbf{x}_{23})]\Big\}\\
&\quad \cdot \Big\{1 - [1 - \chi_{b(0,1)}(\mathbf{x}_{14})][1 - \chi_{b(0,1)}(\mathbf{x}_{24})]\Big\}\, d\mathbf{x}_3\, d\mathbf{x}_4,
\end{aligned} \tag{5.90}$$

where $r_{ij} = |\mathbf{x}_{ij}|$. Since particles cannot overlap, some of the above terms vanish after expanding, and the relation for S_2, and as well as for $S_3(\mathbf{x}_1,\mathbf{x}_2,\mathbf{x}_3)$, can be simplified:

$$S_2^{(0)}(\mathbf{x}_1,\mathbf{x}_2) = 1 - nV_2(\mathbf{x}_1,\mathbf{x}_2) + n^2\nu(1,2), \tag{5.91}$$

$$\begin{aligned} S_3^{(0)}(\mathbf{x}_1,\mathbf{x}_2,\mathbf{x}_3) =& 1 - nV_3(\mathbf{x}_1,\mathbf{x}_2,\mathbf{x}_3) + n^2[\nu(1,2)+\nu(1,3)+\nu(2,3)] \\ &= n^2[w(1,2,3)+w(2,1,3)+w(3,3,2)] - n^3\chi(1,2,3), \end{aligned} \tag{5.92}$$

where

$$\nu(i,j) = \iint \chi_{b(0,1)}(\mathbf{x}_{i4})g(r_{45})\chi_{b(0,1)}(\mathbf{x}_{5j})\,d\mathbf{x}_4\,d\mathbf{x}_5, \tag{5.93}$$

$$w(i,j,k) = \iint \chi_{b(0,1)}(\mathbf{x}_{i4})\chi_{b(0,1)}(\mathbf{x}_{j5})\chi_{b(0,1)}(\mathbf{x}_{k5})g(r_{45})\,d\mathbf{x}_4\,d\mathbf{x}_5, \tag{5.94}$$

$$\chi(i,j,k) = \iiint \chi_{b(0,1)}(\mathbf{x}_{i4})\chi_{b(0,1)}(\mathbf{x}_{j5})\chi_{b(0,1)}(\mathbf{x}_{k6})g(\mathbf{x}_4,\mathbf{x}_5,\mathbf{x}_6)\,d\mathbf{x}_4 d\mathbf{x}_5 d\mathbf{x}_6, \tag{5.95}$$

and $V_n(\mathbf{x}^n)$ are the union volumes of n spheres with configuration $\mathbf{x}^n$:

$$V_2(\mathbf{x}_1,\mathbf{x}_2) = \int \chi_{b(0,1)}(\mathbf{x}_{14})\chi_{b(0,1)}(\mathbf{x}_{24})\,d\mathbf{x}_4, \tag{5.96}$$

$$V_3(\mathbf{x}_1,\mathbf{x}_2,\mathbf{x}_3) = \int \chi_{b(0,1)}(\mathbf{x}_{14})\chi_{b(0,1)}(\mathbf{x}_{24})\chi_{b(0,1)}(\mathbf{x}_{34})\,d\mathbf{x}_4. \tag{5.97}$$

For example, for the first three space dimensions $V_2(\mathbf{x}_1,\mathbf{x}_2) = V_2(r)$ $(r = |\mathbf{x}_1 - \mathbf{x}_2|)$:

$$\begin{aligned} &V_2(r) = (1-\frac{r}{2})\omega_1\chi_{(0,2)}(r), \quad V_2(r) = \frac{2}{\pi}\Big[\cos^{-1}\frac{r}{2} - \frac{r}{2}\Big(1-\frac{r^2}{4}\Big)^{1/2}\Big]\omega_2\chi_{(0,2)}(r), \\ &V_2(r) = \Big[1 - \frac{3r}{4} + \frac{1}{16}r^3\Big]\omega_3\chi_{(0,2)}(r), \end{aligned} \tag{5.98}$$

for $d = 1,2,3$, respectively; here ω_d is the volume of unit sphere in R^d (5.29).

A much more complicated six-tuple integral $\nu(i,j)$ (5.82) was reduced to a simple one-tuple integral in [735] in the case when the distribution of the spheres is statistically isotropic, so that $\varphi(\mathbf{x}_i,\mathbf{x}_j) = g(r)$ becomes a function of $r = |\mathbf{x}_i - \mathbf{x}_j|$. In the framework of the simplest "two-point" level, Markov and Willis [735] demonstrated a simple interconnection expressing $S_2^{(1)}(\mathbf{y}^2)$ for two-phase media as a simple one-tuple integral containing the radial distribution function $g(r)$ (5.30). They obtained the representation for the two-point correlation function $\gamma_2(r)$ (5.85) $\gamma_2(r) = \gamma_2^{\rm ws}(r) + \tilde\gamma_2(r)$. For the well-stirred approximation (5.70), the constituent $\tilde\gamma_2(r) \equiv 0$ vanishes and the representation (5.90) is reduced to the formula $(\rho = r/a)$

$$\gamma_2(r) = \gamma_2^{\rm ws}(r) \equiv \begin{cases} 1 - \frac{3\rho}{4(1-c)} + \frac{(1+3c)\rho^3}{16(1-c)} - \frac{9c\rho^4}{160(1-c)} + \frac{c\rho^6}{2240(1-c)} & (0 \le \rho \le 2), \\ \frac{c}{1-c}\frac{(\rho-4)^2(36-34\rho-16\rho^2-\rho^3)}{2240\rho} & (2 \le \rho \le 4), \\ 0 & (4 \le \rho). \end{cases} \tag{5.99}$$

obtained in [1110]. For an arbitrary $g(r)$,

$$\tilde\gamma_2(r) = \frac{3c}{160\rho(1-c)}\begin{cases} \int_2^{\rho+2} f(\rho-\tau)\tau h(\tau)\,d\tau, \\ \int_2^{\rho+2} f(\rho-\tau)\tau h(\tau)\,d\tau + \int_2^{\rho} f(\rho-\tau)\tau h(\tau)\,d\tau, \\ \int_{\rho-2}^{\rho+2} f(\rho-\tau)\tau h(\tau)\,d\tau + \int_\rho^{\rho+2} f(\rho-\tau)\tau h(\tau)\,d\tau, \end{cases} \tag{5.100}$$

at $0 \le \rho \le 2$, $2 \le \rho \le 4$, and $4 \le \rho$, respectively; here $h(r) = g_2(r) - 1$ and $f(t) = (2+t)^3(4-6t+t^2)$. Thus, the two-point correlation $\gamma_2(r)$ is represented as a simple one-tuple integral containing the radial distribution function of statistically isotropic dispersion of nonoverlapping spheres. Markov and Willis [735] mentioned that due to the well known positive-definiteness of the two-point correlation function $\gamma_2(r)$ for any statistically homogeneous medium, the integral

$$\int_0^{\infty} r^2 \gamma_2^{\rm ws}(r)dr = \frac{(1-8c)}{3(1-c)}a^3 \tag{5.101}$$

(which can obviously estimated through an elementary integration of Eq. (5.94)) should be non-negative which is violated at $c > 1/8$.

Generalization of Eqs. (5.85)–(5.91) was obtained in [915], where the model of impenetrable ellipsoidal aligned particles within correlation spherical holes v_i^0 ("security" spheres) was used to derive $S_2^{(0)}(\mathbf{x}_1, \mathbf{x}_2)$. This assumption is more restrictive than simply requiring the particles to be nonoverlapping, but it made possible the use of well-known statistical models for spherical inclusions (5.85)–(5.91). Quintanilla [915] evaluated numerically the double integral in the generalized Eq. (5.90) by employing the Fourier transform of the density function and the Percus-Yevik-Wertheim approximation for the radial distribution function. Monetto and Drugan [783] obtained an explicite analytical result for the two-point correlation function. In the framework of the same assumption of spherical shape v_i^0 and aligned inclusions, they separated the effects of inclusion shape and their spatial distribution, and incorporated the Verlet-Weis improvement to the standard Percus-Yevick-Wertheim approximation.

5.3.4 Geometrical Description and Averaging of Doubly and Triply Periodic Structures

It is assumed that the representative mesodomain w contains a statistically large number of particles $v_i \subset v^{(1)}$ $(i = 1, 2, \ldots)$ with identical shape, orientation and mechanical properties (i.e., the unmarked point field is analyzed). We now consider a composite medium with particle centers distributed at the nodes of some spatial lattice Λ. Suppose $\mathbf{e}_i$ $(i = 1, 2, 3)$ are linearly independent vectors, so that we can represent any node $\mathbf{m} \in \Lambda$:

$$\mathbf{x_m} = f_1(m_1)\mathbf{e}_1 + f_2(m_2)\mathbf{e}_2 + f_3(m_3)\mathbf{e}_3, \tag{5.102}$$

where $\mathbf{m} = (m_1, m_2, m_3)$ are integer-valued coordinates of the node $\mathbf{m}$ in the basis $\mathbf{e}_i$ which are equal in modulus to $|\mathbf{e}_i|$, and $f_i(m_i) - f_i(m_i+1) \not\equiv \text{const.}$ $(i = 1, 2, 3)$.

For triply periodic structures with linear-independent vectors of the principal period of Λ determining a unit cell Ω of volume $\bar{\Omega} = |\mathbf{e}_1 \cdot (\mathbf{e}_2 \otimes e_3)|$, we can represent any node $\mathbf{m} \in \Lambda$ in the form

$$\mathbf{x_m} = m_1\mathbf{e}_1 + m_2\mathbf{e}_2 + m_3\mathbf{e}_3. \tag{5.103}$$

If, for example, the basis $\mathbf{e}_i$ is orthonormal, and the coefficients $\mathbf{m} = (m_1, m_2, m_3)$ are the integer set Z^3, independent of one other, Λ defines a simple cubic (SC) packing; in the case where the coefficients m_i $(i = 1, 2, 3)$ are either all even or odd, we have a body-centered cubic structure (BCC); a cubic face-centered structure (FCC) is obtained in the case where the coefficients m_i are either all even or two are odd, while the third is even. The method of assigning the lattice Λ is also possible where several nodes are located within the limits of a cell, and the coefficients m_i are the integer set Z^3, independent of one another [636].

For doubly periodical structures

$$\mathbf{x_m} = m_1\mathbf{e}_1 + m_2\mathbf{e}_2 + f_3(m_3)\mathbf{e}_3, \tag{5.104}$$

where $f_3(m_3) - f_3(m_3+1) \neq \text{const}$. In the plane $f(m_3) = \text{const}$. the composite is reinforced by periodic arrays Λ_{m_3} of inclusions in the direction of the $\mathbf{e}_1$ axis and the $\mathbf{e}_2$ axis. The type of lattice Λ_{m_3} is defined by the law governing the variation in the coefficients m_i $(i = 1, 2)$, and also by the magnitude and orientation of the vectors $\mathbf{e}_i$ $(i = 1, 2)$. In the functionally graded direction $\mathbf{e}_3$ the inclusion spacing between adjacent arrays may vary ($f_3(m_3) - f_3(m_3 + 1) \not\equiv \text{const.}$). For a doubly periodic array of inclusions in a finite ply containing $2m^l + 1$ layers of inclusions we have $f(m_3) \equiv 0$ at $|m_3| > m^l$; in the more general case of doubly periodic structures $f(m_3) \not\equiv 0$ at $m_3 \to \pm\infty$. To make the exposition more clear we will assume that the basis $\mathbf{e}_i$ is an orthogonal one and the axes $\mathbf{e}_i$ $(i = 1, 2, 3)$ are directed along axes of the global Cartesian coordinate system (these assumptions are not obligatory).

The statistical description of random structures presented above can also be used for periodic structures by allowing the ensemble distribution function to be expressed in terms of proper delta-functions, when, e.g., the configuration $\mathcal{C}^N$ (5.103) has a distribution function $\varphi(\mathbf{x}) = \sum_{i=1}^N \delta(\mathbf{x} - \mathbf{x}_i)$. As an example, we will consider following [1047] a quadratic point packing with a mesh width a and the number point density $n = a^{-2}$. In such a case a radial distribution function can be defined only in a generalized sense. It is nonzero only for discrete interpoint distances and has long-range order. Indeed, the Ripley's K function (5.26) is given by exact relation:

$$nK(r) = \begin{cases} 0 & (r < a), \\ 4 & (a \le r < \sqrt{2}a), \\ 8 & (\sqrt{2}a \le r < 2a), \\ 12 & (2a \le r < \sqrt{5}a), \\ 20 & (\sqrt{5}a \le r < 2\sqrt{2}a), \\ \dots & \dots \end{cases} \tag{5.105}$$

which makes it possible to formally introduce, according to Eq. (5.30), the generalized radial distribution function $g(r) = \frac{1}{2\pi r}\sum_{i=1}^{\infty} c_i\delta(r - r_i)$, where $r_1 = a$, $r_2 = \sqrt{2}a$, $r_3 = 2a$, $r_4 = \sqrt{5}$, $r_5 = 2\sqrt{2}a, \ldots$, $c_1 = 4a^2$, $c_2 = 4a^2$, $c_3 = 4a^2$, $c_4 = 8a^2$, $c_5 = 4a^2, \ldots$.

The composite material is constructed using the building blocks or cells: $w = \cup\Omega_{\mathbf{m}}$, $v_{\mathbf{m}} \subset \Omega_{\mathbf{m}}$. Hereafter the notation $\mathbf{f}^{\Omega}(\mathbf{x})$ will be used for the average of the function $\mathbf{f}$ over the cell $\mathbf{x} \in \Omega_i$ with the center $\mathbf{x}_i^{\Omega} \in \Omega_i$:

$$\mathbf{f}^{\Omega}(\mathbf{x}) = \mathbf{f}^{\Omega}(\mathbf{x}_i^{\Omega}) \equiv n(\mathbf{x}) \int_{\Omega_i} \mathbf{f}(\mathbf{y})\, d\mathbf{y}, \quad \mathbf{x} \in \Omega_i, \tag{5.106}$$

$n(\mathbf{x}) \equiv 1/\overline{\Omega}_i$ is the number density of inclusions in the cell Ω_i.

Let $\mathcal{V}_{\mathbf{x}}$ be a "moving averaging" cell (or moving-window [395]) with the center $\mathbf{x}$ and characteristic size $a_{\mathcal{V}} = \sqrt[3]{\overline{\mathcal{V}}}$, and let for the sake of definiteness $\boldsymbol{\xi}$ be a random vector uniformly distributed on $\mathcal{V}_{\mathbf{x}}$ whose value at $\mathbf{z} \in \mathcal{V}_{\mathbf{x}}$ is $\varphi_{\boldsymbol{\xi}}(\mathbf{z}) = 1/\overline{\mathcal{V}}_{\mathbf{x}}$ and $\varphi_{\boldsymbol{\xi}}(\mathbf{z}) \equiv 0$ otherwise. Then we can define the average of the function $\mathbf{f}$ with respect to translations of the vector $\boldsymbol{\xi}$:

$$\langle \mathbf{g} \rangle_{\mathbf{x}}(\mathbf{x}-\mathbf{y}) = \frac{1}{\overline{\mathcal{V}}_{\mathbf{x}}} \int_{\mathcal{V}_{\mathbf{x}}} \mathbf{f}(\mathbf{z}-\mathbf{y})\, d\mathbf{z}, \quad \mathbf{x} \in \Omega_i. \tag{5.107}$$

Among other things, "moving averaging" cell $\mathcal{V}_{\mathbf{x}}$ can be obtained by translation of a cell Ω_i and can vary in size and shape during motion from point to point. Clearly, contracting the cell $\mathcal{V}_{\mathbf{x}}$ to the point $\mathbf{x}$ occurs in passing to the limit $\langle \mathbf{f} \rangle_{\mathbf{x}}(\mathbf{x}-\mathbf{y}) \to \mathbf{f}(\mathbf{x}-\mathbf{y})$. To make the exposition more clear we will assume that $\mathcal{V}_{\mathbf{x}}$ results from Ω_i by translation of the vector $\mathbf{x} - \mathbf{x}_i^{\Omega}$; it can be seen, however, that this assumption is not mandatory.

5.3.5 Representations of ODF

To detail the properties of the general density function $\varphi(v_m \mathbf{x}_m|; \mathbf{x}_1, \ldots, \mathbf{x}_n)$ one uses different notions. In the simplest case the one-point function $\varphi(v_m, \mathbf{x})$ depends only on the orientation but not on the position $\mathbf{x}$ of a volume element in the sample. So, the orientation distribution function (ODF) f specifying the volume fraction of crystals with an orientation placed between $\mathbf{g}$ and $\mathbf{g} + d\mathbf{g}$ is $dV/V \equiv f(\mathbf{g})\, d\mathbf{g}$, where $d\mathbf{g}$ is the invariant orientation probability measure (Haar measure) in SO(3) [373] of dV which is the total of all volume elements of the sample with the orientation $\mathbf{g}$, and V is the total sample volume. Arguably, the most representative is volume-fraction distribution (including as a particular case the orientation distribution function) describing the one-point statistics of elements in the structure. The term fundamental zone of crystal orientation expressed by the symbol $SO(3)/G$ refers to the set of all physically distinct orientations of the local crystal that can occur in nature. In the absence of any texture, all rotations of the Euler-angle parametrization belong to the fundamental zone of crystal orientation. If the shape and the size of grains are not altered then $\varphi(v_m) = f$. The ODF describes the probability density for the occurrence of specified crystallographic orientation in a composite material; it contains no information about positional or morphological measures of the microstructure such as the shape and the size of inclusions or grains. However, some other recent studies have combined the morphological and orientational measures of structure to a limited degree. The misorientation distribution function (MDF) gives the probability density for the occurrence of specified intercrystalline misorientation between adjacent grains in the polycrystals. The MDF is closely connected with structure of intercrystalline boundaries, and is not predicted from the ODF. Owing to a significant influence across intercrystalline boundaries which occurs in the consideration of many material properties, microstructural measures containing a positional dependence in conjunction with the orientation, are widely used and the two-point conditional probability function $\varphi(v_j|\mathbf{x}_j; \mathbf{x}_i)$ as well as two-point orientational coherence (or correlation) function (OCF) [6] are exploited.

This function gives the probability density for the simultaneous occurrence of crystallites of specified orientation $\mathbf{g}_i$ and $\mathbf{g}_j$ with the centers in the points $\mathbf{x}_i$ and $\mathbf{x}_j$ independently located in a specified measurement volume separated by a vector $\mathbf{x}_j - \mathbf{x}_i$, and contains morphological information such as inclusion (or crystallite) size and shape as a function of orientation. The OCF is reduced to the ODF upon integration over the domain of one set of orientation variables; the OCF is reduced to $\varphi(v_j|\mathbf{x}_j;\mathbf{x}_i)$ for the fixed shape of the grain and their location centers $\mathbf{x}_j$ and $\mathbf{x}_i$. For the remote centers $|\mathbf{x}_i - \mathbf{x}_j| \to \infty$, the orientation coherence relationship is absent, and the OCF equals the product of the ODF $f(\mathbf{x}_i)f(\mathbf{x}_j)$.

The ODF is used as a "weight function" in the estimation of the average of some function $\mathbf{F}$ over the group O_3:

$$\langle \mathbf{F} \rangle^{\mathbf{g}} = \int \mathbf{F}(\mathbf{g})f(\mathbf{g})\, d\mathbf{g} \tag{5.108}$$

The ODF f is chosen such that for the one-phase polycrystals (when $\langle \cdot \rangle^{\mathbf{g}} \equiv \langle \cdot \rangle$) the integral of the function $\mathbf{F} = \mathbf{I}$ is unity, i.e.,

$$\int f(\mathbf{g})\, d\mathbf{g} = 1. \tag{5.109}$$

The invariant measure for the Euler angle parameters (ϕ_1, θ, ϕ_2) (A.1.4) is $d\mathbf{g} = \sin\theta\, d\phi_1\, d\theta\, d\phi_2$. For the axis-angle parametrization (A.1.2) the invariant measure is [786]: $d\mathbf{g} = \left[\pi(1+\rho^2)/\rho\right]^{-2} d\rho\, d\mathbf{n}$, where $\rho = |\tan(\omega/2)|$ is the magnitude of the Rodrigues vector $\boldsymbol{\rho}$ (see Appendix A.1) and $d\mathbf{n}$ is the invariant measure of an infinitesimal patch lying on the unit sphere S^2 of directions $\mathbf{n} = \sin\theta\cos\psi\mathbf{e}_1 + \sin\theta\sin\psi\mathbf{e}_2 + \cos\theta\mathbf{e}_3$ (see Appendix A.1); for the spherical coordinate representation $d\mathbf{n} = \sin\theta\, d\theta\, d\psi$.

For the fibers, the ODF should also comply with the physical condition of periodicity, because one end of the fiber is indistinguishable from the other $f(\theta, \phi_1, \phi_2) = f(\pi - \theta, \phi_1 + \pi, \phi_2)$ The ODF is a complete and unambiguous description of the fiber orientation state. In particular, for uniformly distributed fiber orientation with $f \equiv$ const. and the Euler angles parametric representation for the rotation group, the condition (5.109) leads to the equality

$$f\, d\mathbf{g} = \frac{1}{8\pi^2}\sin(\theta)\, d\theta\, d\phi_1\, d\phi_2, \tag{5.110}$$

which may be termed as a completely random distribution. Let the preferabed direction be the Ox_1-axis:

$$f\, d\mathbf{g} = \frac{1}{2\pi}\delta(\theta - \pi/2)\delta(\phi_1 - 0)\, d\theta\, d\phi_1\, d\phi_2, \tag{5.111}$$

which represents the situation in which all fibers are aligned and parallel to the Ox_1-axis, and hence fibers are free to rotate only with r espect to the angle ϕ_2. The ODF

$$f\, d\mathbf{g} = \frac{1}{4\pi^2}\delta(\theta - \pi/2) d\theta d\phi_1 d\phi_2, \tag{5.112}$$

represents the case in which the fibers are uniformly lying on the planes parallel to the plane Ox_1x_2, and is hence a two-dimensional in plane distribution. The ODF (5.111), (5.112), and (5.110) will be termed as the $1-D$, $2-D$, and $3-D$, respectively, uniformly random distributions.

The description using three angles (ϕ_1, θ, ϕ_2) is obviously more general than is necessary for fibers with circular cross sections and transversely isotropic symmetry about the fiber axis. In the last case, there is no ϕ_2 dependence to the orientation distribution. In such a case the axis-angle parametrization for the rotation matrix $\mathbf{g} \in SO(3)$ (A.1.2) is more convenient. Owing to the symmetry of fibers, $\mathbf{n}(\theta, \psi) = \mathbf{n}(\pi - \theta, \pi - \psi)$. One usually assumes that the fiber orientations are uniformly distributed with respect to ψ, which is usually the case for injection-molded components. An independence of ODF $f(\theta, \psi)$ on ψ implies transverse isotropy, with Ox_3 being the symmetry axis. With this assumption, a two-parameter exponential function was proposed [728], [1200] to describe the fiber orientation in the injection-molded specimens:

$$f(\theta) = \Big[\int_{\theta_{\min}}^{\theta_{\max}} (\sin\theta)^{2p-1} (\cos\theta)^{2q-1}\, d\theta \Big]^{-1} (\sin\theta)^{2p-1} (\cos\theta)^{2q-1}, \tag{5.113}$$

where $\theta_{\min}$ and $\theta_{\max}$ are the upper and lower limits of the angle θ: $0 \le \theta_{\min} \le \theta \le \theta_{\max} \le \pi/2$, and the parameters $p \ge 0.5$ and $q \ge 0.5$. The ODF (5.113) continuously link the extreme cases of fiber orientations from aligned to uniform random orientation with the most probable fiber orientation angle $\theta_{mod} = \tan^{-1}\{[(2p-1)/(2q-1)]^{1/2}\}$. Among one-parameter axisymmetric ODFs, one can indicate the function $f(\theta) = a \exp\big[-\frac{1}{2}(\theta/s)^2\big]$, where a is a normalized parameter found from the condition (5.109), and s is the standard deviation in a meridional section which can vary from $s \to 0$ (aligned fibers) to $s \to \infty$ (uniform random orientation).

In the general case, the ODF is determined from the complete set of single-orientation measurements called pole figures and is represented as a series expansion [124]:

$$f(\mathbf{g}) = \sum_{l=0}^{\infty} \sum_{m=-l}^{l} \sum_{n=-l}^{l} C_l^{mn} T_l^{mn}(\mathbf{g}) \tag{5.114}$$

in the generalized spherical harmonics T_l^{mn} that form an orthogonal functional system:

$$\int T_{l'}^{m'n'}(\mathbf{g}) T_l^{mn}(\mathbf{g})\, d\mathbf{g} = (2l+1)^{-1} \delta_{ll'} \delta_{mm'} \delta_{nn'}. \tag{5.115}$$

If we denote the indices in Eq. (5.114) by m', n', l', multiply both sides of the equation by $T_l^{mn}(\mathbf{g})$, and integrate over all orientations, we obtain, taking the orthogonality condition (5.115) into account,

$$C_l^{mn} = (2l+1) \int f(\mathbf{g}) T_l^{*mn}(\mathbf{g})\, d\mathbf{g},$$

where the asterisk denotes the complex conjugate property. This relation can be used for determination of the coefficients C_l^{mn}, if the ODF $f(\mathbf{g})$ is known, e.g., from an experiment in a table form. If $f(\mathbf{g})$ is the ODF, then it must be real: $f(\mathbf{g}) = f^*(\mathbf{g})$ that leads to the following relation between the coefficients C_l^{mn}: $C_l^{-m-n} = (-1)^{m+n} C_l^{*mn}$. We shall denote the rotations transforming the crystal from one orientation to a symmetrically equivalent orientation by $\mathbf{g}_B$. These rotations form the group symmetry of the crystal, and the orientations differing by the rotation $\mathbf{g}_B$ should be considered as equivalent:

$$f(\mathbf{g}_B \cdot \mathbf{g}) = f(\mathbf{g}). \tag{5.116}$$

If addition to the crystal symmetry, the sample can also possess statistical symmetry of the orientation distribution described by the group of the sample symmetry with the elements $\mathbf{g}_A$ preserving the ODF:

$$f(\mathbf{g} \cdot \mathbf{g}_A) = f(\mathbf{g}). \tag{5.117}$$

If now $f(\mathbf{g})$ must fulfill both symmetry conditions (5.116) and (5.117), then the coefficients C_l^{mn} should be subjected to some contractions. Some of them must be zero; others must be equal to one another [124].

If one expresses the orientation $\mathbf{g}$ by means of Euler's angles (see Appendix A.1), one thus obtains ($\Phi = \cos\theta$)

$$f(\phi_1, \theta, \phi_2) = \sum_{l=0}^{\infty} \sum_{m=-l}^{l} \sum_{n=-l}^{l} C_l^{mn} e^{im\phi_2} P_l^{mn}(\Phi) e^{in\phi_1}, \tag{5.118}$$

$$P_l^{mn}(\Phi) = \frac{(-1)^{l-m} i^{n-m}}{2^l (l-m)!} \sqrt{\frac{(l-m)!(l+n)!}{(l+m)!(l-n)!}} \cdot (1-\Phi)^{(n-m)/2}$$
$$\cdot\ (1+\Phi)^{-(n+m)/2} \frac{d^{l-n}}{d\Phi^{l-n}} \Big[(1-\Phi)^{l-m} (1+\Phi)^{l+m} \Big], \tag{5.119}$$

where $P_l^{mn}(\Phi)$ are the generalized associated Legendre functions.

If the ODF depends only on two angles ϕ_1, θ, then averaging of Eq. (5.118) over the angle ϕ_2 leads to the expansion:

$$f(\phi_1, \theta) = \sum_{n=0}^{\infty} \sum_{m=-n}^{n} a_n^m P_n^m(\Phi) e^{im\phi_1}, \quad P_n^m(\Phi) = \frac{(1-\Phi^2)^{m/2}}{2^n n!} \frac{d^{m+n}}{d\Phi^{m+n}} (\Phi^2 - 1)^n, \tag{5.120}$$

where $P_n^m(\Phi)$ are the associated Legendre functions. Lastly, if the ODF depends only on angle θ then Eq. (5.120_1) is reduced to the expansion in the Legendre functions $P_n(\Phi)$

$$f(\theta) = \sum_{n=0}^{\infty} a_n P_n(\Phi). \tag{5.121}$$

The representation of the ODF in the form of an expansion in a series of generalized spherical harmonics is useful at the analytical estimations of orientation averages $\langle \mathbf{F} \rangle^{\mathbf{g}}$ (5.108).

5.4 Numerical Simulation of Random Structures

Computer simulation of topologically disordered structures by the random packing of hard spherical particles in 2-D and 3-D cases has a long history originating in the theory of liquids. This problem is closely connected with the known fundamental problem of statistical physics—description of the behavior of the particle system with interaction potential of hard spheres [87]. Random packing of spheres has been studied very extensively because of its technological importance, and the opportunity to model the simple liquid, concentrated suspensions, amorphous, and powder materials.

Computer simulation of packing problems can be classified into three groups of methods: molecular dynamic, Monte Carlo, and dense random packing. Much progress in the theory of the dense random packing was approached by the use of two kinds of methods: the sequential generation models and the collective rearrangement models. In the sequential model in [49], the so-called cluster growth model, a particle being added to the surface of particle cluster which grows outwards is placed sequentially to the point closest to the original such that the new particle established contact with three existing spheres in the cluster. Overlapping is ruled out by checking the center–center distance of the particle. This algorithm was modified [704] by including an artificial input parameter, "coefficient of tetrahedron perfection," k^{tp} to require that a triangular site would be filled only if the resulting tetrahedron satisfied a minimum k^{tp} requirement that ranged from requiring a perfect tetrahedra, $k^{tp} = 1.0$, to placing no requirement on triangular sites except that the added sphere would touch all three of the spheres that made up the site, $k^{tp} = 2.0$. The phenomenological character of the construction algorithm moved the particle cluster from the initial term containing three particles leads to the inhomogeneous and anisotropic inclusion fields with different densities that was demonstrated in [103]. Moreover, the configurations generated do not demonstrate the characteristic split second peak in the radial distribution function observed in experimental packing. Kansal *et al.* [549] proposed an algorithm controlling the degree of order throughout the formation of the packing of growing clusters. In the second type of sequential generation model, called the model of "rigid sphere free fall into a virtual box," one particle is dropped vertically each time from the random point onto the surface of an existing particle cluster growing upwards [201], [366], [583], [815]. The different densities were approached by introducing a phenomenological parameter limiting the number of rotations of each fallen sphere until it becomes a permanent part of the structure. Effects of boundaries of the virtual box are eliminated by introducing conventional cyclic boundary conditions. The algorithms described belong to the class of static methods where the particles are placed at a given time step and cannot thereafter move. For contrast, dynamic methods assume the reorganization of whole packing due to either short–or long-range interactions between particles. In the collective rearrangement model, N points randomly distributed in a virtual box are assigned both radii and random motion vectors. Each sphere is moved until there are no overlaps. Then the radii are increased and the process is repeated until any further increase in radii or any displacement of the spheres create overlaps that cannot be eliminated (the different versions of this method can be found, e.g., in [242], [448], [578], [708], [709], [833], [1235]). More recently, numerical simulations were performed to realize homogeneous and isotropic packing of spheres by various methods, for instance, by assuming hypothetical spheres having dual structure whose inner diameter defines the true density and the outer one a nominal density [510]. An alternative approach (eliminating the boundary effect of the virtual box) is based on the use of spherical boundary conditions instead of periodic ones [418], [1100]. There one simulates hard discs (more exactly a circular cap which can be visualized as a contact lens on the surface of an eyeball) on the surface of the ordinary three-dimensional sphere and hard spheres on the "surface" of a four-dimensional hyposphere. The advantage of this procedure is that there is no preferential direction, and it is impossible to pack particles into perfect regular periodic configurations. In this section we will follow [168].

5.4.1 Materials and Image Analysis Procedures

A digital image processing technique will be considered for measurement of centroid coordinates of fibers with subsequent estimation of statistical parameters and functions describing the stochastic structure of a real fiber composite. As an example, a carbon fiber-reinforced epoxy composite is chosen for microstructural analysis. Ten specimens of this composites produced by different technological regimes are analyzed. Each specimen containing 10 samples was cut using a diamond saw from a unidirectional composite laminate, and each specimen was first sanded and then polished to a 0.5-μm finish on the cross-section area, 20 mm×1.5 mm. A microscopic image with 200× magnitude for microstructural analysis of samples was taken using an optical microscope. Each image of the sample contains approximately 1800 fibers. The fiber volume $c^{(1)}$ and spatial distribution (such as, e.g., $S_1^{(i)}(\mathbf{y})$, $S_2(i)(\mathbf{y}^2)$, $i = 0, 1$) were determined from the image analyzer that is capable of calculating the fiber volume and the coordinate of each fiber [1022], [1215]. To determine $S_1^{(i)}$, a simple Monte-Carlo like simulation can be utilized for a statistically homogeneous medium $S_1^{(i)} = \hat{N}^{(i)}/\hat{N}$, where $\hat{N}$ is a total number of random throws into a sample w and $\hat{N}^{(i)}$ denotes the number of successful hits into the phase $v^{(i)}$. Some sort of the naive *minus-samping* method (5.27) can be employed to determine $S_2^{(r)}(\mathbf{y}^2)$ $(r = 0, 1)$ from a digitized micrograph imagined as a discretization of the characteristic function $V^{(r)}(\mathbf{y})$, usually presented in terms of a $M \cdot N$ bitmap. Denoting the value of $V^{(r)}$ for a pixel located in the ith row and j-column as $V^{(r)}(i, j)$ allows writing the first two statistical descriptors as

$$S_1^{(r)} = \frac{1}{MN} \sum_{i=0}^{M-1} \sum_{j=0}^{N-1} V^{(r)}(i, j), \tag{5.122}$$

$$S_2^{(r,s)}(m, n) = \frac{1}{(i_M - i_m)(j_N - j_n)} \sum_{i=i_m}^{i_M-1} \sum_{j=j_n}^{j_N-1} V^{(r)}(i, j) V^{(s)}(i + m, j + n), \tag{5.123}$$

where $i_m = \max(0, -m)$, $i_M = \min(M, M - m)$, $j_n = \max(0, -n)$, $j_N = \min(N, N - n)$; $r, s = 0, 1$.

Digital images of transverse sections through the fiber composite material were obtained by digitizing high-resolution optical micrographs, using standard methods [989], [1033], [1084], [1205]. Care was taken to maximize the contrast between the fibers and matrix in the original micrograph. One [168] described in detail the analysis of the 1024×1024 digital images using a commercially available desktop software package (Adobe Photoshop 5.5) in conjunction with a plug-in Image Processing Tool Kit (IPTK; see [951]). Once the image is reduced to an array of black pixels corresponding to the centroid positions of each fiber profile, it is a simple matter to extract the coordinates, using the IPTK software or any other suitable computer code. The coordinates are output as integer pairs in the range [1, 1024], and are subsequently renormalized for input into the microstructural models.

Since random packed structures are strongly dependent on the method of their generation, we will consider a few popular algorithms and their combinations, and will compare the statistical parameters of configurations generated by each different method.

5.4.2 Hard-Core Model

The extension of the Poisson distribution process as a static model is the generation of random assemblies of n nonoverlapping discs by the *Hard-Core Model* (HCM) (called also random sequential adsorption model and simple sequential inhibition) [328], [466]: discs with radius a are placed one by one with the center positions $X = (\mathbf{x}_1, \ldots, \mathbf{x}_N)$ being distributed randomly and uniformly over the set of all points in a rectangular region ω of size $[-0.5, 0.5] \times [-0.5, 0.5]$. Although the distribution of discs depends heavily on the shape and size of w, usually researchers assume a homogeneous structure in the whole space which is observed only in w. To avoid this discrepancy, one supposes the periodic boundary conditions, that is, ω and X are periodically replicated in all directions. If the new disc does not overlap already deposited discs, its position is fixed and does not move anymore; otherwise, it is rejected and another random center position is generated. The process is finished when either a preassigned packing fraction is achieved or when no more particles can be added (jamming limit) which occurs at a volume fraction $c \approx 0.55$ (2-D case) or $c = 0.38$ (3-D case) [1076], [703]. The mathematically formalized descriptions of other versions of the HCM using either a birth-and-death process with vanishing death rate or the Matern's thinning rule can be found in [1046].

The HCM provides a more realistic reference model than a Poisson point process, in which arbitrary small distances between points are allowed. The advantage of protocol independence in the HCM is sacrificed in the case of generation of binary or polydisperse structures [265], [449], [1013]. Indeed, after an unacceptable trial one can keep the previous choice of inclusion radius or choose to replace it. The real situation then lies between two limiting cases of jamming limits, the packing density generated by first placing the larger particles and then the smaller particles will be higher than the opposite case, where the small particles are placed before the big ones. The geometrical blocking effect and the process irreversibility leads to packing configurations that are essentially different from the corresponding equilibrium configuration [1176]. To prevent just this kind of low-density jamming in the following model, we will shake the discs.

5.4.3 Hard-Core Shaking Model (HCSM)

Hard-Core Shaking Model (HCSM) is a type of HCM that generates an increasing number of inclusions in a virtual box w accompanied by a shaking process, i.e., giving each disc a small random displacement independent of its neighbors' positions. This makes it possible to unlock the discs from the jamming configuration (which takes place for HCM at $c \approx 0.55$) and allows them to find the most homogeneous and mixed arrangement. Various algorithms have been devised to simulate reordering due to shaking or vibration of dense packing [35] which reduces the volume concentration of the high-density jam configuration. Close random ensembles of spheres have been studied for many years, and the following quantitative parameters are well known. Packing densities for close lattice packing are $\pi/\sqrt{12} \approx 0.9069$ (triangular) in the case of discs packing into the plane, and $\pi/\sqrt{18} \approx 0.7405$ (fcc or hcp) in the case of spheres packed into R^3. Model experiments were performed using steel balls of equal size randomly packed into shaking containers. The measured densities are extrapolated to eliminate finite-size effects. These models provide the conventional value of *random close packing* such as $c^{RCP} = 0.6366 \pm 0.004$ or, as some people believe, $2/\pi = 0.6366$. *Random loose packing* $c^{RLP} = 0.60 \pm 0.2$ is observed at the gentle rolling of steel balls into the packing container without shaking. In two dimensions, the experimental numbers for close and loose random packing are estimated with less accuracy: $c^{RCP} = 0.8225$ and $c^{RLP} = 0.601$ [75], [229]. It should be mentioned that in contradistinction to the periodic packing, random close packing is as ill defined problem that has no unique theoretical definition, and its final states are protocol dependent in both the numerical simulation and experimental sense [1111].

In order to describe the algorithm used in this section, at first one introduces the following definitions. To speed up the calculations it is useful to check the collision partners v_j $(j = j_1^i, \ldots, j_{n_t}^i)$ of v_i only in some restricted neighborhood of $\mathbf{x}_i$ called the testing window $w_t^i = \{\mathbf{x}: \ |\mathbf{x} - \mathbf{x}_i| < R_t\}$ $(R_t = \text{const.})$. In this model, the neighbors of a disc are considered to be the set of discs whose centers lie within some maximum distance called the testing window w_t^i of that disk rather than the "geometric neighbors" which can be determined by the more computationally expensive Delaunay tessellation [836]. The reason for the use of the testing window w_t^i instead of the Delaunay tessellation is that each $n_t\bar{v}_i$ is an estimation of local volume fraction which coincides with c only for an infinitely large testing window. If $n_t < 7$ for $R_t = 3a$ then a new uniformly distributed inclusion is generated in the area $w_t^i \setminus v_i^0$ in a spirit of the HCM described above, where v_i^0 is a spherical "excluded volume" with the center $\mathbf{x}_i$ and the radius $2a$ (since inclusions cannot overlap).

Another way to speed up the calculations is to carry out the shaking process only within a local shakeup window [168] rather than within a whole domain w [225]. Random local shaking is established in a shake up window $w_{sh} = \{\mathbf{x}: \ |\mathbf{x} - \mathbf{x}_i| < R_t - 2a\}$ $(R_t = \text{const.})$ whereby the inclusion center $\mathbf{x}_i$ is randomly moved to a position $\mathbf{x}'_i$ uniformly distributed in w_{sh}. If the particle does not overlap with any other inclusions the shaking is accepted, otherwise the trial shaking is repeated until the number of attempted trial shakings exceeds some limit.

Only the near neighbor set of inclusions v_j $(j = j_1^i, \ldots, j_{n_t}^i)$ which are located in the testing window w_t^i are checked for overlap, which also reduces computer time. One determines the optimal size for the shakeup window $R_t - 2a$ equal to $1.1a$ which provides the minimum average number of trial shaking attempts. This number increases from 5 to 8 when the packing density grows from $c = 0.5$ to $c = 0.8$. This size of shake up window provides the fastest stabilization of statistical parameter estimations. However, their values do not depend on the size of the window and are, furthermore, protocol independent (in contrast with the known methods of random close packing simulations [1111]). The shaking process passes through all inclusions (so called global shaking) with reestimation of the neighbor inclusions in the testing window w_t^i after each local shaking of the inclusion v_i. In so doing, the increasing number of shaking has a twofold effect. On the one hand, the system becomes more homogeneous and well-mixed and on the other hand, the stochastic fluctuations of statistical parameters (such as, e.g., $g(r)$ and $N^{nd}(r)$) estimated by averaging of these structures tend to diminish.

In Fig. 5.1 the histograms of distributions of fibers in a tested window $w_t^i = \{\mathbf{x} : \ |\mathbf{x}_l - \mathbf{x}_{il}| < R_t, \ l = 1, 2\}$ with the size $R_t = 3.1a$ are presented for the averaging over the configurations simulated by the HCSM as well as the averaging over the 100 experimental samples containing 1800 fibers each. As can be seen, the HCSM generates significantly more homogeneous and mixed arrangements than the real slightly clustered structures: the fractions $p(n_t)$ of testing windows containing both the small and large numbers n_t of inclusions decrease. It was detected that the compromise of the shaking procedure with the modified HCM leads to a more homogeneous and well-mixed arrangement than HCM.

5.4.4 Collective Rearrangement Model (CRM)

Just for completeness we will briefly introduce the collective rearrangement model accompanied by the shaking procedure. We start with a unit square $(0, 0)$ and its periodically located neighbors labeled by the douplet of integer numbers $\alpha = (\alpha_1, \alpha_2) \in Z^+$ where $\alpha_1, \alpha_2 = 0, \pm 1$. N random points $\mathbf{x}_i$ $(i = 1, \ldots, N)$ in the central square periodically reflected into the neighboring squares are assigned to the initial velocities $\mathbf{v}_i = (v_{i1}, v_{i2})$ whose components are independently distributed at random between -1 and $+1$ and to the uniformly growing inclusions with the radii $a(t) = a_0 t$. The centers of the inclusions move according to the equations $d\mathbf{x}/dt = \mathbf{v}_i$ with a discontinuous change of the vectors $\mathbf{v}_i$ at the moment the particle exits through a face of a central square as well as during collisions with other inclusions.

The collision time is obtained from the condition that the separation distance is the current diameter, which for the inclusions v_i and v_j is

$$|\mathbf{x}_i + \mathbf{v}_i \Delta t - \mathbf{x}_j - \mathbf{v}_j \Delta t| = (a_0 t + a_0 \Delta t). \tag{5.124}$$

In the case of the collision, the smaller positive root Δt of Eq. (5.124) defines the collision time τ_{ij}. If the collision takes place between the inclusions v_i and v_j from central and neighboring squares (e.g., identified by $(-1, 0)$), respectively, then $\mathbf{x}_j$ in Eq. (5.124) should be replaced by $\mathbf{x}_j + (-1, 0)$ with a subsequent estimation of a collision time τ_{ij}. The exit time τ_i^Γ is estimated as the smallest positive time for exiting of the inclusion v_i being considered through one of the sides of the central square. The determination of $\Delta t_* = \min(\tau_{ij}, \tau_i^\Gamma)$ allows estimation of the new inclusion radii $a_0(t + \Delta t_*)$ and the identification of either colliding (v_{i^*} and v_{j^*}) or exiting (v_{i^*}) inclusions for which the new velocities and locations are reestimated. For all other inclusions the position vectors are updated according to the equation $d\mathbf{x}/dt = \mathbf{v}_i$: $\mathbf{x}_i \to \mathbf{x}_i + \mathbf{v}_i \Delta t$. For these inclusions $\Delta t = \Delta t_*$ and velocities $\mathbf{v}_i$ remain unchanged. In addition, the collision and exit times are corrected $\tau_{ij} \to \tau_{ij} - \Delta t_*$ and $\tau_i^\Gamma \to \tau_i^\Gamma - \Delta t_*$, respectively. Details of the re-estimation of velocities for the inclusions v_{i^*} and v_{j^*} based on the conservation laws of momentum and energy can be found in [709], [578].

Up to this point, a single step of the *Collective Rearrangement Model* (CRM) has been defined. Repeating this step leads to the close packing demonstrated in [709]. However, because our goal is different (i.e., the simulation of well-mixed, random structures over a complete range of inclusion concentrations), further corrections to the algorithm were necessary. The algorithm is modified so that after a few repetitions estimating new velocities for the colliding inclusions, the procedures of (A) generating new inclusions in low-density testing windows and (B) inclusion shaking (described in Subsection 5.5.3) are carried out.

Both experiments and simulations suggest a transition between random and ordered configurations in the vicinity of a density $c \approx 0.8$ [925]. The packing density increases much more slowly beyond this point. An understanding of this transition is more obvious after quantitative analysis given in Figs. 5.2 and 5.3 showing the RDFs $g(r)$ estimated by the modified CRM for different disc concentrations $c = 0.60 - 0.75$ and $c = 0.75 - 0.90$,

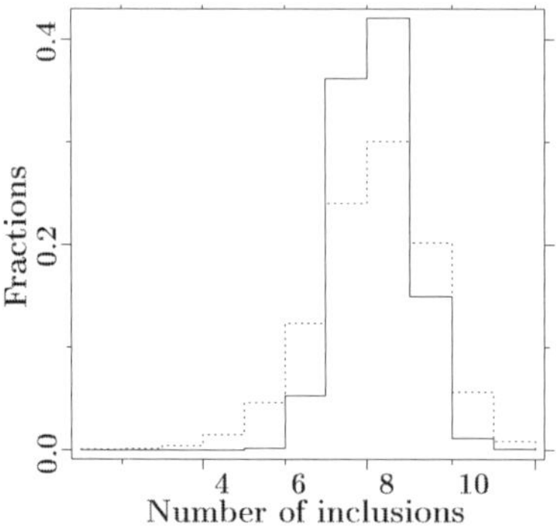

Fig. 5.1. Histogram of fractions $p(n_t)$ of testing window containing n_t fibers: generated by the HCSM at $N = 3700$, $N_t = 3.1a$ (solid curve), averaged over 100 samples (dotted curve).

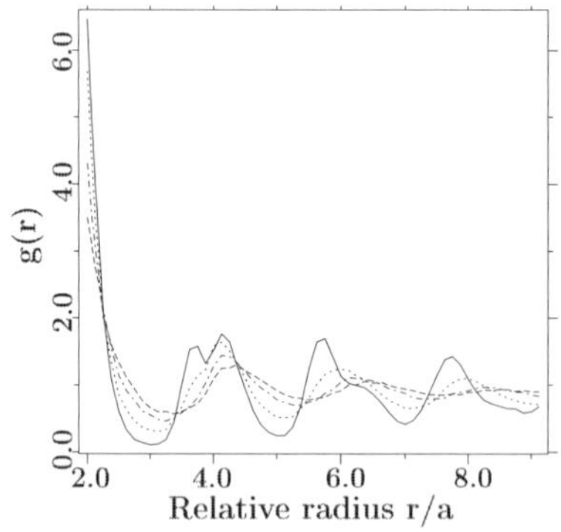

Fig. 5.2. $g(r)$ vs r/a estimated by the modified CRM at $c = 0.60$ (dashed curve), $c = 0.65$ (dot-dashed curve), $c = 0.70$ (dotted curve), $c = 0.75$ (solid curve); N grows from 799 to 811.

As can be seen, for high disc concentrations the plot of the RDF has a second peak that is characteristically split as was observed in the experimental packing [196], [797]. The split peak demonstrates the presence of large clusters with close triangular disc packing [1117]. In order to analyze the plots $g(r) \sim r$ in Fig. 5.2

we will compare the coordinates of their peaks with the analogous peaks for close triangular and square packing. A distinguishing characteristic of the last regular packing is the presence of "fixed" peaks corresponding to $r = 2a, 4a, 6a, \ldots$ and "floating" subpeaks, the locations of which depend on the specific structure of the unit cell. For triangular and square packing the coordinates of the subpeaks are $r = 3.46a, 5.32a, \ldots$ and $r = 2.83a, 4.44a, 6.3a, \ldots$, respectively. Imperfection, both in the identification of the disc centers and in the lattice itself, broaden the peaks and raise the heights of the intervening values. Thus, one can presuppose that the second subpeak in Fig. 5.2 at $r = 3.47a$ is caused by the influence of local ordering in the form of clusters with the triangular structure.

It should be mentioned that the modified CRM is not as optimal as the original CRM for modeling close-packing configurations simply because the added procedure of random shaking is just focused on the "destruction" of dense "locked" local configurations in some testing windows, leading to the generation of highly homogeneous and mixed structures. In relation to the last statement, it should be mentioned that [274] combines various Metropolis-Hastings algorithms to obtain a simulation algorithm with good mixing properties. However, the comparison of Fig. 2 in [274] for the radial distribution function $g(r)$ with $c = 0.65$ and $c = 0.735$ indicates that this $g(r)$ function reflects more order. On the other hand, modification of the known simulation protocols by adding a shaking procedure has some additional benefits. Indeed, the repeating procedure is not necessary in the protocols accompanied by a shaking procedure because a configuration generated by a few global shakings can be regarded as a separate realization.

In Fig. 5.4 we compare the RDF estimated from the experimental fiber centroid data with that from numerical simulation by the CRM, as well as the RDF represented analytically by Eq. (5.56). Figure 5.4 shows a good fit between RDFs estimated from experimental data and a from numerical simulation by the modified CRM and a substantial dissimilarity from the curves (5.56) and (5.57).

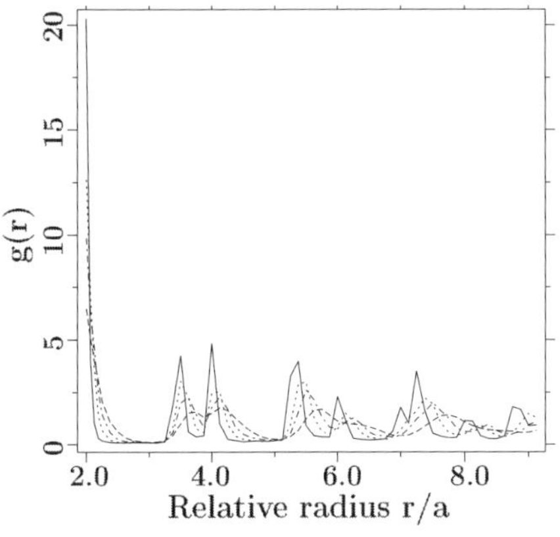

Fig. 5.3. $g(r)$ vs r/a estimated by the modified CRM at $N = 811$ and $c = 0.90$ (solid curve), $c = 0.85$ (dotted curve), $c = 0.8$ (dot-dashed curve), $c = 0.75$ (dashed curve).

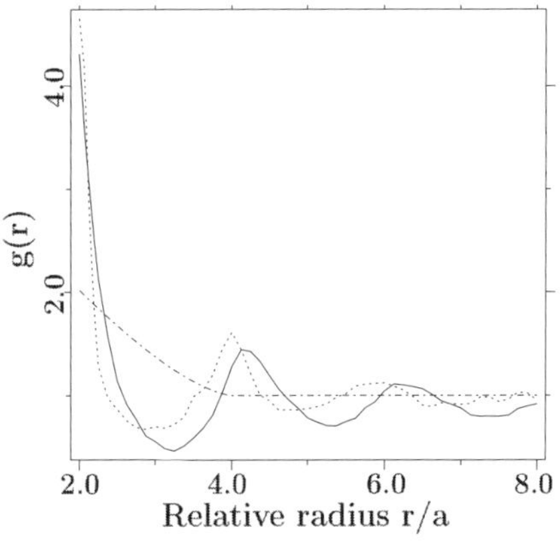

Fig. 5.4. $g(r)$ vs r/a estimated by the numerical simulation (solid curve), from experimental data (dotted curve), by (5.56) (dot-dashed curve).

Initially the periodic shaking model (IPSM) is worthy of notice. The rearrangement of initially periodic structures is conducted by the shaking procedure described in Subsection 5.5.3. The statistical parameters $g(r)$, $N^{nd}(r)$ are estimated by the averaging analysis of a few global shakings allowing inclusion configurations to be generated that are more homogeneous and well-mixed. Stabilization of estimations of $g(r)$, $N^{nd}(r)$ with increased numbers of global shakings leads to convergence of the shaking process [1006].

It is known that taking only one point probability density (volume fraction) into account can provide only a rough estimation of bounds of effective properties and statistical averages of stresses in the constitutive equations of composite materials. More informative characteristics of the point set are obtained using statistical second order quantities (such as two-point probability density, second-order intensity function, and nearest neighbor distribution) which examine the association of points relative to other points. A few contributions have paid attention to the application of these statistical distributions for generation of concrete realizations of the locations of a finite number of interacting inclusions with subsequent analyses [381], [903]. More rigorous estimations of the statistical average of stress fields in the constituents and effective elastic moduli are based on the statistical averaging of random integral equations involved for an infinite number of inclusions whose configurations are described by statistical second-order functions (see Chapters 8–10, and [1106]). In particular, we will demonstrate the strong dependence of effective moduli on the concrete form of the radial distribution function and strong differences between apparently similar distributions.

6

Effective Properties and Energy Methods in Thermoelasticity of Composite Materials

We now summarize the principal formulae of thermoelastostatics of statistically homogeneous composites satisfying the ergodicity condition in a form that is appropriate for our intended application to composites [251], [302], [438], [439], [460], [602], [646], [1180], [1183]. General representations of both effective properties and effective energy functions are presented through the local stress and strain concentrator factors. Some general exact results for both the two-phase composites and polycrystals composed of transversally isotropic crystals are presented. We will consider the variational methods representing the most rigorous trend of micromechanics which generate the bounds of effective properties by the substitution of approximate fields into the strict energy bounds obtained by the volume average of the energy density with the help of lower-order correlation functions.

6.1 Effective Thermoelastic Properties

6.1.1 Hill's Condition and Representative Volume Element

We will consider the local basic equations of thermoelastostatic of composites

$$\nabla\boldsymbol{\sigma}(\mathbf{x}) + \mathbf{f}(\mathbf{x}) = \mathbf{0}, \tag{6.1}$$

$$\boldsymbol{\sigma}(\mathbf{x}) = \mathbf{L}(\mathbf{x})\boldsymbol{\varepsilon}(\mathbf{x}) + \boldsymbol{\alpha}(\mathbf{x}), \quad \text{or} \quad \boldsymbol{\varepsilon}(\mathbf{x}) = \mathbf{M}(\mathbf{x})\boldsymbol{\sigma}(\mathbf{x}) + \boldsymbol{\beta}(\mathbf{x}), \tag{6.2}$$

$$\boldsymbol{\varepsilon}(\mathbf{x}) = \mathrm{Def}\,\mathbf{u}(\mathbf{x}), \quad \mathrm{Inc}\boldsymbol{\varepsilon}(\mathbf{x}) = \mathbf{0}, \tag{6.3}$$

accompanied by one of two homogeneous boundary conditions

$$\mathbf{u}(\mathbf{x}) = \mathbf{u}^{\partial\mathcal{E}}(\mathbf{x}) = \boldsymbol{\varepsilon}^{\partial\mathcal{E}} \cdot \mathbf{x}, \ \boldsymbol{\varepsilon}^{\partial\mathcal{E}} \equiv \text{const.}, \ \mathbf{x} \in \partial\mathcal{E}, \tag{6.4}$$

$$\boldsymbol{\sigma}(\mathbf{x}) \cdot \mathbf{n}^{\partial\mathcal{E}}(\mathbf{x}) = \mathbf{t}^{\partial\mathcal{E}}(\mathbf{x}) \equiv \boldsymbol{\sigma}^{\partial\mathcal{E}} \cdot \mathbf{n}^{\partial\mathcal{E}}(\mathbf{x}), \quad \boldsymbol{\sigma}^{\partial\mathcal{E}} = \text{const.}, \ \mathbf{x} \in \partial\mathcal{E}, \tag{6.5}$$

where $\boldsymbol{\varepsilon}^{\partial\mathcal{E}}(\mathbf{x}) = \frac{1}{2}[\nabla \otimes \mathbf{u}^{\partial\mathcal{E}}(\mathbf{x}) + (\nabla \otimes \mathbf{u}^{\partial\mathcal{E}}(\mathbf{x}))^{\top}]$, $\mathbf{x} \in \partial\mathcal{E}$ and both $\boldsymbol{\varepsilon}^{\partial\mathcal{E}}$ and $\boldsymbol{\sigma}^{\partial\mathcal{E}}$ are symmetric constant given tensors. In the case described by Eq. (6.4), the area $\mathcal{E}$ is subjected to the prescribed surface displacements of the kind that would produce uniform strain $\boldsymbol{\varepsilon}^{\partial\mathcal{E}}$ in a homogeneous material. The surface constraints in the case (6.5) are provided by the surface tractions of the kind that would produce uniform internal stress in a homogeneous material. We assumed that only one of the following two representations of $\partial\mathcal{E}$ takes place: $\partial\mathcal{E} = \partial\mathcal{E}_{\mathbf{u}}$ ($\partial\mathcal{E}_\sigma \equiv 0$) or $\partial\mathcal{E} = \partial\mathcal{E}_\sigma$ ($\partial\mathcal{E}_{\mathbf{u}} \equiv 0$).

We define the *mean value* (or volume average) of an integrable tensor field $\mathbf{g}$ given on the macrodomain $\mathcal{E} \subset R^d$ ($d = 2, 3$) as $\{\mathbf{g}\} = (\bar{\mathcal{E}})^{-1} \int_{\mathcal{E}} \mathbf{g}(\mathbf{x})\, d\mathbf{x}$, where $\bar{\mathcal{E}} = \text{mes}\mathcal{E}$. Then, according to the work-energy (2.104) and Cauchy (6.3_1) relations, we can express the volume averages of stress and strain by integrals over the respective sources and boundary conditions

$$\{\sigma_{ij}\} = \frac{1}{\bar{\mathcal{E}}} \int_{\mathcal{E}} \sigma_{ik} x_{k,j}\, d\mathbf{x} = \frac{1}{\bar{\mathcal{E}}} \Big[\int_{\mathcal{E}} f_{(i}(\mathbf{x}) x_{j)}\, d\mathbf{x} + \int_{\partial\mathcal{E}} t_{(i}(\mathbf{x}) x_{j)}\, d\mathbf{s} \Big], \quad (6.6)$$

$$\{\varepsilon_{ij}\} = \frac{1}{\bar{\mathcal{E}}} \int_{\partial\mathcal{E}} u_{(i}(\mathbf{x}) n_{j)}(\mathbf{x})\, d\mathbf{s}. \quad (6.7)$$

The relation (6.6) yields an interesting conclusion that the average value of residual stresses ($\mathbf{f}, \mathbf{t} \equiv \mathbf{0}$, $\boldsymbol{\beta}_1(\mathbf{x}) \not\equiv \mathbf{0}$) vanishes. In the particular form of the homogeneous boundary conditions (6.4), Eq. (6.7) with $\nabla\mathbf{x} = \mathbf{x}\nabla = \boldsymbol{\delta}$ leads to the conclusion that for any kinematically admissible strain field corresponding to $\boldsymbol{\varepsilon}^{\partial\mathcal{E}} \equiv$ const., the mean value $\{\boldsymbol{\varepsilon}\}$ of $\boldsymbol{\varepsilon}$ is just $\boldsymbol{\varepsilon}^{\partial\mathcal{E}}$:

$$\{\boldsymbol{\varepsilon}\} = \boldsymbol{\varepsilon}^{\partial\mathcal{E}} = \text{const.} \quad (6.8)$$

In a similar manner, the homogeneous traction boundary condition (6.5) incorporated into Eq. (6.6) (at $\mathbf{f} \equiv \mathbf{0}$) leads to the equality for the statically admissible stress field corresponding to $\boldsymbol{\sigma}^{\partial\mathcal{E}}(\mathbf{x})$:

$$\{\boldsymbol{\sigma}\} = \boldsymbol{\sigma}^{\partial\mathcal{E}} = \text{const.} \quad (6.9)$$

Now we will consider the Hill [460] condition which enables one to split a volume average of an energy-like representation into a product of volume averages of stress and strain fields:

$$\{\boldsymbol{\sigma} : \boldsymbol{\varepsilon}\} = \{\boldsymbol{\sigma}\} : \{\boldsymbol{\varepsilon}\} \quad (6.10)$$

for any admissible $\mathbf{u} \in \mathcal{A}^k(\mathbf{u})$ (2.100) ($\boldsymbol{\varepsilon} = \text{Def}\mathbf{u}$) and equilibrium $\boldsymbol{\sigma} \in \mathcal{A}^s(\boldsymbol{\sigma})$ ($\mathbf{f} \equiv \mathbf{0}$) (2.101) compatible with the homogeneous boundary conditions (6.4) and (6.5) , respectively.

Indeed, the next obvious chain of equalities

$$\begin{aligned}
\{\boldsymbol{\sigma} : \boldsymbol{\varepsilon}\} =& \frac{1}{\bar{\mathcal{E}}} \int_{\partial\mathcal{E}} t_i u_i\, d\mathbf{s} = \frac{1}{\bar{\mathcal{E}}} \{\sigma_{ik}\} \int_{\mathcal{E}_\sigma} u_i n_k\, d\mathbf{s} + \frac{1}{\bar{\mathcal{E}}} \{\varepsilon_{ik}\} \int_{\mathcal{E}_{\mathbf{u}}} t_i x_k\, d\mathbf{s} \\
=& \frac{1}{\bar{\mathcal{E}}} \{\sigma_{ik}\} \Big[\int_{\partial\mathcal{E}} u_i n_k\, d\mathbf{s} - \int_{\partial\mathcal{E}_{\mathbf{u}}} u_i n_k\, d\mathbf{s} \Big] + \frac{1}{\bar{\mathcal{E}}} \{\varepsilon_{ik}\} \Big[\int_{\partial\mathcal{E}} t_i x_k\, d\mathbf{s} - \int_{\partial\mathcal{E}_\sigma} t_i x_k\, d\mathbf{s} \Big] \\
=& 2\{\boldsymbol{\sigma}\} : \{\boldsymbol{\varepsilon}\} - \frac{1}{\bar{\mathcal{E}}} \{\sigma_{ik}\}\{\varepsilon_{il}\} \int_{\partial\mathcal{E}_{\mathbf{u}}} u_l n_k\, d\mathbf{s} - \frac{1}{\bar{\mathcal{E}}} \{\sigma_{ik}\}\{\varepsilon_{il}\} \int_{\partial\mathcal{E}_\sigma} t_l x_k\, d\mathbf{s} \\
=& 2\{\boldsymbol{\sigma}\} : \{\boldsymbol{\varepsilon}\} - \frac{1}{\bar{\mathcal{E}}} \{\sigma_{ik}\}\{\varepsilon_{il}\} \int_{\partial\mathcal{E}} u_l n_k\, d\mathbf{s} = \{\boldsymbol{\sigma}\} : \{\boldsymbol{\varepsilon}\} \quad (6.11)
\end{aligned}$$

proves the Hill's condition (6.9). We note that $\boldsymbol{\sigma}$ and $\boldsymbol{\varepsilon}$ need not be connected by a specific stress–strain relation.

We will show that for statistically homogeneous composites satisfying the ergodicity condition, the Hill's condition (6.10) exploiting the volume average operations over the volume containing statistically large number of inclusions can be recast in terms of statistical averages. At the prescribed assumptions, the operations of averaging over the macrovolume and the ensemble averaging are commutative and the next chain of equalities: takes place

$$\langle \boldsymbol{\sigma} : \boldsymbol{\varepsilon} \rangle = \{ \langle \boldsymbol{\sigma} : \boldsymbol{\varepsilon} \rangle \} = \langle \{ \boldsymbol{\sigma} : \boldsymbol{\varepsilon} \} \rangle = \frac{1}{\bar{\mathcal{E}}} \langle \int_{\partial \mathcal{E}} t_i u_i \, d\mathbf{s} \rangle = \frac{1}{\bar{\mathcal{E}}} \int_{\partial \mathcal{E}} \langle t_i u_i \rangle \, d\mathbf{s}. \tag{6.12}$$

A splitting of the displacement field $\mathbf{u}(\mathbf{x}) = \delta \mathbf{u}(\mathbf{x}) + \langle \mathbf{u}(\mathbf{x}) \rangle$ transforms the last integral (6.12) to the form

$$\frac{1}{\bar{\mathcal{E}}} \int_{\partial \mathcal{E}} \langle t_i u_i \rangle \, d\mathbf{s} = \frac{1}{\bar{\mathcal{E}}} \int_{\partial \mathcal{E}} \langle t_i \rangle \langle u_i \rangle \, d\mathbf{s} + \frac{1}{\bar{\mathcal{E}}} \langle \int_{\partial \mathcal{E}_\sigma} t_i \delta u_i \, d\mathbf{s} \rangle + \frac{1}{\bar{\mathcal{E}}} \langle \int_{\partial \mathcal{E}_{\mathbf{u}}} t_i \delta u_i \, d\mathbf{s} \rangle, \tag{6.13}$$

where the last two integrals vanish for prescribed deterministic boundary conditions because the operations $\{\cdot\}$ can be replaced by $\langle \cdot \rangle$ owing to the ergodicity condition. Thus, Eqs. (6.12) and (6.13) leads to the equation

$$\langle \boldsymbol{\sigma} : \boldsymbol{\varepsilon} \rangle = \{ \langle \mathbf{t} \rangle : \langle \mathbf{u} \rangle \}, \tag{6.14}$$

which can be transformed to the desired statistical form of the Hill's condition

$$\langle \boldsymbol{\sigma} : \boldsymbol{\varepsilon} \rangle = \langle \boldsymbol{\sigma} \rangle : \langle \boldsymbol{\varepsilon} \rangle, \tag{6.15}$$

by the use of substitution of the relations

$$\langle \boldsymbol{\varepsilon} \rangle = \text{Def} \langle \mathbf{u} \rangle, \;\; \langle \mathbf{t}(\mathbf{x}) \rangle = \langle \boldsymbol{\sigma} \rangle \cdot \mathbf{n}(\mathbf{x}), \;\; \text{Div} \boldsymbol{\sigma} = \mathbf{0} \tag{6.16}$$

into the right-hand side of Eq. (6.14) and subsequent application of the Gauss theorem (2.23).

Thus, for statistically homogeneous composites satisfying the ergodicity condition, the Hill's condition can be presented in two equivalent forms involving either the volume averages (1.10) or statistical averages (6.15). The Hill's condition holds for any compatible strain field $\boldsymbol{\varepsilon}(\mathbf{x})$ and equilibrium stress field $\boldsymbol{\sigma}(\mathbf{x})$ not necessarily related to each other by a specific stress–strain relation. The result (6.15) does not demand statistical independence of stress and strains fields because it is only a scalar condition.

The basic subject of the theory of statistically homogeneous linear elastic composites is the characterization of the *overall* or *equivalent* elastic behavior of composites. It means that one is asking not for the detailed stress–strain state in each point, but only for the (volume or statistical) averages of the stress-state. This involves finding the *overall* or *effective* moduli as the moduli of some fictitious homogeneous medium with the same average stress–strain state as for the microinhomogeneous one being considered. It means that there exists a *representative volume element Ω* (RVE) of the body under consideration that [460] "(a) is structurally entirely typical of the whole mixture on average, and (b) contains a sufficient number of inclusions for the apparent overall moduli to be effectively independent of the surface values of traction and displacement, so long as these values are 'macroscopically uniform'" (see Eq. (6.4) and (6.5)). These apparent or *effective* moduli fluctuate about a mean value with a wavelength smaller than the dimension l of the sample, which is, however, much larger than the characteristic size if the body inhomogeneity d and much smaller than the characteristic dimension L of the body itself: $d \ll l \ll L$. The effect of such fluctuations becomes insignificant within a few wavelengths from the boundary of Ω. The contribution of this boundary layer effect to any volume averages becomes negligible if the RVE becomes large enough (for details see [281], [843]).

6.1.2 Effective Elastic Moduli

Thus, a notion of RVE is critical to the concrete transition from micro- to macro levels. This transition involves two main problems: (a) to define field and material macrovariables of an effective constitutive equation, and (b) to determine how the macrovariables are related to the microvariables of the local constitutive equations as well to microtopology of a composite material. The term RVE was originally proposed in the analysis of statistically homogeneous linear elastic media ($\mathbf{f}, \boldsymbol{\beta} \equiv \mathbf{0}$) when the answer to the problem (a) is obvious. Indeed, the primary microfields $\boldsymbol{\sigma}(\mathbf{x})$ and $\boldsymbol{\varepsilon}(\mathbf{x})$ correspond to the field macrovariables $\{\boldsymbol{\sigma}\}$ and $\{\boldsymbol{\varepsilon}\}$ dependent only on the homogeneous boundary conditions either (6.8) or (6.9) and related by a material effective tensor of either the elasticity or compliance:

$$\{\boldsymbol{\sigma}\} = \mathbf{L}^*\{\boldsymbol{\varepsilon}\}, \quad \text{or} \quad \{\boldsymbol{\varepsilon}\} = \mathbf{M}^*\{\boldsymbol{\sigma}\}, \tag{6.17}$$

respectively. The problem (b) establishing the constitutive linkage between the suitable chosen micro- and macrofield is significantly more complicated. This problem is reduced to the estimation of the tensor influence functions $\mathbf{A}^*(\mathbf{x})$ and $\mathbf{B}^*(\mathbf{x})$ for the displacement (6.4) and traction (6.5) homogeneous boundary conditions:

$$\boldsymbol{\varepsilon}(\mathbf{x}) = \mathbf{A}^*(\mathbf{x})\{\boldsymbol{\varepsilon}\}, \quad \boldsymbol{\sigma}(\mathbf{x}) = \mathbf{B}^*(\mathbf{x})\{\boldsymbol{\sigma}\}, \tag{6.18}$$

respectively. Due to linearity of the problem these influencing functions exist (but are generally unknown), and have a fundamental role for obtaining the macroscopic elastic response, since they act as weighting functions in averages presenting the overall elastic moduli. Indeed, substitution of the micro-macro linkages (6.18) into the right-hand sides of the averaged constitutive equations $\{\boldsymbol{\sigma}(\mathbf{x})\} = \{\mathbf{L}(\mathbf{x})\boldsymbol{\varepsilon}(\mathbf{x})\}$, and $\{\boldsymbol{\varepsilon}(\mathbf{x})\} = \{\mathbf{M}(\mathbf{x})\boldsymbol{\sigma}(\mathbf{x})\}$ leads to the representations

$$\{\boldsymbol{\sigma}(\mathbf{x})\} = \{\mathbf{L}(\mathbf{x})\mathbf{A}^*(\mathbf{x})\}\{\boldsymbol{\varepsilon}(\mathbf{x})\}, \quad \{\boldsymbol{\varepsilon}(\mathbf{x})\} = \{\mathbf{M}(\mathbf{x})\mathbf{B}^*(\mathbf{x})\}\{\boldsymbol{\sigma}(\mathbf{x})\}, \tag{6.19}$$

respectively. Subsequent substitutions of Eqs. (6.17_1) and (6.17_2) into the left-hand sides of Eqs. (6.19_1) and (6.19_2) yield estimations of the effective stiffness and compliance:

$$\mathbf{L}^* = \{\mathbf{L}(\mathbf{x})\mathbf{A}^*(\mathbf{x})\}, \quad \mathbf{M}^* = \{\mathbf{M}(\mathbf{x})\mathbf{B}^*(\mathbf{x})\}, \tag{6.20}$$

respectively, due to a symmetry of prescribed fields either $\boldsymbol{\varepsilon}^{\partial\mathcal{E}}(\mathbf{x}) = \text{const.}$ or $\boldsymbol{\sigma}^{\partial\mathcal{E}}(\mathbf{x}) = \text{const.}$ at $\mathbf{x} \in \partial\mathcal{E}$, respectively.

Thus, the influence functions $\mathbf{A}^*(\mathbf{x})$ and $\mathbf{B}^*(\mathbf{x})$ act as weighting factors in averages giving the effective stiffness (6.20_1) and compliance (6.20_2). We will consider the properties of these functions in more detail. From the symmetry of the tensors $\{\boldsymbol{\varepsilon}\}$ and $\{\boldsymbol{\sigma}\}$ it follows that the influence tensors A^*_{ijkl} and B^*_{ijkl} are symmetric inside the index pairs ($\mathbf{C} = \mathbf{A}^*,\ \mathbf{B}^*$): $C_{ijkl} = C_{jikl} = C_{ijlk}$. Averaging of Eqs. (6.18) leads to the conclusion that $\{\mathbf{A}^*(\mathbf{x})\} = \{\mathbf{B}^*(\mathbf{x})\} = \mathbf{I}',\ \ I'_{ijkl} = \delta_{ik}\delta_{jl}$. From (6.20), and from the symmetry properties of the microscopic values $\mathbf{L}(\mathbf{x})$, $\mathbf{M}(\mathbf{x})$ and $\mathbf{A}^*(\mathbf{x})$, $\mathbf{B}^*(\mathbf{x})$, the symmetry of macroproperties $\mathbf{L}^*$ and $\mathbf{M}^*$ inside the index pairs follows. The symmetry of $\mathbf{L}^*$ and $\mathbf{M}^*$ between the index pairs (such as, e.g., $L^*_{ijkl} = L^*_{klij}$) can be obtained from energetic analysis. Indeed, for the homogeneous displacement boundary conditions (6.4), substitution of Eqs. (6.2) (at $\boldsymbol{\beta} \equiv \mathbf{0}$) and ($6.18_1$) into Hill's condition (6.10) leads to

$$\{\boldsymbol{\varepsilon} : \boldsymbol{\sigma}\} = \Big\{(\mathbf{A}^*\{\boldsymbol{\varepsilon}\}) : (\mathbf{L}\mathbf{A}^*\{\boldsymbol{\varepsilon}\})\Big\} = \Big\{(\{\boldsymbol{\varepsilon}\}\mathbf{A}^{*\top}) : (\mathbf{L}\mathbf{A}^*\{\boldsymbol{\varepsilon}\})\Big\}, \tag{6.21}$$

where $\mathbf{A}^{*\top}$ denotes the transposition of the tensor $\mathbf{A}^*$. Since the volume average $\{\boldsymbol{\varepsilon}\}$ is a constant tensor, Eq. (6.21) can be recast as

$$\{\boldsymbol{\varepsilon} : \boldsymbol{\sigma}\} = \{\boldsymbol{\varepsilon}\} : \{\mathbf{A}^{*\top}\mathbf{L}\mathbf{A}^*\}\{\boldsymbol{\varepsilon}\}, \tag{6.22}$$

and compared with the result of substitution of the effective constitutive equation (6.17_1) into Hill's condition (6.10):

$$\{\boldsymbol{\varepsilon} : \boldsymbol{\sigma}\} = \{\boldsymbol{\varepsilon}\} : \mathbf{L}^*\{\boldsymbol{\varepsilon}\}, \tag{6.23}$$

which can be considered as a *new energetic definition* of the effective elasticity $\mathbf{L}^*$. The two definitions (6.17) and (6.23) are not in general equivalent to one another for arbitrary inhomogeneous media, and the extent to which they agree in their estimated for $\mathbf{L}^*$ provides a partial check on the validity of the "overall modulus" concept.

Since Eqs. (1.22) and (6.23) coincide for any constant symmetric field $\{\boldsymbol{\varepsilon}\}$, the effective module $\mathbf{L}^*$ can be expressed as

$$\mathbf{L}^* = \{\mathbf{A}^{*\top}\mathbf{L}\mathbf{A}^*\}. \tag{6.24}$$

Owing to the symmetry properties of $\mathbf{L}$, Eq. (6.24) yields the symmetry of $\mathbf{L}^*$: $\mathbf{L}^{*\top} = \mathbf{L}^*$. In a similar manner, for homogeneous traction boundary conditions (6.5), we gen an analogous expression of symmetric effective compliance

$$\mathbf{M}^* = \{\mathbf{B}^{*\top}\mathbf{M}\mathbf{B}^*\}, \quad \mathbf{M}^{*\top} = \mathbf{M}^*. \tag{6.25}$$

Furthermore, from the positive definiteness of local elastic properties $\mathbf{L}(\mathbf{x})$ and $\mathbf{M}(\mathbf{x})$, and Hill's condition (6.23) we can obtain the positive definiteness of macroscopic properties $\{\boldsymbol{\varepsilon}\} : \mathbf{L}^*\{\boldsymbol{\varepsilon}\} \geq 0$, $\{\boldsymbol{\sigma}\} : \mathbf{M}^*\{\boldsymbol{\sigma}\} \geq 0$, and $\{\boldsymbol{\varepsilon}\}:\mathbf{L}^*\{\boldsymbol{\varepsilon}\} = 0$, $\{\boldsymbol{\sigma}\} : \mathbf{M}^*\{\boldsymbol{\sigma}\} = 0$ if and only if $\{\boldsymbol{\varepsilon}\} = \mathbf{0}$ or $\{\boldsymbol{\sigma}\} = \mathbf{0}$, respectively. Then the effective tensors $\mathbf{L}^*$ and $\mathbf{M}^*$ are invertible, and the inverse tensors $(\mathbf{L}^*)^{-1}$ and $(\mathbf{M}^*)^{-1}$ exist.

First, we will analyze the homogeneous displacement boundary condition (6.4) corresponding to $\{\boldsymbol{\varepsilon}\}$, which can be presented as a function of $\{\boldsymbol{\sigma}\}$: $\{\boldsymbol{\varepsilon}\} = (\mathbf{L}^*)^{-1}\{\boldsymbol{\sigma}\}$. Then from the definition of the influence function (6.18$_1$) as well as from the local relation $\boldsymbol{\sigma} = \mathbf{L}(\mathbf{x})\boldsymbol{\varepsilon}$ we can find the symmetric stress influence function $\hat{\mathbf{B}}^*(\mathbf{x})$:

$$\boldsymbol{\sigma}(\mathbf{x}) = \hat{\mathbf{B}}^*(\mathbf{x})\{\boldsymbol{\sigma}\}, \quad \hat{\mathbf{B}}^*(\mathbf{x}) = \mathbf{L}(\mathbf{x})\mathbf{A}^*(\mathbf{x})(\mathbf{M}^*)^{-1}, \tag{6.26}$$

satisfying the averaging condition $\{\hat{\mathbf{B}}^*(\mathbf{x})\} = \mathbf{I}$. In a similar manner, for the homogeneous traction boundary condition (6.5) corresponding to $\{\boldsymbol{\sigma}\}$, we can get the symmetric strain influence function $\hat{\mathbf{A}}^*(\mathbf{x})$:

$$\boldsymbol{\varepsilon}(\mathbf{x}) = \hat{\mathbf{A}}^*(\mathbf{x})\{\boldsymbol{\varepsilon}\}, \quad \hat{\mathbf{A}}^*(\mathbf{x}) = \mathbf{M}(\mathbf{x})\mathbf{B}^*(\mathbf{x})(\mathbf{M}^*)^{-1}, \tag{6.27}$$

with averaging condition $\{\hat{\mathbf{A}}^*(\mathbf{x})\} = \mathbf{I}$.

Thus, we defined two groups of influence functions $\mathbf{A}^*(\mathbf{x})$, $\mathbf{B}^*(\mathbf{x})$ and $\hat{\mathbf{A}}^*(\mathbf{x})$, $\hat{\mathbf{B}}^*(\mathbf{x})$ as well as two effective parameters $\mathbf{L}^*$ and $\mathbf{M}^*$. Up to now we did not establish any connections between the influence tensor functions $\mathbf{A}^*(\mathbf{x})$ and $\hat{\mathbf{A}}^*(\mathbf{x})$ as well as between $\mathbf{B}^*(\mathbf{x})$ and $\hat{\mathbf{B}}^*(\mathbf{x})$ appearing in the different homogeneous boundary conditions (6.4) and (6.5) . We also deed not define a connection between $\mathbf{L}^*$ and $\mathbf{M}^*$. However, the main goal of micromechanical research is to replace a large microinhomogeneous sample (RVE) by an equivalent homogeneous medium. This replacement makes sense if and only if the two effective parameters $\mathbf{L}^*$ and $\mathbf{M}^*$, obtained at two different homogeneous boundary conditions, are reciprocally inverse: $\mathbf{L} = (\mathbf{M}^*)^{-1}$. Therefore, we need to install some additional supplementary conditions providing the equivalence of effective constitutive equations obtained using two different homogeneous boundary conditions. It turns out that such conditions can be formulated in the form of so-called *Hill's weak assumption* (see for details and analyses of another conditions [251]:
If $\{\mathbf{u}, \boldsymbol{\varepsilon}, \boldsymbol{\sigma}\}$ is the solution of the first homogeneous boundary value problem (6.4) corresponding to a prescribed symmetric $\{\boldsymbol{\varepsilon}\}$, if $\{\mathbf{u}_1, \boldsymbol{\varepsilon}_1, \boldsymbol{\sigma}_1\}$ is the solution of the second homogeneous boundary value problem (6.5) corresponding to a prescribed symmetric $\{\boldsymbol{\sigma}_1\}$ and if $\{\boldsymbol{\varepsilon}_1\} = \{\boldsymbol{\varepsilon}\}$, then $\{\boldsymbol{\sigma}\} = \{\boldsymbol{\sigma}_1\}$ and, moreover,

$$\mathbf{A}^*(\mathbf{x}) \equiv \hat{\mathbf{A}}^*(\mathbf{x}), \ \mathbf{B}^*(\mathbf{x}) \equiv \hat{\mathbf{B}}^*(\mathbf{x}), \tag{6.28}$$

$$\mathbf{L}^* = (\mathbf{M}^*)^{-1}. \tag{6.29}$$

Thus, Hill's weak assumption means the compatibility of the influence tensor functions (6.28) as well as the equivalence of the dual effective material laws (6.29). We will assume in a subsequent account that Hill's weak assumption is fulfilled.

We recall that the theory presented above is based on the conception of a RVE for which Hill's condition (6.10) is valid only asymptotically with a tolerance of order d/l: $\{\boldsymbol{\sigma} : \boldsymbol{\varepsilon}\} = \{\boldsymbol{\sigma}\} : \{\boldsymbol{\varepsilon}\} + O(d/l)$ just in the case of the homogeneous boundary conditions, either (6.4) or (6.5). In such a case the volume averages $\{\cdot\}$ appearing in this section can be replaced by statistical averages $\langle \cdot \rangle$ for statistically homogeneous fields satisfying the ergodicity condition. The breakdown of any assumption mentioned above leads to the violation of Hill's condition and a meaningfulness of the RVE's notion is decreased (for details see Chapter 13).

6.1.3 Overall Thermoelastic Properties

In addition to purely mechanical loading $(\mathbf{f}, \boldsymbol{\beta} \equiv \mathbf{0})$, an eigenstress field $\boldsymbol{\alpha}(\mathbf{x})$ and an eigenstrain field $\boldsymbol{\beta}(\mathbf{x})$ may exist in $\mathcal{E}$. They cause an effective eigenstress $\boldsymbol{\alpha}^*$ or eigenstrain $\boldsymbol{\beta}^*$ involved in the overall constitutive relation

$$\langle \boldsymbol{\sigma} \rangle = \mathbf{L}^* \langle \boldsymbol{\varepsilon} \rangle + \boldsymbol{\alpha}^*, \quad \langle \boldsymbol{\varepsilon} \rangle = \mathbf{M}^* \langle \boldsymbol{\sigma} \rangle + \boldsymbol{\beta}^*, \tag{6.30}$$

where $\mathbf{L}^* = (\mathbf{M}^*)^{-1}$ and $\mathbf{M}^*$ are the tensor of effective stiffness and compliance, $\boldsymbol{\beta}^*$ and $\boldsymbol{\alpha}^* \equiv -\mathbf{L}^* \boldsymbol{\beta}^*$ are second-order tensors of macroscopic eigenstrains and eigenstresses, respectively. Without loss of generality, we can assume that the macrodomain (or RVE) $\mathcal{E}$ is divided into subvolumes or local volumes $\mathcal{E}^{(r)}$ ($\mathcal{E} = \cup \mathcal{E}^{(r)}$, $\mathcal{E}^{(r)} \cap \mathcal{E}^{(s)} = \emptyset$, $r, s = 0, 1, \ldots, N$, $r \neq s$) with constant parameters $\mathbf{g}(\mathbf{x}) = \mathbf{g}^{(r)} \equiv$ const. at $\mathbf{x} \in \mathcal{E}^{(r)}$ ($\mathbf{g} = \mathbf{L}, \mathbf{M}, \boldsymbol{\alpha}, \boldsymbol{\beta}$). Then the local constitutive equations can be presented in the form

$$\boldsymbol{\sigma}(\mathbf{x}) = \mathbf{L}^{(r)} \boldsymbol{\varepsilon}(\mathbf{x}) + \boldsymbol{\alpha}^{(r)}, \quad \boldsymbol{\varepsilon}(\mathbf{x}) = \mathbf{M}^{(r)} \boldsymbol{\varepsilon}(\mathbf{x}) + \boldsymbol{\beta}^{(r)}, \tag{6.31}$$

while the extensions of local mechanical concentrator factors (6.18) have the following form ($\mathbf{x} \in \mathcal{E}_s$)

$$\boldsymbol{\varepsilon}(\mathbf{x}) = \mathbf{A}^*(\mathbf{x}) \langle \boldsymbol{\varepsilon} \rangle + \overline{\mathbf{D}}^*(\mathbf{x}), \quad \overline{\mathbf{D}}^*(\mathbf{x}) \equiv \sum_{r=0}^{N} \mathbf{D}_r^*(\mathbf{x}) \boldsymbol{\beta}^{(r)}, \tag{6.32}$$

$$\boldsymbol{\sigma}(\mathbf{x}) = \mathbf{B}^*(\mathbf{x}) \langle \boldsymbol{\sigma} \rangle + \overline{\mathbf{F}}^*(\mathbf{x}), \quad \overline{\mathbf{F}}^*(\mathbf{x}) \equiv \sum_{r=0}^{N} \mathbf{F}_r(\mathbf{x}) \boldsymbol{\alpha}^{(r)}, \tag{6.33}$$

where $\mathbf{D}_r^*(\mathbf{x})$ and $\mathbf{F}_r^*(\mathbf{x})$ are called the eigenstrain influence function and the eigenstress influence function, respectively. Without loss of generality we consider a matrix composite where $\mathcal{E}^{(0)}$ is the matrix.

At first, for the estimation of effective transformation $\boldsymbol{\alpha}^*$ and $\boldsymbol{\beta}^*$, we decompose the overall field as

$$\boldsymbol{\sigma}(\mathbf{x}) = \boldsymbol{\sigma}^I(\mathbf{x}) + \boldsymbol{\sigma}^{II}(\mathbf{x}), \quad \boldsymbol{\varepsilon}(\mathbf{x}) = \boldsymbol{\varepsilon}^I(\mathbf{x}) + \boldsymbol{\varepsilon}^{II}(\mathbf{x}), \tag{6.34}$$

with the sources

$$\boldsymbol{\alpha}^I(\mathbf{x}) = \boldsymbol{\beta}^I(\mathbf{x}) = 0, \quad \text{and} \quad \boldsymbol{\alpha}^{II}(\mathbf{x}) = \boldsymbol{\alpha}(\mathbf{x}), \quad \boldsymbol{\beta}^{II}(\mathbf{x}) = \boldsymbol{\beta}(\mathbf{x}). \tag{6.35}$$

We assume homogeneous traction boundary conditions:

$$\mathbf{t}^I(\mathbf{x}) = \mathbf{t}^{\partial \mathcal{E}_\sigma}(\mathbf{x}) = \langle \boldsymbol{\sigma} \rangle \cdot \mathbf{n}(\mathbf{x}), \quad \mathbf{t}^{II}(\mathbf{x}) = \mathbf{0}, \tag{6.36}$$

$\mathbf{x} \in \partial \mathcal{E}$. The field I is solely created at $\boldsymbol{\beta}(\mathbf{x}) \equiv \mathbf{0}$ by the applied traction compatible with a homogeneous stress field $\langle \boldsymbol{\sigma} \rangle$. The field II is generated by the random stress-free strains $\boldsymbol{\beta}(\mathbf{x})$ in the absence of boundary traction. Because of this, we may call $\boldsymbol{\sigma}^I$ the load stress field and $\boldsymbol{\sigma}^{II}$ the residual stress field. Consequently, the constitutive equations decompose into two kind of equations at the local

$$\boldsymbol{\sigma}^I = \mathbf{L}\boldsymbol{\varepsilon}^I, \text{ or } \boldsymbol{\varepsilon}^I = \mathbf{M}\boldsymbol{\sigma}^I, \tag{6.37}$$

$$\boldsymbol{\sigma}^{II} = \mathbf{L}\boldsymbol{\varepsilon}^{II} + \boldsymbol{\alpha}, \text{ or } \boldsymbol{\varepsilon}^{II} = \mathbf{M}\boldsymbol{\sigma}^{II} + \boldsymbol{\beta} \tag{6.38}$$

and the macroscopic

$$\langle\boldsymbol{\sigma}^I\rangle = \mathbf{L}^*\langle\boldsymbol{\varepsilon}^I\rangle, \text{ or } \langle\boldsymbol{\varepsilon}^I\rangle = \mathbf{M}^*\langle\boldsymbol{\sigma}^I\rangle, \tag{6.39}$$

$$\langle\boldsymbol{\sigma}^{II}\rangle = \mathbf{L}^*\langle\boldsymbol{\varepsilon}^{II}\rangle + \boldsymbol{\alpha}^*, \text{ or } \langle\boldsymbol{\varepsilon}^{II}\rangle = \mathbf{M}^*\langle\boldsymbol{\sigma}^{II}\rangle + \boldsymbol{\beta}^* \tag{6.40}$$

levels, respectively, while the overall constitutive relations are defined by Eq. (6.30).

Taking into account the definitions of the two fields I and II and the boundary conditions, we find the following tensor equations for the average fields

$$\langle\boldsymbol{\sigma}^I\rangle = \langle\boldsymbol{\sigma}\rangle, \quad \langle\boldsymbol{\varepsilon}^I\rangle = \mathbf{M}^*\langle\boldsymbol{\sigma}\rangle, \quad \langle\boldsymbol{\sigma}^{II}\rangle = \mathbf{0}, \quad \langle\boldsymbol{\varepsilon}^{II}\rangle = \boldsymbol{\beta}^*. \tag{6.41}$$

Since $\boldsymbol{\sigma}^I$ and $\boldsymbol{\sigma}^{II}$ are statically admissible stress fields, while $\boldsymbol{\varepsilon}^I$ and $\boldsymbol{\varepsilon}^{II}$ are kinematically admissible strain fields then, according to the Hill's condition, we obtain

$$\langle\boldsymbol{\sigma}^I\boldsymbol{\varepsilon}^I\rangle = \langle\boldsymbol{\sigma}\rangle\mathbf{M}^*\langle\boldsymbol{\sigma}\rangle, \quad \langle\boldsymbol{\varepsilon}^I\boldsymbol{\sigma}^{II}\rangle = 0, \quad \langle\boldsymbol{\varepsilon}^{II}\boldsymbol{\sigma}^I\rangle = \boldsymbol{\beta}^*\langle\boldsymbol{\sigma}\rangle, \quad \langle\boldsymbol{\varepsilon}^{II}\rangle\langle\boldsymbol{\sigma}^{II}\rangle = 0. \tag{6.42}$$

We can recast Eq. (6.42$_3$) by the use of Eqs. (6.30) and (6.38) :

$$\boldsymbol{\beta}^*\langle\boldsymbol{\sigma}\rangle = \langle\boldsymbol{\varepsilon}^{II}\mathbf{L}\boldsymbol{\varepsilon}^I\rangle = \langle(\boldsymbol{\varepsilon}^{II} - \boldsymbol{\beta})\mathbf{L}\boldsymbol{\varepsilon}^I\rangle + \langle\boldsymbol{\beta}\mathbf{L}\boldsymbol{\varepsilon}^I\rangle = \langle\boldsymbol{\varepsilon}^I\boldsymbol{\sigma}^{II}\rangle + \langle\boldsymbol{\beta}\boldsymbol{\sigma}^I\rangle \tag{6.43}$$

Substitution of Eqs. (6.42$_2$) and (6.18$_2$) into Eq. (6.43) yields the representation of the effective eigenstrain through the stress concentrator factors of the pure mechanical loading problem (6.18$_2$):

$$\boldsymbol{\beta}^* = \langle\mathbf{B}^{*\top}\boldsymbol{\beta}\rangle. \tag{6.44}$$

Thus the effective residual strain $\boldsymbol{\beta}^*$ is defined by the averaging of eigenstrains $\boldsymbol{\beta}(\mathbf{x})$, with the weight $\mathbf{B}^{*\top}(\mathbf{x})$ depending only on the elastic properties of the medium.

However, another form of Eq. (6.44) is widely encountered in practice. To obtaining it, we resort to the obvious equalities:

$$\langle\boldsymbol{\varepsilon}^{II}\rangle = \langle V(\mathbf{x})\boldsymbol{\varepsilon}^{II}(\mathbf{x})\rangle + \langle V^{(0)}(\mathbf{x})\boldsymbol{\varepsilon}^{II}(\mathbf{x})\rangle, \quad \mathbf{0} = \langle V(\mathbf{x})\boldsymbol{\sigma}^{II}\rangle + \langle V^{(0)}(\mathbf{x})\boldsymbol{\sigma}^{II}\rangle \tag{6.45}$$

Then averaging of the local constitutive equation over the volume of the matrix and inclusions

$$\langle V^{(0)}(\mathbf{x})\boldsymbol{\varepsilon}^{II}(\mathbf{x})\rangle = \langle V^{(0)}(\mathbf{x})\mathbf{M}(\mathbf{x})\boldsymbol{\sigma}^{II}(\mathbf{x})\rangle + \langle V^{(0)}(\mathbf{x})\boldsymbol{\beta}(\mathbf{x})\rangle, \tag{6.46}$$

$$\langle V(\mathbf{x})\boldsymbol{\varepsilon}^{II}(\mathbf{x})\rangle = \langle V(\mathbf{x})\mathbf{M}(\mathbf{x})\boldsymbol{\sigma}^{II}(\mathbf{x})\rangle + \langle V(\mathbf{x})\boldsymbol{\beta}(\mathbf{x})\rangle \tag{6.47}$$

leads to the representation

$$\boldsymbol{\beta}^* = \langle \boldsymbol{\beta}(\mathbf{x}) \rangle + \langle V(\mathbf{x}) \mathbf{M}_1(\mathbf{x}) \boldsymbol{\sigma}^{II}(\mathbf{x}) \rangle \tag{6.48}$$

at the assumption of homogeneity of matrix elastic properties ($\mathbf{M}(\mathbf{x}) \equiv$ const., $\mathbf{x} \in v^{(0)}$).

The representation (6.48) has some advantages over Eq. (6.44). First of all, we did not use the Hill's condition as well as the assumptions of both the homogeneity of boundary conditions and statistical homogeneity of composite microstructure. Because of this, Eq. (6.45) can be easily generalized to the case of statistically inhomogeneous media and recast in terms of statistical averages:

$$\boldsymbol{\beta}^*(\mathbf{x}) = \langle \boldsymbol{\beta} \rangle(\mathbf{x}) + \langle V \mathbf{M}_1 \boldsymbol{\sigma}^{II} \rangle(\mathbf{x}), \tag{6.49}$$

where the argument $\mathbf{x}$ reflects an opportunity of variation of the effective material and field characteristics along the space of statistically inhomogeneous media.

For the special case of piecewise local uniform eigenstrains and elastic moduli, Eqs. (6.44) and (6.48) can be evaluated in each local volume $\mathcal{E}^{(r)}$ and the results added together to provide the following expressions:

$$\boldsymbol{\beta}^* = \boldsymbol{\beta}^{(0)} + \sum_{r=1}^{N} \langle \mathbf{B}^{*\top} \rangle_r \boldsymbol{\beta}_1^{(r)}, \tag{6.50}$$

$$\boldsymbol{\beta}^* = \langle \boldsymbol{\beta} \rangle - \sum_{r=1}^{N} \sum_{s=1}^{N} c^{(r)} \mathbf{M}_1^{(r)} \langle \mathbf{F}_s^* \rangle_r \mathbf{L}^{(s)} \boldsymbol{\beta}_1^{(s)}. \tag{6.51}$$

In a similar manner, the first homogeneous boundary value problem (6.5) can be analyzed when the following decomposition of the boundary displacement takes place:

$$\mathbf{u}^I(\mathbf{x}) = \boldsymbol{\varepsilon}^{\partial \mathcal{E}_u} \cdot \mathbf{x}, \quad \boldsymbol{\varepsilon}^{\partial \mathcal{E}_u} \equiv \text{const.}, \quad \mathbf{u}^{II}(\mathbf{x}) \equiv \mathbf{0}, \quad \mathbf{x} \in \partial \mathcal{E}_{\mathbf{u}} \equiv \partial \mathcal{E}, \tag{6.52}$$

which leads to the average fields replacing the averages (6.41) and (6.42):

$$\langle \boldsymbol{\varepsilon}^I \rangle = \langle \boldsymbol{\varepsilon} \rangle, \quad \langle \boldsymbol{\sigma}^I \rangle = \mathbf{L}^* \langle \boldsymbol{\varepsilon} \rangle, \quad \langle \boldsymbol{\varepsilon}^{II} \rangle = \mathbf{0}, \quad \langle \boldsymbol{\sigma}^{II} \rangle = \boldsymbol{\alpha}^*, \tag{6.53}$$

$$\langle \boldsymbol{\sigma}^I \boldsymbol{\varepsilon}^I \rangle = \langle \boldsymbol{\varepsilon} \rangle \mathbf{L}^* \langle \boldsymbol{\varepsilon} \rangle, \quad \langle \boldsymbol{\varepsilon}^I \boldsymbol{\sigma}^{II} \rangle = \boldsymbol{\alpha}^* \langle \boldsymbol{\varepsilon} \rangle, \quad \langle \boldsymbol{\varepsilon}^{II} \boldsymbol{\sigma}^I \rangle = 0, \quad \langle \boldsymbol{\varepsilon}^{II} \rangle \langle \boldsymbol{\sigma}^{II} \rangle = 0. \tag{6.54}$$

Consider the loading case $\langle \boldsymbol{\varepsilon} \rangle \equiv \mathbf{0}$, and note from (6.30) there is $\boldsymbol{\alpha}^* = \langle \mathbf{L}(\mathbf{x}) \boldsymbol{\varepsilon}(\mathbf{x}) + \boldsymbol{\alpha}(\mathbf{x}) \rangle$. Then we will get the representations of the effective eigenstress analogous to the representations for the effective eigenstrain (6.44) and (6.48)

$$\boldsymbol{\alpha}^* = \langle \mathbf{A}^{*\top} \boldsymbol{\alpha} \rangle, \quad \boldsymbol{\alpha}^* = \langle \boldsymbol{\alpha} \rangle + \langle \mathbf{L}_1 \boldsymbol{\varepsilon}^{II} \rangle, \tag{6.55}$$

which can be recast for piecewise uniform local properties (such as $\boldsymbol{\alpha}, \mathbf{L}$) as

$$\boldsymbol{\alpha}^* = \boldsymbol{\alpha}^{(0)} + \sum_{r=1}^{N} \langle \mathbf{A}^{*\top} \rangle_r \boldsymbol{\alpha}_1^{(r)}, \tag{6.56}$$

$$\boldsymbol{\alpha}^* = \langle \boldsymbol{\alpha} \rangle - \sum_{r=1}^{N} \sum_{s=1}^{N} c^{(r)} \mathbf{L}_1^{(r)} \langle \mathbf{D}_s^* \rangle_r \mathbf{M}^{(s)} \boldsymbol{\alpha}_1^{(s)}. \tag{6.57}$$

The consistency of the results (6.56), (6.57), and (6.50), (6.51), respectively, are assured as follows. Substitution of Eq. (6.56) into (6.57) at $\boldsymbol{\alpha}^{(r)} = \delta_{rp}\boldsymbol{\alpha}^{(p)}$ ($\boldsymbol{\alpha}^{(p)} \neq \mathbf{0}$) leads to

$$\langle \mathbf{A}^{*\top} \rangle_s = \mathbf{I} - (c^{(s)})^{-1} \sum_{r=1}^{N} \mathbf{L}^{(r)} \langle \mathbf{D}_s \rangle_r \mathbf{M}^{(s)}. \tag{6.58}$$

In a similar manner, we can get from (6.50) and (6.51)

$$\langle \mathbf{B}^{*\top} \rangle_s = \mathbf{I} - (c^{(s)})^{-1} \sum_{r=1}^{N} c^{(r)} \mathbf{M}^{(r)} \langle \mathbf{F}_s \rangle_r \mathbf{L}^{(s)}. \tag{6.59}$$

Summation over s of Eqs. (6.58) and (6.59) then gives another pair of consistency conditions for $\langle \mathbf{D}_s \rangle_s$ and $\langle \mathbf{F}_s \rangle_s$:

$$\sum_{s=1}^{N}\sum_{r=1}^{N} c^{(r)} \mathbf{L}^{(r)} \langle \mathbf{D}_s \rangle_r \mathbf{M}^{(r)} = 0, \quad \sum_{s=1}^{N}\sum_{r=1}^{N} c^{(r)} \mathbf{L}^{(r)} \langle \mathbf{D}_s \rangle_r \mathbf{M}^{(r)} = 0. \tag{6.60}$$

6.2 Effective Energy Functions

We will show that the effective energy functions can be expressed in terms of only the average values of the state variables and the composite effective properties. For this purpose the average energy function densities are presented in terms of the constitutive properties :

$$W = \frac{1}{2\bar{\mathcal{E}}} \int_{\mathcal{E}} \boldsymbol{\sigma} : (\boldsymbol{\varepsilon} - \boldsymbol{\beta}) \, d\mathbf{x}, \tag{6.61}$$

$$\Pi = W - \frac{1}{\bar{\mathcal{E}}} \int_{\mathcal{E}} \rho \mathbf{b} \cdot \mathbf{u} \, d\mathbf{x} - \frac{1}{\bar{\mathcal{E}}} \int_{\partial\mathcal{E}_\sigma} \mathbf{t}^{\partial\mathcal{E}_\sigma} \cdot \mathbf{u} \, ds, \tag{6.62}$$

$$\langle F(\boldsymbol{\varepsilon}, T) \rangle = \frac{1}{\bar{\mathcal{E}}} \int_{\mathcal{E}} \Big[\frac{1}{2} L_{ijkl} \varepsilon_{ij} \varepsilon_{kl} + \alpha_{ij}^T \varepsilon_{ij} \theta + C_\varepsilon \frac{\theta^2}{2T_0} \Big] \, d\mathbf{x}, \tag{6.63}$$

$$\langle G_e(\boldsymbol{\sigma}, T) \rangle = \frac{1}{\bar{\mathcal{E}}} \int_{\mathcal{E}} \Big[-\frac{1}{2} M_{ijkl} \sigma_{ij} \sigma_{kl} - \beta_{ij}^T \sigma_{ij} \theta + C_\sigma \frac{\theta^2}{2T_0} \Big] \, d\mathbf{x}, \tag{6.64}$$

where at the estimation of the effective strain energy W (2.131), potential energy Π (2.133$_1$), Helmholtz free energy (2.114), and Gibbs potential (2.123) for the domain devoid of nonintegrable singularities in the mechanical or physical variables, the simple averaging procedure is applicable. Hereafter we assume without loss of generality that $\bar{\mathcal{E}} = 1$.

At first we will establish some relations for both the potential energy and strain energy for the composite materials. We can consider a strain energy function W presented, according to the principle of virtual work [PVW, see (2.103)], in the form

$$W = \frac{1}{2} \int_{\partial\mathcal{E}_\sigma} \mathbf{t}^{\partial\mathcal{E}_\sigma} \cdot \mathbf{u} \, ds + \frac{1}{2} \int_{\mathcal{E}} \rho \mathbf{b} \cdot \mathbf{u} \, d\mathbf{x} - \frac{1}{2} \int_{\mathcal{E}} \boldsymbol{\sigma} : \boldsymbol{\beta} \, d\mathbf{x}. \tag{6.65}$$

The expression for the potential energy (6.62) can also be recast taking the principle of virtual work into account

$$\begin{aligned}
\Pi &= \int_{\mathcal{E}} \Big[\frac{1}{2}\boldsymbol{\sigma} : (\boldsymbol{\varepsilon} - \boldsymbol{\beta}) - \rho\mathbf{b}\cdot\mathbf{u}\Big]\, d\mathbf{x} - \int_{\partial\mathcal{E}_\sigma} \mathbf{t}^{\partial\mathcal{E}_\sigma}\cdot\mathbf{u}\, ds \\
&= -\frac{1}{2}\Big[\int_{\mathcal{E}} (\boldsymbol{\sigma} : \boldsymbol{\beta} + \rho\mathbf{b}\cdot\mathbf{u})\, d\mathbf{x} + \int_{\mathcal{E}} \mathbf{t}\cdot\mathbf{u}\, ds\Big] + \int_{\partial\mathcal{E}_\mathbf{u}} \mathbf{t}\cdot\mathbf{u}\, ds \\
&= -\frac{1}{2}\int_{\mathcal{E}} \boldsymbol{\sigma}\cdot(\boldsymbol{\varepsilon} + \boldsymbol{\beta})\, d\mathbf{x} + \int_{\partial\mathcal{E}_\mathbf{u}} \mathbf{t}\cdot\mathbf{u}\, ds.
\end{aligned} \tag{6.66}$$

It is interesting to measure the change of potential energy due to the change of elastic constants from $\mathbf{L}^0$ to $\mathbf{L}^1$ at the fixed sources $(\mathbf{u}^{\partial\mathcal{E}}, \mathbf{t}^{\partial\mathcal{E}}, \mathbf{b}, \boldsymbol{\beta})$. Let $\mathbf{f}^0$ and $\mathbf{f}^1$ $(\mathbf{f} = \boldsymbol{\sigma}, \boldsymbol{\varepsilon}, \mathbf{u})$ be the fields related with the material tensors $\mathbf{L}^0$ and $\mathbf{L}^1$, respectively. Then we get from (6.66)

$$\begin{aligned}
2(\Pi^1 - \Pi^0) &= \int_{\partial\mathcal{E}_\mathbf{u}} (\mathbf{t}^1 - \mathbf{t}^0)\cdot\mathbf{u}\, ds - \int_{\partial\mathcal{E}_\sigma} \mathbf{t}\cdot(\mathbf{u}^1 - \mathbf{u}^0)\, ds \\
&= \int_{\mathcal{E}} \Big[\rho\mathbf{b}\cdot(\mathbf{u}^1 - \mathbf{u}^0) + (\boldsymbol{\sigma}^1 - \boldsymbol{\sigma}^0) : \boldsymbol{\beta}\Big]\, d\mathbf{x} \\
&= \int_{\partial\mathcal{E}} (\mathbf{t}^1\cdot\mathbf{u}^0 - \mathbf{t}^0\cdot\mathbf{u}^1)\, ds - \int_{\mathcal{E}} \Big[\rho\mathbf{b}\cdot(\mathbf{u}^1 - \mathbf{u}^0) + (\boldsymbol{\sigma}^1 - \boldsymbol{\sigma}^0) : \boldsymbol{\beta}\Big]\, d\mathbf{x} \\
&= \int_{\mathcal{E}} (\boldsymbol{\sigma}^1 : \boldsymbol{\varepsilon}^0 - \boldsymbol{\sigma}^0 : \boldsymbol{\varepsilon}^1)\, d\mathbf{x} - \int_{\mathcal{E}} (\boldsymbol{\sigma}^1 - \boldsymbol{\sigma}^0) : \boldsymbol{\beta}\, d\mathbf{x} \\
&= \int_{\mathcal{E}} (\boldsymbol{\varepsilon}^1 - \boldsymbol{\beta}):(\mathbf{L}^1 - \mathbf{L}^0):(\boldsymbol{\varepsilon}^0 - \boldsymbol{\beta})\, d\mathbf{x} = \int_{\mathcal{E}} \boldsymbol{\sigma}^1 :(\mathbf{M}^0 - \mathbf{M}^1):\boldsymbol{\sigma}^0\, d\mathbf{x},
\end{aligned} \tag{6.67}$$

where the PVW was again used to obtain the third equality. In particular, the case $\mathbf{L}^0 \equiv$ const. and $\mathbf{L}^1(\mathbf{x}) \equiv \mathbf{L}^0$ at $\mathbf{x} \in v^{(0)}$, where $v^{(0)} \subset \mathcal{E}$ is a simply connected domain called the matrix, models the change of potential energy produced by an "introduction" of inclusions $v^{(1)} \equiv \mathcal{E}\backslash v^{(0)}$ with characteristic function $V(\mathbf{x})$ into the matrix

$$2(\Pi^1 - \Pi^0) = \int_{\mathcal{E}} (\boldsymbol{\varepsilon}^1 - \boldsymbol{\beta}):(\mathbf{L}^1 - \mathbf{L}^0):(\boldsymbol{\varepsilon}^0 - \boldsymbol{\beta}) V(\mathbf{x})\, d\mathbf{x} = \int_{\mathcal{E}} \boldsymbol{\sigma}^1 :(\mathbf{M}^0 - \mathbf{M}^1):\boldsymbol{\sigma}^0 V(\mathbf{x})\, d\mathbf{x}, \tag{6.68}$$

where, in contrast to (6.67), the integration is carried out only over the inclusion volume $v^{(1)}$ rather than over the entire domain $\mathcal{E}$.

Further simplification of Eq. (6.62) can be performed for two particular cases of mixed boundary conditions (2.96) and (2.97):

1) Fixed displacement described by the conditions

$$\mathbf{u}(\mathbf{x}) = \mathbf{u}^{\partial\mathcal{E}}(\mathbf{x}), \quad \mathbf{x} \in \partial\mathcal{E}_\mathbf{u}; \quad \mathbf{b} \equiv \mathbf{0}, \quad \mathbf{t}(\mathbf{y}) = \mathbf{0}, \quad \mathbf{y} \in \mathcal{E}_\sigma \ \text{ or } \ \mathcal{E}_\sigma = \emptyset. \tag{6.69}$$

2) Fixed traction described by the conditions

$$\boldsymbol{\sigma}(\mathbf{x})\cdot\mathbf{n}^{\partial\mathcal{E}}(\mathbf{x}) = \mathbf{t}^{\partial\mathcal{E}}(\mathbf{x}), \quad \mathbf{x} \in \partial\mathcal{E}_\sigma, \quad \boldsymbol{\beta} \equiv \mathbf{0}, \quad \mathbf{u}(\mathbf{y}) = \mathbf{0}, \quad \mathbf{y} \in \mathcal{E}_\mathbf{u} \ \text{ or } \ \mathcal{E}_\mathbf{u} = \emptyset. \tag{6.70}$$

In particular, the boundary conditions (6.69) and (6.71) at $\mathcal{E}_\sigma = \emptyset$ and $\mathcal{E}_{\mathbf{u}} = \emptyset$ yield the displacement and traction boundary conditions, respectively. Moreover, the boundary conditions (6.69) and (6.70) lead to the homogeneous boundary conditions (2.98) and (2.99), respectively, at some obvious additional constraints.

Then for the fixed displacement boundary conditions (6.69), Eqs. (6.65) and (6.67) leads to

$$\Pi = W = \frac{1}{2}\int_{\partial\mathcal{E}_{\mathbf{u}}} \mathbf{t}\cdot\mathbf{u}^{\partial\mathcal{E}_{\mathbf{u}}}\, ds - \frac{1}{2}\int_{\mathcal{E}} \boldsymbol{\sigma}:\boldsymbol{\beta}\, d\mathbf{x} = \frac{1}{2}\int_{\mathcal{E}} \boldsymbol{\sigma}:(\boldsymbol{\varepsilon}^{\partial\mathcal{E}_{\mathbf{u}}} - \boldsymbol{\beta})\, d\mathbf{x}, \quad (6.71)$$

where $\boldsymbol{\varepsilon}^{\partial\mathcal{E}_{\mathbf{u}}}$ is an arbitrary strain field compatible with the given displacement $\mathbf{u}^{\partial\mathcal{E}}(\mathbf{x})$ on $\partial\mathcal{E}_{\mathbf{u}}$. For the fixed traction boundary conditions (6.70), Eqs. (6.65) and (6.67) yield

$$\Pi = -W = -\frac{1}{2}\int_{\partial\mathcal{E}_\sigma} \mathbf{t}^{\partial\mathcal{E}_\sigma}\cdot\mathbf{u}\, ds - \frac{1}{2}\int_{\mathcal{E}} \rho\mathbf{b}\cdot\mathbf{u}\, d\mathbf{x} = -\frac{1}{2}\int_{\mathcal{E}} \boldsymbol{\sigma}^{\partial\mathcal{E}_\sigma}:\boldsymbol{\varepsilon}\, d\mathbf{x}, \quad (6.72)$$

where $\boldsymbol{\sigma}^{\partial\mathcal{E}_\sigma}$ is an arbitrary equilibrium stress field compatible with the given traction $\mathbf{t}^{\partial\mathcal{E}_\sigma}$ on $\partial\mathcal{E}_\sigma$. Thus, the potential energy Π coincides with elastic energy W to within sign for both boundary conditions (6.69) and (6.70).

In the elastic case ($\boldsymbol{\beta}\equiv\mathbf{0}$), applying Gauss's divergence theorem to Eq. (6.68) reduces the volume integrals over the inclusions $v^{(1)}$ to the surface integrals over their surfaces:

$$\Pi^1 - \Pi^0 = \int_{\partial v^{(1)}} \left[\boldsymbol{\sigma}^1:(\mathbf{n}\otimes\mathbf{u}^0) - \boldsymbol{\sigma}^0:(\mathbf{n}\otimes\mathbf{u}^1)\right] ds \quad (6.73)$$

where $\mathbf{n}$ is an outward unit normal to the inclusion surfaces $\partial v^{(1)}$. The relation (6.73) recast in the terms of the elastic energy (6.61) according to the Eqs. (6.71) and (6.72)

$$W^1 - W^0 = \pm\int_{\partial v^{(1)}} \left[\boldsymbol{\sigma}^1:(\mathbf{n}\otimes\mathbf{u}^0) - \boldsymbol{\sigma}^0:(\mathbf{n}\otimes\mathbf{u}^1)\right] ds \quad (6.74)$$

bears the name of Eshelby [317] formula (see also [237]) reducing the usual volume integrations for calculating strain energy to a particular type of surface integrations. Here the signs + and − correspond to the boundary conditions (6.69) and (6.70), respectively.

The relations obtained in the current section have been concerned with deterministic location of inclusions in the macrodomain $\mathcal{E}$ that can be considered as a particular realization of a random inclusion field. For the homogeneous boundary conditions (6.4) and (6.5), due to the assumed statistical homogeneity of the random function involved and realizability of the ergodicity hypothesis, both the volume and surface averages can be replaced by the appropriate statistical averages. However, some interesting results can be obtained directly by the use of a standard technique of random variables that was already explored in previous sections. So, we now turn to the estimation of contributions introduced by the load stress field $\boldsymbol{\sigma}^I$ and residual stress field $\boldsymbol{\sigma}^{II}$ into the elastic energy W for the homogeneous boundary conditions (6.36) and (6.52) at $\mathbf{b}\equiv\mathbf{0}$. We will also use the decomposition of both stress and strain fields (6.34) for the estimation of the average elastic energy (6.61):

$$W \equiv \frac{1}{2}\langle(\boldsymbol{\varepsilon}-\boldsymbol{\beta})\boldsymbol{\sigma}\rangle = \frac{1}{2}\langle\boldsymbol{\varepsilon}^I\boldsymbol{\sigma}^I\rangle + \frac{1}{2}\langle\boldsymbol{\varepsilon}^I\boldsymbol{\sigma}^{II}\rangle + \frac{1}{2}\langle(\boldsymbol{\varepsilon}^{II}-\boldsymbol{\beta})\boldsymbol{\sigma}^I\rangle$$
$$+\frac{1}{2}\langle(\boldsymbol{\varepsilon}^{II}-\boldsymbol{\beta})\boldsymbol{\sigma}^{II}\rangle = W^I + W^{II} + W^{\mathrm{Int}}, \quad W^I = \frac{1}{2}\langle\boldsymbol{\varepsilon}^I\boldsymbol{\sigma}^I\rangle, \tag{6.75}$$
$$W^{II} = \frac{1}{2}\langle(\boldsymbol{\varepsilon}^{II}-\boldsymbol{\beta})\boldsymbol{\sigma}^{II}\rangle, \quad W^{\mathrm{Int}} = \frac{1}{2}\langle\boldsymbol{\varepsilon}^I\boldsymbol{\sigma}^{II}\rangle + \frac{1}{2}\langle(\boldsymbol{\varepsilon}^{II}-\boldsymbol{\beta})\boldsymbol{\sigma}^I\rangle, \tag{6.76}$$

where the superscripts I, II and Int stand for energetic values caused by the fields $\boldsymbol{\sigma}^I$ and $\boldsymbol{\sigma}^{II}$ as well as by the interaction of $\boldsymbol{\sigma}^I$ and $\boldsymbol{\sigma}^{II}$, respectively. We will also introduce the average value of the strain energy density stored in the residual stress field (shortly called stored energy):

$$W^* \equiv \frac{1}{2}\langle\boldsymbol{\sigma} : (\boldsymbol{\varepsilon}-\boldsymbol{\beta})\rangle = -\frac{1}{2}\langle\boldsymbol{\sigma} : \boldsymbol{\beta}\rangle, \tag{6.77}$$

where the Hill's condition (6.15) was used together with the condition $\langle\boldsymbol{\sigma}\rangle = \mathbf{0}$ being assumed. As we will see, the stored energy W^* depends only on $\mathbf{L}(\mathbf{x})$ and $\boldsymbol{\beta}(\mathbf{s})$ and is thus as effective constant of a composite material in parallel with $\mathbf{L}^*$ and $\boldsymbol{\beta}^*$.

According to the average representations (6.41), (6.42), and (6.44) we can write

$$W^I = \frac{1}{2}\langle\boldsymbol{\sigma}^I\rangle\langle\boldsymbol{\varepsilon}^I\rangle, \quad W^{II} = -\frac{1}{2}\langle\boldsymbol{\beta}\boldsymbol{\sigma}^{II}\rangle, \quad W^{\mathrm{Int}} = 0 \tag{6.78}$$

for the traction boundary conditions (6.36). Thus the overall strain energy density W is the sum of the strain energy density W^I due to the applied load and W^{II} stored by the transformation stress field,

$$W = W^I + W^{II}, \quad W^I = \frac{1}{2}\langle\boldsymbol{\sigma}\rangle\mathbf{M}^*\langle\boldsymbol{\sigma}\rangle, \quad W^{II} = -\frac{1}{2}\langle\boldsymbol{\beta}\boldsymbol{\sigma}^{II}\rangle. \tag{6.79}$$

From Eqs. (6.42_2) and (6.43), the energy W^{II} is just the stored energy (6.77) $W^{II} = W^* = -0.5\langle\boldsymbol{\beta} : \boldsymbol{\sigma}^{II}\rangle$. The straight forward substitution of the decompositions (6.77) and (6.78) into the representation for the potential energy density (6.62) at $\mathbf{b} \equiv \mathbf{0}$ and applying the divergence theorem to the last integral in Eq. (6.62) lead to the decomposition of the potential energy at the traction boundary conditions:

$$\Pi = \Pi^I + \Pi^{II} + \Pi^{\mathrm{Int}}, \tag{6.80}$$
$$\Pi^I = W^I - \int_{\partial\mathcal{E}} \mathbf{t}^{\partial\mathcal{E}} \cdot \langle\mathbf{u}^I\rangle\, ds = -W^I, \tag{6.81}$$
$$\Pi^{II} = W^{II}, \quad \Pi^{\mathrm{Int}} = -\boldsymbol{\beta}^* : \langle\boldsymbol{\sigma}^I\rangle. \tag{6.82}$$

Thus, the strain energy W and potential energy Π can be expressed through the average stresses and the effective parameters

$$W = \frac{1}{2}\langle\boldsymbol{\sigma}\rangle\mathbf{M}^*\langle\boldsymbol{\sigma}\rangle + W^*, \quad \Pi = -\frac{1}{2}\langle\boldsymbol{\sigma}\rangle\mathbf{M}^*\langle\boldsymbol{\sigma}\rangle - \boldsymbol{\beta}^*\langle\boldsymbol{\sigma}\rangle + W^*. \tag{6.83}$$

For the displacement boundary conditions (6.52), the decomposition of the strain energy W presented in the form

$$\begin{aligned} W &\equiv \frac{1}{2}\langle(\boldsymbol{\sigma}-\boldsymbol{\alpha})\boldsymbol{\varepsilon}\rangle = \frac{1}{2}\langle\boldsymbol{\sigma}^I\boldsymbol{\varepsilon}^I\rangle + \frac{1}{2}\langle\boldsymbol{\sigma}^I\boldsymbol{\varepsilon}^{II}\rangle + \frac{1}{2}\langle(\boldsymbol{\sigma}^{II}-\boldsymbol{\alpha})\boldsymbol{\varepsilon}^I\rangle \\ &+ \frac{1}{2}\langle(\boldsymbol{\sigma}^{II}-\boldsymbol{\alpha})\boldsymbol{\varepsilon}^{II}\rangle = W^I + W^{II} + W^{\rm Int}, \quad W^I = \frac{1}{2}\langle\boldsymbol{\sigma}^I\rangle\langle\boldsymbol{\varepsilon}^I\rangle, \\ W^{II} &= -\frac{1}{2}\langle(\boldsymbol{\sigma}^{II}-\boldsymbol{\alpha})\boldsymbol{\varepsilon}^{II}\rangle, \quad W^{\rm Int} \equiv \langle\boldsymbol{\sigma}^I\boldsymbol{\varepsilon}^{II}\rangle + \langle(\boldsymbol{\sigma}^{II}-\boldsymbol{\alpha})\boldsymbol{\varepsilon}^I\rangle \end{aligned} \tag{6.84}$$

can be simplified due to the averages (6.53), (6.54), and (6.55_1). Therefore the decompositions (6.78) and (6.80) should be replaced by

$$W^I = \Pi^I = \frac{1}{2}\langle\boldsymbol{\varepsilon}\rangle\mathbf{L}^*\langle\boldsymbol{\varepsilon}\rangle, \quad W^{II} = \Pi^{II} = -\frac{1}{2}\langle\boldsymbol{\beta}\boldsymbol{\sigma}^{II}\rangle, \quad W^{\rm Int} = \Pi^{\rm Int} = \boldsymbol{\alpha}^*\langle\boldsymbol{\varepsilon}\rangle. \tag{6.85}$$

Therefore, the total strain energy and potential energy density become

$$W = \Pi = \frac{1}{2}(\langle\boldsymbol{\varepsilon}\rangle - \boldsymbol{\beta}^*)\mathbf{L}^*(\langle\boldsymbol{\varepsilon}\rangle - \boldsymbol{\beta}^*) + W^* = \frac{1}{2}\langle\boldsymbol{\sigma}\rangle\mathbf{M}^*\langle\boldsymbol{\sigma}\rangle + W^*. \tag{6.86}$$

Thus, the average energy densities (6.61) and (6.62) can be expressed by the effective thermoelastic constants $\mathbf{L}^*, \boldsymbol{\beta}^*$, and W^* for the homogeneous boundary conditions (6.4) and (6.5) .

At the estimation of the effective entropy, the simple averaging procedure is also applicable. With the help of (2.120) the average entropy for the composite medium at the displacement boundary conditions (6.52) is found as

$$S^* = -\frac{1}{\theta}\boldsymbol{\alpha}^*\langle\boldsymbol{\varepsilon}\rangle + \frac{C^*_\varepsilon}{T_0}\theta, \quad \boldsymbol{\alpha}^* = \langle\mathbf{A}^{*\top}\boldsymbol{\alpha}\rangle, \tag{6.87}$$

$$\frac{C^*_\varepsilon}{T_0}\theta = \frac{\langle C_\epsilon\rangle}{T_0}\theta - \frac{1}{\theta}\sum_{s,r=1}^N \langle\boldsymbol{\alpha}^\top\mathbf{D}_r\rangle_s\boldsymbol{\beta}^{(r)}. \tag{6.88}$$

The integrals (6.63) and (6.64) can be evaluated term by term [944]. However, following [646], $\langle F\rangle$ can be easily calculated by adopting Hill's condition (6.10) and taking advantage of the equation $F = (\boldsymbol{\sigma} : \boldsymbol{\varepsilon} - S\theta)/2$ and combining the results with Eqs. (6.30), (6.32), and (6.88)

$$\langle F\rangle = \frac{1}{2}\langle\boldsymbol{\sigma}\rangle\langle\boldsymbol{\varepsilon}\rangle - \frac{1}{2}\langle S\rangle = \frac{1}{2}\langle\boldsymbol{\varepsilon}\rangle\mathbf{L}^*\langle\boldsymbol{\varepsilon}\rangle + \boldsymbol{\alpha}^*\langle\boldsymbol{\varepsilon}\rangle - \frac{C^*_\varepsilon}{2T_0}\theta^2. \tag{6.89}$$

In the case of the traction boundary conditions (6.36), following the same procedure as above yields:

$$S^* = \frac{1}{\theta}\boldsymbol{\beta}^*\langle\boldsymbol{\sigma}\rangle + \frac{C^*_\sigma}{T_0}\theta, \quad \boldsymbol{\beta}^* = \langle\mathbf{B}^{*\top}\boldsymbol{\beta}\rangle, \tag{6.90}$$

$$\frac{C^*_\sigma}{T_0}\theta = \frac{\langle C_\sigma\rangle}{T_0}\theta - \frac{1}{\theta}\sum_{s,r=1}^N \langle\boldsymbol{\beta}^\top\mathbf{F}_r\rangle_s\boldsymbol{\alpha}^{(r)}, \tag{6.91}$$

$$\langle F\rangle = \frac{1}{2}\langle\boldsymbol{\sigma}\rangle\mathbf{M}^*\langle\boldsymbol{\sigma}\rangle + \boldsymbol{\beta}^*\langle\boldsymbol{\sigma}\rangle - \frac{C^*_\sigma}{2T_0}\theta^2. \tag{6.92}$$

Since (6.89) and (6.92) should be the same when the constant traction boundary conditions $\langle\boldsymbol{\sigma}\rangle$ (6.36) correspond with the displacement boundary conditions (6.52) according to Eqs. (6.30), it leads to $C^*_\sigma = C^*_\varepsilon + \boldsymbol{\alpha}^{T*}\mathbf{L}^*\boldsymbol{\alpha}^{T*}T_0$ and, as expected

$$\frac{C^*_\sigma}{T_0}\theta = \frac{\langle C_\sigma\rangle}{T_0}\theta - \frac{1}{\theta}\sum_{s,r=1}^{N}\langle\boldsymbol{\beta}^\top\mathbf{F}_r\rangle_s\boldsymbol{\alpha}^{(r)}. \tag{6.93}$$

Moreover, from Eqs. (6.32), (6.88) and Eqs. (6.33), (6.93) at the boundary conditions (6.52) and (6.36), we obtain

$$2W^{II} = -\langle\boldsymbol{\alpha}\boldsymbol{\varepsilon}^{II}\rangle = (C^*_\varepsilon - \langle C_\varepsilon\rangle)\frac{\theta^2}{T_0},\ \ 2W^{II} = -\langle\boldsymbol{\beta}\boldsymbol{\sigma}^{II}\rangle = (C^*_\sigma - \langle C_\sigma\rangle)\frac{\theta^2}{T_0}, \tag{6.94}$$

respectively.

In a similar manner, the average of the Gibbs thermodynamic potential per unit volume $\langle G_e\rangle$ (6.64) can be treated as

$$\langle G_e\rangle = -\frac{1}{2}\langle\boldsymbol{\sigma}\rangle\mathbf{M}^*\langle\boldsymbol{\sigma}\rangle - \boldsymbol{\beta}^*\langle\boldsymbol{\sigma}\rangle + C^*_\sigma\frac{\theta^2}{2T_0}. \tag{6.95}$$

In so doing, the representations (6.89) and (6.95) can be written in other terms of temperature and either average stresses or average strains by the use of the effective relations (6.30).

Thus the average energy functions (6.89), (6.92), and (6.95) for the statistically homogeneous composite medium subjected to the homogeneous boundary conditions either (6.4) or (6.5) are the same as for a homogeneous continuum with average values of the state variables replacing the local values, and with the effective composite properties replacing the homogeneous properties. The stored energies W^{II} ($6.94_{1,2}$), like $\boldsymbol{\alpha}^*$ and $\boldsymbol{\beta}^*$, depend only on the fields $\mathbf{L}(\mathbf{x}), \mathbf{M}(\mathbf{x})$ and $\boldsymbol{\alpha}(\mathbf{x}), \boldsymbol{\beta}(\mathbf{x})$ and act like effective material constants, which can be observed macroscopically. The stored energies W^{II} can be determined experimentally by measuring the specific heats of the composite according to Eqs. (6.94). The tensors of effective transformations $\boldsymbol{\alpha}^* = \langle\boldsymbol{\sigma}^{II}\rangle$ (6.53_4) and $\boldsymbol{\beta}^* = \langle\boldsymbol{\varepsilon}^{II}\rangle$ (6.41_4) for the displacement (6.52) and traction (6.36) boundary conditions can also be defined though the averages (6.44) and (6.55_1) with the weighting functions in the form of mechanical concentrator factors of stresses $\mathbf{B}^*(\mathbf{x})$ (6.33) and strains $\mathbf{A}^*(\mathbf{x})$ (6.32), respectively. It should be mentioned that the effective thermoelastic constants $\mathbf{L}^*, \boldsymbol{\beta}^*, W^*$ play a key role in micromechanics because of their close relationship to the internal stress fluctuations (see Chapter 13).

6.3 Some General Exact Results

6.3.1 Two-Phase Composites

The binary composites are characterized by piece-wise constant constitutive equations (6.31) at $N = 1$ when the shape, orientation, and mechanical properties in a sample-fixed coordinate system of each element of the phases are identical.

The feature that distinguishes binary composites from all other composites is that the stress concentrator factors (6.32) and (6.33) can be expressed directly in terms of the mechanical properties of both the composite and all constituents.

Indeed, substitution of Eq. (6.33) into Eqs. ($6.41_{1,2}$) leads to

$$c^{(0)}\langle\mathbf{B}^*\rangle_0 + c^{(1)}\langle\mathbf{B}^*\rangle_1 = \mathbf{I}, \quad c^{(0)}\mathbf{M}^{(0)}\langle\mathbf{B}^*\rangle_0 + c^{(1)}\mathbf{M}^{(1)}\langle\mathbf{B}^*\rangle_1 = \mathbf{M}^*, \tag{6.96}$$

which can unequally be solved for $\langle\mathbf{B}^*\rangle_i$, $(i = 0, 1)$ provided that $(\mathbf{M}^{(1)} - \mathbf{M}^{(0)})^{-1}$ exists

$$\begin{aligned}\langle\mathbf{B}^*\rangle_0 &= (c^{(0)})^{-1}(\mathbf{M}^{(0)} - \mathbf{M}^{(1)})^{-1}(\mathbf{M}^* - \mathbf{M}^{(1)}),\\ \langle\mathbf{B}^*\rangle_1 &= (c^{(1)})^{-1}(\mathbf{M}^{(1)} - \mathbf{M}^{(0)})^{-1}(\mathbf{M}^* - \mathbf{M}^{(0)}).\end{aligned} \tag{6.97}$$

Analogous substitution of Eq. (6.32) into Eqs. ($6.53_{1,2}$) leads to the estimation of average strain concentrator factors:

$$\begin{aligned}\langle\mathbf{A}^*\rangle_0 &= (c^{(0)})^{-1}(\mathbf{L}^{(0)} - \mathbf{L}^{(1)})^{-1}(\mathbf{L}^* - \mathbf{L}^{(1)}),\\ \langle\mathbf{A}^*\rangle_1 &= (c^{(1)})^{-1}(\mathbf{L}^{(1)} - \mathbf{L}^{(0)})^{-1}(\mathbf{L}^* - \mathbf{L}^{(0)}).\end{aligned} \tag{6.98}$$

In a similar manner, the transformation concentrator factors can be estimated. So, substitution of Eq. (6.33) and (6.38_2) into Eq. ($6.41_{3,4}$) yields

$$c^{(0)}\langle\overline{\mathbf{F}}^*\rangle_0 + c^{(1)}\langle\overline{\mathbf{F}}^*\rangle_1 = \mathbf{I}, \quad \langle\boldsymbol{\beta}\rangle + c^{(0)}\mathbf{M}^{(0)}\langle\overline{\mathbf{F}}^*\rangle_0 + c^{(1)}\mathbf{M}^{(1)}\langle\overline{\mathbf{F}}^*\rangle_1 = \boldsymbol{\beta}^*. \tag{6.99}$$

Hence

$$\langle\overline{\mathbf{F}}^*\rangle_0 = (\mathbf{M}^{(0)} - \mathbf{M}^{(1)})^{-1}(\boldsymbol{\beta}^* - \langle\boldsymbol{\beta}\rangle), \quad \langle\overline{\mathbf{F}}^*\rangle_1 = (\mathbf{M}^{(1)} - \mathbf{M}^{(0)})^{-1}(\boldsymbol{\beta}^* - \langle\boldsymbol{\beta}\rangle). \tag{6.100}$$

Hereinafter $\langle\mathbf{g}\rangle = c^{(0)}\mathbf{g}^{(0)} + c^{(1)}\mathbf{g}^{(1)}$, $(\mathbf{g} = \boldsymbol{\alpha}, \boldsymbol{\beta}, \mathbf{L}, \mathbf{M})$. Analogous treatment of Eqs. (6.32) and ($6.53_{3,4}$) allows one to obtain the representations of the transformation strain concentrator factors

$$\langle\overline{\mathbf{D}}^*\rangle_0 = (\mathbf{L}^{(0)} - \mathbf{L}^{(1)})^{-1}(\boldsymbol{\alpha}^* - \langle\boldsymbol{\alpha}\rangle), \quad \langle\overline{\mathbf{D}}^*\rangle_1 = (\mathbf{L}^{(1)} - \mathbf{L}^{(0)})^{-1}(\boldsymbol{\alpha}^* - \langle\boldsymbol{\alpha}\rangle). \tag{6.101}$$

Thus, the average strain concentrator factors $\langle\mathbf{A}^*\rangle_i$, $\langle\overline{\mathbf{D}}^*\rangle_i$ and $\langle\mathbf{B}^*\rangle_i$, $\langle\overline{\mathbf{F}}^*\rangle_i$ are completely determined by the effective properties $\mathbf{L}^*$, $\boldsymbol{\alpha}^*$ and $\mathbf{M}^*$, $\boldsymbol{\beta}^*$, respectively, assumed to be known. Moreover, it can be shown that $\boldsymbol{\alpha}^*$ and $\boldsymbol{\beta}^*$ are the unique functions of $\mathbf{L}^*$ and $\mathbf{M}^*$, respectively. Indeed, introducing Eqs. (6.98), (6.97) into Eqs. (6.55_1), (6.44), respectively, leads to the following symmetric connections:

$$\boldsymbol{\alpha}^* = -(\mathbf{L}^* - \mathbf{L}^{(1)})[\mathbf{L}_1^{(1)}]^{-1}\boldsymbol{\alpha}^{(0)} + (\mathbf{L}^* - \mathbf{L}^{(0)})[\mathbf{L}_1^{(1)}]^{-1}\boldsymbol{\alpha}^{(1)}, \tag{6.102}$$

$$\boldsymbol{\beta}^* = -(\mathbf{M}^* - \mathbf{M}^{(1)})[\mathbf{M}_1^{(1)}]^{-1}\boldsymbol{\beta}^{(0)} + (\mathbf{M}^* - \mathbf{M}^{(0)})[\mathbf{M}_1^{(1)}]^{-1}\boldsymbol{\beta}^{(1)} \tag{6.103}$$

as well as to other representations of the effective eigenstresses and eigenstrains:

$$\boldsymbol{\alpha}^* = \boldsymbol{\alpha}^{(0)} + (\mathbf{L}^* - \mathbf{L}^{(0)})(\mathbf{L}^{(1)} - \mathbf{L}^{(0)})^{-1}(\boldsymbol{\alpha}^{(1)} - \boldsymbol{\alpha}^{(0)}), \tag{6.104}$$

$$\boldsymbol{\alpha}^* = \langle\boldsymbol{\alpha}\rangle + (\mathbf{L}^* - \langle\mathbf{L}\rangle)(\mathbf{L}^{(1)} - \mathbf{L}^{(0)})^{-1}(\boldsymbol{\alpha}^{(1)} - \boldsymbol{\alpha}^{(0)}), \tag{6.105}$$

$$\boldsymbol{\beta}^* = \boldsymbol{\beta}^{(0)} + (\mathbf{M}^* - \mathbf{M}^{(0)})(\mathbf{M}^{(1)} - \mathbf{M}^{(0)})^{-1}(\boldsymbol{\beta}^{(1)} - \boldsymbol{\beta}^{(0)}), \tag{6.106}$$

$$\boldsymbol{\beta}^* = \langle\boldsymbol{\beta}\rangle + (\mathbf{M}^* - \langle\mathbf{M}\rangle)(\mathbf{M}^{(1)} - \mathbf{M}^{(0)})^{-1}(\boldsymbol{\beta}^{(1)} - \boldsymbol{\beta}^{(0)}). \tag{6.107}$$

The stored energy W^{II} can also be expressed in terms of effective moduli by the use of substitution of Eqs. (6.33), (6.100), and (6.103) into 6.76_2):

$$W^{II} = \frac{1}{2}\boldsymbol{\beta}_1^{(1)}[\mathbf{M}_1^{(1)}]^{-1}(\langle\mathbf{M}\rangle - \mathbf{M}^*)[\mathbf{M}_1^{(1)}]^{-1}\boldsymbol{\beta}_1^{(1)}. \tag{6.108}$$

with an alternative expression of W^{II} 6.79_3) in terms of the eigenstress $\boldsymbol{\alpha}$ and elastic moduli $\mathbf{L}$. Substitution of Eqs. (6.98), (6.102) and (6.97), (6.103) into Eqs. (6.101) and (6.100), respectively, leads to a unique expression of the transformation concentration functions averaged in the phases through the mechanical concentrator factors and local thermoelastic properties ($i = 0, 1$)

$$\langle\overline{\mathbf{D}}^*\rangle_i = (\mathbf{I} - \langle\mathbf{A}^*\rangle_i)(\mathbf{L}^{(i)} - \mathbf{L}^{(1-i)})^{-1}(\boldsymbol{\alpha}^{(1-i)} - \boldsymbol{\alpha}^{(i)}), \tag{6.109}$$
$$\langle\overline{\mathbf{F}}^*\rangle_i = (\mathbf{I} - \langle\mathbf{B}^*\rangle_i)(\mathbf{M}^{(i)} - \mathbf{M}^{(1-i)})^{-1}(\boldsymbol{\beta}^{(1-i)} - \boldsymbol{\beta}^{(i)}). \tag{6.110}$$

All the results for both the concentration factors and effective transformation fields hold independently in phase topology. The only assumptions are statistical homogeneity and the existence of just two phases. Information about phase topology is completely provided by the effective properties $\mathbf{L}^*$ or $\mathbf{M}^*$.

As an example we will consider the special case of a composite with isotropic phases (2.197): $\mathbf{L}^{(i)} = (3k^{(i)}, 2\mu^{(i)})$, $\boldsymbol{\beta}^{T(i)} = \beta^{T(i)}\boldsymbol{\delta}$. Then the effective eigenstrain (6.102) is

$$\beta_{ij}^{T*} = \langle\beta^T\rangle\delta_{ij} + \frac{(\beta^{T(1)} - \beta^{T(0)})k^{(0)}k^{(1)}}{(k^{(0)} - k^{(1)})}\Big[3M_{kkij}^* - \langle\frac{1}{k}\rangle\delta_{ij}\Big]. \tag{6.111}$$

For transversally isotropic composites of two isotropic phase, Eq. (6.112) can be written as

$$\beta_a^{T*} = \langle\beta^T\rangle + \frac{(\beta^{T(1)} - \beta^{T(0)})k^{(0)}k^{(1)}}{(k^{(0)} - k^{(1)})}\Big[\frac{3(1-\nu_a^*)}{E_a^*} - \langle\frac{1}{k}\rangle\Big],$$
$$\beta_t^{T*} = \langle\beta^T\rangle + \frac{(\beta^{T(1)} - \beta^{T(0)})k^{(0)}k^{(1)}}{(k^{(0)} - k^{(1)})}\Big[\frac{3}{2k_t^*} - \frac{3\nu_a^*(1-2\nu_a^*)}{E_a^*} - \langle\frac{1}{k}\rangle\Big], \tag{6.112}$$

where β_a^{T*} and β_t^{T*} are the effective axial and transversal thermal expansion coefficients, E_a^* and ν_a^* are the effective axial Young's modulus and Poisson's ratio, and k_t^* is the effective plane strain transverse bulk modulus.

If both the composite and its constituents are isotropic, Eq. (6.112) simplifies to the formula of Levin [665]

$$\beta^{T*} = \langle\beta^T\rangle + \frac{(\beta^{T(1)} - \beta^{T(0)})k^{(0)}k^{(1)}}{(k^{(0)} - k^{(1)})}\Big[\frac{1}{k^*} - \langle\frac{1}{k}\rangle\Big] \tag{6.113}$$

with the stress concentrator factors presented in the form (2.197) for the isotropic tensors ($i = 0, 1$):

$$\langle\mathbf{B}^*\rangle_i = \left(\frac{k^{(i)}}{c^{(i)}k^*}\frac{k^{(1-i)} - k^*)}{k^{(1-i)} - k^{(i)}}\,,\ \frac{\mu^{(i)}}{c^{(i)}\mu^*}\frac{\mu^{(1-i)} - \mu^*}{\mu^{(1-i)} - \mu^{(i)}}\right), \tag{6.114}$$

$$\langle\overline{\mathbf{F}}^*\rangle_i = \left(\frac{3(-1)^i}{c^{(i)}}\frac{k^{(0)}k^{(1)}}{k^{(1)} - k^{(0)}}(\beta^* - \langle\beta\rangle)\,,\ 0\right). \tag{6.115}$$

Thus average eigenstress concentrator factors $\langle \overline{\mathbf{F}}^* \rangle_i$ $(i = 0, 1)$ are purely isotropic, and this holds independently of phase geometry. The stored energy (6.108) is given by

$$W^{II} = \frac{9}{2} \frac{k^{(0)} k^{(1)}}{k^{(1)} - k^{(0)}} (\beta^* - \langle \beta \rangle)(\beta^{(1)} - \beta^{(0)}). \tag{6.116}$$

The results of this section have been derived and recovered by several authors [59], [295], [296], [598], [602], [646], [665], [945], [970].

Equations (6.113)–(6.116) show that average phase stresses, effective thermal expansion, and stored energy are monotonic functions of the effective module k^* (concerning (6.114) this holds for the isotropic part; the deviatoric stresses are monotonic functions of μ^*). Therefore, the bounds on the effective moduli lead to bounds on the average phase stresses, effective expansion, and stored energy.

It is interesting that the direct expressions of the transformation concentrator factors through the mechanical ones (6.109), (6.110) were obtained for their averages over the phases. However, Dvorak [295] (see also [59], [302]) has generalized these equalities to the local level by a proposed *uniform field theory.*

Namely, let the homogeneous traction boundary conditions (6.5) be prescribed. The derivation of the local eigenstress concentrator factor $\overline{\mathbf{F}}^*(\mathbf{x})$ from the mechanical one $\mathbf{B}^*(\mathbf{x})$ will be performed via the following scheme. At first, superpose an auxiliary overall stress $\tilde{\boldsymbol{\sigma}}$, so that the total overall stress is given by $\hat{\boldsymbol{\sigma}} = \boldsymbol{\sigma}^{\partial \mathcal{E}} + \tilde{\boldsymbol{\sigma}}$, and that makes both the local stress and strain uniform in $\mathcal{E}$. The goal is to find the overall stress $\hat{\boldsymbol{\sigma}}$ providing the compatibility of strain fields in the phases

$$\hat{\boldsymbol{\varepsilon}} = \mathbf{M}^{(0)} \hat{\boldsymbol{\sigma}} + \boldsymbol{\beta}^{(0)} = \mathbf{M}^{(1)} \hat{\boldsymbol{\sigma}} + \boldsymbol{\beta}^{(1)}, \tag{6.117}$$

$$\hat{\boldsymbol{\sigma}} = (\mathbf{M}^{(1)} - \mathbf{M}^{(0)})^{-1} (\boldsymbol{\beta}^{(0)} - \boldsymbol{\beta}^{(1)}). \tag{6.118}$$

At the last step of the decomposition procedure, we apply the overall traction $-\hat{\boldsymbol{\sigma}} \cdot \mathbf{n}$ to $\partial \mathcal{E}$ to cancel the traction introduced at the previous step. Then we can find from Eq. (6.32) that the eigenstrain problem in two-phase composites can be converted into a solution of a mechanical loading problem $(\mathbf{x} \in \mathcal{E}^{(i)},\ i = 0, 1)$

$$\overline{\mathbf{F}}^*(\mathbf{x}) = (\mathbf{I} - \mathbf{B}^*(\mathbf{x}))(\mathbf{M}^{(i)} - \mathbf{M}^{(1-i)})^{-1} (\boldsymbol{\beta}^{(1-i)} - \boldsymbol{\beta}^{(i)}). \tag{6.119}$$

at the local level which yields the analogous problem at the level of phase averages (6.110) after the averaging of Eq. (6.119) over the phase i.

In a similar manner, a displacement boundary condition (6.4) associated with the homogeneous strain tensor $\boldsymbol{\varepsilon}^{\partial \mathcal{E}}$ may be considered. Then the actual $\boldsymbol{\varepsilon}^{\partial \mathcal{E}}$ is superposed by an auxiliary overall strain $\tilde{\boldsymbol{\varepsilon}}$ generating the overall total strain $\hat{\boldsymbol{\varepsilon}} = \boldsymbol{\sigma}^{\partial \mathcal{E}} + \tilde{\boldsymbol{\varepsilon}}$ and both the local uniform stresses and strains in $\mathcal{E}$

$$\hat{\boldsymbol{\sigma}} = \mathbf{L}^{(0)} \hat{\boldsymbol{\varepsilon}} + \boldsymbol{\alpha}^{(0)} = \mathbf{L}^{(1)} \hat{\boldsymbol{\varepsilon}} + \boldsymbol{\alpha}^{(1)}, \tag{6.120}$$

$$\hat{\boldsymbol{\varepsilon}} = (\mathbf{L}^{(1)} - \mathbf{L}^{(0)})^{-1} (\boldsymbol{\alpha}^{(0)} - \boldsymbol{\alpha}^{(1)}). \tag{6.121}$$

The last step is application of the boundary conditions $\mathbf{u}(\mathbf{x}) = -\hat{\boldsymbol{\varepsilon}} \cdot \mathbf{x}$ that leads to the expression of eigenstrain concentrator factor in Eq. (6.32) by the equation $(\mathbf{x} \in \mathcal{E}^{(i)},\ i = 0, 1)$:

$$\overline{\mathbf{D}}^*(\mathbf{x}) = (\mathbf{I} - \mathbf{A}^*(\mathbf{x}))(\mathbf{L}^{(i)} - \mathbf{L}^{(1-i)})^{-1}(\boldsymbol{\alpha}^{(1-i)} - \boldsymbol{\alpha}^{(i)}). \tag{6.122}$$

Thus, for any two-phase medium, there exists a unique pair of uniform overall strain $\hat{\boldsymbol{\varepsilon}}$ and stress $\hat{\boldsymbol{\sigma}}$ (see Eqs. (6.117), (6.118) and (6.120), (6.121)), which in superposition with transformation fields, make the total local field uniform in the entire volume. Applying the homogeneous boundary conditions associated with either the field $\hat{\boldsymbol{\varepsilon}}$ or $\hat{\boldsymbol{\sigma}}$ leads to the unique explicit expressions of local transformation concentrator factors through the mechanical concentration factors (6.119) and (6.122). The obtained representations are exact ones and hold independently on phase topology. The only assumptions are statistical homogeneity and the existence of just two phases with the ideal contact conditions between them.

Moreover, in two-phase media with isotropic constituents and slipping interfaces, analogous exact relationships are found between mechanical and thermal stress or strain fields in [50], [60], [295], [301] where the emphasis was on evaluation of general thermomechanical connections rather than the formulation of micromechanical models. Following the aforementioned works, we will consider *imperfectly bonded interfaces* allowing nonvanishing relative displacements to exist together with nonzero traction. Let an orthonormalized basis $(\mathbf{n}, \mathbf{q}, \mathbf{s})$ with unit normal $\mathbf{n} \perp \partial\mathcal{E}_{01}$ be considered at each point $\mathbf{x}$ on the interface $\mathbf{x} \in \partial\mathcal{E}^{(01)}$ between the phases. At any generic point $\mathbf{x}$ of the interphase, let us define the decomposition of both the displacement $\mathbf{u} = (\mathbf{u}_n, \mathbf{u}_q, \mathbf{u}_s)^\top$ and traction $\mathbf{t} = (\mathbf{t}_n, \mathbf{t}_q, \mathbf{t}_s)^\top$ vectors in the basis $(\mathbf{n}, \mathbf{q}, \mathbf{s})$; additional indexes $\mathbf{g}^{(ij)}$ ($\mathbf{g} = \mathbf{n}, \mathbf{u}, \mathbf{t}, \partial\mathcal{E}$; $i, j = 0, 1$) denote a direction from phase i to phase j. The interphase $\partial\mathcal{E}^{(01)}$ is modeled by the following equations

$$[\mathbf{u}_n]_{\partial\mathcal{E}^{(01)}} = \mathbf{0}, \quad [\mathbf{t}_n]_{\partial\mathcal{E}^{(01)}} = \mathbf{0}, \tag{6.123}$$

$$[\mathbf{u}_\alpha]_{\partial\mathcal{E}^{(01)}} = M^{\partial}_{\alpha\beta}\mathbf{t}_\beta, \tag{6.124}$$

where $[\mathbf{g}_n]_{\partial\mathcal{E}^{(ij)}} = \mathbf{g}^{(j)}(\mathbf{y}) - \mathbf{g}^{(j)}(\mathbf{z})$ is a jump of the function $\mathbf{g} = \mathbf{u}, \mathbf{t}$ at the transition from phase i to phase j ($i, j = 0, 1$; $i \neq j$; $\mathbf{y} \in \mathcal{E}^{(j)}$, $\mathbf{z} \in \mathcal{E}^{(i)}$; $\mathbf{y}, \mathbf{z} \to \mathbf{x} \in \partial\mathcal{E}^{(ij)}$). $M^{\partial}_{\alpha\beta}$ ($\alpha, \beta = q, s$) are the "compliance" of the interface layer modeling in the limiting cases both the ideal contact ($M^{\partial}_{\alpha\beta} \equiv 0$) and sliding ($M^{\partial}_{\alpha\beta} \equiv \infty$).

Let the field components of the local decompositions $\boldsymbol{\varepsilon}(\mathbf{x}) = \boldsymbol{\varepsilon}^I(\mathbf{x}) + \boldsymbol{\varepsilon}^{II}(\mathbf{x})$ and $\boldsymbol{\sigma}(\mathbf{x}) = \boldsymbol{\sigma}^I(\mathbf{x}) + \boldsymbol{\sigma}^{II}(\mathbf{x})$ be satisfy the equations for the stress–strain concentrator factors in the phases

$$\boldsymbol{\varepsilon}^I(\mathbf{x}) = \mathbf{A}^*(\mathbf{x})\langle\boldsymbol{\varepsilon}\rangle, \quad \boldsymbol{\sigma}^I(\mathbf{x}) = \mathbf{L}(\mathbf{x})\mathbf{A}^*(\mathbf{x})\langle\boldsymbol{\varepsilon}\rangle, \tag{6.125}$$

$$\boldsymbol{\varepsilon}^{II}(\mathbf{x}) = \overline{\mathbf{D}}^*(\mathbf{x}), \quad \boldsymbol{\sigma}^{II}(\mathbf{x}) = \mathbf{L}(\mathbf{x})\overline{\mathbf{D}}^*(\mathbf{x}) + \boldsymbol{\alpha}^*(\mathbf{x}), \tag{6.126}$$

$$\boldsymbol{\sigma}^I(\mathbf{x}) = \mathbf{B}^*(\mathbf{x})\langle\boldsymbol{\sigma}\rangle, \quad \boldsymbol{\varepsilon}^I(\mathbf{x}) = \mathbf{M}(\mathbf{x})\mathbf{B}^*(\mathbf{x})\langle\boldsymbol{\sigma}\rangle, \tag{6.127}$$

$$\boldsymbol{\sigma}^{II}(\mathbf{x}) = \overline{\mathbf{F}}^*(\mathbf{x}), \quad \boldsymbol{\varepsilon}^{II}(\mathbf{x}) = \mathbf{M}(\mathbf{x})\overline{\mathbf{F}}^*(\mathbf{x}) + \boldsymbol{\beta}^*(\mathbf{x}), \tag{6.128}$$

where Eqs (6.125), (6.126) and (6.127), (6.128) correspond to the homogeneous boundary conditions (6.4) and (6.5) , respectively. Furthermore, at the interphase between the phases the following equations for the displacement jumps $\mathbf{u} \equiv \mathbf{u}^I + \mathbf{u}^{II}$ should be added:

$$[\mathbf{u}^I(\mathbf{x})]_{\partial\mathcal{E}^{01}} = \mathbf{d}^I(\mathbf{x})\langle\varepsilon\rangle, \quad [\mathbf{u}^{II}(\mathbf{x})]_{\partial\mathcal{E}^{01}} = \mathbf{d}^{II}(\mathbf{x}), \tag{6.129}$$
$$[\mathbf{u}^I(\mathbf{x})]_{\partial\mathcal{E}^{01}} = \mathbf{f}^I(\mathbf{x})\langle\boldsymbol{\sigma}\rangle, \quad [\mathbf{u}^{II}(\mathbf{x})]_{\partial\mathcal{E}^{01}} = \mathbf{f}^{II}(\mathbf{x}), \tag{6.130}$$

where Eqs. (6.129) and (6.130) are consistent with the boundary conditions (6.4) and (6.5) , respectively, and the local displacement fields (6.129) and (6.130) hold the contact conditions (6.123), (6.124).

We will show that in two-phase systems with isotropic phases and imperfectly bonded interface (6.123) and (6.124), the transformation concentrator factors can be explicitly expressed through the mechanical ones, i.e., the two pairs of transformation tensors $\overline{\mathbf{D}}^*$, $\mathbf{d}^{II}$ and $\overline{\mathbf{F}}^*$, $\mathbf{f}^{II}$ are explicitly expressed through the tensor pairs of pure mechanical loading $\mathbf{A}^*$, $\mathbf{d}^I$ and $\mathbf{B}^*$, $\mathbf{f}^I$, respectively. To accomplish this, we will modify the scheme (6.117)–(6.121) previously proposed for the ideal contact between the phases.

First let the homogeneous traction boundary conditions (6.36) be considered. For imperfectly bonded interface (6.123) and (6.124), Eqs. (6.117)–(6.119) are also valid and define the homogeneous field $\hat{\boldsymbol{\sigma}}$ leading to coinciding of the strain distributions in the phases (6.117). This field $\hat{\boldsymbol{\sigma}}$ provides the continuation of displacement fields in the interface vicinity because the field $\hat{\boldsymbol{\sigma}}$ (6.118) in the microinhomogeneous media with the isotropic constituents is isotropic, the shear stresses are absent at the interface $\partial\mathcal{E}^{(01)}$, and the composite responds as if interfaces were perfectly bounded. Zero tractions at the external interface $\partial\mathcal{E}$ are provided by applying of the traction at the boundary $\partial\mathcal{E}$: $\mathbf{t}(\mathbf{y}) = -\hat{\boldsymbol{\sigma}}\cdot\mathbf{n}$, $\mathbf{y}\in\partial\mathcal{E}$, which produces the local fields inside the medium $\mathbf{x}\in\mathcal{E}$:

$$\boldsymbol{\sigma}^I(\mathbf{x}) = -\mathbf{B}^*(\mathbf{x})\hat{\boldsymbol{\sigma}}, \quad [\mathbf{u}^I(\mathbf{x})]_{\partial\mathcal{E}^{(01)}} = -\mathbf{f}^I(\mathbf{x})\hat{\boldsymbol{\sigma}}. \tag{6.131}$$

Substitution of Eqs. (6.131) into the equality (6.33) yields ($\mathbf{x}\in\mathcal{E}^{(i)}$, $i=0,1$)

$$\boldsymbol{\sigma}^{II}(\mathbf{x}) = (\mathbf{I}-\mathbf{B}^*(\mathbf{x}))(\mathbf{M}^{(i)}-\mathbf{M}^{(1-i)})^{-1}(\boldsymbol{\beta}^{(1-i)}-\boldsymbol{\beta}^{(i)}), \tag{6.132}$$
$$[\mathbf{u}^{II}(\mathbf{x})]_{\partial\mathcal{E}^{(01)}} = -\mathbf{f}^I(\mathbf{x})(\mathbf{M}^{(i)}-\mathbf{M}^{(1-i)})^{-1}(\boldsymbol{\beta}^{(1-i)}-\boldsymbol{\beta}^{(i)}), \tag{6.133}$$

from which the concentrator factors can be found ($\mathbf{x}\in\mathcal{E}^{(i)}$, $i=0,1$)

$$\overline{\mathbf{F}}^*(\mathbf{x}) = (\mathbf{I}-\mathbf{B}^*(\mathbf{x}))(\mathbf{M}^{(i)}-\mathbf{M}^{(1-i)})^{-1}(\boldsymbol{\beta}^{(1-i)}-\boldsymbol{\beta}^{(i)}), \tag{6.134}$$
$$\mathbf{f}^{II}(\mathbf{x}) = -\mathbf{f}^I(\mathbf{x})(\mathbf{M}^{(i)}-\mathbf{M}^{(1-i)})^{-1}(\boldsymbol{\beta}^{(1-i)}-\boldsymbol{\beta}^{(i)}). \tag{6.135}$$

In a similar manner, for the displacement boundary conditions associated with the homogeneous tensor $\boldsymbol{\varepsilon}^{\partial\mathcal{E}_{\mathbf{u}}}$ (6.55), a connection between the local strain concentrator factors can be found ($\mathbf{x}\in\mathcal{E}^{(i)}$, $i=0,1$):

$$\overline{\mathbf{D}}^*(\mathbf{x}) = (\mathbf{I}-\mathbf{A}^*(\mathbf{x}))(\mathbf{L}^{(i)}-\mathbf{L}^{(1-i)})^{-1}(\boldsymbol{\alpha}^{(1-i)}-\boldsymbol{\alpha}^{(i)}), \tag{6.136}$$
$$\mathbf{d}^{II}(\mathbf{x}) = -\mathbf{d}^I(\mathbf{x})(\mathbf{L}^{(i)}-\mathbf{L}^{(1-i)})^{-1}(\boldsymbol{\alpha}^{(1-i)}-\boldsymbol{\alpha}^{(i)}). \tag{6.137}$$

In is notable that the equalities (6.134) and (6.136) coincide with the analogous expression for the ideal interface (6.119) and (6.122), respectively, with the distinguished values of the tensors $\mathbf{A}^*(\mathbf{x}), \mathbf{B}^*(\mathbf{x})$. At the obtaining Eqs. (6.134)–(6.137) the isotropy condition of constituents was essentially explored because in the case of their anisotropy, both the displacement jumps and shear components of interface tractions do not vanish, which makes establishment of the interconnections of the concentration tensors (6.134)–(6.137) difficult.

We now turn our attention to the specification of the link between the overall stresses and strains of microinhomogeneous media. We will show that the effective constitutive equation has the form (6.30) and find the dependences of the effective parameters $\mathbf{L}^*, \mathbf{M}^*, \boldsymbol{\alpha}^*, \boldsymbol{\beta}^*$ on the average concentrator tensors (6.125) and (6.127). We start by writing the average strain in phase r $(r = 0, \ldots, N)$:

$$\langle \varepsilon_{ij} \rangle^{(r)} = \frac{1}{2\bar{\mathcal{E}}^{(r)}} \int_{\partial \mathcal{E}} (u_i n_j + u_j n_i)\, d\mathbf{s} + \frac{1}{2\bar{\mathcal{E}}^{(r)}} \sum_{s=1}^{N} \int_{\partial \mathcal{E}^{(rs)}} (u_i n_j + u_j n_i)\, d\mathbf{s}, \quad (6.138)$$

where Gauss's divergence theorem (2.23) has been applied to the phase r with the outward normal $\mathbf{n}$ to the bounding surfaces $\partial\mathcal{E}$ and $\partial\mathcal{E}^{(rs)}$. Here there are not any limitation on the interface boundary conditions (such as, e.g., (6.123) and (6.124)), and the surface of cracks or cavities that are internal to phase r is denoted by $\partial\mathcal{E}^{(rr)}$. Multiplying Eq. (6.138) by $c^{(r)} \equiv \bar{\mathcal{E}}^{(r)}/\bar{\mathcal{E}}$, and summing over r, results in

$$\langle \boldsymbol{\varepsilon} \rangle = \sum_{r=0}^{N} c^{(r)} \langle \boldsymbol{\varepsilon} \rangle^{(r)} - \sum_{r=0}^{N} \sum_{s=0}^{N} \mathbf{J}^{(rs)}, \quad \mathbf{J}^{(rs)} \equiv \frac{1}{2\bar{\mathcal{E}}} \int_{\partial \mathcal{E}^{(rs)}} (u_i n_j + u_j n_i)\, d\mathbf{s}, \quad (6.139)$$

where Gauss's divergence theorem has again been applied to the total domain $\mathcal{E}$ bounded by the surface $\partial\mathcal{E}$. Equation (6.139$_2$) can also be recast in terms of displacement jumps at the interfaces $\partial\mathcal{E}^{(rs)}$:

$$(\mathbf{J}^{(rs)} + \mathbf{J}^{(sr)})_{ij} = \frac{1}{2\bar{\mathcal{E}}} \int_{\partial \mathcal{E}^{(rs)}} ([u_i] n_j + [u_j] n_i)\, ds \quad (6.140)$$

taking into account that $\mathbf{n}^{(rs)} = -\mathbf{n}^{(sr)}$, $[\mathbf{u}]_{\partial\mathcal{E}^{(rs)}} = \mathbf{u}^{(s)}(\mathbf{y}) - \mathbf{u}^{(r)}(\mathbf{z})$ $(r, s = 0, 1, \ldots N;\ r \neq n;\ \mathbf{y} \in \mathcal{E}^{(s)},\ \mathbf{z} \in \mathcal{E}^{(r)};\ \mathbf{y}, \mathbf{z} \to \mathbf{x} \in \partial\mathcal{E}^{(rs)})$, and that the internal interphase $\partial\mathcal{E}^{(rr)}$ of phase r can be decomposed as $\partial\mathcal{E}^{(rr)} = \partial\mathcal{E}^{+(rr)} + \partial\mathcal{E}^{-(rr)}$, where $\partial\mathcal{E}^{+(rr)}$ and $\partial\mathcal{E}^{-(rr)}$ denote the upper and lower surfaces of the interphase $\partial\mathcal{E}^{(rr)}$. The interface tensors $\mathbf{J}^{(rs)} + \mathbf{J}^{(sr)} = \tilde{\mathbf{d}}^{I(rs)} \langle \boldsymbol{\varepsilon} \rangle + \tilde{\mathbf{d}}^{II(rs)}$ can be decomposed with respect to both the strain concentrator factor $\mathbf{d}^I$ (6.129$_1$) and eigenstrain concentrator factor $\mathbf{d}^{II}$ (6.129$_2$):

$$\tilde{\mathbf{d}}^{I(rs)} = \frac{1}{\bar{\mathcal{E}}} \int_{\partial \mathcal{E}^{(rs)}} d^I_{(ikl}(\mathbf{x}) n_{j)}(\mathbf{x})\, d\mathbf{x}, \quad \tilde{\mathbf{d}}^{II(rs)} = \frac{1}{\bar{\mathcal{E}}} \int_{\partial \mathcal{E}^{(rs)}} d^{II}_{(ikl}(\mathbf{x}) n_{j)}(\mathbf{x})\, d\mathbf{x}, \quad (6.141)$$

where the concentration factors $\tilde{\mathbf{d}}^{I(sr)}$ and $\tilde{\mathbf{d}}^{II(sr)}$ are symmetric with respect to the index interchange $r \leftrightarrow s$. Once the concentrator factors $\langle \mathbf{A}^* \rangle_r$, $\langle \overline{\mathbf{D}}^* \rangle_r$, $\tilde{\mathbf{d}}^{I(rs)}$, $\tilde{\mathbf{d}}^{II(rs)}$ $(r, s = 0, \ldots, N)$ are known, we can obtain the effective properties

$$\mathbf{L}^* = \mathbf{L}^{(0)} + \sum_{r=1}^{N} c^{(r)} \mathbf{L}_1^{(r)} \langle \mathbf{A}^* \rangle_r + \mathbf{L}^{(0)} \sum_{r=0}^{N} \sum_{s=r}^{N} \tilde{\mathbf{d}}^{I(rs)}, \quad (6.142)$$

$$\boldsymbol{\alpha}^* = \boldsymbol{\alpha}^{(0)} + \sum_{r=1}^{N} c^{(r)} \mathbf{L}_1^{(r)} \langle \overline{\mathbf{D}}^* \rangle_r + \mathbf{L}^{(0)} \sum_{r=0}^{N} \sum_{s=r}^{N} \tilde{\mathbf{d}}^{II(rs)} \quad (6.143)$$

with similar representations for $\mathbf{M}^*$ and $\boldsymbol{\beta}^*$. By the use of the theorem of virtual work, Benveniste and Dvorak [60] have proved a generalization of effective eigenvector representations (6.55$_1$) and (6.50) also [as Eqs. (6.142) and (6.143)] for the case of multiphase materials with anisotropic phases and imperfect interfaces of the type described by (6.123) and (6.124):

$$\boldsymbol{\alpha}^* = \boldsymbol{\alpha}^{(0)} + \sum_{r=0}^{N} \langle \mathbf{A}^{*\top} \rangle_r \boldsymbol{\alpha}_1^{(r)}, \quad \boldsymbol{\beta}^* = \boldsymbol{\beta}^{(0)} + \sum_{r=1}^{N} \langle \mathbf{B}^{*\top} \rangle_r \boldsymbol{\beta}_1^{(r)}. \tag{6.144}$$

Initially we will consider N phase composite material with zero boundary conditions (6.52) $\boldsymbol{\varepsilon}^{\partial \mathcal{E}_u} \equiv \mathbf{0}$, when the left hand site of Eq. (6.139$_1$) vanishes $\langle \boldsymbol{\varepsilon} \rangle \equiv \mathbf{0}$, and the average stresses can be expressed as

$$\langle \boldsymbol{\sigma} \rangle = \sum_{r=0}^{N} c^{(r)} (\mathbf{L}^{(r)} \langle \boldsymbol{\varepsilon} \rangle^{(r)} + \boldsymbol{\alpha}^{(r)}) = \boldsymbol{\alpha}^*. \tag{6.145}$$

In subsequent presentations, we will analyze the case of two-phase composites ($r = 0, 1$), when the closed analogy with the calculations for the perfect interphase (6.96)–(6.103) takes place. Expressing the average phase strain $\langle \boldsymbol{\varepsilon} \rangle^{(1)}$ from Eq. (6.139$_1$), and substitution into Eq. (6.145) leads to

$$\boldsymbol{\alpha}^* = \langle \boldsymbol{\alpha} \rangle + c^{(1)} \mathbf{L}_1^{(1)} \langle \overline{\mathbf{D}}^* \rangle_1 + c^{(1)} \mathbf{L}^{(0)} \tilde{\mathbf{d}}^I [\mathbf{L}_1^{(1)}]^{-1} \boldsymbol{\alpha}_1^{(1)} \tag{6.146}$$

where the tensor $\langle \overline{\mathbf{D}}^* \rangle_1$ is the average of $\overline{\mathbf{D}}^*(\mathbf{x})$ (6.32) defined by Eq. (6.109), and the last item in the right-hand side of Eq. (6.146) is obtained by substituting (6.129$_2$), (6.137) into (6.139$_2$) with a subsequent notation $\tilde{\mathbf{d}}^I = \sum_{r=0}^{1} \sum_{s=r}^{1} \tilde{\mathbf{d}}^{I(rs)}$. Hence, due to Eqs. (6.109), Eq. (6.146) can be presented in terms of the constituent phase properties and the mechanical concentrator factors $\langle \mathbf{A}^* \rangle_1$ and $\tilde{\mathbf{d}}^{I(01)}$:

$$\boldsymbol{\alpha}^* = \langle \boldsymbol{\alpha} \rangle - c^{(1)} \mathbf{L}_1^{(1)} (\mathbf{I} - \langle \mathbf{A}^* \rangle_1) [\mathbf{L}_1^{(1)}]^{-1} \boldsymbol{\alpha}_1^{(1)} + c^{(1)} \mathbf{L}^{(0)} \tilde{\mathbf{d}}^I [\mathbf{L}_1^{(1)}]^{-1} \boldsymbol{\alpha}_1^{(1)}. \tag{6.147}$$

The representation (6.147) in turn can be simplified by expressing $\langle \mathbf{A}^* \rangle_1$ through the effective elastic modulus (6.142):

$$\mathbf{L}^* = \mathbf{L}^{(0)} + c^{(1)} \mathbf{L}_1^{(1)} \langle \mathbf{A}^* \rangle_1 + c^{(1)} \mathbf{L}^{(0)} \tilde{\mathbf{d}}^I \tag{6.148}$$

and substituting the result into Eq. (6.147), which leads to the formula

$$\boldsymbol{\alpha}^* = \boldsymbol{\alpha}^{(0)} + (\mathbf{L}^* - \mathbf{L}^{(0)}) [\mathbf{L}_1^{(1)}]^{-1} \boldsymbol{\alpha}_1^{(1)}, \tag{6.149}$$

which formally coincides with Eq. (6.104) for composites with perfect interface where at this point the tensor $\mathbf{L}^*$ is defined by Eq. (6.148) rather than by Eq. (6.20$_1$).

In a similar manner, we can obtain for the zero traction boundary conditions (6.36) $\mathbf{t}^{\partial \mathcal{E}_\sigma}(\mathbf{x}) = \mathbf{0}$, $\mathbf{x} \in \partial \mathcal{E}$ equations analogous to Eqs. (6.149) and (6.148):

$$\boldsymbol{\beta}^* = \boldsymbol{\beta}^{(0)} + (\mathbf{M}^* - \mathbf{M}^{(0)})[\mathbf{M}_1^{(1)}]^{-1}\boldsymbol{\beta}_1^{(1)}, \tag{6.150}$$

$$\mathbf{M}^* = \mathbf{M}^{(0)} + c^{(1)}\mathbf{M}_1^{(1)}\langle\mathbf{B}^*\rangle_1 + c^{(1)}\mathbf{M}^{(0)}\tilde{\mathbf{f}}^I, \tag{6.151}$$

respectively, where $\tilde{f}^I_{ijkl} = \frac{1}{\bar{\mathcal{E}}}\int_{\partial\mathcal{E}^{(01)}} f^I_{(ikl}(\mathbf{x})n_{j)}(\mathbf{x})\, d\mathbf{x}$.

Recall that the effective eigenstrain $\boldsymbol{\beta}^*$ (6.150) and eigenstress $\boldsymbol{\alpha}^*$ (6.149) are obtained for composites with two isotropic phases and slipping interface. For the case of anisotropic constituents, Benveniste and Dvorak [60] have proposed the counterparts of these representations

$$\boldsymbol{\alpha}^* = \boldsymbol{\alpha}^{(0)} + (\mathbf{L}^* - \mathbf{L}^{(0)})[\mathbf{L}_1^{(1)}]^{-1}\boldsymbol{\alpha}_1^{(1)} + c^{(1)}(\tilde{\mathbf{d}}^I)^\top\left\{\boldsymbol{\alpha}^{(0)} - \mathbf{L}^{(0)}[\mathbf{L}_1^{(1)}]^{-1}\boldsymbol{\alpha}_1^{(1)}\right\}, \tag{6.152}$$

$$\boldsymbol{\beta}^* = \boldsymbol{\beta}^{(0)} + (\mathbf{M}^* - \mathbf{M}^{(0)})[\mathbf{M}_1^{(1)}]^{-1}\boldsymbol{\beta}_1^{(1)} + c^{(1)}(\tilde{\mathbf{f}}^I)^\top[\mathbf{M}_1^{(1)}]^{-1}\boldsymbol{\beta}_1^{(1)}, \tag{6.153}$$

where the last terms in Eqs. (6.152) and (6.153) vanish in the special cases of isotropic phases.

6.3.2 Polycrystals Composed of Transversally Isotropic Crystals

It turns out that a similar exact relationship between the thermal expansion coefficient and the bulk modulus holds for statistically isotropic polycrystalline aggregates (PA) composed of crystals of hexagonal, tetragonal, or trigonal symmetry. We will consider one-phase polycrystals composed of one kind of identical crystals with 3-D uniform random orientation of their crystallographic axes described by the orientation distribution function $f(\mathbf{g})$ (ODF) (5.110) with respect to a fixed set of axes $\mathbf{x}_i$. Then the components of the material tensors, e.g, $\mathbf{M}$ and $\boldsymbol{\beta}$ referred to $\mathbf{x}_i$ ($i = 1, 2, 3$) are denoted $\mathbf{M}(\mathbf{g})$ and $\boldsymbol{\beta}(\mathbf{g})$, respectively. Consequently the polycrystal is a microinhomogeneous medium composed of anisotropic crystals of different orientation $\mathbf{g}$. For statistically homogeneous media being considered, $f(\mathbf{g})$ depends only on the orientation $\mathbf{g}$ but not on the position $\mathbf{x}$ of a crystal in the sample and the statistical average $\langle(\cdot)\rangle$ (5.74_1) and orientation average $\langle(\cdot)\rangle^{\mathbf{g}}$ (5.108) coincide: $\langle(\cdot)\rangle = \langle(\cdot)\rangle^{\mathbf{g}}$. It follows that, for such a PA, the general representations for effective properties of microinhomogeneous media (6.20_2), (6.28), and (6.44) can be recast in the form

$$\mathbf{M}^* = \langle\mathbf{M}(\mathbf{g})\mathbf{B}^*(\mathbf{g})\rangle, \quad \mathbf{I} = \langle\mathbf{B}^*(\mathbf{g})\rangle, \quad \boldsymbol{\beta}^* = \langle\mathbf{B}^{*\top}(\mathbf{g})\boldsymbol{\beta}(\mathbf{g})\rangle. \tag{6.154}$$

The tensors $\mathbf{M}^*$ and $\boldsymbol{\beta}^*$ are assumed to be isotropic $\mathbf{M}^* = (1/3k^*, 1/2\mu^*)$ (2.198_1) and $\boldsymbol{\beta}^* = \beta^*\boldsymbol{\delta}$, respectively, where their components can be found by contraction of all free indices in Eqs. (6.154) and analogously Eqs. (2.199):

$$1/k^* = \langle M_{rrkl}(\mathbf{g})B^*_{klii}(\mathbf{g})\rangle, \quad 3 = \langle B^*_{kkii}(\mathbf{g})\rangle, \quad 3\beta^* = \langle\beta_{kl}(\mathbf{g})B^*_{klii}(\mathbf{g})\rangle. \tag{6.155}$$

Since, moreover, each term in angle brackets $\langle(\cdot)\rangle$ is unaltering in the different coordinate systems, Eqs. (6.155) may be written in the crystallographic axes of a crystal:

$$1/k^* = M^c_{rrkl}B^c_{klii}, \quad 3 = B^c_{kkii}, \quad 3\beta^* = \beta^c_{kl}B^c_{klii}, \tag{6.156}$$

where the indexes c denote crystallographic axes.

We will consider a particular case of the traction boundary condition (6.5) restricted to a pure isotropic applied load:

$$\boldsymbol{\sigma}_{ij}^{\partial\mathcal{E}}(\mathbf{x}) = \sigma^0\delta_{ij} \tag{6.157}$$

which allows us to interpret the tensor B^c_{klii} as a stress concentrator factor corresponding to the loading (6.157):

$$\sigma^c_{kl} = B^c_{klii}\sigma^0. \tag{6.158}$$

Furthermore, we will consider all those crystal symmetries for which the second-order tensors β_{ij}, M_{ijmm}, M_{imjm}, $M_{ijkl}\beta_{kl}$ show transversal isotropy where $\mathbf{x}_3$ is the preferred, axial, direction. This includes the hexagonal, tetragonal, and trigonal symmetries [432]. For such an ideal statistically isotropic PA subjected to isotropic loading (6.157), there is no distinction between the statistical averages of the stresses, strains, and eigenstrains in the crystals of the specific direction which must show the same transversal symmetry as the crystals

$$\begin{aligned}&\boldsymbol{\sigma}(\mathbf{g}) = \boldsymbol{\sigma}^c = \mathrm{diag}(\sigma_1,\sigma_2,\sigma_3), \quad \boldsymbol{\varepsilon}(\mathbf{g}) = \boldsymbol{\varepsilon}^c = \mathrm{diag}(\varepsilon_1,\varepsilon_2,\varepsilon_3),\\ &\boldsymbol{\beta}(\mathbf{g}) = \boldsymbol{\beta}^c = \mathrm{diag}(\beta_1,\beta_2,\beta_3);\end{aligned} \tag{6.159}$$

for the present we do not distinguish between fields I and II. The relevant components of the elastic compliance tensor $\mathbf{M}^c$ (and, analogously, the elastic stiffness $\mathbf{L}^c$) are

$$M^c_{1111} = M_{2222},\ M^c_{3333},\ M^c_{1122},\ M^c_{1133} = M^c_{2233},\ M^c_{1212},\ M^c_{1313} = M^c_{2323} \tag{6.160}$$

(the other components vanish or do not enter into the subsequent analysis). Thus, all M_{rrij} with $i \neq j$ vanish and the only surviving groups of $\mathbf{B}$ components are $B_{11ii} = B_{22ii}$ and B_{33ii}. Then, due to the absence of orientation dependence, Eqs. (6.156) are reduced to

$$1/k^* = 2M^c_{rr11}B^c_{11ii} + M^c_{rr33}B^c_{33ii}, \tag{6.161}$$

$$3\beta^* = 2\beta_1 B^c_{11ii} + \beta_3 B^c_{33ii}, \quad 3 = 2B^c_{11ii} + B^c_{33ii}. \tag{6.162}$$

which allows us to find the components B^c_{11ii}, B^c_{33ii} (and, therefore, σ_1^I and σ_3^I (6.158), (6.159) as well as β^*:

$$\sigma_1^I = \frac{1}{2}\frac{1/k^* - 3M_3}{M_1 - M_3}\sigma^0, \quad \sigma_3^I = \frac{3M_1 - 1/k^*}{M_1 - M_3}\sigma^0, \tag{6.163}$$

$$\beta^* = \beta^0 + \frac{\beta_3 - \beta_1}{3(M_3 - M_1)}(1/k^* - 2M_1 - M_3), \tag{6.164}$$

where $\beta^0 = (2\beta_1 + \beta_3)/3$ and

$$M_1 = M^c_{rr11} = M_{11} + M_{12} + M_{13} = \frac{L_{33} - L_{13}}{L_{33}(L_{11} + L_{12}) - 2L_{13}^2}, \tag{6.165}$$

$$M_3 = M^c_{rr33} = 2M_{13} + M_{33} = \frac{L_{11} + L_{12} - 2L_{13}}{L_{33}(L_{11} + L_{12}) - 2L_{13}^2}. \tag{6.166}$$

A more general representation of Eqs. (6.161) and (6.162) was proposed in [978] where the case of a statistically homogeneous, but not isotropic, PA was analyzed.

In a similar manner, the components of $\boldsymbol{\sigma}^{II} = \text{diag}(\sigma_1^{II}, \sigma_1^{II}, \sigma_3^{II})$ can be found from Eq. (6.42_1):

$$\sigma_1^{II} = \frac{3}{2}\frac{\beta^* - \beta^0}{M_1 - M_3}, \quad \sigma_3^{II} = -2\sigma_1^{II}, \tag{6.167}$$

which demonstrates that the statistical average residual stresses in the crystals are purely deviatoric ones because the isotropic part vanishes. Finally, the expression for the stored energy W^{II} can be obtained from 6.77_2) and (6.167)

$$W^{II} = \frac{3}{2}\frac{(\beta^* - \beta^0)(\beta_3 - \beta_1)}{M_1 - M_3}. \tag{6.168}$$

The expressions (6.164) and (6.168) are reminiscent of the corresponding expressions for two-phase materials (6.113) and (6.116), respectively.

Again all effective properties are monotonic functions of the effective bulk modulus k^* and, therefore, the bounds on k^* lead to the bounds on effective eigenstrain, stored energy, as well as on stresses. The basic results of this section were obtained in [432] (see also [598]) where the limiting case of elastic isotropy $(M_1 = M_3,\ \mathbf{M}^c = (3k, 2\mu))$ was also analyzed by the use of a passage to the limit $M_1 - M_3 \to \mathbf{0}$ in Eqs. (6.164), (6.167), and (6.168):

$$\beta^* = \beta^0, \quad W^{II} = \frac{2\mu}{15}\frac{9k + 8\mu}{3k + 4\mu}(\beta_3 - \beta_1)^2, \tag{6.169}$$

$$\sigma_1^{II} = \frac{2\mu}{15}\frac{9k + 8\mu}{3k + 4\mu}(\beta_3 - \beta_1), \quad \sigma_3^{II} = -2\sigma_1^{II}. \tag{6.170}$$

The exact relations (6.169) and (6.170) hold for arbitrary crystal shape. They can also be used as an approximation for PA demonstrating large anisotropic thermal deformation of crystals but weak anisotropy of elastic constants.

6.4 Variational Principle of Hashin and Shtrikman

A famous variational principle of Hashin and Shtrikman [438], [439] is based on both the idea of a homogeneous comparison medium with a stiffness $\mathbf{L}^c$ and compliance $\mathbf{M}^c = (\mathbf{L}^c)^{-1}$ and the use of the notions of the polarization tensors of the stresses $\boldsymbol{\tau}$ and strains $\boldsymbol{\eta}$ introduced by Hill [461]:

$$\boldsymbol{\tau}(\mathbf{x}) = \boldsymbol{\sigma}(\mathbf{x}) - \mathbf{L}^c\boldsymbol{\varepsilon}(\mathbf{x}), \quad \boldsymbol{\sigma}(\mathbf{x}) = \mathbf{L}^c\boldsymbol{\varepsilon}(\mathbf{x}) + \boldsymbol{\tau}(\mathbf{x}) \tag{6.171}$$

$$\boldsymbol{\eta}(\mathbf{x}) = \boldsymbol{\varepsilon}(\mathbf{x}) - \mathbf{M}^c\boldsymbol{\sigma}(\mathbf{x}), \quad \boldsymbol{\varepsilon}(\mathbf{x}) = \mathbf{M}^c\boldsymbol{\sigma}(\mathbf{x}) + \boldsymbol{\eta}(\mathbf{x}) \tag{6.172}$$

instead of the tensors of stresses $\boldsymbol{\sigma}$ and strains $\boldsymbol{\varepsilon}$, respectively. Here the stress polarization and strain polarization tensors are related by the equation $\boldsymbol{\eta}(\mathbf{x}) = -\mathbf{M}^c\boldsymbol{\tau}(\mathbf{x})$.

For the sake of definiteness, we will consider the first boundary-value problem (6.4) with the prescribed surface displacement $\mathbf{u}(\mathbf{x}) = \boldsymbol{\varepsilon}^0 \cdot \mathbf{x}$ compatible with a uniform strain $\boldsymbol{\varepsilon}^0 \equiv \langle\boldsymbol{\varepsilon}\rangle$ which is consequently the overall average strain $\langle\boldsymbol{\varepsilon}\rangle$ (6.8). The general integral equation (see Chapter 7) for statistically homogeneous media can be presented in the following operator form:

$$\boldsymbol{\varepsilon} = \boldsymbol{\varepsilon}^0 + \mathbf{U}\boldsymbol{\tau}, \quad (\mathbf{L}_1^{-1} - \mathbf{U})\boldsymbol{\tau} = \boldsymbol{\varepsilon}^0, \tag{6.173}$$

where the operator $\mathbf{U}\boldsymbol{\tau} = \int \mathbf{U}(\mathbf{x}-\mathbf{y})[\boldsymbol{\tau}(\mathbf{y}) - \langle\boldsymbol{\tau}\rangle]\, d\mathbf{y}$ is defined by the kernel $\mathbf{U}(\mathbf{x})$ (3.19) and $\mathbf{L}_1(\mathbf{x}) \equiv \mathbf{L}(\mathbf{x}) - \mathbf{L}^c$. For a consideration of the properties of the operator $\mathbf{U}$, one introduces the scalar production in the space of square integrable functions $(\mathbf{f}, \mathbf{g}) = \bar{\mathcal{E}}^{-1} \int \mathbf{f}(\mathbf{x})\mathbf{g}(\mathbf{x})\, d\mathbf{x}$. Obviously for the arbitrary constant tensors $\mathbf{b}_1, \mathbf{b}_2 = \text{const}$.

$$\mathbf{U}\boldsymbol{\tau} = \mathbf{U}(\boldsymbol{\tau} + \mathbf{b}_1), \quad (\boldsymbol{\tau}, \mathbf{U}\boldsymbol{\tau}) = (\boldsymbol{\tau} + \mathbf{b}_2, \mathbf{U}\boldsymbol{\tau}). \tag{6.174}$$

To prove self-adjoint of the operator $\mathbf{U}$, recall that $(\tau_1, \mathbf{U}\boldsymbol{\tau}_2) = -(\boldsymbol{\tau}_1, \boldsymbol{\varepsilon}_2) = (\mathbf{L}^c\boldsymbol{\varepsilon}_1 - \boldsymbol{\sigma}_1, \boldsymbol{\varepsilon}_2)$, where $\boldsymbol{\sigma}_1$ and $\boldsymbol{\varepsilon}_1$ are the stress and strain associated with arbitrary $\boldsymbol{\tau}_1$ according to Eqs. (4.1_2) and (6.173_1), respectively, and $\boldsymbol{\varepsilon}_2$ is a strain field similarly consistent with $\boldsymbol{\tau}_2$ and zero boundary conditions (6.4) : $\boldsymbol{\varepsilon}^0 = \mathbf{0}$. The virtual principle (2.104)

$$(\boldsymbol{\sigma}_1, \boldsymbol{\varepsilon}_2) = 0 \tag{6.175}$$

and Eq. (6.173_2) lead to $(\boldsymbol{\tau}_1, \mathbf{U}\boldsymbol{\tau}_2) = -(\mathbf{L}^c\boldsymbol{\varepsilon}_1, \boldsymbol{\varepsilon}_2)$. The operator $\mathbf{U}$ is self-adjoint because the last equality is symmetric with respect to the interchange of indexes due to the symmetry of $\mathbf{L}^c$. It what follows, we see after substitution of $\boldsymbol{\tau}_1 = \boldsymbol{\tau}_2$ into Eq. (6.175) that $\mathbf{U}$ is negative definite and, therefore, $\mathbf{L}_1^{-1} - \mathbf{U}$ is positive definite so long as the tensor $\mathbf{L}_1(\mathbf{x})$ is positive definite $\forall \mathbf{x} \in w$ in the sense of a relevant quadratic form. For analysis of a case of negative definiteness of $\mathbf{L}_1$, we will substitute the tensor of polarization stress (6.171) into Eq. (6.173_1): $-(\boldsymbol{\tau}, \mathbf{U}\boldsymbol{\tau}) = (\boldsymbol{\eta}, \mathbf{L}^c\boldsymbol{\eta}) - (\boldsymbol{\sigma}, \mathbf{M}^c\boldsymbol{\sigma})$. Summation of the last equality with the equation $(\boldsymbol{\tau}, \mathbf{L}_1^{-1}\boldsymbol{\tau}) = (\mathbf{L}^c\boldsymbol{\eta}, \mathbf{L}_1^{-1}\mathbf{L}^c\boldsymbol{\eta}) = -(\boldsymbol{\eta}, \mathbf{M}_1^{-1}\boldsymbol{\eta}) - (\boldsymbol{\eta}, \mathbf{L}^c\boldsymbol{\eta})$, where the equalities $\boldsymbol{\eta} = -\mathbf{M}^c\boldsymbol{\tau}$ and $\mathbf{L}_1^{-1}\mathbf{L}^c = -\mathbf{M}^c\mathbf{M}_1^{-1} - \mathbf{I}$ were used, leads to the conclusion that the value

$$(\boldsymbol{\tau}, \mathbf{L}_1^{-1}\boldsymbol{\tau}) - (\boldsymbol{\tau}, \mathbf{U}\boldsymbol{\tau}) = -(\boldsymbol{\eta}, \mathbf{M}_1^{-1}\boldsymbol{\eta}) - (\boldsymbol{\sigma}, \mathbf{M}^c\boldsymbol{\sigma}) \tag{6.176}$$

is negative definite so long as the tensor $\mathbf{M}_1$ is positive definite or, what is the same, $\mathbf{L}_1$ is negative definite. Thus, we have proved the positive (negative) definiteness of the operator $\mathbf{L}_1^{-1} - \mathbf{U}$ so long as the tensor $\mathbf{L}_1$ is positive (negative) definite at each point $\mathbf{x} \in w$.

From positive (or negative) definiteness of the operator $\mathbf{L}_1^{-1} - \mathbf{U}$, it follows that a functional of an arbitrary "trial" tensor $\boldsymbol{\tau}^*$

$$\mathcal{H}(\boldsymbol{\tau}^*) = (\boldsymbol{\tau}^*, \mathbf{L}_1^{-1}\boldsymbol{\tau}^*) - (\boldsymbol{\tau}^*, \mathbf{U}\boldsymbol{\tau}^*) - 2(\boldsymbol{\tau}^*, \boldsymbol{\varepsilon}^0) \tag{6.177}$$

attains its extremum $\delta\mathcal{H}(\boldsymbol{\tau}) = 0$ at the real values of the field $\boldsymbol{\tau}$ (6.173_2). To find the stationary value of the functional (6.177), let $\boldsymbol{\tau}$, $\boldsymbol{\sigma}$ and $\boldsymbol{\varepsilon}$ be the solution of (6.173_2), and the corresponding stress and strain associated with this $\boldsymbol{\tau}$ according to Eqs. (4.1_2) and (6.173_1), respectively. Then the stationary value of $\mathcal{H}(\boldsymbol{\tau}^*)$ is $-(\boldsymbol{\tau}, \boldsymbol{\varepsilon}^0)$ which is estimated from (4.1_2) as

$$-(\boldsymbol{\tau}, \boldsymbol{\varepsilon}^0) = (\mathbf{L}^c\boldsymbol{\varepsilon} - \boldsymbol{\sigma}, \boldsymbol{\varepsilon}^0) = (\boldsymbol{\varepsilon}^0, \mathbf{L}^c\boldsymbol{\varepsilon}^0) - (\langle\boldsymbol{\sigma}\rangle, \boldsymbol{\varepsilon}^0) = 2(W^0 - W), \tag{6.178}$$

where according to Hill's condition (6.15) $W = \langle \boldsymbol{\sigma}\boldsymbol{\varepsilon} \rangle/2$, $W^0 = \boldsymbol{\varepsilon}^0\mathbf{L}^c\boldsymbol{\varepsilon}^0/2$ are the mean elastic energy densities of the composite material and homogeneous comparison medium, respectively. In so doing the value $\mathcal{H}(\boldsymbol{\tau}^*)$ is minimized (maximized) when the tensor $\mathbf{L}_1(\mathbf{x})$ is positive (negative) definite at each point $\mathbf{x} \in w$. This proved principle is referred to as the Hashin-Shtrikman [438], [439] variational principle which was originally deduced from the field equation rather than directly from Eq. (6.173_2), as reproduced here after [1179].

We will establish the connection between the Hashin-Shtrikman variational principle and classical variational principles following [461], [1186]. For this purpose, one will choose as arbitrary trial field $\boldsymbol{\tau}^*$ of the polarization stress tensor that produces the fields of the strain $\boldsymbol{\varepsilon}^*$ and stress $\boldsymbol{\sigma}^*$ according to Eqs. (6.173_1) and (4.1_2) which, respectively, correspond with the displacement boundary conditions and the equilibrium equation. Then, following the classic minimum potential energy principle (2.106), we will get for the real mean elastic energy density W:

$$2W \leq (\boldsymbol{\varepsilon}^*, \mathbf{L}^c\boldsymbol{\varepsilon}^*) + (\boldsymbol{\varepsilon}^*, \mathbf{L}_1\boldsymbol{\varepsilon}^*). \qquad (6.179)$$

The first item in the right-hand side of Eq. (6.179) can be simplified by the use of Eqs. (6.171) and (6.173_1) and the principle of virtual work (6.175). Then the field $\boldsymbol{\tau}^*$ holds the inequality

$$2(W-W^0) \leq -(\boldsymbol{\tau}^*, \mathbf{U}\boldsymbol{\tau}^*)+(\boldsymbol{\varepsilon}^0, \mathbf{L}_1\boldsymbol{\varepsilon}^0)+2(\boldsymbol{\varepsilon}^0, \mathbf{L}_1\mathbf{U}\boldsymbol{\tau}^*)+(\mathbf{U}\boldsymbol{\tau}^*, \mathbf{L}_1\mathbf{U}\boldsymbol{\tau}^*). \quad (6.180)$$

One can recast the last item in Eq. (6.180) in the form

$$\begin{aligned}(\mathbf{U}\boldsymbol{\tau}^*, \mathbf{L}_1\mathbf{U}\boldsymbol{\tau}^*) &= \Big([\mathbf{L}_1^{-1} - \mathbf{U}]\boldsymbol{\tau}^* - \boldsymbol{\varepsilon}^0, \mathbf{L}_1\{[\mathbf{L}_1^{-1} - \mathbf{U}]\boldsymbol{\tau}^* - \boldsymbol{\varepsilon}^0\}\Big) - (\boldsymbol{\tau}^*, \mathbf{L}_1^{-1}\boldsymbol{\tau}^*)\\ &+ 2(\boldsymbol{\tau}^*, \mathbf{U}\boldsymbol{\tau}^*) + 2(\boldsymbol{\varepsilon}^0, \boldsymbol{\tau}^* - \mathbf{L}_1\mathbf{U}\boldsymbol{\tau}^*) - (\boldsymbol{\varepsilon}^0, \mathbf{L}_1\boldsymbol{\varepsilon}^0). \qquad (6.181)\end{aligned}$$

Inasmuch as the real field $\boldsymbol{\tau}$ obeys Eq. (6.173_1), the first item at the right-hand side of Eq. (6.181) vanishes at $\boldsymbol{\tau}^* = \boldsymbol{\tau}$ and is small if $\boldsymbol{\tau}^*$ is a sufficiently accurate approximation of $\boldsymbol{\tau}$. Then one obtains from (6.180)

$$2(W - W^0) \leq -(\boldsymbol{\tau}^*, \mathbf{L}_1^{-1}\boldsymbol{\tau}^*) + (\boldsymbol{\tau}^*, \mathbf{U}\boldsymbol{\tau}^*) + 2(\boldsymbol{\tau}^*, \boldsymbol{\varepsilon}^0) + (\mathbf{e}^*, \mathbf{L}_1\mathbf{e}^*), \qquad (6.182)$$

where $\mathbf{e}^* \equiv (\mathbf{L}_1^{-1} - \mathbf{U})\boldsymbol{\tau}^* - \boldsymbol{\varepsilon}^0$ is a residual of the trial field. If $\mathbf{L}^c$ is chosen in this manner that $\mathbf{L}_1$ is negative definite at each point $\mathbf{x} \in w$ then dropping of the last item in the right-hand side of the inequality (6.182) yields

$$2(W - W^0) \leq \mathcal{H}(\boldsymbol{\tau}^*). \qquad (6.183)$$

Thus, the value $2(W-W^0)$ is an extremum of the functional $\mathcal{H}(\boldsymbol{\tau}^*)$. The inequality (6.182) demonstrates that the Hashin-Shtrikman principle is weaker than the corresponding classical variational principle due to the negative definiteness of the item $(\mathbf{e}^*, \mathbf{L}_1\mathbf{e}^*)$.

In a similar manner, the minimum principle of the complementary energy (2.108)

$$2(\boldsymbol{\sigma}^*, \boldsymbol{\varepsilon}^0) - (\boldsymbol{\sigma}^*, \mathbf{M}\boldsymbol{\sigma}^*) \leq 2W \qquad (6.184)$$

can be transformed to the form $-\mathcal{H}(\boldsymbol{\tau}^*) - (\mathbf{L}^c\mathbf{e}^*, \mathbf{M}_1^c\mathbf{L}^c\mathbf{e}^*) \leq (W - W^0)$ by the use of an equality $\mathbf{L}^c\mathbf{M}_1 + \mathbf{L}_1(\mathbf{M}^c + \mathbf{L}_1) = \mathbf{0}$. By this means if the tensor $\mathbf{M}_1$ is negative definite (or which is the same, $\mathbf{L}_1$ is positive definite) then the inequality (6.184) leads to $\mathcal{H}(\boldsymbol{\tau}^*) \leq 2(W - W^0)$.

We now turn our attention to the homogeneous second boundary-value (6.5) problem with prescribed tractions $\mathbf{t}^{\partial\mathcal{E}}(\mathbf{x}) \equiv \boldsymbol{\sigma}^0 \cdot \mathbf{n}^{\partial\mathcal{E}}(\mathbf{x})$ $(\mathbf{x} \in \partial\mathcal{E})$ compatible with a uniform stress $\boldsymbol{\sigma}^0 = \langle\boldsymbol{\sigma}\rangle$ which is consequently the overall average stress $\langle\boldsymbol{\sigma}\rangle$ (6.10). If the homogeneous boundary conditions produce a homogeneous stress field $\boldsymbol{\sigma}^0$ in the comparison medium, then the operator representations

$$\boldsymbol{\sigma} = \boldsymbol{\sigma}^0 + \boldsymbol{\Gamma}\boldsymbol{\eta}, \quad (\mathbf{M}_1^{-1} - \boldsymbol{\Gamma})\boldsymbol{\eta} = \boldsymbol{\sigma}^0 \tag{6.185}$$

will be an analog of Eq. (6.173_1) and (6.173_2) respectively. Here the action of the operator $\boldsymbol{\Gamma}$ at the strain polarization tensor $\boldsymbol{\eta}$ (6.172) is described by an equation $\boldsymbol{\Gamma\tau} = \int \boldsymbol{\Gamma}(\mathbf{x}-\mathbf{y})[\boldsymbol{\eta}(\mathbf{y}) - \langle\boldsymbol{\eta}\rangle]\, d\mathbf{y}$, with the kernel $\boldsymbol{\Gamma}(\mathbf{x}) = -\mathbf{L}^c\boldsymbol{\delta}(\mathbf{x}) - \mathbf{L}^c\mathbf{U}(\mathbf{x})\mathbf{L}^c$ (3.37) The following functional can be associated with Eq. (6.185_2):

$$\mathcal{F}(\boldsymbol{\eta}^*) = (\boldsymbol{\eta}^*, [\mathbf{M}_1^{-1} - \boldsymbol{\Gamma}]\boldsymbol{\eta}^*) - 2(\boldsymbol{\eta}^*, \boldsymbol{\sigma}^0). \tag{6.186}$$

The rules for the minimum principles of the elastic energy and complementary energy in the considered case of boundary conditions now change. So, the analog of the principle (6.179), from the complementary energy principle, is

$$2W \leq (\boldsymbol{\sigma}^*, \mathbf{M}^c\boldsymbol{\sigma}^*) + (\boldsymbol{\sigma}^*, \mathbf{M}_1\boldsymbol{\sigma}^*), \tag{6.187}$$

which can be presented in terms of $\boldsymbol{\eta}^*$:

$$2(W - W^0) \leq -(\boldsymbol{\eta}^*, \boldsymbol{\Gamma}\boldsymbol{\eta}^*) + (\boldsymbol{\sigma}^*, \mathbf{M}_1\boldsymbol{\sigma}^*) + 2(\boldsymbol{\sigma}^0, \mathbf{M}_1\boldsymbol{\Gamma}\boldsymbol{\eta}^*) + (\boldsymbol{\Gamma}\boldsymbol{\eta}^*, \mathbf{M}_1\boldsymbol{\Gamma}\boldsymbol{\eta}^*). \tag{6.188}$$

In the final analysis we will obtain similarly to the inequality (6.182) $2(W - W^0) \leq -\mathcal{F}(\boldsymbol{\eta}^*) + (\mathbf{L}^c\mathbf{e}^*, \mathbf{M}_1\mathbf{L}^c\mathbf{e}^*)$, and, therefore, $2(W - W^0) \leq -\mathcal{F}(\boldsymbol{\eta}^*)$ at any choice of $\mathbf{M}^c$ with the negative definite $\mathbf{M}_1(\mathbf{x})$ $(\forall \mathbf{x} \in w)$ (or, what is the same, $\mathbf{L}_1(\mathbf{x})$ is positive definite). The residual in the inequality (6.187) presented in terms of $\boldsymbol{\eta}^*$ is associated with the residual introduced in the inequality (6.182): $\mathbf{L}^c\mathbf{e}^* = (\mathbf{M}_1^{-1} - \boldsymbol{\Gamma})\boldsymbol{\eta}^* - \boldsymbol{\sigma}^0$. Furthermore, one can present the functionals $\mathcal{H}$ and $\mathcal{F}$ in the terms $\boldsymbol{\eta}^*$ and $\boldsymbol{\tau}^*$, respectively, where $\boldsymbol{\eta}^*$ and $\boldsymbol{\tau}^*$ are related by Eq. (6.173). Then it is found to be $\mathcal{H}(\boldsymbol{\tau}^*) = \mathcal{F}(\boldsymbol{\eta}^*)$.

Finally, the following inequalities obtained from the minimum potential energy principle be reciprocal to (6.182) and (6.183) for the second boundary value problem:

$$-\mathcal{F}(\boldsymbol{\eta}^*) - (\mathbf{e}^*, \mathbf{L}_1\mathbf{e}^*) \leq 2(W - W^0), \quad -\mathcal{F}(\boldsymbol{\eta}^*) \leq 2(W - W^0), \tag{6.189}$$

where $\mathbf{L}_1$ is an arbitrary tensor negatively defined at $\forall \mathbf{x} \in w$.

By this means, following [461], [1148], [1149], [1186] we made sure that the variational principle of Hashin-Shtrikman is inferred from the classical ones.

6.5 Bounds of Effective Elastic Moduli

6.5.1 Hill's Bounds

For definiteness sake, we will consider the first homogeneous boundary value problem (6.4). Then the upper bound of the effective modulus $\mathbf{L}^*$ can be estimated by the use of the Hill's condition (6.15) and the classical minimum potential energy principle (6.179):

$$2W \leq \langle \boldsymbol{\varepsilon}^* \mathbf{L} \boldsymbol{\varepsilon}^* \rangle \tag{6.190}$$

where the trial strain field $\boldsymbol{\varepsilon}^*$ is derived from an arbitrary displacement field $\mathbf{u}^*$ consistent with the boundary condition (6.4) . In the simplest case of an approximation of $\boldsymbol{\varepsilon}^*$ by a constant

$$\boldsymbol{\varepsilon}^* = \langle \boldsymbol{\varepsilon} \rangle, \tag{6.191}$$

one obtains an inequality

$$\mathbf{L}^* \leq \mathbf{L}^V, \quad \mathbf{L}^V \equiv \langle \mathbf{L} \rangle, \tag{6.192}$$

proposed by Voight [1142]. The inequality (6.192) is understanded in the sense of an inequality for the corresponding quadratic form when $a_{ij}(L^V_{ijkl} - L^*_{ijkl})a_{kl} \geq 0$ for any symmetric tensor of the second order a_{ij}. In so doing, it is possible that for some components of the tensors L_{ijkl} and L^V_{ijkl}, the inequality (6.192) is violated.

In a similar manner, for the boundary conditions (6.4), the minimum principle of the complementary energy can be written as

$$2(\mathbf{t}^*, \mathbf{u}) - (\boldsymbol{\sigma}^*, \mathbf{M}\boldsymbol{\sigma}^*) \leq W, \tag{6.193}$$

where $\mathbf{t}^* = \boldsymbol{\sigma}^* \mathbf{n}$, $\boldsymbol{\sigma}^*$ is a trial stress field obeying the equilibrium equation, $\mathbf{n}$ is a unit outward normal to the boundary ∂w, and $\mathbf{u} \equiv \langle \boldsymbol{\varepsilon} \rangle \mathbf{x}$. The simplest approximation of a trial stress field $\boldsymbol{\sigma}^*$ is a constant:

$$\boldsymbol{\sigma}^* = \text{const.} \tag{6.194}$$

Then Eq. (6.193) can be transformed according to Gauss's divergence theorem (2.23):

$$2\boldsymbol{\sigma}^* \langle \boldsymbol{\varepsilon} \rangle - \boldsymbol{\sigma}^* \mathbf{M}^R \boldsymbol{\sigma}^* \leq 2W, \tag{6.195}$$

where $\mathbf{M}^R \equiv \langle \mathbf{M} \rangle$ is a definition of a compliance tensor introduced by Reuss [934]. The left-hand side of the inequality (6.195) is reached the maximum at $\boldsymbol{\sigma}^* = \mathbf{L}^R \langle \boldsymbol{\varepsilon} \rangle$, where $\mathbf{L}^R \equiv (\mathbf{M}^R)^{-1}$. Then, in an accompany with the inequalities (6.192) and (6.195), one obtains an inequality

$$\mathbf{L}^R \leq \mathbf{L}^* \leq \mathbf{L}^V, \tag{6.196}$$

referred to as the Hill's [459] (see also [863]) bounds. The inequalities (6.196) can also be deduced in a similar manner if $\mathbf{L}$ is defined for the traction homogeneous boundary conditions (6.5). The upper bound (6.196) is identified as the bound by Voight, while the lower bound is called Reuss's bound. The arithmetic average of the bounds by Voight and Reuss represents the Hill's estimation $\mathbf{L}^H = (\mathbf{L}^R + \mathbf{L}^V)/2$.

For statistically isotropic composites, the bounds by Voight and Reuss will be the isotropic tensors which can be presented in Voight notations (2.197): $\mathbf{L}^* = (3k^*, 2\mu^*)$, $\mathbf{L}^V = (3k^V, 2\mu^V)$, $\mathbf{L}^R = (3k^R, 2\mu^R)$ and the inequality (6.196) acquires the form

$$k^R \le k^* \le k^V, \quad \mu^R \le \mu^* \le \mu^V. \tag{6.197}$$

In the case of a composite material with isotropic constituents, one obtains

$$k^R = \langle k^{-1}\rangle^{-1}, \quad k^V = \langle k\rangle, \quad \mu^R = \langle \mu^{-1}\rangle^{-1}, \quad \mu^V = \langle \mu\rangle. \tag{6.198}$$

For polycrystals, the Hill's bounds (6.196) are estimated as the arithmetic and harmonic means of the local stiffness tensors (5.108):

$$\mathbf{L}^V = \int \mathbf{L}(\mathbf{g})f(\mathbf{g})\,d\mathbf{g}, \quad \mathbf{M}^R = \int \mathbf{M}(\mathbf{g})f(\mathbf{g})\,d\mathbf{g}. \tag{6.199}$$

The arithmetic and harmonic means correspond to the assumption of homogeneous strain and stress field, respectively. Note that $\mathbf{L}^V$ and $\mathbf{L}^R$ differing in general give the upper $\mathbf{L}^V$ (Voigt) and lower $\mathbf{L}^R$ (Reuss) bounds for the effective elastic stiffness $\mathbf{L}^*$, the distance between which describes the degree of elastic anisotropy. For instance, in the case of an aggregate of cubic single crystals this distance can be expressed in terms of the Zener anisotropy ratio Z (2.176): $\| \mathbf{L}^V - \mathbf{L}^R \| = 6\sqrt{5}(Z-1)^2[5(2Z+3)]^{-1}(L_{1111} - L_{1122})$.

For uniformly distributed grain orientation the tensors $\mathbf{L}^V$ and $\mathbf{L}^R \equiv (\mathbf{M}^R)^{-1}$ are isotropic ones with the bulk k^V, k^R and shear μ^V, μ^R moduli defined by the relations $(J = V, R)$:

$$k^J = L^J_{iijj}/9, \quad \mu^J = (L^J_{ijij} - L^J_{iijj}/3)/10 \tag{6.200}$$

which provide the bounds for the effective elastic moduli $\mathbf{L}^* = (3k^*, 2\mu^*)$ (6.197). Due to an equality for the Young modulus $E^{-1} = (3k)^{-1} + (9\mu)^{-1}$, the set of inequalities analogous to (6.197) is valid also for the Young modulus:

$$E^R \le E^* \le E^V. \tag{6.201}$$

The Hill [459] approach combines the upper and lower bounds as both arithmetic and geometric means of the isotropic bounds by Voigt and Reuss, giving a good approximation for the actual effective elastic constant. $k^H = (k^V + k^R)/2$, $\mu^H = (\mu^V + \mu^H)/2$.

Of course, for the uniformly distributed grain orientation, the Voigt-Reuss estimations (6.199) can be obtained by the direct average of representations (2.169) over all possible orientations $\mathbf{g}$. However, the same result can be obtained by the contraction evaluations of (2.170) and (2.171) analogously to (6.200)

$$3k^V = \frac{1}{3}\sum_n(\lambda_n + 6\mu_n + 4\nu_n), \quad 2\mu^V = \frac{2}{15}\sum_n(\lambda_n + 10\nu_n), \tag{6.202}$$

$$(3k^R)^{-1} = \frac{1}{3}\sum_n(p_n + 6q_n + 4r_n), \quad (2\mu^R)^{-1} = \frac{2}{15}\sum_n(p_n + 10r_n). \tag{6.203}$$

Substituting the explicit representations λ_n, μ_n, ν_n and analogous representations for p_n, q_n, r_n for the different symmetric classes (2.170)–(2.177) leads to the representation of the bulk and shear components of the tensors $\mathbf{L}^V$ and $\mathbf{M}^R$:
for orthorhombic symmetry:

$$
\begin{aligned}
3k^V &= \big[L_{11} + L_{22} + L_{33} + 2(L_{12} + L_{23} + L_{13})\big]/3, \\
2\mu^V &= \frac{2}{15}\big[L_{11} + L_{22} + L_{33} - L_{12} - L_{23} - L_{13} + 3(L_{44} + L_{55} + L_{66})\big], \quad (6.204) \\
(3k^R)^{-1} &= \big[M_{11} + M_{23} + M_{33} + 2(M_{12} + M_{23} + M_{13})\big]/3, \\
(2\mu^R)^{-1} &= \frac{2}{15}\big[M_{11} + M_{22} + M_{33} - M_{12} - M_{23} - M_{13} \\
&\quad + \frac{3}{4}(M_{44} + M_{55} + M_{66})\big], \quad (6.205)
\end{aligned}
$$

for tetragonal symmetry:

$$
\begin{aligned}
3k^V &= \big[2L_{11} + L_{33} + 2(L_{12} + 2L_{13})\big]/3, \\
2\mu^V &= \frac{2}{15}\big[2L_{11} + L_{33} - L_{12} - 2L_{13} + 3(2L_{44} + L_{66})\big], \quad (6.206) \\
(3k^R)^{-1} &= \big[2M_{11} + M_{33} + 2(M_{12} + 2M_{13})\big]/3, \\
(2\mu^R)^{-1} &= \frac{2}{15}\big[2M_{11} + M_{33} - M_{12} - 2M_{13} + \frac{3}{4}(2M_{44} + M_{66})\big], \quad (6.207)
\end{aligned}
$$

for hexagonal symmetry:

$$
\begin{aligned}
3k^V &= \big[2L_{11} + L_{33} + 2(L_{12} + 2L_{13})\big]/3, \\
2\mu^V &= \frac{1}{15}\big[7L_{11} + 2L_{33} - 5L_{12} - 4L_{13} + 12L_{44}\big], \quad (6.208) \\
(3k^R)^{-1} &= \big[2M_{11} + M_{33} + 2(M_{12} + 2M_{13})\big]/3, \\
(2\mu^R)^{-1} &= \frac{1}{15}\big[7M_{11} + 2M_{33} - 5M_{12} - 4M_{13} + 3M_{44})\big], \quad (6.209)
\end{aligned}
$$

for cubic symmetry:

$$
3k^V = L_{11} + 2L_{12}, \quad 2\mu^V = \frac{2}{5}(L_{11} - L_{12} + 3L_{44}), \quad (6.210)
$$

$$
(3k^R)^{-1} = M_{11} + 2M_{12}, \quad (2\mu^R)^{-1} = \frac{2}{5}(M_{11} - M_{12} + \frac{3}{4}M_{44}). \quad (6.211)
$$

As expected, $k^V \neq k^R$ and $\mu^V \neq \mu^R$ in general, but for the polycrystals of cubic symmetry $k^V = k^R$ because the local contractions $L_{iijj} =$ const. and $M_{iijj} =$ const. do not vary in the aggregate due to the equality $L_{12} + 2L_{12} = (M_{11} + 2M_{12})^{-1}$. Moreover, the equality $\lambda = k - 2\mu/3$ for the isotropic medium accompanied by the inequality (6.197$_2$) and equality $k^V = k^R$ for the polycrystalline aggregates of cubic crystals leads to $\lambda^V \leq \lambda^* \equiv k^* - 2\mu^* \leq \lambda^R$ [710]. Thus, the upper bound for Lamé constant is the Reuss estimate of this constant, while the lower bound for this constant is given by its Voigt estimate (compare with Eqs. (6.197) and (6.201)).

It should be mentioned that in the contractions (6.200) the tensor containing one of four indexes odd times (such as, e.g., L_{2312}) is absent. Therefore, Eqs. (6.208) and (6.209) obtained for hexagonal crystals are also correct for an aggregate with trigonal crystals, and Eqs. (6.204) and (6.205) describe not only the orthorhombic crystals but also the crystals of monoclinic and triclinic symmetry.

Let us consider the polycrystals of cubic symmetry with the ODF $f(\phi_1, \theta, \phi_2) \equiv f(\theta)$ (5.121) which implies the transverse isotropy with Ox_3 being the symmetry axis. Then the tensor of the average elastic moduli has five independent constants which can be found by averaging of the representation (5.121). Keeping in mind that $f(\theta)$ does not depend on the angles ϕ_1 and ϕ_2, we can easily estimate the averages

$$\langle c_i^2\rangle = \langle s_i^2\rangle = \frac{1}{2}, \ \langle c_i^4\rangle = \langle s_i^4\rangle = \frac{3}{8}, \ \langle c_i^2 s_i^2\rangle = \frac{1}{8}, \ (i = 1, 2). \tag{6.212}$$

In the case of the series expansion of the ODF (5.121), the averages

$$\begin{aligned} \langle c_1^2\rangle &= \frac{1}{3}(1 + \frac{2}{5}a_2), \ \langle c_1^4\rangle = \frac{1}{5}(1 + \frac{4}{7}a_2 + \frac{8}{63}a_4), \\ \langle s_1^2\rangle &= \frac{2}{3}(1 - \frac{1}{5}a_2), \ \langle s_1^4\rangle = \frac{8}{15}(1 - \frac{2}{7}a_2 + \frac{1}{2}a_4) \end{aligned} \tag{6.213}$$

can be found by the use of an explicit representation of the Legendre polynomials in the operation averages (5.108), (6.199), with taking the orthogonality property

$$\int_{-1}^{1} P_l(x)P_{l'}(x)\, dx = \frac{2}{l+1}\delta_{ll'} \tag{6.214}$$

into account. Substitution of the averages (6.212) and (6.213) into the average Eq. (6.199) leads to

$$\begin{aligned} \langle L_{11}\rangle &= L_{11}^0 + \frac{1}{60}a_4 L, \ \langle L_{12}\rangle = L_{12}^0 + \frac{1}{180}a_4 L, \ \langle L_{13}\rangle = L_{12}^0 - \frac{1}{45}a_4 L, \\ \langle L_{33}\rangle &= L_{11}^0 + \frac{2}{45}a_4 L, \ \langle L_{44}\rangle = \frac{1}{2}(L_{11}^0 - L_{12}^0) - \frac{1}{45}a_4 L, \end{aligned} \tag{6.215}$$

where $L = L_{11} - L_{12} - 2L_{44}$, and $L_{11}^0 = 1/15(L_{iijj} + 2L_{ijij})$, $L_{12}^0 = 1/15(2L_{iijj} - L_{ijij})$ are estimated for the uniform random orientation of crystals (6.210) when $a_2 = a_4 = 0$.

The formal generalizations of Hill's [459] estimation $\mathbf{L}^H$ are noteworthy. So an approach [784] exhibiting a natural reciprocity between stiffness and compliance of effective properties takes into account the recursive arithmetic mean by Voigt and Reuss bounds:

$$\begin{aligned} &\mathbf{L}_0^V = \langle \mathbf{L}\rangle, \mathbf{M}_0^R = \langle \mathbf{M}\rangle, \ \mathbf{L}_0^R = \langle \mathbf{L}\rangle^{-1}, \ \mathbf{M}_0^V = \langle \mathbf{M}\rangle^{-1}, \\ &\mathbf{L}_{i+1}^V = (\mathbf{L}_i^V + \mathbf{R}_l^R)/2, \ \mathbf{M}_{i+1}^R = (\mathbf{M}_i^V + \mathbf{M}_i^R)/2, \\ &\mathbf{L}_{i+1}^R = (\mathbf{M}_{i+1}^R)^{-1}, \ \mathbf{M}_{i+1}^V = (\mathbf{L}_{i+1}^V)^{-1}. \end{aligned} \tag{6.216}$$

In the numerical examples being considered, the recursion (6.216) with the initial approximation [459] ($i = 0$) converges in four iterations. This recursive approach has no micromechanical justification other than that effective elastic stiffness and compliance should be reciprocal with one another. A similar level of justification was proposed in a homogenization technique in [11] (see also [746]) using the geometrical mean of local effective properties:

$$\mathbf{L}^A = \exp\Big(\int \ln(\mathbf{L}(\mathbf{g}))f(\mathbf{g})\, d\mathbf{g}\Big),\ \mathbf{M}^A = \exp\Big(\int \ln(\mathbf{M}(\mathbf{g}))f(\mathbf{g})\, d\mathbf{g}\Big) \equiv (\mathbf{L}^A)^{-1}. \tag{6.217}$$

Arithmetic (6.199_1), harmonic (6.199_2), and geometric (6.217) averages were compared by Böhlke and Bertran [92] showing that for cubic crystal aggregates all of these averages depend on the same irreducible fourth-order tensor, which represents the pure anisotropic portion of the effective elasticity tensor. It should be mentioned that the Voiht-Reuss estimations (6.197) for uniformly distributed grain orientation are based on exploiting two invariants (6.200). Another formal approach [10] is based on the use of a determinant of the elastic moduli matrix. For example, for the polycrystals of cubic symmetry with the Zener anisotropy ratio Z (2.176), the desired determinant D equals $D = 12KZ^{-1}L_{44}^5$, $K = (L_{11} + 2L_{12})/3$, while for the equivalent isotropic medium with effective elastic moduli $\mathbf{L}^* = (3k^*, 2\mu^*)$, we have $D = 12k^*(\mu^*)^5$. Taking into account an invariantness of K with respect to rotation of the cubic symmetry crystals, we obtain

$$k^* = K, \quad \mu^* = L_{44}Z^{-2/5} \tag{6.218}$$

with identical representation in the approach based on the use of compliance tensor $1/\mu^* = M_{44}Z^{2/5}$. Thus, even in the case of significant differences of μ^V and μ^R (6.197) obtained by the use of linear invariants of $\mathbf{L}$ and $\mathbf{M}$, respectively, the method (6.218) leads to a natural reciprocity of the shear elasticity and compliance.

6.5.2 Hashin-Shtrikman Bounds

Hill's bounds (6.196) are found to be accurate enough for polycrystals and agree fairly well with experiment [995]. Unfortunately, Hill's bounds (6.196) prove to be so much wider for composites with high elastic mismatch between the phases that they lose their practical meaning. However, it is important to point out that the bounds (6.196) are achieved for some geometrical structures of composites (such solids are simple laminate or fiber media). The question now arises of how to narrow this interval spanned by the plausible effective moduli which collapse to the exact value, in principle, if we know the complete details of the microtopology. The bounds (6.196) depend only on the moduli and phase concentrations and do not consist how the phases are distributed. The construction of better bounds is possible if more information is available. In most applications, it is unlikely that information for higher than two point probabilities densities is avaliable. Therefore, although formal extensions are possible, attention will be focused on the bounds that can be obtained using two-point correlation functions only. This enables one to use more successful approximations of trial fields than assumed in Eqs. (6.191) and (6.194).

Namely, we will assume that the trial strains $\boldsymbol{\varepsilon}^*$ and stresses $\boldsymbol{\sigma}^*$ (and, therefore, the polarization tensors $\boldsymbol{\tau}^*$ (6.171) and $\boldsymbol{\eta}^*$ (6.172)) are homogeneous within the phases only rather than in the whole medium. So, for an n-phase medium

$$\boldsymbol{\tau}^*(\mathbf{x}, \lambda) = \sum_{i=1}^{N} \boldsymbol{\tau}_i V_i(\mathbf{x}, \lambda), \quad \boldsymbol{\tau}_i(\mathbf{x}) = \text{const. at } \mathbf{x} \in v^{(i)} \tag{6.219}$$

where the $\boldsymbol{\tau}_i$ are the functions of only one argument $\mathbf{x}$, independent of the configuration of the field $V(\mathbf{x})$ specified by a random parameter λ: any relaxation of this restriction would introduce more than two-point probability densities into $\mathcal{H}(\boldsymbol{\tau})$ (6.177), so that (6.219) is a most general trial field reducing the required statistical information in the functional $\mathcal{H}(\boldsymbol{\tau})$ to the two-point probability densities.

Substitution of Eq. (6.219) into the expression for the functional $\mathcal{H}(\boldsymbol{\tau}^*)$ yields

$$\mathcal{H}(\boldsymbol{\tau}^*) = \frac{1}{\overline{w}}\sum_{r=1}^{N}\int V_r(\mathbf{x},\lambda)\boldsymbol{\tau}_r(\mathbf{x})[\mathbf{L}_1^{-1}(\mathbf{x})\boldsymbol{\tau}^*(\mathbf{x}) - 2\boldsymbol{\varepsilon}^0]\, d\mathbf{x} = \frac{1}{\overline{w}}\sum_{r=1}^{N}\iint V_r(\mathbf{x},\lambda)$$
$$\times\, \boldsymbol{\tau}_r(\mathbf{x})\mathbf{U}(\mathbf{x}-\mathbf{y})\Big[\sum_{s=0}^{N} V_s(\mathbf{y},\lambda)\boldsymbol{\tau}_s(\mathbf{y}) - \langle\boldsymbol{\tau}\rangle\Big]\, d\mathbf{y}\, d\mathbf{x}. \tag{6.220}$$

Since the polarization stress $\boldsymbol{\tau}^*$ is assumed to be independent of the field realization $V(\mathbf{x},\lambda)$, we can also consider $\mathcal{H}(\boldsymbol{\tau}^*)$ as independent on λ and, therefore, the product $V_r(\mathbf{x},\lambda)V_s(\mathbf{y},\lambda)$ can be replaced by a probability density $S_2^{(rs)}(\mathbf{x},\mathbf{y}) = S_2^{(s|r)}(\mathbf{y},\mathbf{x})S_1^{(r)}(\mathbf{x})$ (see Subsection 5.3.3):

$$\mathcal{H}(\boldsymbol{\tau}^*) = \frac{1}{\overline{w}}\sum_{r=1}^{N}\int S_1^{(r)}(\mathbf{x})\boldsymbol{\tau}_r(\mathbf{x})[\mathbf{L}_1^{-1}(\mathbf{x})\boldsymbol{\tau}_r(\mathbf{x}) - 2\boldsymbol{\varepsilon}^0]\, d\mathbf{x} - \frac{1}{\overline{w}}\sum_{r=1}^{N}\int\int S_1^{(r)}(\mathbf{x})$$
$$\times\, \boldsymbol{\tau}_r(\mathbf{x})\mathbf{U}(\mathbf{x}-\mathbf{y})\Big[\sum_{s=0}^{N} S_2^{(s|r)}(\mathbf{y},\mathbf{x})\boldsymbol{\tau}_s(\mathbf{y}) - \langle\boldsymbol{\tau}\rangle\Big]\, d\mathbf{y}\, d\mathbf{x}. \tag{6.221}$$

where the integrals absolutely converge due to the limit for the conditional probability density for finding the phase s at $\mathbf{y}$: $S_2^{(s|r)}(\mathbf{y},\mathbf{x}) \to S_1^{(s)}(\mathbf{y})$ at $|\mathbf{x}_r - \mathbf{x}_s| \to \infty$. The functional $\mathcal{H}(\boldsymbol{\tau}^*)$ (6.221) gives the extremum value $\mathcal{H}(\boldsymbol{\tau}^*) = -(\langle\boldsymbol{\tau}\rangle, \boldsymbol{\varepsilon}^0)$ when

$$\mathbf{L}_1^{-1}(\mathbf{x})\boldsymbol{\tau}_r(\mathbf{x}) - \sum_{s=0}^{N}\int \mathbf{U}(\mathbf{x}-\mathbf{y})[S_2^{(s|r)}(\mathbf{y},\mathbf{x}) - S_1^{(s)}(\mathbf{y})]\boldsymbol{\tau}_s(\mathbf{y})\, d\mathbf{y} = \boldsymbol{\varepsilon}^0(\mathbf{x}). \tag{6.222}$$

By performing routine matrix algebra, the linear system (6.222) can be solved

$$\boldsymbol{\tau}_s = \mathbf{L}_1^{(s)}\overline{\mathbf{D}}_s\boldsymbol{\varepsilon}^0, \quad \overline{\mathbf{D}}_s = [\mathbf{L}_1^{(s)}]^{-1}\sum_{r=0}^{N}\mathbf{D}_{sr}\mathbf{L}_1^{(r)}, \tag{6.223}$$

and

$$\langle\boldsymbol{\tau}\rangle = \mathbf{L}^c + \sum_{s=0}^{N}\mathbf{L}_1^{(s)}\overline{\mathbf{D}}_s\boldsymbol{\varepsilon}^0, \tag{6.224}$$

where the matrix $\mathbf{D}$ has an inverse matrix $\mathbf{D}^{-1}$ with the submatrix elements

$$\mathbf{D}_{ij}^{-1} = \delta_{ij} - \mathbf{L}_1^{(i)}\int \mathbf{U}(\mathbf{x}-\mathbf{y})[S_2^{(j|i)}(\mathbf{y},\mathbf{x}) - S_1^{(j)}(\mathbf{y})]\, d\mathbf{y}. \tag{6.225}$$

The solution of Eq. (6.222) can be presented in compact form if

$$\int \mathbf{U}(\mathbf{x}-\mathbf{y})[S_2^{(j|i)}(\mathbf{y},\mathbf{x}) - S_1^{(j)}(\mathbf{y})]\, d\mathbf{y} = \mathbf{P}_i(\delta_{ij} - c^{(j)}) = \text{const.}, \quad \forall i,j, \quad (6.226)$$

where $\mathbf{P}_i = \mathbf{P}(v_i^0)$ defined by Eq. (6.226) describes "ellipsoidal" symmetry of the microstructure, such as would be obtained if an isotropic distribution were subjected to an affine transform. Geometrical isotropy usually refers to stronger rotational invariants, namely, the invariance of all correlation functions when the material is rotated. The connection of "ellipsoidal" symmetry of the microstructure with both the shape of a correlation hole v_i^0 considered in Chapters 7 and 9 and the functions $\varphi(v_j|;v_i,\mathbf{x}_i)$ was proposed in [1180], [1183].

Thus, in the case of ellipsoidal microstructures (6.226), the system (6.222) can be simplified $[(\mathbf{L}_1^{(r)})^{-1} + \mathbf{P}_r]\boldsymbol{\tau}_r - \mathbf{P}_r\langle\boldsymbol{\tau}\rangle = \boldsymbol{\varepsilon}^0$ which allows for the average $\langle\boldsymbol{\tau}\rangle$ and the solution $\boldsymbol{\tau}_r$ of Eq. (6.222)

$$\langle\boldsymbol{\tau}\rangle = \Big[\sum_{r=0}^{N} c_r \mathbf{A}_r\Big]^{-1} \sum_{s=0}^{N} c_s \mathbf{A}_s \mathbf{L}_1^{(s)} \boldsymbol{\varepsilon}^0, \quad (6.227)$$

$$\boldsymbol{\tau}_k = \mathbf{A}_k \mathbf{L}_1^{(k)} \Big\{ \mathbf{I} + \mathbf{P}_k \Big[\sum_{p=0}^{N} c_p \mathbf{A}_p\Big]^{-1} \sum_{s=0}^{N} c_s \mathbf{A}_s \mathbf{L}_1^{(s)} \Big\} \boldsymbol{\varepsilon}^0, \quad (6.228)$$

where the tensor $\mathbf{A}_q$ $(q = r,s,p)$: $\mathbf{A}_q = (\mathbf{I} + \mathbf{L}_1^{(q)}\mathbf{P}_q)^{-1}$ introduced by Eq. (3.120) has a sense of a strain concentrator factor at one ellipsoidal inclusion in the infinite matrix with the modulus $\mathbf{L}^c$.

Recall that the bounds of effective moduli follow from the inequality (6.183), (6.178)

$$2(W^0 - W) \le -\langle\boldsymbol{\tau}\rangle\boldsymbol{\varepsilon}^0, \quad (6.229)$$

where the tensor $\langle\boldsymbol{\tau}\rangle$ is defined by either Eq. (6.224) or (6.228). The result (6.229) can be expressed in the form $\boldsymbol{\varepsilon}^0(\mathbf{L}^* - \overline{\mathbf{L}})\boldsymbol{\varepsilon}^0 \ge 0$, whenever $\mathbf{L}_1^{(r)}(\mathbf{x})$ is positive (negative) semi-definite for all r at any $\mathbf{x} \in w$, where

$$\overline{\mathbf{L}} = \Big[\sum_{r=0}^{N} c_r \mathbf{A}_r\Big]^{-1} \sum_{s=0}^{N} c_s \mathbf{A}_s \mathbf{L}_1^{(s)} \quad (6.230)$$

and

$$\overline{\mathbf{L}} = \mathbf{L}^c + \sum_{s=0}^{N} c^{(s)} \mathbf{L}_1^{(s)} \overline{\mathbf{D}}_s \quad (6.231)$$

for the representations of $\langle\boldsymbol{\tau}\rangle$ (6.227) and (6.224), respectively. In other words, $\overline{\mathbf{L}}$ gives the low (upper) bound of the effective modulus $\mathbf{L}^*$ if $\mathbf{L}_1(\mathbf{x})$ is positive (negative) definite at any $\mathbf{x} \in w$.

The bounds $\overline{\mathbf{M}}$ for the effective compliances $\mathbf{M}^*$ can be obtained by inversion of corresponding $\overline{\mathbf{L}}$. These can also be obtained directly from consideration of the traction boundary value problem (6.5) in a similar manner as the bounds $\overline{\mathbf{L}}$ were found for the displacement boundary conditions (6.4). Extremizing $\mathcal{F}(\boldsymbol{\eta}^*)$ (6.186) is performed analogously to extremizing the functional $\mathcal{H}(\boldsymbol{\tau}^*)$ with replacement of $\mathbf{L}_1$ by $\mathbf{M}_1$ and $\mathbf{U}$ by $\boldsymbol{\Gamma}$ taking into account that Eqs. (6.173$_2$) and (6.185$_2$)

must be consistent with $\boldsymbol{\tau} = -\mathbf{L}^c\boldsymbol{\eta}$, $\boldsymbol{\varepsilon}^0 = \langle\boldsymbol{\varepsilon}\rangle$, and $\boldsymbol{\sigma}^0 = \langle\boldsymbol{\sigma}\rangle = \mathbf{L}^c\boldsymbol{\varepsilon}^0 + \langle\boldsymbol{\tau}\rangle$. In such a case the bounds $\overline{\mathbf{M}}$ for a composite with "ellipsoidal" symmetry (6.226) corresponding to the bounds $\overline{\mathbf{L}}$ (6.230) are given by the relations

$$\overline{\mathbf{M}} = \Big[\sum_{r=0}^{N} c_r\mathbf{B}_r\Big]^{-1}\sum_{s=0}^{N} c_s\mathbf{B}_s\mathbf{M}^{(s)}, \tag{6.232}$$

which are inverse to $\overline{\mathbf{L}}$ (6.230). Here the tensor $\mathbf{B}_q$ $(q = r, s)$: $\mathbf{B}_q = (\mathbf{I} + \mathbf{M}_1^{(q)}\mathbf{Q}_q)^{-1}$, $\mathbf{Q}_q = -\mathbf{L}^c(\mathbf{I} - \mathbf{P}_q\mathbf{L}^c)$ introduced by Eq. (3.121) has a sense of a stress concentrator factor at one ellipsoidal inclusion in the infinite matrix with the compliance $\mathbf{M}^c$ (3.119_2).

The estimation (6.230) was originally proposed in [438], [439], [1148], [1149] (see also [958]) by another but equivalent method for the particular case of statistically isotropic composites when the tensors $\mathbf{P}_i$ (6.226) can be assumed to be estimated for the spherical domains v_i^0. In such a case the integral of $\mathbf{U}$ over any sphere v_i^0 is $\mathbf{P}_i \equiv \mathbf{P}^0 =$ const., independently of its radius. A more general case of "ellipsoidal" symmetry was analyzed in [1179], [1180], [1183], [1186] where Willis obtained the formula (6.230) and (6.232) for the particular case $\mathbf{P}_i = \mathbf{P}$ $(\forall i,\ i = 0, \ldots, N)$ with $\mathbf{P}$ taking identical "ellipsoidal" form for all phases which might be appropriate to discuss a composite containing aligned ellipsoidal inclusions, as limiting case, thin needles, and flat cracks. The equations (6.230) and (6.231) enable one to evaluate the bounds of effective moduli of anisotropic media; the case of the transversally isotropic medium was considered in [237], [431]. Simmons and Wang [1012] have summarized in sufficient detail the results of property estimations of cubic symmetry polycrystals. The variation method was also exploited for the analysis of polycrystals with hexagonal, trigonal, tetragonal, and monoclinic symmetry [77], [1162], [1163], [1164], [1165]. Diener *et al.* [270] have obtained a generalization of Hashin-Shtrikman bounds to the nonlocal media with the effective properties described by the nonlocal effective operator rather than by the tensor $\mathbf{L}$. The bounds of effective properties were estimated by the use of the variational principle generalizing the Hashin-Shtrikman principle to the dynamic case in [1066], [1067].

Let us consider two simplest cases of the choice of $\mathbf{L}^c$. At first, $\mathbf{L}_1(\mathbf{x})$ is positive definite for $\forall\mathbf{x} \in w$ if $\mathbf{L}^c \equiv \mathbf{0}$, then passing to the limit $\mathbf{L}^c \to \mathbf{0}$ and $\mathbf{P} \to \infty$ in (6.230) leads to the lower bound of the effective moduli $\overline{\mathbf{L}}^- = \langle\mathbf{M}\rangle^{-1} \equiv \mathbf{L}^R$ which is found to coincide with the lower Reuss bound (6.196). In the opposite case $\mathbf{L}^c \to \infty$, when $\mathbf{P} \to \infty$ and $\mathbf{L}_1(\mathbf{x})$ is negative definite at $\forall\mathbf{x} \in w$. Then the upper bound (6.230) $\overline{\mathbf{L}}^+ = \langle\mathbf{L}\rangle \equiv \mathbf{L}^V$ coincided with the upper Voight bound (6.196) Thus, the Hashin-Shtrikman bounds (6.230) in the considered limiting cases $\mathbf{L}^c \to \mathbf{0}$ and $\mathbf{L}^c \to \infty$ are reduced to the Hill's bounds (6.196), and, moreover, these limiting cases degenerate the Hashin-Shtrikman variational principle into the minimum principles of the potential and complementary energies.

Following [1148], we will consider a statistically isotropic composite with the isotropic tensors $\mathbf{P}_i = \mathbf{P}^0$ (3.167) and with isotropic comparison medium $\mathbf{L}^c = (3k^c, 2\mu^c)$ and phases $\mathbf{L}^{(r)} = (3k^{(r)}, 2\mu^{(r)})$ $(r = 0, \ldots, N)$. Then Eq. (6.230) can be rearranged

$$\overline{\mathbf{L}} = \Big[\sum_{r=0}^{N} c_r(P_0^{-1} - \mathbf{L}^c + \mathbf{L}^{(r)})^{-1}\Big]^{-1} \sum_{s=0}^{N} c_s(P_0^{-1} - \mathbf{L}^c + \mathbf{L}^{(s)})^{-1}\mathbf{L}^{(s)} \tag{6.233}$$

and presented in Voight symbolic notations:

$$\overline{k} = \Big[\sum_{r=0}^{N} c_r(k^{(r)} + \tilde{k})^{-1}\Big]^{-1} - \tilde{k}, \quad \overline{\mu} = \Big[\sum_{r=0}^{N} c_r(\mu^{(r)} + \tilde{\mu})^{-1}\Big]^{-1} - \tilde{\mu}, \tag{6.234}$$

where the components of a symbolic representation of the tensor

$$\mathbf{P}_0^{-1} - \mathbf{L}^c = (3\tilde{k}, 2\tilde{\mu}) = \Big(4\mu^c, \frac{1}{3}\mu^c \frac{9k^c + 8\mu^c}{k^c + 2\mu^c}\Big) \tag{6.235}$$

were introduced by Hill [463]. It can be seen from Eqs. (6.234) that the found values $\overline{k}$ and $\overline{\mu}$ are the monotonically increasing functions of $\tilde{k}$ and $\tilde{\mu}$, which in turn, are monotonically increasing functions of k^c and μ^c. Then the lower and upper bounds of the moduli of the comparison medium defining the lower and upper bounds of effective moduli can be taken as $k^{c-} = \min_r(k^{(r)})$, $\mu^{c-} = \min_r(\mu^{(r)})$, and $k^{c+} = \max_r(k^{(r)})$, $\mu^{c+} = \max_r(\mu^{(r)})$, which define the bounds of effective moduli

$$\Big[\sum_{r=0}^{N} c_r(k^{(r)} + \tilde{k}_-)^{-1}\Big]^{-1} - \tilde{k}_- \le k^* \le \Big[\sum_{r=0}^{N} c_r(k^{(r)} + \tilde{k}_+)^{-1}\Big]^{-1} - \tilde{k}_+, \tag{6.236}$$

$$\Big[\sum_{r=0}^{N} c_r(\mu^{(r)} + \tilde{\mu}_-)^{-1}\Big]^{-1} - \tilde{\mu}_- \le \mu^* \le \Big[\sum_{r=0}^{N} c_r(\mu^{(r)} + \tilde{\mu}_+)^{-1}\Big]^{-1} - \tilde{\mu}_+, \tag{6.237}$$

where $\tilde{k}_- = 4\mu^{c-}/3$, $\tilde{k}_+ = 4\mu^{c+}/3$. The values of $\tilde{\mu}_-$ and $\tilde{\mu}_+$ are chosen depending on the relation between the phase elastic moduli. If $(k^{c+} - k^{c-})(\mu^{c+} - \mu^{c-}) > 0$ then

$$\tilde{\mu}_- = \frac{\mu^{c-}}{6}\frac{9k^{c-} + 8\mu^{c-}}{k^{c-} + 2\mu^{c-}}, \quad \tilde{\mu}_+ = \frac{\mu^{c+}}{6}\frac{9k^{c+} + 8\mu^{c+}}{k^{c+} + 2\mu^{c+}}. \tag{6.238}$$

In the opposite case $(k^{c+} - k^{c-})(\mu^{c+} - \mu^{c-}) < 0$, the values

$$\tilde{\mu}_- = \frac{\mu^{c-}}{6}\frac{9k^{c+} + 8\mu^{c-}}{k^{c+} + 2\mu^{c-}}, \quad \tilde{\mu}_+ = \frac{\mu^{c+}}{6}\frac{9k^{c-} + 8\mu^{c+}}{k^{c-} + 2\mu^{c+}}. \tag{6.239}$$

are chosen for the bound estimations (6.237).

It is interesting to mention that for composites (called Hill's [460] medium) with homogeneous shear modulus $\mu(\mathbf{x}) \equiv \text{const.}$ when $\mu^{c-} = \mu^{c+} = \mu$, the lower and upper bounds (6.236) coincide and, therefore the effective bulk and shear moduli are exactly defined:

$$k^* = \Big[\sum_{r=0}^{N} c_r\big(k^{(r)} + \frac{4}{3}\mu\big)^{-1}\Big]^{-1} - \frac{4}{3}\mu, \quad \mu^* = \mu. \tag{6.240}$$

Equations (6.230) and (6.231) obtained for the composites can be easily generalized to the case of one-phase polycrystals. In such a polycrystal, the identical grains are randomly located with prescribed orientation of local crystallographic axes $\mathbf{e}'_i$ with respect to the global coordinate system $\mathbf{e}_i$ that is described by the rotation matrix $\mathbf{g}$ (see A1). Then the volume concentration of crystals with orientation enclosed by the interval $(\mathbf{g}, \mathbf{g} + d\mathbf{g})$ equals $f(\mathbf{g})d\mathbf{g}$, and the summation sign in Eq. (6.230) should be replaced by averaging over the possible orientations

$$\overline{\mathbf{L}} = \langle \mathbf{A}(\mathbf{g}) \rangle^{-1} \langle \mathbf{A}(\mathbf{g}) \mathbf{L}(\mathbf{g}) \rangle, \tag{6.241}$$

where $\mathbf{A}(\mathbf{g}) = [\mathbf{I} + \mathbf{L}_1(\mathbf{g})\mathbf{P}(\mathbf{g})]^{-1}$ is a strain concentrator factor for one crystal with orientation $\mathbf{g}$ in the infinite matrix with the modulus $\mathbf{L}^c$. In a general case of medium anisotropy, a choice of the comparison medium $\mathbf{L}^{c-} \leq \mathbf{L}^c \leq L^{c+}$ is not trivial. For crystals with a cubic symmetry, the elastic moduli are completely specified by three moduli L_{11}, L_{12} and L_{44} (2.175) in six-by-six matrix notation. The bulk modulus and two shear moduli of such a crystal are defined as

$$k = (L_{11} + 2L_{12})/3, \quad \mu_1 = (L_{11} - L_{12})/2, \; \mu_2 = L_{44} \tag{6.242}$$

where $\mu_1 < \mu_2$. Then for statistically isotropic polycrystals [1012], one can set $k^{c-} = k^{c+} = k$, $\mu^{c-} = \mu_1$, $\mu^{c+} = \mu_2$, and (6.241) yields $k^* = k$ and $(\mu_2^1 = \mu_2 - \mu_1)$:

$$\bar{\mu}^- = \mu_1 + 3\Big[\frac{12(k + 2\mu_1)}{5\mu_1(3k + 4\mu_1)} + \frac{5}{\mu_2^1}\Big]^{-1}, \quad \bar{\mu}^+ = \mu_2 + 2\Big[\frac{18(k + 2\mu_2)}{5\mu_2(3k + 4\mu_2)} - \frac{5}{\mu_2^1}\Big]. \tag{6.243}$$

6.5.3 Bounds of Higher Order

The obtained bound $\overline{\mathbf{L}}$ (6.230) contains one- and two-point phase locations probability. Moreover, for isotropic $P_{j|i}(\mathbf{x})$ the tensor $\overline{\mathbf{L}}$ is expressed only through the volume concentration of phases and does not depend on the two-point probability $P_{j|i}(\mathbf{x})$. Refinement of the bounds (6.230) can be achieved if the microtopology taken into account at the bound estimation contains three-point probabilities. The last condition is accomplished in the case of exploiting classical minimum principles of the potential and complementary energies instead of the weaker Hashin-Shtrikman principle.

Recall that to obtain Voight's bound (6.192), we have substituted the trial field $\boldsymbol{\varepsilon}^* = \langle \boldsymbol{\varepsilon} \rangle$ into the classical minimum principle of potential energy (6.190). Following [266], we will present an improvement of the values $\boldsymbol{\varepsilon}^*$ by the perturbation method. For this purpose, we insert an expression for the local random strain field $\boldsymbol{\varepsilon}(\mathbf{x}) = \mathbf{A}^*(\mathbf{x})\langle \boldsymbol{\varepsilon} \rangle$ (6.179) into the general operator equation (6.173_1): $\mathbf{A}^* = \mathbf{I} + \mathbf{U}(\mathbf{L}_1\mathbf{A}^* - \langle \mathbf{L}_1\mathbf{A}^* \rangle)$, where $\langle \mathbf{A}^* \rangle = \mathbf{I}$, and $\mathbf{L}^* = \langle \mathbf{L}\mathbf{A}^* \rangle$. Expanding the operator $\mathbf{A}^*$ into a series over the powers of $\mathbf{L}_1^V \equiv \mathbf{L} - \langle \mathbf{L} \rangle$ leads to linear item over $\mathbf{L}_1^V$ $\mathbf{A}^* = \mathbf{I} + \mathbf{U}\mathbf{L}_1^V$, which can be used for the representation of the trial field in the form $\boldsymbol{\varepsilon}^* = (\mathbf{I} + \mathbf{U}\mathbf{L}_1^V)\langle \boldsymbol{\varepsilon} \rangle$. Substitution of the proposed $\boldsymbol{\varepsilon}^*$ into the right-hand side of Eq. (6.179) yields

$$\langle \boldsymbol{\varepsilon} \rangle \mathbf{L}^* \langle \boldsymbol{\varepsilon} \rangle \leq \langle \boldsymbol{\varepsilon} \rangle [\langle \mathbf{L} \rangle + 2\langle \mathbf{L}_1^V \mathbf{U} \mathbf{L}_1^V \rangle + \langle \mathbf{L}_1^V \mathbf{U}\mathbf{L}\mathbf{U}\mathbf{L}_1^V \rangle] \langle \boldsymbol{\varepsilon} \rangle. \tag{6.244}$$

where the self-adjoint property of the operator $\mathbf{U}$ was used. From a definition of the Green function (3.3) it follows that $\mathbf{U}\langle\mathbf{L}\rangle\mathbf{U} = -\mathbf{U}$ and, therefore, the inequality (6.244 can be recast as

$$\mathbf{L}^* \leq \langle\mathbf{L}\rangle + \langle\mathbf{L}_1^V\mathbf{U}\mathbf{L}_1^V\rangle + \langle\mathbf{L}_1^V\mathbf{U}\mathbf{L}_1^V\mathbf{U}\mathbf{L}_1^V\rangle. \tag{6.245}$$

An analogous lower bound of $\mathbf{L}^*$ (or the upper bound for $\mathbf{M}^*$) can be deduced by the perturbation method from the minimum principle of the complimentary energy by the use of Eq. (6.185$_1$). The estimation (6.245) has the second order of smallness in the inclusion concentration $c^{(1)}$ and the third order of smallness according to the relative variation of the elastic modulus perturbation: $O\big(c^2[\mathbf{L}_1^V(\mathbf{L}^V)^{-1} - \mathbf{I}]^3\big)$. Higher-order bounds have also been introduced in [611], [750], [752], [776], [1186].

Third-order bounds on k^* and μ^* for statistically isotropic composites were obtained in [71] and simplified in [772]:

$$\Big(\langle\frac{1}{k}\rangle - \frac{4c^{(0)}c^{(1)}(k^{(1)}-k^{(0)})^2}{4(k^{(1)}k^{(0)})^2(\langle\widetilde{1/k}\rangle + 3\langle 1/k\rangle_\zeta)}\Big)^{-1}$$
$$\leq k^* \leq \Big(\langle k\rangle - \frac{3c^{(0)}c^{(1)}(k^{(1)}-k^{(0)})^2}{3\langle\widetilde{k}\rangle + 4\langle k\rangle_\zeta}\Big), \tag{6.246}$$

$$\Big(\langle\frac{1}{\mu}\rangle - \frac{c^{(0)}c^{(1)}(\mu^{(1)}-\mu^{(0)})^2}{(\mu^{(1)}\mu^{(0)})^2(\langle\widetilde{1/\mu}\rangle + 6\Xi)}\Big)^{-1}$$
$$\leq \mu^* \leq \Big(\langle\mu\rangle - \frac{6c^{(0)}c^{(1)}(\mu^{(1)}-\mu^{(0)})^2}{6\langle\widetilde{\mu}\rangle + \Theta}\Big). \tag{6.247}$$

Hereafter one introduces some additional averaging operations for a material parameter Y (e.g. $Y = k, \mu$):

$$\langle\widetilde{Y}\rangle = Y^{(0)} + (1-c^{(1)})(Y^{(1)} - Y^{(0)}), \tag{6.248}$$
$$\langle Y\rangle_\zeta = Y^{(0)} + \zeta(Y^{(1)} - Y^{(0)}), \quad \langle Y\rangle_\eta = Y^{(0)} + \eta(Y^{(1)} - Y^{(0)}), \tag{6.249}$$

and the coefficients Ξ and Θ are defined by the relations

$$\Xi = \Big[10\langle k\rangle^2\langle\frac{1}{k}\rangle_\zeta + 5\langle\mu\rangle\langle 3\mu+2k\rangle\langle\frac{1}{\mu}\rangle_\zeta + \langle 3k+\mu\rangle^2\langle\frac{1}{\mu}\rangle_\eta\Big]/\langle 9k+8\mu\rangle^2, \tag{6.250}$$
$$\Theta = \Big[10\langle\mu\rangle^2\langle k\rangle_\zeta + 5\langle\mu\rangle\langle 3\mu+2k\rangle\langle\mu\rangle_\zeta + \langle 3k+\mu\rangle^2\langle\mu\rangle_\eta\Big]/\langle k+2\mu\rangle^2 \tag{6.251}$$

depending in turn on the microstructural parameters of the medium:

$$\zeta = \frac{9}{2c^{(0)}c^{(1)}}\int_0^\infty\frac{dr}{r}\int_0^\infty\frac{ds}{s}\int_{-1}^1 P_2(x)\Big[S_3^{(1)}(r,s,t) - \frac{S_2^{(1)}(r)S_2^{(1)}(s)}{c^{(1)}}\Big], \tag{6.252}$$
$$\eta = \frac{5}{21}\zeta + \frac{150}{7c^{(0)}c^{(1)}}\int_0^\infty\frac{dr}{r}\int_0^\infty\frac{ds}{s}\int_{-1}^1 P_4(x)\Big[S_3^{(1)}(r,s,t) - \frac{S_2^{(1)}(r)S_2^{(1)}(s)}{c^{(1)}}\Big], \tag{6.253}$$

where $P_2(x)$ and $P_4(x)$ are the Legendre polynomials of degree 2 and 4, respectively. $S_2^{(1)}(r)$ and $S_3^{(1)}(r,s,t)$ (see Subsection 5.3.3) are, respectively, the probability of finding, in phase 1, the end points of a line segment of length r and the vertices of a triangle with sides of length r, s, and t; x is the cosine of the angle opposite the side of length t and so $t^2 = r^2 + s^2 - 2rsx$. Since the square-bracketed expression in (6.252) and (6.253) tends to zero at $r \to 0$, $s \to 0$, $r \to \infty$ or $s \to \infty$, the numerical estimation of these integrals does not present any problems. The central problem with the integral evaluations (6.252), (6.253) is finding the three-point correlation function $S_3^{(1)}(r, s, t)$ which can be measured from cross-sectional photographs of the composite materials. Measurements of the required correlation functions with an uncertain accuracy are tedious and are therefore best done with the aid of numerical simulations (see [246] and Chapter 5), although again, high-order morphological information required is beyond practical reach.

It is interesting to mention that for composites with incompressible phases ($k(\mathbf{x}) \equiv \infty$), the bounds $\mu^{*-} \le \mu^* \le \mu^{*+}$ (6.247) yield

$$\left(\langle\frac{1}{\mu}\rangle - \frac{3c^{(0)}c^{(1)}(\mu^{(1)} - \mu^{(0)})^2}{(\mu^{(1)}\mu^{(0)})^2(\widetilde{\langle 1/\mu\rangle} + 2\langle 1/\mu\rangle_\eta)}\right)^{-1} \le \mu^* \le \left(\langle\mu\rangle - \frac{3c^{(0)}c^{(1)}(\mu^{(1)} - \mu^{(0)})^2}{2\langle\widetilde{\mu}\rangle + 3\langle 1/\mu\rangle_\eta}\right). \tag{6.254}$$

Considering that $\mu^{*-} \le \mu^{*+}$, $\zeta, \eta \in [0, 1]$ and a straightforward check makes sure that in the limiting cases $\eta = 0$ and $\eta = 1$ the bounds (6.254) coincide with the Hashin-Shtrikman bounds. The bounds (6.254) are found to be considerably narrower than the Hashin-Shtrikman bounds at any η.

Due to the completeness of the direct evaluation of the microstructural parameters (6.252) and (6.253), their estimations are carried out via different model representations of the real composite structure. So, Miller [768] constructed symmetric-cell materials by dividing space into nonintersecting cells (possibly of varying shapes and sizes) and then randomly and independently assigning each cell as phase 0 or 1 with probabilities $c^{(0)}$ and $c^{(1)}$, respectively. In the case of a matrix material, the shapes of cells corresponding to the inclusions are taken as the inclusion shapes while a cell shape corresponding to the matrix of the real material is assumed to be arbitrary. For these materials, Milton [772] suggested estimations

$$\zeta = c^{(0)} + \frac{1}{2}(c^{(1)} - c^{(0)})(9a - 1), \quad \eta = c^{(0)} + (c^{(1)} - c^{(0)})[4(5b - 1) - 5(9a - 1)]/6, \tag{6.255}$$

where the parameters $1/9 \le a \le 1/3$ and b depend on the cell shapes and, in particular, for the spherical cells $a = 1/9$, $b = 1/5$; for platelike cells $a = 1/3$, $b = 1$; for needlelike cells $a = 1/6$, $b = 3/8$.

For more realistic matrix composites with the spherical inclusions, the microstructural parameters ζ and η were analytically derived to $O([c^{(1)}]^i)$ in polynomial expansions:

$$\zeta = \sum_{i=0}^{\infty} \zeta_i (c^{(1)})^i, \quad \eta = \sum_{i=0}^{\infty} \eta_i (c^{(1)})^i \tag{6.256}$$

for both impenetrable spheres and spheres with an arbitrary degree of penetrability as well as for equalized spherical particles and polydispersions containing different and widely separated particle sizes (see [1095] for additional references). For the spherical inclusions $\eta_0 = \zeta_0 = 0$, and η_1 and ζ_1 depend only on one- and two-point statistical moments of composite structure. To avoid much algebraic manipulation, we reproduce the representation [1095] for the linear item

$$\zeta_1 = \frac{1}{(c^{(1)})^2}\int_0^\infty\int_0^\infty c_{ab}\Delta_\zeta\Big(\frac{b}{a}\Big)\,da\,db, \quad \eta_1 = \frac{3}{(c^{(1)})^2}\int_0^\infty\int_0^\infty c_{ab}\Delta_\eta\Big(\frac{b}{a}\Big)\,da\,db,$$

where the functions ($c_{ab} = c(a)c(b)$, $\bar\gamma = 1 + 2\gamma$)

$$\begin{aligned}\Delta_\zeta(\gamma) &= \frac{3}{16\gamma^3}\Big[\frac{(1+2\gamma)^4 - 1}{4(1+2\gamma)^2} - \ln(1+2\gamma)\Big],\\ \Delta_\eta(\gamma) &= \frac{1}{16\bar\gamma^3}\Big[\frac{\bar\gamma^2 - 1}{4\bar\gamma^6}(9\bar\gamma^6 + 29\bar\gamma^4 - 10\bar\gamma^2 + 2) - 15\ln\bar\gamma\Big]\end{aligned} \tag{6.257}$$

take into account a size distribution of inclusions described by probability density $c(a)$ of concentration dependence of the inclusion radius a which obeys the normalized condition $c^{(1)} = \int_0^\infty c(a)\,da$. In particular, for a monodispersion

$$\zeta_1 = 3a^3\int_{2a}^\infty \frac{r^2\,dr}{(r^2 - a^2)^3}, \tag{6.258}$$

and $\zeta = 0.21068c^{(1)} + O([c^{(1)}]^2)$, $\eta = 0.48274c^{(1)} + O([c^{(1)}]^2)$. Substitution of these analytical expressions into variational third-order bounds on the effective elastic moduli leads then to rigorous bounds for such model microstructures which are exact to $O([c^{(1)}]^2)$:

$$\begin{aligned}\frac{k^{*+}}{k^{(0)}} &= 1 + L_{k1}c^{(1)} + L_{k2}(c^{(1)})^2\Big[1 + \frac{2}{3}\frac{1-2\nu^{(0)}}{1-\nu^{(0)}}(a_k - 1)\zeta_1\Big] + O((c^{(1)})^3),\\ \frac{k^{*-}}{k^{(0)}} &= 1 + L_{k1}c^{(1)} + L_{k2}(c^{(1)})^2\Big[1 + \frac{2}{3}\frac{1-2\nu^{(0)}}{1-\nu^{(0)}}(1 - a_k^{-1})\zeta_1\Big] + O((c^{(1)})^3),\\ \frac{\mu^{*+}}{\mu^{(0)}} &= 1 + L_{\mu1}c^{(1)} + L_{\mu2}(c^{(1)})^2\Big[1 + \frac{9(a_\mu - 1)\eta_1}{20(1-\nu^{(0)})(4-5\nu^{(0)})} + (1 - 2\nu^{(0)})\\ &\quad\cdot\frac{(13 - 14\nu^{(0)})(a_\mu - 1) + 4(1+\nu^{(0)})(a_k - 1)}{12(1-\nu^{(0)})(4-5\nu^{(0)})}\zeta_1\Big] + O((c^{(1)})^3),\\ \frac{\mu^{*-}}{\mu^{(0)}} &= 1 + L_{\mu1}c^{(1)} + L_{\mu2}(c^{(1)})^2\Big[1 + \frac{9(1 - a_\mu^{-1})\eta_1}{20(1-\nu^{(0)})(4-5\nu^{(0)})} + (1 - 2\nu^{(0)})\\ &\quad\cdot\frac{(13 - 14\nu^{(0)})(1 - a_\mu^{-1}) + 4(1+\nu^{(0)})(1 - a_k^{-1})}{12(1-\nu^{(0)})(4-5\nu^{(0)})}\zeta_1\Big] + O((c^{(1)})^3),\end{aligned} \tag{6.259}$$

where $a_k = k^{(1)}/k^{(0)}$, $a_\mu = \mu^{(1)}\mu^{(0)}$, and

$$\begin{aligned}L_{k1} &= \frac{3(1-\nu^{(0)})(a_k - 1)}{(1+\nu^{(0)})a_k + 2(1-2\nu^{(0)})}, \quad L_{k2} = \frac{3(1-\nu^{(0)})(1+\nu^{(0)})(a_k - 1)^2}{(1+\nu^{(0)})a_k + 2(1-2\nu^{(0)})},\\ L_{\mu1} &= \frac{15(1-\nu^{(0)})(a_\mu - 1)}{2(4-5\nu^{(0)})a_\mu + 7 - 5\nu^{(0)}}, \quad L_{\mu2} = \frac{30(1-\nu^{(0)})(4-5\nu^{(0)})(a_\mu - 1)^2}{[2(4-5\nu^{(0)})a_\mu + 7 - 5\nu^{(0)}]^2},\end{aligned}$$

Thovert *et al.* [1095] have plotted both the third-order and Hashin-Shtrikman bounds on μ^* for the monodisperse and polydisperse composites with compressible phases (the glass spheres in the epoxy matrix). The third-order bounds were found to be considerably narrower than the Hashin-Shtrikman bounds, and the effect of the polydispersivity on the lower bound was not large for the analyzed range of volume fraction. The size distribution of the inclusions in the experimental data [1020] presented in [1095] was not fully characterized, and only the range (1–30 μm) was reported.

Many of the results for ζ and η estimations are summarized in the review article [1103] (see also [503]). Helsing [451] shows how ζ and η can be evaluated from cross-sectional photographs for an arbitrary microgeometry in an efficient manner using interface integral techniques involving polyharmonic Green's functions and the fast multiple method. As a result he was able to evaluate structural parameters ζ and η with an error less than 0.005% as contrasted to the error of 2% in the standard method based on the measurement of the function $S_3^{(1)}$ from cross-sectional photographs of the composite material.

It should be mentioned that the structure parameters ζ and η do not depend on the mechanical properties of the phases, and, therefore, can be found from the model problems. For example, the parameter ζ is involved not only in the bounds of the bulk modulus, but also in the analogous bounds of a transport coefficient. From another perspective, the parameter η defines the effective viscosity of suspensions with incompressible phases.

6.6 Bounds of Effective Conductivity

Substitution of the simplest approximations of the temperature gradient $\nabla T^*(\mathbf{x}) \equiv \mathbf{G}^{\partial w} = \text{const.}$ and flux $\mathbf{q}^*(\mathbf{x}) \equiv \mathbf{q}^{\partial w} = \text{const.}$ at the boundary conditions analogous to (6.4) and (6.5) , respectively, into the classical minimum principles of the potential energy and complementary energy (see Eqs. (2.106) and (2.108)), respectively, lead to as analog of Hill's bounds for the effective conductivity

$$\langle \boldsymbol{\kappa} \rangle \equiv \boldsymbol{\kappa}^V \geq \boldsymbol{\kappa}^* \geq \boldsymbol{\kappa}^R \equiv \langle \boldsymbol{\kappa}^{-1} \rangle^{-1}, \tag{6.260}$$

respectively, proved by Wiener [1175] in the scalar dielectric context obtained by the use of some algebraic arguments; here $\boldsymbol{\kappa}^V$ and $\boldsymbol{\kappa}^R$ are to be attributed to Voigt and Reuss. It should be mentioned that the Voight and Reuss bounds are accessible for some specific composite structures. Indeed, a temperature field with a constant gradient does appear in the fiber or layered material with the prescribed macrogradient $\mathbf{G}^{\partial w}$ along the fibers or layers that provide accessibility of the Voight's bound. If, however, the gradient $\mathbf{G}^{\partial w}$ is perpendicular to the plies in the layered composite then a constant flux field does appear in the same direction that provides the Reuss's bound. Thus, for laminated structures with layers perpendicular to $(1, 0, 0)^\top$ and the local conductivity $\kappa_{ij} \equiv 0$ for all $i \neq j$ we have $\boldsymbol{\kappa}^* = \text{diag}(\langle 1/\kappa_{11} \rangle^{-1}, \langle \kappa_{22} \rangle, \langle \kappa_{33} \rangle)$.

More narrow bounds, analogous to the Hashin-Shtrikman bounds, can be obtained if the conductivity $\boldsymbol{\kappa}^c$ of a comparison medium is chosen either as $\kappa_{\min}$

and $\kappa_{\max}$ from the values $\boldsymbol{\kappa}^{(i)} = \kappa^{(i)}\boldsymbol{\delta}$ $(i = 1, 2)$ for the isotropic composites with two isotropic components or as the maximum κ_1 and minimum κ_3 from the values of the conductivity tensor components $\boldsymbol{\kappa}$ after bringing to the main axes form for the statistically isotropic polycrystals. Then for two-phase composites and one-phase polycrystals, we will get

$$\langle\kappa\rangle - \frac{c^{(0)}c^{(1)}(\kappa^{(1)} - \kappa^{(0)})^2}{\langle\tilde{\kappa}\rangle + 2\kappa_{\min}} \le \kappa^* \le \langle\kappa\rangle - \frac{c^{(0)}c^{(1)}(\kappa^{(1)} - \kappa^{(0)})^2}{\langle\tilde{\kappa}\rangle + 2\kappa_{\max}}, \tag{6.261}$$

$$\kappa_1 \frac{4\kappa_1^2 + 8\kappa_1\kappa_2 + 8\kappa_1\kappa_3 + 7\kappa_2\kappa_3}{16\kappa_1^2 + 5\kappa_1\kappa_2 + 5\kappa_1\kappa_3 + \kappa_2\kappa_3} \le \kappa^* \le \kappa_3 \frac{4\kappa_3^2 + 8\kappa_3\kappa_2 + 8\kappa_1\kappa_3 + 7\kappa_2\kappa_1}{16\kappa_3^2 + 5\kappa_3\kappa_2 + 5\kappa_1\kappa_3 + \kappa_2\kappa_1}, \tag{6.262}$$

respectively. Hashin-Shtrikman bounds, which are essentially narrower Wiener bounds (6.260), cannot be improved if the medium is statistically isotropic and only volume concentrations of constituents are prescribed. For a further narrowing of the bounds it is necessary to use information about the three-point correlation function that was obtained in [67], [772]:

$$\left\{\langle\kappa^{-1}\rangle - \frac{2c^{(0)}c^{(1)}(\kappa^{(1)} - \kappa^{(0)})^2}{(\kappa^{(1)}\kappa^{(0)})^2(\langle\widetilde{1/\kappa}\rangle + \langle 1/\kappa\rangle_\zeta)}\right\}^{-1} \le \kappa^* \le \langle\kappa\rangle - \frac{c^{(0)}c^{(1)}(\kappa^{(1)} - \kappa^{(0)})^2}{\langle\tilde{\kappa}\rangle + 2\langle\kappa\rangle_\zeta}, \tag{6.263}$$

where the notations of Eqs. (6.248) and (6.249$_1$) for the operations $\langle(\tilde{\cdot})\rangle$ and $\langle(\cdot)\rangle_\zeta$ are used and the structural parameter ζ is introduced by Eq. (6.252).

In particular, Thovert *et al.* [1095] obtained the bounds of $\boldsymbol{\kappa}$ exact to $O[(c^{(1)})^2(\kappa^{(1)}/\kappa^{(0)} - 1)^3]$ for composites with spherical inclusions:

$$\begin{aligned} &1 + 3\beta c^{(1)} + 3\beta^2(c^{(1)})^2\left[1 + \frac{2}{3}(1 - a_\kappa^{-1})\zeta_1\right] + O((c^{(1)})^3) \le \kappa^*/\kappa^{(0)} \\ &\le 1 + 3\beta c^{(1)} + 3\beta^2(c^{(1)})^2\left[1 + \frac{2}{3}(a_\kappa - 1)\zeta_1\right] + O((c^{(1)})^3), \end{aligned} \tag{6.264}$$

where $a_\kappa = \kappa^{(1)}/\kappa^{(0)} - 1$, $\beta = (a_\kappa - 1)/(a_\kappa + 2)$ and ζ_1 is a linear coefficient in the decomposition of ζ (6.256).

For the Miller [768] cell material, the Beran-Milton bounds (6.264) can be presented in the form

$$\begin{aligned} &a_\kappa^{1/2}\left[a_\kappa - c^{(1)}(a_\kappa - 1) - \frac{4}{3}\frac{(1 - a_\kappa)^2 c^{(0)}c^{(1)}}{1 + a_\kappa + 3(a_\kappa - 1)[(c^{(1)})^2 a - (c^{(0)})^2 b]}\right]^{-1} \\ &\le \frac{\kappa^*}{(\kappa^{(0)}\kappa^{(1)})^{1/2}} \le a_\kappa^{-1/2}\left[a_\kappa + c^{(1)}(a_\kappa - 1)\right. \\ &\left. - \frac{1}{3}\frac{(1 - a_\kappa)^2 c^{(0)}c^{(1)}}{1 + (a_\kappa - 1)c^{(1)} + 3(a_\kappa - 1)[(c^{(0)})^2 b - (c^{(1)})^2 a]}\right]^{-1}, \end{aligned} \tag{6.265}$$

where a and b are the coefficients characterizing the Miller's cell which were defined by Eq. (6.254).

Beran [67] has also obtained a 2-D analog of the representation (6.265) for fiber composites. For effective conductivity in the transversal direction κ_t^*, the following representations of the bounds hold:

$$a_\kappa^{1/2}\left[1-c^{(1)}(a_\kappa-1)-\frac{1}{2}\frac{(1-a_\kappa)^2c^{(0)}c^{(1)}}{a_\kappa+(1-a_\kappa)c^{(1)}+2(a_\kappa-1)[(c^{(1)})^2b_1-(c^{(0)})^2b_2]}\right]^{-1}$$

$$\le\frac{\kappa_t^*}{(\kappa^{(0)}\kappa^{(1)})^{1/2}}\le a_\kappa^{-1/2}\left[1+c^{(1)}(a_\kappa-1)\right.$$

$$\left.-\frac{1}{2}\frac{(1-a_\kappa)^2c^{(0)}c^{(1)}}{1+(a_\kappa-1)\{c^{(1)}+2(a_\kappa-1)[(c^{(0)})^2b_2-(c^{(1)})^2b_1]\}}\right]^{-1}, \qquad (6.266)$$

where $1/4\le b_i\le 1/2$. Here $b=1/4$ and $b=1/2$ correspond to the circular fibers and parallel plates, respectively. Along the fiber direction the effective constant κ_l^* is equal $\langle\kappa\rangle$.

Dihne [304] has obtained an interesting result for 2-D symmetric composite material with $c^{(0)}=c^{(1)}=0.5$ (see also [557]). The structure of this material will not change if each phase is replaced by another one and it can be either random or deterministic (e.g., a chess-board). It turns out then that the effective conductivity can be estimated exactly $\kappa^*=\sqrt{\kappa^{(0)}\kappa^{(1)}}$. The last result was generalized by Berdichevsky [72], who proved that estimation of the exact physicomechanical parameters is possible if the secondary variational problem is brought to the same form as an initial one. In parallel with conductivity, a plane elasticity problem for composites with the incompressible phases ($\nu^{(0)}=\nu^{(1)}=0.5$) also shows this property and $\mu^*=\sqrt{\mu^{(0)}\mu^{(1)}}$.

6.7 Bounds of Effective Eigenstrain

The problem of the bound estimations of the effective eigenstrains $\boldsymbol\beta^*$ was considered in a fairly small number of papers, possibly because $\boldsymbol\beta^*$ is uniquely determined through the effective moduli for both two phase composites and one phase polycrystal with transversally isotropic grains (see Section 6.3). Because of this we will consider a few results for the multiphase composites following [970], [995].

Consider a constitutive equation for the microinhomogeneous medium (6.2). At first we will obtain the general representations for the bounds of the effective eigenstrain $\boldsymbol\beta^*$ and stored energy W^* (6.232) in the case of the first boundary value problem with prescribed displacement boundary conditions (6.4) generating a homogeneous average field $\langle\boldsymbol\varepsilon\rangle=\boldsymbol\varepsilon^0$. Then for an arbitrary trial field of strains $\boldsymbol\varepsilon^*$ compatible with the boundary conditions ($\langle\boldsymbol\varepsilon^*\rangle=\langle\boldsymbol\varepsilon\rangle=\boldsymbol\varepsilon^0$), the variational principle (2.134_1) can be recast in terms $\boldsymbol\beta^*,\mathbf{L}^*,U^*$:

$$\langle(\boldsymbol\varepsilon^*-\boldsymbol\beta)\mathbf{L}(\boldsymbol\varepsilon^*-\boldsymbol\beta)\rangle\ge(\boldsymbol\varepsilon^0-\boldsymbol\beta^*)\mathbf{L}^*(\boldsymbol\varepsilon^0-\boldsymbol\beta^*)+2W^*. \qquad (6.267)$$

where Eq. (6.86) is taken into account. Analogously to obtaining the Hill bounds (6.196), the trial field $\boldsymbol\varepsilon^*$ is chosen as a simplest approximation $\boldsymbol\varepsilon^*=\boldsymbol\varepsilon^0$ (6.191) that transforms the inequality (6.267) to the form

$$\boldsymbol\varepsilon^0(\langle\mathbf{L}\rangle-\mathbf{L}^*)\boldsymbol\varepsilon^0+2\boldsymbol\varepsilon^0(\mathbf{L}^*\boldsymbol\beta^*-\langle\mathbf{L}\boldsymbol\beta\rangle)-\boldsymbol\beta^*\mathbf{L}^*\boldsymbol\beta^*+\langle\boldsymbol\beta\mathbf{L}\boldsymbol\beta\rangle-2W^*\ge 0. \quad (6.268)$$

This inequality holds at any arbitrary $\boldsymbol\varepsilon^0$. The left-hand side of the inequality (6.268) achieves its minimum

$$2W^* + (\boldsymbol{\beta}^* - \boldsymbol{\beta}_s)[(\mathbf{L}^*)^{-1} - \langle \mathbf{L} \rangle^{-1}]^{-1}(\boldsymbol{\beta}^* - \boldsymbol{\beta}_s) \leq \langle \boldsymbol{\beta}\mathbf{L}\boldsymbol{\beta} \rangle - \boldsymbol{\beta}_s \langle \mathbf{L} \rangle \boldsymbol{\beta}_s \quad (6.269)$$

at the values $\boldsymbol{\varepsilon}^0 = (\langle \mathbf{L} \rangle - \mathbf{L}^*)^{-1}(\mathbf{L}^*\boldsymbol{\beta}^* - \langle \mathbf{L}\boldsymbol{\beta} \rangle)$, where $\boldsymbol{\beta}_s \equiv \langle \mathbf{L} \rangle^{-1} \langle \mathbf{L}\boldsymbol{\beta} \rangle$. The inequality (6.269) can be considered as an upper bound for W^* if $\mathbf{L}^*$ and $\boldsymbol{\beta}^*$ are known. One considers the limiting cases of the inequalities (6.268) and (6.269). Due to both the Hill's bounds $\mathbf{L}^* \leq \langle \mathbf{L} \rangle$ (6.196) and the arbitrariness of $\boldsymbol{\varepsilon}^0$ in (6.268), the second item in (6.268) must vanish if $\mathbf{L}^* = \langle \mathbf{L} \rangle$, and, therefore

$$\boldsymbol{\beta}^* = \boldsymbol{\beta}_s, \text{ at } \mathbf{L}^* = \langle \mathbf{L} \rangle. \quad (6.270)$$

Another degenerate case can be obtained from the inequality (6.269):

$$W^* \leq \frac{1}{2}\Big[\langle \boldsymbol{\beta}\mathbf{L}\boldsymbol{\beta} \rangle - \boldsymbol{\beta}_s \langle \mathbf{L} \rangle \boldsymbol{\beta}_s\Big] \equiv W^+. \quad (6.271)$$

Then, at the attainment of the maximum value $W^* = W^+$, we get from (6.269) and (6.271), similar to (6.270), $\boldsymbol{\beta}^* = \boldsymbol{\beta}_s$ at $W^* = W^+$.

We now turn to the second boundary value problem (6.5) where the traction boundary conditions generate homogeneous average stresses $\langle \boldsymbol{\sigma} \rangle = \boldsymbol{\sigma}^0$. Then for an arbitrary trial stress field $\boldsymbol{\sigma}^*$ obeying the equality equation and compatible with the boundary conditions ($\langle \boldsymbol{\sigma}^* \rangle = \boldsymbol{\sigma}^0$), the minimum principle (2.134_2) can be presented in the form

$$\langle \boldsymbol{\sigma}^* \mathbf{M} \boldsymbol{\sigma}^* \rangle + 2\langle \boldsymbol{\sigma}^* \boldsymbol{\beta} \rangle \geq \langle \boldsymbol{\sigma} \rangle \mathbf{M}^* \langle \boldsymbol{\sigma} \rangle + 2\langle \boldsymbol{\sigma} \rangle \boldsymbol{\beta}^* - 2W^*, \quad (6.272)$$

where Eq. (6.83) is taken into account. We will consider the simplest approximation (6.194) as a trial field $\boldsymbol{\sigma}^* = \langle \boldsymbol{\sigma} \rangle$. Then (6.272) can be simplified as

$$\boldsymbol{\sigma}^0(\langle \mathbf{M} \rangle - \mathbf{M}^*)\boldsymbol{\sigma}^0 - 2\boldsymbol{\sigma}^0(\boldsymbol{\beta}^* - \langle \boldsymbol{\beta} \rangle) + 2W^* \geq 0 \quad (6.273)$$

The minimum value of the left-hand side of (6.273)

$$2W^* - (\boldsymbol{\beta}^* - \boldsymbol{\beta}_e)(\langle \mathbf{M} \rangle - \mathbf{M}^*)^{-1}(\boldsymbol{\beta}^* - \boldsymbol{\beta}_e) \geq 0 \quad (6.274)$$

is achieved at $\boldsymbol{\sigma}^0 = (\langle \mathbf{M} \rangle - \mathbf{M}^*)^{-1}(\boldsymbol{\beta}^* - \langle \boldsymbol{\beta} \rangle)$, where $\boldsymbol{\beta}_e \equiv \langle \boldsymbol{\beta} \rangle$. The inequality (6.274) defines the lower bound for W^* if $\mathbf{M}^*$ and $\boldsymbol{\beta}^*$ are given. If $\mathbf{M}^*$ and $\boldsymbol{\beta}^*$ are not known, the inequality (6.274) yields its limiting cases at either the maximum effective compliance $\mathbf{M}^* = \langle \mathbf{M} \rangle$ or $W^* = 0$ $\boldsymbol{\beta}^* = \langle \boldsymbol{\beta} \rangle$.

Let us consider some particular cases. Specifically, for the homogeneous elastic moduli $\mathbf{L}^* = \mathbf{L}(\mathbf{x}) \equiv$ const. when the stress concentrator factor $\mathbf{B}^*(\mathbf{x}) \equiv \mathbf{I}$ (6.33), we get the exact result [see (6.48)] $\boldsymbol{\beta} = \langle \boldsymbol{\beta} \rangle$, and the inequality

$$0 \leq W^* \leq \Big[\langle \boldsymbol{\beta}\mathbf{L}\boldsymbol{\beta} \rangle - \langle \boldsymbol{\beta} \rangle \mathbf{L} \langle \boldsymbol{\beta} \rangle\Big], \quad (6.275)$$

following from (6.272). However, for some special composites (see Section 6.3) the exact results instead of bounds can also be obtained for W^*.

We will also consider the statistically isotropic composite containing isotropic phases

$$\mathbf{L}^* = (3k^*, 2\mu^*),\ \overline{\mathbf{L}}^{\pm} = (3k^{\pm}, 2\mu^{\pm}),\ \mathbf{L}(\mathbf{x}) = (3k(\mathbf{x}), 2\mu(\mathbf{x})),$$
$$\boldsymbol{\beta}^* = \beta^*\boldsymbol{\delta},\ \boldsymbol{\beta} = \beta\boldsymbol{\delta},\ \boldsymbol{\beta}_s = \beta_s\boldsymbol{\delta},\ \boldsymbol{\beta}_e = \beta_e\boldsymbol{\delta}, \quad (6.276)$$

where $\overline{\mathbf{L}}^{\pm}$ are the Hill's bounds (6.196). The components of isotropic tensors (6.276) can be found by the use of the appropriate tensor convolutions

$$\beta_e = \langle\beta\rangle, \quad \beta_s = \frac{1}{k^+}\langle k\beta\rangle, \quad \beta_w \equiv \sqrt{\frac{1}{k^+}\langle\beta_{ik}L_{iklm}\beta_{lm}\rangle} = \sqrt{\frac{1}{k^+}\langle k\beta^2\rangle}, \qquad (6.277)$$

where a new characteristic strain β_w is introduced. Then the inequalities (6.272), (6.270), and (6.275) leads to

$$0 \leq W^* \leq W^+ = \frac{9}{2}k^+(\beta_w^2 - \beta_s^2), \qquad (6.278)$$

$$(\beta^* - \beta_s)^2 \leq \frac{2}{9}\Big(\frac{1}{k^*} - \frac{1}{k^+}\Big)(W^+ - W^*), \qquad (6.279)$$

$$(\beta^* - \beta_s)^2 \leq \frac{2}{9}\Big(\frac{1}{k^-} - \frac{1}{k^*}\Big)W^*, \qquad (6.280)$$

respectively. The last two inequalities determine a closed domain in the parameter space $(k^*, \boldsymbol{\beta}^*, W^*)$ which does not depend on the effective shear modulus μ^*.

7

General Integral Equations of Micromechanics of Composite Materials

The need to consider the actual microstructure of composite materials subjected to essentially inhomogeneous mechanical, body force, and temperature loading in micromechanics problems is well known. Unfortunately, the starting assumptions made in the majority of studies, namely, that the structure of composite media as well as the random fields of stresses are statistically homogeneous and therefore are invariant with respect to the translation, are incorrect. For example, due to some production technologies, the inclusion concentration may be a function of the coordinates [244], [883], [884]. The accumulation of damage also occurs locally in stress–concentration regions, for example at the tip of a macroscopic crack [529]. Furthermore, in layered composite shells, the location of the fibers is random within the periodic layers, and the micromechanics equations are equations with almost periodic coefficients. Finally, *Functionally Graded Materials* (FGMs) also have a statistically inhomogeneous structure which consists of two or more phases which is fabricated with a spatial variation of its composition that may improve the structural response (see Chapter 12).

In many problems the material may possess a periodic microstructure formed by the spatial repetition of small microstructures, or unit cells. Periodicity of the solution allows one to explore a type of modeling completely different from the average schemes of random media such as asymptotic expansion as well as the Fourier series expansion to field in a solid with periodic microstructure [806]. Such a perfectly regular distribution, of course, does not exist in actual cases, although periodic modeling can be quite useful, since it provides rigorous estimations with *a priori* prescribed accuracy for various material properties. Another problem is that the development of the theory of homogenization of initially periodic structures with the introduced infinite number of imperfections of unit cells randomly distributed in space has an essentially practical meaning. For example, the accumulation of damage, such as the microcracks in the matrix and debonding of inclusions from the matrix, is essentially random in nature.

The final goal of micromechanical research of composites involved in a prediction of both the overall effective properties and statistical moments of stress–strain fields is based on the approximate solution of exact initial integral equations connecting the random stress fields at the point being considered and the surrounding points. This infinite system of coupled integral equations is well known for statistically homogeneous composite materials subjected to homogeneous boundary conditions [561], [668], [995], [1184]. The goal of this chapter is to obtain a generalization of these equations for the case of either the statistically inhomogeneous or doubly periodical structures of composite materials subjected to essentially inhomogeneous loading by fields of the stresses, temperature, and body forces. The case of triply periodic structures with random imperfections is also considered.

7.1 General Integral Equations for Matrix Composites of Any Structure

This chapter discusses a certain representative mesodomain w with a characteristic function W and a boundary Γ containing a set $X = (v_i)$ of inclusions v_i with characteristic functions $V_i \quad (i = 1, 2, \ldots)$. At first no restrictions are imposed on the elastic symmetry of the phases or on, for example, geometry of the inclusions. It is assumed that the inclusions can be grouped into components (phases) $v^{(k)} \quad (k = 1, 2, \ldots, N)$ with identical mechanical and geometrical properties (such as the shape, size, orientation, and microstructure of inclusions). We consider the local basic equations of linear thermoelastostatic of composites (6.1)–(6.3). We introduce a "comparison" body, whose mechanical properties $\mathbf{g}^c$ $(\mathbf{g} = \mathbf{L}, \mathbf{M}, \boldsymbol{\alpha}, \boldsymbol{\beta}, \mathbf{f})$ denoted by the upper index c and $\mathbf{L}^c, \mathbf{M}^c$ will usually be taken as uniform over w, so that the corresponding boundary value problem is easier to solve than that for the original body. All tensors $\mathbf{g}$ $(\mathbf{g} = \mathbf{L}, \mathbf{M}, \boldsymbol{\alpha}, \boldsymbol{\beta})$ of material properties are decomposed as $\mathbf{g} \equiv \mathbf{g}^c + \mathbf{g}_1(\mathbf{x}) = \mathbf{g}^c + \mathbf{g}_1^{(m)}(\mathbf{x})$ (3.67). Here and in the following the upper index $^{(m)}$ indicates the components and the lower index i indicates the individual inclusions; $v^{(0)} = w \backslash v$, $v \equiv \cup v^{(k)} \equiv \cup v_i$, $V(\mathbf{x}) = \sum V^{(k)} = \sum V_i(\mathbf{x})$, and $V^{(k)}(\mathbf{x})$ is a characteristic function of $v^{(k)}$ that equals 1 at $\mathbf{x} \in v^{(k)}$ and 0 otherwise, $(m = 0, k;\ k = 1, 2, \ldots, N; \quad i = 1, 2, \ldots)$.

We assume that the phases are perfectly bounded (3.83), and the boundary conditions at the interface boundary are be considered together with the mixed boundary conditions (4.1) and (4.2) on $\Gamma = \Gamma_u \cup \Gamma_t$ with the unit outward normal $\mathbf{n}^\Gamma$

$$\mathbf{u}(\mathbf{x}) = \mathbf{u}^\Gamma(\mathbf{x}), \quad \mathbf{x} \in \Gamma_u, \quad \text{and} \quad \boldsymbol{\sigma}(\mathbf{x})\mathbf{n}^\Gamma(\mathbf{x}) = \mathbf{t}^\Gamma(\mathbf{x}). \quad \mathbf{x} \in \Gamma_t, \tag{7.1}$$

Of special practical interest are the homogeneous boundary conditions (6.4) and (6.5):

$$\mathbf{u}^\Gamma(\mathbf{x}) = \boldsymbol{\varepsilon}^\Gamma \mathbf{x}, \quad \boldsymbol{\varepsilon}^\Gamma \equiv \text{const.}, \ \mathbf{x} \in \Gamma, \tag{7.2}$$

$$\mathbf{t}^\Gamma(\mathbf{x}) = \boldsymbol{\sigma}^\Gamma \mathbf{n}^\Gamma(\mathbf{x}), \quad \boldsymbol{\sigma}^\Gamma \equiv \text{const.}, \ \mathbf{x} \in \Gamma. \tag{7.3}$$

Hereafter in analysis of integral equations of microinhomogeneous media, the notations of the domain $\mathcal{E}$ and its boundary $\partial\mathcal{E}$ are usually replaced by w and Γ, respectively.

From Eq. (6.1) through (6.3) a general integral equation for $\boldsymbol{\sigma}$ and $\boldsymbol{\varepsilon}$ can be derived. Substituting (6.2) and (6.3) into the equilibrium equation (6.1), we obtain a differential equation with respect to the strain $\boldsymbol{\varepsilon}$:

$$\nabla\{\mathbf{L}(\mathbf{x})[\boldsymbol{\varepsilon}(\mathbf{x})-\boldsymbol{\beta}(\mathbf{x})]\}+\mathbf{f}(\mathbf{x})=\mathbf{0}. \tag{7.4}$$

In the framework of the traditional scheme, we introduce a homogeneous "comparison" body with homogeneous moduli $\mathbf{L}^c$, and with the inhomogeneous deterministic transformation field $\boldsymbol{\beta}^c(\mathbf{x})$ and body force $\mathbf{f}^c(\mathbf{x})$ (and with solution $\boldsymbol{\sigma}^0$, $\boldsymbol{\varepsilon}^0$, $\mathbf{u}^0$ to the same boundary-value problem), and so Eq. (7.4) transforms to the equation:

$$\nabla\mathbf{L}^c\big[\boldsymbol{\varepsilon}(\mathbf{x})-\boldsymbol{\beta}^c(\mathbf{x})\big]+\mathbf{f}^c(\mathbf{x})=-\nabla\{\mathbf{L}_1(\mathbf{x})[\boldsymbol{\varepsilon}(\mathbf{x})-\boldsymbol{\beta}(\mathbf{x})]-\mathbf{L}^c\boldsymbol{\beta}_1(\mathbf{x})\}-\mathbf{f}_1(\mathbf{x}), \tag{7.5}$$

with a fictitious random "body-force" in the right-hand side of the equation, hereafter for all material tensors $\mathbf{g}$ ($\mathbf{L}$, $\mathbf{M}$, $\boldsymbol{\beta}$, $\boldsymbol{\alpha}$, $\mathbf{f}$) the notation $\mathbf{g}_1(\mathbf{x})\equiv\mathbf{g}(\mathbf{x})-\mathbf{g}^c$ is used. Then Eq. (7.5) may be reduced to a symmetrized integral form:

$$\boldsymbol{\varepsilon}(\mathbf{x})=\boldsymbol{\varepsilon}^0(\mathbf{x})+\nabla\int_w \mathbf{G}(\mathbf{x}-\mathbf{y})\nabla\Big\{\mathbf{L}_1(\mathbf{y})[\boldsymbol{\varepsilon}(\mathbf{y})-\boldsymbol{\beta}(\mathbf{x})]-\mathbf{L}^c\boldsymbol{\beta}_1(\mathbf{x})\Big\}\,d\mathbf{y}$$
$$+\nabla\int_w \mathbf{G}(\mathbf{x}-\mathbf{y})\mathbf{f}_1(\mathbf{y})\,d\mathbf{y}, \tag{7.6}$$

where $\mathbf{G}$ (3.3) is the infinite-homogeneous-body Green's function of the Navier equation (3.1) with an elastic modulus tensor $\mathbf{L}^c$. The deterministic function $\boldsymbol{\varepsilon}^0(\mathbf{x})$ is the strain field that would exist in the medium with homogeneous properties $\mathbf{L}^c$ and appropriate boundary conditions:

$$\varepsilon^0_{pq}(\mathbf{x})=\int_\Gamma\Big[G_{i(p,q)}(\mathbf{x}-\mathbf{s})t_i^\Gamma(\mathbf{s})-u_i^\Gamma(\mathbf{s})L^c_{ijkl}G_{k(p,q)l}(\mathbf{x}-\mathbf{s})n_j(\mathbf{s})\Big]\,d\mathbf{s}$$
$$+\int_w G_{i(p,q)}(\mathbf{x}-\mathbf{y})f_i^c(\mathbf{y})\,d\mathbf{y}, \tag{7.7}$$

which conforms with the stress field $\boldsymbol{\sigma}^0(\mathbf{x})=\mathbf{L}^c\boldsymbol{\varepsilon}^0(\mathbf{x})-\boldsymbol{\beta}^c(\mathbf{x})$. In particular, for the condition (7.1_1) with $\Gamma_u=\Gamma$ and $\boldsymbol{\beta}$, $\mathbf{f}\equiv\mathbf{0}$ the right-hand side integral over the external surface (7.7) can be considered as a continuation of $\boldsymbol{\varepsilon}^\Gamma(\mathbf{x})$, $\mathbf{x}\in\Gamma_u$ into w as the strain field that the boundary condition (7.1_1) would generate in the comparison medium with moduli $\mathbf{L}^c$. It should be mentioned that for the constant gravitation loads $f_i^{\mathrm{g}}=\rho^c g_i$ and for a centrifugal load $f_i^c=g_{ij}x_j$, the volume integral in Eq. (7.7) can be transformed into a surface integral [109]. Here we introduced a constant mass density ρ^c, a constant gravitation field g_i, as well as the matrix $g_{ij}=$ const.

Equation (7.7), which is valid for internal points of the body $\mathbf{x}\in w$, should be modified for the treatment of strong singularities in the surface integrals as $\mathbf{x}$ tends to the boundary Γ. It requires either the definition of singular integrals in the sense of a Cauchy principle value or a regularization procedure in the spirit of a subtraction technique to eliminate or reduce the order of strong singularities [29], [109], [1017], [1069]. For simplicity we will consider only internal points $\mathbf{x}\not\to\Gamma$ when the validity of Eq. (7.7) takes place except in some "boundary layer" region close to the surface Γ.

The representation (7.7) is valid for both the general cases of the first and second boundary value problems as well as for the mixed boundary-value problem

[29], [109], [410]. In particular, for the conditions (7.1$_1$), (7.2), and $\boldsymbol{\beta}^c, \mathbf{f}^c \equiv \mathbf{0}$, the right-hand-side integral over the external surface (7.7) can be considered as a continuation of $\boldsymbol{\varepsilon}^{\Gamma}(\mathbf{x})$, $\mathbf{x} \in \Gamma$, that is, (7.2), into w as the strain field that the boundary condition (7.1$_1$), (7.2) would generate in the comparison medium with homogeneous moduli $\mathbf{L}^c$.

After integration of Eq. (7.6) by parts, it is found that

$$\begin{aligned}\boldsymbol{\varepsilon}(\mathbf{x}) = \boldsymbol{\varepsilon}^0(\mathbf{x}) &+ \int_w \mathbf{U}(\mathbf{x}-\mathbf{y})\Big\{\mathbf{L}_1(\mathbf{y})[\boldsymbol{\varepsilon}(\mathbf{y}) - \boldsymbol{\beta}(\mathbf{y})] - \mathbf{L}^c\boldsymbol{\beta}_1(\mathbf{y})\Big\}\, d\mathbf{y} \\ &+ \int_w \nabla\mathbf{G}(\mathbf{x}-\mathbf{y})\mathbf{f}_1(\mathbf{y})\, d\mathbf{y} \\ &+ \int_\Gamma \nabla\mathbf{G}(\mathbf{x}-\mathbf{s})\Big\{\mathbf{L}_1(\mathbf{s})[\boldsymbol{\varepsilon}(\mathbf{s}) - \boldsymbol{\beta}(\mathbf{s})] - \mathbf{L}^c\boldsymbol{\beta}_1(\mathbf{s})\Big\}\mathbf{n}(\mathbf{s})\, ds, \qquad (7.8)\end{aligned}$$

which used that $\nabla_{\mathbf{y}} = -\nabla_{\mathbf{x}}$, and the surface integration is taken over the external surface Γ of the mesodomain w, containing a statistically large number of inclusions, and the integral operator kernel $\mathbf{U}$ is an even homogeneous generalized function of degree -3 defined by Eq. (3.19). For construction of the regularization of a generalized function of the type of derivatives of homogeneous regular function, we will consider a scheme proposed in [374] according to which the tensor $\mathbf{U}(\mathbf{x})$ is split into the singular and regular parts [see Eq. (3.20)]. Both of these terms depend on the shape of an exclusion region prescribed by Eqs. (3.17), while their sum is defined uniquely.

7.2 Random Structure Composites

7.2.1 General Integral Equation for Random Structure Composites

It is assumed that the representative mesodomain w contains a statistically large number of inclusions $v_i \subset v^{(k)}$ $(i = 1, 2, \ldots;\ k = 1, 2, \ldots, N)$; all the random quantities under discussion are described by statistically inhomogeneous random fields. We will use the statistical description presented in Section 5.3.3 in terms of the conditional probability densities $\varphi(v_i, \mathbf{x}_i | v_1, \mathbf{x}_1, \ldots, v_n, \mathbf{x}_n)$. We will consider a general case of statistically inhomogeneous media with the homogeneous matrix (for example, for so-called FGM), when the conditional probability density is not invariant with respect to translation. The average and conditional average taken for the ensemble of a statistically inhomogeneous field $X = (v_i)$ were introduced in Section 5.3.3.

Equation (7.8) is centered, that is, from both sides of Eq. (7.8) their statistical averages are subtracted:

$$\begin{aligned}\boldsymbol{\varepsilon}(\mathbf{x}) = \langle\boldsymbol{\varepsilon}\rangle(\mathbf{x}) &+ \int_w \mathbf{U}(\mathbf{x}-\mathbf{y})\{\boldsymbol{\tau}(\mathbf{y}) - \langle\boldsymbol{\tau}\rangle(\mathbf{y})\}\, d\mathbf{y} \\ &= \int_w \nabla\mathbf{G}(\mathbf{x}-\mathbf{y})\big[\mathbf{f}_1(\mathbf{y}) - \langle\mathbf{f}_1\rangle(\mathbf{y})\big]\, d\mathbf{y} + \boldsymbol{\mathcal{I}}^{\Gamma\epsilon}, \qquad (7.9)\end{aligned}$$

where $\boldsymbol{\tau}(\mathbf{y}) = \mathbf{L}_1(\mathbf{y})[\boldsymbol{\varepsilon}(\mathbf{y}) - \boldsymbol{\beta}(\mathbf{y})] - \mathbf{L}^c\boldsymbol{\beta}_1$ is called the stress polarization tensor and is simply a notation convenience. The subtraction technique used for the transformation of Eq. (7.8) into Eq. (7.9), also called a centering method, is a major stage of a justification of the so-called *locality principle* in micromechanics of composite materials (see Section 7.6). In so doing in the right-hand side of Eq. (7.9) the integral over the external surface Γ

$$\mathcal{I}^{\Gamma\epsilon} \equiv \int_{\Gamma} \nabla \mathbf{G}(\mathbf{x}-\mathbf{s})\Big\{\mathbf{L}_1(\mathbf{s})[\boldsymbol{\varepsilon}(\mathbf{s})-\boldsymbol{\beta}(\mathbf{s})] - \mathbf{L}^c\boldsymbol{\beta}_1(\mathbf{s}) \\ - [\langle \mathbf{L}_1(\boldsymbol{\varepsilon}-\boldsymbol{\beta})\rangle(\mathbf{s}) - \langle \mathbf{L}^c\boldsymbol{\beta}_1\rangle(\mathbf{s})]\Big\}\mathbf{n}(\mathbf{s})\, d\mathbf{s} \tag{7.10}$$

can be dropped out, because this integral vanishes at sufficient distance $\mathbf{x}$ from the boundary Γ. This means that if $|\mathbf{x}-\mathbf{s}|$ is large enough for $\forall \mathbf{s} \in \Gamma$, then at the portion of the smooth surface $d\mathbf{s} \approx |\mathbf{x}-\mathbf{s}|^2\, d\boldsymbol{\omega}^s$ with a small solid angle $d\boldsymbol{\omega}^s$, the tensor $\nabla\mathbf{G}(\mathbf{x}-\mathbf{s})|\mathbf{x}-\mathbf{s}|^2$ depends only on the solid angle $\boldsymbol{\omega}^s$ variables and slowly varies on the portion of the surface $d\mathbf{s}$; in this sense, the tensor $\nabla\mathbf{G}(\mathbf{x}-\mathbf{s})$ is called a "slow" variable of the solid angle $\boldsymbol{\omega}^s$ while the expression in curly brackets on the right-hand-side integral of Eq. (7.10) is a rapidly oscillating function on $d\mathbf{s}$ and is called a "fast" variable. Therefore, we can use a rigorous theory of "separate" integration of "slow" and "fast" variables, according to which (freely speaking) the operation of surface integration may be regarded as averaging, see [337] and its applications [995]. If (as we assume) there is no *long-range* order and the function $\varphi(v_j,\mathbf{x}_j|;v_i,\mathbf{x}_i)-\varphi(v_j,\mathbf{x}_j)$ decays at infinity sufficiently rapidly, then it leads to a degeneration of the surface integral. So, exponential decreasing of this function was obtained in [1180] for spherical inclusions; a more rapidly decreasing function for circle aligned fibers was proposed in [424], [1108].

To express Eq. (7.9) in terms of stresses, we use the identities:

$$\mathbf{L}_1(\boldsymbol{\varepsilon}-\boldsymbol{\beta}) = -\mathbf{L}^c\mathbf{M}_1\boldsymbol{\sigma}, \quad \boldsymbol{\varepsilon} = [\mathbf{M}^c\boldsymbol{\sigma}+\boldsymbol{\beta}^c] + [\mathbf{M}_1\boldsymbol{\sigma}+\boldsymbol{\beta}_1]. \tag{7.11}$$

Substituting (7.11) into the right-hand side and the left-hand side of (7.9), respectively, and contracting with the tensor $\mathbf{L}^c$ gives the general integral equation for stresses:

$$\boldsymbol{\sigma}(\mathbf{x}) = \langle\boldsymbol{\sigma}\rangle(\mathbf{x}) + \int_w \boldsymbol{\Gamma}(\mathbf{x}-\mathbf{y})\{\boldsymbol{\eta}(\mathbf{y})-\langle\boldsymbol{\eta}\rangle(\mathbf{y})\}\, d\mathbf{y} \\ + \int_w \mathbf{L}^c\nabla\mathbf{G}(\mathbf{x}-\mathbf{y})\big[\mathbf{f}_1(\mathbf{y})-\langle\mathbf{f}_1\rangle(\mathbf{y})\big]\, d\mathbf{y} + \mathcal{I}^{\Gamma\sigma}. \tag{7.12}$$

where we define $\mathcal{I}^{\Gamma\sigma} = \int_{\Gamma} \mathbf{L}^c\mathbf{G}(\mathbf{x}-\mathbf{s})\mathbf{L}^c\{\boldsymbol{\eta}(\mathbf{s})-\langle\boldsymbol{\eta}\rangle(\mathbf{s})\}\mathbf{n}(\mathbf{s})\, d\mathbf{s}$ and the strain polarization tensor $\boldsymbol{\eta}(\mathbf{y}) = \mathbf{M}_1(\mathbf{y})\boldsymbol{\sigma}(\mathbf{y})+\boldsymbol{\beta}_1(\mathbf{y})$ (4.8). If we assume no long-range order, then the tensor $\mathcal{I}^{\Gamma\sigma}$ degenerates and can be dropped. In (7.12) $\mathbf{M}_1(\mathbf{y})$ and $\boldsymbol{\beta}_1(\mathbf{y})$ are the jumps of the compliance $\mathbf{M}^{(k)}$ and of the eigenstrain $\boldsymbol{\beta}^{(k)}$ inside the component $v^{(k)}$ $(k=0,\ldots,N)$ with respect to the constant tensors $\mathbf{M}^c$ and $\boldsymbol{\beta}^c$, respectively. The integral operator kernel, $\boldsymbol{\Gamma}(\mathbf{x}-\mathbf{y})$, called the Green stress tensor [611], [613], is defined by Eq. (3.37).

For convenience in the presentation that follows, we will recast Eq. (7.12) in another form, for which we introduce the operation $\boldsymbol{\lambda}^1(\mathbf{x}) = \boldsymbol{\lambda}(\mathbf{x}) - \langle\boldsymbol{\lambda}\rangle_0(\mathbf{x})$ for the random function $\boldsymbol{\lambda}$ (e.g., $\boldsymbol{\lambda} = \boldsymbol{\sigma},\ \boldsymbol{\varepsilon},\ \boldsymbol{\tau},\ \boldsymbol{\eta},\ \mathbf{f}$) with statistical average in the matrix $\langle\boldsymbol{\lambda}\rangle_0(\mathbf{x})$. Then, Eq. (7.12) can be rewritten in the form

$$\boldsymbol{\sigma}(\mathbf{x}) = \langle\boldsymbol{\sigma}\rangle(\mathbf{x}) + \int_w \boldsymbol{\Gamma}(\mathbf{x}-\mathbf{y})\left\{\boldsymbol{\eta}^1(\mathbf{y}) - \langle\boldsymbol{\eta}^1\rangle(\mathbf{y})\right\}\, d\mathbf{y} + \int_w \mathbf{L}^c\nabla\mathbf{G}(\mathbf{x}-\mathbf{y})\left[\mathbf{f}_1^1(\mathbf{y}) - \langle\mathbf{f}_1^1\rangle(\mathbf{y})\right]\, d\mathbf{y}. \tag{7.13}$$

The general integral equations (7.9), (7.12), and (7.13) were proposed in [141] for statistically inhomogeneous media for the general case of inhomogeneity of tensors $\boldsymbol{\beta}^c(\mathbf{x}),\ \boldsymbol{\beta}^{(0)}(\mathbf{x}),\ \mathbf{f}^c(\mathbf{x}),\ \mathbf{f}^{(0)}(\mathbf{x})$ for both the general case of the first and second boundary value problems and for the mixed boundary-value problem.

For a statistically homogeneous ergodic field X, the integral equations (7.9) and (7.12) are equivalent to those known from the literature (see [156], [667], [668], [1186]). Nevertheless, for a statistically inhomogeneous field X the dependence of the statistical averages $\langle(.)\rangle(\mathbf{y})$ (7.9) and (7.12) on the current coordinate $\mathbf{y}$ is of fundamental importance. But even in this case, the expressions in curly brackets Eqs. (7.9), (7.12) are of the order $O(|\mathbf{x}-\mathbf{y}|^{-3})$ as $|\mathbf{x}-\mathbf{y}| \to \infty$, and the integrals in Eqs. (7.9) and (7.12) with the kernels $\mathbf{U}$ and $\boldsymbol{\Gamma}$, respectively, converge absolutely. In a similar manner, the integrals with the body force density converge absolutely. In fact, the kernel $\nabla\mathbf{G}(\mathbf{x}-\mathbf{y})$ is of the order $O(|\mathbf{x}-\mathbf{y}|^{-2})$ as $|\mathbf{x}-\mathbf{y}| \to \infty$. In so doing, the term $\mathbf{f}_1^1(\mathbf{y}) - \langle\mathbf{f}_1^1\rangle$, differing from zero only inside inclusions v_j, $(j = 1, 2, \ldots)$, tends to zero with $|\mathbf{x}-\mathbf{y}| \to \infty$ ($\mathbf{x} \in v_i,\ \mathbf{y} \in v_j$) as $\varphi(v_j, \mathbf{x}_j|; v_i, \mathbf{x}_i) - \varphi(v_j, \mathbf{x}_j)$. For no *long-range* order, an absolute convergence of the integrals involved is assured if the function $\varphi(v_j, \mathbf{x}_j|; v_i, \mathbf{x}_i) - \varphi(v_j, \mathbf{x}_j)$ decays at infinity sufficiently rapidly.

Therefore, for $\mathbf{x} \in w$ considered in Eqs. (7.9), (7.12) and removed far enough from the boundary Γ, the right-hand-side integrals in (7.9) and (7.12) do not depend on the shape and size of the domain w, and they can be replaced by the integrals over the whole space R^3. With this assumption, we hereafter omit explicitly denoting R^3 as the integration domain:

$$\begin{aligned}\boldsymbol{\varepsilon}(\mathbf{x}) &= \langle\boldsymbol{\varepsilon}\rangle(\mathbf{x}) + \int \mathbf{U}(\mathbf{x}-\mathbf{y})\{\boldsymbol{\tau}^1(\mathbf{y}) - \langle\boldsymbol{\tau}^1\rangle(\mathbf{y})\}\, d\mathbf{y} \\ &+ \int \nabla\mathbf{G}(\mathbf{x}-\mathbf{y})\left[\mathbf{f}_1(\mathbf{y}) - \langle\mathbf{f}_1\rangle(\mathbf{y})\right]\, d\mathbf{y}, \\ \boldsymbol{\sigma}(\mathbf{x}) &= \langle\boldsymbol{\sigma}\rangle(\mathbf{x}) + \int \boldsymbol{\Gamma}(\mathbf{x}-\mathbf{y})\left\{\boldsymbol{\eta}^1(\mathbf{y}) - \langle\boldsymbol{\eta}^1\rangle(\mathbf{y})\right\}\, d\mathbf{y} \\ &+ \int \mathbf{L}^c\nabla\mathbf{G}(\mathbf{x}-\mathbf{y})\left[\mathbf{f}_1^1(\mathbf{y}) - \langle\mathbf{f}_1^1\rangle(\mathbf{y})\right]\, d\mathbf{y}.\end{aligned} \tag{7.14}$$

Thus, there are no difficulties connected with the asymptotic behavior of the generalized functions $\nabla\nabla\mathbf{G}$ and $\boldsymbol{\Gamma}$ at infinity (as $|\mathbf{x}-\mathbf{y}|^{-3}$), and there is no need to postulate either the shape or the size of the integration domain w [691] or to resort to either regularization [613], [618] or renormalization [216], [751] of integrals that are divergent at infinity [1184]. The rigorous mathematical analysis of the above-mentioned methods is beyond the purpose of the current presentation. Nevertheless, it should be noted that the disruption of statistical homogeneity of media often leads to additional difficulties, whose resolution by these known mentioned approaches appears to be questionable.

We average Eq. (7.14_2) on sets $X(\cdot|v_1$, $X(\cdot|v_1, v_2)$, and so forth, for fixed sets v_1; v_1, v_2; ... by means of different distribution densities $\varphi(v_m|v_1,\ldots,v_n)$. Considering these conditional statistical averages of the general integral equation (7.14_2) leads to an infinite system of integral equations ($n = 1, 2, \ldots$)

$$\begin{aligned}
&\langle\boldsymbol{\sigma}|v_1,\mathbf{x}_1;\ldots;v_n,\mathbf{x}_n\rangle(\mathbf{x}) - \sum_{i=1}^{n}\int \boldsymbol{\Gamma}(\mathbf{x}-\mathbf{y})\langle V_i(\mathbf{y})\boldsymbol{\eta}^1|v_1,\mathbf{x}_1;\ldots;v_n,\mathbf{x}_n\rangle(\mathbf{y})\,d\mathbf{y}\\
&-\sum_{i=1}^{n}\int \mathbf{L}^c\nabla\mathbf{G}(\mathbf{x}-\mathbf{y})\langle V_i(\mathbf{y})\mathbf{f}^1|v_1,\mathbf{x}_1;\ldots;v_n,\mathbf{x}_n\rangle(\mathbf{y})\,d\mathbf{y}\\
&=\langle\boldsymbol{\sigma}\rangle(\mathbf{x}) + \int \boldsymbol{\Gamma}(\mathbf{x}-\mathbf{y})\{\langle\boldsymbol{\eta}^1|;v_1,\mathbf{x}_1;\ldots;v_n,\mathbf{x}_n\rangle(\mathbf{y}) - \langle\boldsymbol{\eta}^1\rangle(\mathbf{y})\}\,d\mathbf{y}\\
&+\int \mathbf{L}^c\nabla\mathbf{G}(\mathbf{x}-\mathbf{y})\{\langle\mathbf{f}^1|;v_1,\mathbf{x}_1;\ldots;v_n,\mathbf{x}_n\rangle(\mathbf{y}) - \langle\mathbf{f}^1\rangle(\mathbf{y})\}\,d\mathbf{y}. \qquad (7.15)
\end{aligned}$$

Since $\mathbf{x} \in v_1,\ldots,v_n$ in the nth line of the system takes values only in the inclusions $v_1,\ldots,v_n$, the nth line actually contains n equations. In a similar manner, the analogous infinite system of integral equations can be obtained from Eq. (7.9).

7.2.2 Some Particular Cases

Thus, the general integral Eqs. (7.9) and (7.12) over the whole space obtained above are exact and do not depend on particular boundary conditions. The subsequent analysis can be done for the comparison medium with any elastic modulus $\mathbf{L}^c$, which necessarily leads to some additional assumptions for the structure of the strain fields in the matrix. Equations (7.9) and (7.12) are much easier to solve when the stress–strain fields are studied only inside the inclusions. Following [170], we will consider two fundamentally different approaches to ensure that the integration on the right-hand sides of Eqs. (7.9), (7.12), and (7.14) extends only over the volume of the inclusion.

In the first one, we postulate

$$\mathbf{L}^c \equiv \mathbf{L}^{(0)}, \quad \boldsymbol{\beta}^c \equiv \boldsymbol{\beta}^{(0)}. \qquad (7.16)$$

Then, $\mathbf{L}_1(\mathbf{x}) \equiv \mathbf{0}$ and $\boldsymbol{\beta}_1(\mathbf{x}) \equiv \mathbf{0}$ at $\mathbf{x} \in v^{(0)}$ and the integrands with the arguments $\mathbf{y}$ in both Eqs. (7.14) vanish at $\mathbf{y} \in v^{(0)}$. In the second one, we choose $\mathbf{L}^c$ quite arbitrarily, and analyze, e.g., Eq. (7.14_2). Equation (7.14_2) being exact for any $\langle\boldsymbol{\eta}\rangle_0(\mathbf{x})$ can be simplified with the additional assumption that the strain polarization tensor in the matrix $\boldsymbol{\eta}(\mathbf{x})$, ($\mathbf{x} \in v^{(0)}$) coincides with its statistical average in the matrix

$$\boldsymbol{\eta}(\mathbf{x}) \equiv \langle\boldsymbol{\eta}\rangle_0(\mathbf{x}), \ \mathbf{x} \in v^{(0)}. \qquad (7.17)$$

In so doing, the assumption (7.16) is more restricted in the sense that the assumption (7.16) yields the assumption (7.17) (the opposite is not true) and, moreover, in such a case the exact equality $\boldsymbol{\eta}(\mathbf{x}) \equiv \langle\boldsymbol{\eta}\rangle_0(\mathbf{x}) \equiv \mathbf{0}$, $\mathbf{x} \in v^{(0)}$ holds.

Because of this we will consider only that assumption (7.17) ensured the integration of the integrand in the right-hand side of Eqs. (7.14) only by the volume of inclusions.

For both no body forces acting and purely mechanical loading, that is,

$$\mathbf{f}(\mathbf{x}) \equiv \mathbf{0}, \ \boldsymbol{\beta}(\mathbf{x}) \equiv \mathbf{0} \tag{7.18}$$

Equations (7.14) simplify to

$$\boldsymbol{\varepsilon}(\mathbf{x}) = \langle \boldsymbol{\varepsilon} \rangle(\mathbf{x}) + \int \mathbf{U}(\mathbf{x}-\mathbf{y})\Big\{\mathbf{L}_1(\mathbf{y})\boldsymbol{\varepsilon}(\mathbf{y}) - \langle \mathbf{L}_1\boldsymbol{\varepsilon}\rangle(\mathbf{y})\Big\}\, d\mathbf{y}, \tag{7.19}$$

$$\boldsymbol{\sigma}(\mathbf{x}) = \langle \boldsymbol{\sigma} \rangle(\mathbf{x}) + \int \boldsymbol{\Gamma}(\mathbf{x}-\mathbf{y})\Big\{\mathbf{M}_1(\mathbf{y})\boldsymbol{\sigma}(\mathbf{y}) - \langle \mathbf{M}_1\boldsymbol{\sigma}\rangle(\mathbf{y})\Big\}\, d\mathbf{y}. \tag{7.20}$$

For both cases being considered, (7.18) and $\mathbf{M}^c \equiv \mathbf{M}^{(0)}$, Eqs. (7.19) and (7.20) were presented without any justification in [171].

It is interesting that the case of the inhomogeneous transformation field $\boldsymbol{\beta}^{(0)}(\mathbf{x}) \not\equiv$ const. corresponds to a practically important problem of development of a heat-shielding composite material coating for airspace constructions. In so doing in the framework of the assumption (7.17) Eq. (7.14_2) is similar to Eq. (7.20) for which the solution methods are well developed [181], [182] and can be applied for the solution of Eq. (7.14_2).

In what follows for case (7.18) for statistically homogeneous media and homogeneous boundary conditions (7.2) or (7.3), we obtain [1184] $\langle \boldsymbol{\varepsilon} \rangle(\mathbf{x}) \equiv \boldsymbol{\varepsilon}^{\Gamma} =$ const, or $\langle \boldsymbol{\sigma} \rangle(\mathbf{x}) \equiv \boldsymbol{\sigma}^{\Gamma} =$ const., $(\mathbf{x} \in w)$, respectively, and $\langle \mathbf{L}_1\boldsymbol{\varepsilon}\rangle$, $\langle \mathbf{M}_1\boldsymbol{\sigma}\rangle \equiv$ const., Eqs. (7.19) and (7.20) reduce to

$$\boldsymbol{\varepsilon}(\mathbf{x}) = \langle \boldsymbol{\varepsilon}\rangle + \int \mathbf{U}(\mathbf{x}-\mathbf{y})[\boldsymbol{\tau}(\mathbf{y}) - \langle \boldsymbol{\tau}\rangle]\, d\mathbf{y}, \quad \boldsymbol{\sigma}(\mathbf{x}) = \langle \boldsymbol{\sigma}\rangle + \int \boldsymbol{\Gamma}(\mathbf{x}-\mathbf{y})[\boldsymbol{\eta}(\mathbf{y}) - \langle \boldsymbol{\eta}\rangle]\, d\mathbf{y}, \tag{7.21}$$

which are well known in micromechanics; here $\boldsymbol{\tau}(\mathbf{x}) = \mathbf{L}_1(\mathbf{x})\boldsymbol{\varepsilon}(\mathbf{x})$.

Finally, we will consider the field X bounded in one direction such as a laminated structure of some real FGM [883], [884]. Then the surface integral (7.8) over a "cylindrical" surface (with the surface area proportional to $\rho = |\mathbf{x}-\mathbf{s}|$) tends to zero with $|\mathbf{x}-\mathbf{s}| \to \infty$ as ρ^{-1} simply because the generalized function $\nabla \mathbf{G}(\mathbf{x}-\mathbf{s})$ is an even homogeneous function of order -2. Therefore, for infinite media the surface integral (7.8) vanishes, and Eq. (7.8) can be rewritten as

$$\begin{aligned}\boldsymbol{\varepsilon}(\mathbf{x}) &= \boldsymbol{\varepsilon}^0(\mathbf{x}) + \int \nabla \mathbf{G}(\mathbf{x}-\mathbf{y})\mathbf{f}_1(\mathbf{y})\, d\mathbf{y} \\ &+ \int \mathbf{U}(\mathbf{x}-\mathbf{y})\{\mathbf{L}_1(\mathbf{y})[\boldsymbol{\varepsilon}(\mathbf{y}) - \boldsymbol{\beta}(\mathbf{y})] - \mathbf{L}^c\boldsymbol{\beta}_1(\mathbf{y})\}\, d\mathbf{y},\end{aligned} \tag{7.22}$$

or, alternatively, in terms of stresses

$$\boldsymbol{\sigma}(\mathbf{x}) = \boldsymbol{\sigma}^0(\mathbf{x}) + \int \mathbf{L}^c\nabla \mathbf{G}(\mathbf{x}-\mathbf{y})\mathbf{f}_1(\mathbf{y})\, d\mathbf{y} + \int \boldsymbol{\Gamma}(\mathbf{x}-\mathbf{y})\boldsymbol{\eta}(\mathbf{y})\, d\mathbf{y}. \tag{7.23}$$

In so doing the integrals from body forces in Eqs. (7.22) and (7.23) only conditionally converge for a general case of a bounded function $\mathbf{f}_1(\mathbf{y})$; because of this, for the function $\mathbf{f}(\mathbf{y})$ we will assume decay at infinity $|\mathbf{x}-\mathbf{y}| \to \infty$ no less than $O(|\mathbf{x}-\mathbf{y}|^{-\alpha})$, $(\alpha > 0)$ guaranteeing the absolute convergence of body force integrals in Eqs. (7.22) and (7.23). Clearly, in the considered case of X bounded in one direction, Eqs. (7.22) and (7.23) are exact, and the right-hand-side integrals in (7.22) and (7.23) converge absolutely. Equation (7.22) was used in [1105] for the particular case (7.18) with homogeneous boundary conditions (7.2) and for the inclusion field X with a constant concentration of inclusions within an ellipsoidal domain included in the infinite matrix.

The problem of the estimation of effective properties can be simplified if the local thermoelastic properties are functions of only one space coordinate; in so doing the variations in these local properties may be either continuous or discontinuous, as in a layered system. A layered medium is one of the few composite geometries for which one can ascertain exact closed-form expression for the relationship between forces and moments per unit length and the deformations that they produced. So, [847] gave an explicit procedure for determining the effective stiffness coefficients, coupling coefficients, and flexural stiffness for thin laminates with layers of arbitrary material symmetry.

It should be mentioned that Eqs. (7.9) and (7.12), while more complicated than Eq. (7.22) and (7.23), nevertheless provide practical advantages because their integrands decay at infinity faster than the integrands involved in Eqs. (7.22) and (7.23).

7.2.3 Comparison with Related Equations

We now turn our attention to the discussion of related integral equations and consider the case of a statistically homogeneous medium (with $\mathbf{M}^c = \mathbf{M}^{(0)}$ and $\boldsymbol{\beta} \equiv \mathbf{0}$) and homogeneous boundary conditions either (7.2) or (7.3). Kunin [618] uses the equation over the whole space:

$$\boldsymbol{\varepsilon}(\mathbf{x}) = \boldsymbol{\varepsilon}^0 + \int \mathbf{U}(\mathbf{x}-\mathbf{y})\boldsymbol{\tau}(\mathbf{y})\, d\mathbf{y}, \quad \boldsymbol{\sigma}(\mathbf{x}) = \boldsymbol{\sigma}^0 + \int \boldsymbol{\Gamma}(\mathbf{x}-\mathbf{y})\boldsymbol{\eta}(\mathbf{y})\, d\mathbf{y} \tag{7.24}$$

at $\boldsymbol{\varepsilon}^0 \equiv$ const. and $\boldsymbol{\sigma}^0 \equiv$ const., respectively; here $\boldsymbol{\tau} = \mathbf{L}_1\boldsymbol{\varepsilon}$. It is well known that the right-hand-side integrals in (7.24_1) are conditionally convergent, and as a consequence Eqs. (7.24_1) and (7.24_2) are used in association with the regularizations:

$$\int \mathbf{U}(\mathbf{x}-\mathbf{y})\mathbf{h}\, d\mathbf{y} = \mathbf{0}, \quad \text{or} \quad \int \boldsymbol{\Gamma}(\mathbf{x}-\mathbf{y})\mathbf{h}\, d\mathbf{y} = \mathbf{L}^c\mathbf{h}, \tag{7.25}$$

and

$$\int \mathbf{U}(\mathbf{x}-\mathbf{y})\mathbf{h}\, d\mathbf{y} = \mathbf{M}^c\mathbf{h}, \quad \text{or} \quad \int \boldsymbol{\Gamma}(\mathbf{x}-\mathbf{y})\mathbf{h}\, d\mathbf{y} = \mathbf{0}, \tag{7.26}$$

for the first and the second boundary-value problem, respectively; $\mathbf{h}$ is an arbitrary constant symmetric second-order tensor, while, according to (3.37), each of the relations is a consequence of the other from the same pair, either (7.25) or (7.26). In fact, the relations (7.25) and (7.26) introduce an operation of generalized functions $\mathbf{U}$ and $\boldsymbol{\Gamma}$ on a constant symmetric tensor $\mathbf{h} \equiv$ const. (see for comparison [374]).

For completeness of presentation, we will reproduce the verification of Eqs. (7.25) and (7.26) proposed in Kanaun [538], [539] and reproduced in [618]. Let us assume that $\mathbf{L}_1(\mathbf{x})$, $\mathbf{M}_1(\mathbf{x}) \equiv$ const. throughout the whole space. Analyze first the second boundary-value problem $\boldsymbol{\sigma}^{\Gamma} \equiv$ const. (7.3) leading to the homogeneous stress state inside w: $\boldsymbol{\sigma}(\mathbf{x}) \equiv \boldsymbol{\sigma}^{\Gamma}$, $\boldsymbol{\varepsilon}(\mathbf{x}) \equiv \mathbf{M}\boldsymbol{\sigma}^{\Gamma}$. It is natural to require that Eqs. (7.24_1) and (7.24_2) give the same result. This is possible only if $\mathbf{U}$ and $\boldsymbol{\Gamma}$ act on constants as in Eqs. (7.26). Analogously, for the first boundary value problem $\boldsymbol{\varepsilon}^{\Gamma} \equiv$ const. (7.2) the assumptions (7.25) ensure the homogeneous solution $\boldsymbol{\varepsilon}(\mathbf{x}) \equiv \boldsymbol{\varepsilon}^{\Gamma}$, $\boldsymbol{\sigma}(\mathbf{x}) = \mathbf{L}\boldsymbol{\varepsilon}^{\Gamma}$ ($\mathbf{x} \in w$) by Eq. (7.24_1). Thus, the logical scheme of the proof mentioned is as follows: if Eqs. (7.24_1) and (7.24_2) are valid for a particular case of both the boundary conditions and inhomogeneity of the elastic mismatch ($\mathbf{M}_1(\mathbf{x}) \equiv$ const., $\mathbf{x} \in w$), then either equality (7.25) or (7.26) is valid. The extension of this proof for more practically interesting cases is troublesome because the correctness of Eqs. (7.24_1) and (7.24_2) is questionable (compare with Eqs. (7.21_1) and (7.21_2)).

In the considered conditions the "noncanonical regularizations" (7.25) and (7.26) provides an insignificant formal simplification of Eqs. (7.9) and (7.12); however, they offer no promise for analysis of statistically inhomogeneous media, where it becomes necessary in general to determine the convolution of $\boldsymbol{\Gamma}$ with an arbitrary piecewise constant function $\langle \boldsymbol{\eta} \rangle(\mathbf{y})$.

For statistically homogeneous composites and homogeneous boundary conditions Lipinski *et al.* [692] reduced the system (7.7) and (7.8) (at $\mathbf{f}$, $\boldsymbol{\beta} \equiv \mathbf{0}$) to an equation analogous to (7.24_1); the subsequent convergence difficulty was overcome by the use of a self-consistent approach which is equivalent in fact to the termination of the constituent $\widetilde{\mathbf{U}}^f(\mathbf{x})$ (3.20) leading to the known convergence problems. Hori and Kubo [469], Ju and Chen [514], Ju and Tseng [521], [523] also used Eq. (7.24_1) and eliminated the difficulties induced by the dependence of a conditionally convergent integral on the shape of integration domain w by the use of an assumption of this shape. Fassi-Fehri *et al.* [327] postulated the size of integration domain in Eq. (7.24_1). Equation (7.24_1) was also used in [516], [525] with the consequent transformation of the known renormalization scheme that is not fully clear to the present author (see also [748]). The difficulties mentioned above and some others can be avoided easily by using Eqs. (7.9) and (7.12) instead of Eqs. (7.24_1) and (7.24_2), respectively.

For a purely mechanical loading ($\mathbf{f}$, $\boldsymbol{\beta} \equiv \mathbf{0}$) Eq. (7.8) formally coincides with the analogous relationships obtained in [692] by the use of Green's functions for a bounded domain w when the surface integral in the right-hand side (7.8) vanishes. Its implementation is not trivial because the finite-body Green's function $\mathbf{G}^w$ is generally not known, and replacing $\mathbf{G}^w$ by $\mathbf{G}$ (7.7) if w is large enough leads to well known convergence difficulties as discussed in detail by Willis [1184]. For homogeneous boundary conditions (7.2), (7.3), and for $\mathbf{f}$, $\boldsymbol{\beta} \equiv \mathbf{0}$ Eqs. (7.9) and (7.12), and subsequent conversions were considered in [1179], [1184] by a more specific method using essentially the uniformity of the boundary conditions [823]. It was noted, however, "that the argument given is plausible rather than rigorous and would not apply at all for the boundary conditions that would produce stresses and strains that were not uniform "on average" (quoted from

[1184]). The method proposed above, (7.6) through (7.12), is devoid of these limitations.

It should be mentioned that for the case $\mathbf{L}^c \equiv \mathbf{L}^{(0)}$, $\mathbf{f}, \boldsymbol{\beta} \equiv \mathbf{0}$ and homogeneous boundary conditions (7.2), Levin [667], [668] used Eq. (7.24_1) with a subsequent implication of a centrification procedure (see Subsection 7.2.2) leading to the correct Eq. (7.21_1); Levin *et al.* [670] used the same scheme for a thermoelectroelastic problem. Equation (7.8) was used in [561], [562] with a surface integral that can be transformed by utilization of the divergence theorem to the volume integral, providing equivalence with Eq. (7.21_1).

The related integral Eq. (7.9), averaged over the components and the nonlocal constitutive equations involving inhomogeneous $\langle\boldsymbol{\sigma}\rangle(\mathbf{x})$ and $\langle\boldsymbol{\varepsilon}\rangle(\mathbf{x})$ for statistically homogeneous media, was obtained in [279] by considering an infinite body with only some fictitious distribution of deterministic body-force $\mathbf{f}(\mathbf{x})$ applied that decays sufficiently rapidly at infinity ($|\mathbf{x}| \to \infty$). Then, slightly modifying the approach in [279] (proposed for $\boldsymbol{\beta} \equiv \mathbf{0}$) with the transformation field $\boldsymbol{\beta}_1(\mathbf{x})$ decaying sufficiently rapidly at infinity leads to

$$\nabla\{\mathbf{L}(\mathbf{x})[\boldsymbol{\varepsilon}(\mathbf{x}) - \boldsymbol{\beta}(\mathbf{x})]\} + \mathbf{f}(\mathbf{x}) = 0, \tag{7.27}$$

leading to the integral representation over the whole space:

$$\boldsymbol{\varepsilon}(\mathbf{x}) = \boldsymbol{\varepsilon}^{0f}(\mathbf{x}) + \int \mathbf{U}(\mathbf{x}-\mathbf{y})\Big\{\mathbf{L}_1(\mathbf{y})[\boldsymbol{\varepsilon}(\mathbf{y}) - \boldsymbol{\beta}(\mathbf{y})] - \mathbf{L}^c\boldsymbol{\beta}_1(\mathbf{y})\Big\}\, d\mathbf{y}, \tag{7.28}$$

rather than (7.7) and (7.8), where

$$\boldsymbol{\varepsilon}^{0f}(\mathbf{x}) = \int \nabla\mathbf{G}(\mathbf{x}-\mathbf{y})\mathbf{f}(\mathbf{y})\, d\mathbf{y} \tag{7.29}$$

is a solution for the infinite comparison body with moduli $\mathbf{L}^c$ subjected to the same body-force $\mathbf{f}(\mathbf{x})$. Ensemble averaging of Eq. (7.28) and subtracting the result from the initial Eq. (7.28) leads to Eq. (7.9), which yields the approximate Eq. (19) in [279] for $\boldsymbol{\beta}(\mathbf{x}) \equiv \mathbf{0}$. Kunin [619] noted that the surface traction can be included in the body force as a surface δ-function term, and appropriate integration can be extended to the whole space. Although it has no effect on the correctness of Eqs. (7.11_1) and (7.12) obtained by other methods, it should be mentioned that to the author's knowledge, no justification of the introduction of the fictitious body force $\mathbf{f}(\mathbf{x})$ into the present scheme (7.27) through (7.29) has been published before.

7.3 Doubly and Triply Periodical Structure Composites

It is assumed that the representative mesodomain w contains a statistically large number of ellipsoidal inclusions $v_i \subset v^{(1)}$ $(i = 1, 2, \ldots)$ with identical shape, orientation, and mechanical properties. We now consider a composite medium with particle centers distributed at the nodes of some spatial lattice Λ. Suppose $\mathbf{e}_i$ $(i = 1, 2, 3)$ are linearly independent vectors, so that we can represent any node $\mathbf{m} \in \Lambda$: $\mathbf{x}_\mathbf{m} = f_1(m_1)\mathbf{e}_1 + f_2(m_2)\mathbf{e}_2 + f_3(m_3)\mathbf{e}_3$ (5.102), where $\mathbf{m} = (m_1, m_2, m_3)$

are integer-valued coordinates of the node $\mathbf{m}$ in the basis $\mathbf{e}_i$ which are equal in modulus to $|\mathbf{e}_i|$, and $f_i(m_i) - f_i(m_i+1) \not\equiv$ const., $(i = 1,2,3)$. A detailed description of such deterministic structures is presented in Section 5.3.4. The representation (5.102) is reduced in some particular cases to the equations describing both the triply (5.103) and doubly (5.104) periodic structures

$$\mathbf{x}_{\mathbf{m}} = m_1\mathbf{e}_1 + m_2\mathbf{e}_2 + m_3\mathbf{e}_3, \tag{7.30}$$

$$\mathbf{x}_{\mathbf{m}} = m_1\mathbf{e}_1 + m_2\mathbf{e}_2 + f_3(m_3)\mathbf{e}_3, \tag{7.31}$$

respectively. The composite material is constructed using building blocks or cells: $w = \cup\Omega_{\mathbf{m}}$, $v_{\mathbf{m}} \subset \Omega_{\mathbf{m}}$. In Section 5.3.4, the averages over both the cell Ω_i (5.106) and "moving averaging" cell $\mathcal{V}_{\mathbf{x}}$ (or moving-window [395]) (5.107) were defined $(\mathbf{x} \in \Omega_i)$:

$$\mathbf{g}^{\Omega}(\mathbf{x}) = \frac{1}{\bar{\Omega}_i}\int_{\Omega_i} \mathbf{g}(\mathbf{y})\, d\mathbf{y}, \quad \langle \mathbf{g}\rangle_{\mathbf{x}}(\mathbf{x}-\mathbf{y}) = \frac{1}{\bar{\mathcal{V}}_{\mathbf{x}}}\int_{\mathcal{V}_{\mathbf{x}}} \mathbf{g}(\mathbf{z}-\mathbf{y})\, d\mathbf{z}. \tag{7.32}$$

By way of illustration, let us consider the case of triply periodic structures (3.1) under the uniform boundary conditions (7.3) and let $\mathbf{g}$ be governed by the boundary condition (7.3) (for example, $\mathbf{g} \equiv \boldsymbol{\sigma}$). Clearly, for homogeneous boundary conditions (7.3), $\boldsymbol{\sigma}^{\Omega}(\mathbf{x})$ by (3.3_1) is an invariant with respect to the cell number i and $\boldsymbol{\sigma}^{\Omega}(\mathbf{x}) = \langle\boldsymbol{\sigma}\rangle_{\mathbf{x}}(\mathbf{x}) =$ const., $\forall \mathbf{x} \in \Omega_i \subset w$ (if $\mathcal{V}_{\mathbf{x}}$ is a translation of Ω_i). In the general case of inhomogeneous boundary conditions $\boldsymbol{\sigma}^{\Gamma}(\mathbf{x}) \not\equiv$ const. (as well as in the case that the condition (3.1) breaks down) $\boldsymbol{\sigma}^{\Omega}(\mathbf{x})$ is a step function $\boldsymbol{\sigma}^{\Omega}(\mathbf{x}) \neq \boldsymbol{\sigma}^{\Omega}(\mathbf{y})$ at $\mathbf{x} \in \Omega_i$ and $\mathbf{y} \in \Omega_j$ $(i \neq j)$ as well as $\boldsymbol{\sigma}^{\Omega}(\mathbf{x}) \not\equiv \langle\boldsymbol{\sigma}\rangle_{\mathbf{x}}(\mathbf{x})$ at $\mathbf{x} \in \Omega_i$.

Recall that Eq. (7.8) is valid for concrete realization of the inclusion field X which can be doubly periodical (3.2). In such a case, the function of the operation of a statistical average for random structure composites (see Section 7.2) becomes the volume average over the "moving averaging" cell $\mathcal{V}_{\mathbf{x}}$. By doing so, the transformation of Eq. (7.8) in the framework of the centering method is carried out by subtracting from both sides of Eq. (7.8) their average over the "moving averaging" cell $\mathcal{V}_{\mathbf{x}}$ (3.3_2):

$$\begin{aligned}\boldsymbol{\varepsilon}(\mathbf{x}) &= \langle\langle \boldsymbol{\varepsilon}^0\rangle\rangle_{\mathbf{x}}(\mathbf{x}) + \langle\boldsymbol{\varepsilon}\rangle_{\mathbf{x}}(\mathbf{x}) + \int \langle\langle \nabla\mathbf{G}\rangle\rangle_{\mathbf{x}}(\mathbf{x}-\mathbf{y})\mathbf{f}_1(\mathbf{y})\, d\mathbf{y}\\ &+ \int \langle\langle \mathbf{U}\rangle\rangle_{\mathbf{x}}(\mathbf{x}-\mathbf{y})\{\mathbf{L}_1(\mathbf{y})[\boldsymbol{\varepsilon}(\mathbf{y}) - \boldsymbol{\beta}(\mathbf{y})] - \mathbf{L}^c(\mathbf{y})]\boldsymbol{\beta}_1(\mathbf{y})\}\, d\mathbf{y}\\ &+ \oint \langle\langle \nabla\mathbf{G}\rangle\rangle_{\mathbf{x}}(\mathbf{x}-\mathbf{s})\{\mathbf{L}_1(\mathbf{s})[\boldsymbol{\varepsilon}(\mathbf{s}) - \boldsymbol{\beta}(\mathbf{s})] - \mathbf{L}^c(\mathbf{s})]\boldsymbol{\beta}_1(\mathbf{s})\}\mathbf{n}(\mathbf{s})\, ds, \end{aligned}\tag{7.33}$$

where $\mathbf{x} \in \Omega_i$, and one introduces a new centering operation over the "moving averaging" cell $\mathbf{x} \in \mathcal{V}_{\mathbf{x}}$ caused by translation of a cell Ω_i: $\langle\langle \mathbf{g}\rangle\rangle_{\mathbf{x}}(\mathbf{x}-\mathbf{y}) \equiv \mathbf{g}(\mathbf{x}-\mathbf{y}) - \langle\mathbf{g}\rangle_{\mathbf{x}}(\mathbf{x}-\mathbf{y})$. For the analysis of integral convergence in Eq. (7.33), we expand $\mathbf{U}(\mathbf{z}-\mathbf{y})$ $(\mathbf{z} \in \mathcal{V}_{\mathbf{X}})$ in a Taylor series about $\mathbf{x}$ and integrate term by term over the cell $\mathcal{V}_{\mathbf{x}}$ with the center $\mathbf{x}$, then

$$\mathbf{U}(\mathbf{z},\mathbf{y}) = \mathbf{U}(\mathbf{x},\mathbf{y}) + (\mathbf{z}-\mathbf{x})\nabla\mathbf{U}(\mathbf{x},\mathbf{y}) + \frac{1}{2}(\mathbf{z}-\mathbf{x})\otimes(\mathbf{z}-\mathbf{x})\nabla\nabla\mathbf{U}(\mathbf{x},\mathbf{y})\ldots,$$

$$\langle\mathbf{U}\rangle_{\mathbf{x}}(\mathbf{x},\mathbf{y}) = \mathbf{U}(\mathbf{x},\mathbf{y}) + \frac{1}{2\bar{\mathcal{V}}_{\mathbf{x}}}\int_{\mathcal{V}_{\mathbf{x}}} (\mathbf{z}-\mathbf{x})\otimes(\mathbf{z}-\mathbf{x})\, d\mathbf{z}\nabla\nabla\mathbf{U}(\mathbf{x},\mathbf{y})\ldots.$$

A similar expansion can be performed for the tensor $\nabla\mathbf{G}(\mathbf{x}-\mathbf{y})$ that leads to

$$\langle\langle\mathbf{U}\rangle\rangle_{\mathbf{x}}(\mathbf{x}-\mathbf{y}) = -\frac{1}{2\overline{\mathcal{V}}_{\mathbf{x}}}\int_{\mathcal{V}_{\mathbf{x}}}(\mathbf{z}-\mathbf{x})\otimes(\mathbf{z}-\mathbf{x})\,d\mathbf{z}\nabla\nabla\mathbf{U}(\mathbf{x}-\mathbf{y})+\ldots. \quad (7.34)$$

$$\langle\langle\nabla\mathbf{G}\rangle\rangle_{\mathbf{x}}(\mathbf{x}-\mathbf{y}) = -\frac{1}{2\overline{\mathcal{V}}_{\mathbf{x}}}\int_{\mathcal{V}_{\mathbf{x}}}(\mathbf{z}-\mathbf{x})\otimes(\mathbf{z}-\mathbf{x})\,d\mathbf{z}\nabla\nabla\nabla\mathbf{G}(\mathbf{x}-\mathbf{y})+\ldots. \quad (7.35)$$

As is evident from Eq. (7.35), the tensor $\langle\langle\mathbf{U}\rangle\rangle_{\mathbf{x}}(\mathbf{x}-\mathbf{y})$ is of the order $O(a_{\mathcal{V}}^2|\mathbf{x}-\mathbf{y}|^{-5})$ with the dropped terms in Eq. (7.35) being of the order $O(a_{\mathcal{V}}^4|\mathbf{x}-\mathbf{y}|^{-7})$ and higher order terms. Then, the absolute convergence of the volume integral (7.33) is assured because at sufficient distance $\mathbf{x}$ from the boundary Γ and $|\mathbf{x}-\mathbf{y}|\to\infty$, integration over $\mathbf{y}$ can be carried out independently for both $\langle\langle\mathbf{U}\rangle\rangle_{\mathbf{x}}(\mathbf{x}-\mathbf{y})$ (the function of the "slow" variable $\mathbf{x}-\mathbf{y}$) and the expression in curly brackets $\{\mathbf{L}_1(\mathbf{y})[\boldsymbol{\varepsilon}(\mathbf{y})-\boldsymbol{\beta}(\mathbf{y})]-\mathbf{L}^c\boldsymbol{\beta}_1(\mathbf{y})\}$ (the function of "fast" variable $\mathbf{y}$), and therefore the volume integral converges absolutely. In a similar manner the term $\langle\langle\nabla\mathbf{G}\rangle\rangle_{\mathbf{x}}(\mathbf{x}-\mathbf{s})$ in the surface integral (7.33) is of the order $O(a_{\mathcal{V}}^2|\mathbf{x}-\mathbf{y}|^{-4})$, and the surface integral vanishes at $|\mathbf{x}-\mathbf{s}|\to\infty$, $\mathbf{s}\in\Gamma$. For the same reason, the volume integral with the kernel $\langle\langle\nabla\mathbf{G}\rangle\rangle_{\mathbf{x}}(\mathbf{x}-\mathbf{s})$ converges absolutely for bounded functions $\mathbf{f}_1(\mathbf{y})$.

By this means, Eq. (7.33) is reduced to the relation

$$\boldsymbol{\varepsilon}(\mathbf{x}) = \langle\langle\boldsymbol{\varepsilon}^0\rangle\rangle_{\mathbf{x}}(\mathbf{x}) + \langle\boldsymbol{\varepsilon}\rangle_{\mathbf{x}}(\mathbf{x}) + \int\langle\langle\nabla\mathbf{G}\rangle\rangle_{\mathbf{x}}(\mathbf{x}-\mathbf{y})\mathbf{f}_1(\mathbf{y})\,d\mathbf{y}$$
$$+\int\langle\langle\mathbf{U}\rangle\rangle_{\mathbf{x}}(\mathbf{x}-\mathbf{y})\{\mathbf{L}_1(\mathbf{y})[\boldsymbol{\varepsilon}(\mathbf{y})-\boldsymbol{\beta}(\mathbf{y})]-\mathbf{L}^c\boldsymbol{\beta}_1(\mathbf{y})\}\,d\mathbf{y}. \quad (7.36)$$

where the volume integrals converge absolutely.

Expressing Eq. (7.36) in terms of stresses by the use of identities (7.11), the general equation

$$\boldsymbol{\sigma}(\mathbf{x}) = \langle\langle\boldsymbol{\sigma}^0\rangle\rangle_{\mathbf{x}}(\mathbf{x})+\langle\boldsymbol{\sigma}\rangle_{\mathbf{x}}(\mathbf{x})+\int\mathbf{L}^c\langle\langle\nabla\mathbf{G}\rangle\rangle_{\mathbf{x}}(\mathbf{x}-\mathbf{y})\mathbf{f}_1(\mathbf{y})d\mathbf{y}+\int\langle\langle\boldsymbol{\Gamma}\rangle\rangle_{\mathbf{x}}(\mathbf{x}-\mathbf{y})\boldsymbol{\eta}(\mathbf{y})d\mathbf{y} \quad (7.37)$$

is obtained, where the "strain polarization" tensor $\boldsymbol{\eta}$ was utilized.

Evidently, the right-hand-side volume integrals in Eqs. (7.36) and (7.37) converge absolutely, with no restrictions being imposed on the microtopology of the lattice Λ, and Eqs. (7.36) and (7.37) are valid for any deterministic (even nonperiodic) structures. The principal advantages of Eqs. (7.36) and (7.37) as compared with Eq. (7.33) are the lack of the surface integral in Eqs. (7.36) and (7.37), and the local character of Eqs. (7.36) and (7.37). The last-mentioned advantage makes it possible to reduce the analysis of infinite number inclusion problems to the analysis of a finite number of inclusions located in some representative volume element (RVE) [133], [134].

In the interest of obtaining simpler relationships we will consider the case of "slowly varying" fields $\boldsymbol{\varepsilon}^0(\mathbf{x})$ and $\boldsymbol{\sigma}^0(\mathbf{x})$. Then, expanding $\boldsymbol{\varepsilon}^0(\mathbf{z})$ in a Taylor series about $\mathbf{x}$ gives, by analogy with Eq. (7.35), that $\langle\langle\boldsymbol{\varepsilon}^0\rangle\rangle_{\mathbf{x}}(\mathbf{x})$ is of order $O(a_{\mathcal{V}}^2\nabla\nabla\boldsymbol{\varepsilon}^0(\mathbf{x}))$. Therefore, in the interest of obtaining explicit final expressions, we can neglect the term $\langle\langle\boldsymbol{\varepsilon}^0\rangle\rangle_{\mathbf{x}}(\mathbf{x})$ (7.33) as compared with $\langle\boldsymbol{\varepsilon}\rangle_{\mathbf{x}}(\mathbf{x})$ in the "slowly varying" approximation of $\varepsilon^0(\mathbf{x})$. The same simplification is admissible with respect to the term $\langle\langle \boldsymbol{\sigma}^0 \rangle\rangle_{\mathbf{x}}(\mathbf{x})$ of Eq. (7.37). Then Eqs. (7.36) and (7.37) are reduced to the approximate relationships

$$\varepsilon(\mathbf{x}) = \langle \varepsilon \rangle_{\mathbf{x}}(\mathbf{x}) + \int \langle\langle \nabla \mathbf{G} \rangle\rangle_{\mathbf{x}}(\mathbf{x}-\mathbf{y})\mathbf{f}_1(\mathbf{y})\, d\mathbf{y}$$

$$= \int \langle\langle \mathbf{U} \rangle\rangle_{\mathbf{x}}(\mathbf{x}-\mathbf{y})\{\mathbf{L}_1(\mathbf{y})[\varepsilon(\mathbf{y}) - \boldsymbol{\beta}(\mathbf{y})] - \mathbf{L}^c\boldsymbol{\beta}_1(\mathbf{y})\}\, d\mathbf{y}, \tag{7.38}$$

$$\boldsymbol{\sigma}(\mathbf{x}) = \langle \boldsymbol{\sigma} \rangle_{\mathbf{x}}(\mathbf{x}) + \int \mathbf{L}^c \langle\langle \nabla \mathbf{G} \rangle\rangle_{\mathbf{x}}(\mathbf{x}-\mathbf{y})\mathbf{f}_1(\mathbf{y})\, d\mathbf{y} + \int \langle\langle \boldsymbol{\Gamma} \rangle\rangle_{\mathbf{x}}(\mathbf{x}-\mathbf{y})\boldsymbol{\eta}(\mathbf{y})\, d\mathbf{y} \tag{7.39}$$

which are exact for linear functions $\varepsilon^0(\mathbf{x})$. Buryachenko [133], [134] has obtained the particular cases of Eq. (7.38) (triply periodical structures with $\mathbf{f},\ \boldsymbol{\beta} \equiv \mathbf{0}$) and Eq. (7.39) (doubly periodical structures $\mathbf{f} \equiv \mathbf{0}$), respectively.

By way of subsequent simplification, we now consider the case of a triply periodic structure (3.1) under the uniform boundary conditions (7.3). Then, the terms in curly brackets in the right-hand-side integrals of Eqs. (7.36) and (7.37) are invariant with respect to the unit cell number and Eqs. (7.36) and (7.37) can be rewritten in the form

$$\varepsilon(\mathbf{x}) = \varepsilon^{\Omega}(\mathbf{x}) + \int \nabla \mathbf{G}(\mathbf{x}-\mathbf{y})\big[\mathbf{f}_1(\mathbf{y}) - \mathbf{f}_1^{\Omega}(\mathbf{x})\big]\, d\mathbf{y} + \int \mathbf{U}(\mathbf{x}-\mathbf{y})$$

$$\cdot\Big\{\big[\mathbf{L}_1(\varepsilon - \boldsymbol{\beta}) - \mathbf{L}^c\boldsymbol{\beta}_1\big](\mathbf{y}) - \big[\mathbf{L}_1(\varepsilon - \boldsymbol{\beta}) - \mathbf{L}^c\boldsymbol{\beta}_1\big]^{\Omega}(\mathbf{x})\Big\}\, d\mathbf{y}, \tag{7.40}$$

$$\boldsymbol{\sigma}(\mathbf{x}) = \boldsymbol{\sigma}^{\Omega}(\mathbf{x}) + \int \mathbf{L}^c \nabla \mathbf{G}(\mathbf{x}-\mathbf{y})\big[\mathbf{f}_1(\mathbf{y}) - \mathbf{f}_1^{\Omega}(\mathbf{x})\big]\, d\mathbf{y}$$

$$+ \int \boldsymbol{\Gamma}(\mathbf{x}-\mathbf{y})\big\{\boldsymbol{\eta}(\mathbf{y}) - \boldsymbol{\eta}^{\Omega}(\mathbf{x})\big\}\, d\mathbf{y}. \tag{7.41}$$

In the pure elastic case with $\mathbf{M}_1^{(0)}(\mathbf{x}), \mathbf{f}(\mathbf{x}), \boldsymbol{\beta}(\mathbf{x}) \equiv \mathbf{0}$ $(\mathbf{x} \in w)$, exact Eq. (3.11) was previously used in [174] for both $\langle \mathbf{L}_1\varepsilon \rangle_{\mathbf{x}}(\mathbf{x}) = (\mathbf{L}_1\varepsilon)^{\Omega}(\mathbf{x}) \equiv$ const. and $\langle \varepsilon \rangle_{\mathbf{x}}(\mathbf{x}) = \varepsilon^{\Omega}(\mathbf{x}) \equiv$ const. when in Eq. (3.12) $\langle \boldsymbol{\eta} \rangle_{\mathbf{x}}(\mathbf{x}) = \boldsymbol{\eta}^{\Omega}(\mathbf{x}) \equiv$ const. and $\langle \boldsymbol{\sigma} \rangle_{\mathbf{x}}(\mathbf{x}) = \boldsymbol{\sigma}^{\Omega}(\mathbf{x}) \equiv$ const. as well.

It should be emphasized that for the field X bounded in one direction (for example, for the field X (3.2)), Eqs. (7.22) and (7.23) are valid as well as for the deterministic field X. Note that for the elastic analysis ($\mathbf{M}_1^{(0)}(\mathbf{x}),\ \mathbf{f}(\mathbf{x}), \boldsymbol{\beta}(\mathbf{x}) \equiv \mathbf{0}$) of triply periodic structures Fassi Fehri *et al.* [327] used Eq. (7.22) (for $\boldsymbol{\beta} \equiv \mathbf{0}$) without any regularization of the right-hand-side integral in Eq. (7.22) which diverges at infinity for the case of triply periodic composites.

7.4 Random Structure Composites with Long-Range Order

Localized Eqs. (7.9) and (7.12) were obtained in the framework of no long-range order assumption when the integrand in curly brackets decays at infinity $|\mathbf{x} - \mathbf{s}| \to \infty$ sufficiently rapidly. Now, we relax this assumption and, for the sake of definiteness, we will consider some conditional averages of the surface integral of Eq. (7.12)

$$\langle \mathcal{L}^{\Gamma\sigma} | v_1, \mathbf{x}_1; .; v_n, \mathbf{x}_n \rangle(\mathbf{x}) = \int_\Gamma \mathbf{\Gamma}^\Gamma(\mathbf{x}-\mathbf{s}) \{ \langle \boldsymbol{\eta}^1 |; v_1, \mathbf{x}_1; .; v_n, \mathbf{x}_n \rangle(\mathbf{s}) - \langle \boldsymbol{\eta}^1 \rangle(\mathbf{s}) \} \, d\mathbf{s}, \tag{7.42}$$

where $\mathbf{x} \in v_1, \ldots, v_n$, $(n = 1, 2, \ldots)$, $\mathbf{x} \notin \Gamma$ and $\mathbf{\Gamma}^\Gamma(\mathbf{x}-\mathbf{s}) = -\mathbf{L}^c \nabla \mathbf{G}(\mathbf{x}-\mathbf{s}) \mathbf{L}^c$. The asymptotic behavior of the integrand in curly brackets in Eq. (7.42) as $|\mathbf{x}-\mathbf{s}| \to \infty$ can be estimated by the representation of the solution $\langle \boldsymbol{\eta}^1 |; v_1, \mathbf{x}_1; \ldots; v_n, \mathbf{x}_n \rangle(\mathbf{s})$ by the successive approximation method (see [156] and Chapter 9). Then,

$$\begin{aligned} &\langle \boldsymbol{\eta}^1 |; v_1, \mathbf{x}_1; \ldots; v_n, \mathbf{x}_n \rangle(\mathbf{s}) - \langle \boldsymbol{\eta}^1 \rangle(\mathbf{s}) \\ &\quad \to \langle \boldsymbol{\eta}^1 \rangle_i(\mathbf{s}) [\varphi(v_i, \mathbf{x}_i | v_1, \mathbf{x}_1, \ldots, v_n, \mathbf{x}_n) - \varphi(v_i, \mathbf{x}_i)] \\ &\quad - O(r^{-d}) \sum_{j=1}^n \langle \boldsymbol{\eta}^1 \rangle_j(\mathbf{x}_j) \varphi(v_i, \mathbf{x}_i | v_1, \mathbf{x}_1, \ldots, v_n, \mathbf{x}_n), \end{aligned} \tag{7.43}$$

where $\mathbf{s} \in v_i$, $r = \min |\mathbf{x}_j - \mathbf{s}|$, $(j = 1, \ldots, n)$ and the terms in Eq. (7.43) of order $O(1/r^{-2d})$ and higher order terms are dropped. The contribution of terms in (7.43) proportional to $O(r^{-d})$ into the integral (7.42) vanishes at $|\mathbf{x}_j - \mathbf{s}| \to \infty$, and Eq. (7.42) can be simplified:

$$\begin{aligned} \langle \mathcal{L}^{\Gamma\sigma} | v_1, \mathbf{x}_1; \ldots; v_n, \mathbf{x}_n \rangle(\mathbf{x}) = \int_\Gamma & \mathbf{\Gamma}^\Gamma(\mathbf{x}-\mathbf{s}) \langle \boldsymbol{\eta}^1 \rangle_i(\mathbf{s}) [\varphi(v_i, \mathbf{x}_i | v_1, \mathbf{x}_1, \ldots, v_n, \mathbf{x}_n) \\ & - \varphi(v_i, \mathbf{x}_i)] \, d\mathbf{s}. \end{aligned} \tag{7.44}$$

If boundary conditions are applied for which $\langle \boldsymbol{\sigma} \rangle(\mathbf{s})$ and, therefore, $\langle \boldsymbol{\eta}^1 \rangle_j(\mathbf{s})$ vary linearly (or higher) with $\mathbf{s}$ and $[\varphi(v_i, \mathbf{x}_i | v_1, \mathbf{x}_1, \ldots, v_n, \mathbf{x}_n) - \varphi(v_i, \mathbf{x}_i)]$ does not decay sufficiently rapidly as $|\mathbf{x}_j - \mathbf{s}| \to \infty$ $(j = 1, \ldots, n)$ (long-range order), then the integral (7.44) may be divergent. We will consider the interesting practical case of a random structure composite described as either triply or doubly periodical in the broad sense random field X, when the composite material is constructed using the building blocks or cells $w = \cup \Omega_{\mathbf{m}}$ forming either triply or doubly periodical lattice of cell centers Λ (see Section 7.4) with a random transmission of Λ and a random distribution of inclusions in the cells.

The surface integral (7.44) can be eliminated in the equation related to (7.12) by "centrification" achieved by subtracting from both sides of Eq. (7.12) their averages over the moving averaging cell $\mathcal{V}_{\mathbf{x}}$ (3.3$_2$). In so doing, the average operator (3.3$_2$) introduced for a deterministic function $\mathbf{g}(\mathbf{y})$ should be recast for random function $\mathbf{g}(\mathbf{y})$ by the use of a previous estimation of a statistical average $\langle \mathbf{g} \rangle(\mathbf{z}-\mathbf{y})$: $\langle \mathbf{g} \rangle_{\mathbf{x}}(\mathbf{x}-\mathbf{y}) = (\overline{\mathcal{V}}_{\mathbf{x}})^{-1} \int_{\mathcal{V}_{\mathbf{x}}} \langle \mathbf{g} \rangle(\mathbf{z}-\mathbf{y}) \, d\mathbf{z}, \quad \mathbf{x} \in \Omega_i$. Then Eq. (7.12) is reduced to

$$\begin{aligned} \boldsymbol{\sigma}(\mathbf{x}) = \langle \boldsymbol{\sigma} \rangle(\mathbf{x}) &+ \int_w \{ \langle\langle \mathbf{\Gamma}(\mathbf{x}-\mathbf{y}) \boldsymbol{\eta} \rangle\rangle_{\mathbf{x}}(\mathbf{y}) - \langle\langle \mathbf{\Gamma} \rangle\rangle_{\mathbf{x}}(\mathbf{x}-\mathbf{y}) \langle \boldsymbol{\eta} \rangle(\mathbf{y}) \} \, d\mathbf{y} \\ &+ \int_w \mathbf{L}^c \langle\langle \nabla \mathbf{G} \rangle\rangle_{\mathbf{x}}(\mathbf{x}-\mathbf{y}) \big[\mathbf{f}_1(\mathbf{y}) - \langle \mathbf{f}_1 \rangle(\mathbf{y}) \big] \, d\mathbf{y} + \langle\langle \mathcal{I}^{\Gamma\sigma} \rangle\rangle_{\mathbf{x}}, \end{aligned} \tag{7.45}$$

where

$$\langle\langle \mathcal{I}^{\Gamma\sigma} \rangle\rangle_{\mathbf{x}} = \int_\Gamma \langle\langle \mathbf{\Gamma}^\Gamma \rangle\rangle_{\mathbf{x}}(\mathbf{x}-\mathbf{s}) \langle \boldsymbol{\eta}^1 \rangle_i(\mathbf{s}) [\varphi(v_i, \mathbf{x}_i | v_1, \mathbf{x}_1, \ldots, v_n, \mathbf{x}_n) - \varphi(v_i, \mathbf{x}_i)] \, d\mathbf{s}.$$

is a centered surface integral with the kernel $\langle\!\langle \nabla \mathbf{G} \rangle\!\rangle_{\mathbf{x}}(\mathbf{x}-\mathbf{s})$ of order $O(a_{\mathcal{V}}^2|\mathbf{x}-\mathbf{y}|^{-4})$, and the surface integral in the expression $\langle\!\langle \boldsymbol{\mathcal{I}}^{\Gamma\sigma} \rangle\!\rangle_{\mathbf{x}}$ vanishes at $|\mathbf{x}-\mathbf{s}| \to \infty$, $\mathbf{s} \in \Gamma$ if $\langle \boldsymbol{\sigma} \rangle(\mathbf{s})$ and, therefore, $\langle \boldsymbol{\eta}^1 \rangle_j(\mathbf{s})$ grows with $\mathbf{s}$ slower then $O(|\mathbf{x}-\mathbf{s}|^{2-\alpha})$ ($\alpha = \text{const.} > 0$). By this means, the locality principle exists in Eq. (7.45) for the case of long-range order composites being considered if the average stress $\langle \boldsymbol{\sigma} \rangle(\mathbf{s})$ grows with $\mathbf{s}$ slower than $O(|\mathbf{x}-\mathbf{s}|^{2-\alpha})$ ($\alpha = \text{const.} > 0$).

7.5 Triply Periodic Particulate Matrix Composites with Imperfect Unit Cells

We first consider a composite medium with periodic set X of ellipsoidal inclusions with identical shape, orientation, and mechanical properties with particle centers, periodically distributed at the nodes of the same spatial lattice Λ. Suppose $\mathbf{e}_i$ ($i = 1, 2, 3$) are linearly independent vectors of the principal period of Λ, which determine a unit cell Ω of volume $\overline{\Omega} = |\mathbf{e}_1 \cdot (\mathbf{e}_2 \otimes \mathbf{e}_3)|$, so that we can represent any node $\mathbf{m} \in \Lambda$ in the form (3.1).

Now we will consider a *substitutional* sort of the disordered imperfection of the unit cells which connotes a variability in properties per vertex [844]. In particular, this sort of imperfection describes the effect of randomly missing inclusions, when in some randomly distributed cells, the inclusions are absent and replaced by the matrix. In another sort of imperfection, some inclusions are randomly replaced by one or several kinds of inclusions that, perhaps, are the debonding inclusions in the initially periodic structure. In the simplified modeling the completely debonded inclusions are replaced by the voids of the inclusion size [269]. For a description of the random structure of a composite material, let us introduce a probability density $\varphi(v_i, \mathbf{x}_i)$ of location of the inclusion v_i with the center in the point $\mathbf{x}_i$, and a conditional probability density $\varphi(v_j, \mathbf{x}_j | v_i, \mathbf{x}_i)$, which is a probability density to find the jth inclusion with the center $\mathbf{x}_j$ in the domain v_j with fixed inclusion v_i with the centers $\mathbf{x}_i$. The notation $\varphi(v_j, \mathbf{x}_j|; v_i, \mathbf{x}_i)$ denotes the case $\mathbf{x}_j \neq \mathbf{x}_i$. Then

$$\varphi(v_i, \mathbf{x}_i) = \sum_{\mathbf{m}} \lambda^{(k)}(\mathbf{x}_{\mathbf{m}})\delta(\mathbf{x}_i - \mathbf{x}_{\mathbf{m}}), \quad \varphi(v_j, \mathbf{x}_j|; v_i, \mathbf{x}_i) = {\sum_{\mathbf{m}}}' \lambda^{(l)}(\mathbf{x}_{\mathbf{m}})\delta(\mathbf{x}_j - \mathbf{x}_{\mathbf{m}}), \tag{7.46}$$

where $'$ means that the summation symbol does not contain the argument $\mathbf{x}_{\mathbf{m}} = \mathbf{x}_i$, and $v_i \in v^{(k)}$, $v_j \in v^{(l)}$ ($i, j = 1, 2, \ldots,\ k, l = 1, 2, \ldots, N$). The statistical averages of random variables $\lambda(\mathbf{x}_{\mathbf{m}})$: $0 \leq \langle \lambda^{(k)}(\mathbf{x}_{\mathbf{m}}) \rangle \equiv \lambda^{(k)} \leq 1$ and $0 \leq \sum_k \lambda^{(k)} \leq 1$ reflect the measure of the imperfection: $\lambda^{(k)} = 0$ means the absence in the lattice Λ of inclusions of the component k; the case $\lambda^{(k)} = 1$ stands for the perfect periodic structure filled by identical inclusions from the component k. In so doing the volume concentration of the k component $c^{(k)}(\lambda^{(k)}) = \lambda^{(k)}\overline{v}^{(k)}/\overline{\Omega}_{\mathbf{m}}$ which equals $c^{(k)} \equiv c^{(k)}(1)$ for perfect packing.

In Sections 7.1 through 7.4 we decomposed the elastic modulus on two summands $\mathbf{L}(\mathbf{x}) = \mathbf{L}^c + (\mathbf{L}(\mathbf{x}) - \mathbf{L}^c)$. The main goal of this decomposition is a subsequent decomposition of a solution found on two summands. One of them specific to $\mathbf{L}^c$ is known, and, moreover, it is a constant for the homogeneous boundary conditions either (7.2) or (7.3) being considered in this section. The second summand corresponding to the perturbation $(\mathbf{L}(\mathbf{x}) - \mathbf{L}^c)$ is estimated by the use of a Green function for the known first problem. The main goal of this section considered as a generalization of the scheme mentioned above is a decomposition of the desired solution on the known solution for the perfect periodic structure and on the perturbation produced by the imperfections in the perfect periodic structure. In an analogy with the notations of Section 7.1, we represent the tensor of material properties $\mathbf{f}$ ($\mathbf{f} = \mathbf{L}, \mathbf{M}$) as a decomposition $\mathbf{f}(\mathbf{x}) = \mathbf{f}^c + \mathbf{f}_2(\mathbf{x}) + \mathbf{f}_3(\mathbf{x})$, where $\mathbf{f}(\mathbf{x}) = \mathbf{f}^c$ for $\mathbf{x} \in v^{(0)}$, $\mathbf{f}_2(\mathbf{x}) \equiv \mathbf{f}^{\text{per}}(\mathbf{x}) - \mathbf{f}^c$ is a jump of elastic properties of the perfect imaginary periodic structure with respect to the matrix, and $\mathbf{f}_3(\mathbf{x}) \equiv \mathbf{f}(\mathbf{x}) - \mathbf{f}^{\text{per}}(\mathbf{x})$ is a jump of the real elastic properties with respect to the imaginary perfect periodic structure; in the case $\mathbf{f}^c = \mathbf{f}^{(0)} \equiv \text{const.}$, obviously, we have $\mathbf{f}_2(\mathbf{x}) = \mathbf{f}_3(\mathbf{x}) \equiv \mathbf{0}$ at $\mathbf{x} \in v^{(0)}$. Usually for the homogeneous matrix $\mathbf{L}^c = \mathbf{L}^{(0)}$; otherwise, it is reasonable to consider the elastic properties of the comparison medium as the Voight (or Reuss) average of elastic properties of the matrix $\mathbf{L}^c = [v^{(0)}]^{-1}\langle \mathbf{L}(\mathbf{x}) V^{(0)}(\mathbf{x})\rangle$. The comparison medium with the tensor $\mathbf{f}^c$ will be called by a comparison medium of the first level. In a similar manner, the medium with the material tensor $\mathbf{f}^{\text{per}}(\mathbf{x})$ will be called by a comparison medium of the second level.

Then Eq. (7.5) can be recast in the form $\nabla \mathbf{L}\boldsymbol{\varepsilon}(\mathbf{x}) = -\nabla[\mathbf{L}_2(\mathbf{x})\boldsymbol{\varepsilon}(\mathbf{x})] - \nabla[\mathbf{L}_3(\mathbf{x})\boldsymbol{\varepsilon}(\mathbf{x})]$ (at $\boldsymbol{\beta}(\mathbf{x}), \mathbf{f}(\mathbf{x}) \equiv \mathbf{0}$), leading to the representations equivalent to Eq. (7.21_1):

$$\boldsymbol{\varepsilon}(\mathbf{x}) = \langle \boldsymbol{\varepsilon}\rangle + \int \mathbf{U}(\mathbf{x}-\mathbf{y})[\boldsymbol{\eta}_2(\mathbf{y}) - \langle \boldsymbol{\eta}_2\rangle]\, d\mathbf{y} + \int \mathbf{U}(\mathbf{x}-\mathbf{y})[\boldsymbol{\eta}_3(\mathbf{y}) - \langle \boldsymbol{\eta}_3\rangle]\, d\mathbf{y}, \quad (7.47)$$

where one has introduced two other strain polarization tensors $\boldsymbol{\eta}_2(\mathbf{y}) = \mathbf{L}_2(\mathbf{y})\boldsymbol{\varepsilon}(\mathbf{y})$ and $\boldsymbol{\eta}_3(\mathbf{y}) = \mathbf{L}_3(\mathbf{y})\boldsymbol{\varepsilon}(\mathbf{y})$. Multiplication of Eq. (7.47) over the periodic tensor $\mathbf{L}_2(\mathbf{x})$ leads to the following equation represented in the operator form just for the shortening of writings $\boldsymbol{\eta}_2 = \mathbf{L}_2\langle \boldsymbol{\varepsilon}\rangle + \mathbf{L}_2\mathbf{U} * \boldsymbol{\eta}_2 + \mathbf{L}_2\mathbf{U} * \boldsymbol{\eta}_3$, where $\mathbf{U} * \mathbf{f} \equiv \int \mathbf{U}(\mathbf{x}-\mathbf{y})[\mathbf{f}(\mathbf{y}) - \langle \mathbf{f}\rangle]\, d\mathbf{y}$. Following our goal of extraction of the known solution, we introduce the operator $\boldsymbol{\mathcal{A}} \equiv (\mathbf{I} - \mathbf{U}\mathbf{L}_2)^{-1}$, $\boldsymbol{\mathcal{A}} * \mathbf{f} \equiv \int \boldsymbol{\mathcal{A}}(\mathbf{x}-\mathbf{y})\mathbf{f}(\mathbf{y})\, d\mathbf{y}$, reducing to the multiplication over the inhomogeneous tensor $\mathbf{A}^{\text{per}}(\mathbf{x})$ describing the known periodic solution in the case of the action of this operator on the constant tensor $\langle \boldsymbol{\varepsilon}\rangle$:

$$\boldsymbol{\mathcal{A}} * \langle \boldsymbol{\varepsilon}\rangle = \mathbf{A}^{\text{per}}(\mathbf{x})\langle \boldsymbol{\varepsilon}\rangle. \quad (7.48)$$

Introduction of the operator $\boldsymbol{\mathcal{A}}$ allows one to decompose the desired solution in two parts:

$$\boldsymbol{\varepsilon}(\mathbf{x}) = \mathbf{A}^{\text{per}}(\mathbf{x})\langle \boldsymbol{\varepsilon}\rangle + \boldsymbol{\mathcal{A}} * \mathbf{U} * \boldsymbol{\eta}_3, \quad (7.49)$$

where the first tensorial item in the right-hand side presents the periodic solution and is assumed to be known. The second operator item is produced by the random imperfections contributed in the perfect periodic structure and depend also on the periodic solution described by the operator $\boldsymbol{\mathcal{A}}$.

7.6 Conclusion

A generalization of the centering method is proposed for obtaining general integral equations of thermoplasticity for random structure composites with and without long-range order as well as for composites of deterministic structure subjected to essentially inhomogeneous loading by the fields of the stresses, temperature, and body forces. In the limiting case of the statistically homogeneous and periodical media subjected to homogeneous boundary conditions the equations proposed are reduced to known ones. New equations (7.9), (7.12), (7.36), (7.37), and (7.45) are the theoretical basis of the *locality principle* by Sokolkin and Tashkinov [1027], which uses the short-range-order effect in the interactions of the periodic or random problem in the boundary value problem for a domain with a finite number of inhomogeneities. This principle was based on intuitive physical considerations and described for concrete numerical examples. The locality principle will be justified in Chapter 12 for the particular case of simple cubic packing of spherical inclusions. It will be shown that the maximum error caused by using the representative volume (in an excess of three times the distance between the inclusions) is smaller than 2 percent.

The fundamental advantage of equations obtained is a finite size of a region including the inclusions acting on a separate inclusion being analyzed, which determines an essence of the *locality principle.* The locality principle is closely joined with a fundamental intuitive concept of micromechanics such as a representative volume element (RVE) according to which the overall parameters in a point considered are defined by the surrounding inclusions located inside RVE. It should be mentioned, however, that the size of RVE is not a universal parameter of an actual composite being analyzed but depends on both the kind of overall parameter estimated and the method used for the such estimation. For example, the size of RVE used for estimation of effective elastic moduli in two and even more time less than the size of RVE utilized for the estimation of an effective nonlocal differential operator of the second order (see for details Chapter 12). However, more detailed consideration of methods of the solution of Eqs. (7.9), (7.12), (7.36), (7.37), and (7.45) obtained will be done in Chapters 8–12.

8

Multiparticle Effective Field and Related Methods in Micromechanics of Random Structure Composites

The prediction of the behavior composite materials by using of mechanical properties of constituents and their microstructure is a central problem of micromechanics that is eventually reduced to the estimation of stress fields in constituents. The numerous methods in micromechanics can be classified into four broad categories: perturbation methods, self-consistent methods of truncation of a hierarchy, variational methods, and model methods (for details see [1187]). In most details we will consider the self-consistent methods based on some mathematical approximations for solving the infinite systems of integral equations involved. A considerable number of methods are known in the linear theory of composites that yield the effective elastic constants and stress field averages in the components. Appropriate, but by no means exhaustive, references are provided by the reviews [128], [138], [174], [237], [539], [545], [564], [602], [734], [775], [794], [806], [995], [1106], [1184], [1187]. It appears today that variants of the effective medium method [463], [606], and the mean field method [52], [787] are the most popular and widely used methods. The notion of an effective field in which each particle is located is a basic concept of such powerful methods in micromechanics as the methods of self-consistent fields and effective fields (for references see [545], [788]). The concept of the effective field in combination with subsequent assumptions originated in the physics of multiple scattering of waves (see, e.g., [202], [348], [648]) was intensively applied in micromechanics of random and periodic structure composites (for references see, e.g., [139] [539], [545]) as well in micromechanics of multiple interacting cracks under the name traction (for references see, e.g., [529]) or pseudo-load [470]. The "quasi-crystalline" approximation of Lax [649] is often used for truncation of the hierarchy of integral equations involved in leading to neglect of direct multiparticle interactions of inclusions. The last deficiency was overcome recently by the multiparticle effective field method (MEFM), put forward and developed by the author (references may be found in the survey by Buryachenko [138]). The MEFM is based on the theory of functions of random variables and Green's functions. Within this method a hierarchy of statistical moment equations for conditional averages of the stresses in the inclusions is derived. The hierarchy is established by introducing the notion of an effective field. In this way the interaction of different inclusions is taken directly into account.

The outline of this charter is as follows. In Section 8.1 we present the main hypotheses of the proposed MEFM for the approximate solution of a generalization of the infinite system of integral equations under arbitrary choice of comparison medium. In Section 8.2 the MEFM is proposed for the estimation of effective properties (such as compliance, thermal expansion, stored energy) and the first statistical moments of stresses in the components under arbitrary choice of comparison medium. In Section 8.3 one considers in detail the connection of the method proposed with the mean field method. Explicit analytical representations are presented for composites with spheroidal inclusions.

8.1 Definitions of Effective Fields and Effective Field Hypotheses

8.1.1 Effective Fields

From equations (6.1)–(6.3) the general integral equations (7.14) for $\boldsymbol{\varepsilon}$ and $\boldsymbol{\sigma}$ were derived which for no *long-range order* and no body force can be presented in the form

$$\boldsymbol{\varepsilon}(\mathbf{x}) = \langle\boldsymbol{\varepsilon}\rangle(\mathbf{x}) + \int \mathbf{U}(\mathbf{x}-\mathbf{y})\Big\{\mathbf{L}_1(\mathbf{y})[\boldsymbol{\varepsilon}(\mathbf{y}) - \boldsymbol{\beta}(\mathbf{y})] - \mathbf{L}^c\boldsymbol{\beta}_1(\mathbf{y})$$
$$- [\langle\mathbf{L}_1(\boldsymbol{\varepsilon}-\boldsymbol{\beta})\rangle(\mathbf{y}) - \langle\mathbf{L}^c\boldsymbol{\beta}_1\rangle(\mathbf{y})]\Big\}\, d\mathbf{y}, \tag{8.1}$$

$$\boldsymbol{\sigma}(\mathbf{x}) = \langle\boldsymbol{\sigma}\rangle(\mathbf{x}) + \int \boldsymbol{\Gamma}(\mathbf{x}-\mathbf{y})\,\{\boldsymbol{\eta}(\mathbf{y}) - \langle\boldsymbol{\eta}\rangle(\mathbf{y})\}\; d\mathbf{y}, \tag{8.2}$$

where the tensor $\boldsymbol{\eta}(\mathbf{y}) = \mathbf{M}_1(\mathbf{y})\boldsymbol{\sigma}(\mathbf{y}) + \boldsymbol{\beta}_1(\mathbf{y})$ is called the strain polarization tensor and is simply a notational convenience. The general integral equations (8.1) and (8.2) are valid at sufficient distance $\mathbf{x}$ from the boundary ∂w for statistically homogeneous and inhomogeneous media for both the general case of the first and second boundary value problems as well as for the mixed boundary-value problem in the space dimensionality d ($d = 2, 3$). For the sake of definiteness, in 2-D case we will consider a plane-strain problem (see also the footnote at the page 101). We will consider the statistical description presented in Section 5.3.3 (and used in Chapter 7) in terms of the conditional probability densities $\varphi(v_i, \mathbf{x}_i | v_1, \mathbf{x}_1, \ldots, v_n, \mathbf{x}_n)$. The composites being analyzed have statistically homogeneous structure (5.20) and are subjected to the homogeneous boundary conditions either (6.4) or (6.5) so that $\langle\boldsymbol{\sigma}\rangle(\mathbf{x})$, $\langle\boldsymbol{\varepsilon}\rangle(\mathbf{x}) \equiv$ const.

Equations (8.1) and (8.2) are much easier to solve when the stress-strain fields are studied only inside the inclusions. There are two fundamentally different assumptions (7.16) and (7.17) to ensuring the integration on the right-hand sides of Eqs. (8.1) and (8.2) extends only over the volume of the inclusion. We will utilize more general assumption (7.17) providing a transformation of Eq. (8.2) to the form

$$\boldsymbol{\sigma}(\mathbf{x}) = \langle\boldsymbol{\sigma}\rangle(\mathbf{x}) + \int \boldsymbol{\Gamma}(\mathbf{x}-\mathbf{y})\,\{\boldsymbol{\vartheta}(\mathbf{y}) - \langle\boldsymbol{\vartheta}\rangle(\mathbf{y})\}\; d\mathbf{y}, \tag{8.3}$$

ensuring that the integration in the right-hand side of Eq. (8.3) is only taken within the volume of inclusions. Here $\boldsymbol{\vartheta} \equiv \boldsymbol{\eta}^1$ is called the modified strain polarization tensor and one introduced the operation $\boldsymbol{\lambda}^1(\mathbf{x}) \equiv \boldsymbol{\lambda}(\mathbf{x}) - \langle\boldsymbol{\lambda}\rangle^{(0)}$ for the random function $\boldsymbol{\lambda}$ (e.g. $\boldsymbol{\lambda} = \boldsymbol{\sigma}, \boldsymbol{\varepsilon}, \boldsymbol{\eta}$) with statistical average in the matrix $\langle\boldsymbol{\lambda}\rangle^{(0)} \equiv \text{const}$.

Let the inclusions $v_1, \ldots, v_n$ be fixed and we define two sorts of effective fields $\overline{\boldsymbol{\sigma}}_i(\mathbf{x})$ and $\widetilde{\boldsymbol{\sigma}}_{1,\ldots,n}(\mathbf{x})$ ($i = 1, \ldots, n$; $\mathbf{x} \in v_1, \ldots, v_n$) by the use of the rearrangement of Eq. (8.3) in the following form (see the earliest references on related manipulations [151], [182], [184], [545]):

$$\boldsymbol{\sigma}(\mathbf{x}) = \overline{\boldsymbol{\sigma}}_i(\mathbf{x}) + \int \boldsymbol{\Gamma}(\mathbf{x}-\mathbf{y})V_i(\mathbf{y})\boldsymbol{\vartheta}(\mathbf{y})\, d\mathbf{y}, \tag{8.4}$$

$$\overline{\boldsymbol{\sigma}}_i(\mathbf{x}) = \widetilde{\boldsymbol{\sigma}}_{1,\ldots,n}(\mathbf{x}) + \sum_{j\neq i} \int \boldsymbol{\Gamma}(\mathbf{x}-\mathbf{y})V_j(\mathbf{y})\boldsymbol{\vartheta}(\mathbf{y})\, d\mathbf{y}, \tag{8.5}$$

$$\widetilde{\boldsymbol{\sigma}}_{1,\ldots,n}(\mathbf{x}) = \langle\boldsymbol{\sigma}\rangle(\mathbf{x}) + \int \boldsymbol{\Gamma}(\mathbf{x}-\mathbf{y})\{\boldsymbol{\vartheta}(\mathbf{y})V(\mathbf{y}|; v_1, \mathbf{x}_1; \ldots; v_n, \mathbf{x}_n) - \langle\boldsymbol{\vartheta}\rangle(\mathbf{y})\}d\mathbf{y}, \tag{8.6}$$

for $\mathbf{x} \in v_i$, $i = 1, 2, \ldots, n$. Then, considering some conditional statistical averages of the general integral equation (8.3) leads to an infinite system of integral equations that can also be represented in terms of the strain polarization tensor ($n = 1, 2, \ldots$)

$$\begin{aligned}
&\langle\boldsymbol{\sigma}|v_1, \mathbf{x}_1; \ldots; v_n, \mathbf{x}_n\rangle(\mathbf{x}) - \sum_{i=1}^{n} \int \boldsymbol{\Gamma}(\mathbf{x}-\mathbf{y})\langle V_i(\mathbf{y})\boldsymbol{\vartheta}|v_1, \mathbf{x}_1; \ldots; v_n, \mathbf{x}_n\rangle(\mathbf{y})\, d\mathbf{y}\\
&= \langle\boldsymbol{\sigma}\rangle(\mathbf{x}) + \int \boldsymbol{\Gamma}(\mathbf{x}-\mathbf{y})\{\langle\boldsymbol{\vartheta}|; v_1, \mathbf{x}_1; \ldots; v_n, \mathbf{x}_n\rangle(\mathbf{y}) - \langle\boldsymbol{\vartheta}\rangle(\mathbf{y})\}\, d\mathbf{y},\\
&\langle\boldsymbol{\eta}|v_1, \mathbf{x}_1; \ldots; v_n, \mathbf{x}_n\rangle(\mathbf{x}) - \sum_{i=1}^{n} \int \mathbf{M}_1(\mathbf{x})\boldsymbol{\Gamma}(\mathbf{x}-\mathbf{y})\langle V_i(\mathbf{y})\boldsymbol{\vartheta}|v_1, \mathbf{x}_1; \ldots; v_n, \mathbf{x}_n\rangle(\mathbf{y})d\mathbf{y}\\
&= \langle\boldsymbol{\eta}\rangle(\mathbf{x}) + \int \mathbf{M}_1(\mathbf{x})\boldsymbol{\Gamma}(\mathbf{x}-\mathbf{y})\{\langle\boldsymbol{\vartheta}|; v_1, \mathbf{x}_1; \ldots; v_n, \mathbf{x}_n\rangle(\mathbf{y}) - \langle\boldsymbol{\vartheta}\rangle(\mathbf{y})\}\, d\mathbf{y}.
\end{aligned} \tag{8.7}$$

Since $\mathbf{x} \in v_1, \ldots, v_n$ in the nth line of the system can take the values in the inclusions $v_1, \ldots, v_n$, the nth line actually contains n equations.

The definitions of the effective fields $\overline{\boldsymbol{\sigma}}_i(\mathbf{x})$, $\widetilde{\boldsymbol{\sigma}}_{1,2,\ldots,n}(\mathbf{x})$ as well as their statistical averages $\langle\overline{\boldsymbol{\sigma}}_i\rangle(\mathbf{x})$, $\langle\widetilde{\boldsymbol{\sigma}}_{1,2,\ldots,n}\rangle(\mathbf{x})$ are nothing more than notational convenience for different terms of the infinite systems (8.4)–(8.6) and (8.7), respectively. The physical meaning of these fields is the following (graphic illustrations are presented in [182], see also Fig. 8.1). $\widetilde{\boldsymbol{\sigma}}_{1,2,\ldots,n}(\mathbf{x})$ is a stress field in which the chosen fixed inclusions $v_1, \ldots, v_n$ are embedded. This effective field is a random function of all the other positions of the surrounding inhomogeneities, and the average $\langle\widetilde{\boldsymbol{\sigma}}_{1,\ldots,n}\rangle(\mathbf{x})$ of $\widetilde{\boldsymbol{\sigma}}_{1,\ldots,n}(\mathbf{x})$ over a random realization of these inclusions is equal to the right-hand side of the nth line of the system (8.7). Consequently, each inclusion v_i ($i = 1, \ldots, n$) of the chosen fixed set is in a random (generally speaking, nonhomogeneous) field $\overline{\boldsymbol{\sigma}}_i(\mathbf{x})$, ($\mathbf{x} \in v_i$, $i \neq j$, $i, j = 1, 2, \ldots, n$) (8.5), which is the superposition of the effective field $\widetilde{\boldsymbol{\sigma}}_{1,\ldots,n}(\mathbf{x})$ and the distribution caused by the other inclusions v_j, ($j \neq i$, $j = 1, \ldots, n$) of the considered set.

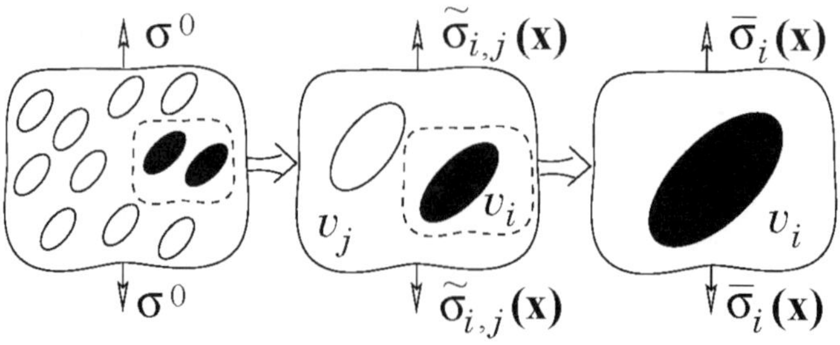

Fig. 8.1. Schematic representation of the effective fields.

In a similar manner, Eqs. (8.4)–(8.7) can be recast in the terms of strains

$$\boldsymbol{\varepsilon}(\mathbf{x}) = \overline{\boldsymbol{\varepsilon}}_i(\mathbf{x}) + \int \mathbf{U}(\mathbf{x}-\mathbf{y})V_i(\mathbf{y})\boldsymbol{\tau}(\mathbf{y})\, d\mathbf{y}, \tag{8.8}$$

$$\overline{\boldsymbol{\varepsilon}}_i(\mathbf{x}) = \widetilde{\boldsymbol{\varepsilon}}_{1,\ldots,n}(\mathbf{x}) + \sum_{j\neq i} \int \mathbf{U}(\mathbf{x}-\mathbf{y})V_j(\mathbf{y})\boldsymbol{\tau}(\mathbf{y})\, d\mathbf{y}, \tag{8.9}$$

$$\widetilde{\boldsymbol{\varepsilon}}_{1,\ldots,n}(\mathbf{x}) = \langle \boldsymbol{\varepsilon}\rangle(\mathbf{x}) + \int \mathbf{U}(\mathbf{x}-\mathbf{y})\big\{\boldsymbol{\tau}(\mathbf{y})V(\mathbf{y}|; v_1, \mathbf{x}_1; \ldots; v_n, \mathbf{x}_n) - \langle\boldsymbol{\tau}\rangle(\mathbf{y})\big\}\, d\mathbf{y}, \tag{8.10}$$

$$\langle \boldsymbol{\varepsilon}|v_1,\mathbf{x}_1;\ldots;v_n,\mathbf{x}_n\rangle(\mathbf{x}) - \sum_{i=1}^{n}\int \mathbf{U}(\mathbf{x}-\mathbf{y})\langle V_i(\mathbf{y})\boldsymbol{\tau}|v_1,\mathbf{x}_1;\ldots;v_n,\mathbf{x}_n\rangle(\mathbf{y})\, d\mathbf{y}$$
$$= \langle\boldsymbol{\varepsilon}\rangle(\mathbf{x}) + \int \mathbf{U}(\mathbf{x}-\mathbf{y})\big\{\langle\boldsymbol{\tau}|; v_1,\mathbf{x}_1;\ldots;v_n,\mathbf{x}_n\rangle(\mathbf{y}) - \langle\boldsymbol{\tau}\rangle(\mathbf{y})\big\}\, d\mathbf{y}. \tag{8.11}$$

for $\mathbf{x} \in v_i$, $i = 1, 2, \ldots, n$ in Eqs. (8.8)–(8.10) and $\mathbf{x} \in v_1, \ldots, v_n$ in Eq. (8.11), respectively. Here $\boldsymbol{\tau}(\mathbf{x}) = \mathbf{L}_1(\mathbf{x})\boldsymbol{\varepsilon}(\mathbf{y})$ is a stress polarization tensor for the pure elastic case $\boldsymbol{\beta}_1 \equiv \mathbf{0}$ being considered.

Both the effective compliance $\mathbf{M}^*$ (6.20) and the effective eigenstrains $\boldsymbol{\beta}^*$ (6.44) governing the overall constitutive relation (6.29) can be rewritten in an equivalent form

$$\begin{aligned}\mathbf{M}^* &= \mathbf{M}^{(0)} + \langle \mathbf{M}_1\mathbf{B}^*V\rangle - \mathbf{M}_1^{(0)}\langle \mathbf{B}^*V\rangle,\\ \boldsymbol{\beta}^* &= \boldsymbol{\beta}^{(0)} + \langle {\mathbf{B}^*}^{\top}\boldsymbol{\beta}_1 V\rangle - \langle {\mathbf{B}^*}^{\top}V\rangle\boldsymbol{\beta}_1^{(0)},\end{aligned} \tag{8.12}$$

where $\mathbf{B}^* = \mathbf{B}^*(\mathbf{x})$ is a local stress concentration tensor obtained under pure mechanical loading (6.18). One uses the decomposition of material properties $\mathbf{g}(\mathbf{x}) = \mathbf{g}^c + \mathbf{g}_1(\mathbf{x}) = \mathbf{g}^c + \mathbf{g}_1^{(m)}(\mathbf{x})$ with respect to the properties $\mathbf{g}^c$ rather than $\mathbf{g}^{(0)}$ (3.67) ($\mathbf{g} = \mathbf{L}, \mathbf{M}, \boldsymbol{\alpha}, \boldsymbol{\beta};\ \ \mathbf{g}^c, \mathbf{g}_1^{(0)} \equiv$ const). Hereafter, as in Chapter 7, the upper index $^{(m)}$ indicates the components and the lower index i indicates the individual inclusions ($m = 0, k;\ k = 1, \ldots, N;\ i = 1, 2, \ldots$). Conversely, the estimation of the residual stresses (for $\langle\boldsymbol{\sigma}\rangle \equiv \mathbf{0}$) can be used for the calculation of the stored energy W^* (6.77) in the transformed stress field as well as of the effective eigenstrains $\boldsymbol{\beta}^*$ (6.48):

$$W^* \equiv \frac{1}{2}\langle \boldsymbol{\sigma}\mathbf{M}\boldsymbol{\sigma}\rangle = -\frac{1}{2}\langle\boldsymbol{\beta}\boldsymbol{\sigma}\rangle,$$

$$\boldsymbol{\beta}^* = \boldsymbol{\beta}^c + \langle \boldsymbol{\eta} \rangle - \mathbf{M}_1^{(0)} \langle \boldsymbol{\sigma} V \rangle. \tag{8.13}$$

8.1.2 Approximate Effective Field Hypothesis

To simplify the exact system (8.7), we now apply the main hypothesis of many micromechanical methods, the approximation called the effective field hypothesis:

H1) *Each inclusion v_i has an ellipsoidal form and is located in the field (8.5) (or (8.9))* $(\mathbf{y} \in v_i)$:

$$\overline{\boldsymbol{\sigma}}_i(\mathbf{y}) \equiv \overline{\boldsymbol{\sigma}}(\mathbf{x}_i), \quad (\text{or} \quad \overline{\boldsymbol{\varepsilon}}_i(\mathbf{y}) \equiv \overline{\boldsymbol{\varepsilon}}(\mathbf{x}_i)) \tag{8.14}$$

which is homogeneous over the inclusion v_i, and the perturbation introduced by the inclusion v_i at the point $\mathbf{y} \notin v_i$ is defined by the relation

$$\int \boldsymbol{\Gamma}(\mathbf{y}-\mathbf{x}) V_i(\mathbf{x}) \boldsymbol{\eta}(\mathbf{x})\, d\mathbf{x} = \bar{v}_i \mathbf{T}_i(\mathbf{y}-\mathbf{x}_i)\boldsymbol{\eta}_i, \quad \text{or}$$
$$\int \mathbf{U}(\mathbf{y}-\mathbf{x}) V_i(\mathbf{x}) \boldsymbol{\tau}(\mathbf{x})\, d\mathbf{x} = \bar{v}_i \mathbf{T}_i^{\epsilon}(\mathbf{y}-\mathbf{x}_i)\boldsymbol{\tau}_i. \tag{8.15}$$

Hereafter, $\boldsymbol{\eta}_i \equiv \langle \boldsymbol{\eta}(\mathbf{x}) V_i(\mathbf{x}) \rangle_{(i)}$ (and $\boldsymbol{\tau}_i$) is an average over the volume of the inclusion v_i (but not over the ensemble), $\langle (.) \rangle_i \equiv \langle \langle (.) \rangle_{(i)} \rangle$, and (see Eqs. (4.56) and Appendix A.3.1)

$$\mathbf{T}_i(\mathbf{y}-\mathbf{x}_i) = \langle \boldsymbol{\Gamma}(\mathbf{y}-\mathbf{x}) \rangle_i, \ \mathbf{T}_{ij}(\mathbf{x}_j-\mathbf{x}_i) = \langle \mathbf{T}_i(\mathbf{y}-\mathbf{x}_i) \rangle_j,$$
$$\mathbf{T}_i^{\epsilon}(\mathbf{y}-\mathbf{x}_i) = \langle \mathbf{U}(\mathbf{y}-\mathbf{x}) \rangle_i, \ \mathbf{T}_{ij}^{\epsilon}(\mathbf{x}_j-\mathbf{x}_i) = \langle \mathbf{T}_i^{\epsilon}(\mathbf{y}-\mathbf{x}_i) \rangle_j \tag{8.16}$$

For a homogeneous ellipsoidal inclusion v_i the standard assumption (8.14) (see, e.g., [138], [545]) yields the assumption (8.15); otherwise, the formula (8.15) defines an additional assumption.

According to hypothesis **H1**, each inclusion v_i being considered is located in a homogeneous field $\overline{\boldsymbol{\sigma}}(\mathbf{x}_i)$ that significantly simplifies the problem (8.4):

$$\boldsymbol{\sigma}(\mathbf{x}) = \overline{\boldsymbol{\sigma}}_i(\mathbf{x}) + \int \boldsymbol{\Gamma}(\mathbf{x}-\mathbf{y}) V_i(\mathbf{y}) \big[\mathbf{M}_1(\mathbf{y})\boldsymbol{\sigma}(\mathbf{y}) + \boldsymbol{\beta}_1(\mathbf{y}) - \langle \boldsymbol{\eta} \rangle^{(0)}\big]\, d\mathbf{y}. \tag{8.17}$$

In view of the linearity of the problem, there exist constant fourth and second-rank tensors $\mathbf{B}^{(i)}(\mathbf{x})$, $\mathbf{R}^{(i)}(\mathbf{x})$, and $\mathbf{C}^{(i)}(\mathbf{x})$, $\mathbf{F}^{(i)}(\mathbf{x})$, such that (see Eqs. (3.179) for comparison):

$$\boldsymbol{\sigma}(\mathbf{x}) = \mathbf{B}^{(i)}(\mathbf{x})\overline{\boldsymbol{\sigma}}(\mathbf{x}_i) + \mathbf{C}^{(i)}(\mathbf{x}),$$
$$\bar{v}_i \boldsymbol{\vartheta}(\mathbf{x}) = \mathbf{R}^{(i)}(\mathbf{x})\overline{\boldsymbol{\sigma}}(\mathbf{x}_i) + \mathbf{F}^{(i)}(\mathbf{x}), \tag{8.18}$$

where $\mathbf{x} \in v_i, \ v_i \subset v^{(i)}$ and

$$\mathbf{C}^{(i)}(\mathbf{x}) \equiv \overline{\mathbf{C}}^{(i)}(\mathbf{x}) + \mathbf{B}_i(\mathbf{x})\mathbf{Q}_i \langle \boldsymbol{\eta} \rangle^{(0)}, \quad \mathbf{R}^{(i)}(\mathbf{x}) = \bar{v}_i \mathbf{M}_1^{(i)}(\mathbf{x})\mathbf{B}^{(i)}(\mathbf{x}), \tag{8.19}$$
$$\mathbf{F}^{(i)}(\mathbf{x}) = \bar{v}_i[\mathbf{M}_1^{(i)}(\mathbf{x})\mathbf{C}^{(i)}(\mathbf{x}) + \boldsymbol{\beta}_1^{(i)}(\mathbf{y}) - \langle \boldsymbol{\eta} \rangle^{(0)}], \tag{8.20}$$

here the tensor $\overline{\mathbf{C}}^{(i)}(\mathbf{x})$ introduced does not depend on $\langle\boldsymbol{\eta}\rangle^{(0)}$. According to Eshelby's theorem ([319], see Chapter 3) the following relations exist between the averaged tensors (8.18_1) and (8.25):

$$\mathbf{R}_i = \overline{v}_i \mathbf{Q}_i^{-1}(\mathbf{I} - \mathbf{B}_i), \quad \mathbf{F}_i = -\overline{v}_i \mathbf{Q}_i^{-1}\mathbf{C}_i \equiv \overline{\mathbf{F}}_i + \widetilde{\mathbf{F}}_i \langle\boldsymbol{\eta}\rangle^{(0)}, \tag{8.21}$$

where $\mathbf{f}_i \equiv \langle \mathbf{f}(\mathbf{x})\rangle_{(i)}$ ($\mathbf{f}$ stands for $\mathbf{B}, \mathbf{C}, \overline{\mathbf{C}}, \mathbf{R}, \mathbf{F}$), the tensor $\mathbf{Q}_i = -\langle \boldsymbol{\Gamma}(\mathbf{x} - \mathbf{y})\rangle_{(i)}$, $\mathbf{x}, \mathbf{y} \in v_i$ (3.38) is associated with the well–known Eshelby tensor by $\mathbf{S}_i = \mathbf{I} - \mathbf{M}^{(0)}\mathbf{Q}_i$, and the tensors $\overline{\mathbf{F}}_i \equiv -\overline{v}_i \mathbf{Q}_i^{-1}\overline{\mathbf{C}}_i$ and $\widetilde{\mathbf{F}}_i \equiv -\overline{v}_i \mathbf{Q}_i^{-1}\mathbf{B}_i\mathbf{Q}_i$ were introduced for separation of influences of tensors $\boldsymbol{\beta}_1^{(i)}$ and $\langle\boldsymbol{\eta}\rangle^{(0)}$ on the tensor $\mathbf{F}_i$. It should be mentioned that the field $\overline{\boldsymbol{\sigma}}(\mathbf{x}_i)$ can vary with variation of the center $\mathbf{x}_i$ of the inclusion considered, but the field $\overline{\boldsymbol{\sigma}}(\mathbf{y})$ ($\mathbf{y} \in v_i$) is homogeneous over the inclusion v_i. Because of this the application of Eshelby's theorem is correct.

For example, for the homogeneous ellipsoidal domain v_i (3.187) we get

$$\mathbf{B}_i = \left(\mathbf{I} + \mathbf{Q}_i \mathbf{M}_1^{(i)}\right)^{-1}, \quad \mathbf{C}_i = -\mathbf{B}_i \mathbf{Q}_i(\boldsymbol{\beta}_1^{(i)} - \langle\boldsymbol{\eta}\rangle^{(0)}), \tag{8.22}$$

$$\mathbf{R}_i = \bar{v}_i \mathbf{M}_1^{(i)}\mathbf{B}_i, \quad \mathbf{F}_i = \bar{v}_i(\mathbf{I} + \mathbf{M}_1^{(i)}\mathbf{Q}_i)^{-1}(\boldsymbol{\beta}_1^{(i)} - \langle\boldsymbol{\eta}\rangle^{(0)}). \tag{8.23}$$

In the general case of inclusions v_i the tensors $\mathbf{B}(\mathbf{x})$ and $\mathbf{C}(\mathbf{x})$ can be found numerically according to the schemes (4.18) and (4.19). For particular cases of coated inclusions different analytical models were considered in Section 3.7.

Averaging (8.7) over the volume of the considered inclusion v_i and using the hypothesis **H1** (8.15) with (8.25) leads to the infinite system of integral equations:

$$\begin{aligned}
&\bar{v}_i \langle \boldsymbol{\vartheta}(\mathbf{x})|\ v_1, \mathbf{x}_1; \ldots; v_n, \mathbf{x}_n\rangle_i - \sum_{j \neq i}^{n} \mathbf{R}_i \mathbf{T}_{ij}(\mathbf{x}_i - \mathbf{x}_j)\langle \boldsymbol{\vartheta}(\mathbf{y})|v_1, \mathbf{x}_1; \ldots; v_n, \mathbf{x}_n\rangle_j \\
&\quad = \bar{v}_i \boldsymbol{\vartheta}_i^0 + \mathbf{R}_i \int \Big[\mathbf{T}_{iq}(\mathbf{x}_i - \mathbf{x}_q)\langle \boldsymbol{\vartheta}(\mathbf{y})|; v_1, \mathbf{x}_1; \ldots; v_n, \mathbf{x}_n\rangle_q \\
&\quad \times \varphi(v_q, \mathbf{x}_q|; v_1, \mathbf{x}_1; \ldots; v_n, \mathbf{x}_n) - \mathbf{T}_i(\mathbf{x}_i - \mathbf{x}_q)\langle \boldsymbol{\vartheta}\rangle n^{(1)}\Big]\, d\mathbf{x}_q,
\end{aligned} \tag{8.24}$$

where $\bar{v}_i \boldsymbol{\vartheta}_i^0 = \mathbf{R}_i\langle\boldsymbol{\sigma}\rangle + \mathbf{F}_i - \bar{v}_i\langle\boldsymbol{\eta}\rangle^{(0)}$ is called the external modified strain polarization tensor in the component $v^{(i)}$.

Using hypothesis **H1**, the system (8.5) for k fixed inclusions with fixed values $\widetilde{\boldsymbol{\sigma}}_{1,\ldots,k}(\mathbf{x})$ ($\mathbf{x} \in v_i$, $i = 1.\ldots,k$) on the right-hand side of the equations becomes algebraic when the solution (8.25) for one inclusion in the field $\overline{\boldsymbol{\sigma}}(\mathbf{x}_i)$ ($i = 1,\ldots,k$) is applied:

$$\mathbf{R}_i \overline{\boldsymbol{\sigma}}(\mathbf{x}_i) + \mathbf{F}_i = \sum_{j=1}^{k} \mathbf{Z}_{ij}\Big\{\mathbf{R}_j \widetilde{\boldsymbol{\sigma}}_{1,\ldots,k}(\mathbf{x}_j) + \mathbf{F}_j\Big\}. \tag{8.25}$$

Here the matrix $\mathbf{Z}^{-1}$ has the elements $(\mathbf{Z}^{-1})_{ij} = \mathbf{I}\delta_{ij} - (1 - \delta_{ij})\mathbf{R}_j \mathbf{T}_{ij}(\mathbf{x}_i - \mathbf{x}_j)$, ($i, j = 1,\ldots,n$) (4.58). Buryachenko considered [140] (see also Section 4.3) the possible refinement of the multiple inclusion interaction problem (8.4) and (8.5) and analyzed the related simplified methods.

8.1.3 Closing Effective Field Hypothesis

For termination of the hierarchy of statistical moment equations (8.7) (or (8.11)) we will use the closing effective field hypothesis:

H2) *For a sufficiently large* n, *we complete the system (8.7) (or (8.11)) by the assumption* $\langle\widetilde{\boldsymbol{\sigma}}_{1,\ldots,j,\ldots,n+1}(\mathbf{x})\rangle_i = \langle\widetilde{\boldsymbol{\sigma}}_{1,\ldots,n}(\mathbf{x})\rangle_i$ (or $\langle\widetilde{\boldsymbol{\varepsilon}}_{1,\ldots,j,\ldots,n+1}(\mathbf{x})\rangle_i = \langle\widetilde{\boldsymbol{\varepsilon}}_{1,\ldots,n}(\mathbf{x})\rangle_i$), *where the right-hand side of the equality does not contain the index* $j \neq i$ $(i = 1,\ldots,n;\ j = 1,\ldots,n+1;\ \mathbf{x} \in v_i)$.

The hypothesis **H2**, rewritten in terms of stresses $\boldsymbol{\sigma}(\mathbf{x}),\ (\mathbf{x} \in v_i)$, is a standard closing assumption (see, e.g., [561], [562], [1184]) degenerating to the "quasi-crystalline" approximation by Lax [649] at $n = 1$ (for analysis see also Subsection 8.3.1).

The solution obtained for one inclusion (8.25) and a finite number of inclusions (8.25), which are located in effective fields $\overline{\boldsymbol{\sigma}}(\mathbf{x}_i)$ and $\widetilde{\boldsymbol{\sigma}}_{1,\ldots,n}(\mathbf{x})$, respectively, and the adoption of hypothesis **H2** make it possible to solve the closing system of coupled integral equations (8.7), which is reduced to

$$\langle\widetilde{\boldsymbol{\sigma}}_{1,\ldots,j}(\mathbf{x})\rangle_i = \langle\boldsymbol{\sigma}\rangle + \int \Big[\mathbf{T}_{iq}(\mathbf{x}_i - \mathbf{x}_q)\varphi(v_q, \mathbf{x}_q|; v_1, \mathbf{x}_1; \ldots; v_j, \mathbf{x}_j)\sum_{l=1}^{j+1}\mathbf{Z}_{ql}$$
$$\times(\mathbf{R}_l\langle\widetilde{\boldsymbol{\sigma}}_{1,\ldots,j+1}(\mathbf{x})\rangle_l + \mathbf{F}_l) - \mathbf{T}_i(\mathbf{x}_i - \mathbf{x}_q)\langle\mathbf{R}\overline{\boldsymbol{\sigma}} + \mathbf{F}\rangle n^{(q)}\Big]\,d\mathbf{x}_q, \quad (8.26)$$

$$\langle\widetilde{\boldsymbol{\sigma}}_{1,\ldots,n}(\mathbf{x})\rangle_i = \langle\boldsymbol{\sigma}\rangle + \int \Big[\mathbf{T}_{iq}(\mathbf{x}_i - \mathbf{x}_q)\varphi(v_q, \mathbf{x}_q|; v_1, \mathbf{x}_1; \ldots; v_j, \mathbf{x}_j)\sum_{l=1}^{n}\mathbf{Z}_{ql}$$
$$\times(\mathbf{R}_l\langle\widetilde{\boldsymbol{\sigma}}_{1,\ldots,n}(\mathbf{x})\rangle_l + \mathbf{F}_l) - \mathbf{T}_i(\mathbf{x}_i - \mathbf{x}_q)\langle\mathbf{R}\overline{\boldsymbol{\sigma}} + \mathbf{F}\rangle n^{(q)}\Big]\,d\mathbf{x}_q. \quad (8.27)$$

The tensor $\widetilde{\boldsymbol{\sigma}}_{1,\ldots,n}(\mathbf{x})$ on the right-hand side of the last equation of (8.26) is formed from the tensor $\widetilde{\boldsymbol{\sigma}}_{1,\ldots,n}(\mathbf{x})$ on the left-hand side by replacing one of the indices by q. System (8.27) is linear with respect to $\langle\widetilde{\boldsymbol{\sigma}}_{1,\ldots,j}(\mathbf{x})\rangle_l$ $(j = 1,\ldots,n; l = 1,\ldots,j)$, and each j-th row with $\langle\widetilde{\boldsymbol{\sigma}}_{1,\ldots,j}(\mathbf{x})\rangle_l$ contains j equations, since $i = 1,\ldots,j$. The value of $\langle\widetilde{\boldsymbol{\sigma}}_{1,\ldots,n}(\mathbf{x})\rangle_i$ $(i = 1,\ldots,n)$ can be estimated from the last row in Eq. (8.27) by the method of successive approximation with all possible positions of the inclusions $v_1,\ldots,v_n$, It is also necessary to take into account the fact that $\langle\widetilde{\boldsymbol{\sigma}}_{1,\ldots,j}(\mathbf{x})\rangle_i \to \langle\overline{\boldsymbol{\sigma}}\rangle_i$ as $|\mathbf{x}_l - \mathbf{x}_i| \to \infty$, $j = 1,\ldots,j$; $l \neq i$. We substitute the value found for $\langle\widetilde{\boldsymbol{\sigma}}_{1,\ldots,n}(\mathbf{x})\rangle_i$ into the right-hand side of the $(n-1)$-th row of system (8.27), determine $\langle\widetilde{\boldsymbol{\sigma}}_{1,\ldots,n-1}(\mathbf{x})\rangle_i$ $(i = 1,\ldots,n-1)$, and so on. After estimating $\langle\overline{\boldsymbol{\sigma}}\rangle_i$ and $\langle\boldsymbol{\eta}^1\rangle_i$ in the relations (8.7) we find from Eq. (1.12_1) the effective compliance with the help of the equality

$$\langle\boldsymbol{\sigma}\rangle^{(0)} = \big(c^{(0)} + \langle V\mathbf{C}\rangle\mathbf{M}_1^{(0)}\big)^{-1}\big(\langle\boldsymbol{\sigma}\rangle - \langle V\mathbf{B}\overline{\boldsymbol{\sigma}}\rangle\big).$$

8.1.4 Effective Field Hypothesis and Composites with One Sort of Inhomogeneities

We will consider in this subsection the case $\mathbf{L}^c = \mathbf{L}^{(0)}$ when $\langle\boldsymbol{\eta}\rangle^{(0)} \equiv \mathbf{0}$ and the system (8.24) has principally the same structure as the system for the pure elastic problem (with $\mathbf{F} \equiv \mathbf{0}$). Therefore, we can apply the traditional analysis procedure of purely elastic composites and represent $\langle \boldsymbol{\eta} \rangle^{(1)}$ as a linear function of the external field $\boldsymbol{\eta}^0$

$$\langle \boldsymbol{\eta} \rangle^{(1)} = \mathbf{Y}\boldsymbol{\eta}^0 \tag{8.28}$$

and therefore

$$\mathbf{R}\langle \overline{\boldsymbol{\sigma}} \rangle^{(1)} + \mathbf{F} = \mathbf{Y}(\mathbf{R}\boldsymbol{\sigma}^0 + \mathbf{F}). \tag{8.29}$$

Comparison of (8.24) with (8.28) leads to the fact that $\mathbf{Y}$ depends only on the tensors $\mathbf{R}$, $\mathbf{T}_{ij}$ and $\mathbf{T}_j$. The tensor $\mathbf{Y}$ is determined by the purely elastic action (with $\mathbf{F} \equiv \mathbf{0}$) of the surrounding inclusions on the separated one. For a dilute concentration of the inclusions, i.e., $c^{(1)} \to 0$, we have $\mathbf{Y} \to \mathbf{I}$. The actual form of the tensor $\mathbf{Y}$, used in the analysis as an approximation, depends on additional assumptions for closing of the infinite system (8.24). In particular, for purely elastic composites (with $\boldsymbol{\beta}_1 \equiv \mathbf{0}$) with fictitious homogeneous inclusions (3.189), (3.190) such relations will be represented later for commonly applied methods of micromechanics, i.e., the effective medium method by Kröner [607] and by Hill [463], Mori–Tanaka [787] method, MEFM.

Taking into account Eqs. (8.25) and (8.28), from the solution of the purely elastic problem for the composite we can calculate the average stresses inside the inclusions by

$$\langle \boldsymbol{\sigma} \rangle^{(1)} = \mathbf{B}\mathbf{R}^{-1}\left[\mathbf{Y}(\mathbf{R}\boldsymbol{\sigma}^0 + \mathbf{F}) - \mathbf{F}\right] + \mathbf{C}. \tag{8.30}$$

The mean matrix stresses follow simply from the condition $\langle \boldsymbol{\sigma} \rangle = \boldsymbol{\sigma}^0$:

$$\langle \boldsymbol{\sigma} \rangle^{(0)} = (c^{(0)})^{-1}\left[\boldsymbol{\sigma}^0 - c^{(1)}(\mathbf{I} - \mathbf{B})^{-1}\mathbf{C} - c^{(1)}\mathbf{B}\mathbf{R}^{-1}\mathbf{Y}(\mathbf{R}\boldsymbol{\sigma}^0 + \mathbf{F})\right]. \tag{8.31}$$

The local stresses inside the inclusion, i.e., in the core and in the coating, respectively, are found by

$$\langle \boldsymbol{\sigma} \rangle_i(\mathbf{x}) = \mathbf{C}(\mathbf{x}) + \mathbf{B}(\mathbf{x})\mathbf{R}^{-1}\left[\mathbf{Y}(\mathbf{R}\boldsymbol{\sigma}^0 + \mathbf{F}) - \mathbf{F}\right], \tag{8.32}$$

where $\langle \boldsymbol{\sigma} \rangle_i(\mathbf{x})$ is the average of the local stress state at $\mathbf{x} \in v_i \subset v^1$ over an ensemble realization of surrounding inclusions (but not over the volume v_i of a particular inclusion, in contrast to $\langle \boldsymbol{\sigma} \rangle^{(1)}$).

Comparing (6.33) with (8.32) leads to the relation for the average local stress concentration tensor inside the inclusions:

$$\langle \boldsymbol{\sigma} \rangle_i(\mathbf{x}) = \langle \mathbf{B}^* \rangle_i(\mathbf{x})\boldsymbol{\sigma}^0 + [\langle \mathbf{B}^* \rangle_i(\mathbf{x}) - \mathbf{B}(\mathbf{x})]\mathbf{R}^{-1}\mathbf{F} + \mathbf{C}(\mathbf{x}), \tag{8.33}$$

where $\langle \mathbf{B}^* \rangle_i(\mathbf{x}) \equiv \mathbf{B}(\mathbf{x})\mathbf{D}$, $(\mathbf{x} \in v_i)$, and the tensor $\mathbf{D} \equiv \mathbf{R}^{-1}\mathbf{Y}\mathbf{R}$ has the simple physical meaning of the action (at $\boldsymbol{\beta}(\mathbf{x}) \equiv \mathbf{0}$) of the surrounding inclusions on the separated one: $\langle \overline{\boldsymbol{\sigma}} \rangle_i = \mathbf{D}\boldsymbol{\sigma}^0$.

Thus the average of local thermal stresses $\langle \boldsymbol{\sigma}^{II} \rangle_i(\mathbf{x})$ (6.34) over the ensemble is defined by the purely elastic solution for the composite medium (the tensor $\langle \mathbf{B}^* \rangle_i(\mathbf{x})$) as well as by the thermoelastic solution for a single inclusion in an infinite matrix (the tensors $\mathbf{B}(\mathbf{x})$, $\mathbf{C}(\mathbf{x})$, $\mathbf{R}$, $\mathbf{F}$). For two-component composites with identical homogeneous inclusions of any shape Benveniste and Dvorak [59] obtained an exact relation for nonaveraged local thermal stresses (6.33) and (6.119)

$$\boldsymbol{\sigma}^{II}(\mathbf{x}) = [\mathbf{B}^*(\mathbf{x}) - \mathbf{I}][\mathbf{M}_1^{(1)}]^{-1}\boldsymbol{\beta}_1^{(1)},$$

from which one can derive Eq. (8.33) in the case of ellipsoidal inclusions.

The matrix stresses in the immediate vicinity of the inclusions v_i, denoted by $\boldsymbol{\sigma}_i^-(\mathbf{n})$, are given by the formula (3.95). Equations (8.32) and (3.95) allow the estimation of the ensemble average of the matrix stresses in the vicinity of the inclusions near a point $\mathbf{x} \in v_i$ with the unit outward normal vector on ∂v_i:

$$\begin{aligned}\langle\boldsymbol{\sigma}^-(\mathbf{n})\rangle_x &= \left[\mathbf{I} + \boldsymbol{\Gamma}^{(0)}(\mathbf{n})\mathbf{M}_1^{(1)}(\mathbf{x})\right]\left\{\mathbf{C}(\mathbf{x}) + \mathbf{B}(\mathbf{x})\mathbf{R}^{-1}\left[\mathbf{Y}(\mathbf{R}\boldsymbol{\sigma}^0 + \mathbf{F}) - \mathbf{F}\right]\right\}\\ &\quad + \boldsymbol{\Gamma}(\mathbf{n})\boldsymbol{\beta}_1^{(1)}(\mathbf{x}).\end{aligned} \tag{8.34}$$

For the estimation of the effective compliance we use relations (6.20) and (8.32) to obtain

$$\mathbf{M}^* = \mathbf{M}^{(0)} + \mathbf{Y}\mathbf{R}n^{(1)}. \tag{8.35}$$

Taking the equality $\boldsymbol{\beta}^* = \boldsymbol{\beta}^{(0)} + \langle\mathbf{M}_1\boldsymbol{\sigma}^{II} + \boldsymbol{\beta}_1\rangle$ (see, e.g., [156] as well as Eq. (6.49)) into account we find the coefficient of thermal expansion from the relations (8.25) and (8.31): $\boldsymbol{\beta}^* = \boldsymbol{\beta}^{(0)} + \mathbf{Y}\mathbf{F}n^{(1)}$. Equations ($8.13_1$) and (8.32) yield the following relation for the stored energy density:

$$2W^* = -\langle\boldsymbol{\beta}_1(\mathbf{x})\mathbf{C}(\mathbf{x})\rangle - \langle\boldsymbol{\beta}_1(\mathbf{x})\mathbf{B}(\mathbf{x})\rangle\mathbf{R}^{-1}(\mathbf{Y} - \mathbf{I})\mathbf{F}.$$

Interestingly, all three effective quantities $\mathbf{M}^*$, $\boldsymbol{\beta}^*$, W^* can be estimated by the use of unique scheme. For example, for the estimation of the effective compliance $\mathbf{M}^*$ as a first step it is necessary to solve the elastic problem for a single inclusion in an infinite matrix (e.g., to find the tensor $\mathbf{B}(\mathbf{x})$, see, e.g., Eqs. (4.18) and (4.22) as well as Section 3.7.2), and in a second step, the single constant tensor $\mathbf{Y}$ is found from the purely elastic problem ($\boldsymbol{\beta}(\mathbf{x}) \equiv 0$) for the composite with ellipsoidal inclusions with the tensor $\mathbf{R}$ (8.19_2). Analogously the problem of evaluating the effective tensors $\boldsymbol{\beta}^*$, W^* and the average stresses inside the components (8.31), (8.33) can be fully solved if one of the two parameters, $\mathbf{Y}$ or $\mathbf{M}^*$, is found (for example, experimentally). For a proof, assume that $\mathbf{M}^*$ is known. Then (8.35) yields $\mathbf{Y} = (\mathbf{M}^* - \mathbf{M}^{(0)})(\mathbf{R}n^{(1)})^{-1}$ and therefore

$$\boldsymbol{\beta}^* = \boldsymbol{\beta}^{(0)} + (\mathbf{M}^* - \mathbf{M}^{(0)})\mathbf{R}^{-1}\mathbf{F}, \tag{8.36}$$

$$\begin{aligned}2W^* &= -\langle\boldsymbol{\beta}_1(\mathbf{x})\mathbf{B}(\mathbf{x})\rangle\mathbf{R}^{-1}\left[\mathbf{M}^* - \mathbf{M}^{(0)} - \mathbf{R}n^{(1)}\right](\mathbf{R}n^{(1)})^{-1}\mathbf{F}\\ &\quad - \langle\boldsymbol{\beta}_1(\mathbf{x})\mathbf{C}(\mathbf{x})\rangle.\end{aligned} \tag{8.37}$$

Equation (8.36) can be rewritten in an alternative elegant word formulation: Theorem. *In the framework of the hypothesis* **H1**, *the ratio of the increment of the effective thermal expansion to the increment of the effective stiffness is a constant tensor defined by the averaging solution for a single coated inclusion in an infinite matrix (8.21) and does not depend on the concrete statistically homogeneous microstructure of the whole composite material and the detailed microstructure of individual inclusions as well:*

$$(\mathbf{M}^* - \mathbf{M}^{(0)})^{-1}(\boldsymbol{\beta}^* - \boldsymbol{\beta}^{(0)}) = \mathbf{R}^{-1}\mathbf{F}. \tag{8.38}$$

In particular, for homogeneous (i.e., noncoated) inclusions the constant tensors $\mathbf{B}$, $\mathbf{C}$, $\mathbf{R}$, and $\mathbf{F}$ according to (8.22) and (8.23) are determined by the Eshelby tensor $\mathbf{S}$ and the jumps of the material property tensors $\mathbf{M}_1^{(1)}$, $\boldsymbol{\beta}_1^{(1)}$. Then from (8.36) and (8.37) the classical results (6.106) and (6.108) for two-phase composites are derived. It is not surprising that the exact relations (6.106) and (6.108) are derived from the approximate ones, i.e., (8.36) and (8.37), since the additional assumption **H1** does not expand the class of the considered materials and homogenization methods. The representations (6.106) and (6.108) are formally invariant with respect to the replacement $v^{(1)} \leftrightarrow v^{(0)}$, although this cannot be said about the relations (8.36) and (8.37), obtained for matrix structure composites with ellipsoidal inclusions.

It should be mentioned that all results of this subsection 8.1.3 where obtained in the framework of hypothesis **H1** only. No restrictions are imposed on the concrete form of the tensor $\mathbf{Y}$ (8.28), on the microtopology of the coated inclusions or on the inhomogeneity of the stress state in the inclusions (8.32). Moreover, the assumption of an ellipsoidal shape of the inclusion in the hypothesis **H1** was used only in order to obtain analytical solutions (8.21). This is because the tensor $\langle \boldsymbol{\Gamma}(\mathbf{x}-\mathbf{y})\rangle_i$ (3.38) is apparently homogeneous for $\mathbf{x}, \mathbf{y} \in v_i$ for an ellipsoid only (see, e.g., [353], [712]). For nonellipsoidal inclusions, one could assume that in some parts of the region $v_i^c \subset v_i$ the properties $\mathbf{M}_1(\mathbf{x}) \equiv \mathbf{0}$, $\boldsymbol{\beta}_1(\mathbf{x}) \equiv \mathbf{0}$ take place, that is, it is sufficient to include a real nonellipsoidal inclusion $v_i \setminus v_i^c$ into an ellipsoid (with smallest possible volume) and call it the inclusion v_i with a "coating" v_i^c. The further scheme for calculating the tensors $\mathbf{B}$, $\mathbf{C}$ (8.18_1) and overall properties is the same, but the prescribed conditional distributions $\varphi(v_m|\mathbf{x}_m;\mathbf{x}_1,\ldots\mathbf{x}_n)$ will have a larger correlation hole $\cup v_{0i}$, $(i=1,\ldots n)$ than in the real composite material. This will result in an underestimation of the computed values of $\mathbf{M}^*$ for inclusions that are softer than the matrix and in an overestimation in the opposite case.

It is noteworthy that we do not pursue the goal of presenting in this chapter all known methods based on hypothesis **H1**. There are, of course, several other methods that are based on hypothesis **H1** and no ranking between them is given here. Our main objective it to prove the general relations (8.36) and (8.37) which are valid in the framework of hypothesis **H1** only. No restrictions are imposed on the concrete statistically homogeneous microstructure of the whole composite material with a single sort of coated inclusions being analyzed as well as on the microtopology of coated inclusions or on the inhomogeneity of the stress state in the inclusions.

8.2 Analytical Representation of Effective Thermoelastic Properties

8.2.1 Average Stresses in the Components

The n-particle approximation solution of problems (8.26) and (8.27) presupposed a solution of systems (8.26) and (8.27) for arbitrary coordinates of the centers of n inclusions and arbitrary orientations of these inclusions. The two-particle approximation and the assumption

$$\langle \widetilde{\boldsymbol{\sigma}}(\mathbf{x})_{1,2} \rangle_i = \langle \overline{\boldsymbol{\sigma}}(\mathbf{x}) \rangle_i = \text{const.} \quad (i = 1, 2). \tag{8.39}$$

greatly simplify the problem. This independence of $\langle \widetilde{\boldsymbol{\sigma}}(\mathbf{x})_{1,2} \rangle$ of the spacing between the inclusions v_1 and v_2 (8.39) occurs for the limiting case $|\mathbf{x}_1 - \mathbf{x}_2| \gg \max a_i^k$, where a_i^k $(k = 1, 2, 3;\ i = 1, 2)$ are the semi-axes of the ellipsoidal inclusions v_1 and v_2, respectively. Then the system (8.26) can be solved by analytical methods, so from (8.26), taking (8.25) and (8.39) into account, we get

$$\overline{v}_i \langle \boldsymbol{\vartheta}_i \rangle = \overline{v}_i \boldsymbol{\vartheta}_i^0 + \mathbf{R}_i \sum_{q=1}^{N} \Big\{ \int \mathbf{T}_{iq}(\mathbf{x}_i - \mathbf{x}_q) \mathbf{Z}_{qi} \varphi(v_q, \mathbf{x}_q |; v_i, \mathbf{x}_i)\, d\mathbf{x}_q \overline{v}_i \langle \boldsymbol{\vartheta}_i \rangle$$
$$= \int \Big[\mathbf{T}_{iq}(\mathbf{x}_i - \mathbf{x}_q) \mathbf{Z}_{qq} \varphi(v_q, \mathbf{x}_q |; v_i, \mathbf{x}_i) - \mathbf{T}_i(\mathbf{x}_i - \mathbf{x}_q) n^{(q)} \Big] \overline{v}_q \langle \boldsymbol{\vartheta}_q \rangle\, d\mathbf{x}_q \Big\}. \tag{8.40}$$

where the matrix elements $\mathbf{Z}_{qi}$, $\mathbf{Z}_{qq}$ are nondiagonal and diagonal elements, respectively, of the binary interaction matrix $\mathbf{Z}$ (4.58), (8.25) for the two inclusions v_q and v_i. The algebraic system can be solved for $\langle \boldsymbol{\vartheta}_i \rangle$:

$$\overline{v}_i \langle \boldsymbol{\vartheta}_i \rangle = \sum_{j=1}^{N} \mathbf{Y}_{ij} \overline{v}_j \boldsymbol{\vartheta}_j^0, \tag{8.41}$$

where the matrix $\mathbf{Y}^{-1}$ has the following elements $(\mathbf{Y}^{-1})_{ij}$ $(i, j = 1, 2, \ldots, N)$:

$$(\mathbf{Y}^{-1})_{ij} = \delta_{ij} \left[\mathbf{I} - \mathbf{R}_i \sum_{q=1}^{N} \int \mathbf{T}_{iq}(\mathbf{x}_i - \mathbf{x}_q) \mathbf{Z}_{qi} \varphi(v_q, \mathbf{x}_q |; v_i, \mathbf{x}_i)\, d\mathbf{x}_q \right]$$
$$- \mathbf{R}_i \int \Big[\mathbf{T}_{ij}(\mathbf{x}_i - \mathbf{x}_j) \mathbf{Z}_{jj} \varphi(v_j, \mathbf{x}_j |; v_i, \mathbf{x}_i) - \mathbf{T}_i(\mathbf{x}_i - \mathbf{x}_j) n^{(j)} \Big]\, d\mathbf{x}_j. \tag{8.42}$$

The statistical averages of the effective and stress fields inside the inclusions are obtained from (8.18_1) and (8.41)

$$\langle \overline{\boldsymbol{\sigma}} \rangle_i = \mathbf{D}_i \boldsymbol{\sigma}^0 + \sum_{j=1}^{N} \mathbf{Y}_{ij} \mathbf{F}_j, \quad \mathbf{D}_i = \mathbf{R}_i^{-1} \sum_{j=1}^{N} \mathbf{Y}_{ij} \mathbf{R}_j, \tag{8.43}$$

$$\langle \boldsymbol{\sigma} \rangle_i(\mathbf{x}) = \mathbf{B}_i(\mathbf{x}) \mathbf{R}_i^{-1} \Big\{ \sum_{j=1}^{N} \mathbf{Y}_{ij} (\mathbf{R}_j \boldsymbol{\sigma}^0 + \mathbf{F}_j) - \mathbf{F}_i \Big\} + \mathbf{C}_i(\mathbf{x}), \tag{8.44}$$

$$\overline{v}_i \langle \boldsymbol{\vartheta}_i \rangle(\mathbf{x}) = \mathbf{R}_i(\mathbf{x}) \mathbf{D}_i \boldsymbol{\sigma}^0 + \mathbf{F}_i(\mathbf{x}) + \mathbf{R}_i(\mathbf{x}) \mathbf{R}_i \sum_{j=1}^{N} \mathbf{Y}_{ij} (\mathbf{F}_j - \mathbf{I}\delta_{ij}) \mathbf{F}_j, \tag{8.45}$$

$$\langle \boldsymbol{\sigma} \rangle_i = \mathbf{B}_i \Big\{ \mathbf{D}_i \boldsymbol{\sigma}^0 + \mathbf{R}_i^{-1} \sum_{j=1}^{N} (\mathbf{Y}_{ij} - \mathbf{I}\delta_{ij}) \mathbf{F}_j \Big\} + \mathbf{C}_i, \tag{8.46}$$

where the tensor $\mathbf{D}_i$ $(i = 1, \ldots, N)$ has a simple physical meaning of the action of surrounding inclusions on the separate one: $\langle \overline{\boldsymbol{\sigma}} \rangle_i = \mathbf{D}_i \boldsymbol{\sigma}^0$ for $\boldsymbol{\beta}(\mathbf{x}) \equiv \mathbf{0}$.

The stress in the matrix in the vicinity of the inhomogeneities v_i, $\sigma_i^-(\mathbf{n})$, is given by the formula (3.101) which holds for any shape, anisotropy, and inhomogeneity of the inclusion v_i. Equations (8.44) and (3.101) make it possible to estimate the ensemble average of the stress in the vicinity of the inhomogeneities near the point $\mathbf{x} \in v_i$ with the unit outward normal vector $\mathbf{n} \perp \partial v_i$

$$\langle \boldsymbol{\sigma}^-(\mathbf{n})\rangle_{\mathbf{x}} = \mathbf{B}_i(\mathbf{n})\mathbf{B}_i(\mathbf{x})\mathbf{D}_i\boldsymbol{\sigma}^0 + \mathbf{B}(\mathbf{n})\mathbf{B}_i(\mathbf{x})\mathbf{R}_i^{-1}\sum_{j=1}^{N}(\mathbf{Y}_{ij} - \mathbf{I}\delta_{ij})\mathbf{F}_j$$
$$+[\mathbf{B}_i(\mathbf{n}) - \mathbf{I}][\mathbf{M}^{(i)}(\mathbf{x}) - \mathbf{M}^{(0)}]^{-1}[\boldsymbol{\beta}^{(i)}(\mathbf{x}) - \boldsymbol{\beta}^{(0)}] + \mathbf{B}_i(\mathbf{n})\mathbf{C}_i(\mathbf{x}), \quad (8.47)$$

where the tensor $\mathbf{B}_i(\mathbf{n})$ (4.69) depends only on the elastic properties of the contacting materials and by the direction of the normal $\mathbf{n}$.

It should be mentioned that according to (8.19$_1$) and (8.21), the right-hand sides of Eqs. (8.44)– (8.46) are the linear functions of tensors $\mathbf{F}_i(\mathbf{x})$, $\mathbf{C}_i(\mathbf{x})$ $(i = 1, \ldots, N)$ depending on $\langle \boldsymbol{\sigma}\rangle^{(0)}$ which can be found by substitution of Eqs. (8.46), (8.19$_1$) and (8.21) into the obvious equality $\langle \boldsymbol{\sigma}\rangle^{(0)} = (\boldsymbol{\sigma}^0 - \langle \boldsymbol{\sigma} V\rangle)/c^{(0)}$:

$$\langle \boldsymbol{\sigma}\rangle^{(0)} = \Big\{c^{(0)}\mathbf{I} + \sum_{i=1}^{N} c^{(i)}\mathbf{B}_i\Big[\mathbf{Q}_i + \mathbf{R}_i^{-1}\sum_{j=1}^{N}\big[\mathbf{Y}_{ij} - \delta_{ij}\mathbf{I}\big]\widetilde{\mathbf{F}}_j\Big]\mathbf{M}_1^{(0)}\Big\}^{-1}$$
$$\times\Big\{\boldsymbol{\sigma}^0 - \sum_{i=1}^{N} c^{(i)}\mathbf{B}_i\mathbf{R}_i^{-1}\sum_{j=1}^{N}\mathbf{Y}_{ij}\big[\mathbf{R}_j\boldsymbol{\sigma}^0 + \overline{\mathbf{F}}_j\big] - \sum_{i=1}^{N} c^{(i)}(\mathbf{I} - \mathbf{B}_i)^{-1}\overline{\mathbf{C}}_i\Big\}. \quad (8.48)$$

8.2.2 Effective Properties of the Composite

After estimating average stresses inside the inclusions, see (8.46), the problem of calculating effective properties becomes trivial and leads, according to (8.3) and (8.4), to the following representations:

$$\mathbf{M}^* = \mathbf{M}^{(0)} + \sum_{i,j=1}^{N}\mathbf{Y}_{ij}\mathbf{R}_j n^{(i)} + \Big\{\mathbf{I} + \sum_{i,j=1}^{N}\mathbf{Y}_{ij}\widetilde{\mathbf{F}}_j n^{(i)}\Big\}\mathbf{M}_1^{(0)}$$
$$\times\Big\{c^{(0)}\mathbf{I} + \sum_{i=1}^{N} c^{(i)}\mathbf{B}_i\Big[\mathbf{Q}_i + \mathbf{R}_i^{-1}\sum_{j=1}^{N}\big[\mathbf{Y}_{ij} - \delta_{ij}\mathbf{I}\big]\widetilde{\mathbf{F}}_j\Big]\mathbf{M}_1^{(0)}\Big\}^{-1}$$
$$\times\Big\{\mathbf{I} - \sum_{i=1}^{N} c^{(i)}\mathbf{B}_i\mathbf{R}_i^{-1}\sum_{j=1}^{N}\mathbf{Y}_{ij}\mathbf{R}_j\Big\}, \quad (8.49)$$

$$\boldsymbol{\beta}^* = \boldsymbol{\beta}^{(0)} + \sum_{i,j=1}^{N}\mathbf{Y}_{ij}\overline{\mathbf{F}}_j n^{(i)} - \Big\{\mathbf{I} + \sum_{i,j=1}^{N}\mathbf{Y}_{ij}\widetilde{\mathbf{F}}_j n^{(i)}\Big\}\mathbf{M}_1^{(0)}$$
$$\times\Big\{c^{(0)}\mathbf{I} + \sum_{i=1}^{N} c^{(i)}\mathbf{B}_i\Big[\mathbf{Q}_i + \mathbf{R}_i^{-1}\sum_{j=1}^{N}\big[\mathbf{Y}_{ij} - \delta_{ij}\mathbf{I}\big]\widetilde{\mathbf{F}}_j\Big]\mathbf{M}_1^{(0)}\Big\}^{-1}$$
$$\times\Big\{\sum_{i=1}^{N} c^{(i)}\mathbf{B}_i\mathbf{R}_i^{-1}\sum_{j=1}^{N}\mathbf{Y}_{ij}\overline{\mathbf{F}}_j + \sum_{i=1}(\mathbf{I} - \mathbf{B}_i)^{-1}\overline{\mathbf{C}}_i\Big\}, \quad (8.50)$$

$$
\begin{aligned}
2W^* = &- \sum_{i,j=1}^{N} \langle \boldsymbol{\beta}_1^{(i)}(\mathbf{x})\mathbf{B}_i(\mathbf{x})\rangle_i \mathbf{R}_i^{-1}(\mathbf{Y}_{ij} - \mathbf{I}\delta_{ij})\overline{\mathbf{F}}_j c^{(i)} \\
&- \sum_{i=1}^{N} \langle \boldsymbol{\beta}_1^{(i)}(\mathbf{x})\mathbf{C}_i(\mathbf{x})\rangle_i c^{(i)} - \sum_{i,j=1}^{N} \langle \boldsymbol{\beta}_1^{(i)}(\mathbf{x})\mathbf{B}_i(\mathbf{x})\rangle_i \mathbf{R}_i^{-1}(\mathbf{Y}_{ij} - \mathbf{I}\delta_{ij})\overline{\mathbf{F}}_j \\
&\times \Big\{ c^{(0)}\mathbf{I} + \sum_{i=1}^{N} c^{(i)}\mathbf{B}_i \Big[\mathbf{Q}_i + \mathbf{R}_i^{-1} \sum_{j=1}^{N} \big[\mathbf{Y}_{ij} - \mathbf{I}\delta_{ij}\big]\widetilde{\mathbf{F}}_j\Big]\mathbf{M}^{(0)} \Big\}^{-1} \\
&\times \Big\{ \sum_{i=1}^{N} c^{(i)}\mathbf{B}_i\mathbf{R}_i^{-1} \sum_{j=1}^{N} \mathbf{Y}_{ij}\overline{\mathbf{F}}_j + \sum_{i=1}^{N} (\mathbf{I} - \mathbf{B}_i)^{-1}\overline{\mathbf{C}}_i \Big\}. \qquad (8.51)
\end{aligned}
$$

In particular, for the case when elastic moduli of comparison medium $\mathbf{M}^c$ coincide with elastic properties of matrix $\mathbf{M}^{(0)}$ (7.16), $\boldsymbol{\eta}(\mathbf{x}) \equiv \mathbf{0}$ ($\mathbf{x} \in v^{(0)}$) and there is no need to postulate that the stress field in the matrix is homogeneous. Then formulae (8.49)–(8.51) are simplified substantially:

$$
\mathbf{M}^* = \mathbf{M}^{(0)} + \sum_{i,j=1}^{N} \mathbf{Y}_{ij}\mathbf{R}_j n^{(i)}, \qquad (8.52)
$$

$$
\boldsymbol{\beta}^* = \boldsymbol{\beta}^{(0)} + \sum_{i,j=1}^{N} \mathbf{Y}_{ij}\mathbf{F}_j n^{(i)}, \qquad (8.53)
$$

$$
\begin{aligned}
2W^* = &- \sum_{i,j=1}^{N} \langle \boldsymbol{\beta}_1^{(i)}(\mathbf{x})\mathbf{B}_i(\mathbf{x})\rangle_i \mathbf{R}_i^{-1}(\mathbf{Y}_{ij} - \mathbf{I}\delta_{ij})\mathbf{F}_j c^{(i)} \\
&- \sum_{i=1}^{N} \langle \boldsymbol{\beta}_1^{(i)}(\mathbf{x})\mathbf{C}_i(\mathbf{x})\rangle_i c^{(i)}. \qquad (8.54)
\end{aligned}
$$

Equations (8.52)–(8.54) were obtained in [182] by a different method. For homogeneous inclusions (8.22) the formulae (8.52)–(8.54) are equivalent to the results derived in [129], [151]. For homogeneous ellipsoidal inclusions and some additional assumptions the relation (8.52) is reduced to the similar representation for effective compliance $\mathbf{M}^*$ obtained by the author [156], [159] from which follows, in turn, a more specific result for the point approximation (4.64) of identical spherical inclusions obtained more recently in [541], [545], and recast in our notation in the following form:

$$
\mathbf{M}^* = \mathbf{M}^{(0)} + \mathbf{R}_1 n^{(1)} \Big\{ \mathbf{I} - \mathbf{Q}_1\mathbf{R}_1 n^{(1)} + \bar{\mathbf{J}} \Big\}^{-1}, \qquad (8.55)
$$

$$
\bar{\mathbf{J}} \equiv \int \boldsymbol{\Gamma}(\mathbf{x}_i - \mathbf{x}_q)\mathbf{R}_1 \Big\{ \big[\mathbf{I} + \boldsymbol{\Gamma}(\mathbf{x}_i - \mathbf{x}_q)\mathbf{R}_1\big]^{-1} - \mathbf{I} \Big\} \varphi(v_q, \mathbf{x}_q|; v_i, \mathbf{x}_i)\, d\mathbf{x}_q. \qquad (8.56)
$$

Further simplifications of the representation of $\mathbf{Y}$ concern the approximative solution of the problem of binary interaction of the inclusions. So the analogous problem for homogeneous ellipsoidal inclusions was solved in [156] by the method of successive approximation (4.67_1) in the framework of point approximation of the inclusions (4.64), when for $\mathbf{Z}_{ji} = \mathbf{I}\delta_{ji} + (1 - \delta_{ji})\mathbf{R}\boldsymbol{\Gamma}(\mathbf{x}_j - \mathbf{x}_i)$, and for

$$\bar{\mathbf{J}} = \int \boldsymbol{\Gamma}(\mathbf{x}_i - \mathbf{x}_q)\mathbf{R}_1\boldsymbol{\Gamma}(\mathbf{x}_i - \mathbf{x}_q)\mathbf{R}_1\varphi(v_q, \mathbf{x}_q|; v_i, \mathbf{x}_i)\, d\mathbf{x}_q,$$

an analytical representation was obtained. The general case of coated inclusions was considered in [183]. Then, for an isotropic matrix $\mathbf{L}^{(0)} = (3k^{(0)}, 2\mu^{(0)})$ and identical spherical inclusions with radius a and isotropic parameters $\mathbf{R}_1 = \frac{4}{3}\pi a^3(3k^R, 2\mu^R)$, $\varphi(v_q|\mathbf{x}_j; \mathbf{x}_i) = ng(|\mathbf{x}_j - \mathbf{x}_i|) \equiv ng(r)$ one can obtain the following representation for the isotropic tensor

$$\bar{\mathbf{J}} = 4\pi a^3 c^{(1)} \int_{2a}^{\infty} \mathbf{J}(r)g(r)dr, \ \ \mathbf{J}(r) = (3k^J, 2\mu^J),$$

where

$$\begin{aligned}
3k^J &= 2\xi^2 3k^R 2\mu^R \big(3k^{(0)}\big)^2 \big(2\mu^{(0)}\big)^2 r^{-4},\\
2\mu^J &= \frac{2}{5r^4}\Big[\xi^2 3k^R 2\mu^R \big(3k^{(0)}\big)^2 + \big(2\mu^R\big)^2 \big(2\mu^{(0)}\big)^2 (7\gamma^2 - \frac{\eta^2}{4} + 2\xi\eta)\Big]\big(2\mu^{(0)}\big)^2,\\
\xi &= \frac{1}{3k^{(0)} + 2\mu^{(0)}}, \quad \eta = \frac{1}{3\mu^{(0)}}, \quad \gamma = -\frac{3k^{(0)} + \mu^{(0)}}{3\mu^{(0)}(3k^{(0)} + 4\mu^{(0)})}.
\end{aligned} \tag{8.57}$$

8.2.3 Some Related Multiparticle Methods

Another alternative version of the multiparticle method deserves more attention. It is the one proposed and developed by Ju and coauthors [513], [514], [515], [516], [525] for a wide class of micromechanical problems. For a clear understanding we will reproduce the general scheme of their method in our notation that allows us to abandon some unnecessary concepts (such as an eigenstrain and a perturbed strain) and use the rigorous techniques of formal mathematics.

At first, one considers two identical spherical homogeneous inclusions (2-D or 3-D cases) in an infinite matrix subjected to the uniform remote stress $\boldsymbol{\sigma}^0$, or to put this another way, the system (8.4) and (8.5) is solved in the framework of hypothesis **H1** for $n = 2$, $\widetilde{\boldsymbol{\sigma}}_{1,2}(\mathbf{x}) \equiv \boldsymbol{\sigma}^0$; $\boldsymbol{\beta} \equiv \mathbf{0}$, $\mathbf{M}^c \equiv \mathbf{M}^{(0)}$ and two identical inclusions, with $a_i^1 = a_i^2 = a_i^3$, $i = 1, 2$. The solution

$$\overline{\boldsymbol{\sigma}}(\mathbf{x}_i) = (\mathbf{I} + \boldsymbol{\mathcal{T}}_i(\mathbf{x}_i - \mathbf{x}_j))\boldsymbol{\sigma}^0, \ i = 1, 2, \ j = 3 - i \tag{8.58}$$

was obtained by an approximate method which is equivalent to the inversion of the matrix $(\mathbf{Z}^{-1})_{ij}$ (8.25) by the method of successive approximation [see [140] and Section 4.3 for detailed comparative analysis of different popular approximate methods of the solution of the system (8.4) and (8.5)]. Independently of the concrete approximate method, the asymptotic behavior at the infinity of the tensor $\boldsymbol{\mathcal{T}}_i(\mathbf{x}_i - \mathbf{x}_j)$ is well known: $\boldsymbol{\mathcal{T}}_i(\mathbf{x}_i - \mathbf{x}_j) = O(|\mathbf{x}_i - \mathbf{x}_j|^{-d})$ as $|\mathbf{x}_i - \mathbf{x}_j| \to \infty$.

To obtain the ensemble-average solution of $\langle\boldsymbol{\sigma}\rangle_i$ in the composite within the context of approximate pairwise inclusion interaction, one needs to integrate (8.48) over all possible positions $\mathbf{x}_j$ for a given location $\mathbf{x}_i$ $(i = 1, 2,\ j = 3 - i)$:

$$\mathbf{R}_i\langle\overline{\boldsymbol{\sigma}}\rangle(\mathbf{x}_i) = \mathbf{R}_i\Big[\mathbf{I} + \int \boldsymbol{\mathcal{T}}_i(\mathbf{x}_i - \mathbf{x}_j)\varphi(v_j, \mathbf{x}_j|v_i, \mathbf{x}_i)\, d\mathbf{x}_j\Big]\boldsymbol{\sigma}^0. \tag{8.59}$$

According to the assumptions considered in [515], [516], [525], the term in the rectangular brackets in Eq. (8.58) is reduced to the tensor $\mathbf{\Gamma}$ defined by Eqs. (4.8) and (36) from [515] and [525], respectively. The passage from Eq. (8.57) to (8.58) as well as the identical passages in related methods is a very popular manipulation in micromechanics (see, e.g., [327], [528], [591], [691], [692], [1227]), which in fact implies the use of one of equations (7.24). A contradiction of the verification of Eqs. (7.24) was considered in detail in Section 7.2. These passages are questionable for the following reasons. In the referenced papers by Ju and coauthors one uses in actual fact the hypothesis $\langle\widetilde{\boldsymbol{\sigma}}(\mathbf{x})_{i,j}\rangle_k = \langle\overline{\boldsymbol{\sigma}}(\mathbf{x})\rangle_k = \boldsymbol{\sigma}^0$, $(k = i, j,\ \mathbf{x} \in v_k)$ leading necessarily to the consistent perturbation method (considered in Section 10.1), differing from Eq. (8.58) by the second summand in the integrands. Then one comes to the known difficulties connected with the absence of absolute convergence of an improper integral in (8.58) since $\boldsymbol{\mathcal{T}}_j(\mathbf{x}_i - \mathbf{x}_j)$ is of the order $|\mathbf{x}_i - \mathbf{x}_j|^{-d}$ (see also Subsection 7.2.2). So, in [515], [516], [521], [523] one eliminated the difficulties produced by the dependence of a conditionally convergent integral on the shape of integration domain w by the use of the postulation of this shape. The size of integration domain in Eq. (8.58) is postulated in [327].

Taking the general Eq. (8.2) into account, for elimination of these errors we recast Eq. (8.58) in the following correct form:

$$\mathbf{R}_i\langle\overline{\boldsymbol{\sigma}}\rangle(\mathbf{x}_i) = \mathbf{R}_i\Big\{\mathbf{I} + \int \big[\boldsymbol{\mathcal{T}}_i(\mathbf{x}_i - \mathbf{x}_j)\varphi(v_j, \mathbf{x}_j|v_i, \mathbf{x}_i) - \mathbf{\Gamma}(\mathbf{x}_i - \mathbf{x}_j)\mathbf{R}_i n^{(1)}\big]\, d\mathbf{x}_j\Big\}\boldsymbol{\sigma}^0, \tag{8.60}$$

To obtain the effective compliance it is enough to find the relation between $\langle\boldsymbol{\sigma}\rangle_i$ and $\langle\boldsymbol{\sigma}\rangle$ and utilize the exact equality:

$$\langle\boldsymbol{\varepsilon}\rangle = \mathbf{M}^{(0)}\langle\boldsymbol{\sigma}\rangle + c^{(1)}\mathbf{M}_1^{(1)}\langle\boldsymbol{\sigma}\rangle_i, \tag{8.61}$$

Nevertheless, Ju and coauthors used an alternative equation:

$$\langle\boldsymbol{\varepsilon}\rangle = \mathbf{M}^{(0)}\langle\boldsymbol{\sigma}\rangle + c^{(1)}\widetilde{\mathbf{S}}\mathbf{M}_1^{(1)}\langle\boldsymbol{\sigma}\rangle_i. \tag{8.62}$$

The consequent substitution of Eq. (8.58) into Eq. (8.62) leading to the representation of the effective compliance $\mathbf{M}^*$ completes the account of Ju and coauthors' approach [513], [515], [516], [525]:

$$\mathbf{M}^* = \mathbf{M}^{(0)} + \langle\mathbf{R}_i^v V\rangle\Big\{\mathbf{I} + \widetilde{\mathbf{S}}\int \boldsymbol{\mathcal{T}}_i(\mathbf{x}_i - \mathbf{x}_j)\varphi(v_j, \mathbf{x}_j|v_i, \mathbf{x}_i)\, d\mathbf{x}_j\Big\}. \tag{8.63}$$

The formulae (8.63), which is equivalent to Eqs. (53) (3-D case) and (57) (2-D case) from [515] and [516], respectively, were applied to a wide class of nonlinear problems such as estimation of the second moment of stresses as well as plasticity and damage phenomena (see [518], [520], [524], [525], [526] where the previous references can be found). However, the internal inconsistency of the mentioned source approach of the estimation of both the effective elastic properties and the second moment of stresses will be discussed in detail in Chapter 14 and, because of this, the correctness of including these approaches into the nonlinear approaches of both plasticity and damage phenomena is questionable.

We will not consider in detail the verification and concrete representation of the so-called depolarization factor tensor $\widetilde{\mathbf{S}}$ (8.62) which for the case being considered coincides with Eshelby tensor: $\widetilde{\mathbf{S}} = \mathbf{S}$. But a simple comparison of Eqs. (8.61) and (8.62) yields the unique conclusion that Eq. (8.62) is valid if and only if $\widetilde{\mathbf{S}} \equiv \mathbf{I}$. In so doing, substitution of Eq. (8.60) into Eq. (8.61) leads to the correct representation of the effective compliance tensor:

$$\mathbf{M}^* = \mathbf{M}^{(0)} + \langle \mathbf{R}_i^v V \rangle \Big\{ \mathbf{I} + \int \big[\boldsymbol{\mathcal{T}}_i(\mathbf{x}_i - \mathbf{x}_j)\varphi(v_j, \mathbf{x}_j | v_i, \mathbf{x}_i) \\ - \boldsymbol{\Gamma}(\mathbf{x}_i - \mathbf{x}_j)\mathbf{R}_i n^{(1)} \big] d\mathbf{x}_j \Big\}^{-1}. \tag{8.64}$$

To put this another way, the obvious correction of two basic faulty Eqs. (8.58) and (8.62) reduces Ju and coauthors' approach to the known perturbation method (8.64) in which the interaction tensor $\boldsymbol{\mathcal{T}}_i(\mathbf{x}_i - \mathbf{x}_j)$ can be considered as a simplified approximation of the matrix $\mathbf{Z}_{ij}$ (8.25).

8.3 One-Particle Approximation of the MEFM and Mori-Tanaka Approach

8.3.1 One-Particle ("Quasi–Crystalline") Approximation of MEFM

The known "quasi–crystalline" approximation by Lax [649] (see also [170], [898]) in our notations has one of three equivalent forms ($\mathbf{x} \in v_i$):

$$\begin{aligned} \langle \boldsymbol{\sigma}(\mathbf{x}) | v_i, \mathbf{x}_i; v_j, \mathbf{x}_j \rangle &= \langle \boldsymbol{\sigma} \rangle_i, \\ \langle \overline{\boldsymbol{\sigma}}_i(\mathbf{x}) | v_i, \mathbf{x}_i; v_j, \mathbf{x}_j \rangle &= \langle \overline{\boldsymbol{\sigma}}_i \rangle, \\ \mathbf{Z}_{ij} &= \mathbf{I}\delta_{ij}. \end{aligned} \tag{8.65}$$

It should be mentioned that the basic closing assumption of the one-particle approximation of the effective field method states: "The random variables $\overline{\boldsymbol{\sigma}}_i$ do not depend statistically on the geometric characteristics and elastic constants of the defect v_i" (the citation from Kanaun [540], see also [545]) that necessarily leads to the "quasi-crystalline" approximation (8.65_2). We will see, however, that $\overline{\boldsymbol{\sigma}}_i$ is explictly depend on the correlation hole shape v_{ij}^0 which in turn depends on the shape of v_i.

Therefore, applying Eq. (8.65_1) immediately for closing of the infinite system (8.7) yields the equation

$$\overline{v}_i \langle \boldsymbol{\vartheta}_i \rangle = \overline{v}_i \boldsymbol{\vartheta}_i^0 + \mathbf{R}_i \sum_{q=1}^{N} \int \Big[\mathbf{T}_{iq}(\mathbf{x}_i - \mathbf{x}_q)\varphi(v_q, \mathbf{x}_q |; v_i, \mathbf{x}_i) \\ - \mathbf{T}_i(\mathbf{x}_i - \mathbf{x}_q) n^{(q)} \Big] \overline{v}_q \langle \boldsymbol{\vartheta}_q \rangle \, d\mathbf{x}_q$$

with the solution presented in the form (8.41) where the matrix $\mathbf{Y}^{-1}$ can be reduced to

$$(\mathbf{Y}^{-1})_{ij} = \mathbf{I}\delta_{ij} - \mathbf{R}_i \int [\mathbf{T}_{ij}(\mathbf{x}_i - \mathbf{x}_j)\varphi(v_j, \mathbf{x}_j|; v_i, \mathbf{x}_i) - \mathbf{T}_i(\mathbf{x}_i - \mathbf{x}_j)n^{(j)}]\, d\mathbf{x}_j. \quad (8.66)$$

To make further progress, use is made the hypothesis of "*ellipsoidal symmetry*" for the distribution of inclusions, introduced by Willis [1179] (see also [170], [898]):

$$\varphi(v_q, \mathbf{x}_q\ |; v_i, \mathbf{x}_i) = h_1(\rho), \quad \rho \equiv |\ (\mathbf{a}^0_{ij})^{-1}(\mathbf{x}_j - \mathbf{x}_i)\ | \quad (8.67)$$

when the conditional probability density function $\varphi(v_q, \mathbf{x}_q\ |; v_i, \mathbf{x}_i)$ depends on $\mathbf{x}_q - \mathbf{x}_i$ only through the combination $\rho = |(\mathbf{a}^0_{ij})^{-1}(\mathbf{x}_j - \mathbf{x}_i)|$, for some matrix $(\mathbf{a}^0_{ij})^{-1}$. A pair distribution function has "ellipsoidal symmetry" but with an ellipsoid shape differing from the one that defines the inclusion shape. Although the assumed statistics may not be exactly realized in any particular composite, the results of effective moduli estimations are explicit and easily to use. It is crucial for the analyst to be aware of their reasonable choice of the shape of "ellipsoidal" spatial correlation of inclusion location (see Section 18.3). For spherical inclusions the relation (8.67) is realized for a statistical isotropy of the composite structure. It is reasonable to assume that $(\mathbf{a}^0_{ij})^{-1}$ identifies a matrix of affine transformation that transfers the ellipsoid v^0_{ij} being the "excluded volume" ("correlation hole") into a unit sphere and, therefore, the representation of the matrix $\mathbf{Y}_{ij}$ can be simplified:

$$(\mathbf{Y}^{-1})_{ij} = \mathbf{I}\delta_{ij} - \mathbf{R}_i\mathbf{Q}^0_{ij}, \quad (8.68)$$

where $\mathbf{Q}^0_{ij} \equiv \mathbf{Q}(v^0_{ij})$ (see Eqs. (3.38) and (3.181)) is a constant for the ellipsoidal domain v^0_{ij}. For the sake of simplicity of the subsequent calculation we will usually assume that the shape of "correlation hole" v^0_{ij} does not depend on the inclusion v_j: $v^0_{ij} = v^0_i$ and $\mathbf{Q}^0_{ij} = \mathbf{Q}^0_i \equiv \mathbf{Q}(v^0_i)$. For spherical inclusions the relation (8.67) is realized for a statistical isotropy of the composite structure.

Thus, the final results may be significantly simplified under the following additional assumptions:

$$\langle V_j(\mathbf{y})\boldsymbol{\vartheta}_j\ |; v_i, \mathbf{x}_i\rangle = \mathbf{h}_2(\langle\boldsymbol{\vartheta}_j\rangle, \rho), \quad (8.69)$$

where the dependence of the function $\mathbf{h}_2$ on the geometrical parameters of the inclusion v_i is defined by the scalar value ρ (8.67). According to relation (8.69) the conditional averaging properties of the composite have level surfaces which are obtained from the ellipsoidal surfaces ∂v^0_i by the use of a homothetic transformation. Under the assumption of (8.65_2) the equality (8.69) is valid under the simplest conditional probability density (8.67). The found tensor $\mathbf{Y}_{ij}$ (8.68) can be used in all formulae of Section 8.2 for estimations of effective thermoelastic properties and stress concentrator factors.

According to Eqs. (3.38) and by virtue of the fact that the generalized function $\boldsymbol{\Gamma}(\mathbf{x})$ is a homogeneous function of order $-d$, we obtain – under the assumptions (8.69) and $v^0_{ij} = v^0_i$ – the following relation:

$$\sum_{q=1}^{N}\int \mathbf{\Gamma}(\mathbf{x}-\mathbf{x}_q)[\langle V_q(\mathbf{x}_q)\boldsymbol{\vartheta}(\mathbf{x}_q)|;v_i,\mathbf{x}_i\rangle - c^{(q)}\langle\boldsymbol{\vartheta}_q\rangle]\,d\mathbf{x}_q = \mathbf{Q}_i^0\sum_{q=1}^{N}\langle\boldsymbol{\vartheta}_q\rangle c^{(q)} \quad (8.70)$$

Taking the assumptions (8.39), (8.65_2), and (8.69) into account, Eq. (8.40) can be combined into a simple equation:

$$\overline{v}_i\langle\boldsymbol{\vartheta}_i\rangle = \overline{v}_i\boldsymbol{\vartheta}_i^0 + \mathbf{R}_i\mathbf{Q}_i^0\sum_{q=1}^{N}\langle\boldsymbol{\vartheta}_q\rangle c^{(q)}, \quad (8.71)$$

leading to the estimation of average effective fields and stresses inside the inclusions

$$\langle\overline{\boldsymbol{\sigma}}\rangle_i = \boldsymbol{\sigma}^0 + \mathbf{Q}_i^0\big[\mathbf{I}-\langle\mathbf{R}^v\mathbf{Q}^0V\rangle\big]^{-1}(\langle\mathbf{R}^vV\rangle\boldsymbol{\sigma}^0+\langle\mathbf{F}^vV\rangle), \quad (8.72)$$

$$\langle\boldsymbol{\sigma}\rangle_i = \mathbf{B}_i\boldsymbol{\sigma}^0 + \mathbf{C}_i + \mathbf{B}_i\mathbf{Q}_i^0\big[\mathbf{I}-\langle\mathbf{R}^v\mathbf{Q}^0V\rangle\big]^{-1}(\langle\mathbf{R}^vV\rangle\boldsymbol{\sigma}^0+\langle\mathbf{F}^vV\rangle), \quad (8.73)$$

where for convenience one introduces the notations $\mathbf{g}(\mathbf{x}) = \mathbf{g}^{(i)}(\mathbf{x})$, $n(\mathbf{x}) \equiv n^{(i)}$ at $\mathbf{x}\in v^{(i)}$, $(i = 1,\dots,N;\quad \mathbf{g} = \mathbf{R},\mathbf{F},\mathbf{B},\mathbf{Q}^0)$ and $\mathbf{f},\mathbf{g},n(\mathbf{x}) \equiv 0$ otherwise. Hereafter, the tensors $\mathbf{C}(\mathbf{x})$, $\mathbf{F}(\mathbf{x})$ depend on the average stresses in the matrix

$$\begin{aligned}\langle\boldsymbol{\sigma}\rangle^{(0)} =& \mathbf{D}^c\Big[(\mathbf{I}-\langle\mathbf{B}V\rangle)\boldsymbol{\sigma}^0 - \langle\overline{\mathbf{C}}V\rangle\\ &- \langle\mathbf{B}\mathbf{Q}^0V\rangle(\mathbf{I}-\langle\mathbf{R}^v\mathbf{Q}^0V\rangle)^{-1}(\langle\mathbf{R}^vV\rangle\boldsymbol{\sigma}^0+\langle\overline{\mathbf{F}}^vV\rangle)\Big],\\ \mathbf{D}^c =& \Big\{c^{(0)}\mathbf{I} + \langle\mathbf{B}\mathbf{Q}V\rangle\mathbf{M}_1^{(0)}\\ &+ \langle\mathbf{B}\mathbf{Q}^0V\rangle(\mathbf{I}-\langle\mathbf{R}^v\mathbf{Q}^0V\rangle)^{-1}\langle\mathbf{Q}^{-1}\mathbf{B}\mathbf{Q}V\rangle\mathbf{M}_1^{(0)}\Big\}^{-1},\end{aligned} \quad (8.74)$$

where the tensor $\mathbf{D}^c \equiv c^{(0)}\mathbf{I}$ if $\mathbf{M}^c = \mathbf{M}^{(0)}$. Equation (8.74) permits the obtaining of an explicit representation of average stresses inside the inclusions:

$$\begin{aligned}\langle\boldsymbol{\sigma}\rangle_i =& \mathbf{B}_i\boldsymbol{\sigma}^0 + \overline{\mathbf{C}}_i + \mathbf{B}_i\mathbf{Q}_i^0\big[\mathbf{I}-\langle\mathbf{R}^v\mathbf{Q}^0V\rangle\big]^{-1}(\langle\mathbf{R}^vV\rangle\boldsymbol{\sigma}^0+\langle\overline{\mathbf{F}}^vV\rangle)\\ &+\Big[-\mathbf{B}_i\mathbf{Q}_i + \mathbf{B}_i\mathbf{Q}_i^0(\mathbf{I}-\langle\mathbf{R}^vV\mathbf{Q}^0\rangle)^{-1}\langle\mathbf{Q}^{-1}\mathbf{B}\mathbf{Q}V\rangle\Big]\mathbf{M}_1^{(0)}\mathbf{D}^c\Big[\boldsymbol{\sigma}^0-\langle\mathbf{B}V\rangle\boldsymbol{\sigma}^0\\ &-\langle\overline{\mathbf{C}}V\rangle - \langle\mathbf{B}\mathbf{Q}^0V\rangle(\mathbf{I}-\langle\mathbf{R}^v\mathbf{Q}^0V\rangle)^{-1}(\langle\mathbf{R}^vV\rangle\boldsymbol{\sigma}^0+\langle\overline{\mathbf{F}}^vV\rangle)\Big].\end{aligned} \quad (8.75)$$

Substitution of Eqs. (8.18_1) and (8.75) into (1.12) and (8.13_1) gives the representation for effective properties:

$$\begin{aligned}\mathbf{M}^* =& \mathbf{M}^c + \langle\mathbf{R}^{0v}V\rangle + \langle\mathbf{R}^{0v}\mathbf{Q}^0V\rangle\big[\mathbf{I}-\langle\mathbf{R}^v\mathbf{Q}^0V\rangle\big]^{-1}\langle\mathbf{R}^vV\rangle\\ &+\Big[-\langle\mathbf{R}^{0v}V\rangle + \langle\mathbf{R}^{0v}\mathbf{Q}^0V\rangle(\mathbf{I}-\langle\mathbf{R}^vV\mathbf{Q}^0\rangle)^{-1}\langle\mathbf{Q}^{-1}\mathbf{B}\mathbf{Q}V\rangle\Big]\mathbf{M}_1^{(0)}\mathbf{D}^c\\ &\times\Big\{\mathbf{I}-\langle\mathbf{B}V\rangle - \langle\mathbf{B}\mathbf{Q}^0V\rangle\big[\mathbf{I}-\langle\mathbf{R}^v\mathbf{Q}^0V\rangle\big]^{-1}\langle\mathbf{R}^vV\rangle\Big\},\end{aligned} \quad (8.76)$$

$$\begin{aligned}\boldsymbol{\beta}^* =& \boldsymbol{\beta}^{(0)} + \big[\mathbf{I}-\mathbf{M}_1^{(0)}\langle\mathbf{R}^v\mathbf{Q}^0V\rangle\big]\big[\mathbf{I}-\langle\mathbf{R}^v\mathbf{Q}^0V\rangle\big]^{-1}\langle\overline{\mathbf{F}}^vV\rangle\\ &-\Big[-\langle\mathbf{R}^{0v}V\rangle + \langle\mathbf{R}^{0v}\mathbf{Q}^0V\rangle(\mathbf{I}-\langle\mathbf{R}^vV\mathbf{Q}^0\rangle)^{-1}\langle\mathbf{Q}^{-1}\mathbf{B}\mathbf{Q}V\rangle\Big]\mathbf{M}_1^{(0)}\mathbf{D}^c\end{aligned}$$

$$\times \left\{ \langle \overline{\mathbf{C}} V \rangle + \langle \mathbf{B} \mathbf{Q}^0 V \rangle \left[\mathbf{I} - \langle \mathbf{R}^v \mathbf{Q}^0 V \rangle \right]^{-1} \langle \overline{\mathbf{F}}^v V \rangle \right\}, \tag{8.77}$$

$$W^* = -\frac{1}{2} \left\{ \langle \boldsymbol{\beta}_1 \overline{\mathbf{C}} V \rangle + \langle \boldsymbol{\beta}_1 \mathbf{B} \mathbf{Q}^0 V \rangle \left[\mathbf{I} - \langle \mathbf{R}^v \mathbf{Q}^0 V \rangle \right]^{-1} \langle \overline{\mathbf{F}}^v V \rangle \right\}$$
$$+ \frac{1}{2} \Big[- \langle \boldsymbol{\beta}_1 \mathbf{B} \mathbf{Q} \rangle + \langle \boldsymbol{\beta}_1 \mathbf{B}_i \mathbf{Q}_i^0 \rangle (\mathbf{I} - \langle \mathbf{R}^v V \mathbf{Q}^0 \rangle)^{-1} \langle \mathbf{Q}^{-1} \mathbf{B} \mathbf{Q} V \rangle \Big] \mathbf{M}_1^{(0)} \mathbf{D}^c$$
$$\times \left\{ \langle \overline{\mathbf{C}} V \rangle + \langle \mathbf{B} \mathbf{Q}^0 V \rangle \left[\mathbf{I} - \langle \mathbf{R}^v \mathbf{Q}^0 V \rangle \right]^{-1} \langle \overline{\mathbf{F}}^v V \rangle \right\}. \tag{8.78}$$

Taking into account the assumption of coincidence of elastic moduli of the comparison medium and matrix (7.16) (when $\mathbf{C}_i(\mathbf{x}) = \overline{\mathbf{C}}_i(\mathbf{x}),\ \mathbf{F}_i(\mathbf{x}) = \overline{\mathbf{F}}_i(\mathbf{x})$) leads to simplification of Eqs (8.73)–(8.78) obtained in [182] by another method:

$$\mathbf{M}^* = \mathbf{M}^{(0)} + \Big[\mathbf{I} - \langle \mathbf{R}^v \mathbf{Q}^0 V \rangle \Big]^{-1} \langle \mathbf{R}^v V \rangle, \tag{8.79}$$

$$\boldsymbol{\beta}^* = \boldsymbol{\beta}^{(0)} + (\mathbf{M}^* - \mathbf{M}^{(0)}) \langle \mathbf{R}^v V \rangle^{-1} \langle \mathbf{F}^v V \rangle, \tag{8.80}$$

$$W^* = -\frac{1}{2} \left\{ \langle \boldsymbol{\beta}_1 \mathbf{C} V \rangle + \langle \boldsymbol{\beta}_1 \mathbf{B} \mathbf{Q}^0 V \rangle (\mathbf{M}^* - \mathbf{M}^{(0)}) \langle \mathbf{R}^v V \rangle^{-1} \langle \mathbf{F}^v V \rangle \right\}. \tag{8.81}$$

Equations for the statistical averages of the stresses (8.74) and related fields inside the heterogeneity v_i are also simplified:

$$\langle \overline{\boldsymbol{\sigma}} \rangle_i = \boldsymbol{\sigma}^0 + \mathbf{Q}_i^0 \left[\mathbf{I} - \langle \mathbf{R}^v \mathbf{Q}^0 V \rangle \right]^{-1} (\langle \mathbf{R}^v V \rangle \boldsymbol{\sigma}^0 + \langle \mathbf{F}^v V \rangle), \tag{8.82}$$

$$\langle \boldsymbol{\sigma} \rangle_i(\mathbf{x}) = \mathbf{B}_i(\mathbf{x}) \boldsymbol{\sigma}^0 + \mathbf{C}_i(\mathbf{x})$$
$$+ \mathbf{B}_i(\mathbf{x}) \mathbf{Q}_i^0 \left[\mathbf{I} - \langle \mathbf{R}^v \mathbf{Q}^0 V \rangle \right]^{-1} (\langle \mathbf{R}^v V \rangle \boldsymbol{\sigma}^0 + \langle \mathbf{F}^v V \rangle), \tag{8.83}$$

$$\bar{v}_i \langle \boldsymbol{\eta}_i \rangle(\mathbf{x}) = \mathbf{R}_i(\mathbf{x}) \boldsymbol{\sigma}^0 + \mathbf{F}_i(\mathbf{x})$$
$$+ \mathbf{R}_i(\mathbf{x}) \mathbf{Q}_i^0 \left[\mathbf{I} - \langle \mathbf{R}^v \mathbf{Q}^0 V \rangle \right]^{-1} (\langle \mathbf{R}^v V \rangle \boldsymbol{\sigma}^0 + \langle \mathbf{F}^v V \rangle), \tag{8.84}$$

The relations (8.80) and (8.81) do not result as a generalization of Eqs. (8.36) and (8.37) to any number of components, but they were obtained under the additional assumptions (8.65_2) and (8.67). For homogeneous inclusions (3.187) Eqs. (8.79) and (8.80) reduce to some results found in [667], [669] and in [898] (see also [545], [1036]).

Thus, if the distribution of inclusions is taken to be the same for all inclusions pairs and, therefore, in Eq. (8.70) $\mathbf{Q}_i \equiv \mathbf{Q}(v^{el}) =$ const. $\forall i$ for some ellipsoidal domain v^{el}, then Eq. (8.79) rewritten in terms of elastic moduli coincides with Eq. (8.52) from [898]. If additionally one assumes that v^{el} has the form a sphere, then Eqs. (8.71) and (8.79) reduce to the results obtained in [669]. For identical homogeneous ellipsoidal inclusions randomly oriented in space, Eq. (8.79) coincides with the result [667]. If one assumes that $\langle \mathbf{R}^v \mathbf{Q}^0 \rangle = \langle \mathbf{R}^v \rangle \langle \mathbf{Q}^0 \rangle$, then Eq. (8.79) reduces to the result [545].

It should be mentioned that for multicomponent media with $N \gg 1$ (for example, for random orientation of inclusions) in the framework of the hypothesis (8.65_2), either Eqs. (8.73)–(8.78) or Eqs. (8.79)–(8.81) are more convenient from the computational point of view than the direct use of more general Eqs. (8.49)–(8.51) or (8.52)–(8.54), respectively. Indeed, the assumption (8.65_2) yields to the termination of integrals numerically integrable in Eq. (8.44), but the difficulties induced by inversion of multidimensional matrix $\mathbf{Y}^{-1}$ (8.44) remain. In so doing the utilization of Eq. (8.73)–(8.78) implies only estimation of statistical averages of a few tensors $\langle \mathbf{B}V\rangle$, $\langle \mathbf{C}V\rangle$, $\langle \mathbf{R}^vV\rangle$, $\langle \mathbf{F}^vV\rangle$, $\langle \mathbf{B}\mathbf{Q}^0V\rangle$, $\langle \boldsymbol{\beta}_1\mathbf{C}V\rangle$, $\langle \boldsymbol{\beta}_1\mathbf{B}\mathbf{Q}^0V\rangle$ that would almost always be preferable for computation.

It should be emphasized that Eqs. (8.80) and (8.81) for one kind of coated inclusions ($N=1$) formally coincide with analogous equations obtained in [184] only by the use of the hypothesis **H1**. Thus, in the framework of the hypothesis **H1** the result [184] can be considered as a generalization of classical results by Levin [665] (see also [945]], which cannot be said about Eqs. (8.80) and (8.81) obtained under the additional assumptions (8.65_2) and (4.7).

Notice that the term one particle for the method considered in this subsection stands for the hypothesis (8.65_2) according to which there is no direct influence of surrounding inclusions v_j on the inclusion v_i being considered. However, such an influence takes place in an average sense via the action of the second summand in the right-hand side of Eq. (8.71). It is interesting that for periodic structure composites considered as a particular realization of a random structure, the hypothesis (8.65_2) is valid exactly. Then the MEFM implies the utilization of the hypothesis **H1** only leading to a high accuracy (in a comparison with exact analytical results) of estimations of both the effective local and nonlocal effective properties (see for details Chapter 13).

A dual structure of effective properties can be derived for a case when the composite is subjected to a boundary displacement corresponding to a uniform strain $\langle \boldsymbol{\varepsilon}\rangle$:

$$\mathbf{L}^* = \mathbf{L}^{(0)} + \big[\mathbf{I} - \langle \mathbf{R}^{\epsilon v}\mathbf{P}^0 V\rangle\big]^{-1}\langle \mathbf{R}^{\epsilon v}V\rangle, \tag{8.85}$$

$$\boldsymbol{\alpha}^* = \boldsymbol{\alpha}^{(0)} + (\mathbf{L}^* - \mathbf{L}^{(0)})\langle \mathbf{R}^{\epsilon v}V\rangle^{-1}\langle \mathbf{F}^{\epsilon v}V\rangle. \tag{8.86}$$

Hereafter $\mathbf{A}$, $\mathbf{C}^{\epsilon}$, $\mathbf{R}^{\epsilon v}$, $\mathbf{F}^{\epsilon v}$ are the concentrator factors analogous to $\mathbf{B}$, $\mathbf{C}$, $\mathbf{R}^v$, $\mathbf{F}^v$ (8.18), respectively ($\mathbf{x}\in v_i\subset v^{(i)}$):

$$\begin{aligned}\boldsymbol{\varepsilon}(\mathbf{x}) &= \mathbf{A}^{(i)}(\mathbf{x})\overline{\boldsymbol{\varepsilon}}(\mathbf{x}_i) + \mathbf{C}^{\epsilon(i)}(\mathbf{x}),\\ \mathbf{L}_1(\mathbf{x})\boldsymbol{\varepsilon}(\mathbf{x}) &= \mathbf{R}^{\epsilon v(i)}(\mathbf{x})\overline{\boldsymbol{\varepsilon}}(\mathbf{x}_i) + \mathbf{F}^{\epsilon v(i)}(\mathbf{x}),\end{aligned} \tag{8.87}$$

It is important that the effective moduli $\mathbf{L}^*$ and compliance $\mathbf{M}^*$ under the different boundary conditions are identical. This can be readily proved by invoking the inter-relations (3.120) and (3.121) leading from Eqs. (8.18_1) and (8.87_1) under nonrestrictive equalities $\overline{\boldsymbol{\sigma}}\equiv\boldsymbol{\sigma}^0=\mathbf{L}^{(0)}\boldsymbol{\varepsilon}^0\equiv\mathbf{L}^{(0)}\overline{\boldsymbol{\varepsilon}}$ and $\boldsymbol{\sigma}(\mathbf{x})=\mathbf{L}^{(i)}(\mathbf{x})\boldsymbol{\varepsilon}(\mathbf{x})$ ($\mathbf{x}\in v_i\subset v^{(i)}$). It then follows that $\mathbf{L}^*$ and $\mathbf{M}^*$ can be recast in the form

$$\mathbf{L}^* = \Big[\mathbf{I} - \langle \mathbf{R}^{\epsilon v}\mathbf{P}^0V\rangle\Big]^{-1}\mathbf{L}^{(0)}\Big[\mathbf{I} - \langle \mathbf{R}^v\mathbf{Q}^0V\rangle\Big], \tag{8.88}$$

$$\mathbf{M}^* = \Big[\mathbf{I} - \langle \mathbf{R}^v\mathbf{Q}^0V\rangle\Big]^{-1}\mathbf{M}^{(0)}\Big[\mathbf{I} - \langle \mathbf{R}^{\epsilon v}\mathbf{P}^0V\rangle\Big], \tag{8.89}$$

providing the desirable reciprocal relation between the two. Now that their equivalency has been proved, one may choose either expression for the determination of the effective properties (8.88) or (8.89).

8.3.2 Mori-Tanaka Approach

As pointed out by Benveniste [52], the essential assumption in the Mori-Tanaka method (MTM) states that each inclusion v_i behaves as an isolated one in the infinite matrix ($\mathbf{M}^c \equiv \mathbf{M}^{(0)}$) and subject to some effective stress field $\overline{\boldsymbol{\sigma}}_i(\mathbf{x})$ (8.4) coinciding with the average stress in the matrix:

$$\overline{\boldsymbol{\sigma}}_i \equiv \langle \boldsymbol{\sigma} \rangle^{(0)}. \tag{8.90}$$

The composite is assumed to be subjected to a boundary traction giving rise to a uniform stress $\langle \boldsymbol{\sigma} \rangle$. The assumption (8.90) allows one to define uniquely the effective thermoelastic properties of multicomponent composite materials. Thus, hypothesis (8.90) is more restricted than hypothesis **H1** which gives an opportunity to use the solution (8.17) for the each inclusion v_i and to find the average stress in the matrix by the use of a representation of the average stress in the whole composite over the average stresses in the separate phases:

$$\langle \boldsymbol{\sigma} \rangle \equiv c^{(0)} \langle \boldsymbol{\sigma} \rangle^{(0)} + \langle \boldsymbol{\sigma} V \rangle = c^{(0)} \langle \boldsymbol{\sigma} \rangle^{(0)} + \langle \mathbf{B} V \rangle \langle \boldsymbol{\sigma} \rangle^{(0)} + \langle \mathbf{C} V \rangle, \tag{8.91}$$

leading to the representations of statistical average of both the stresses in the matrix and the strain polarization tensor $\boldsymbol{\eta}$ in the inclusions

$$\langle \boldsymbol{\sigma} \rangle^{(0)} = (c^{(0)} \mathbf{I} + \langle \mathbf{B} V \rangle)^{-1} (\langle \boldsymbol{\sigma} \rangle - \langle \mathbf{C} V \rangle), \tag{8.92}$$

$$\langle \boldsymbol{\eta} \rangle_i(\mathbf{x}) = \mathbf{R}^{(i)}(\mathbf{x}) (c^{(0)} \mathbf{I} + \langle \mathbf{B} V \rangle)^{-1} (\langle \boldsymbol{\sigma} \rangle - \langle \mathbf{C} V \rangle) + \mathbf{F}^{(i)}(\mathbf{x}). \tag{8.93}$$

Substituting of Eq. (8.93) into the average constitutive equation (6.2) leads to

$$\begin{aligned} \langle \boldsymbol{\varepsilon} \rangle \equiv \mathbf{M}^{(0)} \langle \boldsymbol{\sigma} \rangle + \boldsymbol{\beta}^{(0)} + \langle \boldsymbol{\eta} V \rangle &= \left[\mathbf{M}^{(0)} + \langle \mathbf{R}^v V \rangle (c^{(0)} \mathbf{I} + \langle \mathbf{B} V \rangle)^{-1} \right] \langle \boldsymbol{\sigma} \rangle \\ &+ \boldsymbol{\beta}^{(0)} + \langle \mathbf{F} V \rangle - \langle \mathbf{R}^v V \rangle (c^{(0)} \mathbf{I} + \langle \mathbf{B} V \rangle)^{-1} \langle \mathbf{C} V \rangle. \end{aligned} \tag{8.94}$$

Comparison of the definition of effective parameters $\mathbf{M}^*$ and $\boldsymbol{\beta}^*$ (1.12) with Eq. (8.94) yields to the representation of effective properties and the effective field:

$$\mathbf{M}^* = \mathbf{M}^{(0)} + \langle \mathbf{R}^v V \rangle (c^{(0)} \mathbf{I} + \langle \mathbf{B} V \rangle)^{-1}, \tag{8.95}$$

$$\boldsymbol{\beta}^* = \boldsymbol{\beta}^{(0)} + \langle \mathbf{F} V \rangle - \langle \mathbf{R}^v V \rangle (c^{(0)} \mathbf{I} + \langle \mathbf{B} V \rangle)^{-1} \langle \mathbf{C} V \rangle, \tag{8.96}$$

$$\langle \overline{\boldsymbol{\sigma}}_i \rangle = (c^{(0)} \mathbf{I} + \langle \mathbf{B} V \rangle)^{-1} (\langle \boldsymbol{\sigma} \rangle - \langle \mathbf{C} V \rangle), \tag{8.97}$$

From the definition of the stored energy (8.13_1), in association with Eqs. (8.13_2), and (8.92) we obtain

$$2W^* = \langle \boldsymbol{\beta}_1 V \mathbf{B} \rangle (c^{(0)} \mathbf{I} + \langle \mathbf{B} V \rangle)^{-1} \langle \mathbf{C} V \rangle - \langle \boldsymbol{\beta}_1 V \mathbf{C} \rangle. \tag{8.98}$$

Equations (8.95), (8.96), and (8.93) can be recast in an equivalent form convenient for comparison with the analogous representations obtained by the MEFM:

$$\mathbf{M}^* = \mathbf{M}^{(0)} + \langle \mathbf{R}^v V \rangle (\mathbf{I} - \langle \mathbf{Q} \mathbf{R}^v V \rangle)^{-1}, \tag{8.99}$$

$$\boldsymbol{\beta}^* = \boldsymbol{\beta}^{(0)} + \langle \mathbf{F}^v V \rangle - \langle \mathbf{R}^v V \rangle (\mathbf{I} - \langle \mathbf{Q} \mathbf{R}^v V \rangle)^{-1} \langle \mathbf{C} V \rangle, \tag{8.100}$$

$$\langle \boldsymbol{\eta} \rangle_i(\mathbf{x}) = \mathbf{R}_i^v(\mathbf{x}) (\mathbf{I} - \langle \mathbf{Q} \mathbf{R}^v V \rangle)^{-1} \left[\langle \boldsymbol{\sigma} \rangle - \langle \mathbf{C} V \rangle \right] + \mathbf{F}_i^v(\mathbf{x}). \tag{8.101}$$

Transformation of Eqs. (8.95), (8.96) to Eqs. (8.99), (8.100) was carried out by the use of an equality (8.21) that is valid for ellipsoidal inclusion of any microstructure.

Though once $\mathbf{M}^*$ is found, the displacement at infinity is prescribed to correspond to a uniform strain $\langle\boldsymbol{\varepsilon}\rangle$ that leads to the following representations for the effective thermoelastic properties and effective fields:

$$\mathbf{L}^* = \mathbf{L}^{(0)} + \langle\mathbf{R}^{\varepsilon v}V\rangle(c^{(0)}\mathbf{I} + \langle\mathbf{A}V\rangle)^{-1}, \tag{8.102}$$
$$\boldsymbol{\alpha}^* = \boldsymbol{\alpha}^{(0)} + \langle\mathbf{F}^{\epsilon}V\rangle - \langle\mathbf{R}^{\varepsilon v}V\rangle(c^{(0)}\mathbf{I} + \langle\mathbf{A}V\rangle)^{-1}\langle\mathbf{C}^{\epsilon}V\rangle, \tag{8.103}$$
$$\langle\overline{\boldsymbol{\varepsilon}}_i\rangle = (c^{(0)}\mathbf{I} + \langle\mathbf{A}V\rangle)^{-1}(\langle\boldsymbol{\varepsilon}\rangle - \langle\mathbf{C}^{\epsilon}V\rangle). \tag{8.104}$$

Taking the equalities (8.94) into account, we can recast the $\mathbf{L}^*$ and $\mathbf{M}^*$ in the form

$$\mathbf{L}^* = [c^{(0)}\mathbf{I} + \langle\mathbf{B}V\rangle]\mathbf{L}^{(0)}[c^{(0)}\mathbf{I} + \langle\mathbf{A}V\rangle]^{-1}, \tag{8.105}$$
$$\mathbf{M}^* = [c^{(0)}\mathbf{I} + \langle\mathbf{A}V\rangle]\mathbf{M}^{(0)}[c^{(0)}\mathbf{I} + \langle\mathbf{B}V\rangle]^{-1}, \tag{8.106}$$

providing the reciprocal relation $(\mathbf{M}^*)^{-1} = \mathbf{L}^*$.

It should be mentioned that the stress concentrator factor at the representative inclusion v_i defined by Eq. (8.18_1) depends on two sorts of stress concentrator factors. The first one is a "local" stress concentrator factor defined by the solution for one isolated inclusion in the infinite medium that is described by the tensors $\mathbf{B}_i(\mathbf{x})$ and $\mathbf{C}_i(\mathbf{x})$ and which is the same in both the MEFM and MTM. For example, for a pure elastic problem $\boldsymbol{\beta}_1 \equiv \mathbf{0}$, we have (no summation over i)

$$\langle\boldsymbol{\sigma}\rangle_i(\mathbf{x}) = \mathbf{B}_i(\mathbf{x})\mathbf{B}_i^*\langle\boldsymbol{\sigma}\rangle. \tag{8.107}$$

The second sort of stress concentrator factor $\mathbf{B}_i^*$ is governed by the "global" stress concentrator factors at the effective field either (8.82) or (8.97) and differs in the MEFM and the MTM. Moreover, the effective field (8.97) does not depend on the geometrical parameters of the inclusion v_i, while the effective field (8.82) estimated by the MEFM explicitly depends on both the orientation and shape of the inclusion v_i through the tensor $\mathbf{Q}_i^0$. In particular, for 3D uniform random orientation of inclusions, the tensor $\mathbf{B}_i^*$ estimated by the MTM is an isotropic and does not depend on the orientation and the shape of the inclusion v_i. None of the mentioned properties of the tensor $\mathbf{B}_i^*$ are fulfilled in the case of evaluation of this tensor by the MTM.

The MTM as defined by Eqs. (8.95) has been called the "direct approach" by Benveniste [52], who showed it is identical for composites with the ellipsoidal homogeneous inclusions to the *"equivalent inclusion-average stress"* formalism (see, e.g., [817], [1065], [1170], [1171]) which has played a key role in micromechanics and was originally proposed by Eshelby [318], [319] (see Subsection 3.6.1) for one ellipsoidal homogeneous inclusion in the infinite matrix. Namely, the average strains produced by the homogeneous boundary conditions (6.5) in the composite and in the matrix are given respectively by (following [1171])

$$\langle\boldsymbol{\varepsilon}\rangle = \mathbf{M}^*\langle\boldsymbol{\sigma}\rangle, \quad \langle\boldsymbol{\varepsilon}\rangle^{(0)} = \mathbf{M}^{(0)}\langle\boldsymbol{\sigma}\rangle^{(0)}. \tag{8.108}$$

We will also assume that the heterogeneities are subjected to the stress free strains $\boldsymbol{\beta}_1^{(i)}(\mathbf{x}) \equiv \text{const}$, $(\mathbf{x} \in v^{(i)},\ \boldsymbol{\beta}^{(0)} \equiv \mathbf{0})$. In each heterogeneity v_i the average strain field and stress fields may be written as the superposition of *mean fields* in the matrix with the additional perturbed strain $\boldsymbol{\varepsilon}_i^{\text{pt}}$ and stress $\boldsymbol{\sigma}_i^{\text{pt}}$:

$$\langle \boldsymbol{\varepsilon} \rangle_i = \langle \boldsymbol{\varepsilon} \rangle^{(0)} - \boldsymbol{\beta}_1^{(i)} + \boldsymbol{\varepsilon}_i^{\text{pt}}, \quad \langle \boldsymbol{\sigma} \rangle_i = \langle \boldsymbol{\sigma} \rangle^{(0)} + \boldsymbol{\sigma}_i^{\text{pt}}, \tag{8.109}$$

which is the essence of *Mori-Tanaka mean field theory*. The argument (which is plausible rather than rigorous) given by Mori-Tanaka is as follows. Since there are many heterogeneities, adding a single heterogeneity into the infinite ensemble will not alter the mean field values $\langle \boldsymbol{\varepsilon} \rangle^{(0)}$ and $\langle \boldsymbol{\sigma} \rangle^{(0)}$, and therefore, the average strains and stresses inside an inclusion can be expressed by (8.109_1) and (8.109_2). The average stress inside the inclusion $\langle \boldsymbol{\sigma} \rangle_i$ then will follow from the stress-strain relation as well as from Eshelby's equivalent-inclusion principle:

$$\langle \boldsymbol{\sigma} \rangle_i = \mathbf{L}^{(i)}(\langle \boldsymbol{\varepsilon} \rangle^{(0)} - \boldsymbol{\beta}_1^{(i)} + \boldsymbol{\varepsilon}_i^{\text{pt}}) = \mathbf{L}^{(0)}(\langle \boldsymbol{\varepsilon} \rangle^{(0)} - \boldsymbol{\beta}_1^{(i)} + \boldsymbol{\varepsilon}_i^{\text{pt}} - \boldsymbol{\varepsilon}_i^*), \tag{8.110}$$

differing from the analogous equation for one heterogeneity (3.133) inside the infinite homogeneous matrix by the representation of the average strain inside $v^{(i)}$ in the decomposition form $\langle \boldsymbol{\varepsilon} \rangle_i = \langle \boldsymbol{\varepsilon} \rangle^{(0)} + \boldsymbol{\varepsilon}_i^{\text{pt}}$, where the perturbed part $\boldsymbol{\varepsilon}_i^{\text{pt}}$ with respect to the strain in the matrix is further related to Eshelby's equivalent "stress free strain" $\boldsymbol{\varepsilon}_i^*$ through Eshelby's tensor as (see Eq. (3.140): $\boldsymbol{\varepsilon}_i^{\text{pt}} = \mathbf{S}_i(\boldsymbol{\varepsilon}_i^* + \boldsymbol{\beta}_1^{(i)})$. Equation ($8.109_2$) also provides that the equivalent "eigenstrain" $\boldsymbol{\varepsilon}_i^* = \langle \boldsymbol{\varepsilon} \rangle_i - \mathbf{M}^{(0)}\langle \boldsymbol{\sigma} \rangle_i - \boldsymbol{\beta}_1^{(i)}$, or

$$\boldsymbol{\varepsilon}_i^* = (\mathbf{M}^{(i)} - \mathbf{M}^{(0)})\langle \boldsymbol{\sigma} \rangle_i.$$

It is interesting that for a pure mechanical case ($\boldsymbol{\beta}(\mathbf{x}) \equiv \mathbf{0}$) the equivalent "eigenstrain" $\boldsymbol{\varepsilon}_i^*$ is identical to the average strain polarization tensor $\langle \boldsymbol{\eta} \rangle_i$ inside the inclusion v_i.

We will recast the average stress in the ith phase (8.109_2) by expressing of tensors $\mathbf{S}_i$, $\mathbf{L}_1^{(i)}$, and $\mathbf{L}^{(0)}$ through the tensors $\mathbf{Q}_i$, $\mathbf{M}_1^{(i)}$, and $\mathbf{M}^{(0)}$ (see Section 3.5), respectively, and find

$$\langle \boldsymbol{\sigma} \rangle_i = \mathbf{Q}_i(\boldsymbol{\varepsilon}_i^* + \boldsymbol{\beta}^{(i)}) + \langle \boldsymbol{\sigma} \rangle^{(0)} = \langle \boldsymbol{\sigma} \rangle^{(0)} - \mathbf{Q}_i(\mathbf{M}_i^{(i)}\langle \boldsymbol{\sigma} \rangle_i + \boldsymbol{\beta}_1^{(i)}), \tag{8.111}$$

or rearranging,

$$\langle \boldsymbol{\sigma} \rangle_i = \mathbf{B}_i\langle \boldsymbol{\sigma} \rangle^{(0)} - \mathbf{B}_i\mathbf{Q}_i\boldsymbol{\beta}_1^{(i)}, \tag{8.112}$$

where $\mathbf{B}_i = (\mathbf{I} + \mathbf{Q}_i\mathbf{M}_1^{(i)})^{-1}$ (3.121) is recognizable as the familiar stress-concentrator tensor of the single inclusion (with $\mathbf{L}^{(i)}$) embedded in the infinite matrix (with $\mathbf{L}^{(0)}$) subjected to the uniform stress field $\langle \boldsymbol{\sigma} \rangle^{(0)}$. For the homogeneous ellipsoidal heterogeneities being assumed in Eqs. (8.109)–(8.111), the basic hypothesis (8.90) of the direct MTM is equivalent to Eq. (8.112) due to the Eqs. (8.18_1) and (8.22). In such a case, reducing Eq. (8.112) to (8.90), repeating subsequent mathematics (8.91)–(8.101) results in the conclusion that Eq. (8.112) leads to the representations for the effective thermoelastic properties

(8.99), (8.100) and average strain polarization tensor (8.101) for the composites with homogeneous ellipsoidal heterogeneities.

However, the direct method has a fundamental advantage with respect to the method of equivalent inclusion (8.108)–(8.111) because at the derivative of the initial representation (8.95) no assumptions about both the shape of heterogeneities and their microstructure were used. In so doing, reduction of Eq. (8.95) to Eq. (8.99) does not exclude the possibility that $\mathbf{M}_1(\mathbf{x})$ vanishes at the part of inclusions v_i used for the introduction of the tensors $\mathbf{Q}_i = \mathbf{Q}_i(v_i)$ in Eq. (8.99) (see the comments at the end of Subsection 8.1.3). The direct method is based on a single hypothesis (8.90) which was implicitly used in Eq. (8.109_1) in parallel with the additional assumptions of both the homogeneity of the tensor $\mathbf{L}^{(i)}(\mathbf{x}) = \text{const.}$ ($\mathbf{x} \in v^{(i)}$) and ellipsoidal shape of heterogeneities. Moreover, exploiting a single strong hypothesis (8.90) in the direct method also has a specific merit in shedding new light on the weakness of the direct method that opens up outstanding possibilities to improve the method as was done in the MEFM's hypothesis (8.65_2).

8.3.3 Effective Properties Estimated via the MEF and MTM at $\mathbf{Q}_i \equiv \mathbf{Q}_i^0$

At first, we will recast Eqs. (8.79) and (8.80) accompanied with the representation for the average strain polarization tensor $\langle \boldsymbol{\eta}_i \rangle(\mathbf{x})$ (8.71) obtained by the MEFM in an equivalent form convenient for subsequent analysis

$$\mathbf{M}^* = \mathbf{M}^{(0)} + \langle \mathbf{R}^v V\rangle \big[\mathbf{I} - \langle \mathbf{R}^v V\rangle^{-1} \langle \mathbf{R}^v \mathbf{Q}^0 V\rangle \langle \mathbf{R}^v V\rangle\big]^{-1}, \tag{8.113}$$

$$\begin{aligned}\boldsymbol{\beta}^* = \boldsymbol{\beta}^{(0)} + \langle \mathbf{R}^v V\rangle \big[\mathbf{I} - \langle \mathbf{R}^v V\rangle^{-1} \langle \mathbf{R}^v \mathbf{Q}^0 V\rangle \langle \mathbf{R}^v V\rangle\big]^{-1} \\ \times \langle \mathbf{R}^v V\rangle^{-1} \langle \mathbf{F}^v V\rangle,\end{aligned} \tag{8.114}$$

$$\begin{aligned}\langle \boldsymbol{\eta}_i\rangle(\mathbf{x}) = \mathbf{R}_i^v(\mathbf{x})\boldsymbol{\sigma}^0 + \mathbf{F}_i^v(\mathbf{x}) + \mathbf{R}_i^v(\mathbf{x})\mathbf{Q}_i^0 \langle \mathbf{R}^v V\rangle \\ \times \big[\mathbf{I} - \langle \mathbf{R}^v V\rangle^{-1} \langle \mathbf{R}^v \mathbf{Q}^0 V\rangle \langle \mathbf{R}^v V\rangle\big]^{-1} \Big[\boldsymbol{\sigma}^0 + \langle \mathbf{R}^v V\rangle^{-1} \langle \mathbf{F}^v V\rangle\Big].\end{aligned} \tag{8.115}$$

We will prove the equivalence of representations (8.113), (8.114), and (8.99), (8.100), respectively, in some particular cases. At first, let the tensors $\mathbf{Q}_i^0$ be a constant for all inclusions:

$$\mathbf{Q}_i^0 \equiv \mathbf{Q}(v^{\rm el}) = \text{const.}\ \forall i \tag{8.116}$$

for some ellipsoidal domain v^{el}. Then the tensor $\mathbf{Q}(v^{el})$ can be taken out from the average symbol in Eqs. (8.113) and (8.114) and by trivial manipulation we obtain

$$\langle \mathbf{R}^v V\rangle^{-1} \langle \mathbf{R}^v \mathbf{Q}^0 V\rangle \langle \mathbf{R}^v V\rangle = \mathbf{Q}(v^{\rm el}) \langle \mathbf{R}^v V\rangle. \tag{8.117}$$

Then Eqs. (8.113) and (8.99) are equivalent ($\mathbf{M}^{*\rm MEF} = \mathbf{M}^{*\rm MTM}$) if the ellipsoidal inclusions v_i and correlation holes v_i^0 both have the same shape and fixed orientation:

$$v_i, v_i^0 - \text{ellipsoids;}\quad \frac{a_i^1}{a_i^2} = \frac{a_i^2}{a_i^3} = \frac{a_i^{01}}{a_i^{02}} = \frac{a_i^{02}}{a_i^{03}} \equiv \text{cnst.}\ \forall i, \tag{8.118}$$

when

$$\mathbf{Q}_i = \mathbf{Q}_i^0 \equiv \mathbf{Q}(v^{\text{el}}) = \text{const}, \tag{8.119}$$

which takes place for either the aligned or spherical inclusions. In so doing, the anisotropy of inclusions can be arbitrarily changed from one inclusion to another. For justification of an equivalence of Eqs. (8.114) and (8.100), we will invoke by Eq. (8.21) for obtaining the following chain of equalities:

$$\begin{aligned}\langle \mathbf{R}^v V\rangle^{-1}\langle \mathbf{F}^v V\rangle &= \langle \mathbf{Q}^{-1}(\mathbf{I}-\mathbf{B})V\rangle^{-1}\langle \mathbf{F}^v V\rangle = -\langle(\mathbf{I}-\mathbf{B})V\rangle^{-1}\mathbf{Q}_i\langle \mathbf{Q}^{-1}\mathbf{C}V\rangle\\ &= \langle(\mathbf{I}-\mathbf{B})V\rangle^{-1}\langle \mathbf{C}V\rangle,\end{aligned} \tag{8.120}$$

which is used accompanied with Eq. (8.25) for transformation of Eq. (8.100) to the form

$$\begin{aligned}\boldsymbol{\beta}^* &= \boldsymbol{\beta}^{(0)} - \langle \mathbf{R}^v V\rangle\langle(\mathbf{I}-\mathbf{B})V\rangle^{-1}\langle \mathbf{C}V\rangle - \langle \mathbf{R}^v V\rangle[\mathbf{I}-\langle(\mathbf{I}-\mathbf{B})V\rangle]^{-1}\\ &= \boldsymbol{\beta}^{(0)} - \langle \mathbf{R}^v V\rangle[\mathbf{I}-\langle(\mathbf{I}-\mathbf{B})V\rangle]^{-1}\langle(\mathbf{I}-\mathbf{B})V\rangle\langle \mathbf{C}V\rangle\\ &= \boldsymbol{\beta}^{(0)} + \langle \mathbf{R}^v V\rangle(\mathbf{I}-\mathbf{Q}_i\langle \mathbf{R}^v V\rangle)^{-1}\langle \mathbf{R}^v V\rangle^{-1}\langle \mathbf{F}^v V\rangle\end{aligned} \tag{8.121}$$

coinciding with Eq. (8.114) at the assumption (8.119) ($\boldsymbol{\beta}^{*\text{MEF}} = \boldsymbol{\beta}^{*\text{MTM}}$). In a similar manner, the representation of the statistical average strain polarization tensor (8.84) can be transformed by substitution of Eqs. (8.21), (8.117), (8.118), (8.123) into Eq. (8.84) obtained by the MEF. This leads to the equation

$$\begin{aligned}\bar{v}_i\langle \boldsymbol{\eta}_i\rangle(\mathbf{x}) &= \mathbf{R}_i(\mathbf{x})\Big[c^{(0)}\mathbf{I} + \langle \mathbf{B}V\rangle + \mathbf{Q}_i^0\langle \mathbf{R}^v V\rangle\Big]\Big[c^{(0)}\mathbf{I} + \langle \mathbf{B}V\rangle\Big]^{-1}\boldsymbol{\sigma}^0\\ &\quad + \mathbf{F}_i(\mathbf{x}) + \mathbf{R}_i(\mathbf{x})\mathbf{Q}_i^0\langle \mathbf{R}^v V\rangle\Big[c^{(0)}\mathbf{I} + \langle \mathbf{B}V\rangle\Big]^{-1}\langle(\mathbf{I}-\mathbf{B})V\rangle^{-1}\langle \mathbf{C}V\rangle\\ &= \mathbf{R}_i(\mathbf{x})\Big[c^{(0)}\mathbf{I} + \langle \mathbf{B}V\rangle\Big]^{-1}\boldsymbol{\sigma}^0 + \mathbf{F}_i(\mathbf{x})\\ &\quad + \mathbf{R}_i(\mathbf{x})\langle(\mathbf{I}-\mathbf{B})V\rangle\Big[\mathbf{I}-\langle(\mathbf{I}-\mathbf{B})V\rangle\Big]^{-1}\langle(\mathbf{I}-\mathbf{B})V\rangle^{-1}\langle \mathbf{C}V\rangle.\end{aligned} \tag{8.122}$$

equivalent to the representation (8.101) obtained by the MTM. Thus, the equivalence of Eqs. (8.113), (8.114), (8.84), and (8.99), (8.100), (8.101) ($\mathbf{M}^{*\text{MEF}} = \mathbf{M}^{*\text{MTM}}$, $\boldsymbol{\beta}^{*\text{MEF}} = \boldsymbol{\beta}^{*\text{MTM}}$, $\langle \boldsymbol{\eta}_i\rangle^{*\text{MEF}}(\mathbf{x}) = \langle \boldsymbol{\eta}_i\rangle^{*\text{MTM}}(\mathbf{x})$, respectively, is proved at the condition (8.119) (the *first case*). It should be mentioned that these equivalencies take place in the framework of additional assumptions (8.67) and (8.69) only. No restrictions are imposed on the microtopology of inclusion (the inclusions can be nonellipsoidal and multilayered coated ones) as well as on the stress state in the inclusions. Equations (8.79) and (8.80) are in rather poor agreement with a multicomponent realization of Mori-Tanaka's scheme (see, e.g., [871]) the inconsistency of which is well known: Mori-Tanaka moduli may violate the Hashin-Shtrikman bounds (see [817], [908]) and are nonsymmetric for the general biphase composites (see [63], [334]); the convergence difficulties of the Mori-Tanaka scheme are discussed in [1118].

In the *second case*, let all inclusions be isotropic and homogeneous ellipsoids with the same shape coinciding with the shape of correlation holes (8.118) and thermoelastic properties:

$$\mathbf{M}_1(\mathbf{x}) \equiv \mathbf{M}_1 = \text{const.}, \ \boldsymbol{\beta}_1(\mathbf{x}) \equiv \boldsymbol{\beta}_1 = \text{const.}, \ (\mathbf{x} \in v), \tag{8.123}$$

then the tensor $\mathbf{M}_1$ can be taken out from the average symbols in Eq. (8.113) and (8.114) leading to the sequence of equalities

$$\begin{aligned}\langle \mathbf{R}^v V\rangle^{-1}\langle \mathbf{R}^v V\mathbf{Q}^0\rangle\langle \mathbf{R}^v V\rangle &= \langle \mathbf{B}V\rangle^{-1}\langle \mathbf{B}\mathbf{Q}\mathbf{M}_1 V\rangle\langle \mathbf{B}V\rangle \\ &= \langle \mathbf{B}V\rangle^{-1}\langle(\mathbf{I}-\mathbf{B})V\rangle\langle \mathbf{B}V\rangle = c^{(1)}\mathbf{I} - \langle \mathbf{B}V\rangle, \end{aligned} \tag{8.124}$$

because for homogeneous ellipsoidal inclusions $\mathbf{B}_i\mathbf{Q}_i\mathbf{M}_1^{(i)} = \mathbf{I}-\mathbf{B}_i$ for the homogeneous ellipsoidal inclusion, and, therefore, the representation (8.113) is equivalent to (8.99). For justification of an equivalence of Eqs. (8.114) and (8.100) at the conditions (8.118) and (8.122), we will transform the tensor $\langle \mathbf{R}^v V\rangle^{-1}\langle \mathbf{F}^v V\rangle$ to the form distinguishing from (8.120):

$$\begin{aligned}\langle \mathbf{R}^v V\rangle^{-1}\langle \mathbf{F}^v V\rangle &= \langle \mathbf{B}V\rangle^{-1}(\mathbf{M}_1^{(i)})^{-1}\langle(\mathbf{I}+\mathbf{M}_1\mathbf{Q})^{-1}\boldsymbol{\beta}_1\rangle \\ &= \langle \mathbf{B}V\rangle^{-1}\langle(\mathbf{I}+\mathbf{Q}\mathbf{M}_1)^{-1}(\mathbf{M}_1^{(i)})^{-1}\boldsymbol{\beta}_1\rangle = (\mathbf{M}_1^{(i)})^{-1}\boldsymbol{\beta}_1^{(i)}. \end{aligned} \tag{8.125}$$

The relation (8.125) combined with Eq. (8.21) is used for transformation of Eq. (8.100) to the form equivalent to Eq. (8.114):

$$\begin{aligned}\boldsymbol{\beta}^* &= \boldsymbol{\beta}^{(0)} + \langle \mathbf{R}^v V\rangle\Big\{(\mathbf{M}_1^{(i)})^{-1}\boldsymbol{\beta}_1^{(i)} - [c^{(0)}\mathbf{I}+\langle \mathbf{B}V\rangle]^{-1}\langle \mathbf{C}V\rangle\Big\} \\ &= \boldsymbol{\beta}^{(0)} + \langle \mathbf{R}^v V\rangle[c^{(0)}\mathbf{I}+\langle \mathbf{B}V\rangle]^{-1}\Big[(c^{(0)}\mathbf{I}+\langle \mathbf{B}V\rangle)(\mathbf{M}_1^{(i)})^{-1}\boldsymbol{\beta}_1^{(i)} \\ &+ \langle(\mathbf{I}-\mathbf{B})V\rangle(\mathbf{M}_1^{(i)})^{-1}\boldsymbol{\beta}_1^{(i)}\Big] \\ &= \boldsymbol{\beta}^{(0)} + \langle \mathbf{R}^v V\rangle[c^{(0)}\mathbf{I}+\langle \mathbf{B}V\rangle]^{-1}(\mathbf{M}_1^{(i)})^{-1}\boldsymbol{\beta}_1^{(i)} \\ &= \boldsymbol{\beta}^{(0)} + \langle \mathbf{R}^v V\rangle[c^{(0)}\mathbf{I}+\langle \mathbf{B}V\rangle]^{-1}\langle \mathbf{R}^v V\rangle^{-1}\langle \mathbf{F}^v V\rangle. \end{aligned} \tag{8.126}$$

After obtaining the second line of Eq. (8.126), it was used in the equality (8.124).

Thus, we proved the equivalence of Eqs. (8.113), (8.114) and (8.99), (8.100) ($\mathbf{M}^{*\mathrm{MEF}} = \mathbf{M}^{*\mathrm{MTM}}$, $\boldsymbol{\beta}^{*\mathrm{MEF}} = \boldsymbol{\beta}^{*\mathrm{MTM}}$), respectively, at conditions (8.118) and (8.122) (the *second case*).

Now we will analyze the statistical averages of the strain polarization tensor obtained by both the MEF (8.84) and MTM (8.101). For this purpose we will substitute Eqs. (8.124), (8.125) into Eq. (8.84):

$$\begin{aligned}\bar v_i\langle\boldsymbol{\eta}_i\rangle(\mathbf{x}) &= \mathbf{R}_i\boldsymbol{\sigma}^0 + (\mathbf{I}-\mathbf{R}_i\mathbf{M}_1^{-1})\langle \mathbf{R}^v V\rangle\big[c^{(0)}\mathbf{I}+\langle \mathbf{B}V\rangle\big]^{-1}\boldsymbol{\sigma}^0 \\ &+\mathbf{F}_i + [\langle \mathbf{R}^v V\rangle - \mathbf{R}_i\langle \mathbf{B}V\rangle]\big[c^{(0)}\mathbf{I}+\langle \mathbf{B}V\rangle\big]^{-1}\mathbf{M}_1^{-1}\boldsymbol{\beta}_1 \\ &= \big[c^{(0)}\mathbf{R}_i + \mathbf{R}_i\langle \mathbf{B}V\rangle + \langle \mathbf{R}^v V\rangle - \mathbf{R}_i\langle \mathbf{B}V\rangle\big]\big[c^{(0)}\mathbf{I}+\langle \mathbf{B}V\rangle\big]^{-1}\boldsymbol{\sigma}^0 \\ &+\mathbf{F}_i - \mathbf{R}_i\mathbf{M}_1^{-1}\boldsymbol{\beta}_1 + \big[c^{(0)}\mathbf{R}_i + \langle \mathbf{R}^v V\rangle\big]\big[c^{(0)}\mathbf{I}+\langle \mathbf{B}V\rangle\big]^{-1}\mathbf{M}_1^{-1}\boldsymbol{\beta}_1. \end{aligned} \tag{8.127}$$

At the derivative of the first equality in Eq. (8.127), the equality $\mathbf{R}_i\mathbf{Q}_i = \mathbf{I} - \mathbf{R}_i\mathbf{M}_1^{-1}$ was used. We will also transform Eq. (8.101), obtained by the MTM, by substitution of Eqs. (8.18) into Eq. (8.101)

$$\begin{aligned}\bar{v}_i\langle\boldsymbol{\eta}\rangle_i(\mathbf{x}) &= \mathbf{R}_i[\mathbf{I}+\langle(\mathbf{I}-\mathbf{B})V\rangle)^{-1}\langle\boldsymbol{\sigma}\rangle\\ &\quad+\mathbf{F}_i+\mathbf{R}_i[\mathbf{I}+\langle(\mathbf{I}-\mathbf{B})V\rangle]^{-1}\langle(\mathbf{I}-\mathbf{B})V\rangle\mathbf{M}_1^{-1}\boldsymbol{\beta}_1\\ &= \mathbf{R}_i[c^{(0)}\mathbf{I}+\langle\mathbf{B}V\rangle]^{-1}\langle\boldsymbol{\sigma}\rangle\\ &\quad+\mathbf{F}_i-\mathbf{R}_i\mathbf{M}_1^{-1}\boldsymbol{\beta}_1+\mathbf{R}_i[c^{(0)}\mathbf{I}+\langle\mathbf{B}V\rangle]^{-1}\mathbf{M}_1^{-1}\boldsymbol{\beta}_1.\end{aligned}\tag{8.128}$$

Straightforward comparison of the average strain polarization tensors $\langle\boldsymbol{\eta}\rangle_i^{\mathrm{MEF}}$ and $\langle\boldsymbol{\eta}\rangle_i^{\mathrm{MTM}}$ estimated by the MEF (8.127) and MTM (8.128) leads to the conclusion that

$$\bar{v}_i\langle\boldsymbol{\eta}\rangle_i^{\mathrm{MEF}}=\bar{v}_i\langle\boldsymbol{\eta}\rangle_i^{\mathrm{MTM}}+[\langle\mathbf{R}^vV\rangle-c^{(1)}\mathbf{R}_i][c^{(0)}+\langle\mathbf{B}V\rangle]^{-1}(\boldsymbol{\sigma}^0+\mathbf{M}_1^{-1}\boldsymbol{\beta}_1).\tag{8.129}$$

Thus, in the second case of the conditions (8.118) and (8.123) when the MEF and MTM lead to the identical effective thermoelastic properties (8.113), (8.114) and (8.99), (8.100), respectively, in the general case of inclusion ellipsoidal shape, their estimations of the statistical averages of strain polarization tensors (8.127) and (8.128), respectively, coincide if and only if the ODF satisfies the additional condition of the fiber alignment (8.119); otherwise $\langle\boldsymbol{\eta}\rangle_i^{\mathrm{MEF}}\neq\langle\boldsymbol{\eta}\rangle_i^{\mathrm{MTM}}$. The last inequality immediately yields from a basic assumption according to which the effective field (8.90) acting on the inclusion v_i in the MTM does not depend on the geometrical parameters of the inclusion v_i, in contrast to the effective field estimated by the MEF (8.82), which explicitly depends on the shape of the correlation hole v_i^0 coinciding in the case being considered with the shape of the inclusion v_i. However, angle averaging of Eq. (8.129) yields the equality of total statistical average of strain polarization tensors estimated by both the MEF and MTM $\langle\boldsymbol{\eta}\rangle^{\mathrm{EMF}}=\langle\boldsymbol{\eta}\rangle^{\mathrm{MTM}}$ that in turn lead to the equality of effective thermoelastic properties estimated by these methods

$$\mathbf{M}^{*\mathrm{MEF}}=\mathbf{M}^{*\mathrm{MTM}},\ \boldsymbol{\beta}^{*\mathrm{MEF}}=\boldsymbol{\beta}^{*\mathrm{MTM}}.$$

It is interesting that there is a *third case* when the estimations (8.113) and (8.99) give the symmetric representation for the effective compliance. Namely, for composites with 3-D uniform random orientation of anisotropic fibers in the isotropic matrix, the matrix of effective properties estimated by both the MTM and MEF is symmetric and allows the subsequent simplification of the "correlation hole" v_i^0 homothetic to fiber v_i when $\mathbf{Q}_i^0\equiv\mathbf{Q}_i$. This result is general for the composites with the isotropic matrix and does not depend on the elastic anisotropic properties of the fibers and on their aspect ratio α. Indeed, although the tensor $\mathbf{Q}_i\neq$ const. and varies with the alteration of fiber orientation, the tensors $\langle\mathbf{R}^vV\mathbf{Q}\rangle$ and $\langle\mathbf{R}^vV\rangle$ are isotropic, and because of this, they are commutative $\langle\mathbf{R}^vV\mathbf{Q}\rangle\langle\mathbf{R}^vV\rangle=\langle\mathbf{R}^vV\rangle\langle\mathbf{R}^vV\mathbf{Q}\rangle$ (the last statement is incorrect in both the general case of prescribed random fiber orientation and the anisotropic matrix). Then the effective elastic moduli estimated by the MEF (8.113) and MTM (8.99) are symmetric, isotropic ,and can be presented in the form

$$\begin{aligned}\mathbf{M}^*&=\mathbf{M}^{(0)}+\langle\mathbf{R}^vV\rangle\big[\mathbf{I}-\langle\mathbf{R}^v\mathbf{Q}V\rangle\big]^{-1},\\ \mathbf{M}^*&=\mathbf{M}^{(0)}+\langle\mathbf{R}^vV\rangle\big[\mathbf{I}-\langle\mathbf{Q}\mathbf{R}^vV\rangle\big]^{-1},\end{aligned}\tag{8.130}$$

respectively. In a general case of an inclusion anisotropy, an inequality $\langle\mathbf{R}^v\mathbf{Q}^0V\rangle\neq\langle\mathbf{Q}\mathbf{R}^vV\rangle$ takes place and the representations (8.130_1) and (8.130_2) do not coinside. However, for the rigid inclusions (when $\mathbf{M}(\mathbf{x}) \equiv \mathbf{0},\ \mathbf{x} \notin v^{(0)}$) the relations

$$\langle \mathbf{R}^v \mathbf{Q} V\rangle = c\mathbf{I} - \langle \mathbf{M}^{(0)}\mathbf{B}\mathbf{L}^{(0)} V\rangle, \quad \langle \mathbf{Q}\mathbf{R}^v V\rangle = c\mathbf{I} - \langle \mathbf{B} V\rangle$$

coincide, because $\langle \mathbf{B}\rangle$ and $\mathbf{L}^{(0)}$ are both isotropic and commutative, which leads to the equivalence of the formulae of the effective moduli (8.130_1) and (8.130_2). The last coincidence in the case of rigid inclusions provides confidence that the representations (8.130_1) and (8.130_2) lead to the similar results also if the inclusions are stiff enough.

Thus, we proved the equivalence of Eqs. (8.113), (8.114) and (8.99), (3.36), respectively, in three cases. In so doing, one of the necessary conditions providing this equivalence was coincidence of the sphapes of domains v_i and v_i^0. However, if the shape of the ellipsoidal correlation hole v_i^0 differs from the shape of the inclusion v_i (in particularly, if v_i is not an ellipsoid), then Eqs. (8.99) and (8.113) are not identical in general. Let $v_i^1 \supset v_i$ have a minimum volume among the areas homothetic to v_i^0 with the homothetic center $\mathbf{x}_i$. Then for the stresses averaged over the domain v_i^1 we have $\boldsymbol{\sigma}_i^1 = \overline{\boldsymbol{\sigma}} - \bar{v}_i(\bar{v}_i^1)^{-1}\mathbf{Q}_i^1\langle \mathbf{M}_1\boldsymbol{\sigma}\rangle_i$, where $\mathbf{Q}_i^1 = \mathbf{Q}(v_i^1)$, and, therefore, $\mathbf{R}_i = \bar{v}_i^1(\mathbf{Q}_i^1)^{-1}(\mathbf{I} - \mathbf{B}_i^1)$, where $\mathbf{B}_i^1$ is the average of the stress concentration factor $\mathbf{B}(\mathbf{x})$ over the domain $\mathbf{x} \in v_i^1$. In so doing, $\bar{v}_i^1\mathbf{R}_i^1 = \bar{v}_i\mathbf{R}_i$ because $\mathbf{M}_1^{(i)}(\mathbf{x}) \equiv \mathbf{0}$ at $\mathbf{x} \in v_i^1 \setminus v_i$. Then Eqs. (8.113) and (8.99) can be recast as

$$\mathbf{M}^* = \mathbf{M}^{(0)} + \langle \mathbf{R}^v V\rangle[(1 - c^{1(1)})\mathbf{I} + c^{1(1)}\mathbf{B}_i^1]^{-1}, \tag{8.131}$$

$$\mathbf{M}^* = \mathbf{M}^{(0)} + \langle \mathbf{R}^v V\rangle[(1 - c^{(1)})\mathbf{I} + c^{(1)}\mathbf{B}_i]^{-1}, \tag{8.132}$$

respectively, which are equivalent in general only for $v_i = v_i^1$ when $\mathbf{B}_i \equiv \mathbf{B}_i^1$; here $c^{(1)} = n\bar{v}_i$, $c^{1(1)} = n\bar{v}_i^1$. Otherwise, in general,

$$c^{1(1)}\mathbf{B}_i^1 - c^{1(1)}\mathbf{I} \equiv c^{(1)}\mathbf{B}_i - c^{1(1)}\mathbf{I} + n\int_{v_i^1\setminus v_i} \mathbf{B}(\mathbf{x})\, d\mathbf{x} \neq c^{(1)}\mathbf{B}_i - c^{(1)}\mathbf{I}, \tag{8.133}$$

and, therefore, $\mathbf{M}^{*\mathrm{MEF}} \neq \mathbf{M}^{*\mathrm{MT}}$ even for the identical aligned isotropic fibers if v_i and v_i^0 are not homothetic.

Thus, an alternative MT approach deserving more attention is based on an additional assumption of coinciding of the statistical average of effective fields acting on each inclusion and the volume average of stresses in the matrix. The last assumption leads to the invariance of estimations obtained by the MT approach with respect to the spatial correlation of inclusion locations. Only in the particular case of identical aligned inclusions when the "ellipsoidal symmetry" of both a pair distribution function coincides and inclusion coincide, the MT method yields the Hashin-Shtrikman formula (see [1179]). It may be noted there are certain drawbacks of the extension of Mori-Tanaka method to multiphase composites: Mori-Tanaka moduli may violate the Hashin-Shtrikman bounds [817] and are nonsymmetric for the general diphase composites ([63], [334]). However, Nemat-Nasser and Hori [806] showed that by the energy-based definition, the effective moduli are the symmetric part of the effective elastic moduli defined by the average stress and strain in the composite, and because of this, must always be symmetric. Nevertheless, the mentioned authors proved the symmetry of Mori-Tanaka moduli for biphase composites in two cases. In the first case the composites are reinforced by the aligned identical anisotropic fibers with the conditions (8.116) and (8.118) [or (8.118) and (8.123)], for which the equivalence of Mori-Tanaka and the one-particle approximation of the MEFM approaches take place. The second one is the composite material reinforced by one sort of isotropic fiber with an arbitrary prescribed random orientation. This case of isotropic spheroidal inclusions with different prescribed random orientations was systematically investigated by Rammerstorfer and coworkers (see, e.g., comprehensive reviews in [93], [94], [292], [871]). In both cases the MT approach is equivalent to the one particle approximation of the MEFM. However, to the author's knowledge, no models exist that satisfy all theoretical criteria mentioned above for arbitrary phase anisotropy and fiber-orientation distributions.

8.3.4 Some Methods Related to the One-Particle Approximation of the MEFM

The conditional-moment method proposed and intensity developed by the Kiev school of micromechanicians (see for references Khoroshun [559], [561], [562], [563], [564]) is worthy of critical note. The concrete numerical results were obtained by truncation of the infinite system of integral equations by taking into account only two- and three-point conditional probabilities, and by neglecting fluctuations of stresses within the limits of the components. These are equivalent to acceptance of assumption (7.17), (8.65_2) and consideration of homogeneous inclusions. For analytical estimation of integrals involved similar to the integrals in Eq. (8.70) Khoroshun [561], [562] proposed the analytical representation of the two-point density $\langle V_q(\mathbf{x})|; v_i, \mathbf{x}_i\rangle$ conforming with the shape of the inclusion v_i that with necessity leads to acceptance of Eq. (8.70). The concrete numerical results were obtained for aligned homogeneous ellipsoidal inclusions under different choices of comparison media, either

$$\mathbf{M}^c = \langle\mathbf{M}\rangle \quad \text{or} \quad \mathbf{M}^c = \langle\mathbf{M}^{-1}\rangle^{-1} \tag{8.134}$$

if the compliance of the inclusions is less than that of the matrix or otherwise, respectively. Of course, there is no *a priori* justification for the specific choice of $\mathbf{M}^c$, not counting the condition that the quadratic form $(\mathbf{M}_1\boldsymbol{\sigma})\boldsymbol{\sigma}$, employed in the proof of the Hashin and Shtrikman variational principle, has a constant sign. The only justification for choosing $\mathbf{M}^c$ in the Voight or Reuss estimation of $\mathbf{M}^*$ (see [561], [562]) is the fact that specific experimental data agree with the computing curves. Although the final explicit general representation for effective moduli (similar to (8.73)) was not presented in the conditional-moment method, the equivalence of assumptions admitted leads to the conclusion that the conditional-moment method can be considered as a particular case of the MEFM presented in Section 8.2. In addition, in the conditional-moment method the shape of the inclusions is taken into account via prescribed anisotropy of the conditional probability density $\langle V_q(\mathbf{x})|; v_i, \mathbf{x}_i\rangle$ (or $\varphi(v_q|; v_i, \mathbf{x}_i)$). For equally probable orientation of ellipsoidal inclusions it is possible to obtain an isotropic function $\varphi(v_q|; v_i, \mathbf{x}_i)$ and the estimation of the effective compliance $\mathbf{M}^*$ will be invariant with respect to the shape of the inclusion. This result can be avoided easily by taking into account directly the shape of the inclusions via the tensor $\mathbf{Q}_q$, as done by Willis [1179], [1186] on the basis of a variational principle. For identical oriented ellipsoidal inclusions the estimations of effective moduli obtained in [561], [1179] as well as Eq. (8.76) are equivalent.

We note than in order to solve system (8.7), in any case, at least for homogeneous inclusions, it is not necessary to introduce intermediate concepts – effective fields $\overline{\boldsymbol{\sigma}}_i(\mathbf{x})$ and $\widetilde{\boldsymbol{\sigma}}_{1,\ldots,n}(\mathbf{x})$. System (8.7) is linear in the fields $\langle\boldsymbol{\sigma}|v_1, \mathbf{x}_1; \ldots; v_n, \mathbf{x}_n\rangle(\mathbf{x})$, and on closure [561], [562], [1184], analogous to hypothesis **H2**; it becomes finite and can be solved by the methods of linear algebra; this scheme is implemented by the conditional-moment method by Khoroshun [561], [562]. According to the MEFM, information about geometrical and mechanical characteristics is given by the tensors (8.18_1) and the fields $\langle\widetilde{\boldsymbol{\sigma}}_{1,\ldots,n}(\mathbf{x})\rangle$, in contrast to $\langle\boldsymbol{\sigma}|v_1, \mathbf{x}_1; \ldots; v_n, \mathbf{x}_n\rangle(\mathbf{x})$, are weakly nonuniform. This is why, as noted by in [148], [529], even rough assumptions about the structure of the effective fields (8.39) make it possible to obtain correct results. Different versions of closure assumptions in terms of conditional stress fields analogous to hypothesis **H2** for the effective stress fields are known [561], [562], [1184]. The first-order approximations of these similar approaches are given by Eq. (8.39) and (8.65_2).

Of course, the principal difference between the hypotheses (8.39) and (8.65_2) is beyond the scope of direct substitution of the stress field $\boldsymbol{\sigma}(\mathbf{x})$ for the effective field $\overline{\boldsymbol{\sigma}}(\mathbf{x})$, which leads only to quantitative improvement of the estimates (see for details CI.7). Of more importance is the application of the assumption (8.39) after the consideration of multiparticle inclusion interactions (8.25) rather than before as in hypothesis (8.65_2). What seems to be only a formal trick yields to the discovery of fundamentally new nonlocal effects in the theory of functionally graded materials, which will be demonstrated in Chapter 13. However, even for statistically homogeneous composites being considered in this chapter it will be shown below that the use of assumption (8.39) instead of (8.65_2) can lead to a variation of effective elastic moduli by a factor of two or more, a fact that has been confirmed by classical experimental data [597].

In an other popular generalized singular-approximation method, by Shermergor [995], which is invariant with respect to the shape of the inclusions, the action of the singular integral operator with the kernel $\boldsymbol{\Gamma}$ in the general integral equation (8.3) is replaced by a multiplication on a constant tensor coinciding with the singular component $\boldsymbol{\Gamma}^\delta = -\mathbf{Q}\delta$ for a spherical domain in Eq. (3.38_1). This automatically implies a number of strong assumptions: $\boldsymbol{\Gamma} = \boldsymbol{\Gamma}^\delta$, the field of the stresses in the components are homogeneous, the "quasi-crystalline" approximation (8.65_2) and assumption (8.70) are accepted, and the functions $\varphi(v_q|; v_i, \mathbf{x}_i)$ are isotropic. For this reason, the singular-approximation method is also a consequence of the one-particle approximation of MEFM proposed. The last statements can also be justified by direct reduction of the MEFM to the singular-approximation method if the mentioned assumptions are accepted. Indeed, in the case of coincidence of elastic moduli of the comparison medium and matrix $\mathbf{L}^c = \mathbf{L}^{(0)}$, $\mathbf{Q}_i^0 = \mathbf{Q}_i$, and $c^{(0)} = 0$, Eqs. (8.79) and (8.85) are reduced to

$$\mathbf{M}^* = \langle \mathbf{B}V\rangle^{-1}\langle \mathbf{R}^v V\rangle, \quad \mathbf{L}^* = \langle \mathbf{A}V\rangle^{-1}\langle \mathbf{R}^{\varepsilon v} V\rangle. \tag{8.135}$$

If in additional $\mathbf{P}(\mathbf{x})$, $\mathbf{Q}(\mathbf{x}) \equiv$ const. (for example, for the spherical inclusions) then Eqs. (8.135_1) and (8.135_2) can be recast in the following equivalent forms

$$\begin{aligned} \mathbf{M}^* &= \langle(\mathbf{M}+\mathbf{b}^M)^{-1}\rangle^{-1}\langle \mathbf{M}(\mathbf{M}+\mathbf{b}^M)^{-1}\rangle, \\ (\mathbf{M}^*+\mathbf{b}^M)^{-1} &= \langle(\mathbf{M}+\mathbf{b}^M)^{-1}\rangle, \end{aligned} \tag{8.136}$$

$$\begin{aligned} \mathbf{L}^* &= \langle(\mathbf{L}+\mathbf{b}^L)^{-1}\rangle^{-1}\langle \mathbf{L}(\mathbf{L}+\mathbf{b}^L)^{-1}\rangle, \\ (\mathbf{L}^*+\mathbf{b}^L)^{-1} &= \langle(\mathbf{L}+\mathbf{b}^L)^{-1}\rangle, \end{aligned} \tag{8.137}$$

respectively, where the auxiliary tensors $\mathbf{b}^M$ and $\mathbf{b}^L$ are introduced by the equalities $\mathbf{b}^M = \mathbf{Q}^{-1} - \mathbf{M}^c$, $\mathbf{b}^L = \mathbf{P}^{-1} - \mathbf{L}^c$. The formula (8.137) are exactly the appropriate relations of the generalized singular-approximation method [995] widely used for analyses of both the composite materials and polycrystals. Other forms of these equations were obtained by the different methods attributed to the terms "absolute disorder" [608], "one-point approximation" [559], "strong isotropy" [97], and "singular approximation" [347]. It was shown in [347], [967] that if $\mathbf{L}^c = \mathbf{0}$ and $\mathbf{L}^c = \infty$, we obtain the Reuss [934] and Voight [1143] approximations, respectively. If $\mathbf{L}^c$ is chosen to be equal to the elastic moduli of components with the maximum and minimum rigidities, we get the upper and lower Hashin-Shtrikman [438], [441] bounds. The results of the conditional moment method [559], [561] are obtained if $\mathbf{L}^c$ (or $\mathbf{M}^c$) is taken according to the relations (3.70). Assuming that $\mathbf{L}^c = \mathbf{L}^*$, we arrive at one of the versions of the self-consistent method (see Section 9.2).

It was noted [1184] that for homogeneous ellipsoidal inclusions Eq. (8.85) is identical to the relevant equation (6.231)providing the extremum of some integral functional (6.187) in the Hashin-Shtrikman [441] variational principle. This extremum was found in the trial piecewise-constant function class (6.219) $\boldsymbol{\tau}^*(\mathbf{x}) = \sum_k V^{(k)}(\mathbf{x})\langle\boldsymbol{\tau}\rangle^{(k)}$, $(k = 0,1,\ldots,N)$ that implies automatically satisfaction of both hypothesis **H1** and (8.65_2), and, moreover, of an additional hypothesis of a homogeneity of ellipsoidal inclusions. In such a case Eq. (8.76), derived on the basis of the *ad hoc* assumptions **H1** and (8.65_2), in fact has the status of the boundaries for $\mathbf{M}^*$. In light of the effective field theory developed in this chapter it is natural to find the desirable extremum in other, more broad, classes of trial piecewise-constant functions:

$$\overline{\boldsymbol{\varepsilon}}^*(\mathbf{x}) = \sum_k V^{(k)}(\mathbf{x})\langle\overline{\boldsymbol{\varepsilon}}\rangle^{(k)}, \quad (k = 0,1,\ldots,N). \tag{8.138}$$

With the help of Eqs. (8.87_1) and (8.87_2),this makes it possible to analyze the composite materials filled by inclusions of any shape and microtopology without any restrictions on the inhomogeneity of statistical averages of the stress state in the coated inclusions as well as to take into account immediately the orientation effect of inclusions by the use of tensors $\mathbf{R}_i$ $(i = 1,2,\ldots)$ (for details see [101], [172]). In so doing the approximation (8.138) remains a fundamental disadvantage inherent in the quasi-crystalline approximation (8.65) neglecting a direct binary interaction of inclusions v_i and v_j: $\mathbf{Z}_{ij} = \delta_{ij}$.

In the interesting approach by Dvorak and Srinivas [303] a complete formal similarity of the standard techniques for estimating overall elastic and average stress field in the constitucnts is established for effective medium and Mori-Tanaka methods (see Subsections 8.3.2), and then for Walpole's [1148] formulation of the Hashin-Shtrikman [441] variational bounds for statistically isotropic composites for which the hypotheses **H1** and (8.65_2) are valid. They detected a common analytical regularity of all these methods that allows one to generate any number of related analytical representation for effective properties from other choices of the comparison medium made such that the resulting moduli predictions do not violate the bounds. The generalization of their results for the case of composites with coated inclusions is noteworthy.

8.3.5 Some Analytical Representations for Effective Moduli

Consider first the simple example of a composite consisting of an isotropic matrix containing the identical spherical inclusions of arbitrary stucture with uniform random orientation. Then the representations of the effective elastic moduli $\mathbf{L}^* = (3k^*, 2\mu^*)$ obtained by the quasi-crystalline approximation of the MEFM (8.85) and by the MTM (8.102) coincide:

$$k^* = k^{(0)} + \frac{c^{(1)}k^{R\epsilon}}{1 - c^{(1)}3P_1 3k^{R\epsilon}}, \quad \mu^* = \mu^{(0)} + \frac{c^{(1)}\mu^{R\epsilon}}{1 - c^{(1)}2P_2 2\mu^{R\epsilon}}, \tag{8.139}$$

$$3P_1 = \frac{1}{3k^{(0)} + 4\mu^{(0)}}, \quad 2P_2 = \frac{3(k^{(0)} + 2\mu^{(0)})}{5(3k^{(0)} + 4\mu^{(0)})}, \tag{8.140}$$

where $\mathbf{P}^0 = (3P_1, 2P_2)$, and $(3k^{R\epsilon}, 2\mu^{R\epsilon})$ are the components of the isotropic average polarization tensor $\mathbf{R}^{v\epsilon} = \langle \mathbf{L}_1 \mathbf{A} \rangle$. If the moduli of inclusions are taken as homogeneous and isotropic then

$$3k^{R\epsilon} = \frac{3k_1^{(1)}}{1 + 3P_1 3k_1^{(1)}}, \quad 2\mu^{R\epsilon} = \frac{2\mu_1^{(1)}}{1 + 2P_2 2\mu^{(1)}}, \tag{8.141}$$

and Eq. (8.139) gives

$$k^* = k^{(0)} + \frac{c^{(1)}k_1^{(1)}}{1 + c^{(0)}3P_1 3k_1^{(1)}}, \quad \mu^* = \mu^{(0)} + \frac{c^{(1)}\mu_1^{(1)}}{1 + c^{(0)}2P_2 2\mu_1^{(1)}}. \tag{8.142}$$

The elastic moduli of composites with identical homogeneous alligned ellipsoidal inclusions (with the semi-axes $a_1 = a_2 = a$, a_3, aspect ratio $\gamma = a_3/a$, and a symmetry axis $\mathbf{n}$) also coincide for the MEFM (8.85) and the MTM (8.102) if $\mathbf{P}^0 = \mathbf{P}$

$$\mathbf{L}^* = \mathbf{L}^{(0)} + c^{(1)}\left[\mathbf{I} - c^{(1)}\mathbf{R}^{\epsilon v}\mathbf{P}^0\right]^{-1}\mathbf{R}^{\epsilon v}. \tag{8.143}$$

The tensor $\bar{\mathbf{D}} \equiv \mathbf{I} - n^{(1)}\mathbf{R}^{\epsilon v}\mathbf{P}^0$ presented in Eq. (8.143) can be expressed in the $\mathbf{P}$-basis (A.2.7) $\bar{\mathbf{D}} = (\bar{d}_1, \ldots, \bar{d}_6)$ through the representations $\mathbf{P}^0 = (p_1, \ldots, p_6)$ (3.161) and $\mathbf{R}^{\epsilon v} = (r_1, \ldots, r_6)$ (3.164) as

$$\begin{aligned}
\bar{d}_1 &= 1/2 - c^{(1)}(2r_1p_1 + r_3p_3), \ \bar{d}_2 = 1 - c^{(1)}r_2p_2, \\
\bar{d}_3 &= -c^{(1)}(2r_1p_3 + r_3p_6), \bar{d}_4 = -c^{(1)}(2r_3p_1 + r_6p_3), \\
\bar{d}_5 &= 2 - r_5p_5/2; \ \bar{d}_6 = 1 - c^{(1)}(2r_3p_3 + r_6p_6),
\end{aligned} \tag{8.144}$$

which allows the components of the **P**-basis representation of the effective elastic moduli (see, e.g., [544], [545]):

$$\mathbf{L}^* = (k^*, 2m^*, l^*, l^*, 2\mu^*, n^*), \tag{8.145}$$

$$k^* = \frac{\mu^{(0)}}{1-2\nu^{(0)}} + \frac{r_1\bar{d}_6 - r_3\bar{d}_3}{\Delta}, \ m^* = \mu^{(0)} + \frac{r_2}{2\bar{d}_2}, \ \mu^* = \mu^{(0)} + \frac{r_5}{2\bar{d}_5}, \tag{8.146}$$

$$l^* = \frac{2\mu^{(0)}\nu^{(0)}}{1-2\nu^{(0)}} + \frac{r_3\bar{d}_6 - r_6\bar{d}_3}{\Delta}, \ n^* = \frac{2\mu^{(0)}(1-\nu^{(0)})}{1-2\nu^{(0)}} + \frac{2(r_6\bar{d}_1 - r_3\bar{d}_4)}{\Delta}, \tag{8.147}$$

where $\Delta = 2(\bar{d}_1\bar{d}_6 - \bar{d}_3\bar{d}_4)$. The case of uniform random orientation of spheroids is easily obtained from (8.143) by replacing the tensors $\mathbf{R}^{\epsilon v}\mathbf{P}^0$ and $\mathbf{R}^{\epsilon v}$ with its isotropic form according to the averaging operations (A.2.15):

$$\begin{aligned}
\mathbf{L}^* &= (3k^*, 2\mu^*), \ \ 3k^* = 3k^{(0)} + 3R_1/\tilde{d}_1, \ 2\mu^* = 2\mu^{(0)} + 2R_2/\tilde{d}_2, \\
3R_1 &= \frac{1}{9}[4(r_1 + r_3) + r_6], \ \ 2R_2 = \frac{1}{15}[r_1 + 2r_2 - 2(r_3 - 3r_5/4) + r_6], \\
\tilde{d}_1 &= 1 - \frac{1}{3}[8r_1p_1 + 6r_3p_3 + 4(r_1p_3 + r_3p_1) + 2(r_3p_6 + r_6p_3) + r_6p_6], \\
\tilde{d}_2 &= 1 - \frac{2}{15}[2r_1p_1 + 3(r_3p_3 + r_2p_2) - 2(r_1p_3 + r_3p_1) + \frac{3}{4}r_5p_5 \\
&\quad - (r_3p_6 + r_6p_3) + r_6p_6].
\end{aligned} \tag{8.148}$$

In particular, for the limiting case of coin-shaped microcracks (3.162) Eq. (8.148) is reduced to the representation

$$\begin{aligned}
\frac{k^*}{k^{(0)}} &= \left(1 + \frac{19}{9}\frac{1-(\nu^{(0)})^2}{1-2\nu^{(0)}}\epsilon\right)^{-1}, \\
\frac{\mu^*}{\mu^{(0)}} &= \left(1 + \frac{32}{45}\frac{(1-\nu^{(0)})(5-\nu^{(0)})}{2-\nu^{(0)}}\epsilon\right)^{-1}
\end{aligned} \tag{8.149}$$

allowing subsequent simplification in the dilute case when the crack density parameter $\epsilon = n^{(1)}a^3 \to 0$

$$\begin{aligned}
\frac{k^*}{k^{(0)}} &= 1 - \frac{16}{9}\frac{1-(\nu^{(0)})^2}{1-2\nu^{(0)}}\epsilon, \\
\frac{\mu^*}{\mu^{(0)}} &= 1 - \frac{32}{45}\frac{(1-\nu^{(0)})(5-\nu^{(0)})}{2-\nu^{(0)}}\epsilon.
\end{aligned} \tag{8.150}$$

The representations (8.149) were obtained in [52] by the MTM (8.102) (see also a conductivity problem for the 2-D case [51]) which is equivalent to the MEFM (8.85) in the case $\mathbf{P}_i^0 = \mathbf{P}_i$ being considered. Formulae (8.150) were proposed in [111], [1154].

In a similar manner, for a composite material made from a transversely isotropic matrix and infinite cylindrical fibers with the axes coinciding with the axis of symmetry of the matrix and fiber, the known components of the **P**-basis (A.2.7) representations of the tensors $\mathbf{L}^{(i)}, \mathbf{P}$ (3.175) and (3.176) allow the expansion coefficients of the effective moduli $\mathbf{L}^*$ (8.145) over the **P**-basis:

$$
\begin{aligned}
k^* &= k^{(0)} + c^{(1)} k_1^{(1)} \Big[1 + \frac{c^{(0)} k_1^{(1)}}{k^{(0)} + m^{(0)}} \Big]^{-1}, \\
m^* &= m^{(0)} + c^{(1)} m_1^{(1)} \Big[1 + c^{(0)} \frac{m_1^{(1)} (k^{(0)} + 2m^{(0)})}{2m^{(0)} (k^{(0)} + m^{(0)})} \Big]^{-1}, \\
l^* &= l^{(0)} + c^{(1)} l_1^{(1)} \Big[1 + \frac{c^{(0)} k_1^{(1)}}{k^{(0)} + m^{(0)}} \Big]^{-1}, \\
\mu^* &= \mu^{(0)} + c^{(1)} \mu_1^{(1)} \Big[1 + \frac{c^{(0)} \mu^{(1)}}{2\mu^{(0)}} \Big]^{-1}, \\
n^* &= n^{(0)} + c^{(1)} n_1^{(1)} - \frac{c^{(1)} c^{(0)} l_1^{(1)2}}{k^{(0)} + m^{(0)}} \Big[1 + \frac{c^{(0)} k_1^{(1)}}{k^{(0)} + m^{(0)}} \Big]^{-1}.
\end{aligned}
\tag{8.151}
$$

In closing, we will consider the analytical representations of elastic moduli obtained by the method of the conditional moments [561], [562], [564]. In particular, for a macroisotropic material of granular-matrix structure consisting of two isotropic phases, the effective moduli $\mathbf{L}^* = (3k^*, 2\mu^*)$ are

$$
\begin{aligned}
k^* &= \langle k \rangle - \frac{c^{(0)} c^{(1)} (k_1^{(1)})^2}{c^{(0)} k^{(1)} + c^{(1)} k^{(0)} + \tilde{k}}, \quad \mu^* = \langle \mu \rangle - \frac{c^{(0)} c^{(1)} (\mu_1^{(1)})^2}{c^{(0)} \mu^{(1)} + c^{(1)} \mu^{(0)} + \tilde{\mu}}, \\
\tilde{k} &= \frac{4}{3} \mu^c, \; \tilde{\mu} = \frac{\mu^c (9k^c + 8\mu^c)}{6(k^c + 2\mu^c)},
\end{aligned}
\tag{8.152}
$$

where the moduli k^c and μ^c of the comparison medium can be arbitrarily selected, and $k^c = \langle k \rangle$, $\mu^c = \langle \mu \rangle$ and $k^c = \langle k^{-1} \rangle^{-1}$, $\mu^c = \langle \mu^{-1} \rangle^{-1}$ in the cases (3.70_1) and (3.70_2), respectively. For two-component materials of a unidirectionally fibrous structure, whose effective elastic properties are transversally isotropic, the effective constitutive equation can be presented as

$$
\begin{aligned}
\langle \sigma_{ij} \rangle &= (\lambda_{11}^* - \lambda_{12}^*) \langle \varepsilon_{ij} \rangle + (\lambda_{12}^* \langle \varepsilon_{rr} \rangle + \lambda_{13}^* \langle \varepsilon_{33} \rangle) \delta_{ij}, \\
\langle \sigma_{33} \rangle &= \lambda_{13}^* \langle \varepsilon_{rr} \rangle + \lambda_{33}^* \langle \varepsilon_{33} \rangle, \quad \langle \sigma_{i3} \rangle = 2\lambda_{44}^* \langle \varepsilon_{i3} \rangle,
\end{aligned}
$$

where $i, j, r = 1, 2$, and

$$
\begin{aligned}
\frac{1}{2}(\lambda_{11}^* + \lambda_{12}^*) &= \langle \lambda + \mu \rangle - \frac{c^{(0)} c^{(1)} (\lambda_1^{(1)} + \mu_1^{(1)})^2}{c^{(0)} (\lambda^{(1)} + \mu^{(1)}) + c^{(1)} (\lambda^{(0)} + \mu^{(0)}) + \mu^c}, \\
\frac{1}{2}(\lambda_{11}^* - \lambda_{12}^*) &= \langle \lambda + \mu \rangle - \frac{c^{(0)} c^{(1)} (\mu_1^{(1)})^2 (\lambda^c + 3\mu^c)}{(c^{(0)} \mu^{(1)} + c^{(1)} \mu^{(0)})(\lambda^c + 3\mu^c) + \mu^c (\lambda^c + \mu^c)}, \\
\lambda_{13}^* &= \langle \lambda \rangle - \frac{c^{(0)} c^{(1)} \lambda_1^{(1)} (\lambda_1^{(1)} + \mu_1^{(1)})}{c^{(0)} (\lambda^{(1)} + \mu^{(1)}) + c^{(1)} (\lambda^{(0)} + \mu^{(0)}) + \mu^c}, \\
\lambda_{44}^* &= \langle \mu \rangle - \frac{c^{(0)} c^{(1)} (\mu_1^{(1)})^2}{c^{(0)} \mu^{(1)} + c^{(1)} \mu^{(0)} + \mu^c}
\end{aligned}
\tag{8.153}
$$

Here $f_1^{(1)} = f^{(1)} - f^{(0)}$ $(f = \lambda, \mu)$, and $\lambda^c = \langle \lambda \rangle$, $\mu^c = \langle \mu \rangle$ and $\lambda^c = \langle k^{-1} \rangle^{-1} - 2\langle \mu^{-1} \rangle / 3$, $\mu^c = \langle \mu^{-1} \rangle^{-1}$ in the cases (3.70_1) and (3.70_2).

9

Some Related Methods in Micromechanics of Random Structure Composites

This charter is organized as follows. In Section 9.1 the new combined MEFM-perturbation method is proposed. We demonstrate in Section 9.2 and 9.3 that the MEFM includes, as particular cases the well-known methods of mechanics of strongly heterogeneous media (such as the effective medium and differential methods and some others, see Section 9.4). Finally, in Section 9.5 we employ the proposed explicit representations and some related ones for numerical estimations of effective elastic moduli of composites with inclusions of both spheroidal and nonspheroidal shapes.

9.1 Related Perturbation Methods

9.1.1 Combined MEFM–Perturbation Method

For simplicity, let us consider the problem (8.4) and (8.5) for two ellipsoidal inclusions v_i and v_j subjected to a homogeneous field $\widetilde{\boldsymbol{\sigma}}_{1,2}(\mathbf{x})$:

$$\overline{\boldsymbol{\sigma}}(\mathbf{x}) = \widetilde{\boldsymbol{\sigma}}_{1,2}(\mathbf{x}) + \boldsymbol{\mathcal{T}}^1_{ij}(\mathbf{x}-\mathbf{x}_j)\boldsymbol{\sigma}^0 + \boldsymbol{\mathcal{T}}^2_{ij}(\mathbf{x}-\mathbf{x}_j), \tag{9.1}$$

$$\boldsymbol{\sigma}(\mathbf{x}) = \mathbf{B}_i(\mathbf{x})\widetilde{\boldsymbol{\sigma}}_{1,2}(\mathbf{x}) + \mathbf{C}_i(\mathbf{x}) + \boldsymbol{\mathcal{F}}^1_{ij}(\mathbf{x}-\mathbf{x}_j) + \boldsymbol{\mathcal{F}}^2_{ij}(\mathbf{x}-\mathbf{x}_j), \tag{9.2}$$

where $\mathbf{x} \in v_i,\ i = 1, 2,\ j = 3 - i$. Reference to some exact and approximate analytical methods of estimation of the tensors $\boldsymbol{\mathcal{T}}^1_i(\mathbf{x}-\mathbf{x}_j)$ and $\boldsymbol{\mathcal{T}}^2_i(\mathbf{x}-\mathbf{x}_j)$ can be found in Chapter 4. Independently of the method of the estimation, the tensors $\boldsymbol{\mathcal{T}}^1, \boldsymbol{\mathcal{T}}^2, \boldsymbol{\mathcal{F}}^1, \boldsymbol{\mathcal{F}}^2$ tend to zero with $|\mathbf{x}-\mathbf{x}_j| \to \infty$ as $|\mathbf{x}-\mathbf{x}_j|^{-d}$. In particular, in the framework of the hypothesis **H1** (8.14), taking Eq. (8.25) into account we get (no sum over $i, j;\ i \neq j$):

$$\boldsymbol{\mathcal{T}}^1_{ij}(\mathbf{x}-\mathbf{x}_j) = \mathbf{R}_i^{-1}[\mathbf{Z}_{ii}\mathbf{R}_i + \mathbf{Z}_{ij}\mathbf{R}_j - \mathbf{R}_i], \tag{9.3}$$

$$\boldsymbol{\mathcal{T}}^2_{ij}(\mathbf{x}-\mathbf{x}_j) = \mathbf{R}_i^{-1}[\mathbf{Z}_{ii}\mathbf{F}_i + \mathbf{Z}_{ij}\mathbf{F}_j - \mathbf{F}_i], \tag{9.4}$$

$$\boldsymbol{\mathcal{F}}^1_{ij}(\mathbf{x}-\mathbf{x}_j) = \mathbf{B}_i(\mathbf{x})\mathbf{R}_i^{-1}[\mathbf{Z}_{ii}\mathbf{R}_i + \mathbf{Z}_{ij}\mathbf{R}_j - \mathbf{R}_i], \tag{9.5}$$

$$\boldsymbol{\mathcal{F}}^2_{ij}(\mathbf{x}-\mathbf{x}_j) = \mathbf{B}_i(\mathbf{x})\mathbf{R}_i^{-1}[\mathbf{Z}_{ii}\mathbf{F}_i + \mathbf{Z}_{ij}\mathbf{F}_j - \mathbf{F}_i]. \tag{9.6}$$

For the sake of simplicity of subsequent calculations we will assume the hypothesis (8.67) where the shape of "excluded volume" v_{ij}^0 does not depend on the inclusion v_j: $v_{ij}^0 = v_i^0$. Then substituting Eqs. (9.2) into Eq. (8.7) at $n=1$ leads to

$$\overline{v}_i\langle\boldsymbol{\vartheta}_i\rangle = \overline{v}_i\boldsymbol{\vartheta}_i^0 + \mathbf{R}_i\mathbf{Q}_i^0\sum_{j=1}^N\langle\boldsymbol{\vartheta}_j\rangle c^{(j)} + \boldsymbol{\mathcal{J}}_i, \tag{9.7}$$

where the notation

$$\boldsymbol{\mathcal{J}}_i = \int\Big\{[\langle\mathbf{M}_1(\mathbf{x})\boldsymbol{\mathcal{F}}_{ij}^1(\mathbf{x}-\mathbf{x}_j)\rangle_i\langle\widetilde{\boldsymbol{\sigma}}_{1,2}\rangle + \langle\mathbf{M}_1(\mathbf{x})\boldsymbol{\mathcal{F}}_i^2(\mathbf{x}-\mathbf{x}_j)\rangle_i]\varphi(v_j,\mathbf{x}_j|v_i,\mathbf{x}_i)$$
$$- \mathbf{R}_i\boldsymbol{\Gamma}(\mathbf{x}_i-\mathbf{x}_j)\langle\boldsymbol{\vartheta}_j\rangle c^{(j)}[1-V_i^0(\mathbf{x}_j)]\Big\}\,d\mathbf{x}_j, \tag{9.8}$$

is introduced and the tensor $\mathbf{Q}_i^0$ is estimated by the formula (3.110) with replacement of the volume v_i by the volume v_i^0; in particular, if the shape of "excluded volume" v_i^0 can be produced from the surface ∂v_i by a homothetic transformation then $\mathbf{Q}_i^0 \equiv \mathbf{Q}_i$. At first we assume that the integral (9.8) is known. Then the linear system (9.7) can be solved easily for $\langle\boldsymbol{\sigma}\rangle_i(\mathbf{x})$:

$$\langle\boldsymbol{\sigma}\rangle_i(\mathbf{x}) = \mathbf{B}_i(\mathbf{x})\boldsymbol{\sigma}^0 + \mathbf{C}_i(\mathbf{x})$$
$$+ \mathbf{B}_i(\mathbf{x})\mathbf{Q}_i^0\big[\mathbf{I}-\langle\mathbf{R}^v\mathbf{Q}^0V\rangle\big]^{-1}\Big\{\langle\mathbf{R}^vV\rangle\boldsymbol{\sigma}^0 + \langle\mathbf{F}^vV\rangle + \sum_{i=1}^N c^{(i)}\boldsymbol{\mathcal{J}}_i\Big\}$$
$$+\int\Big\{[\boldsymbol{\mathcal{F}}_{ij}^1(\mathbf{x}-\mathbf{x}_j)\langle\widetilde{\boldsymbol{\sigma}}_{1,2}\rangle + \boldsymbol{\mathcal{F}}_i^2(\mathbf{x}-\mathbf{x}_j)]\varphi(v_j,\mathbf{x}_j|v_i,\mathbf{x}_i)$$
$$- \mathbf{B}_i(\mathbf{x})\boldsymbol{\Gamma}(\mathbf{x}-\mathbf{x}_j)\langle\boldsymbol{\vartheta}_j\rangle c^{(j)}[1-V_i^0(\mathbf{x}_j)]\Big\}\,d\mathbf{x}_j, \tag{9.9}$$

where one introduces the notations $\mathbf{f}^v(\mathbf{x}) = \mathbf{f}^{(i)}(\mathbf{x})/\overline{v}^{(i)}$, at $\mathbf{x}\in v^{(i)}$, $(i = 1,\ldots,N;\ \mathbf{f} = \mathbf{R},\mathbf{F})$. All integrals in Eq. (9.9) are absolutely convergent and do not depend on the shape of the integration region which can be chosen as homothetical to v_i^0. Recall that the stress fields entering in the integrals (9.9) are unknown in actuality and for solution of Eq. (9.9) we will use the standard assumption of the perturbation method

$$\langle\widetilde{\boldsymbol{\sigma}}(\mathbf{x})_{1,2}\rangle_i = \langle\overline{\boldsymbol{\sigma}}(\mathbf{x})\rangle_i = \boldsymbol{\sigma}^0,\ i=1,2,\ \mathbf{x}\in v_i \tag{9.10}$$

Then all integrals in Eq. (9.9) valid for arbitrary $\mathbf{M}^c \equiv$ const. are known that makes it possible to estimate the effective properties from Eqs. (8.1.12) and (8.1.13) in a similar manner to the analogous process with Eqs. (8.49)–(8.51). Nevertheless, for simplicity of calculations one assumes $\mathbf{M}^c = \mathbf{M}^{(0)}$ that yields the following *new representations*:

$$\mathbf{M}^* = \mathbf{M}^0 + \Big[\mathbf{I}-\langle\mathbf{R}^v\mathbf{Q}^0V\rangle\Big]^{-1}\Big\{\langle\mathbf{R}^vV\rangle + \sum_{i=1}^N c^{(i)}\boldsymbol{\mathcal{J}}_i^{1\mathbf{M}}\Big\}, \tag{9.11}$$

$$\boldsymbol{\beta}^* = \boldsymbol{\beta}^{(0)} + \Big[\mathbf{I}-\langle\mathbf{R}^v\mathbf{Q}^0V\rangle\Big]^{-1}\Big\{\langle\mathbf{F}^vV\rangle + \sum_{i=1}^N c^{(i)}\boldsymbol{\mathcal{J}}_i^{2\mathbf{M}}\Big\}, \tag{9.12}$$

$$-2W^* = \langle \boldsymbol{\beta}_1 \mathbf{C} V\rangle + \langle \boldsymbol{\beta}_1 \mathbf{B}\mathbf{Q}^0 V\rangle \left[\mathbf{I} - \langle \mathbf{R}^v \mathbf{Q}^0 V\rangle\right]^{-1} \langle \mathbf{F}^v V\rangle$$
$$+ \langle \boldsymbol{\beta}_1 \mathbf{B}\mathbf{Q}^0 V\rangle \left[\mathbf{I} - \langle \mathbf{R}^v \mathbf{Q}^0 V\rangle\right]^{-1} \sum_{i=1}^{N} c^{(i)} \boldsymbol{\mathcal{J}}_i^{2\mathbf{M}} + \sum_{i=1}^{N} c^{(i)} \boldsymbol{\mathcal{J}}_i^{2\beta}. \quad (9.13)$$

The notation for absolutely convergent integrals is used:

$$\boldsymbol{\mathcal{J}}_i^{m\mathbf{M}} = \int \Big\{ \langle \mathbf{M}(\mathbf{x}) \boldsymbol{\mathcal{F}}_{ij}^m (\mathbf{x} - \mathbf{x}_j)\rangle_i \varphi(v_j, \mathbf{x}_j | v_i, \mathbf{x}_i)$$
$$- \mathbf{B}_i \boldsymbol{\Gamma}(\mathbf{x} - \mathbf{x}_j) \mathbf{R}_i n^{(j)} [1 - V_i^0(\mathbf{x}_j)] \Big\}\, d\mathbf{x}_j, \quad (9.14)$$

$$\boldsymbol{\mathcal{J}}_i^{m\beta} = \int \Big\{ \langle \boldsymbol{\beta}(\mathbf{x}) \boldsymbol{\mathcal{F}}_{ij}^m (\mathbf{x} - \mathbf{x}_j)\rangle_i \varphi(v_j, \mathbf{x}_j | v_i, \mathbf{x}_i)$$
$$- \langle \boldsymbol{\beta} \mathbf{B}_i \rangle \boldsymbol{\Gamma}(\mathbf{x} - \mathbf{x}_j) \mathbf{F}_i n^{(j)} [1 - V_i^0(\mathbf{x}_j)] \Big\}\, d\mathbf{x}_j, \quad (9.15)$$

describing the binary interaction between the inclusions, $m = 1, 2$. If the integration domains of absolutely convergent integrals (9.14) and (9.15) are chosen to be homothetical with the correlation holes v_i^0 $(i = 1, \ldots, N)$ then the second summands in the integrands (9.14) and (9.15) can be dropped out.

It would be interesting to compare (at least in a qualitative sense) the MEFM with some related methods of prediction of effective moduli that account for higher-order microstructural information. So, in the elegant method of Torquato [1104], [1105] the exact series expansions for the effective stiffness tensor were obtained for macroscopically anisotropic, d-dimensional, two-phase composite media in powers of the "elastic polarization." The method departs from previous treatments by introducing an integral equation for the "cavity" strain field which coincides with our notations for the spherical inclusions with the effective field $\overline{\boldsymbol{\varepsilon}}(\mathbf{x}) = \mathbf{A}^{-1}\boldsymbol{\varepsilon}(\mathbf{x})$ [see (8.8) and (8.87$_1$)] introduced analogously to the effective stress field (8.4). At the next step the equation for the "cavity" strain field (or the effective field) was recast in terms of the stress polarization tensor $\boldsymbol{\tau} = \mathbf{L}_1(\mathbf{x})\boldsymbol{\varepsilon}(\mathbf{x})$, where the driving term $\mathbf{R}^{\epsilon}\langle\boldsymbol{\varepsilon}\rangle$ (called the external stress polarization in [138]) was obtained as an extraction of the Eshelby solution for a single inclusion in the infinite matrix subjected to the field $\langle\boldsymbol{\varepsilon}\rangle$. After that the analog of Eq. (8.1) used in [1104] can be solved by the iteration method similarly to the perturbation method or the method of correlation approximation [69]. In so doing the driving term of Eq. (8.1), which is the external stress polarization factor $\mathbf{R}^{\epsilon}\langle\boldsymbol{\varepsilon}\rangle$, is chosen as an initial approximation. The last choice is better than that in the classical method of correlation approximation with the initial approximation $\langle\boldsymbol{\varepsilon}\rangle$. However, Torquato [1104] has proposed a much better choice, and extracted in Eq. (8.1) as a driving term not only the Eshelby solution $\mathbf{R}\langle\boldsymbol{\varepsilon}\rangle$ for a single inclusion but also the solution corresponding to one of the Hashin-Shtrikman [441] bounds for the composite material that makes it possible to analyze any contrasting of the components. The final explicit representations for the effective moduli $\mathbf{L}^*$ depend on the N-point correlation functions $S_N^{(i)}(\mathbf{y}^N)$, and the interactions of different inclusions is not directly taken into account. In essence, this prospective idea [1104] of the extraction of Hashin-Shtrikman's solution as a driving term was developed in [139] (and presented in this subsection) in the combined MEFM-perturbation method in which the scheme [1104] of the iteration approximation of the integral operator involved was directly accompanied by the solution for the binary interacting inclusions explicitly depending on the radial distribution function $g(r)$. In Section 9.4 we will demonstrate the advantage of the standard MEFM over the combined MEFM-perturbation method through comparison with experimental data for the Newtonian suspensions of identical spherical rigid inclusions. It should be mentioned that to obtain concrete numerical estimations of $\mathbf{L}^*$ by the use of exact formulae [1104], it is necessary to know a complete set of the functions $S_N^{(i)}(\mathbf{y}^N)$ $(N \to \infty)$. The last is practically improbable because, to the author's knowledge, systematic estimations even of 3-point functions $S_3^{(i)}(\mathbf{y}^3)$ for the real microstructures were not conducted, although the necessary microstructural parameters were estimated for the hexagonal array of cylinders [307]. But in the case of the limitation of the microstructural knowledge just by 2-point function $S_2^{(i)}(\mathbf{y}^2)$ (or $g(r)$), the estimations [1104] degenerate to one of the Hashin-Shtrikman boundaries or Mori-Tanaka estimations which is worse than the MEFM evaluations (see Chapter 7) using the same information for the function $g(r)$.

It should be mentioned that the argument verifying the combined MEFM-perturbation method proposed is plausible rather than rigorous because the closing assumption (9.10) was applied to Eq. (9.9) rather than to the initial Eq. (8.7) as in the standard perturbation method being considered at the following subsection.

9.1.2 Perturbation Method for Small Concentrations of Inclusions

In the case of a small concentration of inclusions as well as for a weakly inhomogeneous medium

$$c^{(i)} \ll 1, \quad \text{or} \quad ||\mathbf{M}_1^{(i)}\mathbf{L}^c|| \ll 1 \tag{9.16}$$

the perturbation method is appropriate. Then instead of hypothesis **H2** we can apply an assumption (9.10) to initial Eq. (8.7) (in contrast to the combined MEFM-perturbation method), and taking Eq. (9.2) into account, we get

$$\begin{aligned}\langle\boldsymbol{\sigma}\rangle_i(\mathbf{x}) =& \mathbf{B}_i(\mathbf{x})\boldsymbol{\sigma}^0 + \mathbf{C}_i(\mathbf{x})\\ &+\int\Big[(\boldsymbol{\mathcal{F}}^1_{ij}(\mathbf{x}-\mathbf{x}_j)\varphi(v_j,\mathbf{x}_j|v_i,\mathbf{x}_i) - \mathbf{B}_i(\mathbf{x})\boldsymbol{\Gamma}(\mathbf{x}-\mathbf{x}_j)\mathbf{R}_j n^{(j)})\Big]\boldsymbol{\sigma}^0\, d\mathbf{x}_j\\ &+\int\Big[(\boldsymbol{\mathcal{F}}^2_{ij}(\mathbf{x}-\mathbf{x}_j)\varphi(v_j,\mathbf{x}_j|v_i,\mathbf{x}_i) - \mathbf{B}_i(\mathbf{x})\boldsymbol{\Gamma}(\mathbf{x}-\mathbf{x}_j)\mathbf{F}_j n^{(j)})\Big]\, d\mathbf{x}_j \end{aligned}\tag{9.17}$$

differing from the version of the MEFM (8.40) only by the particular solution of the pair-wise inclusion interaction problem (8.25) as well as by the additional assumption (8.67) which uniquely reduces the MEFM to the perturbation method:

$$\mathbf{M}^* = \mathbf{M}^{(0)} + [\mathbf{I} + \langle\mathbf{R}^v\mathbf{Q}^0V\rangle]\langle\mathbf{R}^vV\rangle + \sum_{i=1}^N c^{(i)}\boldsymbol{\mathcal{J}}_i^{1\mathbf{M}}, \tag{9.18}$$

$$\boldsymbol{\beta}^* = \boldsymbol{\beta}^{(0)} + [\mathbf{I} + \langle\mathbf{R}^v\mathbf{Q}^0V\rangle]\langle\mathbf{F}^vV\rangle + \sum_{i=1}^N c^{(i)}\boldsymbol{\mathcal{J}}_i^{2\mathbf{M}}, \tag{9.19}$$

$$-2W^* = \langle \boldsymbol{\beta}_1 \mathbf{C} V \rangle + \langle \boldsymbol{\beta}_1 \mathbf{B} \mathbf{Q}^0 V \rangle \langle \mathbf{F}^v V \rangle + \sum_{i=1}^{N} c^{(i)} \boldsymbol{\mathcal{J}}_i^{2\beta}. \tag{9.20}$$

Equations (9.18)–(9.20) are new. In the case of the exact estimation of the tensor $\boldsymbol{\mathcal{J}}^{1\mathbf{M}}$ (9.15), Eq. (9.18) is reduced to the analogous relation obtained in [216] (see also [573], [1208]) for spherical homogeneous isotropic inclusions of one radius; for an approximate representation of the tensor $\boldsymbol{\mathcal{J}}^{1\mathbf{M}}$ evaluated by the use of Eqs. (4.64), (4.67$_1$), (9.5) and (9.15), Eq. (9.18) is reduced to the result [181] (see also [1189]). Equations (9.18)–(9.20) can be obtained from Eqs. (9.11)–(9.13) by passage to the limit $c \equiv \langle V \rangle \to 0$ and dropping the terms of order $O(c^3)$ and higher-order terms. Passing to the limit $c \equiv \langle V \rangle \to 0$ in Eqs. (8.79)–(8.81) and dropping the terms of order $O(c^3)$ leads to Eqs. (9.18)–(9.20) with the omitted terms $\boldsymbol{\mathcal{J}}^{1\mathbf{M}}$, $\boldsymbol{\mathcal{J}}^{2\mathbf{M}}$, $\boldsymbol{\mathcal{J}}^{2\beta} \equiv \mathbf{0}$; these representations contain the terms of order $O(c^2)$ although their accuracy has the order $O(c)$. Finally, discarding the terms of order $O(c^2)$ and higher order terms in Eqs. (9.18)–(9.20) leads to the dilute approximation of effective properties. Particular cases of this approximation for the effective moduli $\mathbf{L}^*$ of two-phase composites containing homogeneous ellipsoidal inclusions are one of most rederived formulae for a 40-year period (1957-1999) (for references see, e.g., [734]). Very clear common and concise forms of these representations are attributed to Eshelby [318] and to Krivoglaz and Cherevko [604].

The solution (9.17) takes only binary interaction of inclusions into account. A more complete realization of the perturbation method is based on the assumption

$$\langle \widetilde{\boldsymbol{\sigma}}(\mathbf{x})_{1,\ldots,n} \rangle_i - \langle \overline{\boldsymbol{\sigma}}(\mathbf{x}) \rangle_i = \boldsymbol{\sigma}^0, \ \ \langle \widetilde{\boldsymbol{\sigma}}(\mathbf{x})_{1,\ldots,m} \rangle_j \equiv \mathbf{0}, \tag{9.21}$$

($i = 1,\ldots,n;\ \mathbf{x} \in v_i;\ m > n,\ j = 1,\ldots,m,\ \mathbf{y} \in v_j$) with the truncating of the Neumann series involved. The perturbation method has an essential advantage of controlled accuracy: the error of a remainder term truncated by the assumption (9.21) can, in principle, be estimated exactly.

It should be mentioned that effective properties can be estimated by the use of the evaluation of average strains in the components, e.g.,

$$\mathbf{L}^* = \mathbf{L}^{(0)} + [\mathbf{I} + \langle \mathbf{R}^{\epsilon v} \mathbf{P}^0 V \rangle] \langle \mathbf{R}^{\epsilon v} V \rangle + \sum_{i=1}^{N} c^{(i)} \boldsymbol{\mathcal{J}}_i^{\epsilon 1\mathbf{M}}, \tag{9.22}$$

where the tensor $\boldsymbol{\mathcal{J}}_i^{\epsilon 1\mathbf{M}}$ is defined analogously to Eqs. (9.14) and (9.3) with replacement of the tensors $\mathbf{M},\ \mathbf{R},\ \boldsymbol{\Gamma}$ by $\mathbf{L},\ \mathbf{R}^\epsilon,\ \mathbf{U}$, respectively. In so doing, utilization of the perturbation method as well as the combined MEFM-perturbation method reduces to an inequality $\mathbf{L}^* \neq (\mathbf{M}^*)^{-1}$ in contrast to the MEFM (8.49). Because of this Eqs. (9.11)–(9.13) and (9.18)–(9.20) will be employed if $\mathbf{M} - \mathbf{M}^{(0)}$ is positive definite; otherwise a dual scheme based on the estimation of average strains $\langle \boldsymbol{\varepsilon} \rangle_i$ and effective moduli $\mathbf{L}^*$ will be used.

9.1.3 Perturbation Method for Weakly Inhomogeneous Media

Another form of the perturbation method is related to another small parameters (9.16$_2$) such as a slight deviation of the moduli $\mathbf{L}(\mathbf{x})$ from a constant value $\mathbf{L}^c$

$\equiv \langle \mathbf{L} \rangle$: $||\mathbf{L}_1 \mathbf{M}^c|| \ll 1$. To simplify the subsequent derivations, it is convenient to represent Eq. (7.14$_1$) (with $\boldsymbol{\beta}, \mathbf{f} \equiv \mathbf{0}$) in an operator form

$$\mathbf{L}_1 \boldsymbol{\varepsilon} = \mathbf{L}_1 \langle \boldsymbol{\varepsilon} \rangle + \mathbf{L}_1 \boldsymbol{\mathcal{U}} \mathbf{L}_1 \boldsymbol{\varepsilon}, \tag{9.23}$$

where $\boldsymbol{\mathcal{U}}$ denotes the linear integral operator with the kernel $\mathbf{U}(\mathbf{x})$: $(\boldsymbol{\mathcal{U}} \boldsymbol{\tau})(\mathbf{x}) = \int \mathbf{U}(\mathbf{x}-\mathbf{y}) \{ \boldsymbol{\tau}(\mathbf{y}) - \langle \boldsymbol{\tau} \rangle(\mathbf{y}) \} \, d\mathbf{y}$. We construct the solution of Eq. (9.23) by the point Jacobi iterative scheme based on the recursion formula $\mathbf{L}_1 \boldsymbol{\varepsilon}_{[n+1]} = \mathbf{L}_1 \langle \boldsymbol{\varepsilon} \rangle + \mathbf{L}_1 \boldsymbol{\mathcal{U}} \mathbf{L}_1 \boldsymbol{\varepsilon}_{[n]}$, where $\{ \boldsymbol{\varepsilon}_{(n)} \}$ are treated as an approximation of the solution (9.23), which is presented in the Neumann series form

$$\mathbf{L}_1 \boldsymbol{\varepsilon} = \boldsymbol{\mathcal{L}} \mathbf{L}_1 \langle \boldsymbol{\varepsilon} \rangle, \ \boldsymbol{\mathcal{L}} \equiv \sum_{k=0}^{\infty} \boldsymbol{\mathcal{U}}^k, \tag{9.24}$$

where the power $\boldsymbol{\mathcal{U}}^k$ is defined recursively by the condition $\boldsymbol{\mathcal{U}}^1 = \mathbf{L}_1 \boldsymbol{\mathcal{U}}$ and the kernel of $\boldsymbol{\mathcal{U}}^k$ is $\boldsymbol{\mathcal{U}}_k(\mathbf{x},\mathbf{y}) = \mathbf{L}_1(\mathbf{x}) \int \mathbf{U}(\mathbf{x}-\mathbf{z}) \boldsymbol{\mathcal{U}}_{k-1}(\mathbf{z},\mathbf{y}) \, d\mathbf{z}$. Usually the driving term of Eq. (9.23) is used as an initial approximation $\mathbf{L}_1 \boldsymbol{\varepsilon}_{[0]} = \mathbf{L}_1 \langle \boldsymbol{\varepsilon} \rangle$ which, in combination with Eq. (9.24) and the definition of the effective elastic moduli $\mathbf{L}^*$ (6.20), allows a few iterations of $\mathbf{L}^*$:

$$\mathbf{L}^* = \langle \mathbf{L} \rangle, \quad \mathbf{L}^* = \langle \mathbf{L} \rangle + \langle \mathbf{L}_1 \boldsymbol{\mathcal{U}} \mathbf{L}_1 \rangle, \tag{9.25}$$

$$\mathbf{L}^* = \langle \mathbf{L} \rangle + \langle \mathbf{L}_1 \boldsymbol{\mathcal{U}} \mathbf{L}_1 \rangle + \langle \mathbf{L}_1 \boldsymbol{\mathcal{U}} \mathbf{L}_1 \boldsymbol{\mathcal{U}} \mathbf{L}_1 \rangle, \tag{9.26}$$

The representations (9.25) and (9.26) could be obtained from the general Eqs. (8.8)–(8.11) and (6.20) if the integral system (8.8)–(8.11) is truncated at the $n = 1$ and $n = 2$, and in all integral terms we assume that $\boldsymbol{\varepsilon}(\mathbf{y}) \equiv \langle \boldsymbol{\varepsilon} \rangle$ (direct interaction between the inclusions is neglected). Therefore, the perturbation method (9.24) can be considered as a consequence of the MEFM. The perturbation method is attributed to Lifshitz and Rozenzweig [683], [685] and developed for the composite materials and polycrystals by Shermergor and coauthors [257], [258], [259], [345], [346], [347], [995] as well as in [66], [69], [608], [775], [1106]. The sufficient convergence conditions for the Neumann series method (9.24) were obtained in [344] (see also [116]) in the Hilbert space of modified strain fields. To avoid some algebraic manipulation, we reproduce the effective moduli representations from these works.

In particular, the second-order approximation (9.25$_2$) for statistically homogeneous media is

$$L^*_{ijkl} = \langle L_{ijkl} \rangle + \int U_{mnpq}(\mathbf{x}-\mathbf{y}) B^{L|ijmn}_{pqkl}(\mathbf{x}-\mathbf{y}) \, d\mathbf{y}, \tag{9.27}$$

where $\mathbf{U}$ is a second derivative of the Green tensor $\mathbf{G}$ (3.3) with an elastic modulus $\mathbf{L}^c = \langle \mathbf{L} \rangle$: $U_{ijkl} = [\nabla_j \nabla_l G_{ik}(\mathbf{x})]_{(ij)(kl)}$ (3.19), and the binary correlation function of the elasticity tensor $\mathbf{B}^L = \langle \mathbf{L}_1(\mathbf{x}) \otimes \mathbf{L}_1(\mathbf{y}) \rangle$ is related to the binary correlation function $B^V_{ij}(\mathbf{x}) \equiv \langle V^{(i)\prime}(\mathbf{0}) V^{(j)\prime}(\mathbf{x}) \rangle$ $[V^{(i)\prime}(\mathbf{x}) \equiv V^{(i)}(\mathbf{x}) - c^{(i)}]$: $\mathbf{B}^L(\mathbf{x}) = \sum_i \sum_j \mathbf{L}_1^{(i)} \mathbf{L}_1^{(j)} B^V_{ij}(\mathbf{x})$ if the tensor $\mathbf{L}_1$ is described by a piecewise function $\mathbf{L}_1(\mathbf{x}) = \sum_i \mathbf{L}_1^{(i)} V^{(i)}(\mathbf{x})$. In turn, $\mathbf{B}^V_{ii}(\mathbf{x})$ is the autocovariance of phase $v^{(i)}$

(5.84). In particular, for macroscopically isotropic composites $\langle \mathbf{L} \rangle = (dk^c, 2\mu^c)$ with two isotropic phases $\mathbf{L}(\mathbf{x}) = (dk(\mathbf{x}), 2\mu(\mathbf{x}))$ in the d-dimensional space R^d, the integral in Eq. (9.27) can be calculated by the use of the convolution theorem in the Fourier space. Then substitution of the $\mathbf{U}$ decomposition (3.20), (3.167) into (9.27) yields the following relation for the effective elasticity (see for details [1106]): $\mathbf{L}^* = \big(d\langle k \rangle + db_2^k, 2\langle \mu \rangle + 2b_2^{\mu}\big)$, where

$$b_2^k = -\frac{dD_k}{dk^{(2)} + 2(d-1)\mu^{(2)}}, \quad b_2^{\mu} = -\frac{2dD_{\mu}(k^{(2)} + 2\mu^{(2)})}{(d+2)\mu^{(2)}[dk^{(2)} + 2(d-1)\mu^{(2)}]}, \tag{9.28}$$

and $D_g \equiv c^{(1)}c^{(2)}(g^{(2)} - g^{(1)})^2$ is a dispersion of the parameter $g = k, \mu$. Because due to the statistical isotropy of the microstructure, a dependence of the obtained estimations $\mathbf{L}^*$ (9.28) on the two-point average $B_{ij}^V(\mathbf{x})$, concerning the space distribution of the phases, vanishes and $\mathbf{L}^*$ depends only on the volume phase concentrations rather than on the details of microstructure. However, such a dependence is involved for the third-order estimation (9.26): $\mathbf{L}^* = \big(d\langle k \rangle + db_2^k + db_3^k, 2\langle \mu \rangle + 2b_2^{\mu} + 2b_3^{\mu}\big)$, where

$$b_3^k = \frac{dD_k[dc^{(2)}(k^{(1)} - k^{(2)}) + 2(d-1)\zeta(\mu^{(1)} - \mu^{(2)})]}{[dk^{(2)} + 2(d-1)\mu^{(2)}]^2}, \tag{9.29}$$

$$\begin{aligned} b_3^{\mu} = {} & \frac{2dD_{\mu}}{[dk^{(2)} + 2(d-1)\mu^{(2)}]^2} \left\{ \frac{2\zeta}{d+2}(k^{(1)} - k^{(2)}) + \left[\frac{2d(k^{(2)} + 2\mu^{(2)})^2 c^{(2)}}{(d+2)^2(\mu^{(2)})^2} \right. \right. \\ & \left. \left. + \frac{(d-2)(2k^{(2)} + 3\mu^{(2)})\zeta}{(d+2)\mu^{(2)}} + \left[\frac{dk^{(2)} + (d-2)\mu^{(2)}}{(d+2)\mu^{(2)}} \right]^2 \eta \right] (\mu^{(1)} - \mu^{(2)}) \right\}, \end{aligned} \tag{9.30}$$

and the Torquato-Milton microstructural parameters ζ and η are defined by Eqs. (6.252) and (6.253) with the indexes $0, 1$ replaced by $2, 1$, respectively.

Analogously, for nontextured polycrystals the components of the isotropic effective modulus were estimated in [995] for the 3-D case from the algebraic Eq. (9.27) for the crystals of any symmetry. In particular, the effective bulk modulus is found exactly $k^* = k = (L_{11} + 2L_{12})/3$ (6.242) for the cubic polycrystals, while μ^* is estimated with an accuracy of the order $o(||\mathbf{L}_1\mathbf{M}^c||^2)$:

$$\mu^* = \langle \mu \rangle - \frac{L^2}{25\langle \mu \rangle}\left[1 - \frac{2}{5}\frac{2L_{11} + 3L_{12} + L_{44}}{3L_{11} + 2L_{12} + 4L_{44}}\right], \tag{9.31}$$

where $L = L_{11} - L_{12} - 2L_{44}$.

It should be mentioned that zero-order approximation of $\mathbf{L}^*$ (9.25_1) coincides with the Voight bounds (6.196), which is why the estimations (9.25) and (9.26) are called Voight's scheme of the perturbation method. Construction of Reuss's scheme can be performed in a straightforward manner by replacement of the tensors $\mathbf{L}^c$, $\mathbf{L}, \mathbf{U}$ in Eqs. (9.25) and (9.26) by the tensors $\mathbf{M}^c = \langle \mathbf{M} \rangle$, $\mathbf{M}$, $\boldsymbol{\Gamma}$ (3.37), respectively. In so doing, the effective elasticity $\mathbf{L}^*$ and compliance $\mathbf{M}^*$ are not each other's inverse in general: $\mathbf{L}^* \not\equiv (\mathbf{M}^*)^{-1}$.

9.1.4 Elastically Homogeneous Media with Random Field of Residual Microstresses

Driven by modern design concepts such as compactness, reliability, and power density, the constituent phases of composite materials encounter severe conditions with significant thermal expansion, phase transformation, and plastic strains. These strains are commonly called as eigenstrains. Especially interesting results are obtained if the fluctuations of elastic compliance are negligible whereas the stress-free strains fluctuate. Of course, the statistical averages of residual stresses in composite components (8.44) can be estimated with the help of the passage to the zero limit of the elastic mismatch of different components in the corresponding formulae (8.44) and (8.54). Nevertheless, the desired relationships can be found immediately without some assumptions of the MEFM [136].

Namely, the random Eq. (8.2) can be simplified if $\mathbf{M}(\mathbf{x}) \equiv$ const. as

$$\boldsymbol{\sigma}(\mathbf{x}) = \langle \boldsymbol{\sigma} \rangle + \int \boldsymbol{\Gamma}(\mathbf{x}-\mathbf{y})[\boldsymbol{\beta}_1(\mathbf{y}) - \langle \boldsymbol{\beta}_1 \rangle] \, d\mathbf{y}, \tag{9.32}$$

allowing for a representation of a statistical average stresses in the inclusion v_i

$$\langle \boldsymbol{\sigma} \rangle_i(\mathbf{x}) = \langle \boldsymbol{\sigma} \rangle + \mathbf{T}_i^{\beta}(\mathbf{x}-\mathbf{x}_i)\bar{v}_i + \int \Big[\mathbf{T}_q^{\beta}(\mathbf{x}-\mathbf{x}_q)\bar{v}_q \varphi(v_q, \mathbf{x}_q|; v_i, \mathbf{x}_i) - \boldsymbol{\Gamma}(\mathbf{x}-\mathbf{x}_q)\langle \boldsymbol{\beta}_1 \rangle \Big] \, d\mathbf{x}_q, \tag{9.33}$$

where $\mathbf{T}_q^{\beta}(\mathbf{x}-\mathbf{x}_q) = \bar{v}_i^{-1} \int \boldsymbol{\Gamma}(\mathbf{x}-\mathbf{z})\boldsymbol{\beta}_1(\mathbf{z}) V_q(\mathbf{z}) \, d\mathbf{z}$ $(\mathbf{x} \notin v_q)$. In so doing, for ellipsoidal inclusions $\mathbf{x} \in v_i$ we have $\bar{v}_i \langle \mathbf{T}_i^{\beta}(\mathbf{x}-\mathbf{x}_i) \rangle_i = -\mathbf{Q}_i \langle \boldsymbol{\beta}_1 \rangle_i$ and $\bar{v}_i \mathbf{T}_i^{\beta}(\mathbf{x}-\mathbf{x}_i) = -\mathbf{Q}_i \boldsymbol{\beta}_1^{(i)}$ if the inclusion v_i is additionally homogeneous: $\boldsymbol{\beta}_1(\mathbf{x}) = \boldsymbol{\beta}_1^{(i)} \equiv$ const. at $\mathbf{x} \in v_i$. Then for ellipsoidal inclusion $\mathbf{x} \in v_i$ it follows that the relation for average stresses in the inclusion v_i is

$$\langle \boldsymbol{\sigma} \rangle_i = \langle \boldsymbol{\sigma} \rangle - \mathbf{Q}_i \langle \boldsymbol{\beta}_1 \rangle_i + \int \Big[\mathbf{T}_{iq}^{\beta}(\mathbf{x}_i-\mathbf{x}_q)\bar{v}_q \varphi(v_q, \mathbf{x}_q|; v_i, \mathbf{x}_i) - \mathbf{T}_i(\mathbf{x}_i-\mathbf{x}_q)\langle \boldsymbol{\beta}_1 \rangle \Big] \, d\mathbf{x}_q, \tag{9.34}$$

hereafter under $\mathbf{x} \notin v_q$, $v_i \neq v_q$, and $\mathbf{T}_{iq}^{\beta}(\mathbf{x}_i-\mathbf{x}_q) = \langle \mathbf{T}_q^{\beta}(\mathbf{x}-\mathbf{x}_q) \rangle_i$.

The relation for the stored energy from $W^* \equiv -\langle \boldsymbol{\beta}_1 \boldsymbol{\sigma} \rangle / 2$ (for $\langle \boldsymbol{\sigma} \rangle \equiv \mathbf{0}$) follows immediately from Eq. (9.34):

$$W^* = -\frac{1}{2} \sum_{i=1}^{N} \Big\{ c^{(i)} \mathbf{Q}_i \langle \boldsymbol{\beta}_1 \rangle_i + \iint V_i(\mathbf{x}) \boldsymbol{\beta}_1(\mathbf{x}) \Big[\mathbf{T}_q^{\beta}(\mathbf{x}-\mathbf{x}_q)\bar{v}_q \varphi(v_q, \mathbf{x}_q|; v, \mathbf{x}_i) - \boldsymbol{\Gamma}(\mathbf{x}-\mathbf{x}_q)\langle \boldsymbol{\beta}_1 \rangle \Big] \, d\mathbf{x}_q \, d\mathbf{x} \Big\}. \tag{9.35}$$

The formulae (9.34) and (9.35) may be significantly simplified under the assumption (8.69) where in the considered case $\boldsymbol{\vartheta} \equiv \boldsymbol{\beta}_1$. The assumption (8.69) holds for some models of composites. In particular one may consider a grain structure where the correlation of homogeneous transformations $\boldsymbol{\beta}^{(i)}(\mathbf{x}) \equiv$ const.

($\mathbf{x} \in v^{(i)}$) between different grains is lacking [601]. Then the equality (8.69) is valid under the simplest conditional probability density (8.67) which is realized for statistical isotropy of composite structure for spherical inclusions [599]. Then, by virtue of Eq. (8.70), Eq. (9.34) can be combined into a simple equation

$$\langle \boldsymbol{\sigma} \rangle_i(\mathbf{x}) = \langle \boldsymbol{\sigma} \rangle + \mathbf{Q}_i \langle \boldsymbol{\beta}_1 \rangle + \bar{v}_i \mathbf{T}_i^{\beta}(\mathbf{x}, \mathbf{x}_i). \tag{9.36}$$

In so doing for the homogeneous ellipsoidal inclusion v_i the statistical average stress $\langle \boldsymbol{\sigma} \rangle_i(\mathbf{x})$ does not depend on the position $\mathbf{x}$ inside the inclusion being analyzed:

$$\langle \boldsymbol{\sigma} \rangle_i(\mathbf{x}) = \langle \boldsymbol{\sigma} \rangle + \mathbf{Q}_i \big[\langle \boldsymbol{\beta}_1 \rangle - \boldsymbol{\beta}_1^{(i)}\big], \quad \text{if} \quad \boldsymbol{\beta}_1(\mathbf{x}) = \text{const.}, \ \ \mathbf{x} \in v_i. \tag{9.37}$$

In a similar manner the relation for the stored energy (9.35) may be simplified as well:

$$W^* = -\frac{1}{2} \sum_{i=1}^{N} c^{(i)} \Big[\bar{v}_i \langle \boldsymbol{\beta}_1(\mathbf{x}) \mathbf{T}_i^{\beta}(\mathbf{x}, \mathbf{x}_i) \rangle_i + \langle \boldsymbol{\beta}_1 \rangle_i \mathbf{Q}_i \langle \boldsymbol{\beta}_1 \rangle \Big]. \tag{9.38}$$

For homogeneous inclusions Eq. (9.38) admits of further simplification

$$W^* = \frac{1}{2} \sum_{i=1}^{N} c^{(i)} \boldsymbol{\beta}_1^{(i)} \mathbf{Q}_i \Big[\boldsymbol{\beta}_1^{(i)} - \langle \boldsymbol{\beta}_1 \rangle \Big]. \tag{9.39}$$

Under the additional assumption (8.69) the formula (9.39) may be simplified

$$W^* = \frac{1}{2} \langle (\boldsymbol{\beta} - \langle \boldsymbol{\beta} \rangle) \mathbf{Q} (\boldsymbol{\beta} - \langle \boldsymbol{\beta} \rangle) \rangle. \tag{9.40}$$

The relation (9.40) was obtained in [601] by another less formal method.

9.2 Effective Medium Methods

9.2.1 Application to Composite Materials

One of the first schemes of the effective medium method (EMM), also called the self-consistent method, was proposed for both the polycrystals [455], [606] and composite materials [117], [463], [979] and was based on the following hypothesis: each inclusion in the composite material behaves as an isolated one in a homogeneous medium whose properties coincide with the effective properties of the whole composite:

$$\mathbf{M}^c \equiv \mathbf{M}^*. \tag{9.41}$$

The field acting on this inclusion $\overline{\boldsymbol{\sigma}}$ coincides with the average field $\overline{\boldsymbol{\sigma}} \equiv \langle \boldsymbol{\sigma} \rangle$ corresponding to the homogeneous traction boundary condition (6.5) for the macrodomain containing a statistically large number of inclusions. By doing so, the infinite system (8.4)–(8.6) for the matrix composites being considered is truncated after Eq. (7.17) which provides a way of estimating of effective compliance (8.1.12) by the use of the solution (8.18) and (8.87) for an isolated inclusion within an infinite effective medium:

$$\mathbf{M}^* = \mathbf{M}^{(0)} + \langle \bar{\mathbf{R}}^v V\rangle, \quad \mathbf{M}^* = \mathbf{M}^{(0)} + \langle (\mathbf{M} - \mathbf{M}^{(0)})\mathbf{L}\bar{\mathbf{A}}V\rangle\mathbf{M}^*, \tag{9.42}$$

where the tensors $\bar{\mathbf{R}}^v = (\mathbf{M} - \mathbf{M}^{(0)})\bar{\mathbf{B}}$, $\bar{\mathbf{B}} = \bar{\mathbf{B}}(\mathbf{x})$, and $\bar{\mathbf{A}} = \bar{\mathbf{A}}(\mathbf{x})$ implicitly depend on $\mathbf{M}^*$. Here the bar over the material tensor indicate that it is computed using the fact that the matrix properties coincide with those of the effective medium. It was exploited that $\bar{\mathbf{B}}(\mathbf{x}) = \mathbf{L}(\mathbf{x})\bar{\mathbf{A}}(\mathbf{x})\mathbf{M}^*$ and $\langle\boldsymbol{\varepsilon}\rangle = \mathbf{M}^*\langle\boldsymbol{\sigma}\rangle$.

In a similar manner, for the homogeneous displacement boundary conditions (6.4) accompanied by the EMM assumptions $\mathbf{L}^c = \mathbf{L}^*$, $\overline{\boldsymbol{\varepsilon}} \equiv \langle\boldsymbol{\varepsilon}\rangle$, we get two equivalent representations of effective moduli (6.20):

$$\mathbf{L}^* = \mathbf{L}^{(0)} + \langle (\mathbf{L} - \mathbf{L}^{(0)})\bar{\mathbf{A}}V\rangle, \quad \mathbf{L}^* = \mathbf{L}^{(0)} + \langle (\mathbf{L} - \mathbf{L}^{(0)})\mathbf{M}\bar{\mathbf{B}}V\rangle\mathbf{L}^*. \tag{9.43}$$

The obtained effective compliances (9.42_1), (9.42_2) and elastic moduli (9.43_1), (9.43_2) are each other's inverses. Indeed, multiplying Eq. (9.43_1) from the right by $(\mathbf{L}^*)^{-1}$, and from the left by $(\mathbf{L}^{(0)})^{-1} = \mathbf{M}^{(0)}$ leads to

$$\mathbf{M}^{(0)} = (\mathbf{L}^*)^{-1} + \langle \mathbf{M}^{(0)}(\mathbf{L} - \mathbf{L}^{(0)})\bar{\mathbf{A}}\rangle(\mathbf{L}^*)^{-1} = (\mathbf{L}^*)^{-1} - \langle (\mathbf{M} - \mathbf{M}^{(0)})\mathbf{L}\bar{\mathbf{A}}\rangle(\mathbf{L}^*)^{-1}. \tag{9.44}$$

Analogously, multiplying Eq. (9.42_2) by the tensors $(\mathbf{M}^*)^{-1}$ and $(\mathbf{M}^{(0)})^{-1} = \mathbf{L}^{(0)}$ from the right and from the left, respectively, gives

$$\mathbf{L}^{(0)} = (\mathbf{M}^*)^{-1} + \langle \mathbf{L}^{(0)}(\mathbf{M} - \mathbf{M}^{(0)})\bar{\mathbf{B}}\rangle(\mathbf{M}^*)^{-1} = (\mathbf{M}^*)^{-1} - \langle (\mathbf{L} - \mathbf{L}^{(0)})\mathbf{M}\bar{\mathbf{B}}\rangle(\mathbf{M}^*)^{-1}. \tag{9.45}$$

From comparison of (9.44) with (9.42_1), and (9.45) with (9.42_2), it is seen that the effective compliance $\mathbf{M}^*$ (9.42_1), (9.42_2) and the effective moduli (9.43_1), (9.43_2) are mutually inverse $(\mathbf{L}^*)^{-1} = \mathbf{M}^*$, $(\mathbf{M}^*)^{-1} = \mathbf{L}^*$ for the matrix composite which consists of a homogeneous matrix containing a statistically uniform set of inclusions $v = \bigcup v^{(i)}$ $(i = 1, \ldots, N)$ (with the indicator function $V(\mathbf{x}) = \sum V^{(i)}$, $i = 1, \ldots, N$) of any shape and microtopology, and with an arbitrary number of phases.

Moreover, we can assume that the composite has no continuous matrix $c^{(0)} \equiv 0$ as it takes place for polycrystals. Then Eqs. (6.20) and (9.43_1) coupled with $\mathbf{L}^{(0)} = \mathbf{L}^*$ imply the equivalent representations for both the strain concentrator factors

$$\langle \bar{\mathbf{A}}V\rangle = \mathbf{I}, \quad \langle (\mathbf{L} - \mathbf{L}^*)\bar{\mathbf{A}}V\rangle = \mathbf{0}, \tag{9.46}$$

and their counterparts presented in terms of stress concentrator factors:

$$\langle \bar{\mathbf{B}}V\rangle = \mathbf{I}, \quad \langle (\mathbf{M} - \mathbf{M}^*)\bar{\mathbf{B}}V\rangle = \mathbf{0}. \tag{9.47}$$

For concrete estimation of the effective properties, we need to know the tensors $\bar{\mathbf{A}}(\mathbf{x})$ $(\mathbf{x} \in v^{(i)}$, $i = 1, \ldots, N)$. In particular, for the homogeneous ellipsoidal inclusions $\bar{\mathbf{A}}(\mathbf{x}) = [\mathbf{I} + \bar{\mathbf{P}}_i(\mathbf{L}^{(i)} - \mathbf{L}^*)]^{-1}$, $\bar{\mathbf{B}}(\mathbf{x}) = [\mathbf{I} + \bar{\mathbf{Q}}_i(\mathbf{M}^{(i)} - \mathbf{M}^*)]^{-1}$, $(\mathbf{x} \in v^{(i)}$, $i = 1, \ldots, N)$ (3.120), (3.121), and $\bar{\mathbf{Q}}_i$ and $\bar{\mathbf{P}}_i$ are defined by Eqs. (3.110) with replacement of $\mathbf{L}^{(0)}$ by $\mathbf{L}^*$. If, however, $\bar{\mathbf{P}}_i = \bar{\mathbf{P}}$ for all i, then Eqs. (9.43_1), (9.46_1), and (9.46_2) are equivalent to the representation

$$\sum_{i=1}^{N}(\mathbf{L}^{(i)}-\mathbf{L}^{*})[\mathbf{I}+\bar{\mathbf{P}}(\mathbf{L}^{(i)}-\mathbf{L}^{*})]^{-1}=\mathbf{0} \tag{9.48}$$

allowing subsequent simplified decomposition for composites with the spherical isotropic inclusions in R^d-space ($d=2,3$) (see, e.g., [1106]):

$$\sum_{i=1}^{n}\frac{c^{(i)}(k^{(i)}-k^{*})}{k^{(i)}+2(d-1)\mu^{*}/d}=0, \quad \sum_{i=1}^{n}\frac{c^{(i)}(\mu^{(i)}-\mu^{*})}{\mu^{(i)}+g^{*}}=0, \tag{9.49}$$

where $g^{*}=\mu^{*}[dk^{*}/2+(d+1)(d-2)\mu^{*}/d][k^{*}+2\mu^{*}]^{-1}$, and the obtained representations (9.48), (9.49) are symmetric with respect to the simultaneous interchange $\mathbf{L}^{(i)}\leftrightarrow\mathbf{L}^{(j)},\ \ c^{(i)}\leftrightarrow c^{(j)}$ $(i=0,1,\ldots,N.)$

As another limiting case of the inclusion shape, let us consider an isotropic microcracked medium with a uniform distribution of microcrack orientation. In such a case, we can apply the dilute approximation (8.150) with a replacement of the matrix Poisson ratio $\nu^{(0)}$ by the corresponding ratio of the composite medium ν^{*}:

$$\frac{k^{*}}{k^{(0)}}=1-\frac{16}{9}\frac{1-(\nu^{*})^{2}}{1-2\nu^{*}}\epsilon, \qquad \frac{\mu^{*}}{\mu^{(0)}}=1-\frac{32}{45}\frac{(1-\nu^{*})(5-\nu^{*})}{2-\nu^{*}}\epsilon. \tag{9.50}$$

The formulae (9.50) are the solution of the self-consistent approach [119].

It should be mentioned that the representations (9.49) are the particular case of results obtained by the different methods. For example, a choice of the comparison medium as an effective medium (9.41) allows the relations (9.49) by the method of conditional moments (8.152) [561], [967], by the singular-approximation method (see Subsection 8.3.4 and [995]), by the method [97] based on the concept of "strong isotropy," as well as by the variation method [1179], [1184]. Obviously, the relations (9.49) always lie between the Hashin-Shtrikman bounds intentionally obtained at the more "soft" or "stiff" values of $\mathbf{L}^{c}$. Moreover, the EMM estimations (9.49) are realizable in the sense that there is a particular random structure composite whose effective moduli exactly coincide with EM estimations. Such a fractal-like (or self-similar in all-scale) structure justified in [773], [774] consists of two-sort granules $v^{(1)}$ and $v^{(2)}$ of dilute concentration randomly distributed in the matrix consisting in turn of much smaller similar grains, and so on from a finite size of inclusions to infinitesimally small. Realizability of the EM approach provides a location of the estimations (9.49) within Hashin-Shtrikman bounds.

The obtained representations of effective properties [such as, e.g., (9.43_1), (9.48), (9.49)] are the implicit algebraic systems with respect to yet-unknown effective elastic moduli $\mathbf{L}^{*}$ providing the effective moduli only implicitly through a straightforward iterative procedure to the fixed point. For example, Eq. (9.43_1) corresponds to the iterative scheme

$$\mathbf{L}^{*[n+1]}=\mathbf{L}^{(0)}+\langle(\mathbf{L}-\mathbf{L}^{(0)})\bar{\mathbf{A}}^{[n]}V\rangle, \tag{9.51}$$

with an initial condition, e.g., either $\mathbf{L}^{*[0]}=\langle\mathbf{L}\rangle$ or $\mathbf{L}^{*[0]}=\langle\mathbf{M}\rangle^{-1}$ if the rigidity of the inclusions is less than that of the matrix or otherwise, respectively. The scheme (9.51) is known to be stable and unique although this statement is plausible rather than rigorous. The fixed point of the mapping (9.51) is nothing more than a solution of the already analyzed Eq. (9.43_1). Therefore, the described scheme (9.51) is actually an iterative procedure of solving Eq. (9.43_1).

Note that it is possible to get the simple solutions of (9.49) in the limiting cases of incompressible phase $v^{(0)}$ with the phase $v^{(1)}$ formed by either the rigid inclusions or voids:

$$k^* = \infty, \quad \frac{\mu^*}{\mu^{(0)}} = \frac{2 + dc^{(1)}}{2c^{(0)}}, \quad \text{or} \quad \frac{k^*}{\mu^*} = \frac{2(d-1)c^{(0)}}{dc^{(1)}}, \quad \frac{\mu^*}{\mu^{(0)}} = \frac{dc^{(0)}}{d + 2c^{(1)}}, \tag{9.52}$$

respectively. It should be mentioned that the formulae (9.52) predict either infinite or zero effective stiffness if the volume fraction $c^{(1)}$ reaches a so-called *percolation threshold* $c_p^{(1)}$ [957]. Such a behavior is a manifestation of the so-called *percolation phenomenon* base on the fact that the grains of the dispersed phase $v^{(1)}$ form a continuous connected path (called an infinite cluster) through the composite if $c^{(1)} \geq c_p^{(1)}$. For example, for the spherical inclusions with $\mu^{(1)}/\mu^{(0)} = \infty$ (9.52) the percolation threshold $c_p^{(1)} = 2/(d+2)$ which is not sensitive to the microstructure details (such as, e.g., a fractional composition of inclusions and their radial distribution function) and can be quite different from the real percolation threshold of a concrete material. For a porous medium, both the effective bulk and shear moduli have the same percolation threshold $c_p^{(1)} = (d-1)/(d+1)$. The availability of the mentioned percolation thresholds which are invariant with respect to microstructural details is neither a merit nor a fault of the EMM.

In more recent schemes of the effective medium method the composite material outside a certain layer of matrix surrounding each inclusion is replaced by an effective medium [239], [471], [482], [483], [1019], [1221]. In so doing, Eqs. (9.42_1) and (9.43_1) with obvious corrections of tensors $\bar{\mathbf{B}}(\mathbf{x})$ and $\bar{\mathbf{A}}(\mathbf{x})$ are valid. So, following [239], a composite-sphere model (or three-phase model) considers an RVE as a concentric sphere consisting of a matrix $v^{(0)}$ and a single spherical homogeneous inclusion v_i of radii b and a, respectively, which is inserted in the medium with the yet-unknown elastic moduli $\mathbf{L}^* = (3k^*, 2\mu^*)$. The elasticity tensors of $v^{(0)}$ and v_i are $\mathbf{L}^{(0)} = (3k^{(0)}, 2\mu^{(0)})$ and $\mathbf{L}^{(1)} = (3k^{(1)}, 2\mu^{(1)})$, respectively, while the volume fraction of v_i is $c^{(1)} = c = (a/b)^3$. Then the effective elastic moduli k^* and μ^* are computed from the average elastic energy in both the explicit form

$$\frac{k^*}{k^{(0)}} = 1 + \frac{(k^{(1)} - k^{(0)})(3k^{(0)} + 4\mu^{(0)})}{k^{(0)}[3k^{(0)} + 4\mu^{(0)} + 3(1-c)(k^{(1)} - k^{(0)})]} \tag{9.53}$$

and implicit form of a quadratic equation:

$$A\Big(\frac{\mu^*}{\mu^{(0)}}\Big)^2 + 2B\Big(\frac{\mu^*}{\mu^{(0)}}\Big) + C = 0, \tag{9.54}$$

where

$$A = 8c^{10/3}e(4 - 5\nu^{(0)})\eta_1 - 2c^{7/3}(63e\eta_2 + 2\eta_1\eta_3) + 252c^{5/3}e\eta_2$$

$$
\begin{aligned}
&- 25ce(7 - 12\nu^{(0)} + 8(\nu^{(0)})^2)\eta_2 + 4(7 - 10\nu^{(0)})\eta_2\eta_3,\\
B &= -2c^{10/3}e(4 - 5\nu^{(0)})\eta_1 + 2c^{7/3}(63e\eta_2 + 2\eta_1\eta_3) - 252c^{5/3}e\eta_2\\
&+ 75ce(3 - \nu^{(0)})\nu^{(0)}\eta_2 + 3(15\nu^{(0)} - 7)\eta_2\eta_3/2,\\
C &= 4c^{10/3}e(5\nu^{(0)} - 7)\eta_1 - 2c^{7/3}(63e\eta_2 + 2\eta_1\eta_3) - 252c^{5/3}e\eta_2\\
&+ 25ce((\nu^{(0)})^2 - 1)\eta_2 - 4(7 + 5\nu^{(0)})\eta_2\eta_3,
\end{aligned} \tag{9.55}
$$

with $e = \mu^{(1)}/\mu^{(0)} - 1$, and

$$
\begin{aligned}
\eta_1 &= (49 - 50\nu^{(0)}\nu^{(1)})e + 35(e+1)(\nu^{(1)} - 2\nu^{(0)}) + 35(2\nu^{(1)} - \nu^{(0)}),\\
\eta_2 &= 5\nu^{(1)}(e-3) + 7(e+5), \quad \eta_3 = (e+1)(8 - 10\nu^{(0)}) + (7 - 5\nu^{(0)}).
\end{aligned}
$$

Christensen [238] demonstrated that the composite-sphere model (9.53)–(9.55) (also called the *generalized self-consistent method*, GSCM) results correlate better with the experimental data on viscosity of suspensions than the MTM (8.102) and the differential scheme considered in Section 8.3. Siboni and Benveniste [1008] generalized the (GSCM) to the multi-phase composites, while the GSCM was combined with an accumulation of damage in [971]. The main idea of the EMM lies in the substitution of the strain concentrator factor $\bar{\mathbf{A}}(\mathbf{x})$ into the average scheme of the dilute approximation (9.43$_1$), although the tensor $\bar{\mathbf{A}}(\mathbf{x})$ can be incorporated into any other average scheme. For example, Luo and Weng [720] integrated $\bar{\mathbf{A}}(\mathbf{x})$ into the Mori-Tanaka scheme and demonstrated in a concrete example that the effective moduli estimated by the modified GSCM lie between the estimations obtained via the MTM and GSCM. The GSCM is based on the assumption that the particles in the three-phase cells are separated by the layer of a pure matrix which is free of the particle phase. However, a variation of properties of this layer along the radius makes it possible to take into account a probability of entrapment of the particle phase in the matrix layer [121], [1094].

To put this another way, the effective medium method (9.42$_1$) or (9.43$_1$) is equivalent to the perturbation method of the first order over the concentration of inclusions accompanied by the assumption (9.41). An obvious improvement of the method (9.42$_1$) is the use of the perturbation method of the second order described in Subsection 9.1.2 accompanied by the assumption (9.41). Moreover, the combination of the MEFM and the effective medium concept (9.41) is equally trivial in a substitution of the equality (9.41) into either Eqs. (8.49) or (8.76). This direction of a generalization of the EMM is demonstrated in the version of the MEMM proposed in [175] which is based on the solution of Eq. (8.40) where the involved tensors are estimated for $\mathbf{L}^c = \mathbf{L}^*$. It will lead to the representations of effective properties in the form of Eqs. (8.49)–(8.51) where all material tensors are estimated at $\mathbf{L}^c = \mathbf{L}^*$. The mentioned approach is more general than substitution of (9.41) into Eq. (8.76) which, in turn, is more general than the approach [676] (see also [78], [634]) combining the assumption of the Mori-Tanaka (8.90) and effective medium method (9.41) (the last is questionable in the general case of multiphase composites; see Section 9.3). The implicit algebraic relations for effective moduli obtained must be solved numerically, in general, see Eq. (9.51).

To avoid confusion, it should be mentioned that the term *self-consistent methods* was originally proposed merely for the methods related to the assumption (9.41) enabling one to get self-consistent implicit algebraic equations for effective elastic moduli such, as e.g., (9.42_1). We currently use the logical generalization of this term which is widely used to stand for methods based on some closing assumption for effective fields (see, e.g., the hypothesis H2) for obtaining a *self-consistent* final system of either integral or differential equations for statistical moments of local fields. Only in the simplified special cases considered in the current work, this system can be solved explicitly in analytical form.

9.2.2 Analysis of Polycrystal Materials

Let us consider (following [995]) a one-phase polycrystal consisting of homogeneous spherical grains with the same elastic properties and uniform random orientation. Due to the absence of texture, the polycrystal is macroscopically isotropic $\mathbf{L}^* = \mathbf{L}^c = (3k^*, 2\mu^*)$, and the tensor $\mathbf{b}^L$ introduced in Eq. (8.137) is also isotropic:

$$\mathbf{b}^L \equiv (3b_k^L, 2b_\mu^L) = \left(4\mu^*, \frac{\mu^*(9k^* + 8\mu^*)}{3(k^* + 2\mu^*)}\right). \tag{9.56}$$

For averaging in Eq. (8.137_2), we introduce the auxiliary tensors $\mathbf{S}^L$ and $\mathbf{S}^*$:

$$(\mathbf{L} + \mathbf{b}^L)\mathbf{S}^L = \mathbf{I}, \quad (\mathbf{L}^* + \mathbf{b}^L)\mathbf{S}^* = \mathbf{I}, \tag{9.57}$$

allowing us to present Eq. (8.137_2) in the form $\mathbf{S}^* = \langle \mathbf{S}^L \rangle$. Here the components of the isotropic tensor $\mathbf{S}^* \equiv (3S_k^*, 2S_\mu^*) = \left(\frac{1}{3}S_{iikk}^*, \frac{1}{5}(S_{ikik}^* - \frac{1}{3}S_{iikk}^*)\right)$ can be found for the nontextured materials from two scalar equations:

$$\mathbf{N}_1\mathbf{S}^* = \mathbf{N}_1\mathbf{S}^L, \quad \mathbf{N}_2\mathbf{S}^* = \mathbf{N}_2\mathbf{S}^L. \tag{9.58}$$

As an example, we will consider a cubic symmetry of crystals (2.175). Then Eqs. (9.58_1) and (9.58_2) combined with Eq. (6.211) give

$$3k^* = L_{11} + 2L_{12}, \qquad \frac{1}{\mu^* + b_\mu^L} = \frac{2}{5(\mu_1 + b_\mu^L)} + \frac{3}{5(\mu_2 + b_\mu^L)}, \tag{9.59}$$

where $\mu_1 = (L_{11} - L_{12})/2$, $\mu_2 = L_{44}$, and, therefore

$$5\mu^* b_\mu^L - (2\mu_1 + 3\mu_2)b_\mu^L + (3\mu_1 + 2\mu_2)\mu^* - 5\mu_1\mu_2 = 0. \tag{9.60}$$

Substitution of the explicit representation $\mathbf{b}^L$ (9.56) into Eq. (9.60) leads to the final equation:

$$(\mu^*)^3 + a_2(\mu^*)^2 + a_1\mu^* + a_0 = 0, \tag{9.61}$$

with one positive root; here $a_0 = -3\mu_1\mu_2/4$, $a_1 = -3\mu_2(k^* + 4\mu_1)/8$, $a_2 = (9k^* + 4\mu_1)/8$. Eq. (9.61) was obtained by Hershey [455] and by Kröner [606] via the proposed self-consistent method. Some other examples of a similar approach with applications to the polycrystals for different crystal symmetry can be found, e.g., in [77] [612], [784], [785], [995], [1162].

It should be mentioned that a direct analytical solution of the self-consistent Eqs. (9.43_1) or (9.57) is often cumbersome and hides from view the basic ideas of the method. However, Eq. (9.43_1) and (8.135_2) have an advantage of providing an iterative method with the consecutively repeated scheme such as, e.g., for the one-particle approximation of the MEF method (8.135_2):

$$\mathbf{L}^{*[n+1]} = \langle \bar{\mathbf{A}}^{[n]} V \rangle^{-1} \langle \bar{\mathbf{R}}^{\epsilon v[n]} V \rangle, \tag{9.62}$$

which is obvious and easily performed. We will consider this iteration method (9.62) for the previously considered example of quasi-isotropic polycrystals formed by the uniformly oriented crystals of cubic symmetry. Just for concreteness of arbitrary chosen initial conditions of the scheme (9.62), we will use the Reuss $\mathbf{L}_-^{*[0]} = \langle \mathbf{M} \rangle^{-1}$ and Voight $\mathbf{L}_+^{*[0]} = \langle \mathbf{L} \rangle$ bounds. In Table 9.1, the estimated shear moduli of a few polycrystals with different Zener anisotropy ratio Z (2.176) correspond to the Reuss μ^R and Voight μ^V bounds (6.199), to Hashin-Shtrikman bounds μ^{-HS} and μ^{+HS} (6.243), as well as to the first-order approximation μ^{-MEF} and μ^{+MEF} obtained via scheme (9.62) at the initial conditions $\mathbf{L}_-^{*[0]} = \langle \mathbf{M} \rangle^{-1}$ and $\mathbf{L}_+^{*[0]} = \langle \mathbf{L} \rangle$, respectively. As can be seen, even the first iteration of the MEF method (9.62) has the advantage of providing a most accurate estimations in the considered examples. The fast convergence of the iteration method (9.62) is demonstrated in Fig. 9.1.

Table 9.1. Effective shear moduli (GPa) of some cubic polycrystals

Crystals	Z	μ^R	μ^{-HS}	μ^{-MEF}	μ^{+MEF}	μ^{+HS}	μ^V
Na	7.00	0.12	0.17	0.19	0.21	0.23	0.28
Pb	4.08	0.66	0.80	0.86	0.89	0.91	1.03
Th	3.62	2.33	2.73	2.86	2.94	3.01	3.39
Ag	3.01	2.55	2.90	2.99	3.04	3.09	3.38
Nb	1.96	4.11	4.33	4.36	4.37	4.39	4.57

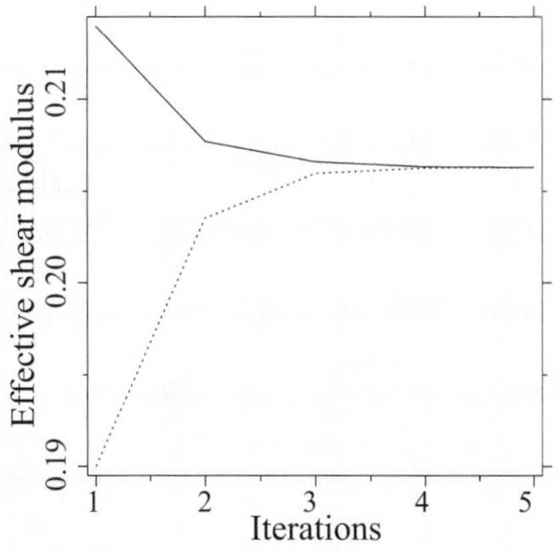

Fig. 9.1. Effective shear moduli μ^{+MEF} (solid curve) and μ^{-MEF} (dotted curve) vs the number of iterations.

It should be mentioned that a significance of the developed models of polycrystalline materials is beyond the scope of its own field of polycrystals and can be applied to the analysis of some skeletal composites (see Subsection 1.1.1). For example, the open-cell carbon foam microstructure is known to possess a tetrahedral structure of the foam ligaments oriented approximately 109° to each other [376], [1009]. The ligaments of foam materials can be modeled as polycrystalline medium (see, e.g., [419], [1161], [1009]) aggregating from the numerous minute crystallites or grains of various sizes and shapes placed in the amorphous matrix and varying along its longitudinal (axial) direction. After evaluation of effective elastic properties of ligament materials by the use of some polycrystalline models such as, e.g., (9.62), the problem of estimation of effective elastic properties of foam materials appears. The application of known methods for the homogenization of random structures to the analyses of the foams and cellular materials has been hampered by their large porosity. Because of this, these materials as well as the waffle structures, are modeled as periodic media using classical homogenization procedures. For example, in approach [1009], because of slenderness, each ligament can be considered as a beam, and the tetrahedral cell microstructure with four ligaments as a structure. The four beams are located in three-dimensional space under arbitrary loading conditions. The cross-section of the beam and their material properties vary along the beam length. The subsequent analysis of the random orientation of the unit cells are performed by the use of either Voight or Reuss volume averaging of effective elastic properties estimated for the perfect periodic structures with prescribed different orientation mentioned above. This popular rough approach can be easily improved by the use of the MEFM generalized in this section for the analysis of polycrystalline materials. In so doing, the role of the individual crystallite is assumed by the homogeneous effective medium corresponding to the involved periodic structure material with the prescribed orientation of unit cells.

9.3 Differential Methods

9.3.1 Scheme of the Differential Method

In addition to effective field (see Section 8.3) and effective medium (see Subsection 9.2.1) methods, the differential effective medium method (EMM) [22], [53], [114], [243], [435], [816], [817], [818], [943] is used in the mechanics of composite materials. This differential scheme of constructing effective elastic moduli also belongs to the class of the effective medium methods. Differential EMM arose as an alternative to the ambiguous assumptions of EMM, where the value of the effective modulus is used as an estimate of the average strain concentration in an isolated inclusion. Differential EMM is considered as a process of consecutive additions of the inclusion phase in a uniform medium with a modulus equal to the effective modulus of the medium with the previous additions of inclusions to the matrix. This process can take place in two ways. In the first, it is assumed that the inclusion phase consists of an infinite number of fractions with infinite size differences, and successive accommodations of the inclusions in the corresponding uniform medium leads to ascending order of inclusion size. In the second, inclusions of the initial size are added to the medium in infinitesimal amounts to reach the final real concentration. Both methods give equivalent results, since at each iteration step, the same solution for the single-particle problem is used to estimate the effective modulus of the medium with an infinitesimal inclusion in the corresponding uniform medium. Buryachenko and Parton [175] proposed two versions of the multiparticle differential methods for multicomponent composites. In the first one the tensors analogous $\bar{\mathbf{R}}_i^v \equiv (\mathbf{M}(\mathbf{x}) - \mathbf{M}^{(0)})\bar{\mathbf{B}}_i$ (9.42_1) are determined from the averaged solution of the problem for n inclusions in a uniform medium with tensor properties $\mathbf{M}^*$ and the field $\boldsymbol{\sigma}^0$ given at infinity. In the second one the tensors $\bar{\mathbf{R}}_i^v$ (9.42_1) are estimated at each iteration step based on n-particle MEFM. Thus both EMM and MEFM are applicable. In this subsection, we do not solve the single-particle problem, but instead solve the multiparticle problem at each iteration step of the differential scheme. To do this we use an algorithm previously proposed in Section 9.1. In comparisons of calculations with experimental data, we show the advantage of the multiparticle differential method over the one-particle method.

Following [175], we will estimate effective stiffness

$$\mathbf{L}^* = \mathbf{L}^{(0)} + \sum_{i=1}^{N} \mathbf{R}_i^{\epsilon *} n^{(i)}, \ \langle \mathbf{L}_1(\mathbf{x})\boldsymbol{\varepsilon}(\mathbf{x})V^{(i)}(\mathbf{x})\rangle \equiv \mathbf{R}_i^{\epsilon *} n^{(i)} \langle \boldsymbol{\varepsilon} \rangle \tag{9.63}$$

in a pure elastic problem ($\boldsymbol{\beta} \equiv \mathbf{0}$). The first equation in (9.63) can be rewritten in the form

$$\mathbf{L}^* = \mathbf{L}^{(0)} + \sum_{i=1}^{N} \mathbf{R}_i^{\epsilon *} \mathbf{A}_0 n^{(i)} \Big[c^{(0)}\mathbf{I} + \sum_{j=1}^{N} c^{(j)} \mathbf{A}_j^0 \Big]^{-1}, \tag{9.64}$$

$$\langle \boldsymbol{\varepsilon} \rangle_i = \mathbf{A}_i^0 \langle \boldsymbol{\varepsilon} \rangle^{(0)}, \ \ \langle \boldsymbol{\varepsilon} \rangle^{(0)} = \mathbf{A}_0 \langle \boldsymbol{\varepsilon} \rangle. \tag{9.65}$$

We will consider a generalization of a differential method scheme [817], [818] in the case of a multicomponent inclusion with complex structure. We take the volume w of the composite medium with some finite concentration of inclusions $i = 1, 2, \ldots, N$ (which in general is different from the concentration $c^{(i)}$). The composite medium is replaced by a uniform volume w with tensor properties $\mathbf{L}$, determined from the equation $\langle \boldsymbol{\sigma} \rangle = \mathbf{L}\langle \boldsymbol{\varepsilon} \rangle$. The infinitesimal discrete volume of the uniform medium is removed and replaced by the sum of the components $i = 1, 2, \ldots, N$. That is, the representative volume $dv^{(i)}$ with properties $\mathbf{L}$ is replaced by the same volume $dv^{(i)}$ with properties of the ith component. If $\tilde{\mathbf{R}}_i^{\epsilon v *}$ is the average value of the coefficient of concentration of the polarization tensor $(\tilde{\mathbf{L}}_1(\mathbf{x}) \equiv \mathbf{L}(\mathbf{x}) - \mathbf{L})$

$$\langle \tilde{\mathbf{L}}_1(\mathbf{x})\boldsymbol{\varepsilon}(\mathbf{x}) \rangle_i \equiv \tilde{\mathbf{R}}_i^{\epsilon v *} \langle \boldsymbol{\varepsilon} \rangle \tag{9.66}$$

of element $dv^{(i)}$, then the increment to the tensor of effective properties $\mathbf{L}$ can be found from (9.63) in the form

$$d\mathbf{L} = \sum_{i=1}^{N} \tilde{\mathbf{R}}_i^{\epsilon v *} \frac{dv^{(i)}}{\bar{w}}. \tag{9.67}$$

Since it is convenient to carry out the calculation with the additional volume portions $dc^{(i)}$, it can be shown that (see [818], [817]) $dv^{(i)}/\bar{w} = \lambda_i dc/(1-c)$, where $c = \sum_{i=1}^{N} c^{(i)}$, $\lambda_i = c^{(i)}/c$, and consequently,

$$d\mathbf{L} = \sum_{i=1}^{N} \tilde{\mathbf{R}}_i^{\epsilon v*} \lambda_i \frac{dc}{1-c}. \tag{9.68}$$

Equation (9.68) is an ordinary differential equation with respect to the unknown tensor $\mathbf{L}$ with initial conditions $\mathbf{L}(0) = \mathbf{L}^{(0)}$, $c^{(i)}(0) = 0$, $c(0) = 0$ ($i = 1, \ldots, N$) and independent variable c ($\lambda_i = c^{(i)}/c = \text{const}$). The basis for (9.68) for a one-component inclusions is given in [243]. To close (9.68), it is necessary to establish a concrete value for $\tilde{\mathbf{R}}_i^{\epsilon v*}$, which can be determined on the basis of one-particle and multiparticle approaches.

9.3.2 One-Particle Differential Method

In the widely used one-particle differential method, it is assumed that the tensor $\tilde{\mathbf{R}}_i^{\epsilon v*} \equiv \tilde{\mathbf{R}}_i^{\epsilon *}/\bar{v}_i$ is determined from the solution to the problem for an isolated inclusion v_i in a homogeneous medium with tensor properties $\mathbf{L}$ and for a uniform field $\langle \varepsilon \rangle$ given at infinity. For the postulated equality of $\mathbf{R}_i^{\epsilon v*}$ (9.63) and $\bar{\mathbf{R}}_i^{\epsilon v}$ (9.42$_1$) ($\mathbf{R}_i^{\epsilon v*} = \bar{\mathbf{R}}_i^{\epsilon v}$), the effective field method is valid. Then the use of $\bar{\mathbf{R}}_i^{\epsilon}$ (9.42$_1$) in the differential equation for the effective modulus (9.68)

$$\tilde{\mathbf{R}}_i^{\epsilon v*} = \bar{\mathbf{R}}_i^{\epsilon v} \tag{9.69}$$

is, in its own way, a combination of the one-particle effective medium method and differential scheme (9.68):

$$d\mathbf{L} = \sum_{i=1}^{N} \bar{\mathbf{R}}_i^{\epsilon v} \lambda_i \frac{dc}{1-c}, \tag{9.70}$$

where $\bar{\mathbf{R}}_i^{\epsilon v} = \bar{\mathbf{P}}_i(\mathbf{I} - \bar{\mathbf{A}}_i)$. For example, a composite material with the isotropic constituents and identical spherical inclusions has the isotropic effective modulus $\mathbf{L} = (3k, 2\mu)$ described by the differential equations

$$\frac{dk}{dc} = \frac{(3k+4\mu)(k^{(1)}-k)}{(3k^{(1)}+4\mu)} \frac{1}{1-c}, \quad \frac{d\mu}{dc} = \frac{5(3k+4\mu)(\mu^{(1)}-\mu)}{6(k+2\mu)\mu^{(1)}/\mu + (9k+8\mu)} \frac{1}{1-c} \tag{9.71}$$

with initial conditions $k(0) = k^{(0)}$, $\mu(0) = \mu^{(0)}$. In the general case, equations (9.70) are coupled and nonlinear, and they admit only a numerical solution. However, in two limiting cases of inclusion moduli [$\mathbf{L}^{(1)} = (\infty, \infty)$, $\mathbf{L}^{(1)} = (0, 0)$] this system was solved in a closed form in [1233] for the rigid inclusions

$$\begin{aligned} \frac{\mu}{\mu^{(0)}} &= \frac{1}{(1-c)^2} \left[\frac{3(1-\nu^{(0)})}{4(1-2\nu) - (1-5\nu^{(0)})(\mu/\mu^{(0)})^{-0.6}} \right]^{1/3}, \\ \frac{k}{\mu^{(0)}} &= \frac{4}{3} - \frac{2}{3} \frac{(1-5\nu^{(0)})}{(1-2\nu^{(0)})} \left(\frac{\mu}{\mu^{(0)}} \right)^{-0.6} \end{aligned} \tag{9.72}$$

and for the pores

$$\frac{\mu}{\mu^{(0)}} = \frac{1}{(1-c)^2}\left[\frac{2(1+\nu^{(0)}) + (1-5\nu^{(0)})(\mu/\mu^{(0)})^{0.6}}{3(1-\nu)}\right]^{1/3},$$
$$\frac{k}{\mu^{(0)}} = \frac{4}{3} + \frac{3}{4}\frac{(1-5\nu^{(0)})}{(1+\nu^{(0)})}\left(\frac{\mu}{\mu^{(0)}}\right)^{0.6}. \tag{9.73}$$

Another version of the one-particle differential method is based on the application of the hypothesis of averaged strains [787], which is a special case of the hypothesis of the method of effective fields (8.14). It follows from just this definition of the tensors $\mathbf{A}_i^0$ $(i = 1, \ldots, N)$ (9.65) that the tensor $\tilde{\mathbf{R}}_i^{\epsilon v*}$ (9.66) can be put in the form

$$\tilde{\mathbf{R}}_i^{\epsilon v*} = \tilde{\mathbf{R}}_i^{\epsilon v*}\mathbf{A}_0^{-1}\left[c^{(0)}\mathbf{I} + \sum_{i=1}^{N} c^{(i)}\mathbf{A}_i^0\right]^{-1}. \tag{9.74}$$

In the Mori-Tanaka hypothesis (8.90), it is assumed that

$$\mathbf{A}_i^0 = \mathbf{A}_i, \tag{9.75}$$

and to obtain an alternate version of the one-particle differential method, it is sufficient to take

$$\tilde{\mathbf{R}}_i^{\epsilon v*}\mathbf{A}_0^{-1} = \mathbf{R}_i^{\epsilon v}, \tag{9.76}$$

then Eqs. (9.68), (9.74)–(9.76) form a closed system for calculation of the effective modulus

$$d\mathbf{L} = \sum_{i=1}^{N}\mathbf{R}_i^{\epsilon v}\left[c^{(0)}\mathbf{I} + \sum_{i=1}^{N} c^{(j)}\mathbf{A}_j\right]^{-1}\lambda_i\frac{dc}{1-c}. \tag{9.77}$$

Hypothesis (9.75) states that the tensor of the mean strain concentration field $\langle\boldsymbol{\varepsilon}\rangle_i$ in the inclusions with respect to the mean strain field in the matrix $\langle\boldsymbol{\varepsilon}\rangle^{(0)}$ is independent of the filler concentration in a real composite. Relation (9.76) additionally confirms the equality of this concentration tensor to the analogous concentration tensor at each step of the iteration process involving the addition of an infinitesimal concentration of inclusions to the uniform medium with tensor properties $\mathbf{L}(c)$. Formula (9.77) is a combination of the one-particle approximation of the MEFM and the differential scheme (9.68). For composite media with one-component homogeneous filler ($c^{(1)} = c$), the following relation holds: $\mathbf{A}_i^0 = (1-c)c^{-1}(\mathbf{L}-\mathbf{L}^{(0)})(\mathbf{L}^{(1)}-\mathbf{L})^{-1}$. By substituting this into (9.64) and using the assumptions (9.42$_1$), (9.63), we obtain $d\mathbf{L}/dc = (\mathbf{L}^{(1)}-\mathbf{L})\mathbf{A}_i(\mathbf{L}^{(1)}-\mathbf{L}^{(0)})^{-1}(\mathbf{L}^{(1)}-\mathbf{L})(1-c)^{-2}$.

9.3.3 Multiparticle Differential Method (Combination with EMM and with MEFM)

In the version of the multiparticle differential method combined with EMM, it is assumed that the tensor $\tilde{\mathbf{R}}_i^{\epsilon v*}$ (9.68) is determined from the averaged solution of the problem for n inclusions in a homogeneous medium with tensor properties $\mathbf{L}$ and the field $\langle\boldsymbol{\varepsilon}\rangle$ given at infinity. The last assumption makes it possible to close the modified system (8.8)–(8.11) where the modulus $\mathbf{L}^{(0)}$ is replaced by $\mathbf{L}$. From the last modified Eq. (8.11), we define the effective field $\langle\widetilde{\varepsilon}_{1,\ldots,n-1}\rangle_i$ with an accuracy to the first-order terms in c. We substitute this value into the preceding equation and so forth. Thus we obtain a representation for $\tilde{\mathbf{R}}_i^{\epsilon v*}$ in the form of a tensor polynomial in c of degree n.

Since the addition of component i $(i = 1, 2, \ldots, N)$ at each iteration step of the differential scheme (9.68) is small $(\lambda_i dc/(1-c) \ll 1)$, then it can be assumed that $\tilde{\mathbf{R}}_i^{\epsilon v*}$ can be represented in the form of a power series in the small parameter $dc/(1-c)$ with the coefficients $\tilde{\mathbf{R}}_i^{\epsilon(k)}$ $(k = 0, n-1)$, and $\tilde{\mathbf{R}}_i^{\epsilon(0)} = \bar{\mathbf{R}}_i^{\epsilon} \equiv \bar{v}_i \bar{\mathbf{R}}_i^{\epsilon v}$. But for each sufficiently smooth function (including $\mathbf{L}$ and $\tilde{\mathbf{R}}_i^{\epsilon v*}$) there is a Taylor expansion in powers of $dc/(1-c)$. Then, from (9.68) and a comparison of the coefficients of like powers of $dc/(1-c)$ in the corresponding Taylor series, we obtain an ordinary differential equation of order n

$$\frac{d^n\mathbf{L}}{dc^n} = \sum_{i=1}^{N} \lambda_i \tilde{\mathbf{R}}_i^{\epsilon(n)} \frac{n!}{(1-c)^{n+1}} \tag{9.78}$$

with initial conditions

$$\mathbf{L}^{(0)}(0) = \mathbf{L}^{(0)},\ \mathbf{L}^{(1)}(0) = \bar{\mathbf{R}}_i^{\epsilon v},\ \ldots,\ \mathbf{L}^{(n-1)}(0) = \sum_{i=1}^{n} \tilde{\mathbf{R}}_i^{(n-1)}(n-1)!, \tag{9.79}$$

where $\mathbf{L}^{(k)}(0) = d^k\mathbf{L}/dc^k(0)$.

In the multiparticle differential method combined with MEFM, the tensor $\tilde{\mathbf{R}}_i^{v*}$ (9.66) is estimated at each iteration step on n-particle MEFM. It is assumed that system (8.26) and (8.27) (expressed in terms of effective strains), which describes the interaction of n inclusions in the medium $\mathbf{L}^c = \mathbf{L}$, is valid for an infinitesimal concentration of the component i $(i = 1, 2, \ldots, N)$, equal to $\lambda_i dc/(1-c)$. This makes it possible to expand $\bar{\mathbf{R}}_i^{\epsilon v*}$ in an infinite series in the small parameter $dc/(1-c)$:

$$\bar{\mathbf{R}}_i^{\epsilon v*} = \sum_{k=0}^{\infty} \mathbf{R}_i^{\epsilon(k)} \left(\frac{dc}{1-c}\right)^k. \tag{9.80}$$

Unlike the multiparticle differential method combined with EMM, series (9.80) is infinite even for the n-particle version of MEFM. Comparing (9.80) with the formal Taylor expansion of the tensor $\tilde{\mathbf{R}}_i^{\epsilon v*}$ in terms $dc/(1-c)$, we find from (9.68)

$$\frac{d^m\mathbf{L}}{dc^m} = \sum_{i=1}^{n} \mathbf{R}_i^{\epsilon(m)} \frac{m!}{(1-c)^{m+1}} \tag{9.81}$$

with Couchy conditions $\mathbf{L}^{(0)}(0) = \mathbf{L}^{(0)}$, $\mathbf{L}^{(1)}(0) = \mathbf{R}_i^{\epsilon v}$, $\ldots$, $\mathbf{L}^{(m-1)}(0) = \mathbf{R}_i^{\epsilon(m-1)}(m-1)!$. In general, the order of (9.81) is larger than the number of interacting inclusions $(m \geq n)$. For $m = n > 2$, the estimates made using the differential methods (EMM) (9.78) and MEFM (9.81) are different.

9.4 Estimation of Effective Properties of Composites with Nonellipsoidal Inclusions

The approach presented is based on such fundamental notions as Green's function and an Eshelby tensor. However, in many cases the matrix of composite materials exhibits systematic anisotropy: the stiffness in the extrusion direction can differ from the stiffness in the out-of-plane direction. The modeling of damage accumulation in the matrix of composite materials is often handled as homogenization of damaged matrix considered as homogeneous anisotropic material [269]. Another common problem dealing with an anisotropic matrix is an incremental micromechanical scheme for composites with the elastoplastic deformation of the matrix estimated by the use of the anisotropic compliance tensor [255], [476], [897]. For all these cases no explicit formula for the Eshelby [318] tensor has been developed. In a widely cited paper [372], Gavazzi and Lagoudas have proposed a numerical scheme for the evaluation of the component of the Eshelby tensor by the use of the surface integral parameterized on the surface of the unit sphere. The solution obtained for one ellipsoidal inclusion in the infinite matrix is further used in the different average schemes such as averaging over the orientation of inclusions [269], [309], [502], [1074]. These methods usually employ the Eshelby theorem [318] about homogeneity of stresses inside a homogeneous ellipsoidal inclusion embedded in the infinite matrix subjected to remote homogeneous stress. As a consequence, these methods cannot be easily extended to composite materials containing inclusions with continuously variable anisotropic properties [217]. For example, Ozmusul and Picu [846] have investigated a polymer matrix composite of periodic structure with nanometer-size filler in which the polymer chains in the matrix are preferentially oriented close to the interface with the relatively rigid fillers, thus adding a graded interfacial layer about each inclusion which can be considered as a thick coating on the inclusion with variable properties.

Kozaczek *et al.* [593] have analyzed a single nonellipsoidal inclusion in an infinite medium which can be considered as a limiting case of a dilute concentration of inclusions. They demonstrated that the shape of the inclusion plays a role in the stress distribution in the grain boundary region; sharp corners raise stress more effectively than rounded edges of oblong-shaped precipitates. At present, the homogenization theory for composites with regular structure is developed in detail (references can be found, for example in [25], [48], [507], [960]. Most of these works are based on the use of multivariable asymptotic techniques with subsequent implementation of different numerical methods. For example, Fan *et al.* [326] analyzed a simple cubic packing of cubic inclusions by the double layer boundary element method for periodic suspensions.

However, a combination of the general anisotropy of the matrix and the general shape of randomly located inclusions presents an impenetrable barrier to the classical approaches using either analytical or numerical representation for the Eshelby tensor for inclusions. The principal analysis schemes of composites with the random arrangement of nonellipsoidal inclusions were considered by many authors [138], [1221]. However, for example, Zheng and Du [1221] have essentially used the homogeneity of inclusions, and presented the numerical results for particular cases of nonelliptic inclusions obtained by the use of the complex potentials method which substantially explores the isotropy of the matrix and 2-D dimensionality of the problem. Thus, systematic numerical analysis of composite materials with the general shape of inclusions and anisotropy of constituents calls for further investigation.

It should be mentioned that there is fundamentally another group of methods based on numerical simulation of random location of inclusions, for example, by the hard-core model, with subsequent analysis of the particle iteration problem via different numerical methods (for references see Chapter 4 and [138], [363], [492], [1236]). Direct consideration of multiparticle interaction problems mentioned in the homogenization schemes involves the complete high-level computer resources of both in terms of CPU time and memory, and usually has limitations in the case of inhomogeneous loading and statistical inhomogeneity of structures, due to exploration of particular realization (by, e.g., Monte-Carlo simulation) of a real random configuration of the inclusion fields. More detailed analysis of these methods is beyond the scope of this section which is dedicated to the more pragmatic goal of an integration of the detailed numerical analysis for one inclusion in the infinite matrix (without any restrictions on the anisotropy of the constituents as well as on the shape and the inhomogeneity of the inclusion) into the known (as well as modified) average scheme of one-particle approximation of the MEFM.

It should be mentioned that the results of this chapter were obtained for any shape and microstructure of inclusions the deformation properties of which are effectively described by the tensors $\mathbf{B}^{(i)}$, $\mathbf{C}^{(i)}$ (8.18). However, the mentioned estimations explicitly depend on the shape of the correlation hole v_i^0 assumed to be ellipsoidal. The last assumption was used only for a statistical description of a composite structure (8.67) and (8.69) allowing us to get an analytical representation of the integral (8.71). This simplification, which seems unessential, is in reality fundamental because of necessity of leads to a homogeneity of the effective field $\overline{\boldsymbol{\sigma}}_i(\mathbf{x})$ which is a basic hypothesis of the MEFM [see Eq. (8.14)]. However, it is interesting that it is possible to generalize both the effective field hypothesis **H1** and the closing one **H2** to the case of a nonellipsoidal shape of the "correlation holes" v_i^0 allowing simplification of subsequent calculations and that are derived from the assumptions **H1** and **H2** for the ellipsoidal v_i^0. We will do so in the framework of a generalized approximation (8.65$_2$), namely:

H1^2) *Each inclusion v_i is located in the field $\overline{\boldsymbol{\sigma}}_i(\mathbf{y})$ which, perhaps, is not a constant and depends on the geometrical parameters of a nonellipsoidal correlation hole v_i^0 but does not depend statistically on the geometric characteristics and elastic constants of the inclusion v_i.*

For termination of statistical moment equations and for their subsequent simplification, we will use the "averaged" analog of the quasi-crystalline approximation by Lax [648] (8.65$_2$):

H2^2) *The equality*

$$\int \boldsymbol{\Gamma}(\mathbf{x}-\mathbf{y})\Big[\langle \boldsymbol{\eta}(\mathbf{y})|; v_i, \mathbf{x}_i\rangle - (1 - V_i^0(\mathbf{y}))\langle \boldsymbol{\eta}\rangle\Big]\, d\mathbf{y} = \mathbf{0} \tag{9.82}$$

holds at $\forall \mathbf{x} \in v_i$.

Obviously, hypotheses (8.14), (8.65_2), and (8.67) are more restrictive than $\mathbf{H1^2}$ and $\mathbf{H2^2}$. Indeed, hypotheses (8.14), (8.65_1), and (8.67) lead to assumption $\mathbf{H1^2}$ for ellipsoidal v_i^0. However, even for ellipsoidal v_i Eq. (8.70) does not yield the hypotheses (8.20), (8.65_2), and (8.67).

Statistical averaging of Eq. (8.39) with the fixed heterogeneity v_i and their extraction from the right-hand side of the averaged equation leads to the representation for statistical average of the effective field

$$\langle\overline{\boldsymbol{\sigma}}_i\rangle(\mathbf{x}) = \langle\boldsymbol{\sigma}\rangle + \sum_{q=1}^{N}\int \boldsymbol{\Gamma}(\mathbf{x}-\mathbf{x}_q)[\langle V_q(\mathbf{x}_q)\boldsymbol{\eta}(\mathbf{x}_q)|; v_i, \mathbf{x}_i\rangle - c^{(q)}\langle\boldsymbol{\eta}_q\rangle]\, d\mathbf{x}_q. \quad (9.83)$$

Due to the hypothesis $\mathbf{H2^2}$ (9.82), the integral on the right-hand side of Eq. (9.83) can be presented in the form

$$\sum_{q=1}^{N}\int \boldsymbol{\Gamma}(\mathbf{x}-\mathbf{x}_q)[\langle V_q(\mathbf{x}_q)\boldsymbol{\eta}(\mathbf{x}_q)|; v_i, \mathbf{x}_i\rangle - c^{(q)}\langle\boldsymbol{\eta}_q\rangle]\, d\mathbf{x}_q = \mathbf{Q}_i^0(\mathbf{x})\sum_{q=1}^{N}\langle\boldsymbol{\eta}_q\rangle c^{(q)}, \quad (9.84)$$

which is also valid at the assumptions (8.14), (8.65_2), and (8.67) for the ellipsoidal correlation hole v_i^0 when $\mathbf{Q}_i^0(\mathbf{x}) \equiv \mathbf{Q}_i^0 =$ const. Thus, Eq. (9.83) can be recast in the form

$$\langle\overline{\boldsymbol{\sigma}}_i\rangle(\mathbf{x}) = \langle\boldsymbol{\sigma}\rangle + \mathbf{Q}_i^0(\mathbf{x})\sum_{q=1}^{N}\langle\boldsymbol{\eta}_q\rangle c^{(q)} \quad (9.85)$$

allowing the linear algebraic equation for $\langle\boldsymbol{\eta}_i\rangle$ to be obtained after acting on Eq. (9.85) by the operator $\boldsymbol{\mathcal{B}} = (\mathbf{I}-\boldsymbol{\Gamma}\mathbf{M}_1)\ {}^1V_i$ (4.16) (see also Eqs. (4.24) and (4.26))

$$\langle\boldsymbol{\eta}\rangle_i(\mathbf{x}) = \mathbf{R}_i(\mathbf{x})\langle\boldsymbol{\sigma}\rangle + \mathbf{F}_i(\mathbf{x}) + \mathbf{R}^{\mathbf{Q}0}(\mathbf{x})\sum_{q=1}^{N}\langle\boldsymbol{\eta}_q\rangle c^{(q)}. \quad (9.86)$$

Volume averaging of Eq. (9.85) over the heterogeneity v_i and summation over the inclusion number i lead to the average strain polarization tensor:

$$\langle\boldsymbol{\eta}\rangle = (\mathbf{I} - \langle\mathbf{R}^{\mathbf{Q}0v}V\rangle)^{-1}(\langle\mathbf{R}^vV\rangle\langle\boldsymbol{\sigma}\rangle + \langle\mathbf{F}^vV\rangle). \quad (9.87)$$

where one introduces the notations $\mathbf{f}^v(\mathbf{x}) = \mathbf{f}^{(i)}(\mathbf{x})/\bar{v}_i$, at $\mathbf{x} \in v^{(i)}$, ($i = 1, 2, \ldots, N$; $\mathbf{f} = \mathbf{R}, \mathbf{F}, \mathbf{R}^{\mathbf{Q}0}$).

The statistical averages of stresses in the heterogeneities can be easily found after substitution of Eq. (9.87) into Eq. (9.86) that also gives the representations for the effective properties substituting of Eq. (9.87) into Eq. (8.12)

$$\mathbf{M}^* = \mathbf{M}^{(0)} + \left[\mathbf{I} - \langle\mathbf{R}^{\mathbf{Q}0v}V\rangle\right]^{-1}\langle\mathbf{R}^vV\rangle, \quad (9.88)$$

$$\boldsymbol{\beta}^* = \boldsymbol{\beta}^{(0)} + \left[\mathbf{I} - \langle\mathbf{R}^{\mathbf{Q}0v}V\rangle\right]^{-1}\langle\mathbf{F}^vV\rangle, \quad (9.89)$$

$$\langle\boldsymbol{\sigma}\rangle(\mathbf{x}) = \mathbf{B}_i(\mathbf{x})\langle\boldsymbol{\sigma}\rangle + \mathbf{C}_i(\mathbf{x}) + \mathbf{B}^{\mathbf{Q}0}(\mathbf{x})\left[(\mathbf{M}^* - \mathbf{M}^{(0)})\langle\boldsymbol{\sigma}\rangle + (\boldsymbol{\beta}^* - \boldsymbol{\beta}^{(0)})\right] \quad (9.90)$$

If (and only if) the correlation hole v_i^0 is chosen as an ellipsoid (coinciding, for example, with the ellipsoid $\tilde{v}_i^0$ described in Subsection 8.3.1) then the tensor

$\mathbf{Q}_i^0(\mathbf{x}) = \mathbf{Q}_i^0 \equiv \text{const.}$, $\mathbf{R}_i^{\mathbf{Q}0}(\mathbf{x}) = \bar{v}_i(\bar{v}_i^0)^{-1}\mathbf{R}_i(\mathbf{x})\mathbf{Q}_i^0$, $\mathbf{B}_i^{\mathbf{Q}0}(\mathbf{x}) = \mathbf{B}_i^0(\mathbf{x})\mathbf{Q}_i^0$, and, therefore, Eqs. (9.88)-(9.90) are reduced to the known representations (8.79), (8.80), (8.82) with the constant tensor $\mathbf{Q}_i^0$ depending on the orientation of the correlation hole v_i^0. As can be seen the effective stiffness (8.79) depends on two tensors $\mathbf{R}_i(\mathbf{x})$ and $\mathbf{Q}_i^0$ related with the heterogeneity v_i, in which only $\mathbf{R}_i(\mathbf{x})$ requires a numerical solution, while the tensor $\mathbf{Q}_i^0 = \mathbf{L}^{(0)}(\mathbf{I} - \mathbf{S}_i)$ is expressed though the Eshelby tensor $\mathbf{S}_i$ having analytical representation, e.g., for the isotropic matrix. However, even for the isotropic matrix, a new expression for the effective stiffness (9.87) has no relation with any Eshelby tensor and depends on an additional tensor $\mathbf{R}_i^{\mathbf{Q}0}(\mathbf{x})$ in which estimation is reduced to the evaluation of strain polarization tensor $\boldsymbol{\eta}(\mathbf{x})$ $(\mathbf{x} \in v_i)$ in the special model problem for a single heterogeneity in the infinite matrix (see Chapter 4). This is a modest sacrifice for the abandonment from any hypotheses about both the shape of a correlation hole v_i^0 and specific structure of effective field $\overline{\boldsymbol{\sigma}}_i(\mathbf{x})$ (compare hypotheses **H1** and **H1^2**).

9.5 Numerical Results

9.5.1 Composites with Spheroidal Inhomogeneities

Let us now demonstrate the application of the theoretical results by considering an isotropic composite made of an incompressible isotropic matrix, filled with rigid spherical inclusions of one size ($n = 1$). This example was chosen deliberately because it provides the maximum difference of predictions of effective shear modulus estimated by various methods and was considered by a number of authors. The following alternative radial distribution functions of inclusions (5.55)–(5.57) will be examined

$$g(r) = H(r - 2a), \tag{9.91}$$

$$g(r) = H(r - 2a)\left[1 + \left(\frac{2+c}{2(1-c)^2} - 1\right)\cos(\frac{\pi r}{a})e^{2(2-r/a)}\right], \tag{9.92}$$

$$g(r) = H(r - 2a)\left\{1 + \frac{4}{\pi}\left[\pi - 2\sin^{-1}(\frac{r}{4a}) - \frac{r}{2a}\sqrt{1 - \frac{r^2}{16a^2}}\right]H(4a - r)\right\} \tag{9.93}$$

for 3-D (9.91), (9.92) and 2-D (9.91), (9.93) cases. Here H denotes the Heaviside step function, $r \equiv |\mathbf{x}_i - \mathbf{x}_j|$ is a distance between the nonintersecting inclusions v_i and v_j, and c is the volume fraction of inclusions. For an isotropic matrix filled by spherical and circle inclusions the tensors $\mathbf{B}$, $\mathbf{Q}$ and $\mathbf{T}_{ij}(\mathbf{x}_i - \mathbf{x}_j)$ are known (see, e.g., Chapter 4).

At first we consider the estimation of the effective shear modulus, giving the analytical representations for a step radial distribution function (9.91) in the 3-D case. In so doing, the perturbation methods (9.11) and (9.18) as well as other methods will be rewritten in the terms of elastic moduli. Then we estimate the effective shear modulus by different methods:

$$\mu^*/\mu^{(0)} = 1 + 2.5c + 4.844c^2, \tag{9.94}$$
$$\mu^*/\mu^{(0)} = 1 + 2.5c + 5.01c^2, \tag{9.95}$$
$$\mu^*/\mu^{(0)} = 1 + 2.5c/(1-c), \tag{9.96}$$
$$\mu^*/\mu^{(0)} = 1 + 2.5c/(1-31c/16), \tag{9.97}$$
$$\mu^*/\mu^{(0)} = 1 + (1-c)^{-1}(2.5c + 2.344c^2), \tag{9.98}$$
$$\mu^*/\mu^{(0)} = 1 + (1-c)^{-1}(2.5c + 2.51c^2), \tag{9.99}$$
$$\mu^*/\mu^{(0)} = (1-2.5c)^{-1}, \tag{9.100}$$
$$\mu^*/\mu^{(0)} = (1-c)^{-2.5}, \tag{9.101}$$
$$\mu^*/\mu^{(0)} = [1-2.5c(1+1.938c)]^{-1}, \tag{9.102}$$

Equations (9.94) and (9.95) are obtained for a small concentration of the inclusions by the perturbation method (4.64), (9.3), (9.18) [with $\boldsymbol{\mathcal{J}}_i^{1\mathbf{M}} = (3J_1^{\mathbf{M}}, 2J_2^{\mathbf{M}})$, $2J_2^{\mathbf{M}} = 31c\mu^{(0)}/16 \approx 1.938c\mu^{(0)}$] and following [216], respectively. In so doing, the isotropic tensors $\boldsymbol{\mathcal{J}}_i^{1\mathbf{M}} = (3J_1^{\mathbf{M}}, 2J_2^{\mathbf{M}})$ used in Eq. (9.94) and (9.95) have the components $2J_2^{\mathbf{M}} = 2.5c \cdot 15/16\mu^{(0)} = 2.344c\mu^{(0)}$ and $2J_2^{\mathbf{M}} = 2.51c\mu^{(0)}$, respectively. Relations (9.96) and (9.97) are calculated via MEFM in the framework of one-particle (or Mori-Tanaka) (8.79) and two-particle (8.55), (4.67), and (4.64) (point approximation) approximations, respectively. Relations (9.98) and (9.99) are estimated by the combined MEFM–perturbation method (9.11) utilizing the tensors $\boldsymbol{\mathcal{J}}_i^{1\mathbf{M}}$ previously evaluated in Eq. (9.94) and (9.95), respectively. The effective medium method (9.42_1) predicts Eq. (9.100) while the differential scheme (9.70) leads to Eq. (9.102); (9.102) is determined using two-particle differential EMM. The differential equation

$$\frac{d^2\mu}{dc^2} = \frac{155}{16}\mu\frac{1}{(1-c)^2}, \quad \mu(0) = \mu^{(0)}, \ \mu^{(1)}(0) = \frac{5}{2}\mu^{(0)} \tag{9.103}$$

corresponds to two-particle differential EMM (9.78) and MEFM (9.81) with $m = 2$ (9.81), while the equation

$$\frac{d^3\mu}{dc^3} = 15\Big(\frac{31}{16}\Big)^2\mu\frac{1}{(1-c)^3}, \quad \mu(0) = \mu^{(0)}, \ \mu^{(1)}(0) = \frac{5}{2}\mu^{(0)}, \ \mu^{(2)}(0) = \frac{155}{16}\mu^{(0)} \tag{9.104}$$

corresponds to two-particle differential MEFM (9.81) with $m = 3$. The numerical results estimated via Eqs. (9.94)—(9.102) are depicted in Figs. 9.2 and 9.3. As can be seen from Fig. 9.2 the use of Eq. (9.96) leads to underestimating evaluation of the effective shear modulus by 2.6 times for $c = 0.43$ compared to the experimental data as well as to the more exact point approximation of weakly interacting inclusions (9.97), which provides the best comparison with experimental data [597] on relative changes of the Newtonian viscosity of a suspension of identical rigid spheres. The results of estimations via Eqs. (9.98) and (9.99) differ from one another by not more than 2%. At the same time the Mori-Tanaka solution (9.96) differs from the nondilute approximations (9.94) and (9.95) by not more than 5% for the concentration of the inclusions $c \leq 0.5$ (see Fig. 9.3). Therefore, even in the limiting case of an infinite elastic mismatch and $c = 0.5$ the perturbation method (9.1) and (9.18) provides the same accuracy as the

Mori-Tanaka method. The relation (9.100) leads to overestimated evaluation of the effective shear modulus.

A comparative analysis of differential methods is presented in Fig. 9.4 which shows the experimental data (points) [597], and curves 1–4 computed from Eqs. (9.100), (9.81) with $m = 30$, (9.103), (9.102), respectively. It is clear that taking the binary interaction of the inclusions into account can increase the accuracy of the differential methods. We now turn to analysis of estimations based on the exact values of the tensors $\mathbf{T}_{ij}(\mathbf{x}_i - \mathbf{x}_j)$ (4.67$_1$). A comparison between the relative shear modulus $\mu^*/\mu^{(0)}$, calculated by the use of the MEFM (8.52), as well as by the point approximation (4.64) of the MEFM (8.55) and (8.56) for different radial functions (9.91) and (9.92), is presented in Fig. 9.5.

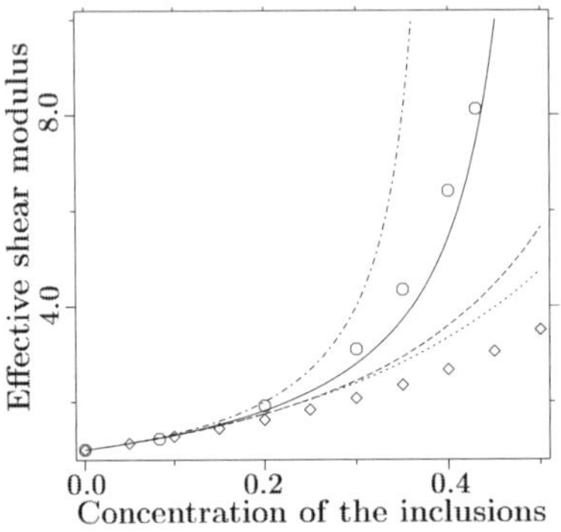

Fig. 9.2. $\mu^*/\mu^{(0)}$ vs c. Experimental data (○) and curves calculated by Eqs. (9.100) (dot-dashed line), (9.97) (solid line), (9.102) (dashed line), (9.98) (dotted line), and (9.96) (◇).

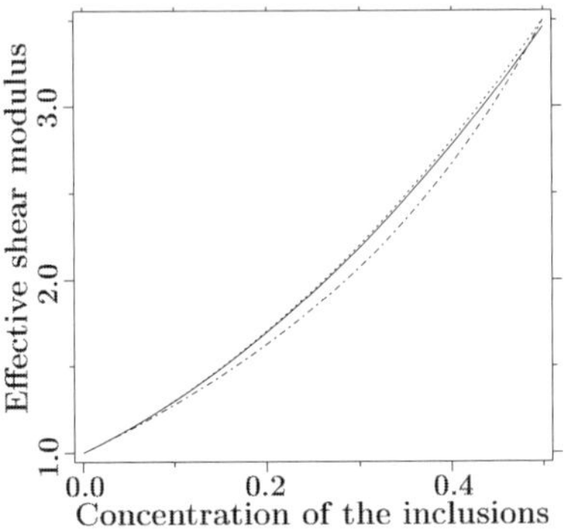

Fig. 9.3. Variation of the relative effective shear modulus $\mu^*/\mu^{(0)}$ calculated by Eqs. (9.95) (dotted line), (9.94) (solid line), and (9.96) (dot-dashed line).

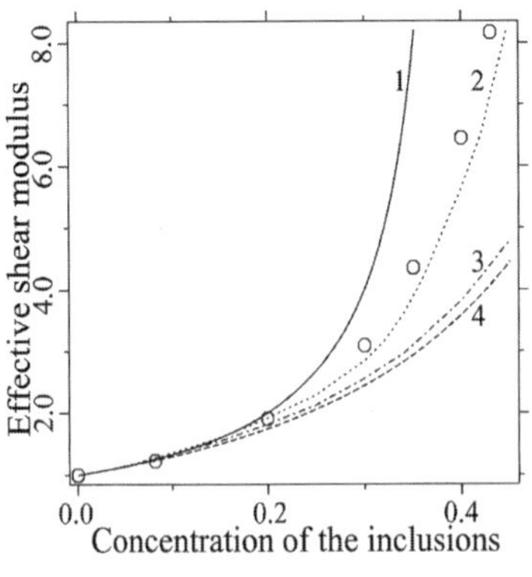

Fig. 9.4. Variation of the relative effective shear modulus $\mu^*/\mu^{(0)}$ calculated by equations (9.100) (curve 1), (9.81) with $m - 30$ (curve 2), (9.103) (curve 3), (9.102) (curve 4).

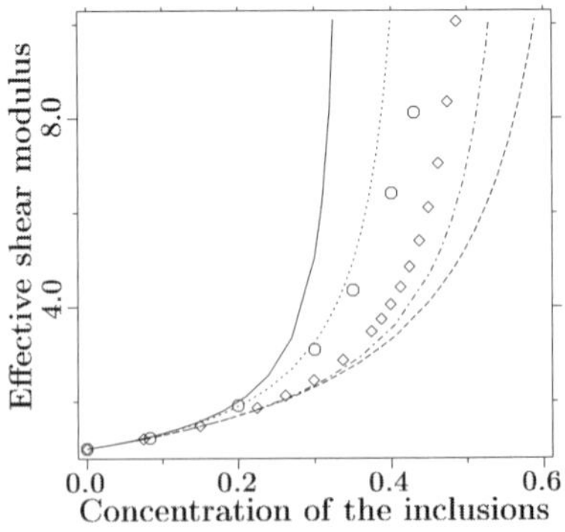

Fig. 9.5. $\mu^*/\mu^{(0)}$ vs c calculated by the MEFM (8.52), (4.64) for RDF (9.92) (solid line) and (9.91) (dotted line) as well as by the use of an approximation (4.67) for RDF (9.92) (dot–dashed line) and (9.91) (dashed line). The line (◇) is estimated by combined method, experimental data (○).

Analysis of Figs. 9.2 and 9.4 leads to the conclusion that the high accuracy of Eq. (9.97) exploited the point approximation (4.64), and the step radial distribution function (9.91) is nothing more than randomness because using both the more accurate radial distribution function (9.92) and the solution (4.58) and (4.64) yields to the considerable sacrifice of a prognostic opportunity of the MEFM (see solid curve in Fig. 9.5); it should be added in this connection that the deficiencies of point approximation (4.64) were analyzed by Buryachenko in [133] also for regular structure composites. In so doing, the MEFM (8.52) exploiting the solution (8.25) and (4.58) of the binary interaction problem admits some improvement based on a more accurate solution of multiple interaction problem (8.5) [140]. For example, in the combined method motivated by [305], after estimation of the matrix $\mathbf{Z}_{ij}$ (4.58) we can replace in Eq. (8.44) the tensor $\mathbf{T}_{ij}$ by the tensor $\mathbf{T}_j$ (8.16); that is, the perturbation of surrounding inclusions v_j are replaced with the perturbation at the inclusion center $\mathbf{x}_i$ (see the curve $\diamond$ in Fig. 9.5).

In order to demonstrate the comparison of the available experimental data with the predicting capability of the proposed method in the 2-D case, we will consider the estimation of the effective elastic moduli $\mathbf{L}^*$. Assume the matrix is epoxy resin ($k^{(0)} = 4.27$ GPa and $\mu^{(0)} = 1.53$ GPa) which contains identical circular glass fibers ($k^{(1)} = 50.89$ GPa and $\mu^{(1)} = 35.04$ GPa). Four different radial distribution functions for the inclusions will be examined. As can be seen from Fig. 9.6 the use of the approach (8.79) based on the quasi-crystalline approximation (8.65_2) (also called Mori-Tanaka (MT) approach) leads to an underestimate of the effective shear modulus by 1.85 times for $c = 0.7$ compared with the experimental data. Much better approximations are given by the MEFM (8.52) which shows good agreement with the experimental data provided in [659]. In the MEFM model, the best fit is obtained using the RDF simulated by the modified CRM.

Let us now demonstrate an application of the theoretical results by considering an isotropic composite made of an incompressible isotropic matrix, filled with rigid disc inclusions of one size ($n = 1$). This example was chosen deliberately because it provides the maximum difference between predictions of effective elastic response, as estimated by the various methods. In Fig. 9.7 the most advanced micromechanical model (8.52) is analyzed for the effect of choosing different RDFs. As can seen, the effective shear moduli can be differ by a factor of two or more depending on the chosen RDF. In so doing, the RDF simulated by the modified CRM (see Chapter 5) provides the estimations of $\mu^*/\mu^{(0)}$ that are very close to those obtained by the real RDF at $c = 0.65$. It is interesting that all RDF lead to infinite values of $\mu^*(c)$ for large values of c, but the simulated RDF provides a limiting upper value of $c = 0.72$.

We now turn to the analysis of another limiting case of porous materials with an incompressible matrix [157], [131]. The formulae

$$\mathbf{L}^* = \mu^{(0)}\left(4\left[\frac{1}{c} - 1\right], 2\left[1 - \frac{5}{3}c\frac{1}{1 + \frac{2c}{3}}\right]\right), \tag{9.105}$$

$$\mathbf{L}^* = \mu^{(0)}\left(4\left[\frac{1}{c} - \frac{29}{24}\right], 2\left[1 - \frac{5}{3}c\frac{1}{1 + \frac{5c}{24}}\right]\right), \tag{9.106}$$

$$\mathbf{L}^* = \mu^{(0)}\left(3\frac{1}{c}, 2\left[1 - \frac{16}{15\pi^2}c\frac{1}{1+\frac{16c}{15\pi^2}}\right]\right), \tag{9.107}$$

$$\mathbf{L}^* = \mu^{(0)}\left(3\left[\frac{1}{c} - \frac{16}{15\pi^2}\right], 2\left[1 - \frac{16}{15\pi^2}c\frac{1}{1+\frac{16c}{125\pi^2}}\right]\right), \tag{9.108}$$

describe the isotropic effective moduli of composites with identical spherical pores ($a_1 = a_2 = a_3 = a$) (9.105), (9.106), and with identical uniformly oriented coin-shaped cracks $a_1 = a_2 = a \gg a_3$) (9.107), (9.108) estimated by the MT (9.105), (9.107) and by the MEFM (9.106), (9.108); the values $c = \frac{4}{3}\pi a^3 n$ in the expressions presented have different physical meaning for the spherical and coin-shaped pores. In the case of the MEFM estimations, RDF (9.91) and two-particle point approximation (8.55), (8.57) presented in the term of stiffness tensors were used. For reduction of computations, an analysis of porous materials with uniformly oriented coin-shaped cracks was also performed via Eq. (8.55), (8.57) with replacement of the tensors $\mathbf{R}_1^\epsilon, \mathbf{P}_1\mathbf{R}_1^\epsilon n^{(1)}$ by the averaged tensors $\langle\mathbf{R}^\epsilon\rangle, (\mathbf{I} - \langle\mathbf{A}\rangle)c$, respectively. Figure 9.8 shows normalized values of $k_{sd} = k^* c/\mu^{(0)}$, and $k_s = 3k^* c/4\mu^{(0)}$ for planar spheroidal and spherical inclusions, as calculated with Eqs. (9.108), (9.106) in curves 1, 2; the curves 3, 4 are values of k_{sd} and k_s estimated by the MTM (9.105), (9.107); curve 5 are values of k_s calculated by the method [561]. We note that for an incompressible matrix ($\nu^{(0)} = 0$) and planar spheroidal pores, according to [119], and to [561] $k^* = k^{(0)}$ for any concentration c, which indicates the invalidity of the mentioned theories in the limiting case considered here.

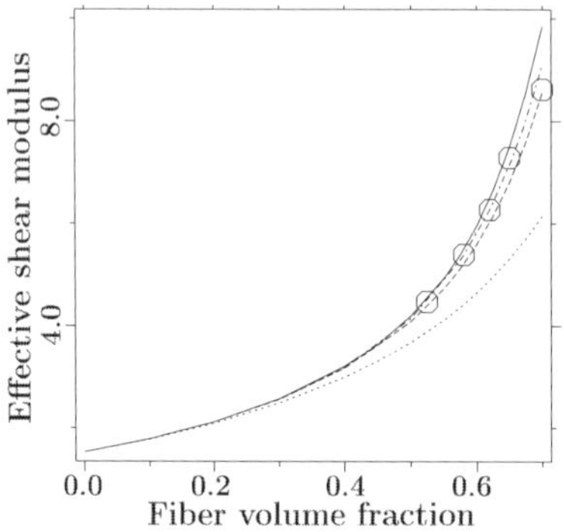

Fig. 9.6. μ^* vs c. Experimental data (○) and curves calculated via Eqs. (8.52) and (9.93) (solid line), via (8.52) with the RDF simulated via the modified CRM (dot-dashed line), by (8.52) and (9.91) (dashed curve), and via the Mori-Tanaka method (dotted line).

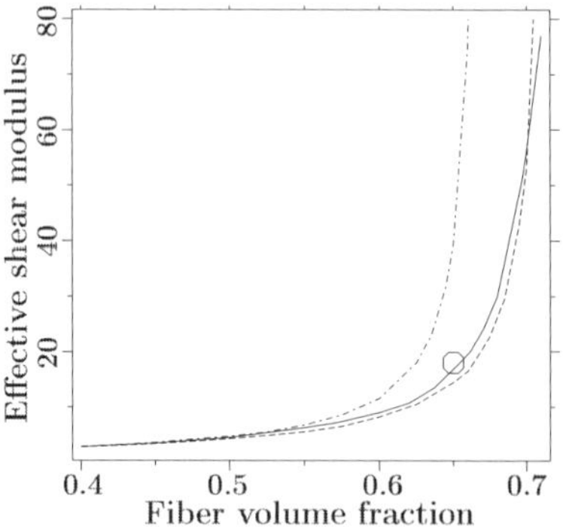

Fig. 9.7. Relative effective shear modulus $\mu^*/\mu^{(0)}$ vs c estimated via Eqs. (8.52) and (9.93) (dot-dashed line), via (8.52) with the RDF simulated by the modified CRM (solid curve), via (8.52) and (9.91) (dashed curve), and via (8.52) with experimentally estimated RDF (○).

It should be mentioned that for $\nu^{(o)} > 0.35$ the influence of the matrix compressibility on the effective moduli $\mathbf{L}^*$ is only moderate. For example, for $\nu^{(0)} = 0.35$ we find in analogy to (9.106)

$$\mathbf{L}^* = \mu^{(0)}\left(4\left[\frac{1}{c} - 1.15\right], 2\left[1 - \frac{5}{3}c\frac{1}{1+0.21c}\right]\right), \tag{9.109}$$

which differs from Eq. (9.106) by less than 5%. Therefore, the matrix of porous materials may be considered as incompressible for $\nu^{(0)} \geq 0.35$.

Let us now study the stress distribution inside the inhomogeneities as a function of the inclusion shape. The elastic isotropic constants for the matrix WC and inclusions Co as usually found in the literature [602] are $k^{(1)} = 389$ GPa, $\mu^{(1)} = 292$ GPa and $k^{(0)} = 167$ GPa, $\mu^{(0)} = 77$ GPa, respectively. We assume spheroidal particles with an aspect ratio a_1^1/a_1^3 for prolate spheroids and a_1^3/a_1^1 for oblate ones ($a_1^1 = a_1^2$). The axes $(\mathbf{x}^1, \mathbf{x}^2, \mathbf{x}^3)$ attached to the crystal lattice are aligned parallel with the global Cartesian reference frame. For the sake of definiteness we will consider the point approximation (4.64) for uniformly oriented inclusions. The binary correlation function has the form $\varphi(v_j, \mathbf{x}_j|; v_i, \mathbf{x}_i) = H(|\mathbf{a}_i^{-1}(\mathbf{x}_j - \mathbf{x}_i)| - 2)n_j$, where $\mathbf{a}_i^{-1} \equiv \operatorname{diag}[(a_i^1)^{-1}, (a_i^2)^{-1}, (a_i^3)^{-1}]$. Fig. 9.9 shows

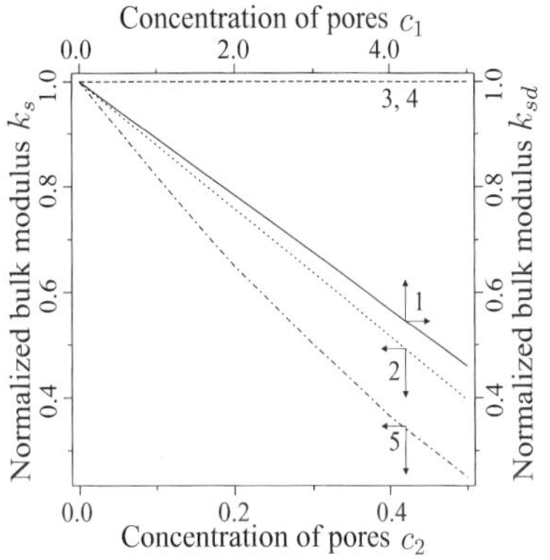

Fig. 9.8. k_s and k_{sd} vs c estimated by Eqs. (9.108) (curve 1), (9.106) (curve 2), (9.105) (curve 3), (9.107) (curve 4), and by the method [561] (curve 5).

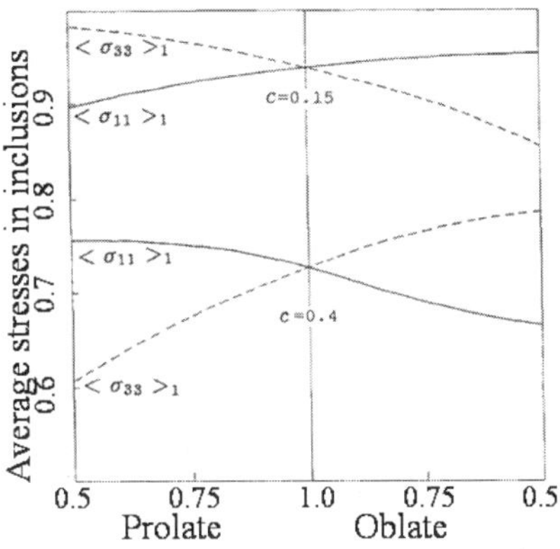

Fig. 9.9. $\langle\sigma_{11}\rangle_1$ (solid lines) and and $\langle\sigma_{33}\rangle_1$ (dashed lines) vs aspect ratio a^1/a^3 (left) and a^3/a^1 (right) for c =0.15 and c =0.4.

the values $\langle\sigma_{11}\rangle_1$ and $\langle\sigma_{33}\rangle_1$ versus shape of inclusions obtained from (8.46). For the small inclusion concentration $c = 0.15$ and prolate (or oblate) inclusion shape the results agree with our expectation that the average stresses $\langle\sigma_{33}\rangle_1$ (or $\langle\sigma_{11}\rangle_1$) are larger than for spherical grains because the stress transfer length is large in the corresponding direction. The relative position of curves are interchanged for the concentration $c = 0.4$ when the binary interaction of inclusions plays an essential part in the generation of stresses inside the inhomogeneities, and a minimum distance between inclusion centers is achieved in the direction of the x_3-axis (x_1-axis) for oblate (prolate) spheroids. The accuracy of the calculations decreases in the limiting cases of spheroids $a^1/a^3 \to 0$ or $\to \infty$ since the point approximation assumption of (8.16) then becomes incorrect.

9.5.2 Composites Reinforced by Nonellipsoidal Inhomogeneities with Ellipsoidal v_i^0

Following [195], we will consider the plane strain state, and first the influence of the noncircular shape of inclusions (see Fig. 4.2.1) on the effective elastic moduli of a composite material with the isotropic components will be analyzed. Let the elastic properties of the components be $E^{(0)} = 1$ GPa, $E^{(1)} = 100$ GPa,

$\nu^{(0)} = \nu^{(1)} = 0.45$. We assume that for all inclusions (4.33) the tensors $\mathbf{Q}_i^0$ are identical and coincide with the tensor $\mathbf{Q}^0$ for the circular inclusion which was estimated both numerically by the use of FEA and analytically by means of the Eshelby tensor. The error of numerical solution is not more than 0.15%, where the ratio $R/r = 40$ is being considered to represent an infinite medium. The variation of relative Young modulus $E_{11}^*/E^{(0)}$ (8.79) is presented in Fig. 9.10 for different concentrations of inclusions, $c \equiv V^{(1)}$, and shape parameter l/r. An estimation of effective compliance $\mathbf{M}^* = \mathbf{M}^*(\phi)$ for the composite with another orientation of inclusions (4.34) was performed via the use of transformation of elastic compliance in the matrix form $\mathbf{M}^*(0)$ (8.79), (4.33) due to rotation of coordinate system on the angle $\phi = \pi/4$ (see, e.g., [933]):

$$\mathbf{M}^*(\phi) = \mathbf{T}^\top \mathbf{M}^*(0)\mathbf{T}, \quad \mathbf{T} = \begin{pmatrix} c^2 & s^2 & 2cs \\ s^2 & c^2 & -2cs \\ -cs & cs & c^2 - s^2 \end{pmatrix}, \tag{9.110}$$

where $\mathbf{T}$ is a transformation matrix depending on a single angle ϕ measured from the global direction X (4.34) to the local X-direction (4.33); $c = \cos(\phi)$, $s = sin(\phi)$. In Fig. 9.11 the estimations of $E_{11}^*/E^{(0)}$ for the composite with the inclusions (4.33) (aligned inclusions 1) and (4.34) (aligned inclusions 2) at $l/r = 0.8$ as well as $l/r = 0.0$ are presented. As can be seen the circular shape of inclusions is intermediate in increasing the rigidity of composites between the inclusions (4.33) and (4.34). The case of uniform random orientation of inclusions is analogously analyzed by the transformation of the tensor $\mathbf{R}_i$ corresponding to the inclusion (4.33) due to all possible rotations of the coordinate system connected with the inclusion (4.33) with subsequent substitution of $\langle \mathbf{R} \rangle_\omega$ found into Eq. (8.79). In doing so the predicted isotropic effective moduli $\mathbf{L}^*$ with uniform random orientation of inclusions (4.33) is stiffer than effective elastic moduli of composites with the circle inclusions ($l/r = 0$; see Fig. 9.11).

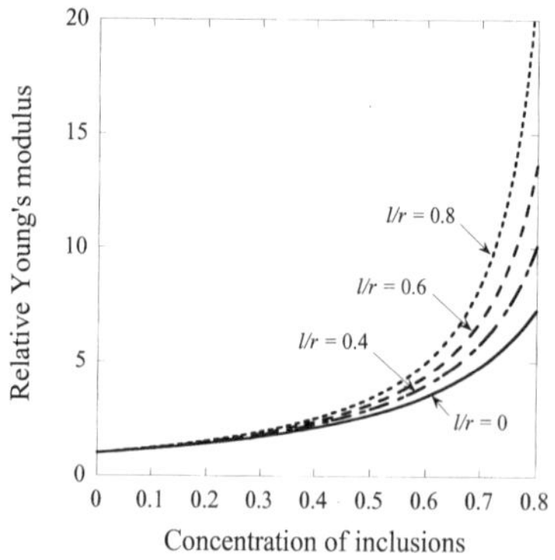

Fig. 9.10. Variation of the relative Young modulus $E_{11}^*/E^{(0)}$ as a function of a concentration of the inclusions c at $l/r = 0.8, 0.6, 0.4, 0.0$.

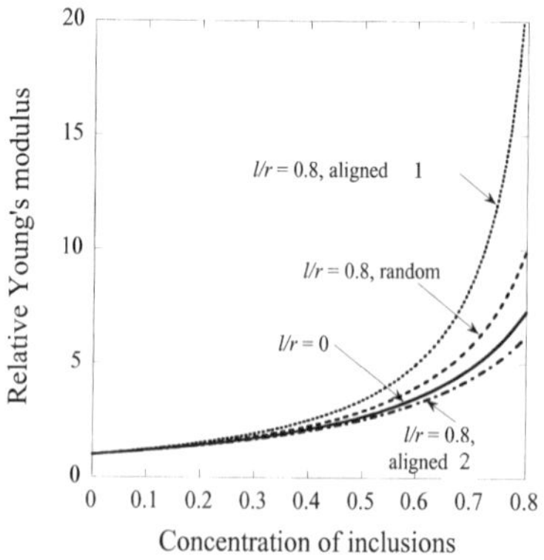

Fig. 9.11. $E_{11}^*/E^{(0)}$ as vs c at $l/r = 0.8$: aligned inclusions 1 and 2 and uniformly random orientation of inclusion (dashed line) as well as at $l/r = 0$ (solid line).

In Fig. 9.12 the distributions of stresses $\langle \sigma_{22} \rangle_i(\mathbf{x})$ along the cross-section $\mathbf{x} = (x_1, 0)^\top$ $(0 \leq x_1 \leq r)$ of the inclusion (4.33) for the external loading $\langle \sigma_{ij} \rangle \equiv$

$\delta_{i1}\delta_{j1}$ are presented for the dilute $c \ll 1$ and nondilute concentration of inclusions $c = 0.5$ and $c = 0.7$. Estimations of stresses for the nondilute concentration of inclusions were carried out via the use of Eq. (8.83) where the tensor $\mathbf{B}_i(\mathbf{x})$ for the inclusion v_i (4.33) was evaluated by the FEA (see Section 4.2.4). The stresses $\langle \boldsymbol{\sigma}^{\rm loc}\rangle_i(\mathbf{x})$ in the local coordinate system (4.33) were transformed into the global coordinate system $\langle \sigma^{\rm glo}\rangle_i(\mathbf{x})$ according to $\langle \sigma^{\rm glo}\rangle_i(\mathbf{x}) = \mathbf{T}^{-1}\langle \boldsymbol{\sigma}^{\rm loc}\rangle_i(\mathbf{x})$. For both random and aligned arrangement of inclusions the statistical averages of stresses in the inclusions $\langle \sigma_{22}\rangle_i(\mathbf{x})$ were estimated over the subset of inclusions with the orientation (4.33). As can be seen, the stress distributions are essentially inhomogeneous along the cross-section of inclusions for any concentration and can even change sign for a larger concentration of inclusions (e.g., at $c = 0.7$).

Next we will consider the plane stress problem for the anisotropic matrix $\mathbf{L}^{(0)\rm anis}$ obtained from the isotropic one $\mathbf{L}^{(0)\rm isot} = (2k^{(0)}, 2\mu^{(0)})$ by the variation of a parameter α in the matrix representation of elastic moduli $\mathbf{L}_{ij}^{(0)\rm anis} = \mathbf{L}^{(0)\rm isot} + 2\mu^{(0)}\delta_{i3}\delta_{j3}(\alpha - 1)$ where $\alpha = 1$ corresponds to the case of an isotropic material. Anisotropic properties of the matrix $\mathbf{L}^{(0)}$ incorporate no modification in the scheme of estimation of effective properties $\mathbf{L}^*$ of composites with isotropic matrix. As can be seen in Fig. 9.13, variation of the range of anisotropy of $\mathbf{L}^{(0)}$ can lead to a change of the effective Young modulus $\mathbf{E}_{11}^*$ by a factor 2.

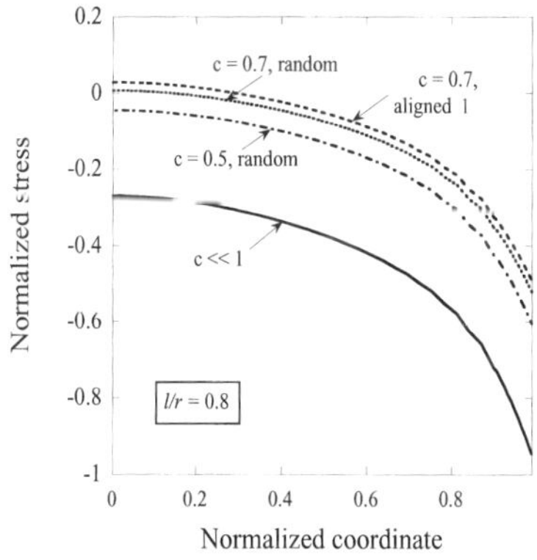

Fig. 9.12. $\langle \sigma_{22}\rangle(\mathbf{x})$ vs x_1/r in the inclusions (4.33) with $l/r = 0.8$: $c \ll 1$ (solid line), $c = 0.5$ and the random orientation of inclusions (dot-dashed line), $c = 0.7$ and the random orientation of inclusions (dotted line), and $c = 0.7$ and the aligned inclusions (4.33) (dashed line).

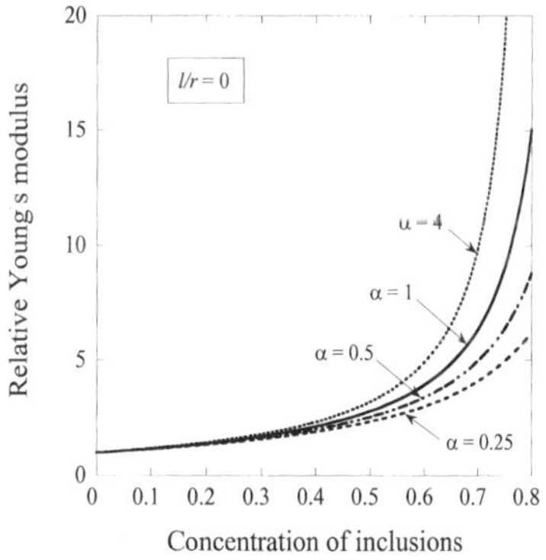

Fig. 9.13. Variation of the relative Young modulus $E_{11}^*/E^{(0)}$ as a function of a concentration of the inclusions c at $l/r = 0$ and $\alpha = 4,\ 1,\ 0.5,\ 0.25$.

It should be mentioned that the one-particle approximation of the MEFM is closely related to the different self-consistent methods that were demonstrated in this chapter. This flexibility points to the possibility that incorporation of the numerical analysis into the known average scheme proposed in the current section can be implemented to other micromechanical models.

9.6 Discussion

The MEFM allows provides the calculation with reasonable accuracy of the effective properties and statistical average of stresses in the components for a whole range of parameters. The method appears to be simple enough in both theoretical and computational aspects and includes plurality of popular average methods based on hypothesis **H1**. The proposed method allows us to consider composites with any number of different components containing inclusions with size distribution, shape, orientation and properties, coated particles and fibers, cracks, etc., that are beyond the scope of the book.

Richer in content is a discussion of the main hypotheses. The assumption **H1** of homogeneity of $\overline{\boldsymbol{\sigma}}(\mathbf{x})$, $(\mathbf{x} \in v_i)$ is a classical hypothesis of micromechanics (see the earliest references [649]) and was required in order to make it easier to solve the algebraic system (8.5) and (8.26). The high accuracy of applications of this hypothesis at $\mathbf{M}^c = \mathbf{M}^{(0)}$ was estimated for two spherical inclusions in [79], [941]. However, the problem (8.5) for k fixed inclusions with fixed values $\widetilde{\boldsymbol{\sigma}}_{1,\ldots,k}(\mathbf{x}) = \text{const.}$ can also be solved for a polynomial function $\overline{\sigma}_i(\mathbf{x})$, $\widetilde{\sigma}(\mathbf{x})_{1,\ldots,n}$ in analogy to [789], [216]. Then in the case of the rejection of the hypothesis **H1**) it is necessary to introduce new concentration tensors of larger dimension in addition to $\mathbf{B}$, $\mathbf{C}$, $\mathbf{F}$, $\mathbf{R}$ (8.18). So, for example, the tensor $\mathbf{B}^{(i)}$ (8.18) should be replaced by differential operator $\sum \mathbf{B}_j^{(i)}(\mathbf{x}^j \otimes \nabla^j)$ (summation over the multi-index j), where the index j takes into account the effect of the term of degree j in the polynomial $\overline{\boldsymbol{\sigma}}_i(\mathbf{x})$. Similarly, in analyzing system (8.5) the kernel $\boldsymbol{\Gamma}(\mathbf{x}_i - \mathbf{y})$ should be expanded in a Taylor series around the center of the inclusion being considered, and the problem should be solved for a finite number of inclusions, as done in [789]. A modification of this scheme was used in [528], [529] for analyses of multiple crack interaction problems. The high accuracy of the effective-field model for stress estimations (called a pseudo-load in [470]) is a result of the fact that the field $\overline{\boldsymbol{\sigma}}(\mathbf{x})$ $(\mathbf{x} \in v_i)$ within an inclusion changes insignificantly [148]. Rejection of the hypothesis **H1** yields to replacement of concentration tensors (8.18) by the operator involved that significantly complicates the calculations and reduces the generality of the obtained formulae (8.49)–(8.51).

The current charpter is dedicated to the estimation of volume averages of stresses in the phases and the related problem of evaluation of effective properties. However, this does not exhaust all possibilities of the MEFM. The main advantage of the method proposed is universal calculation scheme of general integral equations that leaves room for correction of their individual elements if either the initial integral equations and boundary conditions are changed or more improved methods are utilized for the analysis of these individual elements (for details see Chapter 10).

10

Generalization of the MEFM in Random Structure Matrix Composites

It was demonstrated in Chapter 8 that the MEFM does not make use of a number of hypotheses that form the basis of the traditional one-particle methods, and that the MEFM includes in particular cases the well-known methods of mechanics of strongly heterogeneous media. However, it is a fundamental feature of the following general results that all these indicated methods (effective medium, mean field method, MEFM and others) are based on the same so-called effective field hypothesis **H1** (8.14), (8.15), which is merely a zero-order approximation of binary interacting inclusions. This substantial obstacle will be overcome in the generalized version of the MEFM in light of the generalized schemes based on the numerical solution of the problem for both one and two inclusions in an infinite medium. Because of this, the combination of computational micromechanics with analytical micromechanics seems to be very promising. Numerical solutions can be used to construct concentration factors for single and interacting inclusions, which then can be incorporated into the general framework of analytical micromechanics.

The outline of this chapter (where the results in [147] are presented) is as follows. Based on the results of Chapter 4, we present the numerical solution by the volume integral equation (VIE) method of the problem for two inclusions in an infinite medium, subjected to the homogeneous effective field. The iteration the VIE method used is a standard iteration method of the solution of the Fredholm integral equation of the second order involved which led to an impenetrable barrier of computer costs if the number of inclusions with near-to-dense packing is large enough. These solutions fulfill the role of some building blocks used in the subsequent calculations. Just with some additional assumptions (such as an effective field hypothesis) they can be expressed through both the Green function, Eshelby tensor, and external Eshelby tensor. By the use of these building blocks, a *generalization of the MEFM* is proposed for the estimation of effective properties (such as compliance, thermal expansion, stored energy) and the first statistical moments of stresses varying along a cross-section of inclusions. No restriction on the homogeneity of the effective fields (similar to the hypothesis **H1**) acting on the individual inclusions is used. However, we will use the hypothesis of homogeneity of the effective field acting on each pair of inclusions. The last hypothesis is modified taking into account the dependence of these fields on the mutual location of each pair of inclusions estimated. After that the particular cases of the general proposed approach such as the composites with identical aligned inclusions as well as the simplifications produced by both the effective field hypothesis **H1** and quasi-crystalline approximation are analyzed. Finally, we employ the proposed explicit relations and some related ones for numerical estimations of effective elastic moduli in an isotropic composite made of the isotropic matrix and aligned fibers.

10.1 Two Inclusions in an Infinite Matrix

Let us assume that two inclusions v_i and v_j are placed in an infinite homogeneous matrix subjected to the homogeneous stress field $\widetilde{\boldsymbol{\sigma}}_{i,j} = \text{const}$. For $\mathbf{x} \in v_i$, Eqs. (8.4) and (8.5) (at $\mathbf{L}^c \equiv \mathbf{L}^{(0)}$ assumed in this chapter) can be recast in the form

$$\boldsymbol{\eta}(\mathbf{x}) = \overline{\boldsymbol{\eta}}_i(\mathbf{x}) + \int \boldsymbol{\mathcal{K}}_i(\mathbf{x},\mathbf{y})V_i(\mathbf{y})\boldsymbol{\eta}(\mathbf{y})d\mathbf{y}, \quad \overline{\boldsymbol{\eta}}_i(\mathbf{x}) = \widetilde{\boldsymbol{\eta}}_{i,j}(\mathbf{x}) + \int \boldsymbol{\mathcal{K}}_i(\mathbf{x},\mathbf{y})V_j(\mathbf{y})\boldsymbol{\eta}(\mathbf{y})d\mathbf{y}, \tag{10.1}$$

where the kernel $\boldsymbol{\mathcal{K}}_i(\mathbf{x},\mathbf{y})$ is defined by Eq. (4.41), and $\widetilde{\boldsymbol{\eta}}_{i,j}(\mathbf{x}) = \mathbf{E}_i(\mathbf{x})\widetilde{\boldsymbol{\sigma}}_{i,j}(\mathbf{x}) + \mathbf{H}_i(\mathbf{x})$, $(\mathbf{x} \in v_i)$ is another sort of effective strain polarization tensor introduced analogously to the tensor $\boldsymbol{\eta}_i^0$ (4.36)-(4.38). In so doing, $\boldsymbol{\mathcal{K}}_i(\mathbf{x},\mathbf{y}) \equiv \mathbf{K}_i(\mathbf{x},\mathbf{y})$ (4.37) at $\mathbf{x} \in v_i$ and $\mathbf{y} \in v_j \neq v_i$.

Let the integral operator $\boldsymbol{\mathcal{L}}_i$ (4.45) expressing the strain polarization tensor inside the inclusion $\boldsymbol{\eta}(\mathbf{x})$ through the effective strain polarization tensor $\overline{\boldsymbol{\eta}}(\mathbf{x})$ be known. The solution $\boldsymbol{\mathcal{L}}_i$ (4.45) of the singular volume integral equation (4.35) can be solved via the direct quadrature method (called also the Nystron method), via the iteration methods as well as via the Fourier transform method (see, e.g., Chapter 13 where additional references can be found). It should be mentioned that each of these methods has a series of advantages and disadvantages, and it is crucial for the analyst to be aware of their range of application. However, the knowledge of the operator $\boldsymbol{\mathcal{L}}_i$ at the current stage is not necessary and the assumption of the existence of such an operator is enough. At first no restrictions are imposed on the method of estimation of the operator $\boldsymbol{\mathcal{L}}_i$. On the constant stress tensor $\overline{\boldsymbol{\sigma}} \equiv \text{const.}$, according to Eq. (4.53), the linear operator $\boldsymbol{\mathcal{L}}_i$ is degenerated into the inhomogeneous tensors: $\bar{v}_i\boldsymbol{\mathcal{L}}_i\overline{\boldsymbol{\eta}}(\mathbf{x}) = \mathbf{R}_i(\mathbf{x})\overline{\boldsymbol{\sigma}} + \mathbf{F}_i(\mathbf{x})$, $(\mathbf{x} \in v_i)$, where the tensors $\mathbf{R}_i(\mathbf{x})$ and $\mathbf{F}_i(\mathbf{x})$ introduced in Eq. (3.179) are expressed in the case of the homogeneous ellipsoidal inclusion in terms of the Eshelby tensor $\mathbf{S}_i$, see Eqs. (3.188).

Acting on Eq. (10.1_2) by the operator $\boldsymbol{\mathcal{L}}_i$ leads to $(\mathbf{x},\mathbf{y} \in v_i)$:

$$\bar{v}_i\boldsymbol{\eta}(\mathbf{x}) = \mathbf{R}_i(\mathbf{x})\widetilde{\boldsymbol{\sigma}}_{i,j} + \mathbf{F}_i(\mathbf{x}) + \bar{v}_i\boldsymbol{\mathcal{L}}_i \int \mathbf{K}_i(\mathbf{y},\mathbf{z})\boldsymbol{\eta}(\mathbf{z})V_j(\mathbf{z})\, d\mathbf{z}. \tag{10.2}$$

By the use of obvious rearrangement of Eq. (10.2) we will introduce the inhomogeneous tensors $\mathbf{R}_{i,j}(\mathbf{x})$ and $\mathbf{F}_{i,j}(\mathbf{x})$ $(\mathbf{x} \in v_i)$ describing the stress concentrator factor on the one from two inclusions v_i and v_j subjected to the homogeneous field $\widetilde{\boldsymbol{\sigma}}_{i,j}$ at infinity $(\mathbf{x} \in v_i)$:

$$\bar{v}_i\boldsymbol{\mathcal{L}}_i\int \mathbf{K}_i(\mathbf{y},\mathbf{z})\boldsymbol{\eta}(\mathbf{z})V_j(\mathbf{z})\,d\mathbf{z}=\bar{v}_i\boldsymbol{\eta}(\mathbf{x})-\mathbf{R}_i(\mathbf{x})\widetilde{\boldsymbol{\sigma}}_{i,j}-\mathbf{F}_i(\mathbf{x})\equiv\mathbf{R}_{i,j}(\mathbf{x})\widetilde{\boldsymbol{\sigma}}_{i,j}+\mathbf{F}_{i,j}(\mathbf{x}), \tag{10.3}$$

where the tensors $\mathbf{R}_{i,j}(\mathbf{x})$ and $\mathbf{F}_{i,j}(\mathbf{x})$ can be found via any numerical method such as BIE and FEA, complex potential method, and others analogous to the scheme described in Chapter 4.

The stresses inside the inclusion v_i can be estimated from the second Eq. (10.3)

$$\boldsymbol{\sigma}(\mathbf{x})-\mathbf{B}_i(\mathbf{x})\widetilde{\boldsymbol{\sigma}}_{i,j}-\mathbf{C}_i(\mathbf{x})=\mathbf{B}_{i,j}(\mathbf{x})\widetilde{\boldsymbol{\sigma}}_{i,j}+\mathbf{C}_{i,j}(\mathbf{x}) \tag{10.4}$$

where the tensors $\mathbf{B}_{i,j}(\mathbf{x})\equiv\bar{v}_i^{-1}\mathbf{M}_1^{-1}(\mathbf{x})\mathbf{R}_{i,j}(\mathbf{x})$, $\mathbf{C}_{i,j}(\mathbf{x})\equiv\bar{v}_i^{-1}\mathbf{M}_1^{-1}(\mathbf{x})\mathbf{F}_{i,j}(\mathbf{x})$ are defined on the domain $\mathbf{x}\in v_i$. Keeping in mind that the tensors $\mathbf{B}(\mathbf{x})$ and $\mathbf{C}(\mathbf{x})$ can also be defined for points located in the matrix $\mathbf{x}\in v^0$ as $\mathbf{B}(\mathbf{x})\equiv\mathbf{I}$ and $\mathbf{C}(\mathbf{x})=\mathbf{0}$, Eq. (10.4) continued on the matrix ($\mathbf{x}\in v^0=R^d\setminus(v_i\cup v_j)$)

$$\boldsymbol{\sigma}(\mathbf{x})=\widetilde{\boldsymbol{\sigma}}_{i,j}+[\mathbf{B}^I_{i,j}(\mathbf{x})+\mathbf{B}^J_{i,j}(\mathbf{x})]\widetilde{\boldsymbol{\sigma}}_{i,j}+[\mathbf{C}^I_{i,j}(\mathbf{x})+\mathbf{C}^J_{i,j}(\mathbf{x})] \tag{10.5}$$

where the tensors $\mathbf{B}^I_{i,j}(\mathbf{x})$, $\mathbf{C}^I_{i,j}(\mathbf{x})$ and $\mathbf{B}^J_{i,j}(\mathbf{x})$, $\mathbf{C}^J_{i,j}(\mathbf{x})$ describe the perturbations introduced by the inclusions v_i and v_j into the point $\mathbf{x}$:

$$\mathbf{B}^I_{i,j}(\mathbf{x})=\bar{v}_i^{-1}\int\boldsymbol{\Gamma}(\mathbf{x}-\mathbf{y})[\mathbf{R}_i(\mathbf{y})+\mathbf{R}_{i,j}(\mathbf{y})]V_i(\mathbf{y})\,d\mathbf{y}, \tag{10.6}$$

$$\mathbf{B}^J_{i,j}(\mathbf{x})=\bar{v}_j^{-1}\int\boldsymbol{\Gamma}(\mathbf{x}-\mathbf{y})[\mathbf{R}_j(\mathbf{y})+\mathbf{R}_{j,i}(\mathbf{y})]V_j(\mathbf{y})\,d\mathbf{y}, \tag{10.7}$$

$$\mathbf{C}^K_{i,j}(\mathbf{x})=\int\boldsymbol{\Gamma}(\mathbf{x}-\mathbf{y})\boldsymbol{\beta}_1(\mathbf{y})V_K(\mathbf{y})\,d\mathbf{y},\ (K=I,J), \tag{10.8}$$

where one follows Mura's [794] tensorial index notation when uppercase indices always take on the same numbers as the corresponding lowercase ones but are not summed up. Let us replace the inhomogeneity v_j by fictitious inclusion of the matrix elastic properties and the fictitious eigenstrain:

$$\mathbf{L}(\mathbf{y})\equiv\mathbf{L}^{(0)},\quad \bar{v}_j\boldsymbol{\beta}_1^{\rm fict}(\mathbf{y})=\mathbf{R}_j(\mathbf{y})\widetilde{\boldsymbol{\sigma}}_{i,j}+\mathbf{F}_j(\mathbf{y}),\quad(\mathbf{y}\in v_j) \tag{10.9}$$

so that the constitutive equation in the inclusion v_j is $\boldsymbol{\sigma}(\mathbf{y})=\mathbf{L}^{(0)}\boldsymbol{\varepsilon}(\mathbf{y})+\boldsymbol{\beta}_1^{\rm fict}(\mathbf{y})$. Then the strain polarization tensor $\boldsymbol{\eta}(\mathbf{y})=\boldsymbol{\beta}_1^{\rm fict}(\mathbf{y})$ $(\mathbf{y}\in v_j)$ can be assumed to be known and Eq. (10.2) can be reduced to $(\mathbf{x},\mathbf{y}\in v_i)$:

$$\bar{v}_i\boldsymbol{\eta}(\mathbf{x})=\mathbf{R}_i(\mathbf{x})\widetilde{\boldsymbol{\sigma}}_{i,j}+\mathbf{F}_i(\mathbf{x})+\bar{v}_i\boldsymbol{\mathcal{L}}_i\int\mathbf{K}_i(\mathbf{y},\mathbf{z})\boldsymbol{\beta}_1^{\rm fict}(\mathbf{z})V_j(\mathbf{z})\,d\mathbf{z}, \tag{10.10}$$

which can be recast in the form

$$\bar{v}_i\boldsymbol{\mathcal{L}}_i\int\mathbf{K}_i(\mathbf{y},\mathbf{z})\boldsymbol{\beta}_1^{\rm fict}(\mathbf{z})V_j(\mathbf{z})d\mathbf{z}=\bar{v}_i\boldsymbol{\eta}(\mathbf{x})-\mathbf{R}_i(\mathbf{x})\widetilde{\boldsymbol{\sigma}}_{i,j}-\mathbf{F}_i(\mathbf{x})\equiv\mathbf{R}^\infty_{i,j}(\mathbf{x})\widetilde{\boldsymbol{\sigma}}_{i,j}+\mathbf{F}^\infty_{i,j}(\mathbf{x}), \tag{10.11}$$

defining the new concentration factors $\mathbf{R}^\infty_{i,j}(\mathbf{x})$ and $\mathbf{F}^\infty_{i,j}(\mathbf{x})$ $(\mathbf{x}\in v_i)$ which can be estimated analogously to the tensors $\mathbf{R}_{i,j}(\mathbf{x})$ and $\mathbf{F}_{i,j}(\mathbf{x})$ $(\mathbf{x}\in v_i)$ (10.3).

Thus, the tensors $\mathbf{R}_i(\mathbf{x})$ and $\mathbf{F}_i(\mathbf{x})$ $(\mathbf{x}\in v_i)$ describe the strain polarization tensor $\boldsymbol{\eta}_i(\mathbf{x})$ of the isolated inclusion v_i in the matrix subjected at infinity to the homogeneous loading. In so doing, the tensors $\mathbf{R}_{i,j}(\mathbf{x})$ and $\mathbf{F}_{i,j}(\mathbf{x})$ ($\mathbf{x} \in v_i$) define the perturbation of the strain polarization tensor $\boldsymbol{\eta}_i(\mathbf{x})$ introduced by the placement of the inclusion v_j interacting with the inclusion v_i. The tensors $\mathbf{R}^{\infty}_{i,j}(\mathbf{x})$ and $\mathbf{F}^{\infty}_{i,j}(\mathbf{x})$ describe another perturbation of the strain polarization tensor $\boldsymbol{\eta}_i(\mathbf{x})$ at $\mathbf{x} \in v_i$ produced by the inclusion v_j interacting just with the homogeneous external loading but not with the inclusion v_i. Comparison of Eqs. (10.11) defining the tensors $\mathbf{R}^{\infty}_{i,j}(\mathbf{x})$ and $\mathbf{F}^{\infty}_{i,j}(\mathbf{x})$ with the iteration scheme (10.3) leads to a conclusion that the tensors $\mathbf{R}^{\infty}_{i,j}(\mathbf{x})$ and $\mathbf{F}^{\infty}_{i,j}(\mathbf{x})$ construct the first-order approximation of the solution (10.2) by the iteration method (4.43), and are, in fact, the first-order approximations of the tensors $\mathbf{R}_{i,j}(\mathbf{x})$ and $\mathbf{F}_{i,j}(\mathbf{x})$, respectively, describing the solution of Eq. (10.2).

In a similar manner we can define the tensors $\mathbf{T}^{\sigma\infty}_j(\mathbf{x})$ and $\mathbf{T}^{\beta\infty}_j(\mathbf{x})$ describing the perturbation of the effective field $\overline{\boldsymbol{\sigma}}_i(\mathbf{x}) - \widetilde{\boldsymbol{\sigma}}_{i,j}$ introduced by the fictitious inclusion with the elastic modulus $\mathbf{L}^{(0)}$ and eigenstrain $\boldsymbol{\beta}^{\rm fict}_1(\mathbf{y})$ ($\mathbf{y} \in v_j,\ \mathbf{x} \in v_i$) (10.9):

$$\overline{\boldsymbol{\sigma}}_i(\mathbf{x}) - \widetilde{\boldsymbol{\sigma}}_{i,j} = \mathbf{T}^{\sigma\infty}_j(\mathbf{x})\widetilde{\boldsymbol{\sigma}}_{i,j} + \mathbf{T}^{\beta\infty}_j(\mathbf{x}). \tag{10.12}$$

To ascertain the connections between the tensors $\mathbf{R}^{\infty}_{i,j}(\mathbf{x})$, $\mathbf{F}^{\infty}_{i,j}(\mathbf{x})$ and $\mathbf{T}^{\sigma\infty}_j(\mathbf{x})$, $\mathbf{T}^{\beta\infty}_j(\mathbf{x})$, respectively, we recast Eq. (10.12) in terms of the strain polarization tensor

$$\overline{\boldsymbol{\eta}}_i(\mathbf{x}) = \widetilde{\boldsymbol{\eta}}_{i,j} + \mathbf{E}_i(\mathbf{x})\mathbf{T}^{\sigma\infty}_j(\mathbf{x})\widetilde{\boldsymbol{\sigma}}_{i,j} + \mathbf{E}_i(\mathbf{x})\mathbf{T}^{\beta\infty}_j(\mathbf{x}). \tag{10.13}$$

Acting with the operator $\boldsymbol{\mathcal{L}}_i$ on Eq. (10.13) and comparison of the obtained equation with Eq. (10.3) yields the following relationships: $\mathbf{R}^{\infty}_{i,j}(\mathbf{x}) = \boldsymbol{\mathcal{L}}_i\mathbf{E}_i(\mathbf{x})\mathbf{T}^{\sigma\infty}_j(\mathbf{x})$, $\mathbf{F}^{\infty}_{i,j}(\mathbf{x}) = \boldsymbol{\mathcal{L}}_i\mathbf{E}_i(\mathbf{x})\mathbf{T}^{\beta\infty}_j(\mathbf{x})$. It should be mentioned that for the homogeneous ellipsoidal inclusion (3.187), the tensors $\mathbf{T}^{\sigma\infty}_j$ and $\mathbf{T}^{\beta\infty}_j$ have the known representations through the external Eshelby tensor $\mathbf{S}_j(\mathbf{x}-\mathbf{x}_j) \equiv -\langle \mathbf{U}(\mathbf{x}-\mathbf{y})\rangle_{(j)}\mathbf{L}^{(0)}$ ($\mathbf{y} \in v_j,\ \mathbf{x} \notin v_j$) as well as the tensors $\mathbf{R}_j,\ \mathbf{F}_j$ (3.179) ($\mathbf{x} \in v_i$): $\mathbf{T}^{\sigma\infty}_j(\mathbf{x}) = \mathbf{L}^{(0)}\mathbf{S}_j(\mathbf{x}-\mathbf{x}_j)\mathbf{R}_j$, $\mathbf{T}^{\beta\infty}_j(\mathbf{x}) = \mathbf{L}^{(0)}\mathbf{S}_j(\mathbf{x}-\mathbf{x}_j)\mathbf{F}_j$. In a similar manner, averaging of the tensors $\mathbf{B}^{K}_{i,j}(\mathbf{x})$ and $\mathbf{C}^{K}_{i,j}(\mathbf{x})$ over the ellipsoidal matrix area $v_m \subset v^{(0)}$ ($\mathbf{M}_1(\mathbf{x}),\ \boldsymbol{\beta}_1(\mathbf{x}) \equiv \mathbf{0},\ \mathbf{x} \in v_m \subset R^d \setminus (v_i \cup v_j)$) is also represented through the external Eshelby tensors ($\mathbf{y} \in v_K,\ K = I, J$): $\langle \mathbf{B}^{K}_{i,j}\rangle_{(m)} = \mathbf{L}^{(0)}\langle \mathbf{S}_m(\mathbf{x}_m-\mathbf{y})[\mathbf{R}_K + \mathbf{R}_{i,j}(\mathbf{y})]\rangle_{(K)}$, $\langle \mathbf{C}^{K}_{i,j}\rangle_{(m)} = \langle \mathbf{S}_m(\mathbf{x}_m-\mathbf{y})\rangle_K \boldsymbol{\beta}^{(K)}_1$.

Analogously, we can consider the related problems for n inclusions $v_{i_1}, \ldots, v_{i_n}$ with the centers $\mathbf{x}_{i_1}, \ldots, \mathbf{x}_{i_n}$ placed in the infinite matrix subjected to the homogeneous field $\widetilde{\boldsymbol{\sigma}}_{i_1,\ldots,i_n}$ at infinity. For instance, the tensors $\mathbf{R}_{i_1,\ldots,i_n}(\mathbf{x})$ and $\mathbf{F}_{i_1,\ldots,i_n}(\mathbf{x})$ can also be found from the equations analogous to Eq. (10.3) ($\mathbf{x} \in v_{i_j};\ i_j,\ i_k = 1, \ldots, n; i_k \neq i_j$):

$$\begin{aligned}\bar{v}_{i_j}\boldsymbol{\mathcal{L}}_{i_j}\sum_{i_k=1}^{n}{}' \int \boldsymbol{\mathcal{K}}_{i_j}(\mathbf{y},\mathbf{z})\boldsymbol{\eta}(\mathbf{z})V_{i_k}(\mathbf{z})\,d\mathbf{z} &= \bar{v}_{i_j}\boldsymbol{\eta}(\mathbf{x}) - \mathbf{R}_{i_j}(\mathbf{x})\widetilde{\boldsymbol{\sigma}}_{i_1,\ldots,i_n} - \mathbf{F}_{i_j}(\mathbf{x})\\ &\equiv \mathbf{R}_{i_1,\ldots,i_n}(\mathbf{x})\widetilde{\boldsymbol{\sigma}}_{i_1,\ldots,i_n} + \mathbf{F}_{i_1,\ldots,i_n}(\mathbf{x}),\end{aligned} \tag{10.14}$$

via any numerical method such as BIE and FEA, complex potential method, and others. In Eq. (10.14) the prime $'$ at the summation symbol stands for the omission of the summand with the index $i_k = i_j$.

10.2 Composite Material

10.2.1 General Representations

Fixing the inclusion v_i in the composite material produces the random effective field acting on this inclusion $\overline{\boldsymbol{\sigma}}$ $(\mathbf{x})(\mathbf{x} \in v_i)$ (8.5), which can be recast also in terms of strain polarization tensors:

$$\overline{\boldsymbol{\eta}}(\mathbf{x}) = \boldsymbol{\eta}_i^{\infty}(\mathbf{x}) + \int \mathbf{K}_i(\mathbf{x}, \mathbf{y})[\boldsymbol{\eta}(\mathbf{y})V(\mathbf{y}|; v_i, \mathbf{x}_i) - \langle\boldsymbol{\eta}\rangle]\, d\mathbf{y}, \tag{10.15}$$

where the kernel $\mathbf{K}_i(\mathbf{x}, \mathbf{y})$ is defined by Eq. (4.37), and $\boldsymbol{\eta}_i^{\infty}(\mathbf{x}) \equiv \mathbf{E}_i(\mathbf{x})\langle\boldsymbol{\sigma}\rangle + \mathbf{H}_i(\mathbf{x})$ is the external strain polarization tensor generated by the homogeneous loading $\boldsymbol{\sigma}^{\infty} \equiv \langle\boldsymbol{\sigma}\rangle$ and defined by the tensors $\mathbf{E}_i(\mathbf{x})$ and $\mathbf{H}_i(\mathbf{x})$ (4.38).

Acting with the operator $\boldsymbol{\mathcal{L}}_i$ on Eq. (10.15) and taking Eq. (4.45) into account, we obtain

$$\bar{v}_i\boldsymbol{\eta}(\mathbf{x}) = \mathbf{R}_i(\mathbf{x})\langle\boldsymbol{\sigma}\rangle + \mathbf{F}_i(\mathbf{x}) + \bar{v}_i\boldsymbol{\mathcal{L}}_i \int \mathbf{K}_i(\mathbf{z}, \mathbf{y})[\boldsymbol{\eta}(\mathbf{y})V(\mathbf{y}|; v_i, \mathbf{x}_i) - \langle\boldsymbol{\eta}\rangle]\, d\mathbf{y}. \tag{10.16}$$

Hereafter we assume that v_i^0 is the excluded ellipsoidal volume defined in Section 5.3. Extracting of the excluded ellipsoidal volume v_i^0 from the unbounded integration domain in Eq. (10.16) leads Eq. (10.16) to

$$\begin{aligned}\bar{v}_i\boldsymbol{\eta}(\mathbf{x}) &= \mathbf{R}_i(\mathbf{x})\langle\boldsymbol{\sigma}\rangle + \mathbf{F}_i(\mathbf{x}) + \mathbf{R}_i(\mathbf{x})\mathbf{Q}_i^0\langle\boldsymbol{\eta}\rangle \\ &+ \bar{v}_i\boldsymbol{\mathcal{L}}_i \int \mathbf{K}_i(\mathbf{z}, \mathbf{y})\Big\{\boldsymbol{\eta}(\mathbf{y})V(\mathbf{y}|; v_i, \mathbf{x}_i) - [1 - V_i^0(\mathbf{y})]\langle\boldsymbol{\eta}\rangle\Big\}\, d\mathbf{y}.\end{aligned} \tag{10.17}$$

It should be mentioned that the integral in Eq. (10.17) absolutely converges because $\mathbf{K}_i(\mathbf{z}, \mathbf{y}) \sim O(|\mathbf{z} - \mathbf{y}|^{-d})$ and the item in the figure brackets vanishes at infinity $|\mathbf{z} - \mathbf{y}| \to \infty$. Then, it is tempting to avoid consideration of the infinite range improper integral (10.17) over the whole space by truncating the range at a set of some finite domains, say, the set of the homothetic domains $\omega^{R(k)}(\mathbf{x})$ $(k = 1, \ldots, N)$ containing the inscribed spheres with the radius R and with the centers $\mathbf{x}$

$$\begin{aligned}&\int \mathbf{K}_i(\mathbf{z}, \mathbf{y})[\boldsymbol{\eta}(\mathbf{y})V(\mathbf{y}|; v_i, \mathbf{x}_i) - (1 - V_i^0(\mathbf{y}))\langle\boldsymbol{\eta}\rangle]\, d\mathbf{y} = \sum_{k=1}^{N} \lim_{R\to\infty} \int_{\omega^{R(k)}(\mathbf{x})} \mathbf{K}_i(\mathbf{z}, \mathbf{y}) \\ &\times[\boldsymbol{\eta}(\mathbf{y})V^{(k)}(\mathbf{y}|; v_i, \mathbf{x}_i) - (c^{(k)}/c)(1 - V_i^0(\mathbf{y}))\langle\boldsymbol{\eta}\rangle^{(k)}]\, d\mathbf{y}.\end{aligned} \tag{10.18}$$

In so doing the mentioned limit of finite range integrals in the right-hand side of Eq. (10.18) exists and does not depend on the shape of the domains $\omega^{R(k)}(\mathbf{x})$. In particular, the shapes of domains $\omega^{R(k)}(\mathbf{x})$ can be chosen coinciding with the shapes of the ellipsoids v_i^0 $(v_i \subset v^{(k)})$. Then the Eshelby tensor $\mathbf{S}_i^0 \equiv \mathbf{S}(\omega^{R(k)}(\mathbf{x}))$ and, because of this, the item $(1 - V_i^0(\mathbf{y}))\langle\boldsymbol{\eta}\rangle^{(k)}$ can be omitted in the integrals over the mentioned ellipsoidal domains $\omega^{R(k)}(\mathbf{x})$.

Conditional averaging of Eq. (10.17) with fixed inclusion v_i and exploitation of the hypothesis $\widetilde{\boldsymbol{\sigma}}_{i,j} =$ const. (**H**) in the framework of just binary interacting inclusions allows one to reduce Eq. (10.17) to the following:

$$\bar{v}_i\langle\boldsymbol{\eta}\rangle_i(\mathbf{x}) = \mathbf{R}_i(\mathbf{x})\langle\boldsymbol{\sigma}\rangle + \mathbf{F}_i(\mathbf{x}) + \mathbf{R}_i(\mathbf{x})\mathbf{Q}_i^0\langle\boldsymbol{\eta}\rangle$$
$$+ \int \Big\{[\mathbf{R}_{i,j}(\mathbf{x})\langle\widetilde{\boldsymbol{\sigma}}_{i,j}\rangle + \mathbf{F}_{i,j}(\mathbf{x})]\varphi(v_j,\mathbf{x}_j|;v_i,\mathbf{x}_i)$$
$$- (1 - V_i^0(\mathbf{x}_j))n^{(j)}[\mathbf{R}_{i,j}^{\infty}(\mathbf{x})\langle\widetilde{\boldsymbol{\sigma}}_{i,j}\rangle + \mathbf{F}_{i,j}^{\infty}(\mathbf{x})]\Big\}\, d\mathbf{x}_j. \qquad (10.19)$$

The tensors $\mathbf{R}_{i,j}^{\infty}(\mathbf{x})$ and $\mathbf{F}_{i,j}^{\infty}(\mathbf{x})$ and the related tensors produced by the constant average strain polarization tensor $\langle\boldsymbol{\eta}\rangle$ in Eq. (10.19) and in all subsequent integrals can be omitted if the infinite range improper integrals involved are estimated by the passage to the limit in the set of integrals over the domains $\omega^{R(k)}$ (see Eq. (10.18)) which is assumed later on.

We can find the implicit representation for $\langle\boldsymbol{\eta}\rangle$ by averaging Eq. (10.19):

$$\langle\boldsymbol{\eta}\rangle = \Big[\mathbf{I} - \langle\mathbf{R}^v\mathbf{Q}^0V\rangle\Big]^{-1}\Big\{\langle\mathbf{R}^vV\rangle\langle\boldsymbol{\sigma}\rangle + \langle\mathbf{F}^vV\rangle$$
$$+ \int \Big[\sum_i n^{(i)}[\langle\mathbf{R}_{i,j}(\mathbf{x})\rangle_{(i)}\langle\widetilde{\boldsymbol{\sigma}}_{i,j}\rangle + \langle\mathbf{F}_{i,j}(\mathbf{x})\rangle_{(i)}]\varphi(v_j,\mathbf{x}_j|;v_i,\mathbf{x}_i)$$
$$-(1 - V_i^0(\mathbf{x}_j))n^{(j)}\sum_i n^{(i)}[\langle\mathbf{R}_{i,j}^{\infty}(\mathbf{x})\rangle_{(i)}\langle\widetilde{\boldsymbol{\sigma}}_{i,j}\rangle + \langle\mathbf{F}_{i,j}^{\infty}(\mathbf{x})\rangle_{(i)}]\Big]\, d\mathbf{x}_j\Big\}. \qquad (10.20)$$

Therefore

$$\bar{v}_i\langle\boldsymbol{\eta}\rangle_{(i)}(\mathbf{x}) = \mathbf{R}_i(\mathbf{x})\langle\boldsymbol{\sigma}\rangle + \mathbf{F}_i(\mathbf{x}) + \mathbf{R}_i(\mathbf{x})\mathbf{Q}_i^0[\mathbf{I} - \langle\mathbf{R}^v\mathbf{Q}^0V\rangle]^{-1}\Big\{\langle\mathbf{R}^vV\rangle\langle\boldsymbol{\sigma}\rangle$$
$$+ \langle\mathbf{F}_k^vV\rangle + \sum_k n^{(k)}\int\Big[[\langle\mathbf{R}_{k,j}(\mathbf{x})\rangle_{(k)}\langle\widetilde{\boldsymbol{\sigma}}_{k,j}\rangle + \langle\mathbf{F}_{k,j}(\mathbf{x})\rangle_{(k)}]\varphi(v_j,\mathbf{x}_j|;v_k,\mathbf{x}_k)$$
$$- (1 - V_k^0(\mathbf{x}_j))n^{(j)}[\langle\mathbf{R}_{k,j}^{\infty}(\mathbf{x})\rangle_{(k)}\langle\widetilde{\boldsymbol{\sigma}}_{k,j}\rangle + \langle\mathbf{F}_{k,j}^{\infty}(\mathbf{x})\rangle_{(k)}]\Big]\, d\mathbf{x}_j\Big\}$$
$$+ \int \Big[[\mathbf{R}_{i,j}(\mathbf{x})\langle\widetilde{\boldsymbol{\sigma}}_{i,j}\rangle + \mathbf{F}_{i,j}(\mathbf{x})]\varphi(v_j,\mathbf{x}_j|;v_i,\mathbf{x}_i)$$
$$- (1 - V_i^0(\mathbf{x}_j))n^{(j)}[\mathbf{R}_{i,j}^{\infty}(\mathbf{x})\langle\widetilde{\boldsymbol{\sigma}}_{i,j}\rangle + \mathbf{F}_{i,j}^{\infty}(\mathbf{x})]\Big]\, d\mathbf{x}_j. \qquad (10.21)$$

No restrictions are imposed on the microtopology of the microstructure and the shape of inclusions as well as on the inhomogeneity of stress field inside the inclusions. However, the main computational advantage of the proposed Eqs. (10.19)–(10.21) lies in the fact that such fundamental notions of micromechanics as the Green function (Eshelby tensor is used only through the tensor $\mathbf{Q}_i^0$ which can be found numerically, see Eq. (4.20) is not used, and we can analyze any anisotropy of constituents (including the matrix) as well as any shape and any composite structure of inclusions. We constructed Eqs. (10.19)–(10.21) by the use of some building blocks described by the numerical solutions for both one and two inclusions inside the infinite medium subjected to the homogeneous loading at infinity. With just some additional assumptions the tensors mentioned above can be expressed through the Green function, Eshelby, tensor and external Eshelby tensor.

10.2.2 Some Related Integral Equations

Let us ascertain the correlation of Eqs. (10.19) with few in number related equations based on the utilization of the solutions for one and two inclusions rather than the direct use of a Green function. In the framework of the assumption of a perturbation method $\langle\widetilde{\boldsymbol{\sigma}}_{i,j}\rangle = \langle\boldsymbol{\sigma}\rangle$ (9.10), Eq. (10.19) is reduced to:

$$\bar{v}_i\langle\boldsymbol{\eta}\rangle_i(\mathbf{x}) = \mathbf{R}_i(\mathbf{x})\langle\boldsymbol{\sigma}\rangle + \mathbf{F}_i(\mathbf{x}) + \mathbf{R}_i(\mathbf{x})\mathbf{Q}_i^0\langle\boldsymbol{\eta}\rangle + \int \Big\{[\mathbf{R}_{i,j}(\mathbf{x})\langle\boldsymbol{\sigma}\rangle + \mathbf{F}_{i,j}(\mathbf{x})] \\ \times\, \varphi(v_j,\mathbf{x}_j|;v_i,\mathbf{x}_i) - n^{(j)}[1 - V_i^0(\mathbf{x}_j)][\mathbf{R}_{i,j}^{\infty}(\mathbf{x})\langle\boldsymbol{\sigma}\rangle + \mathbf{F}_{i,j}^{\infty}(\mathbf{x})]\Big\}\, d\mathbf{x}_j. \quad (10.22)$$

Chen and Acrivos [216] proposed an equation related to Eq. (10.22) for the pure elastic case of the identical homogeneous spherical inclusions ($\boldsymbol{\beta}_1 \equiv \mathbf{0}$, $N = 1$). In our notations this averaged equation can be represented in the form

$$\bar{v}_i\langle\boldsymbol{\eta}\rangle_i = \mathbf{R}_i\langle\boldsymbol{\sigma}\rangle + \mathbf{R}_i\mathbf{Q}_i^0\langle\boldsymbol{\eta}\rangle + \int \Big[\langle\mathbf{R}_{i,j}(\mathbf{x})\rangle_{(i)}\langle\boldsymbol{\sigma}\rangle\varphi(v_j,\mathbf{x}_j|;v_i,\mathbf{x}_i) \\ - (1 - V_i^0(\mathbf{x}_j))n^{(j)}\{\mathbf{X}\boldsymbol{\sigma}'\}\Big]\, d\mathbf{x}_j, \quad (10.23)$$

where the second summand of the integrand in the brace $\mathbf{X}\boldsymbol{\sigma}'$ where $\boldsymbol{\sigma}' \equiv \boldsymbol{\sigma}'(\mathbf{x}_i) = \boldsymbol{\sigma}(\mathbf{x}_i) - \langle\boldsymbol{\sigma}\rangle$ is a perturbation stress at the $\mathbf{x}_i$ due to a single inclusion at $\mathbf{x}_j$ subjected to the applied strain $\langle\boldsymbol{\sigma}\rangle$ at infinity. The constant tensor $\mathbf{X}$ determined so that the integral will be absolutely convergent has been chosen as $\mathbf{R}_j$. Therefore the item in the brace can be uniquely determined in the term of the solution (10.11): $\mathbf{X}\boldsymbol{\sigma}' = \mathbf{R}_j\langle\mathbf{T}_j^{\sigma\infty}(\mathbf{x})\rangle_{(i)}\langle\boldsymbol{\sigma}\rangle$. The proposed approximation is based on the assumption that the field $\boldsymbol{\sigma}' \equiv$ const. in the area v_i. The last limitation can be easily avoided in the framework of the original scheme [216]. Indeed, the item in the brace should be replaced by the relation $\langle\mathbf{R}_{i,j}^{\infty}\rangle_{(i)}\langle\boldsymbol{\sigma}\rangle$ (10.11), defined through the strain polarization tensor $\langle\boldsymbol{\eta}\rangle_{(i)}$ in the inclusion v_i subjected to both the homogeneous remote field $\langle\boldsymbol{\sigma}\rangle$ and the inhomogeneous field $\mathbf{T}_j^{\sigma\infty}(\mathbf{x})\langle\boldsymbol{\sigma}\rangle$ ($\mathbf{x} \in v_i$). Then Eq. (10.23) has a new representation:

$$\bar{v}_i\langle\boldsymbol{\eta}\rangle_i = \mathbf{R}_i\langle\boldsymbol{\sigma}\rangle + \mathbf{R}_i\mathbf{Q}_i^0\langle\boldsymbol{\eta}\rangle + \int \Big[\langle\mathbf{R}_{i,j}(\mathbf{x})\rangle_{(i)}\langle\boldsymbol{\sigma}\rangle\varphi(v_j,\mathbf{x}_j|;v_i,\mathbf{x}_i) \\ - (1 - V_i^0(\mathbf{x}_j))n^{(j)}\langle\mathbf{R}_{i,j}^{\infty}(\mathbf{x})\rangle_{(i)}\langle\boldsymbol{\sigma}\rangle\Big]\, d\mathbf{x}_j, \quad (10.24)$$

It should be mentioned that the inhomogeneous tensor $\langle\boldsymbol{\sigma}\rangle + \mathbf{T}_j^{\sigma\infty}(\mathbf{x})\langle\boldsymbol{\sigma}\rangle$ ($\mathbf{x} \in v_i$) has a sense of the inhomogeneous effective stress field outside of the inclusion v_j in the area v_i, which can be estimated by different methods, not just by the use of the tensors $\mathbf{T}_j^{\sigma\infty}(\mathbf{x})$. A new representation (10.24) coincides with the old one (10.23) at $\mathbf{X}\boldsymbol{\sigma}' = \mathbf{R}_j\langle\mathbf{T}_j^{\sigma\infty}(\mathbf{x})\rangle_{(i)}\langle\boldsymbol{\sigma}\rangle$ just in the framework of both the case of homogeneous ellipsoidal inclusions and the additional effective field hypothesis **H1** (8.14): each inclusion v_i is located in the homogeneous effective field

$$\overline{\boldsymbol{\sigma}}_i(\mathbf{x}) = \overline{\boldsymbol{\sigma}}_i(\mathbf{x}_i) = \text{const.} \quad (\mathbf{x} \in v_i), \quad (10.25)$$

which is equivalent to the original assumption [216] $\boldsymbol{\sigma}'(\mathbf{x}) \equiv$ const. ($\mathbf{x} \in v_i$). Assumption (10.25) is not a basic one in this chapter and is presented only for explanation of the relationship between the proposed equation (10.19) and the known one (10.23).

10.2.3 Closing Assumption and the Effective Properties

For the closing of Eq. (10.21) we will use the following hypothesis:
H2). Each pair of inclusions v_j and v_j is subjected to the homogeneous field $\widetilde{\boldsymbol{\sigma}}_{i,j} = \text{const.}$, and statistical volume average $\langle \boldsymbol{\eta} \rangle_i$ is defined by the formulae

$$\langle \boldsymbol{\eta} \rangle_i = \mathbf{R}_i \langle \overline{\boldsymbol{\sigma}}_i \rangle + \mathbf{F}_i, \tag{10.26}$$

$$\langle \widetilde{\boldsymbol{\sigma}}_{i,j} \rangle = \langle \overline{\boldsymbol{\sigma}}_i \rangle. \tag{10.27}$$

To keep the calculations down, one introduces the notations for the statistical average of all perturbations introduced by surrounding inclusions v_j ($j = 1, 2, \ldots;\ j \neq i$) into the inclusion v_i:

$$\mathbf{R}^{\text{pert}}_{i,j}(\mathbf{x}) = \int \left[\mathbf{R}_{i,j}(\mathbf{x}) \varphi(v_j, \mathbf{x}_j |; v_i, \mathbf{x}_i) - (1 - V_i^0(\mathbf{x}_j)) n^{(j)} \mathbf{R}^{\infty}_{i,j}(\mathbf{x}) \right] d\mathbf{x}_j, \tag{10.28}$$

$$\mathbf{F}^{\text{pert}}_{i}(\mathbf{x}) = \sum_j \int \left[\mathbf{F}_{i,j}(\mathbf{x})) \varphi(v_j, \mathbf{x}_j |; v_i, \mathbf{x}_i) - (1 - V_i^0(\mathbf{x}_j)) n^{(j)} \mathbf{F}^{\infty}_{i,j}(\mathbf{x}) \right] d\mathbf{x}_j. \tag{10.29}$$

The volume averages of the tensors $\mathbf{R}^{\text{pert}}_{i,j}(\mathbf{x})$ and $\mathbf{F}^{\text{pert}}_{i}(\mathbf{x})$ over the volume of the inclusion v_i are denoted by $\mathbf{R}^{\text{pert}}_{i,j}$ and $\mathbf{F}^{\text{pert}}_{i}$, respectively. For a small concentration of inclusions $c \equiv \langle V \rangle \ll 1$, the tensors $\mathbf{R}^{\text{pert}}_{i,j}(\mathbf{x})$ and $\mathbf{F}^{\text{pert}}_{i}(\mathbf{x})$ are the linear functions of c. For a pure elastic case ($\boldsymbol{\beta}_1 \equiv \mathbf{0}$), the tensor $\mathbf{R}^{\text{pert}}_{i,j}(\mathbf{x})$ has a physical meaning of the perturbation of the statistical average of the strain polarization tensor $\boldsymbol{\eta}$ in the inclusion v_i introduced by the action of surrounding inclusions belonging to the phase $v^{(j)}$. However, the total perturbation introduced by the surrounding inclusions contains also the well known item $\langle \mathbf{R}^v \mathbf{Q}^0 V \rangle$ presented in Eq. (10.20) as well as in the related equations obtained in the framework of quasi-crystallite approximation (see Chapters 8 and 9).

Volume averaging of Eq. (10.19) over the inclusion v_i with subsequent substitution of Eqs. (10.26) and (10.27) into the left-hand side of the equation obtained leads to the closed algebraic equation with the solution:

$$\langle \overline{\boldsymbol{\sigma}}_i \rangle = \sum_m \mathbf{Y}_{im} \Big[\langle \boldsymbol{\sigma} \rangle + \mathbf{Q}^0_m \langle \mathbf{F}^v V \rangle + \mathbf{R}^{-1}_m \mathbf{F}^{\text{pert}}_m \Big], \tag{10.30}$$

$$(\mathbf{Y}^{-1})_{im} = \delta_{im} \Big[\mathbf{I} - \mathbf{R}^{-1}_i \sum_k \mathbf{R}^{\text{pert}}_{i,k} \Big] - n^{(m)} \mathbf{Q}^0_i \mathbf{R}_m. \tag{10.31}$$

Substituting Eq. (10.30) into the right-hand side of Eg. (10.19) leads to the explicit representation for statistical average of strain polarization tensor in the inclusion v_i:

$$\begin{aligned} \bar{v}_i \langle \boldsymbol{\eta} \rangle_i(\mathbf{x}) &= \mathbf{R}_i(\mathbf{x}) \langle \boldsymbol{\sigma} \rangle + \mathbf{F}_i(\mathbf{x}) \\ &+ \mathbf{R}_i(\mathbf{x}) \mathbf{Q}^0_i \sum_{k,m} n^{(k)} \Big\{ \delta_{km} \mathbf{F}_k + \mathbf{R}_k \mathbf{Y}_{km} \big[\langle \boldsymbol{\sigma} \rangle + \langle \mathbf{Q}^0 \mathbf{F}^v V \rangle + \mathbf{R}^{-1}_m \mathbf{F}^{\text{pert}}_m \big] \Big\} \\ &+ \mathbf{F}^{\text{pert}}_i(\mathbf{x}) + \sum_j \mathbf{R}^{\text{pert}}_{i,j}(\mathbf{x}) \sum_m \mathbf{Y}_{im} \big[\langle \boldsymbol{\sigma} \rangle + \langle \mathbf{Q}^0 \mathbf{F}^v V \rangle + \mathbf{R}^{-1}_m \mathbf{F}^{\text{pert}}_m \big]. \end{aligned} \tag{10.32}$$

The mean field inside the inclusions $\langle \boldsymbol{\sigma} \rangle_i(\mathbf{x})$ is obtained from (4.10$_1$) and (10.32):

$$\langle \boldsymbol{\sigma} \rangle_i(\mathbf{x}) = \mathbf{B}_i(\mathbf{x})\langle \boldsymbol{\sigma} \rangle + \mathbf{C}_i(\mathbf{x}) + [\bar{v}_i \mathbf{M}_1(\mathbf{x})]^{-1}\mathbf{F}_i^{\text{pert}}(\mathbf{x})$$
$$+\mathbf{B}_i(\mathbf{x})\mathbf{Q}_i^0 \sum_{k,m} n^{(k)} \Big\{ \delta_{km}\mathbf{F}_k + \mathbf{R}_k \mathbf{Y}_{km} \big[\langle \boldsymbol{\sigma} \rangle + \langle \mathbf{Q}^0 \mathbf{F}^v V \rangle + \mathbf{R}_m^{-1}\mathbf{F}_m^{\text{pert}} \big] \Big\}$$
$$+ [\bar{v}_i \mathbf{M}_1(\mathbf{x})]^{-1} \sum_j \mathbf{R}_{i,j}^{\text{pert}}(\mathbf{x}) \sum_m \mathbf{Y}_{im} \big[\langle \boldsymbol{\sigma} \rangle + \langle \mathbf{Q}^0 \mathbf{F}^v V \rangle + \mathbf{R}_m^{-1}\mathbf{F}_m^{\text{pert}} \big]. \quad (10.33)$$

It should be mentioned that in Eqs. (10.32) and (10.33) one does not assume that the effective field $\overline{\boldsymbol{\sigma}}_i(\mathbf{x})$ is a homogeneous one as proposed in the initial version of the MEFM (see Chapter 8). Because of this the tensors $\langle \boldsymbol{\eta} \rangle_i(\mathbf{x})$, $\langle \boldsymbol{\sigma} \rangle_i(\mathbf{x}) \neq$ const. due to the inhomogeneity of the tensors $\mathbf{R}_{i,j}^{\text{pert}}(\mathbf{x}), \mathbf{F}_i^{\text{pert}}(\mathbf{x}) \neq$ const. even for the homogeneous ellipsoidal inclusions when $\mathbf{R}_i(\mathbf{x}), \mathbf{F}_i(\mathbf{x}), \mathbf{B}_i(\mathbf{x}), \mathbf{C}_i(\mathbf{x}) =$ const. $(\mathbf{x} \in v_i)$.

After estimating average strain polarization tensors inside the inclusions, see Eqs. (10.26), (10.27) and (10.32), the problem of calculating effective properties becomes trivial and leads, according to Sections 6.1 and 6.2, to the following *new representations*:

$$\mathbf{M}^* = \mathbf{M}^{(0)} + \sum_{i,j=1}^{N} n^{(i)} \mathbf{R}_i \mathbf{Y}_{ij}, \quad (10.34)$$

$$\boldsymbol{\beta}^* - \boldsymbol{\beta}^{(0)} + \langle \mathbf{F}^v V \rangle + \sum_{i,m=1}^{N} n^{(i)} \mathbf{R}_i \mathbf{Y}_{im} \Big[\langle \mathbf{F}^v V \rangle + \mathbf{R}_m^{-1}\mathbf{F}_m^{\text{pert}} \Big], \quad (10.35)$$

$$W^* = -\frac{1}{2}\langle \boldsymbol{\beta}_1 \mathbf{C} \rangle - \frac{1}{2}\langle \boldsymbol{\beta}_1 \mathbf{M}_1^{-1}\mathbf{F}^{\text{pert}} \rangle - \frac{1}{2}\langle \boldsymbol{\beta}_1(\mathbf{x})\mathbf{B}(\mathbf{x})\mathbf{Q}^0 \rangle \langle \mathbf{F}^v V \rangle$$
$$-\frac{1}{2} \sum_{i,k,m=1}^{N} n^{(i)} n^{(k)} \langle \boldsymbol{\beta}_1^{(i)}(\mathbf{x})\mathbf{B}_i(\mathbf{x}) \rangle_{(i)} \mathbf{Q}_i^0 \mathbf{R}_k \mathbf{Y}_{km} \Big[\langle \mathbf{F}^v V \rangle + \mathbf{R}_m^{-1}\mathbf{F}_m^{\text{pert}} \Big]$$
$$-\frac{1}{2} \sum_{i,m=1}^{N} n^{(i)} \langle \boldsymbol{\beta}_1^{(i)}(\mathbf{x})\mathbf{M}_1^{-1}(\mathbf{x})\mathbf{R}_i^{\text{pert}}(\mathbf{x}) \rangle_{(i)} \mathbf{Y}_{im} \Big[\langle \mathbf{F}^v V \rangle + \mathbf{R}_m^{-1}\mathbf{F}_m^{\text{pert}} \Big]. \quad (10.36)$$

Exploring the closing assumption

$$\langle \widetilde{\boldsymbol{\sigma}}_{i,j} \rangle = \langle \overline{\boldsymbol{\sigma}}_j \rangle \quad (10.37)$$

instead of Eq. (10.27) makes it possible to obtain Eqs. (10.33)–(10.36) with the tensor

$$(\mathbf{Y}^{-1})_{im} = \delta_{im}\mathbf{I} - \mathbf{R}_i^{-1}\mathbf{R}_{i,m}^{\text{pert}} - n^{(m)}\mathbf{Q}_i^0 \mathbf{R}_m \quad (10.38)$$

differing from the representation (10.31).

10.2.4 Conditional Mean Value of Stresses in the Inclusions

The interaction effects generating the local stresses produced by inclusions are highly sensitive to their locations. In so doing nonlinear processes such as fracture and fatigue are far more sensitive to local stresses and, therefore, to local variations of microstructure than other mechanical phenomena. Because of this, in the deformation and fracture processes, another important statistical parameter of the location of inclusions is a nearest neighbor distribution N^{nd} function [901] used in conjunction with the estimation of conditional mean values of stresses in the inclusion v_i at the condition of the fixed inclusion v_j. It should be mentioned that there are a few models [448]) based on the idea that at high inclusion concentrations, the effective properties are dominated by the interaction forces between neighboring particles that are proportional to δ^{-1}, where δ is the ratio of the mean gap between neighboring particles to particle diameter. Obviously, the use of the average value $\delta \equiv \langle N^{nd}\rangle/(2a)$ instead of random distribution N^{nd} leads to the loss of statistical information concerning the microtopology of the composite and is conceptually questionable because estimating the average of the output parameter (such as effective modulus) by the use of the average of the random input parameter (such as nearest neighbor distance) is essentially a nonlinear problem. The approach [901] based on the elastic solution for two interacting inclusions in an infinite matrix with subsequent averaging by the NND is a distinct improvement because it is more sensitive to the local configuration of inclusions. In so doing, Pyrz [901] has assumed that two inclusions are subjected to the field $\langle\widetilde{\boldsymbol{\sigma}}_{i,j}\rangle \equiv \langle\boldsymbol{\sigma}\rangle$ (see also Eq. (9.10)) leading in our notations (10.3) to

$$\langle\boldsymbol{\sigma}|;v_j,\mathbf{x}_j\rangle(\mathbf{x}) = \mathbf{B}_i(\mathbf{x})\langle\boldsymbol{\sigma}\rangle + \mathbf{C}_i(\mathbf{x}) + [\bar{v}_i\mathbf{M}_1^{(i)}]^{-1}\Big[\mathbf{R}_{i,j}(\mathbf{x})\langle\boldsymbol{\sigma}\rangle + \mathbf{F}_{i,j}(\mathbf{x})\Big], \quad (10.39)$$

where $\mathbf{x} \in v_i$. However, in light of the proposed approach, the crude assumption $\langle\widetilde{\boldsymbol{\sigma}}_{i,j}\rangle \equiv \langle\boldsymbol{\sigma}\rangle$ can be relaxed by use of more accurate estimations (10.27) of the correlation function of the effective field $\langle\widehat{\boldsymbol{\sigma}}_{i,j}\rangle$ leading, in conjunction with Eqs. (10.3) and (10.30), to the representation:

$$\langle\boldsymbol{\sigma}|;v_j,\mathbf{x}_j\rangle(\mathbf{x}) = \mathbf{C}_i(\mathbf{x}) + [\bar{v}_i\mathbf{M}_1^{(i)}]^{-1}\mathbf{F}_{i,j}(\mathbf{x}) + \Big\{\mathbf{B}_i(\mathbf{x})\langle\boldsymbol{\sigma}\rangle + [\bar{v}_i\mathbf{M}_1^{(i)}]^{-1}\mathbf{R}_{i,j}(\mathbf{x})\Big\}$$
$$\times \sum_m \mathbf{Y}_{im}\Big[\langle\boldsymbol{\sigma}\rangle + \langle\mathbf{Q}^0\mathbf{F}^v V\rangle + \mathbf{R}_m^{-1}\mathbf{F}_m^{\text{pert}}\Big]. \quad (10.40)$$

The equation (10.40) is reduced to Eq. (10.39) only in the case $\langle V\rangle \to 0$.

10.3 First-order Approximation of the Closing Assumption and Effective Elastic Moduli

10.3.1 General Equation for the Effective Fields $\langle\overline{\boldsymbol{\sigma}}_{i,j}\rangle$

For statistically homogeneous media, the statistical average $\langle\widetilde{\boldsymbol{\sigma}}_{i,j}\rangle$ in Eq. (10.21) characterizing the pair interactions between the inclusions should in general depend on the difference $\mathbf{x}_i - \mathbf{x}_j$. For an approximate estimation of this dependence, we average Eq. (8.6) under the condition that the inclusions v_i and v_j are fixed:

$$\langle\widetilde{\boldsymbol{\sigma}}_{i,j}\rangle(\mathbf{x}) = \langle\boldsymbol{\sigma}\rangle(\mathbf{x}) + \int \boldsymbol{\Gamma}(\mathbf{x}-\mathbf{y})\Big[\langle\boldsymbol{\eta}(\mathbf{y})|;v_i,\mathbf{x}_i;v_j,\mathbf{x}_j\rangle(\mathbf{y})$$
$$\times\, \varphi(v_k,\mathbf{x}_k|;v_i,\mathbf{x}_i,v_j,\mathbf{x}_j) - \langle\boldsymbol{\eta}\rangle(\mathbf{y})\Big]\, d\mathbf{y}, \quad (10.41)$$

where $\mathbf{y} \in v_k$, and $\langle \boldsymbol{\eta}(\mathbf{y})|; v_i, \mathbf{x}_i; v_j, \mathbf{x}_j\rangle(\mathbf{y})$ is the mean value of the strain polarization tensor for an inclusion v_k located at point $\mathbf{y}_k$ under the condition that the inclusion v_i and v_j are fixed. We combine the three-point averages $\langle \boldsymbol{\eta}(\mathbf{y})|; v_i, \mathbf{x}_i; v_j, \mathbf{x}_j\rangle(\mathbf{y})$ in the integral in Eq. (10.41) with the two-point averages $\langle \boldsymbol{\eta}(\mathbf{y})|; v_i, \mathbf{x}_i\rangle(\mathbf{y})$ presented in the involved equation for the effective field $\langle \overline{\boldsymbol{\sigma}}_i\rangle$ and recast Eg. (10.41) in the form:

$$\begin{aligned}
\langle \widetilde{\boldsymbol{\sigma}}_{i,j}\rangle(\mathbf{x}) &= \langle \boldsymbol{\sigma}\rangle(\mathbf{x}) + \int \boldsymbol{\Gamma}(\mathbf{x}-\mathbf{y})\big\{\langle \boldsymbol{\eta}(\mathbf{y})|; v_i, \mathbf{x}_i\rangle(\mathbf{y})\varphi(v_k, \mathbf{x}_k|; v_i, \mathbf{x}_i) - \langle \boldsymbol{\eta}\rangle\big\}\, d\mathbf{y} \\
&+ \int \boldsymbol{\Gamma}(\mathbf{x}-\mathbf{y})\langle \boldsymbol{\eta}(\mathbf{y})|; v_i, \mathbf{x}_i\rangle(\mathbf{y})[\varphi(v_k, \mathbf{x}_k|; v_i, \mathbf{x}_i, v_j, \mathbf{x}_j) \\
&- (1 - V^{0i}_{0j}(\mathbf{y}))\varphi(v_k, \mathbf{x}_k|; v_i, \mathbf{x}_i)]\, d\mathbf{y} \\
&- \int \boldsymbol{\Gamma}(\mathbf{x}-\mathbf{y})\langle V^{0i}_{0j}(\mathbf{y})\boldsymbol{\eta}(\mathbf{y})|; v_i, \mathbf{x}_i\rangle(\mathbf{y})\varphi(v_k, \mathbf{x}_k|; v_i, \mathbf{x}_i)\, d\mathbf{y} \\
&+ \int \boldsymbol{\Gamma}(\mathbf{x}-\mathbf{y})[\langle \boldsymbol{\eta}(\mathbf{y})|; v_i, \mathbf{x}_i; v_j, \mathbf{x}_j\rangle(\mathbf{y}) - \langle \boldsymbol{\eta}(\mathbf{y})|; v_i, \mathbf{x}_i\rangle(\mathbf{y})] \\
&\times \varphi(v_k, \mathbf{x}_k|; v_i, \mathbf{x}_i, v_j, \mathbf{x}_j)\, d\mathbf{y}, \qquad (10.42)
\end{aligned}$$

where V^{0i}_{0j} is a characteristic function of the domain $v^{0i}_{0j} = v^0_j \backslash (v^0_j \cap v^0_i)$ prohibited for the inclusion centers in the integration area in both Eq. (10.41) and the second integral in Eq. (10.42) and available for the inclusion centers in the integration area in the first integral on the right-hand side of Eq. (10.42). For two circle inclusions of the radii a placed at the distance $|\mathbf{x}_i - \mathbf{x}_j| \equiv r$, the volume of the domain v^{0i}_{0j} takes the value ($z = r - 2a$)

$$\overline{v^{0i}_{0j}} = \begin{cases} 0 & \text{for } r < 2a, \\ 3\pi a^2 + 2a^2 \arcsin\frac{z}{2a} + \frac{z}{2}\sqrt{4a^2 - z^2} & \text{for } 2a \le r \le 4a, \\ 4\pi a^2 & \text{for } r \ge 4a. \end{cases} \qquad (10.43)$$

It is interesting that for well-stirred approximation of statistical descriptors (5.70) the second integral on the right-hand side of Eq. (10.42) vanishes. Different double $\varphi(v_k, \mathbf{x}_k|; v_i, \mathbf{x}_i)$ and triple $\varphi(v_k, \mathbf{x}_k|; v_i, \mathbf{x}_i, v_j, \mathbf{x}_j)$ (e.g., Kirkwood approximation (5.71)) conditional probability densities were analyzed in Sections 5.2 and 5.3.

Comparing Eq. (10.42) and the statistical average of Eq. (8.5) leads to

$$\begin{aligned}
\langle \widetilde{\boldsymbol{\sigma}}_{i,j}\rangle(\mathbf{x}) &= \langle \overline{\boldsymbol{\sigma}}_i\rangle(\mathbf{x}) + \int \boldsymbol{\Gamma}(\mathbf{x}-\mathbf{y})\langle \boldsymbol{\eta}(\mathbf{y})|; v_i, \mathbf{x}_i\rangle(\mathbf{y})\widetilde{\varphi}(v_k, \mathbf{x}_k|; v_i, \mathbf{x}_i, v_j, \mathbf{x}_j)\, d\mathbf{y} \\
&+ \int \boldsymbol{\Gamma}(\mathbf{x}-\mathbf{y})[\langle \boldsymbol{\eta}(\mathbf{y})|; v_i, \mathbf{x}_i; v_j, \mathbf{x}_j\rangle(\mathbf{y}) \\
&- \langle \boldsymbol{\eta}(\mathbf{y})|; v_i, \mathbf{x}_i\rangle(\mathbf{y})]\varphi(v_k, \mathbf{x}_k|; v_i, \mathbf{x}_i, v_j, \mathbf{x}_j)\, d\mathbf{y}, \quad (\mathbf{x} \in v_i), \qquad (10.44)
\end{aligned}$$

where $\widetilde{\varphi}(v_k, \mathbf{x}_k|; v_i, \mathbf{x}_i, v_j, \mathbf{x}_j) = \varphi(v_k, \mathbf{x}_k|; v_i, \mathbf{x}_i, v_j, \mathbf{x}_j) - \varphi(v_k, \mathbf{x}_k|; v_i, \mathbf{x}_i)$. Decaying at infinity as $|\mathbf{x}_i - \mathbf{x}_j|^{-d}$ of the integrals on the right-hand side of Eq. (10.44) leads to both disappearance of the correlations in the arrangement of inclusions when the distance between them increases and degeneration of Eq. (10.44) into the assumption (10.26).

10.3.2 Closing Assumptions for the Strain Polarization Tensor $\langle\boldsymbol{\eta}(\mathbf{y})|; v_i, \mathbf{x}_i; v_j, \mathbf{x}_j\rangle(\mathbf{y})$

Thus, the two-point average $\langle\widetilde{\boldsymbol{\sigma}}_{i,j}\rangle(\mathbf{x})$ depends on the three-point averages of the strain polarization tensor $\langle\boldsymbol{\eta}(\mathbf{y})|; v_i, \mathbf{x}_i; v_j, \mathbf{x}_j\rangle(\mathbf{y})$ that, in turn, depend on the four-point average $\langle\boldsymbol{\eta}(\mathbf{y})|; v_i, \mathbf{x}_i; v_j, \mathbf{x}_j; v_k, \mathbf{x}_k\rangle(\mathbf{y})$ (see Eq. (8.7_2)) and so on. In a limiting case we will get an infinite chain of coupled integral equations connecting all multipoint conditional means of strain polarization tensor. Obtaining a hierarchy of equations connecting the multipoint conditional means of increasing order can be closed by the use of additional assumptions concerning the statistical properties of the strain polarization tensor. The simplest assumptions described by Eqs. (10.27) and (10.37) were applied to Eq. (10.19) which already takes the binary interaction of inclusions into account. The effective fields $\langle\widetilde{\boldsymbol{\sigma}}_{i,j}\rangle$ in the assumptions either (10.27) or (10.37) do not depend explicitly on the distance $|\mathbf{x}_i-\mathbf{x}_j|$ of on the mechanical and geometrical properties of the inclusions v_i and v_j. To overcome this disadvantage, we close Eq. (10.41) by any from the following assumption ($\mathbf{y}\in v_k\neq v_i,\ v_j$):

$$\langle\boldsymbol{\eta}(\mathbf{y})|; v_i, \mathbf{x}_i; v_j, \mathbf{x}_j\rangle(\mathbf{y})=\langle\boldsymbol{\eta}(\mathbf{y})|; v_i, \mathbf{x}_i\rangle(\mathbf{y}), \tag{10.45}$$

$$\langle\boldsymbol{\eta}(\mathbf{y})|; v_i, \mathbf{x}_i; v_j, \mathbf{x}_j\rangle(\mathbf{y})=\langle\boldsymbol{\eta}(\mathbf{y})|; v_j, \mathbf{x}_j\rangle(\mathbf{y}), \tag{10.46}$$

$$\langle\boldsymbol{\eta}(\mathbf{y})|; v_i, \mathbf{x}_i; v_j, \mathbf{x}_j\rangle(\mathbf{y})=\langle\boldsymbol{\eta}(\mathbf{y})|; v_m, \mathbf{x}_m\rangle(\mathbf{y}),\ m=\arg\min_{l=i,j}|\mathbf{x}_k-\mathbf{x}_l| \tag{10.47}$$

with the obvious corrections of Eq. (10.42). For the sake of definiteness, we shall consider the assumption (10.45). Analysis of another closing assumption, either (10.46) or (10.47), causes no complications and can be analogously carried out analogously. Since obtaining Eq. (10.19) assumes homogeneity of the field $\widetilde{\boldsymbol{\sigma}}_{i,j}(\mathbf{x})$ at $\mathbf{x}\in v_i$, we will estimate the desirable field $\widetilde{\boldsymbol{\sigma}}_{i,j}(\mathbf{x})\equiv\widetilde{\boldsymbol{\sigma}}_{i,j}=\text{const.}$ ($\mathbf{x}\in v_i$) by acting by the operator $\boldsymbol{\mathcal{L}}_i$ on Eq. (10.44) (4.42) with subsequent multiplication of the obtained equation on the tensor $\mathbf{R}_i^{-1}(\mathbf{x})$ ($\mathbf{x}\in v_i$):

$$\begin{aligned}\langle\widetilde{\boldsymbol{\sigma}}_{i,j}\rangle &= \langle\overline{\boldsymbol{\sigma}}_i\rangle + \int\Big[\langle\mathbf{R}_i^{-1}(\mathbf{x})\mathbf{R}_{i,k}(\mathbf{x})\rangle_{(i)}\langle\widetilde{\boldsymbol{\sigma}}_{i,k}\rangle + \langle\mathbf{R}_i^{-1}(\mathbf{x})\mathbf{F}_{i,k}(\mathbf{x})\rangle_{(i)}\Big]\\ &\times\widetilde{\varphi}(v_k,\mathbf{x}_k|; v_i,\mathbf{x}_i,v_j,\mathbf{x}_j)\,d\mathbf{x}_k\\ &+\int\mathbf{J}(\mathbf{x}_k)\Big[\langle\mathbf{B}^{\sigma}_{i,j,k}(\mathbf{x})\rangle_{(i)}\langle\widetilde{\boldsymbol{\sigma}}_{j,k}\rangle-\langle\mathbf{R}_i^{-1}(\mathbf{x})\mathbf{R}_{i,k}(\mathbf{x})\rangle_{(i)}\langle\widetilde{\boldsymbol{\sigma}}_{i,k}\rangle\Big]\\ &\times\varphi(v_k,\mathbf{x}_k|; v_i,\mathbf{x}_i,v_j,\mathbf{x}_j)\,d\mathbf{x}_k+\int\mathbf{J}(\mathbf{x}_k)\Big[\langle\mathbf{C}^{\beta}_{i,j,k}(\mathbf{x})\rangle_{(i)}-\langle\mathbf{R}_i^{-1}(\mathbf{x})\mathbf{F}_{i,k}(\mathbf{x})\rangle_{(i)}\Big]\\ &\times\varphi(v_k,\mathbf{x}_k|; v_i,\mathbf{x}_i,v_j,\mathbf{x}_j)\,d\mathbf{x}_k,\end{aligned} \tag{10.48}$$

keeping in mind that the centers of inclusions v_i and $\bar{v}_j$ are fixed in the points $\mathbf{x}_i$ and $\mathbf{x}_j$, respectively. Although the fields $\widetilde{\boldsymbol{\sigma}}_{i,j}$ and $\widetilde{\boldsymbol{\sigma}}_{i,k}$ are constant inside the inclusions v_i, v_j, v_k, they vary with the distance $|\mathbf{x}_i-\mathbf{x}_j|$ and $|\mathbf{x}_i-\mathbf{x}_k|$, respectively. Hereafter the tensors $\mathbf{J}(\mathbf{x}_k)$, $\mathbf{B}^{\sigma}_{i,j,k}(\mathbf{x})$, and $\mathbf{C}^{\beta}_{i,j,k}(\mathbf{x})$ ($\mathbf{x}\in v_i$):

$$\mathbf{J}(\mathbf{x}_k)=\begin{cases}\mathbf{0},\\ \mathbf{I},\\ \mathbf{I}V^+_{i,j}(\mathbf{x}_k),\end{cases}\quad \mathbf{B}^{\sigma}_{i,j,k}(\mathbf{x})=\begin{cases}\mathbf{R}_i^{-1}(\mathbf{x})\mathbf{R}_{i,j}(\mathbf{x}),\\ \mathbf{B}^K_{k,j}(\mathbf{x}),\\ \mathbf{B}^K_{k,j}(\mathbf{x}),\end{cases}\quad \mathbf{C}^{\beta}_{i,j,k}(\mathbf{x})=\begin{cases}\mathbf{R}_i^{-1}(\mathbf{x})\mathbf{F}_{i,j}(\mathbf{x}),\\ \mathbf{C}^K_{k,j}(\mathbf{x}),\\ \mathbf{C}^K_{k,j}(\mathbf{x}),\end{cases} \tag{10.49}$$

take different values, according to which the assumptions (10.45), (10.46) or (10.47), respectively, were admitted. Consistent with Eqs. (10.49), the second and the third integrals in Eq. (10.48) vanish in the case of acceptance of assumption (10.45). If the assumption (10.47) is taken, then the second and the third integrals in Eq. (10.48) are calculated over a half-space $R^+_{i,j} = \{\mathbf{y}|\ |\mathbf{y}-\mathbf{x}_j| < |\mathbf{y}-\mathbf{x}_i|\}$, $\mathbf{y} \in R^d$ with a characteristic function $V^+_{i,j}(\mathbf{y})$.

The integrals in Eq. (10.48) can be estimated by the standard numerical methods mentioned in Chapter 4. To keep the calculations down, one introduces the notations for the constant tensors $\boldsymbol{\mathcal{T}}^{\beta}_{i,j}$ and the integral operator $\boldsymbol{\mathcal{T}}^{\sigma}_{i,j,k,l}$ with the kernel $\boldsymbol{\mathcal{T}}^{\sigma}_{i,j,k,l}(\mathbf{x}_k)$ ($\mathbf{x} \in v_i$):

$$\boldsymbol{\mathcal{T}}^{\sigma}_{i,j,k,l}(\mathbf{x}_k) = \boldsymbol{\mathcal{T}}^{1\sigma}_{i,j,k}(\mathbf{x}_k)\delta_{Il} + \boldsymbol{\mathcal{T}}^{2\sigma}_{i,j,k}(\mathbf{x}_k)\delta_{Jl} - \boldsymbol{\mathcal{T}}^{3\sigma}_{i,j,k}(\mathbf{x}_k)\delta_{Il}, \tag{10.50}$$

$$\boldsymbol{\mathcal{T}}^{1\sigma}_{i,j,k}(\mathbf{x}_k) = \langle \mathbf{R}_i^{-1}(\mathbf{x})\mathbf{R}_{i,k}(\mathbf{x})\rangle_{(i)}\widetilde{\varphi}(v_k, \mathbf{x}_k|; v_i, \mathbf{x}_i; v_j, \mathbf{x}_j), \tag{10.51}$$

$$\boldsymbol{\mathcal{T}}^{2\sigma}_{i,j,k}(\mathbf{x}_k) = \mathbf{J}(\mathbf{x}_k)\langle \mathbf{B}^{\sigma}_{i,j,k}(\mathbf{x})\rangle_{(i)}\varphi(v_k, \mathbf{x}_k|; v_i, \mathbf{x}_i, v_j, \mathbf{x}_j), \tag{10.52}$$

$$\boldsymbol{\mathcal{T}}^{3\sigma}_{i,j,k}(\mathbf{x}_k) = \mathbf{J}(\mathbf{x}_k)\langle \mathbf{R}_i^{-1}(\mathbf{x})\mathbf{R}_{i,k}(\mathbf{x})\rangle_{(i)}\varphi(v_k, \mathbf{x}_k|; v_i, \mathbf{x}_i, v_j, \mathbf{x}_j), \tag{10.53}$$

$$\begin{aligned}\boldsymbol{\mathcal{T}}^{\beta}_{i,j} &= \sum_{k=1}^{N}\int \langle \mathbf{R}_i^{-1}(\mathbf{x})\mathbf{F}_{i,k}(\mathbf{x})\rangle_{(i)}\widetilde{\varphi}(v_k, \mathbf{x}_k|; v_i, \mathbf{x}_i; v_j, \mathbf{x}_j)\, d\mathbf{x}_k \\ &+ \sum_{k=1}^{N}\int \mathbf{J}(\mathbf{x}_k)\Big[\langle \mathbf{C}^{\beta}_{i,j,k}(\mathbf{x})\rangle_{(i)} - \langle \mathbf{R}_i^{-1}(\mathbf{x})\mathbf{F}_{i,k}(\mathbf{x})\rangle_{(i)}\Big] \\ &\times \varphi(v_k, \mathbf{x}_k|; v_i, \mathbf{x}_i, v_j, \mathbf{x}_j)\, d\mathbf{x}_k\end{aligned} \tag{10.54}$$

The action of the integral operators $\boldsymbol{\mathcal{T}}^{\sigma}_{i,j,k,l}$ and $\boldsymbol{\mathcal{T}}^{1\sigma}_{i,j,k}$ at the constant tensors $\mathbf{f} \equiv$ const. is reduced to the product over the constant tensor $\overline{\boldsymbol{\mathcal{T}}}^{\sigma}_{i,j,k,l}$ and $\overline{\boldsymbol{\mathcal{T}}}^{1\sigma}_{i,j,k}$, respectively:

$$\begin{aligned}\boldsymbol{\mathcal{T}}^{\sigma}_{i,j,k,l}(\mathbf{f}) &= \overline{\boldsymbol{\mathcal{T}}}^{\sigma}_{i,j,k,l}\mathbf{f}, \quad \overline{\boldsymbol{\mathcal{T}}}^{\sigma}_{i,j,k,l} \equiv \int \boldsymbol{\mathcal{T}}^{\sigma}_{i,j,k,l}(\mathbf{x}_k)\, d\mathbf{x}_k, \\ \boldsymbol{\mathcal{T}}^{1\sigma}_{i,j,k}(\mathbf{f}) &= \overline{\boldsymbol{\mathcal{T}}}^{1\sigma}_{i,j,k}\mathbf{f}, \quad \overline{\boldsymbol{\mathcal{T}}}^{1\sigma}_{i,j,k} \equiv \int \boldsymbol{\mathcal{T}}^{1\sigma}_{i,j,k}(\mathbf{x}_k)\, d\mathbf{x}_k.\end{aligned} \tag{10.55}$$

Recasting of Eq. (10.48) in a standard matrix form leads to

$$\langle \widetilde{\boldsymbol{\sigma}}_{\alpha}\rangle = \langle \overline{\boldsymbol{\sigma}}_{\alpha}\rangle + \boldsymbol{\mathcal{T}}^{\sigma}_{\alpha\gamma}\langle \widetilde{\boldsymbol{\sigma}}_{\gamma}\rangle + \boldsymbol{\mathcal{T}}^{\beta}_{\alpha}, \tag{10.56}$$

where the Greek indexes $\alpha, \gamma = 1, \ldots, N \times N$ are combined with the Latin indexes $i, j = 1, \ldots, N$ and $k, l = 1, \ldots, N$, respectively: $\alpha = i + N(j-1)$, $\gamma = k + N(l-1)$, and $\langle \widetilde{\boldsymbol{\sigma}}_{\alpha}\rangle = \langle \widetilde{\boldsymbol{\sigma}}_{i,j}\rangle$, $\langle \overline{\boldsymbol{\sigma}}_{\alpha}\rangle = \langle \overline{\boldsymbol{\sigma}}_{i}\rangle$, $\boldsymbol{\mathcal{T}}^{\sigma}_{\alpha\gamma} = \boldsymbol{\mathcal{T}}^{\sigma}_{i,j,k,l}$, $\boldsymbol{\mathcal{T}}^{\beta}_{\alpha} = \boldsymbol{\mathcal{T}}^{\beta}_{i,j}$. By virtue of the fact that between the Greek and Latin indexes there is one-to-one correspondence, we will sometimes use these indexes simultaneously.

We rearrange Eq. (10.56) in the spirit of a subtraction technique for more convenience using the iteration method:

$$\langle \widetilde{\boldsymbol{\sigma}}_{\alpha}\rangle = \langle \overline{\boldsymbol{\sigma}}_{\alpha}\rangle + \overline{\boldsymbol{\mathcal{T}}}^{\sigma}_{\alpha\gamma}\langle \widetilde{\boldsymbol{\sigma}}_{\gamma}\rangle + \boldsymbol{\mathcal{T}}^{\sigma}_{\alpha\gamma}(\langle \widetilde{\boldsymbol{\sigma}}_{\gamma}\rangle') + \boldsymbol{\mathcal{T}}^{\beta}_{\alpha}, \tag{10.57}$$

where $\langle \widetilde{\boldsymbol{\sigma}}_{\gamma}\rangle' \equiv \langle \widetilde{\boldsymbol{\sigma}}_{\gamma}\rangle(\mathbf{s}) - \langle \widetilde{\boldsymbol{\sigma}}_{\gamma}\rangle(\mathbf{r})$, and $\mathbf{r} = \mathbf{x}_i - \mathbf{x}_j$ and $\mathbf{s} = \mathbf{x}_i - \mathbf{x}_k$ are the arguments of the left-hand side of Eq. (10.57) and the integration variable of the operator $\boldsymbol{\mathcal{T}}^{\sigma}_{\alpha\gamma}$ on the right-hand side of Eq. (10.57), respectively. In the following, the assumption (10.45) will be used.

The most descriptive is the first-order iteration of the iteration method based on the substitution of the initial approximation (10.27) into the integral in Eq. (10.57) (no sum on i):

$$\langle \widetilde{\boldsymbol{\sigma}}^{(0)}_{i,j} \rangle = \sum_{k=1}^{N} \left[\mathbf{I}\delta_{ik} + \overline{\boldsymbol{\mathcal{T}}}^{1\sigma}_{i,j,k} \right] \langle \overline{\boldsymbol{\sigma}}_k \rangle + \boldsymbol{\mathcal{T}}^{\beta}_{i,j} \tag{10.58}$$

A more accurate approximation of the solution of Eq. (10.57) is based on the self-consistent assumption:

$$\langle \widetilde{\boldsymbol{\sigma}}_{i,k} \rangle (\mathbf{x}_i - \mathbf{x}_k) \equiv \langle \widetilde{\boldsymbol{\sigma}}_{i,k} \rangle (\mathbf{x}_i - \mathbf{x}_j),\ \mathbf{x}_k \in v^{0i}_{0j} \tag{10.59}$$

substituted into the left-hand side integral of Eq. (10.57) instead of the initial approximation $\langle \widetilde{\boldsymbol{\sigma}}_{i,j} \rangle = \langle \overline{\boldsymbol{\sigma}}_i \rangle$ (10.27). Assumption (10.59) is equivalent to neglecting the item $\boldsymbol{\mathcal{T}}^{\sigma}_{\alpha\gamma}(\langle \widetilde{\boldsymbol{\sigma}}_\gamma \rangle')$ on the right hand side of Eq. (10.57), which leads to refined initial approximation of the solution of Eq. (10.57):

$$\langle \widetilde{\boldsymbol{\sigma}}^{(0)}_{\alpha} \rangle = \sum_{k=1}^{N} \widetilde{\mathbf{Y}}_{\alpha k} \langle \overline{\boldsymbol{\sigma}}_k \rangle + \sum_{\gamma=1}^{N\times N} \widetilde{\mathbf{Y}}_{\alpha\gamma} \boldsymbol{\mathcal{T}}^{\beta}_{\gamma}, \tag{10.60}$$

where $\widetilde{\mathbf{Y}}_{\alpha k} = \sum_{l=1}^{N} \widetilde{\mathbf{Y}}_{\alpha k,l}$, $(\widetilde{\mathbf{Y}}^{-1})_{\alpha\gamma} = \delta_{\alpha\gamma}\mathbf{I} - \overline{\boldsymbol{\mathcal{T}}}^{\sigma}_{\alpha\gamma}$, $\gamma = k + (l-1)N$.

Substitution of the refined initial approximation (10.60) into the operator $\boldsymbol{\mathcal{T}}^{\sigma}_{\alpha\gamma}(\langle \widetilde{\boldsymbol{\sigma}}_\gamma \rangle')$ (10.59) generates the first iteration of the solution of Eq. (10.59). The construction of the next iterations should present no problems ($n = 0, 1, \ldots$):

$$\langle \widetilde{\boldsymbol{\sigma}}^{(n+1)}_{\alpha} \rangle = \sum_{k=1}^{N} \widetilde{\mathbf{Y}}_{\alpha k} \langle \overline{\boldsymbol{\sigma}}_k \rangle + \sum_{\gamma=1}^{N\times N} \widetilde{\mathbf{Y}}_{\alpha\gamma} \boldsymbol{\mathcal{T}}^{\beta}_{\gamma} + \sum_{\gamma,\varsigma=1}^{N\times N} \widetilde{\mathbf{Y}}_{\alpha\gamma} \boldsymbol{\mathcal{T}}^{\sigma}_{\gamma\varsigma}(\langle \widetilde{\boldsymbol{\sigma}}^{(n)}_{\varsigma} \rangle') \tag{10.61}$$

The solution (10.61) can be formally represented as

$$\langle \widetilde{\boldsymbol{\sigma}}_{\alpha} \rangle = \sum_{k=1}^{N} \widetilde{\mathbf{Y}}^{\infty}_{\alpha k} \langle \overline{\boldsymbol{\sigma}}_k \rangle + \sum_{\gamma=1}^{N\times N} \widetilde{\mathbf{Y}}^{\infty}_{\alpha\gamma} \boldsymbol{\mathcal{T}}^{\beta}_{\gamma} \tag{10.62}$$

where the tensors $\widetilde{\mathbf{Y}}^{\infty}_{\alpha k}$ and $\widetilde{\mathbf{Y}}^{\infty}_{\alpha\gamma}$ are constructed via the recursion formula (10.61). The iteration sequence (10.61) converges to a unique solution $\langle \widetilde{\boldsymbol{\sigma}}_{\alpha} \rangle$ if the norm of the integral operator $\widetilde{\mathbf{Y}}\boldsymbol{\mathcal{T}}^{\sigma}$ turns out to be small "enough" (less than 1), and the problem is reduced to computation of the integral involved, the density of which is given. In effect the iteration method transforms the integral equation problem into the linear algebra problem in any case. In so doing, the mentioned iteration method based on the initial approximation (10.60) is equivalent to a point total-step iterative method in a linear algebra [1134]. As can be shown [145] in the concrete examples, only a few iterations of Eq. (10.61) are necessary; however, they refine the estimation of effective moduli based on the initial approximation (10.60) no better than 2%. Because of this, due to the desirability of obtaining a tractable solution, we shall usually restrict our consideration to the initial (10.58) and improved initial (10.60) approximations.

10.3.3 Effective Elastic Properties and Stress Concentrator Factors

Thus, the closing assumption (10.45) leads Eq. (10.44) to Eq. (10.48), which has the first-order polynomial dependence on the concentration of inclusions c as $c \to 0$. The Fredholm integral equation (10.48) was solved under some additional assumptions. So, we can replace the closing assumption for the effective field $\langle\widetilde{\boldsymbol{\sigma}}_{i,j}\rangle = \langle\overline{\boldsymbol{\sigma}}_i\rangle$ (10.27) by either approximation (10.58) or (10.62) explicitly depending on the distance $|\mathbf{x}_i - \mathbf{x}_j|$ between the fixed inclusions and the mechanical and geometrical properties of the inclusions v_i, v_j, v_k.

The improved closing assumptions (10.26) and (10.62) allow us to repeat the scheme of estimations of both the stress concentrator factors and effective properties presented in Subsection 10.2.2 for the simplified assumptions (10.26) and (10.27). The straightforward examination of Eqs. (10.30)–(10.36) corroborates their correctness also for the assumptions (10.26) and the approximation (10.62) if in Eqs. (10.30)–(10.36) all the tensors $\mathbf{R}^{\rm pert}_{i,j}(\mathbf{x})$ and $\mathbf{F}^{\rm pert}_i(\mathbf{x})$ are replaced by the tensors ($\mathbf{x} \in v_i$):

$$
\begin{aligned}
\mathbf{R}^{\rm 1pert}_{i,k}(\mathbf{x}) = \sum_{j=1}^{N} \int \Big[&\mathbf{R}_{i,j}(\mathbf{x})\widetilde{\mathbf{Y}}^{\infty}_{i,j,k}\varphi(v_j, \mathbf{x}_j|; v_i, \mathbf{x}_i) \\
&- (1 - V^0_i(\mathbf{x}_j))n^{(j)}\mathbf{R}^{\infty}_{i,j}(\mathbf{x})\Big]\, d\mathbf{x}_j,
\end{aligned} \tag{10.63}
$$

$$
\begin{aligned}
\mathbf{F}^{\rm 1pert}_{i}(\mathbf{x}) = \sum_{j=1}^{N} \int \Big\{&\Big[\mathbf{F}_{i,j}(\mathbf{x}) - \mathbf{R}_{i,j}(\mathbf{x}) \sum_{\gamma=1}^{N\times N} \widetilde{\mathbf{Y}}^{\infty}_{i,j,\gamma}\boldsymbol{\mathcal{T}}^{\beta}_{\gamma}\Big]\varphi(v_j, \mathbf{x}_j|; v_i, \mathbf{x}_i) \\
&- (1 - V^0_i(\mathbf{x}_j))n^{(j)}\mathbf{F}^{\infty}_{i,j}(\mathbf{x})\Big\}\, d\mathbf{x}_j,
\end{aligned} \tag{10.64}
$$

respectively, and the tensor $\mathbf{Y}$ (10.31) is replaced by the tensor $\mathbf{Y}$ with the elements of the inverse tensor:

$$
(\mathbf{Y}^{-1})_{im} = \delta_{im}\mathbf{I} - \mathbf{R}^{-1}_i\mathbf{R}^{\rm 1pert}_{i,m} - n^{(m)}\mathbf{Q}^0_i\mathbf{R}_m \tag{10.65}
$$

which is analogous to the tensor $\mathbf{Y}^{-1}$ (10.38). Substitution of the relations (10.63)–(10.65) into Eqs. (10.34)–(10.36) yields the effective elastic properties. In a similar manner we can obtain an improved estimation of stress concentrator factors. Analogously, the conditional mean value of stresses in the inclusions (10.40) could be transformed into the representation ($\mathbf{x} \in v_i$):

$$
\begin{aligned}
\langle\boldsymbol{\sigma}|; v_j, \mathbf{x}_j\rangle(\mathbf{x}) = \mathbf{C}_i(\mathbf{x}) + [\mathbf{M}^{(i)}_1]^{-1}\mathbf{F}_{i,j}(\mathbf{x}) + \Big\{&\mathbf{B}_i(\mathbf{x})\langle\boldsymbol{\sigma}\rangle + [\mathbf{M}^{(i)}_1]^{-1}\mathbf{R}_{i,j}(\mathbf{x})\Big\} \\
\times \sum_m \mathbf{Y}_{im}\Big[&\langle\boldsymbol{\sigma}\rangle + \langle\mathbf{Q}^0\mathbf{F}^v V\rangle + \mathbf{R}^{-1}_m\mathbf{F}^{\rm 1pert}_m\Big].
\end{aligned} \tag{10.66}
$$

with the new tensor $\mathbf{Y}$ (10.65).

10.3.4 Symmetric Closing Assumption

Both closing assumptions (10.45) and (10.46) are not symmetric with respect to the numbers of inclusions i and j. Such symmetry takes place for closing assumption (10.47) which is more natural than both Eqs. (10.45) and (10.46), at least for one-component aligned inclusions. Substitution of Eq. (10.47) into the generalization of Eq. (10.41) on N-phase reinforcement leads to the equation ($\mathbf{x} \in v_i,\ \mathbf{y} \in v_k$)

$$\langle \widetilde{\boldsymbol{\sigma}}_{i,j} \rangle(\mathbf{x}) = \langle \overline{\boldsymbol{\sigma}}_i \rangle(\mathbf{x}) + \int \boldsymbol{\Gamma}(\mathbf{x}-\mathbf{y}) \langle \boldsymbol{\eta}(\mathbf{y})|; v_i, \mathbf{x}_i \rangle(\mathbf{y}) \widetilde{\varphi}(v_k, \mathbf{x}_k|; v_i, \mathbf{x}_i, v_j, \mathbf{x}_j)\, d\mathbf{y}$$
$$+ \int V^+_{i,j}(\mathbf{y}) \boldsymbol{\Gamma}(\mathbf{x}-\mathbf{y}) [\langle \boldsymbol{\eta}(\mathbf{y})|; v_j, \mathbf{x}_j \rangle(\mathbf{y}) - \langle \boldsymbol{\eta}(\mathbf{y})|; v_i, \mathbf{x}_i \rangle(\mathbf{y})]$$
$$\times\, \varphi(v_k, \mathbf{x}_k|; v_i, \mathbf{x}_i, v_j, \mathbf{x}_j)\, d\mathbf{y}. \qquad (10.67)$$

Acting on Eq. (10.67) by the operator $\boldsymbol{\mathcal{L}}_i$ (4.42) with subsequent multiplication of the obtained equation on the tensor $\mathbf{R}_i^{-1}(\mathbf{x})$ ($\mathbf{x} \in v_i$) leads to

$$\langle \widetilde{\boldsymbol{\sigma}}_{i,j} \rangle = \langle \overline{\boldsymbol{\sigma}}_i \rangle + \int \Big[\langle \mathbf{R}_i^{-1}(\mathbf{x}) \mathbf{R}_{i,k}(\mathbf{x}) \rangle_{(i)} \langle \widetilde{\boldsymbol{\sigma}}_{i,k} \rangle + \langle \mathbf{R}_i^{-1}(\mathbf{x}) \mathbf{F}_{i,k}(\mathbf{x}) \rangle_{(i)} \Big]$$
$$\times\, \widetilde{\varphi}(v_k, \mathbf{x}_k|; v_i, \mathbf{x}_i, v_j, \mathbf{x}_j)\, d\mathbf{x}_k$$
$$+ \int V^+_{i,j}(\mathbf{x}_k) \Big[\langle \mathbf{B}^K_{k,j}(\mathbf{x}) \rangle_{(i)} \langle \widetilde{\boldsymbol{\sigma}}_{j,k} \rangle - \langle \mathbf{R}_i^{-1}(\mathbf{x}) \mathbf{R}_{i,k}(\mathbf{x}) \rangle_{(i)} \langle \widetilde{\boldsymbol{\sigma}}_{i,k} \rangle \Big]$$
$$\times\, \varphi(v_k, \mathbf{x}_k|; v_i, \mathbf{x}_i, v_j, \mathbf{x}_j)\, d\mathbf{x}_k$$
$$+ \int V^+_{i,j}(\mathbf{x}_k) \Big[\langle \mathbf{C}^K_{k,j}(\mathbf{x}) \rangle_{(i)} - \langle \mathbf{R}_i^{-1}(\mathbf{x}) \mathbf{F}_{i,k}(\mathbf{x}) \rangle_{(i)} \Big]$$
$$\times\, \varphi(v_k, \mathbf{x}_k|; v_i, \mathbf{x}_i, v_j, \mathbf{x}_j)\, d\mathbf{x}_k. \qquad (10.68)$$

Each of the integrals in Eq. (10.68) is absolutely convergent. In particular, the asymptotic behaviors at infinity ($|\mathbf{x}_i - \mathbf{x}_k|,\ |\mathbf{x}_j - \mathbf{x}_k| \to \infty$) of the summands of the integrand in the second integral are $\langle \mathbf{B}^K_{k,j}(\mathbf{x}) \rangle_{(i)} \rangle_{(i)} \sim \boldsymbol{\Gamma}(\mathbf{x}_m - \mathbf{x}_k)\mathbf{R}_k + O(|\mathbf{x}_m - \mathbf{x}_k|^{-2d})$ and $\langle \mathbf{R}_i^{-1}(\mathbf{x}) \mathbf{R}_{i,k}(\mathbf{x}) \rangle_{(i)} \sim \boldsymbol{\Gamma}(\mathbf{x}_m - \mathbf{x}_k)\mathbf{R}_k + O(|\mathbf{x}_m - \mathbf{x}_k|^{-2d})$, ($\mathbf{x}_m = (\mathbf{x}_i + \mathbf{x}_j)/2$) as well as $\langle \widetilde{\boldsymbol{\sigma}}_{i,k} \rangle - \langle \widetilde{\boldsymbol{\sigma}}_{j,k} \rangle \sim O(1)$. Therefore, the relation in the square brackets in the second integral (10.68) decays at infinity as $O(|\mathbf{x}_m - \mathbf{x}_k|^{-2d})$ and the second integral in Eq. (10.68) is absolutely convergent.

Analogously to the notations (10.52) and (10.54), we introduce the notations for the constant tensors $\boldsymbol{\mathcal{T}}^{\beta}_{\alpha}$ and the integral operator $\boldsymbol{\mathcal{T}}^{\sigma}_{\alpha\gamma}$ with the kernel $\boldsymbol{\mathcal{T}}^{\sigma}_{\alpha\gamma}(\mathbf{x}_k)$, where $\mathbf{J} = \mathbf{I}V^+_{i,j}(\mathbf{x}_k)$, $\mathbf{B}^{\sigma}_{i,j,k}(\mathbf{x}) = \mathbf{B}^K_{i,j,k}(\mathbf{x})$, $\mathbf{C}^{\beta}_{i,j,k}(\mathbf{x}) = \mathbf{C}^K_{k,j}(\mathbf{x})$ ($\mathbf{x} \in v_i$). Recasting of Eq. (10.68) in a standard matrix form leads to Eq. (10.56) which can be also solved by the iteration method (10.62) with the initial approximation analogous to either (10.58) or (10.60).

Thus, the approximate closing assumption (10.47) can be replaced by the modified representation (10.62) with obvious corrections of Eqs. (10.63)–(10.66) describing the involved formulae for the effective elastic properties and stress concentrator factors.

10.3.5 Closing Assumptions for the Effective Fields $\langle \overline{\boldsymbol{\sigma}}_{i,j,k} \rangle$

The logical scheme of the solution of Eq. (10.44) was the following. At first, the three-point statistical moments $\langle \boldsymbol{\eta}(\mathbf{y})|; v_i, \mathbf{x}_i; v_j, \mathbf{x}_j \rangle(\mathbf{y})$ ($\mathbf{y} \in v_k$) were approximated by one of the two-point statistical moments (10.45)–(10.47). After that, the two-point statistical moment of strain polarization tensor $\langle\boldsymbol{\eta}(\mathbf{y})|; v_i, \mathbf{x}_i\rangle(\mathbf{y})$ $(\mathbf{y} \in v_m)$ was expressed through the two-point statistical moments of the effective field $\langle\overline{\boldsymbol{\sigma}}_{i,m}\rangle$ (10.3) and (10.5) that led to closed equations with respect to the two-point statistical moments of the effective field (see either equation (10.48) or (10.68)). The same goal of obtaining the closed equations could be reached by use of the modified scheme. Namely, we can, at first, express the three-point statistical moments $\langle\boldsymbol{\eta}(\mathbf{y})|; v_i, \mathbf{x}_i; v_j, \mathbf{x}_j\rangle(\mathbf{y})$ $(\mathbf{y} \in v_k)$ through the three-point statistical moment of the effective field $\langle\overline{\boldsymbol{\sigma}}_{i,j,k}\rangle$, and subsequently we can approximate the three-point statistical moment $\langle\overline{\boldsymbol{\sigma}}_{i,j,k}\rangle$ by any of the following two-point statistical moments of the effective fields:

$$\langle\overline{\boldsymbol{\sigma}}_{i,j,k}\rangle = \langle\overline{\boldsymbol{\sigma}}_{i,j}\rangle, \quad \langle\overline{\boldsymbol{\sigma}}_{i,j,k}\rangle = \langle\overline{\boldsymbol{\sigma}}_{i,k}\rangle, \quad \langle\overline{\boldsymbol{\sigma}}_{i,j,k}\rangle = \langle\overline{\boldsymbol{\sigma}}_{i,m}\rangle, \tag{10.69}$$

where $m = \arg\min_{l=i,j} |\mathbf{x}_k - \mathbf{x}_l|$.The closing assumptions (10.69) are analogous to the closing ones (10.45)–(10.47). The conditions (10.69_1) and (10.69_2) are nonsymmetric, while the assumption (10.69_3) is symmetric.

Thus, substituting of the representations (10.14) into the general Eq. (10.44) leads to exclusion of the strain polarization tensor from Eq. (10.44):

$$\begin{aligned}
\langle\widetilde{\boldsymbol{\sigma}}_{i,j}\rangle &= \langle\overline{\boldsymbol{\sigma}}_i\rangle + \int \Big[\langle\mathbf{R}_i^{-1}(\mathbf{x})\mathbf{R}_{i,k}(\mathbf{x})\rangle_{(i)}\langle\widetilde{\boldsymbol{\sigma}}_{i,k}\rangle + \langle\mathbf{R}_i^{-1}(\mathbf{x})\mathbf{F}_{i,k}(\mathbf{x})\rangle_{(i)}\Big] \\
&\quad\times \widetilde{\varphi}(v_k, \mathbf{x}_k|; v_i, \mathbf{x}_i, v_j, \mathbf{x}_j)\, d\mathbf{x}_k \\
&\quad+ \int \Big[\langle\mathbf{R}_i^{-1}(\mathbf{x})\mathbf{R}_{i,j,k}(\mathbf{x})\rangle_{(i)}\langle\widetilde{\boldsymbol{\sigma}}_{i,j,k}\rangle - \langle\mathbf{R}_i^{-1}(\mathbf{x})\mathbf{R}_{i,k}(\mathbf{x})\rangle_{(i)}\langle\widetilde{\boldsymbol{\sigma}}_{i,k}\rangle\Big] \\
&\quad\times \varphi(v_k, \mathbf{x}_k|; v_i, \mathbf{x}_i, v_j, \mathbf{x}_j)\, d\mathbf{x}_k \\
&\quad+ \int \Big[\langle\mathbf{R}_i^{-1}(\mathbf{x})\mathbf{F}_{i,j,k}(\mathbf{x})\rangle_{(i)} - \langle\mathbf{R}_i^{-1}(\mathbf{x})\mathbf{F}_{i,k}(\mathbf{x})\rangle_{(i)}\Big] \\
&\quad\times \varphi(v_k, \mathbf{x}_k|; v_i, \mathbf{x}_i, v_j, \mathbf{x}_j)\, d\mathbf{x}_k.
\end{aligned} \tag{10.70}$$

Utilizing the closing assumptions (10.69) in Eq. (10.70) yields the closed equation:

$$\begin{aligned}
\langle\widetilde{\boldsymbol{\sigma}}_{i,j}\rangle &= \langle\overline{\boldsymbol{\sigma}}_i\rangle + \int \Big[\langle\mathbf{R}_i^{-1}(\mathbf{x})\mathbf{R}_{i,k}(\mathbf{x})\rangle_{(i)}\langle\widetilde{\boldsymbol{\sigma}}_{i,k}\rangle + \langle\mathbf{R}_i^{-1}(\mathbf{x})\mathbf{F}_{i,k}(\mathbf{x})\rangle_{(i)}\Big] \\
&\quad\times \widetilde{\varphi}(v_k, \mathbf{x}_k|; v_i, \mathbf{x}_i, v_j, \mathbf{x}_j)\, d\mathbf{x}_k \\
&\quad+ \int \Big[\langle\mathbf{R}_i^{-1}(\mathbf{x})\mathbf{R}_{i,j,k}(\mathbf{x})\rangle_{(i)}\langle\widetilde{\boldsymbol{\sigma}}_{i,l}\rangle - \langle\mathbf{R}_i^{-1}(\mathbf{x})\mathbf{R}_{i,k}(\mathbf{x})\rangle_{(i)}\langle\widetilde{\boldsymbol{\sigma}}_{i,k}\rangle\Big] \\
&\quad\times \varphi(v_k, \mathbf{x}_k|; v_i, \mathbf{x}_i, v_j, \mathbf{x}_j)\, d\mathbf{x}_k \\
&\quad+ \int \Big[\langle\mathbf{R}_i^{-1}(\mathbf{x})\mathbf{F}_{i,j,k}(\mathbf{x})\rangle_{(i)} - \langle\mathbf{R}_i^{-1}(\mathbf{x})\mathbf{F}_{i,k}(\mathbf{x})\rangle_{(i)}\Big] \\
&\quad\times \varphi(v_k, \mathbf{x}_k|; v_i, \mathbf{x}_i, v_j, \mathbf{x}_j)\, d\mathbf{x}_k,
\end{aligned} \tag{10.71}$$

where either $l = i,\ j$, or m in the case of using of closing assumptions (10.69_1), (10.69_2) or (10.69_3), respectively.

Analogously to the notations (10.49) and (10.55), we introduce the notations for the constant tensors $\boldsymbol{\mathcal{T}}_{\alpha}^{\beta}$ and the integral operator $\boldsymbol{\mathcal{T}}_{\alpha\gamma}^{2\sigma}$ with the kernel $\boldsymbol{\mathcal{T}}_{\alpha\gamma}^{2\sigma}(\mathbf{x}_k)$ $(\mathbf{x} \in v_i)$:

$$\boldsymbol{\mathcal{T}}^{2\sigma}_{i,j,k,l}(\mathbf{x}_k) = \boldsymbol{\mathcal{T}}^{1\sigma}_{i,j,k}(\mathbf{x}_k)\delta_{Il} + \boldsymbol{\mathcal{T}}^{B\sigma}_{i,j,k}(\mathbf{x}_k)\delta_{Pl} - \boldsymbol{\mathcal{T}}^{R\sigma}_{i,j,k}(\mathbf{x}_k)\delta_{Il}, \tag{10.72}$$

$$\boldsymbol{\mathcal{T}}^{B\sigma}_{i,j,k}(\mathbf{x}_k) = \langle \mathbf{R}_i^{-1}(\mathbf{x})\mathbf{R}_{i,j,k}(\mathbf{x})\rangle_{(i)}\varphi(v_k, \mathbf{x}_k|; v_i, \mathbf{x}_i, v_j, \mathbf{x}_j), \tag{10.73}$$

$$\boldsymbol{\mathcal{T}}^{R\sigma}_{i,j,k}(\mathbf{x}_k) = \langle \mathbf{R}_i^{-1}(\mathbf{x})\mathbf{R}_{i,k}(\mathbf{x})\rangle_{(i)}\varphi(v_k, \mathbf{x}_k|; v_i, \mathbf{x}_i, v_j, \mathbf{x}_j), \tag{10.74}$$

$$\begin{aligned}\boldsymbol{\mathcal{T}}^{2\beta}_{i,j} = &\sum_{k=1}^{N}\int \langle \mathbf{R}_i^{-1}(\mathbf{x})\mathbf{F}_{i,k}(\mathbf{x})\rangle_{(i)}\widetilde{\varphi}(v_k, \mathbf{x}_k|; v_i, \mathbf{x}_i; v_j, \mathbf{x}_j)\, d\mathbf{x}_k \\ &+ \sum_{k=1}^{N}\int \langle \mathbf{R}_i^{-1}(\mathbf{x})[\mathbf{F}_{i,j,k}(\mathbf{x}) - \mathbf{F}_{i,k}(\mathbf{x})]\rangle_{(i)} \\ &\times \varphi(v_k, \mathbf{x}_k|; v_i, \mathbf{x}_i, v_j, \mathbf{x}_j)\, d\mathbf{x}_k, \end{aligned} \tag{10.75}$$

where the tensor $\boldsymbol{\mathcal{T}}^{1\sigma}_{i,j,k}(\mathbf{x}_k)$ was defined via Eq. (10.51), and the index P is chosen according to sampling of either closing assumption (10.69_1), (10.69_2), or (10.69_3): $P = I$, $P = J$ or $P = M$, respectively, where $M = \arg\min_{l=i,j} |\mathbf{x}_k - \mathbf{x}_l|$. The introduced tensors $\boldsymbol{\mathcal{T}}^{2\sigma}_{i,j,k,l}(\mathbf{x}_k)$ (10.72) and $\boldsymbol{\mathcal{T}}^{2\beta}_{i,j}$ (10.75) will be exploited in Eq. (10.56) (instead of $\boldsymbol{\mathcal{T}}^{\sigma}_{i,j,k,l}(\mathbf{x}_k)$, $\boldsymbol{\mathcal{T}}^{\beta}_{i,j}$, respectively) for the estimation of the effective field $\langle \widetilde{\boldsymbol{\sigma}}_\alpha\rangle$ by the use of Eq. (10.62). Solution (10.62) in turn will be substituted into Eqs. (10.63)–(10.66) for the subsequent evaluation of the effective elastic properties (10.34)–(10.36) and stress concentrator (10.33).

It should be mentioned that after some simplifications, the representations (10.72) and (10.75) are reduced to the involved relations preliminarily obtained for the closing assumptions for the strain polarization tensors (10.45)–(10.47). For instance, the acceptance of both assumption (10.69_1) and the equality $\mathbf{R}_{i,j,k}(\mathbf{x}) = \mathbf{R}_{i,k}(\mathbf{x})$ leads to the elimination of the remainder $\boldsymbol{\mathcal{T}}^{B\sigma}_{i,j,k}(\mathbf{x}_k)\delta_{Pl} - \boldsymbol{\mathcal{T}}^{R\sigma}_{i,j,k}(\mathbf{x}_k)\delta_{Il} \equiv \mathbf{0}$ as well as to the equivalence of the representations (10.50), (10.55) and (10.72), (10.75), respectively. The effective elastic properties (10.34)–(10.36) and stress concentrator factors (10.33) obtained in the framework of the mentioned assumptions are also equivalent. Furthermore, the equivalence of the results obtained at the assumptions (10.47) and (10.69_3) takes place in the approximative representation of the tensor $\mathbf{R}_{i,j,k}(\mathbf{x})$ describing the triple interaction of the inclusions $\mathbf{R}_i^{-1}(\mathbf{x})\mathbf{R}_{i,j,k}(\mathbf{x}) = \mathbf{B}^K_{k,j}(\mathbf{x})$ $(\mathbf{x} \in v_i)$. The last equality is based on neglecting of the influence of the inclusion v_i on the stress distribution in the interacting inclusions v_j and v_k.

10.4 Abandonment from the Approximative Hypothesis (10.26)

In Section 10.3 we refined the closing assumption (10.27) by the improved estimation (10.62). We are now coming to the critical analysis of the approximative assumption (10.26) which is exact only for the homogeneous fields $\overline{\boldsymbol{\sigma}}(\mathbf{x})$ $(\mathbf{x} \in v_i)$. It should be mentioned, however, that the assumption (10.26) was used after substitution of the accurate solutions $\mathbf{R}_{i,j}(\mathbf{x})$, $\mathbf{F}_{i,j}(\mathbf{x})$ for the binary interacting inclusions into Eq. (10.19) rather than before as was assumed in the one-particle approximation of the MEFM (for details see Chapter 8): $\overline{\boldsymbol{\sigma}}(\mathbf{x}) \equiv \text{const}$ $(\mathbf{x} \in v_i)$.

Because of this, hypothesis (10.26), applying only in the average sense, is more accurate than effective field hypothesis **H1** $\overline{\boldsymbol{\sigma}}(\mathbf{x}) \equiv \text{const}$ ($\mathbf{x} \in v_i$) (10.25). Nevertheless, exploiting the accurate numerical solutions for the binary interacting inclusions described by the tensors $\mathbf{R}_{i,j}(\mathbf{x})$, $\mathbf{F}_{i,j}$, $\mathbf{R}^{\infty}_{i,j}(\mathbf{x})$, $\mathbf{F}^{\infty}_{i,j}$ allows us to abandon the use of the approximative assumption (10.26).

Indeed, substituting the refined estimation of the effective field $\widetilde{\boldsymbol{\sigma}}_{i,j}$ (10.62) into Eqs. (10.20) and (10.21) yields

$$\langle \boldsymbol{\eta} \rangle = \big[\mathbf{I} - \langle \mathbf{R}^v \mathbf{Q}^0 V \rangle\big]^{-1} \Big\{ \langle \mathbf{R}^v V \rangle \langle \boldsymbol{\sigma} \rangle + \langle \mathbf{F}^v V \rangle + \sum_{k,j} n^{(k)} \Big[\mathbf{R}^{1\text{pert}}_{k,j} \langle \overline{\boldsymbol{\sigma}}_j \rangle + \delta_{kj} \mathbf{F}^{1\text{pert}}_k \Big] \Big\}, \tag{10.76}$$

$$\begin{aligned} \bar{v}_i \langle \boldsymbol{\eta} \rangle_i(\mathbf{x}) &= \mathbf{R}_i(\mathbf{x}) \langle \boldsymbol{\sigma} \rangle + \mathbf{F}_i(\mathbf{x}) + \mathbf{R}_i(\mathbf{x}) \mathbf{Q}^0_i \big[\mathbf{I} - \langle \mathbf{R}^v \mathbf{Q}^0 V \rangle\big]^{-1} \\ &\times \Big\{ \langle \mathbf{R}^v V \rangle \langle \boldsymbol{\sigma} \rangle + \langle \mathbf{F}^v V \rangle + \sum_{k,j} n^{(k)} \Big[\mathbf{R}^{1\text{pert}}_{k,j} \langle \overline{\boldsymbol{\sigma}}_j \rangle + \delta_{kj} \mathbf{F}^{1\text{pert}}_k \Big] \Big\} \\ &+ \sum_j \Big[\mathbf{R}^{1\text{pert}}_{i,j}(\mathbf{x}) \langle \overline{\boldsymbol{\sigma}}_j \rangle + \delta_{ij} \mathbf{F}^{1\text{pert}}_j(\mathbf{x}) \Big], \end{aligned} \tag{10.77}$$

respectively. The system of equations (10.77) contains N equations with $2N$ unknowns $\langle \boldsymbol{\eta} \rangle_i(\mathbf{x})$ and $\langle \overline{\boldsymbol{\sigma}}_j \rangle$. In Section 10.5 the closing of the system analogous to Eq. (10.77) will be affected by the supplementary N equations (10.25). However, the necessary additional equations can be carried out immediately from Eq. (8.5) in a similar manner to how Eq. (10.19) was obtained. With this in mind, we will perform the statistical average of Eq. (8.5) over the fixed inclusion v_i:

$$\langle \overline{\boldsymbol{\sigma}}_i \rangle = \langle \boldsymbol{\sigma} \rangle + \mathbf{Q}^0_i \langle \boldsymbol{\eta} \rangle + \int \mathbf{T}_i(\mathbf{x}_i - \mathbf{y}) [\boldsymbol{\eta}(\mathbf{y}) V(\mathbf{y} |; v_i, \mathbf{x}_i) - (1 - V^0_i(\mathbf{y})) \langle \boldsymbol{\eta} \rangle] \, d\mathbf{y}. \tag{10.78}$$

Substituting into the averaged Eq. (10.78) the solution for the binary interacting inclusions (10.3), (10.11) accompanied by both the approximation of the effective field $\langle \widetilde{\boldsymbol{\sigma}}_{i,j} \rangle$ (10.62) and the representation for the average strain polarization tensor (10.76) leads to the closed equation

$$\begin{aligned} \langle \overline{\boldsymbol{\sigma}}_i \rangle &= \langle \boldsymbol{\sigma} \rangle + \mathbf{Q}^0_i \big[\mathbf{I} - \langle \mathbf{R}^v \mathbf{Q}^0 V \rangle\big]^{-1} \Big\{ \langle \mathbf{R}^v V \rangle \langle \boldsymbol{\sigma} \rangle + \langle \mathbf{F}^v V \rangle \\ &+ \sum_{k,j} n^{(k)} \Big[\mathbf{R}^{1\text{pert}}_{k,j} \langle \overline{\boldsymbol{\sigma}}_j \rangle + \delta_{kj} \mathbf{F}^{1\text{pert}}_k \Big] \Big\} + \sum_k \Big[\mathbf{R}^{2\text{pert}}_{i,k} \langle \overline{\boldsymbol{\sigma}}_k \rangle + \delta_{ik} \mathbf{F}^{2\text{pert}}_i \Big], \end{aligned} \tag{10.79}$$

where one introduces the constant tensors ($\mathbf{y} \in v_j$):

$$\begin{aligned} \mathbf{R}^{2\text{pert}}_{i,k} &= \sum_{j=1}^{N} \int \Big\{ \mathbf{T}_i(\mathbf{x}_i - \mathbf{y}) \big[\mathbf{R}_j(\mathbf{y}) + \mathbf{R}_{j,i}(\mathbf{y}) \big] \widetilde{\mathbf{Y}}^{\infty}_{j,i,k} \varphi(v_j, \mathbf{x}_j |; v_i, \mathbf{x}_i) \\ &- (1 - V^0_i(\mathbf{y})) n^{(j)} \langle \mathbf{T}^{\sigma\infty}_j(\mathbf{x}) \rangle_{(i)} \Big\} \, d\mathbf{y}, \end{aligned} \tag{10.80}$$

$$\begin{aligned} \mathbf{F}^{2\text{pert}}_i &= \sum_{j=1}^{N} \int \Big\{ \mathbf{T}_i(\mathbf{x}_i - \mathbf{y}) \Big[\mathbf{F}_j(\mathbf{y}) + \mathbf{F}_{j,i}(\mathbf{y}) - \mathbf{R}_{j,i}(\mathbf{y}) \sum_{\gamma=1}^{N \times N} \widetilde{\mathbf{Y}}^{\infty}_{j,i,\gamma} \boldsymbol{\mathcal{T}}^{\beta}_{\gamma} \Big] \\ &\times \varphi(v_j, \mathbf{x}_j |; v_i, \mathbf{x}_i) - (1 - V^0_i(\mathbf{y})) n^{(j)} \langle \mathbf{T}^{\beta\infty}_j(\mathbf{x}) \rangle_{(i)} \Big\} \, d\mathbf{y} \end{aligned} \tag{10.81}$$

constructed from already known tensors described in Section 10.1.

The solution of Eq. (10.79) is

$$\langle\overline{\sigma}_i\rangle = \sum_m \mathbf{Y}_{im}\Big\{\big[(\mathbf{I} - \langle\mathbf{R}^v\mathbf{Q}^0 V\rangle)(\mathbf{Q}^0_m)^{-1} + \langle\mathbf{R}^v V\rangle\big]\langle\boldsymbol{\sigma}\rangle + \langle\mathbf{F}^v V\rangle + (\mathbf{I} - \langle\mathbf{R}^v\mathbf{Q}^0 V\rangle)(\mathbf{Q}^0_m)^{-1}\mathbf{F}^{2\mathrm{pert}}_m + \sum_k n^{(k)}\mathbf{F}^{1\mathrm{pert}}_k\Big\}, \quad (10.82)$$

$$(\mathbf{Y}^{-1})_{im} = (\mathbf{I} - \langle\mathbf{R}^v\mathbf{Q}^0 V\rangle)(\mathbf{Q}^0_m)^{-1}\big[\delta_{im}\mathbf{I} - \mathbf{R}^{2\mathrm{pert}}_{i,m}\big] - \sum_k n^{(k)}\mathbf{R}^{1\mathrm{pert}}_{k,m}. \quad (10.83)$$

10.5 Some Particular Cases

10.5.1 Identical Aligned Inclusions

Let us consider simplifications of the equations obtained in case of the closing assumptions (10.25) and (10.27) applied to the involved equations for the composites with the identical inclusions ($N = 1$) when the matrix $\mathbf{Y}_{ij}$ ($i, j = 1, \ldots, N$) is reduced to the tensor, and Eq. (10.30) for the average effective field can be recast in the form

$$\langle\overline{\sigma}_1\rangle = \mathbf{Y}_{11}\big[\langle\boldsymbol{\sigma}\rangle + n^{(1)}\mathbf{Q}^0_1\mathbf{F}_1 + \mathbf{R}_1^{-1}\mathbf{F}^{\mathrm{pert}}_1\big], \quad (10.84)$$

$$(\mathbf{Y}^{-1})_{11} = \mathbf{I} - n^{(1)}\mathbf{Q}^0_1\mathbf{R}_1 - \mathbf{R}_1^{-1}\mathbf{R}^{\mathrm{pert}}_{1,1}. \quad (10.85)$$

Substitution of Eqs. (10.84) into the right-hand side of Eq. (10.19) enables one to get the representations for the inhomogeneous statistical average of both the strain polarization tensor and stresses:

$$\begin{aligned}\bar{v}_i\langle\boldsymbol{\eta}\rangle_1(\mathbf{x}) &= \mathbf{R}_1(\mathbf{x})\langle\boldsymbol{\sigma}\rangle + \mathbf{F}_1(\mathbf{x}) + \mathbf{F}^{\mathrm{pert}}_1(\mathbf{x}) + n^{(1)}\mathbf{R}_1(\mathbf{x})\mathbf{Q}^0_1\mathbf{F}_1\\ &+ \Big[n^{(1)}\mathbf{R}_1(\mathbf{x})\mathbf{Q}^0_1\mathbf{R}_1 + \mathbf{R}^{\mathrm{pert}}_{1,1}(\mathbf{x})\Big]\\ &\times \mathbf{Y}_{11}\Big[\langle\boldsymbol{\sigma}\rangle + n^{(1)}\mathbf{Q}^0_1\mathbf{F}_1 + \mathbf{R}_1^{-1}\mathbf{F}^{\mathrm{pert}}_1\Big], \end{aligned} \quad (10.86)$$

$$\begin{aligned}\langle\boldsymbol{\sigma}\rangle_1(\mathbf{x}) &= \mathbf{B}_1(\mathbf{x})\langle\boldsymbol{\sigma}\rangle + \mathbf{C}_1(\mathbf{x}) + [\bar{v}_i\mathbf{M}^{(1)}_1(\mathbf{x})]^{-1}\mathbf{F}^{\mathrm{pert}}_1(\mathbf{x}) + n^{(1)}\mathbf{B}_1(\mathbf{x})\mathbf{Q}^0_1\mathbf{F}_1\\ &+ \Big\{n^{(1)}\mathbf{B}_1(\mathbf{x})\mathbf{Q}^0_1\mathbf{R}_1 + [\bar{v}_i\mathbf{M}^{(1)}_1(\mathbf{x})]^{-1}\mathbf{R}^{\mathrm{pert}}_{1,1}(\mathbf{x})\Big\}\\ &\times \mathbf{Y}_{11}\Big[\langle\boldsymbol{\sigma}\rangle + n^{(1)}\mathbf{Q}^0\mathbf{F}_1 + \mathbf{R}_1^{-1}\mathbf{F}^{\mathrm{pert}}_1\Big], \end{aligned} \quad (10.87)$$

which can be recast after the substitution of Eq. (10.85) in the following form:

$$\begin{aligned}\langle\boldsymbol{\eta}\rangle_1(\mathbf{x}) &= \mathbf{F}_1(\mathbf{x}) + \mathbf{F}^{\mathrm{pert}}_1(\mathbf{x}) - \mathbf{R}_1(\mathbf{x})\mathbf{R}_1^{-1}\mathbf{F}^{\mathrm{pert}}_1\\ &+ \Big[\mathbf{R}_1(\mathbf{x}) + \mathbf{R}^{\mathrm{pert}}_{1,1}(\mathbf{x}) - \mathbf{R}_1(\mathbf{x})\mathbf{R}_1^{-1}\mathbf{R}^{\mathrm{pert}}_{1,1}\Big]\\ &\times \mathbf{Y}_{11}\Big[\langle\boldsymbol{\sigma}\rangle + n^{(1)}\mathbf{Q}^0_1\mathbf{F} + \mathbf{R}_1^{-1}\mathbf{F}^{\mathrm{pert}}_1\Big], \end{aligned} \quad (10.88)$$

$$\begin{aligned}\langle\boldsymbol{\sigma}\rangle_1(\mathbf{x}) &= \mathbf{C}_1(\mathbf{x}) + [\bar{v}_i\mathbf{M}^{(1)}_1(\mathbf{x})]^{-1}\mathbf{F}^{\mathrm{pert}}_1(\mathbf{x}) - \mathbf{B}_1(\mathbf{x})\mathbf{R}_1^{-1}\mathbf{F}^{\mathrm{pert}}_1\\ &+ \mathbf{B}_1(\mathbf{x})\Big[\mathbf{I} + \mathbf{R}_1^{-1}(\mathbf{x})\mathbf{R}^{\mathrm{pert}}_{1,1}(\mathbf{x}) - \mathbf{R}_1^{-1}\mathbf{R}^{\mathrm{pert}}_{1,1}\Big]\\ &\times \mathbf{Y}_{11}\Big[\langle\boldsymbol{\sigma}\rangle + n^{(1)}\mathbf{Q}^0_1\mathbf{F}_1 + \mathbf{R}_1^{-1}\mathbf{F}^{\mathrm{pert}}_1\Big]. \end{aligned} \quad (10.89)$$

The volume average of Eqs. (10.88) and (10.89) over the representative inclusion v_1 leading to the representations:

$$\langle \boldsymbol{\sigma} \rangle_1 = \mathbf{C}_1 + \mathbf{B}_1 \mathbf{Y}_{11} \Big[\langle \boldsymbol{\sigma} \rangle + n^{(1)} \mathbf{Q}^0 \mathbf{F}_1 + \mathbf{R}_1^{-1} \mathbf{F}_1^{\text{pert}} \Big] \tag{10.90}$$

are substituted in turn into Eq. (6.20), (6.48) and (6.77) for the estimation of effective properties:

$$\mathbf{M}^* = \mathbf{M}^{(0)} + n^{(1)} \mathbf{R}_1 \mathbf{Y}_{11}, \tag{10.91}$$

$$\boldsymbol{\beta}^* = \boldsymbol{\beta}^{(0)} + n^{(1)} \mathbf{F}_1 + n^{(1)} \mathbf{R}_1 \mathbf{Y}_{11} [n^{(1)} \mathbf{Q}_1^0 \mathbf{F}_1 + \mathbf{R}_1^{-1} \mathbf{F}_1^{\text{pert}}], \tag{10.92}$$

$$\begin{aligned} -2W^* &= \langle \boldsymbol{\beta}_1(\mathbf{x}) \mathbf{C}_1(\mathbf{x}) \rangle + \langle \boldsymbol{\beta}_1(\mathbf{x}) \mathbf{B}_1(\mathbf{x}) \rangle \mathbf{Q}_1^0 \mathbf{F}_1 \\ &+ \langle \boldsymbol{\beta}_1(\mathbf{x}) [\mathbf{M}_1^{(1)}(\mathbf{x})]^{-1} \mathbf{F}_1^{\text{pert}}(\mathbf{x}) V(\mathbf{x}) \rangle \\ &+ \Big[\langle \boldsymbol{\beta}_1(\mathbf{x}) \mathbf{B}(\mathbf{x}) V(\mathbf{x}) \rangle \mathbf{Q}_1^0 \mathbf{R}_1^v + \langle \boldsymbol{\beta}_1(\mathbf{x}) [\mathbf{M}_1^{(1)}(\mathbf{x})]^{-1} \mathbf{R}_{1,1}^{\text{pert}}(\mathbf{x}) V(\mathbf{x}) \rangle \Big] \\ &\times \mathbf{Y}_{11} [n^{(1)} \mathbf{Q}_1^0 \mathbf{F}_1 + \mathbf{R}_1^{-1} \mathbf{F}_1^{\text{pert}}]. \end{aligned} \tag{10.93}$$

Combining Eq. (10.89) with (10.90) allows us to extract the explicit dependence of the statistical average of stresses in the inclusions on their volume averages:

$$\begin{aligned} \langle \boldsymbol{\sigma} \rangle_1(\mathbf{x}) &= \mathbf{C}_1(\mathbf{x}) + [\bar{v}_i \mathbf{M}_1^{(1)}(\mathbf{x})]^{-1} \mathbf{F}_1^{\text{pert}}(\mathbf{x}) - \mathbf{B}_1(\mathbf{x}) \mathbf{R}_1^{-1} \mathbf{F}_1^{\text{pert}} \\ &+ \mathbf{B}_1(\mathbf{x}) \Big[\mathbf{I} + \mathbf{R}_1^{-1}(\mathbf{x}) \mathbf{R}_{1,1}^{\text{pert}}(\mathbf{x}) - \mathbf{R}_1^{-1} \mathbf{R}_{1,1}^{\text{pert}} \Big] \mathbf{B}_1^{-1} (\langle \boldsymbol{\sigma} \rangle_1 - \mathbf{C}_1). \end{aligned} \tag{10.94}$$

It is interesting that for homogeneous ellipsoidal inclusions (3.187) when the $\mathbf{M}_1(\mathbf{x})$, $\mathbf{B}_1(\mathbf{x})$, $\mathbf{C}_1(\mathbf{x})$, $\mathbf{R}_1(\mathbf{x}) \equiv$ const. ($\mathbf{x} \in v^{(1)}$), the inhomogeneity of statistical average of stresses in the inclusions also takes place and is defined only by the binary interaction of inclusions described by the inhomogeneous tensors $\mathbf{R}_{1,1}^{\text{pert}}(\mathbf{x}), \mathbf{F}_1^{\text{pert}}(\mathbf{x})$.

$$\begin{aligned} \langle \boldsymbol{\sigma} \rangle_1(\mathbf{x}) - \langle \boldsymbol{\sigma} \rangle_1 &= [\bar{v}_1 \mathbf{M}_1^{(1)}]^{-1} \Big[\mathbf{F}_1^{\text{pert}}(\mathbf{x}) - \mathbf{F}_1^{\text{pert}} \Big] \\ &+ [\bar{v}_1 \mathbf{M}_1^{(1)}]^{-1} \Big[\mathbf{R}_{1,1}^{\text{pert}}(\mathbf{x}) - \mathbf{R}_{1,1}^{\text{pert}} \Big] \mathbf{B}_1^{-1} (\langle \boldsymbol{\sigma} \rangle_1 - \mathbf{C}_1). \end{aligned} \tag{10.95}$$

In so doing, for the pure elastic problem ($\boldsymbol{\beta}_1 \equiv \mathbf{0}$) with the well-stirred approximation of the RDF (5.70_2), the relative variation of the statistical average of stresses in the inclusions $\langle \boldsymbol{\sigma} \rangle_1(\mathbf{x})$ with respect the volume average $\langle \boldsymbol{\sigma} \rangle_1$ is a linear function of the volume concentration of the inclusions $c^{(1)}$.

It should be mentioned that a standard so-called computational micromechanics usually limited by estimation of effective elastic moduli is based on periodization of random media [954] reducing to the modeling of periodic boundary conditions at the unit cell containing just a few tens n of inclusions with the centers $\mathbf{x}_{\alpha\beta}$ in *beta's* Monte Carlo realization ($\alpha = 1, \ldots n$, $\beta = 1, \ldots N$) (see, e.g., [93], [412], [414], [537], [631], [698], [973], [982], [1237]). These approaches are reduced to evaluation of the volume stress average $\langle \boldsymbol{\sigma} \rangle_1$ through the estimation of average stresses over both n inclusions in a unite cell and N random generations of sets of n inclusions by the Monte-Carlo simulations: $\langle \boldsymbol{\sigma} \rangle_1 \approx (nN)^{-1} \sum_\alpha \sum_\beta \langle \boldsymbol{\sigma}(\mathbf{x}_{\alpha\beta}, \mathbf{x}) \rangle_{(i_{\alpha\beta})}$. where $\boldsymbol{\sigma}(\mathbf{x}_{\alpha\beta}, \mathbf{x})$ is a stress distribution in the local coordinate system in the inclusion $\mathbf{x} \in v_{i_{\alpha\beta}}$ with the center $\mathbf{x}_{\alpha\beta}$. Obviously, an estimation of the desired statistical stress averages [similar to Eq. (10.95)] can be obtained by more simple formula $\langle \boldsymbol{\sigma} \rangle_1(\mathbf{x}) \approx (nN)^{-1} \sum_\alpha \sum_\beta \boldsymbol{\sigma}(\mathbf{x}_{\alpha\beta}, \mathbf{x})$ (see, e.g., [101]), which was not, nevertheless, investigated in detail (although the histograms of stress distributions in the matrix were found [1113]).

The next simplification of Eqs. (10.91)–(10.93) is connected with the assumption $c^{(1)} \to 0$. Then, restricting the Taylor extension over the powers of c of the tensors $\mathbf{Y}_{11}$, $\langle \boldsymbol{\eta} \rangle_1$, $\langle \boldsymbol{\sigma} \rangle_1(\mathbf{x})$ by the first-order $O(c)$ approximation leads to $O(c^2)$ estimations of effective properties:

$$\mathbf{M}^* = \mathbf{M}^{(0)} + n^{(1)}\mathbf{R}_1 + (n^{(1)})^2 \mathbf{R}_1 \mathbf{Q}_1^0 \mathbf{R}_1 + n^{(1)} \mathbf{R}_{1,1}^{\text{pert}}, \tag{10.96}$$

$$\boldsymbol{\beta}^* = \boldsymbol{\beta}^{(0)} + n^{(1)}\mathbf{F}_1 + (n^{(1)})^2 \mathbf{R}_1 \mathbf{Q}_1^0 \mathbf{F}_1 + n^{(1)} \mathbf{F}_1^{\text{pert}}, \tag{10.97}$$

$$\begin{aligned} -2W^* &= \langle \boldsymbol{\beta}_1(\mathbf{x}) \mathbf{C}_1(\mathbf{x}) \rangle + \langle \boldsymbol{\beta}_1(\mathbf{x}) [\mathbf{M}_1^{-1}]^{-1}(\mathbf{x}) \mathbf{F}_1^{\text{pert}}(\mathbf{x}) V \rangle \\ &+ \Big[\langle \boldsymbol{\beta}_1(\mathbf{x}) \mathbf{B}(\mathbf{x}) V(\mathbf{x}) \rangle \mathbf{Q}_1^0 \mathbf{R}_1^v + \langle \boldsymbol{\beta}_1(\mathbf{x}) [\mathbf{M}_1^{(1)}(\mathbf{x})]^{-1} \mathbf{R}_{1,1}^{\text{pert}}(\mathbf{x}) V(\mathbf{x}) \rangle \Big] \\ &\times (n^{(1)} \mathbf{Q}_1^0 \mathbf{F}_1 + \mathbf{R}_1^{-1} \mathbf{F}_1^{\text{pert}}) + \langle \boldsymbol{\beta}_1(\mathbf{x}) \mathbf{B}_1(\mathbf{x}) \rangle \mathbf{Q}_1^0 \mathbf{F}. \end{aligned} \tag{10.98}$$

It should be mentioned that a natural reciprocity between stiffness and compliance of effective properties $\mathbf{L}^* = (\mathbf{M}^*)^{-1}$ is not valid for the representation (10.96) and, because of this, we estimate the effective stiffness $\mathbf{L}^*$ by the use of Eq. (10.96) recast in the terms of stiffnesses [the same can be performed with respect to Eqs. (10.97) and (10.98)]:

$$\mathbf{L}^* = \mathbf{L}^{(0)} + n^{(1)}\mathbf{R}_1^\epsilon + (n^{(1)})^2 \mathbf{R}_1^\epsilon \mathbf{Q}_1^0 \mathbf{R}_1^\epsilon + n^{(1)} \mathbf{R}_{1,1}^{\epsilon\text{pert}}, \tag{10.99}$$

where the tensors $\mathbf{R}_1$ (3.179), $\mathbf{R}_{i,j}$ (10.3), $\mathbf{R}_{i,j}^{\text{pert}}$ (10.51) describing the properties of the strain polarization tensor $\boldsymbol{\eta}$ are replaced by the tensors $\mathbf{R}_1^\epsilon$, $\mathbf{R}_{i,j}^\epsilon$, $\mathbf{R}_{i,j}^{\epsilon\text{pert}}$ introduced in a similar manner for describing analogous properties of a stress polarization tensor $\boldsymbol{\xi}$ (see Eq. (4.12)). The integral terms $\mathbf{R}_{i,j}^{\text{pert}}$ and $\mathbf{R}_{i,j}^{\epsilon\text{pert}}$ in Eqs. (10.96) and (10.99) are absolutely convergent and, because of this, do not depend on the shape of the integration domains. However, the choice "ellipsoidal symmetry" (8.67) for the distribution of inclusions leads to the vanishing of a contribution produced by the second summands $(1 - V_i^0(\mathbf{x}_j))n^{(j)}\mathbf{R}_{i,j}^\infty(\mathbf{x})$ and $(1 - V_i^0(\mathbf{x}_j))n^{(j)}\mathbf{R}_{i,j}^{\epsilon\infty}(\mathbf{x})$ in $\mathbf{R}_{i,j}^{\text{pert}}$ and $\mathbf{R}_{i,j}^{\epsilon\text{pert}}$, respectively, and then the terms $(1 - V_i^0(\mathbf{x}_j))n^{(j)}\mathbf{R}_{i,j}^\infty(\mathbf{x})$ and $(1 - V_i^0(\mathbf{x}_j))n^{(j)}\mathbf{R}_{i,j}^{\epsilon\infty}(\mathbf{x})$ can be omitted.

Let us compare Eqs. (10.46) and (10.47) with estimations by Mori-Tanaka method (MTM) (8.102) which coincide with the variational lower bounds [428] (see Chapters 6 and 8) when the fibers are stiffer than the matrix, and with the upper bounds when the fibers are weaker than the matrix. To $O(c^2)$, the relevant representations via the MTM (called $O(c^2)$ approximation of the MTM) for the effective stiffness and compliance coincide with Eqs. (10.99) and (10.96), respectively, if in the last representation the integral terms are neglected. The absence of integral terms in Eqs. (10.96) and (10.99) is equivalent to ignoring binary interaction of inclusions provided by the one-particle approximation $\langle \boldsymbol{\eta}(\mathbf{x}) |; v_j, \mathbf{y}_j \rangle_i \equiv \langle \boldsymbol{\eta} \rangle_i$ and $\langle \boldsymbol{\xi}(\mathbf{x}) |; v_j, \mathbf{y}_j \rangle_i \equiv \langle \boldsymbol{\xi} \rangle_i$ (called also quasi-crystalline approximation, see Eqs. (8.65)). The effective elastic moduli to $O(c^2)$ were estimated in [1189] for the particular matrix composites through the use of an integral equation method in the framework of the far-field solution $O((a/R)^6)$ accompanied by the effective field hypothesis $\overline{\boldsymbol{\sigma}}(\mathbf{x}) \equiv$ const. (or $\overline{\boldsymbol{\varepsilon}}(\mathbf{x}) \equiv$ const. at $\mathbf{x} \in v_i$ (see Eq. (8.14)), where $R = |\mathbf{x}_i - \mathbf{y}_j|$ is a distance between the centers of inclusions v_i and v_j. These estimations follow as a limiting case from more general representations considered in Chapter 9.1.2 at $c \to 0$ (the analogous results for the fiber composites were obtained in [168]):

$$\begin{aligned}\mathbf{L}^* &= \mathbf{L}^{(0)} + n\mathbf{R}^{\varepsilon} + n^2\mathbf{R}^{\varepsilon}\mathbf{P}(v_i^0)\mathbf{R}^{\varepsilon} \\ &+ n\int \mathbf{R}^{\varepsilon}\mathbf{T}^{\varepsilon}_{ij}(\mathbf{x}_i - \mathbf{y}_j)\mathbf{R}^{\varepsilon}\mathbf{T}^{\varepsilon}_{ij}(\mathbf{x}_i - \mathbf{y}_j)\mathbf{R}^{\varepsilon}\phi(v_j, \mathbf{y}_j|; v_i, \mathbf{x}_i)\, d\mathbf{y}_j, \qquad (10.100)\\ \mathbf{M}^* &= \mathbf{M}^{(0)} + n\mathbf{R} + n^2\mathbf{R}\mathbf{Q}(v_i^0)\mathbf{R} \\ &+ n\int \mathbf{R}\mathbf{T}_{ij}(\mathbf{x}_i - \mathbf{y}_j)\mathbf{R}\mathbf{T}_{ij}(\mathbf{x}_i - \mathbf{y}_j)\mathbf{R}\phi(v_j, \mathbf{y}_j|; v_i, \mathbf{x}_i)\, d\mathbf{y}_j, \qquad (10.101)\end{aligned}$$

where the tensor $\mathbf{T}^{\varepsilon}_{ij}(\mathbf{x}_i - \mathbf{y}_j)$ is defined by Eq. (4.56_4) and the terms analogous to the second ones in the integrand in Eqs. (10.28) were omitted due to choice of both the ellipsoidal symmetry (8.67) and appropriate ellipsoidal integration areas in Eqs. (10.100) and (10.101). It should be mentioned that natural reciprocity between stiffness and compliance of effective properties $\mathbf{L}^* = (\mathbf{M}^*)^{-1}$ is not valid for the representations for (10.96) and (10.99) as well as for Eqs. (10.100) and (10.101).

The classical related method of estimation of effective properties of composites with the circle fibers proposed in [234] is noteworthy. They used the equation analogous to Eq. (10.96) accompanied by an approximative method of reflection in a two-cylinder system in which the disturbance of the field around a first cylinder is compensated by an additional homogeneous field produced by the second cylinder in the center of the first cylinder if the first cylinder is absent. This assumption in our notations is equivalent to the equality $\mathbf{T}_{ij}(\mathbf{x}_i - \mathbf{y}_j) \equiv \mathbf{T}_j(\mathbf{x}_i - \mathbf{y}_j)$ in Eq. (10.101) and is known as an asymptotic approximation which, along with another simplified assumptions of the approximative solutions of multiple inclusion interaction problems, was analyzed in detail in Chapter 4.

10.5.2 Improved Analysis of Composites with Identical Aligned Fibers

All relations (10.84)–(10.98) in this section were obtained in the framework of the zero-order approximations (10.25) and (10.27) of the closing assumption. However, the straightforward examination of Eqs. (10.84)–(10.98) corroborates their correctness also for assumptions (10.25) and (10.59) if in Eqs. (10.84)–(10.98) all the tensors $\mathbf{R}^{\text{pert}}_{i,j}(\mathbf{x})$ (10.28) and $\mathbf{F}^{\text{pert}}_{i}(\mathbf{x})$ (10.29) are replaced by the tensors $\mathbf{R}^{1\text{pert}}_{i,j}(\mathbf{x})$ (10.63) and $\mathbf{F}^{1\text{pert}}_{i}(\mathbf{x})$ (10.64), respectively, which in turn are reduced to the following tensors in the case of identical inclusions

$$\begin{aligned}\mathbf{R}^{1\text{pert}}_{i}(\mathbf{x}) = \int \Big\{&\mathbf{R}_{i,j}(\mathbf{x})\big(\mathbf{I} - \mathcal{T}^{1\sigma}_{i,j,1}\big)^{-1}\varphi(v_j, \mathbf{x}_j|; v_i, \mathbf{x}_i) \\ &-(1 - V^0_i(\mathbf{x}_j))n^{(j)}\mathbf{R}^{\infty}_{i,j}(\mathbf{x})\Big\}\, d\mathbf{y}, \qquad (10.102)\end{aligned}$$

$$\mathbf{F}_i^{1\text{pert}}(\mathbf{x}) = \int \Big\{ \Big[\mathbf{F}_{i,j}(\mathbf{x}) - \mathbf{R}_{i,j}(\mathbf{x}) \big(\mathbf{I} - \boldsymbol{\mathcal{T}}^{1\sigma}_{i,j,1} \big)^{-1} \boldsymbol{\mathcal{T}}^{\beta}_{i,j} \Big] \varphi(v_j, \mathbf{x}_j|; v_i, \mathbf{x}_i)$$
$$-(1 - V_i^0(\mathbf{x}_j)) n^{(j)} \mathbf{F}^{\infty}_{i,j}(\mathbf{x}) \Big\} \, d\mathbf{x}_j, \tag{10.103}$$

respectively; here $i, j, N = 1$, $\mathbf{x} \in v_i$. In a similar manner, the tensors (10.80) and (10.81) are reduced to ($i, j, N = 1$, $\mathbf{y} \in v_j$):

$$\mathbf{R}_i^{2\text{pert}} = \int \Big\{ \mathbf{T}_i(\mathbf{x}_i - \mathbf{y}) \mathbf{R}_{j,i}(\mathbf{y}) \big[\mathbf{I} - \boldsymbol{\mathcal{T}}^{1\sigma}_{j,i,1} \big]^{-1} \varphi(v_j, \mathbf{x}_j|; v_i, \mathbf{x}_i)$$
$$-(1 - V_i^0(\mathbf{y})) n^{(j)} \langle \mathbf{T}_j^{\sigma\infty}(\mathbf{x}) \rangle_{(i)} \Big\} \, d\mathbf{x}_j, \tag{10.104}$$

$$\mathbf{F}_i^{2\text{pert}} = \int \mathbf{T}_i(\mathbf{x}_i - \mathbf{y}) \Big[\{ \mathbf{F}_{j,i}(\mathbf{y}) - \mathbf{R}_{j,i}(\mathbf{y}) \big[\mathbf{I} - \boldsymbol{\mathcal{T}}^{1\sigma}_{j,i,1} \big]^{-1} \boldsymbol{\mathcal{T}}^{\beta}_{j,i} \}$$
$$\times \varphi(v_j, \mathbf{x}_j|; v_i, \mathbf{x}_i) - (1 - V_i^0(\mathbf{y})) n^{(j)} \langle \mathbf{T}_j^{\beta\infty}(\mathbf{x}) \rangle_{(i)} \Big] \, d\mathbf{y} \tag{10.105}$$

The matrix representation for both the average strain polarization tensor (10.77) and the effective field (10.82) inside the inclusions is reduced to the tensor one:

$$\bar{v}_1 \langle \boldsymbol{\eta} \rangle_1 = \big[\mathbf{I} - \langle \mathbf{R}^v \mathbf{Q}^0 V \rangle \big]^{-1} \Big[\mathbf{R}_1 \langle \boldsymbol{\sigma} \rangle + \mathbf{F}_1 + \mathbf{R}_1^{1\text{pert}} \langle \overline{\boldsymbol{\sigma}}_1 \rangle + \mathbf{F}_1^{1\text{pert}} \Big], \tag{10.106}$$

$$\langle \overline{\boldsymbol{\sigma}}_1 \rangle = \mathbf{Y}_{11} \Big[\langle \boldsymbol{\sigma} \rangle + n^{(1)} \mathbf{Q}_1^0 \mathbf{F}_1 + n^{(1)} \mathbf{Q}_1^0 \mathbf{F}_1^{1\text{pert}} + (\mathbf{I} - n^{(1)} \mathbf{Q}_1^0 \mathbf{R}_1) \mathbf{F}_1^{2\text{pert}} \Big], \tag{10.107}$$

$$\mathbf{Y}_{11}^{-1} = \mathbf{I} - n^{(1)} \mathbf{Q}_1^0 \mathbf{R}_1 - n^{(1)} \mathbf{Q}_1^0 \mathbf{R}_1^{1\text{pert}} - (\mathbf{I} - n^{(1)} \mathbf{Q}_1^0 \mathbf{R}_1) \mathbf{R}_1^{2\text{pert}}. \tag{10.108}$$

Substitution of Eqs. (10.106)–(10.107) into Eqs. (6.20) and (6.48) enables one to obtain the representation for the effective properties:

$$\mathbf{M}^* = \mathbf{M}^{(0)} + n^{(1)} [\mathbf{R}_1 + \mathbf{R}_1^{1\text{pert}} - \mathbf{R}_1 \mathbf{R}_1^{2\text{pert}}] \mathbf{Y}_{11}, \tag{10.109}$$

$$\boldsymbol{\beta}^* = \boldsymbol{\beta}^{(0)} + n^{(1)} \mathbf{F}_1 + n^{(1)} \mathbf{R}_1 \mathbf{Y}_{11} (n^{(1)} \mathbf{Q}_1^0 \mathbf{F}_1 + \mathbf{R}_1^{-1} \mathbf{F}_1^{\text{pert}}), \tag{10.110}$$

For analysis of stress distribution in the inclusion v_1, we multiply Eq. (10.77) over the tensor $\mathbf{M}_1^{-1}(\mathbf{x})$ and make some obvious transformation ($\mathbf{x} \in v_1$):

$$\langle \boldsymbol{\sigma} \rangle(\mathbf{x}) = \mathbf{B}_1(\mathbf{x}) \big[\mathbf{I} - n^{(1)} \mathbf{Q}_1^0 \mathbf{R}_1 \big]^{-1} \Big\{ \langle \boldsymbol{\sigma} \rangle + n^{(1)} \mathbf{Q}_1^0 \big[\mathbf{F}_1 + \mathbf{R}_1^{1\text{pert}} \langle \overline{\boldsymbol{\sigma}}_1 \rangle + \mathbf{F}_1^{1\text{pert}} \big]$$
$$+ (\mathbf{I} - n^{(1)} \mathbf{Q}_1^0 \mathbf{R}_1) \mathbf{R}_1^{-1}(\mathbf{x}) \big[\mathbf{R}_1^{1\text{pert}}(\mathbf{x}) \langle \overline{\boldsymbol{\sigma}}_1 \rangle + \mathbf{F}_1^{1\text{pert}}(\mathbf{x}) \big] \Big\}, \tag{10.111}$$

$$\langle \boldsymbol{\sigma} \rangle_1 = \mathbf{B}_1 \big[\mathbf{I} - n^{(1)} \mathbf{Q}_1^0 \mathbf{R}_1 \big]^{-1} \Big\{ \langle \boldsymbol{\sigma} \rangle + n^{(1)} \mathbf{Q}_1^0 \mathbf{F}_1 + \mathbf{R}_1^{-1} \big[\mathbf{R}_1^{1\text{pert}} \langle \overline{\boldsymbol{\sigma}}_1 \rangle + \mathbf{F}_1^{1\text{pert}} \big] \Big\}, \tag{10.112}$$

where Eq. (10.111) for the simplification of calculations was obtained for the homogeneous ellipsoidal inclusions (3.187) that will thereafter be assumed. Subtracting Eqs. (10.111) and (10.112) yields

$$\langle \boldsymbol{\sigma} \rangle_1(\mathbf{x}) - \langle \boldsymbol{\sigma} \rangle_1 = [\bar{v}_1 \mathbf{M}_1^{(1)}]^{-1} \Big\{ [\mathbf{R}_{1,1}^{\text{pert}}(\mathbf{x}) - \mathbf{R}_{1,1}^{\text{pert}}] \langle \overline{\boldsymbol{\sigma}}_1 \rangle + \mathbf{F}_1^{\text{pert}}(\mathbf{x}) - \mathbf{F}_1^{\text{pert}} \Big\}. \tag{10.113}$$

It is interesting that the measure of the stress inhomogeneity inside the homogeneous ellipsoidal inclusions $\mathbf{x} \in v_i$ (at $\boldsymbol{\beta}_1(\mathbf{x}) \equiv \mathbf{0}$): $\langle \boldsymbol{\sigma} \rangle_1(\mathbf{x}) - \langle \boldsymbol{\sigma} \rangle_1 = \big[\bar{v}_i \mathbf{M}_1^{(1)} \big]^{-1} \big[\mathbf{R}_1^{1\text{pert}}(\mathbf{x}) - \mathbf{R}_1^{1\text{pert}} \big] \mathbf{Y}_{11} \langle \boldsymbol{\sigma} \rangle$ is reduced to Eq. (10.95) at $\boldsymbol{\mathcal{T}}^{1\sigma}_{j,i,1} \to \mathbf{0}$.

It should be mentioned that the differences between Eqs. (10.84) and (10.106) are explained by exploiting the transformation of the approximating hypothesis (10.25) which is not necessary and is not used in obtaining of Eq. (10.106). Admission of the hypothesis (10.25) in Eq. (10.19) can be considered as an effect of a more inflexible assumption of homogeneity of the perturbation of the effective field $\overline{\boldsymbol{\sigma}}(\mathbf{x})$ ($\mathbf{x} \in v_i$) produced by the stress field $\boldsymbol{\sigma}(\mathbf{y})$ ($\mathbf{y} \in v_j$) evaluated in the inclusion v_j via the accurate Eq. (10.3). In such a case, $\mathbf{R}_{i,j}^{2\text{pert}} = \mathbf{R}_i^{-1}\mathbf{R}_{i,j}^{1\text{pert}}$, $\mathbf{F}_i^{2\text{pert}} = \mathbf{R}_i^{-1}\mathbf{F}_i^{1\text{pert}}$, and Eqs. (10.106) and (10.107) are reduced to Eqs. (10.84) and (10.85), respectively, because, e.g., $n^{(1)}\mathbf{Q}_1^0\mathbf{R}_1^{1\text{pert}} + (\mathbf{I} - n^{(1)}\mathbf{Q}_1^0\mathbf{R}_1)\mathbf{R}_1^{2\text{pert}} = \mathbf{R}_1^{2\text{pert}}$.

Thus, the proposed method allows us to estimate not only the effective elastic moduli and some related parameters, such as the volume average of the stress concentrator factor inside the inclusion, but also the statistical average of stresses varying along the cross-section of inclusions. Such a variation can be easily estimated for nonellipsoidal shape of inclusion even for a much simplified popular assumption as the homogeneity of the effective field $\overline{\boldsymbol{\sigma}}(\mathbf{x}) = \text{const.}$ ($\mathbf{x} \in v_i$) (10.26) as well as for the quasicrystallite approximation (see for detailed analysis Chapter 8). However, for statistically homogeneous composites subjected to the homogeneous loading and containing the homogeneous ellipsoidal inclusions, the acceptance of the assumptions mentioned above leads necessity to the homogeneity of statistical averages of stresses inside the inclusions. The surprising thing is that exploiting the proposed approach discovers a fundamentally new effect in the stress distributions inside the inclusions. Precisely, the statistical average of stresses is essentially inhomogeneous even inside the homogeneous ellipsoidal inclusions.

10.5.3 Effective Field Hypothesis

Usually the effective field hypothesis (10.25) is accompanied by the assumption (8.15) $\mathbf{T}_j^{\sigma\infty}(\mathbf{x}) = \mathbf{T}_j(\mathbf{x} - \mathbf{x}_j)\mathbf{R}_j$, $\mathbf{T}_j^{\beta\infty}(\mathbf{x}) = \mathbf{T}_j(\mathbf{x} - \mathbf{x}_j)\mathbf{F}_j$, where the tensors $\mathbf{T}_j(\mathbf{x} - \mathbf{x}_j)$ and $\mathbf{T}_{ij}(\mathbf{x}_i - \mathbf{x}_j)$ were defined via Eq. (8.16). For a homogeneous ellipsoidal inclusion v_i, the well known assumption (10.25) yields the equality (8.15); otherwise, Eq. (8.15) should be considered as an additional assumption.

The acceptance of the effective field hypothesis (10.25) for both interacting inclusions v_i and v_i reduces the integral Eqs. (10.2) and (10.10) describing the solution for the binary interacting inclusions to the algebraic ones. This allows us to significantly simplify the estimation of the tensors $\mathbf{R}_{i,k}(\mathbf{x})$, $\mathbf{F}_{i,k}(\mathbf{x})$ (10.3), and $\mathbf{R}_{i,k}^{\infty}(\mathbf{x})$, $\mathbf{F}_{i,k}^{\infty}(\mathbf{x})$ (10.11) $\mathbf{R}_{i,k}^{\infty}(\mathbf{x}) = \mathbf{R}_i(\mathbf{x})\mathbf{T}_{ij}(\mathbf{x}_i - \mathbf{x}_j)\mathbf{R}_j$, $\mathbf{F}_{i,k}^{\infty}(\mathbf{x}) = \mathbf{R}_i(\mathbf{x})\mathbf{T}_{ij}(\mathbf{x}_i - \mathbf{x}_j)\mathbf{F}_j$. The tensors $\mathbf{R}_{i,k}(\mathbf{x})$ and $\mathbf{F}_{i,k}(\mathbf{x})$ (10.3) can also easily be estimated in the framework of the effective field hypothesis (10.25) and (8.15) from the matrix representation of the solution of Eq. (10.2) ($\mathbf{x} \in v_i$):

$$\boldsymbol{\eta}(\mathbf{x}) = \mathbf{F}_i(\mathbf{x}) - \mathbf{R}_i(\mathbf{x})\mathbf{R}_i^{-1}\mathbf{F}_i + \mathbf{R}_i(\mathbf{x})\mathbf{R}_i^{-1}\sum_{k=1}^{2}\mathbf{Z}_{ik}[\mathbf{R}_k\widetilde{\boldsymbol{\sigma}}_{i,j} + \mathbf{F}_k], \quad (10.114)$$

where for simplicity $i, j = 1, 2$ are taken and the matrix $\mathbf{Z}^{-1}$ was defined by Eq. (4.58). Substitution of the solution (10.114) into Eq. (10.3) leads to the following desired representation ($\mathbf{x} \in v_i$):

$$\mathbf{R}_{i,j}(\mathbf{x}) = \mathbf{R}_i(\mathbf{x})\mathbf{R}_i^{-1}\sum_{k=1}^{2}(\mathbf{Z}_{ik} - \mathbf{I}\delta_{ik})\mathbf{R}_k, \tag{10.115}$$

$$\mathbf{F}_{i,j}(\mathbf{x}) = \mathbf{R}_i(\mathbf{x})\mathbf{R}_i^{-1}\sum_{k=1}^{2}(\mathbf{Z}_{ik} - \mathbf{I}\delta_{ik})\mathbf{F}_k. \tag{10.116}$$

Buryachenko [139] has analyzed the representations for the effective parameters (10.34)–(10.36) and some related problems using the tensors $\mathbf{R}_{i,j}(\mathbf{x})$ and $\mathbf{F}_{i,j}(\mathbf{x})$. In so doing, he utilized the effective field hypotheses (10.25) and (8.15), which, however, were not applied for the first-order approximation of the effective field $\widetilde{\boldsymbol{\sigma}}_{i,j}$ (10.48).

Substitution of the tensors $\mathbf{R}_i(\mathbf{x})$, $\mathbf{F}_i(\mathbf{x})$ and $\mathbf{R}_{i,j}(\mathbf{x})$ and $\mathbf{F}_{i,j}(\mathbf{x})$ into Eqs. (10.50) and (10.55) gives the representation of the tensors $\boldsymbol{\mathcal{T}}^{1\sigma}_{i,j,k}$ and $\boldsymbol{\mathcal{T}}^{\beta}_{i,j}$ used for the refined estimation of the effective field $\widetilde{\boldsymbol{\sigma}}_{i,j}$ (10.48). It should be mentioned, however, that instead of the tensors $\mathbf{R}_{i,k}(\mathbf{x})$ and $\mathbf{F}_{i,k}(\mathbf{x})$ ($\mathbf{x} \in v_i$) (10.3) in Eqs. (10.48)–(10.64), one can use their first-order approximation $\mathbf{R}^{\infty}_{i,k}(\mathbf{x})$ and $\mathbf{F}^{\infty}_{i,k}(\mathbf{x})$ (10.11), which in the framework of the effective field hypothesis, in turn, can be represented through the tensor $\mathbf{T}_k(\mathbf{x}-\mathbf{x}_k)$. Then the tensors $\boldsymbol{\mathcal{T}}^{1\sigma}_{i,j,k}(\mathbf{x}_k)$ and $\boldsymbol{\mathcal{T}}^{\beta}_{i,j}$ in Eqs. (10.50)–(10.64) can be replaced by the tensors:

$$\boldsymbol{\mathcal{T}}^{1\sigma\infty}_{i,j,k}(\mathbf{x}_k) = V^{0i}_{0j}(\mathbf{x}_k)\mathbf{T}_{ik}(\mathbf{x}_i - \mathbf{x}_k)\mathbf{R}_k\varphi(v_k, \mathbf{x}_k|; v_i, \mathbf{x}_i), \tag{10.117}$$

$$\boldsymbol{\mathcal{T}}^{\beta\infty}_{i,j} = \sum_{k=1}^{N}\int V^{0i}_{0j}(\mathbf{x}_k)\mathbf{T}_{ik}(\mathbf{x}_i - \mathbf{x}_k)\mathbf{F}_k\varphi(v_k, \mathbf{x}_k|; v_i, \mathbf{x}_i)\, d\mathbf{x}_k, \tag{10.118}$$

respectively. In the case of identical inclusions ($N = 1$), the relations (10.115), (10.116) and (10.28), (10.29) can be simplified (see Chapter 8) ($\mathbf{x} \in v_i$):

$$\mathbf{R}_{i,j}(\mathbf{x}) = \mathbf{R}_1(\mathbf{x})\mathbf{T}_{ij}(\mathbf{x}_i - \mathbf{x}_j)[\mathbf{I} - \mathbf{R}_1\mathbf{T}_{ij}(\mathbf{x}_i - \mathbf{x}_j)]^{-1}\mathbf{R}_1, \tag{10.119}$$

$$\mathbf{F}_{i,j}(\mathbf{x}) = \mathbf{R}_1(\mathbf{x})\mathbf{T}_{ij}(\mathbf{x}_i - \mathbf{x}_j)[\mathbf{I} - \mathbf{R}_1\mathbf{T}_{ij}(\mathbf{x}_i - \mathbf{x}_j)]^{-1}\mathbf{F}_1, \tag{10.120}$$

$$\mathbf{R}^{\rm pert}_{i,j}(\mathbf{x}) = \mathbf{R}_1(\mathbf{x})\int \mathbf{T}_{ij}(\mathbf{x}_i - \mathbf{x}_j)\Big\{[\mathbf{I} - \mathbf{R}_1\mathbf{T}_{ij}(\mathbf{x}_i - \mathbf{x}_j)]^{-1} \times \varphi(v_j, \mathbf{x}_j|; v_i, \mathbf{x}_i) - (1 - V^0_i(\mathbf{x}_j))n^{(j)}\Big\}\, d\mathbf{x}_j\mathbf{R}_1, \tag{10.121}$$

$$\mathbf{F}^{\rm pert}_{i}(\mathbf{x}) = \mathbf{R}_1(\mathbf{x})\int \mathbf{T}_{ij}(\mathbf{x}_i - \mathbf{x}_j)\Big\{[\mathbf{I} - \mathbf{R}_1\mathbf{T}_{ij}(\mathbf{x}_i - \mathbf{x}_j)]^{-1} \times \varphi(v_j, \mathbf{x}_j|; v_i, \mathbf{x}_i) - (1 - V^0_i(\mathbf{x}_j))n^{(j)}\Big\}\, d\mathbf{x}_j\mathbf{F}_1. \tag{10.122}$$

Substitution of Eqs. (10.119)-(10.122) into Eqs (10.91)-(10.93) and (10.89) leads to the representations of effective parameters and stress concentrator factors, respectively. It should be mentioned that for the identical spherical homogeneous inclusions with the RDF $g(r) \equiv \varphi(v_j, \mathbf{x}_j|; v_i, \mathbf{x}_i)$ ($r = |\mathbf{x}_i - \mathbf{x}_j|$) (when the term $(1 - V^0_i(\mathbf{x}_j))n^{(j)}$ (10.121) can be omitted due to the choice of the spherical integration area in Eq. (10.121)), the tensor $\mathbf{R}^{\epsilon\rm pert}_{i,j}$ (10.99) written in the form (10.121) coincides with an analogous tensor proposed in [1207] in the framework of the effective field hypothesis. Then the result [1207] can be rephrased in our notations (10.19) (at $\boldsymbol{\beta}_1 \equiv \mathbf{0}$) as $\bar{v}_i\langle\boldsymbol{\eta}\rangle_i = \mathbf{R}_i\langle\boldsymbol{\sigma}\rangle^{(0)} + \mathbf{R}^{\text{pert}}_{i,j}\langle\boldsymbol{\sigma}\rangle^{(0)}$. In doing so, Eq. (10.19) is reduced to the last one in the case of acceptance of two postulates: $\mathbf{R}_i\langle\boldsymbol{\sigma}\rangle + \mathbf{R}_i\mathbf{Q}^0_i\langle\boldsymbol{\eta}\rangle = \mathbf{R}_i\langle\boldsymbol{\sigma}\rangle^{(0)}$ (which is the same as $\langle\boldsymbol{\sigma}\rangle^{(1)} = \mathbf{B}_i\langle\boldsymbol{\sigma}\rangle^{(0)}$) and $\mathbf{R}_{i,j}\langle\widetilde{\boldsymbol{\sigma}}_{i,j}\rangle = \mathbf{R}^{\text{pert}}_{i,j}\langle\boldsymbol{\sigma}\rangle^{(0)}(\mathbf{x}_j)$ (which is appropriate for statistically homogeneous media), which are more restricted than the closing assumptions (10.26) and (10.27). Both postulates are similar to the Mori-Tanaka hypothesis (8.90) automatically neglecting the binary interaction of inclusions ($\mathbf{R}^{\text{pert}}_{i,j} \equiv \mathbf{0}$, see Chapter 8).

The refined estimations corresponding to the initial approximation (10.60) of parameters mentioned above, can be achieved by the replacement of the values $\mathbf{R}^{\text{pert}}_{i,j}(\mathbf{x})$ and $\mathbf{F}^{\text{pert}}_i(\mathbf{x})$ (10.121) and (10.122) by the related tensors

$$\mathbf{R}^{1\text{pert}}_{i,j}(\mathbf{x}) = \mathbf{R}_1(\mathbf{x}) \int \mathbf{T}_{NJ}(\mathbf{x}_i - \mathbf{x}_j)\Big\{[\mathbf{I} - \mathbf{R}_1\mathbf{T}_{NJ}(\mathbf{x}_i - \mathbf{x}_j)]^{-1} \times \varphi(v_j, \mathbf{x}_j|; v_i, \mathbf{x}_i) - (1 - V^0_i(\mathbf{x}_j))n^{(j)}\Big\}[\mathbf{I} - \overline{\boldsymbol{\mathcal{T}}}^{1\sigma}_{i,j,1}]^{-1}\, d\mathbf{x}_j \mathbf{R}_1, \tag{10.123}$$

$$\mathbf{F}^{1\text{pert}}_{i}(\mathbf{x}) = \mathbf{R}_1(\mathbf{x}) \int \mathbf{T}_{NJ}(\mathbf{x}_i - \mathbf{x}_j)\Big\{[\mathbf{I} + \mathbf{R}_1\mathbf{T}_{NJ}(\mathbf{x}_i - \mathbf{x}_j)]^{-1} \times \varphi(v_j, \mathbf{x}_j|; v_i, \mathbf{x}_i) - (1 - V^0_i(\mathbf{x}_j))n^{(j)}\Big\}[\mathbf{I} - \overline{\boldsymbol{\mathcal{T}}}^{1\sigma}_{i,j,1}]^{-1}\, d\mathbf{x}_j \mathbf{F}_1, \tag{10.124}$$

where $\overline{\boldsymbol{\mathcal{T}}}^{1\sigma}_{i,j,1} = \mathbf{R}_1 \int V^{0i}_{0j}(\mathbf{x}_1)\mathbf{T}_{ij}(\mathbf{x}_i - \mathbf{x}_j)[\mathbf{I} - \mathbf{R}_1\mathbf{T}_{ij}(\mathbf{x}_i - \mathbf{x}_j)]^{-1}\varphi(v_1, \mathbf{x}_1|; v_i, \mathbf{x}_i)d\mathbf{x}_1$, and the equality $\widetilde{\mathbf{Y}}_{i,j,1,1} = (\mathbf{I} - \overline{\boldsymbol{\mathcal{T}}}^{1\sigma}_{i,j,1})^{-1}$ was used. The subsequent iterations of the effective field $\widetilde{\boldsymbol{\sigma}}_{i,j}$ (10.61) can be used for the corrections of the tensors $\overline{\boldsymbol{\mathcal{T}}}^{\sigma}_{i,j,1}$, $\widetilde{\mathbf{Y}}_{i,j,1,1}$ in an obvious manner.

In particular, substitution of Eqs. (10.123) and (10.124) into the representations (10.91) and (10.92) yields the formulae for the effective properties:

$$\mathbf{M}^* = \mathbf{M}^{(0)} + n^{(1)}\Big\{\mathbf{I} - n^{(1)}\mathbf{R}_1\mathbf{Q}^0_1 - \int \mathbf{R}_1\mathbf{T}_{NJ}(x_i - \mathbf{x}_j)\{[\mathbf{I} - \mathbf{R}_1\mathbf{T}_{NJ}(\mathbf{x}_i - \mathbf{x}_j)]^{-1} \times \varphi(v_j, \mathbf{x}_j|; v_i, \mathbf{x}_i) - (1 - V^0_i(\mathbf{x}_j))n^{(j)}\}[\mathbf{I} - \overline{\boldsymbol{\mathcal{T}}}^{1\sigma}_{i,j,1}]^{-1}\, d\mathbf{x}_j\Big\}^{-1}\mathbf{R}_1, \tag{10.125}$$

$$\boldsymbol{\beta}^* = \boldsymbol{\beta}^{(0)} + n^{(1)}\Big\{\mathbf{I} - n^{(1)}\mathbf{R}_1\mathbf{Q}^0_1 - \int \mathbf{R}_1\mathbf{T}_{NJ}(x_i - \mathbf{x}_j)\{[\mathbf{I} - \mathbf{R}_1\mathbf{T}_{NJ}(\mathbf{x}_i - \mathbf{x}_j)]^{-1} \times \varphi(v_j, \mathbf{x}_j|; v_i, \mathbf{x}_i) - (1 - V^0_i(\mathbf{x}_j))n^{(j)}\}[\mathbf{I} - \overline{\boldsymbol{\mathcal{T}}}^{1\sigma}_{i,j,1}]^{-1}\, d\mathbf{x}_j\Big\}^{-1}\mathbf{F}_1. \tag{10.126}$$

A conditional mean value of stresses in the inclusions can be found analogously:

$$\langle\boldsymbol{\sigma}|; v_j, \mathbf{x}_j\rangle_i = \mathbf{B}_1[\mathbf{I} - \mathbf{R}_1\mathbf{T}_{NJ}(\mathbf{x}_i - \mathbf{x}_j)]^{-1}(\mathbf{I} - \boldsymbol{\mathcal{T}}^{1\sigma}_{i,j,1})^{-1}\mathbf{B}^{-1}_1\langle\boldsymbol{\sigma}\rangle_i. \tag{10.127}$$

10.5.4 Quasi-crystalline Approximation

Another sort of simplification of Eqs. (10.34)-(10.36) is generated by neglecting of binary interaction of inclusions $\mathbf{R}^{\text{pert}}_{i,j}(\mathbf{x}), \mathbf{F}^{\text{pert}}_i(\mathbf{x}) \equiv \mathbf{0}$, $(\mathbf{x} \in v_i)$ that is equivalent to the quasi-crystalline approximation (8.65). Then we get the next representations of both the effective parameters:

$$\mathbf{M}^* = \mathbf{M}^{(0)} + \Big[\mathbf{I} - \langle \mathbf{R}^v \mathbf{Q}^0 V \rangle\Big]^{-1} \langle \mathbf{R}^v V \rangle, \tag{10.128}$$

$$\boldsymbol{\beta}^* = \boldsymbol{\beta}^{(0)} + (\mathbf{M}^* - \mathbf{M}^{(0)}) \langle \mathbf{R}^v V \rangle^{-1} \langle \mathbf{F}^v V \rangle, \tag{10.129}$$

$$W^* = -\frac{1}{2}\Big\{ \langle \boldsymbol{\beta}_1 \mathbf{C} V \rangle + \langle \boldsymbol{\beta}_1 \mathbf{B} \mathbf{Q}^0 V \rangle (\mathbf{M}^* - \mathbf{M}^{(0)}) \langle \mathbf{R}^v V \rangle^{-1} \langle \mathbf{F}^v V \rangle \Big\}, \tag{10.130}$$

and the stress concentrator factors on the inclusions:

$$\langle \boldsymbol{\sigma} \rangle_i(\mathbf{x}) = \mathbf{B}_i(\mathbf{x}) \langle \boldsymbol{\sigma} \rangle + \mathbf{C}_i(\mathbf{x}) + \mathbf{B}_i(\mathbf{x}) \mathbf{Q}_i^0 \Big[\mathbf{I} - \langle \mathbf{R}^v \mathbf{Q}^0 V \rangle\Big]^{-1} \Big[\langle \mathbf{R}^v V \rangle \langle \boldsymbol{\sigma} \rangle + \langle \mathbf{F}^v V \rangle\Big] \tag{10.131}$$

for an arbitrary number of the inclusion constituents N. The representations of the involved tensors $\mathbf{B}_i(\mathbf{x}), \mathbf{C}_i(\mathbf{x}), \mathbf{Q}_i^0$ were obtained in Chapter 4, where the related particular equations for the homogeneous ellipsoidal inclusions were also considered. The tensors $\mathbf{R}_i(\mathbf{x})$ were estimated in Subsection 4.2.4 via the FEA for the different noncanonical shape of inclusions and anisotropic matrix. After that we evaluated in Section 9.5 the effective elastic moduli as well as the statistical averages of stress concentrator factors inhomogeneous along a cross-section of inclusions. We considered both aligned inclusions and inclusions uniformly distributed over the orientation. Buryechenko and Roy [189] (see Chapter 18) performed a detailed comparative analysis of Eq. (10.128)–(10.131) with analogous formulae of the Mori-Tanaka approach ($\langle \overline{\boldsymbol{\sigma}}_i \rangle \equiv (c^{(0)})^{-1} \langle (1-V) \boldsymbol{\sigma} \rangle$, (see [52] and Chapter 8) for the composites reinforced by the spheroidal anisotropic inclusions with a prescribed random orientation.

10.6 Some Particular Numerical Results

For the sake of definiteness, we will consider a plane strain 2-D problem for the random structure composites containing the circle inclusions (the fibers). The estimation of effective parameters (10.125)–(10.127) will be performed for the different radial distribution functions (RDF) for the inclusions [RDF simulated by the modified collective rearrangement model (CRM), see Section 5.4, and the well-stirred RDF ((5.70$_2$)]. At moderate concentration of inclusions and elastic mismatch between the matrix and the inclusions, the difference between the predicted curves (10.34), (10.121) and (10.125) is small. Because of this, let us demonstrate an application of the theoretical results by considering an isotropic composite made of an incompressible isotropic matrix, filled with rigid fiber inclusions of one radius. This example was chosen deliberately because it provides the maximum difference between predictions of effective elastic response, as estimated by the various methods. At first we will analyze the refined factor $\widetilde{\mathbf{y}} \equiv [\mathbf{I} - \overline{\boldsymbol{\mathcal{T}}}^{1\sigma}_{i,j,1}]^{-1}$ describing refinement of estimations (10.125)–(10.127) due to the use of the improved Eq. (10.125) instead of the initial representation (10.34) and (10.121). The plot $\widetilde{y}_{1111}$ vs the relative coordinates of the inclusion locations $r/a \equiv |\mathbf{x}_i - \mathbf{x}_j|/a$ is pictured in Fig. 10.1 at the inclusion centers $\mathbf{x}_i$ and $\mathbf{x}_j$ placed at the axis Oe_1: $\mathbf{x}_i - \mathbf{x}_j = |\mathbf{x}_i - \mathbf{x}_j|(1,0)^\top$. As can be seen the variation of the component $\widetilde{y}_{1111}$ is significantly more than the variation of $\widetilde{y}_{2211}$. The variation of the component $\widetilde{y}_{1111}$ is significantly greater than the variation of $\widetilde{y}_{2211}$, in the range $-0.08 < \widetilde{y}_{2211} < 0$; in so doing, the function $\widetilde{y}_{1111}$ estimated for the step RDF (5.70_2) varies in the range $0.94 < \widetilde{y}_{1111} < 1.0$ (see Fig. 10.2). The variation of the function $\widetilde{y}_{klmn} \sim r/a$ is nonmonotonic even at the RDF in the form of a step function. The dramatic variation of the function $\widetilde{y}_{klmn} \sim r/a$ in the vicinity of the point $\widetilde{y}_{klmn} \sim r/a = 4$ is explained by the geometric limitations of surrounding inclusion locations due to the fact that an arbitrary surrounding inclusion can be placed between the fixed inclusions v_i and v_j just in the case $|\mathbf{x}_i - \mathbf{x}_j|/a \geq 4$. The distinguishing characteristics of the functions $\widetilde{y}_{1111} \sim r/a$ estimated by the simulated RDF (Fig. 10.1) and the step RDF (Fig. 10.2) is explained by the presence of "fixed" peaks at the simulated RDF (see Fig. 5.4) corresponding to $r = 2a, 4a, 6a, \ldots$.

The numerical estimations of the relative effective shear modulus $\mu^*/\mu^{(0)}$ are depicted in Fig. 10.3 for both the simulated and step RDFs as well as for both the known representation (10.34), (10.121) and improved one (10.125). As can be seen the significant difference between the curves reached 50% and even more are observed only for the large concentration of fibers. In so doing, the limiting case $\mu^* = \infty$ are attained at $c = 0.71$ and $c = 0.70$ during the estimation by Eqs. (10.34), (10.121) for the step RDF and simulated RDF, respectively. The use of refined Eq. (10.125) leads to the extension of the prediction area till $c = 0.725$ and $c = 0.75$ for the step RDF and simulated RDF, respectively, which is circumstantial evidence of the improvement of the approach proposed.

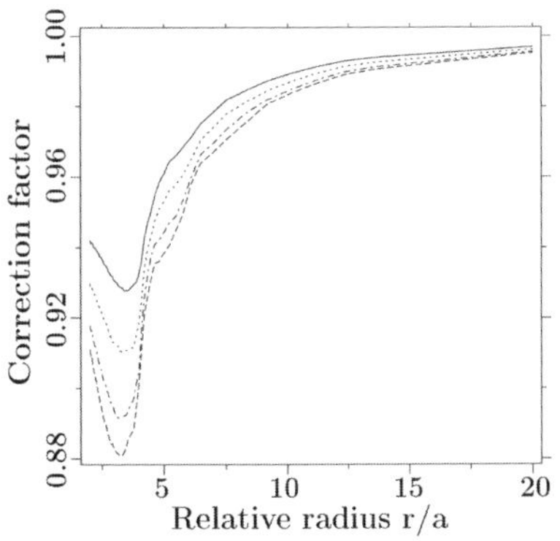

Fig. 10.1. Function $\widetilde{y}_{1111}$ vs r/a: c=0.45 (solid curve), 0.55 (dotted curve), 0.65 (dot-dashed curve), 0.70 (dashed curve).

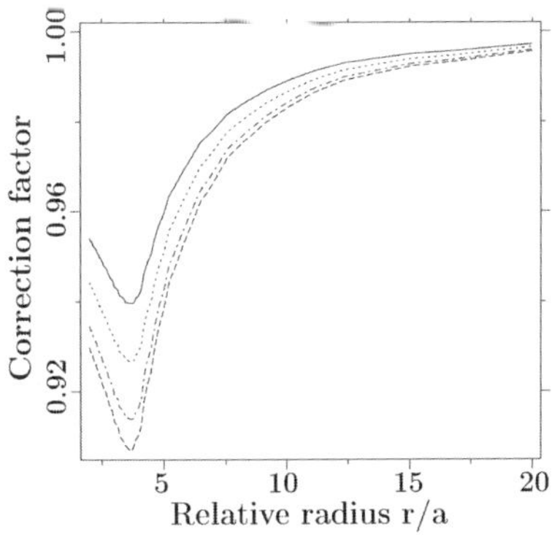

Fig. 10.2. Function $\widetilde{y}_{1111}$ vs r/a: c=0.45 (solid curve), 0.55 (dotted curve), 0.65 (dot-dashed curve), 0.70 (dashed curve).

The numerical estimations of the relative effective shear modulus $\mu^*/\mu^{(0)}$ are depicted in Fig. 10.3 for both the simulated and step RDFs as well as for both the known representation (10.34), (10.121) and improved one (10.125). As can be seen, the significant difference between the curves reached 50% and even more are observed only for the large concentration of fibers. In so doing, the limiting case $\mu^* = \infty$ are attained at $c = 0.71$ and $c = 0.70$ during the estimation by Eqs. (10.34), (10.121) for the step RDF and simulated RDF, respectively. The use of refined Eq. (10.125) leads to the extension of the prediction area until $c = 0.725$ and $c = 0.75$ for the step RDF and simulated RDF, respectively, which is circumstantial evidence of the improvement of the approach proposed.

The interaction effects generating the local stresses produced by inclusions are highly sensitive to their locations. The volume average of the conditional mean values of stresses $\langle\boldsymbol{\sigma}|; v_j, \mathbf{x}_j\rangle_i$ are presented in Fig. 10.4 as functions of the relative distance r/a between the fiber centers $\mathbf{x}_i - \mathbf{x}_j = r(1,0)^\top$. The external loading $\langle\sigma_{ij}\rangle = \delta_{1i}\delta_{1j}$ and the simulated RDF (see Section 5.4) were used for the analysis of the different concentration of fibers $c = 0.0,\ 0.55,\ 0.65,\ 0.70$. Obviously $\langle\boldsymbol{\sigma}|; v_j, \mathbf{x}_j\rangle_i \to \langle\boldsymbol{\sigma}\rangle_1$ as $r \to \infty$. The slight distinguishing characteristics of the function $\langle\boldsymbol{\sigma}|; v_j, \mathbf{x}_j\rangle_i \sim r$ estimated by the simulated RDF (Figs. 5.2–5.4) in the vicinity of the point $r = 4a$ are explained by the presence of a pick at the simulated RDF in the point $r = 4a$. In so doing, the crude assumption [901] that two inclusions are subject to the field $\langle\widetilde{\boldsymbol{\sigma}}_{i,j}\rangle \equiv \langle\boldsymbol{\sigma}\rangle$ (10.2) can be relaxed by the use of more accurate estimations (10.61) of the correlation function of the effective field $\langle\widetilde{\boldsymbol{\sigma}}_{i,j}\rangle$ depending on the vector $(\mathbf{x}_i - \mathbf{x}_j)$ and acting on two inclusions v_i and v_j fixed in the composite material rather than in the infinite matrix.

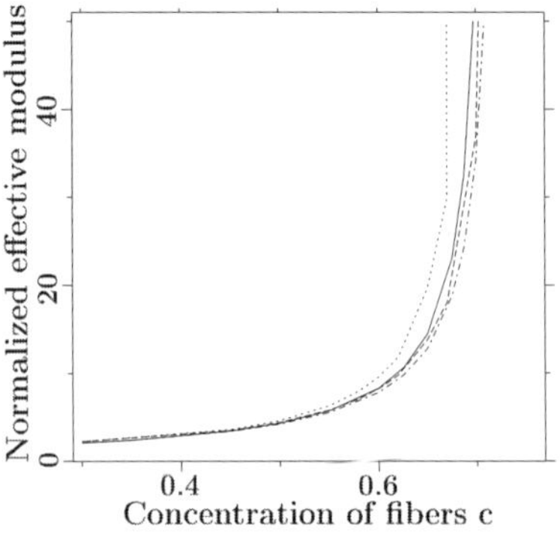

Fig. 10.3. $\mu^*/\mu^{(0)}$ vs c estimated by: Eqs. (10.34), (10.121) for the simulated and step RDFs (dotted and solid curves), Eq. (10.125) for the simulated and step RDFs (dashed and dot-dashed curves).

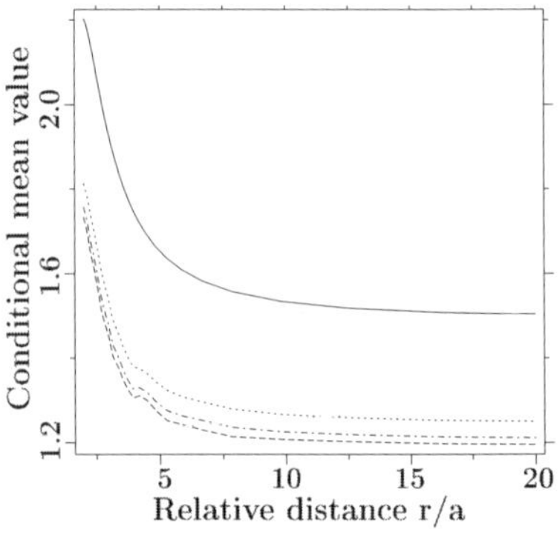

Fig. 10.4. Conditional mean value $\langle\boldsymbol{\sigma}|; v_j, \mathbf{x}_j\rangle_i$ vs the relative distance r/a estimated at: c=0.0 (solid curve), 0.55 (dotted curve), 0.65 (dot-dashed curve), 0.70 (dashed curve).

Up to now, the numerical results were obtained in the framework of the effective field hypothesis (10.25). Now we consider a transversal loading of a linearly elastic isotropic media containing the identical isotropic aligned circular fibers at non-dilute concentration c. The fiber–fiber interaction is estimated via the Kolosov-Muskhelishvili complex potential method which is a simple and powerful tool for solving a variety of plane linear elastic problems. The error less 0.1% is provided by the number of harmonics $n_h = 70$ for the distance between the fiber centers $D = |\mathbf{x}_i - \mathbf{x}_j| = 2.01a$. By the use of solution obtained by the potential method for two interacting circles subjected to three different applied stresses at infinity, and exact integral representations for both the stress and strain distributions in a microinhomogeneous medium, one estimates the effective moduli of the composite accurately to order c^2 by the use of Eq. (10.96) (see for details [152]. Although the present numerical estimation for the local stress field in the fiber loses accuracy (which can be protected by increasing the number n_h of harmonics) when the fibers almost touch, this does not affect the accuracy of the estimated coefficient of the $O(c^2)$ term which, being dependent on the smoothing integral operator, is very insensitive to the exact local stress distribution in the fibers when $D \to 2$. For the well-stirred RDF (5.70_2), the isotropic effective elastic moduli $\mathbf{L}^* = (2k^*, 2\mu^*)$ (10.99) have polynomial dependence on the fiber concentration $k^*/k^{(0)} = 1 + a_1^k c + a_2^k c^2$ and $\mu^*/\mu^{(0)} = 1 + a_1^\mu c + a_2^\mu c^2$. Similarly to [216] (see [42]) we present the coefficient a_2^μ in the form $a_2^\mu = a_1^\mu H$ where the parameter H was estimated in Fig. 10.5 by different methods as a function of the Poisson ratio $\nu^{(0)}$ for two different values $\mu^{(1)} = 1000$ and $\mu^{(1)} = 3$ (other elastic moduli are fixed: $\mu^{(0)} = 1$, $\nu^{(1)} = 0.5$). In particular, we considered an approximation of the tensor (10.121):

$$\mathbf{R}_{i,j}^{\text{pert}} = \mathbf{R}_1 \int \mathbf{T}_{ij}(\mathbf{x}_i - \mathbf{x}_j)\mathbf{R}_1 \mathbf{T}_{ij}(\mathbf{x}_i - \mathbf{x}_j)\varphi(v_j, \mathbf{x}_j|; v_i, \mathbf{x}_i)\, d\mathbf{x}_j \mathbf{R}_1 \qquad (10.132)$$

obtained in [1189] in the framework of the effective field hypothesis (10.25). Methods (10.96) as well as the $O(c^2)$ approximation of the MT method were also used when $\mathbf{R}_{i,j}^{\text{pert}} \equiv \mathbf{0}$. As can be seen, using the accurate solution (10.99) leads to significant refinement of the results obtained. The coefficients H estimated for the different varying numbers $n_h = n_h(|\mathbf{x}_i - \mathbf{y}_j|)$ corresponding to the errors 0.1% and 1.0% differ from one another by 0.002%.

In order to compare of the available experimental data (see Fig. 10.6) with the predictive capability of the proposed method, we will consider the estimation of the effective elastic moduli $\mathbf{L}^*$ (10.99). Assume the matrix is epoxy resin ($k^{(0)} = 4.27$ GPa and $\mu^{(0)} = 1.53$ GPa) which contains identical circular glass fibers ($k^{(1)} = 50.89$ GPa and $\mu^{(1)} = 35.04$ GPa). In Fig. 10.6 the experimental data [659] are compared with the estimations of the effective shear modulus μ^* obtained by the use of Eqs. (10.99), (10.100), by the $O(c^2)$ approximation of the MT method, which is equivalent to the vanishing of the integral term in Eq. (10.100) as well as by the linear approximation $O(c)$ of Eq. (10.99). As can be seen from Fig. 10.6, the use of method (10.99), and especially method (10.100), slightly improves the approximation of experimental data in comparison with the $O(c^2)$ approximation of the MT approach. Only an insignificant difference of the mentioned curved is explained by a slight variation of the considered composite material of the parameter $H = 0.932,\ 1.073,\ 1.200$ estimated via the $O(c^2)$ approximation of the MT method, via (10.99) and by (10.100) methods, respectively (compared with the analogous parameter H in Fig. 10.5 for the rigid fibers in the incompressible matrix). It sould be mentioned that all analyzed methods lead to an underestimate of the effective shear modulus compared with the experimental data. Better approximations are given by method (10.100) and, especially by method (10.99). Thus, taking of binary interaction effects into account even in the framework of the assumption $\overline{\boldsymbol{\varepsilon}}_{i,j} \equiv \langle \boldsymbol{\varepsilon} \rangle$ (or $\overline{\boldsymbol{\sigma}}_{i,j} \equiv \langle \boldsymbol{\sigma} \rangle$) (and even for the step RDF) leads to more significant refinement of estimated effective properties than one-particle approximation $\langle \boldsymbol{\xi}(\mathbf{x})|; v_j, \mathbf{y}_j \rangle_i = \langle \boldsymbol{\xi} \rangle_i$ ($\mathbf{x} \in v_i \neq v_j$)

accompanied by some sort of self-consistent correction of the field $\overline{\varepsilon}_i = \langle\varepsilon\rangle_0$ (or $\overline{\boldsymbol{\sigma}}_i = \langle\boldsymbol{\sigma}\rangle_0$) as in Mori-Tanaka method. The next improvement of the proposed method (10.96) is possible in the framework of the update version of the MEFM proposed in this Chapter with a non-well-stirred RDF $g(|\mathbf{x}_i - \mathbf{y}_j|)$ presented in Fig. 5.4.

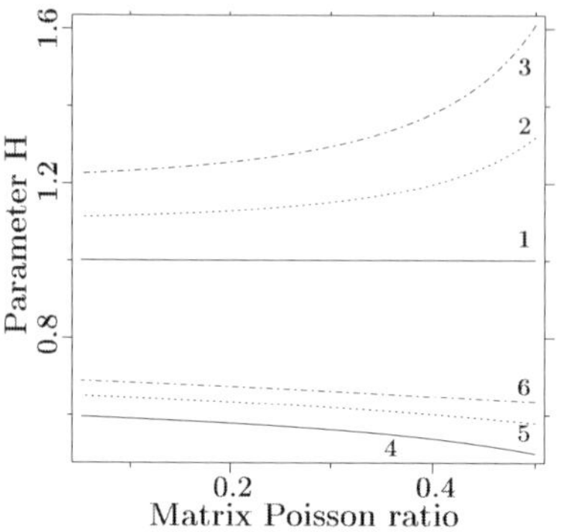

Fig. 10.5. The estimation of H vs $\nu^{(0)}$: $O(c^2)$ approximation of the MT method (solid curves 1, 4), Eqs. (10.96) and (10.132) (dotted curves 2, 5), Eq. (10.96) (dot-dashed curve 3, 6). The curves 1, 2, 3 are estimated for $\mu^{(1)} = 1000$, and the curves 4, 5, 6 for $\mu^{(1)} = 3$.

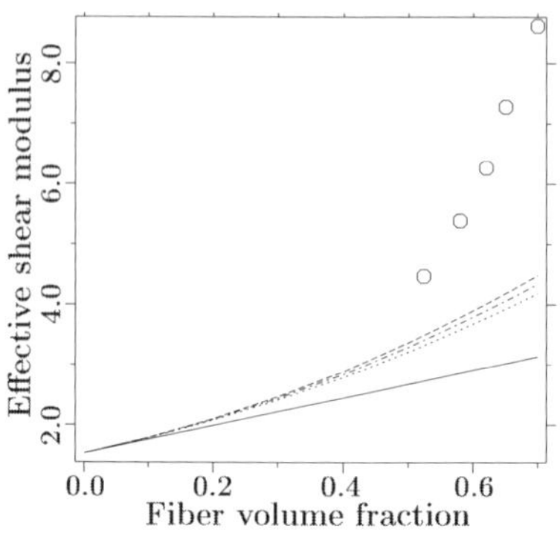

Fig. 10.6. Estimation of $\mu^*/\mu^{(0)}$ vs c. Experimental data ○ and curves calculated via the linear approximation (solid line), via $O(c^2)$ approximation of Mori-Tanaka method (dotted curve), via Eq. (10.100) (dot-dashed line), (10.99) (dashed line).

Thus, we can conclude that the assumption **H1** (10.25) of homogeneity of $\overline{\boldsymbol{\sigma}}(\mathbf{x})$, $(\mathbf{x} \in v_i)$ is a classical hypothesis of micromechanics and was accepted to make it easier to estimate the tensors $\mathbf{R}_i(\mathbf{x})$, $\mathbf{F}_i(\mathbf{x})$, $\mathbf{R}_{i,j}(\mathbf{x})$, $\mathbf{F}_{i,j}(\mathbf{x})$, $\mathbf{R}^{\infty}_{i,j}(\mathbf{x})$, $\mathbf{F}^{\infty}_{i,j}(\mathbf{x})$. In the framework of this assumption, the updated version of the MEFM is degenerated into the classical MEFM (see the references in Chapters 8 and 9). However, hypothesis **H1** is merely a zero-order approximation of binary interacting inclusions that results in a significant shortcoming of the MEFM. This substantial obstacle is overcome in the upgraded version of the MEFM in light of the generalized schemes based on the numerical solution of the problem for both one and two inclusions in the infinite media. No restrictions are imposed on the microtopology of the microstructure and the shape of inclusions as well as on the inhomogeneity of stress field inside the inclusions. However, the main computational advantage of the generalized MEFM proposed in this chapter (and its possible development in light of both Section 9.4 and Subsection 14.4.4) lies in the fact that such fundamental notions of micromechanics as a Green function and Eshelby tensor are not used, and we can analyze any anisotropy of constituents (including the matrix) as well as any shape and any composite structure of inclusions. We constructed Eqs. (10.19)–(10.21) by the use of some building blocks described by the numerical solutions for both one and two inclusions inside the infinite medium subjected to the homogeneous loading at infinity. Just with some additional assumptions considered in Chapter 8 the tensors mentioned above can be expressed through the Green function, Eshelby tensor, and external Eshelby tensor.

11

Periodic Structures and Periodic Structures with Random Imperfections

11.1 General Analysis of Periodic Structures and Periodic Structures with Random Imperfections

In many problems the material may possess a periodic microstructure formed by the spatial repetition of small microstructures, or unit cells. The periodic structure of composites is very attractive because it provides an estimation of interaction effects for an infinite number of inclusions. This interaction greatly influences composite elastic properties, especially for strongly heterogeneous materials with a high inclusion concentration. The periodicity of structures offers the possibility of finding an analytical (or numerical) solution of the corresponding periodic boundary–value problem with controlled accuracy.

For homogeneous boundary conditions periodicity of a composite structure leads to a periodicity of the solution for the stress and strain distributions with the same period as that of the microstructure. Periodicity of the solution allows one to explore a type of modeling completely different from the average schemes of random media (considered in Chapters 8–10) such as asymptotic expansion as well as the Fourier series expansion to field in a solid with periodic microstructure [806]. This asymptotic homogenization modeling coupled the finite element analysis (FEA) has been supposed to be most effective tool especially for the investigation of the microstructural effect on the overall properties (see, e.g., [25], [48], [507], [532], [960]). Such a perfectly regular distribution, of course, does not exist in actual cases, although the periodic modeling can be quite useful, since it provides rigorous estimations with *a priori* prescribed accuracy for various material properties.

For periodic structure composites there are also known semianalytical methods for solving the cell problem. The method of Eshelby transformation strain, taking account of the variability of the field of transformation strain within the inclusion, has been proposed in [807]. However, because they used very slowly converging series, their approach is hardly applicable. Using the periodic fundamental solution for an isotropic medium [442] (see also [1147]) numerical values of the effective characteristics of dispersionally reinforced composites with isotropic components were obtained in [962] by multiple expansions. Those authors extended a collocation technique employed in [820]. Media with arbitrary elastic anisotropy were considered in [636]. Kushch [626] analyzed the elastic isotropic medium containing several triply periodic lattices of aligned spheroidal isotropic inclusions with different size, shape, and properties.

By virtue of the fact that a periodic structure is a particular case of random structure, the MEFM was applied to the analysis of periodic structures in [174]. They used the main hypothesis of many micromechanical methods, according to which each inclusion is located inside a homogeneous so–called effective field (see also Chapter 8). Note that in the framework of the popular simplification, the surrounding inclusions are simulated by the singular sources of polarization strains located at the centers of inclusions [618]. In a proposed method [156], the approximation is considered in the spirit of the Saint–Venant principle so that the effective field is uniformly distributed inside the inclusion. At the same time a similar approach for a particular case of ellipsoidal cavities (microcracks) was proposed in [528]. Buryachenko and Parton [174] reduced the system of integral equations to a linear algebraic system of equations with respect to effective fields; the number of unknowns is finite in the case of a periodic structure. The final solution was obtained for the general case of coated inclusions and any ellipsoidal shape of a representative volume element (RVE) (providing homogeneity of the effective field $\overline{\boldsymbol{\sigma}}(\mathbf{x})$, see Section 9.4).

The assumption usually used is that all characteristic lengths associated with the spatial variations of the mean field quantities are large compared to all characteristic lengths associated with the spatial variations in the material properties. Then the governing equations for the mean field are identical in form to the familiar equations for a homogeneous solid with material properties replaced by effective properties. If the material is periodic but the solution is not periodic, such as in the case of inhomogeneous loading, the overall constitutive equation becomes nonlocal, which was analyzed in detail in [1023]. The breakdown of the condition afore-mentioned leads to a *nonlocal* coupling between statistical average for random structures (or average over the cell for periodic structures) of stresses and strains which is represented more easily by the integral or by the differential operators. The method of Fourier transform has been investigated in nonlocal micromechanics of random structure composites and was used with slight modifications in [69], [132], [163], [279], [563]. As will be shown in this chapter the same approach can be employed for the analysis of triply periodic structures.

The doubly periodic structure of composite materials is very attractive both because it provides an estimation of interaction effects for an infinite number of inclusions and because the breakdown of the periodicity in one direction leading to such structures can be considered as a model of so-called Functionally Graded Materials (FGMs). FGMs are materials that feature gradual compositional or microstructural transitions, being designed to deliver in an optimal way certain functional performance requirements that vary with location within a part. The FGMs approach is intrinsically far more multidisciplinary than almost any other undertaking in material research. In particular cases of the above problems it is possible to use different generalizations of the known methods for triply periodic structures. In particular, for numerical solutions of homogenization theory the number of the inclusions in the unit cell can be increased significantly by increasing the size of unit cell as compared with inclusion size; so Nakamura and Suresh [803] considered 30 and 60 fibers in one unit cell. In particular, for modeling of FGM, a unit cell was considered [1169] as a column containing an inclusion set in one side of the column only. Pindera *et al.* [879] used an analogous scheme with an approximate representation of the local stress states by second-order polynomials in the neighborhood of each inclusion. The elastic response of selected plane models of graded materials under both uniform and linearly varying boundary conditions was estimated by means of detailed finite element studies of large domains containing up to several thousand inclusions that offer very good possibilities [931], [932] (see also [256], [406], [780], [880]). In practice, however, components may be subjected to nonuniform stress states, such as in a circumferential reinforced ring subjected to radial turbine blade loads and centrifugal inertial loads [282]. The breakdown of the periodicity condition mentioned above leads to a *nonlocal* coupling between statistical average for random structures (or average over the cell for periodic structures) of stresses and strains that makes intuitive sense since the stress at any point will depend on the arrangement of the surrounding inclusions. This is especially true if the inclusion number density varies over distances that are comparable to the particle size. As previously mentioned, the method of Fourier transforms has been much investigated in nonlocal micromechanics of both random structure composites and triply periodic structures. Nevertheless, using the Fourier transform method in the case of doubly periodic structures leads to the principal difficulties because the general integral equation being analyzed in this chapter is not a convolution equation, and therefore the standard Fourier transform properties for the transform of both convolution integrals and derivatives cannot be used. For elimination of these difficulties we will propose modified version of the iteration method.

If the material is locally nonperiodic, such as in the case of graded materials, the asymptotic expansion technique generally leads to a poor approximation of the local fields. Kalamkarov [532], [534] has developed the scheme of "matching rules" for the analysis of a rectilinear crack in a composite material of doubly periodic structure. Fish and Wagiman [340] have developed the method originally proposed by Fish [338] under the name of s-version FEA, for the analysis of isolated imperfections in periodic structure (cutouts, cracks, etc). Compatibility between the outer and inner regions was explicitly enforced. The proposed method was applied to resolve the structure of the microscopic field in the single-ply composite plates with a centered hole and a centered crack and in the laminated plate. A boundary-layer method based on the use of the asymptotic method of averaging periodic structures, taking additional solutions of boundary-layer type [25], [1064] into account to allow the edge effect that occurs near the boundary of the crack outline to be considered, was proposed for the analysis of the stress field in the vicinity of the macrocrack.

However, the development of the theory of homogenization of initially periodic structures with the introduced infinite number of imperfections of unit cells randomly distributed in space has an essentially practical meaning. For example, the accumulation of damage, such as the microcracks in the matrix and debonding of inclusions from the matrix, is essentially random in nature. Because of this the popular modeling of failure by the use of periodic placement of microcracks and debonding with identical locations in all unit cells appears to be a poorly justified assumption. The adequate approach of the real process (e.g., failure) is a modeling of imperfections (e.g., cracks and debondings) existing in the finite locations in the unit cell with prescribed probability. Among sparse works dedicated to the mathematical homogenization of periodic structure with the imperfect unit cells, the principle of splitting an average operator for random structure in [855] (see also [25]) is worthy of notice. The final representations were obtained for the averaged problem of percolation of a viscous fluid in a connected system of indefinite capillaries stretched in three coordinate directions. The generator axes of cylindrical capillaries combine into the periodic lattice, while their radiuses are the independent random variables, taking the finite values (including zero) with the prescribed probability. The method of averaging of the differential operator involved is limited by just infinitesimal porosity.

The application of known methods for the homogenization of random structures to the analyses of the foam and cellular materials has been hampered by their large porosity. Because of this, these materials as well as the waffle structures, are modeled as periodic media with subsequent standard homogenization procedures [25], [48], [507], [532], [960]. The most common imperfections for these lightweight materials are the random distribution of fractured cell walls, the missing cells, as well as the availability of stiff inclusions [210]. Then the method of modeling that first comes to mind is Monte Carlo simulation (associated with the term *computational micromechanics*) to generate a random distribution of broken walls with the subsequent FEA of periodically distributed mesocells containing a reasonably large number of both perfect and imperfect unit cells [211], [975], [1011]. The usual shortcomings of the Monte Carlo simulation are the prohibitive computer costs for the reasonably large mesocells, especially for estimation of the first and second statistical moments of stresses in the walls (such estimations are absent) that are more sensitive to the defects of microstructure than the effective elastic moduli. Because of this, combining of opportunities of computational micromechanics with basic assumptions of analytical micromechanics is very promising. This makes possible the replacement of some analytical solutions for single and interacting inclusions by their numerical representations with subsequent incorporation of results into the one from the general schemes of analytical micromechanics. The challenge of modern micromechanics [143] is a development of the general method incorporating the solution for multiply interacting inhomogeneities obtained by highly accurate numerical methods into the most general scheme of analytical micromechanics.

Triply periodic particulate matrix composites with imperfect unit cells are analyzed in this chapter by three methods. The first one is a generalization of the version of the MEFM proposed for the analysis of the perfect periodic particulate composites and based on the choice of a comparison medium coinciding with the matrix. The second one is a Monte Carlo simulation exploring an analytical approximate solution for the binary interacting inclusions obtained in the framework of the hypothesis H1. The third method uses a decomposition of the desired solution on the solution for the perfect periodic structure and on the perturbation produced by the imperfections in the perfect periodic structure. The general scheme of the last method is presented with a subsequent simplification in the framework of the MEFM. All three methods lead to close results in the considered examples; however, the CPU time expended for the solution estimation by Monte Carlo simulation differs by a factor of 1000.

11.2 Triply Periodical Particular Matrix Composites in Varying External Stress Field

11.2.1 Basic Equation and Approximative Effective Field Hypothesis

To simplify the exact systems (7.38) and (7.39) at $\mathbf{f}(\mathbf{x}) \equiv \mathbf{0}$ we will introduce some definitions (analogously to Chapters 4 and 8) and apply the effective field hypothesis. We define the effective field $\overline{\boldsymbol{\varepsilon}}_i(\mathbf{x})$ (or $\overline{\boldsymbol{\sigma}}_i(\mathbf{x})$) ($\mathbf{x} \in v_i,\ i = 1,\ldots$) (generally speaking nonhomogeneous) as a strain (or stress) field in which the chosen inclusion v_i is embedded:

$$\overline{\boldsymbol{\varepsilon}}(\mathbf{x}) = \langle\boldsymbol{\varepsilon}\rangle_{\mathbf{x}}(\mathbf{x}) + \int \Big[\mathbf{U}(\mathbf{x}-\mathbf{y})\boldsymbol{\tau}(\mathbf{y})(1 - V_i(\mathbf{y})) - \langle\mathbf{U}\rangle_{\mathbf{x}}(\mathbf{x}-\mathbf{y})\boldsymbol{\tau}(\mathbf{y})\Big]\, d\mathbf{y},$$
$$\overline{\boldsymbol{\sigma}}(\mathbf{x}) = \langle\boldsymbol{\sigma}\rangle_{\mathbf{x}}(\mathbf{x}) + \int \Big[\boldsymbol{\Gamma}(\mathbf{x}-\mathbf{y})\boldsymbol{\eta}(\mathbf{y})(1 - V_i(\mathbf{y})) - \langle\boldsymbol{\Gamma}\rangle_{\mathbf{x}}(\mathbf{x}-\mathbf{y})\boldsymbol{\eta}(\mathbf{y})\Big]\, d\mathbf{y}, \quad (11.1)$$

where $\boldsymbol{\tau}(\mathbf{y}) = \mathbf{L}_1(\mathbf{y})\boldsymbol{\varepsilon}(\mathbf{y}) + \boldsymbol{\alpha}_1(\mathbf{y})$ and $\boldsymbol{\eta}(\mathbf{y}) = \mathbf{M}_1(\mathbf{y})\boldsymbol{\eta}(\mathbf{y}) + \boldsymbol{\beta}_1(\mathbf{y})$ are the stress and strain polarization tensors, respectively.

In order to simplify the systems (7.38) and (7.39)

$$\boldsymbol{\varepsilon}(\mathbf{x}) = \langle\boldsymbol{\varepsilon}\rangle_{\mathbf{x}}(\mathbf{x}) + \int \langle\langle\mathbf{U}(\mathbf{x}-\mathbf{y})\rangle\rangle_{\mathbf{x}}\boldsymbol{\tau}(\mathbf{y})\, d\mathbf{y},$$
$$\boldsymbol{\sigma}(\mathbf{x}) = \langle\boldsymbol{\sigma}\rangle_{\mathbf{x}}(\mathbf{x}) + \int \langle\langle\boldsymbol{\Gamma}(\mathbf{x}-\mathbf{y})\rangle\rangle_{\mathbf{x}}\boldsymbol{\eta}(\mathbf{y})\, d\mathbf{y}, \quad (11.2)$$

we now apply the main hypothesis of many micromechanical methods, the so-called effective field hypothesis (8.14) and (8.15):

H1) Each inclusion v_i has an ellipsoidal form and is located in the field $\overline{\boldsymbol{\varepsilon}}_i = \overline{\boldsymbol{\varepsilon}}(\mathbf{y}) \equiv$ const. (or $\overline{\boldsymbol{\sigma}}_i = \overline{\boldsymbol{\sigma}}(\mathbf{y}) \equiv$ const.) ($\mathbf{x} \in v_i$) which is homogeneous over the inclusion v_i. The perturbations introduced by the inclusion v_i in the point $\mathbf{x}$ are defined by the relations

$$\int \mathbf{U}(\mathbf{x}-\mathbf{y})V_i(\mathbf{y})\boldsymbol{\tau}(\mathbf{y})\, d\mathbf{y} = \overline{v}_i \mathbf{T}_i^{\epsilon}(\mathbf{x}-\mathbf{x}_i)\boldsymbol{\tau}_i,$$
$$\int \boldsymbol{\Gamma}(\mathbf{x}-\mathbf{y})V_i(\mathbf{y})\boldsymbol{\eta}(\mathbf{y})\, d\mathbf{y} = \overline{v}_i \mathbf{T}_i(\mathbf{x}-\mathbf{x}_i)\boldsymbol{\eta}_i, \quad (11.3)$$

where $\boldsymbol{\tau}_i = \langle\boldsymbol{\tau}(\mathbf{x})V_i(\mathbf{x})\rangle_{(i)} \equiv \overline{v}_i^{-1}\int \boldsymbol{\tau}(\mathbf{x})V_i(\mathbf{x})\, d\mathbf{x}$, $\boldsymbol{\eta}_i = \langle\boldsymbol{\eta}(\mathbf{x})V_i(\mathbf{x})\rangle_{(i)}$, and hereafter one introduced the tensors

$$\mathbf{T}_i^{\epsilon}(\mathbf{x}-\mathbf{x}_i) = \begin{cases} \overline{v}_i^{-1}\int \mathbf{U}(\mathbf{x}-\mathbf{y})V_i(\mathbf{y})\, d\mathbf{y}, \\ -\mathbf{P}, \end{cases} \quad \mathbf{T}_i(\mathbf{x}-\mathbf{x}_i) = \begin{cases} \overline{v}_i^{-1}\int \boldsymbol{\Gamma}(\mathbf{x}-\mathbf{y})V_i(\mathbf{y})\, d\mathbf{y}, \\ -\mathbf{Q} \end{cases} \quad (11.4)$$

for $\mathbf{x} \notin v_i$ (upper line) and $\mathbf{x} \in v_i$ (lower line), respectively. Here the tensors $\mathbf{P}$, $\mathbf{Q}$ are defined by Eqs. (3.110).

Then in the framework of hypothesis **H1** and in view of the linearity of the problem there exist constant fourth-rank tensors $\mathbf{A}(\mathbf{x}), \mathbf{C}^{\epsilon}(\mathbf{x})$, $\mathbf{R}^{\epsilon}(\mathbf{x})$, $\mathbf{F}^{\epsilon}(\mathbf{x})$, and $\mathbf{B}(\mathbf{x}), \mathbf{C}(\mathbf{x})$, $\mathbf{R}(\mathbf{x})$, and $\mathbf{F}(\mathbf{x})$ defined by Eqs. (4.10)–(4.12) and relating the stress and strain distributions inside the inclusion $\mathbf{x} \in v_i$ with the effective fields $\overline{\boldsymbol{\varepsilon}}$ and $\overline{\boldsymbol{\sigma}}$. Strictly speaking the hypothesis **H1** cannot be satisfied, nevertheless it will be showed that using it provides a high precision analysis of some regular structures under the homogeneous external loading. Because of this we will employ the hypothesis **H1**.

The equations, notions, and notations mentioned in 11.2.1 are valid for any deterministic (periodic and nonperiodic) structures. By way of subsequent simplification, we consider in Subsections 11.2.2–11.2.4 the pure elastic case ($\boldsymbol{\beta}(\mathbf{x}) \equiv \mathbf{0}$) of triply periodic structures (5.103).

11.2.2 The Fourier Transform Method

In the framework of the hypothesis **H1** the system (11.2) for the periodic structure is reduced to

$$\overline{\boldsymbol{\varepsilon}}(\mathbf{x}_i) = \langle \boldsymbol{\varepsilon} \rangle_{\mathbf{x}_i}(\mathbf{x}_i) + \int \boldsymbol{\mathcal{F}}(\mathbf{x}_i - \mathbf{y}) \overline{\boldsymbol{\varepsilon}}(\mathbf{y}) \, d\mathbf{y}, \tag{11.5}$$

the tensor $\boldsymbol{\mathcal{F}}(\mathbf{x}_i - \mathbf{y})$ introduced in Eq. (11.5) is defined as

$$\boldsymbol{\mathcal{F}}(\mathbf{x}_i - \mathbf{y}) = \sum_{\mathbf{m}} \Big[\mathbf{T}^{\epsilon}_{i\mathbf{m}}(\mathbf{x}_i - \mathbf{x}_{\mathbf{m}})(1 - V_i(\mathbf{y})) - \langle \mathbf{T}^{\epsilon}_{\mathbf{m}} \rangle_{\mathbf{x}_i}(\mathbf{x}_i - \mathbf{x}_{\mathbf{m}}) \Big] \mathbf{R}^{\epsilon} \delta(\mathbf{y} - \mathbf{x}_{\mathbf{m}}), \tag{11.6}$$

and $\mathbf{T}^{\epsilon}_{i\mathbf{m}}(\mathbf{x}_i - \mathbf{x}_{\mathbf{m}}) = (\overline{v}_i \overline{v}_{\mathbf{m}})^{-1} \int \mathbf{U}(\mathbf{x} - \mathbf{y}) V_i(\mathbf{x}) V_{\mathbf{m}}(\mathbf{y}) \, d\mathbf{x} \, d\mathbf{y}$.

According to the definition the fields $\overline{\boldsymbol{\eta}}(\mathbf{x}_i)$ are invariants with respect to the inclusion number i and depend on the argument $\mathbf{x}$. Since we desire an explicit representation for $\overline{\boldsymbol{\varepsilon}}(\mathbf{x})$ we will approximate $\overline{\boldsymbol{\varepsilon}}(\mathbf{y})$ by the first three terms of its Taylor expansion about $\mathbf{x}$:

$$\overline{\boldsymbol{\varepsilon}}(\mathbf{y}) \approx \overline{\boldsymbol{\varepsilon}}(\mathbf{x}) + (\mathbf{y} - \mathbf{x}) \nabla \overline{\boldsymbol{\varepsilon}}(\mathbf{x}) + \frac{1}{2} (\mathbf{y} - \mathbf{x}) \otimes (\mathbf{y} - \mathbf{x}) \nabla \nabla \overline{\boldsymbol{\varepsilon}}(\mathbf{x}). \tag{11.7}$$

Substituting (11.7) into Eq. (11.5) gives

$$\begin{aligned} \overline{\boldsymbol{\varepsilon}}(\mathbf{x}_i) &= \langle \boldsymbol{\varepsilon} \rangle_{\mathbf{x}_i}(\mathbf{x}_i) + \int \boldsymbol{\mathcal{F}}(\mathbf{x}_i - \mathbf{y}) \, d\mathbf{y} \overline{\boldsymbol{\varepsilon}}(\mathbf{x}_i) \\ &+ \int \boldsymbol{\mathcal{F}}_1(\mathbf{x}_i - \mathbf{y}) \, d\mathbf{y} \nabla \overline{\boldsymbol{\varepsilon}}(\mathbf{x}_i) + \int \boldsymbol{\mathcal{F}}_2(\mathbf{x}_i - \mathbf{y}) \, d\mathbf{y} \nabla \nabla \overline{\boldsymbol{\varepsilon}}(\mathbf{x}_i), \end{aligned} \tag{11.8}$$

where the integral operator kernels $\boldsymbol{\mathcal{F}}_1(\mathbf{x}_i - \mathbf{y})$ and $\boldsymbol{\mathcal{F}}_2(\mathbf{x}_i - \mathbf{y})$ are defined by the tensor $\boldsymbol{\mathcal{F}}(\mathbf{x}_i - \mathbf{y})$ (11.6): $\boldsymbol{\mathcal{F}}_1(\mathbf{x}_i - \mathbf{y}) = \boldsymbol{\mathcal{F}}(\mathbf{x}_i - \mathbf{y}) \otimes (\mathbf{y} - \mathbf{x}_i)$, $\boldsymbol{\mathcal{F}}_2(\mathbf{x}_i - \mathbf{y}) = \frac{1}{2} \boldsymbol{\mathcal{F}}(\mathbf{x}_i - \mathbf{y}) \otimes (\mathbf{y} - \mathbf{x}_i) \otimes (\mathbf{y} - \mathbf{x}_i)$. For triply periodic structures all integrals in Eq. (11.8) are constant tensors. Moreover, the second right-hand-side integral in Eq. (11.8) vanishes, because the tensor $\boldsymbol{\mathcal{F}}_1(\mathbf{x}_i - \mathbf{y})$ is an odd function: $\boldsymbol{\mathcal{F}}_1(\mathbf{x}_i - \mathbf{y}) = -\boldsymbol{\mathcal{F}}_1(\mathbf{y} - \mathbf{x}_i)$.

Considering that Eq. (11.8) is a differential equation with constant coefficients, the method of solution that first comes to mind is using of Fourier transform to transform the differential problem of solving (11.8) into the division problem of solving the multiplicative equation [1112] $\boldsymbol{\mathcal{P}}(i\boldsymbol{\xi})\widetilde{\overline{\boldsymbol{\varepsilon}}}(\boldsymbol{\xi}) = \langle\widetilde{\boldsymbol{\varepsilon}}\rangle(\boldsymbol{\xi})$, where a symbol $\boldsymbol{\mathcal{P}}(i\boldsymbol{\xi}) = \mathbf{I} - \int \left[\boldsymbol{\mathcal{F}}(\mathbf{x}_i - \mathbf{y}) - \boldsymbol{\mathcal{F}}_2(\mathbf{x}_i - \mathbf{y})\boldsymbol{\xi}\otimes\boldsymbol{\xi}\right] d\mathbf{y}$ of the differential operator (11.8) is a polynomial with real constant coefficients in 3 real transform variable $\boldsymbol{\xi} = (\xi_1, \xi_2, \xi_3)^\top$. Here the Fourier transformation $\widetilde{\mathbf{g}}(\boldsymbol{\xi})$ of a function $\mathbf{g}(\mathbf{x})$ and its inverse are defined by the formulae (3.4) provided, of course, that the integrals on the right–hand sides of the equations are convergent.

Therefore $\overline{\boldsymbol{\varepsilon}}(\mathbf{x}) = (2\pi)^{-3}\int e^{i\boldsymbol{\xi}\cdot\mathbf{x}}\boldsymbol{\mathcal{P}}^{-1}(i\boldsymbol{\xi})\langle\widetilde{\boldsymbol{\varepsilon}}\rangle(\boldsymbol{\xi})\, d\boldsymbol{\xi}$ should be a solution of (11.8) in view of (3.4). Substituting as explicit expression for $\boldsymbol{\mathcal{P}}(i\boldsymbol{\xi})$ into the representation for $\overline{\boldsymbol{\varepsilon}}(\mathbf{x})$ and restricting the result to terms of no greater than second order in the expansion

$$\boldsymbol{\mathcal{P}}^{-1}(i\boldsymbol{\xi}) \approx \mathbf{Y}^{\epsilon} - \boldsymbol{\mathcal{Y}}\otimes\boldsymbol{\xi}\otimes\boldsymbol{\xi}, \tag{11.9}$$

we get

$$\overline{\boldsymbol{\varepsilon}}(\mathbf{x}_i) = \mathbf{Y}^{\epsilon}\langle\boldsymbol{\varepsilon}\rangle_{\mathbf{x}_i}(\mathbf{x}_i) + \boldsymbol{\mathcal{Y}}\nabla\nabla\langle\boldsymbol{\varepsilon}\rangle_{\mathbf{x}_i}(\mathbf{x}_i), \tag{11.10}$$

where the local part $\mathbf{Y}^{\epsilon}$ and the nonlocal part $\boldsymbol{\mathcal{Y}}$ of the differential operator (11.10) have the forms

$$\mathbf{Y}^{\epsilon} = \left[\mathbf{I} - \int \boldsymbol{\mathcal{F}}(\mathbf{x}_i - \mathbf{y})\, d\mathbf{y}\right]^{-1}, \quad \boldsymbol{\mathcal{Y}} = \mathbf{Y}^{\epsilon}\int \boldsymbol{\mathcal{F}}_2(\mathbf{x}_i - \mathbf{y})\, d\mathbf{y}\mathbf{Y}^{\epsilon}, \tag{11.11}$$

respectively. To summarize, we have found a nonlocal effective field representation for triply periodic composites having periodic distribution of arbitrarily shape coated inclusions and arbitrarily anisotropic phases.

11.2.3 Iteration Method

Strictly speaking Eq. (11.10) is defined at both sufficiently slowly varying fields $\langle\boldsymbol{\varepsilon}\rangle_{\mathbf{x}}(\mathbf{x})$ and $\overline{\boldsymbol{\varepsilon}}(\mathbf{x})$, and should be considered as an approximation of the real nonlocal operator $\boldsymbol{\mathcal{P}}^{-1}$ by the differential operator of the second order. For elimination of these limitations we will represent the right-hand side of Eq. (11.10) in the form of the integral operator. With this aim it should be mentioned that in additional to the method of Fourier transforms used above there are other approximate methods that can be employed for solving Fredholm integral equations of the second kind such as Eq. (11.5) (see, e.g, [268]). For instance, if the kernel of a Fredholm integral equation of the second kind is small enough then the answer for $\overline{\boldsymbol{\varepsilon}}(\mathbf{x}_i)$ can be constructed by the method of Liouville–Neumann series, proceeding by the recurrence formula

$$\overline{\boldsymbol{\varepsilon}}(\mathbf{x}_i)^{(n+1)} = \langle\boldsymbol{\varepsilon}\rangle_{\mathbf{x}_i}(\mathbf{x}_i) + \int \boldsymbol{\mathcal{F}}(\mathbf{x}_i - \mathbf{y})\overline{\boldsymbol{\varepsilon}}(\mathbf{y})^{(n)}\, d\mathbf{y}. \tag{11.12}$$

Usually the driving term of this equation is used as an initial approximation: $\overline{\boldsymbol{\varepsilon}}(\mathbf{x}_i)^{(0)} = \langle\boldsymbol{\varepsilon}\rangle_{\mathbf{x}_i}(\mathbf{x}_i)$. However, if we assume a sufficiently slowly varying average field $\langle\varepsilon\rangle_{\mathbf{x}_i}(\mathbf{x}_i)$, and for $\langle\varepsilon\rangle_{\mathbf{x}_i}(\mathbf{x}_i) \equiv$ const. the exact solution of Eq. (11.12) is given by

$$\overline{\varepsilon}(\mathbf{x}_i) = \mathbf{Y}^\epsilon \langle\varepsilon\rangle_{\mathbf{x}_i}(\mathbf{x}_i) \equiv \text{const.} \tag{11.13}$$

Then for slowly varying average field $\langle\varepsilon\rangle_{\mathbf{x}_i}(\mathbf{x}_i)$ it would appear reasonable to employ the value $\overline{\varepsilon}(\mathbf{x}_i)^{(0)} = \mathbf{Y}\langle\varepsilon\rangle_{\mathbf{x}_i}(\mathbf{x}_i) \not\equiv$ const. as the zero-order approximation, so that the first-order approximation can be obtained from the modified recurrence relation (11.12) as follows:

$$\overline{\varepsilon}(\mathbf{x}_i) = \mathbf{Y}^\epsilon \langle\varepsilon\rangle_{\mathbf{x}_i}(\mathbf{x}_i) + \int \boldsymbol{\mathcal{Z}}(\mathbf{x}_i - \mathbf{y})\big[\langle\varepsilon\rangle_{\mathbf{y}}(\mathbf{y}) - \langle\varepsilon\rangle_{\mathbf{x}_i}(\mathbf{x}_i)\big]\, d\mathbf{y}, \tag{11.14}$$

where the integral operator kernel $\boldsymbol{\mathcal{Z}}(\mathbf{x}_i - \mathbf{y})$ (11.14) is defined by the relation $\boldsymbol{\mathcal{Z}}(\mathbf{x}_i - \mathbf{y}) = \mathbf{Y}^\epsilon \boldsymbol{\mathcal{F}}(\mathbf{x}_i - \mathbf{y})\mathbf{Y}^\epsilon$. It is obvious that for the homogeneous boundary conditions $\boldsymbol{\varepsilon}^0(\mathbf{x}) \equiv$ const. carries into $\langle\varepsilon\rangle_{\mathbf{y}}(\mathbf{y}) \equiv$ const., the right-hand-side integral in Eq. (11.14) vanishes, and the solutions (11.13) and (11.14) coincide.

Now we prove that for sufficiently smooth average strain fields and some additional assumptions both the Fourier transform method and the iteration method lead to the same results. It has been assumed previously that the average field $\langle\varepsilon\rangle_{\mathbf{y}}(\mathbf{y})$ is slowly varying enough. Then we can approximate $\langle\varepsilon\rangle(\mathbf{y})$ by the first three terms of its Taylor expansion about $\mathbf{x}_i$: $\langle\varepsilon\rangle_{\mathbf{y}}(\mathbf{y}) \approx \langle\varepsilon\rangle_{\mathbf{x}_i}(\mathbf{x}_i) + (\mathbf{y} - \mathbf{x}_i)\nabla\langle\varepsilon\rangle_{\mathbf{x}_i}(\mathbf{x}_i) + \frac{1}{2}(\mathbf{y} - \mathbf{x}_i) \otimes (\mathbf{y} - \mathbf{x}_i)\nabla\nabla\langle\varepsilon\rangle_{\mathbf{x}_i}(\mathbf{x}_i)$. Substituting the last equation into (11.14) and since $\boldsymbol{\mathcal{Z}}$ (11.14) is an even function, (11.14) finally reduces to (11.10). By this means for sufficiently slowly varying average fields $\langle\varepsilon\rangle_{\mathbf{y}}(\mathbf{y})$ the first few steps of both successive iterations and Taylor expansions in the iteration method and in the Fourier transform method, respectively, lead to the same relation (11.10). Nevertheless, since the Fourier transform method employed Taylor's expansion twice (11.7) and (11.9) one should expect that the iteration method is a better choice between these two methods.

The method of Fourier transform used above (11.10) has been investigated in nonlocal micromechanics of random structure composites and was used with slight modifications in [69], [132], [163], [279], [563]. Nevertheless the differential nonlocal relation (11.10) has the disadvantage that is uses the concrete polynomial approximations (11.7) and (11.9) in a sufficiently large neighborhood, which is sometimes violated in practice. In contrast, the integral nonlocal Eq. (11.14) does not use the concrete representations (11.7) and (11.9), and can be applied with controlled accuracy for the analysis of a wider class of average fields $\langle\varepsilon\rangle_{\mathbf{y}}(\mathbf{y})$, since integration is a smoothing operation and the right-hand-side integral (11.14) is likely to be a rather smooth function even when $\langle\varepsilon\rangle_{\mathbf{y}}(\mathbf{y})$ is very jagged. However, more detailed consideration of convergence rates and an estimation of its accuracy for the procedures of nonlocal operators used here are beyond the scope of the current section (this line of research will be pursued in the Section 11.2).

11.2.4 Average Strains in the Components

The strain field inside the inclusions $\boldsymbol{\varepsilon}(\mathbf{y})$ $(\mathbf{z} \in v_i)$ is obtained from (4.12) and (11.10)

$$\boldsymbol{\varepsilon}(\mathbf{x}_i, \mathbf{y}) = \mathbf{A}(\mathbf{y})\mathbf{Y}^{\epsilon}\langle\boldsymbol{\varepsilon}\rangle_{\mathbf{x}_i}(\mathbf{x}_i) + \mathbf{A}(\mathbf{y})\boldsymbol{\mathcal{Y}}\nabla\nabla\langle\boldsymbol{\varepsilon}\rangle_{\mathbf{x}_i}(\mathbf{x}_i), \tag{11.15}$$

from which the representation for the average strains inside the inclusion v_i follows

$$\langle\boldsymbol{\varepsilon}\rangle_{(i)} = \mathbf{A}\mathbf{Y}^{\epsilon}\langle\boldsymbol{\varepsilon}\rangle_{\mathbf{x}_i}(\mathbf{x}_i) + \mathbf{A}\boldsymbol{\mathcal{Y}}\nabla\nabla\langle\boldsymbol{\varepsilon}\rangle_{\mathbf{x}_i}(\mathbf{x}_i), \tag{11.16}$$

here the "fast" independent variable $\mathbf{y} \in v_i$ characterizing the strain state is defined in local coordinate system connected with the semi-axes of the ellipsoid v_i. There is a connection between the "slow" $\mathbf{x}$ and "fast" $\mathbf{z} \in v_i$ variables: $\mathbf{x} = \sum m_j \mathbf{e}_j + \mathbf{y}$.

The mean matrix strains follow simply from Eq. (11.16) and the relation $\langle\boldsymbol{\varepsilon}\rangle_0(\mathbf{x}) = (\langle\boldsymbol{\varepsilon}\rangle_{\mathbf{x}}(\mathbf{x}) - c^{(1)}\langle\boldsymbol{\varepsilon}\rangle_{(i)})/c^{(0)}$, where $\mathbf{x} \in \Omega_i \backslash v_i$. Substituting (11.16) into (11.2) gives the local strains in the matrix $\mathbf{z} \in \Omega_i \backslash v_i$ in the form of an integro–differential equation

$$\boldsymbol{\varepsilon}(\mathbf{x}_i, \mathbf{z}) = \langle\boldsymbol{\varepsilon}\rangle_{\mathbf{x}_i}(\mathbf{x}_i) + \int \sum_{\mathbf{m}} \Big[\mathbf{T}^{\epsilon}_{i\mathbf{m}}(\mathbf{z}-\mathbf{y}) - \langle\mathbf{T}^{\epsilon}_{\mathbf{m}}\rangle_{\mathbf{z}}(\mathbf{z}-\mathbf{y})\Big]\mathbf{R}^{\epsilon}\delta(\mathbf{y}-\mathbf{x}_{\mathbf{m}}) \times \Big[\mathbf{Y}^{\epsilon}\langle\boldsymbol{\varepsilon}\rangle_{\mathbf{y}}(\mathbf{y}) + \boldsymbol{\mathcal{Y}}\nabla\nabla\langle\boldsymbol{\varepsilon}\rangle_{\mathbf{y}}\Big]\, d\mathbf{y}. \tag{11.17}$$

When using the integral equation (11.14) rather than the differential dependence form of the effective strain $\overline{\boldsymbol{\varepsilon}}(\mathbf{x}_i)$ on the mean strain $\boldsymbol{\varepsilon}^{\Omega}(\mathbf{x}_i)$, Eqs. (11.15), (11.16), and (11.17) should be replaced by ($\mathbf{r} \in v_i \subset \Omega_i,\ \mathbf{z} \in \Omega_i \backslash v_i$):

$$\begin{aligned}
\boldsymbol{\varepsilon}(\mathbf{x}_i, \mathbf{r}) &= \mathbf{A}(\mathbf{r})\Big\{\mathbf{Y}^{\epsilon}\langle\boldsymbol{\varepsilon}\rangle_{\mathbf{x}_i}(\mathbf{x}_i) + \int \boldsymbol{\mathcal{Z}}(\mathbf{x}_i - \mathbf{y})\big[\langle\boldsymbol{\varepsilon}\rangle_{\mathbf{y}}(\mathbf{y}) - \langle\boldsymbol{\varepsilon}\rangle_{\mathbf{x}_i}(\mathbf{x}_i)\big]\, d\mathbf{y}\Big\}, \\
\langle\boldsymbol{\varepsilon}\rangle_{(i)} &= \mathbf{A}\Big\{\mathbf{Y}^{\epsilon}\langle\boldsymbol{\varepsilon}\rangle_{\mathbf{x}_i}(\mathbf{x}_i) + \int \boldsymbol{\mathcal{Z}}(\mathbf{x}_i - \mathbf{y})\big[\langle\boldsymbol{\varepsilon}\rangle_{\mathbf{y}}(\mathbf{y}) - \langle\boldsymbol{\varepsilon}\rangle_{\mathbf{x}_i}(\mathbf{x}_i)\big]\, d\mathbf{y}\Big\}, \\
\boldsymbol{\varepsilon}(\mathbf{x}_i, \mathbf{z}) &= \langle\boldsymbol{\varepsilon}\rangle_{\mathbf{x}_i}(\mathbf{x}_i) + \int \sum_{\mathbf{m}} \Big[\mathbf{T}^{\epsilon}_{i\mathbf{m}}(\mathbf{z}-\mathbf{y}) - \langle\mathbf{T}_{\mathbf{m}}{}^{\epsilon}\rangle_{\mathbf{z}}(\mathbf{z}-\mathbf{y})\Big]\mathbf{R}^{\epsilon}\delta(\mathbf{y}-\mathbf{x}_{\mathbf{m}}) \\
&\quad \times \Big\{\mathbf{Y}^{\epsilon}\langle\boldsymbol{\varepsilon}\rangle_{\mathbf{x}_{\mathbf{m}}}(\mathbf{x}_{\mathbf{m}}) + \int \boldsymbol{\mathcal{Z}}(\mathbf{x}_{\mathbf{m}} - \mathbf{t})\big[\langle\boldsymbol{\varepsilon}\rangle_{\mathbf{t}}(\mathbf{t}) - \langle\boldsymbol{\varepsilon}\rangle_{\mathbf{x}_{\mathbf{m}}}(\mathbf{x}_{\mathbf{m}})\big]\, d\mathbf{t}\Big\}\, d\mathbf{y},
\end{aligned} \tag{11.18}$$

respectively.

11.2.5 Effective Properties of Composites

Taking the average strain in the inclusions (11.16) gives a macroscopic constitutive equation that relates $\langle\boldsymbol{\sigma}\rangle(\mathbf{x})$ and $\langle\boldsymbol{\varepsilon}\rangle(\mathbf{x})$:

$$\langle\boldsymbol{\sigma}\rangle(\mathbf{x}) = \mathbf{L}^*\langle\boldsymbol{\varepsilon}\rangle(\mathbf{x}) + \boldsymbol{\mathcal{L}}^*\nabla\nabla\langle\boldsymbol{\varepsilon}\rangle(\mathbf{x}), \quad \mathbf{L}^* = \mathbf{L}^{(0)} + \mathbf{R}^{\epsilon}\mathbf{Y}^{\epsilon}n, \quad \boldsymbol{\mathcal{L}}^* = \mathbf{R}^{\epsilon}\boldsymbol{\mathcal{Y}}n. \tag{11.19}$$

The treatment of the integral form of the differential macroscopic constitutive Eq. (11.19_1) leads to the nonlocal relation:

$$\langle\boldsymbol{\sigma}\rangle(\mathbf{x}) = \mathbf{L}^*\langle\boldsymbol{\varepsilon}\rangle(\mathbf{x}) + n\mathbf{R}^{\epsilon}\int \boldsymbol{\mathcal{Z}}(\mathbf{x}-\mathbf{y})\big[\langle\boldsymbol{\varepsilon}\rangle_{\mathbf{y}}(\mathbf{y}) - \langle\boldsymbol{\varepsilon}\rangle_{\mathbf{x}}(\mathbf{x})\big]\, d\mathbf{y}. \tag{11.20}$$

For the sake of definiteness, let the composite material be subjected to a strain gradient along the direction $\mathbf{e}_3$ of the orthogonal basis $\mathbf{e}_1,\ \mathbf{e}_2,\ \mathbf{e}_3$. Then

$$\langle \boldsymbol{\varepsilon} \rangle_{\mathbf{x}}(\mathbf{x}) = f(x_3)\boldsymbol{\varepsilon}^{\text{con}}, \quad \mathbf{x} = (x_1, x_2, x_3)^\top, \tag{11.21}$$

where $\boldsymbol{\varepsilon}^{\text{con}} \equiv$ const. and $f(x_3) \not\equiv$ const., and a three-dimensional integral (11.20) can be reduced to the one-dimensional

$$\langle \boldsymbol{\sigma} \rangle_{\mathbf{x}}(\mathbf{x}) = \mathbf{L}^* \boldsymbol{\varepsilon}^{\text{con}} f(x_3) + \int \boldsymbol{\mathcal{Z}}^1(x_3 - y_3)\big[f(y_3) - f(x_3)\big] dy_3 \; \boldsymbol{\varepsilon}^{\text{con}}, \tag{11.22}$$

where the integral operator kernel $\boldsymbol{\mathcal{Z}}^1(x_3 - y_3) \equiv n\mathbf{R}^\epsilon \int\int \boldsymbol{\mathcal{Z}}(\mathbf{x} - \mathbf{y}) \; dy_1 \; dy_2$.

The preceding relation for the local effective properties $\mathbf{L}^*$ may be simplified by means of addition assumptions. For example, an expression for $\mathbf{L}^*$ for the ellipsoidal RVE w^{el} (providing homogeneity of the effective field, see Section 9.4) was obtained [174]. In the case of triply periodic structures (5.103) under the uniform boundary conditions, we have

$$\boldsymbol{\sigma}^0(\mathbf{x}) = \boldsymbol{\sigma}^0 \equiv \text{const}. \tag{11.23}$$

Then the tensor $\mathbf{L}^*$ now has a form

$$\mathbf{L}^* = \mathbf{L}^{(0)} + \mathbf{R}^\epsilon n \Big\{ \mathbf{I} - \mathbf{P}(w^{\text{el}})\mathbf{R}^\epsilon n - \sum_{\mathbf{m} \neq \mathbf{0}} \mathbf{T}^\epsilon_{i\mathbf{m}}(\mathbf{x}_i - \mathbf{x}_{\mathbf{m}})\mathbf{R}^\epsilon \Big\}^{-1}, \tag{11.24}$$

where $\mathbf{x}_{\mathbf{m}} \in w^{el}$, and for the sake of definiteness $\mathbf{x}_i = \mathbf{x}_0 = \mathbf{0}$ and the index $\mathbf{m}$ is a triplet: $\mathbf{m} = (m_1, m_2, m_3)$. It is assumed that $\mathbf{x}_i$ coincides with the center of the region w^{el}, containing a quite large number of inclusions $\mathbf{x}_{\mathbf{m}} \in w^{\text{el}}$, and $\mathbf{P}(w^{\text{el}})$ is defined by Eq. (3.110) for the ellipsoidal domain w^{el}.

Alternatively in (11.24) one may use a so-called point approximation [618] $\mathbf{T}^\epsilon_{ij}(\mathbf{x}_i - \mathbf{x}_j) = \mathbf{U}(\mathbf{x}_i - \mathbf{x}_j)$, which is exact for infinitely spaced heterogeneities. Then Eq. (11.24) is reduced to

$$\mathbf{L}^* = \mathbf{L}^{(0)} + \mathbf{R}^\epsilon n \Big\{ \mathbf{I} - \mathbf{P}(w^{\text{el}})\mathbf{R}^\epsilon n - \sum_{\mathbf{m} \neq \mathbf{0}} \mathbf{U}(\mathbf{x}_i - \mathbf{x}_{\mathbf{m}})\mathbf{R}^\epsilon \Big\}^{-1}, \quad \mathbf{x}_{\mathbf{m}} \in w^{el}. \tag{11.25}$$

A significant error of the relation (11.25) as compared to (11.24) was demonstrated in [174] (see also Subsection 11.2.6).

11.2.6 Numerical Results

Let us consider as an example a composite consisting of isotropic homogeneous components and having identical spherical inclusions $\mathbf{L}^{(i)} = (3k^{(i)}, 2\mu^{(i)}) \equiv 3k^{(i)}\mathbf{N}_1 + 2\mu^{(i)}\mathbf{N}_2$, $(\mathbf{N}_1 = \delta \otimes \delta/3,\ \mathbf{N}_2 = \mathbf{I} - \mathbf{N}_1)$. For a simple cubic (SC) lattice of spherical inclusions the tensor of effective moduli $\mathbf{L}^*$ (11.19$_2$) is characterized by three elastic moduli:

$$k^* = (L^*_{1111} + 2L^*_{1122})/3, \quad \mu^* = L^*_{1212}, \quad \widetilde{\mu}^* = (L^*_{1111} - L^*_{1122})/2, \tag{11.26}$$

where the stiffness components are given with respect to a coordinate system whose base vectors are normal to the faces of the unit cell. In the interest of obtaining maximum difference between the effective properties, estimated by the different methods we will consider the examples for hard inclusions ($\nu^{(0)} = \nu^{(1)} = 0.3$, $\mu^{(1)}/\mu^{(0)} = 1000$) as well as for the voids ($\mathbf{L}^{(1)} \equiv \mathbf{0}$, see for details [133], and a number of values of the volume concentration of inclusions. The local elastic moduli (11.26) are computed by analytical methods in [624], [820], as well as by the formulae (11.24) and (11.25) (see Table 11.1)); here $\nu^{(i)} \equiv (3k^{(i)} - 2\mu^{(i)})/(6k^{(i)} - 2\mu^{(i)})$, $(i = 0, 1)$ is a Poisson ratio.

It was shown that the error of Eq. (11.24) is maximum for $c = 0.5$ and does not exceed 30% and 23% for the rigid inclusions and for the voids, respectively; similar errors for $c = 0.4$ do not exceed 19% and 14%, respectively. The method (11.24) performs quite well (at least for $\mathbf{L}^{(1)} \equiv \mathbf{0}$) even if c is quite close to limiting packing coefficient for SC: $c_{\max} = \pi/6 \cong 0.52$. The calculation by the approximate variant (11.25) gives contradictory results for $c > 0.35$: the component L^*_{1111} oscillates around zero as c increases. Table 11.1 give the values $\mathbf{L}^*$ calculated from formulae (11.24) and (11.25) for spherical RVE with radius $r_{25} = 25|\mathbf{e}_1|$, containing 25 layers of inclusions around a considered inclusion v_i. Similar results for elastic moduli $\mathbf{L}^*$ (11.24) estimated for different radii of RVE are represented in Table 11.2. The RVE with $r_1 = |\mathbf{e}_1|$, $r_3 = 3|\mathbf{e}_1|$, and $r_6 = 6|\mathbf{e}_1|$ have one, three, and six layers of surrounding inclusions around the considered one, respectively. As can be seen from Table 11.2, the estimations of $\mathbf{L}^*$ (11.24) for the RVE with the radii r_3 and r_6 (at $c = 0.5$) differ from those presented in Table 11.1 by 10% and 3% as maximum, respectively; i.e., the RVE with three layers of inclusions can already be considered as representative and the principle of locality [1027] holds. The use of a 6-layers spherical RVE guarantees at least three-digit accuracy for the voids and two-digit accuracy for the rigid inclusions.

Table 11.1. Overall elastic constants of SC arrays of rigid inclusions: (N) Nunan and Keller [820] for $c = 0.1$–0.4, (K) Kushch [624] for $c = 0.5$, (P) point approximation (11.25), (H1) the proposed method (11.24).

	$k^*/k^{(0)}$			$\mu^*/\mu^{(0)}$			$\widetilde{\mu}^*/\mu^{(0)}$		
c	N/K	P	H1	N/K	P	H1	N/K	P	H1
0.10	1.180	1.179	1.179	1.216	1.207	1.213	1.274	1.286	1.269
0.20	1.405	1.403	1.403	1.455	1.410	1.451	1.704	1.897	1.690
0.30	1.706	1.691	1.691	1.766	1.608	1.740	2.35	4.132	2.319
0.40	2.173	2.074	2.074	2.25	1.802	2.120	3.74	–11.8	3.207
0.50	3.503	2.610	2.610	3.14	1.999	2.674	6.49	–2.15	4.334

Table 11.2. Overall elastic constants of SC arrays of rigid inclusions estimated by the Eq. (11.24) for RVE with the radii r_1, r_3, and r_6.

	$k^*/k^{(0)}$			$\mu^*/\mu^{(0)}$			$\widetilde{\mu}^*/\mu^{(0)}$		
c	r_1	r_3	r_6	r_1	r_3	r_6	r_1	r_3	r_6
0.10	1.179	1.179	1.179	1.207	1.215	1.214	1.285	1.266	1.268
0.20	1.403	1.403	1.403	1.427	1.458	1.453	1.791	1.667	1.682
0.30	1.691	1.691	1.691	1.682	1.759	1.746	2.708	2.236	2.291
0.40	2.074	2.074	2.074	2.000	2.164	2.134	4.423	2.985	3.131
0.50	2.610	2.610	2.610	2.434	2.774	2.705	7.686	3.858	4.165

The analysis of Tables 11.1 and 11.2 is an extension summary of [174], where the stated problem was investigated. More recently the same problem has also been considered by use of some additional unnecessary assumptions. So Rodin [940] proposed a different equivalent approach based on the eigenstrain method; the convergence of integral representations was justified for media with an isotropic matrix containing homogeneous spherical inclusions. The analogous problem for the spherical RVE was analyzed in [782]. This example is considered deliberately for the demonstration of high accuracy of the proposed method (11.24) based on the use of hypothesis **H1**. Still, one could argue that if hypothesis **H1** is used for the estimation of effective nonlocal properties then the accuracy is satisfactory for many purposes, although to our knowledge, the exact analytical methods has never been actually implemented in the estimation of nonlocal effective properties, which we will consider now by the approximate method (11.11_1) and (11.19_3).

Now let us compare the different components of the normalized tensor $\boldsymbol{\mathcal{L}}^{*nor} \equiv 10\boldsymbol{\mathcal{L}}^*/(\mu^{(0)}|\mathbf{e}_1|^2)$ describing the nonlocal properties and obtained by the use of Eqs. (11.10_2) and (11.19_3) for different radii of the RVE: $r_6 = 6|\mathbf{e}_1|$, $r_9 = 9|\mathbf{e}_1|$, $r_{25} = 25|\mathbf{e}_1|$. As can be seen from Table 11.3 for rigid inclusions the maximum errors are 18% at $r = r_6$ and 7% at $r = r_9$ compared to the result at $r = r_{25}$. Therefore for good accuracy (i.e., 7% error) of a constitutive nonlocal model, the minimum RVE size is relatively large: $r = 9|\mathbf{e}_1|$. In the case of rigid inclusions the minimum RVE size required for the same accuracy of estimated nonlocal parameters is always substantially larger as compared to the case of voids considered in [133]. Table 11.3 shows that for the systems considered, the minimum RVE size increases with increasing inclusion volume fraction. Comparing the analysis of Tables 11.1 to 11.3 shows that for equal accuracy of effective elastic properties for local and nonlocal response, the RVE must be larger in the case of nonlocal properties. This fact is explained by different behavior at infinity of integrand functions $\boldsymbol{\mathcal{F}}(\mathbf{x}_i - \mathbf{y})$ and $\boldsymbol{\mathcal{F}}_2(\mathbf{x}_i - \mathbf{y})$ in the integral representations of the local (11.11_1), (11.19_2) and nonlocal (11.11_2), (11.19_3) operators, respectively.

Table 11.3. Overall normalized nonlocal elastic constants of SC arrays of rigid inclusions estimated by the Eq. (11.19_3) for RVE with the radii r_6, r_9, and r_{25}.

	$\mathcal{L}^{*nor}_{111111}$			$\mathcal{L}^{*nor}_{112211}$			$\mathcal{L}^{*nor}_{121211}$		
c	r_6	r_9	r_{25}	r_6	r_9	r_{25}	r_6	r_9	r_{25}
0.10	0.164	0.166	0.167	-0.059	-0.060	-0.060	-0.068	-0.068	-0.069
0.20	0.875	0.894	0.904	-0.303	-0.312	-0.318	-0.233	-0.235	-0.236
0.30	2.588	2.678	2.725	-0.859	-0.898	-0.918	-0.468	-0.471	-0.473
0.40	5.771	6.067	6.226	-1.771	-1.893	-1.958	-0.749	-0.752	-0.754
0.50	10.37	11.10	11.50	-2.700	-2.981	-3.133	-0.997	-1.003	-1.005

As mentioned above, the differential representation of the nonlocal operator (11.19_1) has the disadvantage that is uses the Taylor expansions (11.7) and (11.9) in the RVE about the center $\mathbf{x}$; it is sometimes violated in practice. For instance an infinite or nonexisting derivative of some finite order in $\langle\varepsilon\rangle_{\mathbf{x}}(\mathbf{x})$ can take place; the radius of convergence of Taylor series (11.7) and (11.9) can be less than the radius of the RVE, which may cause slow convergence at an infinite range for integral operators. So in our case, the integral (11.11_2) is conditionally convergent, i.e., it depends on the shape of the RVE. Similar circumstances can lower the feasibility of the Fourier transform method which is considered in the current section in sufficient detail deliberately for demonstration of disadvantages of this popular method. For later use we derived the integral form of the nonlocal operator (11.14) which does not suffer from this limitation, and which will be considered now.

For the sake of definiteness the composite materials is subjected to a strain gradient along the direction $\mathbf{e}_3$ and only one component $\varepsilon_{ij}^{\rm con}$ (11.21) differs from zero: $\varepsilon_{ij}^{\rm con} \neq 0$; all other $\varepsilon_{kl}^{\rm con} \equiv 0 \ \ (kl \neq ij)$. Finally, lest it be thought that all nonlocal operators with smooth $\langle \boldsymbol{\varepsilon} \rangle_{\mathbf{x}}(\mathbf{x})$ can straightforwardly be solved by the method (11.10) and (11.19_1), we give two counterexamples:
a monotonical smooth function

$$f(x_3) = f_1(x_3) \equiv H(x_3)\frac{(x_3/a)^4}{1+(x_3/a)^4} \tag{11.27}$$

and an even infinitely differentiable function

$$f(x_3) = f_2(x_3) \equiv 1 - e^{-(x_3/a)^4}, \tag{11.28}$$

where a is a positive length parameter. Clearly $f_3(x_3),\ f_1''(x_3) \equiv 0$ and $f_2(x_3),\ f_2''(x_3) = 0$ at $x_3 \leq 0$ and $x_3 = 0$, respectively. Therefore the differential approach (11.19_1) leads to degenerate results: $\langle \boldsymbol{\sigma} \rangle_{\mathbf{x}}(\mathbf{x}) \equiv \mathbf{0}$ for $x_3 \leq 0$, and $\langle \boldsymbol{\sigma} \rangle_{\mathbf{x}}(\mathbf{x}) = \mathbf{0}$ for $x_3 = 0$, for the functions f_1 (11.27) and f_2 (11.28), respectively.

Thus the approximation of a nonlocal operator by the second-order differential operator might be too crude even if the driving function $\langle \boldsymbol{\varepsilon} \rangle_{\mathbf{x}}(\mathbf{x})$ is smooth enough (11.21), (11.27) and (11.28). At the same time, the treatment of indicated functions within the framework of the method of successive approximations (11.14), (11.22) is quite efficient. We illustrate the statement made above with the function $f_1(x_3)$ (11.27) for rigid inclusions (when the estimations obtained by different methods will have the maximum dissimilarity from one another), and a number of values of a. Let the tensors $\boldsymbol{\sigma}^{\rm loc}(\mathbf{x}) \equiv \mathbf{L}^* \langle \boldsymbol{\varepsilon} \rangle_{\mathbf{x}}(\mathbf{x})$ and $\boldsymbol{\sigma}^{\rm non}(\mathbf{x}) \equiv \langle \boldsymbol{\sigma} \rangle_{\mathbf{x}}(\mathbf{x}) - \mathbf{L}^* \langle \boldsymbol{\varepsilon} \rangle_{\mathbf{x}}(\mathbf{x})$ be named the local and nonlocal stresses, respectively. In Fig. 11.1 the 33 components of normalized nonlocal stresses $\sigma_{33}^{\rm non}(\mathbf{x})/(\mu^{(0)}\varepsilon_{33}^{\rm con})$ are plotted as the functions of dimensionless parameters $a^{\rm nor} \equiv a/|\mathbf{e}_3|$ for different values of normalized coordinates $x^{\rm nor} \equiv x_3/|\mathbf{e}_3| = 0, 1, 2, 3$ (layer number). As can be seen from Fig. 11.1, the nonlocal stress $\sigma_{33}^{\rm non}(0)/(\mu^{(0)}\varepsilon_{33}^{\rm con}) \neq 0$ and decreases with increasing $a^{\rm nor}$, notwithstanding the fact, that according to the formulae (11.19_1) and (11.27), $\boldsymbol{\sigma}^{\rm loc}(\mathbf{0}) = \boldsymbol{\sigma}^{\rm non}(\mathbf{0}) \equiv \mathbf{0}$, $\forall a > 0$. At the points $x_3 = |\mathbf{e}_3|$ the nonlocal stress $\sigma_{33}^{\rm non}(\mathbf{x})/(\mu^{(0)}\varepsilon_{33}^{\rm con})$ reaches its maximum at $a^{\rm nor} = 1.65$ and equals 8% of the local stress $\sigma_{33}^{\rm loc}(\mathbf{x})/(\mu^{(0)}\varepsilon_{33}^{\rm con})$ $(x_3 = |\mathbf{e}_3|)$. It should be mentioned that $f_1(x_3)$ (11.27) is a monotonic function; however, the nonlocal stresses may reverse sign, which is compatible with changing of the sign of the second derivation $f_1''(x_3)$ in the framework of the differential approach (11.19_1). For the function $f_2(x_3)$ one obtains a similar dependence of nonvanished nonlocal stresses $\sigma_{ij}^{\rm non}(\mathbf{x})/(\mu^{(0)}\varepsilon_{ij}^{\rm con})$ $(ij = 11,\ 33,\ 13,\ 12)$ (which increase with increasing reinforcement volume fraction c) as the functions of parameters $a^{\rm nor}$ at $x_3 = 0$, which is in contrast to the results obtained by conventional differential approach (11.19$_1$): $\boldsymbol{\sigma}^{\rm non}(\mathbf{0}) \equiv \mathbf{0}$.

Let us now compare the 33 components of nonlocal normalized stresses $\sigma_{33}^{\rm non}(x_3)\ /(\mu^{(0)}\varepsilon_{33}^{\rm con})$ estimated by both the Fourier transform method (11.19$_1$) and by the iteration method (11.20). Figure 11.2 shows the normalized stresses $\sigma_{33}^{\rm non}(x_3)/(\mu^{(0)}\varepsilon_{33}^{\rm con})$ as the functions of layer numbers $x_n^{\rm nor} \equiv x_{n3}/|\mathbf{e}_3| = 0, \pm 1, \pm 2, \ldots$ $(n = 0, \pm 1, \pm 2, \ldots)$ for the function $f_2(x_3)$ (11.28) with dimensionless parameter values $a^{\rm nor} = 1$ and $a^{\rm nor} = 2$, and $c = 0.5$. It is evident from Fig. 11.2 that for strongly varying average strains $\langle \boldsymbol{\varepsilon} \rangle_{\mathbf{x}}(\mathbf{x})$ (for $a^{\rm nor} = 1$) a qualitative difference between the results obtained by dissimilar methods occurs. For smoother fields $\langle \boldsymbol{\varepsilon} \rangle_{\mathbf{x}}(\mathbf{x})$ (for $a^{\rm nor} = 2$) the analogous curves are distinguished from one another, not nearly so much as in a case $a^{\rm nor} = 1$. In so doing at $a^{\rm nor} = 2$ the use of the iteration method leads to the values $\sigma_{33}^{\rm non}(x_{13}) = 1.16\sigma_{33}^{\rm loc}(x_{13})$, $\sigma_{33}^{\rm non}(x_{23}) = 0.15\sigma_{33}^{\rm loc}(x_{23})$, and $\sigma_{33}^{\rm non}(x_{03}) = 0.03\sigma_{33}^{\rm loc}(x_{23})$, although $\sigma_{33}^{\rm loc}(x_{03}) = \sigma_{33}^{\rm non}(x_{03}) \equiv 0$, $\forall a > 0$ with the use of the Fourier transform method.

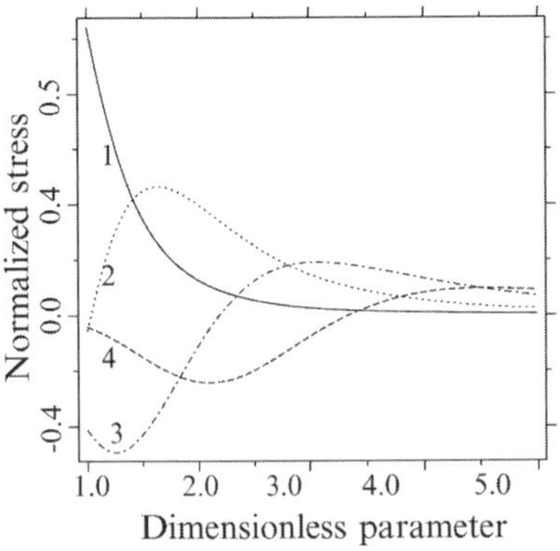

Fig. 11.1. Normalized nonlocal stresses $\sigma_{33}^{\rm non}(x_3)/(\mu^{(0)}\varepsilon_{33}^{\rm con})$ as functions of the dimensionless parameter $a^{\rm nor}$: $x_3/|\mathbf{e}_3| = 0$ (curve 1), 1 (curve 2), 2 (curve 3), 3 (curve 4).

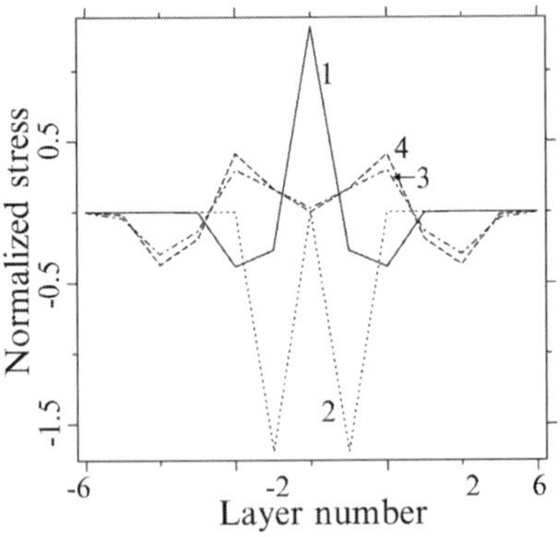

Fig. 11.2. $\sigma_{33}^{\rm non}(\mathbf{x})/(\mu^{(0)}\varepsilon_{33}^{\rm con})$ vs $x^{\rm nor}$ calculated via the Fourier method (11.19$_1$) (dotted and dashed curves) and via the iteration method (11.20) (curves 1,3) for $a^{\rm nor} = 1$ (curves 1, 2) and $a^{\rm nor} = 2$ (curves 4, 3).

In conclusion, it may be said that the relations obtained depend on the values associated with the mean distance between inclusions and do not depend on the other characteristic size, i.e., the mean inclusion diameter. This fact may be explained by the initial acceptance of hypothesis **H1** dealing with the homogeneity of the field $\overline{\boldsymbol{\sigma}}(\mathbf{x})$ inside each inclusion. In the case of a variable representation of $\overline{\boldsymbol{\sigma}}(\mathbf{x})$ $(\mathbf{x} \in v_i)$, for instance in polynomial form, the mean size of the inclusions will be contained in the nonlocal dependence of microstresses on the average stress $\langle \boldsymbol{\sigma} \rangle_{\mathbf{x}}(\mathbf{x})$.

It should be mentioned that the effective constitutive Eq. (11.18) was derived for points $\mathbf{x}_i$ located sufficiently far from the boundary of the body ∂w. In so doing the relations developed have been obtained by the use of the whole-space Green's function. Then use of nonlocal constitutive relations (11.18) requires more complicated boundary conditions [69], and [279], and [621]; this question is beyond the scope of the current study.

11.3 Graded Doubly Periodical Particular Matrix Composites in Varying External Stress Field

We will consider now (see Fig. 11.3) a graded doubly periodic set X of ellipsoidal inclusions with identical shape, orientation, and mechanical properties when the inclusion centers $\mathbf{x_m}$ are distributed at the nodes of some spatial lattice $\mathbf{m} \in \Lambda$ (5.104): $\mathbf{x_m} = m_1\mathbf{e}_1 + m_2\mathbf{e}_2 + f(m_3)\mathbf{e}_3$, where $\mathbf{m} = (m_1, m_2, m_3)$ are integer–valued coordinates of the node $\mathbf{m}$ in the basis $\mathbf{e}_i$ which are equal in modulus to $|\mathbf{e}_i|$, and $f(m_3) - f(m_3+1) \not\equiv \text{const}$. In the functionally graded direction $\mathbf{e}_3$ the inclusion spacing between adjacent arrays may vary ($f(m_3) - f(m_3+1) \not\equiv$ const.). For a doubly periodic array of inclusions in a finite ply containing $2m^l + 1$ layers of inclusions we have $f(m_3) \equiv 0$ at $|m_3| > m^l$; in the more general case of doubly periodic structures $f(m_3) \not\equiv 0$ at $m_3 \to \pm\infty$. To make exposition more clear we will assume that the basis $\mathbf{e}_i$ is an orthogonal one and the axes $\mathbf{e}_i$ ($i = 1, 2, 3$) are directed along axes of the global Cartesian coordinate system (these assumptions are not obligatory).

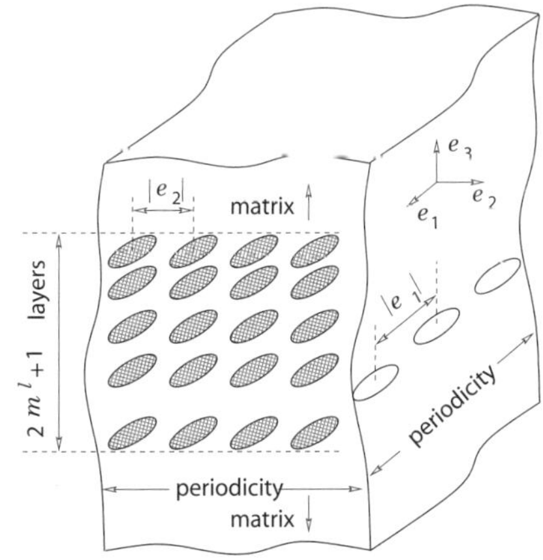

Fig. 11.3. Schematic representation of the doubly periodic inclusion ply.

11.3.1 Local Approximation of Effective Stresses

In the framework of hypothesis **H1**, from (11.2), taking (4.10) into account we get

$$\overline{\boldsymbol{\eta}}(\mathbf{x}_i) = \boldsymbol{\eta}^{av}(\mathbf{x}_i) + \int \mathbf{K}(\mathbf{x}_i, \mathbf{y})\overline{\boldsymbol{\eta}}(\mathbf{y})\, d\mathbf{y}, \tag{11.29}$$

where $\overline{\boldsymbol{\eta}}(\mathbf{x}) \equiv \boldsymbol{\eta}_i$, and $\overline{v}_i\boldsymbol{\eta}^{av}(\mathbf{x}) \equiv \mathbf{R}\langle\boldsymbol{\sigma}\rangle_{\mathbf{x}}(\mathbf{x}) + \mathbf{F}$ (at $\mathbf{x} \in v_i$) will be named the modified strain polarization tensor of average stresses. The integral operator kernel introduced in Eq. (11.29) is defined on the lattice Λ:

$$\mathbf{K}(\mathbf{x}_i,\mathbf{y}) = \mathbf{R}\sum_{\mathbf{m}}\Big[\mathbf{T}_{i\mathbf{m}}(\mathbf{x}_i-\mathbf{x}_\mathbf{m})(1-V_i(\mathbf{y}))-\langle\mathbf{T}_\mathbf{m}\rangle_{\mathbf{x}_i}(\mathbf{x}_i-\mathbf{x}_\mathbf{m})\Big]\delta(\mathbf{y}-\mathbf{x}_\mathbf{m}), \tag{11.30}$$

and $\mathbf{T}_{i\mathbf{m}}(\mathbf{x}_i-\mathbf{x}_\mathbf{m}) = (\overline{v}_i\overline{v}_\mathbf{m})^{-1}\int \boldsymbol{\Gamma}(\mathbf{x}-\mathbf{y})V_i(\mathbf{x})V_\mathbf{m}(\mathbf{y})\,d\mathbf{x}\,d\mathbf{y}$. Rewriting Eq. (11.29) in the spirit of a subtraction technique gives

$$\overline{\boldsymbol{\eta}}(\mathbf{x}_i) = \mathbf{Y}(\mathbf{x}_i)\boldsymbol{\eta}^{av}(\mathbf{x}_i) + \int \boldsymbol{\mathcal{K}}(\mathbf{x}_i,\mathbf{y})\overline{\boldsymbol{\eta}}(\mathbf{y})\,d\mathbf{y}, \tag{11.31}$$

where

$$\mathbf{Y}(\mathbf{x}_i) \equiv \Big(\mathbf{I} - \int \mathbf{K}(\mathbf{x}_i,\mathbf{y})\,d\mathbf{y}\Big)^{-1}, \tag{11.32}$$

$$\boldsymbol{\mathcal{K}}(\mathbf{x}_i,\mathbf{y}) \equiv \mathbf{Y}(\mathbf{x}_i)\Big[\mathbf{K}(\mathbf{x}_i,\mathbf{y}) - \delta(\mathbf{x}_i-\mathbf{y})\int \mathbf{K}(\mathbf{x}_i,\mathbf{z})\,d\mathbf{z}\Big]. \tag{11.33}$$

The matrix $\mathbf{Y}(\mathbf{x}_i)$ determines the "local" action of the surrounding inclusions on the separated one, while the integral operator kernel $\boldsymbol{\mathcal{K}}(\mathbf{x}_i,\mathbf{y})$ describes a "nonlocal" action of these inclusions. For the purpose of clarifying the above statement we consider, as an example, a particular problem for triply periodic structures (5.103) under the homogeneous boundary conditions (11.23). In such a case $\mathbf{Y}(\mathbf{x}_i) \equiv \mathbf{Y}^{\text{tri}} = \text{const.}$, and for an ellipsoidal representative volume element w^{el} (RVE) containing a statistically large number of inclusions the result can be rewritten in the current notation as

$$\mathbf{Y}^{\text{tri}} = \Big(\mathbf{I} - \mathbf{R}\mathbf{Q}(w^{\text{el}}) - \mathbf{R}\sum_{\mathbf{m}\neq i}\mathbf{T}_{i\mathbf{m}}(\mathbf{x}_i-\mathbf{x}_\mathbf{m})\Big)^{-1}, \quad \mathbf{x}_\mathbf{m}\in w^{\text{el}}, \tag{11.34}$$

where the tensor $\mathbf{Q}(w^{\text{el}})$ is defined for the domain w^{el} in a similar manner to Eq. (3.110). Clearly for triply periodic structures (5.103) Eqs. (11.32) and (11.34) coincide. If furthermore, $\boldsymbol{\sigma}^0(\mathbf{x}) \equiv \text{const.}$ then $\overline{\boldsymbol{\eta}}(\mathbf{y})$ is insensitive to translations, and the right-hand-side integral in Eq. (11.31) vanishes, and Eq. (11.31) is local.

When one of two (or both) assumptions (5.103) and (11.23) breaks down, Eq. (11.31) is nonlocal on two counts. So, if only the assumption (11.23) breaks down then $\overline{\boldsymbol{\eta}}(\mathbf{y}) \not\equiv \text{const.}$, and the integral in Eq. (11.31) does not vanish. Nevertheless the current kind of the nonlocalization can usually be easily analyzed. So, in this case the kernel $\boldsymbol{\mathcal{K}}(\mathbf{x}_i,\mathbf{y})$ is a translation kernel: $\boldsymbol{\mathcal{K}}(\mathbf{x}_i,\mathbf{y}) = \boldsymbol{\mathcal{K}}(\mathbf{x}_i-\mathbf{y})$, and the current problem is akin to the estimation of nonlocal effects in statistically homogeneous random structure composites. So, for slowly varying functions $\overline{\boldsymbol{\eta}}(\mathbf{y})$ Taylor expansion of $\overline{\boldsymbol{\eta}}(\mathbf{y})$ about $\mathbf{x}_i$ reduces Eq. (11.31) to a differential equation with constant coefficients. The method of solving this that first comes to mind is using the Fourier transformation to transform the differential version of solving (11.31) into the inverse problem of solving the multiplicative equation [132], [133], [134], [181].

The breakdown of the assumption (5.103) is more common in practice, because it leads to the inequality $\mathbf{Y}(\mathbf{x}_i) \not\equiv \text{const.}$ Then the average stresses $\langle\boldsymbol{\sigma}\rangle_\mathbf{x}(\mathbf{x}) \not\equiv \text{const.}$ and hence $\overline{\boldsymbol{\eta}}(\mathbf{y}) \not\equiv \text{const.}$ To put this another way, we will have a nonlocal Eq. (11.31) even at the homogeneous boundary conditions (11.23), which is difficult to solve by the Fourier transform method insofar as

$\mathcal{K}(\mathbf{x}_i, \mathbf{y}) \neq \mathcal{K}(\mathbf{x}_i - \mathbf{y})$. Moreover, let us consider $\mathbf{Y}(\mathbf{x}_i)$ (11.32) as an approximation of the corresponding right-hand-side nonlocal operator in Eq. (11.31) by constant tensor (zero-order approximation). From such a consideration it can be concluded that "local" part $\mathbf{Y}(\mathbf{x}_i)$ of the nonlocal operator (11.31) depends explicitly not only on the local parameters of the inclusion distribution at the point $\mathbf{x}_i$, but also in a certain neighborhood of that point. This sort of a so-called nonlocal effect was identified in [181] for statistically inhomogeneous random structure composites, and takes place for doubly periodic structure as well (see also Section 12.5).

11.3.2 Estimation of the Nonlocal Operator via the Iteration Method

As mentioned, the method of Fourier transform has been much investigated in nonlocal micromechanics, but its use is difficult if $\mathcal{K}(\mathbf{x}_i, \mathbf{y}) \neq \mathcal{K}(\mathbf{x}_i - \mathbf{y})$. This inconsistency can be avoided if the method of successive approximations, which is also called the Neumann series method, is used. With this in mind we initially define the function $(\mathcal{K}\overline{\boldsymbol{\eta}})(\mathbf{x}_i) = \int \mathcal{K}(\mathbf{x}_i, \mathbf{y})\overline{\boldsymbol{\eta}}(\mathbf{y})\, d\mathbf{y}$. Then Eq. (11.31) can be abbreviated as

$$\overline{\boldsymbol{\eta}} = \mathbf{Y}\boldsymbol{\eta}^{av} + \mathcal{K}\overline{\boldsymbol{\eta}}. \tag{11.35}$$

The iteration method proceeds by using the recursion formula $\overline{\boldsymbol{\eta}}_{(k+1)} = \mathbf{Y}\boldsymbol{\eta}^{av} + \mathcal{K}\overline{\boldsymbol{\eta}}_{(k)}$ to construct a sequence of functions $\{\overline{\boldsymbol{\eta}}_{(k)}\}$ that can be treated as an approximation of the solution of Eq. (11.35). Usually the driving term of this equation is used as an initial approximation: $\overline{\boldsymbol{\eta}}_{(0)}(\mathbf{x}_i) = \mathbf{Y}(\mathbf{x}_i)\boldsymbol{\eta}^{av}(\mathbf{x}_i)$, which is a local approximation of the effective stress in terms of the Subsection 11.3.1. The next approximations are presented by the Neumann series for the solution $\overline{\boldsymbol{\eta}}$ of (11.35):

$$\overline{\boldsymbol{\eta}} = \sum_{k=0}^{\infty} \mathcal{K}^k \mathbf{Y}\boldsymbol{\eta}^{av}, \tag{11.36}$$

where the power $\mathcal{K}^k$ is defined recursively by the condition $\mathcal{K}^1 = \mathcal{K}$ and the kernel of $\mathcal{K}^k$ is $\mathcal{K}_k(\mathbf{x}, \mathbf{y}) = \int \mathcal{K}(\mathbf{x}, \mathbf{z})\mathcal{K}_{k-1}(\mathbf{z}, \mathbf{y})\, d\mathbf{z}$, see [882]. In effect the iteration method transforms the integral equation problem (11.31) into the linear algebra problem (11.36) in any case. The sequence $\{\overline{\boldsymbol{\eta}}_{(k)}\}$ converges to a solution $\overline{\boldsymbol{\eta}}$ (11.31) for the kernel of $\mathcal{K}$ "small" enough. The classical mathematical existence and uniqueness problems (and what is meant by saying "small" or "slowly-varying" enough), as are usually assumed in micromechanics, are beyond the scope of the current presentation. Nevertheless, it should be mentioned that $\mathbf{Y}(\mathbf{x}_i) \to \mathbf{I}$ and $\int \mathcal{K}(\mathbf{x}_i, \mathbf{y})\, d\mathbf{y} \to \mathbf{0}$ as $n(\mathbf{x}_j)\overline{v}_j \to 0\ \ \forall j$. Therefore, the iteration method is appropriate at least for a dilute concentration of inclusions. However, as we shall see in Subsection 11.3.7, it is easy to compute the approximate solution by the constructed procedure (11.36) used with the acceptable controlled accuracy.

11.3.3 General Relations for Average Stresses and Effective Thermoelastic Properties

Substituting (11.36) into (4.10) and combining terms, (4.10) finally gives the stress field inside the inclusions, $\boldsymbol{\sigma}(\mathbf{z})$ ($\mathbf{z} \in v_i$):

$$\boldsymbol{\sigma}(\mathbf{x}_i, \mathbf{z}) = \mathbf{B}(\mathbf{z})\mathbf{R}^{-1}\Big\{ -\mathbf{F} + \overline{v}_i \sum_{n=0}^{\infty} (\mathcal{K}^n \mathbf{Y} \boldsymbol{\eta}^{av})(\mathbf{x}_i)\Big\} + \mathbf{C}(\mathbf{z}), \tag{11.37}$$

from which the representation for the average stress inside the inclusion v_i follows:

$$\langle \boldsymbol{\sigma} \rangle_i = \mathbf{B}\mathbf{R}^{-1}\Big\{ -\mathbf{F} + \overline{v}_i \sum_{n=0}^{\infty} (\mathcal{K}^n \mathbf{Y} \boldsymbol{\eta}^{av})(\mathbf{x}_i)\Big\} + \mathbf{C}, \tag{11.38}$$

where the "fast" variable $\mathbf{z} \in v_i$ characterizing the stress state is defined in the local coordinate system connected with the semi-axes of the ellipsoid v_i. There is a connection between the "slow" $\mathbf{x}$ and "fast" $\mathbf{z} \in v_i$ variables: $\mathbf{x} = \sum m_j \mathbf{e}_j + \mathbf{z}$.

The mean stress in the matrix of the cell Ω_i follows simply from Eq. (11.38) and the relation $\langle \boldsymbol{\sigma} \rangle_0(\mathbf{x}) = (\langle \boldsymbol{\sigma} \rangle_{\mathbf{x}}(\mathbf{x}) - c^{(1)} \langle \boldsymbol{\sigma} \rangle_i)/c^{(0)}$, where $\mathbf{x} \in \Omega_i \backslash v_i$. Substituting (11.38) into (11.2) gives the local stress in the matrix $\mathbf{x} \in \Omega_i \backslash v_i$ $\boldsymbol{\sigma}(\mathbf{x}) = \langle \boldsymbol{\sigma} \rangle_{\mathbf{x}}(\mathbf{x}) + \int \langle\!\langle \boldsymbol{\Gamma} \rangle\!\rangle_{\mathbf{x}}(\mathbf{x} - \mathbf{y}) \sum_{n=0}^{\infty} (\mathcal{K}^n \mathbf{Y} \boldsymbol{\eta}^{av})(\mathbf{y})\, d\mathbf{y}$.

Our goal is to find a constitutive equation relating $\langle \boldsymbol{\varepsilon} \rangle_{\mathbf{x}_i}(\mathbf{x})$ to $\langle \boldsymbol{\sigma} \rangle_{\mathbf{x}_i}(\mathbf{x})$, which is valid when these vary with $\mathbf{x}_i$. After estimating local stresses inside the inclusions, see (11.37), this problem becomes trivial, and, taking the average of (11.37), leads to

$$\langle \boldsymbol{\varepsilon} \rangle_{\mathbf{x}_i}(\mathbf{x}_i) = (\boldsymbol{\mathcal{M}}^* \langle \boldsymbol{\sigma} \rangle_{\mathbf{x}_i})(\mathbf{x}_i) + \boldsymbol{\mathcal{B}}^*(\mathbf{x}_i), \tag{11.39}$$

where the integral operator $\boldsymbol{\mathcal{M}}^*$ and the tensor $\boldsymbol{\mathcal{B}}^*$ admit the representation:

$$\boldsymbol{\mathcal{M}}^* = \mathbf{M}^{(0)} + \boldsymbol{\mathcal{Y}}\mathbf{R}, \quad \boldsymbol{\mathcal{B}}^* = \boldsymbol{\beta}^0 + \boldsymbol{\mathcal{Y}}\mathbf{F}, \tag{11.40}$$

and the operator $\boldsymbol{\mathcal{Y}}$ is defined by the Neumann series $\boldsymbol{\mathcal{Y}} = \sum_{k=0}^{\infty} n \mathcal{K}^k \mathbf{Y}$. The quantities $\boldsymbol{\mathcal{M}}^*$ and $\boldsymbol{\mathcal{B}}^*$ are called the effective compliance operator and the effective eigenstrains, respectively, and are simply a notational convenience. It should be emphasized that both $\boldsymbol{\mathcal{M}}^*$ (11.40_1) and $\boldsymbol{\mathcal{B}}^*$ (11.40_2) are linear functions of the operator $\boldsymbol{\mathcal{Y}}$. Therefore the effective eigenstrains $\boldsymbol{\mathcal{B}}^*$ can be expressed in terms of the effective compliance operator $\boldsymbol{\mathcal{M}}^*$: $\boldsymbol{\mathcal{B}}^* = \boldsymbol{\beta}^{(0)} - \mathbf{M}^{(0)}\mathbf{R}^{-1}\mathbf{F} + \boldsymbol{\mathcal{M}}^*\mathbf{R}^{-1}\mathbf{F}$.

11.3.4 Some Particular Cases for Effective Properties Representations

For example, the zeroth order approximation of Eq. (11.39) is contained in the familiar constitutive equations for a homogeneous solid with material properties replaced by effective properties

$$\boldsymbol{\mathcal{M}}^*_{[0]}(\mathbf{x}_i) = \mathbf{M}^{(0)} + \mathbf{Y}(\mathbf{x}_i)\mathbf{R}n(\mathbf{x}_i), \quad \boldsymbol{\mathcal{B}}^*_{[0]}(\mathbf{x}_i) = \boldsymbol{\beta}^{(0)} + \mathbf{Y}(\mathbf{x}_i)\mathbf{F}n(\mathbf{x}_i). \tag{11.41}$$

In the first-order approximation of the constitutive equation (11.39) additional integral terms involving average strains and stresses arise:

$$\langle\varepsilon\rangle_{\mathbf{x}_i}(\mathbf{x}_i) = (\boldsymbol{\mathcal{M}}^*_{(1)}\langle\boldsymbol{\sigma}\rangle_{\mathbf{x}_i})(\mathbf{x}_i) + \boldsymbol{\mathcal{B}}^*_{(1)}(\mathbf{x}_i), \tag{11.42}$$

where

$$(\boldsymbol{\mathcal{M}}^*_{(1)}\langle\boldsymbol{\sigma}\rangle_{\mathbf{x}_i})(\mathbf{x}_i) = \mathbf{M}^*(\mathbf{x}_i)\langle\boldsymbol{\sigma}\rangle_{\mathbf{x}_i}(\mathbf{x}_i) + n(\mathbf{x}_i)\mathbf{Y}(\mathbf{x}_i)\int \mathbf{K}(\mathbf{x}_i,\mathbf{y})\langle\boldsymbol{\sigma}\rangle_{\mathbf{y}}(\mathbf{y}) \times \left[\mathbf{Y}(\mathbf{y}) - \mathbf{Y}(\mathbf{x}_i)\langle\boldsymbol{\sigma}\rangle_{\mathbf{x}_i}(\mathbf{x}_i)\right]\, d\mathbf{y}\mathbf{R}, \tag{11.43}$$

$$\boldsymbol{\mathcal{B}}^*_{(1)}(\mathbf{x}_i) = \boldsymbol{\beta}^*(\mathbf{x}_i) + n(\mathbf{x}_i)\mathbf{Y}(\mathbf{x}_i)\int \mathbf{K}(\mathbf{x}_i,\mathbf{y})\left[\mathbf{Y}(\mathbf{y}) - \mathbf{Y}(\mathbf{x}_i)\right]\, d\mathbf{y}\mathbf{F}. \tag{11.44}$$

Therefore the average strains at a point are related to the average stresses at every point.

Substituting the zero-order approximation of both $\boldsymbol{\mathcal{M}}^*_{[0]} \equiv \mathbf{M}^*(\mathbf{x}_i)$ (11.41_1) and $\boldsymbol{\mathcal{B}}^*_{[0]}$ (11.41_2) into (11.40) leads to $\boldsymbol{\beta}^*(\mathbf{x}_i) = \boldsymbol{\beta}^{(0)} + (\mathbf{M}^*(\mathbf{x}_i) - \mathbf{M}^{(0)})\mathbf{R}^{-1}\mathbf{F}$. In particular, for triply periodic composites (5.103) (when $\boldsymbol{\beta}^*(\mathbf{x}) = \boldsymbol{\beta}^* =$ const. and $\mathbf{M}^*(\mathbf{x}) = \mathbf{M}^* =$ const.) with ellipsoidal homogeneous inclusions the classical formula (6.106) for two-phase statistically homogeneous composites by Levin [665] (see also [945]) follows from the last relation for $\boldsymbol{\beta}^*(\mathbf{x}_i)$. However, in contrast to the case of statistically homogeneous composites (6.106), the last formula for $\boldsymbol{\beta}^*(\mathbf{x}_i)$ is not an exact one, because the initial relation (11.29) is obtained under some additional assumptions.

11.3.5 Doubly Periodic Inclusion Field in a Finite Stringer

For the general case of doubly periodic structures we obtained the representations of nonlocal operators (11.38) and (11.39) in terms of the Neumann series. Nevertheless, for a doubly periodic array of inclusions (11.29) in a finite ply containing $2m^l + 1$ layers of the inclusions ($f(m_3) \equiv 0$ at $|m_3| > m^l$) the problem can be solved immediately if the composite material is subjected to a stress gradient along the functionally graded direction $\mathbf{e}_3$. Then in the framework of the effective field hypothesis, the general exact Eq. (7.23) is reduced to the linear system of $2m^l + 1$ algebraic equations for $\overline{\boldsymbol{\eta}}(\mathbf{x_m})$ in terms of $\boldsymbol{\eta}^0(\mathbf{x_n})$:

$$\overline{v}_{\mathbf{n}}\overline{\boldsymbol{\eta}}(\mathbf{x_n}) = \overline{v}_{\mathbf{n}}\boldsymbol{\eta}^0(\mathbf{x_n}) + \mathbf{R}\sum_{\mathbf{m}\neq\mathbf{n}} \mathbf{T_{nm}}(\mathbf{x_n} - \mathbf{x_m})\overline{v}_{\mathbf{m}}\overline{\boldsymbol{\eta}}(\mathbf{x_m}), \tag{11.45}$$

where $\mathbf{x_m} = (x^1_{\mathbf{m}}, x^2_{\mathbf{m}}, x^3_{\mathbf{m}})$ and $\overline{\boldsymbol{\eta}}(\mathbf{x_m}) \equiv \overline{\boldsymbol{\eta}}(x^3_{\mathbf{m}})$; $\overline{v}_{\mathbf{n}}\boldsymbol{\eta}^0(\mathbf{x_n}) \equiv \overline{v}_{\mathbf{n}}\boldsymbol{\eta}^0(x^3_{\mathbf{n}}) \equiv \mathbf{R}\langle\boldsymbol{\sigma}^0(\mathbf{x})\rangle_{\mathbf{n}} + \mathbf{F}$ is called the external strain polarization tensor.

Recognizing that in each inclusion layer $x_3 =$ const. $\overline{\boldsymbol{\eta}}(\mathbf{x_n}) \equiv \overline{\boldsymbol{\eta}}(\mathbf{x_m})$ for $\forall\mathbf{n},\mathbf{m}$ the system (11.45) a the finite number of unknowns, and we may express a solution of Eq. (11.45) in the following compact form:

$$\overline{v}_{\mathbf{n}}\overline{\boldsymbol{\eta}}(\mathbf{x_n}) = \sum_{m_3=-m^l}^{m_3=m^l} \overline{v}_{\mathbf{m}}\mathbf{D}_{n_3m_3}\boldsymbol{\eta}^0(\mathbf{x_m}). \tag{11.46}$$

Here the inverse matrix $\mathbf{D}^{-1}$

$$(\mathbf{D}^{-1})_{n_3m_3} = \delta_{n_3m_3}\Big\{\mathbf{I} - \mathbf{R}\sum_{m_1,m_2\neq 0}\mathbf{T}_{\mathbf{nm}}(\mathbf{x_n}-\mathbf{x_m})\Big\} - (1-\delta_{n_3m_3})\mathbf{R}\sum_{\mathbf{m},|m_3|\leq m^l}\mathbf{T}_{\mathbf{nm}}(\mathbf{x_n}-\mathbf{x_m}), \qquad (11.47)$$

where $n_3, m_3 \doteq -m^l, m^l$. By virtue of the fact that $\overline{\boldsymbol{\eta}}(\mathbf{x_n})$ and $\boldsymbol{\eta}^0(\mathbf{x_n})$ do not vary in the layer $x^3_{n_3}$ = const. one bears in mind, for the sake of definiteness, that $n_1, n_2 = 0$ (11.46).

Substituting (4.10) into (11.46) gives the explicit relations for the stress:

$$\boldsymbol{\sigma}(\mathbf{x_n},\mathbf{z}) = \mathbf{B}(\mathbf{z})\mathbf{R}^{-1}\Big\{-\mathbf{F} + \sum_{m_3=-m^l}^{m^l}\mathbf{D}_{n_3m_3}\big[\mathbf{R}\langle\boldsymbol{\sigma}^0\rangle_{(\mathbf{m})} + \mathbf{F}\big]\Big\} + \mathbf{C}(\mathbf{z}), \quad (11.48)$$

$$\boldsymbol{\sigma}(\mathbf{x_n},\mathbf{y}) = \boldsymbol{\sigma}^0(\mathbf{y}) + \sum_{\mathbf{m}}\mathbf{T_m}(\mathbf{y}-\mathbf{x_m})\sum_{k_3=-m^l}^{m^l}\mathbf{D}_{m_3k_3}\big[\mathbf{R}\langle\boldsymbol{\sigma}^0\rangle_{(\mathbf{k})}(\mathbf{x_k}) + \mathbf{F}\big]. \quad (11.49)$$

in the inclusions $\mathbf{z}\in v_\mathbf{n}$ and in the matrix $\mathbf{y}\in\Omega_\mathbf{n}\backslash v_\mathbf{n}$, respectively.

The obtained formulae (11.48) and (11.49) make possible detailed estimations of local stresses in the cell $\Omega_\mathbf{n}$. However, the explicit relations between the average values $\langle\boldsymbol{\sigma}\rangle_{\mathbf{x_n}}(\mathbf{x_n})$ and $\langle\boldsymbol{\varepsilon}\rangle_{\mathbf{x_n}}(\mathbf{x_n})$ are more convenient to use for analysis of the overall response of the inclusion ply. With this in mind we will average Eqs. $\boldsymbol{\varepsilon}(\mathbf{x}) = \mathbf{M}(\mathbf{x})\boldsymbol{\sigma}(\mathbf{x}) + \boldsymbol{\beta}(\mathbf{x})$ and (7.22) over the cell $\Omega_\mathbf{n}$ taking Eqs. (5.107) and (11.3) into account:

$$\langle\boldsymbol{\varepsilon}\rangle_{\mathbf{x_n}}(\mathbf{x_n}) = \mathbf{M}^{(0)}\langle\boldsymbol{\sigma}\rangle_{\mathbf{x_n}}(\mathbf{x_n}) + \boldsymbol{\beta}^{(0)} + n(\mathbf{x_n})\overline{v}_\mathbf{n}\overline{\boldsymbol{\eta}}(\mathbf{x_n}), \qquad (11.50)$$

$$\langle\boldsymbol{\sigma}\rangle_{\mathbf{x_n}}(\mathbf{x_n}) = \langle\boldsymbol{\sigma}^0\rangle_{\mathbf{x_n}}(\mathbf{x_n}) + \sum_{\mathbf{m}}\langle\mathbf{T_m}\rangle_{\mathbf{x_n}}(\mathbf{x_n}-\mathbf{x_m})\overline{v}_\mathbf{m}\overline{\boldsymbol{\eta}}(\mathbf{x_m}). \qquad (11.51)$$

For simplicity we will approximate $\langle\boldsymbol{\sigma}^0\rangle_{\mathbf{x_n}}(\mathbf{x_n}) = \langle\boldsymbol{\sigma}^0(\mathbf{x})\rangle_{(\mathbf{n})}$ $(\mathbf{x}\in v_\mathbf{n})$. Then, substitution of (11.45) into (11.51) gives

$$\mathbf{R}\langle\boldsymbol{\sigma}\rangle_{\mathbf{x_n}}(\mathbf{x_n}) + \mathbf{F} = \overline{v}_\mathbf{n}\overline{\boldsymbol{\eta}}(\mathbf{x_n}) - \mathbf{R}\sum_{\mathbf{m}\neq\mathbf{n}}\mathbf{T}_{\mathbf{nm}}(\mathbf{x_n}-\mathbf{x_m})\overline{v}_\mathbf{m}\overline{\boldsymbol{\eta}}(\mathbf{x_m}) + \mathbf{R}\sum_{\mathbf{m}}\langle\mathbf{T_m}\rangle_{\mathbf{x_n}}(\mathbf{x_n}-\mathbf{x_m})\overline{v}_\mathbf{m}\overline{\boldsymbol{\eta}}(\mathbf{x_m}). \qquad (11.52)$$

Solving the algebraic system (11.52) in terms of $\langle\boldsymbol{\sigma}\rangle_{\mathbf{x_n}}(\mathbf{x_n})$ and substituting this solution into (11.50) gives

$$\langle\boldsymbol{\varepsilon}\rangle_{\mathbf{x_n}}(\mathbf{x_n}) = \mathbf{M}^{(0)}\langle\boldsymbol{\sigma}\rangle_{\mathbf{x_n}}(\mathbf{x_n}) + \boldsymbol{\beta}^0 + n(\mathbf{x_n})\sum_{m_3=-m_l}^{m_l}\mathbf{Z}_{n_3m_3}\big[\mathbf{R}\langle\boldsymbol{\sigma}\rangle_{\mathbf{x_m}}(\mathbf{x_m}) + \mathbf{F}\big], \qquad (11.53)$$

where the elements of $\mathbf{Z}^{-1}$ are: $(\mathbf{Z}^{-1})_{n_3m_3} = (\mathbf{D}^{-1})_{n_3m_3} + \sum_{m_1,m_2}\mathbf{R}\langle\mathbf{T_m}\rangle_{\mathbf{x_n}}(\mathbf{x_n}-\mathbf{x_m})$. The nonlocal Eq. (11.53) may be represented in a standard integral form:

$$\langle\boldsymbol{\varepsilon}\rangle_{\mathbf{x_n}}(\mathbf{x_n}) = \int\boldsymbol{\mathcal{M}}^*(\mathbf{x_n},\mathbf{y})\langle\boldsymbol{\sigma}\rangle_\mathbf{y}(\mathbf{y})dy^3 + \boldsymbol{\mathcal{B}}^*(\mathbf{x_n}), \qquad (11.54)$$

where the kernel $\boldsymbol{\mathcal{M}}^*(\mathbf{x_n},\mathbf{y})$ of the nonlocal operator and the effective eigenstrain $\boldsymbol{\mathcal{B}}^*(\mathbf{x_n})$, introduced in (11.54), are defined as

$$\boldsymbol{\mathcal{M}}^*(\mathbf{x_n},\mathbf{y}) = \mathbf{M}^{(0)}\delta(\mathbf{x_n}-\mathbf{y}) + n(\mathbf{x_n})\sum_{m_3=-m_l}^{m_l}\delta(\mathbf{x_m}-\mathbf{y})\mathbf{Z}_{n_3m_3}\mathbf{R}, \quad (11.55)$$

$$\boldsymbol{\mathcal{B}}^*(\mathbf{x_n}) = \boldsymbol{\beta}^0 + n(\mathbf{x_n})\sum_{m_3=-m_l}^{m_l}\mathbf{Y}_{n_3m_3}\mathbf{F}. \quad (11.56)$$

For triply periodic structures (5.103) and the homogeneous boundary conditions (11.23) the overall constitutive equations (11.54)–(11.56) are reduced to Eq. (11.39) with the effective material tensors (11.41_1) and (11.41_2). The same equivalence takes place also for central cells of the ply thick enough: $m^l \gg 1$, $|m_3| \ll m^l$. Nevertheless, in the general case of the finite inclusion ply we have $\langle\boldsymbol{\sigma}\rangle_{\mathbf{x}}(\mathbf{x}) \not\equiv$ const. even for homogeneous boundary conditions (11.23). So, from Eqs. (11.46) and (11.51) we find

$$\langle\boldsymbol{\sigma}\rangle_{\mathbf{x_n}}(\mathbf{x_n}) = \boldsymbol{\sigma}^0 + \sum_{\mathbf{m}}\langle\mathbf{T_m}\rangle_{\mathbf{x_n}}(\mathbf{x_n}-\mathbf{x_m})\sum_{l_3=-m^l}^{m^l}\mathbf{D}_{m_3l_3}\big[\mathbf{R}\boldsymbol{\sigma}^0+\mathbf{F}\big], \quad (11.57)$$

and therefore $\langle\boldsymbol{\sigma}\rangle_{\mathbf{x_n}}(\mathbf{x_n}) \not\equiv$ const. even for $\boldsymbol{\sigma}^0 \equiv$ const.

It should be pointed out that the long-range effect takes place at the estimation of average stresses (11.57). This fact is explained by different behaviors at the infinity of summed functions $\mathbf{K}(\mathbf{x}_i,\mathbf{x}_j)$ (as $O(|\mathbf{x}_i-\mathbf{x}_j|^{-5})$) (11.30) and $\mathbf{T}_{ij}(\mathbf{x}_i-\mathbf{x}_j)$ (as $O(|\mathbf{x}_i-\mathbf{x}_j|^{-3})$) in the representations of effective properties (11.43) and the stress concentrator tensor (11.57), respectively. Although it is well known that the integral of the function $\mathbf{T}_{ij}(\mathbf{x}_i-\mathbf{x}_j)$ is not absolutely convergent in the whole Euclidean space R^3, we have no convergence problems for any finite thickness of the ply, since in the case considered the domain of integration is bounded in the direction $\mathbf{e}_3$.

By this means the initial problem with infinite number of the spherical inclusions (micro level) is reduced to the problem with $2m^l+1$ layers (meso level). At the microlevel, each layer is treated as a particulate composite of appropriate inclusion volume fraction, the matrix-inclusion microtopology of which is explicitly accounted for and the interactions of inclusions from different layers are considered as well. At the meso level the composite is viewed as consisting of alternating homogenized layers. As this takes place, even for identical inclusion layers the effective local and nonlocal parameters of homogenized layers change from layer to layer (see for comparison Chapter 12). These nontrivial dependencies are explained by the interactions of inclusions from different layers and therefore the coupling of micro and meso levels is established explicitly.

11.3.6 Numerical Results for Three-Dimensional Fields

Let us consider as an example a composite consisting of isotropic homogeneous components and having identical spherical inclusions $\mathbf{L}^{(i)} = (3k^{(i)}, 2\mu^{(i)}) \equiv$

$3k^{(i)}\mathbf{N}_1 + 2\mu^{(i)}\mathbf{N}_2$, $(\mathbf{N}_1 = \boldsymbol{\delta}\otimes\boldsymbol{\delta}/3,\ \mathbf{N}_2 = \mathbf{I}-\mathbf{N}_1)$. Let an inclusion ply (see Fig. 11.3) have a simple cubic (SC) lattice containing $2m^l+1$ layers of inclusions. For central inclusion layers of the thick ply, the local effective properties $\mathbf{M}^*(\mathbf{x}_i) = \mathbf{M}^* \equiv$ const. (11.41$_1$) coincide with the properties of triply periodic structures and the tensor of effective moduli $\mathbf{L}^* \equiv (\mathbf{M}^*)^{-1}$ is characterized by three elastic moduli: $k^*_{13} = (L^*_{3333} + 2L^*_{1133})/3$, $\mu^*_{13} = L^*_{1313}$, $\widetilde{\mu}^*_{13} = (L^*_{3333} - L^*_{1133})/2$, where the stiffness components are given with respect to a coordinate system whose base vectors are normal to the faces of the unit cell. In the interest of obtaining maximum difference between the effective properties estimated by the different methods we will consider the examples for hard inclusions ($\nu^{(0)} = \nu^{(1)} = 0.3$, $\mu^{(1)}/\mu^{(0)} = 1000$) as well as for the voids ($\mathbf{L}^{(1)} \equiv \mathbf{0}$), and a number of values of the volume concentration of inclusions. For triply periodic SC arrays the local elastic moduli (11.26) are computed via analytical methods in [820], and [624] as well as via the formulae (11.34) and (11.41$_1$) (see Table 11.4); in addition to the relations (11.26) the parameters k^*_{12}, μ^*_{12} and $\widetilde{\mu}^*_{12}$ obtained by replacement of the index 3 by the index 2 were estimated for the boundary layer $m_3 = m^l$, $m^l = 50$ (see the Table 11.4; the case of voids was considered in [134]. For the boundary layer $m_3 = m^l$ the tensor of local effective moduli $\mathbf{L}^*(\mathbf{x})$ shows hexagonal symmetry.

Table 11.4. Overall elastic constants of SC arrays of rigid inclusions in the thick ply: the components (ij) =(13) and (ij) =(12) for the boundary layer (11.32), (11.41$_1$); (N) Nunan and Keller [820] for c =0.1–0.4, (K) Kushch [624] for c =0.5, (H1) the proposed method (11.34), (11.41$_1$).

	Boundary layer						Central layer					
	$k^*_{ij}/k^{(0)}$		$\mu^*_{ij}/\mu^{(0)}$		$\widetilde{\mu}^*_{ij}/\mu^{(0)}$		$k^*_{13}/k^{(0)}$		$\mu^*_{13}/\mu^{(0)}$		$\widetilde{\mu}^*_{13}/\mu^{(0)}$	
c	13	12	13	12	13	12	N/K	H1	N/K	H1	N/K	H1
0.10	1.18	1.18	1.22	1.21	1.26	1.27	1.18	1.18	1.22	1.21	1.27	1.27
0.20	1.40	1.40	1.46	1.46	1.64	1.68	1.41	1.40	1.46	1.45	1.70	1.69
0.30	1.68	1.69	1.76	1.75	2.17	2.29	1.71	1.69	1.77	1.74	2.35	2.32
0.40	2.05	2.07	2.15	2.15	2.91	3.19	2.17	2.07	2.25	2.12	3.74	3.21
0.50	2.54	2.61	2.69	2.73	3.77	4.27	3.50	2.61	3.14	2.67	6.49	4.33

It was demonstrated in [134] (see also Table 11.4) that in the interior of a thick layer ply ($m^l \to \infty$), sufficiently far away from its boundary, $\mathbf{L}^*(\mathbf{x})$ coincides with the effective moduli $\mathbf{L}^*$ for the triply periodic structure. Near the boundary of the ply the tensors of the effective moduli $\mathbf{L}^*(\mathbf{x})$ vary significantly within the boundary layer $x^3 = \pm m^l|\mathbf{e}_3|$ (*boundary layer effect*). There is a slight dependence of local overall effective moduli $\mathbf{L}^*(\mathbf{x})$ on the ply size m^l (*scale effect*).

Buryachenko [134] analyzed also the nonlocal constitutive equations (11.39), (11.40$_1$), and (11.42). For the sake of definiteness, let the finite ply of rigid spherical inclusions be subjected to a known average stress $\langle\boldsymbol{\sigma}\rangle_{\mathbf{x}}$ along the functionally graded direction e_3 and only one component of $\langle\sigma_{ij}\rangle_{\mathbf{x}}$ $(i,j=1,2,3)$ differs from zero: $\langle\boldsymbol{\sigma}\rangle_{\mathbf{x}}(\mathbf{x}) = f(x^3)\boldsymbol{\sigma}^{\rm con}$, $\sigma^{\rm con}_{ij} = {\rm const.} \neq 0$, all other $\sigma^{\rm con}_{kl} = 0$, $(kl \neq ij)$. Here $f(x^3) = 1 - \cos(\pi x^3/a)$, and a is a positive length parameter. Let the tensors

$\boldsymbol{\varepsilon}^{\rm loc}(\mathbf{x}) \equiv \mathbf{M}^*(\mathbf{x})\langle\boldsymbol{\sigma}\rangle_{\mathbf{x}}(\mathbf{x})$ and $\boldsymbol{\varepsilon}^{\rm non}(\mathbf{x}) \equiv \langle\boldsymbol{\varepsilon}\rangle_{\mathbf{x}}(\mathbf{x}) - \mathbf{M}^*(\mathbf{x})\langle\boldsymbol{\sigma}\rangle_{\mathbf{x}}(\mathbf{x})$ be named the local and nonlocal average strains, respectively. Clearly $f(2am) = 0$ if m is the integer set Z, and therefore $\boldsymbol{\varepsilon}^{\rm loc}(\mathbf{x}) = \mathbf{0}$ at $x^3 = 2am$, whereas $\boldsymbol{\varepsilon}^{\rm non}(\mathbf{x}) \neq \mathbf{0}$ at the mentioned points. The boundary layer effect shows up most vividly for nonlocal components of strains; so $\varepsilon_{33}^{\rm non}(\mathbf{x})$ in the boundary and central layers can differ from one another by a factor of two or even more, whereas the corresponding local components of strains differ by only 13%.

11.3.7 Numerical Results for Two-Dimensional Fields

The method being proposed for the analysis of stress fields within doubly periodic structures (11.57) can be used for the consideration of some singly periodic structures if we assume that $|\mathbf{e}_2| \gg |\mathbf{e}_1|, |\mathbf{e}_3|$. Then the problem being analyzed is reduced to a problem of a single layer of inclusions periodical in the direction $|\mathbf{e}_1|$. The two-dimensional analog of this arrangement is a nonperiodic inclusion field located in one line. For the purpose of an evaluation of the accuracy of the proposed method we will consider an example that has an analytical solution obtained in [949] by the method of the theory of functions of complex variables.

Namely, let us consider the plane problem of a semi-infinite regular grid of straight cuts (cracks) of the length $2l$ on line L ($x^2 = 0$) at nodes of a semi-infinite regular grid $x_n^1 = nh$ ($n \in Z^+ \equiv \{0, 1, \ldots\}$; $h > 2l$). In the case considered the method analyzed in Subsection 11.3.5 is reduced to the method in [528], and we will use the analytical solution for two cracks presented in the paper mentioned above. The external field $\boldsymbol{\sigma}^0$ is uniaxial tension in the direction of the normal $\mathbf{n}\perp L$ and has the form $\sigma_{\alpha\beta}^0 = \sigma_0^0 n_\alpha n_\beta$, σ_0^0 is a scalar, $\alpha, \beta = 1, 2$. Then the state of each defect is determined by the field $\overline{\boldsymbol{\sigma}}$ and $\overline{\sigma}_{\alpha\beta} n_\beta = \overline{\sigma}_0^0 n_\alpha$, where $\overline{\sigma}_0^0$ is a scalar. We estimate the relative change in the stress intensity factor (SIF) $k_I = K_I/K_I^0$ versus h, where $K_I^0 = \sigma_0^0\sqrt{\pi l}$ is the SIF for an isolated crack in an unbounded plane. Then in the framework of the hypothesis **H1** we get

$$D_{nm}^{\alpha\beta} = D_{nm}^0 n^\alpha n^\beta, \ \left(D^0\right)_{nm}^{-1} \equiv \delta_{nm} - (1 - \delta_{nm})T_{nm}, \tag{11.58}$$

$$T_{nm}(x_n^1 - x_m^1) = \frac{\sqrt{|n-m|h}}{2l}\left(\sqrt{|n-m|h + 2l} - \sqrt{|n-m|h - 2l}\right), \tag{11.59}$$

$$T_m(x^1 - x_m^1) = \frac{|(n-m)h - x^1|}{\sqrt{[(n-m)h - x^1]^2 - l^2}} - 1, \tag{11.60}$$

where $n \neq m$, and $x^1 \notin [x_m^1 - l, x_m^1 + l]$.

In the simple case we adopt the estimation $k_I = \overline{\sigma}_0^0/\sigma_0^0$, following from Eq. (11.46)

$$k_I(x_n^1 \pm l) = \sum_{m=0}^{\infty} D_{nm}^0. \tag{11.61}$$

A more accurate expression for k_I can be constructed with consideration of the fact that the field $\overline{\sigma}_0^0 = \overline{\sigma}_0^0(x^1)$ $(x^1 \in [x_n - l, x_n + l])$ is inhomogeneous in the neighborhood of the defect and equal to the superposition of the fields induced by the surrounding cracks:

$$k_I(x_n^3 \pm l) = 1 + \frac{1}{\sqrt{\pi l}} \sum_{m \in Z_n^+} \int_{-l}^{l} \sqrt{\frac{l \pm \xi}{l \mp \xi}} T_m(\xi + x_n^1 - x_m^1) d\xi \sum_{k=0}^{\infty} D_{mk}^0, \quad (11.62)$$

where Z_n^+ denotes the set $Z^+ \setminus n$. For a finite number of cracks as well as for the infinite regular grid of cracks ($x_n^1 = 0, \pm 1, \pm 2, \ldots$), Eq. (11.62) is reduced to the relations analyzed previously in [528]. Buryachenko and Parton [171] obtained the relations similar to (11.62) by a more approximative method based on the consideration of triply interactive effects of cracks.

In Table 11.5 the exact solution by Koiter [580] for the infinite regular grid of cracks ($x_n^1 = 0, \pm 1, \pm 2, \ldots$): $k_I(\pm l) = \sqrt{\gamma^{-1} \tan \gamma}$, ($\gamma = \pi l/h$), as well as the accurate solution for the boundary crack $k_I(0 \pm l)$ obtained in [949] are compared with the approximate solutions (11.61) and (11.62). Agreement of the approximate Eq. (11.62) with the exact ones is satisfactory. It was shown that the boundary layer effect has a short range of influence: its impact is practically confined to the nearest four cracks. It should be mentioned that SIF depends essentially on the nonhomogeneity of the effective field $\overline{\boldsymbol{\sigma}}_0^0(\mathbf{x})$ leading to a significant difference of SIFs estimated by the formulae (11.61) and (11.62). In the estimation of both average stresses and effective properties of composites with ellipsoidal inclusions this dependence appears only slightly, and higher accuracy should be expected than using the effective field hypothesis **H1** (at least it was shown in the examples presented in Table 11.4).

Table 11.5. Normalized SIF k_I in a semi-infinite periodic collinear row of cracks: (H1) effective field hypothesis method (11.61); (I) the improved method (11.62); (R) solution by Rubinstein [949], (K) the solution Koiter [580]

	$k_I(-l)$			$k_I(l)$			$k_I(\infty \pm l)$		
$h/(2l)-1$	H1	I	R	H1	I	R	H1	I	K
0.30	1.209	1.148	1.093	1.209	1.307	1.243	1.443	1.482	1.477
0.20	1.289	1.194	1.172	1.289	1.454	1.425	1.628	1.703	1.689
0.10	1.467	1.290	1.276	1.467	1.830	1.794	2.064	2.271	2.207

11.3.8 Conclusion

Joint solution of the equilibrium equation, boundary conditions, and effective constitutive relations using either (11.53) or (11.54) leads to the estimation of average stresses $\langle \boldsymbol{\sigma} \rangle_{\mathbf{x}}(\mathbf{x})$ and the average strains $\langle \boldsymbol{\varepsilon} \rangle_{\mathbf{x}}(\mathbf{x})$. Of course, the mentioned scheme can be generalized easily to the case where instead of each inclusion layer one considers an individual ply consisting of few inclusion layers and different plies can be distinguished by the type of lattice periodicity as well as by the mechanical and geometrical parameters of the inclusions. In any case the effective properties of either the inclusion layers or the plies depend not only on the individual structure of the layer considered (as usually one assumes [883]) but on the parameters of others layers.

The obtained relations depend on the values associated with the mean distance between inclusions, and do not depend on the other characteristic size, i.e., the mean inclusion diameter. This fact may be explained by the initial use of the hypothesis **H1** dealing with the homogeneity of the field $\overline{\boldsymbol{\sigma}}(\mathbf{x})$ inside each inclusion. In the case of a variable representation of $\overline{\boldsymbol{\sigma}}(\mathbf{x}) \quad (\mathbf{x} \in v_i)$, for instance in polynomial form, the mean size of the inclusions will be contained in the nonlocal dependence of microstresses on the average stress $\langle\boldsymbol{\sigma}\rangle_{\mathbf{x}}(\mathbf{x})$. Such an improvement was done in Eq. (11.62) in comparison with Eq. (11.61).

It should be mentioned that the effective constitutive Eq. (11.39) was derived for points $\mathbf{x}_i$ located sufficiently far from the boundary of the body ∂w. In so doing the relations developed have been obtained by the use of the whole-space Green's function. Then use of nonlocal constitutive relations (11.39) requires more complicated boundary conditions [69], [279]; this question is beyond the scope of the current study.

11.4 Triply Periodic Particulate Matrix Composites with Imperfect Unit Cells

In this section we will consider the pure elastic problems ($\boldsymbol{\beta}(\mathbf{x}) \equiv \mathbf{0}$) for triply periodic particulate matrix composites with imperfect unit cells the geometrical random structure of which is described by the probability density $\varphi(v_i, \mathbf{x}_i)$ (7.46$_1$) and conditional probability density $\varphi(v_j, \mathbf{x}_j|; v_i, \mathbf{x}_i)$ (7.46$_2$). This is a *substitutional* sort of the disordered imperfection of the unit cells that connotes a variability in properties per vertex (see e.g. [844]).

11.4.1 Choice of the Homogeneous Comparison Medium

In the framework of the traditional scheme, we introduce a homogeneous "comparison" body with homogeneous moduli $\mathbf{L}^c$ (and with solution $\boldsymbol{\sigma}^0$, $\boldsymbol{\varepsilon}^0$, $\mathbf{u}^0$ to the same boundary-value problem), and so the equilibrium equation presented in the terms of the strains transforms to the equation

$$\nabla\mathbf{L}^c\boldsymbol{\varepsilon}(\mathbf{x}) = -\nabla\big\{[\mathbf{L}(\mathbf{x}) - \mathbf{L}^c]\boldsymbol{\varepsilon}(\mathbf{x})\big\}, \tag{11.63}$$

with a fictitious random "body-force" on the right-hand side of the equation. The equation (11.63) is simplified for $\mathbf{L}^c = \mathbf{L}^{(0)} \equiv$ const. (that will be assumed hereafter), when the right-hand side of Eq. (11.63) differs from zero just inside the inclusions. In a more general case of the inhomogeneous elastic properties of the matrix, it is reasonable to consider the elastic properties of the comparison medium as the Voight (or Reuss) average of elastic properties of the matrix $\mathbf{L}^c = (v^{(0)})^{-1}\langle\mathbf{L}(\mathbf{x})V^{(0)}(\mathbf{x})\rangle$.

The stochastic equation (11.63) can be reduced to the system (8.8)–(8.11) with their subsequent approximative solution in the framework of the effective field hypotheses **H1** and **H2** (see Section 8.1). Analogously Eq. (8.46), we will get analytical representation for a strain concentrator factor

$$\langle\boldsymbol{\varepsilon}\rangle_i = \mathbf{A}_i\mathbf{D}^{\epsilon}_i\langle\boldsymbol{\varepsilon}\rangle, \quad \mathbf{D}^{\epsilon}_i = \mathbf{R}^{\epsilon\,-1}_i\sum_{j=1}^{N}\mathbf{Y}^{\epsilon}_{ij}\mathbf{R}^{\epsilon}_j, \tag{11.64}$$

where the tensor $\mathbf{D}_i^\epsilon$ $(i = 1, \ldots, N)$ has a simple physical meaning of the action of surrounding inclusions on the separate one: $\langle \overline{\varepsilon} \rangle_i = \mathbf{D}_i^\epsilon \langle \varepsilon \rangle$, and the matrix $(\mathbf{Y}\epsilon)^{-1}$ has the following elements $(\mathbf{Y}^{\epsilon^{-1}})_{ij}$ $(i, j = 1, 2, \ldots, N)$:

$$(\mathbf{Y}^{\epsilon^{-1}})_{ij} = \delta_{ij} \left[\mathbf{I} - \mathbf{R}_i^\epsilon \sum_{q=1}^{N} \int \mathbf{T}_{iq}^\epsilon (\mathbf{x}_i - \mathbf{x}_q) \mathbf{Z}_{qi}^\epsilon \varphi(v_q, \mathbf{x}_q | ; v_i, \mathbf{x}_i) \, d\mathbf{x}_q \right]$$
$$- \mathbf{R}_i^\epsilon \int \left[\mathbf{T}_{ij}^\epsilon (\mathbf{x}_i - \mathbf{x}_j) \mathbf{Z}_{jj}^\epsilon \varphi(v_j, \mathbf{x}_j | ; v_i, \mathbf{x}_i) - \mathbf{T}_i^\epsilon (\mathbf{x}_i - \mathbf{x}_j) \lambda^{(j)} n^{(j)} \right] \, d\mathbf{x}_j, \quad (11.65)$$

where the matrix $\mathbf{Z}_{jj}^\epsilon$ is defined analogously to Eq. (4.58) by replacement of the tensors $\mathbf{R}_j$, $\mathbf{T}_{ij}$ by $\mathbf{R}_j^\epsilon$, $\mathbf{T}_{ij}^\epsilon$. After estimating average strains inside the inclusions, see (11.64), the problem of calculating effective properties becomes trivial and leads to

$$\mathbf{L}^* = \mathbf{L}^{(0)} + \sum_{i,j=1}^{N} \mathbf{Y}_{ij}^\epsilon \mathbf{R}_j^\epsilon \lambda^{(i)} n^{(i)}, \quad (11.66)$$

where the statistical averages of random variables $\lambda(\mathbf{x_m})$: $0 \leq \langle \lambda^{(k)}(\mathbf{x_m}) \rangle \equiv \lambda^{(k)} \leq 1$ and $0 \leq \sum_k \lambda^{(k)} \leq 1$ reflect the measure of the imperfection: $\lambda^{(k)} = 0$ means the absence in the lattice Λ of inclusions of the component k, the case $\lambda^{(k)} = 1$ stands for the perfect periodic structure filled by identical inclusions from the component k. Equations (11.64)–(11.66) comprise a slight modification of the analogous equations for the random statistically homogeneous media (see Chapter 8) in which the parameters describing the geometrical structures had a different sense and were obtained in a similar manner. The known "quasi-crystalline" approximation **H1** (11.3) in our notations has the form $\mathbf{Z}_{ij}^\epsilon = \mathbf{I}\delta_{ij}$. In such a case for the perfect periodic structures with the identical ellipsoidal inclusions $N = 1$, $\lambda^{(1)} = 1$, the Eqs. (11.64)–(11.66) are reduced to the known one (11.24).

Let us consider as an example (demonstrating the accuracy and efficiency of estimations obtained in the framework of the hypothesis **H1**) a composite consisting of isotropic homogeneous components and having identical spherical inclusions $\mathbf{L}^{(i)} = (3k^{(i)}, 2\mu^{(i)})$. For SC lattice of spherical inclusions, the tensor of effective moduli $\mathbf{L}^*$ is characterized by three elastic moduli (11.26), where the stiffness components are given with respect to a coordinate system whose base vectors are normal to the faces of the unit cell. In the interest of obtaining the maximum difference between the effective properties, estimated by the different methods, we will consider the examples for hard inclusions ($\nu^{(0)} = \nu^{(1)} = 0.3$, $\mu^{(1)}/\mu^{(0)} = 1000$) and a number of values of the volume concentration of inclusions (see Table 11.6). As can be seen for the nondilute

Table 11.6. The overall elastic constants of SC arrays of rigid inclusions: (N) Nunan and Keller [820] for $c = 0.1$–0.4, (K) Kushch [624] for $c = 0.5$, (TP) Two-particle approximation (11.65), (11.66), (OP) One-particle approximation (11.24).

	$k^*/k^{(0)}$			$\mu^*/\mu^{(0)}$			$\widetilde{\mu}^*/\mu^{(0)}$		
c	N/K	TP	OP	N/K	TP	OP	N/K	TP	OP
0.10	1.180	1.188	1.179	1.216	1.224	1.213	1.274	1.292	1.269
0.20	1.405	1.411	1.403	1.455	1.462	1.451	1.704	1.776	1.690
0.30	1.706	1.732	1.691	1.766	1.778	1.740	2.35	2.819	2.319
0.40	2.173	2.224	2.074	2.25	2.233	2.120	3.74	5.773	3.207
0.50	3.503	3.095	2.610	3.14	2.986	2.674	6.49	29.263	4.334

concentration of hard inclusions, a two-particle approximation (11.65) and (11.66) gives more accurate results than a one-particle approximation (1.40) at the estimation of effective parameters k^* and μ^* (but not $\widetilde{\mu}^*$).

Let us consider an imperfect SC packing of hard inclusions described by a perfect parameter $\lambda^{(1)}$ and analyzed by the two-particle and one-particle approximations. In Fig. 11.4 the plots of the relative effective moduli $\mu^*/\mu^{(0)}$ and $\widetilde{\mu}^*/\mu^{(0)}$ as functions of the perfect parameter $\lambda^{(1)}$ at $c^{(1)} = 0.3$ are displayed in comparison with the analytical estimations [820].

In Fig. 11.5 the analogous results obtained via two-particle approximation (11.65) and (11.66) are presented for the initially perfect packing of hard spheres $v^{(1)}$ ($\lambda^{(1)} = 1,\ \lambda^{(2)} = 0$) disordered by the random replacement of some hard spheres by the holes $v^{(2)}$. At the variation of the concentration of the hard inclusions $c^{(1)}(\lambda^{(1)})$ and the holes $c^{(2)}(\lambda^{(2)})$, their complete concentration is fixed $c^{(1)}(\lambda^{(1)}) + c^{(2)}(\lambda^{(2)}) = 0.3$. Effective elastic moduli presented in Fig. 11.5 are consistent with a continuous transition from the perfect packing of hard spheres ($\lambda^{(1)} = 1,\ \lambda^{(2)} = 0$) to the perfect packing of holes ($\lambda^{(1)} = 0,\ \lambda^{(2)} = 1$).

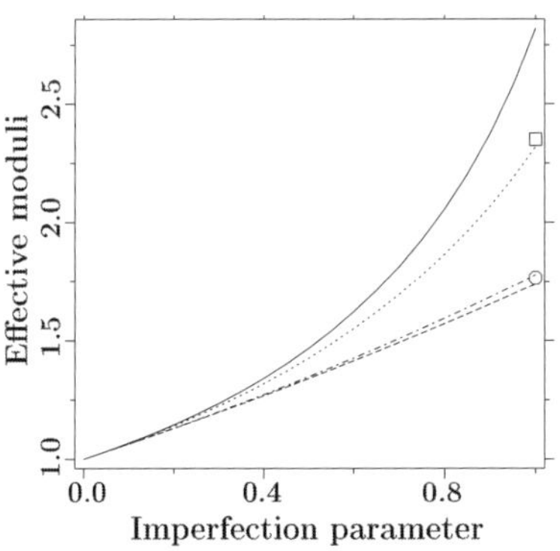

Fig. 11.4. Normalized effective elastic moduli vs $\lambda^{(1)}$: $\widetilde{\mu}^*/\mu^{(0)}$ (TP, solid line; OP, dotted line [820], □) and $\mu^*/\mu^{(0)}$ (TP, dot-dashed line; OP, dashed line; [820],○).

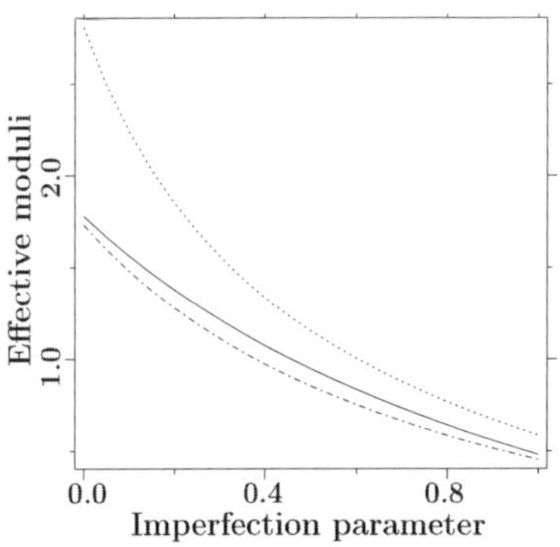

Fig. 11.5. Normalized effective elastic moduli as the functions of $\lambda^{(2)}$: $\widetilde{\mu}^*/\mu^{(0)}$ (dotted line), $\mu^*/\mu^{(0)}$ (solid line), $k^*/k^{(0)}$ (dashed line).

The example in Fig. 11.5 can be considered as a simplified damage model (see, e.g., [269]) where the broken inclusions are replaced by some microdiscontinuities, that is the stiffness tensor of one broken inclusion is reduced to zero. This inclusion is removed from the whole inclusion, and one void is introduced inside the matrix. This void has the same orientation and size as the broken inclusion.

Contrary to the model [425], when a broken inclusion occurs, it is replaced by the void. More detailed analysis of damage models considering different statistical treatments of composite strength [235] is beyond the scope of the current presentation.

11.4.2 MEFM Accompanied by Monte Carlo Simulation

The Monte Carlo technique is a standard method for solving probabilistic problems, by sampling from random distributions utilizing concepts of probability theory. The popular method of the analysis of imperfect periodic structures is based on the Monte Carlo simulation of randomly distributed imperfections in a mesocell $\Omega^{\rm meso}$ containing 100 and more unit cells $\Omega^{\rm meso} = \cup\Omega_i$ $(i = 1,\ldots,N^{\rm meso})$ with subsequent application of periodic boundary conditions to the mesocell $\Omega^{\rm meso}$ with the random set of inclusions $X^{\rm meso}$ [210], [1011]. An immediate realization of this scheme in the MEFM presents problems due to the necessity of considering an infinite medium. However, a slight modification of the MEFM adapted to the exploration of the Monte Carlo simulation can be proposed in light of an alternative *toroidal edge correction* used in a statistical description of random packing [939] in which rectangular region $\Omega^{\rm meso}$ can be regarded as a torus, so that inclusions on the opposite edges are considered to be closed (the 3-D case is considered in a similar manner). Then $\Omega^{\rm meso}$ can be considered to be part of a grid of identical rectangles, forming a border around the pattern inside $\Omega^{\rm meso}$. Distances are then measured from the point in the central rectangle $\Omega^{\rm meso}$ to points in the surrounding periodic rectangles. Then Eqs. (11.65) and (11.66) admit straightforward generalization when the strain concentrator factors (11.64) are estimated for each inclusion $v_i \subset \Omega^{\rm meso}$ with the centers $\mathbf{x}_i$ $(i = 1,\ldots,N^{\rm meso})$ and random elastic moduli $\mathbf{L}$ described by independent random variables which takes values $\{\mathbf{L}^{(1)},\ldots,\mathbf{L}^{(N)}\}$ with the probabilities $P\{\mathbf{L} = \mathbf{L}^{(m)}\} = \lambda^{(m)}$ $(m = 1,\ldots,N)$, and the index $m = 1$ is assumed to be assigned to the elastic properties of the inclusions in the perfect structures. Thus, our configuration space is discrete, and all rates at which they occur can be determined, and we can choose and execute a single change to the system from the list of all possible changes at each Monte Carlo step. Let the tensors $\mathbf{R}_i^\epsilon$ in the point $\mathbf{x}_i$ takes the values $\mathbf{R}_{(i_r)}^\epsilon \equiv \mathbf{R}^{\epsilon(m)}$ and, therefore for a concrete realization of a random process we will have the equations for $N^{\rm meso}$ inclusions:

$$\overline{\varepsilon}(\mathbf{x}_i) = \langle\varepsilon\rangle + \sum_{j=1}^{N^{\rm meso}} \mathbf{T}_{i[j]}^\epsilon(\mathbf{x}_i - \mathbf{x}_{[j]})\mathbf{R}_{j_r}^\epsilon\overline{\varepsilon}(\mathbf{x}_j) + \mathbf{P}(w_i^{\rm el})\langle\mathbf{R}^\epsilon\overline{\varepsilon}n\rangle, \tag{11.67}$$

where the points $\mathbf{x}_j$ pass through all grid nodes in both the mesocell $\Omega^{\rm meso}$ and in the moving region $w^{\rm el}$ with the center $\mathbf{x}_i$ defined in a similar way to Eq. (11.24) where the tensor $\mathbf{P}(w_i^{\rm el})$ was defined as in Eq. (11.24). $\langle\mathbf{R}^\epsilon\overline{\varepsilon}n\rangle$ is a statistical average over the statistically large number of realization, and the tensor $\mathbf{T}_{i[j]}^\epsilon$ is made of from the tensors $\mathbf{T}_{iq}^\epsilon(\mathbf{x}_i-\mathbf{x}_q)$ of inclusions v_i and $v_q \subset w_i^{\rm el}$ periodical to v_j $(\mathbf{x}_q = \mathbf{x}_j + \mathbf{l},\ \mathbf{l} = (l_1e_1, l_2e_2, l_3e_3),\ l_1, l_2, l_3 = 0, \pm1, \pm2, \ldots)$: $\mathbf{T}_{i[j]}^\epsilon(\mathbf{x}_i - \mathbf{x}_{[j]}) = \sum_q \mathbf{T}_{iq}^\epsilon(\mathbf{x}_i - \mathbf{x}_q)$.

Therefore an implicit representation for the effective fields is

$$\overline{\varepsilon}(\mathbf{x}_i) = \sum_{j=1}^{N^{\text{meso}}} \widetilde{\mathbf{Y}}^{\epsilon}_{ij} \Big[\langle \varepsilon \rangle + \mathbf{P}(w_i^{\text{el}}) \langle \mathbf{R}^{\epsilon} \overline{\varepsilon} n \rangle \Big], \tag{11.68}$$

where the matrix $(\widetilde{\mathbf{Y}}^{\epsilon})^{-1}$ has the elements $(i, j = 1, \ldots, N^{\text{meso}})$: $(\widetilde{\mathbf{Y}}^{\epsilon})^{-1}_{ij} = \delta_{ij}\mathbf{I} - \sum_{j=1}^{N^{\text{meso}}} \mathbf{T}^{\epsilon}_{i[j]}(\mathbf{x}_i - \mathbf{x}_{[j]})\mathbf{R}^{\epsilon}_{j_r}$. The operation of averaging is performed over a number of Monte Carlo simulations; in so doing the tensors $\mathbf{R}^{\epsilon}_{i_r}, \widetilde{\mathbf{Y}}^{\epsilon}_{ij}$ are estimated for the each concrete realization of Monte Carlo simulation: in each point $\mathbf{x}_i \subset \Omega^{\text{meso}}$ the elastic properties of the inclusion $\mathbf{L}(\mathbf{x}_i)$ are taken as $\mathbf{L}^{(m)}$ if the generated random variable uniformly distributed on the interval $[0, 1]$ takes the value ξ in the interval $\sum_{k=0}^{m-1} \lambda^{(k)} < \xi < \sum_{k=0}^{m} \lambda^{(k)}$ $(\lambda^{(0)} = 0)$. The representation of the effective fields (11.68) for a single realization admits of the estimation $\langle \mathbf{R}^{\epsilon} \overline{\varepsilon} n \rangle$ averaged over the statistically large number n^{MC} of the Monte Carlo simulation:

$$\langle \mathbf{R}^{\epsilon} \overline{\varepsilon} n \rangle = (\mathbf{I} - \mathbf{B}\mathbf{P}(w_i^{\text{el}}))^{-1} \mathbf{B} \langle \varepsilon \rangle, \tag{11.69}$$

$$\mathbf{B} = \frac{n}{N^{\text{meso}}} \langle \sum_{j=1}^{N^{\text{meso}}} \mathbf{R}^{\epsilon}_{i_r} \widetilde{\mathbf{Y}}^{\epsilon}_{ij} \rangle \equiv \lim_{n^{MC} \to \infty} \frac{n}{n^{MC} N^{\text{meso}}} \sum_{m=1}^{n^{MC}} \sum_{j=1}^{N^{\text{meso}}} \mathbf{R}^{\epsilon}_{i_r} \widetilde{\mathbf{Y}}^{\epsilon}_{ij} \tag{11.70}$$

as well as the estimation of the effective elastic modulus:

$$\mathbf{L}^* = \mathbf{L}^{(0)} + (\mathbf{I} - \mathbf{B}\mathbf{P}(w_i^{\text{el}}))^{-1} \mathbf{B}, \tag{11.71}$$

Substitution of Eqs. (4.12) and (11.69) into Eq. (11.68) leads to estimation of the stress field in the inclusion v_i in a concrete realization of Monte Carlo simulation $\varepsilon(\mathbf{x}_i) = \sum_{j=1}^{N^{\text{meso}}} \mathbf{A}_{i_r} \widetilde{\mathbf{Y}}^{\epsilon}_{ij} [\mathbf{I} - \mathbf{P}(w_i^{\text{el}})\mathbf{B}]^{-1} \langle \varepsilon \rangle$. Statistical averaging of the last equation over the number of realization $r = 1, \ldots, n^{MC}$ yields the estimation of the statistical averages of the stresses in the inclusions belonging to the component $m = 1, \ldots, N$: $\langle \varepsilon(\mathbf{x}_i) \rangle = \langle \sum_{j=1}^{N^{\text{meso}}} \mathbf{A}_{i_r} \widetilde{\mathbf{Y}}^{\epsilon}_{ij} \rangle [\mathbf{I} - \mathbf{P}(w_i^{\text{el}})\mathbf{B}]^{-1} \langle \varepsilon \rangle$.

Now we will obtain the numerical results associated with Eq. (11.71) as well as execute their comparison with the results of Subsection 11.4.1. The advantage of Eq. (11.71) obtained without restrictions produced by the closing assumptions **H2** or (11.3) is accomplished sacrificing the computing costs. For example even for a single Monte Carlo realization ($n^{MC} = 1$), the dimensions of the matrix $\widetilde{\mathbf{Y}}^{\epsilon}_{ij}$ in Eqs. (11.65) and (11.71) differ by a factor N^{meso}/N, and the CPU time for a PC with a 2.4 GHz processor expended for the estimation of $\mathbf{L}^*$ (11.66) for $N = 1$ equals 0.1 sec, while the CPU time exploited for evaluation of just a single realization ($n^{MC} = 1$) in Eq. (11.69) equals 5 sec, 40 sec, and 310 sec for a simple cubic packing of the mesocell Ω^{meso} containing 3 ($N^{\text{meso}} = 27$), 5 ($N^{\text{meso}} = 125$) and 7 ($N^{\text{meso}} = 343$) inclusions in the each direction, respectively.

It should be mentioned that the representation of effective properties (11.71) was obtained by the average over the realization number n^{MC} of periodic structure with mesocell Ω^{meso} containing the random set of inclusions X^{meso}. In the theory of periodic structures [25], [48], [507]), [532], [960], this problem is usually solved on the mesocell $\Omega^{\rm meso}$ with periodic boundary conditions. In the current approach, we analyzed all the inclusions from the whole space R^3 with the boundary conditions at infinity. However, due to the absolute convergence of the right-hand side of Eq. (11.67), the real number of analyzed inclusions is limited by the inclusions belonging to the domain $w_i^{\rm el}$ with linear size exceeding in a few times the linear size of $\Omega^{\rm meso}$ (see for details Section 11.2). Thus, although the periodic boundary conditions were not used in Eq. (11.67); nevertheless, the periodicity condition was explored for the solution $\overline{\boldsymbol{\varepsilon}}(\mathbf{x}_i)$ (11.67), and, because of this, Eq. (11.67) contains only $N^{\rm meso}$ unknown strains associated with the inclusions embedded in the mesocell $\Omega^{\rm meso}$.

As an example, we will consider an imperfect SC packing of identical hard spherical inclusions (the phase $v^{(1)}$) randomly replaced by the spherical voids of the same radius (the phase $v^{(2)}$). The statistical descriptors of the imperfect packing are fixed: $c^{(1)}(\lambda^{(1)})+c^{(2)}(\lambda^{(2)})=0.4,\ \lambda^{(1)}=0.9,\ \lambda^{(2)}=0.1$. In Fig. 11.6 the convergence of the estimations (11.69) and (11.71) is demonstrated. Stabilization of the estimations of $\mathbf{L}^*$ with as increase of the numbers of inclusions $N^{\rm meso}$ in the mesocell indicates the stabilization of $\mathbf{L}^*$ at $N^{\rm meso}\to\infty$ and $n^{MC}\to\infty$. For instance, the values $\mu^*/\mu^{(0)}$ estimated for $400<n^{MC}<500$ distinguish from their estimations for $n^{MC}=2000$, $n^{MC}=600$ no more than 1.0% and 0.05% for $N^{\rm meso}=27$ and $N^{\rm meso}=343$, respectively. In so doing, the values $\mu^*/\mu^{(0)}$ estimated for $N^{\rm meso}=27$ and $N^{\rm meso}=343$ at $n^{MC}=2000$ and $n^{MC}=600$, respectively, can be distinguished by 0.35%. In a similar manner, the case of missing stiff inclusions ($c^{(1)}(1)=0.4,\ \lambda^{(1)}=0.9,\ \lambda^{(2)}\equiv 0,\ \mathbf{L}^{(2)}\equiv\mathbf{L}^{(0)}$) was considered in Fig. 11.7.

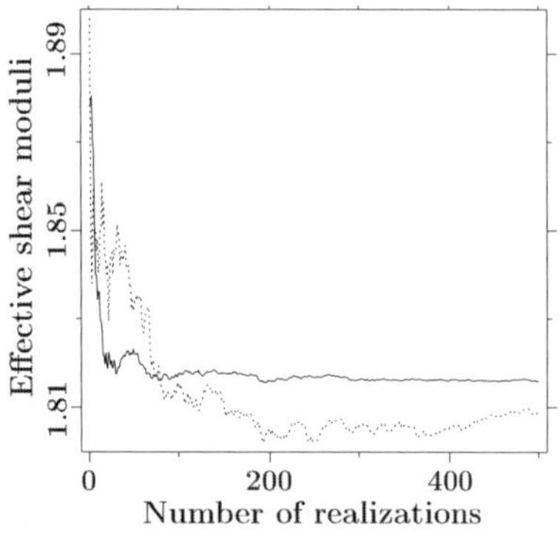

Fig. 11.6. $\mu^*/\mu^{(0)}$ vs the number of Monte Carlo realization n^{MC}: for $N^{\rm meso}=343$ (solid line), $N^{\rm meso}=27$ (dotted line).

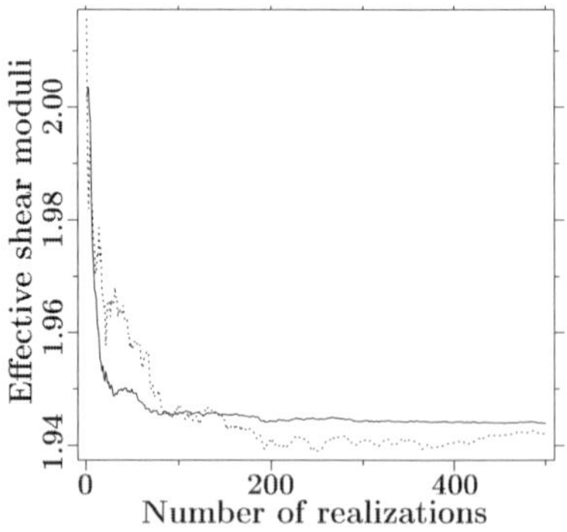

Fig. 11.7. $\mu^*/\mu^{(0)}$ vs the number of Monte Carlo realization n^{MC}: for $N^{\rm meso}=343$ (solid line), $N^{\rm meso}=27$ (dotted line).

11.4.3 Choice of the Periodic Comparison Medium. General Scheme

The general integral equations for the periodic structures with random imperfections were obtained in Section 7.5. Thus, the initial problem (7.47) was reduced to the estimation of statistical average of the second item on the right-hand side of Eq. (7.49). We will begin the study of this operator item from the problem for one imperfection $v_i^{\rm imp} \subset \Omega_i$ with a characteristic function $V_i^{\rm imp}$ in an infinite perfect periodic structure loaded by a homogeneous field $\overline{\boldsymbol{\varepsilon}}$ at infinity $\boldsymbol{\varepsilon}(\mathbf{x}) = \mathbf{A}^{\rm per}(\mathbf{x})\overline{\boldsymbol{\varepsilon}} + \boldsymbol{\mathcal{A}} * \int \mathbf{U}(\mathbf{y}-\mathbf{z})\boldsymbol{\eta}_3(\mathbf{z})V_i^{\rm imp}(\mathbf{z})\, d\mathbf{z}$. Then the operator term can be presented in the form ($\mathbf{x} \in \Omega_k,\ k = 1, 2, \ldots$)

$$\boldsymbol{\mathcal{A}} * \Big\{ \int \mathbf{U}(\mathbf{y}-\mathbf{z})\boldsymbol{\eta}_3(\mathbf{z})V_i^{\rm imp}(\mathbf{z})\, d\mathbf{z} \Big\}(\mathbf{x}) = \boldsymbol{\varepsilon}(\mathbf{x}) - \mathbf{A}^{\rm per}(\mathbf{x})\overline{\boldsymbol{\varepsilon}} \equiv \mathbf{A}_i^{\rm imp}(\mathbf{x})\overline{\boldsymbol{\varepsilon}}, \quad (11.72)$$

where the introduction of the inhomogeneous tensor $\mathbf{A}_i^{\rm imp}(\mathbf{x})$ is admissible due to the linearity of the problem and to the homogeneity of the strain $\overline{\boldsymbol{\varepsilon}}$. The rigorous solution $\boldsymbol{\varepsilon}(\mathbf{x})$ ($\mathbf{x} \in \Omega_k,\ k = 1, 2, \ldots$) (11.72) of the problem for a single imperfection (microcrack) in an infinite periodic medium that is equivalent to the estimation of the tensor $\mathbf{A}_i^{\rm imp}(\mathbf{x})$ (11.72) was considered in [340], [534]. Estimation of $\mathbf{A}_i^{\rm imp}(\mathbf{x})$ (11.72) by the MEFM will be considered in Subsection 11.4.4.

In a similar manner we can define the tensor $\mathbf{A}_{i,j}^{\rm imp}(\mathbf{x})$ describing the solution for two imperfections $v_i^{\rm imp} \subset \Omega_i$ and $v_j^{\rm imp} \subset \Omega_j$ with characteristic functions $V_i^{\rm imp}(\mathbf{z})$ and $V_j^{\rm imp}(\mathbf{z})$, respectively, in the infinite periodic structure ($\mathbf{x} \in \Omega_k$):

$$\boldsymbol{\mathcal{A}} * \Big\{ \int \mathbf{U}(\mathbf{y}-\mathbf{z})\boldsymbol{\eta}_3(\mathbf{z})(V_i^{\rm imp}(\mathbf{z}) + V_j^{\rm imp}(\mathbf{z}))\, d\mathbf{z} \Big\}(\mathbf{x}) = \boldsymbol{\varepsilon}(\mathbf{x}) - \mathbf{A}^{\rm per}(\mathbf{x})\overline{\boldsymbol{\varepsilon}} \equiv \mathbf{A}_{i,j}^{\rm imp}(\mathbf{x})\overline{\boldsymbol{\varepsilon}},$$

where the decomposition of the tensor $\mathbf{A}_{i,j}^{\rm imp}(\mathbf{x}) = \check{\mathbf{A}}_{i,j}^{\rm imp}(\mathbf{x}) + \hat{\mathbf{A}}_{i,j}^{\rm imp}(\mathbf{x})$ describes the perturbations introduced by the imperfection $v_i^{\rm imp}$ and $v_j^{\rm imp}$, respectively. The tensors $\mathbf{A}_i^{\rm imp}(\mathbf{x})\overline{\boldsymbol{\varepsilon}}$ and $\mathbf{A}_{i,j}^{\rm imp}(\mathbf{x})\overline{\boldsymbol{\varepsilon}}$ have a trivial sense of the perturbation introduced in the point $\mathbf{x} \in \Omega_k$ by one ($v_i^{\rm imp}$) and two ($v_i^{\rm imp}$ and $v_j^{\rm imp}$) imperfections, respectively.

After conditional statistical averaging, Eq. (5.7) turns into an infinite system of integral equations that can be closed and approximately solved in the framework of the modified version of the MEFM hypotheses (compare with Subsection 11.4.1).

H1^2) *Each imperfection $v_i^{\rm imp} \subset \Omega_i$ is considered as an isolated one and located in the infinite periodic structure loaded by the homogeneous effective field at infinity $\overline{\boldsymbol{\varepsilon}}_i(\mathbf{y}) \equiv \overline{\boldsymbol{\varepsilon}}_i$, ($\mathbf{y} \in \Omega_i$).*

H2^2) *Each pair of the imperfections $v_i^{\rm imp} \subset \Omega_i$ and $v_j^{\rm imp} \subset \Omega_j$ is located in an effective field $\widehat{\boldsymbol{\varepsilon}}(\mathbf{x})_{i,j}$ and: $\langle \widehat{\boldsymbol{\varepsilon}}(\mathbf{x})_{i,j} \rangle_k = \langle \overline{\boldsymbol{\varepsilon}}_k \rangle(\mathbf{x}) = {\rm const.}$ ($\mathbf{x} \in \Omega_k,\ k = i, j$).*

As in Section 11.4.1 the hypothesis **H2^2)** is applied for the closing of the infinite integral equation system after the substitution of the solution for two imperfections in the infinite periodic structure ($v_i^{\rm imp}, v_j^{\rm imp} \subset w_i^{\rm el},\ \mathbf{x} \subset \Omega_i$):

$$\begin{aligned} \mathbf{A}^{\rm per}(\mathbf{x})\langle \overline{\boldsymbol{\varepsilon}}_i \rangle &= \mathbf{A}^{\rm per}(\mathbf{x})\langle \boldsymbol{\varepsilon} \rangle + \mathbf{A}^{\rm per}(\mathbf{x})\mathbf{P}(w_i^{\rm el}) \sum_{j=1}^{N} \lambda^{(j)}\mathbf{R}_3^{\epsilon(j)} \langle \overline{\boldsymbol{\varepsilon}}_j \rangle \\ &+ \int \check{\mathbf{A}}_{i,j}^{\rm imp}(\mathbf{x})\varphi(v_j^{\rm imp}, \mathbf{x}_j | ; v_i^{\rm imp}, \mathbf{x}_i)V_i^{\rm el}(\mathbf{x}_j)\, d\mathbf{x}_j \langle \overline{\boldsymbol{\varepsilon}}_i \rangle \\ &+ \int \hat{\mathbf{A}}_{i,j}^{\rm imp}(\mathbf{x})\varphi(v_j^{\rm imp}, \mathbf{x}_j | ; v_i^{\rm imp}, \mathbf{x}_i)V_i^{\rm el}(\mathbf{x}_j)\langle \overline{\boldsymbol{\varepsilon}}_j \rangle\, d\mathbf{x}_j, \quad (11.73) \end{aligned}$$

where the tensor $\mathbf{R}_3^{\epsilon(j)}$ is obtained by the volume average over the unit cell $\langle(\cdot)\rangle_{(i)}^{\Omega}$ with the imperfection $\mathbf{R}_3^{\epsilon(j)} = \langle \mathbf{L}_3(\mathbf{x})[\mathbf{A}_j^{\rm imp}(\mathbf{x}) + A^{\rm per}(\mathbf{x})]\rangle_{(j)}^{\Omega}$, and $\mathbf{R}_3^{\epsilon(1)} \equiv \mathbf{0}$ for the perfect inclusion. In (11.73) one takes into account that the item with the tensor $\mathbf{A}_{i,i}^{\rm imp}(\mathbf{x})$ is extracted from the integral sign, and $\mathbf{A}_{i,i}^{\rm imp}(\mathbf{x}) = \mathbf{A}_i^{\rm imp}(\mathbf{x})$.

The estimation of statistical averages can be easily found from Eq. (11.73) of both effective fields and strain fields:

$$\langle\overline{\varepsilon}_i\rangle = \sum_{j=1}^{N} \mathbf{D}_{ij}^{\epsilon}\langle\varepsilon\rangle, \quad \mathbf{D}_{ij}^{\epsilon-1} = -\lambda^{(j)}\mathbf{P}(V_i^{\rm el})\mathbf{R}_3^{\epsilon(j)}$$
$$+\delta_{ij}\Big[\mathbf{I} - \langle(\mathbf{A}^{\rm per}(\mathbf{x}))^{-1}\int \check{\mathbf{A}}_{i,j}^{\rm imp}(\mathbf{x})\varphi(v_j^{\rm imp},\mathbf{x}_j|;v_i^{\rm imp},\mathbf{x}_i)W_i^{\rm el}(\mathbf{x}_j)\,d\mathbf{x}_j\rangle_{(i)}^{\Omega}\Big]$$
$$-\langle(\mathbf{A}^{\rm per}(\mathbf{x}))^{-1}\int \hat{\mathbf{A}}_{i,j}^{\rm imp}(\mathbf{x})\varphi(v_j^{\rm imp},\mathbf{x}_j|;v_i^{\rm imp},\mathbf{x}_i)W_i^{\rm el}(\mathbf{x}_j)\,d\mathbf{x}_j\rangle_{(i)}^{\Omega}, \qquad (11.74)$$

$$\langle\varepsilon\rangle(\mathbf{x}) = [\mathbf{A}_i^{\rm imp}(\mathbf{x}) + \mathbf{A}^{\rm per}(\mathbf{x})]\sum_{j=1}^{N}\mathbf{D}_{ij}^{\epsilon}\langle\varepsilon\rangle, \qquad (11.75)$$

The conditional statistical average of strains in the unit cell Ω_i in the case of the absence of the imperfection can be estimated as

$$\langle\varepsilon\rangle(\mathbf{x}) = \mathbf{A}^{\rm per}(\mathbf{x})\langle\varepsilon\rangle + \mathbf{A}^{\rm per}(\mathbf{x})\mathbf{P}(w_i^{\rm el})\sum_{j=1}^{N}\lambda^{(j)}\mathbf{R}_3^{\epsilon(j)}\langle\overline{\varepsilon}_j\rangle$$
$$+\int \mathbf{A}_j^{\rm imp}(\mathbf{x})\varphi(v_j^{\rm imp},\mathbf{x}_j|;\Omega_i)W_i^{\rm el}(\mathbf{x}_j)\langle\overline{\varepsilon}_j\rangle\,d\mathbf{x}_j, \qquad (11.76)$$

where the statistical averages of effective fields $\langle\overline{\varepsilon}_j\rangle$ were obtained via Eq. (11.74_1):

$$\langle\varepsilon\rangle(\mathbf{x}) = \mathbf{A}^{\rm per}(\mathbf{x})\mathbf{D}_0^{\epsilon}(\mathbf{x})\langle\varepsilon\rangle, \quad \mathbf{D}_0^{\epsilon}(\mathbf{x}) = \mathbf{I} + \mathbf{P}(w_i^{\rm el})\sum_{j,k=1}^{N}\lambda^{(j)}\mathbf{R}_3^{\epsilon(j)}\mathbf{D}_{jk}^{\epsilon}$$
$$+[\mathbf{A}^{\rm per}(\mathbf{x})]^{-1}\sum_{k=1}^{N}\int \mathbf{A}_j^{\rm imp}(\mathbf{x})\varphi(v_j^{\rm imp},\mathbf{x}_j|;\Omega_i)W_i^{\rm el}(\mathbf{x}_j)\mathbf{D}_{jk}^{\epsilon}\,d\mathbf{x}_j. \qquad (11.77)$$

Effective elastic properties are defined from the average constitutive low $\langle\boldsymbol{\sigma}\rangle = \langle\mathbf{L}\varepsilon\rangle$ accompanied by Eqs. (11.75) and (11.77_1) for the strain concentrator factor:

$$\mathbf{L}^* = \mathbf{L}^{(0)} + \sum_{i,j=2}^{N}\lambda^{(i)}\mathbf{R}_3^{\epsilon(i)}\mathbf{D}_{ij}^{\epsilon} + \lambda^{(1)}\langle\mathbf{L}_2(\mathbf{x})\mathbf{A}^{\rm per}(\mathbf{x})\mathbf{D}_0^{\epsilon}(\mathbf{x})\rangle_{(i)}^{\Omega}, \qquad (11.78)$$

from which the effective elastic modulus for the perfect periodic structure $\mathbf{L}^{per*}$ assumed to be known can be extracted at $\lambda^{(1)} = 1,\ \lambda^{(i)} = 0,\ i = 2,\ldots,N$:

$$\mathbf{L}^{\rm per*} = \mathbf{L}^{(0)} + \langle\mathbf{L}_2(\mathbf{x})\mathbf{A}^{\rm per}(\mathbf{x})\rangle_{(i)}^{\Omega}. \qquad (11.79)$$

In so doing, the estimations of averages over the unit cell $\langle(\cdot)\rangle_i^{\Omega}$ in Eqs. (11.78) and (11.79) are performed over the volume of inclusions because $\mathbf{L}_2(\mathbf{x}), \mathbf{L}_3(\mathbf{x}) \equiv$

const. in the matrix. The method proposed is general; it is not limited by a concrete numerical scheme and can use particular known solutions for particular problems mentioned above (for example, the solutions for the perfect periodic structure $\mathbf{A}^{\rm per}(\mathbf{x})$ as well as the solution for one crack in doubly periodic structure described by the tensor $\mathbf{A}_i^{\rm imp}(\mathbf{x})$, see [534]. The main idea of the proposed method is a decomposition of the desired solution on the known solution for the perfect periodic structure ($\mathbf{L}(\mathbf{x}) = \mathbf{L}^{\rm per}(\mathbf{x})$), and on the perturbation produced by the imperfections in the perfect periodic structure. Analogously to the random structure matrix composites (RSMC), the effective moduli depend on three particular solutions: and strain concentration factor $\mathbf{A}^{\rm per}(\mathbf{x})$ in the perfect periodic structure (analog of homogeneous strains in the RSMC without defects), strain concentration factor in the infinite periodic medium with one imperfection $\mathbf{A}_i^{\rm imp}(\mathbf{x})$ (analog of Eshelby solution in the RSMC), strain concentration factor in the infinite periodic medium with two arbitrary located imperfections $\mathbf{A}_{i,j}^{\rm imp}(\mathbf{x})$ (analog of the solution for two inclusions inside an infinite matrix; see Chapter 4).

The obtained representations for the strain concentrators can be simplified in the framework of the generalized quasi-crystalline approximation by Lax [649], which is interpreted in our cases (11.73), (11.74_2) and (11.77_1) that perturbation produced by the imperfection $v_j^{\rm imp}$ does not depend on the imperfection $v_i^{\rm imp}$:

$$\mathbf{A}_{i,j}^{\rm imp}(\mathbf{x}) = \mathbf{A}_j^{\rm imp}(\mathbf{x}), \ \ (\mathbf{x} \in \Omega_k, \ \ i,j,k = 1,2,\ldots). \tag{11.80}$$

Assumption (11.80) allows the further simplification at $N = 2$ (a single sort of imperfections):

$$\langle \boldsymbol{\varepsilon}\rangle(\mathbf{x}) = [\mathbf{A}_i^{\rm imp}(\mathbf{x}) + \mathbf{A}^{\rm per}(\mathbf{x})]\mathbf{D}_1^{\epsilon}\langle\boldsymbol{\varepsilon}\rangle, \quad \mathbf{x} \in v_i^{\rm imp} \subset \Omega_i, \tag{11.81}$$

$$\langle \boldsymbol{\varepsilon}\rangle(\mathbf{y}) = \mathbf{A}^{\rm per}(\mathbf{y})\mathbf{D}_0^{\epsilon}(\mathbf{x})\langle\boldsymbol{\varepsilon}\rangle, \quad \mathbf{y} \in \Omega_j \not\supset v_i^{\rm imp} \tag{11.82}$$

$$\begin{aligned}\mathbf{D}_1^{\epsilon-1} &= \mathbf{I} - \lambda^{(2)}\mathbf{P}(V_i^{\rm el})\mathbf{R}_3^{\epsilon(2)} \\ &\quad - \langle(\mathbf{A}^{\rm per}(\mathbf{x}))^{-1}\int \mathbf{A}_j^{\rm imp}(\mathbf{x})\varphi(v_j^{\rm imp},\mathbf{x}_j|; v_i^{\rm imp},\mathbf{x}_i)W_i^{\rm el}(\mathbf{x}_j)\, d\mathbf{x}_j\rangle_{(i)}^{\Omega},\end{aligned} \tag{11.83}$$

$$\begin{aligned}\mathbf{D}_0(\mathbf{x}) &= \mathbf{I} + \lambda^{(2)}\mathbf{P}(w_i^{\rm el})\mathbf{R}_3^{\epsilon(2)}\mathbf{D}_1^{\epsilon} \\ &\quad + [\mathbf{A}^{\rm per}(\mathbf{x})]^{-1}\int \mathbf{A}_j^{\rm imp}(\mathbf{x})\varphi(v_j^{\rm imp},\mathbf{x}_j|;\Omega_i)W_i^{\rm el}(\mathbf{x}_j)\, d\mathbf{x}_j\mathbf{D}_1^{\epsilon}.\end{aligned} \tag{11.84}$$

It should be mentioned that the so-called Green's function model [252] considering the finite number of broken fibers in the periodic mesocell $\Omega^{\rm meso}$ is in actual truth an approximate method of the estimation of the tensors $\mathbf{A}_{i,j}^{\rm imp}(\mathbf{x})$ by the use of the tensor $\mathbf{A}_i^{\rm imp}(\mathbf{x})$ for a single broken fiber in the periodic mesocell $\Omega^{\rm meso}$. In light of the current approach proposed, the incorporation of the tensors $\mathbf{A}_i^{\rm imp}(\mathbf{x})$ and $\mathbf{A}_{i,j}^{\rm imp}(\mathbf{x})$ estimated by such a manner into Eqs. (11.74_2)–(11.78) presents no difficulties.

Broadly speaking, for the case where the defect embedded within a very large system, can be made small compared to the system size, Green's functions are a very powerful method for studying the physics of defects and their interaction. In light of this, the proposed method is related to a hierarchical method [1088] (see also [1203]) in which Green's functions are calculated for the perfect lattice, for increasingly complicated defect lattices. The authors suggested that the motivation for using this method, rather than the direct simulation methods that are more widely used today, is the prospect for studying a very large system in 3-D, with the order of 10^7–10^8 atoms. At this point in the analysis, a set of perfect lattice Green's functions has become available in real-space coordinates. Imperfections are treated in the lattice via changes in the coupling between the atoms. When the number of altered sites is a minority of the sites in the lattice, a straightforward approach offers itself through the Dyson equation.

11.4.4 The Version of MEFM Using the Periodic Comparison Medium

We will consider the realization of the MEFM in the framework of the scheme presented in Subsection 11.4.3. Up to now the hypothesis **H1** was not used. The harnessing of this hypothesis makes it possible to obtain the representation of the tensor $\mathbf{A}^{\rm imp}_{i,j}(\mathbf{x})$ just through the tensor $\mathbf{A}^{\rm imp}_{i}(\mathbf{x})$ in much the same manner as in Subsection 11.4.1. We will first estimate the tensor $\mathbf{A}^{\rm per}(\mathbf{x})$ (5.6) for the perfectly periodic structure at the MEFM's assumptions **H1** and **H2** (a single sort of inclusions). Then, taking (11.24) into account, we get the following representation for the stress concentrator factor for the points both inside $\mathbf{x} \in v_i \subset \Omega_i$ and outside $\mathbf{y} \in \Omega_i \setminus v_i$ of the inclusion in the unit cell:

$$\mathbf{A}^{\rm per}(\mathbf{x}) = \begin{cases} \mathbf{A}_1 \mathbf{D}^{\epsilon \rm per} & \text{for } \mathbf{x} \in v_i \subset \Omega_i, \\ \mathbf{D}^{\epsilon \rm per}_0(\mathbf{x}) & \text{for } \mathbf{x} \in \Omega_i \setminus v_i, \end{cases} \tag{11.85}$$

where $(\mathbf{D}^{\epsilon \rm per})^{-1} = \mathbf{I} - \mathbf{P}(w^{\rm el}_i)\mathbf{R}^{\epsilon} n^{(1)} - \sum_{\mathbf{m}\neq \mathbf{0}} \mathbf{T}^{\epsilon}_{i\mathbf{m}}(\mathbf{x}_i - \mathbf{x}_{\mathbf{m}})\mathbf{R}^{\epsilon}$, $\mathbf{D}^{\epsilon \rm per}_0(\mathbf{x}) = \mathbf{I} + \mathbf{P}(w^{\rm el}_i)\mathbf{R}^{\epsilon} n^{(1)} \mathbf{D}^{\epsilon \rm per} + \sum_{\mathbf{m}} \mathbf{T}^{\epsilon}_{\mathbf{m}}(\mathbf{x} - \mathbf{x}_{\mathbf{m}})\mathbf{R}^{\epsilon}\mathbf{D}^{\epsilon \rm per}$, and $\mathbf{x}_{\mathbf{m}} \in w^{\rm el}_i$. Estimation of the tensors $\mathbf{A}^{\rm imp}_i(\mathbf{x})$ and $\mathbf{A}^{\rm imp}_{i,j}(\mathbf{x})$ can be performed in a manner like the solutions (4.12) and $\mathbf{Z}^{\epsilon}_{ij}$ with the simplest case corresponding to the single sort of inclusions and imperfections $N = 2$. In so doing the *toroidal edge correction* used for the periodic structures in Subsection 11.4.2 should be replaced by another sort of the edge correction [939], and the strain concentrator factors in the inclusions v_k ($k = 1, \ldots, N^{\rm meso}$) at the fixed imperfection $v^{\rm imp}_i$ in the center of meso element $\Omega^{\rm meso}$ are defined by the relations ($\mathbf{x} \in v_k$):

$$\mathbf{A}^{\rm imp}_i(\mathbf{x}) = \mathbf{A}_k \mathbf{D}^{\epsilon \rm imp}_k - \mathbf{A}^{\rm per}(\mathbf{x}), \ \ \mathbf{D}^{\epsilon \rm imp}_k = \sum_{k=1}^{N^{\rm meso}} \overline{\mathbf{Y}}_{kl}[\mathbf{I} + \mathbf{P}(w^{\rm el}_k)\mathbf{R}^{\epsilon}\mathbf{D}^{\epsilon \rm per} n], \tag{11.86}$$

where the matrix $\overline{\mathbf{Y}}^{-1}$ has the elements ($k, l = 1, \ldots, N^{\rm meso}$) analogous to $(\widetilde{\mathbf{Y}}^{\epsilon})^{-1}_{ij}$ (11.68)

$$(\overline{\mathbf{Y}}^{-1})_{kl} = \delta_{kl}\mathbf{I} - \sum_{\mathbf{m}\neq k} \mathbf{T}^{\epsilon}_{k\mathbf{m}}(\mathbf{x}_k - \mathbf{x}_{\mathbf{m}})\mathbf{R}^{\epsilon}_{\mathbf{m}_r}, \ \ \mathbf{x}_{\mathbf{m}} \in w^{\rm el}_k \tag{11.87}$$

where the single imperfect cell is assumed to be located in the center of $\Omega^{\rm meso}$ and just a single $\mathbf{R}^{\epsilon}_{\mathbf{m}_r} = \mathbf{R}^{\epsilon \rm imp}$ at $\mathbf{m} = i$, whereas other $\mathbf{R}^{\epsilon}_{\mathbf{m}_r}$ are identical and appropriate to the perfect inclusions $\mathbf{R}$. Obviously, that in the absent of the imperfection $\mathbf{D}_k^{\epsilon\text{imp}} \equiv \mathbf{D}^{\epsilon\text{per}}$ and $\mathbf{A}_i^{\text{imp}}(\mathbf{x}) \equiv \mathbf{0}$.

The strain concentrator factor in the matrix can be found in much the same way as Eq. (11.85) ($\mathbf{x} \in \Omega_k \setminus v_k,\ \mathbf{x_m} \in w_k^{\text{el}}$):

$$\mathbf{A}_i^{\text{imp}}(\mathbf{x}) = \mathbf{A}^{\text{per}}(\mathbf{x})[\mathbf{D}_{0k}^{\epsilon\text{imp}}(\mathbf{x}) - \mathbf{I}], \quad \mathbf{D}_{0k}^{\epsilon\text{imp}}(\mathbf{x}) = \mathbf{I} + \mathbf{P}(w_i^{\text{el}})\mathbf{R}^{\epsilon}n^{(1)}\mathbf{D}^{\epsilon\text{per}} + \sum_{\mathbf{m}} \mathbf{T}_{\mathbf{m}}^{\epsilon}(\mathbf{x} - \mathbf{x_m})\mathbf{R}_{\mathbf{m}_r}^{\epsilon}\mathbf{D}_{\mathbf{m}}^{\epsilon\text{imp}}. \quad (11.88)$$

As in (11.87), the imperfect cell is assumed to be located in the center of Ω^{meso} and in Eq. (11.88_1) just a single $\mathbf{R}_k^{\epsilon} = \mathbf{R}_i^{\epsilon\text{imp}}$ at $k = i$, whereas other $\mathbf{R}_j^{\epsilon} \equiv \mathbf{R}^{\epsilon}$ are identical and appropriate to the perfect inclusions.

The tensor $\mathbf{A}_{i,j}^{\text{imp}}(\mathbf{x})$ can be found by the use of Eqs. (11.86) and (11.88_1) taking into account that in Eqs. (11.87) and (11.88_2), respectively, the summation is performed over the inclusions containing two imperfections v_i^{imp} and v_j^{imp} that yields the equality $\mathbf{R}_{\mathbf{m}_r}^{\epsilon} = \mathbf{R}^{\epsilon\text{imp}}$ at $\mathbf{m} = i, j$, whereas other $\mathbf{R}_{\mathbf{m}_r}^{\epsilon}$ are identical $\mathbf{R}^{\epsilon}$ for the perfect inclusions.

In the case of two imperfections v_i^{imp} and v_j^{imp} subjected to the remote homogeneous loading $\overline{\boldsymbol{\varepsilon}}$, the tensor $\mathbf{A}_{i,j}^{\text{imp}}(\mathbf{x})$ can be estimated by the use of the tensor $\mathbf{A}_i^{\text{imp}}(\mathbf{x})$ (11.86). Indeed, analogously to (4.57) we can get $\overline{\boldsymbol{\varepsilon}}_i = \sum_{j=1}^{2} \mathbf{Z}_{ij}^{\text{imp}}\overline{\boldsymbol{\varepsilon}}$, where the matrix $(\mathbf{Z}^{\text{imp}})^{-1}$ has the elements $[(\mathbf{Z}^{\text{imp}})^{-1}]_{ij} = \mathbf{I}\delta_{ij} - (1 - \delta_{ij})(\mathbf{A}^{\text{per}}(\mathbf{x}_i))^{-1}\mathbf{A}_j^{\text{imp}}(\mathbf{x}_i)$, $(i = 1, 2;\ j = 3 - i)$. The equation for $\overline{\boldsymbol{\varepsilon}}_i$ allows one to find the perturbation introduced by the imperfections v_i^{imp} and v_j^{imp} in the point $\mathbf{x} \in \Omega_k$:

$$\check{\mathbf{A}}_{i,j}^{\text{imp}}(\mathbf{x}) = \mathbf{A}_i^{\text{imp}}(\mathbf{x}) \sum_{l=1}^{2} \mathbf{Z}_{il}, \quad \text{and} \quad \hat{\mathbf{A}}_{i,j}^{\text{imp}}(\mathbf{x}) = \mathbf{A}_j^{\text{imp}}(\mathbf{x}) \sum_{l=1}^{2} \mathbf{Z}_{jl}, \quad (11.89)$$

respectively, which can be simplified for the identical imperfections ($\mathbf{x} \in v_i^{\text{imp}}$):

$$\mathbf{A}_{i,j}^{\text{imp}}(\mathbf{x}) = \mathbf{A}_j^{\text{imp}}(\mathbf{x})[\mathbf{I} - \mathbf{A}_i^{\text{imp}}(\mathbf{x}_j)]^{-1}. \quad (11.90)$$

After estimation of the tensors $\mathbf{A}^{\text{per}}(\mathbf{x})$ as well as $\mathbf{A}_i^{\text{imp}}(\mathbf{x})$, $\mathbf{A}_{i,j}^{\text{imp}}(\mathbf{x})$, $\mathbf{D}_1^{\epsilon\text{imp}}$, $\mathbf{D}_0^{\epsilon\text{imp}}(\mathbf{x})$ for one and two imperfections in the infinite periodic structure, the evaluation of the strain concentrator factors analogous to Eqs. (11.75) and (11.77_1) as well as of the effective elastic moduli (11.78) presents no special problems. All these equations are simplified because $\mathbf{A}^{\text{per}}(\mathbf{x}), \mathbf{A}_i^{\text{imp}}(\mathbf{x}), \mathbf{A}_{i,j}^{\text{imp}}(\mathbf{x}) \equiv \text{const.}$ in the inclusions.

The obtained representations for the strain concentrators can be simplified in the framework of the generalized quasi-crystalline approximation by Lax [649], which is interpreted in our cases as the equality (11.80). In the framework of the generalized quasi-crystalline approximation (11.80), by a simple substitution we can verify that in Eqs. (11.81)–(11.84) $\mathbf{D}_0^{\epsilon} = \mathbf{D}_1^{\epsilon}$, and, therefore,

$$\mathbf{L}^* = \mathbf{L}^{(0)} + c_1\langle(\lambda^{(2)}\mathbf{L}_3^{(i)} + \lambda^{(1)}\mathbf{L}_2^{(i)})\mathbf{A}^{\text{per}}(\mathbf{x})\rangle_{(i)}^{\Omega}\mathbf{D}_1^{\epsilon} + c_1\lambda^{(2)}\langle\mathbf{L}_3^{(i)}\mathbf{A}_i^{\text{imp}}(\mathbf{x})\rangle_{(i)}^{\Omega}\mathbf{D}_1^{\epsilon}. \quad (11.91)$$

As an example we will demonstrate the previous results obtained in the framework of hypothesis **H1** for a simple cubic packing of stiff spherical inclusions with elastic properties $\mathbf{L}^{(1)} = 100\mathbf{L}^{(0)}$ randomly replaced by the holes of the same size with the probability $\lambda^{(2)} = 1 - \lambda^{(1)}$, $c = 0.4$. In Fig. 11.8 the normalized effective shear moduli $\mu^*/\mu^{(0)}$ are estimated by the use of two-point approximation (11.88_1) and (11.90), one-particle approximation (11.80) and (11.91), one particle approximation (11.3) using the homogeneous comparison medium $\mathbf{L}^c = \mathbf{L}^{(0)}$ as well as via the Monte Carlo simulation described in Subsection 11.4.2 with $n^{MC} = 250$, $N^{\rm meso} = 343$. Figure 11.8 shows the higher accuracy of the two-point approximation described by hypothesis $\mathbf{H2}^2$ and by the quasi-crystalline, one-point approximation (11.91) that was confirmed by comparison with the Monte-Carlo simulation, which does not use any closing assumptions of either the **H2** or $\mathbf{H2}^2$ types. The CPU times expended for estimation of each point obtained by the Monte Carlo simulation and by Eq. (11.78) differ by a factor of 1000.

Thus, we proposed a general explicit analytical representation of effective moduli of periodic structures with the random field of imperfections. This representation explicitly depends on three tensors describing the solutions for the perfect periodic medium, for the periodic medium with one imperfection, and for the periodic medium with all possible relative locations of two imperfections. No restrictions were assumed on the method used for obtaining these solutions. It should be mentioned that the known numerical methods (such as, e.g., finite element analysis and boundary element method) have a series of advantages and disadvantages (see, e.g., Chapter 4), and it is crucial for the analyst to be aware of their range of applications. However, such an analysis is beyond the scope of the current presentation where we consider the application of the proposed method in the framework of the additional hypothesis **H1** significantly simplifying the analytical representations of the tensors $\mathbf{A}^{\rm per}(\mathbf{x})$, $\mathbf{A}_i^{\rm imp}(\mathbf{x})$, $\mathbf{A}_{i,j}^{\rm imp}(\mathbf{x})$.

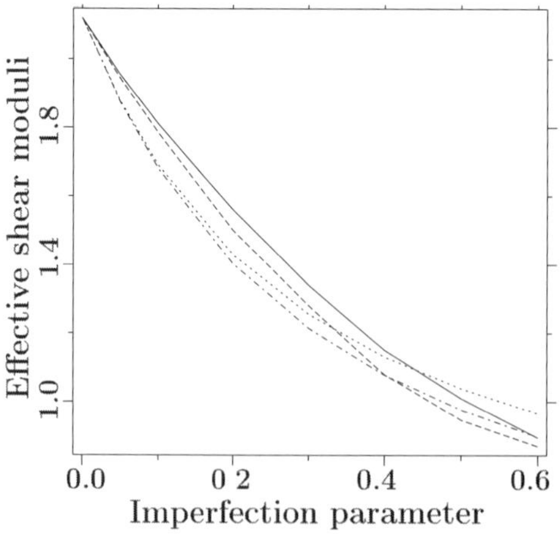

Fig. 11.8. $\mu^*/\mu^{(0)}$ vs $\lambda^{(2)}$: Monte Carlo random simulation (solid line), one-point approximation at $\mathbf{L}^c = \mathbf{L}^{(0)}$ (dashed line), one-point approximation (11.80) and (11.91) (dotted line), two-point approximation (11.88_1) and (11.90) (dot-dashed line).

11.4.5 Concluding Remarks

We will shortly present some of the specific peculiarities of analysis of periodic media with randomly distributed imperfections. Richer in content is a discussion of the main hypotheses as well as the limitations of the proposed estimations and their possible generalizations.

Choice of the comparison medium. It should be mentioned that the choice of the elastic properties of a comparison medium as elastic properties of either the matrix or the periodic medium has a series of advantages and disadvantages, and it is crucial for the analyst to be aware of their range of applications. For example, for pure random arrangement of inclusions, the assumption (11.3) is the only possible one among those considered. It is reasonable to suggest that the elastic properties of the comparison medium and the matrix coincide at the estimation of their induced perturbations for the case of a large concentration of imperfections (prescribed by $\lambda^{(k)}$, $k = 2, 3, \ldots$) and moderate concentration of inclusions. In so doing, the coinciding of the elastic properties of both the matrix and comparison medium implies homogeneity of the matrix that leads to the necessity of estimations of stress fields just inside the inhomogeneities. This simplification explains the effectiveness of the analysis of the microinhomogeneous media by the boundary integral equation (BIE) method or by the volume integral equation method (VIE, see, e.g., Chapter 4). In the general case of the inhomogeneity of the matrix in the unit cells, the choice of the homogeneous comparison medium is based on the intuition of the analyst; for the case considered, for instance, the elastic properties of the comparison medium can coincide with the Voight or Reuss averages of the matrix elastic properties. For the small concentration of imperfections ($\lambda^{(k)} \ll 1$, $k = 2, 3, \ldots$) in the media with evidently expressed periodic structure, the choice of a comparison medium as a perfect periodic one is preferable. Acceptance of hypothesis **H1** allows one to withhold from the assumptions the homogeneity and ellipsoidal shape of inclusions that can be done in the framework of the notion of the fictitious homogeneous ellipsoidal inclusion analyzed in Chapters 4 and 8. In such a case the statistical averages of stresses inside the inclusions will be, in general, inhomogeneous (see, e.g., [195] and Chapter 8).

Geometrical imperfections. Up to now the term imperfection implied constitutive imperfection with material properties $\mathbf{f}^{(k)}(\mathbf{x})$ ($\mathbf{f} = \mathbf{L}, \mathbf{M}$; $k = 2, 3, \ldots N$) differing from the properties of the periodic properties $\mathbf{f}^{\rm per}(\mathbf{x})$. However, the geometrical imperfections should also present no difficulties. For example, identical inclusions may be randomly distributed near the centers of unit cells of the periodic lattice, or, alternatively, the wall thickness of the honeycomb structure [210] is a random variable. In such a case the choice of the comparison medium of the second level $\mathbf{f}^{\rm per}(\mathbf{x})$ is obvious, and the number N of imperfections is defined by the digitization of possible locations of imperfection in the unit cell.

Lightweight materials. For lightweight materials with essentially anisotropic and inhomogeneous cell walls, it is reasonable to consider the elastic properties of the comparison medium of the first level as the Voight average of elastic properties of the solid phase $\mathbf{L}^c = (v^{(0)})^{-1}\langle \mathbf{L}(\mathbf{x})V^{(0)}(\mathbf{x})\rangle$. The comparison medium of the second level is chosen as a perfect periodic structure. The standard imperfections for these lightweight structures are the random distribution of fractured cell walls and missing cells. In such a case the imperfections of the same sort (e.g., broken ligament), with both different orientation and location inside the unit cell, are recognized as belonging to different components of imperfections. Thus, the problem of the estimation of effective compliance (11.78) is reduced to the solution of at least N problems (11.72) for a single imperfection in the infinite periodic medium.

Thus, the method proposed provides the calculation, with reasonable accuracy, of the effective properties and statistical average of stresses in the components for a whole range of parameters. The method appears to be simple enough in both theoretical and computational aspects, and contains in particular limiting cases, the known solutions for both the perfect periodic structures and the statistically homogeneous random structures (for example, Mori-Tanaka approach and others; for details see Chapter 8). In the limiting case of the perfect periodic structures ($\mathbf{L}(\mathbf{x}) \equiv \mathbf{L}^{\rm per}(\mathbf{x})$), the proposed method reduced to the well developed approaches analyzing periodic systems [25], [48], [507], [532], [960]. In the case of the standard random structure composites ($\mathbf{L}^{\rm per} \equiv \mathbf{L}^c = \text{const.}$), the proposed approach is reduced to the known version of MEFM. The accuracy of this version was verified by comparison with the accurate numerical analyses of periodic particulate and microcracked composites as well as by comparison with the experimental data for a wide class of linear and nonlinear problems (of conductivity, elasticity, thermoelastoplasticity, strength, and fracture) (numerous references can be found in Chapters 8–16). The main result of the proposed approach for the periodic structures with random imperfections is presented by Eqs. (11.74_2)–(11.78) which depends on three solutions: the known solution for the perfect periodic structure $\mathbf{A}^{\rm per}(\mathbf{x})$, the solution for the infinite periodic medium with one imperfection $\mathbf{A}_i^{\rm imp}(\mathbf{x})$, and the solution for the infinite periodic medium with two arbitrary located imperfections $\mathbf{A}_{i,j}^{\rm imp}(\mathbf{x})$. No restrictions were assumed for the methods of estimation of these tensors. However, the numerical results were obtained by the use of approximative analytical representations of mentioned tensors in the framework of hypothesis **H1** (the accuracy of the last hypothesis was analyzed for some particular micromechanical problems, see, e.g., Chapters 8 and 10). The numerical results obtained by the approach based on the choice of the periodic comparison medium (see Subsections 11.4.3 and 11.4.4) demonstrated the closed comparison with results based on both the choice of the homogeneous comparison medium (Section 4.1) and Monte Carlo simulation (Subsection 11.4.2). The proposed methods allow us to consider composites with any number of different components containing inclusions with size distribution, shape, orientation and properties, coated particles and fibers, cracks, etc. However, more detailed consideration of composites with anisotropic components, especially with the continuously variable anisotropic and nonlinear properties of components, is beyond the scope of the current study.

12

Nonlocal Effects in Statistically Homogeneous and Inhomogeneous Random Structure composites

12.1 General Analysis of Approaches in Nonlocal Micromechanics of Random Structure Composites

When the statistical average stress in a composite medium varies sufficiently slowly compared to the size scale of the microstructure (i.e., roughly speaking the separation of the external and internal scales takes place), the material macroscopically behaves as a homogeneous body with some effective moduli which can be estimated by different methods referred to earlier. If the condition of the separation of scales summarized earlier is not valid, the material's response cannot be adequately described in the framework of the local theory of elasticity for the homogeneous medium.

For analysis of the widely separated scales (but not "too widely"), a number of *phenomenological* approaches have been proposed to enhance the continuum model by nonlocal terms, introduced either through an integral equation, or through an additional gradient equation including one or several intrinsic length scales and based on the assumption that the forces between material points can be long-range in character (see [9], [45], [315], [343], [368], [942]) and further references therein). The integral formulation may be reduced to a gradient form by truncating the series expansion of the non-local kernel. Although an approximation, the gradient representation usually leads to problems that are more tractable than those obtained within the integral formulation. The different modifications of these phenomenological constitutive relations were used for the analyses of various size effects such as strain gradient hardening, size effect in torsion and indentation, strain gradient plasticity, and nonlocal damage mechanics.

Micromechanical models are used to answer a fundamental question of how length scales in the effective constitutive equations could be directly derived from the geometrical and mechanical properties of constituent phases. The mechanics of generalized continua has actively developed from the pioneer works [312], [313], [314], [315], [609], [610], [615], [618] establishing nonlocal elastic theories based on the lattice theory, on generalized Cosserat continua approximately considering independent micro-displacemens and a scale parameter in a medium, as well as on idea that cohesive forces have a long range, and are in fact nonlocal.

Kröner [610] showed that the eventual abandonment of the hypothesis of statistically homogeneous fields leads to a *nonlocal* coupling between statistical averages of the stress and strain tensors when the statistical average stress is given by an integral of the field quantity weighted by some tensorial kernel, i.e. the *nonlocal* effective elastic operator $\boldsymbol{\mathcal{L}}^*$: $\langle \boldsymbol{\sigma} \rangle(\mathbf{x}) = \int \boldsymbol{\mathcal{L}}^*(\mathbf{x},\mathbf{y})\langle \boldsymbol{\varepsilon} \rangle(\mathbf{y})d\mathbf{y}$. It leads to the violation of Hill's condition and a meaningfulness of the RVE notion is lowered. The reason for this violation is that in micromechanics of random structure composites the assumption (see Chapter 6) is usually made that all characteristic lengths associated with the spatial variations of the statistical mean field quantities are significantly large $d \ll l \ll L$ compared to all characteristic lengths associated with the spatial variations in the material's properties (such as the size of the inclusions, the mean distance between them and the characteristic length of the variation of the inclusion number density). If this assumption is valid, the governing equations for the ensemble-averaged fields are identical in form with the usual local elastic equations for homogeneous media. Otherwise, the effective constitutive law is nonlocal. In the consideration of dispersed media this nonlocal approach makes intuitive sense since the stress at any point will depend on the arrangement of the surrounding inclusions. Therefore the value of the statistical average field will locally depend on its value at the other points in its vicinity. This is especially true if the inclusion number density varies over distances that are comparable to the particle size. Thus, even if RVE is introduced as a minimum size of a microinhomogeneous medium providing the same (in some sense) effective elastic response as an infinite microinhomogeneous medium, this notion cannot be consistent. Indeed, the size of such as RVE will depend on the kind of effective elastic parameters (both local and nonlocal) involved in *a priori* unknown type of effective constitutive equation, and, moreover, it will depend on the method used for estimation of this effective constitutive equation (for details see Chapters 9, 11). Indeed, estimation of an effective elastic moduli by the one-particle effective field method (EFM) (see [170], [540]) does not depend on the size of RVE, in contrast with the evaluation of $\mathbf{L}^*$ by the MEFM (see Chapter 9).

If the field $\langle \boldsymbol{\varepsilon} \rangle(\mathbf{y})$ is varying sufficiently slowly in the neighborhood of an arbitrary point $\mathbf{y}$, then it can be expanded about $\mathbf{x}$ in a Taylor series and, therefore, the integral operator of nonlocal effective elastic properties can be considered as a differential one. Beran and McCoy [69], and Levin [666] considered a weakly inhomogeneous medium when the perturbation method of solution of integral equations involved with Green's functions kernels is appropriate; they found nonlocal dependence of statistical averages of the stresses and strains in the form of either integral or differential operators. Analysis of highly contrasted statistically homogeneous media is simplified for sufficiently smooth external loading permitting the use of the Taylor expansion for the statistical average of a stress field and the application of a Fourier transform method. In so doing the initial integral equation is reduced to an algebraic polynomial equation with constant coefficients in a Fourier space with subsequent solution and the implementation of the inverse Fourier transform. The scheme summarized informally above is usually based on the hypothesis of the homogeneity of the effective field in which each particle is located. The "quasi-crystalline" approximation by Lax [649] is often used for truncation of the hierarchy of integral equations involved leading to neglect of direct multiparticle interactions of inclusions. This reduces the analysis to the use of statistical information of up to two-point correlation functions and allows one to derive explicit relations for the nonlocal overall differential operator by the different methods: via the effective field method [545], or via the method of conditional moments [563]. Diener *et al.* [270] derived bounds on the Fourier transform of the nonlocal overall operator in the special case of "cell-structure" models. The nonlocal influence of density variations in an otherwise homogeneous medium was analyzed in [1188] assuming an exponential form for the two-point correlation function. An advantage of the rigorous approach [278], [279] is that it is based on variational principles, providing the bounds of approximations; Drugan and Willis obtained an elegant explicit representation of the nonlocal effective differential operators of the second and fourth order and systematically estimated long-range action of a nonlocal effect and the necessary sizes of representative volume elements providing *a'priori* prescribed accuracy of the evaluation of nonlocal effects. A similar Hashin-Shtrikman formalism was exploited in [715], [716] for the analysis of nonlocal effects produced by random fluctuations in the applied body force. Besides the usual (nonlocal) effective modulus operator, Luciano and Willis [715], [716] estimated another operator that provides a contribution to the mean stress field with itself.

The main disadvantage of "quasi-crystalline" approximation consisting of neglecting of direct multiparticle interactions of inclusions was overcome recently by the multiparticle effective field method (MEFM) (references may be found in Chapter 8). The MEFM is based on the theory of functions of random variables and Green's functions. Within this method a hierarchy of statistical moment equations for conditional averages of the stresses in the inclusions is derived. The hierarchy is established by introducing the notion of an effective field. In this way the interaction of different inclusions is taken directly into account in the framework of the homogeneity hypothesis of the effective field. Combining the MEFM with the standard scheme of the Fourier transform method usually used for the analysis of the statistically inhomogeneous stress fields allowed [132], [163], [181] to obtain the explicit representation of a nonlocal overall operator in the form of the second-order differential operator. Buryachenko [134] estimated the nonlocal integral operator of the overall constitutive equation for periodic structure composites by the use of the first iteration of the iteration method and showed the reduction of the integral operator obtained to the differential operator involved derived by the Fourier transform method.

More accurate analysis of the periodic structures is based on the well developed method of asymptotic expansions (see, e.g., [25], [532], [533], [885]) by construction of a "two-scale" asymptotic expansion of the solution with respect to the small parameter ε which is the ratio of the scale of the internal "micro" (the size of unit cell) and external "macro" (the size of the body) scales. Substitution of the expansion of the displacement in the asymptotic series over ε into the original system of equations and boundary conditions leads to so-called problems on a cell. Taking into account higher order terms ε, one can see that the stresses in the effective medium depend on higher order derivatives of displacements with respect to the coordinates considered, in particular, in couple stress elasticity

[104], [350], [367]. The variational-asymptotic approach proposed in [1023] may provide a better approximation to the real solution than the "purely" asymptotic approach in the sequential powers of the small parameter ε. The reason for this is that the variational construction is intrinsically such that it tends to minimize the error in a certain variational sense for the final values ε, while the asymptotic approach simply constructs the perturbations assuming that ε is very small. The effective material parameters were determined from the response of a unit cell under either displacement, displacement-periodic, or traction boundary conditions [105]. It was found that the three boundary conditions result in hierarchies of both couple-stress moduli and characteristic length. Smyshlyaev and Cherednichenko [1023] (see also [105], [851]) proposed a combination of the effective medium and micromechanical theories leading to reasonably accurate interface stresses. Vanin [1129] obtained an analytical representation for couple stress effective elastic properties of doubly periodic fiber reinforced medium; in the framework of moment elasticity for the effective medium he estimated the stress concentration around a circular hole in a composite plate.

The approach presented in Section 11.3 is generalized in Section 12.2 at first for the analysis of statistically homogeneous random structure composites. By the use of the iteration and Fourier transform methods one obtains the explicit representations for the nonlocal integral and differential operators, respectively, of any order describing overall effective properties as well as the stress concentration factor in the components. We show the reduction of the integral operator to the differential one for sufficiently smooth statistically average stress fields. With some additional assumptions the proposed method is reduced to the perturbation method as well as to the "quasi-crystalline" approach. With concrete numerical examples one demonstrates the advantage of the iteration method over the Fourier transform method.

Up to now we considered statistically homogeneous media. However, it is well known that the possible starting assumptions may be wrong: namely that the composite medium is statistically homogeneous and, specifically, that the distribution function of inclusions is invariant with respect to the translation are well known. For example, due to some production technologies, the inclusion concentration may be a function of the coordinates (see [244], [674], [883], [884], and also Fig. 12.1).

The accumulation of damage also occurs locally in stress-concentration regions [652], for example, at the tip of a macroscopic crack [529]. Furthermore, in layered composite shells the location of the fibers is random within the periodic layers (see Fig. 12.2), and the micromechanics equations are equations with almost periodic coefficients. Finally, *Functionally Graded Materials* (FGMs) have been the subject of intense research efforts from mid-1980s when this term was originated in Japan in the framework of a national project to develop heat-shielding structural materials for the future Japanese space program. FGMs are composites consisting of two or more phases that are fabricated with a spatial variation of composition which may improve the structural response (see, e.g., [736], [813], [1059],) In particular cases of the above problems it is possible to use different generalizations of the known methods for random structures and periodic ones.

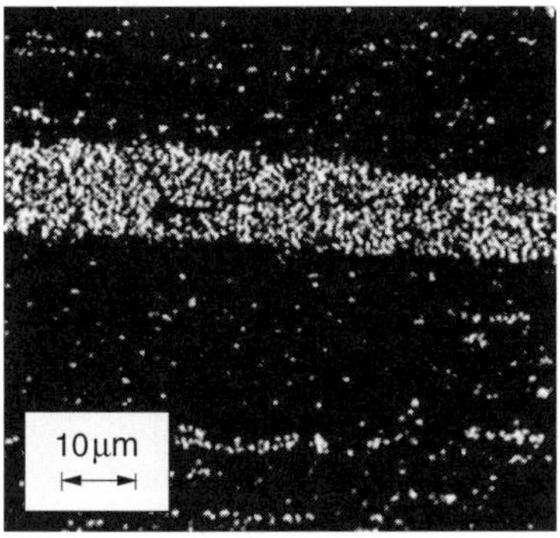

Fig. 12.1. The metallograph of high speed steel reinforced by carbide inclusion sand produced by electro-slag remelting by [883], [884]).

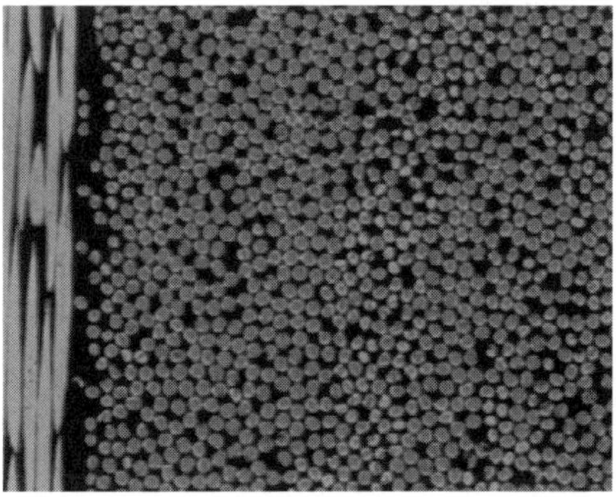

Fig. 12.2. Fragment of the original micrograph of fiber composite material microstructure near the boundary between two plies with the different orientation of aligned fibers.

FGMs of deterministic structure (e.g., doubly periodic composites) were considered in Chapter 10. Analysis of random structure composites is a significantly more complicated problem. A considerable number of methods is known in the linear theory of statistically homogeneous composites, which yield the effective thermoelastic constants and stress field averages in the components. Classification and analysis of these methods were considered in Chapter 8. The most popular so-called one-particle methods (such as, e.g., effective medium and Mori-Tanaka methods) provide relatively simple analytical estimations. However, they have a principal limitation following immediately from the basic hypothesis of these methods about statistical homogeneity of the composite structure. Therefore the use of the referenced methods for the analysis of FGMs is very limited and (as will be shown in the current chapter) can even lead to a qualitatively incorrect prediction of effective properties. In so doing, the MEFM does not make use of a number of hypotheses that form the basis of the traditional one-particle methods and, moreover, the MEFM includes as particular cases the afore-mentioned methods. However, MEFM has not only quantitative advantages (in comparison with experimental data and exact analytical solutions for some regular structures) but MEFM also qualitative benefits following immediately from the consideration of multiparticle interactions. From such considerations it can be concluded that the final relations for effective properties depend explicitly not only on the local concentration of the inclusions but also on at least binary correlation functions of the inclusions that will also be considered in this chapter. Therefore, effective properties as well as local statistical average stresses in the components are nonlocal functions of the inclusion concentration; this was shown in [171] under some additional assumptions for pure mechanical loading of FGM with homogeneous inclusions.

12.2 The Nonlocal Integral Equation

We will consider the solution of a two-particle approximation ($n = 1$, $\mathbf{L}^c = \mathbf{L}^{(0)}$) of nonlocal integral equation (8.27) which is valid for both statistically homogeneous and statistically inhomogeneous (so-called graded) media. Averaging the result obtained over the inclusion v_i gives

$$\begin{aligned}\bar v_i\langle\boldsymbol{\eta}_i\rangle(\mathbf{x}_i) &= \bar v_i\langle\boldsymbol{\eta}_i^{av}\rangle(\mathbf{x}_i)\\ &+\mathbf{R}_i\sum_{q=1}^{N}\Big\{\int \mathbf{T}_{iq}(\mathbf{x}_i-\mathbf{x}_q)\mathbf{Z}_{qi}\varphi(v_q,\mathbf{x}_q|;v_i,\mathbf{x}_i)d\mathbf{x}_q\big[\mathbf{R}_i\langle\widetilde{\boldsymbol{\sigma}}_{i,q}(\mathbf{x}_i)\rangle_i+\mathbf{F}_i\big]\\ &+\int\Big[\mathbf{T}_{iq}(\mathbf{x}_i-\mathbf{x}_q)\mathbf{Z}_{qq}\varphi(v_q,\mathbf{x}_q|;v_i,\mathbf{x}_i)\big[\mathbf{R}_q\langle\widetilde{\boldsymbol{\sigma}}_{i,q}(\mathbf{x}_q)\rangle_q+\mathbf{F}_q\big]\\ &-\mathbf{T}_i(\mathbf{x}_i-\mathbf{x}_q)n^{(q)}(\mathbf{x}_q)\bar v_q\langle\boldsymbol{\eta}_q\rangle(\mathbf{x}_q)\Big] \qquad (12.1)\end{aligned}$$

where N is a number of inclusion components, and $\bar v_i\langle\boldsymbol{\eta}_i^{av}\rangle(\mathbf{x}_i) = \mathbf{R}_i\langle\boldsymbol{\sigma}\rangle(\mathbf{x}_i)+\mathbf{F}_i$ is called the strain polarization tensor of the average stress $\langle\boldsymbol{\sigma}\rangle(\mathbf{x}_i)$ in the component $v^{(i)}$, and the matrix elements $\mathbf{Z}_{qi}$, $\mathbf{Z}_{qq}$ (no sum on q) are nondiagonal elements and diagonal ones, respectively, of the binary interaction matrix $\mathbf{Z}$ (8.25) for the two inclusions v_q and v_i.

Equation (12.1) can be solved using the effective field hypothesis **H2** with the first order approximation (8.39): $\langle\widetilde{\boldsymbol{\sigma}}_{i,q}(\mathbf{x})\rangle_j = \langle\overline{\boldsymbol{\sigma}}(\mathbf{x})\rangle_j = \text{const.}$ $(j = i, q)$. Then from (12.1), taking (8.39) into account we get

$$\begin{aligned}\bar v_i\langle\boldsymbol{\eta}_i\rangle(\mathbf{x}_i) = \bar v_i\langle\boldsymbol{\eta}_i^{av}\rangle(\mathbf{x}_i)+\sum_{q=1}^{N}\Big\{\int \boldsymbol{\mathcal{T}}_{iq}(\mathbf{x}_i,\mathbf{x}_q)\; d\mathbf{x}_q\bar v_i\langle\boldsymbol{\eta}_i\rangle(\mathbf{x}_i)\\ +\int \boldsymbol{\mathcal{F}}_{iq}(\mathbf{x}_i,\mathbf{x}_q)\bar v_q\langle\boldsymbol{\eta}_q\rangle(\mathbf{x}_q)\; d\mathbf{x}_q\Big\}, \qquad (12.2)\end{aligned}$$

where

$$\boldsymbol{\mathcal{T}}_{iq}(\mathbf{x}_i,\mathbf{x}_q)=\mathbf{R}_i\mathbf{T}_{iq}(\mathbf{x}_i-\mathbf{x}_q)\mathbf{Z}_{qi}\varphi(v_q,\mathbf{x}_q|;v_i,\mathbf{x}_i), \qquad (12.3)$$

$$\boldsymbol{\mathcal{F}}_{iq}(\mathbf{x}_i,\mathbf{x}_q)=\mathbf{R}_i\big[\mathbf{T}_{iq}(\mathbf{x}_i-\mathbf{x}_q)\mathbf{Z}_{qq}\varphi(v_q,\mathbf{x}_q|;v_i,\mathbf{x}_i)-\mathbf{T}_i(\mathbf{x}_i-\mathbf{x}_q)n^{(q)}(\mathbf{x}_q)\big]. \qquad (12.4)$$

The hypothesis **H2** (see Subsection 8.1.3) is a standard closing assumption degenerating to the "quasi-crystalline" approximation (8.65): $\langle\overline{\boldsymbol{\sigma}}_i(\mathbf{x})|v_i,\mathbf{x}_i;v_j,\mathbf{x}_j\rangle = \langle\overline{\boldsymbol{\sigma}}_i\rangle$, $\mathbf{x}\in v_i$. It should be mentioned that the assumption (8.39) is similar but not equivalent to the quasi-crystalline approximation (8.65) usually applied for closing of the infinite system (8.7), immediately, leading to a trivial result:

$$\bar v_i\langle\boldsymbol{\eta}_i\rangle(\mathbf{x}_i) = \bar v_i\langle\boldsymbol{\eta}_i^{av}\rangle(\mathbf{x}_i)+\sum_{q=1}^{N}\int \boldsymbol{\mathcal{F}}_{iq}(\mathbf{x}_i,\mathbf{x}_q)\bar v_q\langle\boldsymbol{\eta}_q\rangle(\mathbf{x}_q)\; d\mathbf{x}_q, \qquad (12.5)$$

coinciding with Eq. (12.2) under the additional assumption (8.65_3) $\mathbf{Z}_{iq} = \mathbf{I}\delta_{ij}$, which is in fact equivalent to then quasi-crystalline approximation (8.65_2). Equations (12.2) and (12.5) for statistically homogeneous media and homogeneous boundary conditions were considered in Chapter 8. However, even for this case it was shown that the use of Eq. (12.2) instead of Eq. (12.5) can lead to a variation of effective elastic moduli by a factor of two or more, a fact that has been confirmed by classical experimental data [597]. Of course, the principal difference between the hypotheses (8.39) and (8.65_2) is beyond the scope of direct substitution of the stress field $\boldsymbol{\sigma}(\mathbf{x})$ for the effective field $\overline{\boldsymbol{\sigma}}(\mathbf{x})$, which leads only to quantitative improvement of the estimates; see Chapter 8. Of more importance is the application of the assumption (8.39) after the consideration of multiparticle inclusion interactions (8.25) rather than before as in hypothesis (8.65_2). What seems to be only a formal trick yields to the discovery of fundamentally new effects, which are demonstrated below.

For statistically homogeneous media (5.20), when the probability densities $\varphi(v_i, \mathbf{x}_i)$ and $\varphi(v_q, \mathbf{x}_q | v_i, \mathbf{x}_i)$ are insensitive to translations:

$$\varphi(v_i, \mathbf{x}_i) = n^{(i)} \equiv \text{const.}, \ \ \varphi(v_q, \mathbf{x}_q + \mathbf{x} | v_i, \mathbf{x}_i + \mathbf{x}) \equiv \varphi(v_q, \mathbf{x}_q | v_i, \mathbf{x}_i) \qquad (12.6)$$

for any $\mathbf{x}$ and the integral kernels (12.3) and (12.4) are the convolution ones: $\boldsymbol{\mathcal{T}}_{iq}(\mathbf{x}_i, \mathbf{x}_q) = \boldsymbol{\mathcal{T}}_{iq}(\mathbf{x}_i - \mathbf{x}_q)$, $\boldsymbol{\mathcal{F}}_{iq}(\mathbf{x}_i, \mathbf{x}_q) = \boldsymbol{\mathcal{F}}_{iq}(\mathbf{x}_i - \mathbf{x}_q)$.

Transforming of the second integral in Eq. (12.2) in the spirit of a subtraction technique reduces Eq. (12.2) to the representation

$$\bar{v}_i \langle \boldsymbol{\eta}_i \rangle (\mathbf{x}_i) = \sum_{j=1}^{N} \mathbf{Y}_{ij} \Big\{ \bar{v}_j \langle \boldsymbol{\eta}_j^{av} \rangle (\mathbf{x}_j) + \sum_{q=1}^{N} \int \boldsymbol{\mathcal{F}}_{jq}(\mathbf{x}_j, \mathbf{x}_q) \bar{v}_q \left[\langle \boldsymbol{\eta}_q \rangle (\mathbf{x}_q) - \langle \boldsymbol{\eta}_q \rangle (\mathbf{x}_j) \right] \, d\mathbf{x}_q \Big\} \qquad (12.7)$$

where the matrix $\mathbf{Y}^{-1}$ has the following elements:

$$(\mathbf{Y}^{-1})_{ij} = \delta_{ij} \left[\mathbf{I} - \sum_{q=1}^{N} \int \boldsymbol{\mathcal{T}}_{iq}(\mathbf{x}_i, \mathbf{x}_q) \, d\mathbf{x}_q \right] - \int \boldsymbol{\mathcal{F}}_{ij}(\mathbf{x}_i, \mathbf{x}_j) \, d\mathbf{x}_j, \qquad (12.8)$$

for $i, j = 1, 2, \ldots, N$. This matrix $\mathbf{Y}$ determines the action of the surrounding inclusions on the isolated one and is defined simply by the solution of the problem for purely mechanical loading (with $\boldsymbol{\beta} \equiv 0$).

Then Eq. (12.2) can be rewritten in compact form:

$$\boldsymbol{\mathcal{E}} = \mathbf{Y} \boldsymbol{\mathcal{E}}^{av} + \boldsymbol{\mathcal{K}} \boldsymbol{\mathcal{E}}, \quad \mathbf{x} \in v_i, \qquad (12.9)$$

where $\boldsymbol{\mathcal{E}}(\mathbf{x}) \equiv (\overline{v}_1 \boldsymbol{\eta}_1(\mathbf{x}), \ldots, \overline{v}_N \boldsymbol{\eta}_N(\mathbf{x}))^\top$, $\boldsymbol{\mathcal{E}}^{av}(\mathbf{x}) \equiv (\overline{v}_1 \boldsymbol{\eta}_1^{av}(\mathbf{x}), \ldots, \overline{v}_N \boldsymbol{\eta}_N^{av}(\mathbf{x}))^\top$ and $(\boldsymbol{\mathcal{K}} \boldsymbol{\mathcal{E}})(\mathbf{x}) = \int \mathbf{K}(\mathbf{x}, \mathbf{y}) [\boldsymbol{\mathcal{E}}(\mathbf{y}) - \boldsymbol{\mathcal{E}}(\mathbf{x})] \, d\mathbf{y}$ defines the integral operator $\boldsymbol{\mathcal{K}}$ with the matrix kernel $\mathbf{K} \equiv \mathbf{Y} \boldsymbol{\mathcal{F}}(\mathbf{x}_j, \mathbf{x}_q)$ with elements formally represented as

$$\mathbf{K}_{iq}(\mathbf{x}_i, \mathbf{x}_q) = \sum_{j=1}^{n} \mathbf{Y}_{ij} \boldsymbol{\mathcal{F}}_{jq}(\mathbf{x}_j, \mathbf{x}_q). \qquad (12.10)$$

Thus we reduced the integral equation (12.2) to the standard form of the operator equation (12.9) with the regular integral kernel of convolution type that can be solved by such known methods as the method of mechanical quadratures, successive approximations, and Fourier transform methods. We formally write the solution of Eq. (12.9) as $\boldsymbol{\mathcal{E}} = \boldsymbol{\mathcal{L}} \boldsymbol{\mathcal{E}}^{av}$, where the inverse operator $\boldsymbol{\mathcal{L}} = (\mathbf{I} - \boldsymbol{\mathcal{K}})^{-1}$ will be constructed via three methods mentioned in the next section.

12.3 Methods for the Solution of the Nonlocal Integral Equation

12.3.1 Direct Quadrature Method

Application of the method of mechanical quadratures (called also Nystrom method) is very popular [656], [657] although it is necessary to solve linear systems of high order even for smoothly varying load. In effect the Nystron method transforms the integral equation problem (12.9) with regular kernel into the linear algebra problem in any case $\boldsymbol{\mathcal{E}}_k = \mathbf{Y}\boldsymbol{\mathcal{E}}_k^{av} + \sum_l \boldsymbol{\mathcal{K}}_{kl}\boldsymbol{\mathcal{E}}_l$, where $\boldsymbol{\mathcal{E}}_k = \boldsymbol{\mathcal{E}}(\mathbf{x}_k)$, $\boldsymbol{\mathcal{E}}_l = \boldsymbol{\mathcal{E}}(\mathbf{x}_l)$, $\boldsymbol{\mathcal{E}}_k^{av} = \boldsymbol{\mathcal{E}}^{av}(\mathbf{x}_k)$, $\boldsymbol{\mathcal{K}}_{kl} = w_l\boldsymbol{\mathcal{K}}(\mathbf{x}_k, \mathbf{x}_l)$, and w_l are the weights $(k, l = 1, 2, \ldots)$ of some known quadrature rule [268]. For periodic function $\boldsymbol{\mathcal{E}}^{av}$, the integral (12.9) with the infinite range is in fact reduced to the integral over the unit cell. Since the kernel $\boldsymbol{\mathcal{K}}(\mathbf{x}, \mathbf{y})$ decays at infinity fast enough (as $O(|\mathbf{x}-\mathbf{y}|^{-2d})$, then for $\boldsymbol{\mathcal{E}}^{av}$ stabilizing at infinity ($\boldsymbol{\mathcal{E}}^{av}(\mathbf{x}) \to$ const. as $|\mathbf{x}| \to \infty$), it is tempting to avoid consideration of the infinite range problem by truncating the range at some finite domain, say, the sphere with the radius R and the characteristic function describing by the Heaviside step function $H(R - |\mathbf{x}-\mathbf{y}|)$: $\int \mathbf{K}(\mathbf{x}, \mathbf{y})[\boldsymbol{\mathcal{E}}(\mathbf{y}) - \boldsymbol{\mathcal{E}}(\mathbf{x})]\, d\mathbf{y} = \lim_{R\to\infty} \int \mathbf{K}(\mathbf{x}, \mathbf{y})[\boldsymbol{\mathcal{E}}(\mathbf{y}) - \boldsymbol{\mathcal{E}}(\mathbf{x})]H(R - |\mathbf{x}-\mathbf{y}|)\, d\mathbf{y}$. The effectiveness of this scheme in some concrete numerical examples will be demonstrated in Subsection 12.4.4. Analysis of more general cases of the behavior of the function $\boldsymbol{\mathcal{E}}^{av}$ based on the use of a rule developed directly for the infinite range (such as Gauss-Languerre and Gauss-Hermite rules [268] is performed in a straightforward manner and will not be considered in more detail in the current presentation.

12.3.2 The Iteration Method

Although the direct quadrature method usually causes no problems of accuracy, for a large number of unknown variables N its $O(N^3)$ cost dependence can lead to surprisingly long computing time. The obvious way of reducing this cost is to construct an iterative scheme for Eq. (12.9). Because of this, at first we will solve Eq. (12.9) by the method of successive approximations, which is also called the Neumann series method. The connection with the Fourier transform method will be considered in Subsection 12.3.3.

For simplicity we will consider the point Jacoby iterative method based on the recursion formula

$$\boldsymbol{\mathcal{E}}_{[k+1]} = \mathbf{Y}\boldsymbol{\mathcal{E}}^{av} + \boldsymbol{\mathcal{K}}\boldsymbol{\mathcal{E}}_{[k]} \tag{12.11}$$

to construct a sequence of functions $\{\boldsymbol{\mathcal{E}}_{(k)}\}$ that can be treated as an approximation of the solution of Eq. (12.9). We will not analyze other methods, for example the accelerated Liebmann method (called also extrapolated Gauss-Siedel method) which is usually "faster" that the point Jacobi method, and has the computational advantage than it does not require simultaneous storage of the two iterations $\boldsymbol{\mathcal{E}}_{(k+1)}$ and $\boldsymbol{\mathcal{E}}_{(k)}$ [1134].

Usually the driving term of Eq. (12.11) is used as an initial approximation:

$$\boldsymbol{\mathcal{E}}_{[0]}(\mathbf{x}) = \mathbf{Y}\boldsymbol{\mathcal{E}}^{av}(\mathbf{x}), \quad \bar{v}_i\langle\boldsymbol{\eta}_i\rangle(\mathbf{x}_i) = \sum_{j=1}^{N} \mathbf{Y}_{ij}\bar{v}_j\langle\boldsymbol{\eta}_j^{av}\rangle(\mathbf{x}_j) \tag{12.12}$$

which is exact (in the framework of hypotheses **H1** and **H2**) for a statistically homogeneous media subjected to a homogeneous boundary conditions either (7.2) or (7.3). The matrix $\mathbf{Y}$ determines the "local" action of the surrounding inclusions on the one v_i, while the integral operator kernel $\boldsymbol{\mathcal{K}}$ describes a "non-local" action of these inclusions. The next two iterations have the form

$$\boldsymbol{\mathcal{E}}_{[1]}(\mathbf{x}) = \mathbf{Y}\boldsymbol{\mathcal{E}}^{av}(\mathbf{x}) + \int \mathbf{K}(\mathbf{x},\mathbf{y})\mathbf{Y}[\boldsymbol{\mathcal{E}}^{av}(\mathbf{y}) - \boldsymbol{\mathcal{E}}^{av}(\mathbf{x})]\, d\mathbf{y}, \tag{12.13}$$

$$\begin{aligned}\boldsymbol{\mathcal{E}}_{[2]}(\mathbf{x}) &= \mathbf{Y}\boldsymbol{\mathcal{E}}^{av}(\mathbf{x}) + \int \mathbf{K}(\mathbf{x},\mathbf{y})\mathbf{Y}[\boldsymbol{\mathcal{E}}^{av}(\mathbf{y}) - \boldsymbol{\mathcal{E}}^{av}(\mathbf{x})]\, d\mathbf{y} \\ &+ \int \mathbf{K}(\mathbf{x},\mathbf{y})\mathbf{Y}\Big\{\int \mathbf{K}(\mathbf{y},\mathbf{z})\mathbf{Y}\big[\boldsymbol{\mathcal{E}}^{av}(\mathbf{z}) - \boldsymbol{\mathcal{E}}^{av}(\mathbf{y})\big]\, d\mathbf{z} \\ &- \int \mathbf{K}(\mathbf{x},\mathbf{z})\mathbf{Y}\big[\boldsymbol{\mathcal{E}}^{av}(\mathbf{z}) - \boldsymbol{\mathcal{E}}^{av}(\mathbf{x})\big]\, d\mathbf{z}\Big\}\, d\mathbf{y},\end{aligned} \tag{12.14}$$

and again proceeding formally, it suggests the Neumann series form for the solution $\boldsymbol{\mathcal{E}}$ of (12.8)

$$\boldsymbol{\mathcal{E}} = \boldsymbol{\mathcal{L}}\boldsymbol{\mathcal{E}}^{av}, \quad \boldsymbol{\mathcal{L}} \equiv \sum_{k=0}^{\infty} \boldsymbol{\mathcal{K}}^k, \tag{12.15}$$

where $\boldsymbol{\mathcal{L}}$ is the inverse integral operator with the kernel $\boldsymbol{\mathcal{L}}(\mathbf{x},\mathbf{y})$, and the power $\boldsymbol{\mathcal{K}}^k$ is defined recursively by the condition $\boldsymbol{\mathcal{K}}^1 = \boldsymbol{\mathcal{K}}$ and the kernel of $\boldsymbol{\mathcal{K}}^k$ is (see, e.g., [882]): $\boldsymbol{\mathcal{K}}_k(\mathbf{x},\mathbf{y}) = \int \boldsymbol{\mathcal{K}}(\mathbf{x},\mathbf{z})\boldsymbol{\mathcal{K}}_{k-1}(\mathbf{z},\mathbf{y})\, d\mathbf{z}$. In effect the iteration method (12.11) transforms the integral equation problem (12.7) into the linear algebra problem (12.13) and (12.15) in any case. The sequence $\{\boldsymbol{\mathcal{E}}_{(k)}\}$ (12.11) with arbitrary continuous $\boldsymbol{\mathcal{E}}_{av}(\mathbf{x})$ converges to a unique solution $\boldsymbol{\mathcal{E}}$ (12.9) for the kernel of $\boldsymbol{\mathcal{K}}$ "small" enough, which is to say that

$$\|\boldsymbol{\mathcal{K}}\|_{\infty,v_i} \equiv \max_{\mathbf{x}\in v_i} \int |\boldsymbol{\mathcal{K}}(\mathbf{x},\mathbf{y})|\, d\mathbf{y} < 1. \tag{12.16}$$

As will be shown, only a few iterations of Eq. (12.11) are necessary; these iterations prove very much faster than a direct inversion of the operator $\mathbf{I} - \boldsymbol{\mathcal{K}}$ by the quadrature method. In so doing, condition (12.16) is fulfilled in all examples considered.

12.3.3 The Fourier Transform Method for Statistically Homogeneous Media

One assumes that the statistical average of strain polarization tensor $\langle\boldsymbol{\eta}_q\rangle(\mathbf{x})_q$ belongs to the class of m-times continuously differential functions $C^m(w)$. Since we desire an explicit representation for the stress concentration factor, we will approximate $\langle\boldsymbol{\eta}_q\rangle(\mathbf{x})_q$ by the first terms of its Taylor expansion

$$\langle \boldsymbol{\eta}_q \rangle(\mathbf{x}_q) \approx \sum_{|\alpha|=0}^{m} \frac{1}{\alpha!}[\otimes(\mathbf{x}_q - \mathbf{x}_i)]^\alpha \nabla^\alpha \langle \boldsymbol{\eta}_q \rangle(\mathbf{x}_i), \tag{12.17}$$

where for the multi-indices of non-negative integers $\alpha = (\alpha_1, \ldots, \alpha_d) \in Z^d_+$, and $\beta = (\beta_1, \ldots, \beta_d) \in Z^d_+$ $(d = 2, 3)$ the following notations were used:

$$\alpha + \beta = (\alpha_1 + \beta_1, \ldots, \alpha_d + \beta_d), \quad |\alpha| \equiv \alpha_1 + \ldots + \alpha_d,$$
$$\nabla^\alpha = \frac{\partial^{\alpha_1}}{\partial x_1^{\alpha_1}} \cdots \frac{\partial^{\alpha_d}}{\partial x_d^{\alpha_d}}, \quad \sum_{|\alpha|=0}^{m} = \sum_{|\alpha|=0}^{m} \sum_{\alpha_1+\ldots+\alpha_d=|\alpha|},$$
$$(\otimes \mathbf{x})^\alpha = (x_1)^{\alpha_1} \ldots (x_d)^{\alpha_d}, \quad \alpha! = \alpha_1! \ldots \alpha_d!. \tag{12.18}$$

Taking the expansions (12.17) into account, Eq. (12.2) can be rewritten in compact form

$$\boldsymbol{\mathcal{P}}(\mathbf{x}_i, \nabla)\boldsymbol{\mathcal{E}}(\mathbf{x}_i) = \boldsymbol{\mathcal{E}}^{av}(\mathbf{x}_i), \tag{12.19}$$

where the matrix of the linear partial differential operator $\boldsymbol{\mathcal{P}}(\mathbf{x}_i, \nabla) = [\boldsymbol{\mathcal{P}}(\mathbf{x}_i, \nabla)_{iq}]$ $(i, q = 1, \ldots, N)$ has the operator elements

$$\boldsymbol{\mathcal{P}}(\mathbf{x}_i,\nabla)_{iq} = \delta_{iq}\Big[\mathbf{I} - \sum_{q=1}^{N} \int \boldsymbol{\mathcal{T}}_{iq}(\mathbf{x}_i, \mathbf{x}_q)\, d\mathbf{x}_q\Big]$$
$$- \sum_{|\alpha|=0}^{m} \frac{1}{\alpha!} \int \boldsymbol{\mathcal{F}}_{iq}(\mathbf{x}_i, \mathbf{x}_q) \otimes (\mathbf{x}_i - \mathbf{x}_q)^\alpha \, d\mathbf{x}_q \nabla^\alpha.$$

It should be mentioned that Eq. (12.19) as well as the results of Sections 12.2 and 3.1, 3.2 were obtained for the general cases of statistically inhomogeneous media. The analytical solution of Eq. (12.19) can be obtained with the additional assumption of statistical homogeneity (12.6) of random structures being considered. In such a case the kernels (12.3) and (12.4) of involved integral operators are the convolution ones. Then for statistically homogeneous media (12.6), the differential operator $\boldsymbol{\mathcal{P}}(\mathbf{x}_i, \nabla)$ (12.19) is a linear partial differential operator with constant coefficients: $\boldsymbol{\mathcal{P}}(\mathbf{x}_i, \nabla) \equiv \boldsymbol{\mathcal{P}}(\nabla)$.

Considering that Eq. (12.19) is a differential equation with constant coefficients, the method of solution that first comes to mind is using the Fourier transform method to transform the differential problem of solving (12.19) into the division problem of solving the multiplicative equation [1003], [1112]. The Fourier transformation $\widetilde{\mathbf{g}}(\boldsymbol{\xi})$ of a function $\mathbf{g}(\mathbf{x})$ and its inverse are defined by the formulae (3.4) provided, of course, that the integrals on the right-hand sides of Eqs. (3.4) are convergent. Using the properties of Fourier transforms of fundamental functions $F(\nabla^\alpha \mathbf{g}(\mathbf{x})) = (i\boldsymbol{\xi})^\alpha \widetilde{\mathbf{g}}(\boldsymbol{\xi})$, $F^{-1}((i\boldsymbol{\xi})^\alpha \widetilde{\mathbf{g}}(\boldsymbol{\xi})) = \nabla^\alpha \mathbf{g}(\mathbf{x})$ enables one to transform the linear differential equation with the constant coefficient (12.19) to the algebraic multiplicative equation $\boldsymbol{\mathcal{P}}(i\boldsymbol{\xi})\widetilde{\boldsymbol{\mathcal{E}}}(\boldsymbol{\xi}) = \widetilde{\boldsymbol{\mathcal{E}}^{av}}(\boldsymbol{\xi})$, where the symbol $\boldsymbol{\mathcal{P}}(i\boldsymbol{\xi})$ of the operator $\boldsymbol{\mathcal{P}}(\nabla)$ is a polynomial with complex coefficients in R^d of real transform variable $\boldsymbol{\xi} = (\xi_1, \ldots, \xi_d)^\top$. Taking advantage of (3.4_2) into account, we can write

$$\boldsymbol{\mathcal{E}}(\mathbf{x}) = (2\pi)^{-d} \int e^{i\xi\cdot\mathbf{x}} \boldsymbol{\mathcal{P}}^{-1}(i\boldsymbol{\xi})\widetilde{\boldsymbol{\mathcal{E}}^{av}}(\boldsymbol{\xi})\, d\boldsymbol{\xi} \tag{12.20}$$

which should be a solution of (12.19) in view of (3.4). Equation (12.20) provides a convenient way of calculating the average polarization tensor $\boldsymbol{\mathcal{E}}(\mathbf{x})$ for given average $\widetilde{\boldsymbol{\mathcal{E}}^{av}}(\mathbf{x})$ if we know the transformed properties $\boldsymbol{\mathcal{P}}^{-1}(i\boldsymbol{\xi})$. To facilitate explicit results we restrict attention to "long-wave" approximations and approximate $\boldsymbol{\mathcal{P}}^{-1}(i\boldsymbol{\xi})$ by its Taylor expansion about $\boldsymbol{\xi} = \mathbf{0}$:

$$\boldsymbol{\mathcal{P}}^{-1}(i\boldsymbol{\xi}) = \sum_{k=0}^{m} \boldsymbol{\mathcal{Y}}_{(k)}(i\boldsymbol{\xi}), \tag{12.21}$$

where the functions $\boldsymbol{\mathcal{Y}}_{(k)} = [\boldsymbol{\mathcal{Y}}_{(k)ij}] = \sum_{|\alpha|=k} \boldsymbol{\mathcal{Y}}^{\alpha}(i\boldsymbol{\xi})$ are constructed by the functions $\boldsymbol{\mathcal{Y}}^{\alpha}(i\boldsymbol{\xi})$ proportional to $(i\boldsymbol{\xi})^{\alpha}$. Then (12.20) and (12.21) lead to the representation of concentration of strain polarization by differential operator:

$$\boldsymbol{\mathcal{E}}(\mathbf{x}) = \sum_{k=0}^{m} \boldsymbol{\mathcal{Y}}_{(k)}(\nabla)\boldsymbol{\mathcal{E}}^{av}(\mathbf{x}). \tag{12.22}$$

The local part of the differential operator (12.22) coincides with the local part of the integral operator (12.12): $\boldsymbol{\mathcal{Y}}_{(0)}(\nabla) \equiv \mathbf{Y} = \left[\mathbf{Y}_{ij}\right]$. Truncating Eq. (12.22) after the first four sets of terms, making use of Eq. (12.21) gives the following equations on $\boldsymbol{\mathcal{Y}}_{(m)}(\nabla)$ $(m = 1, \ldots, 4)$:

$$\boldsymbol{\mathcal{Y}}_{(1)}(\nabla) = \mathbf{Y}\boldsymbol{\mathcal{B}}_1(\nabla)\mathbf{Y}, \tag{12.23}$$

$$\boldsymbol{\mathcal{Y}}_{(2)}(\nabla) = \mathbf{Y}\boldsymbol{\mathcal{B}}_2(\nabla)\mathbf{Y} + \mathbf{Y}\boldsymbol{\mathcal{B}}_1(\nabla)\mathbf{Y}\boldsymbol{\mathcal{B}}_1(\nabla)\mathbf{Y}, \tag{12.24}$$

$$\begin{aligned}\boldsymbol{\mathcal{Y}}_{(3)}(\nabla) - \mathbf{Y}\boldsymbol{\mathcal{B}}_3(\nabla)\mathbf{Y} &+ \mathbf{Y}\boldsymbol{\mathcal{B}}_1(\nabla)\mathbf{Y}\boldsymbol{\mathcal{B}}_2(\nabla)\mathbf{Y}\\ &+ \mathbf{Y}\boldsymbol{\mathcal{B}}_2(\nabla)\mathbf{Y}\boldsymbol{\mathcal{B}}_1(\nabla)\mathbf{Y} + \mathbf{Y}\boldsymbol{\mathcal{B}}_1(\nabla)\mathbf{Y}\boldsymbol{\mathcal{B}}_1(\nabla)\mathbf{Y}\boldsymbol{\mathcal{B}}_1(\nabla)\mathbf{Y},\end{aligned} \tag{12.25}$$

$$\begin{aligned}\boldsymbol{\mathcal{Y}}_{(4)}(\nabla) = \mathbf{Y}\boldsymbol{\mathcal{B}}_4(\nabla)\mathbf{Y} &+ \mathbf{Y}\boldsymbol{\mathcal{B}}_2(\nabla)\mathbf{Y}\boldsymbol{\mathcal{B}}_2(\nabla)\mathbf{Y} + \mathbf{Y}\boldsymbol{\mathcal{B}}_1(\nabla)\mathbf{Y}\boldsymbol{\mathcal{B}}_3(\nabla)\mathbf{Y}\\ &+ \mathbf{Y}\boldsymbol{\mathcal{B}}_3(\nabla)\mathbf{Y}\boldsymbol{\mathcal{B}}_1(\nabla)\mathbf{Y} + \mathbf{Y}\boldsymbol{\mathcal{B}}_1(\nabla)\mathbf{Y}\boldsymbol{\mathcal{B}}_1(\nabla)\mathbf{Y}\boldsymbol{\mathcal{B}}_2(\nabla)\mathbf{Y}\\ &+ \mathbf{Y}\boldsymbol{\mathcal{B}}_1(\nabla)\mathbf{Y}\boldsymbol{\mathcal{B}}_2(\nabla)\mathbf{Y}\boldsymbol{\mathcal{B}}_1(\nabla)\mathbf{Y} + \mathbf{Y}\boldsymbol{\mathcal{B}}_2(\nabla)\mathbf{Y}\boldsymbol{\mathcal{B}}_1(\nabla)\mathbf{Y}\boldsymbol{\mathcal{B}}_1(\nabla)\mathbf{Y}\\ &+ \mathbf{Y}\boldsymbol{\mathcal{B}}_1(\nabla)\mathbf{Y}\boldsymbol{\mathcal{B}}_1(\nabla)\mathbf{Y}\boldsymbol{\mathcal{B}}_1(\nabla)\mathbf{Y}\boldsymbol{\mathcal{B}}_1(\nabla)\mathbf{Y}.\end{aligned} \tag{12.26}$$

The next iterations have the series form for the solution $\boldsymbol{\mathcal{Y}}_{(m)}(\nabla)$ $(m = 1, \ldots)$:

$$\boldsymbol{\mathcal{Y}}_{(m)}(\nabla) = \mathbf{Y}\sum_{n=1}^{m} \sum_{k_1+\ldots+k_n=m} \boldsymbol{\mathcal{B}}_{k_1}(\nabla)\mathbf{Y}\ldots\boldsymbol{\mathcal{B}}_{k_n}(\nabla)\mathbf{Y}, \tag{12.27}$$

where

$$\boldsymbol{\mathcal{B}}_k(\nabla) = \sum_{|\alpha|=k} \boldsymbol{\mathcal{B}}^{\alpha}(\nabla), \quad \boldsymbol{\mathcal{B}}^{\alpha}(\nabla) = [\boldsymbol{\mathcal{B}}^{\alpha}_{ij}(\nabla)], \tag{12.28}$$

$$\boldsymbol{\mathcal{B}}^{\alpha}_{ij}(\nabla) = \frac{1}{\alpha!}\int \boldsymbol{\mathcal{F}}_{ij}(\mathbf{x}_i - \mathbf{x}_j)\left[(\mathbf{x}_j - \mathbf{x}_i)^{\alpha} \cdot \nabla^{\alpha}\right] d\mathbf{x}_j, \tag{12.29}$$

$\alpha \in Z^d_+$, $k, k_1, \ldots, k_n \in Z^1_+$, $n = 1, \ldots, m$. For statistically isotropic composites the differential operators $\boldsymbol{\mathcal{B}}_k(\nabla)$ of odd order $k = 2n - 1$ $(n = 1, \ldots)$ equal zero identically, simply because, for considered composites, the generalized functions $\boldsymbol{\mathcal{F}}_{ij}(\mathbf{x}_i - \mathbf{x}_j)$ will be even homogeneous functions.

12.4 Average Stresses in the Components and Effective Properties for Statistically Homogeneous Media

12.4.1 Differential Representations

The mean field of elastic stresses inside the inclusions $\langle\boldsymbol{\sigma}\rangle_i(\mathbf{z})$ ($\mathbf{z} \in v_i$) is obtained from (8.18) and (12.22):

$$\langle\boldsymbol{\sigma}\rangle_i(\mathbf{z}) = \mathbf{B}_i(\mathbf{z})\mathbf{R}_i^{-1}\Big\{\sum_{j=1}^{N}\mathbf{Y}_{ij}(\mathbf{R}_j\langle\boldsymbol{\sigma}\rangle(\mathbf{x}_i)+\mathbf{F}_j)-\mathbf{F}_i\Big\}+\mathbf{C}_i(\mathbf{z})$$
$$+\mathbf{B}_i(\mathbf{z})\mathbf{R}_i^{-1}\sum_{j=1}^{N}\sum_{k=1}^{m}\boldsymbol{\mathcal{Y}}_{(k)ij}(\nabla)\mathbf{R}_j\langle\boldsymbol{\sigma}\rangle(\mathbf{x}_i), \tag{12.30}$$

and therefore

$$\langle\boldsymbol{\sigma}\rangle_i(\mathbf{x}_i) = \mathbf{B}_i\Big\{\mathbf{D}_i\langle\boldsymbol{\sigma}\rangle(\mathbf{x}_i)+\mathbf{R}_i^{-1}\sum_{j=1}^{N}(\mathbf{Y}_{ij}-\mathbf{I}\delta_{ij})\mathbf{F}_j\Big\}+\mathbf{C}_i$$
$$+\mathbf{B}_i\mathbf{R}_i^{-1}\sum_{j=1}^{N}\sum_{k=1}^{m}\boldsymbol{\mathcal{Y}}_{(k)ij}(\nabla)\Big[\mathbf{R}_j\langle\boldsymbol{\sigma}\rangle(\mathbf{x}_i)+\mathbf{F}_j\Big], \tag{12.31}$$

where the tensor $\mathbf{D}_i$ was defined by Eq. (8.43_2), and the variable $\mathbf{z} \in v_i$ is defined in the local coordinate system connected with the semi-axis of the ellipsoid v_i. The mean matrix stress follows simply from the relation $\langle\boldsymbol{\sigma}\rangle_0(\mathbf{x}) = (\langle\boldsymbol{\sigma}\rangle(\mathbf{x})-\langle\boldsymbol{\sigma}V\rangle(\mathbf{x}))/c^{(0)}$. In a similar manner the statistical average stresses in the components in the form of integral operators can be obtained from Eq. (12.15).

Taking the ensemble average of a local constitutive equation combined with Eq. (12.22) give a macroscopic constitutive equation that relates $\langle\boldsymbol{\varepsilon}\rangle(\mathbf{x})$ and $\langle\boldsymbol{\sigma}\rangle(\mathbf{x})$:

$$\langle\boldsymbol{\varepsilon}\rangle(\mathbf{x}) = \mathbf{M}^*\langle\boldsymbol{\sigma}\rangle(\mathbf{x})+\boldsymbol{\beta}^*+\boldsymbol{\mathcal{M}}^*(\langle\boldsymbol{\sigma}\rangle)(\mathbf{x}), \tag{12.32}$$

$$\mathbf{M}^* = \mathbf{M}^{(0)}+\sum_{i,j=1}^{N}\mathbf{Y}_{ij}\mathbf{R}_j n^{(i)}, \tag{12.33}$$

$$\boldsymbol{\beta}^* = \boldsymbol{\beta}^{(0)}+\sum_{i,j=1}^{N}\mathbf{Y}_{ij}\mathbf{F}_j n^{(i)}, \tag{12.34}$$

$$\boldsymbol{\mathcal{M}}^*(\langle\boldsymbol{\sigma}\rangle)(\mathbf{x}) = \sum_{i,j=1}^{N}\sum_{k=1}^{m}n^{(i)}\boldsymbol{\mathcal{Y}}_{(k)ij}(\nabla)\mathbf{R}_j\langle\boldsymbol{\sigma}\rangle(\mathbf{x}). \tag{12.35}$$

The differential operator $\boldsymbol{\mathcal{M}}^*(\langle\boldsymbol{\sigma}\rangle)$ of the second order ($m = 2$) (12.35) is reduced to the analogous relation proposed in [132] for the identical homogeneous inclusions (3.187) ($N = 1$).

12.4.2 The Reduction of Integral Overall Constitutive Equations to Differential Ones

If $\boldsymbol{\mathcal{E}}^{av}(\mathbf{x})$ belongs to $C^m(w)$, then substituting its Taylor expansion analogous to (12.17) into the first order iteration for the average strain polarization tensor $\boldsymbol{\mathcal{E}}_{(1)}(\mathbf{x})$ (12.13) reduces this integral equation to the differential one:

$$\boldsymbol{\mathcal{E}}_{(1)}(\mathbf{x}) = [\boldsymbol{\mathcal{Z}}_{(0)}(\nabla) + \boldsymbol{\mathcal{Z}}_{(1)}(\nabla)]\boldsymbol{\mathcal{E}}^{av}(\mathbf{x}), \tag{12.36}$$

where $\boldsymbol{\mathcal{Z}}_{(0)}(\nabla) \equiv \mathbf{Y}$ and the differential operators with the constant coefficients

$$\boldsymbol{\mathcal{Z}}_{(1)}(\nabla) = \sum_{k=1}^{m} \boldsymbol{\mathcal{Z}}^k_{(1)}(\nabla),$$
$$\boldsymbol{\mathcal{Z}}^k_{(1)}(\nabla) = \sum_{|\alpha|=k} \frac{1}{\alpha!} \int \mathbf{K}(\mathbf{x},\mathbf{y})\mathbf{Y}[\otimes(\mathbf{y}-\mathbf{x})^{\alpha}]\, d\mathbf{y}\nabla^{\alpha}, \tag{12.37}$$

can be recast, according to the notations (12.10) and (12.28), in the form $\boldsymbol{\mathcal{Z}}^k_{(1)}(\nabla) = \mathbf{Y}\boldsymbol{\mathcal{B}}_k(\nabla)\mathbf{Y}$. For statistically isotropic composites all odd operators $\boldsymbol{\mathcal{B}}^{\alpha}(\nabla) \equiv \mathbf{0}$ at $|\alpha| = 2n-1,\ n = 1, \ldots$.

In a similar manner, substitution of the Taylor expansion $\boldsymbol{\mathcal{E}}^{av}(\mathbf{x})$ for (12.17) into the representation for the second iteration (12.14) leads to the differential equation

$$\boldsymbol{\mathcal{E}}_{(2)}(\mathbf{x}) = [\boldsymbol{\mathcal{Z}}_{(0)}(\nabla) + \boldsymbol{\mathcal{Z}}_{(1)}(\nabla) + \boldsymbol{\mathcal{Z}}_{(2)}(\nabla)]\boldsymbol{\mathcal{E}}^{av}(\mathbf{x}), \tag{12.38}$$

where for the representation of $\boldsymbol{\mathcal{Z}}_{(2)}(\nabla)$ we transform the iterated integral in Eq. (12.14) to the following form:

$$\begin{aligned}
&\int \mathbf{K}(\mathbf{x},\mathbf{y})\mathbf{Y}\Big\{\int \mathbf{K}(\mathbf{y},\mathbf{z})\mathbf{Y}\big[\boldsymbol{\mathcal{E}}^{av}(\mathbf{z}) - \boldsymbol{\mathcal{E}}^{av}(\mathbf{y})\big]\, d\mathbf{z} \\
&-\int \mathbf{K}(\mathbf{x},\mathbf{z})\mathbf{Y}\big[\boldsymbol{\mathcal{E}}^{av}(\mathbf{z}) - \boldsymbol{\mathcal{E}}^{av}(\mathbf{x})\big]\, d\mathbf{z}\Big\}\, d\mathbf{y} \\
&= \int \mathbf{Y}\boldsymbol{\mathcal{F}}(\mathbf{x}-\mathbf{y})\mathbf{Y}\boldsymbol{\mathcal{B}}^{\alpha}(\nabla)\mathbf{Y}[\boldsymbol{\mathcal{E}}^{av}(\mathbf{y}) - \boldsymbol{\mathcal{E}}^{av}(\mathbf{x})]\, d\mathbf{y}.
\end{aligned} \tag{12.39}$$

Then the repeated use of the Taylor expansion for the function $\boldsymbol{\mathcal{E}}^{av}(\mathbf{y})$ leads to the final representation $\boldsymbol{\mathcal{Z}}_{(2)}(\nabla) = \sum_{k=2}^{m} \boldsymbol{\mathcal{Z}}^k_{(2)}(\nabla)$, hereafter $\boldsymbol{\mathcal{Z}}^k_{(m)}(\nabla) = \sum_{|\alpha^1+\ldots\alpha^m|=k} \mathbf{Y}\boldsymbol{\mathcal{B}}^{\alpha^1}(\nabla) \ldots \mathbf{Y}\boldsymbol{\mathcal{B}}^{\alpha^m}(\nabla)\mathbf{Y}$ and $\alpha^i = (\alpha^i_1, \ldots, \alpha^i_d) \in Z^d_+;\ i = 1, \ldots, m;\ k = m, m+1, \ldots;\ k, m \in Z^1_+$. The solutions (12.36) and (12.38) coincide with the first- (12.13) and the second- (12.14) order approximation, respectively, only for both $\boldsymbol{\mathcal{E}}^{av}(\mathbf{x}) \in C^{\infty}(w)$ and taking into account infinite number of terms of the Taylor expansion of $\boldsymbol{\mathcal{E}}^{av}(\mathbf{x})$, $m = \infty$ (12.17). The construction of the following differential analogs $\boldsymbol{\mathcal{Z}}_{(n)}$, $(n = 3, 4, \ldots)$ of integral iterations $\boldsymbol{\mathcal{K}}^n$ (12.15) is obvious. In so doing, the differential operators $\boldsymbol{\mathcal{Y}}_{(n)}$ $(n = 0, 1, \ldots)$ (12.22) can be obtained by truncations of differential operators $\boldsymbol{\mathcal{Z}}_{(m)}$ $(m = 0, \ldots, n)$:

$$\boldsymbol{\mathcal{Y}}_{(1)}(\nabla) = \boldsymbol{\mathcal{Z}}^1_{(1)}, \quad \boldsymbol{\mathcal{Y}}_{(2)}(\nabla) = \boldsymbol{\mathcal{Z}}^2_{(1)} + \boldsymbol{\mathcal{Z}}^2_{(2)}, \tag{12.40}$$
$$\boldsymbol{\mathcal{Y}}_{(3)}(\nabla) = \boldsymbol{\mathcal{Z}}^3_{(1)} + \boldsymbol{\mathcal{Z}}^3_{(2)} + \boldsymbol{\mathcal{Z}}^3_{(3)}, \tag{12.41}$$
$$\boldsymbol{\mathcal{Y}}_{(4)}(\nabla) = \boldsymbol{\mathcal{Z}}^4_{(1)} + \boldsymbol{\mathcal{Z}}^4_{(2)} + \boldsymbol{\mathcal{Z}}^4_{(3)} + \boldsymbol{\mathcal{Z}}^4_{(4)}, \tag{12.42}$$

and so on. For instance, for the estimation of the operator $\boldsymbol{\mathcal{Y}}_{(2)}(\nabla)$ it is enough to apply the first-order approximation (12.13) to the quadratic polynomial approximation of the Taylor expansion (12.17) of the function $\boldsymbol{\mathcal{E}}^{av}(\mathbf{x})$. The operator $\boldsymbol{\mathcal{Y}}_{(m)}$ can be estimated by application of the $(m-1)$ order approximation of integral operator $\boldsymbol{\mathcal{K}}^{(m-1)}$ (12.15) to the m's polynomial approximation of the Taylor expansion (12.17) of the function $\boldsymbol{\mathcal{E}}^{av}(\mathbf{x})$. Again, all operators $\boldsymbol{\mathcal{Z}}^{2n-1}_{(m)}$ $(n = 1,2,\ldots,\ m = 1,\ldots,2n-1)$ as well as the operators $\boldsymbol{\mathcal{Z}}^{2n}_{(m)}$ $(n = 1,2,\ldots,\ m = 1,\ldots,2n)$ constructed by the operators $\boldsymbol{\mathcal{B}}_{2k-1}(\nabla)$ $(k < n)$ of odd order vanish for statistically isotropic media.

12.4.3 "Quasi-crystalline" Approximation

In the framework of the "quasi-crystalline" approximation (8.65$_2$), the matrix $\mathbf{Y}^{-1}$ can be reduced to the representation (8.66) which can be significantly simplified under the following additional assumptions (8.67) and (8.69). Then, by virtue of the fact that the generalized function $\boldsymbol{\Gamma}(\mathbf{x})$ is an even homogeneous function of order -3, we obtain – under the assumption (8.67) and (8.69) – the following relation:

$$\begin{aligned}\textstyle\sum_{q=1}^{N}\int\int \boldsymbol{\Gamma}(\mathbf{x}-\mathbf{x}_q)[\langle V_q(\mathbf{x}_q)\boldsymbol{\eta}(\mathbf{x}_q)|;v_i,\mathbf{x}_i\rangle - c^{(q)}\langle\boldsymbol{\eta}_q\rangle(\mathbf{x}_q)]V_i(\mathbf{x})\,d\mathbf{x}_q\,d\mathbf{x}\\ = \sum_{k=0}^{m}\boldsymbol{\mathcal{Q}}_i^k(\nabla)\sum_{q=1}^{N}\langle\boldsymbol{\eta}_q\rangle(\mathbf{x}_i)c^{(q)},\end{aligned} \tag{12.43}$$

where

$$\begin{aligned}\boldsymbol{\mathcal{Q}}_i^k(\nabla) = \sum_{|\alpha|=k}\frac{1}{\alpha!}\int[\mathbf{T}_{iq}(\mathbf{x}_i-\mathbf{x}_q)\varphi(v_q,\mathbf{x}_q|;v_i,\mathbf{x}_i)\frac{1}{n^{(q)}}\\ -\mathbf{T}_i(\mathbf{x}_i-\mathbf{x}_q)](\mathbf{x}_q-\mathbf{x}_i)^\alpha\,d\mathbf{x}_q\nabla^\alpha\end{aligned}$$

for $k = 1,\ldots$ and $\boldsymbol{\mathcal{Q}}_i^0(\nabla) \equiv \mathbf{Q}_i$.

In the framework of "quasi-crystallite approximation" (8.65$_2$), the operator $\boldsymbol{\mathcal{Q}}_i^k(\nabla)$ is associated with the operator $\boldsymbol{\mathcal{B}}_{kij}(\nabla)$ (12.28): $\boldsymbol{\mathcal{B}}_{kij}(\mathbf{x}_i,\nabla) = \mathbf{R}_i\boldsymbol{\mathcal{Q}}_i^k(\nabla)n^{(j)}$. Taking the assumptions (8.65$_2$) into account, Eq. (12.31) can be combined into a simple equation:

$$\langle\boldsymbol{\sigma}\rangle_i = \mathbf{B}_i\langle\boldsymbol{\sigma}\rangle + \mathbf{C}_i + \mathbf{B}_i\mathbf{Q}_i\sum_{k=0}^{m}\boldsymbol{\mathcal{Y}}^Q_{(k)}(\nabla)\sum_{q=1}^{N}\mathbf{R}_q\langle\boldsymbol{\sigma}\rangle(\mathbf{x})n^{(q)}, \tag{12.44}$$

where the tensor $\boldsymbol{\mathcal{Y}}^Q_{(0)}$, introduced in (12.44), is defined as

$$\boldsymbol{\mathcal{Y}}^Q_{(0)}(\nabla) \equiv \mathbf{Y}^Q = \left[\mathbf{I} - \sum_{q=1}^{N}n^{(q)}\mathbf{R}_q\mathbf{Q}_q\right]^{-1}, \tag{12.45}$$

and for statistically isotropic media the differential operators of odd order (12.44) vanish identically $\boldsymbol{\mathcal{Y}}^Q_{(k)}(\nabla) \equiv \mathbf{0}$; $(k = 2n-1,\ n = 1,2,\ldots)$. In such a case the first nonzero operators $\boldsymbol{\mathcal{Y}}^Q_{(k)}(\nabla)$, $(k = 2,4)$ (12.44) are represented by the formulae

$$\mathcal{Y}^Q_{(2)}(\nabla)=\mathbf{Y}^Q\Big[\sum_{q=1}^N n^{(q)}\mathbf{R}_q\mathcal{Q}^2_q(\nabla)\Big]\mathbf{Y}^Q,$$

$$\mathcal{Y}^Q_{(4)}(\nabla)=\mathbf{Y}^Q\Big[\sum_{q=1}^N n^{(q)}\mathbf{R}_q\mathcal{Q}^4_q(\nabla)\Big]\mathbf{Y}^Q$$
$$+\mathbf{Y}^Q\Big[\sum_{q=1}^N n^{(q)}\mathbf{R}_q\mathcal{Q}^2_q(\nabla)\Big]\mathbf{Y}^Q\Big[\sum_{q=1}^N n^{(q)}\mathbf{R}_q\mathcal{Q}^2_q(\nabla)\Big]\mathbf{Y}^Q. \quad (12.46)$$

Substitution of (12.46) into (12.37) results in

$$\mathbf{M}^* = \mathbf{M}^{(0)} + \mathbf{Y}^Q\sum_{i=1}^N \mathbf{R}_i n^{(i)}, \quad \boldsymbol{\beta}^* = \langle\boldsymbol{\beta}\rangle + \mathbf{Y}^Q\sum_{i=1}^N \mathbf{F}_i n^{(i)}, \quad (12.47)$$

$$\mathcal{M}^*(\langle\boldsymbol{\sigma}\rangle)(\mathbf{x}) = \Big[\mathbf{I}-(\mathbf{Y}^Q)^{-1}\Big]\Big[\sum_{k=2}^m \mathcal{Y}^Q_{(k)}(\nabla)\Big]\sum_{q=1}^N \mathbf{R}_q\langle\boldsymbol{\sigma}\rangle(\mathbf{x})n^{(q)}. \quad (12.48)$$

The relations (12.47) and (12.48) are obtained for N-component inclusions, which can differ from one another by their thermoelastic properties and internal microtopologies. For identical unidirectionally aligned inclusions ($N=1$), when $\mathcal{Q}^k_q(\nabla)\equiv\mathcal{Q}^k(\nabla)$ and $\varphi(v_q,\mathbf{x}_q|;v_i,\mathbf{x}_i)/n^{(q)}\equiv\varphi(v_p,\mathbf{x}_p|;v_i,\mathbf{x}_i)/n^{(p)}$ for $\forall p,q\neq i$ and $\forall k=0,1,\ldots$, the relation (12.44) is simplified ($i=1$):

$$\langle\boldsymbol{\sigma}\rangle_i = \Big[\mathbf{I}-\mathbf{Q}\mathbf{R}_i n^{(i)}\Big]^{-1}(\mathbf{B}_i\langle\boldsymbol{\sigma}\rangle(\mathbf{x})+\mathbf{C}_i)+\mathbf{B}_i\mathbf{Q}\sum_{k=2}^m \mathcal{Y}^Q_{(k)}(\nabla)\mathbf{R}_i\langle\boldsymbol{\sigma}\rangle(\mathbf{x}_i)n^{(i)}, \quad (12.49)$$

and the nonlocal operators $\mathcal{Y}^Q_{(2)}(\nabla)$ and $\mathcal{Y}^Q_{(4)}(\nabla)$ (12.46) are defined by the effective elastic compliance:

$$\mathcal{Y}^Q_{(2)}(\nabla)\mathbf{R}_i n^{(i)}=(\mathbf{M}^*-\mathbf{M}^{(0)})\mathcal{Q}^2(\nabla)(\mathbf{M}^*-\mathbf{M}^{(0)}), \quad (12.50)$$

$$\mathcal{Y}^Q_{(4)}(\nabla)\mathbf{R}_i n^{(i)}=(\mathbf{M}^*-\mathbf{M}^{(0)})\mathcal{Q}^4(\nabla)(\mathbf{M}^*-\mathbf{M}^{(0)})$$
$$+(\mathbf{M}^*-\mathbf{M}^{(0)})\mathcal{Q}^2(\nabla)(\mathbf{M}^*-\mathbf{M}^{(0)})\mathcal{Q}^2(\nabla)(\mathbf{M}^*-\mathbf{M}^{(0)}). \quad (12.51)$$

A "quasi-crystallite" approximation (8.65_2), which is equivalent to the assumption $\mathbf{Z}_{ij}=\mathbf{I}\delta_{ij}$, when $\mathcal{T}\equiv\mathbf{0}$ (12.3) and $\mathbf{Z}_{qq}=\mathbf{I}$ (12.4) was used in [279], [563], [1024], [1239], where the authors obtained the differential operator $\mathcal{M}^*$ (12.35) of the second and fourth orders for an arbitrary comparison moduli and for another conditional probability density $\langle V^{(i)}(\mathbf{x})V^{(q)}(\mathbf{y})\rangle$ describing the random structure of composites instead of $\varphi(v_q,\mathbf{x}_q|v_i,\mathbf{x}_i)$. Drugan [278] has proposed a similar approach for the estimation of the fourth-order differential operator.

12.4.4 Numerical Analysis of Nonlocal Effects for Statistically Homogeneous Composites

We will estimate effective nonlocal properties of statistically homogeneous composite material with an isotropic incompressible matrix containing spherical identical rigid inclusions [132]. This example with the infinite contrast between the two phases was chosen deliberately because it provides the maximum difference of predictions of effective elastic response estimated by various methods, and was considered by a number of authors. We will consider a well-stirred RDF $g(r) \equiv \varphi(v_i, \mathbf{x}_i|; v_q, \mathbf{x}_q)/n^{(1)} = H(r - 2a)$ where H denotes the Heaviside step function, $r \equiv |\mathbf{x}_i - \mathbf{x}_q|$ is the distance between the nonintersecting inclusions v_i and v_q, and $c(\mathbf{x}) \equiv 0.4$. Let $\boldsymbol{\mathcal{L}}^*$ be the nonlocal effective elastic operator in the governing equation $\langle \varepsilon \rangle(\mathbf{x}) = \mathbf{L}^*(\mathbf{x})\langle \varepsilon \rangle(\mathbf{x}) + \boldsymbol{\mathcal{L}}^*(\langle \varepsilon \rangle)(\mathbf{x})$, where

$$\boldsymbol{\mathcal{L}}^*(\langle \varepsilon \rangle)(\mathbf{x}) = b_1 \delta_{kl} \Delta \langle \varepsilon_0 \rangle(\mathbf{x}) + b_2 \partial_{kl} \langle \varepsilon_0 \rangle(\mathbf{x}) + b_3 \Delta \langle \mathrm{e}_{kl} \rangle(\mathbf{x}),$$

and $\varepsilon_0 \equiv \boldsymbol{\varepsilon\delta}/3$, $\mathbf{e} \equiv \boldsymbol{\varepsilon} - \varepsilon_0 \boldsymbol{\delta}$, $\Delta \equiv \partial_i \partial_i$. In Fig. 12.3 a comparison between the normalized parameters $b_3^\mu \equiv b_3(2\mu^{(0)}a)^{-2}$ calculated by the use of far-field $\mathbf{Z}_{ij} = \mathbf{I}\delta_{ij} + (1-\delta_{ij})\mathbf{U}(\mathbf{x}_i - \mathbf{x}_j)\mathbf{R}$ and the "quasi-crystalline" $\mathbf{Z}_{ij} = \mathbf{I}\delta_{ij}$ approximations is presented. For sufficiently large concentrations of the inclusions the values b_3^μ (solid and dotted curves) differ by a factor of six or even more.

Now we will consider the 2-D case (plane strain problem) for the two-phase composites consisting of an isotropic incompressible matrix reinforced by a statistically homogeneous field of rigid identical circle fibers of radius a. We will exam two alternative RDFs of inclusions (9.91) and (9.93). Estimation of effective moduli $\mathbf{L}^*$ of composites with RDFs (9.91) and (9.93) were already considered in Chapter 8. Our approach to addressing this in Subsection 12.4.4 will be to employ the nonlocal equations for stress concentrator tensors we derived in the previous sections by the iteration and Fourier transform methods. We will consider ensemble-averaged stress $\langle \boldsymbol{\sigma} \rangle(\mathbf{x})$, and determine the stress concentration factors $\langle \boldsymbol{\sigma} \rangle_i(\mathbf{x})$ at the inclusions by the quadrature method (QM), by the iteration method (IM) and, in some particular cases, by the Fourier transform methods (FTM). The quantitative results we obtain will be for the 2-D case (plane strain problem) for the two-phase composites consisting of an isotropic matrix reinforced by a random dispersion of isotropic identical circle particles.

Let us now demonstrate the application of the theoretical results by considering an isotropic highly filled composite ($c = 0.6$) made of an incompressible isotropic matrix $\mathbf{L}^{(0)} = (\infty, 2\mu^{(0)})$, filled with rigid circle inclusions $\mathbf{L}^{(1)} = (\infty, \infty)$ of one size $a = 1$, $N = 1$. Again, infinite elastic mismatch of constituents was chosen deliberately. We shall consider the response for a normal loading that varies with position in its loading direction:

$$\langle \boldsymbol{\sigma}_{ij} \rangle(\mathbf{x}) = f(x_1)\delta_{1i}\delta_{1j}. \tag{12.52}$$

with four specific cases of the functions

$$f_1(x_1) = \sin(\frac{\pi}{4}x_1), \quad f_2(x_1) = 0.6579|x_1|^{2.001}e^{-0.2422x_1^2}, \tag{12.53}$$

$$f_3(x_1) = 0.6584|x_1|^{1.999}e^{-0.2422x_1^2}, \; f_4(x_1) = 0.6580x_1^2 e^{-0.2420x_1^2}. \tag{12.54}$$

The loading (12.52), (12.53_1) with $f_1(x_1) \in C^\infty(R)$ was analyzed by the FTM in detail in [278], [279], for the different arguments and the different concentrations of the spherical inclusions. We can only define the mth derivatives of the functions $f_j(0)$ ($j = 2, 3$) if $f_j^{(m)}(-0) = f_j^{(m)}(+0)$. Now $f_2 \in C^2$ and $f_3 \in C^1$ and the third and the second derivatives do not exist for the functions f_2 and f_3, respectively. The functions f_j $(j = 2, 3, 4)$ have approximately the same max and the same max of their first derivatives as the function f_1: $|\max f_j(x_1) - 1| < 10^{-4}$, $|\max f'_j(x_1) - \max f'_2(y_1)| < 10^{-4}$, $(j = 2, 3, 4;\ x_1, y_1 \in R)$ (see Fig. 12.4).

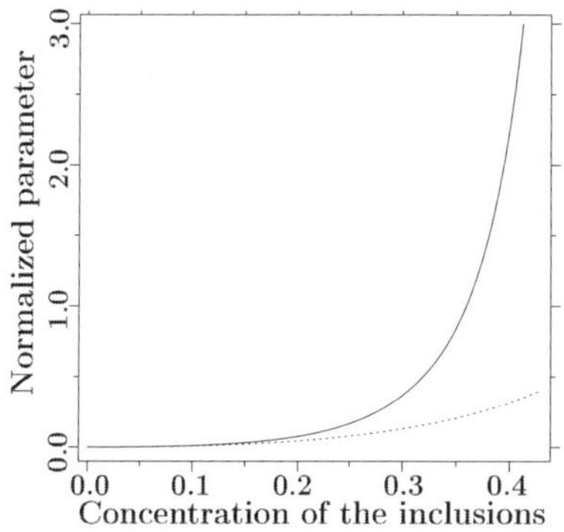

Fig. 12.3 Normalized parameters b_3^μ vs c: the far field (solid curve) and "quasi-crystalline" (doted curve) approximation.

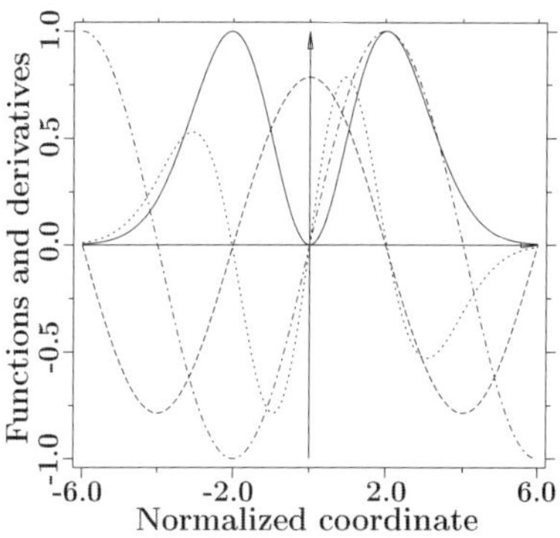

Fig. 12.4 The functions and their first derivatives vs argument x_1: $f_2(x_1)$ (solid line), $f_1(x_1)$ (dot-dashed line), $f'_1(x_1)$ (dashed line), $f'_2(x_1)$ (dotted line).

In so doing $f''_2(0) = 0$ and, therefore, $\boldsymbol{\mathcal{Y}}_{(2)}\langle\boldsymbol{\sigma}\rangle(\mathbf{0}) = \mathbf{0}$ and the actions of the next order differential operators (12.24) $\boldsymbol{\mathcal{Y}}_{(2)}\langle\boldsymbol{\sigma}\rangle(\mathbf{x})$ cannot be defined for the function $f_2(x_1)$ due to non-differentiability of the function $f''_2(x_1)$. Thus, independently from the concrete micromechanical average scheme, the FTM will predict zero nonlocal effects at the point $\mathbf{0}$ for the function $f_2(x_1)$. Analogous analysis leads to a more dramatic conclusion for the function $f_3(x_1)$, that is, the FTM cannot be applied in principle to the field (12.52), (12.54$_1$) because the differentiable function $f_3(x_1)$ which is very close to $f_2(x_1) \in C^2$ and $f_4(x_1) \in C^\infty$ is not twice differentiable at $x_1 = 0$.

The comparative results estimated by the MEFM for the radial distribution function (9.93) $(c = 0.6)$ and for different functions $f_j(x_1)$ $(j = 1, \ldots, 4)$ for 0th and 7th iterations are presented in Table 12.1; the results for functions f_2, f_3, f_4 differ from one another by less than 1%. Table 12.1 also gives the quantitative analysis of nonlocal effects presented in the terms of the stress variations $\Delta_1(\%) \equiv \langle\sigma_{11}\rangle_{i(7)}(\mathbf{0})/\max\langle\sigma_{11}\rangle_{i(0)}$; $\Delta_2(\%) \equiv \langle\max\sigma_{11}\rangle_{i(7)}/\max\langle\sigma_{11}\rangle_{i(0)}$. Table 12.2 presents nth iterations of the statistical average of stresses at the origin $\mathbf{0}$ estimated by the iteration method $\langle\sigma_{11}\rangle_{i(n)}(\mathbf{0})$ $(n = 0, 7, 20)$ and by the quadrature method (12.17) $\langle\sigma_{11}^{QM}\rangle_i(\mathbf{0})$. The 7th and 20th iterations differ from one other by 3%; the difference of the estimations by the quadrature method and the 20th iterations is not over 0.2%. In Fig. 12.5 one presents the stresses $\langle\sigma_{11}^{QM}\rangle_i(x_1)$ for the function f_2 (12.53$_2$), radial distribution function $g(\mathbf{x}_i - \mathbf{x}_j)$ (9.93), and $c = 0.15,\ 0.3,\ 0.45,\ 0.6$ estimated by the QM which differ from the estimations of 20th iterations by the IM less than 0.2%. Comparison of Figs. 9.2 with 12.3 and 12.5 leads to the conclusion that the values of both the nonlocal stresses $\langle\sigma_{11}^{QM}\rangle_i(x_1) - \langle\sigma_{11}\rangle_{i(0)}(x_1)$ and nonlocal differential operator $\boldsymbol{\mathcal{L}}^*$ are more sensitive then the local effective modulus $\mathbf{L}^*$ to the value of the volume fiber concentration c. The first few iterations of estimations of stresses in the

Table 12.1. Comparative analysis of different functions (12.53), (12.54)

	$\max\langle\sigma_{11}\rangle_{i(0)}$	$\langle\sigma_{11}\rangle_{i(7)}(\mathbf{0})$	$\max\langle\sigma_{11}\rangle_{i(7)}$	$\Delta_1(\%)$	$\Delta_2(\%)$
f_1	1.2446	0.0	1.3106	0	5.3
f_2	1.2439	−0.4295	1.5664	34.5	25.9
f_3	1.2430	−0.4300	1.5580	34.6	25.3
f_4	1.2441	−0.4296	1.5592	35.0	26.2

Table 12.2. Comparative analysis of estimations by the IM and QM

	$\langle\sigma_{11}\rangle_{i(0)}(\mathbf{0})$	$\langle\sigma_{11}\rangle_{i(7)}(\mathbf{0})$	$\langle\sigma_{11}\rangle_{i(20)}(\mathbf{0})$	$\langle\sigma_{11}^{QM}\rangle_i(\mathbf{0})$
f_1	0.0	0.0	0.0	0.0
f_2	0.0	−0.4295	−0.4419	−0.4426
f_3	0.0	−0.4300	−0.4433	−0.4440
f_4	0.0	−0.4296	−0.4429	−0.4427

fibers $\langle\sigma_{11}\rangle_i(x_1)$ for the function f_2 are presented in Fig. 12.6. The first-order approximation of the IM leads to $\langle\sigma_{11}\rangle_{i(1)}(\mathbf{0})/\max\langle\sigma_{11}\rangle_{i(0)}(x_1) = 0.2134$ provided, according to Eqs. (12.36) and (12.37), by the derivatives $f_2^{(m)}$, $m = 4, 6, \ldots$ (if they exist), whereas the second-order approximation of the FTM leads to the degenerate result $\boldsymbol{\mathcal{Y}}_{(2)}(\nabla)\langle\boldsymbol{\sigma}\rangle_i(\mathbf{0}) = \mathbf{0}$ (nonlocal effect is absent at the point $\mathbf{x} = \mathbf{0}$).

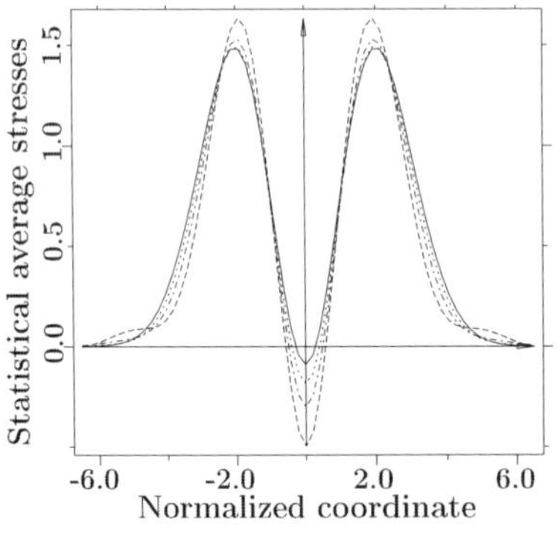

Fig. 12.5. The statistical averages $\langle\sigma_{11}^{QM}\rangle_i(x_1)$ vs x_1 estimated for the functions $f_2(x_i)$ (12.53$_2$) and $g_2(r)$ (9.93) by the MEFM: $c = 0.15$ (solid line), 0.30 (dotted line), 0.45 (dot-dashed line), 0.60 (dashed line).

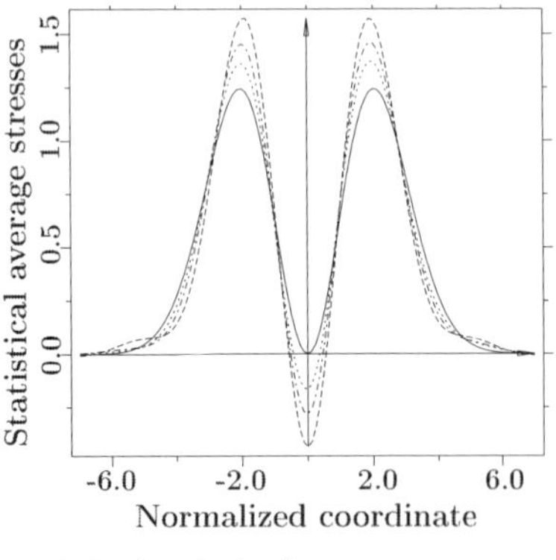

Fig. 12.6. $\langle\sigma_{11}\rangle_i(x_1)$ vs x_1 estimated for $c = 0.6$, the functions $f_2(x_i)$ (12.53$_2$) and $g_2(r)$ (9.93) via the MEFM. Zero-order (solid line), first-order (dotted line), second order (dot-dashed line), seventh order (dashed line) approximations.

In Table 12.3 only the function f_2 is analyzed in the framework of the iteration scheme. A comparative analysis of the MEFM (12.11), Mori-Tanaka method (MT) (12.44), and the perturbation method (PM) (see [166] and (12.66)) for the different radial correlation functions (9.91) and (9.93) ($c = 0.6$) is presented. As can be seen, the MEFM is most sensitive to the choice of radial distribution function and leads to the maximum nonlocal effects predicted. It is interesting that nonlocal response, in contrast to the local one, estimated by the Mori-Tanaka method, depends on the radial distribution function $g(r)$. The difference of estimations for the different average methods analyzed in Table. 12.3 by the quadrature method (12.17) and 7th iteration are usually not over 3% as in Table 12.1.

Table 12.3. Comparative analysis of different methods and the functions $g(r)$ (9.91)–(9.93)

Method	$g(r)$	$\max\langle\sigma_{11}\rangle_{i(0)}$	$\langle\sigma_{11}\rangle_{i(7)}(\mathbf{0})$	$\max\langle\sigma_{11}\rangle_{i(7)}$	$\Delta_1(\%)$	$\Delta_2(\%)$
MEFM	(9.91)	1.2146	−0.3564	1.4938	29.3	23.0
MEFM	(9.93)	1.2439	−0.4295	1.5664	34.5	25.9
MT	(9.91)	1.1247	−0.2311	1.2955	20.5	15.2
MT	(9.93)	1.1247	−0.2412	1.3163	21.4	17.0
PM	(9.91)	1.1631	−0.1542	1.2830	13.3	10.3
PM	(9.93)	1.1336	−0.1995	1.2976	17.6	14.7

Thus, in some specific numerical examples the feasibility, efficiency, and accuracy of all three numerical methods of solution of nonlocal integral equation (12.9), quadrature method (QM), iteration method (IM), and Fourier transform method (FTM) were emphatically demonstrated. Their qualitative comparative analyses will be presented in Section 12.6.

12.5 Effective Properties of Statistically Inhomogeneous Media

12.5.1 Local Effective Properties of FGMs

Equation (12.19) was obtained for a statistically inhomogeneous field of inclusions as well. For the sake of simplicity we will assume that the elastic compliances of the comparison medium and matrix coincide and we will consider "slowly varying" effective fields providing the approximation $\langle\boldsymbol{\eta}_q\rangle(\mathbf{x}_q)$ (12.19) by the first term of its Taylor expansion about $\mathbf{x}_i$. Then Eq. (12.19) is reduced to the differential one

$$\bar{v}_i\langle\boldsymbol{\eta}_i\rangle(\mathbf{x}_i)=\sum_{j=1}^{N}\mathbf{Y}_{ij}(\mathbf{x}_i)\bar{v}_j\langle\boldsymbol{\eta}_j^e\rangle(\mathbf{x}_i)$$
$$+\sum_{j,q=1}^{N}\bar{v}_q\mathbf{Y}_{ij}(\mathbf{x}_i)\sum_{|\alpha|=1}^{\infty}\boldsymbol{\mathcal{Y}}_{jq}^{\boldsymbol{\alpha}}(\mathbf{x}_i)\nabla^{\boldsymbol{\alpha}}\langle\boldsymbol{\eta}_q\rangle(\mathbf{x}_i), \qquad (12.55)$$

where the matrix $\mathbf{Y}^{-1}(\mathbf{x}_i)$ is defined previously by Eq. (12.8). Here $\boldsymbol{\alpha} = (\alpha_1,\alpha_2,\alpha_3)$ is the multi-index (12.18) and the tensors

$$\boldsymbol{\mathcal{Y}}_{jq}^{\boldsymbol{\alpha}}(\mathbf{x}_i)=(\boldsymbol{\alpha}!)^{-1}\mathbf{R}_j\int\boldsymbol{\mathcal{F}}_{jq}[\otimes(\mathbf{x}_q-\mathbf{x}_i)]^{\boldsymbol{\alpha}}\,d\mathbf{x}_q$$

vary in space. Obtaining the analytical solution (12.55) in the traditional framework of Fourier transform method poses major problems because Eq. (12.55) has variable coefficients. Since we desire an explicit representation for $\langle\boldsymbol{\eta}_i\rangle(\mathbf{x}_i)$, we will omit the second right-hand-side sum in Eq. (12.55): $\boldsymbol{\mathcal{Y}}^{\boldsymbol{\alpha}}_{jq} \equiv \mathbf{0} \quad (\forall\boldsymbol{\alpha} > \mathbf{0})$. This means that all subsequent estimations of both average effective fields and local effective properties should be considered as zero-order approximations of appropriate nonlocal operators. For the zero-order approximations we have the ordinary theory of thermoelasticity at the macrolevel; for the following approximations a weak dispersion theory takes place. In the light of aforementioned approximation the equalities (12.55) add no error to the estimation of local effective properties which are the principal focus of Subsection 12.5.1.

The conditional statistical average of the stresses $\langle\boldsymbol{\sigma}|v_i,\mathbf{x}_i\rangle(\mathbf{x}_i)$ inside the inclusions is obtained from (12.55)

$$\langle\boldsymbol{\sigma}|v_i,\mathbf{x}_i\rangle(\mathbf{x}_i) = \mathbf{B}_i\Big\{\mathbf{D}_i(\mathbf{x}_i)\langle\boldsymbol{\sigma}\rangle(\mathbf{x}_i) + \mathbf{R}_i^{-1}\sum_{j=1}^{N}(\mathbf{Y}_{ij}(\mathbf{x}_i) - \mathbf{I}\delta_{ij})\mathbf{F}_j\Big\} + \mathbf{C}_i, \quad (12.56)$$

where the tensor $\mathbf{D}_i = \mathbf{R}_i^{-1}\sum_{j=1}^{N}\mathbf{Y}_{ij}(\mathbf{x}_i)\mathbf{R}_j$ $(i = 1,\ldots,N)$ is defined analogously to the representation (8.43_2) for statistically homogeneous media. In (12.56) it is necessary to distinguish between the conditional average $\langle\boldsymbol{\sigma}|v_i,\mathbf{x}_i\rangle(\mathbf{x}_i)$ at the point $\mathbf{x}_i \in v_i$ and the statistical average $\langle\boldsymbol{\sigma}\rangle(\mathbf{x}_i)$.

The statistical average of the stresses in the matrix $\mathbf{x} \in v^{(0)}$ is found simply from the relations $\langle\boldsymbol{\sigma}|v^{(0)},\mathbf{x}\rangle(\mathbf{x}) = [\langle\boldsymbol{\sigma}\rangle(\mathbf{x}) - \langle\boldsymbol{\sigma}V|v^{\text{incl}},\mathbf{x}\rangle(\mathbf{x})]/c^{(0)}$. On the right side of the last equation the statistical average $\langle\boldsymbol{\sigma}\rangle(\mathbf{x})$ and the conditional statistical average $\langle\boldsymbol{\sigma}V|v^{\text{incl}},\mathbf{x}\rangle(\mathbf{x})$ (under the condition that there is an inclusion at the point $\mathbf{x} \in v^{\text{incl}}$, $v^{\text{incl}} \equiv \cup v^{(k)}$, $k = 1,\ldots,N$) are used.

From (12.56) the representation for the statistical average of the strain polarization tensor follows:

$$\langle\boldsymbol{\eta}\rangle(\mathbf{x}) = \boldsymbol{\mathcal{R}}^*(\mathbf{x})\langle\boldsymbol{\sigma}\rangle(\mathbf{x}) + \boldsymbol{\mathcal{F}}^*(\mathbf{x}), \quad (12.57)$$

where the tensors $\boldsymbol{\mathcal{R}}^*(\mathbf{x})$ and $\boldsymbol{\mathcal{F}}^*(\mathbf{x})$ have explicit representations according to Eq. (12.56) and can be used for the estimation of the local effective parameters

$$\mathbf{M}^*(\mathbf{x}) = \mathbf{M}^{(0)} + \boldsymbol{\mathcal{R}}^*(\mathbf{x}), \quad \boldsymbol{\beta}^*(\mathbf{x}) = \boldsymbol{\beta}^{(0)} + \boldsymbol{\mathcal{F}}^*(\mathbf{x}), \quad (12.58)$$

which governs the statistically averaged constitutive relation $\langle\boldsymbol{\varepsilon}\rangle(\mathbf{x}) = \mathbf{M}^*(\mathbf{x})\langle\boldsymbol{\sigma}\rangle(\mathbf{x}) + \boldsymbol{\beta}^*(\mathbf{x})$. Only for a statistically homogeneous ergodic field X and uniform boundary conditions, the local effective parameters coincide with the global ones: $\mathbf{M}^*(\mathbf{x}),\ \boldsymbol{\beta}^*(\mathbf{x}) \equiv \text{const}$.

After estimating average strain polarization tensors, see (12.56) and (12.57), the problem of calculating effective properties becomes trivial and leads, according to (12.56) and (12.58), to:

$$\mathbf{M}^*(\mathbf{x}) = \mathbf{M}^{(0)} + \sum_{i,j=1}^{N}\langle\mathbf{R}^{(i)}\rangle(\mathbf{x})\mathbf{R}_i^{-1}\mathbf{Y}_{ij}(\mathbf{x})\mathbf{R}_j n^{(i)}(\mathbf{x}), \quad (12.59)$$

$$\boldsymbol{\beta}^*(\mathbf{x}) = \langle\mathbf{M}_1\mathbf{C} + \boldsymbol{\beta}\rangle(\mathbf{x}) + \sum_{i,j=1}^{N}\langle\mathbf{R}^{(i)}\rangle(\mathbf{x})\mathbf{R}_i^{-1}\Big[\mathbf{Y}_{ij}(\mathbf{x}) - \mathbf{I}\delta_{ij}\Big]\mathbf{F}_j n^{(i)}(\mathbf{x}). \quad (12.60)$$

For a statistically homogeneous field of homogeneous inclusions the formulae (12.59) and (12.60) are equivalent to the results (for global effective properties) derived in [156] (see also Chapter 8), when $\langle \mathbf{R}^{(i)} \rangle(\mathbf{x}) \equiv \mathbf{R}_i$, $\mathbf{Y}_{ij}(\mathbf{x}) \equiv \mathbf{Y}_{ij} =$ const. Only at first glance are the relations (12.60) equivalent to (12.33) for global effective properties. The main difference is that $\mathbf{M}^*(\mathbf{x})$ and $\boldsymbol{\beta}^*(\mathbf{x})$ depend on the parameters of the inclusion distribution not only at the point $\mathbf{x}$, but also in a certain neighborhood of that point leading to a so-called *nonlocal effect*, though, of course, the effective parameter $\mathbf{M}^*(\mathbf{x})$ (12.60) is the local one in the sense of "nonlocal elasticity" theory (for details see Subsection 12.5.2-12.5.5). The diameter of this region mentioned above is estimated as three times the characteristic dimension of the inclusions (see also [132], [134], [181]. As a result, a statistically inhomogeneous composite medium behaves like a macroscopically inhomogeneous medium with local effective parameter $\mathbf{M}^*(\mathbf{x})$ determined for a nonlocal distribution of the inclusions in a certain neighborhood of the point considered. Relations (12.59) and (12.60) can be used to find the stress and strain distributions $\langle \boldsymbol{\sigma} \rangle(\mathbf{x})$, $\langle \boldsymbol{\varepsilon} \rangle(\mathbf{x})$ in the mesodomain w for prescribed boundary conditions, and a conditional statistical average stress field inside the inclusions (12.56) can be estimated. Therefore, the present approach explicitly couples the microstructural details with the global analysis.

A particular case of the statistically inhomogeneous field X is its representation in the form of a stationary random process that is periodic in the broad sense (in practice one has such a situation, for example, for laminated composites of periodic structure, when each layer contains a random field of inclusions). Since $\varphi(v_q, \mathbf{x}_i)$ and $\varphi(v_q, \mathbf{x}_q|; v_i, \mathbf{x}_i)$ are periodic functions of $\mathbf{x}_i$, it follows from (12.55), (12.59) and (12.60) that the local effective properties $\mathbf{M}^*(\mathbf{x})$ and $\boldsymbol{\beta}^*(\mathbf{x})$ are deterministic periodic functions of the coordinates; if the characteristic dimension of the inclusions is comparable to the dimension of the periodicity of the cell, the integration region in (12.55) contains several neighboring cells. After finding the local effective properties (12.59) and (12.60) the estimation of the stress and strain state $\langle \boldsymbol{\sigma} \rangle(\mathbf{x})$, $\langle \boldsymbol{\varepsilon} \rangle(\mathbf{x})$ can be performed by standard methods of periodic structure analysis (see, e.g., [25], [48], [507] [532], [960] and the references in Chapter 11).

Of course, the explicit relations for local effective properties (12.59) and (12.60) can be applied easily to the cases where instead of inclusion clusters (clouds) that form either half-space or sphere one considers a finite or infinite (random or periodical structure) number of individual clouds which can be distinguished by type of probability densities $\varphi(v_q, \mathbf{x}_q|; v_i, \mathbf{x}_i)$ and $\varphi(v_q, \mathbf{x}_q)$ as well as by the mechanical and geometrical parameters of the inclusions. In any case, the effective properties of the inclusion cluster depend not only on the local concentration of inclusions at the point considered but also on a certain neighborhood of that point as well as on parameters of other neighboring clusters.

A popular macroscopic approach is the modeling of FGMs as macroscopically elastic materials, in which the material properties are graded but continuous and are described by a local effective constitutive equation (see, e.g., [311], [899]). Obviously, for the one-particle approximation of MEFM (12.44) we obtain a trivial dependence of $\mathbf{M}^*(\mathbf{x})$ on the point $\mathbf{x}$, from which estimations by the Mori–Tanaka method can be derived for identical aligned inclusions. Mori–Tanaka methods are widely used in the analysis of graded materials [800], [1146], [1238], although in the case of statistically inhomogeneous materials the Mori–Tanaka hypothesis is particularly questionable (see also [1207] analyzed in Subsection 10.5.3). We can say the same about other one-particle methods considered in [1238] (see also [931], [932], [987]) based on the hypothesis of a statistically homogeneous structure of the composites. Nevertheless, these theories provide an acceptable accuracy if the gradients of statistical averages of stress fields are small enough.

12.5.2 Elastically Homogeneous Composites

In this section specializations of the general formulae given in the preceding sections are derived. Especially interesting results are obtained if the fluctuations of the elastic compliance vanish, i.e., $\mathbf{M}(\mathbf{x}) \equiv \mathbf{M}^{(0)} = \text{const}$, and the stress–free strains $\boldsymbol{\beta}$ fluctuate. Of course, the average stresses in the components can be estimated with the help of the passage to the limit $\mathbf{M}^{(i)} \to \mathbf{0}$ in the corresponding formulae. However, some relations can be found immediately without the assumptions of MEFM.

If we have $\boldsymbol{\eta}(\mathbf{x}) \equiv \boldsymbol{\beta}_1(\mathbf{x})$, the general integral equation (7.20) for statistically inhomogeneous media yields an exact formula for the average stresses in the component $v^{(i)}$ $(i = 1, \ldots, N)$:

$$\langle \boldsymbol{\sigma} | v_i, \mathbf{x}_i \rangle(\mathbf{x}) = \langle \boldsymbol{\sigma} \rangle(\mathbf{x}) + \int \boldsymbol{\Gamma}(\mathbf{x} - \mathbf{y})[\langle \boldsymbol{\beta}_1(\mathbf{y}) | v_i, \mathbf{x}_i \rangle - \langle \boldsymbol{\beta}_1 \rangle(\mathbf{y})] \, d\mathbf{y} \qquad (12.61)$$

for $\mathbf{x} \in v^{(i)}$. From Eq. (12.58) the trivial relation for local effective eigenstrains follows: $\boldsymbol{\beta}^*(\mathbf{x}) = \boldsymbol{\beta}^{(0)} + \langle \boldsymbol{\beta}_1 \rangle(\mathbf{x})$. For the estimation of the statistical average $\langle \boldsymbol{\sigma} \rangle(\mathbf{x})$ it is sufficient to analyze the linear problem for elastically homogeneous material with macroscopically inhomogeneous eigenstrains $\boldsymbol{\beta}^*(\mathbf{x})$. The evaluation of conditional statistical eigenstrains $\langle \boldsymbol{\sigma} | v_i, \mathbf{x}_i \rangle(\mathbf{x})$ (12.61) is more complicated.

Let us consider different cases of simplifications of the relation (12.61). The final results may be significantly simplified for statistically homogeneous composites under the additional assumptions (8.67) and (8.69). In such a case the exact relations for the average stresses (9.36) follow from (12.61) for statistically homogeneous composites.

For statistically inhomogeneous composites a significant simplification can be obtained if there is a domain $w^0 \subset w$ such that $\boldsymbol{\beta}_1(\mathbf{x}) \equiv 0$ for $\mathbf{x} \in w \setminus w^0$ and $w^0 \cap \partial w = 0$. We will assume also that $\boldsymbol{\beta}_1(\mathbf{x}) = \boldsymbol{\beta}_1^{(k)} = \text{const.}$ for $\mathbf{x} \in v^{(k)}$. Then the surface integral in Eqs. (7.9), (7.10) becomes zero and (7.9) yields

$$\langle \boldsymbol{\sigma} | v_i, \mathbf{x}_i \rangle(\mathbf{x}) = \boldsymbol{\sigma}^0 - \mathbf{Q}_i \boldsymbol{\beta}_{1i} + \int \mathbf{T}_q(\mathbf{x} - \mathbf{x}_q) \boldsymbol{\beta}_{1q} \overline{v}_q \varphi(v_q, \mathbf{x}_q | ; v_i, \mathbf{x}_i) \, d\mathbf{x}_q, \quad (12.62)$$

where $\mathbf{x} \in v_i \subset w^0$. Since the integration on the right-hand side of (12.62) is performed over the domain $\mathbf{x} \notin w^0$, outside of the domain w^0 we obtain for the matrix $\mathbf{x} \in v^{(0)}$ (see also [1155]: $\langle \boldsymbol{\sigma} | v^{(0)}, \mathbf{x} \rangle(\mathbf{x}) = \boldsymbol{\sigma}^0 + \int \mathbf{T}_q(\mathbf{x} - \mathbf{x}_q) \langle \boldsymbol{\beta}_1 \rangle(\mathbf{x}_q) \, d\mathbf{x}_q$. An analogous equation can be applied for the estimation of the statistical average of the stresses in the domain w^0

$$\langle \boldsymbol{\sigma} \rangle(\mathbf{x}) = \boldsymbol{\sigma}^0 + \int \boldsymbol{\Gamma}(\mathbf{x}-\mathbf{y})\langle \boldsymbol{\beta}_1 \rangle(\mathbf{y})\, d\mathbf{y}, \quad \mathbf{x} \in w^0. \tag{12.63}$$

Consideration of conditional statistical averages of the stresses $\langle \boldsymbol{\sigma}|v_i, \mathbf{x}_i\rangle(\mathbf{x})$, $\mathbf{x} \in v_i$ necessitates the use of the more complicated Eq. (12.62). Then, at least near the boundary ∂w^0 the estimations $\langle \boldsymbol{\sigma}|v_i, \mathbf{x}_i\rangle(\mathbf{x})$ for $\mathbf{x} \in v_i \subset w^0$ will depend on the conditional probability density (8.67) (in contrast to (9.37)) as well as on the distance between $\mathbf{x}_i$ and ∂w^0.

It should be mentioned that Eq. (12.63) is equivalent to those used in the popular problem of a macrocrack in the cluster (or the cloud) of randomly distributed inclusions with eigenstrains, and the estimations of statistical averages of the stresses (12.63) are exploited for the estimation of statistical averages of stress intensity factors at the crack tip (see, e.g., [118], [136], [641], [755]). However, the local distribution of the stresses near the crack tip plays a crucial role in fracture mechanics. As will be shown in Subsection 12.5.3 in a concrete numerical example, the conditional average $\langle \boldsymbol{\sigma}|v_i, \mathbf{x}_i\rangle(\mathbf{x})$ for $\mathbf{x} \in v_i \subset w^0$ can differ significantly from the statistical average $\langle \boldsymbol{\sigma} \rangle(\mathbf{x})$, that must be applied for the estimation of the conditional statistical average of the stress intensity factor (this question is not discussed in more detail in this chapter).

12.5.3 Numerical Results of Estimation of Effective Properties of FGMs

Let examine relations (12.56) for the statistical average of the stresses inside the inclusions for the case of pure mechanical loading ($\boldsymbol{\beta}(\mathbf{x}) \equiv 0$). We consider a finite ellipsoidal cluster w^0 with semi-axes a_w^k $(k = 1, 2, 3;\ a_w^1 \geq a_w^2 \geq a_w^3)$ with a homogeneous distribution of identical homogeneous ellipsoidal inclusions within the cluster (see Fig. 12.7a)

$$\varphi(v_i, \mathbf{x}_i) = \begin{cases} n & \text{for } \mathbf{x} \in w^0, \\ 0 & \text{for } \mathbf{x} \notin w^0, \end{cases}, \quad \varphi(v_q, \mathbf{x}_q|v_i, \mathbf{x}_i) = \begin{cases} n & \text{for } \mathbf{x} \in w^0 \setminus v^0_{qi}, \\ 0 & \text{for } \mathbf{x} \in (w \setminus w^0) \cup v^0_{qi}. \end{cases} \tag{12.64}$$

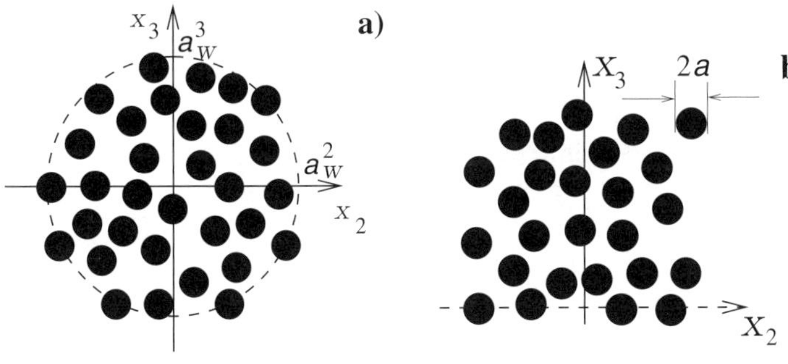

Fig. 12.7. Schematic representations of the inclusion clusters: a) the ellipsoidal cluster, b) the cluster of the form of an upper half space.

It should be mentioned that the final explicit representation for local thermoelastic properties (12.59) and (12.60) were obtained for unrestricted probability densities $\varphi(v_i, \mathbf{x}_i)$ and $\varphi(v_q, \mathbf{x}_q|v_i, \mathbf{x}_i)$. For statistically homogeneous composites the conditional probability density $\varphi(v_q, \mathbf{x}_q|v_i, \mathbf{x}_i)$ was investigated in detail by theoretical and experimental methods (at least for spherical inclusions). However, for graded composites ergodicity is lost, i.e., one cannot equal ensemble and volume average, and a rigorous theoretical and experimental justification of analytical representations for the functions $\varphi(v_q, \mathbf{x}_q|v_i, \mathbf{x}_i)$ is required. For this purpose we choose simple step functions (12.64_1) and (12.64_2) in order to reduce the influence of undesirable side effects on the interesting particular results obtained. The step functions are the limiting cases of real distribution functions of the concentration of inclusions in FGM described in [916] in the framework of the theory of a general Poisson processes.

If $a_w^3 \gg a_i^k$, $(k = 1, 2, 3)$, the local effective compliance of the cluster (12.59) is equal to the global effective compliance $\mathbf{M}^*$ of the corresponding statistically homogeneous composite except in a certain boundary layer, the thickness of which will be evaluated later. Then the average stresses in the interior of the cluster are determined by the relations (8.18) and (8.22) $\langle\boldsymbol{\sigma}\rangle(\mathbf{x}) = \mathbf{B}(w^0)\boldsymbol{\sigma}^0$, where for the calculation of the tensor $\mathbf{B}(w^0)$ by the formula (8.22) the tensor $\mathbf{M}^{(i)}$ must be replaced by the global effective compliance $\mathbf{M}^*$. Then the average value of the stresses in the inclusions can be found from Eqs. (12.56) and (12.59): $\langle\boldsymbol{\sigma}|v_i, \mathbf{x}_i\rangle = (c^{(1)}\mathbf{M}_1^{(1)})^{-1}(\mathbf{M}^* - \mathbf{M}^{(0)})\mathbf{B}(w^0)\boldsymbol{\sigma}^0$, which differs from the expression for the average stresses in the inclusions of a statistically homogeneous composite: $\langle\boldsymbol{\sigma}\rangle_i = (c^{(1)}\mathbf{M}_1^{(1)})^{-1}(\mathbf{M}^* - \mathbf{M}^{(0)})\boldsymbol{\sigma}^0$ by the factor $\mathbf{B}(w^0) \neq \mathbf{I}$. This difference depends on the shape but not on the dimensions of cloud w^0. In particular, for an incompressible matrix $\mathbf{L}^{(0)} = (\infty,\ 2\mu^{(0)})$ and a spherical cloud w^0 with rigid spherical inclusions we obtain

$$\mathbf{L}^* \equiv (\mathbf{M}^*)^{-1} = \left(\infty, 2\mu^{(0)}\frac{16 + 9c^{(1)}}{16 - 31c^{(1)}}\right), \quad \mathbf{B}(w^0) = \left(1, \frac{16 + 9c^{(1)}}{16 - 15c^{(1)}}\right), \tag{12.65}$$

where $c^{(1)}$ is a volume concentration of inclusions in the cluster. The relations (12.65) were obtained by use of Eqs. (8.22), (8.55), (8.57) and (12.59) (see also [156]); Hill's symbolic notations (2.197) for isotropic tensors were used.

Therefore, the use of any large but finite number of inclusions for estimating the statistical average stresses in the inclusions is generally illegitimate. However, according to (12.59) a finite number of inclusions can be used for the estimation of the local effective compliance (12.59) which is equal to the global effective compliance of the corresponding statistically homogeneous composite except in a certain boundary layer ∂w^0. For evaluation of the thickness of this layer let us calculate the local effective modulus $\mathbf{L}^*(\mathbf{x}) = \mathbf{L}^*(x_3)$ (12.59) as a function of the distance of the point $\mathbf{x} = (x_1, x_2, x_3)$ from the flat boundary $x_3 = 0$ of an inclusion cluster of the form of an upper half space (see Fig. 12.7b) simulating the real structure of some graded composites (Fig. 12.1). Estimations of the normalized effective moduli $\mathbf{L}^{\mu*}(x_3) = \mathbf{L}^{\mu*}(x_3, c^{(1)}) = \mathbf{L}^*(x_3, c^{1)})/2\mu^{(0)}$ by Eq. (12.59) are presented in Fig. 12.8 for the concentration of the rigid spherical inclusions (with radius a) inside the cloud $c^{(1)} \equiv \overline{v}_i n = 0.4$; the matrix is assumed to be incompressible. In the cluster's interior (i.e., when $x_3 \to \infty$) $\mathbf{L}^*(\infty)$ coincides with the isotropic effective moduli $\mathbf{L}^*$ for the statistically homogeneous medium. Near the cluster boundary the tensor of the effective moduli $\mathbf{L}^*(x_3)$ is transversally isotropic. As can be seen from Fig. 12.8 the thickness of the boundary layer with significant variation of the effective elastic moduli is not in excess of two times the size (diameter) of the inclusions. It is obvious that in the layer $-0.5a < x_3 < 0$ the number density $\varphi(v_i, \mathbf{x}) \equiv 0$ and $0 < \langle c^{(1)}\rangle(\mathbf{x}) < c^{(1)}$ and, therefore, $L^{\mu(0)}_{3333} < L^{\mu*}_{3333}(x_3, c^{(1)}) < L^{\mu*}_{3333}(0, c^{(1)})$. For simplicity, in the evaluation of $L^*_{3333}(x_3, c^{(1)})$ one can assume that $\mathbf{Y}(x_3) = \mathbf{Y}(0)$ for $-0.5a < x_3 < 0$. For the two positions: $x_3 = 0$ and ∞, numerical estimations for $L^{\mu*}_{3333}(x_3, c^{(1)})$ as well as experimental data [597] for statistically homogeneous media (Newtonian viscosity of a suspension of identical rigid spheres) are presented in Fig. 12.9 for the range of concentration $0 < c^{(1)} < 0.5$. As can be seen, for sufficiently large concentrations of the inclusions the values of the local effective moduli L^*_{3333} near the boundary ($x_3 = 0$) and in the interior of the inclusion cluster ($x_3 \to \infty$) differ by a factor of two or even more.

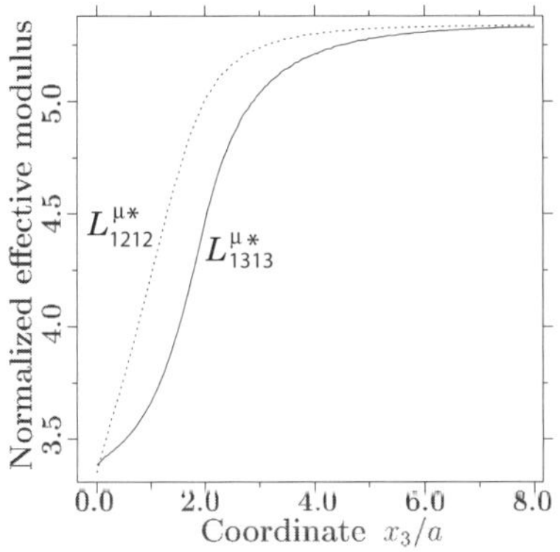

Fig. 12.8. Normalized local effective moduli $\mathbf{L}^{\mu*}(x_3)$ as functions of the distance x_3/a from the cloud boundary for the concentration of the inclusions $c^{(1)} = 0.4$: $L^{\mu*}_{1212}(x_3)$ (dotted curve) and $L^{\mu*}_{1313}(x_3)$ (solid curve).

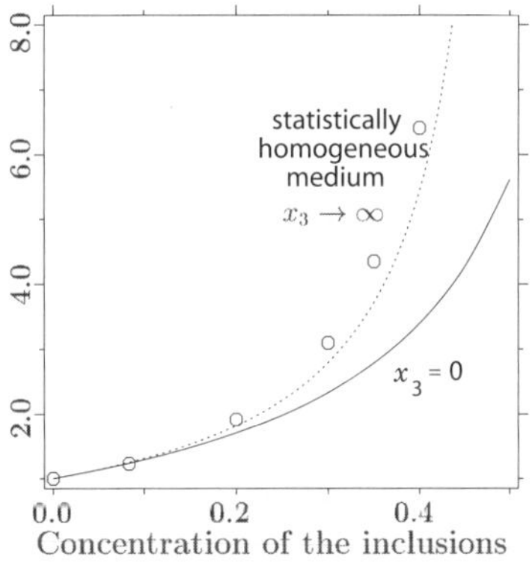

Fig. 12.9. Normalized effective shear moduli $L^{\mu*}_{1212}(x_3, c)$ vs $c^{(1)}$, calculated for $x_3 = 0$ (solid curve) and $x_3 \to \infty$ (dotted curve) as well as experimental data by [597] ((○), $x_3 = \infty$).

In an opposite case let us consider the estimation of the conditional average of the residual stresses ($\boldsymbol{\sigma}^0 \equiv 0$) in spherical clouds w^0 with spherical inclusions (with the radius a), when the fluctuations of the elastic moduli vanish, i.e. $\mathbf{L}(\mathbf{x}) \equiv \mathbf{L}^{(0)} = \text{const}$, and the stress-free strains $\boldsymbol{\beta}$ fluctuate. For the sake of definiteness, we will analyze a composite with isotropic components $\mathbf{L}^{(0)} = (3k^{(0)}, 2\mu^{(0)})$, $\boldsymbol{\beta}^{(1)}_1 = \beta_{01}\boldsymbol{\delta}$ and homogeneous distributions of identical spherical inclusions (12.64_1) within spherical clouds with the center $\mathbf{0}$ and with the radius a_w (see Fig. 12.7a). For the concentration of the inclusions inside the cloud $c(\mathbf{x}) = 0.5$ ($\mathbf{x} \in w^0$) and Poisson's ratio $\nu^{(0)} = 0.3$ conditional statistical averages $\langle \boldsymbol{\sigma} | v_i, \mathbf{x}_i\rangle(\mathbf{x}_i)/(\mu^{(0)}\beta_{01})$ (12.62) and the isotropic average $\langle \boldsymbol{\sigma}\rangle_i/(\mu^{(0)}\beta_{01})$ (9.37) of the corresponding statistically homogeneous composite as well as the isotropic average $\langle \boldsymbol{\sigma}\rangle(\mathbf{x}_i)/(\mu^{(0)}\beta_{01})$ (12.63) are presented in Fig. 12.10. From this figure it can be seen that these stresses at the boundary of the cluster and at its center differ by a factor of more than two; $\mathbf{x}_i = (0, 0, x_3)$ and $a_w = 10a$. The conditional average inside the cluster $\langle \boldsymbol{\sigma}_{33} | v_i, \mathbf{x}_i\rangle(\mathbf{x}_i)$ changes more than 25% in the boundary layer with thickness $2a$. As this takes place in the boundary layer the tensor $\langle\boldsymbol{\sigma}|v_i,\mathbf{x}_i\rangle(\mathbf{x}_i)$ is anisotropic and the absolute magnitude of its third component is larger than that of the first one: $|\langle\sigma_{33}|v_i,\mathbf{x}_i\rangle(\mathbf{x}_i)| > |\langle\sigma_{11}|v_i,\mathbf{x}_i\rangle(\mathbf{x}_i)|$.

A similar spherical cloud w^0 of the rigid spherical inclusions in the incompressible matrix was analyzed by Buryachenko and Rammerstorfer (1998a). It was assumed that $n(\mathbf{x}) = 0.4/\bar{v}_i$ at $\mathbf{x} \in w^0$ and $n(\mathbf{x}) \equiv 0$ at $\mathbf{x} \notin w^0$. In Fig. 12.11 the estimations

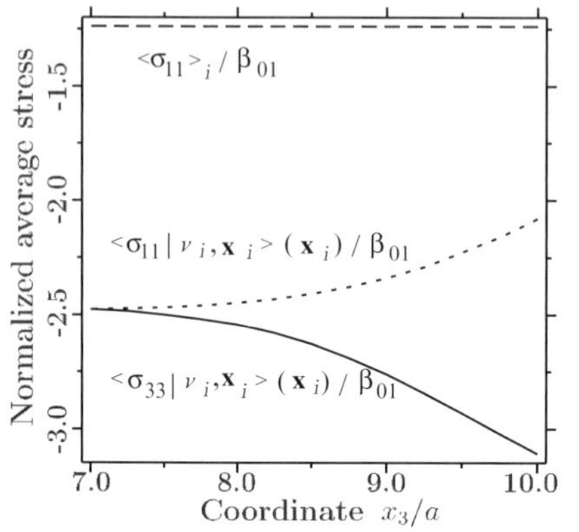

Fig. 12.10. Normalized average stresses vs x_3/a for $c^{(1)}$=0.5: $\langle\sigma_{33}|v_i,\mathbf{x}_i\rangle(\mathbf{x}_i)/(\mu^{(0)}\beta_{01})$ (solid curve), $\langle\sigma_{11}|v_i,\mathbf{x}_i\rangle(\mathbf{x}_i)/(\mu^{(0)}\beta_{01})$ (dotted curve), $\langle\sigma_{11}\rangle_i/(\mu^{(0)}\beta_{01}) \equiv \langle\sigma_{11}\rangle(\mathbf{x}_i)/(\mu^{(0)}\beta_{01})$ (dashed curve).

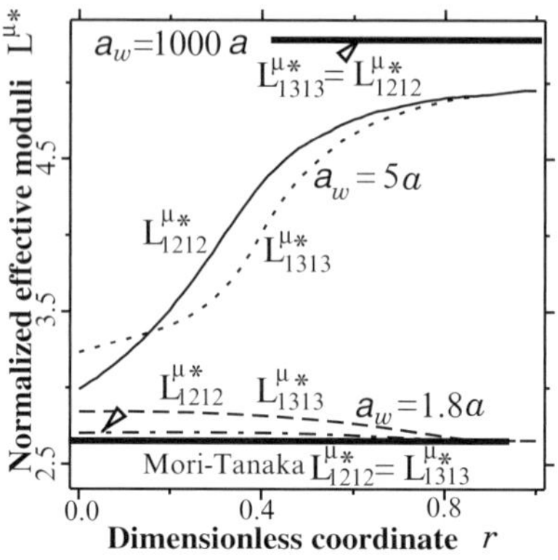

Fig. 12.11. Normalized effective shear moduli $L^{\mu*}_{1212}(r)$ and $L^{\mu*}_{1313}(r)$ vs r at $c^{(1)} = 0.4$ for the different sizes of clusters.

of normalized effective moduli $\mathbf{L}^{\mu*} \equiv \mathbf{L}^*(\mathbf{x})/(2\mu^{(0)})$, $\mathbf{L}^*(\mathbf{x}) \equiv [\mathbf{M}^*(\mathbf{x})]^{-1}$ as functions of the dimensionless coordinate r are represented for $\mathbf{x} = (0,0,a_w(1-r))$, where r represents the relative distance from the cloud boundary. In the interior of a large cloud ($a_w \to \infty$), sufficiently far away from its boundary, $\mathbf{L}^*(\infty)$ coincides with the isotropic effective moduli $\mathbf{L}^*$ for the statistically homogeneous medium. Near the boundary of the cloud, the tensors of the effective moduli $\mathbf{L}^{\mu*}(\mathbf{x})$ are transversally isotropic and vary significantly within the boundary layer $a_w^2 \geq \mathbf{x}^2 > (a_w - 4a)^2$ (nonlocal *boundary layer effect*). The character of the dependence $\mathbf{L}^*(\mathbf{x}) = \mathbf{L}^*(r)$ varies (increases or decreases monotonically or nonmonotonically with $r = |\mathbf{x}|$) with the variation of the cloud size a_w (*scale effect*). For $a_w = 2.2a$ the curves $L^{\mu*}_{1212}(r)$ and $L^{\mu*}_{1313}(r)$ are nonmonotonical functions of r (see Fig. 12.12 and [144]). For $a_w < 2a$ the values $\mathbf{L}^*(\mathbf{0})$ (12.59) coincide with the Mori–Tanaka approach with degeneration of binary interaction effect [52]. It is interesting that the scaler effect analogous to the one described above for elastic graded composite was estimated in [874] numerically for effective conductivity over the cross-section of the composite layer of finite thickness between two homogeneous half-spaces.

In order to demonstrate a similar effect in 2-D case, let us consider a strip model of ideal fiber cluster (see Subsection 5.2.3) with probability densities $\varphi(v_i,\mathbf{x}_i) = n$ and $\varphi(v_j,\mathbf{x}_j|v_i,\mathbf{x}_i) = ng(|\mathbf{x}_i - \mathbf{x}_j|)$ inside the thick ply

$|x_{i|2}| < a_w = 16a$ ($a = 1$ is the inclusion radius) and 0 otherwise. We will consider the volume fiber fraction inside the ply $c^{(1)} = 0.65$, and two radial distribution functions $g(r)$ (step (9.91) and nonstep (9.93) functions). The neglect of the binary interaction of inclusions for statistically homogeneous medium $n(\mathbf{x}) = \text{const.}$ reduces the formula for the effective elastic moduli to the analogous relation obtained by the Mori-Tanaka method which is invariant to the $g(r)$. Assume the matrix is epoxy resin which contains identical circular glass fibers with experimental data corresponding to Fig. 9.6. As can be seen from Fig. 12.13 the use of the approach based on the quasi-crystalline approximation (8.65_2) (for brevity called the MT approach, see also [718]) leads to an underestimate of the effective moduli L^*_{2222} compared to the more exact approximation of the MEFM which provides a good comparison with experimental data for statistically homogeneous media (see Chapter 9). A drastic change of $\mathbf{L}^*(\mathbf{x})$ in the boundary layers $a_w < x_{i|2} < a_w + a$ and $-a_w - a < x_{i|2} < -a_w$ of the fiber ply is explained by the variation of the glass concentration $0 < c(\mathbf{x}) < n\bar{v}_i$ while $c(\mathbf{x}) \equiv n\bar{v} = \text{const.}$ at $|x_{i|2}| \leq a_w$. As can be see in Fig. 12.13, the quasi-crystalline approximation (8.65_2) for the step function $g(r)$ (9.91) significantly reduces the nonlocal effects lie in the decreasing of both the length scale $x^{\text{bl}} = a$ of boundary layer v^{bl} effect and variation of the effective elastic modulus in this layer ($|L^*_{2222}(\mathbf{x}) - L^*_{2222}(\mathbf{0})| \leq 0.99L^*_{2222}(\mathbf{0})$ for $\forall\mathbf{x} \in v^{\text{bl}} \equiv \{\mathbf{x}|0 \leq a_w - |x_{i|2}| \leq x^{\text{bl}}\}$). In so doing, taking the binary interaction of inclusions into account in the MEFM leads to more significant manifestation of the nonlocal effect: $x^{\text{bl}} = 1.7a$ and $|L^*_{2222}(\mathbf{x}) - L^*_{2222}(\mathbf{0})| \geq 0.95L^*_{2222}(\mathbf{0})$ for $\forall\mathbf{x} \in v^{\text{bl}}$.

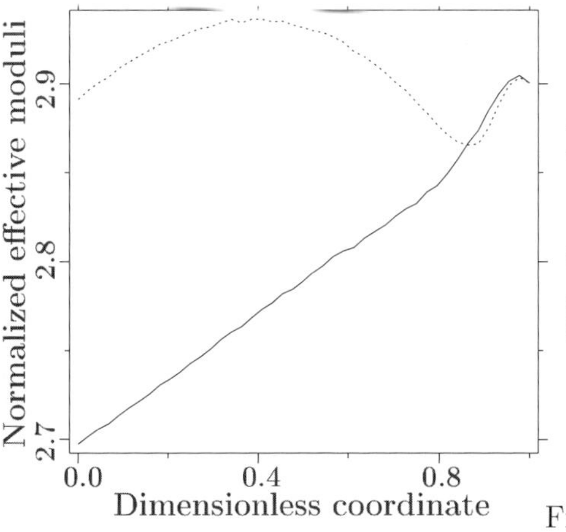

Fig. 12.12. Normalized effective moduli $L^{*\mu}_{1212}$ (solid line), $L^{*\mu}_{133}$ (dotted line) vs r at $a_w = 2.2a$.

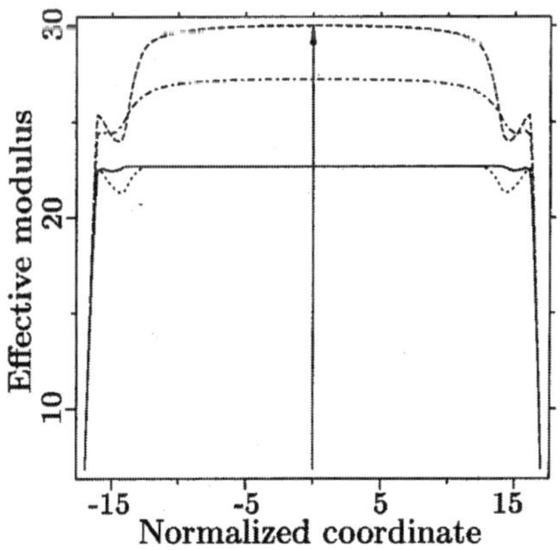

Fig. 12.13. $L^{*\mu}_{2222}$ vs x_2 estimated by: MEFM and nonstep $g(r)$ (dashed line), MEFM and step $g(r)$ (dot-dashed line), MT and nonstep $g(r)$ (dotted line), MT and step $g(r)$ (solid line).

It should be mentioned that the numerical results presented in Figs. 12.8 and 12.9 are approximate, although the reasonable accuracy of the method proposed was demonstrated in some particular examples having analytical solution or explored experimentally (see, e.g., Chapters 8, 9, 10). The subsequent improvement of the method proposed (for example abandonment of the hypothesis **H1** (see

Chapter 10) can lead to some corrections of results presented in Figs. 12.8 and 12.9. At the same time the results presented in Fig. 12.10 are exact ones for the microtopology of the composite considered. No restrictions are imposed on the stress state in the inclusions (hypothesis **H1**) as well as on the number of interacting inclusions (hypothesis **H2**. Nevertheless, the boundary layer effect is made evident in this example as well. Thus we uncover a fundamentally new effect of strong nonlocal dependence of local effective thermoelastic properties as well as the conditional averages of the stresses in the random structure components on the varying inclusion concentration. In the author's opinion, this effect cannot be detected by other popular methods referenced in Chapter 8. Moreover, experimental data confirming (or disproving) the nonlocal uncovered effect are unknown to the author.

12.5.4 Perturbation Method

In the case of a small concentration of the inclusions $c \ll 1$ (9.16_1) as well as for weakly inhomogeneous medium or $||\mathbf{M}_1^{(i)}\mathbf{L}^{(0)}|| \ll 1$ (9.16_2), the perturbation method is appropriate. Then instead of hypothesis **H2** (8.39) we can use the assumption $\langle\widetilde{\boldsymbol{\sigma}}_{i,q}(\mathbf{x})\rangle_j = \langle\boldsymbol{\sigma}\rangle(\mathbf{x}_j)$. $(j = i, q)$, and Eq. (12.2) is reduced to

$$\bar{v}_i\langle\boldsymbol{\eta}_i\rangle(\mathbf{x}_i) = \bar{v}_i\langle\boldsymbol{\eta}_i^{av}\rangle(\mathbf{x}_i) + \sum_{q=1}^{N}\Big\{\int \boldsymbol{\mathcal{T}}_{iq}(\mathbf{x}_i-\mathbf{x}_q)\,d\mathbf{x}_q\bar{v}_i\langle\boldsymbol{\eta}_i^{av}\rangle(\mathbf{x}_i) + \int \boldsymbol{\mathcal{F}}_{iq}(\mathbf{x}_i-\mathbf{x}_q)\bar{v}_q\langle\boldsymbol{\eta}_q^{av}\rangle(\mathbf{x}_q)\,d\mathbf{x}_q\Big\} \tag{12.66}$$

leading to the overall constitutive equations (12.32) with effective properties

$$\mathbf{M}^{*\mathrm{per}} = \mathbf{M}^{(0)} + \sum_{i=1}^{N}\Big\{\mathbf{I} + \sum_{q=1}^{N}\int[\boldsymbol{\mathcal{T}}_{iq}(\mathbf{x}_i-\mathbf{x}_q) + \boldsymbol{\mathcal{F}}_{iq}(\mathbf{x}_i-\mathbf{x}_q)]d\mathbf{x}_q\Big\}n^{(i)}\mathbf{R}_q, \tag{12.67}$$

$$\boldsymbol{\beta}^{*\mathrm{per}} = \boldsymbol{\beta}^{(0)} + \sum_{i=1}^{N}\Big\{\mathbf{I} + \sum_{q=1}^{N}\int[\boldsymbol{\mathcal{T}}_{iq}(\mathbf{x}_i-\mathbf{x}_q) + \boldsymbol{\mathcal{F}}_{iq}(\mathbf{x}_i-\mathbf{x}_q)]d\mathbf{x}_q\Big\}n^{(i)}\mathbf{F}_q, \tag{12.68}$$

$$\boldsymbol{\mathcal{M}}^{*\mathrm{per}}(\langle\boldsymbol{\sigma}\rangle)(\mathbf{x}) = \sum_{i,q=1}^{N} n^{(i)}\int \boldsymbol{\mathcal{F}}_{iq}(\mathbf{x}_i-\mathbf{x}_q)\mathbf{R}_q[\langle\boldsymbol{\sigma}\rangle(\mathbf{x}_q) - \langle\boldsymbol{\sigma}\rangle(\mathbf{x}_i)]\,d\mathbf{x}_q. \tag{12.69}$$

For representation of the integral operator (12.69) in the differential form, we expand $\langle\boldsymbol{\sigma}\rangle(\mathbf{x}_q)$ about $\mathbf{x}_i$ in a Taylor series and integrate term by term over the whole space

$$\boldsymbol{\mathcal{M}}^{*\mathrm{per}}(\langle\boldsymbol{\sigma}\rangle)(\mathbf{x}) = \sum_{i,q=1}^{N}\sum_{|\alpha|=1}^{\infty} n^{(i)}\boldsymbol{\mathcal{B}}_{iq}^{\alpha}(\nabla)\mathbf{R}_q\langle\boldsymbol{\sigma}\rangle(\mathbf{x}). \tag{12.70}$$

Equations (12.67)–(12.70) are reduced to the analogous relations proposed in [132] for identical homogeneous inclusions (3.187) $(N = 1)$; a simplified approach

[181] used a point approximation $\mathbf{T}_{ij}(\mathbf{x}_i-\mathbf{x}_j)=\mathbf{\Gamma}(\mathbf{x}_i-\mathbf{x}_j)$ of binary interactions of the inclusions described by the matrix $\mathbf{Z}_{ij}$ (4.67) obtained from (4.58) by a few iterations of a successive approximation method.

It should be mentioned that effective properties can be estimated by the use of the evaluation of average strains in the components. In so doing, utilization of the perturbation method as well as the combined MEFM-perturbation method reduce to an inequality $\mathbf{L}^*\neq(\mathbf{M}^*)^{-1}$ in contrast to the MEFM (12.36). Because of this Eqs. (12.67)–(12.69) will be employed if $\mathbf{M}_1(\mathbf{x})$ is positive definite; otherwise a dual scheme based on the estimation of average strains $\langle\boldsymbol{\varepsilon}\rangle_i$ and effective moduli $\mathbf{L}^*$ will be used.

12.5.5 Combined MEFM-Perturbation Method

Instead of a standard perturbation method considered in Subsection 12.5.4, we will analyze one possible version of the combined MEFM-perturbation method described in Subsection 9.1.1 and will apply the hypotheses $\langle\widetilde{\boldsymbol{\sigma}}(\mathbf{x})_{1,2}\rangle_i=\langle\overline{\boldsymbol{\sigma}}(\mathbf{x})\rangle_i=\boldsymbol{\sigma}^0,\ i=1,2,\ \mathbf{x}\in v_i$ (9.10) to Eq. (12.55) (before the Taylor expansion of $\langle\boldsymbol{\eta}_q\rangle(\mathbf{x}_q)$ about $\mathbf{x}_i$) only for estimation of the second sum on the right-hand side of Eq. (12.55) rather than to the initial system (8.24). Then from (12.55), taking Eq. (8.18) into account, we get the overall constitutive equation:

$$\langle\boldsymbol{\varepsilon}\rangle(\mathbf{x})=\mathbf{M}^*(\mathbf{x})\langle\boldsymbol{\sigma}\rangle(\mathbf{x})+\boldsymbol{\beta}^*(\mathbf{x})+\boldsymbol{\mathcal{M}}^{*\mathrm{per}}(\langle\boldsymbol{\sigma}\rangle)(\mathbf{x}), \tag{12.71}$$

where the local thermoelastic parameters described by the tensors $\mathbf{M}^*(\mathbf{x}_i)$, $\boldsymbol{\beta}^*(\mathbf{x}_i)$ coincide with (12.59), (12.60) and the nonlocal part of effective compliance is defined by

$$\boldsymbol{\mathcal{M}}^{*\mathrm{per}}(\langle\boldsymbol{\sigma}\rangle)(\mathbf{x}_i)=\sum_{i,j,q=1}^{N} n^{(i)}\mathbf{Y}_{ij}(\mathbf{x}_i)\mathbf{R}_j\int\overline{\boldsymbol{\mathcal{T}}}_{jq}\mathbf{R}_q[\langle\boldsymbol{\sigma}\rangle(\mathbf{x}_q)-\langle\boldsymbol{\sigma}\rangle(\mathbf{x}_i)]\,d\mathbf{x}_q. \tag{12.72}$$

For representation of the integral operator (12.72) in differential form we expand $\langle\boldsymbol{\sigma}\rangle(\mathbf{x}_q)$ about $\mathbf{x}_i$ in a Taylor series and integrate term by term over the whole space:

$$\boldsymbol{\mathcal{M}}^{*\mathrm{per}}(\langle\boldsymbol{\sigma}\rangle)(\mathbf{x}_i)=\sum_{\alpha_s=1}^{\infty}\boldsymbol{\mathcal{B}}^{\mathrm{per}}_{\alpha}(\mathbf{x}_i)\nabla^{\alpha}\langle\boldsymbol{\sigma}\rangle(\mathbf{x}_i), \tag{12.73}$$

$$\boldsymbol{\mathcal{B}}^{\mathrm{per}}_{\alpha}(\mathbf{x}_i)\equiv\frac{1}{\alpha!}\sum_{i,j,q=1}^{N} n^{(i)}\mathbf{Y}_{ij}(\mathbf{x}_i)\mathbf{R}_j\int\overline{\boldsymbol{\mathcal{T}}}_{jq}\mathbf{R}_q[\otimes(\mathbf{x}_q-\mathbf{x}_i)]^{\alpha}\,d\mathbf{x}_q. \tag{12.74}$$

One [181] (see also [69]) used a standard perturbation method implying the application of assumptions (9.10) to the initial system (8.24); in such a case the representation (12.73) is valid if in Eq. (12.74) one postulates $\mathbf{Y}_{ij}(\mathbf{x})\equiv\mathbf{I}\delta_{ij}$.

12.5.6 The MEF Method

In a general case of composite materials when the assumption (12.6) is not valid, it is possible to obtain the representation of nonlocal operator $\boldsymbol{\mathcal{M}}^*$ in series form.

Recasting the second integral on the right-hand side of Eq. (12.31) in a spirit of a subtraction technique and applying the standard scheme of the iteration method [882] leads to the Neumann series form for the solution

$$\mathcal{M}^* = \sum_{k=1}^{\infty} n\mathcal{K}^k \mathbf{Y}, \tag{12.75}$$

where the power $\mathcal{K}^k$ is defined recursively with the following representation of the kernels of $\mathcal{K}^k$ and the initial approximation $\mathcal{K}^1$:

$$\mathcal{K}_k(\mathbf{x},\mathbf{y}) = \int \mathcal{K}(\mathbf{x},\mathbf{z})\mathcal{K}_{k-1}(\mathbf{z},\mathbf{y})d\mathbf{z},$$

$$\mathcal{K}_1(\mathbf{x},\mathbf{y}) = \mathcal{K}(\mathbf{x},\mathbf{y}) \equiv \mathbf{Y}(\mathbf{x})\mathbf{R}\big[\overline{\mathcal{T}}(\mathbf{x},\mathbf{y}) - \delta(\mathbf{x}-\mathbf{y})\int \overline{\mathcal{T}}(\mathbf{x},\mathbf{z})\, d\mathbf{z}\big],$$

respectively. In Eq. (12.75) for simplicity sake, $N = 1$, $\boldsymbol{\beta}(\mathbf{x}) \equiv \mathbf{0}$ was considered and the tensor indices enumerating the constituents are omitted. In effect the iteration method (12.75) transforms the integral equation problem (12.55) into the linear algebra problem in any case. The series (12.75) converges to a solution $\mathcal{M}^*$ (12.71) for the kernel of $\mathcal{K}$ "small" enough. Buryachenko [134] showed in particular examples the high effectiveness of the constructed procedure (12.75) of the approximate solution with acceptable controlled accuracy. However, the classical mathematical existence and uniqueness problems (and what is meant by saying "small" or "slowly varying" enough), were not considered. Nevertheless, it should by mentioned that $\mathbf{Y}(\mathbf{x}) \to \mathbf{I}$ and $\int \mathcal{K}(\mathbf{x},\mathbf{y})\, d\mathbf{y} \to \mathbf{0}$ as $n(\mathbf{x})\overline{v}_i \to 0$. Therefore, the iteration method is appropriate at least for dilute concentration of inclusions.

12.6 Concluding Remarks

Let us discuss the main scheme as well as short sketch of limitations and of possible generalization and application of the methods proposed.

The obtained relations depend on the values associated with the mean distance between inclusions, and do not depend on the other characteristic size, i.e., the mean inclusion diameter. This fact may be explained by the initial use of the hypothesis **H1** dealing with the homogeneity of the field $\overline{\boldsymbol{\sigma}}(\mathbf{x})$ inside each inclusion. In the case of a variable representation of $\overline{\boldsymbol{\sigma}}(\mathbf{x})$ $(\mathbf{x} \in v_i)$, for instance in polynomial form, the mean size of the inclusions will be contained in the nonlocal dependence of microstresses on the average stress $\langle\boldsymbol{\sigma}\rangle(\mathbf{x})$. Such an improvement based on the abandonment from hypothesis **H1** was considered in Chapters 9 and 10 where the comparison with available analytical solution for semi-infinite periodic collinear row of cracks is presented, and there are the numerous references confirming the estimation of accuracy of the hypotheses **H1** and **H2** in connection with relevant experimental data and analytical solutions for the linear problems for statistically homogeneous and periodic structure 2-D and 3-D composites.

It should be mentioned that QM, IM, and FTM have a series of advantages and disadvantages, and it is crucial for the analyst to be aware of their range of application. The QM caused no problems of accuracy but can become less effective in the CPU time sense than the IM if the number of variables increases. The IM (12.15) has two known drawbacks. First, the Neumann series ensures the existence of solutions to integral equations of the second kind only for sufficiently small kernels (12.16), and second, in general, it cannot be summed in closed form. Of course Eq. (12.15) can be solved directly by the quadrature method even if the condition (12.16) is not valid. However, strongly inhomogeneous problems may lead to much larger numbers of quadrature points, making iteration potentially worthwhile. Moreover, increasing the problem dimensionality (from 2-D to 3-D) raises the number of nodes to the dimensional power and the situation changes radically. As was shown, only a few iterations of Eq. (12.15) are necessary; these iterations prove very much faster than a direct inversion of the operator $\mathbf{I}-\boldsymbol{\mathcal{K}}$ by the quadrature method. In so doing, the condition (12.16) is fulfilled in all examples considered for the highly filled composites with an infinite contrast in moduli between the phases.

The reduction of integral operators to the differential ones allows an understanding of drawbacks of the FTM. The first one is that for obtaining an m order differential operator, it is necessary that $\boldsymbol{\mathcal{E}}^{av}$ belongs to $C^m(w)$. In so doing, the IM providing the accuracy of differential operator of an infinite order does not have even continuity of $\boldsymbol{\mathcal{E}}^{av}$ since integration is a smoothing operation and the right-hand-side integral (12.9) is likely to be a rather smooth function even when $\boldsymbol{\mathcal{E}}^{av}(\mathbf{x})$ is very jagged. But even the m'th approximation of the IM contains the differential operators of infinite orders (although not all, see Eq. (12.37)) that are lost in the FTM. The question of the convenience of using one method over another is solved also in favor of the IM because in the FTM, it is necessary to calculate the cumbersome tensors $\boldsymbol{\mathcal{Y}}_{(m)}$ (12.22) and $\boldsymbol{\mathcal{B}}_m$ (12.28), completeness of estimation of which increase dramatically with m, while in the IM it is enough to estimate a single tensor $\mathbf{K}(\mathbf{x}_i-\mathbf{x}_q)$ (12.10) and consecutively to apply the recursion scheme (12.15) the completeness of which does not depend on the iteration number. Thus, the advantage of the FTM comprised of obtaining analytical explicit relations dwindles in light of the disadvantages mentioned above such as a requirement of smoothness of $\boldsymbol{\mathcal{E}}^{av}\in C^m(w)$ and an intricacy of analytical calculations. Nevertheless, there is some advantage of the FTM naturally connected with the explicit representation of the differential operators when tensors $\boldsymbol{\mathcal{Y}}_{(m)}$ (12.22) and $\boldsymbol{\mathcal{B}}_m$ (12.28) need to be estimated just one time and can be used for subsequent analysis of any $\boldsymbol{\mathcal{E}}^{av}\in C^m(w)$ smooth enough (although, of course, the question of which order of differential operator (12.22), (12.27) provides *a priori* prescribed accuracy of the solution of Eq. (12.9) and how it is connected with the smoothness of the function $\boldsymbol{\mathcal{E}}^{av}$, remains valid).

We indicated only mathematical and computational difficulties in the use of the FTM which can be solved, at least in principle, at the cost of great effort if an analytical solution is necessary. However, it should be mentioned that there is an extremely important class of micromechanical problems for statistically inhomogeneous media (such as functionally graded and clustered materials considered in the current chapter, analysis of which by FTM is questionable). The breakdown of the assumption of statistical homogeneity leads to the inequality $\mathbf{Y}(\mathbf{x}_i) \not\equiv$ const. Then the average stresses $\langle \boldsymbol{\sigma} \rangle(\mathbf{x}) \not\equiv$ const. and $\boldsymbol{\mathcal{K}}\boldsymbol{\mathcal{E}} \not\equiv \mathbf{0}$ even at the homogeneous boundary conditions either (7.2) or (7.3). However, even this simplest case of homogeneous boundary conditions leads to a fundamental prohibitive obstacle against the use of the FTM. Indeed, the inhomogeneity $\langle \boldsymbol{\sigma} \rangle(\mathbf{x}) \not\equiv$ const. yields the inequality $\boldsymbol{\mathcal{K}}(\mathbf{x}_i, \mathbf{x}_q) \neq \boldsymbol{\mathcal{K}}(\mathbf{x}_i - \mathbf{x}_q)$ and, therefore, the linear differential operator (12.19) has variable coefficients. Then the jump from Eqs. (12.19) to Eqs. (9.12) and (9.13), respectively, based on the properties of the Fourier transform, is difficult, and applicability of the FTM is questionable. In so doing, the use of the IM for analyis of statistically inhomogeneous media (see Subsections 12.5.5 and 12.5.6) inserts requires a slight modification of the scheme presented for application of the iteration method for research of periodic graded composites (see Section 11.3).

13

Stress Fluctuations in Random Structure Composites

Although the effective behavior of the composite is traditionally the main focus of micromechanics, it is also essential to supply insight into the statistical description of the local strains and stresses, such as their statistical moments of different order in each phase and at interphase. Estimation of these local fields are extremely useful for understanding the evolution of nonlinear phenomena such as plasticity, creep, and damage. When one tries to estimate the equivalent stress in the strength theories as well as in nonlinear creep theory, or when the yield function in plasticity theory is considered, squares of the first invariant or the second invariant of the deviator of local stresses are frequently used. Several papers have already been written on the problem of estimation of values of invariants averaged over the volume of the components, which involve particular assumptions or simplifications as, for instance, the two-dimensional model [362], special correlation function [840]. A very prospective idea of a perturbation method proposed by Bergman [74] is based on the estimation of the perturbation of an energetic function due to a variation of the material properties such as e.g. conductivity and elasticity modulus. This idea was developed for estimation of stress fluctuations in the case of isotropy of materials constants [89], [90], [599] and isotropy of fluctuations [602].

Shermergor [995] investigated this problem by the use of a perturbation method with small fluctuations of components' moduli. Recursive formulae for computing a probability density function of stresses are derived in [469] from the perturbation analysis of field described by Eq. (7.24); this equation is questionable in the general case of unbounded inclusion fields; see Section 7.2. A general scheme for calculating the second-order moments of the random elastic field in the case of a composite of inclusion-matrix type is presented in [125]. The final results are obtained for the volume averaged inclusions of moment functions by means of a particular realization of the MEFM. Stress fluctuations in the components of random structure composites represent a measure of inhomogeneity of stress fields in the components; numerical statistical analysis at the inclusion scale level was performed, e.g., in [24]. The fundamental roles of such inhomogeneities described by the stress fluctuations are discussed in detail in [131], [897] for a wide class of nonlinear problems of micromechanics such as plasticity, damage, viscosity, or creeping.

The perturbation method was developed for the exact estimation of all components of the second moment tensor of the pure elastic [860] and internal residual stresses [151], averaged over the volume of components, is obtained. It is found that the second moment of the stress field is constant within the inclusions if a homogeneity of some random effective stress fields in the neighborhood of each ellipsoidal inclusion was additionally assumed. Furthermore an expression for the second moment of the stress tensor in the matrix in the vicinity of the inclusion is derived and a method for the construction of the correlation function of the internal stresses in separate inclusions is developed. The applicability of the proposed methods for estimating second moments of stresses in the matrix is significantly limited, because they require evaluation of the effective properties of the composite for the general case of matrix anisotropy (even for the case of actually isotropic matrix). This disadvantage was eliminated via the method of integral equations [179], [182] for the estimation of second moments of stresses in the components. Considering both binary and triple interaction of the inclusions, explicit relations for second moments of stresses are obtained. The inhomogeneity of the second moment of stresses in coated inclusions is shown. A comparison between the second invariants of deviator stresses in the matrix estimated via the perturbation method [151] and that estimated via the integral equation method [179], [182] will be conducted in Chapter 15.

The exact solution is possible for deterministic structure by means of numerical methods. Composites with regular structure [322], [362], [1120] and modifications of self-consistent scheme models [647] among them are only two-dimensional, and often the elastic anisotropy is neglected. Although these regularization theories allow detailed stress calculation (for instance local stress intensity factors for flaws in the real interface can be derived), they do not seem suited to describe the real three-dimensional and stochastic microstructures.

A significant advance has been made in the study of residual stresses by the Monte-Carlo method of numerical simulation of the random structure of materials [803], [689], [842]. Because of long computing time two-dimensional models of composites with a special random structure are used. The numerical simulation points to the fact that spatial distribution of components has a significant effect on the local stresses. Therefore these methods do not seem suited to prediction strength of three-dimensional random structures.

The study of a macrocrack propagation in composite materials is of immediate interest to industry. The strength calculations of the composites serve as examples of a nonlinear problem. We will consider the different nonlinear phenomena and related problems of estimation of overall properties. However, fracture-related parameters such as the stress intensity factors (SIF) are sensitive to the stress distribution, while the overall properties are not. The problem of interaction between a crack tip and a source of internal stress (which is given an Eshelby transformation of its stress-free state) is discussed by a number authors (see for references, e.g., [935], [937], [1054]). A considerable number of papers are concerned with a study of crack-microcracks or crack-microinclusions interactions (appropriate, but by no means exhaustive, references are provided by the review [529], see also [106], [240], [254], [806], [1080]). As this takes place a variety of simplifications are used: two-dimensional problem, regular structure of the microdefect field, approximation of microcrack array by a "soft inclusion," weak interaction between microinclusions, and numerical simulation of inclusion distribution with subsequent solution of the determinate problem. But in the last case, the immense computations render a direct numerical analysis of macrocrack-microcrack interaction impractical.

Previous analyses in fracture mechanics usually have not focused on the influence of statistical effects of microstructure of composite materials. But it becomes obvious that, because of random microstructure, the energy release rate at the tip of a macro crack is also random. With the simplifying assumption of an elastically homogeneous plane with a macrocrack, the random finite set of inclusions is modeled as an elastic continuum within which the deterministic stress-free transformation has a magnitude equal to a statistical average of the random transformed field [118], [641], [755].

For estimation of second moments of stresses averaged over the volume of constituents we will consider such general methods as perturbation method and the method of integral equations.

13.1 Perturbation Method

13.1.1 Exact Representation for First and Second Moments of Stresses Averaged over the Phase Volumes

In this section we derive exact representations for all components for second-order moments of the stress and strain fields averaged over the volume of a component by utilizing the functional dependence of effective properties on the thermoelastic constants of the component being considered [74], [89], [90], [602], [860]. This functional dependence can be obtained only by employing of special theoretical micromechanical model rather than from an experimental measurement of effective properties.

Just for concreteness, we will consider homogeneous traction boundary conditions (6.36) and decompose the overall field as $\boldsymbol{\sigma}(\mathbf{x}) = \boldsymbol{\sigma}^I(\mathbf{x}) + \boldsymbol{\sigma}^{II}(\mathbf{x})$, with the sources (6.35) and (6.36) $\boldsymbol{\beta}^I(\mathbf{x}) = \mathbf{0}$, $\mathbf{t}^I(\mathbf{y}) = \mathbf{t}^{\partial\mathcal{E}_\sigma}(\mathbf{y})$ and $\boldsymbol{\beta}^{II}(\mathbf{x}) = \boldsymbol{\beta}(\mathbf{x})$, $\mathbf{t}^{II}(\mathbf{y}) = \mathbf{0}$ for the field $\boldsymbol{\sigma}^I$ and $\boldsymbol{\sigma}^{II}$, respectively, where $\mathbf{t}^{\partial\mathcal{E}_\sigma}$ $(\mathbf{x} \in w,\ \mathbf{y} \in \partial\mathcal{E}_\sigma)$ is a prescribed traction in the boundary conditions (6.36). We now summarize the results of Section 6.2 in a form appropriated to our anticipated purpose:

$$\langle \boldsymbol{\sigma}^I \rangle = \langle \boldsymbol{\sigma} \rangle, \quad \langle \boldsymbol{\sigma}^{II} \rangle = \mathbf{0}, \tag{13.1}$$

$$\langle \boldsymbol{\sigma}^I \mathbf{M} \boldsymbol{\sigma}^I \rangle = \langle \boldsymbol{\sigma} \rangle \mathbf{M}^* \langle \boldsymbol{\sigma} \rangle, \quad \langle \boldsymbol{\sigma}^{II} \mathbf{M} \boldsymbol{\sigma}^{II} \rangle = 2W^*, \quad \langle \boldsymbol{\sigma}^I \mathbf{M} \boldsymbol{\sigma}^{II} \rangle = 0, \tag{13.2}$$

$$\langle \boldsymbol{\sigma}^I \boldsymbol{\beta} \rangle = \langle \boldsymbol{\sigma} \rangle \boldsymbol{\beta}^*, \quad \langle \boldsymbol{\sigma}^{II} \boldsymbol{\beta} \rangle = -2W^*, \tag{13.3}$$

where the right-hand side of the mentioned scalar equations only depend on the effective constant $\mathbf{M}^*$, $\boldsymbol{\beta}^*$, W^* assumed to be known in the following.

For a fixed boundary conditions (6.36), we first consider a small variation of the local compliance $\delta\mathbf{M}(\mathbf{x})$ as compared to $\mathbf{M}(\mathbf{x})$: $\mathbf{M}(\mathbf{x}) \to \mathbf{M}(\mathbf{x}) + \delta\mathbf{M}(\mathbf{x})$ implying corresponding variations of the stress and strain fields: $\boldsymbol{\sigma}(\mathbf{x}) \to \boldsymbol{\sigma}(\mathbf{x}) + \delta\boldsymbol{\sigma}(\mathbf{x})$ and $\boldsymbol{\varepsilon}(\mathbf{x}) \to \boldsymbol{\varepsilon}(\mathbf{x}) + \delta\boldsymbol{\varepsilon}(\mathbf{x})$, respectively. Since both $\boldsymbol{\sigma}(\mathbf{x})$ and $\boldsymbol{\sigma}(\mathbf{x}) +$

$\delta\boldsymbol{\sigma}(\mathbf{x})$ satisfy the same boundary condition, we obtain from (13.1) $\langle\delta\boldsymbol{\sigma}(\mathbf{x})\rangle = \mathbf{0}$. Moreover, $\delta\boldsymbol{\sigma}(\mathbf{x})$ and $\delta\boldsymbol{\varepsilon}(\mathbf{x})$ accomplish the requirements of the Hill's condition (6.15) leading to the scale equalities

$$\langle\delta\boldsymbol{\sigma}\boldsymbol{\varepsilon}\rangle = \langle\delta\boldsymbol{\sigma}\rangle\langle\boldsymbol{\varepsilon}\rangle = 0, \qquad \langle\boldsymbol{\sigma}\delta\boldsymbol{\varepsilon}\rangle = \langle\boldsymbol{\sigma}\rangle\langle\delta\boldsymbol{\varepsilon}\rangle = 0, \tag{13.4}$$

which are fulfilled for both $\boldsymbol{\varepsilon}^I$ and $\boldsymbol{\varepsilon}^{II}$ combined with both fields $\boldsymbol{\sigma}^I$ and $\boldsymbol{\sigma}^{II}$.

Varying (13.2_1) in an accompany with Eqs. (13.4_1) and (6.37) leads to

$$\langle\boldsymbol{\sigma}\rangle\delta\mathbf{M}^*\langle\boldsymbol{\sigma}\rangle = \langle\boldsymbol{\sigma}^I\delta\mathbf{M}\boldsymbol{\sigma}^I\rangle + 2\langle\boldsymbol{\sigma}^I\mathbf{M}\delta\boldsymbol{\sigma}^I\rangle = \langle\boldsymbol{\sigma}^I\delta\mathbf{M}\boldsymbol{\sigma}\rangle. \tag{13.5}$$

In a similar manner, variation of Eq. (13.2_2) by varying of $\mathbf{M}(\mathbf{x})$ at the fixed $\boldsymbol{\beta}(\mathbf{x})$ yields

$$2\delta W^* = \langle\boldsymbol{\sigma}^{II}\delta\mathbf{M}\boldsymbol{\sigma}^{II}\rangle + 2\langle\boldsymbol{\sigma}^{II}\mathbf{M}\delta\boldsymbol{\sigma}^{II}\rangle = \langle\boldsymbol{\sigma}^{II}\delta\mathbf{M}\boldsymbol{\sigma}^{II}\rangle + 2\langle\boldsymbol{\varepsilon}^{II}\delta\boldsymbol{\sigma}^{II}\rangle - \langle\boldsymbol{\beta}\delta\boldsymbol{\sigma}^{II}\rangle \tag{13.6}$$

where the second term in the right-hand side of Eq. (13.6) vanishes in virtue of (13.4_1). Combining Eqs. (13.3_2) and (13.6) leads to

$$\langle\boldsymbol{\sigma}^{II}\delta\mathbf{M}\boldsymbol{\sigma}^{II}\rangle = -2\delta W^*. \tag{13.7}$$

Finally, variation of (13.2_3) with varying $\mathbf{M}(\mathbf{x})$ and fixed $\boldsymbol{\beta}(\mathbf{x})$ gives

$$\begin{aligned} 0 = \langle\boldsymbol{\sigma}^I\delta\mathbf{M}\boldsymbol{\sigma}^{II}\rangle + \langle\delta\boldsymbol{\sigma}^I\mathbf{M}\boldsymbol{\sigma}^{II}\rangle + \langle\boldsymbol{\sigma}^I\mathbf{M}\delta\boldsymbol{\sigma}^{II}\rangle = \langle\boldsymbol{\sigma}^I\delta\mathbf{M}\boldsymbol{\sigma}^{II}\rangle \\ + \langle\delta\boldsymbol{\sigma}^I\boldsymbol{\varepsilon}^{II}\rangle - \langle\delta\boldsymbol{\sigma}^I\boldsymbol{\beta}\rangle + \langle\boldsymbol{\sigma}^I\delta\boldsymbol{\sigma}^{II}\rangle = \langle\boldsymbol{\sigma}^I\delta\mathbf{M}\boldsymbol{\sigma}^{II}\rangle + \langle\boldsymbol{\sigma}\rangle\delta\boldsymbol{\beta}^*, \end{aligned} \tag{13.8}$$

where the last equality was obtained utilizing Eqs. (13.3_1), (13.4_1), and (13.4_2).

Here one may use a special variation of the compliance tensor $(k = 0, 1, \ldots)$:

$$\delta\mathbf{M}(\mathbf{x}) \equiv \delta\mathbf{M}^{(k)}V^{(k)}(\mathbf{x}), \ \ \delta\mathbf{M}^{(k)} = \text{const}, \quad \mathbf{x} \in v^{(k)}. \tag{13.9}$$

In this case $\delta\mathbf{M}$ may be taken outside the averaging sign over the volume of component $v^{(k)}$ $(k = 0, 1, \ldots, N)$. Furthermore we have

$$\delta\mathbf{M}^* = \frac{\partial\mathbf{M}^*}{\partial\mathbf{M}^{(k)}}\delta\mathbf{M}^{(k)}, \quad \delta W^* = \frac{\partial W^*}{\partial\mathbf{M}^{(k)}}\delta\mathbf{M}^{(k)}, \quad \delta\boldsymbol{\beta}^* = \frac{\partial\boldsymbol{\beta}^*}{\partial\mathbf{M}^{(\mathbf{k})}}\delta\mathbf{M}^{(k)} \tag{13.10}$$

which together with Eqs. (13.5), (13.7), (13.8), and (13.9) lead to the final expressions for the second moments of stresses

$$\langle\boldsymbol{\sigma}^I \otimes \boldsymbol{\sigma}^I\rangle^{(k)} = \frac{1}{c^{(k)}}\frac{\partial\mathbf{M}^*}{\partial\mathbf{M}^{(k)}}\langle\boldsymbol{\sigma}\rangle \otimes \langle\boldsymbol{\sigma}\rangle, \tag{13.11}$$

$$\langle\boldsymbol{\sigma}^{II} \otimes \boldsymbol{\sigma}^{II}\rangle^{(k)} = -\frac{2}{c^{(k)}}\frac{\partial W^*}{\partial\mathbf{M}^{(k)}}, \quad \langle\boldsymbol{\sigma}^I \otimes \boldsymbol{\sigma}^{II}\rangle^{(k)} = \frac{2}{c^{(k)}}\frac{\partial\boldsymbol{\beta}^*}{\partial\mathbf{M}^{(k)}}\langle\boldsymbol{\sigma}\rangle \tag{13.12}$$

or in index form

$$\langle\sigma^I_{pq}\sigma^I_{rs}\rangle^{(k)} = \frac{1}{c^{(k)}}\frac{\partial M^*_{ijmn}}{\partial M^{(k)}_{pqrs}}\langle\sigma_{ij}\rangle\langle\sigma_{mn}\rangle, \tag{13.13}$$

$$\langle\sigma^{II}_{pq}\sigma^{II}_{rs}\rangle^{(k)} = -\frac{2}{c^{(k)}}\frac{\partial W^*}{\partial M^{(k)}_{pqrs}}, \quad \langle\sigma^I_{pq}\sigma^{II}_{rs}\rangle^{(k)} = \frac{2}{c^{(k)}}\frac{\partial\beta^*_{ij}}{\partial M^{(k)}_{pqrs}}\langle\sigma_{ij}\rangle. \tag{13.14}$$

Summation of Eqs. (13.11) and (13.12) yields the representation for the second moment of the overall stresses $\boldsymbol{\sigma} = \boldsymbol{\sigma}^I + \boldsymbol{\sigma}^{II}$

$$\langle \boldsymbol{\sigma} \otimes \boldsymbol{\sigma} \rangle^{(k)} = \frac{1}{c^{(k)}} \frac{\partial \mathbf{M}^*}{\partial \mathbf{M}^{(k)}} \langle \boldsymbol{\sigma} \rangle \otimes \langle \boldsymbol{\sigma} \rangle - \frac{2}{c^{(k)}} \frac{\partial W^*}{\partial \mathbf{M}^{(k)}} + \frac{2}{c^{(k)}} \frac{\partial \boldsymbol{\beta}^*}{\partial \mathbf{M}^{(k)}} \langle \boldsymbol{\sigma} \rangle, \quad (13.15)$$

where the partial derivatives are calculated under the assumption of fixed transformation fields $\boldsymbol{\beta}(\mathbf{x})$, $\mathbf{x} \in w$. Invariants $\langle \sigma_0^2 \rangle^{(k)}$ and $\langle s_{ij} s_{ij} \rangle^{(k)}$ ($\sigma_0 = \sigma_{ii}/3$, $s_{ij} = \sigma_{ij} - \sigma_0$) are estimated employing the convolutions of Eqs. (13.13) and (13.14). For this purpose, we will assume that $\mathbf{M}^{(k)} = 3\bar{p}^{(k)}\mathbf{N}_1 + 2\bar{q}^{(k)}\mathbf{N}_2$, $3\bar{p}^{(k)} = M_{iijj}^{(k)}/3$, $2\bar{q}^{(k)} = (M_{ijij}^{(k)} - 3\bar{p}^{(k)})/5$, ($\mathbf{N}_1 = \boldsymbol{\delta} \otimes \boldsymbol{\delta}/3$, $\mathbf{N}_2 = \mathbf{I} - \mathbf{N}_1$, (see Eq. (2.200)) and take into account a differential rule of a composite function such as, e.g.,

$$\begin{aligned}
\langle \boldsymbol{\sigma} \rangle \frac{\partial \mathbf{M}^*}{\partial \bar{p}^{(k)}} \langle \boldsymbol{\sigma} \rangle &= \langle \boldsymbol{\sigma} \rangle \frac{\partial \mathbf{M}^*}{\partial M_{ijmn}^{(k)}} \langle \boldsymbol{\sigma} \rangle \frac{\partial M_{ijmn}^{(k)}}{\partial \bar{p}^{(k)}} \\
&= c^{(k)} \langle \sigma_{ij} \sigma_{mn} \rangle^{(k)} 3N_{1|ijmn} = 9 \langle \sigma_0^2 \rangle^{(k)} \qquad (13.16) \\
\langle \boldsymbol{\sigma} \rangle \frac{\partial \mathbf{M}^*}{\partial \bar{q}^{(k)}} \langle \boldsymbol{\sigma} \rangle &= \langle \boldsymbol{\sigma} \rangle \frac{\partial \mathbf{M}^*}{\partial M_{ijmn}^{(k)}} \langle \boldsymbol{\sigma} \rangle \frac{\partial M_{ijmn}^{(k)}}{\partial \bar{q}^{(k)}} \\
&= c^{(k)} \langle \sigma_{ij} \sigma_{mn} \rangle^{(k)} 2N_{2|ijmn} = 2 \langle s_{ij} s_{ij} \rangle^{(k)}. \qquad (13.17)
\end{aligned}$$

Then, since the variations $\delta\bar{p}^{(k)}$ and $\delta\bar{q}^{(k)}$ are mutually independent, we can get from (13.13) and (13.14) the representations for the stress invariants averaged over the k's phase

$$\langle \sigma_0^2 \rangle^{(k)} = \frac{1}{9c^{(k)}} \frac{\partial \mathbf{M}^*}{\partial \bar{p}^{(k)}} \langle \boldsymbol{\sigma} \rangle \otimes \langle \boldsymbol{\sigma} \rangle - \frac{2}{9c^{(k)}} \frac{\partial W^*}{\partial \bar{p}^{(k)}} + \frac{2}{9c^{(k)}} \frac{\partial \boldsymbol{\beta}^*}{\partial \bar{p}^{(k)}} \langle \boldsymbol{\sigma} \rangle, \quad (13.18)$$

$$\langle s_{ij} s_{ij} \rangle^{(k)} = \frac{1}{2c^{(k)}} \frac{\partial \mathbf{M}^*}{\partial \bar{q}^{(k)}} \langle \boldsymbol{\sigma} \rangle \otimes \langle \boldsymbol{\sigma} \rangle - \frac{1}{c^{(k)}} \frac{\partial W^*}{\partial \bar{q}^{(k)}} + \frac{1}{c^{(k)}} \frac{\partial \boldsymbol{\beta}^*}{\partial \bar{q}^{(k)}} \langle \boldsymbol{\sigma} \rangle, \quad (13.19)$$

where the first, second, and third terms in the right-hand sides of Eqs. (13.18) and (13.19) comprise the appropriate representations for the stress invariants of the corresponding fourth-rank tensors of stresses $\langle \boldsymbol{\sigma}^I \otimes \boldsymbol{\sigma}^I \rangle^{(k)}$, $\langle \boldsymbol{\sigma}^{II} \otimes \boldsymbol{\sigma}^{II} \rangle^{(k)}$, and $\langle \boldsymbol{\sigma}^I \otimes \boldsymbol{\sigma}^{II} \rangle^{(k)}$, respectively, averaged over the volume of the k's phase.

Of course, analogous relations can be derived for the second moment of strains, either through the transformations of Eq. (13.1)–(13.3) into $\boldsymbol{\varepsilon}$-expressions or by recasting of Eqs. (13.11) and (13.12) in $\boldsymbol{\varepsilon}$-terms by the use of the constituent equations (6.37) and (6.38). For example, for pure mechanical stresses $\boldsymbol{\beta}(\mathbf{x}) \equiv \mathbf{0}$), we get

$$\langle \boldsymbol{\varepsilon}^I \otimes \boldsymbol{\varepsilon}^I \rangle^{(k)} = \frac{1}{c^{(k)}} \frac{\partial \mathbf{L}^*}{\partial \mathbf{L}^{(k)}} \langle \boldsymbol{\varepsilon} \rangle \otimes \langle \boldsymbol{\varepsilon} \rangle, \quad (13.20)$$

$$\langle (\varepsilon_0^I)^2 \rangle^{(k)} = \frac{1}{9c^{(k)}} \frac{\partial \mathbf{L}^*}{\partial \bar{k}^{(k)}} \langle \boldsymbol{\varepsilon} \rangle \otimes \langle \boldsymbol{\varepsilon} \rangle, \quad \langle \varepsilon_{\rm eq}^2 \rangle^{(k)} = \frac{1}{3c^{(k)}} \frac{\partial \mathbf{L}^*}{\partial \bar{\mu}^{(k)}} \langle \boldsymbol{\varepsilon} \rangle \otimes \langle \boldsymbol{\varepsilon} \rangle, \quad (13.21)$$

where $\varepsilon_0 = \varepsilon_{ii}/3$, $\varepsilon_{\rm eq}^2 = 2{\rm e}_{ij}{\rm e}_{ij}/3$, $\mathbf{e} = \boldsymbol{\varepsilon} - \varepsilon_0 \boldsymbol{\delta}$, and $3\bar{k}^{(k)} = L_{iijj}^{(k)}/3$, $2\bar{\mu}^{(k)} = (L_{ijij}^{(k)} - 3\bar{k}^{(k)})/5$. It should be mentioned that we did not assume isotropy of the tensors $\mathbf{M}^{(k)}$, $\mathbf{L}^{(k)}$, and, because of this, $3\bar{p}^{(k)} \neq (3\bar{k}^{(k)})^{-1}$, $2\bar{q}^{(k)} \neq (2\bar{\mu}^{(k)})^{-1}$ in general. If the elasticity tensor of the k's phase is isotropic $\mathbf{L}^{(k)} = 3k^{(k)}\mathbf{N}_1 + 2\mu^{(k)}\mathbf{N}_2$ then $\bar{p}^{(k)} = p^{(k)}$, $\bar{q}^{(k)} = q^{(k)}$, $\bar{k}^{(k)} = k^{(k)}$, $\bar{\mu}^{(k)} = \mu^{(k)}$, and $3p^{(k)} = (3k^{(k)})^{-1}$, $2q^{(k)} = (2\mu^{(k)})^{-1}$.

Some particular results of relationships (13.15) and (13.20) were obtained in [860] (for $\boldsymbol{\beta} \equiv 0$) and [151] (for $\boldsymbol{\sigma}^0 \equiv \mathbf{0}$). A more tractable particular case of Eq. (13.15) is proposed in [477]. Buryachenko and Shermergor [194] considered the generalization of Eqs. (13.15) for thermo–elastic–electric fields. Relations (13.15) have been obtained for any degree of anisotropy of $\mathbf{M}^*$, $\mathbf{M}^{(i)}$ $(i = 0, 1, \ldots, N)$. For isotropic tensors $\mathbf{M}^{(i)}$ the relations (13.15) reduce to the results [90] and [602] for macroisotropic composites.

In Chapter 6 the exact representation of stress concentrator factors averaged over the volume of the constituents of the two-phase composites and statistically isotropic polycrystalline aggregates composed of crystals of some symmetries was obtained. It was found that exactly analogous results can be obtained in a general case of multiphase composites if the functional dependencies of effective constants on the constituent compliance are known. Indeed, following [602], we will vary a transformation field $\delta\boldsymbol{\beta}(\mathbf{x})$ at the fixed $\mathbf{M}(\mathbf{x})$ instead of variation of $\delta\mathbf{M}(\mathbf{x})$ at the fixed $\boldsymbol{\beta}(\mathbf{x})$. Then varying Eq. (13.3_1) leads to

$$\begin{aligned}\langle\boldsymbol{\sigma}\rangle\delta\boldsymbol{\beta}^* &= \langle\boldsymbol{\sigma}^I\delta\boldsymbol{\beta}\rangle + \langle\delta\boldsymbol{\sigma}^I\boldsymbol{\beta}\rangle = \langle\boldsymbol{\sigma}^I\delta\boldsymbol{\beta}\rangle + \langle\delta\boldsymbol{\sigma}^I\boldsymbol{\varepsilon}^{II}\rangle - \langle\delta\boldsymbol{\sigma}^I(\boldsymbol{\varepsilon}^{II} - \boldsymbol{\beta})\rangle = \langle\boldsymbol{\sigma}^I\delta\boldsymbol{\beta}\rangle \\ &+ \langle\delta\boldsymbol{\sigma}^I\boldsymbol{\varepsilon}^{II}\rangle - \langle\delta\boldsymbol{\varepsilon}^I\mathbf{L}(\boldsymbol{\varepsilon}^{II} - \boldsymbol{\beta})\rangle = \langle\boldsymbol{\sigma}^I\delta\boldsymbol{\beta}\rangle + \langle\delta\boldsymbol{\sigma}^I\boldsymbol{\varepsilon}^{II}\rangle - \langle\delta\boldsymbol{\varepsilon}^I\boldsymbol{\sigma}^{II}\rangle, \quad (13.22)\end{aligned}$$

where the second and the third term in the right-hand side vanish in virtue of (13.1) and (13.4_2) that reduce Eq. (13.22) to $\langle\boldsymbol{\sigma}^I\delta\boldsymbol{\beta}\rangle = \langle\boldsymbol{\sigma}\rangle\delta\boldsymbol{\beta}^*$. In a similar manner, variation of (13.3_2) by varying $\boldsymbol{\beta}(\mathbf{x})$ yields

$$\begin{aligned}-\delta W^* &= \langle\boldsymbol{\sigma}^{II}\delta\boldsymbol{\beta}\rangle + \langle\delta\boldsymbol{\sigma}^{II}\boldsymbol{\beta}\rangle = \langle\boldsymbol{\sigma}^{II}\delta\boldsymbol{\beta}\rangle + \langle\delta\boldsymbol{\sigma}^{II}\boldsymbol{\varepsilon}^{II}\rangle - \langle\delta\boldsymbol{\sigma}^{II}(\boldsymbol{\varepsilon}^{II} - \boldsymbol{\beta})\rangle \\ &= \langle\boldsymbol{\sigma}^{II}\delta\boldsymbol{\beta}\rangle + \langle\delta\boldsymbol{\sigma}^{II}\boldsymbol{\varepsilon}^{II}\rangle - \langle\delta(\boldsymbol{\varepsilon}^{II} - \boldsymbol{\beta})\boldsymbol{\sigma}^{II}\rangle = 2\langle\boldsymbol{\sigma}^{II}\delta\boldsymbol{\beta}\rangle + \langle\delta\boldsymbol{\sigma}^{II}\boldsymbol{\varepsilon}^{II}\rangle - \langle\boldsymbol{\sigma}^{II}\delta\boldsymbol{\varepsilon}^{II}\rangle,\end{aligned}$$

which can be simplified due to Eqs. (13.1) and (13.12_1) $\langle\boldsymbol{\sigma}^{II}\delta\boldsymbol{\beta}\rangle = -\delta W^*$. A special choice of the varying eigenstrains $(k = 0, 1, \ldots) : \delta\boldsymbol{\beta}(\mathbf{x}) \equiv \delta\boldsymbol{\beta}^{(k)}V^{(k)}(\mathbf{x})$, $\delta\boldsymbol{\beta}^{(k)} = \text{const}$ $(\mathbf{x} \in v^{(k)})$ analogously to (13.9) allows the explicit representations for the average stresses in the k's phase

$$\langle\boldsymbol{\sigma}^I\rangle = \frac{1}{c^{(k)}}\langle\boldsymbol{\sigma}\rangle\frac{\partial\boldsymbol{\beta}^*}{\partial\boldsymbol{\beta}^{(k)}}, \quad \langle\boldsymbol{\sigma}^{II}\rangle = -\frac{1}{c^{(k)}}\frac{\partial W^*}{\partial\boldsymbol{\beta}^{(k)}}, \tag{13.23}$$

under the assumption of fixed compliance $\mathbf{M}(\mathbf{x})$, $\mathbf{x} \in w$.

The preceding considerations show that in order to obtain a quantitative assessment of $\langle\boldsymbol{\sigma}\rangle^{(k)}$ and $\langle\boldsymbol{\sigma}\otimes\boldsymbol{\sigma}\rangle^{(k)}$ it is necessary to know the functional dependence of $\mathbf{M}^*$, W^*, $\boldsymbol{\beta}^*$ on both the compliance tensors $\mathbf{M}^{(k)}$ and eigenstrain $\boldsymbol{\beta}_1^{(k)}$. The dependencies can be calculated only approximately, except for some degenerate structures. Therefore one may think that the relationships are not indeed strictly exact. However, all of them can be measured experimentally, and thus the exact relationships obtained in this work have an immediate physical sense. Therefore eventually the error of numerical results for the first and second field moments is governed by the error of determining the dependencies of $\mathbb{M}^*$ and $\langle W \rangle$ upon $\mathbb{M}(\mathbf{x})$ and This dependence will be determined below with the help of MEFM, which is physically consistent and requires only weak assumptions.

13.1.2 Local Fluctuation of Stresses

From the relation (13.15) one may generally estimate the stress fluctuations at any point $\mathbf{x} \in v_i \quad (i = 1, 2, \ldots)$. For this purpose it is sufficient to choose $v_i' \subset v_i \quad (\mathbf{x} \in v_i', \quad i = 1, 2, \ldots)$ in such a manner, that $\overline{v}_i' \to 0$ and the variation $\delta\mathbf{M}$ differs from zero only in the v_i' region. So it is necessary to evaluate W^* with a compliance $\mathbf{M}(\mathbf{x})$ being changed into $\mathbf{M}(\mathbf{x}) \to \mathbf{M}(\mathbf{x}) + \delta\mathbf{M}(\mathbf{x})$. This problem may be solved approximately by using the MEFM.

Formulae (8.52)–(8.54) were derived on the basis of the two-particle approximation, where only binary interactions of inhomogeneities are taken into account. However, [184] have shown that irrespective of the order of approximation, the influence of the elastic properties of inhomogeneities on the effective parameter W^* is governed only by the tensors $\mathbf{R}_i$ and $\mathbf{F}_i$. Thus, in order to calculate $\partial W^*/\partial\mathbf{M}^{(i)}$, it is enough to determine $\partial\mathbf{R}^{(i)}/\partial\mathbf{M}^{(i)}, \quad \partial\mathbf{F}^{(i)}/\partial\mathbf{M}^{(i)}$, for which the perturbation method can be used. A variation of compliance tensor $\mathbf{M}^{(i)}(\mathbf{x}) \to \mathbf{M}^{(i)}(\mathbf{x}) + \delta\mathbf{M}^{(i)}(\mathbf{x})$ only in the region $\mathbf{x} \in v_i' \subset v_i$ (i.e. $\delta\mathbf{M}^{(i)}(\mathbf{x}) = 0$ for $\mathbf{x} \notin v_i'$) is found, with consequent variations in the stresses $\delta\boldsymbol{\sigma}(\mathbf{y}) \to \boldsymbol{\sigma}(\mathbf{y}) + \delta\boldsymbol{\sigma}(\mathbf{y}), \ \mathbf{y} \in v_i$. According to this we obtain from (8.18): $\langle\boldsymbol{\sigma} + \delta\boldsymbol{\sigma}\rangle_i = (\mathbf{I} - \mathbf{B}_i\mathbf{Q}_i\langle\delta\mathbf{M}^{(i)}\rangle_i)\mathbf{B}_i(\langle\overline{\boldsymbol{\sigma}}\rangle_i - \mathbf{Q}_i\beta_1^{(i)})$ with an accuracy of up to the fist-order term in $\delta\mathbf{M}^{(i)}$. Thus, the changes of $\mathbf{R}_i$ and $\mathbf{F}_i$ do not depend on the location of the variation region v_i', and within the MEFM the second moments of stress and strain fields are homogeneous over the inclusion volume. The last statement makes it possible to estimate not only the value $\langle\boldsymbol{\sigma}\otimes\boldsymbol{\sigma}\rangle_0$ but also the second moment of stress in the matrix in the vicinity of the inclusion boundary. According to (3.95) we find $(\mathbf{x} \in \partial v_i)$:

$$\begin{aligned}\langle\boldsymbol{\sigma}^-(\mathbf{n}) \otimes \boldsymbol{\sigma}^-(\mathbf{x})\rangle_x &= \langle[\mathbf{B}(\mathbf{n})\boldsymbol{\sigma}] \otimes [\mathbf{B}(\mathbf{n})\boldsymbol{\sigma}]\rangle_i + \langle\boldsymbol{\sigma}^-(\mathbf{n})\rangle_x \otimes \Big\{[\mathbf{B}(\mathbf{n}) - \mathbf{I}] \\ &\times (\mathbf{M}_1^{(i)})^{-1}\beta_1^{(i)}\Big\} + \Big\{[\mathbf{B}(\mathbf{n}) - \mathbf{I}](\mathbf{M}_1^{(i)})^{-1}\beta_1^{(i)}\Big\} \otimes [\mathbf{B}(\mathbf{n})\langle\boldsymbol{\sigma}\rangle_i], \qquad (13.24)\end{aligned}$$

where the first term on the right-hand side can be obtained by using Eq. (13.15). Expressions for $\langle\boldsymbol{\sigma}\rangle_i$ and $\langle\boldsymbol{\sigma}^-(\mathbf{n})\rangle_x$ are given by formulae (8.46) and (8.47), respectively. Relation (13.24) was proposed in [860] by a less general and more cumbersome method. It should be mentioned that for the calculation of the invariants $\langle\boldsymbol{\sigma}^-(\mathbf{n})\mathbf{N}_1\boldsymbol{\sigma}^-(\mathbf{n})\rangle_x$ and $\langle\boldsymbol{\sigma}^-(\mathbf{n})\mathbf{N}_2\boldsymbol{\sigma}^-(\mathbf{n})\rangle_x$ it is necessary to estimate all components of the averaged second moment of the stresses inside the inclusion v_i by Eq. (13.15) (in contrast to Eqs. (13.18) and (13.19)).

13.1.3 Correlation Function of Stresses

A general logic pattern for the calculation of n-point elastic stress moments $\langle\boldsymbol{\sigma}(\mathbf{x})_1\otimes\ldots\otimes\boldsymbol{\sigma}(\mathbf{x}_n)|v_1, \mathbf{x}_1; \ldots; v_n, \mathbf{x}_n; v_{n+1}, \mathbf{x}_{n+1}; \ldots; v_m, \mathbf{x}_m\rangle$ has been obtained in [125] by using MEFM. The closing of the infinite system of integral equations in two-particle approximation was made with the help of the following assumptions $(i = 1, 2; \quad j = 3 - i)$:

$$\langle \widetilde{\boldsymbol{\sigma}}_{i,j}(\mathbf{x}_i) \otimes \widetilde{\boldsymbol{\sigma}}_{i,j}(\mathbf{x}_j) \rangle = \langle \overline{\boldsymbol{\sigma}} \rangle_i \otimes \langle \overline{\boldsymbol{\sigma}} \rangle_j, \quad \langle \widetilde{\boldsymbol{\sigma}}_{i,j}(\mathbf{x}_i) \otimes \widetilde{\boldsymbol{\sigma}}_{i,j}(\mathbf{x}_i) \rangle = \langle \overline{\boldsymbol{\sigma}} \otimes \overline{\boldsymbol{\sigma}} \rangle_i. \quad (13.25)$$

The assumption of no correlation between the values of the field $\widetilde{\boldsymbol{\sigma}}(\mathbf{x})_{ij}$ at different points (13.25_1) does not mean a lack of correlation between the stresses $\boldsymbol{\sigma}(\mathbf{x}_i)$ and $\boldsymbol{\sigma}(\mathbf{x}_j)$. The average inclusion stresses $\langle \boldsymbol{\sigma} \rangle_i$ may be obtained by using formula (8.44). For the calculation of a second one-point effective stress moment $\langle \boldsymbol{\sigma} \otimes \boldsymbol{\sigma} \rangle_i$ [125] applied, a cumbersome approximate method. Now we can use the exact relation (13.15) for this purpose.

For two fixed inhomogeneities $v_i, \quad v_j$ we have according to formulae (8.18) and (8.25):

$$\boldsymbol{\sigma}(\mathbf{x}_i) = \mathbf{B}_i \Big\{ \mathbf{R}_j^{-1} \sum_{j=1}^{2} \mathbf{Z}_{ij} \left[\mathbf{R}_j \widetilde{\boldsymbol{\sigma}}_{i,j}(\mathbf{x}_j) + \mathbf{F}_j \right] - \mathbf{F}_j \Big\} + \mathbf{C}_i. \quad (13.26)$$

Here the $\boldsymbol{\sigma}(\mathbf{x}_i), \quad \widetilde{\boldsymbol{\sigma}}_{i,j}(\mathbf{x}_j) \quad (i = 1, 2; \quad j = 3 - i)$ are random fields. The statistical moment of the stresses inside the inhomogeneity v_i is obtained by taking the tensor product of (13.26)

$$\begin{aligned} \langle \boldsymbol{\sigma}(\mathbf{x}_1) \otimes \boldsymbol{\sigma}(\mathbf{x}_j) \rangle = \Big\langle \Big\{ \mathbf{B}_i \Big[\mathbf{R}_i^{-1} \sum_{j=1}^{2} \mathbf{Z}_{ij} (\mathbf{R}_j \widetilde{\boldsymbol{\sigma}}_{i,j}(\mathbf{x}_j) + \mathbf{F}_j) - \mathbf{F}_i \Big] + \mathbf{C}_i \Big\} \\ \otimes \Big\{ \mathbf{B}_j \Big[\mathbf{R}_j^{-1} \sum_{i=1}^{2} \mathbf{Z}_{ij} (\mathbf{R}_i \widetilde{\boldsymbol{\sigma}}_{i,j}(\mathbf{x}_i) + \mathbf{F}_i) - \mathbf{F}_j \Big] + \mathbf{C}_j \Big\} \Big\rangle \quad (13.27) \end{aligned}$$

$(i = 1, 2; \quad j = |i - 2| + 1)$. According to (8.18) and (13.25) we use the following assumptions to modify (13.27):

$$\begin{aligned} [\mathbf{R}_i \widetilde{\boldsymbol{\sigma}}_{i,j}(\mathbf{x}_i) + \mathbf{F}_i] &\otimes [\mathbf{R}_i \widetilde{\boldsymbol{\sigma}}_{i,j}(\mathbf{x}_i) + \mathbf{F}_i] \\ &= \overline{v}_i^2 \langle [\mathbf{M}_1^{(i)} \boldsymbol{\sigma}(\mathbf{x}) + \boldsymbol{\beta}_1^{(i)}] \otimes [\mathbf{M}_1^{(i)} \boldsymbol{\sigma}(\mathbf{x}) + \boldsymbol{\beta}_1^{(i)}] \rangle_i, \\ [\mathbf{R}_i \widetilde{\boldsymbol{\sigma}}(\mathbf{x}_i)_{ij} + \mathbf{F}_i] &\otimes [\mathbf{R}_j \widetilde{\boldsymbol{\sigma}}(\mathbf{x}_j)_{ij} + \mathbf{F}_j] \\ &= \overline{v}_i \overline{v}_j [\mathbf{M}_1^{(i)} \langle \boldsymbol{\sigma} \rangle_i + \boldsymbol{\beta}_1^{(i)}] \otimes [\mathbf{M}_1^{(j)} \langle \boldsymbol{\sigma} \rangle_j + \boldsymbol{\beta}_1^{(j)}]. \quad (13.28) \end{aligned}$$

These equations are exact for the limiting case $|\mathbf{x}_i - \mathbf{x}_j| \gg \max a_m^k \quad (k = 1, 2, 3; m = i, j)$. Then from (13.27) with (13.28) we find

$$\begin{aligned} \langle \boldsymbol{\sigma}(\mathbf{x}_i) \otimes \boldsymbol{\sigma}(\mathbf{x}_j) \rangle = {} & \mathbf{S}_{ii}^{\mathrm{h}} \otimes \mathbf{S}_{ji}^{\mathrm{h}} [\langle \boldsymbol{\sigma} \otimes \boldsymbol{\sigma} \rangle_i + \langle \boldsymbol{\sigma} \rangle_i \otimes \boldsymbol{\Gamma}_i^{\mathrm{h}} + \boldsymbol{\Gamma}_i^{\mathrm{h}} \otimes \langle \boldsymbol{\sigma} \rangle_i + \boldsymbol{\Gamma}_i^{\mathrm{h}} \otimes \boldsymbol{\Gamma}_j^{\mathrm{h}}] \\ & + \mathbf{S}_{ij}^{\mathrm{h}} \otimes \mathbf{S}_{jj}^{\mathrm{h}} [\langle \boldsymbol{\sigma} \otimes \boldsymbol{\sigma} \rangle_j + \langle \boldsymbol{\sigma} \rangle_j \otimes \boldsymbol{\Gamma}_j^{\mathrm{h}} + \boldsymbol{\Gamma}_j^{\mathrm{h}} \otimes \langle \boldsymbol{\sigma} \rangle_j + \boldsymbol{\Gamma}_j^{\mathrm{h}} \otimes \boldsymbol{\Gamma}_j^{\mathrm{h}}] \\ & + \mathbf{S}_{ii}^{\mathrm{h}} \otimes \mathbf{S}_{jj}^{\mathrm{h}} [\langle \boldsymbol{\sigma} \rangle_i \otimes \langle \boldsymbol{\sigma} \rangle_j + \langle \boldsymbol{\sigma} \rangle_i \otimes \boldsymbol{\Gamma}_j^{\mathrm{h}} + \boldsymbol{\Gamma}_i^{\mathrm{h}} \otimes \langle \boldsymbol{\sigma} \rangle_j + \boldsymbol{\Gamma}_i^{\mathrm{h}} \otimes \boldsymbol{\Gamma}_j^{\mathrm{h}}] \\ & + \mathbf{S}_{ij}^{\mathrm{h}} \otimes \mathbf{S}_{ji}^{\mathrm{h}} [\langle \boldsymbol{\sigma} \rangle_j \otimes \langle \boldsymbol{\sigma} \rangle_i + \langle \boldsymbol{\sigma} \rangle_j \otimes \boldsymbol{\Gamma}_i^{\mathrm{h}} + \boldsymbol{\Gamma}_j^{\mathrm{h}} \otimes \langle \boldsymbol{\sigma} \rangle_i + \boldsymbol{\Gamma}_j^{\mathrm{h}} \otimes \boldsymbol{\Gamma}_i^{\mathrm{h}}] \\ & - \boldsymbol{\Gamma}_i^{\mathrm{h}} \otimes \mathbf{S}_{ji}^{\mathrm{h}} (\langle \boldsymbol{\sigma} \rangle_i + \boldsymbol{\Gamma}_i^{\mathrm{h}}) - \boldsymbol{\Gamma}_i^{\mathrm{h}} \otimes \mathbf{S}_{jj}^{\mathrm{h}} (\langle \boldsymbol{\sigma} \rangle_j + \boldsymbol{\Gamma}_j^{\mathrm{h}}) + \boldsymbol{\Gamma}_i^{\mathrm{h}} \otimes \boldsymbol{\Gamma}_j^{\mathrm{h}} \\ & - [\mathbf{S}_{ii}^{\mathrm{h}} (\langle \boldsymbol{\sigma} \rangle_i + \boldsymbol{\Gamma}_i^{\mathrm{h}})] \otimes \boldsymbol{\Gamma}_j^{\mathrm{h}} - [\mathbf{S}_{ij}^{\mathrm{h}} (\langle \boldsymbol{\sigma} \rangle_j + \boldsymbol{\Gamma}_j^{\mathrm{h}})] \otimes \boldsymbol{\Gamma}_j^{\mathrm{h}}, \quad (13.29) \end{aligned}$$

where we have used the notation $\mathbf{S}^{\rm h}_{ij} \equiv [\overline{v}_i \mathbf{M}_1^{(i)}]^{-1} \mathbf{Z}_{ij} \mathbf{M}_1^{(j)} \overline{v}_j$, $\boldsymbol{\Gamma}^{\rm h}_i \equiv [\mathbf{M}_1^{(i)}]^{-1} \boldsymbol{\beta}_1^{(i)}$, $(i,j=1,2)$, with the superscript $^{\rm h}$ indicated an estimation of the tensors $\mathbf{S}^{\rm h}_{ij}$ and $\boldsymbol{\Gamma}^{\rm h}_i$ for the homogeneous inclusions. Te following tensor product notation has been applied

$$[(\mathbf{B}_1 \otimes \mathbf{B}_2)(\mathbf{b}_1 \otimes \mathbf{b}_2)]_{pqrs} = B_{1|pqkl} b_{1|kl} B_{2|rsmn} b_{2|mn},$$
$$[\mathbf{b}_1 \otimes \mathbf{B}_2 \mathbf{b}_2]_{pqrs} = b_{1|pq} B_{2|rskl} b_{2|kl}, \quad [(\mathbf{B}_1 \mathbf{b}_1) \otimes \mathbf{b}_2]_{pqrs} = B_{1|pqkl} b_{1|kl} b_{2|rs}, \quad (13.30)$$

where $\mathbf{B}_1$, $\mathbf{B}_2$ and $\mathbf{b}_1$, $\mathbf{b}_2$ stand for any fourth- and second-order tensor, respectively. On the right-hand side of equation (13.29) the values $\langle \boldsymbol{\sigma} \otimes \boldsymbol{\sigma} \rangle_i$ and $\langle \boldsymbol{\sigma} \rangle_i$ $(i=1,2)$ are obtained from (13.15) with (8.52)–(8.54) and from (8.44).

At $|\mathbf{x}_i - \mathbf{x}_j| \to \infty$ $(i=1,2;\ j=3-i)$ the limiting relations $\mathbf{S}^{\rm h}_{ii} \to \mathbf{I}$, $\mathbf{S}^{\rm h}_{ij} \to 0$ hold, and from (13.29) we obtain an obvious result:

$$\langle \boldsymbol{\sigma}(\mathbf{x}_i) \otimes \boldsymbol{\sigma}(\mathbf{x}_j) \rangle \to \langle \boldsymbol{\sigma} \rangle_i \otimes \langle \boldsymbol{\sigma} \rangle_j, \quad |\mathbf{x}_i - \mathbf{x}_j| \to \infty. \quad (13.31)$$

13.1.4 Numerical Results and Discussions

If analytical representations for the effective moduli are known, it's very easy to find the invariants of the fourth-order tensor $\langle \boldsymbol{\sigma} \otimes \boldsymbol{\sigma} \rangle^{(k)}$. For example, for the porous matrix materials with identical spherical pores and uniformly oriented identical microcracks, we will get the representations of the second invariants of deviator stresses in the matrix utilizing the appropriate estimations of $\mathbf{M}^* = (\mathbf{L}^*)^{-1}$ via MEFM (9.106) and (9.108):

$$\langle s_{ij} s_{ij} \rangle^{(0)} = \frac{9c}{2(1-c)} \left(1 - \frac{29}{24} c\right)^{-1} \langle \sigma_0^2 \rangle + \left(1 - \frac{35}{24} c\right)\left(1 + \frac{5}{24}\right) \langle s_{ij} \rangle \langle s_{ij} \rangle, \quad (13.32)$$
$$\langle s_{ij} s_{ij} \rangle^{(0)} = 2c \left(1 - \frac{16}{15\pi^2} c\right)^{-1} \langle \sigma_0^2 \rangle + \left(1 - \frac{448}{375\pi^2} c\right)\left(1 - \frac{848}{375\pi^2}\right) \langle s_{ij} \rangle \langle s_{ij} \rangle, \quad (13.33)$$

respectively. Here $c = 4\pi a^3 n/3$ are defined by the maximum semi-axis a of both the spherical pores and coin-shaped microcracks.

Let us now demonstrate the application of the theoretical results by considering a more general isotropic composite made of WC and Co and subjected only residual stresses $\langle \boldsymbol{\sigma} \rangle \equiv \mathbf{0}$. Both components are described via isotropic elastic properties to $\mathbf{L}(\mathbf{x}) = 3k(\mathbf{x})\mathbf{N}_1 + 2\mu(\mathbf{x})\mathbf{N}_2$, $\boldsymbol{\beta}(\mathbf{x}) = \beta_0(\mathbf{x})\boldsymbol{\delta}$. The thermoelastic constants as usually found in literature [602] are for the WC matrix $k^{(0)} = 389$ GPa, $\mu^{(0)} = 292$ GPa, $\beta_0^{(0)} = -0.003$, and for Co inclusions $k^{(1)} = 167$ GPa, $\mu^{(1)} = 77$ GPa, $\beta_0^{(1)} = -0.006$. The thermal strains have been calculated with the assumption that internal stresses do not relax by creep of Co below $T_0 = 800K$, i.e. we have put $T - T_0 = -500K$. For a low volume fraction of Co one may describe the composite as a WC matrix with Co particles embedded in it. The Co particles are assumed to be approximately spheres with radius a. Two alternative radial distribution functions of inclusion will be examined (9.91) and (9.92). For an isotropic matrix and spherical inclusions the tensors $\mathbf{Q}$ (3.168) and $\mathbf{B}$ (3.170) are known. The expression for the tensor $\mathbf{T}_{ij}(\mathbf{x}_i - \mathbf{x}_j)$ is presented through the tensor $\mathbf{T}^S_{ij}(\mathbf{x}_i - \mathbf{x}_j)$ by the formula (A3.1) of Appendix A.3.1.

At first we calculated the average residual stress inside the inclusions as a function of the concentration c. It turns out that the estimations via the one-particle approximation of MEFM (8.83) (coinciding in the considered case with the Mori-Tanaka method (MTM) (8.95)) differs from the calculations according to relation (8.87) with (9.91) and (9.92) on 3.3% and 6.2%, respectively, at $c = 0.5$. Thus, there is a weak influence of the binary interaction of the inclusions on average stresses inside particles. The situation changes qualitatively when one considers the stress fluctuations inside inclusions $\Delta\boldsymbol{\sigma}_1^2 \equiv \langle\boldsymbol{\sigma}\otimes\boldsymbol{\sigma}\rangle_1 - \langle\boldsymbol{\sigma}\rangle_1\otimes\langle\boldsymbol{\sigma}\rangle_1$. In this case $\Delta\boldsymbol{\sigma}_1^2 \equiv \mathbf{0}$ if the binary interaction is not considered (8.95). This general statement was proved in [602]. If the pair interaction is taken into account, we find $\Delta\boldsymbol{\sigma}_1^2 \not\equiv \mathbf{0}$ and moreover $\Delta\sigma^2_{1|ijkl} \neq 0$ $(i = k \neq l,\ j = l)$. The values $(\Delta\sigma^2_{1|1212})^{1/2}$ and $(\Delta\sigma^2_{1|1111})^{1/2}$ shown in Fig. 13.1 are calculated according to the relations (13.14) with (8.54) and (9.92) or (9.91). Thus we draw the conclusion that MTM (8.95) is suitable mainly for approximate estimation of effective constants but is debatable in the case of field fluctuation estimations.

Let us now study the stress distribution inside the inhomogeneities as a function of the inclusion shape. The average stresses inside the inclusions are presented in Fig. 9.9 of Chapter 9. We assume spheroidal particles with an aspect ratio a_1^1/a_1^3 for prolate spheroids and a_1^3/a_1^1 for oblate ones $(a_1^1 = a_1^2)$. The axes $(\mathbf{x}^1, \mathbf{x}^2, \mathbf{x}^3)$ attached to the crystal lattice are aligned parallel with the global Cartesian reference frame. Fig. 13.2 shows the values $\Delta\sigma^2_{1|3333}$ and $\Delta\sigma^2_{1|1111}$ (determining by the binary interactions of inclusions only) versus shape of

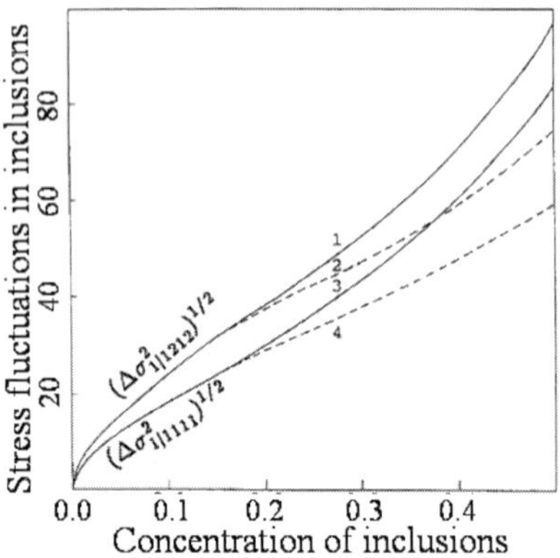

13.1. Stress fluctuations inside inclusions (MPa) $(\Delta\sigma^2_{1|1212})^{1/2}$ (1, 2) and $(\Delta\sigma^2_{1|1111})^{1/2}$ (3, 4) for RDF (9.91) (dashed curves) (9.92) (solid curves).

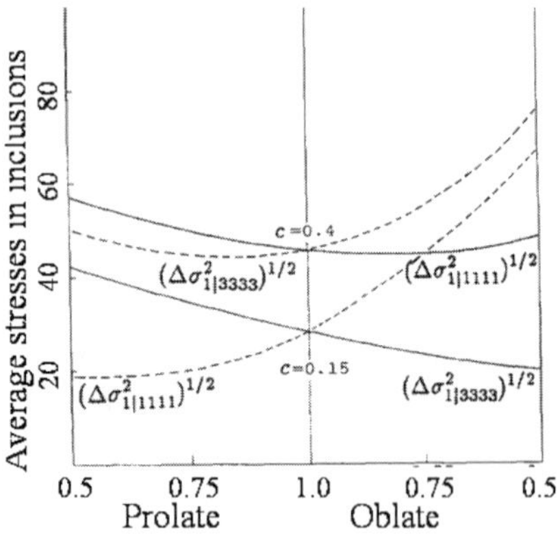

Fig. 13.2. Stress fluctuations (MPa) inside inclusions as a function of particle shape. $(\Delta\sigma^2_{1|1111})^{1/2}$ (solid lines) and $(\Delta\sigma^2_{1|3333})^{1/2}$ (dashed lines) were plotted for c=0.4 and c=0.15.

inclusions for the two particle concentration $c = 0.4$ and $c = 0.15$. In all cases stress fluctuations are larger in the direction of smallest inclusion size than in the other directions. The accuracy of the calculations decreases in the limiting cases of spheroids $a^1/a^3 \to 0$ or $\to \infty$ since the point approximation assumption of (4.64) then becomes uncorrected.

For spherical inclusions the average values $\langle\sigma_{11}^{-}(n)\rangle_x$, $\langle\sigma_{22}^{-}(n)\rangle_x$, $\langle\sigma_{33}^{-}(n)\rangle_x$ were found according to the formulae statistical average $\langle\boldsymbol{\sigma}^{-}(\mathbf{n})\rangle_x$ (8.47), (9.91) as a function of the polar angle θ with $\mathbf{n} = (\sin\theta, 0, \cos\theta)$ and $c = 0.15$. Figure 13.3 presents the fluctuations $\Delta\boldsymbol{\sigma}^{-}(\mathbf{n})^2 \equiv \langle\boldsymbol{\sigma}^{-}(\mathbf{n}) \otimes \boldsymbol{\sigma}^{-}(\mathbf{n})\rangle_x - \langle\boldsymbol{\sigma}^{-}(\mathbf{n})\rangle_x \otimes \langle\boldsymbol{\sigma}^{-}(\mathbf{n})\rangle_x$ of associated stress components.

Now we pass to the examination of the correlation function of internal stresses (13.29) by using the formulae (13.12_1), (8.54), (8.44), (9.91) and (A3.1). Let us define $\Delta\sigma_{ij}(\mathbf{x}_1)\sigma_{kl}(\mathbf{x}_2) \equiv \langle\sigma_{ij}(\mathbf{x}_1)\sigma_{kl}(\mathbf{x}_2)\rangle - \langle\sigma_{ij}(\mathbf{x}_1)\rangle_1\langle\sigma_{kl}(\mathbf{x}_2)\rangle_2$, $\mathbf{x}_1 \in v_1$, $\mathbf{x}_2 \in v_2$. Fig. 13.4 shows three components of $\Delta\boldsymbol{\sigma}(\mathbf{x}_1) \otimes \boldsymbol{\sigma}(\mathbf{x}_2)$ as a function of the distance between the spherical inclusions. From this figure we notice that the equality (13.31) is accurate to 15% when the spacing between inhomogeneities is as large as 6 times inclusion radius.

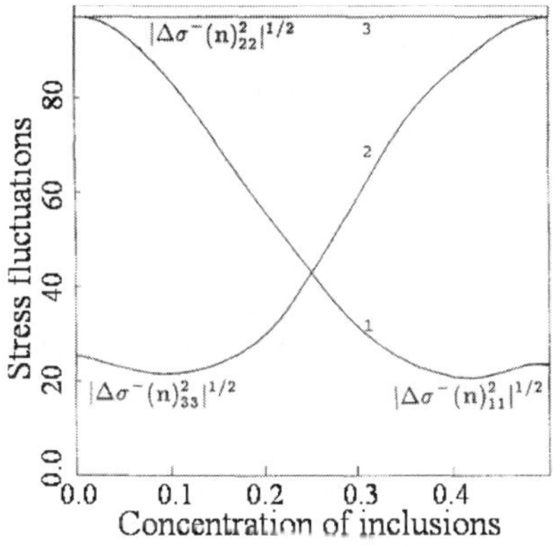

Fig. 13.3. The components of stress fluctuations (MPa) $|\Delta\sigma_{11}^{-}(n)^2|^{1/2}$ (1), $|\Delta\sigma_{33}^{-}(n)^2|^{1/2}$ (2), $|\Delta\sigma_{22}^{-}(n)^2|^{1/2}$ (3) versus the polar angle $\theta(\pi)$.

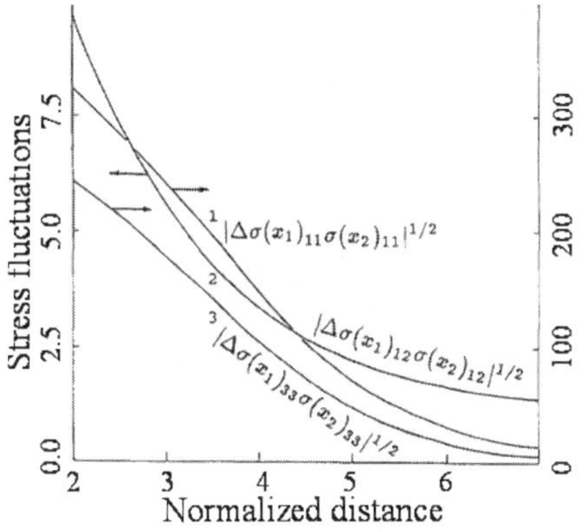

Fig. 13.4. $|\Delta\sigma_{11}(x_1)\sigma_{11}(x_2)|^{1/2}$(1), $|\Delta\sigma_{12}(x_1)\sigma_{12}(x_2)|^{1/2}$(2), $|\Delta\sigma_{33}(x_1)\sigma_{33}(x_2)|^{1/2}$ (3) (MPa) vs r/a.

13.2 Method of Integral Equations

It was demonstrated in Section 13.1 that the perturbation method allows the expressions for the second moment of stresses averaged over the phase volumes. However, if for example the perturbation method is used leading to Eq. (13.15) even in simplest of the considered cases of actually isotropic matrix materials it is necessary to estimate the effective compliance $\mathbf{M}^*$ for the general case of anisotropy of the matrix. This is not required for the method of integral equations considered below, which also makes it possible to estimate inhomogeneity of stress fluctuations inside the inclusions.

13.2.1 Estimation of the Second Moment of Effective Stresses

To obtain the second moments of effective stresses in the component $v^{(i)}$ of the inclusions ($i = 1, 2, \ldots.$) it is necessary to put the tensor product of (8.5) at $n = 1$ into $\overline{\boldsymbol{\sigma}}(\mathbf{x})$, $\quad \mathbf{x} \in v_i$:

$$\overline{\boldsymbol{\sigma}}(\mathbf{x}) \otimes \overline{\boldsymbol{\sigma}}(\mathbf{x}) = \boldsymbol{\sigma}^0 \otimes \boldsymbol{\sigma}^0 + \boldsymbol{\sigma}^0 \otimes \int \boldsymbol{\Gamma}(\mathbf{x}-\mathbf{x}_p)[\boldsymbol{\eta}(\mathbf{x}_p)V_p(\mathbf{x}_p) - \langle \boldsymbol{\eta} \rangle]\, d\mathbf{x}_p$$
$$+ \int \boldsymbol{\Gamma}(\mathbf{x}-\mathbf{x}_q)[\boldsymbol{\eta}(\mathbf{x}_q)V_q(\mathbf{x}_q) - \langle \boldsymbol{\eta} \rangle]\, d\mathbf{x}_q \otimes \boldsymbol{\sigma}^0 + \int\int \boldsymbol{\Gamma}(\mathbf{x}-\mathbf{x}_p)$$
$$\times [\boldsymbol{\eta}(\mathbf{x}_p)V_p(\mathbf{x}_p) - \langle \boldsymbol{\eta} \rangle] \otimes \Gamma(\mathbf{x}-\mathbf{x}_q)[\boldsymbol{\eta}(\mathbf{x}_q)V_q(\mathbf{x}_q) - \langle \boldsymbol{\eta} \rangle]\, d\mathbf{x}_p\, d\mathbf{x}_q. \quad (13.34)$$

The right-hand side of Eq. (13.34) is a random function of the arrangements of surrounding inclusions $v_p,\ v_q \neq v_i \quad (p,q = 1,2,\ldots)$. Averaging (13.34) over a realization ensemble leads to the equation containing double-point and triple-point conditional probability densities in which the terms with $\mathbf{x}_p = \mathbf{x}_q$ under the condition $\mathbf{x}_p \neq \mathbf{x}_i, \quad \mathbf{x}_q \neq \mathbf{x}_i$ may be isolated with the help of the equality $\varphi(v_p, \mathbf{x}_p, v_q, \mathbf{x}_q \mid; v_i, \mathbf{x}_i) = \delta(\mathbf{x}_p - \mathbf{x}_q)\varphi(v_p, \mathbf{x}_p \mid; v_i, \mathbf{x}_i) + \varphi(v_p, \mathbf{x}_p \mid; v_i, \mathbf{x}_i)\varphi(v_q, \mathbf{x}_q \mid ; v_p, \mathbf{x}_p; v_i, \mathbf{x}_i)$. Then the obtained equation can be simplified by the use of the assumption **H1** (8.14) and the known tensors (8.16) that allows us to get the expression for the averaged effective stress in the components:

$$\langle \overline{\boldsymbol{\sigma}} \otimes \overline{\boldsymbol{\sigma}} \rangle_i = \langle \overline{\boldsymbol{\sigma}} \rangle_i \otimes \langle \overline{\boldsymbol{\sigma}} \rangle_i + \int \langle [\mathbf{T}_{ip}(\mathbf{x}_i - \mathbf{x}_p)\boldsymbol{\eta}_p \bar{v}_p]$$
$$\otimes [\mathbf{T}_{ip}(\mathbf{x}_i - \mathbf{x}_p)\boldsymbol{\eta}_p \bar{v}_p] | v_p, \mathbf{x}_p; v_i, \mathbf{x}_i \rangle \varphi(v_p, \mathbf{x}_p \mid; v_i, \mathbf{x}_i)\, d\mathbf{x}_p$$
$$+ \int\int \Big\{ \langle [\mathbf{T}_{ip}(\mathbf{x}_i - \mathbf{x}_p)\boldsymbol{\eta}_p] \otimes [\mathbf{T}_{iq}(\mathbf{x}_i - \mathbf{x}_q)\boldsymbol{\eta}_q] | v_p, \mathbf{x}_p; v_q, \mathbf{x}_q; v_i, \mathbf{x}_i \rangle$$
$$\times \varphi(v_p, \mathbf{x}_p \mid; v_i, \mathbf{x}_i)\varphi(v_q, \mathbf{x}_q \mid; v_p, \mathbf{x}_p; v_i, \mathbf{x}_i)$$
$$- [\mathbf{T}_{ip}(\mathbf{x}_i - \mathbf{x}_p)\langle \boldsymbol{\eta}_p | v_p, \mathbf{x}_p; v_i, \mathbf{x}_i \rangle] \otimes [\mathbf{T}_{iq}(\mathbf{x}_i - \mathbf{x}_q)\langle \boldsymbol{\eta}_q | v_q, \mathbf{x}_q; v_i, \mathbf{x}_i \rangle]$$
$$\times \varphi(v_p, \mathbf{x}_p \mid; v_i, \mathbf{x}_i)\varphi(v_q, \mathbf{x}_q \mid; v_i, \mathbf{x}_i) \Big\} \bar{v}_p \bar{v}_q\, d\mathbf{x}_q\, d\mathbf{x}_p. \quad (13.35)$$

From equation (13.35) one can see that the effective stress dispersion

$$\Delta \overline{\sigma}^{(i)2}_{klmn} \equiv \langle \overline{\sigma}_{kl} \overline{\sigma}_{mn} \rangle_i - \langle \overline{\sigma}_{kl} \rangle_i \langle \overline{\sigma}_{mn} \rangle_i \quad (13.36)$$

presents a determined homogeneous function inside the components $v_i, \quad (i = 1,\ldots,N)$.

13.2.2 Implicit Representations for the Second Moment of Stresses

The effective stress dispersion (13.36) will be used for the estimation of the stress dispersion:

$$\Delta \boldsymbol{\sigma}^{(i)2}(\mathbf{x}) \equiv \langle \boldsymbol{\sigma} \otimes \boldsymbol{\sigma} \rangle_i(\mathbf{x}) - \langle \boldsymbol{\sigma} \rangle_i(\mathbf{x}) \otimes \langle \boldsymbol{\sigma} \rangle_i(\mathbf{x}) = \mathbf{B}_i(\mathbf{x}) \Delta \overline{\boldsymbol{\sigma}}^{(i)2} \mathbf{B}_i^\top(\mathbf{x}), \quad (13.37)$$

which is an inhomogeneous function of the coordinates in the components $v^{(i)}$. The formula (13.37) allows estimation of the averaged stress dispersion in the inclusion volume $\Delta \boldsymbol{\sigma}^{(i)2} = \langle \Delta^{(i)2}(\mathbf{x}) \rangle_i$. By the use of the known properties of the second statistical moments, (13.37) and (13.35), a relation for the average second moment of stresses in the component v_i is achieved:

$$\langle \boldsymbol{\sigma} \otimes \boldsymbol{\sigma} \rangle_i(\mathbf{x}) = \langle \boldsymbol{\sigma} \rangle_i(\mathbf{x}) \otimes \langle \boldsymbol{\sigma} \rangle_i(\mathbf{x})$$

$$
\begin{aligned}
&+\int \langle [\mathbf{B}_i(\mathbf{x})\mathbf{T}_{ip}(\mathbf{x}_i-\mathbf{x}_p)\boldsymbol{\eta}_p]\otimes[\mathbf{B}_i(\mathbf{x})\mathbf{T}_{ip}(\mathbf{x}_i-\mathbf{x}_p)\boldsymbol{\eta}_p]|v_p,\mathbf{x}_p;v_i,\mathbf{x}_i\rangle\\
&\times\bar{v}_p^2\varphi(v_p,\mathbf{x}_p\ |;v_i,\mathbf{x}_i)\ d\mathbf{x}_p+\bar{v}_p\bar{v}_q\int\int\Big\{\langle [\mathbf{B}_i(\mathbf{x})\mathbf{T}_{ip}(\mathbf{x}_i-\mathbf{x}_p)\boldsymbol{\eta}_p]\\
&\otimes[\mathbf{B}_i(\mathbf{x})\mathbf{T}_{iq}(\mathbf{x}_i-\mathbf{x}_q)\boldsymbol{\eta}_q]|v_p,\mathbf{x}_p;v_q,\mathbf{x}_q;v_i,\mathbf{x}_i\rangle\\
&\times\ \varphi(v_p,\mathbf{x}_p\ |;v_i,\mathbf{x}_i)\varphi(v_q,\mathbf{x}_q\ |;v_p,\mathbf{x}_p;v_i,\mathbf{x}_i)\\
&-\ [\mathbf{B}_i(\mathbf{x})\mathbf{T}_{ip}(\mathbf{x}_i-\mathbf{x}_p)\langle\boldsymbol{\eta}_p|v_p,\mathbf{x}_p;v_i,\mathbf{x}_i\rangle\varphi(v_p,\mathbf{x}_p\ |;v_i,\mathbf{x}_i)]\\
&\otimes[\mathbf{B}_i(\mathbf{x})\mathbf{T}_{iq}(\mathbf{x}_i-\mathbf{x}_q)\langle\boldsymbol{\eta}_q|v_q,\mathbf{x}_q;v_i,\mathbf{x}_i\rangle\varphi(v_q,\mathbf{x}_q\ |;v_i,\mathbf{x}_i)]\Big\}\ d\mathbf{x}_q\ d\mathbf{x}_p. \quad (13.38)
\end{aligned}
$$

The relation (13.38) is derived by the use of triple interaction of the inclusions. As may be seen from Eq. (13.38), the neglection of binary interaction is tantamount to assuming homogeneity of the stresses inside the component v_i $\langle\boldsymbol{\sigma}\otimes\boldsymbol{\sigma}\rangle_i(\mathbf{x}) = \langle\boldsymbol{\sigma}\rangle_i(\mathbf{x})\otimes\langle\boldsymbol{\sigma}\rangle_i(\mathbf{x})$. The following approximation for the second moment can be obtained by taking only binary interaction of the inclusions into account:

$$
\begin{aligned}
\langle\boldsymbol{\sigma}\otimes\boldsymbol{\sigma}\rangle_i(\mathbf{x}) &= \langle\boldsymbol{\sigma}\rangle_i(\mathbf{x})\otimes\langle\boldsymbol{\sigma}\rangle_i(\mathbf{x})+\int\langle[\mathbf{B}_i(\mathbf{x})\mathbf{T}_{ip}(\mathbf{x}_i-\mathbf{x}_p)\boldsymbol{\eta}_p\bar{v}_p]\\
&\otimes\ [\mathbf{B}_i(\mathbf{x})\mathbf{T}_{ip}(\mathbf{x}_i-\mathbf{x}_p)\boldsymbol{\eta}_q\bar{v}_q]|v_p,\mathbf{x}_p;v_i,\mathbf{x}_i\rangle\varphi(v_p,\mathbf{x}_p\ |;v_i,\mathbf{x}_i)\ d\mathbf{x}_p. \quad (13.39)
\end{aligned}
$$

Ignoring effective stress fluctuations in the inclusions v_p in the integral term (13.39) leads to

$$
\begin{aligned}
\langle\boldsymbol{\sigma}\otimes\boldsymbol{\sigma}\rangle_i(\mathbf{x}) &= \langle\boldsymbol{\sigma}\rangle_i(\mathbf{x})\otimes\langle\boldsymbol{\sigma}\rangle_i(\mathbf{x})+\int[\mathbf{B}_i(\mathbf{x})\mathbf{T}_{ip}(\mathbf{x}_i-\mathbf{x}_p)\langle\boldsymbol{\eta}\rangle_p\bar{v}_p]\\
&\otimes\ [\mathbf{B}_i(\mathbf{x})\mathbf{T}_{ip}(\mathbf{x}_i-\mathbf{x}_p)\langle\boldsymbol{\eta}\rangle_p\bar{v}_q]\varphi(v_p,\mathbf{x}_p|;v_i,\mathbf{x}_i)\ d\mathbf{x}_p \qquad (13.40)
\end{aligned}
$$

The relation (13.40) is known as "far field approximation".

In a similar manner it is possible to obtain the estimation of the second stress moment averaged over a volume of the matrix. At first we define the conditional probability density $\varphi(v_p,\mathbf{x}_p\ |;v_q,\mathbf{x}_q;\ldots;v_0,\mathbf{x}_0)$ under the condition that the inclusions $v_q,\ldots$ are located at points $\mathbf{x}_q,\ldots$, whereas the matrix material appears at the point $\mathbf{x}_0$. In analogy to the above considerations we obtain

$$
\begin{aligned}
\langle\boldsymbol{\sigma}\otimes\boldsymbol{\sigma}\rangle_0 &= \langle\boldsymbol{\sigma}\rangle_0\otimes\langle\boldsymbol{\sigma}\rangle_0+\int\langle[\mathbf{T}_p(\mathbf{x}_0-\mathbf{x}_p)\boldsymbol{\eta}_p\bar{v}_p]\\
&\otimes[\mathbf{T}_p(\mathbf{x}_0-\mathbf{x}_p)\boldsymbol{\eta}_p\bar{v}_p]|v_p,\mathbf{x}_p;v_0,\mathbf{x}_0\rangle\varphi(v_p,\mathbf{x}_p\ |;v_0,\mathbf{x}_0)\ d\mathbf{x}_p\\
&+\int\int\Big\{\langle[\mathbf{T}_p(\mathbf{x}_0-\mathbf{x}_p)\boldsymbol{\eta}_p]\otimes[\mathbf{T}_q(\mathbf{x}_0-\mathbf{x}_q)\boldsymbol{\eta}_q]|v_p,\mathbf{x}_p;v_q,\mathbf{x}_q;v_0,\mathbf{x}_0\rangle\\
&\times\varphi(v_p,\mathbf{x}_p\ |;v_0,\mathbf{x}_0)\varphi(v_q,\mathbf{x}_q\ |;v_p,\mathbf{x}_p;v_0,\mathbf{x}_0)\\
&-[\mathbf{T}_p(\mathbf{x}_0-\mathbf{x}_p)\langle\boldsymbol{\eta}_p|v_p,\mathbf{x}_p;v_i,\mathbf{x}_i\rangle\varphi(v_p,\mathbf{x}_p\ |;v_0,\mathbf{x}_0)]\\
&\otimes[\mathbf{T}_q(\mathbf{x}_0-\mathbf{x}_q)\langle\boldsymbol{\eta}_q|v_q,\mathbf{x}_q;v_i,\mathbf{x}_i\rangle\varphi(v_q,\mathbf{x}_q\ |;v_0,\mathbf{x}_0)]\Big\}\bar{v}_p\bar{v}_q\ d\mathbf{x}_q\ d\mathbf{x}_p. \quad (13.41)
\end{aligned}
$$

In contrast to (13.38) the second stress moment in the matrix, (13.41), does not depend on $\mathbf{x}_0\in v_0$. A more approximative estimation of the second moment $\langle\boldsymbol{\sigma}\otimes\boldsymbol{\sigma}\rangle_0$ can be obtained in direct analogy to (13.39), (13.40) by replacing v_i, $\mathbf{x}_i$ by v_0, $\mathbf{x}_0$.

13.2.3 Explicit Estimation of Second Moments of Stresses Inside the Phases

A general logic pattern for the calculation of n-point elastic stress moments $\langle \boldsymbol{\sigma}(\mathbf{x}_1)\otimes\ldots\otimes\boldsymbol{\sigma}(\mathbf{x}_n)|v_1,\mathbf{x}_1;\ldots;v_n,\mathbf{x}_n;v_{n+1},\mathbf{x}_{n+1};\ldots;v_m,\mathbf{x}_m\rangle$ has been obtained via the MEFM [125] by the use of a cumbersome approximation method for the calculation of a second one-point effective stress moment $\langle\boldsymbol{\sigma}\otimes\boldsymbol{\sigma}\rangle_i$. This method was improved in [151] (see Subsection 13.1.3) via the exact relation (13.15). For the same purpose we now use the approach based on the approximative integral equation (13.35) as well as on the effective field approximation (8.14) of the binary interaction of the inclusions.

For two fixed inhomogeneities $v_i,\ v_j$ we have, according to (8.18) and (8.25),

$$\overline{\boldsymbol{\sigma}}(\mathbf{x}_i)=\mathbf{R}_i^{-1}\Big[\sum_{j=1}^{2}\mathbf{Z}_{ij}\big(\mathbf{R}_j\widetilde{\boldsymbol{\sigma}}(\mathbf{x}_j)_{1,2}+\mathbf{F}_j\big)-\mathbf{F}_i\Big]. \tag{13.42}$$

Here $\overline{\boldsymbol{\sigma}}(\mathbf{x}_i)$ and $\widetilde{\boldsymbol{\sigma}}(\mathbf{x}_j)_{1,2}$ $(\mathbf{x}_i\in v_i,\ \ \mathbf{x}_j\in v_j,\ \ i,j=1,2)$ are homogeneous random fields in the inclusions $v_i,\ v_j$. The statistical moment of the effective stresses inside the inhomogeneity v_i is obtained by taking the tensor product of (13.42): $\langle\overline{\boldsymbol{\sigma}}\otimes\overline{\boldsymbol{\sigma}}\rangle_i$. According to (8.14) and (13.25) we use the following assumption to modify (13.35) $(i\neq j)$:

$$\langle\boldsymbol{\eta}_i\otimes\boldsymbol{\eta}_j\rangle=\langle\boldsymbol{\eta}_i\rangle\otimes\langle\boldsymbol{\eta}_j\rangle,\quad \langle\boldsymbol{\eta}_i\otimes\boldsymbol{\eta}_i\rangle=\Big\langle\Big\{\boldsymbol{\Gamma}_i+\sum_{k=1}^{2}\mathbf{S}_{ik}\overline{\boldsymbol{\sigma}}_k\Big\}\otimes\Big\{\boldsymbol{\Gamma}_i+\sum_{k=1}^{2}\mathbf{S}_{ik}\overline{\boldsymbol{\sigma}}_k\Big\}\Big\rangle, \tag{13.43}$$

where $\mathbf{S}_{ik}=\mathbf{Z}_{ik}\mathbf{R}_k,\ \boldsymbol{\Gamma}_i=\sum_{k=1}^{2}\mathbf{Z}_{ik}\mathbf{F}_k$. Equations (13.43) are exact for the limiting case $|\mathbf{x}-\mathbf{y}|\gg\max a_m^k,\quad \mathbf{x}\in v_i,\ \mathbf{y}\in v_j\quad(k=1,2,3;\quad m=i,j)$.

Then from (13.35) with (13.43) we find

$$\begin{aligned}\langle\overline{\boldsymbol{\sigma}}\otimes\overline{\boldsymbol{\sigma}}\rangle_i&=\langle\overline{\boldsymbol{\sigma}}\rangle_i\otimes\langle\overline{\boldsymbol{\sigma}}\rangle_i+\sum_{p=1}^{N}\Big\{\mathbf{H}^1_{ip}\langle\overline{\boldsymbol{\sigma}}\otimes\overline{\boldsymbol{\sigma}}\rangle_i+\mathbf{H}^2_{ip}\langle\overline{\boldsymbol{\sigma}}\otimes\overline{\boldsymbol{\sigma}}\rangle_p+\mathbf{H}^3_{ip}\langle\overline{\boldsymbol{\sigma}}\rangle_p\otimes\langle\overline{\boldsymbol{\sigma}}\rangle_i\\&+\mathbf{H}^4_{ip}\langle\overline{\boldsymbol{\sigma}}\rangle_i\otimes\langle\overline{\boldsymbol{\sigma}}\rangle_p+\mathbf{P}^1_{ip}\langle\overline{\boldsymbol{\sigma}}\rangle_p+\mathbf{P}^2_{ip}\langle\overline{\boldsymbol{\sigma}}\rangle_i+\mathbf{P}^3_{ip}\Big\},\end{aligned} \tag{13.44}$$

where the tensors $\mathbf{H}^k_{ip},\ \mathbf{P}^l_{ip}$ $(k=1,2,3,4;\ l=1,2,3)$ depend on the binary probability density $\varphi(v_p,\mathbf{x}_p|;v_i,\mathbf{x}_i)$ and are represented in Appendix A.3.2. The tensor product notation has been applied via Eq. (13.30) and $[(B_1\otimes b_1)b_2)]_{pqrs}=B_{1pqkl}b_{2kl}b_{1rs}$, where $\mathbf{B}_1,\ \mathbf{B}_2$ and $\mathbf{b}_1,\ \mathbf{b}_2$ stand for any fourth and second order tensor, respectively.

Finally, from (13.44) we get the relation for second moments of stresses in the inclusions:

$$\begin{aligned}\langle\overline{\boldsymbol{\sigma}}\otimes\overline{\boldsymbol{\sigma}}\rangle_i=&\sum_{j=1}^{N}\mathbf{X}_{ij}\sum_{p=1}^{N}\{(\delta_{pj}+\mathbf{H}^3_{jp})\langle\overline{\boldsymbol{\sigma}}\rangle_p\otimes\langle\overline{\boldsymbol{\sigma}}\rangle_j\\&+\mathbf{H}^4_{jp}\langle\overline{\boldsymbol{\sigma}}\rangle_j\otimes\langle\overline{\boldsymbol{\sigma}}\rangle_p+\mathbf{P}^1_{jp}\langle\overline{\boldsymbol{\sigma}}\rangle_p+\mathbf{P}^2_{jp}\langle\overline{\boldsymbol{\sigma}}\rangle_j+\mathbf{P}^3_{jp}\},\end{aligned} \tag{13.45}$$

where the inverse matrix $\mathbf{X}^{-1}$ of $\mathbf{X}$ has the following elements: $(\mathbf{X}^{-1})_{ij} = \delta_{ij}\{\mathbf{I}\otimes\mathbf{I}-\sum_{q=1}^{N}\mathbf{H}^1_{iq}\}-\mathbf{H}^2_{ij}$, $(i,j=1,\ldots,N)$. The inversion of the eighth-order tensor $\mathbf{X}$ was carried out by the use of the tensor analogy of the Taylor expansion $(1-x)^{-1}=\sum x^n, \quad (n=0,1,\ldots)$; in the considered examples of composites ten terms of the expansion provide an error that is less than 1%. On the right-hand side of Eq. (13.45) the values $\langle\overline{\boldsymbol{\sigma}}\rangle_i \quad (i=1,\ldots,N)$ are obtained from (8.44). An estimation of the stress dispersion (13.37) can be obtained from Eq. (13.37) by the use of Eq. (13.45). A simplification of Eq. (13.45) can be achieved by accepting the quasi–crystalline approximation (8.65_3), when $\mathbf{H}^1_{ij},\ \mathbf{H}^3_{ij},\ \mathbf{H}^4_{ij},\ \mathbf{P}^2_{ij}\equiv 0$.

An estimation of the second stress moment averaged over the matrix can be found in analogy to the derivation of (13.37) and (13.45). If we restrict our consideration to binary interaction of the inclusions and use the assumption $\langle\overline{\boldsymbol{\sigma}}\otimes\overline{\boldsymbol{\sigma}}|v_i,\mathbf{x}_i;v_0,\mathbf{x}_0\rangle_i\equiv\langle\overline{\boldsymbol{\sigma}}\otimes\overline{\boldsymbol{\sigma}}\rangle_i$, we obtain

$$\langle\boldsymbol{\sigma}\otimes\boldsymbol{\sigma}\rangle^{(0)}=\langle\boldsymbol{\sigma}\rangle^{(0)}\langle\boldsymbol{\sigma}\rangle^{(0)}+\sum_{p=1}^{N}\{\mathbf{H}^{10}_p\langle\overline{\boldsymbol{\sigma}}\otimes\overline{\boldsymbol{\sigma}}\rangle_p+\mathbf{P}^{10}_p\langle\overline{\boldsymbol{\sigma}}\rangle_p+\mathbf{P}^{30}_p\}, \qquad (13.46)$$

where the tensors $\mathbf{H}^{10}_p,\ \mathbf{P}^{10}_p,\ \mathbf{P}^{30}_p$ are represented in Appendix A.3.2. The quantities $\langle\overline{\boldsymbol{\sigma}}\otimes\overline{\boldsymbol{\sigma}}\rangle_p$ and $\langle\overline{\boldsymbol{\sigma}}\rangle_p$ can be estimated by the use of Eqs. (13.45) and (8.41), respectively. Ignoring stress fluctuations in the inclusions v_p $\langle\overline{\boldsymbol{\sigma}}\otimes\overline{\boldsymbol{\sigma}}\rangle=\langle\overline{\boldsymbol{\sigma}}\rangle\otimes\langle\overline{\boldsymbol{\sigma}}\rangle$, from relation (13.46) we obtain analogously (13.40)

$$\begin{aligned}\langle\boldsymbol{\sigma}\otimes\boldsymbol{\sigma}\rangle^{(0)}&=\langle\boldsymbol{\sigma}\rangle^{(0)}\otimes\langle\boldsymbol{\sigma}\rangle^{(0)}+\int[\mathbf{T}_{ip}(\mathbf{x}_i-\mathbf{x}_p)\langle\boldsymbol{\eta}\rangle_p\bar{v}_p]\\&\otimes[\mathbf{T}_{ip}(\mathbf{x}_i-\mathbf{x}_p)\langle\boldsymbol{\eta}\rangle_p\bar{v}_q]\times\varphi(v_p,\mathbf{x}_p|;v_i,\mathbf{x}_i)\,d\mathbf{x}_p. \qquad (13.47)\end{aligned}$$

For homogeneous inclusions (3.187) (considered in Section 13.1) without any transformation field $\boldsymbol{\beta}\equiv\mathbf{0}$ the relations (13.45) and (13.46) reduce to the results [179]. It should be mentioned that in a number of practical cases some simplified methods (such as Mori-Tanaka approach, see Subsection 8.3.2) lead to reasonably accurate estimations of the effective parameters $\mathbf{M}^*,\ \boldsymbol{\beta}^*,\ W^*$. However, in the case of estimating central stress moments inside the inclusions, one obtains the trivial result $\Delta\boldsymbol{\sigma}^{(i)2}(\mathbf{x})\equiv 0, \quad \mathbf{x}\in v^{(i)} \quad (i=1,\ldots)$.

13.2.4 Numerical Estimation of the Second Moments of Stresses in the Phases

Let us now demonstrate the application of the integral equation method by considering a problem of the estimation of all components of the second moment of stresses averaged over the matrix. The solution of such problems by perturbation methods (13.15) requires determination of the effective properties of composites with a general anisotropy of the matrix (even for an actually isotropic matrix). However, by using the proposed method (13.46) these difficulties do not appear. For composites considered in this section some normalized components of the stress fluctuation $\Delta\sigma^{(0)}/\langle\sigma_{11}\rangle$ $(\Delta\sigma^{(0)}\equiv\sqrt{\Delta\sigma^{(0)2}})$ (13.46) under uniaxial tension $\langle\sigma_{ij}\rangle=\delta_{i1}\delta_{j1}$ are plotted in Fig. 13.5. From this figure it becomes obvious that for sufficiently large void concentrations the non-axial components of the stress fluctuations are of the same order of magnitude as the uniaxial average stresses in the matrix.

If the case of identical rigid spherical inclusions in an incompressible matrix is considered, one can see from Fig. 13.6 that for sufficiently large inclusion concentrations the stress fluctuations are also of considerable importance in the formation of the stress state of the matrix.

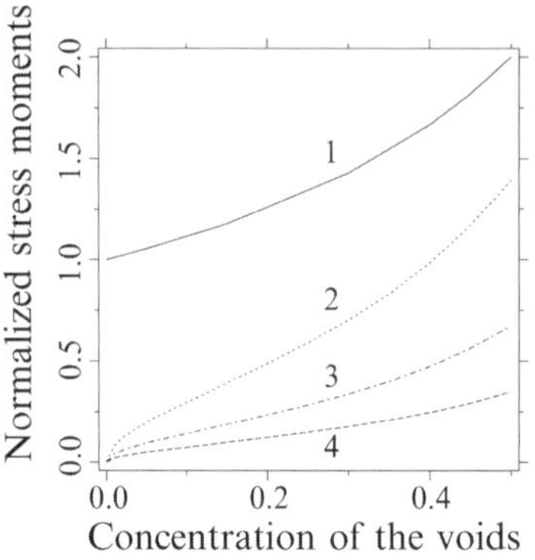

Fig. 13.5. Normalized average stress $\langle\sigma_{11}\rangle^{(0)}/\langle\sigma_{11}\rangle$ (1) and stress fluctuations $\Delta\sigma^{(0)}_{1111}/\langle\sigma_{11}\rangle$ (2), $\Delta\sigma^{(0)}_{2222}/\langle\sigma_{11}\rangle$ (3) and $\Delta\sigma^{(0)}_{2323}/\langle\sigma_{11}\rangle$ (4) as a functions of porosity.

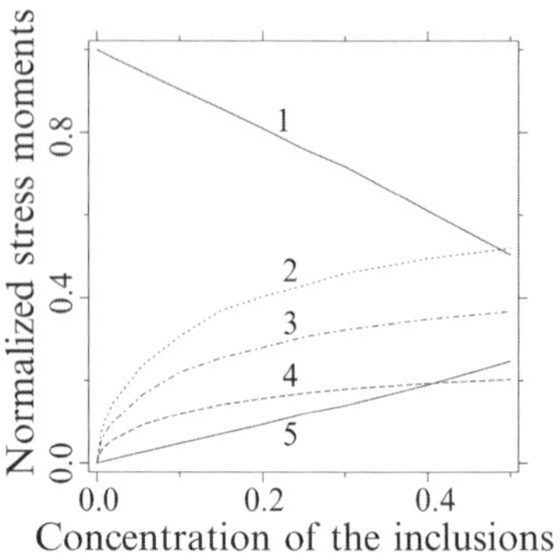

Fig. 13.6 $\langle\sigma_{11}\rangle^{(0)}/\langle\sigma_{11}\rangle$ (1), $\langle\sigma_{22}\rangle^{(0)}/\langle\sigma_{11}\rangle$ (5) $\Delta\sigma^{(0)}_{1111}/\langle\sigma_{11}\rangle$ (2), $\Delta\sigma^{(0)}_{2222}/\sigma^0_{11}$ (3), and $\Delta\sigma^{(0)}_{2323}/\langle\sigma_{11}\rangle$ (4) vc c.

We will now demonstrate the application of the method of integral equations for the estimation of inhomogeneous stress fluctuations inside the inhomogeneous inclusions. To obtain concrete numerical results we choose the Poisson's ratio of the matrix $\nu = 0.3$, and the spherical coated inclusions v_i $(i = 1, 2, \ldots)$ of the radius a have a thin coating $v^c \subset v_i$ of the thickness ha $(h \ll 1)$ with elastic properties of the matrix and homogeneous mismatch properties: $\mathbf{M}_1(\mathbf{x}) = 0$, $\boldsymbol{\beta}_1(\mathbf{x}) = \boldsymbol{\beta}^c_1 \equiv \beta^c_{10}\boldsymbol{\delta}$ at $\mathbf{x} \in V^c$, and the spherical rigid cores $v^i \equiv v_i \setminus v^c$ have thermoelastic parameters $\mathbf{M}_1(\mathbf{x}) = \mathbf{M}^{(i)}_1 = (\infty, \infty)$, $\boldsymbol{\beta}_1(\mathbf{x}) \equiv 0$ at $\mathbf{x} \in v^i$. The conditional probability densities are the step functions (9.91) and $\varphi(v_p, \mathbf{x}_p|; v_0, \mathbf{x}_0) = H(|\mathbf{x}_p - \mathbf{x}_0| - a)n^{(1)}$.

The solution for a single coated ellipsoidal inclusion in an infinite matrix is represented in Subsection 3.7.2: $\mathbf{B}(\mathbf{x}) = \mathbf{B}^i \equiv (\mathbf{I}+\mathbf{Q}^i\mathbf{M}^{(i)}_1)^{-1}$, $\mathbf{C}(\mathbf{x}) \equiv \mathbf{0}$ for the core $\mathbf{x} \in v^i$, and $\mathbf{B}(\mathbf{x}) = \mathbf{I}+\overline{v}^i\mathbf{T}^i(\mathbf{x}_i-\mathbf{x})\mathbf{M}^{(i)}_1\mathbf{B}^i$, $\mathbf{C}(\mathbf{x}) = -[\mathbf{Q}^i+\overline{v}^i\mathbf{T}^i(\mathbf{x}_i-\mathbf{x})]\boldsymbol{\beta}^c_1$ for the coating $\mathbf{x} \in v^c$. Here the upper index i for the tensors $\mathbf{Q}^i$, $\mathbf{B}^i$ and $\mathbf{T}^i(\mathbf{x}_i - \mathbf{x})$ stands on the calculation of these tensors by the use of the formulae (8.21), (8.22) and (8.15), respectively, by replacing of v_i by v^i. Then the averaged concentration tensors of the coated inclusion (8.18) can be obtained by the use of the analogous tensors for homogeneous inclusions $\mathbf{B}_i = \mathbf{I}+\overline{v}^i(\overline{v}_i)^{-1}(\mathbf{B}^i-\mathbf{I})$, $\mathbf{C}_i = -\overline{v}^c(\overline{v}_i)^{-1}\mathbf{Q}^i\boldsymbol{\beta}^c_1$, $\mathbf{R}_i = \mathbf{R}^i \equiv \overline{v}^i\mathbf{M}^{(i)}_1\mathbf{B}^i$, $\mathbf{F}_i = \overline{v}^c\boldsymbol{\beta}^c_1$. After that we can find the dispersion of the effective field $\Delta\overline{\boldsymbol{\sigma}}^2_i$ (13.36), (13.45) as well as an inhomogeneous stress dispersion $\Delta\boldsymbol{\sigma}^2_i(\mathbf{x}) = \mathbf{B}(\mathbf{x})\Delta\overline{\boldsymbol{\sigma}}^2_i\mathbf{B}^\top(\mathbf{x})$ (13.37) inside the coated inclusion $\mathbf{x} \in v_i$; $\langle\sigma_{ij}\rangle \equiv \sigma^0_{33}\delta_{i3}\delta_{j3}$. Let us define the normalized fluctuation of the effective v.Mises stress $\tau^{Mis}(\mathbf{x}) \equiv 1.5|\mathbf{N}_2 : \Delta\boldsymbol{\sigma}^{(i)2}(\mathbf{x})|^{1/2}/\ |\sigma^0_{33}|$. Figure 13.7 shows the values $\tau^{Mis}(\mathbf{x})$ in a spherical coordinate system.

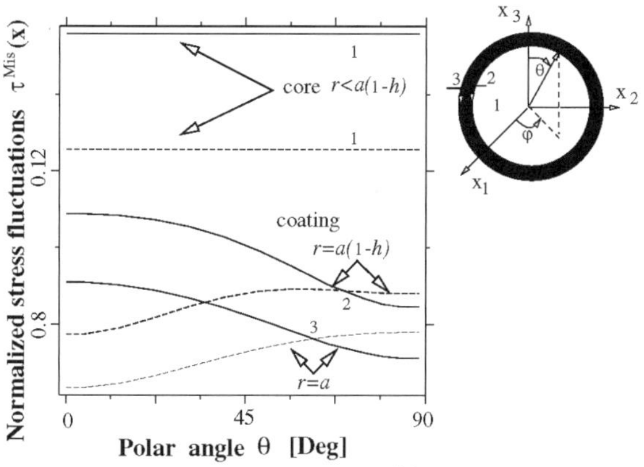

Fig. 13.7. Normalized von Mises stress fluctuations $\tau^{Mis}(\mathbf{x})$ vs θ in the core $r < a(1-h)$ (lines 1) as well as in the coating for $r = a(1-h)+0$ (lines 2) and $r = a-0$ (lines 3) for unit axial loading (dashed lines) and thermo-mechanical loading (solid lines).

$\mathbf{x} = r(\cos\varphi\sin\theta, \sin\varphi\sin\theta, \cos\theta)^\top$ as functions of the polar angle θ at $\varphi \equiv 0$ in the core $r < a(1-h)$ (lines 1) as well as in the coating for $r = a(1-h)$ (lines 2) and $r = a$ (lines 3). A relative thickness $h = 0.1$ and a concentration of inclusions $c \equiv c^{(1)} = \langle V^{(1)}\rangle = 4\pi a^3 n/3 = 0.4$ were assumed. In Fig. 13.7 two cases of loading, $\sigma^0_{33} \neq 0$, $\beta^c_{10} \equiv 0$ (dashed lines) and $\sigma^0_{33} \neq 0$, $\beta^c_{10} = \sigma^0_{33}/(3h3Q^k_i)$ (solid lines), are represented. We see that $\tau^{Mis}(\mathbf{x})$ is a homogeneous function in the core v^i, but it is inhomogeneous in the coating v^c. At the same time for purely thermal loading, $\boldsymbol{\sigma}^0 \equiv \mathbf{0}$, $\beta^c_{10} \neq 0$, the effective von Mises stress in the core v^i, calculated by the use of statistical average stresses $\langle\boldsymbol{\sigma}\rangle_i(\mathbf{x})$ (8.44), is zero.

13.2.5 Related Method of Estimations of the Second Moments of Stresses

It should be mentioned that for the pure elastic case ($\boldsymbol{\beta}_1(\mathbf{x}) \equiv \mathbf{0}$) an alternative formula for the estimation of the second moment $\langle\mathbf{ss}\rangle_0$ was proposed in [517] and [524] without any justification

$$\langle\mathbf{ss}\rangle_0 = H^0 + \int \left[\langle\mathbf{ss}|v_p, \mathbf{x}_p; v^{(0)}, \mathbf{x}_0\rangle - H^0\right]\varphi(v_p, \mathbf{x}_p|; v^{(0)}, \mathbf{x}_0)\, d\mathbf{x}_p, \qquad (13.48)$$

where $H^0 = (\mathbf{B}_1\boldsymbol{\sigma}^0)(\mathbf{B}_1\boldsymbol{\sigma}^0)$, and the tensor $\mathbf{B}_1 = $ const. is defined by an average scheme. The estimation of the second conditional moment $\langle\mathbf{ss}|v_p, \mathbf{x}_p; v^{(0)}, \mathbf{x}_0\rangle$ was fulfilled on the basis of some additional assumptions which are similar (but not equivalent) to the "far field approximation" (13.47)

$$\langle\mathbf{ss}|v_p, \mathbf{x}_p; v^{(0)}, \mathbf{x}_0\rangle = (\mathbf{B}_1\boldsymbol{\sigma}^0 + \mathbf{T}_p(\mathbf{x}_0, \mathbf{x}_p)\mathbf{B}_2\boldsymbol{\sigma}^0)(\mathbf{B}_1\boldsymbol{\sigma}^0 + \mathbf{T}_p(\mathbf{x}_0, \mathbf{x}_p)\mathbf{B}_2\boldsymbol{\sigma}^0), \qquad (13.49)$$

where the tensor $\mathbf{A}_2 = $ const. is defined again by an average scheme. More correct estimations of second conditional moment of stresses were proposed by Eq. (13.47) (see also [179], [182]). It is well known that integrals such as in (13.48) with (13.49) are not absolutely convergent simply because $\mathbf{T}_p(\mathbf{x}_0-\mathbf{x}_p)$ is of order $|\mathbf{x}_0-\mathbf{x}_p|^{-d}$. For dealing with this difficulty one [517], [524] reduced a volume improper integral to a particular case of a triple integral, a procedure that is questionable. Moreover, formulae (13.48) and (13.49) were used by the aforementioned authors for the estimation of second moments of stresses $\langle \mathbf{ss}\rangle_0$ in the elastoplastic analysis, when $\boldsymbol{\beta}_1(\mathbf{x}) \not\equiv \mathbf{0}$ or, in other words, the authors mentioned assumed that the second $\langle \mathbf{ss}\rangle_0$ and first $\langle \mathbf{s}\rangle_0$ moments are invariant with respect to the transformation strain tensor (such as e.g. $\boldsymbol{\beta}_1$) (a mathematically consistent consideration of this problem that is more exact is presented in [151], [179]).

Buryachenko [126], [127], and [159], [160] proposed an approximate method of the estimation of the second moment of stresses only inside the matrix from the elastic energy which is exact for the particular case of the composite with incompressible matrix containing rigid inclusions or pores. They used the additional assumption

$$\langle \sigma_0^2\rangle_0 = \langle \sigma_0\rangle_0^2 \tag{13.50}$$

(i.e., perturbed hydrostatic fields in the matrix are neglected). Later, the assumption (13.50) was used [83], [852], [911], [912], [913], and equivalent results found by a different method based on the energy approach (see also [733]), rendering a physical meaning for the adoption of the second stress moment in a homogenization scheme. Without such a connection the concept of second stress moment would remain purely mathematical. Buryachenko [131] gave estimates of the error caused by the assumptions (13.50) by using Eq. (13.15), which for porous materials gives

$$\langle \sigma_0^2\rangle_0 - \langle \sigma_0\rangle_0^2 = \frac{1}{9(1-c)}\left(\frac{\partial \mathbf{M}^*}{\partial p^{(0)}} - \frac{3\mathbf{N}_1}{1-c}\right)(\langle \sigma\rangle \otimes \langle \sigma\rangle), \tag{13.51}$$

where the right-hand side of (13.51) equals zero only for an incompressible matrix.

13.3 Elastically Homogeneous Composites with Randomly Distributed Residual Microstresses

In this section specializations of the general formulae given in the preceding sections are derived. Especially interesting results are obtained if the mismatch in the elastic compliance is negligible, i.e., $\mathbf{M}(\mathbf{x}) \equiv \mathbf{M}^{(0)} = \text{const}$, and the stress-free strains $\boldsymbol{\beta}$ fluctuate. Some relations represented below were found immediately [136] without of the assumptions of MEFM.

The relations for the stored energy of elastically homogeneous composites (9.38)–(9.40) were obtained previously. Utilizing Eq. (13.7) with a special variation of the compliance tensor $\mathbf{M} \to \mathbf{M}^{(0)} + \delta\mathbf{M}$ ($\mathbf{M}, \delta\mathbf{M} \equiv \text{const.}$) leads to the exact representation for the second moment of residual stresses averaged over both the whole composite and each phase (13.15)

$$\langle \boldsymbol{\sigma} \otimes \boldsymbol{\sigma}\rangle = -2\frac{\partial W^*}{\partial \mathbf{M}^{(0)}}, \quad \langle \boldsymbol{\sigma} \otimes \boldsymbol{\sigma}\rangle_i = -\frac{2}{c^{(i)}}\frac{\partial W^*}{\partial \mathbf{M}^{(i)}}\bigg|_{\boldsymbol{\beta}}, \tag{13.52}$$

respectively. Unfortunately, Eq. (13.52_2) is less simple and requires a consideration of multiparticle interaction of the inclusions. This can be done by the use of MEFM. Nevertheless relation (3.1_1) is useful for testing of the exactness of the integral equation method proposed in the present chapter for the estimation of second moments of stresses in the components.

13.3.1 The Conditional Average of the Stresses Inside the Components

Let $\langle \boldsymbol{\sigma} | v_1, \mathbf{x}_1; v_2, \mathbf{x}_2 \rangle(\mathbf{x})$ denote conditional average stresses over ensemble realization in the point $\mathbf{x} \in v_1$ under condition that there are fixed inclusions $v_1 \neq v_2$ in the points $\mathbf{x}_1$, $\mathbf{x}_2$. This average can be found by the use of the general equation (9.32) ($\mathbf{x} \in v_1 \neq v_2$)

$$\boldsymbol{\sigma}(\mathbf{x}|v_1, \mathbf{x}_1; v_2, \mathbf{x}_2) = \boldsymbol{\sigma}^0 + \int \boldsymbol{\Gamma}(\mathbf{x}-\mathbf{y})[\boldsymbol{\beta}_1(\mathbf{y}) V(\mathbf{y}|v_1, \mathbf{x}_1; v_2, \mathbf{x}_2) - \langle \boldsymbol{\beta}_1 \rangle] \, d\mathbf{y}, \quad (13.53)$$

where $V(\mathbf{y}|v_1, \mathbf{x}_1; v_2, \mathbf{x}_2)$ is a random characteristic function of inclusions $\mathbf{y}$ under the condition that the inclusions $v_1 \neq v_2$ are located in the domains with the centers $\mathbf{x}_1$ and $\mathbf{x}_2$. The terms with $\mathbf{y} \in v_1$ and $\mathbf{y} \in v_2$ may be isolated in the right-hand-side integral (13.53) with the help of the equalities (5.72) and (5.73). Then averaging (13.53) by the use (5.73) and taking the relation for average stresses (9.33) into account leads to

$$\begin{aligned} \langle \boldsymbol{\sigma} \mid v_1, \mathbf{x}_1; v_2, \mathbf{x}_2 \rangle(\mathbf{x}) &= \langle \boldsymbol{\sigma} \rangle_1(\mathbf{x}) + \mathbf{T}_2^{\beta}(\mathbf{x} - \mathbf{x}_2) \\ &+ \int \mathbf{T}_q^{\beta}(\mathbf{x} - \mathbf{x}_q) v_q \left[\varphi(v_q, \mathbf{x}_q |; v_1, \mathbf{x}_1; v_2, \mathbf{x}_2) - \varphi(v_q, \mathbf{x}_q |; v_1, \mathbf{x}_1) \right] d\mathbf{y}, \quad (13.54) \end{aligned}$$

where $\mathbf{x} \in v_1$ and the tensors $\mathbf{T}_i^{\beta}$ and $\mathbf{T}_{ij}^{\beta}$ were introduced by Eqs. (9.33) and (9.34). Proper allowance must be made for calculation of the integral in (13.54) that the triple conditional probability density $\varphi(v_q, \mathbf{x}_q |; v_1, \mathbf{x}_1; v_2, \mathbf{x}_2)$ is unknown. In a similar manner the n-point conditional average of stresses in the inclusions $\langle \boldsymbol{\sigma} | v_1, \mathbf{x}_1; v_1, \mathbf{x}_2; \ldots; v_n, \mathbf{x}_n \rangle_1(\mathbf{x})$ ($\mathbf{x} \in v_1$) may be derived under known ($n-1$)–point conditional average ones

$$\begin{aligned} &\langle \boldsymbol{\sigma} | v_1, \mathbf{x}_1; v_2, \mathbf{x}_2; \ldots; v_n, \mathbf{x}_n \rangle(\mathbf{x}) \\ &= \langle \boldsymbol{\sigma} | v_1, \mathbf{x}_1; v_2, \mathbf{x}_2; \ldots; v_{n-1}, \mathbf{x}_{n-1} \rangle(\mathbf{x}) + \mathbf{T}_n^{\beta}(\mathbf{x} - \mathbf{x}_n) + \int \mathbf{T}_q^{\beta}(\mathbf{x} - \mathbf{x}_q) \bar{v}_q \\ &\times [\varphi(v_q, \mathbf{x}_q |; v_1, \mathbf{x}_1; \ldots ..; v_n, \mathbf{x}_n) - \varphi(v_q, \mathbf{x}_q |; v_1, \mathbf{x}_1; \ldots; v_{n-1}, \mathbf{x}_{n-1})] d\mathbf{x}_q. \quad (13.55) \end{aligned}$$

The right-hand-side integrals in Eqs. (13.54) and (13.55) are of the first order of the inclusion concentration c. Therefore for dilute concentration of inclusions their summations are small compared with $\mathbf{T}_n^{\beta}(\mathbf{x} - \mathbf{x}_n)$ ($n = 2, \ldots$).

13.3.2 The Second Moment Stresses Inside the Phases

To obtain the second moment of stresses in the component $v^{(i)}$ of the inclusions ($i = 1, 2, \ldots.$) it is necessary to take the tensor product of (13.35) into $\boldsymbol{\sigma}(\mathbf{x})$ ($\mathbf{x} \in v_i$) and to average over realization ensemble

$$\langle \boldsymbol{\sigma} \otimes \boldsymbol{\sigma} \rangle_i(\mathbf{x}) = \boldsymbol{\sigma}^0 \otimes \boldsymbol{\sigma}^0 + \boldsymbol{\sigma}^0 \otimes \int [\mathbf{T}_q^\beta(\mathbf{x}-\mathbf{x}_q)\varphi(v_q, \mathbf{x}_q|v_i, \mathbf{x}_i) - \boldsymbol{\Gamma}(\mathbf{x}-\mathbf{x}_q)\langle \boldsymbol{\beta}_1 \rangle] \, d\mathbf{x}_q$$
$$+ \int [\mathbf{T}_p^\beta(\mathbf{x}-\mathbf{x}_p)\varphi(v_p, \mathbf{x}_p|v_i, \mathbf{x}_i) - \boldsymbol{\Gamma}(\mathbf{x}-\mathbf{x}_p)\langle \boldsymbol{\beta}_1 \rangle] \, d\mathbf{x}_p \otimes \boldsymbol{\sigma}^0$$
$$+ \int\int \Big\{ \mathbf{T}_p(\mathbf{x}-\mathbf{x}_p, \boldsymbol{\beta}_1) \otimes \mathbf{T}_q^\beta(\mathbf{x}-\mathbf{x}_q,)\varphi(v_p, \mathbf{x}_p, v_q, \mathbf{x}_q|v_i, \mathbf{x}_i)$$
$$- \mathbf{T}_p^\beta(\mathbf{x}-\mathbf{x}_p) \otimes \boldsymbol{\Gamma}(\mathbf{x}-\mathbf{x}_q)\langle \boldsymbol{\beta}_1 \rangle \varphi(v_p, \mathbf{x}_p|v_i, \mathbf{x}_i)$$
$$- [\boldsymbol{\Gamma}(\mathbf{x}-\mathbf{x}_p)\langle \boldsymbol{\beta}_1 \rangle] \otimes \mathbf{T}_q^\beta(\mathbf{x}-\mathbf{x}_q)\varphi(v_q, \mathbf{x}_q|v_i, \mathbf{x}_i)$$
$$+ [\boldsymbol{\Gamma}(\mathbf{x}-\mathbf{x}_p)\langle \boldsymbol{\beta}_1 \rangle] \otimes [\boldsymbol{\Gamma}(\mathbf{x}-\mathbf{x}_q)\langle \boldsymbol{\beta}_1 \rangle] \Big\} \, d\mathbf{x}_p \, d\mathbf{x}_q. \tag{13.56}$$

The right-hand side of Eq. (13.56) includes one-point and double-point conditional probability densities in which the terms with $\mathbf{x}_p = \mathbf{x}_i$, $\mathbf{x}_q = \mathbf{x}_i$ and $\mathbf{x}_p = \mathbf{x}_q$ may be isolated with the help of the equalities

$$\varphi(v_p, \mathbf{x}_p|v_1, \mathbf{x}_i) = \delta(\mathbf{x}_p - \mathbf{x}_i) + \varphi(v_p, \mathbf{x}_p|; v_i, \mathbf{x}_i),$$
$$\varphi(v_p, \mathbf{x}_p, v_q, \mathbf{x}_q|v_i, \mathbf{x}_i) = \delta(\mathbf{x}_p - \mathbf{x}_q)\delta(\mathbf{x}_q - \mathbf{x}_i) + \delta(\mathbf{x}_p - \mathbf{x}_q)\varphi(v_p, \mathbf{x}_p|; v_i, \mathbf{x}_i)$$
$$+ \delta(\mathbf{x}_p - \mathbf{x}_i)\varphi(v_q, \mathbf{x}_q|; v_i, \mathbf{x}_i) + \delta(\mathbf{x}_q - \mathbf{x}_i)\varphi(v_p, \mathbf{x}_p|; v_i, \mathbf{x}_i)$$
$$+ \varphi(v_p, \mathbf{x}_p|; v_i, \mathbf{x}_i)\varphi(v_q, \mathbf{x}_q|; v_p, \mathbf{x}_p; v_i, \mathbf{x}_i). \tag{13.57}$$

Then the relations (13.56) may be simplified to

$$\langle \boldsymbol{\sigma} \otimes \boldsymbol{\sigma} \rangle_i(\mathbf{x}) = \langle \boldsymbol{\sigma} \rangle_i(\mathbf{x}) \otimes \langle \boldsymbol{\sigma} \rangle_i(\mathbf{x})$$
$$+ \int \mathbf{T}_p^\beta(\mathbf{x}-\mathbf{x}_p)\bar{v}_p \otimes \mathbf{T}_p^\beta(\mathbf{x}-\mathbf{x}_p)\bar{v}_p \varphi(v_p, \mathbf{x}_p|; v_i, \mathbf{x}_i) \, d\mathbf{x}_p$$
$$+ \int\int \mathbf{T}_p^\beta(\mathbf{x}-\mathbf{x}_p)\bar{v}_p \otimes \mathbf{T}_q^\beta(\mathbf{x}-\mathbf{x}_q)\bar{v}_q$$
$$\times \varphi(v_p, \mathbf{x}_p|; v_i, \mathbf{x}_i)[\varphi(v_q, \mathbf{x}_q|; v_p, \mathbf{x}_p; v_i, \mathbf{x}_i) - \varphi(v_q, \mathbf{x}_q|; v_i, \mathbf{x}_i)] d\mathbf{x}_q d\mathbf{x}_p, \tag{13.58}$$

where the expression for the average stress (9.31) was used. The new exact relation (13.58) is derived by the use of triple interaction of the inclusions. As may be seen from Eq. (13.58), neglect of binary interaction is tantamount to assuming homogeneity of stresses inside component v_i: $\langle \boldsymbol{\sigma} \otimes \boldsymbol{\sigma} \rangle_i(\mathbf{x}) = \langle \boldsymbol{\sigma} \rangle_i(\mathbf{x}) \otimes \langle \boldsymbol{\sigma} \rangle_i(\mathbf{x})$.

The following approximation of second moment estimation can be obtained by taking into account only binary interaction of inclusions

$$\langle \boldsymbol{\sigma} \otimes \boldsymbol{\sigma} \rangle_i(\mathbf{x}) = \langle \boldsymbol{\sigma} \rangle_i(\mathbf{x}) \otimes \langle \boldsymbol{\sigma} \rangle_i(\mathbf{x})$$
$$+ \int \mathbf{T}_p^\beta(\mathbf{x}-\mathbf{x}_p)\bar{v}_p \otimes \mathbf{T}_p^\beta(\mathbf{x}-\mathbf{x}_p)\bar{v}_p \varphi(v_p, \mathbf{x}_p|; v_i, \mathbf{x}_i) \, d\mathbf{x}. \tag{13.59}$$

In Subsection 13.3.3 it will be shown that formula (13.59) provides a sufficiently good approximation of the exact solution. One can see from the equation (13.59) that covariance matrix $\Delta\boldsymbol{\sigma}_i^2(\mathbf{x})$ (13.37$_1$) presents a determined nonhomogeneous function of coordinate $\mathbf{x}$ inside inclusion v_i, in contrast with $\langle \boldsymbol{\sigma} \rangle_i(\mathbf{x}) = \text{const.}$ Formulae (13.37$_1$) and (13.59) make possible the estimation of average fluctuations over the inclusion volume $\langle \Delta\boldsymbol{\sigma}^2 \rangle_i = \langle \Delta\boldsymbol{\sigma}_i^2(\mathbf{x}) \rangle_i$.

The previous method of the estimation of stress second moment can be used for calculation of conditional moments of stresses. But it is more convenient to directly employ conditional average (13.55) for this purpose. In a similar spirit it is possible to obtain the estimation of the second stress moment averaging over a volume of the matrix $\langle\boldsymbol{\sigma}\otimes\boldsymbol{\sigma}\rangle_0$, which does not depend on $\mathbf{x}_0\in v_0$ in contradict to (13.58). An estimation of the second moment $\langle\boldsymbol{\sigma}\otimes\boldsymbol{\sigma}\rangle_0$ and conditional one $\langle\boldsymbol{\sigma}\otimes\boldsymbol{\sigma}|;v_i,\mathbf{x}_i\rangle_0$ can be obtained in perfect analogy to (13.59) by the use of the replacement of values $v_i,\ \mathbf{x}_i,\ \mathbf{x}$ by $v_0,\ \mathbf{x}_0,\ \mathbf{x}_0$.

The resultant evaluations of stress moments in individual components $\langle\boldsymbol{\sigma}\otimes\boldsymbol{\sigma}\rangle_i$ $(i=1,\ldots,N)$ (13.58) and $\langle\boldsymbol{\sigma}\otimes\boldsymbol{\sigma}\rangle_0$ enable one to estimate the average over the whole volume of the composite:

$$
\begin{aligned}
\langle\boldsymbol{\sigma}\otimes\boldsymbol{\sigma}\rangle &= \sum_{j=0}^{N} c^{(j)}\langle\boldsymbol{\sigma}\rangle_j\otimes\langle\boldsymbol{\sigma}\rangle_j\\
&+\sum_{i=1}^{N} n^{(i)}\int\!\!\int[\mathbf{T}_p^{\beta}(\mathbf{x}-\mathbf{x}_p)\bar{v}_p\otimes\mathbf{T}_p^{\beta}(\mathbf{x}-\mathbf{x}_p)\bar{v}_p\varphi(v_p,\mathbf{x}_p|;v_i,\mathbf{x}_i)V_i(\mathbf{x})\,d\mathbf{x}_p\,d\mathbf{x}\\
&+\sum_{i=1}^{N} n^{(i)}\int\!\!\int\!\!\int\mathbf{T}_p^{\beta}(\mathbf{x}-\mathbf{x}_p)\bar{v}_p\otimes\mathbf{T}_q^{\beta}(\mathbf{x}-\mathbf{x}_q))\bar{v}_q\\
&\times\varphi(v_p,\mathbf{x}_p|;v_i,\mathbf{x}_i)[\varphi(v_q,\mathbf{x}_q|;v_p,\mathbf{x}_p;v_i,\mathbf{x}_i)-\varphi(v_q,\mathbf{x}_q|;v_i,\mathbf{x}_i)]V_i(\mathbf{x})\,d\mathbf{x}_q\,d\mathbf{x}_p\,d\mathbf{x}\\
&+c^{(0)}\int\mathbf{T}_p^{\beta}(\mathbf{x}_0-\mathbf{x}_p)\bar{v}_p\otimes\mathbf{T}_p^{\beta}(\mathbf{x}_0-\mathbf{x}_p)\bar{v}_p\varphi(v_p,\mathbf{x}_p|;v_0,\mathbf{x}_0)\,d\mathbf{x}_p\\
&+c^{(0)}\int\!\!\int\mathbf{T}_p^{\beta}(\mathbf{x}_0-\mathbf{x}_p)\bar{v}_p\otimes\mathbf{T}_q^{\beta}(\mathbf{x}_0-\mathbf{x}_q)\bar{v}_q\\
&\times\varphi(v_p,\mathbf{x}_p|;v_0,\mathbf{x}_0)[\varphi(v_q,\mathbf{x}_q|;v_p,\mathbf{x}_p;v_0,\mathbf{x}_0)-\varphi(v_q,\mathbf{x}_q|;v_0,\mathbf{x}_0)]\,d\mathbf{x}_q\,d\mathbf{x}_p. \qquad (13.60)
\end{aligned}
$$

Therefore the exact estimation $\langle\boldsymbol{\sigma}\otimes\boldsymbol{\sigma}\rangle$ (9.40) and (13.52_1) offers a test of the accuracy of evaluation (13.60) which uses dissimilar approximate conditional probability densities. It should be noted in connection with this that the second stress moment in the composite $\langle\boldsymbol{\sigma}\otimes\boldsymbol{\sigma}\rangle$ can be represented in the form $\langle\boldsymbol{\sigma}\otimes\boldsymbol{\sigma}\rangle=\sum_{j=0}^{N}c_j\langle\boldsymbol{\sigma}\rangle_j\otimes\langle\boldsymbol{\sigma}\rangle_j+\overline{\Delta\boldsymbol{\sigma}}^2$, where the first term in the right-hand side of the last equation is defined by average stresses inside the components and the second one $\overline{\Delta\boldsymbol{\sigma}}^2$ depends on the stress fluctuation inside each components; because $\langle\boldsymbol{\sigma}\rangle\equiv\boldsymbol{\sigma}^0$, the second moment of stresses is in fact the stress fluctuations in the composite $\langle\boldsymbol{\sigma}\otimes\boldsymbol{\sigma}\rangle\equiv\Delta\boldsymbol{\sigma}^2$ at $\boldsymbol{\sigma}^0=\mathbf{0}$. By virtue of the fact that the average stresses in the components are calculated exactly by Eq. (9.36) the difference $\overline{\Delta\boldsymbol{\sigma}}^2=\langle\boldsymbol{\sigma}\otimes\boldsymbol{\sigma}\rangle-\sum_{j=0}^{N}c_j\langle\boldsymbol{\sigma}\rangle_j\otimes\langle\boldsymbol{\sigma}\rangle_j$, can be used for the estimation of the accuracy of calculated stress fluctuations inside the components (13.58), (13.60).

It is pertinent to note that in a number of example cases simplified methods (such as the Mori-Tanaka one) permit one to get reasonable accuracy of effective parameters $\mathbf{M}^*,\boldsymbol{\beta}^*,U^*$. But for estimation of central stress moments inside component inclusions one obtains a trivial result $\Delta\boldsymbol{\sigma}_i^2(\mathbf{x})\equiv 0,\ \mathbf{x}\in v^{(i)}$ $(i=0,1,\ldots)$.

13.3.3 Numerical Evaluation of Statistical Residual Stress Distribution in Elastically Homogeneous Media

As an example we consider a Si_3N_4 composite with isotropic components $\mathbf{L} = (3k, 2\mu) \equiv 3k\mathbf{N}_1 + 2\mu\mathbf{N}_2$, $\mathbf{N}_1 \equiv \boldsymbol{\delta}\otimes\boldsymbol{\delta}/3$, $\mathbf{N}_2 \equiv \mathbf{I} - \mathbf{N}_1$ containing the identical SiC spherical inclusion. We will use the following elastic constants and thermal expansion coefficients as usually found in the literature [600]: for the matrix (Si_3N_4) $k^{(0)} = 236.4$ GPa, $\mu^{(0)} = 121.9$ GPa, $a^{T(0)} = 3.4 \cdot 10^{-6}(K)$, and for the inclusions (SiC) $k^{(1)} = 208.3$ GPa, $\mu^{(1)} = 169.5$ GPa, $a^{T(1)} = 4.4 \cdot 10^{-6}(K)$. The thermal strains have been calculated with the assumption that internal stresses do not relax by creep of components below T_0. For an assessment of stresses a temperature difference $(T - T_0)$ of about 1000 between room temperature and the stress-free state at the elevated temperature is assumed ($\boldsymbol{\beta} \equiv \mathbf{a}^T(T - T_0)$, $\mathbf{a}^T = a^T\delta_{ij}$, $\boldsymbol{\beta}_1^{(1)} \equiv \beta_{10}\boldsymbol{\beta}$, $\beta_{10} = 10^{-3}$.

For a low volume fraction of SiC one may describe the composite by Si_3N_4 matrix with SiC particles embedded in it and $\mathbf{M}(\mathbf{x}) \equiv \mathbf{M}^{(0)}$. The SiC particles are assumed to be approximately spheres with radius a. Two alternative radial functions (9.91) and (9.92) of inclusion distribution will be examined. For representation of numerical results in dimensionless form we define normalizing coefficient $\tau \equiv -3Q^k\beta_{10}$, where $3Q^k$ equals the bulk component of the tensor $\mathbf{Q} = (3Q^k, 2Q^\mu)$. The physical meaning of τ follows from equation (9.37), according to which τ equals the component of hydrostatic stress inside a single isolated inclusion in an infinite homogeneous matrix (at $\mathbf{M}(\mathbf{x}) \equiv$ const.); for our specific composite SiC–Si_3N_4 we have $\tau = 289$ (MPa) and Poisson's ratio $\nu = 0.28$. To carry out the needed numerical estimates we will use the expressions of the tensors $\mathbf{Q}_p$, $\mathbf{T}_p(\mathbf{x} - \mathbf{x}_p)$ $(p = 1, 2, \ldots)$ which are presented by Eqs. (3.168), and (A3.1), respectively.

At first we will estimate the effect of the assumption of elastic homogeneity of the materials ($\mathbf{M}(\mathbf{x}) \equiv \mathbf{M}^{(0)}$). Buryachenko [136] calculated $\langle\boldsymbol{\sigma}_{11}\rangle_1 \sim c$ by both the exact relation (9.34) ($\mathbf{M}(\mathbf{x}) \equiv \mathbf{M}^{(0)}$) and by the approximate MEFM (8.44) with $\mathbf{M}(\mathbf{x}) \neq$ const. He also compared the estimations of stress fluctuations estimated by Eq. (3.1_2) for both $\mathbf{M}(\mathbf{x}) \equiv \mathbf{M}^{(0)}$ and $\mathbf{M}(\mathbf{x}) \neq$ const. It was noted that the error of the assumption of homogeneity of elastic properties of the ceramic is about error caused by ignoring of the radial distribution function (9.92) and less than 5%. Because of this, below we will consider only the case $\mathbf{M}(\mathbf{x}) \equiv \mathbf{M}^{(0)}$.

Let us evaluate a perturbation $\delta\boldsymbol{\sigma}_1(\mathbf{x}) \equiv \langle\boldsymbol{\sigma}|v_1, \mathbf{x}_1; v_2, \mathbf{x}_2\rangle(\mathbf{x}) - \langle\boldsymbol{\sigma}\rangle_1(\mathbf{x})$, $\mathbf{x} = \mathbf{x}_1$ caused by the inclusion v_2 at the center of the inclusion v_1; $\mathbf{x}_1 = (0, 0, 0)$, $\mathbf{x}_2 = (R, 0, 0)$. We will consider triple conditional probability densities (5.71) and binary ones (9.91), (9.92). A neighboring order in the triple point distribution (13.54) and (5.71) is constructed by the use of the binary probability density; a selection of more complicated distribution functions will not be considered. Figure 13.8 shows the curves of the normalized perturbation $\delta\sigma_{1|11}(\mathbf{x})/\tau \sim R/a$ ($\mathbf{x} = \mathbf{0}$) produced by the inclusion v_2 into a center of the inclusion v_1. The results are calculated for $c = 0$, 0.1 and 0.4 by the use of formulae (13.54), (5.71) for radial distribution functions (9.91) and (9.92). It is evident from this curves, that considering only the principal part in the action of second inclusion $\mathbf{T}_{12}(\mathbf{x}_1 - \mathbf{x}_2)\boldsymbol{\beta}_1^{(2)}\bar{v}_2$ leads to significant errors under nondilute inclusion concentration. The non-monotonical character of the curves 3, 4 near the point $R = 4a$ is explained by the occurrence of the nonzero probability of locating of some inclusion v_p between fixed inclusions v_1, v_2 under $R/a > 4$ only.

We come now to the estimation of the nonhomogeneity of stress fluctuations $\Delta\boldsymbol{\sigma}_1^2(\mathbf{x})$ inside the inclusions. The curves in Fig. 13.9 were calculated by use of formulae (13.37$_1$) and (13.59) with $c = 0.4$ for radial distribution functions (9.92) and (9.91), respectively; $\mathbf{x} = (r, 0, 0)$. The normalized components $\Delta\sigma^2_{1|1111}(\mathbf{x})$ and $\Delta\sigma^2_{1|3333}(\mathbf{x})$ are displayed in the Figure 13.9 by solid curves and dashed ones, respectively. For step correlation function (9.91) the calculated curves are invariant with respect to the inclusion concentration.

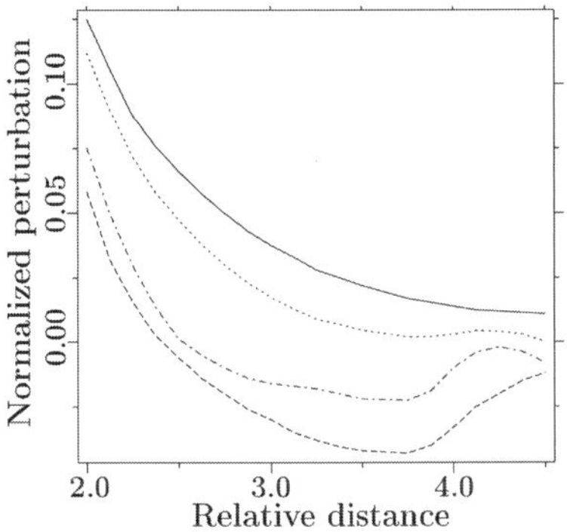

Fig. 13.8. $\delta\sigma_{1|11}(\mathbf{x})/\tau$ vs R/a for $c = 0$ (solid line), 0.1 (dotted line) and 0.4 (dot-dashed and dashed curves). Dashed line is calculated for (9.91), and dot-dashed line is obtained for (9.92).

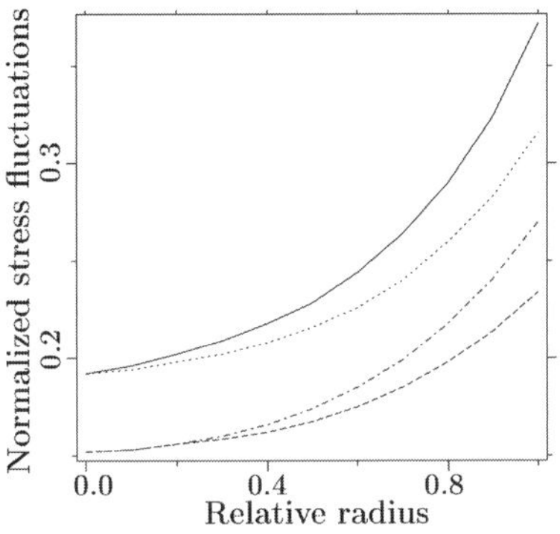

Fig. 13.9. $|\Delta\sigma^2_{1|1111}(\mathbf{x})|^{0.5}/\tau$ vs r (solid and dot-dashed lines) and $|\Delta\sigma_{1|3333}(\mathbf{x})|^{0.5}/\tau$ (dot and dashed lines) both for (9.92) (solid and dotted lines) and (9.91) (dot-dashed and dashed lines).

Now we evaluate the accuracy of stress fluctuation in the components $\overline{\Delta\boldsymbol{\sigma}}^2$ by use of both the three-point approximation (13.60) and the two-point approximation one (13.59) and $\langle\boldsymbol{\sigma}\otimes\boldsymbol{\sigma}\rangle_0$; it turns out that the contributions from stress fluctuation inside the components $\langle\Delta\sigma^2_{1111}\rangle_i$ $(i = 0, 1)$ is less than the exact relation for stress fluctuations in the whole composite $\Delta\boldsymbol{\sigma}^2 \equiv \langle\boldsymbol{\sigma}\otimes\boldsymbol{\sigma}\rangle - \langle\boldsymbol{\sigma}\rangle\otimes\langle\boldsymbol{\sigma}\rangle$ (defined in [840]) by a factor 5. In Fig. 13.10 the curves of the two-point approximations were calculated by using the principal part of the expansion for $\langle\Delta\boldsymbol{\sigma}^2\rangle_1$ (13.59) and $\langle\Delta\boldsymbol{\sigma}^2\rangle_0$ (which are proportional to c under $c \to 0$) for radial distribution function (9.92) and (9.91), respectively. A step function

$$\varphi(v_p, \mathbf{x}_p|; v_0, \mathbf{x}_0) = H(|\mathbf{x}_p - \mathbf{x}_0| - a)n_1 \tag{13.61}$$

has been applied for calculation of these curves. For comparison, the exact result described by the Eqs. (9.36) and (9.40) as well as two curves of three-point approximations calculated by Eq. (13.60) are also plotted in the same figure. One uses triple conditional probability densities (5.71) and $\varphi(v_q, \mathbf{x}_q|; v_p, \mathbf{x}_p; v_0, \mathbf{x}_0) = n_q H(|\mathbf{x}_q - \mathbf{x}_0| - a)H(|\mathbf{x}_p - \mathbf{x}_i| - a)$, and the binary probability density (9.91), (9.92), (13.61). Curves of three-point approximation are computed with the radial distribution functions (9.92) and (9.91) as well. As may be seen from Fig.

13.10 the accuracy of the estimations may be substantially extended by taking into account a threefold interaction of the inclusions. The error of the calculation by the exact formula (13.60) is dictated by the inaccuracy of the determination of different conditional probability densities.

The degree of dissimilarity between the different models should be expected from the calculation of stress fluctuations inside inclusions. The results $(\langle\Delta\sigma^2_{1111}\rangle_1)^{0.5} \sim c$ outlying at Fig. 13.11 are calculated in an exact three-point relation (13.58) and approximate two-point formula (13.59) when used with correlation functions (5.71) with radial distribution functions (9.92) (solid lines) and (9.91) (dashed lines). From this figure we notice that the approximate formula (13.59) under the binary-step function (9.91) can be used for the estimation of stress fluctuations $\langle\Delta\boldsymbol{\sigma}^2\rangle_i$ with sufficient accuracy.

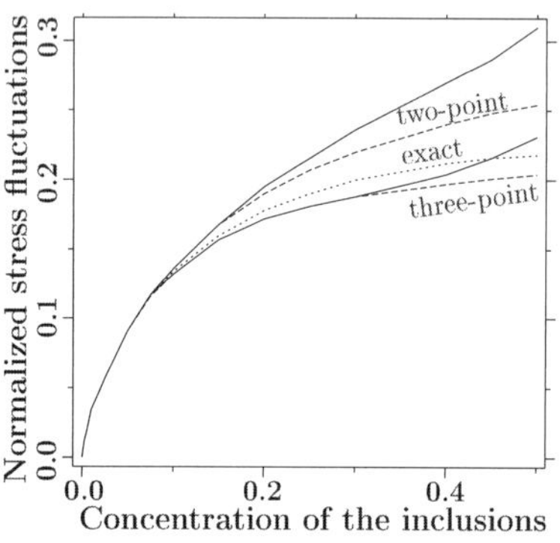

Fig. 13.10. $|\overline{\Delta\sigma^2_{1111}}|^{0.5}/\tau$ vs c estimated for both (9.91) (dashed lines) and (9.92) (solid lines) which are obtained under the two-point and three-point approximations. The exact solution is plotted in the dotted line.

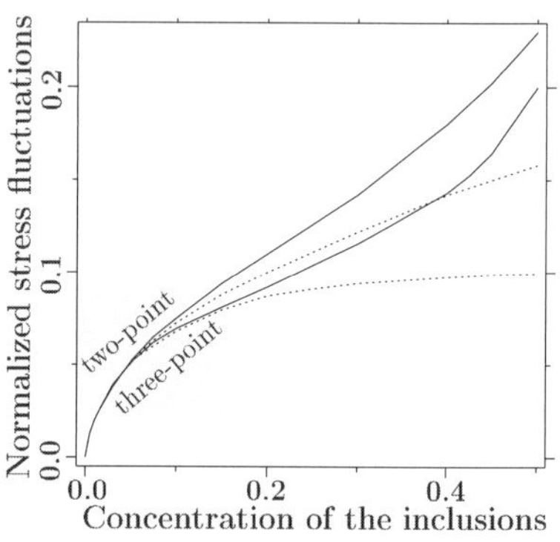

Fig. 13.11. $|\langle\Delta\sigma^2_{1111}\rangle_1|^{0.5}/\tau$ vs c calculated by the use of two-point and three-point approximations for both the step correlation function (9.91) (dashed lines) and the real one (9.92) (solid lines).

13.4 Stress Fluctuations Near a Crack Tip in Elastically Homogeneous Materials with Randomly Distributed Residual Microstresses

A scatter fracture related properties are inherent to brittle materials, and a statistical approach would be a natural way to deal with the problem. We will consider the model [137] in more details for reasons of simplicity of initial assumptions in conjunction with the possible generalizations. Namely, one considers a linearly elastic composite medium, which consists of a homogeneous matrix containing a coin-shaped crack and a homogeneous and statistically uniform random set of ellipsoidal inclusions; elastic properties of the matrix and the inclusions are the same, but stress-free strains fluctuate. One obtains the estimation of both the average and conditional average of stress intensity factors. One can show that at least for an infinite statistically homogeneous field of transforming inclusions the proposed method leads to a valid result (average stress intensity factor equals zero), which cannot be applied for practical purposes. The relations for statistical second moments of stresses in the vicinity of crack tip averaged over ensemble realization will be proposed as well. One can show a fundamental difference between the estimations of statistical moments of stresses for an infinite inclusion field and for an arbitrary large (but finite) inclusion cloud.

13.4.1 The Average and Conditional Mean Values of SIF for Isolated Crack in a Composite Material

Let us consider a coin-shaped crack with radius R^c, center $\mathbf{x}^c = (0,0,0)$ and unit normal $\mathbf{n} = (1,0,0)$ to the crack surface S in an infinite homogeneous elastic body. It is well known that the problem of a linear elastic solid with a crack under remote loading $\boldsymbol{\sigma}(\mathbf{x})$ has the same stress intensity factors as the problem with the crack faces loaded by traction $\mathbf{t}(\mathbf{s}) = -\mathbf{n}(\mathbf{s})\boldsymbol{\sigma}(\mathbf{s})$ ($\mathbf{s} \in S$) and stresses vanishing at infinity. This loading generates a singular stress component $\boldsymbol{\sigma}$ ahead of the crack tip $\mathbf{z} \in \Gamma^c$: $\boldsymbol{\sigma}(\mathbf{x}) \sim \mathbf{J}(\theta,\mathbf{z})/\sqrt{\rho}$, where $\rho \equiv \min_{\mathbf{s}\in\Gamma^c} |\mathbf{x}-\mathbf{s}|$ is a minimum distance between $\mathbf{x}$ and crack tip Γ^c, $\mathbf{z} \equiv \arg\min_{\mathbf{s}\in\Gamma^c} |\mathbf{x}-\mathbf{s}|$; (ρ,θ) are the polar coordinates of the point $\mathbf{x}$ with the origin of the polar coordinates located at the point $\mathbf{z}$. $\mathbf{J}(\theta,\mathbf{z})$ is a tensor stress intensity factor, which is connected with the usual stress intensity factors (SIF) $\mathbf{K}(\mathbf{z})$ $K_j(\mathbf{z}) = J_{1j}(0,\mathbf{z})\sqrt{2\pi}$, $(j = 1,2,3)$. In so doing axes of the local system are labeled to agree with mode number designations for stress intensity factors K_j $(j = 1,2,3)$ [936], and $\mathbf{K}(\mathbf{z}) = (K_1(\mathbf{z}), K_2(\mathbf{z}), K_3(\mathbf{z}))^\top \equiv (K_I(\mathbf{z}), K_{II}(\mathbf{z}), K_{III}(\mathbf{z}))^\top$. Then we can define $\mathbf{K}(\mathbf{z})$ and $\mathbf{J}(\theta,\mathbf{z})$ by the integrals

$$\mathbf{K}(\mathbf{z}) = \int_S \mathbf{k}(\mathbf{z},\mathbf{s})\mathbf{t}(\mathbf{s})\, ds, \quad \mathbf{J}(\theta,\mathbf{z}) = \int_S \mathbf{g}(\theta,\mathbf{z},\mathbf{s})\mathbf{t}(\mathbf{s})\, ds, \tag{13.62}$$

respectively, over surface $\mathbf{s} \in S$ from both the crack-face weight functions $\mathbf{k}(\mathbf{z},\mathbf{s})$ and tensor crack-face weight vector-function $\mathbf{g}(\theta,\mathbf{z},\mathbf{s})$. One may observe that symmetry requires $k_{12} = k_{13} = k_{21} = k_{23} = k_{31} = k_{32} = 0$; the tensile mode crack face weight functions k_{11} and the shear ones k_{22}, k_{33} are given in [936], [370]. By virtue of the fact that in linear fracture mechanics usually only SIF (13.62_1) plays an important role, it is not necessary to know all components of third rank tensor $\mathbf{g}(\theta,\mathbf{z},\mathbf{s})$, which do not coincide with the corresponding components of $\mathbf{k}(\mathbf{z},\mathbf{s})$ (13.62_1).

For a coin-shaped crack in an isotropic medium the relations of $\mathbf{K}(\mathbf{z})$ (13.62_1) may be significantly simplified. We will use the solution [324] according to which the representation (13.62_1) can actually be expressed in the convenient complex form. Kachanov and Laures [530] eliminated the singularities in proposed integrals by way of transfer to the new coordinate system; Buryachenko Buryachenko'2000b corrected some misprints made in [530] and [529] in these representations (see Appendix A.3.2).

At first, the problem for macrocrack S interacting with a single inclusion was considered (an analogous 2-D case was analyzed in [793] by the use of a standard integral representation for SIF $\mathbf{K}(\mathbf{z})$ at the crack tip $\mathbf{z} \in \Gamma^c$ (13.62_1) utilizing weight functions and exactly defined tractions $\mathbf{t}(\mathbf{s})$ ($\mathbf{s} \in S$) induced by the external loading and residual stresses. After that, let a coin-shaped crack of radius R^c with center $\mathbf{x}^c = (0, 0, 0)$ and unit normal $\mathbf{n} = (1, 0, 0)$ be located in an elastically homogeneous composite with statistically homogeneous field of transformed inclusions v_i $(i = 1, 2, \ldots)$. In the subsequent presentation we will make use of following definition for some functions $\mathbf{f}$ $(\boldsymbol{\sigma},\ \mathbf{t},\ \mathbf{J},\ \mathbf{K})$. The notations $\mathbf{f}(\mathbf{x}|v_i, \mathbf{x}_i)$ and $\mathbf{f}(\mathbf{x}|S, v_i, \mathbf{x}_i)$ denote the values of the random function $\mathbf{f}$ of surrounding inclusions in the point $\mathbf{x}$ under the condition that the inclusion v_i with center $\mathbf{x}_i$ is fixed (in the absence of a crack) and under the condition that the inclusion v_i and the crack S are fixed, respectively. The notations $\langle \mathbf{f}(\mathbf{x})|v_i, \mathbf{x}_i\rangle$ and $\langle \mathbf{f}(\mathbf{x})|S, v_i, \mathbf{x}_i\rangle$ are used for the conditional averages of appropriate random variables taken for the ensemble of a statistically homogeneous ergodic field $X = (v_j)$. An added sign ";" in the conditional average $\langle \mathbf{f}(\mathbf{x})|; v_i, \mathbf{x}_i\rangle$ denotes the case $\mathbf{x} \notin v_i$.

The stress field in any point $\mathbf{x}$ is defined by the superposition of the external loading $\boldsymbol{\sigma}^0$ and by the perturbation generated by both the transformed inclusions and by the crack

$$\boldsymbol{\sigma}(\mathbf{x}) = \boldsymbol{\sigma}^0 + \int \boldsymbol{\Gamma}(\mathbf{x} - \mathbf{y})[\boldsymbol{\beta}_1(\mathbf{y}) - \langle \boldsymbol{\beta}_1 \rangle]\, d\mathbf{y} + \mathbf{J}(\theta, \mathbf{z})/\sqrt{\rho}. \tag{13.63}$$

The expected value of the right-hand-side integral in Eq. (13.63) over ensemble realization vanishes and the expected value of the tensor $\mathbf{K}$ is defined by relation (13.62$_1$): $\langle \mathbf{K}(\mathbf{z})\rangle = \int \mathbf{k}(\mathbf{z}, \mathbf{s})\langle \mathbf{t}(\mathbf{s})\rangle\, ds$, where the expected value of the traction $\langle \mathbf{t}(\mathbf{s})\rangle$ may be found as $\langle \mathbf{t}(\mathbf{s})\rangle = -\mathbf{n}(\mathbf{s})\boldsymbol{\sigma}^0$. Thus the expectation value of SIFs $\mathbf{K}$ coincides with the stress intensity factor $\mathbf{K}^0(\mathbf{z})$ for an isolated crack inside the infinite medium without residual stresses:

$$\langle \mathbf{K}(\mathbf{z})\rangle \equiv \mathbf{K}^0(\mathbf{z}). \tag{13.64}$$

Hereafter the superscript "0" denotes SIFs (e.g. $\mathbf{K}^0, \mathbf{J}^0$) in the untransformed body ($\boldsymbol{\beta} \equiv \mathbf{0}$) subjected to the given external loading $\boldsymbol{\sigma}^0$, and the superscript "1" stands for the residual stress generated SIFs (called also the internal SIFs) (e.g. $\mathbf{K}^1, \mathbf{J}^1$). This result (13.64) is an exact one and does not depend on the microtopology of statistically homogeneous composites, this is a consequence of a self-equilibrium of internal residual stresses $\langle \boldsymbol{\sigma}\rangle(\mathbf{x}) \equiv \boldsymbol{\sigma}^0$. It is obvious that the use of this estimation makes no sense for strength calculation. The applications of Eqs. (13.64) do not seem suited to strength calculation, because the fracture takes place at a particular point in the vicinity of a crack tip. This point may be located inside either the matrix or in an inclusion. Therefore, for the fracture analysis the estimations of conditional averages either $\langle \mathbf{K}(\mathbf{z})|v_i, \mathbf{x}_i\rangle$ or $\langle \mathbf{K}(\mathbf{z})|v_0, \mathbf{x}_0\rangle$ under conditions that either the center of the inclusion v_i or the matrix v_0 are located in the point $\mathbf{x}_i$ or $\mathbf{x}_0$, respectively, are preferred over the average $\langle \mathbf{K}(\mathbf{z})\rangle$.

Let us locate an arbitrary inclusion v_i with the center $\mathbf{x}_i$ alongside the crack. Then the stress field $\boldsymbol{\sigma}(\mathbf{x}|S, v_i, \mathbf{x}_i)$ in any point $\mathbf{x}$ can be decomposed into

$$\boldsymbol{\sigma}(\mathbf{x}|S, v_i, \mathbf{x}_i) = \boldsymbol{\sigma}^{\mathrm{ac}}(\mathbf{x}|v_i, \mathbf{x}_i) + \mathbf{J}(\theta, \mathbf{z}|v_i, \mathbf{x}_i)/\sqrt{\rho}, \tag{13.65}$$

where $\boldsymbol{\sigma}^{\mathrm{ac}}(\mathbf{x}|v_i, \mathbf{x}_i)$ and $\mathbf{J}(\theta, \mathbf{z}|v_i, \mathbf{x}_i)$ are determined by the actions of the inclusions and the remote loading (in the absence of a crack), and by the crack provided that there is a fixed inclusion v_i with the center $\mathbf{x}_i$. To estimate of conditional average of (13.65) $\langle\boldsymbol{\sigma}(\mathbf{x}|S,v_i,\mathbf{x}_i)\rangle = \langle\boldsymbol{\sigma}^{\mathrm{ac}}(\mathbf{x})|v_i,\mathbf{x}_i\rangle + \langle\mathbf{J}(\theta,\mathbf{z}|v_i,\mathbf{x}_i)\rangle/\sqrt{\rho}$, over the ensemble realization of surrounding inclusions we start with an evaluation of the conditional average $\langle\boldsymbol{\sigma}^{\mathrm{ac}}(\mathbf{x})|v_i,\mathbf{x}_i\rangle$:

$$\langle\boldsymbol{\sigma}^{\mathrm{ac}}(\mathbf{x}) \mid v_i,\mathbf{x}_i\rangle = \boldsymbol{\sigma}^0 + \mathbf{T}_i^{\beta}(\mathbf{x}-\mathbf{x}_i)\bar{v}_i + \int \Big[\mathbf{T}_p^{\beta}(\mathbf{x}-\mathbf{x}_p)\bar{v}_p\varphi(v_p,\mathbf{x}_p|;v_i,\mathbf{x}_i) - \boldsymbol{\Gamma}(\mathbf{x}-\mathbf{x}_p)\langle\boldsymbol{\beta}_1\rangle\Big]\,d\mathbf{x}_p. \quad (13.66)$$

Here it is necessary to recognize two cases of the location $\mathbf{x}$: $\mathbf{x}\in v_i$ and $\mathbf{x}\notin v_i$. Hereafter for formula simplifications we will use the assumption of special composite structure, see (8.67), which includes the case of statistically isotropy of composites. Then in the first case when $\mathbf{x}\in v_i$ we obtain $\langle\boldsymbol{\sigma}^{\mathrm{ac}}(\mathbf{x})|v_i,\mathbf{x}_i\rangle = \boldsymbol{\sigma}^0 + \mathbf{Q}_i\langle\boldsymbol{\beta}_1\rangle + \mathbf{T}_i^{\beta}(\mathbf{x})$. If the relevant point $\mathbf{x}$ lies outside of the fixed inclusion v_i, Eq. (13.66) can be used. As can be seen from (13.66) the range of the action of surrounding inclusions v_p $(p\neq i)$ is localized in the neighborhood of fixed inclusion v_i, that is the so-called locality principle (see, e.g, Chapter 7).

We now turn our attention to the analysis of the case when the point considered $\mathbf{x}_0$ is located inside the matrix v_0. Analogously with (13.66) we find

$$\langle\boldsymbol{\sigma}^{\mathrm{ac}}(\mathbf{x})|v_0,\mathbf{x}_0\rangle = \boldsymbol{\sigma}^0 + \int \Big[\mathbf{T}_p^{\beta}(\mathbf{x}-\mathbf{x}_p)\bar{v}_p\varphi(v_p,\mathbf{x}_p|;v_0,\mathbf{x}_0) - \boldsymbol{\Gamma}(\mathbf{x}-\mathbf{x}_p)\langle\boldsymbol{\beta}_1\rangle\Big]\,d\mathbf{x}_p. \quad (13.67)$$

In a similar manner the estimation of conditional expectation of values $\langle\mathbf{t}(\mathbf{s})|v_i,\mathbf{x}_i\rangle$ and $\langle\mathbf{t}(\mathbf{s})|v_0,\mathbf{x}_0\rangle$ may be derived:

$$\langle\mathbf{t}(\mathbf{s}) \mid v_J,\mathbf{x}_J\rangle = -\mathbf{n}(\mathbf{s})\Big\{\boldsymbol{\sigma}^0 + \mathbf{T}_J^{\beta}(\mathbf{x}-\mathbf{x}_J)\bar{v}_J + \int \Big[\mathbf{T}_p^{\beta}(\mathbf{x}-\mathbf{x}_p)\bar{v}_p\varphi(v_p,\mathbf{x}_p|;v_J,\mathbf{x}_J) - \boldsymbol{\Gamma}(\mathbf{x}-\mathbf{x}_p)\langle\boldsymbol{\beta}_1\rangle\Big]\,d\mathbf{x}_p\Big\}, \quad (13.68)$$

Hereafter for shortening of representations the subscript $J=0,i$ $(i=1,2,\ldots)$ indicates the location of the point $\mathbf{x}_J$ inside either the matrix v_0 $(J=0)$ or the inclusion v_i $(J=i;\ i=1,2,\ldots)$; in so doing $\mathbf{T}_0^{\beta}(\mathbf{x}-\mathbf{x}_0)$ is taken as zero. When calculating the right-hand-side integrals in Eqs. (13.68) are carried out it is necessary to test the possible locations $\mathbf{s}$ with respect to the inclusion v_p : $\mathbf{s}\in v_p$ or $\mathbf{s}\notin v_p$ that can be calculated with ease.

After finding the average conditional tractions on the crack face described by Eqs. (13.68) we obtain the final result for the conditional expectation of values of both SIFs and the tensor SIFs:

$$\langle\mathbf{K}(\mathbf{z})|v_J,\mathbf{x}_J\rangle = \int_S \mathbf{k}(\mathbf{z},\mathbf{s})\langle\mathbf{t}(\mathbf{s})|v_J,\mathbf{x}_J\rangle\,ds, \quad (13.69)$$

$$\langle\mathbf{J}(\theta,\mathbf{z})|v_J,\mathbf{x}_J\rangle = \int_S \mathbf{g}(\theta,\mathbf{z},\mathbf{s})\langle\mathbf{t}(\mathbf{s})|v_J,\mathbf{x}_J\rangle\,ds, \quad (13.70)$$

respectively.

It is interesting to note that conditional average stresses inside each component with the absence of a crack $(i=1,2,\ldots)$:

$$\langle \boldsymbol{\sigma}^{\rm ac}\rangle_i = \boldsymbol{\sigma}^0 + \mathbf{Q}_i\Big(\langle \boldsymbol{\beta}_1\rangle - \langle \boldsymbol{\beta}_1\rangle_i\Big), \quad \langle \boldsymbol{\sigma}^{\rm ac}\rangle_0 = \frac{1}{c^{(0)}}\Big[\boldsymbol{\sigma}^0 - \sum_{i=1}^{N} c^{(i)}\langle \boldsymbol{\sigma}^{\rm ac}\rangle_i\Big] \quad (13.71)$$

do not depend on either of the conditional probability densities $\varphi(v_p, \mathbf{x}_p|; v_i, \mathbf{x}_i)$ or $\varphi(v_p, \mathbf{x}_p|; v_0, \mathbf{x}_0)$ of the inclusion arrangement. At the same time according to Eqs. (13.68) the conditional averages of SIFs (13.69) and (13.70) are explicitly expressed in terms of $\varphi(v_p, \mathbf{x}_p, |; v_1, \mathbf{x}_i)$ and $\varphi(v_p, \mathbf{x}_p|; v_0, \mathbf{x}_0)$. For $\mathbf{x}_i, \mathbf{x}_0$ at infinity far from the point at the crack tip, the formulae (13.69) and (13.70) lead to the vanishing results $\langle \mathbf{K}^1(\mathbf{z})|v_i, \mathbf{x}_i\rangle = \langle \mathbf{K}^1(\mathbf{z})|v_0, \mathbf{x}_0\rangle = \langle \mathbf{K}^1(\mathbf{z})\rangle \equiv \mathbf{0}$, and the mean stresses inside each component are defined by the relations (13.71).

13.4.2 Conditional Dispersion of SIF for a Crack in the Composite Medium

To obtain the conditional second moment of the SIF and, consequently, stresses in the neighborhood $\mathbf{x} \in v_i$ of the crack tip it is necessary to take the tensor product of (13.65) into $\boldsymbol{\sigma}(\mathbf{x}|S, v_i, \mathbf{x}_i)$: $\boldsymbol{\sigma}(\mathbf{x}|S, v_i, \mathbf{x}_i) \otimes \boldsymbol{\sigma}(\mathbf{x}|S, v_i, \mathbf{x}_i) = [\boldsymbol{\sigma}^{\rm ac}(\mathbf{x}|v_i, \mathbf{x}_i) + \mathbf{J}(\theta, \mathbf{z}|v_i, \mathbf{x}_i)/\sqrt{\rho}] \otimes [\boldsymbol{\sigma}^{\rm ac}(\mathbf{x}|v_i, \mathbf{x}_i) + \mathbf{J}(\theta, \mathbf{z}|v_i, \mathbf{z}_i)/\sqrt{\rho}]$. Averaging the last equation over ensemble realization in terms of Eqs. (13.67) and (13.68) leads to the conditional expectation composed of three terms:

$$\langle \boldsymbol{\sigma}(\mathbf{x}|S, v_i, \mathbf{x}_i) \otimes \boldsymbol{\sigma}(\mathbf{x}|S, v_i, \mathbf{x}_i)\rangle = \mathbf{I}_{20} + \mathbf{I}_{21} + \mathbf{I}_{22}. \quad (13.72)$$

The first term $\mathbf{I}_{20}$ is determined by the second moment of stresses in the point $\mathbf{x} \in v_i$ in the absence of a crack (13.58). $\mathbf{I}_{2i}$ $(i = 0, 1, 2)$ have the singularities $\rho^{-i/2}$ which are explicitly presented in [137]. The third term $\mathbf{I}_{22}$ (13.72) is most important for applications to fracture mechanics, because the expression (13.72) is a conditional elastic energy, and the contribution of $\mathbf{I}_{20}$, $\mathbf{I}_{21}$ to Rice's integral [935] equals zero.

If the matrix is located in the vicinity of a crack tip then the covariance matrix of the principal part of stresses (13.63) can be derived in much the same way as Eq. (13.72):

$$\begin{aligned}
&\langle\, \boldsymbol{\sigma}(\mathbf{x}|S, v_0, \mathbf{x}_0) \otimes \boldsymbol{\sigma}(\mathbf{x}|S, v_0, \mathbf{x}_0)\rangle \rho = \langle \mathbf{J}(\theta, \mathbf{z})|v_0, \mathbf{x}_0\rangle \otimes \langle \mathbf{J}(\theta, \mathbf{z})|v_0, \mathbf{x}_0\rangle \\
&+ \int \mathbf{J}^1(\theta, \mathbf{z}, v_p, \mathbf{x}_p) \otimes \mathbf{J}^1(\theta, \mathbf{z}, v_q, \mathbf{x}_q)\phi(v_p, \mathbf{x}_p|; v_0, \mathbf{x}_0)\, d\mathbf{x}_p \\
&+ \int\!\!\int \mathbf{J}^1(\theta, \mathbf{z}, v_p, \mathbf{x}_p) \otimes \mathbf{J}^1(\theta, \mathbf{z}, v_q, x_q)\varphi(v_p, x_p, |; v_0, \mathbf{x}_0) \\
&\times [\varphi(v_q, \mathbf{x}_q, |; v_p, \mathbf{x}_p; v_0, \mathbf{x}_0) - \varphi(v_q, \mathbf{x}_q|; v_0, \mathbf{x}_0)]\, d\mathbf{x}_q\, d\mathbf{x}_p, \qquad (13.73)
\end{aligned}$$

where $\langle \mathbf{J}(\theta, \mathbf{z})|v_0, \mathbf{x}_0\rangle$ can be found in a manner like (13.62_2), (13.68).

After the production of expressions (13.72) and (13.73) it is an easy matter to obtain the conditional second moment of SIF $\langle \mathbf{K}(\mathbf{z}|v_i, \mathbf{x}_i) \otimes \mathbf{K}(\mathbf{z}|v_i, \mathbf{x}_i)\rangle$, $\langle \mathbf{K}(\mathbf{z}|v_0, \mathbf{x}_0) \otimes \mathbf{K}(\mathbf{z}|v_0, \mathbf{x}_0)\rangle$. If $\mathbf{x}_1$ and $\mathbf{x}_0$ are infinitely far from the crack tip, evidently, SIF does not depend on the coordinate of the fixed point $\langle \mathbf{K}(\mathbf{z}|v_i, \mathbf{x}_i) \otimes \mathbf{K}(\mathbf{z}|v_i, \mathbf{x}_i)\rangle \to \langle \mathbf{K}(\mathbf{z}) \otimes \mathbf{K}(\mathbf{z})\rangle = \text{const.} \neq \mathbf{K}^0(\mathbf{z}) \otimes \mathbf{K}^0(\mathbf{z})$ and $\langle \mathbf{K}(\mathbf{z}|v_0, \mathbf{x}_0) \otimes \mathbf{K}(\mathbf{z}|v_0, \mathbf{x}_0)\rangle \to \langle \mathbf{K}(\mathbf{z}) \otimes \mathbf{K}(\mathbf{z})\rangle = \text{const.} \neq \mathbf{K}^0(\mathbf{z}) \otimes \mathbf{K}^0(\mathbf{z})$, respectively. The estimation method of the second moments of the tensor SIF were used in [137] to evaluate the statistical moments of any order of the tensor SIF. In the framework of binary interactions of inclusions, the principal part of a correction of nth order was estimated to the Gaussian approximation of the random variable $\langle \mathbf{J}(\theta, \mathbf{z}|v_i, \mathbf{x}_i)$.

13.4.3 Crack in a Finite Inclusion Cloud

In a wealth of practical problems there is a need for an analysis of finite inclusion field. So the high stresses in the vicinity of macroscopic crack induce a transformation toughening of zirconia (ZrO_2) inclusions embedded in a matrix of nontransforming ceramic. When ZrO_2 particles are unconstrained by a surrounding matrix, their transformation from tetragonal to a monoclinic crystal structure can be decomposed into a volume expansion of 4% and a shear of about 16% [118], [641]. The transformation in turn alters the stress distribution near the crack tip. In this connection we will consider the case of statistically inhomogeneous inclusion field, when numerical probability density is a function of current coordinate and equals zero outside of some domain $w^{\rm fin}$ with characteristic function $W^{\rm fin}$.

Then in much the same manner as [1155] (see also Eq. (12.61)) one can obtain a general equation for random stresses at any point in the absence of a crack (denoted by the superscript $^{\rm ac}$):

$$\boldsymbol{\sigma}^{\rm ac}(\mathbf{x}) = \boldsymbol{\sigma}^0 + \int \boldsymbol{\Gamma}(\mathbf{x}-\mathbf{y})\boldsymbol{\beta}_1(\mathbf{y})\, d\mathbf{y}, \tag{13.74}$$

$$\boldsymbol{\sigma}^{\rm ac}(\mathbf{x}) = \langle \boldsymbol{\sigma}^{\rm ac}\rangle(\mathbf{x}) + \int \boldsymbol{\Gamma}(\mathbf{x}-\mathbf{y})[\boldsymbol{\beta}_1(\mathbf{y}) - \langle \boldsymbol{\beta}_1\rangle(\mathbf{y})]\, d\mathbf{y}. \tag{13.75}$$

Here the average stress $\langle \boldsymbol{\sigma}^{\rm ac}\rangle(\mathbf{x}) \not\equiv \boldsymbol{\sigma}^0$ in contradistinction to the statistically homogeneous structure and can be defined by the relation

$$\langle \boldsymbol{\sigma}^{\rm ac}\rangle(\mathbf{x}) = \boldsymbol{\sigma}^0 + \int \boldsymbol{\Gamma}(\mathbf{x}-\mathbf{y})\langle \boldsymbol{\beta}_1\rangle(\mathbf{y})\, d\mathbf{y}. \tag{13.76}$$

Hereafter the notation $\langle(\cdot)\rangle(\mathbf{x})$ denotes statistical average in the point $\mathbf{x}$ over the ensemble realization of a the statistically inhomogeneous inclusion field. If $\langle \boldsymbol{\beta}_1\rangle(\mathbf{y}) \equiv \langle \boldsymbol{\beta}_1\rangle = \text{const.}$ in some finite domain $w^{\rm fin}$, the representation (13.76) may be simplified to $\langle \boldsymbol{\sigma}^{\rm ac}\rangle(\mathbf{x}) = \boldsymbol{\sigma}^0 + \int \boldsymbol{\Gamma}(\mathbf{x}-\mathbf{y})W^{\rm fin}(\mathbf{y})\, d\mathbf{y}\langle \boldsymbol{\beta}_1\rangle$. If besides $w^{\rm fin}$ is an ellipsoid with the center $\mathbf{x}^{\rm fin}$ then

$$\langle \boldsymbol{\sigma}^{\rm ac}\rangle(\mathbf{x}) = \begin{cases} \boldsymbol{\sigma}^0 - \mathbf{Q}(w^{\rm fin})\langle \boldsymbol{\beta}_1\rangle, & \mathbf{x} \in w^{\rm fin}, \\ \boldsymbol{\sigma}^0 + \mathbf{T}^{\rm fin}(\mathbf{x}-\mathbf{x}^{\rm fin})\langle \boldsymbol{\beta}_1\rangle \overline{w}^{\rm fin}, & \mathbf{x} \notin w^{\rm fin}, \end{cases} \tag{13.77}$$

where $\mathbf{T}^{\rm fin}(\mathbf{x}-\mathbf{x}^{\rm fin})$ and $\mathbf{Q}(w^{\rm fin})$ are defined analogously to $\mathbf{T}_i$ and $\mathbf{Q}_i$, respectively, with replacement of v_i by $w^{\rm fin}$.

From Eq. (13.75) we see that the analysis of the finite inclusion cloud $w^{\rm fin}$ is formally reduced to the replacement of $\boldsymbol{\sigma}^0$ on the determinate terms $\langle \boldsymbol{\sigma}^{\rm ac}\rangle(\mathbf{x})$ in the equation (13.75). Of course, as this takes place, the number probability density $\phi(v_i, \mathbf{x}_i)$ and conditional one $\phi(v_p, \mathbf{x}_p|; v_i, \mathbf{x}_i)$ determining the values of right-hand-side integrals in Eqs. (13.75) and (13.76), are functions of a coordinate $\mathbf{x}_i$ and are not invariants with respect to translations $\mathbf{x}_i \to \mathbf{x}_i + \mathbf{y}$.

Thus we obtain the estimations for both the average SIFs and the average tensor SIFs

$$\langle \mathbf{K}(\mathbf{z})\rangle = \mathbf{K}(\langle \boldsymbol{\sigma}^{\rm ac}\rangle, \mathbf{z}) \equiv -\int_S \mathbf{k}(\mathbf{z},\mathbf{s})\mathbf{n}(\mathbf{s})\langle \boldsymbol{\sigma}^{\rm ac}\rangle(\mathbf{s})\, ds, \tag{13.78}$$

$$\langle \mathbf{J}(\theta,\mathbf{z})\rangle = \mathbf{J}(\langle \boldsymbol{\sigma}^{\rm ac}\rangle, \theta, \mathbf{z}) \equiv -\int_S \mathbf{g}(\theta,\mathbf{z},\mathbf{s})\mathbf{n}(\mathbf{s})\langle \boldsymbol{\sigma}^{\rm ac}\rangle(\mathbf{s})\, ds. \tag{13.79}$$

For example, for the crack tip inside the statistically homogeneous inclusion cloud $w^{\rm fin}$ having an ellipsoidal shape, we have

$$\langle \mathbf{K}(\mathbf{z})\rangle = \mathbf{K}^0(\mathbf{z}) - \int_S \mathbf{k}(\mathbf{z},\mathbf{s})\mathbf{n}(\mathbf{s})W(\mathbf{s})\, ds\, \mathbf{Q}(w^{\rm fin})\langle \boldsymbol{\beta}_1\rangle. \tag{13.80}$$

This reproduces the known results from the two-dimensional models developed in the literature on transformation toughening of ceramics [641], [755], [1054]. The estimation (13.80) does not depend on the microgeometric structure of composite and is correct for the case when the distance between the crack tip and the cloud boundary is large enough.

However, the local distribution of the stresses near the crack tip plays a crucial role in linear fracture mechanics. Because of this let us now turn to the estimations of conditional expectation values of both the SIF and the tensor SIF. With this aim in view the conditional expectation of values of stresses can be found in much the same way as Eqs. (13.66) and (13.67):

$$\langle \boldsymbol{\sigma}^{\rm ac}(\mathbf{x})|v_J,\mathbf{x}_J\rangle = \langle \boldsymbol{\sigma}^{\rm ac}\rangle(\mathbf{x}) + \mathbf{T}^\beta_J(\mathbf{x}-\mathbf{x}_J)\bar v_J + \int \Big[\mathbf{T}^\beta_p(\mathbf{x}-\mathbf{x}_p) \\ \times \bar v_p \varphi(v_p,\mathbf{x}_p|;v_J,\mathbf{x}_J) - \boldsymbol{\Gamma}(\mathbf{x}-\mathbf{x}_p)\langle \boldsymbol{\beta}_1\rangle(\mathbf{x}_p)\Big]\, d\mathbf{x}_p, \tag{13.81}$$

where the subscript $J = 0, i$ $(i = 1,2,\ldots)$ indicates the location of the point $\mathbf{x}_J$ inside either the matrix v_0 $(J=0)$ or the inclusion v_i $(J=i;\ i=1,2,\ldots)$. Contrary to Eqs. (13.66) and (13.67) the right-hand-side of Eq. (13.81) consists of both the average stresses $\langle \boldsymbol{\sigma}^{\rm ac}(\mathbf{x})\rangle \not\equiv \boldsymbol{\sigma}^0$ estimated above by the relation (13.76) and the macroscopically inhomogeneous deterministic tensor $\langle \boldsymbol{\beta}_1\rangle(\mathbf{x}_p)$ describing the statistical inhomogeneous transformed field. Substituting Eq. (13.81) into (13.62_2) gives the conditional expectation of the value of the tensor SIF:

$$\langle \mathbf{J}(\theta,\mathbf{z})|v_J,\mathbf{x}_J\rangle = \langle \mathbf{J}^{\rm fluct}(\theta,\mathbf{z})|v_J,\mathbf{x}_J\rangle + \langle \mathbf{J}(\langle \boldsymbol{\sigma}^{\rm ac}\rangle,\theta,\mathbf{z})\rangle, \tag{13.82}$$

where $J=0,i;\ i=1,2,\ldots$, and the average tensor SIF $\langle \mathbf{J}(\langle \boldsymbol{\sigma}^{\rm ac}\rangle,\theta,\mathbf{z})\rangle \not\equiv \mathbf{J}^0(\theta,\mathbf{z})$ is estimated by the relations (13.76) and (13.79) for the macroscopically inhomogeneous cloud $w^{\rm fin}$ with the deterministic transformed field $\langle \boldsymbol{\beta}_1\rangle(\mathbf{x})$. The fluctuating constituent term $\langle \mathbf{J}^{\rm fluct}(\theta,\mathbf{z})|v_J,\mathbf{x}_J\rangle$ $(J=0,i;\ i=1,2\ldots)$ is defined by the formula

$$\langle \mathbf{J}^{\rm fluct}(\theta,\mathbf{z}) \mid v_J,\mathbf{x}_J\rangle = -\int_S \mathbf{g}(\theta,\mathbf{z},\mathbf{s})\mathbf{n}(\mathbf{s})\Big\{\mathbf{T}^\beta_J(\mathbf{x}-\mathbf{x}_J)\bar v_J + \int \Big[\mathbf{T}^\beta_p(\mathbf{x}-\mathbf{x}_p) \\ \times \bar v_p\varphi(v_p,\mathbf{x}_p|;v_J,\mathbf{x}_J) - \boldsymbol{\Gamma}(\mathbf{x}-\mathbf{x}_p)\langle \boldsymbol{\beta}_1\rangle(\mathbf{x}_p)\Big]\, d\mathbf{x}_p\Big\}\, ds, \tag{13.83}$$

and formally coincides with a similar representation for statistically homogeneous infinite inclusion field, although at the considered juncture the probability densities $\phi(v_i, \mathbf{x}_i)$ and $\phi(v_p, \mathbf{x}_p|; v_J, \mathbf{x}_J)$ $(J = 0, i;\ i = 1, 2, \ldots)$ (13.83) are sensitive to translations. Repeating the derivation of Eq. (13.73) gives the covariance matrix of the principal part of conditional average stresses in the components $(J = 0, i)$ $\langle \boldsymbol{\sigma}(\mathbf{x}|S, v_J, \mathbf{x}_J) \otimes \boldsymbol{\sigma}(\mathbf{x}|S, v_J, \mathbf{x}_J)\rangle$ which depends on both $\langle \mathbf{J}(\theta, \mathbf{z})|v_J, \mathbf{x}_J\rangle$ defined by Eq. (13.82) and $\mathbf{J}^1(\theta, \mathbf{z}, v_p, \mathbf{x}_p)$. As can be shown in a concrete numerical example, the conditional average $\langle \boldsymbol{\sigma}|v_i, \mathbf{x}_i\rangle(\mathbf{x})$ for $\mathbf{x} \in v_i \subset w^{\text{fin}}$ can differ significantly from the statistical average $\langle \boldsymbol{\sigma}\rangle(\mathbf{x})$, that leads to all the more difference of estimation obtained by the use of the popular formula (13.79) as compared to more accurate relations, see, e.g., Eq. (13.82).

13.4.4 Numerical Estimation of the First and Second Statistical Moments of Stress Intensity Factors

As an example we consider Si_3N_4 composite (already considered in Subsection 13.3.3) containing the identical SiC spherical inclusions with the radius $a = 10^{-5}$m. For the representation of numerical results in dimensionless form we define the normalizing coefficient $\vartheta \equiv -3Q^k\beta_{10}\sqrt{a}$, where $3Q^k = \mathbf{Q}_i : \mathbf{N}_1$; for our concrete composite SiC/Si_3N_4: $\vartheta = 0.914$ MPa$\sqrt{\text{m}}$, $\beta_{10} = -10^{-3}$. We will estimate the action of the alternative radial distribution function for an inclusion (9.91) and (9.93).

At first we consider (see [137]) one fixed inclusion v_1 with the center $\mathbf{x}_1 = (0, R^c + r, 0)$ near the crack tip $\mathbf{z} = (0, R^c, 0) \in \boldsymbol{\Gamma}^c$, $S \perp \mathbf{n} = (1, 0, 0)$, and $\boldsymbol{\sigma}^0 \equiv \mathbf{0}$. Figure 13.12 shows the residual stress generated normalized mode I SIF $K_I^1(\mathbf{z}, v_1, \mathbf{x}_1)/\vartheta$ as a function of normalized distance r/a from the inclusion center $\mathbf{x}_1 = (0, R^c + r, 0)$ to the crack tip $\mathbf{z}$ for the different relative sizes of the crack $R^c = 1000a$, $R^c = 10a$ and $R^c = 3a$. It can be seen from Fig. 13.12 that the variation between calculated values $K_I^1(\mathbf{z}, v_1, \mathbf{x}_1)/\vartheta$ is not more than 4% for $R^c = 1000a$ and $R^c = 3a$ under $|r| < a$; for $R^c = 1000a$ and $R^c = 10a$ this quantity is less than 2% $\forall r$. Therefore for $R^c > 10a$ the crack may be considered as a semi-infinite crack; in the following, unless otherwise specified, we will consider the case $R^c = 100$a. It is interesting that if inclusion is located in the point $r/a \cong 0.93$ (as read approximately from the graphs) it does not initiate a stress singularity near crack tip $K_I^1(\mathbf{z}, v_1, \mathbf{x}_1) = 0$ (in view of the problem symmetry $K_{II}^1(\mathbf{z}, v_1, \mathbf{x}_1) = K_{III}^1(\mathbf{z}, v_1, \mathbf{x}_1) \equiv 0\ \forall r$). Hereafter instances of negative $K_I^1(\mathbf{z}, v_1, \mathbf{x}_1) = 0$ are simply numerical results related to the compression, and have no physical meaning in the sense of material overlap or penetration. Addition of remote loading generating $K_I^0(\mathbf{z})$ will increase the total stress intensity factor $K_I(\mathbf{z}, v_1, \mathbf{x}_1) = K_I^0(\mathbf{z}) + K_I^1(\mathbf{z}, v_1, \mathbf{x}_1)$ as started previously. If in so doing $K_I(\mathbf{z}, v_1, \mathbf{x}_1) > 0$ then the negative residual stress generated SIF $K_I^1(\mathbf{z}, v_1, \mathbf{x}_1)$ has a physical meaning of the shielding (or unloading) effect. For example, at the spacing $r/a < -1$ the crack-inclusion interaction results in shielding: $K_I(\mathbf{z}, v_1, \mathbf{x}_1) < K_I^0(\mathbf{z})$. Impact of the inclusion on the crack is highly localized in the region $|r| < 2a$ and rapidly becomes negligible at further points. This indicates, that, in the case of a crack interacting with many inclusions, one can expect a short-range interaction effect. This seems to imply that the size of so-called representative volume element (RVE) will be small enough.

In order to illustrate the method and to examine the influence of the spatial arrangement of inclusions we come now to the evaluation of conditional averages $\langle \mathbf{K}^1(\mathbf{z})|v_J, \mathbf{x}_J\rangle$ $(J = 0, 1)$ (13.70). For the fixed inclusion the normalized curves $\langle K_I^1(\mathbf{z})|v_1, \mathbf{x}_1\rangle/\vartheta$ are calculated in Fig. 13.13 by the use of the step radial distribution function (9.91) and the real one (9.92) for $c = 0.4$. For the matrix located in the point $\mathbf{x}_0 = (0, R^c + r, 0)$ the relevant curve is calculated under the step correlation function (13.61).

Notwithstanding the fact that the conditional average stresses inside each component in the absence of a crack $\langle \boldsymbol{\sigma}^{\rm ac}\rangle_1$ (13.71$_1$) and $\langle \boldsymbol{\sigma}^{\rm ac}\rangle_0$ (13.71$_2$) do not depend on the radial distribution function g, the significant influence of neighboring order in localization of inclusions on the values of $\langle K_I^1(\mathbf{z})|v_J, \mathbf{x}_J\rangle/\vartheta$ $(J = 0, 1)$ is evident from Fig. 13.13. Such a strong influence of g on the conditional average SIF is explained by essentially nonlinear dependence of SIF on the local stresses near the crack tip. In this connection it should be mentioned that the analogous two-dimensional model representation of a two-component material containing randomly fluctuating residual stresses was analyzed in [689] by the use of the Monte Carlo simulations of the microstructure with a wide distribution of inclusion sizes. For $c = 0.3$ and a single case of the fixed inclusion v_1 with the center $\mathbf{x}_1 = (0, R^c + a, 0)$ Lipetzky and Kreher evaluated $\langle K_{II}^1(\mathbf{z})|v_1, \mathbf{x}_1\rangle = 0$ and $\langle K_I^1(\mathbf{z})|v_1, \mathbf{x}_1\rangle = -0.5\text{MPa}\sqrt{\text{m}}$ in conformity with Fig. 13.13: $\langle K_I^1(\mathbf{z})|v_1, \mathbf{x}_1\rangle = -0.7\text{MPa}\sqrt{\text{m}}$.

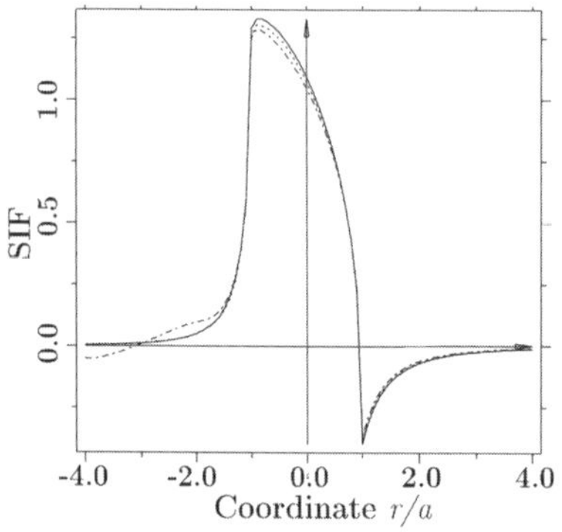

Fig. 13.12. Normalized modes I SIF $K_I^1(\mathbf{z}, v_1, \mathbf{x}_1)/\vartheta$ vs r/a for different crack sizes $R^c = 1000a$ (solid curve), $R^c = 10a$ (dotted curve), and $R^c = 3a$ (dot-dashed curve).

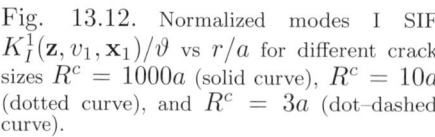

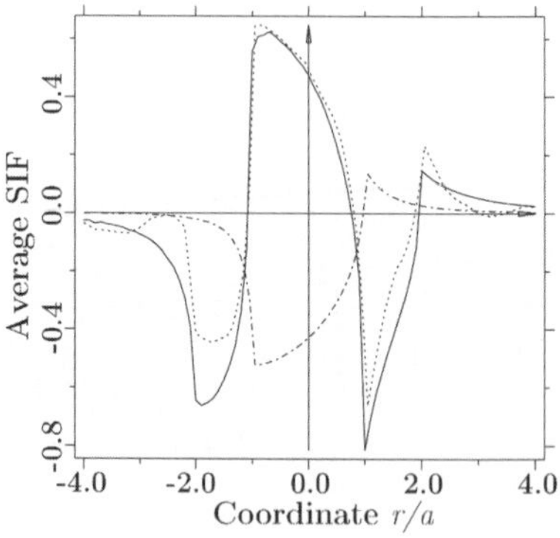

Fig. 13.13. Normalized averages $\langle K_I^1(\mathbf{z})|v_0, \mathbf{x}_0\rangle/\vartheta$ (dot-dashed curve) and $\langle K_I^1(\mathbf{z})|v_1, \mathbf{x}_1\rangle/\vartheta$ (solid and dotted curves) estimated for either (9.92) (dotted curve) or (9.91), (13.61) (solid and dot-dashed curve).

For zero remote loading ($\boldsymbol{\sigma}^0 \equiv \mathbf{0}$) in Figs. 13.14 and 13.15 the normalized curves $[\Delta K_i^{1|2}(\mathbf{z}|v_k, \mathbf{x}_k)]^{1/2}/\vartheta \sim r/a$ ($i =$I, II; $k = 0, 1$; $\Delta K_i^{1|2}(\mathbf{z}|v_k, \mathbf{x}_k) \equiv \langle (K_i^1)^2(\mathbf{z}|v_k, \mathbf{x}_k)\rangle - \langle K_i^1(\mathbf{z}|v_k, \mathbf{x}_k)\rangle^2$) are plotted for the step radial function (9.91) and the real one (9.92) (similar results take place for $i = III$, $k = 0, 1$). Analogous results for the matrix located in the point $\mathbf{x}_0 = (0, R^c + r, 0)$ are represented for the case of a step correlation function (13.61) only. We see that the fluctuation of the mode II SIF (and the mode III SIF as well) are not equal to zero, although there is symmetry with the problem being analyzed, and, therefore, $\langle K^1_{II}(\mathbf{z}|v_k, \mathbf{x}_k)\rangle = \langle K^1_{III}(\mathbf{z}|v_k, \mathbf{x}_k)\rangle \equiv \mathbf{0}$ at $\boldsymbol{\sigma} = 0$. The reason why the conditional fluctuations of SIF under matrix cracks are materially greater than fluctuations of SIF with the fixed inclusion is clear. Really, these values defined by the action of surrounding inclusions v_p $(p = 2, 3, \ldots)$, but the distances from the center $\mathbf{x}_p$ of the nearest inclusion v_p $(p = 2, 3, ...)$ to the fixed point $\mathbf{x}_J$ $(J = 0, 1)$ differ among themselves: $|\mathbf{x}_p - \mathbf{x}_1| = 2|\mathbf{x}_p - \mathbf{x}_0|$. The minimum of $\Delta K_i^{1|2}(\mathbf{z}|v_J, \mathbf{x}_J)$ $(i = I, II, III;\ J = 0, 1)$ occurs for the location of the fixed point near a crack tip $\mathbf{z}$. For the remote field points $\mathbf{K}^{1|2}(\mathbf{z}|v_J, \mathbf{x}_J) \to \text{const.} \neq \mathbf{0}$ under $|\mathbf{x}_J - \mathbf{z}| \to \infty$ $(J = 0, 1)$. For the step radial functions (9.91) and (13.61) $\Delta\mathbf{K}^{1|2}(\mathbf{z}|v_J, \mathbf{x}_J)/c$ does not depend on the inclusion concentration. The fluctuations of SIFs $\Delta\mathbf{K}^{1|2}(\mathbf{z}) \neq \mathbf{0}$ in spite of the exact result $\langle\mathbf{K}(\mathbf{z})\rangle \equiv \mathbf{0}$ at $\boldsymbol{\sigma}^0 \equiv \mathbf{0}$.

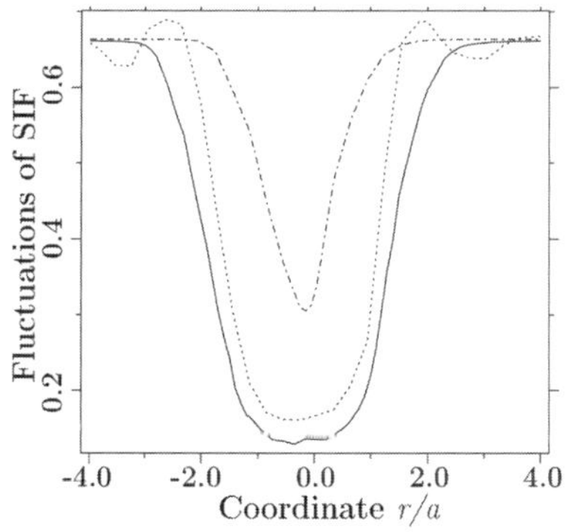

Fig. 13.14. $[\Delta K_I^{1|2}(\mathbf{z}|v_k, \mathbf{x}_k)]^{1/2}/\vartheta$ vs r/a (solid and dotted curves, $k = 1$) and (dot-dashed curve, $k = 0$) calculated for (9.92) (dotted curve) and (9.91) and (13.61) (solid and dot-dashed curve, respectively).

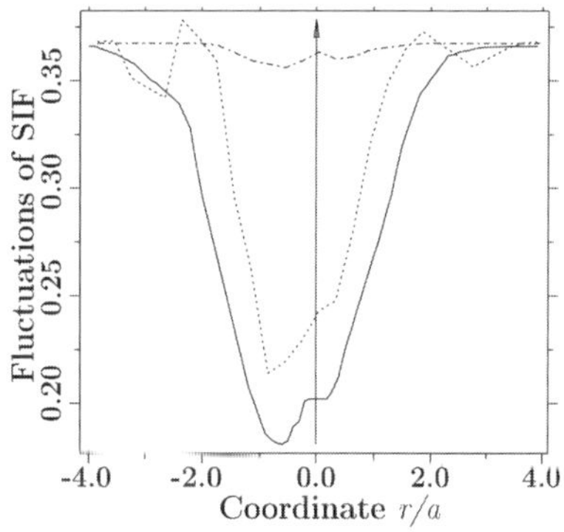

Fig. 13.15. Normalized SIF fluctuation of mode II $[\Delta K_{II}^{1|2}(\mathbf{z}|v_k, \mathbf{x}_k)]^{1/2}/\vartheta$ (the notation of Fig. 13.14).

The curves in Figs. 13.12–13.15 depend only on Poisson's ratio ν and are invariants with respect to both the other thermoelastic properties of components and the inclusion size. But for the range of values $0.2 \leq \nu \leq 0.4$ the calculated curves for both the conditional averages and fluctuations of SIF vary at most 0.5% for mode I and 4% for modes II, III.

13.5 Concluding Remarks

It should be mentioned that the proposed method of integral equations (13.37), (13.45), and (13.46) is just one among others that can be applied for estimating the second moment of the stresses $\langle\boldsymbol{\sigma} \otimes \boldsymbol{\sigma}\rangle^{(k)}$ $(k = 0, 1, \ldots, N)$. However, if for example the perturbation method is used leading to Eq. (13.15) even in the simplest of the considered cases of actually isotropic matrix materials it is necessary to estimate the effective compliance $\mathbf{M}^*$ for the general case of anisotropy of the matrix (which is not required for the proposed method). This estimate could be derived by methods with simple explicit representation of $\mathbf{M}^* = \mathbf{M}^*(\mathbf{M}^{(0)})$. However, we demonstrated in Chapter 8 the advantage of the MEFM in comparison with some popular methods (as, e.g., effective medium [607] and Mori–Tanaka method [787]) with respect to the estimation of the effective compliance $\mathbf{M}^*$. Furthermore, if these more approximative methods are used for estimation $\langle \boldsymbol{\sigma} \otimes \boldsymbol{\sigma} \rangle^{(k)}$ $(k = 0, 1, \ldots, N)$ the results would be invariant with respect to the conditional probability densities $\varphi(v_j, \mathbf{x}_j|; v_i, \mathbf{x}_i)$. In contrast to this, the estimations $\langle \boldsymbol{\sigma} \otimes \boldsymbol{\sigma} \rangle^{(k)}$ $(k = 0, 1, \ldots, N)$ obtained by the proposed method (13.37), (13.45), and (13.46) depend explicitly on the functions $\varphi(v_j, \mathbf{x}_j|; v_i, \mathbf{x}_i)$ as well as on the average stresses in the components.

For elastoplastic analyses based on estimations of some nonlinear functions of local stresses, e.g., the yield condition, taking stress inhomogeneities in the components into account, it is very popular to use secant and tangent moduli concepts (see the references in Chapter 16). This way the nonlinear problem at each solution increment reduces to the averaging linear elastic problem with $\boldsymbol{\beta} \equiv \mathbf{0}$. The use of the secant modulus concept creates known complications, since generally the local stress state is not monotonical and proportional even with monotonical and proportional external loading. The tangent moduli concept leads to the necessity of considerating also the matrix material as being anisotropic at each solution step. This would not lead to any problem in the framework of the "quasi-crystalline" approximation (8.65), but it leads to some computing difficulties at the realization of the MEFM for which advantages in comparison with some popular methods were justified (see Chapter 8). However, the integral representations for stress fluctuations (13.37), (13.45), and (13.46) permit the use of the incremental method with fixed elastic properties of the components and with accumulating plastic strains ($\boldsymbol{\beta}^{(i)} \neq$ const.) (see Chapter 16) without the assumption of homogeneity of plastic strains in the matrix (see Chapter 16). With some additional assumptions considered in Chapter 12 the method of integral equations can be used for the analysis of statistically inhomogeneous composites, i.e., when the concentration of the inclusions is a function of the coordinates. Moreover, we will demonstrate an applicability of this method for analysis of stress fluctuations in a microinhomogeneous half-space (see Chapter 14). The solution of the indicated problems of estimating all components of the second moments $\langle \boldsymbol{\sigma} \otimes \boldsymbol{\sigma} \rangle^{(k)}$ $(k = 0, 1, \ldots, N)$ by the perturbation method (13.15) as well as by variational methods [897] might be a challenging idea.

Finally, the representations for elastically homogeneous composites (9.33), (13.54), (13.58), and (13.60) obtained are exact. For some additional assumptions in the concrete examples it was shown the second moment of stresses inside inclusions are essentially inhomogeneous functions and changed by a factor 2 or more along the inclusion radius (see Fig. 13.11). Moreover even under the evaluation of average statistical moments of stresses inside the constituents the method proposed makes it possible to improve the estimation by 34% over other methods based on hypothesis **H1** (8.14).

14

Random Structure Matrix Composites in a Half-Space

14.1 General Analysis of Approaches in Micromechanics of Random Structure Composites in a Half-space

As analysis of publications shows (see, e.g., the references in Chapters 8 and 9), most micromechanical research done relates to the composite's bulk, whereas relatively few publications can be found that apply to a study of the edge effects in composites. At the same time, the mentioned problem is of major practical importance. A well-known fact is that in many cases strength of a solid is governed by its near-surface area being a zone of initiating and accumulating the cracks. The first reason for it is accumulation near the surface of the defects resulting from fabrication, service and environment. The second one is that the near-surface stress field differs substantially from that in the bulk and often appears to be favorable for the crack initiation and propagation. Therefore, to get a reliable estimate of a composite product's strength, the edge effects should be taken into account. The investigation of the free edge effect in micromechanics of composites has a long history initiated in pioneer papers by Pagano (see [850], [928], [953]). Significant progress in this field was achieved for the composites of regular structures (for references see [2], [851], and Chapter 12). If the material is locally nonperiodic, such as for the case of periodic media in the vicinity of the free edge, the asymptotic expansion technique generally leads to the poor approximation of the local fields. These difficulties have been partially overcome by employing an asymptotic expansion of a different nature in the boundary layer region and matching the solutions in the outer and inner regions using one of the matching rules [961]. The theoretical foundation of this boundary layer and edge effects in homogenization theory were developed in [961]; the appropriate analyses were carried out for laminated composites [39], [286], [848], as well as for composites reinforced by fibers [340] and by particulates [827]. Pagano and Yuan [851] (where additional references can be found) proposed a combination of the effective medium and micromechanical theories leading to reasonably accurate interface stresses. The scheme of matching rules for the analysis of a rectilinear crack in a composite material of doubly periodic structure [534] (see also [532]) was developed. However, the analogous results for the random structure composites are significantly less developed. The perturbation method for weakly inhomogeneous media was proposed for the free-edge analysis of microinhomogeneous half-space by Podalkov and Romanov [886] (see also [699], [700]), who constructed dispersions of the strains by this method in a first-order approximation. Luciano and Willis [717] proposed a methodology for obtaining a nonlocal effective constitutive operator in the vicinity of the free edge of composites with any elastic contrast between the constituent properties; they numerically analyzed a thermal conductivity problem for a random 2-D model structure specified by a two-point exponential correlation function. However, to the author's knowledge, the analogous elastic problem for an arbitrary elastic contrast between the constituent properties is solved only for simplest random structures, such as a dilute concentration of spherical inclusions [633]. The last classical problem for the 2-D case will be considered in this chapter.

It should be mentioned that self-assembled quantum dots (QDs) (and quantum wires, QWs) have attached substantial attention in recent years [86], [426]. A semiconductor QD is a small volume of one semiconductor material, with dimensions of usually a few nanometers, buried within a second semiconductor material called a matrix. QD and QW structures are fabricated by spontaneous growth of small islands (including zero-dimensional dots and one-dimensional wires) from a wetting layer due to its mismatch strain to the substrate according to the Stranski-Krastanov (SK) mechanism. The QD and QW ordering are driven by the elastic field of subsurface stressors, which affects both a datom diffusion and SK island nucleation rates. The stress fields caused by QDs and QWs strongly affect the electronic properties in the vicinity of the dots. To understanding both ordering and electronic properties formation, it is important to determine the statistical residual elastic stress distribution in both the dots and matrix. In the framework of a continuum mechanics approach, QDs and QWs can be modeled as the inclusions of some prescribed shape embedded in a dissimilar matrix half-space. Due to compositional differences of the dots and matrix, the inclusions may possess a lattice parameter mismatch or possibly a thermal expansion mismatch as well as different elastic moduli from the matrix material. However, a common simplification assumed by many researchers is that the thermal or lattice mismatch plays the dominant role, and an elastic mismatch can be neglected. In such a case among the most prominent continuum techniques for stress analysis (the finite elements, Green's function, and the Fourier series; see, e.g., [308], [794], the Green's function method acquires additional benefits. It is explained by the fact that the stress fields acting on the considered inclusion from the surrounding ones does not depend on the stress distribution inside this inclusion and can be obtained by direct superposition.

Independent of practical applications, the general inclusion problem (with zero elastic mismatch) initiated in a pioneer paper by Eshelby [318] was extensively developed (see for references, e.g., [167], [845]), and we consider this approach now for the inclusion with a homogeneous eigenstrain (and zero elastic mismatch) in an isotropic half-space. The solution for an ellipsoidal inclusion with a dilatational eigenstrain. The solution for a semispherical inclusion in a half-space has been obtained in [1196] was [985] (see also [1209]) analyzed. A facetted (parallepipedic) inclusion in a half-space was analyzed in [386] (see also [478]) by the use of the expressions for the elastic displacements of a platelike dilatational inclusion whose eigenstrain varied in one direction in an oscillatory manner. As a limiting case of the parallepipedic inclusion, Glas [386] also estimated the stress field of an infinite wire inclusion with a rectangular cross section (the case of an array of these inclusions was considered in [531]), parallel to a free surface. Glas [387] has also generalized the mentioned approach to the inclusions with the shapes of a truncated pyramid and infinite trapezoidal wire that were modeled by continuous distributions of infinitely thin parallepipedic inclusions. The same approach was used in [388] for the estimation of strains induced by a finite length circular cylindrical inclusion parallel to a flat free surface; this solution amplifies the previous results for the finite-length cylinder perpendicular to a free surface [447], [1197]. The solutions of many problems in the half-space were extended to the case of a two-phase infinite medium with a flat interface boundary [590], [1153], [1209]. This class of problems can effectively be solved by the Green function method through integration of the elastic field created by a unit point force over the inclusion volume. Yang and Pan [1202] have also used the Green's function method in the form of the boundary element method for the analysis of finite numbers of cuboidal inclusions in anisotropic and linearly elastic multilayers, derived within the framework of generalized Stroh formalism and Fourier transforms, in conjunction with the Bettis reciprocal theorem. For a cuboidal inclusion with uniform eigenstrains, analytical formulae for influence coefficients in terms of derivatives of four key integrals were [695] obtained. Arbitrary distributed eigenstrains in a half-space were analyzed by the discrete correlation and fast Fourier transform algorithm, along with the discrete convolution and fast Fourier transform algorithm.

Along with 3-D problems, the solutions of 2-D problems for the inclusions of arbitrary shape in a two-phase plane were also derived [948], [1057] using the conformal mapping approach. The method proposed in [948] is based on the use of an auxiliary function constructed from the conformal mapping of the exterior of the inclusion onto the exterior of the unit circle and analytical continuation. The solution obtained is exact provided that the expansion of the mapping function reduces to only a finite number of terms. The solutions in isotropic 2-D solids with elliptical inclusion were generalized to the case of elastic anisotropic half-space [744] and two-phase anisotropic medium [1218]. An inclusion of arbitrary shape with the uniform eigenstrain and vanishing elastic mismatch was also analyzed by the conformal-mapping technique in [946]. However, the use of this technique is troublesome in the case of an elliptical inclusion with elastic mismatch because the exterior and interior of the heterogeneity should be simultaneously mapped onto a plane with a simple interface. In such a case, Kushch *et al.* [633] combined the Muskhelishvili [798] method of complex potentials with the Fourier integral transform technique reducing a primary boundary-value elastostatics problem for a piece-homogeneous domain to an ordinary well-posed set of linear algebraic equations. A similar problem was also solved by the boundary integral equations method [662] as well as by the volume integral equation method (see [277] and Chapter 4). A fundamental role of both the free edge effect and boundary conditions yielding the redistribution of local stresses in the heterogeneity was [13] demonstrated. Buryachenko and Pagano [167] (see Chapter 4) considered the hybrid BIE and VIE method for multiscale analysis of multiple interacting inclusions of essentially different scale with its distinctive in 10^7 times. The stress distribution near either the large hole or large stiff inclusion provides a way to model the problems for small inclusions in a half-space with either the free or fixed edge, respectively. The generalization of the scheme presented for the analysis of microinhomogeneous media with fixed macroinclusion is well worth further investigation. In so doing, the general integral equations (7.20) and (7.37) should take into account the surface integral over the fixed macroinclusion. In addition to the surface integral over the macroinclusion, the general scheme of the MEFM can be modified by entering the dependence of effective fields on the distance from the macroboundary. So, in a simplest case of a dilute approximation (see Subsection 12.5.4) the effective field $\overline{\boldsymbol{\sigma}}(\mathbf{x})$ ($\mathbf{x} \in v_i$) will depend on the absolute coordinate (not only on $\mathbf{x}_i - \mathbf{x}$) but not on the other inclusions. The last dependence may be estimated in the framework of either the one-particle or multiparticle approximation of the MEFM, when the matrix of binary interactions $\boldsymbol{\mathcal{T}}_{iq}$, $\boldsymbol{\mathcal{F}}_{iq}$ will depend on their absolute positions.

The related phenomena of a size effect in microinhomogeneous media considered in [446], [484], [485], [843], [865], [1198], should be mentioned. The authors detected the dependence of effective elastic properties obtained by the volume average of both the local strains and stresses in the mesosample on the size of the samples being analyzed as well as on the boundary conditions (Dirichlet or Neumann type) acting either on this sample or on the subdomain in the domain decomposition method [1236]. In view of the aforementioned these effects can be explained by the multiple interactions between inclusions as well as by the interactions of inclusions with the boundary of the sample. Because of this the effective constitutive equation produced via the ensemble averaging procedure will be a nonlocal one and both local and nonlocal parts of the effective operator will depend on the coordinate of the macropoint being considered.

At first, in the present chapter, the general methods of estimation of both the stress concentrator factors and nonlocal effective operators will be presented for the matrix composites with an arbitrary elastic and thermal mismatch of constituents in a half-space. At the end of the chapter following [153], a generalization of the method of integral equations [136] (see Section 14.3) is proposed for the estimation of the first and second moments of random residual microstresses in the constituents of elastically homogeneous composites in a half-space with a free edge. Explicit relations for these statistical moments are obtained taking the binary interactions of the inclusions into account which are expressed through the numerical solution for one inclusion in the half-space. The statistical averages of stress fluctuations varying along the inclusion cross sections are completely defined by the random locations of surrounding inclusions. The numerical results are presented for a half-plane containing a random distribution of circular identical inclusions modeling a random array of aligned quantum wires parallel to a free surface. The solution for one inclusion is obtained in [632].

14.2 General Integral Equation, Definitions of the Nonlocal Effective Properties, and Averaging Operations

At first we will present a slight modification of general integral equations for both the stresses and effective stresses proposed in Chapter 7 for both statistically homogeneous and inhomogeneous composites in whole space. Adaptation of the mentioned equations to the composite materials in a half-space is a straightforward and based on replacement of a Green function for a whole space by a Green function for a half-space.

Indeed, substituting of the constitutive (6.2) and Cauchy (6.3) equations into the equilibrium equation (6.1) (at $\mathbf{f} \equiv \mathbf{0}$), we obtain a differential equation with respect to the displacement $\mathbf{u}$ which may be reduced to a symmetrized integral form for the stresses

$$\boldsymbol{\sigma}(\mathbf{x}) = \langle\boldsymbol{\sigma}\rangle(\mathbf{x}) + \int \boldsymbol{\Gamma}(\mathbf{x},\mathbf{y})\left\{\boldsymbol{\eta}(\mathbf{y}) - \langle\boldsymbol{\eta}\rangle(\mathbf{y})\right\}\, d\mathbf{y} \tag{14.1}$$

which can be justified in a similar manner as the analogous Eqs. (7.12) and (7.20). Here the integral operator kernels $\mathbf{U}$ and $\boldsymbol{\Gamma}$ are the generalized functions of the order $-d$ defined by the Green tensor $\mathbf{G}$: $U_{ijkl}(\mathbf{x},\mathbf{y}) = \left[\nabla_j \nabla_l G_{ik}(\mathbf{x},\mathbf{x})\right]_{(ij)(kl)}$, $\boldsymbol{\Gamma}(\mathbf{x},\mathbf{y}) = -\mathbf{L}^{(0)}\left[\mathbf{I}\delta(\mathbf{x}-\mathbf{y}) + \mathbf{U}(\mathbf{x},\mathbf{y})\mathbf{L}^{(0)}\right]$, where $\mathbf{G}$ is the Green's function of the Navier equation with an elastic modulus tensor $\mathbf{L}^{(0)}$: $\nabla\{\mathbf{L}^{(0)}[\nabla\otimes\mathbf{G}(\mathbf{x},\mathbf{y}) + (\nabla\otimes\mathbf{G}(\mathbf{x},\mathbf{y}))^\top]/2\} = -\boldsymbol{\delta}\delta(\mathbf{x}-\mathbf{y})$, defined for the semi-infinite body w and the given boundary conditions (see e.g. [28], [717]).

Let the inclusions $v_1,\ldots,v_n$ be fixed, and we define two sorts of effective fields $\overline{\boldsymbol{\sigma}}_i(\mathbf{x})$ and $\widetilde{\boldsymbol{\sigma}}_{1,\ldots,n}(\mathbf{x})$ ($i = 1,\ldots,n$; $\mathbf{x} \in v_1,\ldots,v_n$) by the use of the rearrangement of Eq. (14.1) in the following form (see the earliest references of related manipulations Chapters 8 and 12):

$$\boldsymbol{\sigma}(\mathbf{x}) = \overline{\boldsymbol{\sigma}}_i(\mathbf{x}) + \int \boldsymbol{\Gamma}(\mathbf{x},\mathbf{y})V_i(\mathbf{y})\boldsymbol{\eta}(\mathbf{y})\, d\mathbf{y}, \tag{14.2}$$

$$\overline{\boldsymbol{\sigma}}_i(\mathbf{x}) = \widetilde{\boldsymbol{\sigma}}_{1,\ldots,n}(\mathbf{x}) + \sum_{j\neq i}\int \boldsymbol{\Gamma}(\mathbf{x},\mathbf{y})V_j(\mathbf{y})\boldsymbol{\eta}(\mathbf{y})\, d\mathbf{y}, \tag{14.3}$$

$$\widetilde{\boldsymbol{\sigma}}_{1,\ldots,n}(\mathbf{x}) = \langle\boldsymbol{\sigma}\rangle(\mathbf{x}) + \int \boldsymbol{\Gamma}(\mathbf{x},\mathbf{y})\left\{\boldsymbol{\eta}(\mathbf{y})V(\mathbf{y}|;v_1,\mathbf{x}_1;\ldots;v_n,\mathbf{x}_n) - \langle\boldsymbol{\eta}\rangle(\mathbf{y})\right\}\, d\mathbf{y}, \tag{14.4}$$

for $\mathbf{x} \in v_i$, $i = 1,2,\ldots,n$. Then, considering some conditional statistical averages of the general integral Eq. (14.1) leads to an infinite system of integral equations ($n = 1,2,\ldots$)

$$\langle\boldsymbol{\sigma}|v_1,\mathbf{x}_1;\ldots;v_n,\mathbf{x}_n\rangle(\mathbf{x}) - \sum_{i=1}^{n}\int \boldsymbol{\Gamma}(\mathbf{x},\mathbf{y})\langle V_i(\mathbf{y})\boldsymbol{\eta}|v_1,\mathbf{x}_1;\ldots;v_n,\mathbf{x}_n\rangle(\mathbf{y})\, d\mathbf{y}$$
$$= \langle\boldsymbol{\sigma}\rangle(\mathbf{x}) + \int \boldsymbol{\Gamma}(\mathbf{x},\mathbf{y})\left\{\langle\boldsymbol{\eta}|;v_1,\mathbf{x}_1;\ldots;v_n,\mathbf{x}_n\rangle(\mathbf{y}) - \langle\boldsymbol{\eta}\rangle(\mathbf{y})\right\}\, d\mathbf{y}, \tag{14.5}$$

which can be also represented in the terms of the strain polarization tensor

$$\langle \boldsymbol{\eta}|v_1,\mathbf{x}_1;\ldots;v_n,\mathbf{x}_n\rangle(\mathbf{x}) - \sum_{i=1}^{n}\int \mathbf{M}_1(\mathbf{x})\boldsymbol{\Gamma}(\mathbf{x},\mathbf{y})\langle V_i(\mathbf{y})\boldsymbol{\eta}|v_1,\mathbf{x}_1;\ldots;v_n,\mathbf{x}_n\rangle(\mathbf{y})\,d\mathbf{y}$$

$$=\langle \boldsymbol{\eta}\rangle(\mathbf{x}) + \int \mathbf{M}_1(\mathbf{x})\boldsymbol{\Gamma}(\mathbf{x},\mathbf{y})\{\langle \boldsymbol{\eta}|;v_1,\mathbf{x}_1;\ldots;v_n,\mathbf{x}_n\rangle(\mathbf{y}) - \langle \boldsymbol{\eta}\rangle(\mathbf{y})\}\,d\mathbf{y}. \tag{14.6}$$

Since $\mathbf{x} \in v_1,\ldots,v_n$ in the nth line of the systems (14.5) and (14.6) can take the values in the inclusions $v_1,\ldots,v_n$, the nth line actually contains n equations. The definitions of the effective fields $\overline{\boldsymbol{\sigma}}_i(\mathbf{x})$, $\widetilde{\boldsymbol{\sigma}}_{1,2,\ldots,n}(\mathbf{x})$ as well as their statistical averages $\langle\overline{\boldsymbol{\sigma}}_i\rangle(\mathbf{x})$, $\langle\widetilde{\boldsymbol{\sigma}}_{1,2,\ldots,n}\rangle(\mathbf{x})$ are identical to the appropriate definitions of the effective stresses in a whole space introduced in Chapter 8.

A constitutive equation for microinhomogeneous medium can be presented in the form

$$\langle \boldsymbol{\sigma}\rangle(\mathbf{x}) = \mathbf{L}^{(0)}\langle \boldsymbol{\varepsilon}\rangle(\mathbf{x}) + \boldsymbol{\alpha}^{(0)} + \langle \boldsymbol{\tau}\rangle(\mathbf{x}), \quad \langle \boldsymbol{\varepsilon}\rangle(\mathbf{x}) = \mathbf{M}^{(0)}\langle \boldsymbol{\sigma}\rangle(\mathbf{x}) + \boldsymbol{\beta}^{(0)} + \langle \boldsymbol{\eta}\rangle(\mathbf{x}), \tag{14.7}$$

where the tensors $\boldsymbol{\tau}(\mathbf{y}) = \mathbf{L}_1(\mathbf{y})\boldsymbol{\varepsilon}(\mathbf{y}) + \boldsymbol{\alpha}_1(\mathbf{y})$ and $\boldsymbol{\eta}(\mathbf{y}) = \mathbf{M}_1(\mathbf{y})\boldsymbol{\sigma}(\mathbf{y}) + \boldsymbol{\beta}_1(\mathbf{y})$ are called the stress polarization tensor and the strain polarization tensor, respectively, and are simply a notational convenience. Averaged constitutive equations (14.7) leads to macroscopic constitutive equations relating $\langle \boldsymbol{\varepsilon}\rangle(\mathbf{x})$ and $\langle \boldsymbol{\sigma}\rangle(\mathbf{x})$

$$\langle \boldsymbol{\sigma}\rangle(\mathbf{x}) = \mathbf{L}^*(\mathbf{x})\langle \boldsymbol{\varepsilon}\rangle(\mathbf{x}) + \boldsymbol{\alpha}^*(\mathbf{x}) + \boldsymbol{\mathcal{L}}^*(\langle \boldsymbol{\varepsilon}\rangle)(\mathbf{x}), \tag{14.8}$$

$$\langle \boldsymbol{\varepsilon}\rangle(\mathbf{x}) = \mathbf{M}^*(\mathbf{x})\langle \boldsymbol{\sigma}\rangle(\mathbf{x}) + \boldsymbol{\beta}^*(\mathbf{x}) + \boldsymbol{\mathcal{M}}^*(\langle \boldsymbol{\sigma}\rangle)(\mathbf{x}), \tag{14.9}$$

where nonlocal operators of the stiffness and compliance:

$$\boldsymbol{\mathcal{L}}^*(\langle \boldsymbol{\varepsilon}\rangle)(\mathbf{x}) = \int \boldsymbol{\mathcal{L}}^*(\mathbf{x},\mathbf{y})[\langle \boldsymbol{\varepsilon}\rangle(\mathbf{y}) - \langle \boldsymbol{\varepsilon}\rangle(\mathbf{x})]\,d\mathbf{y}, \tag{14.10}$$

$$\boldsymbol{\mathcal{M}}^*(\langle \boldsymbol{\sigma}\rangle)(\mathbf{x}) = \int \boldsymbol{\mathcal{M}}^*(\mathbf{x},\mathbf{y})[\langle \boldsymbol{\sigma}\rangle(\mathbf{y}) - \langle \boldsymbol{\sigma}\rangle(\mathbf{x})]\,d\mathbf{y}, \tag{14.11}$$

respectively, are presented in the form vanishing at the constants, and

$$\mathbf{L}^*(\mathbf{x}) = \int \boldsymbol{\mathcal{L}}^*(\mathbf{x},\mathbf{y})\,d\mathbf{y}, \quad \mathbf{M}^*(\mathbf{x}) = \int \boldsymbol{\mathcal{M}}^*(\mathbf{x},\mathbf{y})\,d\mathbf{y}. \tag{14.12}$$

Local effective parameters $\mathbf{L}^*(\mathbf{x})$, $\mathbf{M}^*(\mathbf{x})$ and $\boldsymbol{\alpha}^*(\mathbf{x})$, $\boldsymbol{\beta}^*(\mathbf{x})$ are defined by

$$\mathbf{L}^*(\mathbf{x}) = \mathbf{L}^{(0)} + \langle \mathbf{L}_1\mathbf{A}^*\rangle(\mathbf{x}), \quad \mathbf{M}^*(\mathbf{x}) = \mathbf{M}^{(0)} + \langle \mathbf{M}_1\mathbf{B}^*\rangle(\mathbf{x}), \tag{14.13}$$

$$\boldsymbol{\alpha}^*(\mathbf{x}) = \boldsymbol{\alpha}^{(0)} + \langle {\mathbf{A}^*}^\top\boldsymbol{\alpha}\rangle(\mathbf{x}), \quad \boldsymbol{\beta}^*(\mathbf{x}) = \boldsymbol{\beta}^{(0)} + \langle {\mathbf{B}^*}^\top\boldsymbol{\beta}\rangle(\mathbf{x}), \tag{14.14}$$

where $\mathbf{A}^* = \mathbf{A}^*(\mathbf{x})$ and $\mathbf{B}^* = \mathbf{B}^*(\mathbf{x})$ are the local strain and stress concentration tensors obtained under pure mechanical loading ($\boldsymbol{\alpha}_1(\mathbf{x}),\boldsymbol{\beta}_1(\mathbf{x}) \equiv \mathbf{0}$): $\boldsymbol{\varepsilon}(\mathbf{x}) = \mathbf{A}^*(\mathbf{x})\langle \boldsymbol{\varepsilon}\rangle(\mathbf{x})$, $\boldsymbol{\sigma}(\mathbf{x}) = \mathbf{B}^*(\mathbf{x})\langle \boldsymbol{\sigma}\rangle(\mathbf{x})$. Conversely, the estimation of the residual stresses (for $\boldsymbol{\sigma}^0 \equiv 0$) can be used for the calculation of both the effective eigenstresses and eigenstrains $\boldsymbol{\beta}^*$: $\boldsymbol{\alpha}^*(\mathbf{x}) = \boldsymbol{\alpha}^{(0)} + \langle \boldsymbol{\tau}\rangle(\mathbf{x})$, $\boldsymbol{\beta}^*(\mathbf{x}) = \boldsymbol{\beta}^{(0)} + \langle \boldsymbol{\eta}\rangle(\mathbf{x})$.

Since the number density $n(\mathbf{x})$ depends only on the coordinate $x_d \le 0$, the statistical averages $\langle(\cdot)\rangle(\mathbf{x})$ in the point $\mathbf{x}$ can be reduced to the estimation of the average over the plane (or line) $x_d = \text{const}$. The equations obtained above involve the statistical average $\langle \mathbf{f}V^{(k)}\rangle(\mathbf{x})$ in the inclusion phase $v^{(k)}$ of the different tensors $\mathbf{f}$ [e.g., $\mathbf{f} = \boldsymbol{\sigma}(\mathbf{x})$, $\boldsymbol{\sigma}\otimes\boldsymbol{\sigma}$, $c(\mathbf{x})$]. In particular, the random characteristic function of inclusions $V(\mathbf{x})$ makes it possible to define a concentration of inclusions $c(\mathbf{x}) = \langle V\rangle(\mathbf{x})$ (in so doing, $\langle V|v_i\rangle(\mathbf{x}) = 1$ if $\mathbf{x} \in v_i$, otherwise $\langle V|v_i\rangle(\mathbf{x}) = 0$) as well as the conditional stresses inside the inclusion phase $\langle V\boldsymbol{\sigma}\rangle(\mathbf{x})$ which depend only on coordinate x_1. The conditional statistical average in the inclusion phase $\langle \mathbf{f}V\rangle^{(k)}(\mathbf{x})$ at the condition that the point $\mathbf{x}$ is located in the inclusion phase $\mathbf{x} \in v^{(k)}$ can be found as $\langle \mathbf{f}V\rangle^{(k)}(\mathbf{x}) = \langle V^{(k)}(\mathbf{x})\rangle^{-1}\langle \mathbf{f}V^{(k)}\rangle(\mathbf{x})$. Usually, it is simpler to estimate the condition averages of these tensors in the concrete point $\mathbf{x}$ of the fixed inclusion $\mathbf{x} \in v_i$: $\langle \mathbf{f}|v_i\rangle(\mathbf{x})$.

It can be easy to establish a straightforward relation between these averages. Indeed, at first we built some auxiliary set $v_i^{\oplus}(\mathbf{x}, \mathbf{a}_i) \equiv \mathbf{x} \oplus v_i(\mathbf{0}, \mathbf{a}_i) = \cup_{\mathbf{y}\in v_i(\mathbf{0},\mathbf{a}_i)}\{\mathbf{x} + \mathbf{y}\}$ obtaining by Mincowski addition $\oplus$ (5.18) to the fixed point $\mathbf{x}$ of a translated ellipsoid $v_i(\mathbf{0}, \mathbf{a})$ with the center $\mathbf{0}$ and semi-axes $\mathbf{a}_i = (a_i^1, \dots, a_i^d)^\top$. Subsequently we construct subset $v_i^0(\mathbf{x}, \mathbf{a}_i) \subset v_i^{\oplus}(\mathbf{x}, \mathbf{a}_i)$ with the boundary $\partial v_i^0(\mathbf{x}, \mathbf{a}_i)$ formed by the centers of translated ellipsoids $v_i(\mathbf{0}, \mathbf{a})$ around the fixed point $\mathbf{x}$. Obviously, $v_{ki}^0 \to v_i^0(\mathbf{x}, \mathbf{a}_i)$ if a fixed ellipsoid v_k is shrinking to the point $\mathbf{x}$. Then we can get a relation between the mentioned averages $[\mathbf{x} = (x_1, \dots, x_d)^\top]$:

$$\langle V^{(i)}\mathbf{f}\rangle(\mathbf{x}) = \int_{v_i^0(\mathbf{x},\mathbf{a}_i)\cap R_-^{d0}} n(\mathbf{y})\langle \mathbf{f}|v_i\rangle(\mathbf{x})\, d\mathbf{y}, \tag{14.15}$$

where $R_-^{d0} \equiv \{\mathbf{y}|y_d \le -a_i^d\}$. Formula (14.15) is valid for any material inhomogeneity of inclusions of any concentration in a mesodomain w of any shape (e.g., $w \not\equiv R_-^{d0}$) and can easily be generalized to nonelliptical inclusions.

In particular, the relation (14.15) can be concretized for aligned ellipsoidals in the form:

$$\langle V\mathbf{f}\rangle(\mathbf{x}) = \int_{x_2-a_i^2}^{\min(-a_i^2, x_2+a_i^2)} n(y_2) \int_{-h_1(x_2,y_2)}^{h_1(x_2,y_2)} \langle \mathbf{f}|v_i\rangle(\mathbf{x})dy_1dy_2, \tag{14.16}$$

$$\langle V\mathbf{f}\rangle(\mathbf{x}) = \int_{x_3-a_i^3}^{\min(-a_i^3, x_3+a_i^3)} n(y_3) \int_{-h_1(x_3,y_3)}^{h_1(x_3,y_3)} \int_{-h_2(x_3,y_3,y_2)}^{h_2(x_3,y_3,y_2)} \times \langle \mathbf{f}|v_i\rangle(\mathbf{x})\, dy_1\, dy_2\, dy_3, \tag{14.17}$$

for 2-D and 3-D cases, respectively, where the moving ellipsoids v_i and $v_i^0(\mathbf{x}, \mathbf{a}_i)$ have the centers $\mathbf{y} \in v_i^0(\mathbf{x}, \mathbf{a}_i)$ and $\mathbf{y}_i = (0, \dots, y_d)^\top$, respectively. The second integration in Eq. (14.16) is performed over a chord $[-h_1(x_2, y_2), h_1(x_2, y_2)]$ with $h_1(x_2, y_2) = (a_i^1/a_i^2)\sqrt{(a_i^2)^2 - (y_2 - x_2)^2}$ defined by an intersection of the fixed line $x_2 = \text{const.}$ and the moving ellipse with the center $\mathbf{y}_i = (0, y_2)^\top$. We introduced for the 3-D case in Eq. (14.17), the chord $[-h_1(x_3, y_3), h_1(x_3, y_3)]$ with $h_1(x_3, y_3) = (a_i^2/a_i^3)\sqrt{(a_i^3)^2 - (y_3 - x_3)^2}$ as well as another chord $[-h_2(x_3, y_3, y_2), h_2(x_3, y_3, y_2)]$ with $h_2(x_3, y_3, y_2) = a_i^1\sqrt{1 - (y_3 - x_3)^2/(a_i^3)^2 - y_2^2/(a_i^2)^2}$, respectively. If $\langle \mathbf{f}|v_i\rangle(\mathbf{y}) \equiv \langle \mathbf{f}|v_i\rangle(y_3)$ for $\forall y \in v_i$, the integral representations (14.16) (or (14.17)) can be simplified, e.g.

$$\langle V\mathbf{f}\rangle(\mathbf{x}) = 2\int_{x_3-a_i^3}^{\min(-a_i^3, x_3+a_i^3)} n(y_3)\mathbf{f}(y_3)h_1(x_3, y_3)\, dy_3. \tag{14.18}$$

14.3 Finite Number of Inclusions in a Half-Space

14.3.1 A Single Inclusion Subjected to Inhomogeneous Effective Stress

We will consider a half-space $w = \{\mathbf{x}|x_3 \leq 0\}$ containing one and two ellipsoidal inclusions (the case of the finite number of inclusions is analogous). In the case of one inclusion, we have

$$\boldsymbol{\sigma}(\mathbf{x}) = \overline{\boldsymbol{\sigma}}_i(\mathbf{x}) + \int V_i(\mathbf{y})\boldsymbol{\Gamma}(\mathbf{x},\mathbf{y})[\mathbf{M}_1(\mathbf{y})\boldsymbol{\sigma}(\mathbf{y}) + \boldsymbol{\beta}_1(\mathbf{y})]\, d\mathbf{y}, \tag{14.19}$$

where $\overline{\boldsymbol{\sigma}}_i(\mathbf{x})$ is an effective stress field (14.2) existing in the homogeneous half-space without the inclusion and corresponding to the non-random body force $\mathbf{f}$ and to boundary conditions that can be on the displacements or the tractions or mixed. Then the solution of Eq. (14.19) can be formally presented in the form of the integral operators (no summation over i):

$$\boldsymbol{\sigma}(\mathbf{x}) = \boldsymbol{\mathcal{B}}_i(\overline{\boldsymbol{\sigma}}_i)(\mathbf{x}) + \mathbf{C}_i(\mathbf{x}), \quad \mathbf{C}_i(\mathbf{x}) = \boldsymbol{\mathcal{B}}_i\Big(\int \boldsymbol{\Gamma}(\mathbf{z},\mathbf{y})\boldsymbol{\beta}_1(\mathbf{y})\, d\mathbf{y}\Big)(\mathbf{x}), \tag{14.20}$$

where the integral operator $\boldsymbol{\mathcal{B}}_i = (\mathbf{I} - \boldsymbol{\Gamma}\mathbf{M}_1 V_i)^{-1}$. Let the integral operator $\boldsymbol{\mathcal{B}}_i$ (14.20) express the stress tensor inside the inclusion $\boldsymbol{\sigma}(\mathbf{x})$ through the effective field $\overline{\boldsymbol{\sigma}}_i(\mathbf{x})$ known. The solution $\boldsymbol{\mathcal{B}}_i$ (14.20) of the singular volume integral equation (14.19) can be solved by the direct quadrature method (also called the Nystron method), by the iteration method as well as by the Fourier transform method (see, e.g., Chapter 12 where additional references can be found). However, the knowledge of the operator $\boldsymbol{\mathcal{B}}_i$ at the current stage is not necessary and the assumption of the existence of such an operator is enough. At first no restrictions are imposed on the method of estimation of the operator $\boldsymbol{\mathcal{B}}_i$.

In particular, for the constant effective stress field $\overline{\boldsymbol{\sigma}}_i = \boldsymbol{\sigma}^0 \equiv$ const., the integral operator $\boldsymbol{\mathcal{B}}_i$ is reduced to the inhomogeneous tensor $\mathbf{B}_i(\mathbf{x}) = \int V_i(\mathbf{y})\boldsymbol{\mathcal{B}}_i(\mathbf{x},\mathbf{y})\, d\mathbf{y}$ assumed to be known. So-called local approximation of the field $\overline{\boldsymbol{\sigma}}_i(\mathbf{x})$:

$$\boldsymbol{\sigma}(\mathbf{x}) = \mathbf{B}_i(\mathbf{x})\boldsymbol{\sigma}^0(\mathbf{x}) + \mathbf{C}_i(\mathbf{x}) \tag{14.21}$$

neglects inhomogeneity of the stress field $\overline{\boldsymbol{\sigma}}_i(\mathbf{x})$ inside the area $\mathbf{x} \in v_i$ and allows an equivalent representation of Eq. (14.20):

$$\boldsymbol{\sigma}(\mathbf{x}) = \mathbf{B}_i(\mathbf{x})\overline{\boldsymbol{\sigma}}_i(\mathbf{x}) + \mathbf{C}_i(\mathbf{x}) + \boldsymbol{\mathcal{B}}_i^1(\overline{\boldsymbol{\sigma}}_i)(\mathbf{x}), \tag{14.22}$$

where an integral operator $\boldsymbol{\mathcal{B}}_i^1(\overline{\boldsymbol{\sigma}}_i)(\mathbf{x}) \equiv \boldsymbol{\mathcal{B}}_i(\overline{\boldsymbol{\sigma}}_i)(\mathbf{x}) - \mathbf{B}_i(\mathbf{x})\overline{\boldsymbol{\sigma}}_i(\mathbf{x})$ vanishes at the constant $\overline{\boldsymbol{\sigma}}_i(\mathbf{x}) = \boldsymbol{\sigma}^0 \equiv$ const. We can also estimate the strain polarization tensor and its local approximation

$$\bar{v}_i\boldsymbol{\eta}_i(\mathbf{x}) = \boldsymbol{\mathcal{R}}_i(\mathbf{x})(\overline{\boldsymbol{\sigma}}_i) + \mathbf{F}_i(\mathbf{x}), \qquad \bar{v}_i\boldsymbol{\eta}_i(\mathbf{x}) = \mathbf{R}_i(\mathbf{x})\overline{\boldsymbol{\sigma}}_i(\mathbf{x}) + \mathbf{F}_i(\mathbf{x}), \tag{14.23}$$

where $\boldsymbol{\mathcal{R}}_i = \bar{v}_i\mathbf{M}_1(\mathbf{x})\boldsymbol{\mathcal{B}}_i$, $\mathbf{R}_i(\mathbf{x}) = \bar{v}_i\mathbf{M}_1(\mathbf{x})\mathbf{B}_i(\mathbf{x})$, $\mathbf{F}_i(\mathbf{x}) = \bar{v}_i[\mathbf{M}_1(\mathbf{x})\mathbf{C}_i(\mathbf{x}) + \boldsymbol{\beta}_1^{(i)}]$.

The approximation (14.23_2) can be used for the iterative solution of Eq. (14.19) ($\mathbf{x} \in v_i,\ k = 0, 1, \ldots$) presented in symbolic form in terms of the strain polarization tensor $\boldsymbol{\eta}(\mathbf{x})$:

$$\boldsymbol{\eta} = \overline{\boldsymbol{\eta}}_i + \boldsymbol{\mathcal{K}}_i\boldsymbol{\eta}, \tag{14.24}$$

where $\bar{v}_i\overline{\boldsymbol{\eta}}_i(\mathbf{x}) = \mathbf{R}_i(\mathbf{x})\overline{\boldsymbol{\sigma}}_i(\mathbf{x}) + \mathbf{F}_i(\mathbf{x})$ and $(\boldsymbol{\mathcal{K}}_i\boldsymbol{\eta})(\mathbf{x}) = \int \boldsymbol{\mathcal{K}}_i(\mathbf{x},\mathbf{y})V_i(\mathbf{y})\boldsymbol{\eta}(\mathbf{y})\, d\mathbf{y}$ defines the integral operator $\boldsymbol{\mathcal{K}}_i$ with the kernel formally represented as $\boldsymbol{\mathcal{K}}_i(\mathbf{x},\mathbf{y}) = \mathbf{K}_i(\mathbf{x},\mathbf{y}) - \delta(\mathbf{x}-\mathbf{y})\int V_i(\mathbf{z})\mathbf{K}_i(\mathbf{x},\mathbf{z})\, d\mathbf{z}$ with $\mathbf{K}_i(\mathbf{x},\mathbf{y}) = \mathbf{R}_i(\mathbf{x})\boldsymbol{\Gamma}(\mathbf{x},\mathbf{y})$.

We formally write the solution of Eq. (14.24) as $\boldsymbol{\eta} = \boldsymbol{\mathcal{L}}_i\overline{\boldsymbol{\eta}}_i$, where the inverse operator $\boldsymbol{\mathcal{L}}_i = (\mathbf{I} - \boldsymbol{\mathcal{K}}_i)^{-1}$ will be constructed by the iteration method based on the recursion formula

$$\boldsymbol{\eta}_{[k+1]} = \overline{\boldsymbol{\eta}}_i + \boldsymbol{\mathcal{K}}_i\boldsymbol{\eta}_{[k]} \tag{14.25}$$

to construct a sequence of functions $\{\boldsymbol{\eta}_{[k]}\}$ that can be treated as an approximation of the solution of Eq. (14.24). A more detailed iteration scheme in terms of the stress field can be presented as

$$\boldsymbol{\sigma}_{[k+1]}(\mathbf{x}) = \mathbf{B}_i(\mathbf{x})\overline{\boldsymbol{\sigma}}_i(\mathbf{x}) + \mathbf{C}_i(\mathbf{x}) + \mathbf{B}_i(\mathbf{x})\int V_i(\mathbf{y})\boldsymbol{\Gamma}(\mathbf{x},\mathbf{y})\mathbf{M}_1(\mathbf{y})[\boldsymbol{\sigma}_{[k]}(\mathbf{y}) - \boldsymbol{\sigma}_{[k]}(\mathbf{x})]\, d\mathbf{y}$$

with an initial approximation $\boldsymbol{\sigma}_{[0]}(\mathbf{x}) = \mathbf{B}_i(\mathbf{x})\overline{\boldsymbol{\sigma}}_i(\mathbf{x}) + \mathbf{C}_i(\mathbf{x})$. We presented a point Jacobi (called also Richardson and point total-step) iterative scheme for ease of calculations. Other iteration methods used for the solution of Eq. (14.24) are considered in [1134]. Usually the driving term of this equation is used as an initial approximation:

$$\boldsymbol{\eta}_{[0]}(\mathbf{x}) = \overline{\boldsymbol{\eta}}_i(\mathbf{x}), \tag{14.26}$$

which is exact for a homogeneous ellipsoidal inclusion subjected to remote homogeneous stress field $\boldsymbol{\sigma}^0(\mathbf{x}) \equiv \boldsymbol{\sigma}^0 = \text{const}$. The sequence $\{\boldsymbol{\eta}_{[k]}\}$ (14.25) with arbitrary continuous $\boldsymbol{\eta}_{[0]}(\mathbf{x})$ converges to a unique solution $\boldsymbol{\eta}$ if the norm of the integral operator $\boldsymbol{\mathcal{K}}_i$ turns out to be small "enough" (less than 1), and the problem is reduced to computation of the integrals involved, the density of which is given. In effect the iteration method (14.25) transforms the integral equation problem (14.25) into the linear algebra problem in any case. The fast convergence $\boldsymbol{\sigma}_{[k]}(\mathbf{x}) \to \boldsymbol{\sigma}(\mathbf{x})$ $(k \to \infty)$ of the scheme (14.25) with a removed singularity at $\mathbf{y} \to \mathbf{x}$ was demonstrated in [167] (see Chapter 12) by the volume integral equation method for some particular problems in the 2-D case.

Explicit representations of the tensors $\mathbf{B}_i(\mathbf{x})$ and $\mathbf{C}_i(\mathbf{x})$ for both inside and outside the inclusion v_i are obtained in [633] for the 2-D case for an isotropic circular inclusion inside the isotropic half-plane. At infinity $x_3, y_3 \to -\infty$, the tensor $\boldsymbol{\Gamma}(\mathbf{x},\mathbf{y})$ tends to be an appropriate representation for the infinite-homogeneous-body Green's stress function $\boldsymbol{\Gamma}^\infty(\mathbf{x}-\mathbf{y})$, which is the even homogeneous generalized function of the order $-d$. In such a case the tensors $\mathbf{B}_i(\mathbf{x}), \mathbf{C}_i(\mathbf{x})$ are the constants at $\mathbf{x} \in v_i$ and expressed through the Eshelby tensor $\mathbf{S}_i^{\text{Esh}} = \mathbf{I} - \mathbf{M}^{(0)}\mathbf{Q}_i^{\text{Esh}}$, $\mathbf{Q}_i^{\text{Esh}} \equiv -\langle\boldsymbol{\Gamma}^\infty(\mathbf{x}-\mathbf{y})\rangle_{(i)}$ $(\mathbf{x},\mathbf{y} \in v_i)$:

$$\mathbf{B}_i(\mathbf{x}) \equiv \mathbf{B}_i^{\text{Esh}} = \left(\mathbf{I} + \mathbf{Q}_i^{\text{Esh}}\mathbf{M}_1^{(i)}\right)^{-1}, \quad \mathbf{C}_i(\mathbf{x}) \equiv \mathbf{C}_i^{\text{Esh}} = -\mathbf{B}_i^{\text{Esh}}\mathbf{Q}_i^{\text{Esh}}\boldsymbol{\beta}_1^{(i)}, \tag{14.27}$$

14.3.2 Two Inclusions

For two inhomogeneities v_i and v_j subjected to the homogeneous stress field $\widetilde{\boldsymbol{\sigma}}_{i,j}(\mathbf{x}) = \boldsymbol{\sigma}^0 \equiv \text{const.}$, we have an equation

$$\boldsymbol{\sigma}(\mathbf{x}) = \boldsymbol{\sigma}^0 + \int [V_i(\mathbf{y}) + V_j(\mathbf{y})]\boldsymbol{\Gamma}(\mathbf{x},\mathbf{y})[\mathbf{M}_1(\mathbf{y})\boldsymbol{\sigma}(\mathbf{y}) + \boldsymbol{\beta}_1(\mathbf{y})]\, d\mathbf{y}, \tag{14.28}$$

instead of Eq. (14.19). Acting on Eq. (14.28) by the operator $\boldsymbol{\mathcal{B}}_i$ and taking Eq. (14.21) into account transforms Eq. (14.28) into the following: $\boldsymbol{\sigma}(\mathbf{x}) = \mathbf{B}_i(\mathbf{x})\boldsymbol{\sigma}^0 + \mathbf{C}_i(\mathbf{x}) + \boldsymbol{\mathcal{B}}_i \int \boldsymbol{\Gamma}(\mathbf{y},\mathbf{z})\boldsymbol{\eta}(\mathbf{z})V_j(\mathbf{z})\, d\mathbf{z},\ \mathbf{x},\mathbf{y} \in v_i$. By the use of obvious rearrangement of the last equation, we will introduce the inhomogeneous tensors $\mathbf{B}_{i,j}(\mathbf{x})$ and $\mathbf{C}_{i,j}(\mathbf{x})$ ($\mathbf{x} \in v_i$) describing the stress concentrator factor on the inclusions v_i:

$$\boldsymbol{\mathcal{B}}_i \int \boldsymbol{\Gamma}(\mathbf{y},\mathbf{z})\boldsymbol{\eta}(\mathbf{z})V_j(\mathbf{z})\, d\mathbf{z} = \boldsymbol{\sigma}(\mathbf{x}) - \mathbf{B}_i(\mathbf{x})\boldsymbol{\sigma}^0 - \mathbf{C}_i(\mathbf{x}) \equiv \mathbf{B}_{i,j}(\mathbf{x})\boldsymbol{\sigma}^0 + \mathbf{C}_{i,j}(\mathbf{x}), \tag{14.29}$$

where the tensors $\mathbf{B}_{i,j}(\mathbf{x})$ and $\mathbf{C}_{i,j}(\mathbf{x})$ can be found via any numerical method such as BIE and FEA, complex potential method (see Chapter 4) and others analogously to the scheme described in Subsection 4.2.2.

Generalization of Eq. (14.29) to the case of the inhomogeneous field $\widetilde{\boldsymbol{\sigma}}_{i,j}(\mathbf{x}) \not\equiv$ const. is ($\mathbf{x} \in v_i$)

$$\begin{aligned}\boldsymbol{\mathcal{B}}_i \int \boldsymbol{\Gamma}(\mathbf{y},\mathbf{z})\boldsymbol{\eta}(\mathbf{z})V_j(\mathbf{z})\, d\mathbf{z} &= \boldsymbol{\sigma}(\mathbf{x}) - \boldsymbol{\mathcal{B}}_i(\boldsymbol{\sigma}^0)(\mathbf{x}) - \mathbf{C}_i(\mathbf{x}) \\ &\equiv [\boldsymbol{\mathcal{B}}^I_{i,j}(\boldsymbol{\sigma}^{0I})(\mathbf{x}) + \boldsymbol{\mathcal{B}}^J_{i,j}(\boldsymbol{\sigma}^{0J})(\mathbf{x})] + \mathbf{C}_{i,j}(\mathbf{x}),\end{aligned} \tag{14.30}$$

where the operators $\boldsymbol{\mathcal{B}}^I_{i,j}$ and $\boldsymbol{\mathcal{B}}^J_{i,j}$ are the integral operators over the domains v_i and v_j subjected to the remote stresses $\boldsymbol{\sigma}^{0I}(\mathbf{x}) = \widetilde{\boldsymbol{\sigma}}_{i,j}(\mathbf{x})$ ($\mathbf{x} \in v_i$) and $\boldsymbol{\sigma}^{0J}(\mathbf{x}) = \widetilde{\boldsymbol{\sigma}}_{i,j}(\mathbf{x})$ ($\mathbf{x} \in v_j$), respectively.

Let us replace the inhomogeneity v_j by the fictitious inclusion with the matrix elastic properties and the fictitious eigenstrain ($\mathbf{y} \in v_j$, $\boldsymbol{\sigma}^0 \equiv$ const.)

$$\mathbf{L}(\mathbf{y}) \equiv \mathbf{L}^{(0)}, \quad \bar{v}_j\boldsymbol{\beta}_1^{\text{fict}}(\mathbf{y}) = \mathbf{R}_j(\mathbf{y})\boldsymbol{\sigma}^0 + \mathbf{F}_j(\mathbf{y}), \tag{14.31}$$

so that the constitutive equation in the inclusion $\mathbf{y} \in v_j$ is $\boldsymbol{\sigma}(\mathbf{y}) = \mathbf{L}^{(0)}\boldsymbol{\varepsilon}(\mathbf{y}) + \boldsymbol{\beta}_1^{\text{fict}}(\mathbf{y})$; here the tensors $\mathbf{R}_j(\mathbf{x})$ and $\mathbf{F}_j(\mathbf{x})$ are defined by the stress concentrator factors for one isolated inclusion (14.23$_2$). Then the strain polarization tensor $\boldsymbol{\eta}(\mathbf{y}) = \boldsymbol{\beta}_1^{\text{fict}}(\mathbf{y})$ ($\mathbf{y} \in v_j$) can be assumed to be known, and Eqs. (14.28) and (14.29) can be reduced to ($\mathbf{x},\mathbf{y} \in v_i$)

$$\overline{\boldsymbol{\sigma}}(\mathbf{x}) = \boldsymbol{\sigma}^0 + \int \boldsymbol{\Gamma}(\mathbf{x},\mathbf{z})\boldsymbol{\beta}_1^{\text{fict}}(\mathbf{z})V_j(\mathbf{z})\, d\mathbf{z}, \tag{14.32}$$

$$\boldsymbol{\sigma}(\mathbf{x}) = \mathbf{B}_i(\mathbf{x})\boldsymbol{\sigma}^0 + \mathbf{C}_i(\mathbf{x}) + \boldsymbol{\mathcal{B}}_i \int \boldsymbol{\Gamma}(\mathbf{y},\mathbf{z})\boldsymbol{\beta}_1^{\text{fict}}(\mathbf{z})V_j(\mathbf{z})\, d\mathbf{z}, \tag{14.33}$$

respectively, which can be recast in the form

$$\int \boldsymbol{\Gamma}(\mathbf{x},\mathbf{z})\boldsymbol{\beta}_1^{\text{fict}}(\mathbf{z})V_j(\mathbf{z})\, d\mathbf{z} = \overline{\boldsymbol{\sigma}}(\mathbf{x}) - \boldsymbol{\sigma}^0 \equiv \mathbf{E}^{\infty}_{i,j}(\mathbf{x})\boldsymbol{\sigma}^0 + \mathbf{E}^{\beta\infty}_{i,j}(\mathbf{x}), \tag{14.34}$$

$$\begin{aligned}\boldsymbol{\mathcal{B}}_i \int \boldsymbol{\Gamma}(\mathbf{y},\mathbf{z})\boldsymbol{\beta}_1^{\text{fict}}(\mathbf{z})V_j(\mathbf{z})\, d\mathbf{z} &= \boldsymbol{\sigma}(\mathbf{x}) - \mathbf{B}_i(\mathbf{x})\boldsymbol{\sigma}^0 - \mathbf{C}_i(\mathbf{x}) \\ &\equiv \mathbf{B}^{\infty}_{i,j}(\mathbf{x})\boldsymbol{\sigma}^0 + \mathbf{C}^{\infty}_{i,j}(\mathbf{x})\end{aligned} \tag{14.35}$$

defining the new concentrator factors $\mathbf{E}^{\infty}_{i,j}(\mathbf{x})$, $\mathbf{B}^{\infty}_{i,j}(\mathbf{x})$ and $\mathbf{E}^{\beta\infty}_{i,j}(\mathbf{x})$, $\mathbf{C}^{\infty}_{i,j}(\mathbf{x})$ $(\mathbf{x}\in v_i)$ which can be estimated analogously to the tensors $\mathbf{B}_{i,j}(\mathbf{x})$ and $\mathbf{C}_{i,j}(\mathbf{x})$ $(\mathbf{x}\in v_i)$ (14.29). In so doing, $\overline{\boldsymbol{\sigma}}_i(\mathbf{x})$ is an effective field acting on the inclusion v_i does not depend on the properties of the inclusion v_i and can be defined by Eq. (14.32) in the case of the absence of the inclusion v_i. The tensors $\mathbf{E}^{\infty}_{i,j}(\mathbf{x})$, $\mathbf{B}^{\infty}_{i,j}(\mathbf{x})$ were introduced by Eqs. (14.34) and (14.35) for the case of the homogeneous effective field $\boldsymbol{\sigma}^0(\mathbf{x}) = \text{const}$. In the case of inhomogeneous $\widetilde{\boldsymbol{\sigma}}_{i,j}(\mathbf{x})$, the mentioned tensors should be replaced by the appropriate operators $(\mathbf{x}\in v_i)$:

$$\int \boldsymbol{\Gamma}(\mathbf{x},\mathbf{z})\boldsymbol{\beta}^{\text{fict}}_1(\mathbf{z})V_j(\mathbf{z})\,d\mathbf{z} = \overline{\boldsymbol{\sigma}}(\mathbf{x}) - \widetilde{\boldsymbol{\sigma}}_{i,j}(\mathbf{x}) \equiv \boldsymbol{\mathcal{E}}^{\infty}_{i,j}(\widetilde{\boldsymbol{\sigma}}_{i,j})(\mathbf{x}) + \mathbf{E}^{\beta\infty}_{i,j}(\mathbf{x}),$$

$$\boldsymbol{\mathcal{B}}_i\!\int \boldsymbol{\Gamma}(\mathbf{y},\mathbf{z})\boldsymbol{\beta}^{\text{fict}}_1(\mathbf{z})V_j(\mathbf{z})\,d\mathbf{z} = \boldsymbol{\sigma}(\mathbf{x}) - \boldsymbol{\mathcal{B}}_i(\widetilde{\boldsymbol{\sigma}}_{i,j})(\mathbf{x}) - \mathbf{C}_i(\mathbf{x}) \equiv \boldsymbol{\mathcal{B}}^{\infty}_{i,j}(\widetilde{\boldsymbol{\sigma}}_{i,j})(\mathbf{x}) + \mathbf{C}^{\infty}_{i,j}(\mathbf{x}),$$

respectively, where the fictitious eigenstrain $\boldsymbol{\beta}^{\text{fict}}_1(\mathbf{z})$ is defined by a relation $\bar v_j\boldsymbol{\beta}^{\text{fict}}_1(\mathbf{y}) = \boldsymbol{\mathcal{R}}_j(\boldsymbol{\sigma}^0)(\mathbf{y}) + \mathbf{F}_j(\mathbf{y})$ rather than by Eq. (14.31).

The operators (14.30) have the local approximation analogous to Eq. (14.20) $(\mathbf{x}\in v_i)$

$$\begin{aligned}\boldsymbol{\mathcal{B}}_i\!\int \boldsymbol{\Gamma}(\mathbf{y},\mathbf{z})\boldsymbol{\eta}(\mathbf{z})V_j(\mathbf{z})\,d\mathbf{z} &= \boldsymbol{\sigma}(\mathbf{x}) - \mathbf{B}_i(\mathbf{x})\widetilde{\boldsymbol{\sigma}}_{i,j}(\mathbf{x}_i) - \mathbf{C}_i(\mathbf{x})\\ &\equiv [\mathbf{B}^I_{i,j}(\mathbf{x})\widetilde{\boldsymbol{\sigma}}_{i,j}(\mathbf{x}_i) + \mathbf{B}^J_{i,j}(\mathbf{x})\widetilde{\boldsymbol{\sigma}}_{i,j}(\mathbf{x}_j)] + \mathbf{C}_{i,j}(\mathbf{x}). \quad (14.36)\end{aligned}$$

Here, at the evaluations of the tensors $\mathbf{B}^I_{i,j}(\mathbf{x}) = \langle \boldsymbol{\mathcal{B}}^I_{i,j}(\mathbf{x},\mathbf{y})\rangle_{(i)}(\mathbf{x})$ $(\mathbf{y}\in v_i)$ and $\mathbf{B}^J_{i,j}(\mathbf{x}) = \langle \mathbf{B}^I_{i,j}(\mathbf{x},\mathbf{z})\rangle_{(j)}(\mathbf{x})$ $(\mathbf{z}\in v_j)$, $[\mathbf{B}^I_{i,j}(\mathbf{x}) + \mathbf{B}^J_{i,j}(\mathbf{x}) = \mathbf{B}_{i,j}(\mathbf{x})]$, one does not assume homogeneity of perturbation stress fields produced by the inclusion v_i (or v_j) in the vicinity of the inclusion v_j (or v_i). In the case of this assumption

$$\mathbf{B}^{\infty}_{i,j}(\mathbf{x}) = \mathbf{B}_i(\mathbf{x})\mathbf{T}^{\mathbf{R}}_{ij}(\mathbf{x}_i,\mathbf{y}_j) \quad \mathbf{C}^{\infty}_{i,j}(\mathbf{x}) = \mathbf{B}_i(\mathbf{x})\mathbf{T}^{\mathbf{F}}_{ij}(\mathbf{x}_i,\mathbf{y}_j), \qquad (14.37)$$

where the notations $\mathbf{T}^{\mathbf{g}}_j = \mathbf{T}^{\mathbf{g}}_j(\mathbf{x},\mathbf{y}_j) \equiv \langle \boldsymbol{\Gamma}(\mathbf{x},\mathbf{y})\mathbf{g}(\mathbf{y})\rangle_j$, $\mathbf{T}^{\mathbf{g}}_{ij} = \mathbf{T}^{\mathbf{g}}_{ij}(\mathbf{x}_i,\mathbf{y}_j) \equiv \langle \mathbf{T}^{\mathbf{g}}(\mathbf{x},\mathbf{y}_j)\rangle_i$ are introduced for $\mathbf{x}\in v_i$, $\mathbf{y}\in v_j$, and $\mathbf{g}(\mathbf{y}) = \boldsymbol{\beta}(\mathbf{y})$, $\mathbf{B}(\mathbf{y})$, $\mathbf{C}(\mathbf{y})$, $\mathbf{R}(\mathbf{y})$, $\mathbf{F}_i(\mathbf{y})$. Hereafter, one introduces the operators and tensors ($K = I, J$; $\mathbf{x}\in v_i$)

$$\begin{aligned}&\boldsymbol{\mathcal{R}}^K_{i,j}(\mathbf{x},\mathbf{y}) = \bar v_i\mathbf{M}^{(i)}_1\boldsymbol{\mathcal{B}}^K_{i,j}(\mathbf{x},\mathbf{y}),\ \mathbf{R}^K_{i,j}(\mathbf{x}) = \bar v_i\mathbf{M}^{(i)}_1\mathbf{B}^K_{i,j}(\mathbf{x}),\ \mathbf{R}^{\infty}_{i,j}(\mathbf{x}) = \bar v_i\mathbf{M}^{(i)}_1\mathbf{B}^{\infty}_{i,j}(\mathbf{x}),\\ &\mathbf{F}_{i,j}(\mathbf{x}) = \bar v_i[\mathbf{M}^{(i)}_1\mathbf{C}_{i,j}(\mathbf{x}) + \boldsymbol{\beta}^{(i)}_1],\ \ \mathbf{F}^{\infty}_{i,j}(\mathbf{x}) = \bar v_i[\mathbf{M}^{(i)}_1\mathbf{C}^{\infty}_{i,j}(\mathbf{x}) + \boldsymbol{\beta}^{(i)}_1] \qquad (14.38)\end{aligned}$$

as well as a notation $\mathbf{g}^v_i = \mathbf{g}_i/\bar v_i$ ($\mathbf{g}_i = \mathbf{R}_i(\mathbf{x}),\mathbf{F}_i(\mathbf{x})$). The tensors $\mathbf{B}^I_{i,j}(\mathbf{x})$, $\mathbf{B}^J_{i,j}(\mathbf{x})$, and $\mathbf{C}_{i,j}(\mathbf{x})$ describing a binary interaction of inclusions, can be expressed through the solutions for a single inclusion inside the half-space in the case of assumption (14.37) ($\mathbf{x}\in v_i$, no sum on i,j) $\mathbf{B}^I_{i,j}(\mathbf{x}) = \mathbf{B}_i(\mathbf{x})(\mathbf{Z}_{ii} - \mathbf{I})$, $\mathbf{B}^J_{i,j}(\mathbf{x}) = \mathbf{B}_i(\mathbf{x})\mathbf{Z}_{ij}$, $\mathbf{C}_{i,j}(\mathbf{x}) = \mathbf{B}_i(\mathbf{x})[\mathbf{Z}_{ii}\mathbf{T}^{\mathbf{F}}_{i,j} + \mathbf{Z}_{ij}\mathbf{T}^{\mathbf{F}}_{ji}]$, where for simplicity $i,j = 1,2$ are taken, and the matrix elements $\mathbf{Z}_{ij}$, $\mathbf{Z}_{jj}$ are nondiagonal and diagonal elements, respectively, of the binary interaction matrix $\mathbf{Z}$ for the two inclusions v_i and v_j; $\mathbf{Z}^{-1}$ has the elements $(\mathbf{Z}^{-1})_{ij} = \mathbf{I}\delta_{ij} - (1-\delta_{ij})\mathbf{T}^{\mathbf{B}}_{ij}$.

Thus, the tensors $\mathbf{B}_i(\mathbf{x})$ and $\mathbf{C}_i(\mathbf{x})$ ($\mathbf{x} \in v_i$) describe the stress tensor $\boldsymbol{\sigma}(\mathbf{x})$ in the isolated inclusion v_i in the matrix subjected to the homogeneous loading. In so doing, the tensors $\mathbf{B}_{i,j}(\mathbf{x})$ and $\mathbf{C}_{i,j}(\mathbf{x})$ ($\mathbf{x} \in v_i$) define the perturbation of the stress tensor $\boldsymbol{\sigma}(\mathbf{x})$ introduced by the placement of the inclusion v_j interacting with the inclusion v_i. The tensors $\mathbf{B}^{\infty}_{i,j}(\mathbf{x})$ and $\mathbf{C}^{\infty}_{i,j}(\mathbf{x})$ describe another perturbation of the stress tensor $\boldsymbol{\sigma}(\mathbf{x})$ at $\mathbf{x} \in v_i$ produced by the inclusion v_j interacting just with the homogeneous external loading but not with the inclusion v_i. Comparison of Eqs. (14.31) defining the tensors $\mathbf{B}^{\infty}_{i,j}(\mathbf{x})$ and $\mathbf{C}^{\infty}_{i,j}(\mathbf{x})$ with the iteration scheme of the solution of Eq. (14.29) leads to a conclusion that the tensors $\mathbf{B}^{\infty}_{i,j}(\mathbf{x})$ and $\mathbf{C}^{\infty}_{i,j}(\mathbf{x})$ construct the first-order approximation of the solution (14.29) by the analogous iteration method (14.24), and are, in fact, the first-order approximation of the tensors $\mathbf{B}_{i,j}(\mathbf{x})$ and $\mathbf{C}_{i,j}(\mathbf{x})$, respectively, describing the solution of Eq. (14.29).

The schemes of estimation of the tensors $\mathbf{B}_i(\mathbf{x})$ and $\mathbf{C}_i(\mathbf{x})$, as well as $\mathbf{B}_{i,j}(\mathbf{x})$, $\mathbf{C}_{i,j}(\mathbf{x})$ and $\mathbf{B}^{\infty}_{i,j}(\mathbf{x})$, $\mathbf{C}^{\infty}_{i,j}(\mathbf{x})(\mathbf{x} \in v_i)$ are identical to the appropriate schemes for the case of an infinite matrix (see Chapter 4). The estimation of the tensor $\mathbf{T}^{\beta}_i(\mathbf{x},\mathbf{x}_i) = \int \boldsymbol{\Gamma}(\mathbf{x},\mathbf{z})\boldsymbol{\beta}_1(\mathbf{z})V_i(\mathbf{z})\,d\mathbf{z}$ is performed in a similar manner. Indeed, let, instead of the real eigenstrain $\boldsymbol{\beta}_1(\mathbf{x})$, the inclusion be subjected to the fictitious eigenstrain $\boldsymbol{\beta}^{\,\mathrm{f}}$ such that $\beta^{\,\mathrm{f}}_j = \beta_{1|j}(\mathbf{y})$ ($\mathbf{y} \in v_i$) ($j = 1,\ldots,3d-3$), otherwise $\beta^{\mathrm{f}}_k \equiv 0$ ($k = 1,\ldots,3d-3; k \neq j$), and $\overline{\boldsymbol{\sigma}} \equiv \mathbf{0}$. Then we obtain the explicit representation of the tensor $\mathbf{T}^{\beta}_i(\mathbf{x},\mathbf{x}_i)$ over the known stress field $\boldsymbol{\sigma}(\mathbf{x})$ inside and outside the inclusion v_i ($\mathbf{x} \in R^d$):

$$\mathbf{T}^{\beta}_{i|mj}(\mathbf{x},\mathbf{x}_i) = \sigma_m(\mathbf{x}) \quad \text{for } \overline{\boldsymbol{\sigma}} \equiv \mathbf{0},\ \beta^{\mathrm{f}}_j(\mathbf{x}) = \beta_{1|j}(\mathbf{x}),\ \ \beta^{\mathrm{f}}_k = 0\ (j \neq k), \tag{14.39}$$

where $j = 1,\ldots,d$; $k,m = 1,\ldots,3d-3$ and $i = 1,\ldots$. If $\boldsymbol{\beta}_1(\mathbf{y}) =$ const. at $\mathbf{y} \in v_i$, we can estimate the tensor $\mathbf{T}_i(\mathbf{x},\mathbf{x}_i)$ in a similar manner through the tensor $\mathbf{T}^{\beta^{\mathrm{f}}}_i(\mathbf{x},\mathbf{x}_i)$ by substitution of $\beta^{\,\mathrm{f}}_j(\mathbf{y}) = 1$ ($\mathbf{y} \in v_i$) ($j = 1,\ldots,3d-3$), otherwise $\beta^{\mathrm{f}}_k(\mathbf{y}) \equiv 0$ ($k = 1,\ldots,3d-3; k \neq j$).

Thus, the tensor $\mathbf{T}^{\beta}_i(\mathbf{x},\mathbf{x}_i)$ is found if the stress distribution both outside and inside the inclusion v_i with the fictitious eigenstrain $\boldsymbol{\beta}^{\,\mathrm{f}}$ (14.39) in a half-space is estimated. With non-essential restriction on space dimensionality d and the shape of inclusion we will consider 2-D problems for circular inclusions. To evaluate $\mathbf{T}^{\beta^{\mathrm{f}}}_i(\mathbf{x},\mathbf{x}_i)$ and other mentioned tensors the method [632] combining the Muskelishvili's method of complex potentials with the Fourier integral transform technique will be utilized.

14.4 Nonlocal Effective Operators of Thermoelastic Properties of Microinhomogeneous Half-Space

14.4.1 Dilute Concentration of Inclusions

In the case of a dilute concentration of the inclusions (9.16_1) as well as for a weakly inhomogeneous medium (9.16_2) the standard assumption $\langle\widetilde{\boldsymbol{\sigma}}(\mathbf{x})_{1,2}\rangle_i = \langle\overline{\boldsymbol{\sigma}}(\mathbf{x})\rangle_i = \boldsymbol{\sigma}^0,\ i = 1,2,\ \mathbf{x} \in v_i$ (9.10) of the perturbation method closing the system (14.2)-(14.5) is appropriate. This permits obtaining in Subsections 14.4.1 and 14.4.2 a few representations for the effective elastic properties.

We will consider in this subsection 14.4.1 only the dilute concentration of inclusions, when the interactions between the inclusions are neglected, and the effective thermoelastic properties $\mathbf{B}^*(\mathbf{x})$, $\boldsymbol{\beta}^*(\mathbf{x})$, $\boldsymbol{\mathcal{M}}^*(\mathbf{x},\mathbf{y})$ (and analogously $\mathbf{A}^*(\mathbf{x})$, $\boldsymbol{\alpha}^*(\mathbf{x})$, $\boldsymbol{\mathcal{L}}^*(\mathbf{x},\mathbf{y})$) can be estimated from the solution series for one elliptical inclusion with the center moving along the line $(0,\ldots,y_d)$ ($y_d \leq -1.05a_i^2$) by the use of an averaging operation (14.15). In such a case, the average $\langle\boldsymbol{\eta}\rangle(\mathbf{x})$ can be approximated by the operator $\langle V\mathbf{M}_1\boldsymbol{\mathcal{B}}_i^v\rangle(\mathbf{x})$ and the tensor $\langle\mathbf{F}_i^v\rangle(\mathbf{x})$ obtained from the single-inclusion solution with applied inhomogeneous stress $\langle\boldsymbol{\sigma}\rangle(\mathbf{x})$. Then the local thermoelastic properties can be obtained in the form

$$\mathbf{M}^*(\mathbf{x}) = \mathbf{M}^{(0)} + \langle V\mathbf{R}_i^v\rangle(\mathbf{x}), \quad \boldsymbol{\beta}^*(\mathbf{x}) = \langle\boldsymbol{\beta}\rangle(\mathbf{x}) + \langle V\mathbf{F}_i^v\rangle(\mathbf{x}), \tag{14.40}$$

while the nonlocal part of the stiffness tensor

$$\boldsymbol{\mathcal{M}}^*(\langle\boldsymbol{\sigma}\rangle)(\mathbf{x}) = \langle\mathbf{M}_1(\mathbf{y})\boldsymbol{\mathcal{B}}_i^1(\langle\boldsymbol{\sigma}\rangle)(\mathbf{y})\rangle(\mathbf{x}) \tag{14.41}$$

can also be presented in the differential form obtained for the first order approximation (14.24) of operators $\boldsymbol{\mathcal{B}}$.

$$\boldsymbol{\mathcal{M}}^*(\nabla)(\mathbf{x}) = \sum_{|\alpha|=1}^{m} \frac{1}{\alpha!}\langle\boldsymbol{\mathcal{M}}_\alpha(\nabla)(\mathbf{y})\rangle(\mathbf{x}), \tag{14.42}$$

$$\boldsymbol{\mathcal{M}}_\alpha(\nabla)(\mathbf{y}) = \mathbf{R}_i^v(\mathbf{y})\int V_i(\mathbf{z})\boldsymbol{\Gamma}(\mathbf{y},\mathbf{z})\mathbf{R}_i^v(\mathbf{z}) \otimes (\mathbf{y}-\mathbf{z})^\alpha d\mathbf{z}\nabla^\alpha, \tag{14.43}$$

where the argument $\mathbf{y} \in v_i$ in Eqs. (14.41) and (14.42) paths through the volume of the moving ellipsoid introduced in the definition of the averaged operator $\langle(\cdot)\rangle(\mathbf{x})$ (14.17), and $\alpha = (\alpha_1,\ldots,\alpha_d) \in Z_+^d$ is a multi-index of non-negative integers with the notations (12.18). At the derivative of Eq. (14.42) and (14.43), we assumed that the statistical average of stresses $\langle\boldsymbol{\sigma}\rangle(\mathbf{x})$ belongs to the class of m-times continuous differential functions $C^m(w)$. Since we desire an explicit representation for stress concentration factor, we approximated $\langle\boldsymbol{\sigma}\rangle(\mathbf{z})$ by the first $m+1$ terms of its Taylor expansion in the vicinity of the point $\mathbf{y}$.

Representations (14.40) are reduced to the Eshelby [318] solution for the remote points $\mathbf{x}^\infty = (x_1,\ldots,-\infty)^\top$: $\mathbf{M}^{*\infty} = \mathbf{M}^{(0)} + c^\infty\mathbf{M}_1^{(1)}\mathbf{B}^{\rm Esh}$, $\boldsymbol{\beta}^{*\infty} = \boldsymbol{\beta}^{(0)} + c^\infty(\mathbf{I}-\mathbf{B}^{\rm Esh}\mathbf{Q}^{\rm Esh})\boldsymbol{\beta}_1^{(1)}$, where $c^\infty = c(\mathbf{x}^\infty)$, and the tensors $\mathbf{B}^{\rm Esh}$ and $\mathbf{Q}^{\rm Esh}$ are defined by Eqs. (14.27).

14.4.2 c^2 Order Accurate Estimation of Effective Thermoelastic Properties

To obtain c^2 order of accuracy of the effective properties estimation, it is necessary to take into account a direct binary interaction of inclusions v_i and v_j at the assumption that the statistical average of the effective field $\widetilde{\boldsymbol{\sigma}}_{i,j}(\mathbf{x})$ acting on each pair of inclusions coincides with the statistical average of a stress field in the considered domain $\langle\widetilde{\boldsymbol{\sigma}}_{i,j}\rangle(\mathbf{x}) = \langle\boldsymbol{\sigma}\rangle(\mathbf{x})$ ($\mathbf{x} \in v_i, v_j$). The conditional average stresses in a concrete point $\mathbf{x} \in v_i$ of the fixed inclusion v_i can be expressed as

$$\langle\boldsymbol{\sigma}|v_i\rangle(\mathbf{x})=\mathbf{B}_i(\mathbf{x})\langle\boldsymbol{\sigma}\rangle(\mathbf{x}_i)+\mathbf{C}_i(\mathbf{x})+\int[\mathbf{C}_{i,j}(\mathbf{x})\varphi(v_j,\mathbf{z}|;v_i,\mathbf{x}_i)-\mathbf{C}^{\infty}_{i,j}(\mathbf{x})\varphi(v_j,\mathbf{z})]d\mathbf{z}$$
$$+\int[\mathbf{B}_{i,j}(\mathbf{x})\varphi(v_j,\mathbf{z}|;v_i,\mathbf{x}_i)-\mathbf{B}^{\infty}_{i,j}(\mathbf{x})\varphi(v_j,\mathbf{z})]\,d\mathbf{z}\langle\boldsymbol{\sigma}\rangle(\mathbf{x}_i)$$
$$+\int[\mathbf{B}^{J}_{i,j}(\mathbf{x})\varphi(v_j,\mathbf{z}|;v_i,\mathbf{x}_i)-\mathbf{B}^{\infty}_{i,j}(\mathbf{x})\varphi(v_j,\mathbf{z})][\langle\boldsymbol{\sigma}\rangle(\mathbf{z})-\langle\boldsymbol{\sigma}\rangle(\mathbf{x}_i)]\,d\mathbf{z}. \qquad (14.44)$$

To obtain affordable numerical results, we used in Eq. (14.44) a local approximation (14.21) of the effective field $\overline{\boldsymbol{\sigma}}_i(\mathbf{x})$ acting on fixed inclusion v_i. Obvious generalization of Eq. (14.44) to the analysis of inhomogeneity of statistical average stresses $\langle\boldsymbol{\sigma}\rangle(\mathbf{x})$ in the vicinity of inclusions v_i, and v_j is

$$\langle\boldsymbol{\sigma}|v_i\rangle(\mathbf{x})=\langle\boldsymbol{\sigma}|v_i\rangle^{(14.44)}(\mathbf{x})+\int[\boldsymbol{\mathcal{B}}^{1}_{i,j}(\langle\boldsymbol{\sigma}\rangle^{I})(\mathbf{x})\varphi(v_j,\mathbf{z}|;v_i,\mathbf{x}_i)$$
$$-\boldsymbol{\mathcal{B}}^{\infty 1}_{i,j}(\langle\boldsymbol{\sigma}\rangle^{I})(\mathbf{x})\varphi(v_j,\mathbf{z})]\,d\mathbf{z}+\int[\boldsymbol{\mathcal{B}}^{J1}_{i,j}(\langle\boldsymbol{\sigma}\rangle^{J}-\langle\boldsymbol{\sigma}\rangle^{I})(\mathbf{x})$$
$$\times\varphi(v_j,\mathbf{z}|;v_i,\mathbf{x}_i)-\boldsymbol{\mathcal{B}}^{\infty 1}_{i,j}(\langle\boldsymbol{\sigma}\rangle^{J}-\langle\boldsymbol{\sigma}\rangle^{I})(\mathbf{x})\varphi(v_j,\mathbf{z})]\,d\mathbf{z}, \qquad (14.45)$$

where $\langle\boldsymbol{\sigma}|v_i\rangle^{(14.44)}(\mathbf{x})$ is the conditional stress $\langle\boldsymbol{\sigma}|v_i\rangle(\mathbf{x})$ estimated by the right-hand side of Eq. (14.44), and the upper index 1 defines a transformation of involved integral operators analogously to the transformation of the integral operator $\boldsymbol{\mathcal{B}}_i$ (14.22).

It is interesting that even for a remote point $x_d\to-\infty$ and constant average stresses $\langle\boldsymbol{\sigma}\rangle(\mathbf{x})=\langle\boldsymbol{\sigma}\rangle\equiv$ const. (at $x_d\to-\infty$), the statistical average stresses in the fixed inclusion are not homogeneous ($\mathbf{x}\in v_i$)

$$\langle\boldsymbol{\sigma}|v_i\rangle(\mathbf{x})=\mathbf{B}^{\rm Esh}_i\langle\boldsymbol{\sigma}\rangle+\mathbf{C}^{\rm Esh}_i+\int[\mathbf{C}_{i,j}(\mathbf{x})\varphi(v_j,\mathbf{z}|;v_i,\mathbf{x}_i)-\mathbf{C}^{\infty}_{i,j}(\mathbf{x})\varphi(v_j,\mathbf{z})]\,d\mathbf{z}$$
$$+\int[\mathbf{B}_{i,j}(\mathbf{x})\varphi(v_j,\mathbf{z}|;v_i,\mathbf{x}_i)-\mathbf{B}^{\infty}_{i,j}(\mathbf{x})\varphi(v_j,\mathbf{z})]\,d\mathbf{z}\langle\boldsymbol{\sigma}\rangle. \qquad (14.46)$$

Substituting the representation $\langle\boldsymbol{\sigma}|v_i\rangle(\mathbf{x})$ (14.44) into the average operation either (14.16) or (14.17) allows us to find the conditional average of stresses in the inclusion phase

$$\langle V\boldsymbol{\sigma}\rangle(\mathbf{x})=\langle\mathbf{B}_i\rangle(\mathbf{x})\langle\boldsymbol{\sigma}\rangle(\mathbf{x}_i)+\langle\mathbf{C}_i\rangle(\mathbf{x})$$
$$+\Big\langle\int[\mathbf{B}_{i,j}(\mathbf{y})\varphi(v_j,\mathbf{z}|;v_i,\mathbf{x}_i)-\mathbf{B}^{\infty}_{i,j}(\mathbf{y})\varphi(v_j,\mathbf{z})]\,d\mathbf{z}\Big\rangle(\mathbf{x})\langle\boldsymbol{\sigma}\rangle(\mathbf{x}_i)$$
$$+\Big\langle\int[\mathbf{C}_{i,j}(\mathbf{y})\varphi(v_j,\mathbf{z}|;v_i,\mathbf{x}_i)-\mathbf{C}^{\infty}_{i,j}(\mathbf{y})\varphi(v_j,\mathbf{z})]\,d\mathbf{z}\Big\rangle(\mathbf{x})$$
$$+\Big\langle\int[\mathbf{B}^{J}_{i,j}(\mathbf{y})\varphi(v_j,\mathbf{z}|;v_i,\mathbf{x}_i)-\mathbf{B}^{\infty}_{i,j}(\mathbf{y})\varphi(v_j,\mathbf{z})]$$
$$\times[\langle\boldsymbol{\sigma}\rangle(\mathbf{z})-\langle\boldsymbol{\sigma}\rangle(\mathbf{y})]\,d\mathbf{z}\Big\rangle(\mathbf{x}), \qquad (14.47)$$

where the last integral of Eq. (14.47) describes nonlocal effects of the binary interaction of inclusions; the variable $\mathbf{y}$ in the averaged integral path through the volume of a moving ellipsoid v_i as in Eq. (14.17).

Substituting the conditional average of the strain polarization tensor $\langle\boldsymbol{\eta}\rangle(\mathbf{x}) = \langle V(\mathbf{M}_1\boldsymbol{\sigma} + \boldsymbol{\beta}_1)\rangle(\mathbf{x})$ [found by the use of Eq. (14.47)] into Eq. (14.7$_2$) allows the effective thermoelastic properties:

$$\mathbf{M}^*(\mathbf{x}) = \mathbf{M}^{(0)} + \langle V\mathbf{R}_i^v\rangle(\mathbf{x}) + \Big\langle \int [\mathbf{R}_{i,j}(\mathbf{y})\varphi(v_j, \mathbf{z}|; v_i, \mathbf{x}_i) \\ - \mathbf{R}_{i,j}^\infty(\mathbf{y})\varphi(v_j, \mathbf{z})]\, d\mathbf{z}\Big\rangle(\mathbf{x}), \tag{14.48}$$

$$\boldsymbol{\beta}^*(\mathbf{x}) = \boldsymbol{\beta}^{(0)} + \langle V\mathbf{F}_i^v\rangle(\mathbf{x}) + \Big\langle \int [\mathbf{C}_{i,j}(\mathbf{y})\varphi(v_j, \mathbf{z}|; v_i, \mathbf{y}) \\ - \mathbf{C}_{i,j}^\infty(\mathbf{y})\varphi(v_j, \mathbf{z})]\, d\mathbf{z}\Big\rangle(\mathbf{x}), \tag{14.49}$$

$$\boldsymbol{\mathcal{M}}^*(\langle\boldsymbol{\sigma}\rangle)(\mathbf{x}) = \Big\langle \int [\mathbf{R}_{i,j}^J(\mathbf{y})\varphi(v_j, \mathbf{z}|; v_i, \mathbf{y}) \\ - \mathbf{R}_{i,j}^\infty(\mathbf{y})\varphi(v_j, \mathbf{z})][\langle\boldsymbol{\sigma}\rangle(\mathbf{z}) - \langle\boldsymbol{\sigma}\rangle(\mathbf{y})]\, d\mathbf{z}\Big\rangle(\mathbf{x}). \tag{14.50}$$

The nonlocal integral operator (14.50) can also be presented in the differential form (14.42) where at this point ($\mathbf{y} \in v_i$)

$$\boldsymbol{\mathcal{M}}_\alpha(\nabla)(\mathbf{y}) = \int [\mathbf{R}_{i,j}^J(\mathbf{y})\varphi(v_j, \mathbf{z}|; v_i, \mathbf{x}_i) - \mathbf{R}_{i,j}^\infty(\mathbf{y})\varphi(v_j, \mathbf{z})] \otimes (\mathbf{y} - \mathbf{z})^\alpha\, d\mathbf{z}\nabla^\alpha. \tag{14.51}$$

Representations (14.48)–(14.51) were obtained in the framework of the local approximation for the average stress $\langle\boldsymbol{\sigma}\rangle(\mathbf{x})$ analogous to Eq. (14.21). They can be obviously recast without this hypothesis as it was performed in Eq. (14.45).

Solutions (14.45)–(14.48) take only binary interaction of inclusions into account. More complete realization of the perturbation method is based on the assumption $\langle\widetilde{\boldsymbol{\sigma}}(\mathbf{x})_{1,\ldots,n}\rangle_i = \langle\overline{\boldsymbol{\sigma}}(\mathbf{x})\rangle_i = \langle\boldsymbol{\sigma}\rangle(\mathbf{x})$, $\langle\widetilde{\boldsymbol{\sigma}}(\mathbf{x})_{1,\ldots,m}\rangle_j \equiv \mathbf{0}$, ($i = 1, \ldots, n$; $\mathbf{x} \in v_i$; $m > n$, $j = 1, \ldots, m$, $\mathbf{y} \in v_j$) with truncating of the Neumann series involved. The perturbation method has an essential advantage of controlled accuracy: the error of a remainder term truncated can, in principle, be estimated exactly.

It should be mentioned that effective properties can be estimated by the use of the evaluation of average strains in the components. In so doing, utilization of the perturbation method presented in Subsections 14.4.1 and 14.4.2 reduces to the inequality $\mathbf{L}^* \neq (\mathbf{M}^*)^{-1}$ in contrast to the MEFM (see Subsection 14.4.3). Because of this Eqs. (14.48) through (14.51) will be employed if $\mathbf{M}(\mathbf{x}) - \mathbf{M}^{(0)}$ is positively definite, otherwise a dual scheme based on the estimation of average strains $\langle\boldsymbol{\varepsilon}\rangle_i$ and effective moduli $\mathbf{L}^*(\mathbf{x})$ will be used.

14.4.3 Quasi-crystalline Approximation

In the framework of the "quasi-crystalline" approximation by Lax [649] (8.65) ($\mathbf{y} \in v_j \neq v_i$)

$$\langle\overline{\boldsymbol{\sigma}}_j(\mathbf{y})|v_j, \mathbf{x}_j; v_i, \mathbf{x}_i\rangle(\mathbf{y}) = \langle\overline{\boldsymbol{\sigma}}_j(\mathbf{y})|v_j, \mathbf{x}_j\rangle(\mathbf{y}) \tag{14.52}$$

the tensors describing the binary interaction of inclusions are simplified ($\mathbf{X} = \mathbf{B}, \mathbf{C}, \mathbf{R}, \mathbf{F}$; $\mathbf{x} \in v_i$) $\mathbf{X}_{i,j}(\mathbf{x}) = \mathbf{X}_{i,j}^\infty(\mathbf{x})$. Then instead of Eq. (14.44), we will get a nonlocal equation for the effective field $\langle\overline{\boldsymbol{\sigma}}\rangle(\mathbf{x})$:

$$\langle\overline{\boldsymbol{\sigma}}\rangle(\mathbf{x}) = \langle\boldsymbol{\sigma}\rangle(\mathbf{x}) + \int \mathbf{E}^{\infty}_{i,j}(\mathbf{x})[\varphi(v_j,\mathbf{z}|;v_i,\mathbf{x}_i) - \varphi(v_j,\mathbf{z})]\, d\mathbf{z}\langle\overline{\boldsymbol{\sigma}}\rangle(\mathbf{x})$$
$$+ \int \mathbf{E}^{\infty}_{i,j}(\mathbf{x})[\varphi(v_j,\mathbf{z}|;v_i,\mathbf{x}_i) - \varphi(v_j,\mathbf{z})][\langle\overline{\boldsymbol{\sigma}}\rangle(\mathbf{z}) - \langle\overline{\boldsymbol{\sigma}}\rangle(\mathbf{x})]\, d\mathbf{z}$$
$$+ \int \mathbf{E}^{\beta\infty}_{i,j}(\mathbf{x})[\varphi(v_j,\mathbf{z}|;v_i,\mathbf{x}_i) - \varphi(v_j,\mathbf{z})]\, d\mathbf{z}, \tag{14.53}$$

which was obtained at the local approximation of effective field $\langle\overline{\boldsymbol{\sigma}}\rangle(\mathbf{z})$ (14.34) acting on the surrounding inclusions v_j; it is assumed that all inclusions are aligned and identical: $v_i,\ v_j \subset v^{(1)}$. Equation (14.53) can be recast in the form

$$\langle\overline{\boldsymbol{\sigma}}\rangle(\mathbf{x}) = \mathbf{D}_i(\mathbf{x})\langle\boldsymbol{\sigma}\rangle(\mathbf{x}) + \mathbf{D}_i(\mathbf{x}) \int \mathbf{E}^{\beta\infty}_{i,j}(\mathbf{x})[\varphi(v_j,\mathbf{z}|;v_i,\mathbf{x}_i) - \varphi(v_j,\mathbf{z})]\, d\mathbf{z}$$
$$+ \mathbf{D}_i(\mathbf{x}) \int \mathbf{E}^{\infty}_{i,j}(\mathbf{x})[\varphi(v_j,\mathbf{z}|;v_i,\mathbf{x}_i) - \varphi(v_j,\mathbf{z})][\langle\overline{\boldsymbol{\sigma}}\rangle(\mathbf{z}) - \langle\overline{\boldsymbol{\sigma}}\rangle(\mathbf{x})]\, d\mathbf{z}, \tag{14.54}$$

where $\mathbf{D}_i^{-1}(\mathbf{x}) = \mathbf{I} - \int \mathbf{E}^{\infty}_{i,j}(\mathbf{x})[\varphi(v_j,\mathbf{z}|;v_i,\mathbf{x}_i) - \varphi(v_j,\mathbf{z})]\, d\mathbf{z}$, $(\mathbf{x}\in v_i)$ is the inversion of the local concentration factor. Substitution of Eq. (14.54) into Eq. (14.20) yields a local (zero-order) approximation of the statistical average of stresses in the fixed inclusion v_i

$$\langle\boldsymbol{\sigma}|v_i\rangle(\mathbf{x}) = \boldsymbol{\mathcal{B}}_i(\mathbf{D}_i)(\mathbf{x})\langle\boldsymbol{\sigma}\rangle(\mathbf{x}) + \mathbf{C}_i(\mathbf{x})$$
$$+ \boldsymbol{\mathcal{B}}_i(\mathbf{D}_i \int \mathbf{E}^{\beta\infty}_{i,j}(\mathbf{x})[\varphi(v_j,\mathbf{z}|;v_i,\mathbf{x}_i) - \varphi(v_j,\mathbf{z})]\, d\mathbf{z})(\mathbf{x}), \tag{14.55}$$

which allows the following representation for the local effective properties by the use of Eq. (14.13_2)

$$\mathbf{M}^*(\mathbf{x}) = \mathbf{M}^{(0)} + \langle V\boldsymbol{\mathcal{R}}^v_i(\mathbf{D})\rangle(\mathbf{x}), \tag{14.56}$$
$$\boldsymbol{\beta}^*(\mathbf{x}) = \boldsymbol{\beta}^{(0)} + \langle V\mathbf{F}^v_i\rangle(\mathbf{x})$$
$$+ \Big\langle V\boldsymbol{\mathcal{R}}^v_i\Big(\mathbf{D}\int \mathbf{C}^{\infty}_{i,j}(\mathbf{x})[\varphi(v_j,\mathbf{z}|;v_i,\mathbf{x}_i) - \varphi(v_j,\mathbf{z})]\, d\mathbf{z}\Big)\Big\rangle(\mathbf{x}), \tag{14.57}$$

respectively. In doing so, the averages of two kinds in Eqs. (14.56), (14.57), and (14.55) are related by the formulae (7.9) and (14.1).

For statistically homogeneous media, we solved in Chapter 13 the nonlocal integral equation analogous to (14.54) by three different methods: by the direct quadrature method (called also the Nystron method), the iteration method (or the Neumann method), and the Fourier transform method. Although the direct quadrature method usually causes no problems of accuracy, for large number of unknown variables N its $O(N^3)$ cost dependence can lead to surprisingly long computing time. The use of the Fourier transform method is made difficult by the fact the coefficients of Eq. (14.54) are variable. Because of this, we will solve Eq. (14.54) by the method of successive approximations providing explicit representations for the nonlocal integral and differential operators. For simplicity, we will consider the point Jacoby iterative method for the pure elastic version ($\boldsymbol{\beta}_1(\mathbf{x}) \equiv \mathbf{0}$) of Eq. (14.54) with the driving term used as an initial approximation $\langle\overline{\boldsymbol{\sigma}}_{[0]}\rangle(\mathbf{x}) = \mathbf{D}_i(\mathbf{x})\langle\boldsymbol{\sigma}\rangle(\mathbf{x})$. The next two iterations have the form

$$\langle\overline{\boldsymbol{\sigma}}_{[1]}\rangle(\mathbf{x}) = \mathbf{D}_i(\mathbf{x})\langle\boldsymbol{\sigma}\rangle(\mathbf{x}) + \mathbf{D}_i(\mathbf{x})\int \mathbf{K}(\mathbf{x},\mathbf{z})[\mathbf{D}_i(\mathbf{z})\langle\boldsymbol{\sigma}\rangle(\mathbf{z}) - \mathbf{D}_i(\mathbf{x})\langle\boldsymbol{\sigma}\rangle(\mathbf{x})]\, d\mathbf{z},$$

$$\begin{aligned}\langle\overline{\boldsymbol{\sigma}}_{[2]}\rangle(\mathbf{x}) &= \mathbf{D}_i(\mathbf{x})\langle\boldsymbol{\sigma}\rangle(\mathbf{x}) + \mathbf{D}_i(\mathbf{x})\int \mathbf{K}(\mathbf{x},\mathbf{z})[\mathbf{D}_i(\mathbf{z})\langle\boldsymbol{\sigma}\rangle(\mathbf{z}) - \mathbf{D}_i(\mathbf{x})\langle\boldsymbol{\sigma}\rangle(\mathbf{x})]\, d\mathbf{z}\\ &\quad + \mathbf{D}_i(\mathbf{x})\int \mathbf{K}(\mathbf{x},\mathbf{z})\Big\{\int \mathbf{K}(\mathbf{z},\mathbf{y})[\mathbf{D}_i(\mathbf{y})\langle\boldsymbol{\sigma}\rangle(\mathbf{y}) - \mathbf{D}_i(\mathbf{z})\langle\boldsymbol{\sigma}\rangle(\mathbf{z})]\, d\mathbf{y}\\ &\quad - \int \mathbf{K}(\mathbf{x},\mathbf{y})[\mathbf{D}_i(\mathbf{y})\langle\boldsymbol{\sigma}\rangle(\mathbf{y}) - \mathbf{D}_i(\mathbf{x})\langle\boldsymbol{\sigma}\rangle(\mathbf{x})]\, d\mathbf{y}\Big\}\, d\mathbf{z}, \end{aligned} \tag{14.58}$$

where $\mathbf{x} \in v_i$ and the kernel $\mathbf{K}(\mathbf{x},\mathbf{z}) \equiv \mathbf{E}^{\infty}_{i,j}(\mathbf{x})[\varphi(v_j,\mathbf{z}|;v_i,\mathbf{x}_i) - \varphi(v_j,\mathbf{z})]$. Again proceeding formally, the Neumann series form for the solution $\langle\overline{\boldsymbol{\sigma}}\rangle(\mathbf{x}) = \lim_{n\to\infty}\langle\overline{\boldsymbol{\sigma}}_{[n]}\rangle(\mathbf{x})$ $(n = 0,\dots)$ is suggested:

$$\langle\overline{\boldsymbol{\sigma}}_{[n]}\rangle(\mathbf{x}) = \mathbf{D}_i(\mathbf{x})\langle\boldsymbol{\sigma}\rangle(\mathbf{x}) + \mathbf{D}_i(\mathbf{x})\int \mathbf{K}(\mathbf{x},\mathbf{z})[\langle\overline{\boldsymbol{\sigma}}_{[n-1]}\rangle(\mathbf{z}) - \langle\overline{\boldsymbol{\sigma}}_{[n-1]}\rangle(\mathbf{x})]\, d\mathbf{z}, \tag{14.59}$$

Substituting Eqs. (14.54)–(14.58) into Eq. (14.21) recast in terms of the effective field $\langle\overline{\boldsymbol{\sigma}}\rangle(\mathbf{x})$ allows estimation of the statistical averages of stresses inside both the fixed inclusion v_i: $\langle\boldsymbol{\sigma}|v_i\rangle(\mathbf{x})$ and the phase $v^{(1)}$: $\langle V\boldsymbol{\sigma}\rangle(\mathbf{x})$ that can be exploited for the estimation of the effective properties. For example, in the case of the first-order approximation (14.58$_1$), the effective constitutive law for the pure elastic problem ($\boldsymbol{\beta}_1(\mathbf{x}) \equiv \mathbf{0}$) is

$$\begin{aligned}\langle\varepsilon\rangle(\mathbf{x}) &= \mathbf{M}^0\langle\boldsymbol{\sigma}\rangle(\mathbf{x}) + \langle V\mathcal{R}_i(\mathbf{D}_i)\rangle(\mathbf{x})\langle\boldsymbol{\sigma}\rangle(\mathbf{x})\\ &\quad + \Big\langle V\mathcal{R}_i\Big(\mathbf{D}_i\int \mathbf{K}(\mathbf{x},\mathbf{z})[\mathbf{D}_i(\mathbf{z})\langle\boldsymbol{\sigma}\rangle(\mathbf{z}) - \mathbf{D}_i(\mathbf{x})\langle\boldsymbol{\sigma}\rangle(\mathbf{x})]\, d\mathbf{z}\Big)\Big\rangle(\mathbf{x}). \end{aligned} \tag{14.60}$$

A differential form of Eq. (14.59) can be obtained analogously to the representation (14.51) by the use of the Taylor expansion of $\mathbf{D}_i(\mathbf{z})\langle\boldsymbol{\sigma}\rangle(\mathbf{z})$ in the vicinity $\mathbf{x}_i$. In so doing, the last integral in Eq. (14.60) does not in general vanish at $\langle\boldsymbol{\sigma}\rangle(\mathbf{x}) \equiv \mathbf{0}$ in opposition to Eq. (14.50) and contains, therefore, a local constituent of the effective compliance operator.

The nonlocal representations for the effective fields (14.58) were derived from Eq. (14.53) obtained in turn at the local approximation of effective field $\langle\overline{\boldsymbol{\sigma}}\rangle(\mathbf{z})$ (14.34) acting on the surrounding inclusions v_j. In the case of arbitrary inhomogeneous remote loading $\langle\boldsymbol{\sigma}\rangle(\mathbf{x})$, Eqs. (14.58) should be recast:

$$\langle\overline{\boldsymbol{\sigma}}_{[1]}\rangle(\mathbf{x}) = \mathbf{D}_i(\mathbf{x})\langle\boldsymbol{\sigma}\rangle(\mathbf{x}) + \mathbf{D}_i(\mathbf{x})\int [\mathcal{K}(\mathbf{D}_i\langle\boldsymbol{\sigma}\rangle)(\mathbf{x},\mathbf{z}) - \mathbf{K}(\mathbf{x},\mathbf{z})\mathbf{D}_i(\mathbf{x})\langle\boldsymbol{\sigma}\rangle(\mathbf{x})]\, d\mathbf{z},$$

$$\begin{aligned}\langle\overline{\boldsymbol{\sigma}}_{[2]}\rangle(\mathbf{x}) &= \mathbf{D}_i(\mathbf{x})\langle\boldsymbol{\sigma}\rangle(\mathbf{x}) + \mathbf{D}_i(\mathbf{x})\int [\mathcal{K}\Big(\mathbf{D}_i\langle\boldsymbol{\sigma}\rangle)(\mathbf{x},\mathbf{z}) - \mathbf{K}(\mathbf{x},\mathbf{z})\mathbf{D}_i(\mathbf{x})\langle\boldsymbol{\sigma}\rangle(\mathbf{x})]\, d\mathbf{z}\\ &\quad + \mathbf{D}_i(\mathbf{x})\int\Big\{\mathcal{K}\Big(\int [\mathcal{K}(\mathbf{D}_i\langle\boldsymbol{\sigma}\rangle)(\mathbf{z},\mathbf{y}) - \mathbf{K}(\mathbf{z},\mathbf{y})\mathbf{D}_i(\mathbf{z})\langle\boldsymbol{\sigma}\rangle(\mathbf{z})]\, d\mathbf{y}\Big)(\mathbf{x},\mathbf{z})\\ &\quad - \mathbf{K}(\mathbf{x},\mathbf{z})\int [\mathcal{K}(\mathbf{D}_i\langle\boldsymbol{\sigma}\rangle)(\mathbf{x},\mathbf{y}) - \mathbf{K}(\mathbf{x},\mathbf{y})\mathbf{D}_i(\mathbf{x})\langle\boldsymbol{\sigma}\rangle(\mathbf{x})]\, d\mathbf{y}\Big\}\, d\mathbf{z}, \end{aligned} \tag{14.61}$$

respectively, through the operator kernel $\mathcal{K}(\mathbf{x},\mathbf{z}) \equiv \mathcal{E}^{\infty}_{i,j}(\mathbf{x})[\varphi(v_j,\mathbf{z}|;v_i,\mathbf{x}_i) - \varphi(v_j,\mathbf{z})]$. In a similar manner, the constitutive Eq. (14.60) can be recast in a nonlocal form

$$\langle \varepsilon \rangle(\mathbf{x}) = \mathbf{M}^0 \langle \boldsymbol{\sigma} \rangle(\mathbf{x}) + \langle V \boldsymbol{\mathcal{R}}_i(\mathbf{D}_i) \rangle(\mathbf{x}) \langle \boldsymbol{\sigma} \rangle(\mathbf{x})$$
$$+ \Big\langle V \boldsymbol{\mathcal{R}}_i \Big(\mathbf{D}_i \int \Big[\boldsymbol{\mathcal{K}}(\mathbf{D}_i \langle \boldsymbol{\sigma} \rangle)(\mathbf{y}, \mathbf{z}) - \mathbf{K}(\mathbf{y}, \mathbf{z}) \mathbf{D}_i(\mathbf{y}) \langle \boldsymbol{\sigma} \rangle(\mathbf{y}) \Big] d\mathbf{z} \Big) \Big\rangle(\mathbf{x}). \quad (14.62)$$

14.4.4 Influence of a Correlation Hole v^0_{ij}

The integration domains R^d $(d = 2, 3)$ of the integrals in Eqs. (14.44) and (14.53) can be decomposed as $R^d = (R^d \setminus v^0_{ij}) \cup v^0_{ij}$. The evaluations of integrals over the domains $R^d \setminus v^0_{ij}$ are reduced to the estimation of perturbations around the inclusion v_i introduced by the surrounding inclusions v_j that can be performed in a straightforward manner. A more interesting case takes place for the integrals over the domain v^0_{ij} such as e.g.

$$\int_{v^0_{ij}} \mathbf{B}^{\infty}_{i,j}(\mathbf{x}) \varphi(v_j, \mathbf{z}) \; d\mathbf{z} \quad \text{and} \quad \int_{v^0_{ij}} \mathbf{E}^{\infty}_{i,j}(\mathbf{x}) \varphi(v_j, \mathbf{z}) [\langle \overline{\boldsymbol{\sigma}} \rangle(\mathbf{z}) - \langle \overline{\boldsymbol{\sigma}} \rangle(\mathbf{x})] \; d\mathbf{z} \quad (14.63)$$

where the perturbations $\mathbf{B}^{\infty}_{i,j}(\mathbf{x})$ and $\mathbf{E}^{\infty}_{i,j}(\mathbf{x})$ are produced by the heterogeneities v_j in the area $\mathbf{x} \in v_i$ in the absence of the inclusion v_i when $\mathbf{M}_1(\mathbf{x}), \boldsymbol{\beta}_1(\mathbf{x}) \equiv \mathbf{0}$ if $\mathbf{x} \in v_i \cap (R^d \setminus v_j)$ and $v_i \cap v_j \neq \emptyset$. Put another way, the tensors $\mathbf{B}^{\infty}_{i,j}(\mathbf{x})$ and $\mathbf{E}^{\infty}_{i,j}(\mathbf{x})$ in the integrals (14.63) should be estimated both outside and inside the inclusion v_j rather than only outside the inclusion v_j in the case if $\mathbf{x}_j \in (R^d \setminus v^0_{ij})$. In so doing, no restriction on the shape of the correlation hole v^0_{ij} is assumed. The importance of the last statement is beyond the scope of the concrete considered method.

Indeed, for statistically homogeneous infinite microinhomogeneous media, the integral (14.63) is equivalent to the integral

$$\int_{v^0_{ij}} \mathbf{E}^{\infty}_{i,j}(\mathbf{x}) \varphi(v_j, \mathbf{z}) \; d\mathbf{z} = \int_{v^0_{ij}} \boldsymbol{\Gamma}(\mathbf{x} - \mathbf{z}) \; d\mathbf{z} \langle \mathbf{R}_j \rangle n^{(j)}, \quad (14.64)$$

which is significantly simplified in the case of an ellipsoidal domain v^0_{ij}: $\int_{v^0_{ij}} \boldsymbol{\Gamma}(\mathbf{x} - \mathbf{z}) \; d\mathbf{z} \langle \mathbf{R}_j \rangle \, n^{(j)} = -\mathbf{Q}^0_{ij} \langle \mathbf{R}_j \rangle n^{(j)}$, where $\mathbf{Q}^0_{ij} \equiv \mathbf{Q}(v^0_{ij}) \equiv$ const. It should be mentioned that Eq. (14.64) are valid for any nonellipsoidal shape of inclusions v_j taking by the tensors $\mathbf{R}_j$ into account (for details see Chapter 8). However, a choice of ellipsoidal shape for the correlation hole v^0_{ij} has significantly more fundamental sense than only simplification caused by replacement of an inhomogeneous tensor $\mathbf{Q}^0_{ij}(\mathbf{x})$ for nonellipsoidal domain v^0_{ij} by a homogeneous tensor $\mathbf{Q}^0_{ij} =$ const. for the ellipsoidal v^0_{ij}. We will demonstrate it, just for simplicity, in the framework of the "quasi-crystalline" approximation (14.52) reducing Eq. (14.53) for statistically homogeneous medium with aligned identical inclusions $(i = j, \; \boldsymbol{\beta}_1(\mathbf{x}) \equiv \mathbf{0})$ at $\langle \boldsymbol{\sigma} \rangle =$ const. to

$$\langle \overline{\boldsymbol{\sigma}} \rangle = \langle \boldsymbol{\sigma} \rangle + \mathbf{Q}^0_{ij} \langle \mathbf{R}_j \rangle n^{(j)} \langle \overline{\boldsymbol{\sigma}} \rangle, \quad (14.65)$$

where the tensor $\mathbf{R}_j$ (14.23_2) could be introduced only for the homogeneous effective field $\langle \overline{\boldsymbol{\sigma}} \rangle$. A linear algebraic Eq. (14.65) allows a well known representation of the effective compliance (see Chapter 8):

$$\mathbf{M}^* = \mathbf{M}^{(0)} + \langle \mathbf{R}_i \rangle n^{(i)} [\mathbf{I} - \mathbf{Q}^0_{ij} \langle \mathbf{R}_j \rangle n^{(j)}]^{-1}. \tag{14.66}$$

Nevertheless, in the case of the nonellipsoidal domain v^0_{ij}, the tensor $\mathbf{Q}^0_{ij}(\mathbf{x}) \not\equiv$ const. and the effective field $\langle \overline{\boldsymbol{\sigma}} \rangle(\mathbf{x})$ on the right-hand side of Eq. (14.65) should be inhomogeneous at $\mathbf{x} \in v_i$ with inhomogeneity depending on the inhomogeneity v_i of the tensor $\mathbf{Q}^0_{ij}(\mathbf{x})$. Then the tensor $\mathbf{R}_j$ must be replaced by the operator $\boldsymbol{\mathcal{R}}_i$ (14.23$_1$) that transforms linear algebraic Eq. (14.65) into the integral operator equation with respect to $\langle \overline{\boldsymbol{\sigma}} \rangle(\mathbf{x})$: $\langle \overline{\boldsymbol{\sigma}} \rangle(\mathbf{x}) = \langle \boldsymbol{\sigma} \rangle + \mathbf{Q}^0_{ij}(\mathbf{x}) \langle \boldsymbol{\mathcal{R}}_j(\langle \overline{\boldsymbol{\sigma}} \rangle)(\mathbf{y}) \rangle_{(j)} n^{(j)}$.

Thus, a violation of the ellipsoidal shape assumption for the correlation hole v^0_{ij} raises unsurmountable obstacles in the standard scheme (14.64)–(14.66) even for statistically homogeneous media. However, a straightforward evaluation of the integrals (14.63) by the use of the tensors $\mathbf{B}^{\infty}_{i,j}(\mathbf{x})$ and $\mathbf{E}^{\infty}_{i,j}(\mathbf{x})$ presents no complications in the case of replacement of the ellipsoidal v^0_{ij} on the nonellipsoidal one even for a microinhomogeneous half-space that allows an explicit representation of the local effective compliance (14.56).

14.5 Statistical Properties of Local Residual Microstresses in Elastically Homogeneous Half-Space

14.5.1 First Moment of Stresses in the Inclusions

In this section we will consider an elastically homogeneous half-space when the elastic properties of the matrix and the randomly distributed inclusions are the same, but the stress-free strains are different. Of course, the statistical moments of residual stresses in composite components can be estimated with the help of the passage to the zero limit of the elastic mismatch of different components in the corresponding formulae of Section 14.4. Nevertheless, the desired relationships can be found immediately without some assumptions of previous results.

Indeed, a constitutive equation for microinhomogeneous medium (14.8) and (14.9) can be presented in the form

$$\langle \boldsymbol{\sigma} \rangle(\mathbf{x}) = \mathbf{L}^{(0)} \langle \boldsymbol{\varepsilon} \rangle(\mathbf{x}) + \boldsymbol{\alpha}^*(\mathbf{x}), \quad \langle \boldsymbol{\varepsilon} \rangle(\mathbf{x}) = \mathbf{M}^{(0)} \langle \boldsymbol{\sigma} \rangle(\mathbf{x}) + \boldsymbol{\beta}^*(\mathbf{x}), \tag{14.67}$$

where the tensors of the effective eigenvectors are defined by simple geometrical averaging $\boldsymbol{\alpha}^*(\mathbf{x}) = \langle \boldsymbol{\alpha} \rangle(\mathbf{x})$, $\boldsymbol{\beta}^*(\mathbf{x}) = \langle \boldsymbol{\beta} \rangle(\mathbf{x})$. In a similar manner, the general integral Eq. (14.1) is also simplified:

$$\boldsymbol{\sigma}(\mathbf{x}) = \langle \boldsymbol{\sigma} \rangle(\mathbf{x}) + \int [\boldsymbol{\Gamma}(\mathbf{x},\mathbf{y}) \boldsymbol{\beta}(\mathbf{y}) V(\mathbf{y}) - \mathbf{T}^{\beta}_k(\mathbf{x},\mathbf{y}) \varphi(v_k, \mathbf{y})] \, d\mathbf{y}. \tag{14.68}$$

The tensor $\mathbf{T}^{\beta}_k(\mathbf{x},\mathbf{y}) = \int \boldsymbol{\Gamma}(\mathbf{x},\mathbf{z}) \boldsymbol{\beta}_1(\mathbf{z}) V_k(\mathbf{z}) \, d\mathbf{z}$ describing a perturbation of inclusion v_k with the varying center $\mathbf{y}$ is defined though the tensor $\mathbf{T}_k(\mathbf{x},\mathbf{y})$ if $\boldsymbol{\beta}_1(\mathbf{x}) = \boldsymbol{\beta}^{(k)}_1 \equiv$ const. at $\mathbf{x} \not\in v_k$ as ($\mathbf{z} \in v_k$): $\mathbf{T}^{\beta}_k(\mathbf{x},\mathbf{y}) = \mathbf{T}_k(\mathbf{x},\mathbf{y}) \boldsymbol{\beta}^{(k)}_1$, $\mathbf{T}_k(\mathbf{x},\mathbf{y}) = \bar{v}_k \langle \boldsymbol{\Gamma}(\mathbf{x},\mathbf{z}) \rangle_k$. Estimations of the tensors $\mathbf{T}^{\beta}_i(\mathbf{x},\mathbf{x}_i)$ both inside and outside the inclusion v_i are reduced to the model problem for residual stresses produced by an inclusion inside a half-space (14.39). It should be mentioned that the statistical average of stresses $\langle\boldsymbol{\sigma}\rangle(\mathbf{x})$ can be obtained from the solution of the macroproblem (6.1), (6.3), and (14.67_2) with prescribed boundary conditions. In so doing, the macroscopic constitutive law (14.67_2) contains effective eigenstrain $\boldsymbol{\beta}^*(\mathbf{x}) \equiv \langle\boldsymbol{\beta}\rangle(\mathbf{x})$ easily estimated via geometrical averaging either (14.17) or (14.18). Because of this, the deterministic field $\langle\boldsymbol{\sigma}\rangle(\mathbf{x})$ is assumed to be known.

Let us consider an arbitrary fixed inclusion v_i; then for $\mathbf{x} \in v_i$ from Eq. (14.68) we obtain the relation for the stresses in the inclusion v_i:

$$\boldsymbol{\sigma}(\mathbf{x}) = \overline{\boldsymbol{\sigma}}_i(\mathbf{x}) + \mathbf{T}_i^{\beta}(\mathbf{x}, \mathbf{x}_i), \tag{14.69}$$

which is the superposition of the disturbance $\mathbf{T}_i(\mathbf{x}, \mathbf{x}_i)\boldsymbol{\beta}_1^i$ caused by the transformation field in the inclusion v_i considered and the effective field $\overline{\boldsymbol{\sigma}}_i(\mathbf{x})$ produced by the external loading $\boldsymbol{\sigma}^0$ and by the surrounding inhomogeneities:

$$\overline{\boldsymbol{\sigma}}_i(\mathbf{x}) = \langle\boldsymbol{\sigma}\rangle(\mathbf{x}) + \int \Big[\boldsymbol{\Gamma}(\mathbf{x}, \mathbf{y})\boldsymbol{\beta}_1(\mathbf{y})[V(\mathbf{y}) - V_i(\mathbf{y})] - \mathbf{T}_k^{\beta}(\mathbf{x}, \mathbf{y})\varphi(v_k, \mathbf{y})\Big]\, d\mathbf{y}. \tag{14.70}$$

This effective field is a random function of all the other positions of the surrounding inhomogeneities, which is the superposition of the average field $\langle\boldsymbol{\sigma}\rangle(\mathbf{x})$ and the distribution caused by the other inclusions v_j, $(j \neq i,\ j = 1, \ldots)$ of the considered set.

Averaging Eqs. (14.69) and (14.70) over a random realization of surrounding inclusions $v_q \neq v_i$ gives

$$\begin{aligned}\langle\boldsymbol{\sigma}\rangle_i(\mathbf{x}) &= \langle\boldsymbol{\sigma}\rangle(\mathbf{x}) + \mathbf{T}_i^{\beta}(\mathbf{x}, \mathbf{x}_i)\\ &+ \int \Big[\mathbf{T}_q^{\beta}(\mathbf{x}, \mathbf{x}_q)\bar{v}_q\varphi(v_q, \mathbf{x}_q|; v_i, \mathbf{x}_i) - \mathbf{T}^{\beta}(\mathbf{x}, \mathbf{x}_q)\varphi(v_q, \mathbf{x}_q)\Big]\, d\mathbf{x}_q.\end{aligned} \tag{14.71}$$

Here the integral item can be decomposed on two integrals

$$\begin{aligned}\langle\boldsymbol{\sigma}\rangle_i(\mathbf{x}) &= \langle\boldsymbol{\sigma}\rangle(\mathbf{x}) + \mathbf{T}_i^{\beta}(\mathbf{x}, \mathbf{x}_i) - \int V_i^0(\mathbf{x}_k)\mathbf{T}_k^{\beta}(\mathbf{x}, \mathbf{x}_k)\varphi(v_k, \mathbf{x}_k)\, d\mathbf{x}_k\\ &+ \int \mathbf{T}_q^{\beta}(\mathbf{x}, \mathbf{x}_q)\{\varphi(v_q, \mathbf{x}_q|; v_i, \mathbf{x}_i) - \varphi(v_q, \mathbf{x}_q)[1 - V_i^0(\mathbf{x}_q)]\}\, d\mathbf{x}_q,\end{aligned} \tag{14.72}$$

where the first integral describes the stress perturbation generated by averaged inclusions in the correlation hole v_i^0, while the second integral vanishing for step correlation function $\varphi(v_q, \mathbf{x}_q|v_i, \mathbf{x}_i) = \varphi(v_q, \mathbf{x}_q)$ expresses the influence of inclusions outside of the correlation hole v_i^0.

It should be mentioned that the perturbations produced by the different inclusions do not interacted with one another and can be evaluated by a simple summation. However, correctness of the mentioned simple superposition technique for a finite inclusion number does not provide an appropriateness of their direct utilization for an infinite number of inclusions in the following formula (compare with Eq. (14.71))

$$\langle\boldsymbol{\sigma}\rangle_i(\mathbf{x}) = \langle\boldsymbol{\sigma}\rangle(\mathbf{x}) + \int \mathbf{T}_q^{\beta}(\mathbf{x}, \mathbf{x}_q)\bar{v}_q\varphi(v_q, \mathbf{x}_q|v_i, \mathbf{x}_i)\, d\mathbf{x}_q. \tag{14.73}$$

Fundamental incorrectness of Eq. (14.73) was demonstrated in Subsection 7.2.3 for the infinite composite media. However, the modified superposition scheme used at the obtaining of Eqs. (14.71) is prevented from this drawback.

14.5.2 Limiting Case for a Statistically Homogeneous Medium

At the infinity $x_3, y_3 \to -\infty$, the tensor $\mathbf{\Gamma}(\mathbf{x},\mathbf{y})$ tends to an appropriate representation for the infinite-homogeneous-body Green's stress function $\mathbf{\Gamma}^\infty(\mathbf{x}-\mathbf{y})$, which is the even homogeneous generalized function of the order $-d$. In such a case the tensor $\mathbf{T}_i(\mathbf{x},\mathbf{x}_i)$ is defined by either the external ($\mathbf{x}\notin v_i$) or internal ($\mathbf{x}\in v_i$) Eshelby tensor, which is a constant at $\mathbf{x}\in v_i$: $\mathbf{S}_i^{\rm Esh}=\mathbf{I}-\mathbf{M}^{(0)}\mathbf{Q}_i^{\rm Esh}\equiv\text{const}$, $\mathbf{Q}_i^{\rm Esh}\equiv-\langle\mathbf{\Gamma}^\infty(\mathbf{x}-\mathbf{y})\rangle_{(i)}$ $(\mathbf{x},\mathbf{y}\in v_i)$ and $\mathbf{T}_i(\mathbf{x},\mathbf{x}_i)\equiv\mathbf{T}_i^{\rm Esh}=-\mathbf{Q}_i^{\rm Esh}$. Because of this, the tensor $\mathbf{T}_k(\mathbf{x},\mathbf{x}_k)$ is called a generalized Eshelby tensor (internal or external at $\mathbf{x}\in v_k$ or $\mathbf{x}\notin v_k$, respectively).

For statistically homogeneous media, Eq. (14.68) is reduced to the equivalent representation $\boldsymbol{\sigma}(\mathbf{x})=\langle\boldsymbol{\sigma}\rangle(\mathbf{x})+\int\mathbf{\Gamma}^\infty(\mathbf{x}-\mathbf{y})[\boldsymbol{\beta}_1(\mathbf{y})V(\mathbf{y})-\langle\boldsymbol{\beta}_1\rangle]\,d\mathbf{y}$, which leads to the analog of Eq. (14.72):

$$\begin{aligned}\langle\boldsymbol{\sigma}\rangle_i(\mathbf{x})&=\langle\boldsymbol{\sigma}\rangle(\mathbf{x})+\mathbf{T}_i^\beta(\mathbf{x},\mathbf{x}_i)\\&+\int\Big[\mathbf{T}_q^\beta(\mathbf{x},\mathbf{x}_q)\varphi(v_q,\mathbf{x}_q|;v_i,\mathbf{x}_i)-\mathbf{\Gamma}(\mathbf{x}-\mathbf{x}_q)\langle\boldsymbol{\beta}_1\rangle\Big]\,d\mathbf{x}_q.\end{aligned}\tag{14.74}$$

The formula (14.74) may be significantly simplified under the assumption

$$\langle V_q(\mathbf{y})\boldsymbol{\beta}_1^{(q)}(\mathbf{y})|;v_i,\mathbf{x}_i\rangle=\mathbf{f}_1(\langle\boldsymbol{\beta}_1^{(q)}\rangle,\rho),\tag{14.75}$$

where $\rho\equiv|(\mathbf{a}_i^0)^{-1}(\mathbf{x}_q-\mathbf{x}_i)|$. The assumption (14.75) was considered in detail in Chapter 8; see Eq. (8.69).

By virtue of the fact that generalized function $\mathbf{\Gamma}^\infty(\mathbf{x})$ is an even homogeneous function of order $-d$, we have a relation (see Eq. (8.70)):

$$\int\Big[\mathbf{T}_q^\beta(\mathbf{x},\mathbf{x}_q)\bar v_q\varphi(v_q,\mathbf{x}_q|;v_i,\mathbf{x}_i)-\mathbf{\Gamma}(\mathbf{x}-\mathbf{x}_q)\langle\boldsymbol{\beta}_1\rangle\Big]\,d\mathbf{x}_q=\mathbf{Q}_i^{0\rm Esh}\langle\boldsymbol{\beta}_1\rangle\tag{14.76}$$

under the assumption (14.75). Then Eq. (14.74) can be combined into a simple equation ($\mathbf{x}\in v_i$):

$$\langle\boldsymbol{\sigma}\rangle_i(\mathbf{x})=\boldsymbol{\sigma}^0+\mathbf{Q}_i^{0\rm Esh}\langle\boldsymbol{\beta}_1\rangle+\mathbf{T}_i^\beta(\mathbf{x},\mathbf{x}_i).\tag{14.77}$$

In so doing for the homogeneous ellipsoidal inclusion v_i with $\mathbf{Q}_i^{\rm Esh}=\mathbf{Q}_i^{0\rm Esh}$, the statistical average stress $\langle\boldsymbol{\sigma}\rangle_i(\mathbf{x})$ does not depend on the position $\mathbf{x}$ inside the inclusion being analyzed $\langle\boldsymbol{\sigma}\rangle_i(\mathbf{x})=\boldsymbol{\sigma}^0+\mathbf{Q}_i^{\rm Esh}[\langle\boldsymbol{\beta}_1\rangle-\boldsymbol{\beta}_1^{(i)}]$, if $\boldsymbol{\beta}_1(\mathbf{x})=\text{const.}$, $\mathbf{x}\in v_i$.

It should be mentioned that the representations (14.72) and (14.77) are fundamentally different although both of them are exact. First of all, the second integral in Eq. (14.72) vanishes for statistically homogeneous media at the assumption (14.75) that defines the first simplification of Eq. (14.72) in the case of statistically homogeneous media. The second simplification of the assumption (14.75) is an opportunity to use a merit of ellipsoidal shape of the correlation hole v_i^0 consisting of the analytical representation of the tensor $\mathbf{Q}_i^{0\rm Esh}$ (14.76). However, consideration of a microinhomogeneous medium in a half-space essentially complicates the problem in two directions. First, the statistical average of stresses $\langle\boldsymbol{\sigma}\rangle_i(\mathbf{x})$ (14.72) will depend on the conditional probability density $\varphi(v_q,\mathbf{x}_q|;v_i,\mathbf{x}_i)$ in contrast to $\langle\boldsymbol{\sigma}\rangle_i(\mathbf{x})$ (14.77) invariant with respect to $\varphi(v_q,\mathbf{x}_q|;v_i,\mathbf{x}_i)$. Nevertheless, even for the step pair distribution function $\varphi(v_q,\mathbf{x}_q|;v_i,\mathbf{x}_i)=(1-V_i^0(\mathbf{x}_q))\varphi(v_q,\mathbf{x}_q)$ vanishing the second integral in Eq. (14.72), the fundamental difference between Eqs. (14.72) and (14.77) remains. Indeed, the first integral in Eq. (14.72) serving as the tensor $\mathbf{Q}_i^{0\mathrm{Esh}}\langle\boldsymbol{\beta}_1\rangle$ in Eq. (14.77) is defined only by the tensors $\mathbf{T}_q^{\beta}(\mathbf{x},\mathbf{x}_q)$ and does not exploit any assumptions about both the specific arrangement of inclusion (14.75) and the ellipsoidal shape of the correlation hole v_i^0 (14.76). This generalization of Eq. (14.72) with respect to Eq. (14.77) gives some benefit for the homogeneous ellipsoidal inclusions v_i $[\boldsymbol{\beta}_1(\mathbf{x})=\text{const. at }\mathbf{x}\in v_i,\ i=1,\ldots]$ when Eq. (14.72) depends only on the generalized Eshelby tensor $\mathbf{T}_i(\mathbf{x},\mathbf{x}_i)$. As this takes place, a direct attempt to generalize Eq. (14.74) to the case of microinhomogeneous half-space will lead to the necessity of an integral estimation $\mathbf{T}_i^{0\beta}(\mathbf{x},\mathbf{y},\langle\boldsymbol{\beta}_1\rangle)=\int\boldsymbol{\Gamma}(\mathbf{x},\mathbf{z})\langle\boldsymbol{\beta}_1\rangle(\mathbf{z})V_i^0(\mathbf{z})\,d\mathbf{z}$ for inhomogeneous field $\langle\boldsymbol{\beta}_1\rangle(\mathbf{z})$ $(\mathbf{z}\in v_i^0)$ that is significantly more complicated than the utilization of the generalized Eshelby tensors $\mathbf{T}_i(\mathbf{x},\mathbf{x}_i)$ in Eq. (14.72).

14.5.3 Stress Fluctuations Inside the Inclusions

To obtain the second moment of stresses in the component $v^{(i)}$ of the inclusions $(i=1,2,\ldots.)$, it is necessary to take the tensor product of (14.68) into $\boldsymbol{\sigma}(\mathbf{x}),\ \mathbf{x}\in v_i$:

$$\begin{aligned}
\boldsymbol{\sigma}(\mathbf{x})\otimes\boldsymbol{\sigma}(\mathbf{x}) &= \langle\boldsymbol{\sigma}\rangle(\mathbf{x})\otimes\langle\boldsymbol{\sigma}\rangle(\mathbf{x})\\
&+\langle\boldsymbol{\sigma}\rangle(\mathbf{x})\otimes\int[\boldsymbol{\Gamma}(\mathbf{x},\mathbf{z})\boldsymbol{\beta}_1(\mathbf{z})V_q(\mathbf{z})-\mathbf{T}_k^{\beta}(\mathbf{x},\mathbf{z})\varphi(v_k,\mathbf{z})]\,d\mathbf{z}\\
&+\int[\boldsymbol{\Gamma}(\mathbf{x},\mathbf{y})\boldsymbol{\beta}_1(\mathbf{y})V_p(\mathbf{y})-\mathbf{T}_k^{\beta}(\mathbf{x},\mathbf{y})\varphi(v_k,\mathbf{y})]\,d\mathbf{y}\otimes\langle\boldsymbol{\sigma}\rangle(\mathbf{x})\\
&+\iint[\boldsymbol{\Gamma}(\mathbf{x},\mathbf{y})\boldsymbol{\beta}_1(\mathbf{y})V_p(\mathbf{y})-\mathbf{T}_k^{\beta}(\mathbf{x},\mathbf{y})\varphi(v_k,\mathbf{y})]\\
&\otimes[\boldsymbol{\Gamma}(\mathbf{x},\mathbf{z})\boldsymbol{\beta}_1(\mathbf{z})V_q(\mathbf{z})-\mathbf{T}_k^{\beta}(\mathbf{x},\mathbf{z})\varphi(v_k,\mathbf{z})]\,d\mathbf{y}\,d\mathbf{z}.
\end{aligned}\qquad(14.78)$$

The right-hand side of Eq. (14.78) is a random function of arrangements of surrounding inclusions $v_p,\ v_q$ $(p,q=1,2,\ldots)$. Equation (14.78) is a generalization of the corresponding equation suggested for statistically homogeneous media (see Chapter 13) and can be analyzed in a similar manner. Indeed, the right-hand side of the averaged Eq. (14.78) includes two-point and three-point conditional probability densities which can be expressed through the Eqs. (13.57). Then Eq. (14.78) is reduced to

$$\begin{aligned}
\langle\boldsymbol{\sigma}\otimes\boldsymbol{\sigma}\rangle_i(\mathbf{x}) &= \langle\boldsymbol{\sigma}\rangle_i(\mathbf{x})\otimes\langle\boldsymbol{\sigma}\rangle_i(\mathbf{x})\\
&+\int\mathbf{T}_p^{\beta}(\mathbf{x},\mathbf{x}_p)\bar{v}_p\otimes\mathbf{T}_p^{\beta}(\mathbf{x},\mathbf{x}_p)\bar{v}_p\varphi(v_p,\mathbf{x}_p|;v_i,\mathbf{x}_i)\,d\mathbf{x}_p\\
&+\int\int\mathbf{T}_p^{\beta}(\mathbf{x},\mathbf{x}_p)\bar{v}_p\otimes\mathbf{T}_q^{\beta}(\mathbf{x},\mathbf{x}_q)\bar{v}_q\varphi(v_p,\mathbf{x}_p|;v_i,\mathbf{x}_i)\\
&\times[\varphi(v_q,\mathbf{x}_q|;v_p,\mathbf{x}_p;v_i,\mathbf{x}_i)-\varphi(v_q,\mathbf{x}_q|;v_i,\mathbf{x}_i)]\,d\mathbf{x}_q\,d\mathbf{x}_p.
\end{aligned}\qquad(14.79)$$

The new exact relation (14.79) is derived by the use of triple interaction of the inclusions. As may be seen from Eq. (14.79), neglect of binary interaction is tantamount to assuming homogeneity of stresses inside component v_i $\langle\boldsymbol{\sigma}\otimes\boldsymbol{\sigma}\rangle_i(\mathbf{x}) = \langle\boldsymbol{\sigma}\rangle_i(\mathbf{x})\otimes\langle\boldsymbol{\sigma}\rangle_i(\mathbf{x})$. The following approximation of second moment estimation can be obtained by taking into account only binary interaction of inclusions:

$$\langle\boldsymbol{\sigma}\otimes\boldsymbol{\sigma}\rangle_i(\mathbf{x}) = \langle\boldsymbol{\sigma}\rangle_i(\mathbf{x})\otimes\langle\boldsymbol{\sigma}\rangle_i(\mathbf{x}) + \int \mathbf{T}_p^{\beta}(\mathbf{x},\mathbf{x}_p)\bar{v}_p\otimes\mathbf{T}_p^{\beta}(\mathbf{x},\mathbf{x}_p)\bar{v}_p\varphi(v_p,\mathbf{x}_p|;v_i,\mathbf{x}_i)\,d\mathbf{x}. \qquad (14.80)$$

We proved in Section 13.3 that the formula (14.80) provides a sufficiently good approximation of the exact solution for statistically homogeneous infinite media. One can see from the equation (14.80) that covariance matrix $\Delta\boldsymbol{\sigma}_i^2(\mathbf{x}) \equiv \langle\boldsymbol{\sigma}\otimes\boldsymbol{\sigma}\rangle_i(\mathbf{x}) - \langle\boldsymbol{\sigma}\rangle_i(\mathbf{x})\otimes\langle\boldsymbol{\sigma}\rangle_i(\mathbf{x})$ presents a determined nonhomogeneous function of coordinate $\mathbf{x}$ inside inclusion v_i (even at $\mathbf{x}_{i|2} = -\infty$ or $\mathbf{x}_{i|3} = -\infty$ for 2-D or 3-D problems, respectively). It is significant that in deriving (14.79) one did not used the hypothesis **H1** of the so-called MEFM (see for details Chapter 8).

In a similar manner we can obtain the representations of the first and second statistical moment of stresses inside the matrix

$$\langle\boldsymbol{\sigma}\rangle_0(\mathbf{x}) = \langle\boldsymbol{\sigma}\rangle(\mathbf{x}) + \int \mathbf{T}_q^{\beta}(\mathbf{x},\mathbf{x}_q)\bar{v}_q[\varphi(v_q,\mathbf{x}_q|;v_0,\mathbf{x}_0) - \varphi(v_k,\mathbf{x}_q)]\,d\mathbf{x}_q, \qquad (14.81)$$

$$\begin{aligned}\langle\boldsymbol{\sigma}\otimes\boldsymbol{\sigma}\rangle_0(\mathbf{x}) &= \langle\boldsymbol{\sigma}\rangle_0(\mathbf{x})\otimes\langle\boldsymbol{\sigma}\rangle_0(\mathbf{x}) \\ &+ \int \mathbf{T}_p^{\beta}(\mathbf{x},\mathbf{x}_p)\bar{v}_p\otimes\mathbf{T}_p^{\beta}(\mathbf{x},\mathbf{x}_p)\bar{v}_p\varphi(v_p,\mathbf{x}_p|;v_0,\mathbf{x}_0)\,d\mathbf{x}_p \\ &+ \int\int \mathbf{T}_p^{\beta}(\mathbf{x},\mathbf{x}_p)\bar{v}_p\otimes\mathbf{T}_q^{\beta}(\mathbf{x},\mathbf{x}_q)\bar{v}_q\varphi(v_p,\mathbf{x}_p|;v_0,\mathbf{x}_0) \\ &\times[\varphi(v_q,\mathbf{x}_q|;v_p,\mathbf{x}_p;v_0,\mathbf{x}_0) - \varphi(v_q,\mathbf{x}_q|;v_0,\mathbf{x}_0)]\,d\mathbf{x}_q\,d\mathbf{x}_p. \end{aligned} \qquad (14.82)$$

Thus, we can see from Eqs. (14.72) and (14.79) that the first and the second conditional moment of stresses inside the inclusions are defined by the tensors $\mathbf{T}_i^{\beta}(\mathbf{x},\mathbf{x}_i)$ describing solution for a single inclusion inside a homogeneous half-space as well as by the one-point, two-point, and three-point probability densities $\varphi(v_q,\mathbf{x}_q)$, $\varphi(v_q,\mathbf{x}_q|;v_i,\mathbf{x}_i)$, $\varphi(v_q,\mathbf{x}_q|;v_p,\mathbf{x}_p;v_i,\mathbf{x}_i)$ [or $\varphi(v_q,\mathbf{x}_q|;v_0,\mathbf{x}_0)$, $\varphi(v_q,\mathbf{x}_q|;v_p,\mathbf{x}_p;v_0,\mathbf{x}_0)$ in the case of a stress analysis inside the matrix, see Eqs. (14.81) and (14.82)], respectively. Hence for elastically homogeneous media, the perturbations produced by the different inclusions do not interact one with another and can be evaluated by a simple summation. However, correctness of the mentioned simple superposition technique for a finite inclusion number does not provide an appropriateness of their direct utilization for an infinite number of inclusions in the following formula (compare with Eqs. (14.72) and (14.81), $I = 0, i$) $\langle\boldsymbol{\sigma}\rangle_I(\mathbf{x}) = \langle\boldsymbol{\sigma}\rangle(\mathbf{x}) + \int \mathbf{T}_q^{\beta}(\mathbf{x},\mathbf{x}_q)\bar{v}_q\varphi(v_q,\mathbf{x}_q|;v_I,\mathbf{x}_I)d\mathbf{x}_q$. Fundamental incorrectness of the last equation was demonstrated in Chapter 7 for the infinite composite media. However, the modified superposition scheme used at the obtaining of Eqs. (14.72) and (14.81) is prevented from this drawback.

14.6 Numerical Results

We will consider the circular domains Ω_i $(i = 2,\ldots,n)$ discretized along the polar angle and the radius in the local polar coordinate system with centers $\mathbf{x}_i$. Then the points

$$\Big\{(r,\varphi) \mid (p-1)\frac{2\pi}{l} < \varphi < p\frac{2\pi}{l},\ (q-1)\frac{a_i}{m} < r < q\frac{a_i}{m}\Big\} \qquad (14.83)$$

$(p = 1,2\ldots,l;\ q = 1,2,\ldots,m)$ represent the elements of Γ_i^{pq} of the meshes Ω_i $(i = 2,\ldots,n)$ that is not optimized but is efficient. Moreover, the square meshes

$$\Big\{(x_l^1,x_2^1)^\top \mid (p-1)\frac{a_i}{l} < x_l^1 < p\frac{a_i}{l},\ (q-1)\frac{a_i}{l} < x_2^1 < q\frac{a_i}{l}\Big\}, \qquad (14.84)$$

where $\mathbf{x}^l \equiv (x_l^1,\ x_2^1)^\top$ are local coordinates with origins at the fiber centers, will be analyzed. We will use piecewise-constant elements of the meshes which are not very cost-efficient but are very easy for computer programming, and the discretization (14.84) permits the analysis of nonregular inclusion shapes. For simplicity in estimation of the integrals involved, we will utilize the basic numerical integration formulas of Simpson's rule for the uniform (14.83), (14.84) meshes. The mesh (14.84) is used, e.g., at the estimation of first integral over the v_i^0 in the representation of the first moment of stresses (14.72) (and (14.81)). The second integral in Eq. (14.72) as well as the integral in the representation of the second moment of stresses (14.80) are estimated in the polar coordinate system (14.83). At last, evaluation of statistical averages $\langle(\cdot)\rangle(\mathbf{x})$ (14.17) of the different values (such as, e.g., $V(\mathbf{x})$, $\langle\boldsymbol{\sigma}\rangle_i(\mathbf{x})$, $\langle\boldsymbol{\sigma}\otimes\boldsymbol{\sigma}\rangle_i(\mathbf{x})$) based on the averaging over the lines $x_2 =$ const. can be performed more easily by the use of the rectangular mesh (14.84).

For a half-plane containing identical circles with radius $a_1 = a_2 = a$, we will consider one probability density $\varphi(v_q,\mathbf{x}_q) = n(1 - H(x_{q|2} + a))$ and two conditional probability density functions for the inclusions will be examined (compare with (9.91) and (9.93)):

$$\varphi(v_q,\mathbf{x}_q|;v_i,\mathbf{x}_i) \equiv n(1 - H(x_{q|2} + a))(1 - H(x_{i|2} + a))H(r - 2a), \qquad (14.85)$$

$$\varphi(v_q,\mathbf{x}_q|;v_i,\mathbf{x}_i) = n(1 - H(x_{q|2} + a))(1 - H(x_{i|2} + a))H(r - 2a)$$
$$\times\left\{1+\frac{4c}{\pi}\left[\pi-2\sin^{-1}(\frac{r}{4a})-\frac{r}{2a}\sqrt{1-\frac{r^2}{16a^2}}\right]H(4a-r)\right\}, \qquad (14.86)$$

where H denotes the Heaviside step function, $r \equiv |\mathbf{x}_i - \mathbf{x}_q|$ is the distance between the nonintersecting inclusions v_i and v_q, and $c = \pi a^2 n$ is the area fraction of circle inclusions with the radius a. For composites with free edge, ergodicity is lost and functions $\varphi(v_q,\mathbf{x}_q|;v_i,\mathbf{x}_i)$ is required. Due to the absence of these functions, we use the step functions for the construction of the densities 14.85 and 14.86 (the last one is not a step function) in order to reduce the influence of undesirable size effects on the interesting particular results obtained. Thus, we use some sort of an assumption by Luciano and Willis [717]: "the body under analysis has been cut out of a medium which is statistically stationary". For the function

$\varphi(v_q, \mathbf{x}_q|; v_i, \mathbf{x}_i)$ (14.86), the integrable function in the second integral in the representation of the first moment of stresses (14.72) vanishes at $|\mathbf{x}_q - \mathbf{x}_i| \geq 4a$ and vanishes for $\varphi(v_q, \mathbf{x}_q|; v_i, \mathbf{x}_i)$ (14.85) at $\forall \mathbf{x}_q$.

At first we will consider a microinhomogeneous half plane with dilute concentration of randomly dispersed identical circular inclusions ($a_1 = a_2 = a$). We decompose the overall field as $\boldsymbol{\sigma}(\mathbf{x}) = \boldsymbol{\sigma}^I(\mathbf{x}) + \boldsymbol{\sigma}^{II}(\mathbf{x})$, with the sources $\boldsymbol{\beta}^I(\mathbf{x}) = \mathbf{0}$, $\mathbf{t}^{\Gamma I}(\mathbf{y}) = \mathbf{t}^{\Gamma}(\mathbf{y})$ and $\boldsymbol{\beta}^{II}(\mathbf{x}) = \boldsymbol{\beta}(\mathbf{x})$, $\mathbf{t}^{\Gamma II}(\mathbf{y}) = \mathbf{0}$ for the field $\boldsymbol{\sigma}^I$ and $\boldsymbol{\sigma}^{II}$, respectively, where $\mathbf{t}^{\Gamma}(\mathbf{y})$ ($\mathbf{x} \in R^d_-$, $\mathbf{y} \in \Gamma_t$) is a prescribed traction in the boundary conditions (7.1_2). Let the phase thermal expansion coefficients (CTEs) $\boldsymbol{\beta}^{(0)} = (\beta_0, \beta_0, 0)^\top$ and $\boldsymbol{\beta}^{(1)} = (\beta^1, \beta^1, 0)^\top$ be isotropic which will be convenient for result presentations in a normalized form with respect to the CTEs. Just for concreteness, elastic properties are taken corresponding to the properties of silicate nanocomposites constituents [190]: Young's modulus $E^{(1)} = 300$ GPa and Poisson's ratio $\nu^{(1)} = 0.4$ of inclusions $E^{(0)} = 3.01$ GPa and $\nu^{(0)} = 0.41$ of *Epox*862 matrix. The normalized surface concentration of inclusions $c(\mathbf{x})/c^\infty$ is related to the normalized average coefficient of thermal expansion (CTE) $(\langle V\boldsymbol{\beta}\rangle(\mathbf{x}) - \boldsymbol{\beta}^{(0)})/(\beta_1 c^\infty) = \boldsymbol{\delta} c(\mathbf{x})/c^\infty$, $\beta_1 \equiv \beta^1 - \beta^0$. The normalized effective CTE $\widetilde{\beta}^*_{ij} \equiv (\beta^*_{ij}(\mathbf{x}) - \langle \beta_{ij}\rangle(\mathbf{x}))/(\beta^{*\infty}_{ij} - \langle \beta_{ij}\rangle^\infty)$ (see Fig. 14.1), depending on the elastic mismatch of the matrix and inclusions, is found to be anisotropic due to anisotropic stress distribution $\boldsymbol{\sigma}^{II}(\mathbf{x})$ inside the inclusions $\mathbf{x} = (x_1, x_2) \in \Omega^{(1)}$ in the boundary layer $-5 \leq x_2/a \leq 0$; here $\langle \boldsymbol{\beta}\rangle^\infty = \langle \boldsymbol{\beta}\rangle(\mathbf{x})$ ($\mathbf{x} = (0, -\infty)^\top$). It should be mentioned that the boundary layer effect for the inclusion surface concentration $c(\mathbf{x})$ is limited by the thin area $-1 \leq x_2/a \leq 0$ due to the trivial geometrical constraints on the possible location of inclusion centers $y_2 \leq -1.05a$, while the nature of the free edge effect for the effective CTE appearing in a significantly wider strip $-5 \leq x_2/a \leq 0$ is defined by more complicated reasons of inhomogeneity of residual stress $\sigma^{II}(\mathbf{x})$ inside the inclusion in this strip (see the components of the tensor $\widetilde{\boldsymbol{\beta}}^*(\mathbf{x})/c(\mathbf{x})$ in Fig. 14.1). In Fig. 14.2 are shown

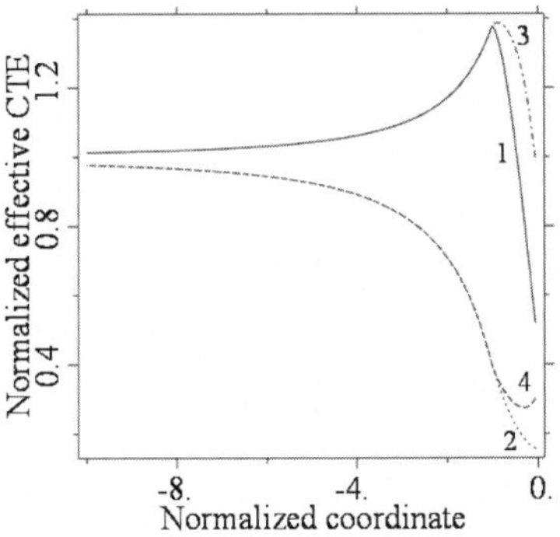

Fig. 14.1. The normalized effective CTE $\widetilde{\boldsymbol{\beta}}^*$ vs coordinate x_2/a: $\widetilde{\beta}^*_{11}$ (1), $\widetilde{\beta}^*_{22}$ (2), $\widetilde{\beta}^*_{11}/c(\mathbf{x})$ (3) and $\widetilde{\beta}^*_{22}/c(\mathbf{x})$ (4).

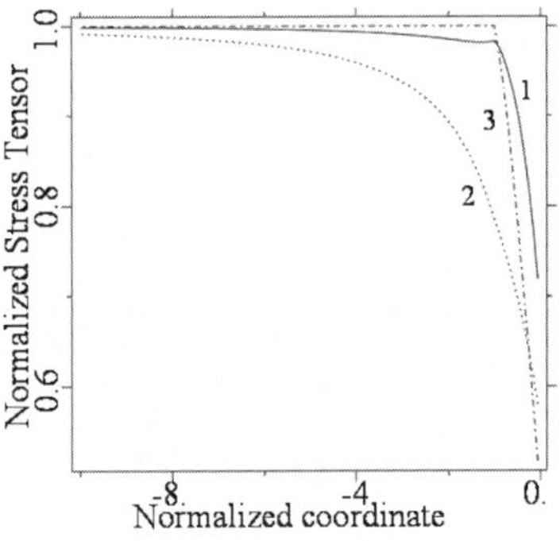

Fig. 14.2. The components of the normalized stress tensor $\widetilde{\boldsymbol{\sigma}}^{II}(\mathbf{x})$ vs coordinate x_2/a: $\widetilde{\sigma}^{II}_{11}(\mathbf{x})$ (1), $\widetilde{\sigma}^{II}_{22}(\mathbf{x})$ (2), $c(\mathbf{x})$ (3)

the components of the normalized conditional statistical averages of residual stresses inside the inclusions $\tilde{\boldsymbol{\sigma}}^{II}(\mathbf{x}) \equiv \langle V\boldsymbol{\sigma}^{II}\rangle(\mathbf{x})/(c(\mathbf{x})\sigma_{i0}^{II\infty})$ $(\mathbf{x} \in \Omega^{(1)})$, where $\boldsymbol{\sigma}_i^{II\infty} \equiv \sigma_{i0}^{II\infty}\boldsymbol{\delta} = -\beta_1\mathbf{BQ}\boldsymbol{\delta}$ is homogeneous isotropic residual stresses (defined by the Eshelby solution) inside the inclusions ($x_2 = -\infty$) remote from the free edge. The use of the conditional statistical average of stresses inside inclusions instead of $\langle V\boldsymbol{\sigma}^{II}\rangle(\mathbf{x})/(c^\infty\sigma_0^{II\infty})$ eliminates the variation of the inclusion concentration $c(\mathbf{x})$ in the boundary layer $-a < \mathbf{x} < 0$.

In a similar manner, the components of the normalized effective compliance $\widetilde{M}^*_{ijkl} \equiv (M^*_{ijkl}(\mathbf{x}) - M^{(0)}_{ijkl})(M^{*\infty}_{ijkl} - M^{(0)}_{ijkl})^{-1}$ (no summation over $i, j, k, l = 1, 2$) are depicted in Fig. 14.3; here $\mathbf{M}^{*\infty} \equiv \mathbf{M}^*(0, -\infty)$ is the effective compliance in the remote area $x_2 = -\infty$ obtained from the Eshelby solution. As can be seen in Fig. 14.3, the effective compliance $\mathbf{M}^*(\mathbf{x})$ is slightly anisotropic: $\widetilde{M}^*_{1111}(\mathbf{x}) \neq \widetilde{M}^*_{2222}(\mathbf{x})$ at $-5a < x_2 < 0$; the component $\widetilde{M}^*_{1212}$ exceeding $\widetilde{M}^*_{2222}$ no more than 0.2% is not depicted. Stress concentrator factors $\tilde{\sigma}^I_{\alpha\beta}(\mathbf{x}) = \langle V\sigma_{\alpha\beta}\rangle(\mathbf{x})/(c(\mathbf{x})\sigma^\infty_{i|\alpha\beta})$ (no summation over $i, j = 1, 2$) for the different unit loadings $\sigma_{\alpha\beta} = \delta_{1\alpha}\delta_{1\beta}, \delta_{2\alpha}\delta_{2\beta}$, and $\delta_{1\alpha}\delta_{2\beta}$ are presented in Fig. 14.4; here $\boldsymbol{\sigma}_i^\infty = \mathbf{B}\boldsymbol{\sigma}^\infty$ is the Eshelby solution for the homogeneous stress distribution inside the inclusion remote from the free edge and subjected to the homogeneous loading $\boldsymbol{\sigma}^\infty$ at infinity. As can be seen, the free edge effects for both $\widetilde{\mathbf{M}}^*(\mathbf{x})$ and $\tilde{\boldsymbol{\sigma}}^I(\mathbf{x})$ are also manifested in a wide strip $-5 \leq x_2/a \leq 0$ that reflects inhomogeneity of the stresses inside the inclusions (distinct from the Eshelby solution) in the vicinity of the free edge. In addition, a strong inhomogeneity of $\widetilde{\mathbf{M}}^*(\mathbf{x})$ in the thin strip $-a < x_2 < 0$ is also defined by the inhomogeneity of $c(\mathbf{x})$ in this area.

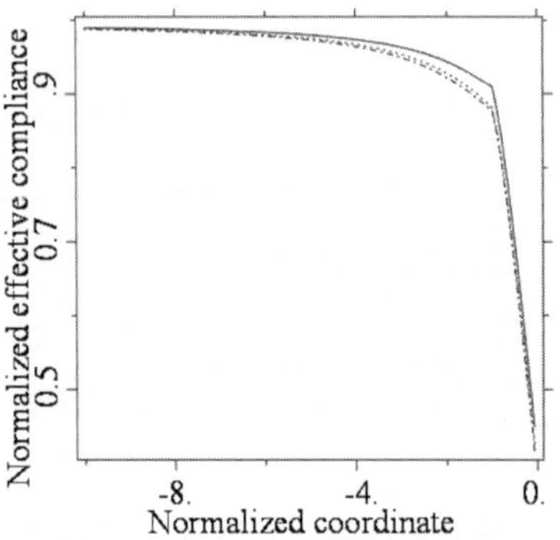

Fig. 14.3. The components of the normalized effective compliance vs the coordinate x_2/a: $\widetilde{M}^*_{1111}(\mathbf{x})$ (solid line), $\widetilde{M}^*_{2222}(\mathbf{x})$ (dotted line), and $\widetilde{M}^*_{1122}(\mathbf{x})$ (dot-dashed line).

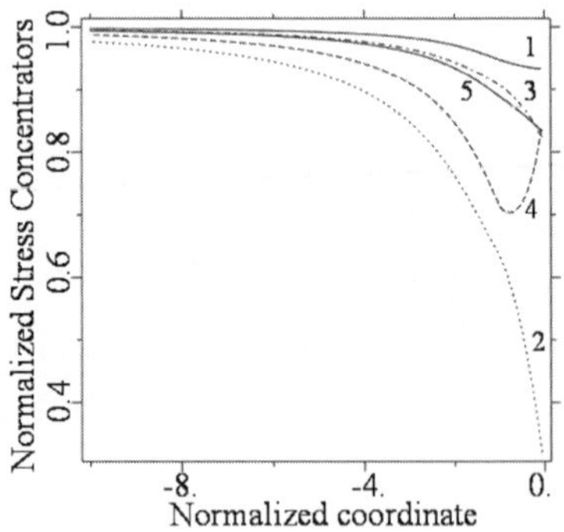

Fig. 14.4. The normalized stress concentration factors vs x_2/a: $\tilde{\sigma}^I_{11}(\mathbf{x})$ (1) and $\tilde{\sigma}^I_{22}(\mathbf{x})$ (2) for $\sigma^\infty_{\alpha\beta} = \delta_{1\alpha}\delta_{1\beta}$; $\tilde{\sigma}^I_{22}(\mathbf{x})$ (3) and $\tilde{\sigma}^I_{11}(\mathbf{x})$ (4) for $\sigma^\infty_{\alpha\beta} = \delta_{2\alpha}\delta_{2\beta}$; $\tilde{\sigma}^I_{12}(\mathbf{x})$ (5) for $\sigma^\infty_{\alpha\beta} = \delta_{1\alpha}\delta_{2\beta}$

Thus, two length scales in the microinhomogeneous half-space were detected. The first length scale $a_1^c = a$ is defined by the geometrical (constraints) on the location of the inclusion centers ($y_2 \leq -a_1^c$). The second length scale a_2^c ($a_1^c \leq a_2^c \leq 5a$) equals the distance of the inclusion to the free edge generating inhomogeneous stresses in the considered inclusion. However, the free edge problem has a nondetected third length scale a_3^c defined by long distance interaction between the inclusions. The influence of a_3^c on $\mathbf{M}^*$ negligible at the dilute concentration of inclusion being considered was analyzed for the functionally graded material (see Chapter 13) when the second kind of length scale effect is absent. It was found that the effective elastic moduli in the large spherical cluster with the constant concentration of spherical inclusions vary in the boundary layer of the cluster with the thickness $5a$. This effect is expected for nondilute concentration of inclusions in the vicinity of the free edge will lead to the inhomogeneity of statistically averaged stresses $\langle \boldsymbol{\sigma} \rangle(\mathbf{x})$ that in turn will appear in the nonlocal nature of the effective constitutive equation; for references see Chapter 12. However this sort of problem is beyond the scope of the current presentation.

We now turn our attention to the analysis of the residual stresses inside fixed inclusions at the different distance x_2 of their centers from the free surface of the elastically homogeneous composite material at $c = 0.5$ (it is assumed that the known $\langle \boldsymbol{\sigma} \rangle(\mathbf{x}) \equiv \mathbf{0}$ and $\mathbf{M}_1(\mathbf{x}) \equiv \mathbf{0}$). We will assume a step correlation function (14.85) vanishing the second integral of the used formula (14.72). In Figs. 14.5a and 14.5b the normalized conditional stress concentrator

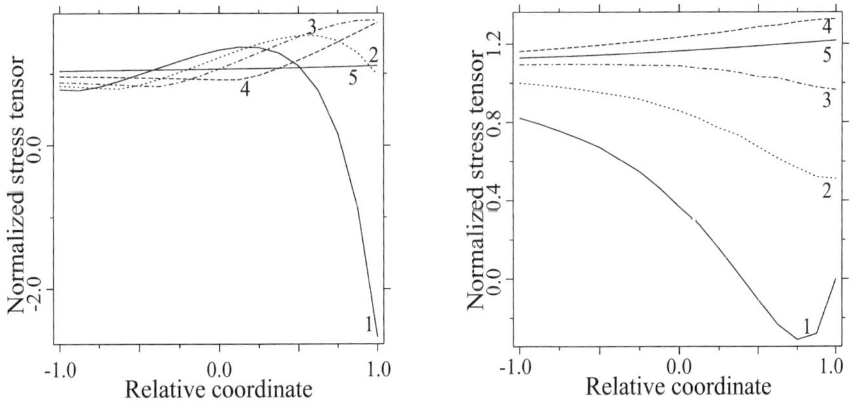

a) b)

Fig. 14.5. $\langle \sigma_{11} \rangle_i(\mathbf{x})/\langle \sigma_{11} \rangle_i^\infty$ (Fig. a) and $\langle \sigma_{22} \rangle_i(\mathbf{x})/\langle \sigma_{22} \rangle_i^\infty$ (Fig. b) vs x_2^l/a for $x_{i|2}/a = -1.0, -1.25, -1.5, -2.0, -6.0$ (curves 1–5, respectively) estimated for $\varphi(v_q, \mathbf{x}_q | v_i, \mathbf{x}_i)$ (14.85).

factors $\langle \sigma_{11} \rangle_i(\mathbf{x})/\langle \sigma_{11} \rangle_i^\infty$ and $\langle \sigma_{22} \rangle_i(\mathbf{x})/\langle \sigma_{22} \rangle_i^\infty$, respectively, are represented as the functions of the relative local coordinate x_2^l/a ($x_1^l \equiv 0$) for the different coordinates of the inclusion centers $x_{i|2}/a$. Here $\langle \sigma_{22} \rangle_i^\infty$ is the stress distribution inside the fixed inclusion v_i of a statistically homogeneous inclusion field estimated through the Eshelby solution (14.77). The normalized tensors of stresses $\langle \sigma_{kl} \rangle_i(\mathbf{x})/\langle \sigma_{kl} \rangle_i^\infty$ (no summation over k, l) do not depend on the eigenstrain $\boldsymbol{\beta}_1(\mathbf{x})$ and the shear modulus μ. As can be seen from Fig. 14.5a, the normalized stresses $\langle \sigma_{11} \rangle_i(\mathbf{x})/\langle \sigma_{11} \rangle_i^\infty$ ($\mathbf{x} \in v_i$) vary less than 10% if the distance between the center of the fixed inclusions x_2, and the free surface exceeds $2.6a$ that defines the so-called boundary-layer effect reflecting inhomogeneity of the stresses inside the inclusions (distinct from the Eshelby solution) in the vicinity of the free edge. In so doing the scale of the free edge effect (defined with a tolerance 10%) for the stresses $\langle\sigma_{22}\rangle_i(\mathbf{x})/\langle\sigma_{22}\rangle_i^\infty$ (see Fig. 14.5b) is manifested in a wider strip $-4.6 \le x_2/a \le 0$.

Let us consider now the normalized conditional stress concentrator factors (no summation over $K = 1,2$) $\langle\sigma_{KK}\rangle^{(1)}(\mathbf{x})/\langle\sigma_{KK}\rangle_i^\infty$ for the different volume concentrations c^∞ varying in the range from 0 to 0.5. The variation of the inclusion concentration $c(\mathbf{x})$ in the boundary layer $-a < \mathbf{x} < 0$ is eliminated for $\langle\sigma_{KK}\rangle^{(1)}(\mathbf{x})/\langle\sigma_{KK}\rangle_i^\infty$ [opposite to $\langle V\sigma_{KK}\rangle(\mathbf{x})/\langle\sigma_{KK}\rangle_i^\infty$ (14.17)]. As can be seen from Figs. 14.6a and 14.6b, increasing of c^∞ leads to the variation of the characters of the curves $\langle\sigma_{KK}\rangle^{(1)}(\mathbf{x})/\langle\sigma_{KK}\rangle_i^\infty \sim x_2/a$ ($K = 1,2$, respectively) from monotonic (at the small $c^\infty \le 0.1$) to nonmonotonic ones (at $c^\infty \ge 0.2$) with two distinguishing points $x_2 = -a$ (minimum possible distance of the inclusion center from the free edge) and $x_2 = -2a$ (the radius of the correlation hole v_i^0). In so doing the free edge effects for $\langle\sigma_{KK}\rangle^{(1)}(\mathbf{x})$ (Fig. 14.7) with $K = 1$ and $K = 2$ are exhibited in the strips $-3 \le x_2/a \le 0$ and $-5 \le x_2/a \le 0$, respectively.

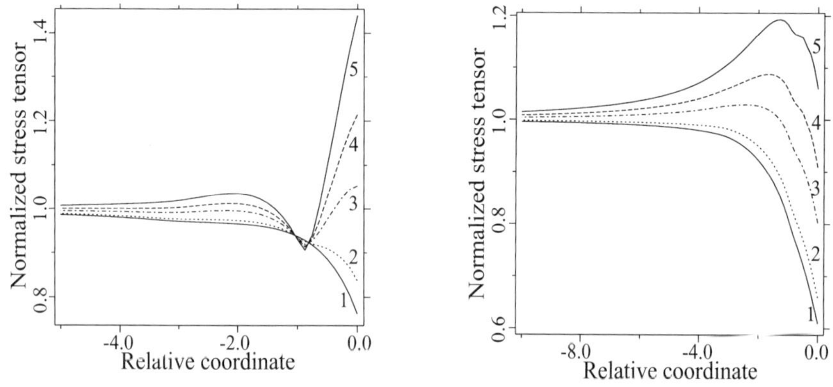

a) **b)**

Fig. 14.6. $\langle\sigma_{11}\rangle^{(1)}(\mathbf{x})/\langle\sigma_{11}\rangle_i^\infty$ (Fig. a) and $\langle\sigma_{22}\rangle^{(1)}(\mathbf{x})/\langle\sigma_{22}\rangle_i^\infty$ (Fig. b) vs relative coordinate x_2/a at $c^\infty = 0.001$, 0.1, 0.3, 0.4, 0.5 (curves 1–5, respectively) estimated for $\varphi(v_q, \mathbf{x}_q|v_i, \mathbf{x}_i)$ (14.85).

In Fig. 14.7 the stress distributions $\langle\sigma_{KK}\rangle^{(1)}(\mathbf{x})/\langle\sigma_{KK}\rangle_i^\infty \sim r$ ($K = 1,2; c^\infty = 0.5$) are estimated for the correlation functions (14.85) and (14.86). In the last case, the second integral in Eq. (14.72) is not vanished that leads to the enhancement of thickness of a domain with a distinct free-edge effect to $5a$ and $10a$ for $K = 1$ and $K = 2$, respectively.

Influence of the correlation functions (14.85) and (14.86) on the local stress distribution is shown in Fig. 14.8 by estimation of $\langle\sigma_{KK}\rangle_i(\mathbf{x})/\langle\sigma_{KK}\rangle_i^\infty$ ($\mathbf{x}^l = (0, x_2^l)^\top \in v_i$, no summation over $K = 1,2$; $c^\infty = 0.5$). Comparison of Figs. 14.8a and 14.8b with Figs. 14.5a and 14.5b, respectively, estimated for the step correlation function (14.85) demonstrates a more pronounced free edge effect. So, the thicknesses of boundary layers providing variation of local stresses with a tolerance of 10% are $5.6a$ and $7.6a$ for the components $\langle\sigma_{11}\rangle_i(\mathbf{x}^l)$ and $\langle\sigma_{22}\rangle_i(\mathbf{x}^l)$, respectively, that significantly exceed the relevant estimations ($2.6a$ and $4.6a$, respectively) obtained for the step correlation function (14.85).

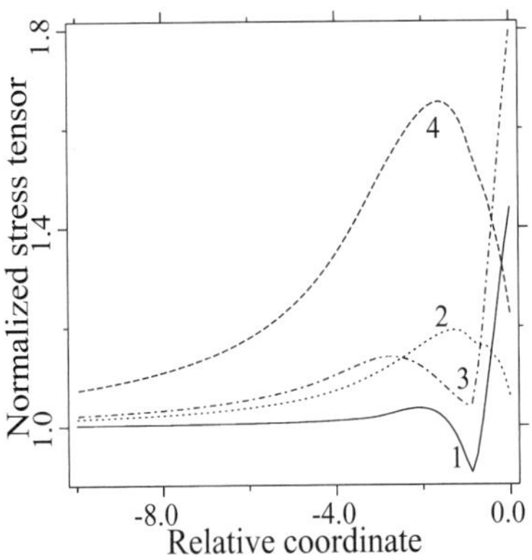

Fig. 14.7. $\langle\sigma_{11}\rangle^{(1)}(\mathbf{x})/\langle\sigma_{11}\rangle_i^{\infty}$ (curves 1 and 3) and $\langle\sigma_{22}\rangle^{(1)}(\mathbf{x})/\langle\sigma_{22}\rangle_i^{\infty}$ (curves 2 and 4) vs x_2/a at $c^{\infty} = 0.5$ estimated for the correlation functions (14.85) (curves 1, 2) and (14.86) (curves 3, 4).

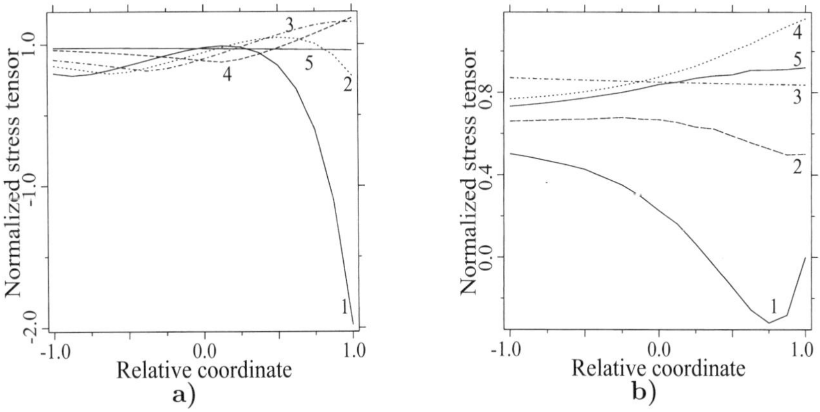

Fig. 14.8. $\langle\sigma_{11}\rangle_i(\mathbf{x}^l)/\langle\sigma_{11}\rangle_i^{\infty}$ (Fig. a) and $\langle\sigma_{22}\rangle_i(\mathbf{x}^l)/\langle\sigma_{22}\rangle_i^{\infty}$ (Fig. b) vs x_2^l/a for the different coordinates of the inclusion centers $x_{i|2}/a$ =-1.0, -1.25, -1.5, -2.0, -6.0 (curves 1-5, respectively) estimated for $\varphi(v_q, \mathbf{x}_q|v_i, \mathbf{x}_i)$ (14.86).

Thus, we analyzed two sorts of the conditional averages of stresses inside the inclusion phase $\langle\boldsymbol{\sigma}\rangle_i(\mathbf{x}^l)$ ($\mathbf{x}^l \in v_i \subset w$) and $\langle\boldsymbol{\sigma}\rangle^{(1)}(\mathbf{x})$ ($\mathbf{x} \in w$). In so doing, the estimations of $\langle\boldsymbol{\sigma}\rangle_i(\mathbf{x}^l)$ are more informative and contain detailed stress distributions. Indeed, the stresses $\langle\sigma_{11}\rangle_i(\mathbf{x}^l)$ vary in the range $-2.63 \le \langle\sigma_{11}\rangle_i(\mathbf{x}^l)/\langle\sigma_{11}\rangle_i^{\infty} \le 1.80$ with extreme values archived at $x_{i|2}/a = -1.0$ and $x_{i|2}/a = -1.75$ while $0.91 \le \langle\boldsymbol{\sigma}_{11}\rangle^{(1)}(\mathbf{x})/\langle\sigma_{11}\rangle_i^{\infty} \le 1.44$. Obviously, at the strength analysis of composite materials, it should be used precisely $\langle\boldsymbol{\sigma}\rangle_i(\mathbf{x}^l)$ ($\mathbf{x}^l \in v_i \subset w$), rather than widely exploited $\langle\boldsymbol{\sigma}\rangle^{(1)}(\mathbf{x})$, ($\mathbf{x} \in w$) (see, e.g., Eq. (15.10) and [719]) which is a too rough descriptor of the stress distribution. This conclusion having a critical practical meaning for a composite half-space being analyzed has no sense at $x_{i|2}, x_2 \to -\infty$ when the statistical averages of stresses inside the inclusions are homogeneous: $\langle\boldsymbol{\sigma}\rangle_i(\mathbf{x}^l) \equiv \langle\boldsymbol{\sigma}\rangle^{(1)}(\mathbf{x}) \equiv \langle\boldsymbol{\sigma}\rangle_i^{\infty}$.

We now turn to the estimation of normalized stress fluctuations (relative stress dispersion) $\bar{\Delta}\sigma_{i|KK}(\mathbf{x}^l) \equiv (\Delta\sigma^2_{i|KK}(\mathbf{x}^l))^{1/2}/|\langle\sigma_{KK}\rangle_i^{\infty}|$ (no sum over $K = 1,2$) (14.80). It is interesting that $\bar{\Delta}\sigma_{i|KK}(\mathbf{x})$ does not vanish even for the remote inclusions with $x_{i|2} = -\infty$ and for the step function $\varphi(v_q, \mathbf{x}_q|v_i, \mathbf{x}_i)$ (14.85). In so doing, $\bar{\Delta}\sigma_{i|11}(\mathbf{x}^l) \to \bar{\Delta}\sigma_{i|22}(\mathbf{x}^l) \not\equiv 0$ at $x_{i|2} \to -\infty$. That is to say, the stress fluctuation is defined, first of all, by the random location of surrounding inclusions (that also takes place for statistically homogeneous inclusion field) which is disrupted by the free edge. This defines a few distinguishing points in the variations of the relative stress dispersion $\bar{\Delta}\sigma_{i|KK}(\mathbf{x}^l) \sim x_2/a$ (see Fig. 14.9): $x_2/a = -1$ (minimum possible distance of the inclusion center from the free edge) $x_2/a = -2$ (the radius of the correlation hole v_i^0), $x_2 = -3/a$ (minimum possible distance of the inclusion center x_2 providing a falling of surrounding inclusions between the fixed inclusion and the free edge). As can be seen, the relative stress dispersion for the first component of stresses $\bar{\Delta}\sigma_{i|11}(\mathbf{x}^l)$ is larger than for the second component $\bar{\Delta}\sigma_{i|22}(\mathbf{x}^l)$, and it varies in the range from 1.70 ($x_{i|2}/a = -1,\ x_2^l/a = 1$) to 0.77 ($x_{i|2}/a = -3,\ x_2^l/a = 0.5$).

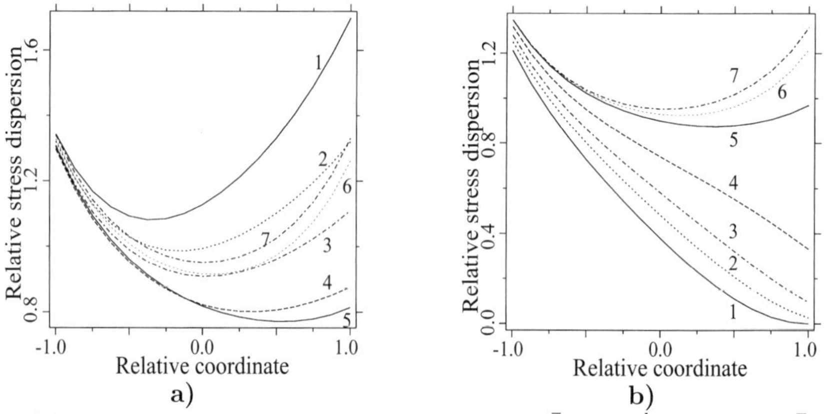

Fig. 14.9. Relative stress dispersion of concentrator factors $\bar{\Delta}\boldsymbol{\sigma}_{i|11}(\mathbf{x}^l)$ (Fig. a) and $\bar{\Delta}\boldsymbol{\sigma}_{i|22}$ $(\mathbf{x}^l)$ (Fig. b) vs relative coordinate $x_2^l/a \in [-1, 1]$ at c^{∞} =0.5 for the different coordinates of the inclusion centers $x_{i|2}/a = -1.0, -1.25, -1.5, -2.0, -3.0, -4.0, -6.0$ (curves 1-7, respectively) estimate for $\varphi(v_q, \mathbf{x}_q|v_i, \mathbf{x}_i)$ (14.86).

Finally, it should be particularly emphasized that the representations for statistical moments of stresses (Eqs. (14.72), (14.80), (14.82)) obtained are new, and (14.69) is exact. The statistical moments of stresses depend explicitly not only on the local concentration of the inclusions, but also on at least binary correlation functions of the inclusions. Therefore, for statistically inhomogeneous composites, including the composites with free edge, the local statistical moments of stresses in the components are nonlocal functions of the inclusion concentration. Thus, the method of integral equations proposed enables one to discover the new nonlocal effects and promises large benefits in analyses of a wide class of nonlinear problems for composite materials (see Chapters 15 and 16).

15

Effective Limiting Surfaces in the Theory of Nonlinear Composites

In contrast to the extensive work on the elastic behavior of composites, only a limited number of studies for the nonlinear range have been published (for references see [299], [897]). Standard methods of successive approximation may be used in nonlinear problems of composites with random structure, linearized problems being solved at each stage. The well-known concept of secant moduli has been combined with the hypothesis of the homogeneity of the increments of the plastic or creep strains [80], [161], [162], [247], [298], [1075], [1146]. Since the widely used method of average strains is capable of estimating only the average stresses in the components, its use for the linearization of functions describing nonlinear effects, e.g., strength ([19], [930]), yielding [687], damage accumulation ([390], [596]; see also the theories combining the damage accumulation and life-fraction rule [150], [1042]), hardening [910], creep [1229], and dynamic viscoplasticity [1231] may be problematical. Due to the significant inhomogeneity of the stress fields in the components (especially in the matrix), such linearization entails physical inconsistencies, which was discussed in detail in [164], [180] within the context of predicting the flow behavior of porous media. Alternatively, for linearizing nonlinear functions (such as the yield or strength criterion, dissipative function) physically consistent assumptions for the dependence of these functions on the second moment of stresses are employed. In this way, the results of Chapter 13 for the volume averages of the stress second moments will be used. The fundamental role of the statistical averages of the second moments of stress concentration factors in nonlinear analysis is explained by the fact that both the yield surface, fiber/matrix interface failure criterion and the energy release rate are the quadratic functions of the local stress distributions. In so doing, estimation of the effective limiting surfaces separating the linear and nonlinear behavior domains of the mentioned nonlinear phenomena is of profound importance for the practical applications and can be done by the methods describing the linear deformations of composite materials. It should be mentioned that the estimation of the effective elastic moduli is a linear problem, with respect to the stress field analyzed which is less sensitive to the local stress distribution than nonlinear micromechanical problems of elastoplastic deformation, fracture, and fatigue of composite materials depending, at least, on mean-square stress fluctuations in the constituents (see, e.g., [138], [693], [897]). The estimations of second moment of stresses are defined by both the random stress fluctuations in the components and the inhomogeneity of the statistical average of stress fields in the constituents; the separation of these stresses presents a real challenge in the framework of the perturbation method considered in Section 13.1. However, the method of integral equations easy allows one to estimate the inhomogeneous second moment of stresses in the inclusions and at each point on the interface between the matrix and inclusions. The dispersion of these interface stresses, defined only by stress fluctuations, will be used in this chapter for the prediction of the failure initiation. The failure initiation is dependent on the size and the volume fraction of inclusions, their surface treatment, matrix, and inclusion properties. A change of the adhesion properties of the interface, defined by the surface treatment, has a smaller effect on modulus than on strength. Indeed, even poor adhesion between the constituents does not appear to be an important factor as long as the frictional forces between the phases are not exceeded by the interface stress. Because of this, the estimation of the failure initiation envelope of the interface is of practical interest.

15.1 Local Limiting Surface

15.1.1 Local Limiting Surface for Bulk Stresses

The failure analysis of composite materials considers the initiation and accumulation of damage occuring in each phase of the material and involves several types of local degradation processes including matrix microcracking (type I), interfacial debonding (type II), and fiber breakages (type III), etc. Generally, these failure mechanisms may initiate concurrently in an early loading stage and progressively accumulate inside the materials (see for references and detail [269], [322], [323], [558], [635], [760], [928]).

The first type of damage processes are those that relate to matrix degradation. They include matrix microcracking and pseudo-delamination. The second type of degradation models describe interfacial decohesion and related mechanisms, such as fiber matrix friction and fiber pull-out processes. Let us assume that the well known tensor-polynomial strength criterion (see, e.g., [1115]) describes the initiation of failure mechanisms of the types I and III for each component, i.e. the equivalent stress is given by

$$\boldsymbol{\Pi}^{(i)}(\boldsymbol{\sigma}) = \boldsymbol{\Pi}^{1(i)}\boldsymbol{\sigma} + \boldsymbol{\Pi}^{2(i)}(\boldsymbol{\sigma}\otimes\boldsymbol{\sigma}) + \boldsymbol{\Pi}^{3(i)}(\boldsymbol{\sigma}\otimes\boldsymbol{\sigma}\otimes\boldsymbol{\sigma}) + \ldots = 1, \tag{15.1}$$

where $i = 0, 1, \ldots$, and the second-, fourth- and sixth-rank tensors of strength $\boldsymbol{\Pi}^1$, $\boldsymbol{\Pi}^2$, $\boldsymbol{\Pi}^3$ are expressed through technical strength parameters for different classes of material symmetry [1086], [1222]. It should be mentioned that in Eq. (15.1), the tensor $\boldsymbol{\sigma} = \boldsymbol{\sigma}(\mathbf{x})$ stands the local stresses in the composite material. Because of this, the criterion (15.1) drown his name Tsai-Wu criterion used macroscopic stresses $\langle\boldsymbol{\sigma}\rangle$ from just a notation convenience.

Similarly to (15.1) for onset of yielding in the case of elastic plastic composites a quadratic yield condition

$$\mathbf{\Pi}^{(i)}(\sigma) \equiv \Pi^{4(i)}_{ijmn}\sigma_{ij}\sigma_{mn} = 1, \quad (i = 0, 1, \ldots, N) \tag{15.2}$$

is used [73] which generalizes some classical criteria, as for example the von Mises equivalent stress criterion:

$$\frac{3}{2}\mathbf{ss} = \tau_i^2, \tag{15.3}$$

for which we have $\mathbf{\Pi}^{4(i)} = 3\mathbf{N}_2/(2\tau_i^2)$ and $\mathbf{\Pi}^{4(i)}(\boldsymbol{\sigma}) = \frac{1}{2\tau_i^2}\{(\sigma_{11}-\sigma_{22})^2 + (\sigma_{22} - \sigma_{33})^2 + (\sigma_{11}-\sigma_{33})^2 + 3[(\sigma_{12})^2 + (\sigma_{13})^2 + (\sigma_{23})^2]\} = 1$, where τ_i stands for the yield stress of the component $v^{(i)}$.

15.1.2 Local Limiting Surface for Interface Stresses

The modeling of global composite behavior necessarily requires the consideration of interface degradation in addition to the matrix and fiber phase degradation (15.1). This requires not only the calculation of the stress at the interface, but equally the identification and the application of a local failure criterion. In a similar manner with Eq. (15.1) we can present a tensor-polynomial failure criterion for the interface failure initiation:

$$\mathbf{\Pi}^{(i)}_a(\mathbf{n},\boldsymbol{\sigma}) = \mathbf{\Pi}^{1(i)}_a(\mathbf{n})\boldsymbol{\sigma}^-(\mathbf{n}) + \mathbf{\Pi}^{2(i)}_a(\mathbf{n})[\boldsymbol{\sigma}^-(\mathbf{n})\otimes\boldsymbol{\sigma}^-(\mathbf{n})] + \ldots = 1, \tag{15.4}$$

where $\boldsymbol{\sigma}^-(\mathbf{n})$ is the limiting stress within the matrix near the inclusion boundary $\mathbf{x} \in \partial v_i$ with the unit outward normal vector $\mathbf{n}$. Generally speaking, adhesion strength parameters $\mathbf{\Pi}^{1(i)}_a(\mathbf{n})$, $\mathbf{\Pi}^{2(i)}_a(\mathbf{n})$, $\mathbf{\Pi}^{3(i)}_a(\mathbf{n})$, which convey the normal and the shear debonding at the interface between the matrix and the fibers, differ from $\mathbf{\Pi}^{1(i)}$, $\mathbf{\Pi}^{2(i)}$, $\mathbf{\Pi}^{3(i)}$.

We will present now popular local criteria that convey the normal and the shear debonding at the interface between the matrix and the fibers. Determination of the failure characteristics of the interface is carried out through a mechanical characterization of the interfacial resistance by means of specific tests such as fiber pull-out, fiber push-out, etc (see for references and details). The interface stresses $\boldsymbol{\sigma}^-_{\mathbf{n}} \equiv \boldsymbol{\sigma}^-(\mathbf{n})\mathbf{n}$ can be partitioned as $\boldsymbol{\sigma}^-_{\mathbf{n}} = \mathbf{N}^{\mathbf{n}}\boldsymbol{\sigma}^-(\mathbf{n}) + \mathbf{T}^{\mathbf{n}}\boldsymbol{\sigma}^-(\mathbf{n})$, where $\mathbf{N}^{\mathbf{n}}$ and $\mathbf{T}^{\mathbf{n}}$ are the three-rank functions of the normal $\mathbf{n}$ such that $N^{\mathbf{n}}_{ikl} = n_i n_k n_l$, $N^{\mathbf{n}}_{ikl} = (\delta_{ik}n_l + \delta_{il}n_k)/2 - n_i n_k n_l$, where the tensors $N^{\mathbf{n}}_{ikl}$ and $T^{\mathbf{n}}_{ikl}$ symmetrical under the interchanges $k \leftrightarrow l$ generate the normal $\boldsymbol{\sigma}^-_n = \mathbf{N}^{\mathbf{n}}\boldsymbol{\sigma}^-(\mathbf{n})$ and tangential components $\boldsymbol{\sigma}^-_\tau = \mathbf{T}^{\mathbf{n}}\boldsymbol{\sigma}^-(\mathbf{n})$ of the traction $\boldsymbol{\sigma}^-\mathbf{n}$ with the magnitudes $\sigma_n \equiv \|\boldsymbol{\sigma}^-_n\| = \sigma^-_{kl}n_k n_l$ and $\sigma_\tau \equiv \|\boldsymbol{\sigma}^-_\tau\| = \sqrt{\sigma^-_{kl}\sigma^-_{kl} - (\sigma_n)^2}$, respectively.

In a simple maximum stress criterion, the normal σ_n and tangential σ_τ components are compared to maximum values $\sigma_n^{\max}$ and $\sigma_\tau^{\max}$ characterizing the interface, and the failure tensors $\mathbf{\Pi}^{2(i)}_a(\mathbf{n})$ have a form

$$\bar{\mathbf{\Pi}}^{2(i)}_a(\mathbf{n}) = \max\left[\frac{\|\mathbf{N}^{\mathbf{n}}\boldsymbol{\sigma}^-(\mathbf{n})\|}{\sigma_n^{\max}}, \frac{\|\mathbf{T}^{\mathbf{n}}\boldsymbol{\sigma}^-(\mathbf{n})\|}{\sigma_\tau^{\max}}\right] = \max\left[\frac{\sigma_n}{\sigma_n^{\max}}, \frac{\sigma_\tau}{\sigma_\tau^{\max}}\right] = 1. \tag{15.5}$$

Other type of criteria taking into account the friction problem were considered by a number of authors (for references see [795]). The Coulomb form [17] permits the introduction of the friction coefficient at the interface by the use of a linear combination between the normal and the shear interface stresses

$$\check{\mathbf{\Pi}}_a^{2(i)}(\mathbf{n}) = \frac{\|\mathbf{N}^{\mathbf{n}}\boldsymbol{\sigma}^-(\mathbf{n})\|}{\sigma_n^{\max}} + \frac{\|\mathbf{T}^{\mathbf{n}}\boldsymbol{\sigma}^-(\mathbf{n})\|}{\sigma_\tau^{\max}} = 1. \quad (15.6)$$

Logical generalization of tensor-polynomial criteria (15.1) to the interface failure initiation was proposed in [1056] (see also [638]) for the cylindrical fibers, which in our more general notations has the form

$$\hat{\mathbf{\Pi}}_a^{2(i)}(\mathbf{n}) = \frac{(\mathbf{N}^{\mathbf{n}}\boldsymbol{\sigma}^-(\mathbf{n}))(\mathbf{N}^{\mathbf{n}}\boldsymbol{\sigma}^-(\mathbf{n}))}{(\sigma_n^{\max})^2} + \frac{(\mathbf{T}^{\mathbf{n}}\boldsymbol{\sigma}^-(\mathbf{n}))(\mathbf{T}^{\mathbf{n}}\boldsymbol{\sigma}^-(\mathbf{n}))}{(\sigma_\tau^{\max})^2} = 1. \quad (15.7)$$

The criterion (15.7) can be recast in the tensor-polynomial form (15.4) with the tensors $\mathbf{\Pi}_a^{1(i)}(\mathbf{n}) \equiv \mathbf{0}$ and $\Pi_{a|ijkl}^{2(i)}(\mathbf{n}) = \left[(\sigma_n^{\max})^{-2} - (\sigma_\tau^{\max})^{-2}\right] n_i n_j n_k n_l + (2\sigma_\tau^{\max})^{-2}(\delta_{ik}n_{jl} + \delta_{il}n_{jk} + \delta_{jl}n_{ik} + \delta_{jk}n_{il})$.

Because the criteria (15.5)–(15.7) should predict the identical stresses of the failure initiation at the pure local normal and shear stresses the empirical interfacial strengths corresponding to tension and shear should be the same in criteria (15.5)–(15.7), and, therefore, the failure envelope (15.7) is inserted between the failure surfaces (15.5) and (15.6) $\check{\mathbf{\Pi}}_a^{2(i)}(\mathbf{n}) < \hat{\mathbf{\Pi}}_a^{2(i)}(\mathbf{n}) < \bar{\mathbf{\Pi}}_a^{2(i)}(\mathbf{n})$, where the equalities

$$\check{\mathbf{\Pi}}_a^{2(i)}(\mathbf{n}) = \hat{\mathbf{\Pi}}_a^{2(i)}(\mathbf{n}) = \bar{\mathbf{\Pi}}_a^{2(i)}(\mathbf{n}) \quad (15.8)$$

in general hold just for the normal $\mathbf{n}$ with either the pure normal or pure tangential local loading $\boldsymbol{\sigma}^-(\mathbf{n}) \equiv \mathbf{N}^{\mathbf{n}}\boldsymbol{\sigma}^-(\mathbf{n})$, or $\boldsymbol{\sigma}^-(\mathbf{n}) \equiv \mathbf{T}^{\mathbf{n}}\boldsymbol{\sigma}^-(\mathbf{n})$. The equalities also hold in some particular cases of correlations between the elastic and strength properties of constitutives. For example, the equalities (15.8) are valid for any $\mathbf{n}$ for the limiting type of soft fibers (hole, $\mathbf{L}^{(1)} \equiv \mathbf{0}$) when $\boldsymbol{\sigma}_{\mathbf{n}}^- \equiv \mathbf{0}$ and the interface failure is degenerated into the failure of the matrix in the vicinity of the interface. In another limiting case of perfect sliding $\sigma_\tau^{\max} = 0$ (for references see [795]) the equality is also valid for any $\mathbf{n}$. Moreover, the last statement also holds if under the failure initiation one understands the normal debonding $(\mathbf{u}^+ - \mathbf{u}^-)\mathbf{n} > 0$.

It should be mentioned that the transverse strength of the reinforced fiber composites is usually significantly higher than that of the matrix. The popular assumption is that the strengths of the interface and the matrix are equal (i.e., the bonding between fiber and matrix is assumed to be perfect; see, e.g., [377]). As an approximation, $\sigma_\tau^{\max}$ is also taken to be a half of $\sigma_n^{\max}$ as usually assumed in a homogeneous isotropic material (see e.g. [638]). In light of the heuristic level of justification, the importance of the fundamental experimental work [1071] can scarcely be exaggerated. The single-fiber cruciform test was used to characterize the initiation of fiber-matrix interface failure in a model composite with the interface subjected to a combined state of transverse and shear stress at a location away from a crack tip or free edge. The elimination of the free-edge effect that requires modeling of a stress singularity was accomplished by utilizing a cruciform specimen geometry with the arms containing the embedded fiber inclined at the different angles with respect to unit axial tension. The ratio of the normal and shear loading at the interface was governed by the amount of off-axis angle the fiber made with the loading direction.

15.1.3 Fracture Criterion for an Isolated Crack

Let us consider a coin-shaped crack of radius R^c with center $\mathbf{x}^c = (0, 0, 0)$ and unit normal $\mathbf{n} = (1, 0, 0)$ to the crack surface S in an infinite elastic homogeneous medium. The energy release rate may be defined by means of $\mathbf{K} \equiv (K_1, K_2, K_3)^\top$ as $\mathcal{J} = \mathbf{\Lambda}(\mathbf{K} \otimes \mathbf{K})$, where for the general anisotropic material the matrix Λ_{ij} $(i, j = 1, 2, 3)$ is symmetric [37], [936]. For an isotropic material the matrix Λ_{ij} is diagonal: $\mathbf{\Lambda} = (2\mu)^{-1}\mathrm{diag}(1 - \nu, 1 - \nu, 1)$. The energy release rate $\mathcal{J}$ could alternatively be expressed as a path-independent line J^r Rice's integral which is invariant with respect to the integration along an arbitrary path encircling the crack tip (see, e.g., [1192])

The energy release rate $\mathcal{J}$ provides a means to introduce a crack propagation criterion on a physical basis: a crack can propagate if the potential energy released per unit area of newly created crack surface exceeds the work that is consumed in creating this new amount of surface. The fracture criterion that will be used for the remainder of this work therefore based on the equality $\mathcal{J} = \mathcal{J}^c \equiv 2\gamma$, where γ called fracture surface energy. For homogeneous external loading the equation $\mathcal{J} = 2\gamma$ defines the surface of second order in six-dimensional space of stresses $\boldsymbol{\sigma}$. In particular, for coin-shaped crack with radius R^c the last formula turns into

$$\frac{2(1-\nu)R^c}{\pi\mu}\left(\sigma_{11}^2 + \left[\frac{2}{2-\nu}\right]^2\sigma_{12}^2 + \left[\frac{2(1-\nu)}{2-\nu}^2\right]\sigma_{13}^2\right) = 2\gamma, \qquad (15.9)$$

(see, e.g., [796]). Thus we have obtained the strength criterion (15.9) in terms of (15.2) in which the strength tensor $\mathbf{\Pi}^4$ is a function of the size and orientation of the crack. This analogy permits the use of strength calculation method of composites developed in the next section.

15.2 Effective Limiting Surface

15.2.1 Utilizing Fluctuations of Bulk Stresses Inside the Phases

The limiting surfaces $\mathbf{\Pi}^{(i)}(\boldsymbol{\sigma}) = 1$ (15.1), (15.2) represent nonlinear functions of $\boldsymbol{\sigma}$. Usually in the treatment of random structure composites homogeneity of the stress or strain fields are used even for these nonlinear functions. Thus, a common way to produce an effective failure envelope for the composite materials is substitution of the component average stress values into the formula (15.1) ([19], [298], [569], [930]) $(i = 0, 1, \ldots)$

$$\mathbf{\Pi}^*(\boldsymbol{\sigma}) \equiv \max_i \Big[\mathbf{\Pi}^{1(i)}\langle\boldsymbol{\sigma}\rangle^{(i)} + \mathbf{\Pi}^{2(i)}(\langle\boldsymbol{\sigma}\rangle^{(i)} \otimes \langle\boldsymbol{\sigma}\rangle^{(i)})$$
$$+ \mathbf{\Pi}^{3(i)}(\langle\boldsymbol{\sigma}\rangle^{(i)} \otimes \langle\boldsymbol{\sigma}\rangle^{(i)} \otimes \langle\boldsymbol{\sigma}\rangle^{(i)}) + \ldots\Big] = 1. \qquad (15.10)$$

This hypothesis is, however, in some sense inconsistent. For example, let us consider an isotropic porous material with matrix yielding properties described by the von Mises criterion (15.3). In this case, under hydrostatic loading the condition $\langle\sigma_{ij}\rangle = \langle\sigma_0\rangle\delta_{ij}$ leads to $\langle\mathbf{s}\rangle^{(0)} \equiv \mathbf{0}$ irrespective of the microstructure of the pores and the method of calculation of $\langle\boldsymbol{\sigma}\rangle^{(0)}$. This would mean that such a porous material has an infinite yield stress under hydrostatic compression as mentioned, e.g., in [164], [910].

It is believed that the following definition of effective limiting surface proposed in [159], [160] and based on fewer assumptions is more correct (see, e.g., [138], [897])

$$\widetilde{\mathbf{\Pi}}^*(\boldsymbol{\sigma}) \equiv \max_i \Big[\mathbf{\Pi}^{2(i)}\langle\boldsymbol{\sigma}\rangle^{(i)} + \mathbf{\Pi}^{4(i)}\langle\boldsymbol{\sigma}\otimes\boldsymbol{\sigma}\rangle^{(i)} + \mathbf{\Pi}^{6(i)}\langle\boldsymbol{\sigma}\otimes\boldsymbol{\sigma}\otimes\boldsymbol{\sigma}\rangle^{(i)} + \ldots\Big] = 1, \tag{15.11}$$

where the estimations of average stress moments of different orders $\langle\boldsymbol{\sigma}\rangle^{(i)}$, $\langle\boldsymbol{\sigma}\otimes\boldsymbol{\sigma}\rangle^{(i)}$, $\langle\boldsymbol{\sigma}\otimes\boldsymbol{\sigma}\otimes\boldsymbol{\sigma}\rangle^{(i)}$ $(i = 0, 1, \ldots)$ can be found by the use of the relevant formulae of Chapter 13.

Let us show the physical consistency of the effective strength criterion (15.11) (in contrast to (15.10)) subjected to the thermal loading ($\langle\boldsymbol{\sigma}\rangle \equiv \mathbf{0}$). In fact, let us consider a two-component isotropic composite with isotropic phases. In this case one may observe that symmetry requires that average stresses inside both components will be hydrostator, one $\langle\sigma_{kl}\rangle^{(1)} \equiv \langle\sigma_{kl}\rangle^{(0)}(c-1)/c \equiv \sigma^0_{11}\delta_{kl}$; in so doing the microstructure of and the method of calculation of average stresses inside the components (for example, (8.46) or any other formula) influence the value of scalar σ^0_{11}, but have no effect on the tensor structure of the fields $\langle\boldsymbol{\sigma}\rangle^{(0)}$, $\langle\boldsymbol{\sigma}\rangle^{(1)}$. Then the composite strength is dictated by the strength of the component which is to be found under conditions of hydrostatic tension and is not determined by the strength of second component. If the strength of the second component falls far short of the strength of the first one, we will obtain an improper prediction of composite strength. In fact, according to Chapter 13, the average values of second deviator invariant inside each component $\langle\mathbf{ss}\rangle^{(i)} \neq 0$ $(s_{kl} \equiv \sigma_{kl} - \sigma_{nn}\delta_{kl}/3;\ i = 0, 1, \ldots)$. Therefore the composite strength is defined by the strength of second more weak component at the cost of the fluctuations of the stress deviator.

The above mentioned inconsistency can be also overcome if the local maximum criterion

$$\mathbf{\Pi}_{\max}(\boldsymbol{\sigma}) \equiv \max_i \max_{\mathbf{x}\in v_i} \Big[\mathbf{\Pi}^{2(i)}\boldsymbol{\sigma}(\mathbf{x}) + \mathbf{\Pi}^{4(i)}\boldsymbol{\sigma}(\mathbf{x})\otimes\boldsymbol{\sigma}(\mathbf{x}) + \ldots\Big] = 1 \tag{15.12}$$

in the phases is considered. In particular, for local limiting surfaces of von Mises's type (15.3), where we have

$$\langle\widetilde{\mathbf{\Pi}}^*(\langle\sigma\rangle) \equiv \max_i \Big\{\frac{3}{2\tau_i^2}\mathbf{N}_2(\langle\sigma\otimes\sigma\rangle^{(i)})\Big\} = 1, \tag{15.13}$$

and according to Holder's inequality $\langle\mathbf{fg}\rangle \leq \langle\mathbf{f}^p\rangle^{1/p}\langle\mathbf{g}^q\rangle^{1/q}$, $1/p + 1/q = 1,\ p, q =$ const. we obtain

$$\mathbf{\Pi}^*_{\max}(\langle\boldsymbol{\sigma}\rangle) \geq \widetilde{\mathbf{\Pi}}^*(\langle\boldsymbol{\sigma}\rangle) \geq \mathbf{\Pi}^*(\langle\boldsymbol{\sigma}\rangle) \geq \max_i \langle\mathbf{\Pi}(\boldsymbol{\sigma})\rangle^{(i)}. \tag{15.14}$$

Here, the equalities are obtained only for homogeneous stress fields in the components, which holds for the mixture theory models with either parallel or series arrangements of the components (for instance, the schemes of Voight and Reuss). Thus, according to Eqs. (15.14) the application of the component volume averaging of the second stress moment in the yield criterion Eq. (15.11) leads to higher values of the calculated estimates for the onset of yielding in comparison with Eq. (15.10); however, they are lower than achieved from the local criterion Eq. (15.12).

Since the main interest in elastoplastic analysis is the consideration of the elastic-plastic deformations of the whole of the components $v^{(i)}$ $(i = 0, 1, \ldots, N)$, for the sake of simplicity we are using the yield criterion Eq. (15.11), instead of Eqs. (15.10) and (15.12). We demonstrated that this assumption is physically consistent in distinction to Eq. (15.10). From the above discussion one should not conclude that for all real composites the criterion (15.11) provides a qualitatively better approximation for the actual onset of yielding at any applied stress state. The particular case of criterion (15.10) was applied in [1075] to determinate the tensile behavior of a silica/epoxy composite and found good agreement with experimental results. Such an approach is reasonably acceptable for a particle–reinforced solid under a stress state with only low triaxiality (e.g., pure tension or shear; see [83], [910]. However, for the criterion Eq. (15.13) under a pure hydrostatic loading the macroisotropic composite would never yield. To remove this undesired feature the different versions of criterion (15.11) can be used.

The composite starts yielding when $\max_i \widetilde{\mathbf{\Pi}}^{*(i)} = 1$ $(i = 0, 1, \ldots, N)$ is fulfilled. Up to the moment when the yielding begins, the problem of estimating $\widetilde{\mathbf{\Pi}}^{*(i)}$ is linear elastic, which allows the use of closed relations for calculating the yield surface $(i = 0, 1, \ldots, N)$, e.g.,

$$F_y(\langle\boldsymbol{\sigma}\rangle) \equiv \max_i \frac{1}{2\tau_i}\sqrt{\frac{3}{c^{(i)}}\frac{\partial \mathbf{M}^*}{\partial q^{(i)}}\langle\boldsymbol{\sigma}\rangle \otimes \langle\boldsymbol{\sigma}\rangle} = 1. \tag{15.15}$$

Equation (15.11) determines the second-order surface $\widetilde{\mathbf{\Pi}}^*(\langle\sigma\rangle) = 1$ for the onset of yielding within the six-dimensional stress space which, in the general case, depends on $\langle\sigma_0\rangle$. The yield surface of the components $\widetilde{\mathbf{\Pi}}^{*(i)}$ may be embedded within each other or they may intersect in the space of macrostrains $\langle\boldsymbol{\sigma}\rangle$.

15.2.2 Utilizing Interface Stress Fluctuations

If the possibility of interfacial fracture is taken into account, the macrostrength criterion can be express in the following forms, analogous to Eqs. (15.10) and (15.11) $(i = 0, 1, \ldots)$

$$\mathbf{\Pi}^*_a(\boldsymbol{\sigma}) = \max\Big\{\mathbf{\Pi}^*(\langle\boldsymbol{\sigma}\rangle),\ \max_i \max_{\mathbf{n}} \Big[\mathbf{\Pi}^{1(i)}_a(\mathbf{n})\langle\boldsymbol{\sigma}^-(\mathbf{n})\rangle_x + \mathbf{\Pi}^{2(i)}_a(\mathbf{n})\langle\boldsymbol{\sigma}^-(\mathbf{n})\rangle_x \otimes \langle\boldsymbol{\sigma}^-(\mathbf{n})\rangle_x + \ldots\Big]\Big\} = 1, \tag{15.16}$$

$$\widetilde{\mathbf{\Pi}}^*_a(\boldsymbol{\sigma}) = \max\Big\{\widetilde{\mathbf{\Pi}}^*(\langle\boldsymbol{\sigma}\rangle),\ \max_i \max_{\mathbf{n}} \Big[\mathbf{\Pi}^{1(i)}_a(\mathbf{n})\langle\boldsymbol{\sigma}^-(\mathbf{n})\rangle_x + \mathbf{\Pi}^{2(i)}_a(\mathbf{n})\langle\boldsymbol{\sigma}^-(\mathbf{n}) \otimes \boldsymbol{\sigma}^-(\mathbf{n})\rangle_x + \ldots\Big]\Big\} = 1, \tag{15.17}$$

respectively, where $\langle\boldsymbol{\sigma}^-(\mathbf{n})\rangle_x$, $\langle\boldsymbol{\sigma}^-(\mathbf{n})\otimes\boldsymbol{\sigma}^-(\mathbf{n})\rangle_x$, and are the statistical moments of limiting stresses within the matrix near the inclusion boundary $\mathbf{x}\in\partial v_i$ with the unit outward normal vector $\mathbf{n}$, see, e.g., (8.47) and (13.24). These criteria can be viewed as intermediate statements between the local (15.12) and averaged (15.10), (15.11) criteria. The inequality sequence can be proved analogous to Eq. (15.14)

$$\boldsymbol{\Pi}^*_{\max}(\langle\boldsymbol{\sigma}\rangle)\geq\widetilde{\boldsymbol{\Pi}}^*_a(\langle\boldsymbol{\sigma}\rangle)\geq\boldsymbol{\Pi}^*_a(\langle\boldsymbol{\sigma}\rangle), \tag{15.18}$$

where the equalities are obtained only for homogeneous stress fields in the components.

In particular, exploring the local failure envelope (15.7) for a pure elastic problem ($\boldsymbol{\beta}(\mathbf{x})\equiv\mathbf{0}$) for a two-phase composite yields the effective failure criterion

$$\widetilde{\boldsymbol{\Pi}}^*_a(\boldsymbol{\sigma})=\max\Big\{\widetilde{\boldsymbol{\Pi}}^*(\langle\boldsymbol{\sigma}\rangle),\ \max_{\mathbf{n}}\Big[\frac{\mathbf{N}^{\mathbf{n}}\boldsymbol{\mathcal{B}}^*(\mathbf{n})\otimes\mathbf{N}^{\mathbf{n}}\boldsymbol{\mathcal{B}}^*(\mathbf{n})}{(\sigma_n^{\max})^2} + \frac{\mathbf{T}^{\mathbf{n}}\boldsymbol{\mathcal{B}}^*(\mathbf{n})\otimes\mathbf{T}^{\mathbf{n}}\boldsymbol{\mathcal{B}}^*(\mathbf{n})}{(\sigma_\tau^{\max})^2}\Big]\langle\boldsymbol{\sigma}\rangle\otimes\langle\boldsymbol{\sigma}\rangle\Big\}. \tag{15.19}$$

Hereafter one introduced the interface stress concentrator factors of the second moment of these stresses:

$$\langle\boldsymbol{\sigma}_i^-(\mathbf{n})\otimes\boldsymbol{\sigma}_i^-(\mathbf{n})\rangle_{\mathbf{x}}=[\boldsymbol{\mathcal{B}}^*(\mathbf{n})\otimes\boldsymbol{\mathcal{B}}^*(\mathbf{n})]\langle\boldsymbol{\sigma}\rangle\otimes\langle\boldsymbol{\sigma}\rangle, \tag{15.20}$$

Neglecting of stress fluctuations in the inclusions $\langle\boldsymbol{\sigma}\otimes\boldsymbol{\sigma}\rangle^{(1)}\equiv\langle\boldsymbol{\sigma}\rangle^{(1)}\otimes\langle\boldsymbol{\sigma}\rangle^{(1)}$ leads to the simplification of Eq. (15.20):

$$[\boldsymbol{\mathcal{B}}^*(\mathbf{n})\otimes\boldsymbol{\mathcal{B}}^*(\mathbf{n})]=\mathbf{B}(\mathbf{n})\mathbf{BD}\otimes\mathbf{B}(\mathbf{n})\mathbf{BD},\quad \langle\boldsymbol{\sigma}_i^-(\mathbf{n})\rangle_{\mathbf{x}}=\mathbf{B}(\mathbf{n})\mathbf{BD}\langle\boldsymbol{\sigma}\rangle, \tag{15.21}$$

where one uses the estimation (8.47) of statistical averages of stresses in the matrix in the vicinity of the inclusion near a point $\mathbf{x}\in\partial v_i$. Taking the stress fluctuations into account, the interface stress concentrator factors can be found by either the perturbation method (13.15) or integral equation method (13.41), (8.47):

$$[\boldsymbol{\mathcal{B}}^{*\mathrm{per}}(\mathbf{n})\otimes\boldsymbol{\mathcal{B}}^{*\mathrm{per}}(\mathbf{n})]=\frac{1}{c^{(1)}}[\mathbf{B}(\mathbf{n})\otimes\mathbf{B}(\mathbf{n})]\frac{\partial\mathbf{M}^*}{\partial\mathbf{M}^{(1)}}, \tag{15.22}$$

$$[\boldsymbol{\mathcal{B}}^{*\mathrm{int}}(\mathbf{n})\otimes\boldsymbol{\mathcal{B}}^{*\mathrm{int}}(\mathbf{n})]=\mathbf{B}(\mathbf{n})\mathbf{BD}\otimes\mathbf{B}(\mathbf{n})\mathbf{BD}+\int\mathbf{B}(\mathbf{n})\mathbf{BT}_{ip}(\mathbf{x}_i-\mathbf{x}_p)\mathbf{R} \otimes\mathbf{B}(\mathbf{n})\mathbf{BT}_{ip}(\mathbf{x}_i-\mathbf{x}_p)\mathbf{R}\varphi(v_p,\mathbf{x}_p|;v_i,\mathbf{x}_i)\,d\mathbf{x}_p, \tag{15.23}$$

respectively.

It should be mentioned that special so-called localized models of plasticity of particulate composites have been developed. Herve and Zaoui [457] proposed a generalization of the elastic three-phase spherical model by Hashin [427], [432] toward the elastoplastic case in the framework of a secant modulus concept. We will not discuss here in general the known advantages of flow theory over deformation theory of plasticity in the case of nonradial loading, that takes place usually at the local level near the inclusions even if the overall stress path is radial. We just look at the method of calculation of onset of yielding in the layer model [457], which used the criterion (in our notations) $\sqrt{1.5\langle\boldsymbol{\sigma}^-(\mathbf{n})\rangle_{s_i} : \mathbf{N}_2 : \langle\boldsymbol{\sigma}^-(\mathbf{n})\rangle_{s_i}} - \tau_0^{(0)} = 0$, where $\langle\boldsymbol{\sigma}^-(\mathbf{n})\rangle_{s_i}$ is the surface average of the stresses $\boldsymbol{\sigma}^-(\mathbf{n})$ over the surface of the inclusion s_i. As a consequence of the properties of the interface operator $\boldsymbol{\Gamma}(\mathbf{n})$ (3.95) this average is given by $\langle\boldsymbol{\sigma}^-(\mathbf{n})\rangle_{s_i} \equiv \langle\boldsymbol{\sigma}\rangle^{(1)}$ (at least for an ellipsoidal inclusion in the framework of the hypothesis **H1**, see Chapter 8) and, therefore, for porous materials one would obtain an infinite overall initial yield stress for pure hydrostatic loading σ_0^0. More recently a yield criterion [100] was proposed $\sqrt{1.5\langle\{\langle\boldsymbol{\sigma}^-(\mathbf{n})\rangle_x : \mathbf{N}_2 : \langle\boldsymbol{\sigma}^-(\mathbf{n})\rangle_x\}\rangle_{s_i}} - \tau_0^{(0)} = 0$, which contains inside the generalized yield surface (15.17) in a dimensionless coordinate system $X_\tau = \sigma_0^0/\tau_0^{(0)}$, $Y_\tau = (1.5\mathrm{s}^0 : \mathrm{s}^0)^{1/2}/\tau_0^{(0)}$, where $\sigma_0^0 \equiv \boldsymbol{\delta} : \boldsymbol{\sigma}^0/3$, $\mathrm{s}^0 \equiv \mathbf{N}_2\boldsymbol{\sigma}^0$.

15.2.3 Effective Fracture Surface for an Isolated Crack in the Elastically Homogeneous Medium with Random Residual Microstresses

Let us consider a crack within elastically homogeneous medium containing a statistically homogeneous inclusion field with the eigenstrains. At first glance it would seem that it is possible to define the first approximation to an estimated effective energy release rate $\mathcal{J}^*(\mathbf{z})$ by means of SIF

$$\mathcal{J}^*(\mathbf{z}) = \boldsymbol{\Lambda}(\langle\mathbf{K}(\mathbf{z})\rangle \otimes \langle\mathbf{K}(\mathbf{z})\rangle) \equiv \boldsymbol{\Lambda}(\mathbf{K}^0(\mathbf{z}) \otimes \mathbf{K}^0(\mathbf{z})), \tag{15.24}$$

where the equality follows from (13.64) and indicates that residual stress generated $\mathcal{J}^*(\mathbf{z})$ (at $\boldsymbol{\sigma}^0 \equiv \mathbf{0}$) is determined extremely by stress fluctuation in the vicinity of crack tip.

It is expected to be more correct the use of a conditional SIF ($k = 0, 1, \ldots$)

$$\mathcal{J}^*(\mathbf{z}) = \max_k \max_{\mathbf{x}_k} \mathcal{J}(\langle\mathbf{K}(\mathbf{z})|v_k, \mathbf{x}_k\rangle)\langle\gamma\rangle/\gamma^{(k)} = 2\langle\gamma\rangle, \tag{15.25}$$

$$\mathcal{J}(\langle\mathbf{K}(\mathbf{z})|v_k, \mathbf{x}_k\rangle) \equiv \boldsymbol{\Lambda}\Big[\langle\mathbf{K}(\mathbf{z})|v_k, \mathbf{x}_k\rangle \otimes \langle\mathbf{K}(\mathbf{z})|v_k, \mathbf{x}_k\rangle\Big], \tag{15.26}$$

and the multiplier $\langle\gamma\rangle$ is introduced for the purpose of preserving the conservation of the dimensionality of $\mathcal{J}^*$. This multiplier is used for the case $\gamma^{(0)} > \gamma^{(i)}$ ($\forall i = 1, 2, \ldots, N$); otherwise the multiplier $\langle\gamma\rangle$ should be replaced by the factor $\langle 1/\gamma\rangle^{-1}$.

The criterion (15.25) is based on the concept of weakest link. The mixture rule can be realized by the use of total probability formula

$$\mathcal{J}^*(\mathbf{z}) = (1-c)\mathcal{J}(\langle\mathbf{K}(\mathbf{z})|v_0, \mathbf{x}_0\rangle)\langle\gamma\rangle/\gamma^{(0)} + \sum_{m=1}^{N} \frac{\langle\gamma\rangle n^{(m)}}{\gamma^{(m)}} \int \boldsymbol{\Lambda}\langle\mathbf{K}(\mathbf{z})|v_m, \mathbf{x}_m\rangle$$
$$\otimes \langle\mathbf{K}(\mathbf{z})|v_m, \mathbf{x}_m\rangle V_m^{\mathbf{z}}(\mathbf{x}_m)\, d\mathbf{x}_m = 2\langle\gamma\rangle, \tag{15.27}$$

where $V_m^{\mathbf{z}}$ designates a characteristic function of the inclusion v_m with the center $\mathbf{z}$; the formula (15.27) has been outlined for spherical inclusions and can be used as an approximation in the general case.

However, at this time we can calculate the second conditional moments of SIF and therefore it would appear reasonable that the generalization of (15.25), (15.27) is ($k = 0, 1, \ldots$)

$$\mathcal{J}^*(\mathbf{z}) = \max_k \max_{\mathbf{x}_k} \langle \mathcal{J}(\mathbf{z})|v_k, \mathbf{x}_k\rangle \langle\gamma\rangle/\gamma^{(k)} = 2\langle\gamma\rangle, \tag{15.28}$$

$$\mathcal{J}^*(\mathbf{z}) = (1-c)\langle \mathcal{J}(\mathbf{z})|v_0, , \mathbf{x}_0\rangle \langle\gamma\rangle/\gamma^{(0)}$$
$$+ \sum_{m=1}^{N} \langle\gamma\rangle n^{(k)}/\gamma^{(m)} \int \langle \mathcal{J}(\mathbf{z})|v_m, \mathbf{x}_m\rangle V_m^{\mathbf{z}}(\mathbf{x}_m)\, d\mathbf{x}_m = 2\langle\gamma\rangle, \tag{15.29}$$

respectively, where $k = 0, 1, \ldots$ and $\langle \mathcal{J}(\mathbf{z})|v_k, \mathbf{x}_k\rangle \equiv \boldsymbol{\Lambda}\langle \mathbf{K}(\mathbf{z}|v_k, \mathbf{x}_k) \otimes \mathbf{K}(\mathbf{z}|v_k, \mathbf{x}_k)\rangle$ can be found from the formulae for the second stress moment in the vicinity of the crack tip (13.72); in so doing only singularity terms proportional to degree -1 in ρ are taken into account in the calculation of $\langle \mathcal{J}(\mathbf{z})|v_k, \mathbf{x}_k\rangle$ because according to [1192] the contribution of terms with the lesser singularity equals zero. A schematic representation of the definition (15.25) and (15.26) is based on the assumption $\langle \mathcal{J}|v_J, \mathbf{x}_J\rangle = \mathcal{J}(\langle \mathbf{K}|v_J, \mathbf{x}_J\rangle)$ ($J = 0, i;\ i = 1, 2, \ldots$) which produces a large error as the SIF dispersion increases. For analyses of stress fluctuation effects we will define a fluctuation part of both the expectation and conditional expectation of values of energy release rate ($J = 0, i;\ i = 1, 2, \ldots$)

$$\Delta\mathcal{J}(\mathbf{z}) \equiv \langle \mathcal{J}(\mathbf{z})\rangle - \mathcal{J}(\langle \mathbf{K}(\mathbf{z})\rangle), \tag{15.30}$$

$$\Delta\mathcal{J}(\mathbf{z}|v_J, \mathbf{x}_J) \equiv \langle \mathcal{J}(\mathbf{z})|v_J, \mathbf{x}_J\rangle - \mathcal{J}(\langle \mathbf{K}(\mathbf{z})|v_J, \mathbf{x}_J\rangle), \tag{15.31}$$

where the terms in the right-hand side of Eq. (15.31) were determined previously. Taking Eq. (15.24) into account, the second term of the right-hand-side of Eq. (15.30) $\mathcal{J}(\langle \mathbf{K}(\mathbf{z})\rangle)$ is defined only by the remote stresses: $\mathcal{J}(\langle \mathbf{K}(\mathbf{z})\rangle) = \mathcal{J}(\mathbf{K}^0(\mathbf{z}))$. Moreover, in conformity with the Jensen inequality for the convex function $\mathcal{J} = \mathcal{J}(\mathbf{K})$ we have the following inequalities $\Delta\mathcal{J}(\mathbf{z}) \geq 0,\ \Delta\mathcal{J}(\mathbf{z}|v_J, \mathbf{x}_J) \geq 0$. Here, the equalities take place if and only if the inclusion fields are deterministic.

It should be mentioned that the criteria (15.28) and (15.29) proposed for a special case of a single microcrack loaded by the random residual stress field can be easily generalized to the cases of both the composites and microcracked media. Indeed, e.g., a fracture criterion for media with identical aligned coin-shaped cracks is based on the estimation of $\mathcal{J}^*(\mathbf{z}) = \boldsymbol{\Lambda}(\langle \mathbf{K}(\mathbf{z}) \otimes \mathbf{K}(\mathbf{z})\rangle_i)$ for a representative crack v_i. In the framework of the effective field hypothesis **H1)**, $\mathbf{K}(\mathbf{z}) = \mathbf{K}_i^\sigma(\mathbf{z})\overline{\boldsymbol{\sigma}}_i$ where $\mathbf{K}_i^\sigma \equiv$ const. is assumed to be known (see, e.g., [324], [796]). Then $\mathcal{J}^*(\mathbf{z}) = \boldsymbol{\Lambda}(\mathbf{K}_i^\sigma(\mathbf{z}) \otimes \mathbf{K}_i^\sigma(\mathbf{z}))(\langle \overline{\boldsymbol{\sigma}} \otimes \overline{\boldsymbol{\sigma}}\rangle_i)$ are found after estimation of the second moment of the effective field $\langle \overline{\boldsymbol{\sigma}} \otimes \overline{\boldsymbol{\sigma}}\rangle_i$ obtained, e.g., by Eq. (13.45). The proposed criterion is reduced to the simplified formulae obtained under the additional assumptions such as, e.g., the dilute approximation $\langle \overline{\boldsymbol{\sigma}} \otimes \overline{\boldsymbol{\sigma}}\rangle_i \equiv \langle \boldsymbol{\sigma}\rangle \otimes \langle \boldsymbol{\sigma}\rangle$ [271], [655], [1240] and the effective medium approach $\langle \overline{\boldsymbol{\sigma}} \otimes \overline{\boldsymbol{\sigma}}\rangle_i \equiv \langle \boldsymbol{\sigma}\rangle \otimes \langle \boldsymbol{\sigma}\rangle$ and $\boldsymbol{\Lambda} = \boldsymbol{\Lambda}(\mathbf{L}^*)$, $\mathbf{K}_i^\sigma = \mathbf{K}_i^\sigma(\mathbf{L}^*)$ [923].

15.2.4 Scheme of Simple Probability Model of Composite Fracture

The proposed strength and fracture criteria (15.17) and (15.27)–(15.29) are based on the determination of conditional averages $\langle \boldsymbol{\Pi}^{(k)}(\boldsymbol{\sigma})\rangle$ and $\langle \mathcal{J}(\mathbf{z})|v_J, \mathbf{x}_J\rangle$ ($k =$

$0, 1, \ldots, N$; $J = 0, i$; $i = 0, 1, \ldots$), respectively. In fracture mechanics for random loading another approach is known.

One of these methods is based on the calculation of distribution $F_Y(y)$ of some fracture parameter Y [96]. Such a parameter can be identified with different nonlinear functions of local stresses. For example, for $Y = \mathbf{\Pi}^{(k)}(\boldsymbol{\sigma})$ (8.1) we have a critical value $y^{\text{crit}} = 1$; in a similar manner in fracture mechanics for $Y = \mathcal{J}(\mathbf{K})$ we have $y^{\text{crit}} = 2\gamma$. Thereafter the first-order estimation of the fraction of fractured component f (or grain faces) is calculated as $f = 1 - F_Y(y)$. Usually the approach assumes that the damage density is small enough so that the interaction of fractured elements is negligible small.

For simplicity the stress distribution within each component is assumed as six-dimensional Gaussian one with distribution function $\Phi_\sigma^{(k)}(\boldsymbol{\sigma})$, $\boldsymbol{\sigma} = (\sigma_1, \ldots, \sigma_6)^\top$. Then pertinent damages (or fracture probability of the component) can be defined by the relation [96] ($k = 0, 1, \ldots N$)

$$f^{(k)} = 1 - \int d\Phi^{(k)}(\boldsymbol{\sigma}), \ (k = 0, 1, \ldots, N). \tag{15.32}$$

In a similar manner, the probability of a crack propagation in the point $\mathbf{z} \in \Gamma^c$ located either the matrix or inclusion v_k ($k = 0, 1, \ldots$) can be calculated as

$$f^{(k)}(\mathbf{z}|v_k, \mathbf{x}_k) = 1 - \int d\Phi^{(k)}(\mathbf{K}|v_k, \mathbf{x}_k), \tag{15.33}$$

where one assumes a three-dimensional Gaussian distribution of SIF with the expectation value $\langle \mathbf{K}(\mathbf{z})|v_k, \mathbf{x}_k\rangle$ (4.17), and the covariance matrix

$$\Delta\mathbf{K}^2(\mathbf{z}|v_k, \mathbf{x}_k) \equiv \langle \mathbf{K}(\mathbf{z}|v_k, \mathbf{x}_k) \otimes \mathbf{K}(\mathbf{z}|v_k, \mathbf{x}_k)\rangle - \langle \mathbf{K}(\mathbf{z})|v_k, \mathbf{x}_k\rangle \otimes \langle \mathbf{K}(\mathbf{z})|v_k, \mathbf{x}_k\rangle, \tag{15.34}$$

which can be found by the formulae either (4.22) or (4.26).

In the right-hand-side integrals of Eqs. (15.32) and (15.33) the integral domains are determined by one of two inequalities ($k = 0, 1, \ldots, N$) $\mathbf{\Pi}^{(k)}(\boldsymbol{\sigma}) \leq 1$ for Eq. (15.32), and

$$\boldsymbol{\Lambda}^{(k)}(\mathbf{K} \otimes \mathbf{K}) \leq 2\gamma^{(k)} \quad \text{or} \quad K_1 < 0. \tag{15.35}$$

for Eq. (15.33). By the use of second inequality (15.35) one assumes the impossibility of fracture under compressive normal loading. The boundaries of the mentioned domains are described by the surface of second order in stress space in the simple cases of a quadratic strength criterion (15.11) and an elliptic plane crack within a homogeneous stress field (15.9).

The most simple phenomenological ways of fracture probability calculation for composites [1025] are based either on the total probability formula or extreme value distribution:

$$f = \sum_{k=0}^{N} c^{(k)} f^{(k)}, \quad f = 1 - \prod_{k=0}^{N} (1 - f^{(k)}), \tag{15.36}$$

respectively. Ortiz and Molinari [840] were the first to consider the particular case of this scheme with an application to random structure composites (see also

[689]). By the use of Fourier's method they obtained as estimation for the second moment of stresses averaging over all volume of a composite with elastically homogeneous properties, but with stress-free strain fluctuations. Because the residual stresses are self equilibrating we have $\delta\boldsymbol{\sigma} \equiv \boldsymbol{\sigma} - \langle\boldsymbol{\sigma}\rangle = \boldsymbol{\sigma}$. Thereupon Ortiz and Molinari [840] (see also [727]) have assumed that the residual stresses are normally distributed with a zero expectation in each point of the composite. In such a case the probability density function of $\delta\boldsymbol{\sigma}$ is entirely defined by the covariance matrix $\mathbf{K}^\sigma = \langle\delta\boldsymbol{\sigma} \otimes \delta\boldsymbol{\sigma}\rangle$ and takes the form (5.13):

$$d\Phi_\sigma(\delta\boldsymbol{\sigma}) = \frac{1}{(2\pi)^3\sqrt{\det(\mathbf{K}^\sigma)}}\exp^{-\frac{1}{2}(\delta\boldsymbol{\sigma})^\top(\mathbf{K}^\sigma)^{-1}\delta\boldsymbol{\sigma}}, \qquad (15.37)$$

where $\det(\mathbf{K}^\sigma)$ is the determinant of the matrix $\mathbf{K}^\sigma$. The indicated approximation of the real process $\delta\boldsymbol{\sigma}$ by Gaussian distributions with zero average stress $\langle\boldsymbol{\sigma}\rangle \equiv \mathbf{0}$ can lead to too crude estimations if conditional average stresses $\langle\boldsymbol{\sigma}\rangle^{(i)} \neq \mathbf{0}$ $(i = 0, \ldots, N)$. However, a slight modification of this approach takes into account the average stresses in the components which can be estimated by using the exact formula (13.71). Therefore, the definition of the conditional covariance matrix $K^{\sigma(i)}_{klmn} = \langle\sigma_{kl}\sigma_{mn}\rangle^{(i)}(\mathbf{x}) - \langle\sigma_{kl}\rangle^{(i)}\langle\sigma_{mn}\rangle^{(i)}$ and the probability density function by

$$d\Phi_\sigma(\boldsymbol{\sigma}) = \frac{1}{\sqrt{(2\pi)^N\det\mathbf{K}^{\sigma(i)}}}\exp^{-\frac{1}{2}(\sigma_{kl}-\langle\sigma_{kl}\rangle^{(i)})(K^{\sigma(i)})^{-1}_{klmn}(\sigma_{mn}-\langle\sigma_{mn}\rangle^{(i)})} \qquad (15.38)$$

seem to be more correct. According to previously obtained estimations [151] the random residual stress fluctuations inside the components are not in excess of 10% of the average values $|\langle\sigma_{11}\rangle^{(i)}|$. Thus, the probability that the stresses inside the component $v^{(i)}$ exceed the value $1.3|\langle\sigma_{11}\rangle^{(i)}|$ is much smaller than predicted by using the formula (15.37).

15.3 Numerical Results

15.3.1 Utilizing Fluctuations of Bulk Stresses Inside the Phases

At first let us consider as an example a composite material consisting of an incompressible matrix satisfying the von Mises yield criterion (15.3) and containing either the rigid inclusions or voids. Then the onset of yielding of the composite is defined by

$$\frac{1}{2c^{(0)}}\frac{\partial M^*_{mnpq}}{\partial q^{(0)}}\langle\sigma_{mn}\rangle\langle\sigma_{pq}\rangle = \frac{2}{3}\tau_0^2. \qquad (15.39)$$

For a macroisotropic composite we will obtain an elliptical criterion for the onset of yielding

$$\frac{X_\tau^2}{a_\tau^2} + \frac{Y_\tau^2}{b_\tau^2} = 1, \qquad (15.40)$$

in the coordinate system of normalized overall stresses: $X_\tau = \langle\sigma_0\rangle/\tau_0$, $Y_\tau = \langle\tau\rangle/\tau_0$, $\langle\tau\rangle = (3\langle\mathbf{s}\rangle\langle\mathbf{s}\rangle/2)^{1/2}$. The effective moduli (9.97), (9.106), (9.108) estimated by the MEFM are consistent with the yields

$$a_\tau^2 = 0, \quad b_\tau^2 = (1-c)\left(1 + \frac{5}{2}c\frac{1}{1-31c/16}\right), \tag{15.41}$$

$$a_\tau^2 = \frac{4}{9c}\left(1 - \frac{29}{24}c\right), \quad b_\tau^2 = (1-c)\left(1 - \frac{5}{3}c\frac{1}{1+5c/24}\right), \tag{15.42}$$

$$a_\tau^2 = \frac{1}{3c}\left(1 - \frac{16}{15\pi^2}c\right), \quad b_\tau^2 = \left(1 - \frac{16}{15\pi^2}c\frac{1}{1-16c/125\pi^2}\right), \tag{15.43}$$

for the composites with identical rigid spherical inclusions ($a_1 = a_2 = a_3 = a$) (15.41), with identical spherical pores ($a_1 = a_2 = a_3 = a$) (15.42), and with identical uniformly oriented coin-shaped cracks ($a_1 = a_2 = a \gg a_3$) (15.43); the values $c = \frac{4}{3}\pi a^3 n$ (where n in a number density) in the expressions presented have different physical meaning for the spherical and coin-shaped pours. The relations (15.42) and (15.43) show that yielding will also take place under purely hydrostatic loading, which is in contradiction to the results for onset of yielding obtained by conventional yielding analysis of composites based on the assumption of homogeneity of the microstress fields in the matrix, Eq. (15.10).

For the "quasi-crystalline" approximation (8.65), (8.79), Eqs. (15.39) and (15.40) give

$$a_\tau^2 = 0, \quad b_\tau^2 = (1-c)\left(1 + \frac{5}{2}c\frac{1}{1-c}\right), \tag{15.44}$$

$$a_\tau^2 = \frac{4}{9c}(1-c), \quad b_\tau^2 = (1-c)\left(1 - \frac{5}{3}c\frac{1}{1+2c/3}\right), \tag{15.45}$$

$$a_\tau^2 = \frac{1}{3c}, \quad b_\tau^2 = \left(1 - \frac{16}{15\pi^2}c\frac{1}{1+16c/15\pi^2}\right). \tag{15.46}$$

The representations (15.41), (15.42), (15.44), (15.45) were obtained in [159] for the case of perfect plasticity of composite materials. More recently the relations (15.44)–(15.46) were also obtained in [764], [895], [896] using variational methods; the relations (15.42) and (15.43) were obtained in [130] for linear elastic composites.

Figure 15.1 shows a comparison of the above surfaces in terms of normalized overall stresses X_τ, Y_τ for a chosen spherical void volume fraction of $c = 0.25$. As expected, the yield surface (15.42) is enclosed inside the yield surface (15.45) because, according to [895], [896], [1190], the yield surface (15.45) is a rigorous upper bound for all other yield surfaces. Of course these yield surfaces approach each other for $c \to 0$.

For comparison with yield surfaces (15.42) and (15.45) we now consider the application of the integral equation method (IEM) for estimating the second moment of stresses (13.43), when the second moment of stresses inside the inclusions $\langle \boldsymbol{\sigma} \otimes \boldsymbol{\sigma} \rangle^{(i)}$ is estimated by the integral equation method (13.46) (first case) and by the relation (13.47) ignoring the stress fluctuations in the inclusions (second case). The small difference between the calculated yield curves is explained by the fact that the stress fluctuations in the inclusions are very small in comparison with their average stresses. In Fig. 15.1 we see that in both cases the yield surfaces are located inside the rigorous upper bound (15.45).

It should be noted that Ju and coauthors [517], [522], [524] have used the estimation of the averaged second moment of the stress deviator $\langle \mathbf{ss} \rangle^{(0)}$ for the determination of yield surfaces. For this purpose they proposed an approximate integral equation method (13.48), (13.49) ignoring the stress fluctuations inside the inclusions; they suggested to evaluate the effective yield surface by the formula $(1-c)^2\langle \mathbf{ss} \rangle^{(0)} = 2\tau_0^2/3$, where a factor $(1-c)^2$ is added without any justification. However, their results differ from Eqs. (13.47) and do not fall within the rigorous upper bound (15.45), see Fig. 15.1 and [180], [182].

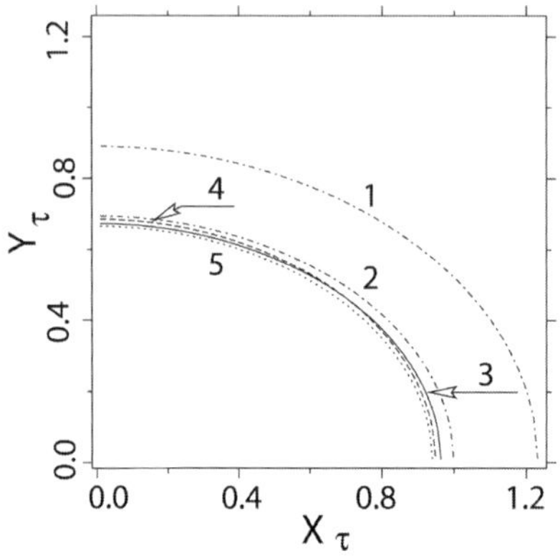

Fig. 15.1. Plots of the yield surface calculated via the method [517] (line 1), via the MEFM (15.45)(line 2), via the MEF (15.41) (solid line 3), via the IEM (13.47) (dashed line 4), via the refined IEM (13.46) (dotted line 5).

For regular porous structures the criteria of Gurson [409] ($\beta = 1$) as well as Koplik and Needleman [586] ($\beta = 1.25$):

$$Y_\tau^2 + 2c\beta \cosh\left(\frac{3X_\tau}{2\tau_0}\right) = 1 - \beta^2 c^2 \tag{15.47}$$

(see also [98], Tvergard'N'1997) construct the limiting surface by the numerical averaging of the dissipative function inside the matrix. Analogously to [352] it is interesting to note that for sufficiently small values X_τ/τ_0 we find from (15.47) by using a Taylor expansion of $\cosh(3X_\tau/2\tau_0)$ that for second degree surfaces for the onset of yielding, Eq. (15.40) has parameters

$$a_\tau^2 = \frac{4}{9c}(1-\beta c)^2, \quad b_\tau^2 = (1-\beta c)^2. \tag{15.48}$$

The predictions of the MEFM, Eq. (15.42), the results of Koplik and Needleman [586], Eqn. (15.48), of Ponte Castañeda, Eq. (15.45), and of Gurson, Eq. (15.48), respectively, are related by the inequalities $b_\tau^{\rm MEFM} < b_\tau^{\rm KN} < b_\tau^{\rm P-C} < b_\tau^{\rm G}$ for $c > 0.2$. The difference between $b_\tau^{\rm MEFM}$ and $b_\tau^{\rm KN}$ is less than 5% for $c < 0.5$. The proximity of estimates by the methods mentioned was recognized [131], [192].

A comparison between the yielding parameters b_τ^2 calculated by the use of two- (15.41) and one-particle (15.44) approximations of the MEFM is presented in Fig. 15.2. As can be seen, a difference between the curves calculated via two- and one-particle approximations of the MEFM in the case of rigid inclusions is high than for porous materials. Buryachenko and Lipanov [159] showed that the two-particle approximation of the MEFM is superior to the one-particle approximation in representing the experimental data (see Fig. 15.3) for uniaxial compressive loading (pressure p) of porous sintered electrolytic nickel subjected to uniaxial compression inside a die mold. The predicted dependences of the pressure leading to onset of yielding for $c \to 0$ are asymptotically similar to the calculated curves. At $c = 0.45$, however, the results of the various methods differ from each other by about 30% with respect to p. More accurate curve 2 estimated by two-particle approximation of the MEFM (15.42) will shift to the left on axis p the more ellipsoidal inclusions differ from spheroidal inclusions adopted in calculations of (15.42).

The example of a medium with incompressible matrix is considered deliberately, because comparison of yield surfaces (15.44)–(15.46) with other analytical models and numerical studies on unit cell models as well as with experimental data are well known [numerous references can be found, e.g., in [83], [131], [895]. The purpose of such consideration is to show that the proposed method of integral equations (13.45)–(13.47) (for a more general estimation of all component of second moment of the stresses $\langle \boldsymbol{\sigma} \otimes \boldsymbol{\sigma} \rangle^{(k)}$ $(k = 0, 1, \ldots, N)$) offers high exactness precisely in the particular classical example.

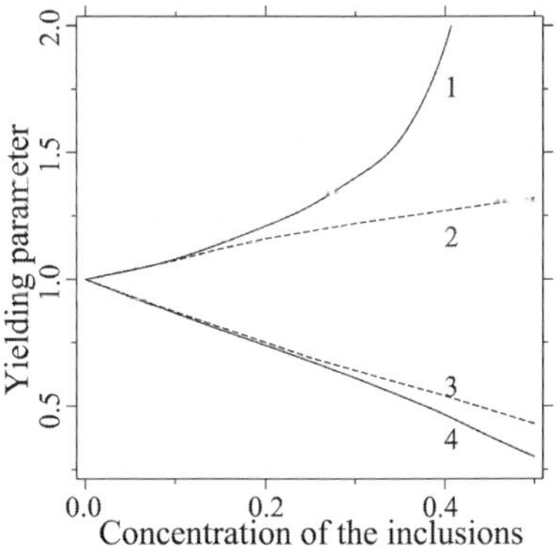

Fig. 15.2. Yielding parameter b_τ for materials with rigid inclusions (1,2) and pores (3,4) calculated by the two-particle (solid line) and one-particle (dashed line) approximations of the MEFM.

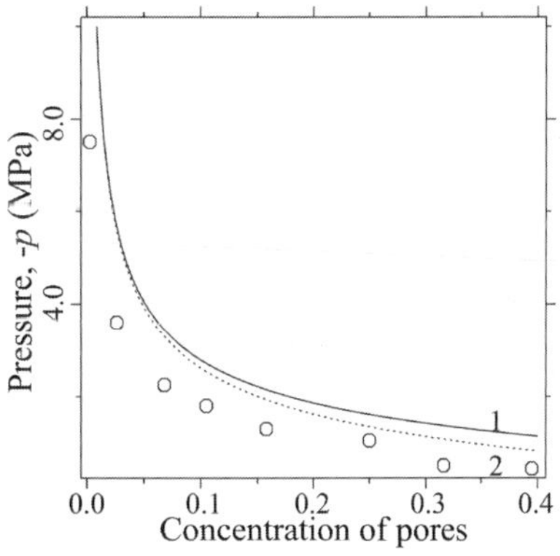

Fig. 15.3. Onset of yielding of porous nickel vs concentration of pores. Estimations by the MEFM (15.45) (curve 1), and (15.41) (curve 2).

Let us analyze an influence of residual stresses on the limiting surface estimated. First we will consider elastically homogeneous isotropic medium containing statistically homogeneous of identical homogeneous spherical inclusions with radius a and eigenstrains $\boldsymbol{\beta}_1(\mathbf{x}) \equiv \beta_1^{(1)}\boldsymbol{\delta} = \text{const.}$ $(\mathbf{x} \in v^{(1)})$. The step functions (9.91) and (13.61) are applied for the calculation of the stress fluctuations inside the inclusions and in the matrix, respectively. For the representation of numerical results in dimensionless form we define a slight modification of a normalizing coefficient $\eta \equiv -3Q^k\beta_{10}c^{1/2}$, where $3Q^k$ represents the bulk component of the tensor $\mathbf{Q}_1 = (3Q^k, 2Q^\mu)$ and $\boldsymbol{\beta}(\mathbf{x}) \equiv \beta_{10}\boldsymbol{\delta}$ $(\mathbf{x} \in v^{(1)})$. The physical meaning of η was also explained in Section 13.3.3. In addition we define the dimensionless variables $Y^{(i)} \equiv (3\langle\mathbf{s}\rangle:\langle\mathbf{s}\rangle/2)^{1/2}/\tau_0^{(i)}$, $X^{(i)} \equiv \eta/\tau_0^{(i)}$ for the component $v^{(i)}$ $(i = 0,1)$. Then the criterion for onset of yielding (15.11) of the component $v^{(i)}$ is circumscribed by an ellipsoidal curve (do not sum to i)

$$\frac{(X^{(i)})^2}{(a_\tau^{(i)})^2} + \frac{(Y^{(i)})^2}{(b_\tau^{(i)})^2} = 1 \qquad (15.49)$$

in the coordinates $X^{(i)}$, $Y^{(i)}$. The semi-axis $b^{(i)} = 1$ does not depend on the phase transformation strains $\boldsymbol{\beta}_1(1)$ and on the microtopology of the composite. The semi-axes

$$\begin{aligned} a_\tau^{(0)} &= \Bigg(\frac{3}{2\eta^2}\int \left[\mathbf{T}_p(\mathbf{x}_0-\mathbf{x}_p)\boldsymbol{\beta}_1^{(1)}\overline{v}_p\right]\mathbf{N}_2\left[\mathbf{T}_p(\mathbf{x}_0-\mathbf{x}_p)\boldsymbol{\beta}_1^{(1)}\overline{v}_p\right] \\ &\quad\times \varphi(v_p,\mathbf{x}_p|;v_0,\mathbf{x}_0)\,d\mathbf{x}_p\,d\mathbf{x}\Bigg)^{-\frac{1}{2}}, \qquad (15.50) \\ a_\tau^{(i)} &= \Bigg(\frac{3}{2\eta^2\overline{v}^{(i)}}\int\!\!\int \left[\mathbf{T}_p(\mathbf{x}-\mathbf{x}_p)\boldsymbol{\beta}_1^{(1)}\overline{v}_p\right]\mathbf{N}_2\left[\mathbf{T}_p(\mathbf{x}-\mathbf{x}_p)\boldsymbol{\beta}_1^{(1)}\overline{v}_p\right] \\ &\quad\times V_i(\mathbf{x})\varphi(v_p,\mathbf{x}_p|;v_i,\mathbf{x}_i)\,d\mathbf{x}\Bigg)^{-\frac{1}{2}} \qquad (15.51) \end{aligned}$$

determined according to Eq. (13.60) are identically zero if binary interaction of the inclusions is neglected or, what amounts to the same thing, if the assumption of homogeneity of the stresses inside each component (15.10) is used. The value of $a_\tau^{(i)}$ does not depend on the inclusion concentration c if the step radius distribution functions (9.91) and (13.61) are assumed. The obtained criterion (15.49) is invariant with respect to the hydrostatic portion of the external stresses.

For the mean field criterion (15.10) we obtain a degenerated criterion for onset of yielding of the component $v^{(i)}$ that is invariant with respect to transformation strain $\boldsymbol{\beta}_1$, when $b_\tau^{(i)} = 1$, $a_\tau^{(i)} = \infty$ in (15.49):

$$(Y^{(i)})^2 = 1. \qquad (15.52)$$

Figure 15.4 shows the curves calculated by using the criteria (15.50), (15.51) and (15.52), respectively, for the matrix and the inclusions ($\nu = 0.3$). The curve of yielding of the matrix (15.50) is enclosed by the curve for yielding of the inclusions (15.51). Hence, if both components have the same initial yield stress, $\tau_0^{(0)} = \tau_0^{(1)}$, the yielding of the composite due to a stress-free strain $\boldsymbol{\beta}_1 \neq 0$ (e.g., caused by a thermal mismatch) will begin from the matrix. This statement is a simple consequence of the difference in conditional probability densities (9.91) and (13.61), because the stress fluctuations are caused by perturbations due to the surrounding inclusions in the vicinity of the considered point $\mathbf{x}_i$ (9.91) or $\mathbf{x}_0$ (13.61), respectively. It should be noted that the normalized numerical results shown in Fig. 15.4 are obtained for $\nu = 0.3$, they depend only on Poisson's ratio and are invariant with respect to other material properties of the components. Buryachenko [135] demonstrated the dependence of the parameters $a_\tau^{(i)}$ $(i = 0,1)$ on Poisson's ratio in the range $0.1 \le \nu \le 0.499$.

Of course the criteria (15.49) and (15.52) apply only for statistically isotropic microstructure with spherical inclusions. For statistically anisotropic composites, for example with unidirectionally aligned identically shaped inclusions, the mean field criterion (15.10) will depend on the transformation strains ε^t. However, also in this case the yield surface (15.11) will be located inside the yield surface (15.10), which can be seen by taking the inequality (15.14) into account.

Let us consider as an example a composite consisting of isotropic homogeneous components with both the thermal and elastic mismatch and having identical spherical inclusions with yield properties according to (15.3). At first we will use the integral equation method for the estimation of the second moments (13.45) and (13.46), accompanied by the determination of the yield surface (15.3) and (15.11). According to (13.45) and (13.46) all tensors used for the calculation of second moments of stresses (such as, e.g., $\mathbf{H}^j_{ip}$, $\mathbf{P}^l_{ip}$; $j = 1, \ldots, 4$, $l = 1, 2, 3$, $i, p = 1, \ldots, N$) are isotropic ones. Then the yield surface (15.11) has ellipsoidal form in the space of nondimensional coordinates

$$X_{\tau 1} = \frac{\langle \overline{\sigma}_0 \rangle^{(i)}}{\tau_i}, \quad Y_{\tau 1} = \frac{\sqrt{3 \langle \overline{\mathbf{s}} \rangle^{(i)} \langle \overline{\mathbf{s}} \rangle^{(i)} / 2}}{\tau_i}, \quad Z_{\tau 1} = \frac{\sigma^{\mathrm{hyd}} \boldsymbol{\beta}^{(i)}_{10}}{\tau_i} \tag{15.53}$$

for $i = 0, 1, \ldots, N$ and, therefore, this yield surface is described by

$$\frac{X^2_\tau}{a^2_\tau} + \frac{Y^2_\tau}{b^2_\tau} + \frac{Z^2_\tau}{c^2_\tau} - \frac{X_\tau Z_\tau}{d^2_\tau} = 1, \tag{15.54}$$

in the coordinates $X_\tau = \langle \sigma^0_0 \rangle / \tau_i$, $Y_\tau = \sqrt{3 \langle \mathbf{s}^0 \rangle \langle \mathbf{s}^0 \rangle / 2} / \tau_i$, $Z_\tau = \sigma^{\mathrm{hyd}} / \tau_i$, because, according to Eqs. (15.41) and (15.53), there is a linear dependence between the effective stresses $\langle \overline{\boldsymbol{\sigma}} \rangle^{(i)}$ and the external loading $\boldsymbol{\sigma}^0$, $\boldsymbol{\beta}^{(i)}_1$. Here the nondimensional coefficients a_τ, b_τ, c_τ, d_τ are expressed by means of the tensors $\mathbf{H}^{10}_p$, $\mathbf{P}^{10}_p$, $\mathbf{P}^{30}$, $\mathbf{H}^j_{ip}$, $\mathbf{P}^l_{ip}$ ($j = 1, \ldots, 4$; $l = 1, 2, 3$; $i, p = 1, \ldots, N$) and the quantity $\sigma^{\mathrm{hyd}} = -3B^k_i 3Q^k_i \beta^{(i)}_{10}$ has the simple physical meaning of the hydrostatic component of the residual stresses inside the single isolated inclusion in an infinite homogeneous matrix: $\beta^{(i)}_{10} = \boldsymbol{\beta}^{(i)}_1 \boldsymbol{\delta}/3$, $\mathbf{B}_i = (3B^k_i, 2B^\mu_i)$, $\mathbf{Q}_i = (3Q^k_i, 2Q^\mu_i)$. The half-ellipsoidal surface (15.54) (with $\mathbf{Y}_{\tau 1} \geq 0$) can be transformed into the canonical form by a rotation in the three-dimensional space $\{X_\tau,\ Y_\tau,\ Z_\tau\}$ around the axis Y_τ. It is interesting that, according to Eqs. (13.39), the stress fluctuation $\Delta \boldsymbol{\sigma}^{(i)2}$ (13.37) is identically zero for $\langle \boldsymbol{\eta}_i \rangle \equiv \mathbf{0}$. The last equation defines the straight line $X_\tau = e_\tau Z_\tau$, $Y_\tau = 0$ ($e_\tau = \mathrm{const}$) in the coordinates $\{X_\tau,\ Y_\tau,\ Z_\tau\}$. Therefore, in reality the ellipsoidal surface is degenerated into an elliptical cylinder with a symmetry axis coinciding with the indicated straight line and $d^2_\tau = a_\tau c_\tau / 2$. By consideration of Eq. (15.54) one can find that this symmetry axis does not depend on the concentration of the inclusions. It should be mentioned, that the widely used yield surface described by (15.3), (15.10) $Y^2_\tau / a^2_{\tau 2} = 1$ reduces to a plane parallel to the coordinate plane $Y_\tau \equiv 0$ with $a_{\tau 2} > a_\tau$. Therefore, the hydrostatic component due to the external loading σ^0_0 as well as an eigenstrain $\beta^{(i)}_{10}$ do not influence the composite's yielding in traditional calculation methods.

Analogously we consider the exact relation for the calculation of the second invariant of stress deviator (15.23). Then the proposed yield surface (15.11) of the component $v^{(i)}$, $(i = 0, 1, \ldots, N)$ is defined by the equation

$$\frac{\partial \mathbf{M}^*}{\partial q^{(i)}} \boldsymbol{\sigma}^0 \otimes \boldsymbol{\sigma}^0 - 2\frac{\partial U^*}{\partial q^{(i)}} + 2\frac{\partial \boldsymbol{\beta}^*}{\partial q^{(i)}} \boldsymbol{\sigma}^0 = \frac{4}{3} c^{(i)} \tau_i^2. \tag{15.55}$$

However, according to (8.52)–(8.54) the tensors used for the calculation of the effective parameters $\mathbf{M}^*$, U^*, $\boldsymbol{\beta}^*$ are isotropic and constant, which are derived from the tensors $\mathbf{Y}_{ij}$, $\mathbf{B}_i$, $\mathbf{R}_j$ $(i, j = 1, \ldots, N)$, defined by the solution of the purely elastic problem ($\boldsymbol{\beta} \equiv \mathbf{0}$) by means of particular derivatives with respect to $q^{(i)}$.

It should be mentioned that yield surfaces in the form of elliptical cylinders (15.54) are obtained if the averages of the second moment of the stresses in the component (15.11) are used. Buryachenko [131] has shown that using the average second moment of the stresses in the matrix in the vicinity of the inclusion leads to a new yield surface which is inserted in the old one. The cross-section of the elliptical yield cylinder in the plane $Z_\tau = 0$ (purely elastic case, $\boldsymbol{\beta} \equiv \mathbf{0}$) was already studied comprehensively in this section (see also [131], [180]), where numerous references and the comparison with known results can be found. Because of this, in the following we consider only the cross-section $Y_\tau = 0$ of the yield surface.

To obtain concrete numerical results we choose the Poisson's ratio of the matrix $\nu = 0.3$, and the concentration of rigid identical spherical inclusions of the radius a is denoted by $c^{(1)}$; $\tau_0 = \tau_1$. The conditional probability densities are the step functions (9.91) and (13.61). Figure 15.5 gives the corresponding cross-sections $\mathbf{Y}_\tau = 0$ of the

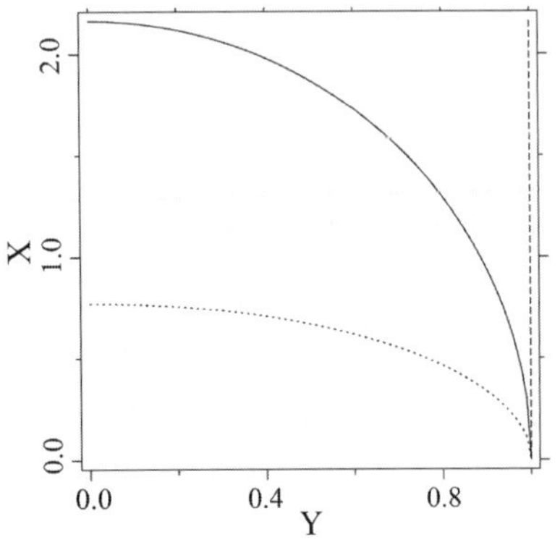

Fig. 15.4. Normalized coordinates $X^{(i)}$ as the functions of $Y^{(i)}$. The curves for onset of yielding of the matrix ($i = 0$, dotted curve) and the inclusions ($i = 1$, solid curve) are calculated by using the criterion (15.50) and (15.51). The criterion (15.52) leads to the dashed line.

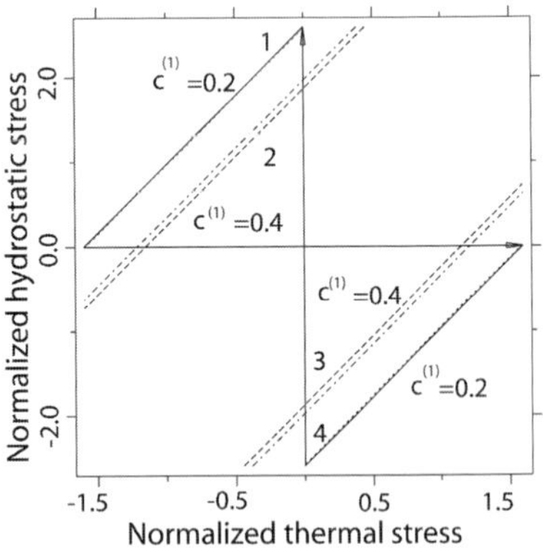

Fig. 15.5. Cross–section $\mathbf{Y}_\tau = 0$ of the yield surfaces of the matrix, calculated by (13.46) (solid and dot-dashed lines) and by (15.55) (dotted and dashed lines) for $c^{(1)} = 0.2$ (lines 1, 4) and $c^{(1)} = 0.4$ (lines 2, 3).

yield surface of the matrix calculated by either the integral equation method (13.46), (15.54) or by the exact relation (15.55) for $c^{(1)} = 0.2$ as well as $c^{(1)} = 0.4$. One can see that both methods, (13.46), (15.54), and (15.55), lead to very similar results. The yield surface (15.55) has the form of an elliptical cylinder (15.54) with dimensionless coefficients $a_{\tau 1}$, $b_{\tau 1}$, $c_{\tau 1}$, $d_{\tau 1}$. The surface (15.55) is inserted inside the other one (13.46), (15.54): $a_{\tau 1} < a_{\tau}$, $b_{\tau 1} < b_{\tau}$, $c_{\tau 1} < c_{\tau}$. Obviously, for $c^{(1)} \to 0$ we have $a^2_{\tau 1} \to a^2_{\tau} \sim 1/c^{(1)}$, $b^2_{\tau 1}, b^2_{\tau} \to 1$, $c^2_{\tau 1}, c^2_{\tau} \sim 1/c^{(1)}$, and the yield surface converges to a plane. Thus, we have again proved that the method of integral equations provides the closed results with the perturbation method in some particular problems (see Figs. 15.1 and 15.5) and offers additional advantages and other interesting aspects when we need to estimate all components of the second moment of stresses varying over the phases (see, e.g., Fig. 13.2.3).

15.3.2 Utilizing Interface Stress Fluctuations

Let us compare the estimates obtained by local averaging, criteria (15.16) and (15.17) involving the interface stresses, with Eq. (15.11) based on utilizing of bulk stresses inside the phases. In Fig. 15.6 the calculated curves for the onset of yielding under hydrostatic loading $\langle \sigma_0 \rangle$ according to criteria (15.11), (15.16) and (15.17) are shown in normalized coordinates $\langle \sigma_0 \rangle / \tau_0$ for a material with the identical spherical pores already considered in Fig. 15.1. As was to be expected (see Eq. (15.18)) the local averaged criterion (15.17) predicts the lowest value for the onset of yielding, but for estimating the elastoplastic behavior of all components (and therefore of the composite) the averaged criterion (15.11) is to be preferred.

In the general case of loading we plot in Fig. 15.7 the effective yield surface describing by Eq. (15.16) of the porous material considered in Fig. 15.1. Just for comparison, we reproduce with the same notations the yield surfaces estimated via Eq. (15.10), (15.45), and (15.41). As can be seen the yield surface corresponding to the yield condition (15.16) accompanied by Eq. (15.21$_2$) is not elliptic and plotted for the particular case of external loading $\langle \sigma_{ij} \rangle = \alpha \delta_{i3} \delta_{j3} + \beta \delta_{ij}$, where α and β are arbitrary variables with $\alpha > 0$. As indicated in Fig. 15.7, the criterion (15.16) as well as (15.45) and (15.41) show that yielding will also take place under purely hydrostatic loading.

We attempt now to quantitatively investigate the performance of the present approach to the failure analysis of fiber composites [191]. The results are directly compared with solutions extracted from simplified assumptions (such as Mori-Tanaka approach as well as hypothesis of a homogeneity of stresses in the constituents) and they are presented in order to place the advantages and limitations of the refined approach in evidence.

For failure analysis, let us consider an isotropic composite made from the epoxy matrix and SCS-0 fibers. Both components are described by isotropic elastic properties with the mechanical constants as usually found in the literature (see [1071]): $k^{(0)} = 3.82$ GPa, $\mu^{(0)} = 1.74$ GPa, $k^{(1)} = 190.47$ GPa, $\mu^{(1)} = 173.91$ GPa, $\sigma_n^{\max} = 34.8$ MPa, $\sigma_{\tau}^{\max} = 32.5$ MPa. At first we will consider well-stirred approximation of the RDF (9.91). In general, interface failure occurs much more easily under a tensile normal than a compressive normal stress. That is, the normal strength $\sigma_n^{\max}$ is much greater for compression than for tension. In Fig. 15.8 the failure envelopes are plotted in the first quadrant of a coordinates system

$X = \langle\sigma_{11}\rangle \geq 0$ and $Y = \langle\sigma_{12}\rangle \geq 0$. The nonelliptical curve 1 corresponds to the dilute concentration of fibers $c^{(1)} \ll 1$.

Curves 2 and 3 were estimated by the MEFM method (15.13), (15.22) with the RDF (9.93) and (9.91), respectively. Neglect of stress fluctuation (15.21_1) transforms curve 3 into curve 4. Ignoring of the binary interaction of fibers (8.65_3) automatically leads to the neglect of stress fluctuations and tends to increase the failure prediction as described by curve 5. It should be mentioned that all curves 1–5 are not elliptical in the global coordinate system of macrostresses $\langle\boldsymbol{\sigma}\rangle$ although the failure envelope (15.7) is described by a quadratic function of the traction $\langle\boldsymbol{\sigma}_{\mathbf{n}}^-\rangle^{(i)}$ in the local coordinate system connected with the fiber surface. The nonelliptical shape of the effective failure envelope is demonstrated by comparison of curve 5 with the elliptical curve 6 with the same semi-axes as the curve 5. It's interesting that the nonelliptical shape of the effective limiting surface for the porous materials was demonstrated in [183] (see Fig. 15.7) in the related problem of onset of yielding at the porous materials.

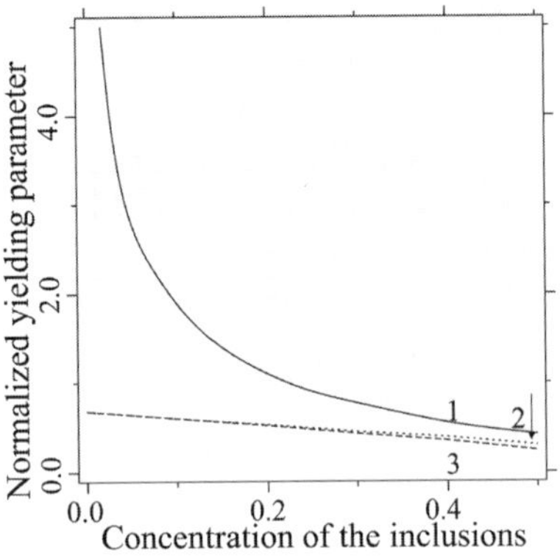

Fig. 15.6. Onset of yielding calculated from Eqs. (15.11), (15.39) (curve 1), Eq. (15.16) (curve 2), Eq. (15.17) (curve 3).

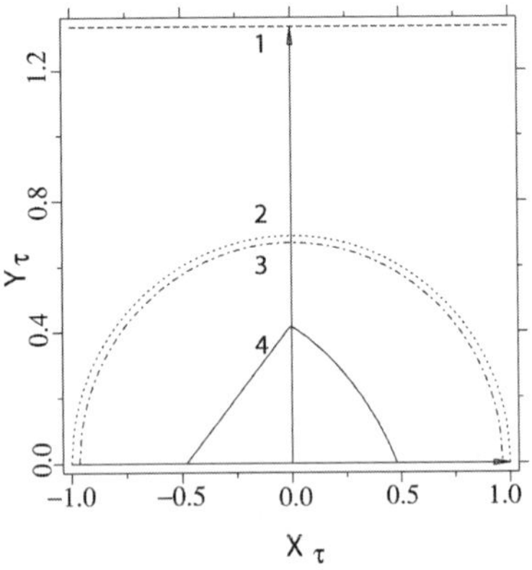

Fig. 15.7. Plots of the yield surfaces calculated by Eq. (15.10) and (8.48) (curve 1), by Eq. (15.45) (curve 2), by (15.41) (curve 3), and by Eqs. (15.11) and (15.21_2) (curve 4).

The popular engineering simplification is based on the neglect of shearing failure in comparison with the failure initiated by the normal component of the traction $\boldsymbol{\sigma}_n^-$. That is equivalent to the assumption

$$\sigma_\tau^{\max} = \infty, \tag{15.56}$$

the error of which we will estimate now by the example of the comparison of the curves 2 and 4 in Fig. 15.8 with the corresponding curves of effective failure envelopes plotted in the framework of assumption (15.56). As can be seen in Fig. 15.9, the significant differences of effective failure envelopes estimated for the real failure parameters $\sigma_\tau^{\max}$ and $\sigma_n^{\max}$ as well as for the assumed one (15.56) are observed just at the small values σ_{11}. In the case of the well stirred RDF (9.91) accompanied by the disregard of stress fluctuations in the fibers (15.21_1), the influence of the assumption (15.56) can be neglected in the area of moderate tension loading $\langle\sigma_{11}\rangle > 0.2\sigma_n^{\max}$.

The influence of the RDF on the effective failure envelopes estimated for the analytical representations (9.91) and (9.93) as well as for the numerical simulation by the CRM accompanied by the random shaking procedure (see Chapter 5) was analyzed. The difference between the estimations increasing with the rise of the fiber concentration vary from 1% until 6% at $c = 0.45$ and $c = 0.75$, respectively. In so doing the difference between the estimations obtained for the RDF (9.93) and for the simulated RDF does not exceed 0.8%. The maximum of the difference of the effective failure envelopes does not exceed 1.7% in the cases of two different methods of the estimation of the second moments of interface stresses such as the perturbation method (15.20) and the method of integral equations (15.23).

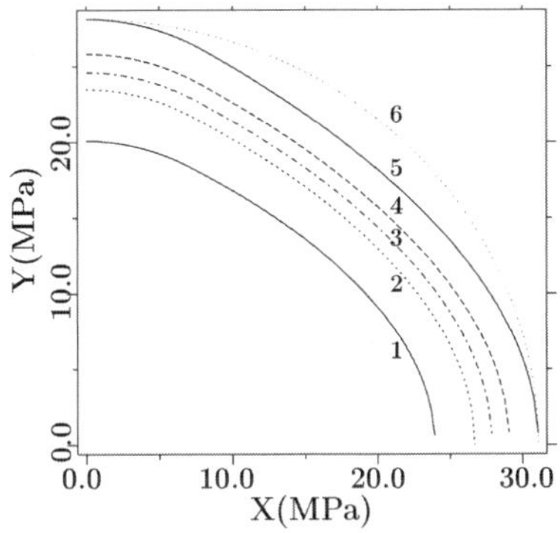

Fig. 15.8. Effective failure envelopes estimated by the different methods; dilute concentration of fibers (curve 1), the MEFM (15.13), (15.22) with the RDF (9.93) (curve 2); the MEFM (15.13), (15.22) with the RDF (9.91) (curve 3); the MEFM (15.13), (15.21$_1$) (curve 4); Mori-Tanaka approach (curve 5); elliptical approximation of Mori-Tanaka approach (curve 6).

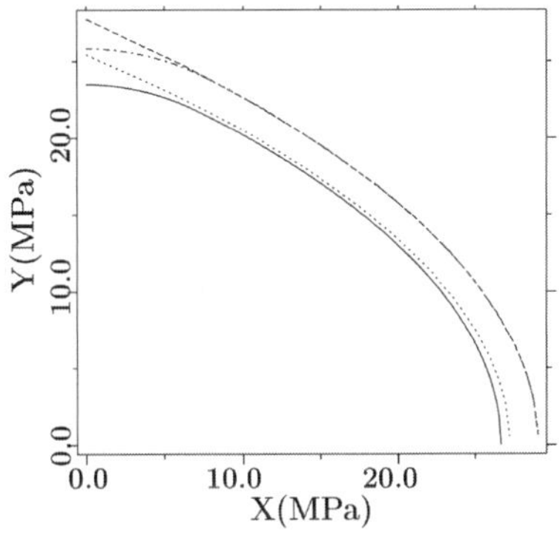

Fig. 15.9. Effect of shearing stresses on effective failure envelopes. Estimation by the MEFM (15.13), (15.11), (9.93) and (15.13), (15.12), (9.91) for the real failure parameters (solid and dot-dashed curves, respectively). Analogous estimations for the assumed failure parameter (15.56) (dotted and dashed curves, respectively).

15.3.3 Effective Energy Release Rate and Fracture Probability

Following Buryachenko [137], [142], let us consider a crack within elastically homogeneous medium containing a statistically homogeneous spherical inclusion field with the constant eigenstrains. For representation of numerical estimations of $\langle \mathcal{J}(\mathbf{z})|v_J, \mathbf{x}_J\rangle$ and $\Delta\mathcal{J}(\mathbf{z}|v_J, \mathbf{x}_J)$ ($J = 0, i;\ i = 1, 2, \ldots$) in dimensionless form we define normalized coefficient $\theta \equiv (1-\nu)\vartheta^2/(2\mu)$ and a dimensionless parameter $\xi = 2\gamma/\theta$, where the parameter $\vartheta = -Q^k\beta_{10}\sqrt{a}$ was introduced in Subsection 13.4.4. According to Eq. (15.9) the coefficient θ is proportional to the energy release rate for a single coin-shaped microcrack with radius $R^c = a$ within an isolated spherical inclusion in the infinite homogeneous matrix. For example, in line with subsection 13.4.4 the composite material Si_3N_4–SiC with SiC spherical inclusions may be considered as an elastically homogeneous material with the parameters $k = 236.4$GPa, $\mu = 121.9$GPa, $\beta_{10} = -10^{-3}$, $a = 10^{-5}$m, $\gamma(\mathbf{x}) \equiv \gamma = 7\cdot 10^{-2}$MPa·m; therefore in this concrete case $\theta = 2.47\cdot 10^{-6}$MPa·m, $\xi = 5.67 \cdot 10^4$. Below, the numerical estimations will be obtained in the form of normalized curves (having a weak dependence on ν) by the use of coefficient θ. We will represent the calculation only for $\nu = 0.28$ (composite Si_3N_4–SiC) and will estimate the boundaries of its variation under the change of ν. For simplicity's sake only binary interaction of inclusions will be taken into account.

First we analyze $\mathcal{J}(\langle \mathbf{K}(\mathbf{z})|v_k, \mathbf{x}_k\rangle)$ ($k = 1, 0$; $\boldsymbol{\sigma}^0 \equiv \mathbf{0}$) (15.25) as a function of the location of either the inclusion ($k = 1$) or the matrix ($k = 0$) in the point $\mathbf{x}_k = (0, R^c + r, 0)$. As a consequence of problem symmetry $\langle K^1_{II}(\mathbf{z})|v_k, \mathbf{x}_k\rangle = \langle K^1_{III}(\mathbf{z})|v_k, \mathbf{x}_k\rangle \equiv 0$ for $\forall \mathbf{x}_k = (0, R^c + r, 0)$ ($k = 0, 1$) and therefore $\mathcal{J}(\langle \mathbf{K}(\mathbf{z})|v_k, \mathbf{x}_k\rangle) \equiv \mathcal{J}(\langle K_I(\mathbf{z})|v_k, \mathbf{x}_k\rangle)$ is defined only by mode I SIF. Then, for example, at $r/a < -1$ we have $\langle K^1_I(\mathbf{z})|v_1\mathbf{x}_1\rangle < 0$ (see Fig. 13.13) and the location of the inclusion leads to the reduction of $\mathcal{J}(\langle K(\mathbf{z})|v_1, \mathbf{x}_1\rangle)$ (shielding effect).

It is of interest to estimate the contributions of normal $\Delta\mathcal{J}^{no}(\mathbf{z}|v_k, \mathbf{x}_k)$ and shear $\Delta\mathcal{J}^{sh}(\mathbf{z}|v_k, \mathbf{x})$ stress fluctuations to $\Delta\mathcal{J}(\mathbf{z}|v_k, \mathbf{x}_k) \equiv \Delta\mathcal{J}^{no}(\mathbf{z}|v_k, \mathbf{x}_k) + \Delta\mathcal{J}^{sh}(\mathbf{z}|v_k, \mathbf{x}_k)$. The normal component $\Delta\mathcal{J}^{no}(\mathbf{z}|v_k, \mathbf{x}_k)/\theta$ is congruent with the shear one $\Delta\mathcal{J}^{sh}(\mathbf{z}|v_k, \mathbf{x}_k)/\theta$ for the case of step radial functions (9.91) ($k = 1$) and (13.61) ($k = 0$) (see Fig. 15.10). Moreover $\Delta\mathcal{J}^{sh}(\mathbf{z}|v_k, \mathbf{x}_k) > \Delta\mathcal{J}^{no}(\mathbf{z}|v_k, \mathbf{x}_k)$ over the regions $-1 < r/a < 1$ ($k = 1$) and $r = 0$ ($k = 0$), which arouse considerable interest in actual practice. The curves plotted for $\nu = 0.28$ vary in value no greater than 0.5% for the normal components and 4% for the shear one according to changes in ν through a range $0.2 \leq \nu \leq 0.4$; the curves calculated exchange places within $0.05 \leq \nu \leq 0.49$ by, at most, 1% and 10%, respectively.

We turn our attention to prediction of fracture probability of separate components near macrocrack $S \perp \mathbf{n} = (1, 0, 0)^\top$ with radius $R^c = 100a$ and center $\mathbf{x}^c = (0, 0, 0)$ under $c = 0.4$ and step radial function (9.91). In view of the large expenditure of computational time for the case of fixed inclusion we will consider only the point $\mathbf{x}_1 = (0, R^c - 0.7a, 0)^\top$ with maximum conditional SIF $\langle K^1_I(\mathbf{z})|v_1, \mathbf{x}_1\rangle = \max$, taking into account that SIF fluctuation is a weak function of $\mathbf{x}_1$ under $|\mathbf{x}_1| < a$. We assume a Gaussian distribution of SIF with conditional average $\langle \mathbf{K}(\mathbf{z})|v_k, \mathbf{x}_k\rangle$ ($k = 0, 1$) (13.69) and the covariance matrix (15.34), which has diagonal form for our case being analyzed. The integration domain in (15.33) (the region of safe loading) constitutes the union of an ellipsoidal domain and a half-space (15.35). Figure 15.11 shows the plots of fracture probability $f^{(k)}(\mathbf{z}|v_k, \mathbf{x})$ (15.33), (15.35) (for $k = 1$ and for $k = 0$) as a function of dimensionless number ξ for $c = 0.4$ and step conditional probability densities ((9.91) and (13.61)). At first glance it would seem strange that for sufficiently strong material $\xi > 1.36$ the fracture probability of the matrix is more than fracture probability of the inclusion notwithstanding the fact that $\langle K_I(\mathbf{z})|v_1, \mathbf{x}_1\rangle > 0 > \langle K_I(\mathbf{z})|v_0, \mathbf{x}_0\rangle$. Such a result stands clear when one takes into account the greater level of SIF fluctuation inside the matrix (see Fig. 13.13). The fracture probability will be determined fundamentally by the maximum SIF fluctuations (other than conditional average SIF) over the region $\xi \gg 1$, which arouses considerable interest in actual practice. For zero strength of the materials $f^{(k)}(\mathbf{z}|v_k, \mathbf{x}_k) \to 1 - P\{K_I(\mathbf{z}|v_k, \mathbf{x}_k) < 0\}$ and $f^{(1)}(\mathbf{z}|v_1, \mathbf{x}_1) \to 0.97$, $f^{(0)}(\mathbf{z}|v_0, \mathbf{x}_0) \to 0.28$, which correlates well with our simplifying assumption, that the material will not fail under compression.

It is interesting to compare the numerical results just obtained by the use of a probabilistic model with a deterministic model of fracture (15.9) near macrocrack tip ($R^c = 100a$) $\langle \mathcal{J}(\mathbf{z})|v_k, \mathbf{x}_k\rangle = 2\gamma$, when the unit probability of the fracture occurs at $\xi \leq 0.25$ ($k = 1$) and $\xi \leq 0.50$ ($k = 0$). With the use of the conditional average SIF (15.25) one should expect fracture of the inclusion at $\xi \leq 0.12$, while fracture of the matrix is impossible. We see that the estimation of fracture probability may be defined more exactly by the application of more correct model (15.23).

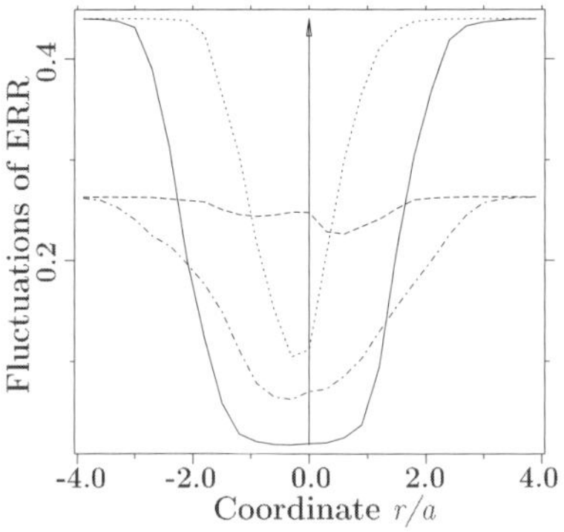

Fig. 15.10. Normal $\Delta\mathcal{J}^{no}(\mathbf{z}|v_k, \mathbf{x}_k)/\theta$ (solid and dotted lines) and shear $\Delta\mathcal{J}^{sh}(\mathbf{z}|v_k, \mathbf{x})/\theta$ (dashed and dot-dashed lines) components of energy release rate $\Delta\mathcal{J}(\mathbf{z}|v_1, \mathbf{x}_1)/\theta$ for fixed matrix (dotted and dashed lines) and inclusions (solid and dot-dashed lines).

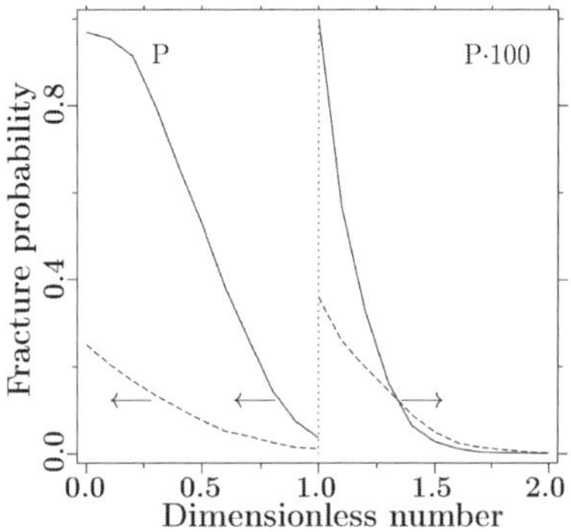

Fig. 15.11. Fracture probability $P \equiv f^{(k)}(\mathbf{z}, v_k, \mathbf{x}_k)$ ($k = 0, 1$) as a function of dimensionless number ξ for located inclusions in the point of maximum conditional average SIF (solid line) and for fixed matrix (dashed line).

It should be noted that all these values ξ significantly below the quantity $\xi = 5.67 \cdot 10^4$ for the Si_3N_4–SiC composite, therefore internal thermal stresses alone can not break down this specific composite. In the case of additional action of external loading the stress fluctuation can involve decreasing the level of fracture loading (see as an example [689]), but the consideration of this problem is beyond the scope of present analysis.

15.4 Concluding Remarks

The scheme presented in Sections 13.4 and 15.3 can be generalized easily for the case of the elastic mismatch of inclusions in the framework of the method of successive approximations with the realization of the MEFM at each iteration. So, the initial problem for multiple interacting inclusions (8.86), analogous to the macrocrack-microcracks interaction problem considered in [529], should be accompanied by additional equation for the effective field for the crack $\overline{\boldsymbol{\sigma}}^{\text{crack}}(\mathbf{s})$, or, what is the same – for the traction $\mathbf{t}(\mathbf{s})$, coupled with the SIF by the use of explicit integral representation of the tensor $\boldsymbol{\mathcal{L}}^{\text{crack}}$ (4.42). The further realization of the MEF method (12.2) should be accompanied by as additional equation for an isolated single crack subjected to an unknown effective field $\overline{\boldsymbol{\sigma}}^{\text{crack}}(\mathbf{s})$, that makes the problem essentially nonlocal one even for statistically homogeneous inclusion field. In so doing, the application of the iteration method that is simplified in the framework of "quasi-crystalline" approximation (8.65) is preferable.

A problem of damage processes in brittle matrix composites under tension parallel to the fiber direction [32], [205], [571], [749], [791], [792], [849] is of considerable interest in practice. After crack nucleation, two models of damage evaluation are possible: multiple cracking and crack expansion. The first model can be described in a spirit of the standard models for multicomponent composites. The second model is determination of an appropriate force-separation law that characterized the restraining traction provided by the fibers. The popular simplified models for overall responses of composites with bridged cracks have the following drawbacks: two-dimensional bridging analysis leading to the neglect of the spatial variation of crack tip behavior, replacement of surrounding fiber reinforced composites by a homogeneous isotropic material with effective properties (see, e.g., [205], [341]), a continuous bridging modeling replacing the action of the fibers by a smeared bridging stress distribution [792]. The exact and approximate solution for the one and deterministic field of fibers bridging the crack [205], [336], [341] are known. The stochastic aspects of damage processes are valid ones for further investigation.

It should be mentioned that there is some similarity between Eq. (12.2) for microinhomogeneous medium and equations involved for bridged crack model that coincide between one to other except the notations in the framework of simplified fiber-bridged crack model [205] assuming a coincidence of elastic properties of the matrix and fibers as well as a small variation of bridging force inside each fiber (analog of hypothesis **H1**). Then in Eq. (12.2) the integral over the whole space should be replaced by the integral over the crack surface $\mathbf{s} \in S$. The tensor $\mathbf{B}_i(\mathbf{s}_i)$ (8.18) is estimated from the solution for a single fiber bridging a crack and the matrix of binary interactions $\mathbf{Z}_{ij}(\mathbf{s}_i, \mathbf{s}_j)$ are also known. The knowledge of tensors mentioned makes it possible to utilize the scheme of a two-particle approximation of the MEFM for finding of the nonlocal dependence of bridging forces $\boldsymbol{\sigma}(\mathbf{s})$ on the applied stress field $\boldsymbol{\sigma}^0(\mathbf{s})$. The subsequent estimation of effective parameters $\langle \mathbf{K}(\mathbf{z})\rangle$ and $\langle \mathcal{J}(\mathbf{z})\rangle$ ($\mathbf{z} \in \Gamma^{\rm crack}$) is obvious. The scheme presented was in actuality realized in [791] in the framework of the Mori-Tanaka method for the homogeneous $\boldsymbol{\sigma}^0 \equiv$ const. and a limiting case of a completely failed matrix material (i.e., two half-spaces connected by fibers are considered). These assumptions allow one to eliminate from consideration all nonlocal effects induced in the general case by the nonhomogeneity of the effective field $\overline{\boldsymbol{\sigma}}(\mathbf{x})$.

16

Nonlinear Composites

In contrast to the extensive work on the elastic behavior of composites, only a limited number of studies for the nonlinear range, in particular for the elastoplastic regime, have been published. Although significant advances have been achieved with respect to disorder and nonlinearity separately, the situation becomes quite complex and interesting when both are important. When plastic deformations occur, the homogeneity of the mechanical properties of the components is lost and the local properties of the phases become position dependent. Exact solutions via analytical or numerical methods can be obtained only for model composites having deterministic phase arrangements, such as composites with regular structure (see, e.g., [3], [95], [298], [803], [805], [804], [1060], [1168]) Alternatively, nonlinear multiphase materials may be described by self-consistent schemes. For example, a special case of infinitely small concentrations of heterogeneities in nonlinear matrix composites was studied [294], [653] and then differential self-consistent schemes were used to obtain approximate constitutive relations applicable for a large range of inclusion concentrations. In addition, effective medium methods were employed for carrying out analysis of elastoplastic polycrystals [487], [808]. The three-phase spherical model [432] was used in [1230] to predict the creep behavior of metal–matrix composites (see also [100], [457]) and in [837] for elastoplastic analysis of thermal cycling of particulate composites. A principal role of the concentration of plastic strains in the vicinity of the inclusions in the overall deformation properties of composite materials was shown [454], [1063]. The method [74] was extended in [1216] for linear problem and made a representation of the effective nonlinear dielectric constant and cubic susceptibility which effectively used a perturbation solution. Agarwal and Dutta Gupta [8] proposed a single-grain T-matrix approach for the calculation of the third- and fifth-order nonlinear susceptibilities of a heterogeneous medium. A determination of the third-order elastic constant by using the perturbation approximation was obtained in [27].

Other noteworthy achievements in the theory of nonlinear composite are connected with generalizations of the Hashin–Shtrikman [441] variational principles to nonlinear materials, as proposed in [1070] (for references see [897]). Bounds for the effective properties of a broad class of nonlinear composites were determined in [764], [1190]. A new variational procedure that can be used to obtain general types of bounds as well as estimates for the effective properties of nonlinear composites was developed [893], [894], [895].

Standard methods of successive approximation may be used in nonlinear problems of composites with random structure, linearized problems being solved at each stage. So, in a classical method of successive approximation the secant tensor is defined by the average strains in the constituent (see, e.g., [27], [161], [162], [568]). The well-known concept of secant moduli has been combined with the hypothesis of the homogeneity of the increments of the plastic or creep strains [80], [162], [298], [1075]. Since the widely used method of average strains is capable of estimating only the average stresses in the components, its use for the linearization of functions describing nonlinear effects, e.g., strength ([19], [930], yielding [687], hardening [910], creep [1230], and dynamic viscoplasticity [1231] may be problematical. Owing to the significant inhomogeneity of the stress fields in the components (especially in the matrix), such linearization entails physical inconsistencies, which will be discussed within the context of predicting the flow behavior of porous media.

When the investigations are conducted with the aim of determining effective nonlinear elastic parameters, a concept of secant modulus is used. But at each step the linearized problem is solved with the help of the MEFM which has none of the disadvantages of Mori-Tanaka's method (for details see Chapter 8). In the modified approach the nonlinear dependence of elastic properties on the strain field is taken into account and one assumes that the effective strain for the phase being considered equals the second moment of strains averaged over this phase [129], [130], [1058]. In doing so the exact estimates [602], [860] (as distinct from those contained in [80], [910], [1075]) in relation to the average in volume of anisotropic components single-point second moments of the stress fields are used. A similar approach was employed previously in [129], [130] for estimating the overall moduli of nonlinear elastic composites. Finally, the application of the proposed theory to the case of different problems will be considered (following Buryachenko [131]): the effective moduli, the yield surface, the power-law creep response and the elastoplastic deformation of composites with compressible and incompressible matrix are predicted. The approach described here has been applied for simulating the monotonic and cyclic thermomechanical behavior of a multiphase model material. Onset of plastic deformations, overall hardening behavior, shake down, and ratchetting are studied.

16.1 Nonlinear Elastic Composites

16.1.1 Popular Linearization Scheme

We consider statistically homogeneous matrix composites consisting of a homogeneous matrix containing a statistically uniform set of inclusions. The local strain and stress tensors satisfy the linearized strain-displacement (6.3) relations and the equilibrium equations (6.1) (with no body force acting), respectively. We consider a mesodomain w containing a statistically large number of inclusions and subjected to the homogeneous boundary conditions (6.5) generating a homogeneous macroscopic stress $\boldsymbol{\sigma}^0 = \langle\boldsymbol{\sigma}\rangle$; the phases are assumed to be perfectly connected.

We will consider a constitutive equation

$$\boldsymbol{\sigma}(\mathbf{x}) = \mathbf{L}(\boldsymbol{\varepsilon}, \mathbf{x}) + \boldsymbol{\mathcal{L}}(\boldsymbol{\varepsilon}, \mathbf{x})\boldsymbol{\varepsilon} \otimes \boldsymbol{\varepsilon}, \tag{16.1}$$

where $\otimes$ stands for a tensor product. $\mathbf{L}(\boldsymbol{\varepsilon}, \mathbf{x})$ and $\boldsymbol{\mathcal{L}}(\boldsymbol{\varepsilon}, \mathbf{x})$ are the second-order elastic compliance and the third order elastic stiffness, respectively. Hereafter we assume that $\mathbf{L}(\boldsymbol{\varepsilon}, \mathbf{x})$, $\boldsymbol{\mathcal{L}}(\boldsymbol{\varepsilon}, \mathbf{x})$ on depend on $\boldsymbol{\varepsilon}$ only through the equivalent strains $\varepsilon_k^e(\mathbf{x}) : \mathbf{L}(\boldsymbol{\varepsilon}, \mathbf{x}) \equiv \mathbf{L}(\varepsilon_1^e, \mathbf{x})$, $\boldsymbol{\mathcal{L}}(\boldsymbol{\varepsilon}, \mathbf{x}) \equiv \boldsymbol{\mathcal{L}}(\varepsilon_2^e, \mathbf{x})$ $(k = 1, 2)$ and: $\varepsilon_k^e = \mathbf{E}_k\boldsymbol{\varepsilon} + \boldsymbol{\mathcal{E}}_k\boldsymbol{\varepsilon}\otimes\boldsymbol{\varepsilon}$, where $\mathbf{E}_k$, $\boldsymbol{\mathcal{E}}_k$ $(k = 1, 2)$ are the some second-order and fourth-order tensors, respectively.

In particular, for the Murnaghan stain potential w (2.78) we obtain from expression $\boldsymbol{\sigma} = \partial w/\partial \boldsymbol{\varepsilon}$

$$\begin{aligned} &L_{ijkl}(\boldsymbol{\varepsilon}) = (3k, 2\mu) \equiv 3k\mathbf{N}_1 + 2\mu\mathbf{N}_2, \quad \mathbf{E}_1 = \mathbf{E}_2 = \mathbf{0}, \quad \boldsymbol{\mathcal{E}}_1 = \boldsymbol{\mathcal{E}}_2 = \mathbf{0}, \\ &\mathcal{L}_{ijklmn} = a\delta_{ij}\delta_{kl}\delta_{mn} + b(\delta_{ij}I_{klmn} + \delta_{mn}I_{ijkl} + \delta_{kl}I_{mnij}) + dJ_{ijklmn}, \end{aligned} \tag{16.2}$$

where $J_{ijklmn} = (I_{ipmn}I_{jpkl} + I_{ipkl}I_{jpmn})/2$. For the Kauderer strain potential w (2.83) we have

$$\mathbf{L}(\boldsymbol{\varepsilon}) = \mathbf{L}^{[0]} + \psi(\boldsymbol{\varepsilon}), \quad \boldsymbol{\mathcal{L}} \equiv \mathbf{0}, \quad \mathbf{L}^{[0]} = (3k, 2\mu), \quad \psi(\boldsymbol{\varepsilon}) = 2\gamma\varepsilon_{\text{eq}}^2\mathbf{N}_2, \tag{16.3}$$

where $\varepsilon_{\text{eq}}^2 \equiv 2\mathbf{e} : \mathbf{e}/3$, $\mathbf{e} \equiv \boldsymbol{\varepsilon} - \varepsilon_0\boldsymbol{\delta}$, $\varepsilon_0 \equiv \varepsilon_{ii}/3$, and γ is a nonlinear parameter.

The tensor of material properties $\mathbf{f} \equiv \mathbf{f}^{(0)} + \mathbf{f}_1(\mathbf{x})$ $(\mathbf{f} = \mathbf{L}, \mathbf{M}, \boldsymbol{\mathcal{L}}, \mathbf{M} \equiv \mathbf{L}^{-1})$ is assumed to be constant in the matrix $v^{(0)} = w\backslash v$ $(v \equiv \cup v_i,\ i = 1, 2, \ldots)$ $\mathbf{f}(\mathbf{x}) \equiv \mathbf{f}^{(0)}$ and in each inclusion phase v_i $(i = 1, 2, \ldots)$ $\mathbf{f}(\mathbf{x}) = \mathbf{f}^{(0)} + \mathbf{f}_1^{(i)}$; hereinafter the upper index of materials properties tensor, given in the parentheses, shows the number of the respective components. The rule of calculation of these piecewise constant tensors is described below. We recast the constitutive equation in the form

$$\boldsymbol{\sigma}(\mathbf{x}) = \mathbf{L}(\boldsymbol{\varepsilon}, \mathbf{x})\boldsymbol{\varepsilon}(\mathbf{x}) + \boldsymbol{\alpha}(\boldsymbol{\varepsilon}, \mathbf{x}), \quad \boldsymbol{\alpha}(\boldsymbol{\varepsilon}, \mathbf{x}) = \boldsymbol{\mathcal{L}}(\boldsymbol{\varepsilon}, \mathbf{x})\boldsymbol{\varepsilon} \otimes \boldsymbol{\varepsilon}, \tag{16.4}$$

where $k = 0, 1, \ldots, N$. In such a case Eq. (16.4) formally coincides with an appropriate equation of thermoelasticity (6.2). The effective moduli of the second and third orders in the relation $\langle\boldsymbol{\sigma}\rangle = \mathbf{L}^*\langle\boldsymbol{\varepsilon}\rangle + \boldsymbol{\mathcal{L}}^*\langle\boldsymbol{\varepsilon}\rangle \otimes \langle\boldsymbol{\varepsilon}\rangle$ can be found by averaging local equation (16.4):

$$\mathbf{L}^* = \mathbf{L}^{(0)} + \langle\mathbf{L}_1\mathbf{A}^*\rangle, \quad \boldsymbol{\mathcal{L}}^* = \sum_{i=1}^N \langle\mathbf{L}_1^{(i)}\boldsymbol{\mathcal{F}}_1^{(i)}\rangle^{(i)} + \sum_{k=0}^N \langle\boldsymbol{\mathcal{L}}^{(k)}\rangle\boldsymbol{\mathcal{F}}_2^{(k)}, \tag{16.5}$$

where the tensors of the fourth rank $\mathbf{A}^*$, the sixth rank $\boldsymbol{\mathcal{F}}_1$, and the eighth rank $\boldsymbol{\mathcal{F}}_2$ define the average strain concentrations in a component $\mathbf{x} \in v^{(k)}$ $(k = 0, 1, \ldots, N)$ $\langle\boldsymbol{\varepsilon}\rangle^{(k)} = \mathbf{A}^*(\mathbf{x}) + \boldsymbol{\mathcal{F}}_1(\mathbf{x})\langle\boldsymbol{\varepsilon}\rangle \otimes \langle\boldsymbol{\varepsilon}\rangle$, $\langle\boldsymbol{\varepsilon} \otimes \boldsymbol{\varepsilon}\rangle^{(k)} = \boldsymbol{\mathcal{F}}_2(\mathbf{x})\langle\boldsymbol{\varepsilon}\rangle \otimes \langle\boldsymbol{\varepsilon}\rangle$.

To obtain final results that can be visualized, we adopt linearization of the Eqs. (16.1), (16.4) and subsequent equations, which assumes the homogeneity of $\boldsymbol{\varepsilon}(\mathbf{x}) \otimes \boldsymbol{\varepsilon}(\mathbf{x})$, $\mathbf{L}(\boldsymbol{\varepsilon}, \mathbf{x})$, and $\boldsymbol{\mathcal{L}}(\boldsymbol{\varepsilon}, \mathbf{x})$ within the phase $v^{(k)}$ $(k = 0, 1, \ldots, N)$: $\boldsymbol{\varepsilon}(\mathbf{x}) \otimes$

$\varepsilon(\mathbf{x}) \equiv \langle \boldsymbol{\varepsilon}(\mathbf{x}) \otimes \boldsymbol{\varepsilon}(\mathbf{x}) \rangle^{(k)}$, $\mathbf{L}(\boldsymbol{\varepsilon}, \mathbf{x}) = \mathbf{L}(\langle \varepsilon_1^e \rangle^{(k)}, \mathbf{x})$, and $\boldsymbol{\mathcal{L}}(\boldsymbol{\varepsilon}, \mathbf{x}) \equiv \boldsymbol{\mathcal{L}}(\langle \varepsilon_2^e \rangle^{(k)}, \mathbf{x})$ at $\mathbf{x} \in v^{(k)}$. This method of linearization assumes homogeneity of material parameters within the phases $\mathbf{x} \in v^{(k)},\ k = 0, 1, \ldots, N)$:

$$\mathbf{L}(\boldsymbol{\varepsilon}, \mathbf{x}) \equiv \mathbf{L}^{s(k)} = \mathbf{L}(\langle \varepsilon_1^e \rangle^{(k)}),\ \boldsymbol{\mathcal{L}}(\boldsymbol{\varepsilon}, \mathbf{x}) \equiv \boldsymbol{\mathcal{L}}^{s(i)} = \boldsymbol{\mathcal{L}}(\langle \varepsilon_2^e \rangle^{(k)}). \tag{16.6}$$

Hereafter the superscript ^{s} indicates the calculation of the values the help of L^s . The method of estimation of secant modulus $\mathbf{L}^{s(k)},\ \boldsymbol{\mathcal{L}}^{s(k)},\ (k = 0, 1, \ldots)$ is based on application of a method of successive approximation for the solution of nonlinear problems (see, e.g., [831], [397]). The constitutive equation (16.4) then yields

$$\boldsymbol{\sigma}^{[1]} = \mathbf{L}^{s[1]} \boldsymbol{\varepsilon}^{[1]}, \tag{16.7}$$

$$\boldsymbol{\sigma}^{[2]} = \mathbf{L}^{s[2]} \boldsymbol{\varepsilon}^{[2]} + \boldsymbol{\alpha}^{[2]},\ \boldsymbol{\alpha}^{[2]} \equiv \boldsymbol{\mathcal{L}}^{s[2]} \boldsymbol{\varepsilon}^{[1]} \otimes \boldsymbol{\varepsilon}^{[1]}, \ldots \tag{16.8}$$

$$\boldsymbol{\sigma}^{[n+1]} = \mathbf{L}^{s[n+1]} \boldsymbol{\varepsilon}^{[n+1]} + \boldsymbol{\alpha}^{[n+1]},\ \ \boldsymbol{\alpha}^{[n+1]} \equiv \boldsymbol{\mathcal{L}}^{s[n+1]} \boldsymbol{\varepsilon}^{[n]} \otimes \boldsymbol{\varepsilon}^{[n]}, \tag{16.9}$$

where the upper index in parentheses indicates the iteration number of the tensors.

The tensors $\mathbf{L}^s$ and $\boldsymbol{\mathcal{L}}^s$ are not known in general spiking. The problem of their estimation even at a local level is a substantially nonlinear one because of nonlinear dependence between strains and stresses. Since the secant modulus $\mathbf{L}^{s(i)}$ and $\boldsymbol{\mathcal{L}}^{s(i)}$ (and consequently $\mathbf{L}^{s*}$ and $\boldsymbol{\mathcal{L}}^{s*}$) are not known *a priori*, an iterative procedure is usually required. We may start by assuming $(\mathbf{x} \in v^{(k)},\ k = 0, 1, \ldots)$: $\mathbf{L}^{s(k)[0]}(\mathbf{x}) \equiv \mathbf{L}^{(k)}(\mathbf{0}, \mathbf{x}),\ \boldsymbol{\mathcal{L}}^{s(k)[0]}(\mathbf{x}) \equiv \boldsymbol{\mathcal{L}}^{(k)}(\mathbf{0}, \mathbf{x}),\ \boldsymbol{\alpha}^{[0]} = \mathbf{0}$, and calculate the corresponding $\mathbf{L}^{*[0]}$ and $\boldsymbol{\mathcal{L}}^{*[0]}$ by the appropriate equations of Chapter 8. The other secant modulus then follows from equations $(\mathbf{x} \in v^{(k)};\ k = 0, 1, \ldots;\ n = 0, 1, \ldots)$:

$$\mathbf{L}^{s(k)[n+1]}(\mathbf{x}) = \mathbf{L}(\langle \varepsilon_1^{e[n]} \rangle^{(i)}, \mathbf{x}),\ \boldsymbol{\mathcal{L}}^{s(k)[n+1]}(\mathbf{x}) = \boldsymbol{\mathcal{L}}(\langle \varepsilon_2^{e[n]} \rangle^{(i)}, \mathbf{x}), \tag{16.10}$$

$$\boldsymbol{\varepsilon}^{(k)[n+1]}(\mathbf{x}) \otimes \boldsymbol{\varepsilon}^{(k)[n+1]}(\mathbf{x}) = \langle \boldsymbol{\varepsilon} \otimes \boldsymbol{\varepsilon} \rangle^{(k)[n]} \tag{16.11}$$

where $[n]$ is the number of an iteration; $\langle \varepsilon_1^{e[n]} \rangle^{(k)},\ \langle \varepsilon_2^{e[n]} \rangle^{(k)}$ are defined by $\mathbf{L}^{s*} = \mathbf{L}^{s*[n]},\ \boldsymbol{\mathcal{L}}^{s(k)} = \boldsymbol{\mathcal{L}}^{s(k)[n]}$. If $||\mathbf{I} - \mathbf{L}^{s*[n]}/\mathbf{L}^{s*[n]}|| \equiv \xi \ll 1$ then the solution is found; usually $n \leq 5$ at $\xi = 10^{-2}$.

A widely distributed method of linearization in nonlinear elasticity problems assumes homogeneity the field of stresses or strains (contrary to Eq. (1/6)) when computing material tensors, which describe nonlinear effects. One usually equates to terms $(\mathbf{x} \in v^{(k)};\ k = 0, 1, \ldots)$:

$$\mathbf{L}(\mathbf{x}) = \mathbf{L}^{s(k)} \equiv \mathbf{L}(\mathbf{E}_1 \langle \boldsymbol{\varepsilon} \rangle^{(k)} + \boldsymbol{\mathcal{E}}_1 \langle \boldsymbol{\varepsilon} \rangle^{(k)} \otimes \langle \boldsymbol{\varepsilon} \rangle^{(k)}), \tag{16.12}$$

$$\boldsymbol{\mathcal{L}}(\mathbf{x}) = \boldsymbol{\mathcal{L}}^{s(k)}(\mathbf{x}) \equiv \boldsymbol{\mathcal{L}}(\mathbf{E}_2 \langle \boldsymbol{\varepsilon} \rangle^{(k)} + \boldsymbol{\mathcal{E}}_2 \langle \boldsymbol{\varepsilon} \rangle^{(k)} \otimes \langle \boldsymbol{\varepsilon} \rangle^{(k)}). \tag{16.13}$$

Then for the Murnaghan strain potential w (2.78) with the elastic moduli (16.2) we get for the second-order approximation (16.8): $\boldsymbol{\mathcal{L}}^* = \sum_{k=0}^{N} c_i \boldsymbol{\mathcal{L}}^{(k)}(\mathbf{A}_k^* \otimes \mathbf{A}_k^* \otimes \mathbf{A}_k^*)$, where the strain concentrator factors $\mathbf{A}_k^*$ are found for the first-order approximation (16.7) by some averaging method prescribed in Chapter 8. In particularly for the isotropic composites with the components which are described by elastic potential (2.78) we can write [161]

$$\mathcal{L}^*_{ijklmn} = a^*\delta_{ij}\delta_{lk}\delta_{mn} + b^*(\delta_{ij}I_{klmn} + \delta_{mn}I_{ijkl} + \delta_{kl}I_{mnij}) + d^*J_{ijklmn}, \quad (16.14)$$

where

$$a^* = \sum_{k=0}^{N} 3c^{(k)}\left[9a^{(k)}r_k^3 + 3b^{(k)}g_k r_k(3r_k + 2s_k) + d^{(k)}g_k^2(g_k + 2s_k)\right], \quad (16.15)$$

$$b^* = \sum_{k=0}^{N} c^{(k)}(2s_k)^2(3b^{(k)}r_k + d^{(k)}g_k), \quad d^* = \sum_{k=0}^{N} c^{(k)}d^{(k)}(2s_k)^3, \quad (16.16)$$

and $\mathbf{A}^*_k \equiv (3r_k, 2s_k)$, $3g_k \equiv 3r_k - 2s_k$, $\mathcal{L}^{(k)}_{ijmnpq} = a^{(k)}\delta_{ij}\delta_{pq}\delta_{mn} + b^{(k)}(\delta_{ij}I_{mnpq} + \delta_{pq}I_{ijmn} + \delta_{mn}I_{ijpq}) + d^{(k)}J_{ijmnpq}$ $(k = 0, 1, \ldots)$. The expression (16.14) is a generalization of the corresponding relation [740] to an arbitrary number of components with a different estimation method of $\mathbf{A}^*_i$. Figure 16.1 plots $d^* \sim c^{(1)}$ (curve 1) calculated from (16.14) with $\mathbf{A}^*_k$ estimated by the MEFM (8.48) for 09G2S steel with identical spherical pores and the following parameters (GPa) $\lambda^{(0)} = 94.4$, $\mu^{(0)} = 79$, $a^{(0)} = -825$, $b^{(0)} = -309$, $d^{(0)} = -799$. The values of $d^* \sim c^{(1)}$ at $c^{(1)} = 0.4$ on curve 1 in Fig. 16.1 are greater by 20% than the estimate by the method of conditional moments [740]. We should note that for a porous medium the ratio of d^* values based on (16.14) to those calculated by the method in [740] is equal to the cube of s_0. Therefore, the difference in estimates of d^* by (16.14) and by method [740] will grow as $k^{(0)}$ increases and as the shape of the inclusions approaches a flat spheroid. Indeed, or spherical pores and $k^{0)} = \infty$, we show in Fig. 16.1 the values of $d^* \sim c^{(1)}$ (curves 2 and 3) calculated from formulae [740] and from (16.14), (8.48), respectively.

Let us consider a porous material with the matrix described by the Kauderer potential (16.3) in the framework of the linearization scheme (16.12) with $\boldsymbol{\alpha} \equiv \mathbf{0}$. As computations of model examples showed, 5–7 iterations ($n = 5 \div 7$, see (16.9)) ensure less than 1% error of the estimate ψ^*. Shown in Fig. 16.2 are curves 1 and 2

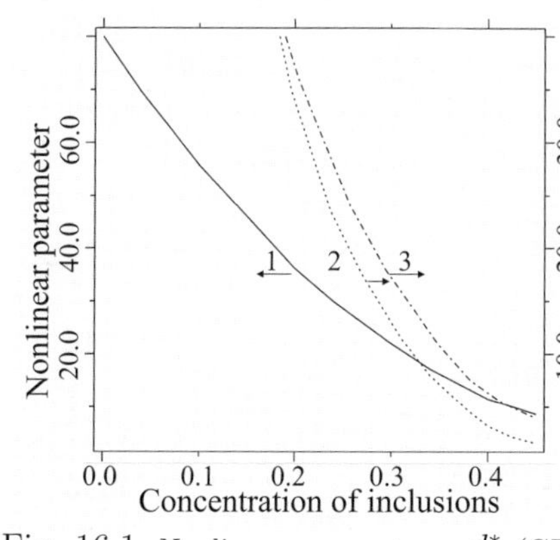

Fig. 16.1. Nonlinear parameter $-d^*$ (GPa) vs $c^{(1)}$: use of the MEFM for porous steel (curve 1), use of the MEFM (curve 2) and the method of condition moments (curve 3) for the model material.

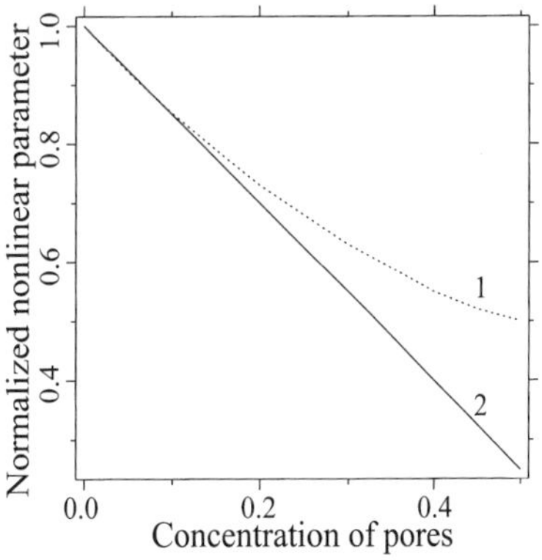

Fig. 16.2. Variation on the $\gamma^*/\gamma^{(0)}$ vs $c^{(1)}$ estimated by the method of conditional moments (curve 1) and by the MEFM (curve 2).

with respect to the change in the nonlinearity parameter $\gamma^*/\gamma^{(0)}$ of copper with spherical pores computed by means of the method of conditional moments [743] and by the point approximation (4.64) of the MEFM (8.48).

Hypothesis (16.12) is inconsistent for potential (2.78). In fact, let us consider an isotropical porous material with a matrix, the elastic properties of which are described by the stress potential (2.78). In such a case $\langle\varepsilon_{ij}\rangle = \langle\varepsilon_0\rangle\delta_{ij}$ at the hydrostatic loading condition irrespective of microstructure of pores and the method of calculation of $\langle\varepsilon\rangle^{(0)}$ (for example, by the formula (8.48) or by any other formula) we obtain that $\langle\mathbf{e}\rangle^{(0)} \equiv \mathbf{0}$ ($\mathbf{e} \equiv \varepsilon - \varepsilon_0\boldsymbol{\delta}$). It means that this porous material with incompressible nonlinear elastic modulus has only linear elastic deformations at hydrostatic compression.

16.1.2 Modified Linearization Scheme

The above mentioned inconsistency can be easily avoided if the linearizations (16.12) and (16.13) are replaced by the linearizations

$$\mathbf{L}(\mathbf{x}) = \mathbf{L}^{s(k)} \equiv \mathbf{L}(\mathbf{E}_1\langle\varepsilon\rangle^{(k)} + \boldsymbol{\mathcal{E}}_1\langle\varepsilon\otimes\varepsilon\rangle^{(k)}), \tag{16.17}$$

$$\boldsymbol{\mathcal{L}}(\mathbf{x}) = \boldsymbol{\mathcal{L}}^{s(k)}(\mathbf{x}) \equiv \boldsymbol{\mathcal{L}}(\mathbf{E}_2\langle\varepsilon\rangle^{(k)} + \boldsymbol{\mathcal{E}}_2\langle\varepsilon\otimes\varepsilon\rangle^{(k)}). \tag{16.18}$$

where $k = 0, 1, \ldots, N$, and the second moment of strains $\langle\varepsilon\otimes\varepsilon\rangle^{(k)}$ are estimated by one of the methods considered in Chapter 13. In particular, at the second-order approximation (16.8) this value can be estimated by the perturbation method (13.20) for the purely elastic case ($\boldsymbol{\alpha} \equiv \mathbf{0}$):

$$\langle\varepsilon\otimes\varepsilon\rangle^{(k)[1]} = \frac{1}{c^{(k)}}\frac{\partial\mathbf{L}^{s*[0]}}{\partial\mathbf{L}^{s(k)[0]}}\langle\varepsilon\rangle\otimes\langle\varepsilon\rangle, \tag{16.19}$$

while the subsequent approximations $\langle\varepsilon\otimes\varepsilon\rangle^{(k)[n]}$ should be estimated through consideration of an appropriate thermoelastic problem (like Eq. (13.15)).

Let us consider, as an example, a porous material containing the spherical identical voids in the incompressible matrix with the elastic properties described by the Kauderer potential (16.3). Then we have an initial approximation of the effective moduli $\mathbf{L}^{*[0]} = \mu^{(0)}(4f_1, 2f_2)$, where

$$f_1 = \frac{1}{c} - \frac{29}{24}, \quad f_2 = (1 - \frac{35c}{24})\Big(1 + \frac{5c}{24}\Big)^{-1}, \tag{16.20}$$

$$f_1 = \frac{1}{3}\Big(\frac{1}{c} - \frac{16}{15\pi^2}\Big), \quad f_2 = \Big(1 - \frac{448c}{375\pi}\Big)\Big(1 - \frac{848c}{375\pi^2}\Big)^{-1}, \tag{16.21}$$

for the spherical pores ($a_1 = a_2 = a_3 = a$, $c = 4/3\pi a^3 n$) and coin-shaped cracks ($a_1 = a_2 = a \gg a_3$, $c = 4/3\pi a^3 n$), respectively, are found by the point approximation of the MEFM (9.106), (9.108). Then the effective isotropic tensor $\boldsymbol{\psi}^*$ (16.3) obtained from linearization conditions (16.12) and (16.17), (16.19) are given in [130] $\boldsymbol{\psi}^* = \gamma^{(0)}(0, f_2^2\langle\varepsilon\rangle^2_{\rm eq})$, and $\boldsymbol{\psi}^* = \gamma^{(0)}(1-c)^{-1}(16f_1^2\langle\varepsilon_0\rangle^2 + 2f_1f_2\langle\varepsilon\rangle^2_{\rm eq}, 8f_1f_2\langle\varepsilon_0\rangle^2 + f_2^2\langle\varepsilon\rangle^2_{\rm eq})$, respectively, where $\langle\varepsilon\rangle^2_{\rm eq} = 2\langle\mathbf{e}\rangle\langle\mathbf{e}\rangle/3$, $\varepsilon_0 =$

$\varepsilon_{ii}/3$, $\mathbf{e} = \boldsymbol{\varepsilon} - \varepsilon_0\boldsymbol{\delta}$. Due to the incompressibility of the matrix the obtained representations are consistent with an infinite number of iterations (16.10).

Now let us consider a composite material with only one phase demonstrating nonlinear properties $\boldsymbol{\sigma}(\mathbf{x}) = \mathbf{L}^{s(i_1)}\boldsymbol{\varepsilon}(\mathbf{x})$, $\mathbf{L}^{s(i_1)} = \mathbf{L}(\mathbf{x}, \varepsilon_1^e)$, $\mathbf{x} \in v^{(i_1)}$, while the linear law $\boldsymbol{\sigma}(\mathbf{x}) = \mathbf{L}^{s(j)[0]}\boldsymbol{\varepsilon}(\mathbf{x})$, $\mathbf{x} \in v^{(j)}$, $(j = 1, \ldots, N;\ j \neq i_1)$ holds good for other phases. In such a case the general iteration scheme (16.10) can be simplified. Indeed, we assume that a monotonically increasing proportional loading

$$\langle \varepsilon_{ij}(t) \rangle = \xi_{ij}\varepsilon^t(t) \tag{16.22}$$

takes place, where ξ_{ij} is a desired proportional constant and $\varepsilon^t(t)$ is a monotonically increasing scalar function depending on a time parameter t. At the first step we will give a small increment of the second strain invariant $\langle \varepsilon \rangle_{\text{eq}}^{(i_1)2}$ in the phase $v^{(i_1)}$. Thereafter we will estimate the secant modulus values according to Eqs. (16.17), (16.19) and find $\mathbf{L}^{s*}$ by the use of the determined $\mathbf{L}^{s(i_1)}$. For the desired proportional loading (16.22), the value $\varepsilon^t(t)$ is given by

$$\varepsilon^t(t) = \left[\frac{1}{3c^{(i_1)}}\frac{\partial \mathbf{L}^{s*}}{\partial \mu^{s(i_1)}}\langle \boldsymbol{\xi} \rangle \otimes \langle \boldsymbol{\xi} \rangle\right]^{-1/2}, \tag{16.23}$$

which provides a way of estimating the components $\langle \varepsilon_{ij} \rangle$ from (16.22). The average stresses $\langle \boldsymbol{\sigma} \rangle$ can be found from the equation $\langle \boldsymbol{\sigma} \rangle = \mathbf{L}^{s*}\langle \boldsymbol{\varepsilon} \rangle$ through the known tensors $\mathbf{L}^{s*}$ and $\langle \boldsymbol{\varepsilon} \rangle$. The second step of deformations at the stress–strain curve can be performed by increasing the values $\langle \varepsilon \rangle_{\text{eq}}^{(i_1)2}$. This procedure continues up to a value of $\varepsilon^t(t) = 1$.

It should be mentioned that in contrast, the problem involving geometrically nonlinear materials has hardly been investigated, which is understandable given the inherent mathematical difficulties that lead to some simplifications of the models proposed. So, Hashin [434] used the composite spheres assemblage model, taking the large strains of the spherical element into account. To derive the second-order elasticity equations one uses the method of successive approximation [81] with a power series expansion of the displacements, stresses, and their gradients in a certain small parameter. In so doing, the case of dilute concentration of spherical inclusions was considered in [491], [714], and [831]. The case of a finite concentration of ellipsoidal inclusions was analyzed in [741], [742] (see also [566] by the method of conditional moments (linear versions of this method were described in Chapter 8) at each step of successive approximation; the average strains in the components estimated at the first step were used at the second step or, in other words, one used the linearization (16.9). A similar approach with the construction of infinite number of recursions in the scheme of successive approximation method was realized in [614] by the perturbation method combined with the assumption $\boldsymbol{\Gamma}(\mathbf{x}) = -\mathbf{Q}\delta(\mathbf{x})$.

In more detail we will consider the model [158] generalizing the Hashin's [434] model which presumes equivalent strain problems for a porous medium and a thick-walled spherical shell and equality of the ratio of the volumes of the pore and spherical element to the porosity of the composite medium being modeled. The model by [158] uses the advantage of the MEFM and places the thick-walled spherical elements $v^{\rm sp}$ in the matrix with prescribed effective stress field at infinity. The internal R_1 and external R_2 radii of the shell element in the undeformed state transform into $r_1 = \lambda_1 R_1$ and $r_2 = \lambda_2 R_2$, where the extension parameter λ determines the relation describing the distances of points of the element to the center of the shell before and after deformation $R = \lambda^{-1} r$. The elastic properties of the matrix are described by the Mooney potential with constants C_1 and C_2 (2.74): $w^M = C_1(I_\epsilon - 3) + C_2(II_\epsilon - 3)$, where $I_\epsilon = 3\lambda^2$, $II_\epsilon = 3\lambda^4$, and $III_\epsilon \equiv 1$ are the invariants associated with the Lagrangian strain tensor under conditions of central-symmetric deformation. In the case of loading of the shell by internal and external pressures p_1 and p_2, the solution for large strains of the spherical element is known [398], [434]:

$$-p_2 = C_1[1/\lambda_2^4 + 4/\lambda_2 - (1/\lambda_1^4 + 4/\lambda_1)] + 2C_2[1/\lambda_2^2 - 2\lambda_2 - (1/\lambda_1^2 - 2\lambda_1)] - p_1. \tag{16.24}$$

For incompressible materials fitting the equality $r_1^3 - R_1^3 = r_2 - R_2^3$, we find $\lambda_1^3 = (\lambda_2^3 - 1)/\gamma + 1$, where the parameter $\gamma = R_1^3/R_2^3$ characterizes the relative fraction of the pore volume in the undeformed spherical elements $v^{\rm sp}$. Equation (16.24) allows us to use the assigned values of p_1 and p_2 to find λ_1 and λ_2. We henceforth choose γ to be small enough so that ξ is small in the expression $\lambda_2 = 1 + \xi$ and $\lambda_1^3 = 3\xi/\gamma + 1$. Then the spherical element can be replaced by a linear elastic sphere whose strain properties are described by Eq. (6.2) $\boldsymbol{\sigma}(\mathbf{x}) = \mathbf{L}(\mathbf{x})[\boldsymbol{\varepsilon}(\mathbf{x}) - \boldsymbol{\beta}(\mathbf{x})]$ with the parameters

$$\mathbf{L}(\mathbf{x}) = \mathbf{L}^{\rm e} = (3k^{\rm e}, 2\mu^{\rm e}), \quad \boldsymbol{\beta}(\mathbf{x}) = \boldsymbol{\beta}^{\rm e} = 3\xi^{\rm e}\boldsymbol{\delta}, \tag{16.25}$$

where $\xi^{\rm e}$ is found from the solution of Eq. (16.24) with $p_2 = 0$ and $k^{\rm e} = p_2/(1 + \xi^{\rm e} - \lambda_2)$; due to the small effect of porosity on μ^* (9.106) with small concentration of pores c in undeformed state , we used the representation (9.106), obtained with small deformation of the pores, for $\mu^{\rm e} = \mu^*$.

In accordance with the physical model of a gas-saturated medium, we will assume that its isotropic deformation of a composite medium consisting of a matrix with the modulus $\mathbf{L}^{(0)} = (3k^{(0)}, 2\mu^{(0)})$, $k^{(0)} = \infty$, $\mu^{(0)} = 2(C_1 + C_2)$ and a random set of spherical inclusions with the parameters $\mathbf{L}^{\rm e}$ and $\boldsymbol{\beta}^{\rm e}$. For the sake of definiteness, we will examine inclusions of one size with the degree of fullness $c^{\rm e} = c/\gamma$. Then the equation of isotropic deformation is described by Eq. (6.61) with the tensors $\mathbf{L}^*$ and $\boldsymbol{\beta}^*$ expressed in a known manner (8.52) and (8.53) in terms of the quantities $\mathbf{L}^{(0)}$, $\mathbf{L}^{\rm e}$, $\boldsymbol{\beta}^{\rm e}$, $c^{\rm e}$. The parameter $\boldsymbol{\beta}^*$ depends on the gas pressure p_1, which in turn is determined by the deformation of the pore phase. In the case where the empirically established mean-value concentration of gas $c^{\rm gas}$ in the macro-region is assumed, then in accordance with the Henry and Mendeleev-Clapeyron laws $p_1 = c^{\rm gas}[(1 - c\lambda_1^3)\Gamma + c\lambda_1^3\mu'/(R'T)]$. Here, the first term in brackets with the Henry constant Γ describes the contribution of the mean concentration of gas dissolved in the solid phase. The second term accounts for the presence of gas with a molecular weight μ' at the temperature T with the gas constant R'.

As an example, we will find at first the average strains $\langle\boldsymbol{\varepsilon}\rangle$ with different values p_1 at $\langle\boldsymbol{\sigma}\rangle = \mathbf{0}$ for the raw rubber with the parameters $C_1 = 0.1$ MPa, $C_2 = 0.01$ MPa [434] and $c = 10^{-3}$. Curves 2 and 3 in Fig. 16.3 (for $\gamma = 10^{-2}$ and $2\cdot 10^{-2}$)

were calculated from nonlinear theory (8.52), (8.53), and (16.25), while curve 1 was calculated from linear relations (8.52), (8.53) with $\boldsymbol{\beta}^* = p_1(\mathbf{M}^* - \mathbf{M}^{(0)})\boldsymbol{\delta}$ and $\mathbf{L}^*$ calculated by Eq. (9.106). Figure 16.3 makes it possible to find the value of c^{gas} necessary for the given loading regime. Curves 1, 4 and 2, 3 in Fig. 16.4 show the bulk strains of the medium at $p_1 = 0$, $c = 10^{-3}$ and $c = 5 \cdot 10^{-3}$, respectively. With a negative hydrostatic stress, the curves $p_2 \sim \langle \varepsilon_{ii} \rangle$ a vertical asymptote $\langle \varepsilon_{ii} \rangle = c$ at small c, while there is no vertical asymptote with isotropic expansion, and even at $c = 10^{-3}$ isotropic expansion of the medium can exceed the value $\langle \varepsilon_{ii} \rangle = 0.1$. Since a material always contains a certain number of pores, it follows that perfectly isotropic inextensible materials do not exist. Moreover, the parameters of different elastic potentials of isotropically deformable materials are generally determined in hydrostatic compression (see [231], [232]). In light of the above analysis, the use of these parameters in the region of large hydrostatic tension would appear to be incorrect.

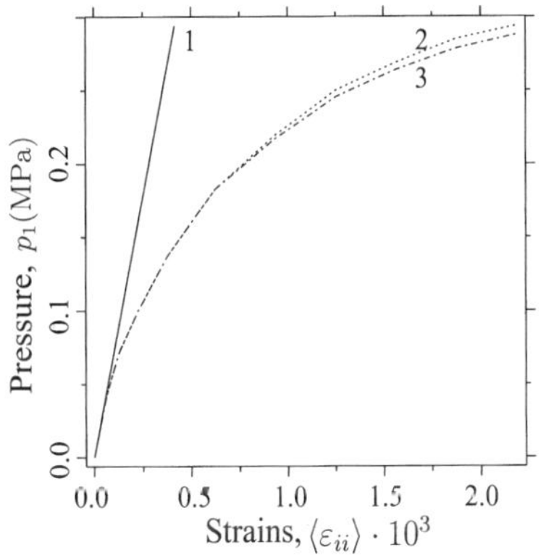

Fig. 16.3. p_1 vs $\langle \varepsilon_{ii} \rangle$ estimated from nonlinear (curves 2, 3), and linear theory (curve 1)

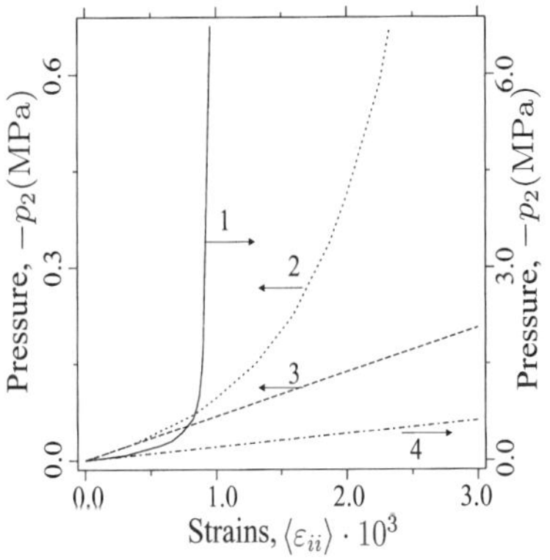

Fig. 16.4. p_2 vs $\langle \varepsilon_{ii} \rangle$ calculated from the linear (curves 3, 4) and nonlinear theory (curves 1, 2)

Other methods of analysis of nonlinear properties are based on the *incremental method* or *tangent modulus concept* (for details see [897]). Constitutive law is written in the incremental form $\dot{\boldsymbol{\sigma}} = \mathbf{L}_t(x, \boldsymbol{\varepsilon})\dot{\boldsymbol{\varepsilon}}$ where the tangent tensors $\mathbf{L}_t(\mathbf{x}, \boldsymbol{\varepsilon})$ are replaced by tensors $\mathbf{L}_t^{(i)}(\varepsilon_1^e)$, which are constant in each phase. The most obvious choice for the effective strain of phase $v^{(i)}$ is the average strain over this phase $\varepsilon(\mathbf{x}) = \langle \boldsymbol{\varepsilon} \rangle^{(i)}$ $(\mathbf{x} \in v^{(i)})$ which can be modified taking the second moment of strains in account. $\mathbf{L}_t^{(i)}$ is generally anisotropic (even for isotropic phases) that yields to known difficulties in the estimation of the effective tangent stiffness $\mathbf{L}^*$.

16.2 Deformation Plasticity Theory of Composite Materials

16.2.1 General Scheme

It is assumed that the rheological properties of composite media with isotropic components are described by the theory of small elastoplastic strains under monotonic, proportional loading. We will not discuss here in general the known advantages of flow theory over deformation theory of plasticity in the case of nonradial loading, which takes place usually at the local level even if the overall stress path is radial.

Specifically, the total strain $\boldsymbol{\varepsilon}$ is written as the sum of elastic $\boldsymbol{\varepsilon}^e$ and plastic $\boldsymbol{\varepsilon}^p$ contributions $\boldsymbol{\varepsilon} = \boldsymbol{\varepsilon}^e + \boldsymbol{\varepsilon}^p$ (2.226). The local equation for the elastic material state, which relates the stress tensor $\boldsymbol{\sigma}(\mathbf{x})$ and the elastic strain tensor $\boldsymbol{\varepsilon}^e(\mathbf{x})$, is given in the form (2.197)

$$\boldsymbol{\sigma}(\mathbf{x}) = \mathbf{L}(\mathbf{x})\boldsymbol{\varepsilon}^e(\mathbf{x}),\ \mathbf{L}(\mathbf{x}) = (3k^{(j)}, 2\mu^{(j)}), \tag{16.26}$$

where $\mathbf{x} \in v^{(j)}$, $j = 0, 1, \ldots, N$. The relation between flow stress and plastic strains is represented by the expression (2.227): $\tau = \tau_0 + f(\varepsilon^p_{\rm eq})$, $f(0) = 0$ in terms of the von Mises effective stress τ and the effective strain $\epsilon^p_{\rm eq}$ (2.228); τ_0 is the initial yield stress and f is a nonlinear function (2.229) describing the material's hardening behavior.

One may determine for each phase j ($j = 0, 1, \ldots, N$) the Young's secant modulus $E^{s(j)}$ (2.232) as well as (due to the plastic incompressibility) the secant Poisson's ratio $\nu^{s(j)}$ and shear modulus $\mu^{s(j)}$ (2.233). Hereafter the superscript ^{s} indicates the calculation of the parameter under consideration with the help of the secant modulus E^s. The effective compliances $\mathbf{M}^{s*}$ and $\mathbf{M}^*$ in the macroscopic equations of state,

$$\mathbf{M}^{s*}\langle\boldsymbol{\sigma}\rangle = \langle\boldsymbol{\varepsilon}\rangle, \quad \mathbf{M}^*\langle\boldsymbol{\sigma}\rangle = \langle\boldsymbol{\varepsilon}^e\rangle, \tag{16.27}$$

are defined by the relations $\mathbf{M}^{s*} = \langle\mathbf{M}^s\mathbf{B}^{s*}\rangle$, $\mathbf{M}^* = \langle\mathbf{M}\mathbf{B}^*\rangle$, where $\mathbf{M}(\mathbf{x}) = \mathbf{L}(\mathbf{x})^{-1}$ is the position dependent compliance tensor. The tensors $\mathbf{B}^{s*}$ and $\mathbf{B}^*$ are found from the solutions of the elastoplastic problem $\boldsymbol{\sigma}(\mathbf{x}) = \mathbf{B}^{s*}(\mathbf{x})\langle\boldsymbol{\sigma}\rangle$ and the elastic problem $\boldsymbol{\sigma}(\mathbf{x}) = \mathbf{B}^*(\mathbf{x})\langle\boldsymbol{\sigma}\rangle$ (at $E^s = E$), respectively. Once $\mathbf{M}^{s*}$ and $\mathbf{M}^*$ have been found, the plastic strain of the composite can be determined from the unloading (elastic) process, i.e.,

$$\epsilon^p_* = (\mathbf{M}^{s*} - \mathbf{M}^*)\langle\boldsymbol{\sigma}\rangle. \tag{16.28}$$

The composite begins its plastic deformation, when

$$\max_i \frac{\widetilde{\tau}^{(i)}}{\tau_0^{(i)}} = 1,\ \ \widetilde{\tau}^{(i)} \equiv \sqrt{\frac{3}{2}\langle s_{kl}s_{kl}\rangle^{(i)}} \tag{16.29}$$

($i = 0, 1, \ldots, N$) is fulfilled. Up to the moment when yielding begins, the problem of estimating $\widetilde{\tau}^{(i)}$ is linear, which allows the use of closed relations for calculating the yield surface ($i = 0, 1, \ldots, N$)

$$F_y(\langle\boldsymbol{\sigma}\rangle) \equiv \max_i \frac{1}{2\tau_0^{(i)}}\left[\frac{3}{c^{(i)}}\frac{\partial \mathbf{M}^*}{\partial q^{(i)}}\langle\boldsymbol{\sigma}\rangle \otimes \langle\boldsymbol{\sigma}\rangle\right]^{1/2} = 1. \tag{16.30}$$

Equation (16.30) determines the second order surface $F_y(\langle\boldsymbol{\sigma}\rangle) = 1$ for the onset of yielding within the six-dimensional stress space which depends on $\langle\sigma_0\rangle$. The yield surfaces of the components $\widetilde{\tau}^{(i)} = \tau_0^{(i)}$ may be embedded within each other or they may intersect in the space of macrostresses $\langle \boldsymbol{\sigma} \rangle$.

Now we shall deal with the elastoplastic state, when for the sake of definiteness, the surface of i_1 is embedded within the others, or we may examine those loading regimes for which the yield surface i_1 is reached first. In this case the strain law (2.227) is true for a component i_1, whereas Eq. (16.26) holds for the other components. Then the macroscopic equation of state relating the tensors $\boldsymbol{\sigma}$ and $\boldsymbol{\varepsilon}$ may be found by the following simple algorithm.

In the first step we add a small increment to the effective plastic strain $\epsilon_{\rm eq}^{p(i_1)}$ in the component i_1. Then we will estimate $\widetilde{\tau}^{(i_1)}$ and the values of the secant modulus in accordance with the formulae (2.227) and (2.232)

$$\widetilde{\tau}^{(i_1)} = \tau_0^{(i_1)} + f(\varepsilon_{\rm eq}^{p(i_1)}), \quad E^{s(i_1)} = \left[\frac{1}{E} + \frac{\varepsilon_{\rm eq}^{p(i_1)}}{\widetilde{\tau}^{(i_1)}} \right]^{-1}. \tag{16.31}$$

Using E^s we obtain $\mathbf{M}^{s*}$ from Eq. (8.52). For a desired proportional loading $\langle \boldsymbol{\sigma} \rangle = \boldsymbol{\xi} \sigma^t(t)$ ($\sigma^t(t)$ is a monotonically increasing scalar function) the value of $\sigma^t(t)$ is given by

$$\sigma^t(t) = 2\widetilde{\tau}^{(i_1)} \left[\frac{3}{c^{(i_1)}} \frac{\partial \mathbf{M}^{s*}}{\partial q^{s(i_1)}} \boldsymbol{\xi} \otimes \boldsymbol{\xi} \right]^{-1/2}, \tag{16.32}$$

which allows the components σ_{ij}^0 to be determined. By using $\mathbf{M}^{s*}$, $\boldsymbol{\sigma}^0$ and Eqs. (16.27) and (16.28) we find the total strain $\langle \varepsilon \rangle$ and the overall plastic strain ϵ_*^p. By increasing the value $\langle \varepsilon_{\rm eq}^{p(i_1)} \rangle$, the stress–strain curves of the composite during the second step of deformation can be obtained.

This procedure is continued up to the yielding of the next component i_2 (where applicable). This occurs when condition $\widetilde{\tau}^{(i)} = \tau_0^{(i)}$ is met for $i = i_2$:

$$\sigma^t(t) = 2\widetilde{\tau}^{(i_2)} \left[\frac{3}{c^{(i_2)}} \frac{\partial \mathbf{M}^{s*}}{\partial q^{s(i_2)}} \boldsymbol{\xi} \otimes \boldsymbol{\xi} \right]^{-1/2}. \tag{16.33}$$

Let us assume that at the moment of interest the components $i_1, \ldots, i_m$ ($m > 1$) are in the plastic state. In that event the problem of estimating the macro parameters $\mathbf{M}^{s*}$ and ϵ_*^p on the basis of the given values $\varepsilon_{\rm eq}^{p(i)}$ ($i = i_1, \ldots, i_m$) becomes more difficult and is equivalent to a search for a minimum of the m-dimensional function with the components $|\widetilde{\tau}^{(j)} - f(\epsilon_{\rm eq}^{p(i)})|$ ($j = i_1, \ldots, i_m$), which depends on m arguments of $\varepsilon_{\rm eq}^{p(j)}$ ($i = i_1, \ldots, i_m$), where $\widetilde{\tau}^{(j)}$ is defined by Eq. $\widetilde{\tau}^{(i)} = \tau_0^{(i)}$. With small increments of the loading parameter at each iteration, this problem can be solved by standard gradient methods.

Note that, as a consequence of the heterogeneity of the field of stresses in each component, the value $\varepsilon_{\rm eq}^{p(j)}$ ($j = i_1, \ldots, i_m$) depends in a complex manner on the local values of the effective plastic strain $\epsilon_{\rm eq}^p(\mathbf{x})$ ($\mathbf{x} \in v_j$) and the average plastic deformations $\langle \epsilon^p \rangle_j$ in the component j. The values $\epsilon^p(\mathbf{x})$ and $\langle \epsilon^p \rangle_j$ ($\mathbf{x} \in v_j$; $j = i_1, \ldots, i_m$) cannot be found from the macroscopic deformation parameters because the problem is itself nonlinear. With this general principle in mind, we will estimate the components of an average plastic deformation on the basis of the secant modulus concept $\epsilon^{p(j)} \equiv [\mathbf{M}^{s(j)} - \mathbf{M}^{(j)}]\langle\boldsymbol{\sigma}\rangle^{(j)} \neq \langle\epsilon^p\rangle^{(j)}$, where the average stresses $\langle\boldsymbol{\sigma}\rangle_j$ in a component are calculated from Eq. (8.46). It is worth noting that

$$\varepsilon^{p(j)}_{\text{eq}} \neq \left[\frac{2}{3}\epsilon^{p(j)}_{kl}\epsilon^{p(j)}_{kl}\right]^{1/2}, \quad \varepsilon^{p(j)}_{\text{eq}} \neq \left[\frac{2}{3}\langle\epsilon^{p}_{kl}\rangle^{(j)}\langle\epsilon^{p}_{kl}\rangle^{(j)}\right]^{1/2} \tag{16.34}$$

and, in consequence, the flow law (2.230) written in terms of $\epsilon^{p(j)}$ and $\epsilon^{p*(j)}$ and $\langle\boldsymbol{\sigma}\rangle^{(j)}$, $\langle\widetilde{\tau}\rangle^{(j)}$ is not satisfied.

Finally, it should be noted that generally the local stress state is not monotonical and proportional even with monotonical and proportional external loading. However, using the linearization $\widetilde{\tau}^{(i)} = \tau_0^{(i)}$, one can see from Eqs. (16.31), (16.32) that the effective stress and the plastic strain are homogeneous, monotonical, and proportional functions inside each component for at least a particular case of the composites consisting of an incompressible matrix containing rigid inclusions or pores and, therefore, the above secant modulus concept can be used.

16.2.2 Elastoplastic Deformation of Composites with an Incompressible Matrix

Now we consider the general case of an incompressible matrix with finite values of the parameters $\mu^{(0)} < \infty$, $\tau_0 > 0$. In this case the representation for the secant Young's modulus of the matrix must take the general form (2.232)

$$E^{s(0)} = \left(\frac{1}{E^{(0)}} + \frac{\varepsilon^{p(0)}_{\text{eq}}}{\widetilde{\tau}^{(0)}}\right)^{-1}, \tag{16.35}$$

where the ratio $\varepsilon^{p(0)}_{\text{eq}}/\widetilde{\tau}^{(0)}$ can be expressed in terms of the macrostresses $\langle\boldsymbol{\sigma}\rangle$ as

$$\frac{\epsilon^{p(0)}_{\text{eq}}}{\widetilde{\tau}^{(0)}} = \frac{g(\widetilde{\tau}^{(0)} - \tau_0^{(0)})}{\widetilde{\tau}^{(0)}}, \quad \widetilde{\tau}^{(0)} = \left[\frac{3}{2}\mathbf{F}(\langle\boldsymbol{\sigma}\rangle\otimes\langle\boldsymbol{\sigma}\rangle)\right]^{1/2}, \quad \mathbf{F} = \frac{1}{2(1-c)}\frac{\partial\mathbf{M}^{s*}}{\partial q^{s(0)}}. \tag{16.36}$$

Here the function g is introduced, which is the inverse of the nonlinear hardening function (2.227). It is assumed that $g \equiv 0$ for $\widetilde{\tau}^{(0)} < \tau_0^{(0)}$.

For either the pores or rigid inclusions, taking into account the relations (16.36) and $\mathbf{F} = 2\mu^{s(0)}(1-c)^{-1}\mathbf{M}^{s*}$ we obtain the following expressions for the overall stress potential

$$w^{c*} = \frac{3}{4}(1-c)\left[\frac{1}{E^{(0)}} + \frac{g(\widetilde{\tau}^{(0)} - \tau_0^{(0)})}{\widetilde{\tau}^{(0)}}\right]\mathbf{F}(\langle\boldsymbol{\sigma}\rangle\otimes\langle\boldsymbol{\sigma}\rangle) \tag{16.37}$$

and for the overall plastic strain $\epsilon^p_* = 1.5(1-c)g(\widetilde{\tau}^{(0)} - \tau_0^{(0)})[\widetilde{\tau}^{(0)}]^{-1}\mathbf{F}\langle\boldsymbol{\sigma}\rangle$. It is interesting that the overall plastic deformation of the composite ϵ^p_* is independent of the initial modulus of the matrix $\mu^{(0)}$.

16.2.3 General Case of Elastoplastic Deformation

At present, it is impossible to construct sufficiently simple analytical solutions and we will use numerical calculations on the basis of Eqs. (16.28), (16.30), and (8.48), (8.52). Let us consider a material containing spherical pores. The elastic and plastic properties of an aluminum matrix were given in [910] as $E^{(0)} = 68.3\,\text{GPa}$, $\nu^{(0)} = 0.33$; $\tau_0^{(0)} = 250\,\text{MPa}$, $h^{(0)} = 173\,\text{MPa}$, $n^{(0)} = 0.455$. Figure 16.5 gives the corresponding

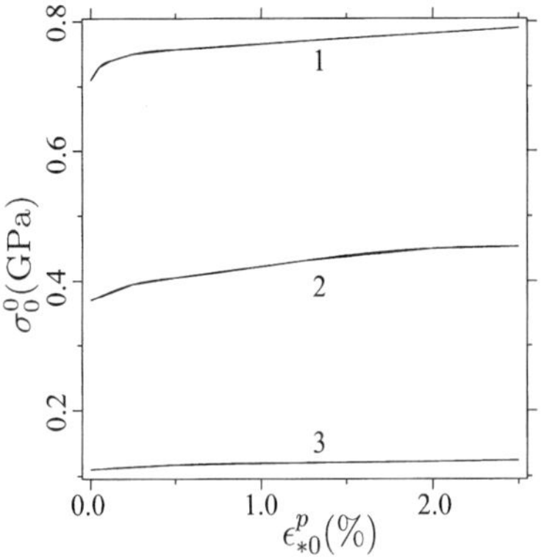

Fig. 16.5. Calculated curves of the relations between hydrostatic stress and plastic strain for porous aluminum.

calculated plastic deformations $\epsilon^p_{*ij} = \epsilon^p_{*0}\delta_{ij}$ for different values of c in dependence of the hydrostatic stress $\sigma^0_{ij} = \sigma^0\delta_{ij}$, taking into account the solution of problem (16.28). As expected from Eq. (9.109), there is only a moderate influence of the matrix compressibility on the elastoplastic deformation of porous materials over a wide range of values of $\nu^{(0)}$. Accordingly, the assumption of an incompressible matrix $\nu^{(0)} = 0.5$ leads to an increase of 1–2 % of the calculated plastic deformations in Fig. 16.5. In Fig. 16.5 the values σ_0^0 for $\epsilon_0^p = 0$ indicate the onset of yielding. Note that according to criterion (15.10) we will obtain $\epsilon^p_{*0} \equiv 0$ for any σ_0^0 or, in other words, the porous material has infinite values of initial yield stress under hydrostatic compression.

The modified secant modulus model of plasticity described above was proposed in [131], [149] and is equivalent to the model proposed independently in [474], [475]. The related approach [83], [852], [912], [913], is based on the approximative method of estimation of second moments of stresses (as opposed to (13.15) used the assumption (13.50) which was utilized also in the flow model of plasticity in [127], [160]. The deformation theory of plasticity in the case of nonradial loading, which takes place usually at the local level, even if the overall stress path is radial, has the known difficulties avoided easily by an incremental plasticity model described below in Sections 16.4 and 16.5.

16.3 Power-Law Creep

In the case of a matrix with zero initial yield stress

$$\mu^{(0)}, \frac{k^{(0)}}{\mu^{(0)}} \to \infty, \quad \tau_0^{(0)} = 0 \tag{16.38}$$

the relation between flow stress and plastic strain, Eq. (16.29), takes the form ($\varepsilon^p_{ii} \equiv 0$)

$$\tau = h(\varepsilon^p_{\rm eq})^n, \quad \sigma_{ij} = \frac{2}{3} h(\varepsilon^p_{\rm eq})^{n-1} \epsilon^p_{ij}. \tag{16.39}$$

The local stress potential corresponding to the constitutive relation (16.39) can be expressed as

$$w^c = \frac{n}{n+1} h^{-\frac{1}{n}} \left[\frac{3}{2}\mathbf{N}_2(\sigma \otimes \sigma)\right]^{\frac{1}{2}+\frac{1}{2n}} = \frac{n}{n+1} h^{-\frac{1}{n}} (\tau)^{1+\frac{1}{n}}, \tag{16.40}$$

where the tensor $\mathbf{N}_2 = (0,1)$ is isotropic and contains no hydrostatic contributions.

If ϵ^p is considered as a rate-of-deformation (strain rate) tensor, Eqs. (16.39) and (16.40) describe a power-law creep of the matrix. In the theory of non-Newtonian fluids it is customary to express the second Eq. (16.39) in the form

$$s_{ij} = 2\eta^{(0)} \epsilon^p_{ij}, \quad 2\eta^{(0)} \equiv \frac{2}{3} h \left(\varepsilon^p_{\rm eq}\right)^{n-1}, \tag{16.41}$$

where $\eta^{(0)}$ is a nonlinear viscosity of the matrix.

In order to estimate the second moment of the stress inside the matrix of the composites with the incompressible matrix containing either the pores or rigid inclusions we will make use of Eqs. (13.15) and (9.97), (9.106), (9.108) with consequent passage to the limit (16.38). Insofar our concern is only with the plastic deformation ϵ^p of a single component it is clearly advantageous to use the iterative scheme (16.31), (16.32). For the initial increment $\varepsilon^{p(0)}_{\rm eq}$ we find $\widetilde{\tau}^{(0)}$ according to Eq. (16.31$_2$), the secant modulus of the matrix takes the form $E^{s(0)} = \widetilde{\tau}^{(0)}/\varepsilon^{p(0)}_{\rm eq}$, and the second moment of the stress inside the matrix is obtained as (13.15)

$$\langle s_{ij} s_{ij}\rangle^{(0)} = F_{mnpq}\langle \sigma_{mn}\rangle\langle\sigma_{pq}\rangle. \tag{16.42}$$

For pores or rigid inclusions the tensor

$$\mathbf{F} \equiv \frac{1}{2(1-c)} \frac{\partial \mathbf{M}^{s*}}{\partial q^{s(0)}} \tag{16.43}$$

depends only on the microtopology of the composite and is independent of the mechanical properties of the matrix. From the relations (9.97), (9.106), and (9.108) it can be shown that

$$\mathbf{F} \equiv \frac{1}{(1-c)} 2\mu^{s(0)} \mathbf{M}^{s*} \tag{16.44}$$

for the considered configurations. It can be proved that the identity (16.44) is valid either for rigid inclusions or for pores in an incompressible matrix. Actually, according to Hill's condition (see Section 6.1) we obtain for rigid inclusions or for pores $\mathbf{M}^{s*}(\langle\boldsymbol{\sigma}\rangle \otimes \langle\boldsymbol{\sigma}\rangle) = (1-c)\mathbf{M}^{s(0)}\langle\boldsymbol{\sigma}\otimes\boldsymbol{\sigma}\rangle^{(0)}$, from which Eq. (16.44) follows under the assumption of an incompressible matrix. Therefore, the actual value of the effective stress in the matrix $\widetilde{\tau}^{(0)}$ can be obtained directly from (16.42) without employing the iterative process (16.31), (16.32).

From (16.42) and (16.44) together with the principle of minimum complementary energy the expression for the overall stress potential is found as

$$w^{c*} = (1-c)\langle w^c\rangle^{(0)} = \frac{n}{n+1}(1-c)h^{-\frac{1}{n}}\left[\frac{3}{2}\mathbf{F}(\langle\boldsymbol{\sigma}\rangle \otimes \langle\boldsymbol{\sigma}\rangle)\right]^{\frac{1}{2}+\frac{1}{2n}} \tag{16.45}$$

and the overall plastic strain is obtained as

$$\epsilon^p_{\text{eq}} = \frac{3}{2}(1-c)h^{-\frac{1}{n}}\left[\frac{3}{2}\mathbf{F}(\langle\boldsymbol{\sigma}\rangle \otimes \langle\boldsymbol{\sigma}\rangle)\right]^{\frac{1}{2n}-\frac{1}{2}}. \tag{16.46}$$

Rather than employing relations (16.45) and (16.46), we will present our result in a more standard manner making use of a "reference tensor"

$$\widetilde{\mathbf{F}} = (1-c)^n\mathbf{F}^{\frac{1}{2}+\frac{n}{2}}. \tag{16.47}$$

As can be seen below, the tensor $\widetilde{\mathbf{F}}$ shows the effect of the inclusions on the variation of the rheological properties of the composite. Obviously, we have $\widetilde{\mathbf{F}} \to \mathbf{N}_2$ for $c \to 0$. For $n = 1$ the tensor $\widetilde{\mathbf{F}} \equiv 2\mu^{s(0)}\mathbf{M}^{s*}$ agrees with the well-known linear results on the relative change of the effective compliance with respect to the matrix properties.

The expression for the tensor $\widetilde{\mathbf{F}}$ for the two-particle approximation of the MEFM was obtained in [160] by a special method for spherical inclusions and by the use of the representations (9.97) and (9.106). General Eq. (16.45) for rigid inclusion and its generalization for the viscoplastic matrix was obtained in [126], [193] in the framework of the same method of the estimation of the second moment of stresses in the incompressible matrix used the assumption (13.50). From relation (16.45), which was obtained using the one-particle approximation of the MEFM, the analogous results [764], [894], [895], [1070] follow, which were obtained by variational methods. It was shown that the upper bound for linear elastic comparison composite $\mathbf{L}^*$ (for example, for porous materials) generate an upper bound for w^{c*}. However, linear estimated lower bounds (for example for composites with rigid inclusions) cannot be used, in general, for generation of lower bounds for the nonlinear composite by the variational procedure proposed which can be used nevertheless for obtaining estimations for nonlinear composites by application of a fairly accurate estimation for linear comparison composites described in Chapter 8.

Let us now consider in more detail the case of rigid inclusions (9.97) (porous materials (9.106), (9.108) can be investigated in the same manner) by the use of the terminology of the theory of non-Newtonian suspensions in a fluid matrix (16.41). Then the tensor $\mathbf{F}$, Eq. (16.42), has the structure of the tensor $\mathbf{N}_2$, with zero hydrostatic contributions, and for a suspension of the particles we have a flow law related to (16.41)

$$\langle s_{ij} \rangle = 2\eta^* \epsilon_{ij}^p, \quad 2\eta^* = \frac{2}{3} h (\varepsilon_{\rm eq}^p)^{n-1}, \tag{16.48}$$

where $\varepsilon_{\rm eq}^p = (2\epsilon_{ij}^p \epsilon_{ij}^p/3)^{1/2}$ is an effective overall plastic strain and $\eta^* = \eta^{(0)}$ for $c = 0$. Therefore, for the same rate of shear strain ϵ_*^p we find from (16.41) and (16.46)– (16.48) the relative change of the effective viscosity as $\eta^*/\eta^{(0)} = (2\widetilde{f}^\mu)^{-1}$, $\eta^{(0)} = (\varepsilon_{\rm eq}^p)^{n-1}/3$, where $2\widetilde{f}^\mu$ is the shear component of the Voight representation of the tensor $\widetilde{\mathbf{F}} = (3\widetilde{f}^k, 2\widetilde{f}^\mu)$.

For the two-particle and one-particle approximations of the MEFM we find from Eqs. (9.97), (16.47) and Eqs. (9.96), (16.47), respectively,

$$\frac{\eta^*}{\eta^{(0)}} = \left[1 + \frac{5}{2} c \left(1 - \frac{31}{16} c \right)^{-1} \right]^{(n+1)/2} (1-c)^{(1-n)/2}, \tag{16.49}$$

$$\frac{\eta^*}{\eta^{(0)}} = \left(1 + \frac{3}{2} c \right)^{(n+1)/2} (1-c)^{-n}. \tag{16.50}$$

Relations (16.49) and (16.50) were obtained in [160] via the MEFM. An expression analogous to Eq. (16.50) was subsequently proposed and analyzed in detail in [894], [895] via a variational method. For dilute concentrations of the inclusions ($c \to 0$) the limits of Eqs. (16.49) and (16.50) take the form

$$\eta^*/\eta^{(0)} = 1 + \gamma c, \quad \gamma = (3 + 7n)/4, \tag{16.51}$$

which agree with the Einstein linear result $\gamma = 5/2$ for $n = 1$. In Fig. 16.6 the experimental data [603] are compared with the results of Eq. (16.51) and with predictions obtained in [33] and [653] from numerical studies of a sphere in an infinite matrix of a power-law material (dilute approximation). For nondilute concentrations of the inclusions, Fig. 16.7 shows a comparison of the experimental data [123] ($n = 0.41$) with the results of formulae (16.49), (16.50) and the dilute approximation (of first degree with respect to c) [653].

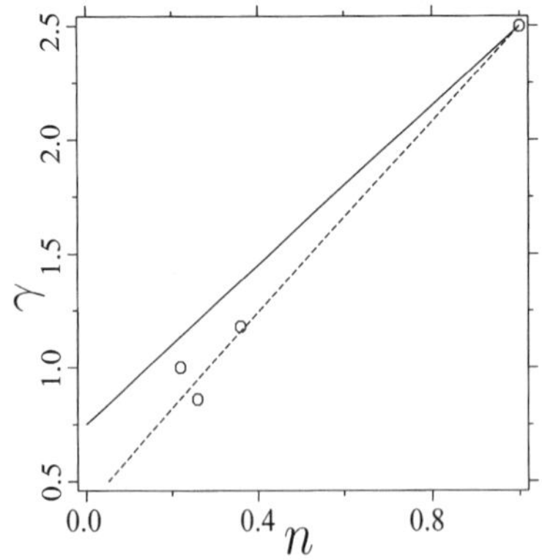

Fig. 16.6. Parameter γ (16.51) vs n. Experimental data (○) and calculations via the MEFM (solid line) and dilute approximation [653] (dotted line).

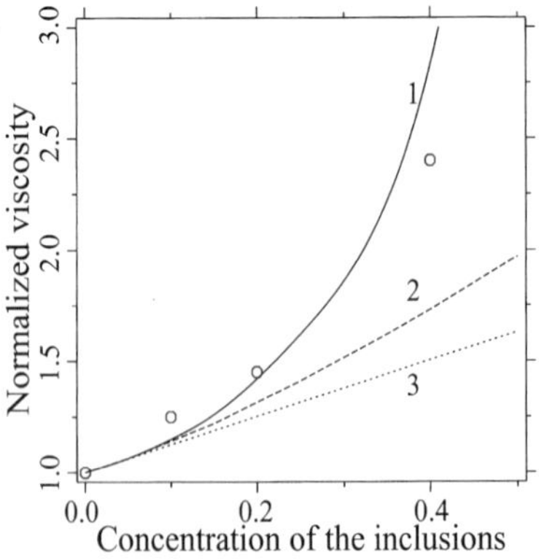

Fig. 16.7. $\eta^*/\eta^{(0)}$ vs c. Experimental data (○) and predictions from Eqs. (16.49) (solid line), (16.50) (dashed line) and from the dilute approximation (dotted line).

16.4 Elastic–Plastic Behavior of Elastically Homogeneous Materials with Random Residual Microstresses

16.4.1 Leading Equations and Elastoplastic Deformations

We assume that the rheological properties of the composite medium with, generally speaking, anisotropic components are described by the theory of small elastic-plastic strains under arbitrarily varying external loading. Additive decomposition of the incremental of total strain tensor $d\boldsymbol{\varepsilon}$ is used in each component of the composite (2.213), i.e., $\boldsymbol{\varepsilon} = \boldsymbol{\varepsilon}^e + \boldsymbol{\varepsilon}^t + \boldsymbol{\varepsilon}^p$, with the elastic strains $\boldsymbol{\varepsilon}^e$, the so-called transformation strains $\boldsymbol{\varepsilon}^t$ including thermal strains, too, and plastic strains $\boldsymbol{\varepsilon}^p$; $\boldsymbol{\beta} \equiv \boldsymbol{\varepsilon}^t + \boldsymbol{\varepsilon}^p$. The local basic Eqs. (6.1)–(6.3) are assumed to be accomplished (with no body forces acting). The uniform traction boundary condition at the surface ∂w of the sample domain w (6.5) generating a homogeneous macroscopic stress $\boldsymbol{\sigma}^0$ is given, and the phases are assumed to be perfectly connected.

For the description of the behavior of the matrix and the elastic–plastic coating the so-called J_2-flow theory with combined isotropic-kinematic hardening is used; see Eqs. (2.215)–(2.217) for details. $\mathbf{s}^a$ is the deviator of the active stresses; $\mathbf{a}^p$ is a symmetric second-order tensor corresponding to the "back-stress" defining the location of the center of the yield surface in the deviatoric stress space. For the evaluation of the back stress $\mathbf{a}^p$ Ziegler's hardening rule (2.218) is used. The material behaves elastically if $f < 0$, or if $f = 0$ and $(\partial f/\partial \boldsymbol{\sigma}) : d\boldsymbol{\sigma} \leq 0$; elastic–plastic deformations take place under active loading, when $f = 0$ and $(\partial f/\partial \boldsymbol{\sigma}) : d\boldsymbol{\sigma} > 0$; see Eq. (2.220).

With non-essential restriction on the number of phases we will consider (following [135]) two-component statistically homogeneous composites containing identical aligned inclusions. The elastic properties of the matrix and the inclusions are the same, but the so-called "stress-free strains," i.e., the strain contributions due to temperature loading, phase transformations, and the plastic strains, fluctuate.

At first we will recast the representations obtained in Chapters 8 and 13 for the statistical moments of residual stresses in the phases in the form convenience for subsequent calculations. So, for two-component composites the relation (9.37) for the average stresses inside inclusions can be represented in unit form for the inclusions ($i = 1$) and the matrix ($i = 0$):

$$\langle \boldsymbol{\sigma} \rangle^{(i)} = \boldsymbol{\sigma}^0 + \mathbf{C}^{(i)} \boldsymbol{\beta}_1^{(1)}, \quad \mathbf{C}^{(i)} = (-1)^i (1 - c^{(i)}) \mathbf{Q}_i, \tag{16.52}$$

where $\boldsymbol{\sigma}^0 = \langle \boldsymbol{\sigma} \rangle$. The representations for the second moments of stresses inside the phases (13.59) and (13.60) taking Eq. (16.52) and only binary interaction of inclusions into account can be recast in terms of the average second invariant of deviator stress stresses inside the inclusions (do not sum i):

$$\langle \boldsymbol{\sigma} \mathbf{N}_2 \boldsymbol{\sigma} \rangle^{(i)} - \langle \boldsymbol{\sigma} \rangle^{(i)} \mathbf{N}_2 \langle \boldsymbol{\sigma} \rangle^{(i)} = \widetilde{\mathbf{B}}^{(i)} (\langle \boldsymbol{\sigma} \rangle^{(i)} - \boldsymbol{\sigma}^0) \otimes (\langle \boldsymbol{\sigma} \rangle^{(i)} - \boldsymbol{\sigma}^0), \tag{16.53}$$

where

$$\widetilde{\mathbf{B}}^{(i)} = \mathbf{N}_2 : \widetilde{\mathbf{C}}^{(i)} \left[\left(\mathbf{C}^{(i)}\right)^{-1} \otimes \left(\mathbf{C}^{(i)}\right)^{-1} \right],$$

$$\widetilde{\mathbf{C}}^{(0)} = \int [\mathbf{T}_p(\mathbf{x}_0 - \mathbf{x}_p)\overline{v}_p] \otimes [\mathbf{T}_p(\mathbf{x}_0 - \mathbf{x}_p)\overline{v}_p] \, \varphi(v_p|\mathbf{x}_p; \mathbf{x}_0) \, d\mathbf{x}_p,$$

$$\widetilde{\mathbf{C}}^{(1)} = \overline{v}_1^{-1} \int\int [\mathbf{T}_p(\mathbf{x} - \mathbf{x}_p)\overline{v}_p] \otimes [\mathbf{T}_p(\mathbf{x} - \mathbf{x}_p)\overline{v}_p] \, V_1(\mathbf{x})\varphi(v_p|\mathbf{x}_p; \mathbf{x}_1) \, d\mathbf{x} \, d\mathbf{x}_p.$$

Since our main interest here is the consideration of the elastic–plastic deformations of the whole of the components $v^{(i)}$ $(i = 0, 1)$ and for the sake of simplicity we are using the yield criterion (15.11), instead of (15.20). In Section 15.1 we demonstrated that this assumption is physically consistent, in contrast with (15.10).

The composite starts yielding when

$$f^* \equiv \max_i [\widetilde{\tau}^{(i)}/\tau_0^{(i)}] - 1 = 0, \quad (i = 0, 1). \tag{16.54}$$

Of course, if only the first onset of yielding is concerned, the problem of estimating $\widetilde{\tau}^{(i)}$ is an elastic one.

At first let us consider situations in which at zero time $\boldsymbol{\sigma}^0 = \mathbf{0}$ and $f^* < 0$. Under monotonically increased external active loading elastic–plastic deformations will start to appear, when $f^* = 0$, according to (16.54), and $[(\partial f^*/\partial \boldsymbol{\sigma}^0) : d\boldsymbol{\sigma}^0 + (\partial f^*/\partial \theta) d\theta] > 0$. The initial yield surfaces of the components (cf. (15.11)) may be embedded in each other or they may intersect in the space spanned by the macrostresses $\boldsymbol{\sigma}^0$ and temperature θ. We define the state of the composite as being elastic–plastic if at least one component $v^{(s)}$ is in an elastic–plastic state. In this case the macroscopic constitutive law can be found as follows.

Taking into account the estimation for the yield surface (15.11) and (16.54) the increments of homogeneous plastic strains in the component $v^{(s)}$,

$$d\varepsilon^{p(s)} = d\lambda^{(s)} \frac{\partial \widetilde{f}^{(s)}}{\partial \langle \boldsymbol{\sigma} \rangle^{(s)}}, \tag{16.55}$$

are defined by the use of the functions $\widetilde{f}^{(s)} \equiv \widetilde{\tau}^{(s)} - F^{(s)}(\gamma^{(s)}, \theta) = 0$, $\widetilde{\tau}^{(s)} = \left(1.5\langle \mathbf{s}^a : \mathbf{s}^a \rangle^{(s)}\right)^{1/2}$. At each incremental step of the external stresses and of the temperature, respectively, we assume homogeneous plastic strains $\varepsilon^{p(s)}$ and increments of the hardening parameters inside the component $v^{(s)}$. If Ziegler's rule (2.218) is used, we have

$$d\mathbf{a}^{p(s)} = d\gamma^{(s)} A(\gamma^{(s)}) \langle \mathbf{s}^{a(s)} \rangle^{(s)}, \quad A(\gamma^{(s)}) = H^{(s)}/F^{(s)}(\gamma^{(s)}, \theta), \tag{16.56}$$

where $H^{(s)}$ is the current plastic tangent modulus of the uniaxial stress–plastic strain curve:

$$H^{(s)} = \frac{\partial \widetilde{\tau}^{a(s)}}{\partial \gamma^{(s)}}, \quad d\widetilde{\tau}^{a(s)} \equiv \left(\frac{3}{2} \langle \, d\mathbf{s}^a : d\mathbf{s}^a \rangle^{(s)}\right)^{1/2}, \quad d\gamma^{(s)} = \left(\frac{2}{3} \, d\varepsilon^{p(s)} : d\varepsilon^{p(s)}\right)^{1/2}.$$

Here the estimations (13.58) and (13.60) can be used for the evaluation of the second moment $\langle \, d\mathbf{s}^a : d\mathbf{s}^a \rangle^{(s)} \equiv \langle \, d\mathbf{s} : d\mathbf{s} \rangle^{(s)} - 2\langle d\mathbf{s} \rangle^{(s)} : \mathbf{a}^{(s)} + \mathbf{a}^{p(s)} : \mathbf{a}^{p(s)}$.

It should be noted that the assumption of homogeneity of the stress and strain fields inside each component, as used in (15.10) for mean field theory, leads to the following relation for the increments of the plastic strains inside the component $v^{(s)}$ [298], [300]:

$$d\varepsilon^{p(s)} = d\lambda^{(s)} \frac{\partial f^{(s)}}{\partial \langle \boldsymbol{\sigma} \rangle^{(s)}}, \tag{16.57}$$

where the yield function, $f^{(s)} \equiv \tau^{(s)} - F^{(s)}(\gamma^{(s)}, \theta) = 0$, is defined by the average stresses inside the component $\tau^{(s)} = \big(1.5\langle \mathbf{s}^a \rangle^{(s)} : \langle \mathbf{s}^a \rangle^{(s)}\big)^{1/2}$.

Of course, if the formulation (16.57) (based on estimation of $\tau^{(s)}$) is used it is not necessary to estimate the second stress moments inside the component $v^{(s)}$. In reality for statistically isotropic composites containing spherical inclusions we see from (16.52) that the deviator of the active stresses inside the component $\langle \mathbf{s}^a \rangle^{(s)} \equiv \mathbf{N}_2\big(\boldsymbol{\sigma}^0 - \mathbf{a}^{p(s)}\big)$ does not depend on the stress-free strain field $\boldsymbol{\beta}$ and the internal microtopology of the composite, and, therefore, residual stresses does not affect the elastic–plastic strains of the composite under consideration. However, in subsequent discussion it will be shown that residual stresses are of fundamental importance.

We now turn our attention to the discussion of the yield condition (16.57_1) under consideration of residual stresses. In view of some further simplifications of the relations (16.55) to (16.56_1) we restrict ourselves to two-phase materials. Taking Eq (16.53) into account we obtain from (16.57_2):

$$\widetilde{\tau}^{(s)} = \left[\frac{3}{2}\langle \mathbf{s}^a \rangle^{(s)} : \langle \mathbf{s}^a \rangle^{(s)} + \frac{3}{2}\widetilde{\mathbf{B}}^{(s)}\Big((\langle \boldsymbol{\sigma} \rangle^{(s)} - \boldsymbol{\sigma}^0) \otimes (\langle \boldsymbol{\sigma} \rangle^{(s)} - \boldsymbol{\sigma}^0)\Big)\right]^{1/2}. \tag{16.58}$$

For consistency of the plastic deformation process the following holds in analogy to (2.221):

$$d\widetilde{f}^{(s)} \equiv \frac{\partial \widetilde{f}^{(s)}}{\partial \langle \boldsymbol{\sigma} \rangle^{(s)}} : d\langle \boldsymbol{\sigma} \rangle^{(s)} + \frac{\partial \widetilde{f}^{(s)}}{\partial \boldsymbol{\sigma}^0} : d\boldsymbol{\sigma}^0 + \frac{\partial \widetilde{f}^{(s)}}{\partial \mathbf{a}^{p(s)}} : d\mathbf{a}^{p(s)} + \frac{\partial \widetilde{f}^{(s)}}{\partial \gamma^{(s)}} d\gamma^{(s)} + \frac{\partial \widetilde{f}^{(s)}}{\partial \theta} d\theta = 0. \tag{16.59}$$

At first we will calculate some partial derivatives in (16.59) by using (no s sum)

$$\frac{\partial \widetilde{f}^{(s)}}{\partial \langle \boldsymbol{\sigma} \rangle^{(s)}} = \frac{3}{2F^{(s)}}\left[\langle \mathbf{s}^a \rangle^{(s)} + \widetilde{\mathbf{B}}^{(s)}(\langle \boldsymbol{\sigma} \rangle^{(s)} - \boldsymbol{\sigma}^0)\right], \tag{16.60}$$

$$\frac{\partial \widetilde{f}^{(s)}}{\partial \boldsymbol{\sigma}^0} = -\frac{3}{2F^{(s)}}\widetilde{\mathbf{B}}^{(s)}(\langle \boldsymbol{\sigma} \rangle^{(s)} - \boldsymbol{\sigma}^0), \tag{16.61}$$

$$\frac{\partial \widetilde{f}^{(s)}}{\partial \mathbf{a}^{p(s)}} = -\frac{3}{2F^{(s)}}\langle \mathbf{s}^a \rangle^{(s)}, \quad \frac{\partial \widetilde{f}^{(s)}}{\partial \gamma^{(s)}} = -\frac{\partial F^{(s)}}{\partial \gamma^{(s)}}, \quad \frac{\partial \widetilde{f}^{(s)}}{\partial \theta} = -\frac{\partial F^{(s)}}{\partial \theta}, \tag{16.62}$$

From (16.52), (16.55), and (16.56_1) the relations for the differential of the average stresses and hardening parameters inside component $v^{(s)}$ are obtained:

$$d\langle \boldsymbol{\sigma} \rangle^{(s)} = d\boldsymbol{\sigma}^0 + \mathbf{C}^{(s)}\bigg(d\varepsilon_1^{t(s)} + (-1)^{(s+1)} d\lambda^{(s)} \frac{\partial \widetilde{f}^{(s)}}{\partial \langle \boldsymbol{\sigma} \rangle^{(s)}}\bigg), \tag{16.63}$$

$$d\gamma^{(s)} = d\lambda^{(s)} \left(\frac{2}{3} \frac{\partial \widetilde{f}^s}{\partial \langle \boldsymbol{\sigma} \rangle^{(s)}} : \frac{\partial \widetilde{f}^{(s)}}{\partial \langle \boldsymbol{\sigma} \rangle^{(s)}} \right)^{1/2}, \tag{16.64}$$

$$d\mathbf{a}^{p(s)} = d\lambda^{(s)} A(\gamma^{(s)}) \langle \mathbf{s}^a \rangle^{(s)} \left(\frac{2}{3} \frac{\partial \widetilde{f}^s}{\partial \langle \boldsymbol{\sigma} \rangle^{(s)}} : \frac{\partial \widetilde{f}^{(s)}}{\partial \langle \boldsymbol{\sigma} \rangle^{(s)}} \right)^{1/2}. \tag{16.65}$$

Substitution of the equations (16.60)–(16.65) in (16.59) leads to the relation of proportionality factor (do not sum to s, $s = 0, 1$): $d\lambda^{(s)} = b_s / \boldsymbol{\beta}_{ss}$, where

$$b_s = -\left(\frac{3}{2} \langle \mathbf{s}^a \rangle^{(s)} : d\boldsymbol{\sigma}^0 + \frac{\partial \widetilde{f}^{(s)}}{\partial \langle \boldsymbol{\sigma} \rangle^{(s)}} \mathbf{C}^{(s)} \, d\boldsymbol{\varepsilon}_1^{t(1)} - \frac{\partial F^{(s)}}{\partial \theta} d\theta \right), \tag{16.66}$$

$$\begin{aligned} \boldsymbol{\beta}_{ss} &= (-1)^s \frac{\partial \widetilde{f}^{(s)}}{\partial \langle \boldsymbol{\sigma} \rangle^{(s)}} \mathbf{C}^{(s)} \frac{\partial \widetilde{f}^{(s)}}{\partial \langle \boldsymbol{\sigma} \rangle^{(s)}} \\ &- \left[\left(\frac{\widetilde{\tau}^{(s)}}{F^{(s)}} \right)^2 H^{(s)} + \frac{\partial F^{(s)}}{\partial \gamma^{(s)}} \right] \left[\frac{2}{3} \frac{\partial \widetilde{f}^{(s)}}{\partial \langle \boldsymbol{\sigma} \rangle^{(s)}} \mathbf{N}_2 \frac{\partial \widetilde{f}^{(s)}}{\partial \langle \boldsymbol{\sigma} \rangle^{(s)}} \right]^{1/2}. \end{aligned} \tag{16.67}$$

Let us consider the process of the deformation, when for each component $v^{(i)}$ $(i = 0, 1)$ the yield conditions are fulfilled

$$\widetilde{f}^{(i)} = 0, \quad \frac{\partial \widetilde{f}^{(i)}}{\partial \langle \boldsymbol{\sigma} \rangle^{(i)}} : d\langle \boldsymbol{\sigma} \rangle^{(i)} + \frac{\partial \widetilde{f}^{(i)}}{\partial \theta} \, d\theta > 0. \tag{16.68}$$

Then from reduced formulae (16.59)–(16.65), only Eq. (16.63) will be changed

$$d\langle \boldsymbol{\sigma} \rangle^{(s)} = d\boldsymbol{\sigma}^0 + \mathbf{C}^{(s)} \left(d\boldsymbol{\varepsilon}_1^{t(1)} + d\lambda^{(1)} \frac{\partial \widetilde{f}^{(1)}}{\partial \langle \boldsymbol{\sigma} \rangle^{(1)}} - d\lambda^{(0)} \frac{\partial \widetilde{f}^{(0)}}{\partial \langle \boldsymbol{\sigma} \rangle^{(0)}} \right). \tag{16.69}$$

By substituting the relations (16.60)–(16.62), (16.64), (16.65), (16.69) in (16.59) we will obtain the system of two linear equations for the proportionality factors $d\lambda^{(i)}$ $(i = 0, 1)$: $\sum_{j=0}^{1} \boldsymbol{\beta}_{ij} d\lambda^{(j)} = b_i$, where b_i, $\boldsymbol{\beta}_{ii}$ $(i = 0, 1)$ are determined by (16.66) and (16.67), and the off-diagonal matrix elements $\boldsymbol{\beta}_{ij}$ $(i = 0, 1;\ j = 2 - i)$ are defined by the relations

$$\boldsymbol{\beta}_{ij} = (-1)^j \frac{\partial \widetilde{f}^{(i)}}{\partial \langle \boldsymbol{\sigma} \rangle^{(i)}} \mathbf{C}^{(i)} \frac{\partial \widetilde{f}^{(j)}}{\partial \langle \boldsymbol{\sigma} \rangle^{(j)}}. \tag{16.70}$$

Then we get $d\lambda^{(i)} = \sum_{j=0}^{1} h_{ij} b_j$, where h_{ij} can be considered functions of hardening $h_{00} = \boldsymbol{\beta}_{11}/\Delta$, $h_{01} = -\boldsymbol{\beta}_{01}/\Delta$, $h_{10} = -\boldsymbol{\beta}_{10}/\Delta$, $h_{11} = \boldsymbol{\beta}_{00}/\Delta$, $\Delta = \boldsymbol{\beta}_{00}\boldsymbol{\beta}_{11} - \boldsymbol{\beta}_{10}\boldsymbol{\beta}_{01}$.

This way for prescribed paths of loading $\boldsymbol{\sigma}(t)$, $\theta(t)$ (where t is a monotonically varying parameter) the system of equations (16.57) becomes a system of six (or twelve, respectively, for both phases yielding) ordinary differential equations for the determination of all components of $d\boldsymbol{\varepsilon}^{p(s)}$. In so doing only five or ten, respectively, these differential equations from ones are independent of each other by virtue of the yield condition (16.57_1) or (16.68), respectively. After integration of this system we will find the plastic strains inside each component and find the total strains: $\langle \varepsilon \rangle^{(s)} = \mathbf{M}\langle \boldsymbol{\sigma} \rangle^{(s)} - \boldsymbol{\varepsilon}^{t(s)} - \boldsymbol{\varepsilon}^{p(s)}$. The overall plastic strains $\boldsymbol{\varepsilon}^{po}$ and the average stresses are defined by the relations

$$\boldsymbol{\varepsilon}^{po} = \langle \boldsymbol{\varepsilon}^{p} \rangle, \quad \langle \boldsymbol{\varepsilon} \rangle = \mathbf{M}\boldsymbol{\sigma}^{0} + \langle \boldsymbol{\varepsilon}^{t} \rangle + \boldsymbol{\varepsilon}^{po}. \tag{16.71}$$

The known numerical schemes [857], [640], [300] are modified due to the additional concentration factors $\widetilde{\mathbf{B}}^{(i)}$ $(i = 0, 1)$ (16.53) which are calculated only once from the elastic problem. Here the backward difference method with a variable integration step is used for integrating Eq. (16.55). A Newton-type iterative procedure is used, starting with the elastic predictor which provides a first estimate of the stress increment in each component.

16.4.2 Numerical Results for Temperature-Independent Properties

In the general case of an elastic–plastic deformation process there are too many constitutive parameters, and normalized curves hardly can be constructed. We consider a particular composite with the thermomechanical properties of a duplex steel, however with much simpler microtopology. Namely, we consider a composite with ferritic matrix containing identical spherical austenite inclusions $(c = 0.5, \tau_0 = \tau_0(°C))$. The thermoelastoplastic constants as usually found in the literature (see [1173]) are displayed in Table 16.1.

Table 16.1. *Thermoelastoplastic constants*

Phase	τ_0[GPa]	$m_0 \cdot 10^6[°C^{-1}]$	μ[GPa]	ν
Ferrite	0.32	12.5	76.9	0.3
Austenite	0.20	17.1	76.9	0.3

Here the thermal expansion coefficient $\mathbf{m} \equiv m_0\boldsymbol{\delta}$ defines the transformation strains (cf. Eq (2.213)) inside the components $\boldsymbol{\varepsilon}^t_{10} = (m_0^{(1)} - m_0^{(0)})\theta = (m_0^{(1)} - m_0^{(0)})(T - T^{\max})$ due to a temperature difference $(T - T^{\max})$ between the stress-free reference temperature $T^{\max} = 900°C$ (from which the cooldown starts) and a current temperature T.

Just for the sake of theoretical interest, let us consider the components with temperature independent parameters $\tau_0, h, n =$ const and small kinematic hardening $H^{(i)} = 0.1\tau^{(i)}$, $h^{(i)} = 0$ $(i = 0, 1)$. Let the composite being loaded by a constant uniaxial external stress $\sigma^0_{ij} = \sigma^0_1\delta_{i1}\delta_{j1}$ and by temperature cycles with one cycle described by $T = T^{\min} + (T^{\max} - T^{\min})(t-1)[H(t-1) - H(1-t)], \quad t \in (0, 2)$, where H is the Heaviside step function and $T^{\min}$ is a minimum temperature.

In Fig. 16.8 typical calculated curves for the variation of the first component ε^{po}_{11} of the overall plastic strain tensor (16.71_1) under the action of different external stresses $\sigma^0_1 = 0$, 0.125, 0.135 GPa and four temperature cycles $900°C \to 0°C \to 900°C$ are shown. Analogous results are presented in Fig. 16.9 for 400 temperature cycles. The curves show the growth of the longitudinal overall plastic strain ε^{po}_{11} accumulated from cycle to cycle. This means that only the plastic strain after completion of each cycle is represented in Fig. 16.9 as a function of cycle number $l = 0, \ldots, 400$. From Figs. 16.8 and 16.9 we see that for $\sigma^0_1 \leq 0.125$ GPa the overall plastic strain after a few temperature cycles $l \leq 4$ converges to a steady-state response (or steady cycle), i.e., a closed loop is received. In structural analysis such behavior is specified as plastic shakedown (or cyclic collapse) (see references [834], [890], [1037]). For the sake of brevity we use this terminology also.

The results show the following behavior: For sufficiently small inclusion concentrations c or/and temperature amplitudes $(T^{\max} - T^{\min})/2$ or /and external loading σ_1^0 the process will be situated in the domain of purely elastic cycling with $f^* \leq 0$ according to (16.54). Increase of c or /and $(T^{\max} - T^{\min})/2$ or /and σ_1^0 leads to elastic shakedown with $\varepsilon^{po}(t) = \text{const.}$ at $t > t_e$, with t_e sufficiently large. A further increase of the above indicated parameters leads to plastic shakedown (curves 1,2 in Fig. 16.8 and curve 1 Fig. 16.9), and, finally, ratchetting (or incremental collapse) appears – see curve 3 in Fig. 16.8 and curves 2, 3 in Fig. 16.9 – where the composite undergoes continued plastic strain growth. In the considered example the transition from plastic shakedown to ratchetting appears approximately at $\sigma_1^0 = 0.126$ GPa.

To obtain a ratchetting effect it is necessary that not only one but also the other component shows plastic deformations in the course of the cycle. Such an onset of yielding of the second component, i.e., of the inclusions in the considered example, appears at point A of the curve 3 in Fig. 16.8. If the noncyclic external loading $\sigma_1^0 = \text{const.}$ is changed, the point A transfers along the deformation path, and it is possible that point A does not appear on the first but a later cycle. It is interesting to notice that in the process with $\sigma_1^0 = 0.135$ GPa the transversal components of overall plastic strain tensor decrease and become negative after the eighth cycle.

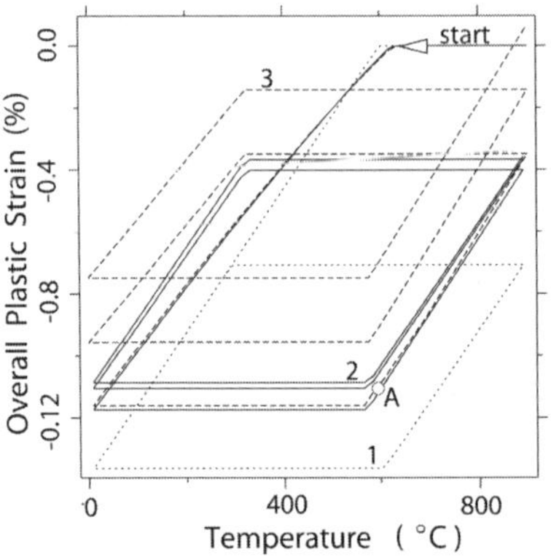

Fig. 16.8. Development of the overall axial plastic strain ε_{11}^{po} as a function of current temperature during the first four temperature cycles 900°C→0°C→900°C at fixed external loading σ_1^0=0 (dotted curve 1), 0.125 GPa (solid curve 2), 0.135 GPa (dashed curve 3).

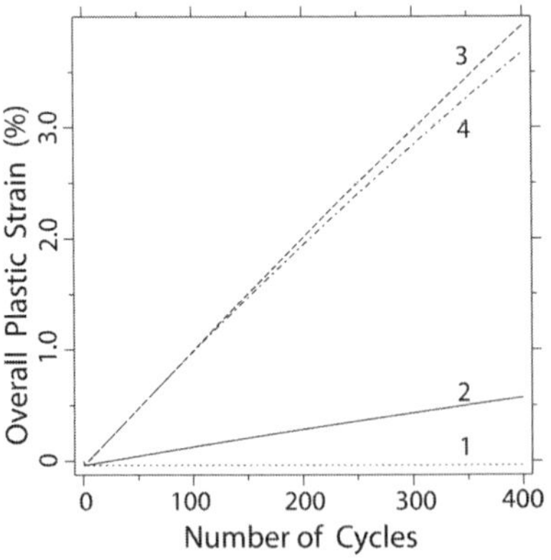

Fig. 16.9. Accumulation of ε_{11}^{po} at σ_1^0= 0.125 GPa (dotted curve 1), 0.127 GPa (solid curve 2), 0.135 GPa (dashed curve 3, dot-dashed line 4). Curves 1, 2 and 3 are obtained under the assumption of kinematic hardening, curve 4 is for mixed kinematic-isotropic hardening.

If an additional small isotropic hardening, $h^{(i)} = 0.1\tau_0^{(i)}$, $n^{(i)} = 1$ $(i = 0, 1)$, is introduced it can be observed, that the progressive plastic longitudinal strain in the "ratchetting" domain, which accumulates cycle-by-cycle, becomes bounded, i.e. it approaches an asymptotic value, leading finally to plastic or elastic "shakedown." The dot-dashed curve 4 in Fig. 16.9 shows the beginning of such a stabilization, when the curve ε^{po} vs. l deviates from the curve 3 obtained for $h^{(i)} = 0$ $(i = 0, 1)$. Similar observations were made in experimental investigations with austenitic–ferritic duplex steel specimens under pure thermal cycles [43], [1173].

16.5 A Local Theory of Elastoplastic Deformations of Metal Matrix Composites

It should be mentioned that the results presented in Section 16.4 were obtained by the use of assumption (16.55) which might be too approximate in particular cases. When plastic deformations occur, the homogeneity of the mechanical properties of the components is lost and the local properties of the phases become position dependent. Appropriate, but by no means exhaustive, references about particulate MMCs are provided in the review [489] (see also [778], [1060]). We will consider (following [186], [187]) a local theory of elastoplastic deformation of random structure composites in the framework of flow theory and small elastoplastic strains. In this model each inclusion consists of an elastic core and a layer of thin coatings. The mechanical properties of the coatings are the same as that of the matrix. Homogeneity of the plastic strains is assumed inside the matrix and in individual subdomains of the coatings, which are considered as the individual components. By this means the proposed method can be considered as a logical extension of the transformation field analysis by Dvorak [297] to random arrangement, when the phases with inhomogeneous stress states are subdivided into a finite number of subdomains with homogeneous stress states. At each step of the incremental macroscopic stresses changes of the average stresses within the components are estimated with the help of the MEFM considered in Chapter 8 for linear problems. For a single inclusion the micromechanical approach is based on the Green function technique as well as on the Hill's interfacial operators; the thin-layer hypothesis and the assumption of a homogeneous stress state in the core are used (see Chapter 4). For a dilute concentration of the inclusions the proposed model is tested by the use of finite element analysis. The employment of the proposed theory for predicting the elastoplastic deformation behavior of a model material with a finite concentration of inclusions is shown.

16.5.1 Geometrical Structure of the Components

At first, due to the analyzed model, we consider a slight modification of a statistical description of statistically homogeneous composites presented in Subsection 5.3.1. Namely, we discusses a mesodomain w with a characteristic function W containing a set $X = (V_i, \mathbf{x}_i, \omega_i)$, $(i = 1, 2, \ldots)$ of coated ellipsoidal inclusions v_i with characteristic functions V_i, centers x_i (that forms a Poisson set), semi-axes a_j^c $(j = 1, 2, 3)$, and aggregate of Euler angles ω. It is assumed that the inclusions v_i have identical mechanical and geometrical properties. Each inclusion consists of an elastic ellipsoidal core $v_i^s \subset v_i$ with semi-axes a_j^s $(j = 1, 2, 3)$, characteristic function $V_i^s(\mathbf{x})$, and a thin coating $v_i^c \equiv v_i \backslash v_i^s$ bounded by a homothetic ellipsoidal surface ∂v_i^c with semi-axes $a_j^c = a_j^s(1+\xi)$ $(j = 1,2,3,\ 0 < \xi < 1)$ and a characteristic function $V_i^c(\mathbf{x}) \equiv V_i(\mathbf{x}) - V_i^s$; here ξ is the relative thickness of the coating (see Fig. 16.10). The mechanical properties $\mathbf{g}(\mathbf{x}) = \mathbf{g}^{(0)}$ $(\mathbf{g} = \mathbf{L}, \tau^0, h, n)$ are the same for both the matrix $v^{(0)} \equiv w \backslash \cup v_i$ and the coating v_i^c. The plastic strains are constant in the matrix and are an inhomogeneous function along the surface ∂v_i^s. The volume v_i^c is subdivided along ∂v_i^s into several local volumes v_{ij}^c $(mes\ v_{ij}^c \ll mes\ v_i^c,\ j = 1,\ldots n^c)$, such that the plastic strains $\boldsymbol{\varepsilon}^p(\mathbf{x})$, $(\mathbf{x} \in v_{ij}^c)$ are constant inside the each individual subdomain v_{ij}^c. In the core of the inclusions we assume $\mathbf{g}(\mathbf{x}) = \mathbf{g}^{(0)} + \mathbf{g}_1(\mathbf{x}) = \mathbf{g}^{(0)} + \mathbf{g}_1$; $\mathbf{g}_1(\mathbf{x}) = \text{const}$ for $\mathbf{x} \in v^{(1)}$ and any elastic mismatch of the matrix and inclusion core can take place. The upper index of the material properties tensor, written in parentheses, shows the number of the respective component: $^{(0)}$ correspond to the matrix, $^{(1)}$ to the core of the inclusions and $^{(k)}$ $(k = 2, n^c+1)$ to the coating.

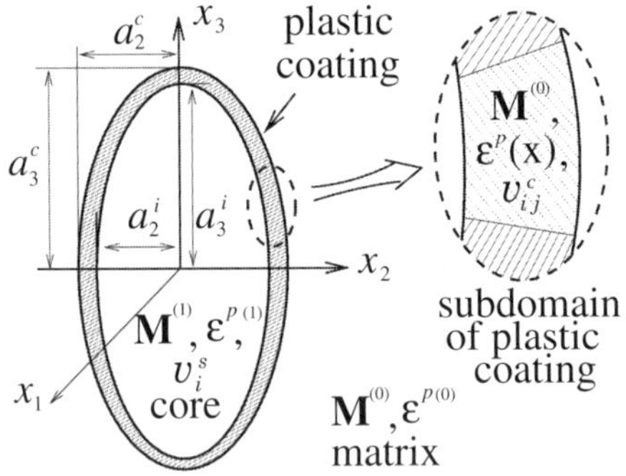

Fig. 16.10. Local model of elastoplastic deformation

It is assumed that the representative domain w contains a statistically large number of inclusions v_k; all random quantities under discussion are described by statistically homogeneous ergodic random fields and, hence, the ensemble averaging could be replaced by volume averaging $\langle(.)\rangle$ and $\langle(.)\rangle^{(k)}$ $(k = 0,\ldots,n^c+1)$ for the over-all average and for phase averages, respectively. The bar appearing above the region represents its measure, $\overline{v} \equiv \text{mes } v$. $V^{(k)}$ is the characteristic function of $v^{(k)}$. The average over an individual inclusion $v_i \in v^{(1)}$ $(i = 1,2,\ldots)$: is defined by $\langle(.)\rangle^{(i)} = \langle(.)\rangle^{(1)}$ if $v_i \in v^{(1)}$. $c^{(k)}$ is the concentration, i. e. volume fraction, of component $v^{(k)}$: $c^{(k)} = \langle V^{(k)}\rangle$; $c^{(1)} = \overline{v}_m^s n^{(1)}$, $c^{(0)} = 1 - \sum c^{(l)}$ $(k = 0,\ldots,n^c+1;\ l = 1,\ldots,n^c+1,\ m = 1,2,\ldots)$. In the following the notation $\langle(.)(\mathbf{x})|\mathbf{x}_1;\ldots;\mathbf{x}_m\rangle$ is used for the conditional average taken for the ensemble of a statistically homogeneous ergodic field $X = (v_i)$, on the condition that there are inclusions at the points $\mathbf{x}_1,\ldots,\mathbf{x}_m$ and $\mathbf{x}_1 \neq \ldots \neq \mathbf{x}_m$. The notation $\langle(.)(\mathbf{y})|;\mathbf{x}_1;\ldots;\mathbf{x}_m\rangle$ means the case $\mathbf{y} \notin v_1,\ldots,v_m$.

16.5.2 Average Stresses Inside the Components and

Overall Elastic Moduli

The average stresses within the components are estimated for linear intermediate problems with the help of the MEFM considered in Chapter 8. In view of the linearity of the problem there exist constant fourth- and second-rank tensors $\mathbf{B}^{(j)}$, and $\mathbf{C}^{(jk)}$, respectively, connecting the stresses inside the component $\mathbf{x} \in v^{(j)}$ $(j = 1, \dots n^c + 1)$ and effective field acting on the inclusion v_i being considered

$$\boldsymbol{\sigma}(\mathbf{x}) = \mathbf{B}^{(j)}\overline{\boldsymbol{\sigma}}_i + \sum_{k=1}^{n^c+1} \mathbf{C}^{(jk)}\boldsymbol{\varepsilon}_1^{p(k)}, \tag{16.72}$$

$(\mathbf{x} \in v^{(j)})$, where $\mathbf{C}^{(j1)} \equiv \mathbf{0}$. The tensors $\mathbf{B}^{(j)}$ and $\mathbf{C}^{(jk)}$ can be found in the framework of the thin-layer hypothesis considered in Subsection 3.7.2. Namely, one obtains an approximate solution (3.205) for a single ellipsoidal inclusion with thin coating under the approximate assumption of a homogeneous stress state in the core, $\boldsymbol{\sigma}^i$, and at the infinity $\overline{\boldsymbol{\sigma}}_i$: $\boldsymbol{\sigma}(\mathbf{x}) \equiv \boldsymbol{\sigma}^i = \text{const}, \quad \mathbf{x} \in v_i^s$. Then the stresses in the coating $\boldsymbol{\sigma}^c(\mathbf{s})$ are found by the relation $\boldsymbol{\sigma}^c(\mathbf{s}) = \boldsymbol{\sigma}^i + \boldsymbol{\Gamma}(\mathbf{n})\left[\mathbf{M}_1^{(1)}(\mathbf{s})\boldsymbol{\sigma}^i - \boldsymbol{\varepsilon}^p(\mathbf{s})\right]$, where the interface operator $\boldsymbol{\Gamma}(\mathbf{n})$ (3.95) is defined by Eq. (3.88); $\mathbf{n}$ is the unit outward normal vector on s_-^i in the point $\mathbf{s} \in s_-^i$, s_-^i is the outer surface of the boundary ∂v_i^s. Let the volume v_i^c be subdivided into one layer of n^c individual local volumes v_{ij}^c ($mes\ v_{ij}^c \ll mes\ v_i^c,\ j = 1, \dots n^c$), such that the plastic strain $\boldsymbol{\varepsilon}^p(\mathbf{x})$ and $\boldsymbol{\sigma}^c(\mathbf{x})$ $(\mathbf{x} \in v_{ij}^c)$ are constant inside each subdomain v_{ij}^c. For example, in a spherical coordinates system (θ, φ, r) coinciding with the semi-axes a_j^s this subdivision can be done by the use of cutting the coating v_i^c along surfaces $\theta = \text{const}$, $\varphi = \text{const}$. Estimations for the tensors $\mathbf{B}^{(j)}$, $\mathbf{C}^{(jk)}$ $(j, k = 1, \dots, n^c + 1)$ can be obtained

$$\begin{aligned}
\mathbf{B}^{(1)} &= (\mathbf{I} + \mathbf{Q}\mathbf{M}_1^{(1)})^{-1},\ \mathbf{B}^{(l)} = \left[\mathbf{I} + \tilde{\boldsymbol{\Gamma}}(\mathbf{n}_l)\mathbf{M}_1^{(1)}\right]\mathbf{B}^{(1)},\\
\mathbf{C}^{(1l)} &= -\mathbf{B}^{(1)}\mathbf{Q}\left[\delta_{1l} + \frac{\overline{v}_{il}}{\overline{v}_i}(1 - \delta_{1l})\right] + \frac{\overline{v}_{il}}{\overline{v}_i}(1 - \delta_{1l})\mathbf{B}^{(1)}\boldsymbol{\Gamma}(\mathbf{x}_i - \mathbf{y}_{il}),\\
\mathbf{C}^{(ml)} &= \left[\mathbf{I} + \tilde{\boldsymbol{\Gamma}}(\mathbf{n}_m)\mathbf{M}_1^{(1)}\right]\mathbf{C}^{(1l)} + \tilde{\boldsymbol{\Gamma}}(\mathbf{n}_m)(\delta_{1l} - \delta_{ml}),
\end{aligned} \tag{16.73}$$

The representations for both the averaged stresses and strain polarization tensors can be found from (16.72) and (16.73)

$$\begin{aligned}
\langle\boldsymbol{\sigma}\rangle^{(i)} &= \mathbf{B}\langle\overline{\boldsymbol{\sigma}}(x)\rangle^{(i)} + \mathbf{C},\quad \bar{v}_1\langle\boldsymbol{\eta}\rangle^{(i)} = \mathbf{R}\langle\overline{\boldsymbol{\sigma}}\rangle^{(i)} + \mathbf{F},\\
\mathbf{B} &\equiv \frac{1}{1 - c^{(0)}}\sum_{j=1}^{n^c+1} c^{(j)}\mathbf{B}^{(j)} = \mathbf{I} + \frac{\overline{v}_i^s}{\overline{v}_i}(\mathbf{B}^{(1)} - \mathbf{I}),\\
\mathbf{C} &= \frac{1}{1 - c^{(0)}}\sum_{j,k=1}^{n^c+1} c^{(j)}\mathbf{C}^{(jk)}\boldsymbol{\varepsilon}_1^{p(k)},
\end{aligned} \tag{16.74}$$

where $l, m = 2, \dots, n^c + 1$; $\mathbf{x}_i$ and $\mathbf{y}_{il}$ are the centers of the domains v_i and v_{il}^c, respectively; $\mathbf{n}_l$ is the unit outward normal vector on s_-^i in the point $\mathbf{y}_{il}$. The tensors $\mathbf{R}$ and $\mathbf{F}$ are found by the use of the Eshelby theorem $\mathbf{R} = \overline{v}_i\mathbf{Q}^{-1}(\mathbf{I} - \mathbf{B})$,

$\mathbf{F} = -\overline{v}_i \mathbf{Q}^{-1}\mathbf{C}$ (see Eq. (3.185)); the tensor $\mathbf{Q}$ is associated with the well-known Eshelby tensor $\mathbf{S}$ by $\mathbf{S} = \mathbf{I} - \mathbf{M}^{(0)}\mathbf{Q}$ (3.136).

In Section 8.1.4 we proved that in the framework of the effective field hypothesis **H1**) only the effective parameters $\mathbf{M}^*, \boldsymbol{\varepsilon}^p_*$ governing the overall constitutive relation $\langle \boldsymbol{\varepsilon}\rangle = \mathbf{M}^*\langle \boldsymbol{\sigma}\rangle + \boldsymbol{\varepsilon}^p_*$ as well as statistical average of the local stresses inside the inclusions have the general representation. In particular, from Eq. (16.72) we get

$$\mathbf{M}^* = \mathbf{M}^{(0)} + \mathbf{Y}\mathbf{R}n^{(1)}, \ \boldsymbol{\varepsilon}^p_* = \boldsymbol{\varepsilon}^{p(0)} + \mathbf{Y}\mathbf{F}n^{(1)}, \tag{16.75}$$

$$\langle \boldsymbol{\sigma}\rangle^{(j)} = \mathbf{B}^{*(j)}\boldsymbol{\sigma}^0 + \sum_{k=0}^{n^c+1} \mathbf{C}^{*(jk)}\boldsymbol{\varepsilon}_1^{p(k)}, \tag{16.76}$$

where $\mathbf{C}^{*(j0)}, \mathbf{C}^{*(j1)} \equiv 0$, $(j = 0, \ldots, n^c + 1)$ and $(l = 1, \ldots n^c + 1)$:

$$\mathbf{B}^{*(l)} = \mathbf{B}^{(l)}\mathbf{R}^{-1}\mathbf{Y}\mathbf{R}, \quad \mathbf{B}^{*(0)} = \frac{\mathbf{I}}{c^{(0)}} - \sum_{l=1}^{n^c+1} \frac{c^{(l)}}{c^{(0)}}\mathbf{B}^{*(l)}, \quad \mathbf{C}^{*(0l)} = -\sum_{k=1}^{n^c+1} \frac{c^{(k)}}{c^{(0)}}\mathbf{C}^{*(kl)},$$

$$\mathbf{C}^{*(jl)} = \mathbf{C}^{(jl)} + \mathbf{B}^{(j)}\mathbf{R}^{-1}(\mathbf{Y} - \mathbf{I})\left[\overline{v}_i^s \mathbf{M}_1^{(1)}\mathbf{C}^{(1l)} + \frac{c^{(l)}\mathbf{I}}{n^{(1)}}\right], \tag{16.77}$$

The tensor $\mathbf{Y}$ is determined by the purely elastic action (with $\mathbf{F} \equiv \mathbf{0}$) of the surrounding inclusions on the separated one. For a dilute concentration of the inclusions, i.e., $c^{(1)} \to 0$, we have $\mathbf{Y} \to \mathbf{I}$. The actual form of the tensor $\mathbf{Y}$, used in the analysis as an approximation, depends on additional assumptions for closing of the infinite system (8.7). For the purely elastic case (with $\mathbf{F} \equiv 0$) such relations are represented in Chapter 8. Concrete numerical analysis in Subsection 16.5.4 will be performed for $\mathbf{Y}$ obtained by the MEFM (8.52).

16.5.3 Elastoplastic Deformation

In our case for varied loading onset of yielding appears in the individual subdomain of the coating v^c_{ij} or in the remaining matrix $v^{(0)}$, respectively, if the corresponding yield criterion

$$\widetilde{\tau}^{(j)} = \tau_0^{(0)}, \quad \widetilde{\tau}^{(j)} \equiv \sqrt{1.5\langle \boldsymbol{\sigma}\rangle^{(j)} : \mathbf{N}_2 : \langle \boldsymbol{\sigma}\rangle^{(j)}}, \tag{16.78}$$

$(j = 0, 1, \ldots, n^c + 1)$ is used. The composite starts to deform plastically, when $\max_j \widetilde{\tau}^{(j)} - \tau_0^{(0)} = 0$ $(j = 0, \ldots, n^c + 1)$ is fulfilled. Up to the moment when yielding starts, the problem of estimating $\widetilde{\tau}^{(j)}$ is linear, which allows the use of the relations (16.76) for calculating the yield surface

$$\widetilde{f}(\langle \boldsymbol{\sigma}\rangle) = \max_j \sqrt{1.5(\mathbf{B}^{*(j)}\boldsymbol{\sigma}^0)^\top : \mathbf{N}_2 : (\mathbf{B}^{*(j)}\boldsymbol{\sigma}^0)} - \tau_0^{(0)} = 0, \tag{16.79}$$

$(j = 0, \ldots, n^c + 1)$. The initial yield surfaces of the components (i.e., the individual subdomains) (16.78) may be embedded within each other or they may intersect in the space of macrostresses $\langle \boldsymbol{\sigma}\rangle$. It should be mentioned that Zhu and

Weng [1230] used a criterion, equivalent to (16.78), in that approach at each increment of external loading the increment of homogeneous inelastic strains in the matrix was assumed to be the average of the local inelastic strain increments over the matrix volume. This, however, would lead to zero increments of inelastic strains in the case of hydrostatic loading of macroisotropic media.

Now we shall deal with the elastoplastic state, when under successive external loading elastic–plastic deformations will take place for the components $i_1, \ldots, i_r$

$$\widetilde{f}^{(i_q)} = 0, \quad \frac{\partial \widetilde{f}^{(i_q)}}{\partial \langle \boldsymbol{\sigma} \rangle^{(i_q)}} : d\langle \boldsymbol{\sigma} \rangle^{(i_q)} > 0, \tag{16.80}$$

where the index i_q, $(q = 1, \ldots, r)$ passes through the number $(i_1, \ldots, i_r)$ of plastically deformed components $v^{(i_q)}$, $(0 \leq i_q \leq n^c + 1,\ i_q \neq 1)$ and where a homogeneous yield criterion inside the volume $v^{(i_q)}$ is assumed:

$$\widetilde{f}^{(i_q)} \equiv \widetilde{\tau}^{(i_q)} - F^{(i_q)}(\gamma^{(i_q)}) = 0, \quad \widetilde{\tau}^{(i_q)} = \sqrt{1.5 \langle \mathbf{s}^a \rangle^{(i_q)} : \langle \mathbf{s}^a \rangle^{(i_q)}}, \tag{16.81}$$

where $\mathbf{s}^a = \mathbf{N}_2(\boldsymbol{\sigma} - \mathbf{a}^p)$, and $\mathbf{a}^p$ is a back-stresses. By using the associated flow rule, the yield function $\widetilde{f}^{(i_q)}$ is taken as plastic potential function of the matrix from which the incremental plastic strains inside the matrix can be determined as

$$d\boldsymbol{\varepsilon}^{p(i_q)} = d\lambda^{(i_q)} \frac{\partial \widetilde{f}^{(i_q)}}{\partial \langle \boldsymbol{\sigma} \rangle^{(i_q)}}. \tag{16.82}$$

At each incremental step of the external stresses homogeneous plastic strains $\boldsymbol{\varepsilon}^{p(i_q)}$ and increments of hardening parameters are assumed within each subdomain. With Ziegler's rule (2.218) we obtain

$$d\mathbf{a}^{p(i_q)} = d\gamma^{(i_q)} A \langle \mathbf{s}^a \rangle^{(i_q)}, \quad A = \frac{H^{(i_q)}}{F^{(i_q)}(\gamma^{(i_q)})}, \quad H^{(i_q)} = \frac{\partial \widetilde{\tau}^{a(i_q)}}{\partial \gamma^{(i_q)}}, \tag{16.83}$$

$$d\gamma^{(i_q)} = \sqrt{2\, d\boldsymbol{\varepsilon}^{p(i_q)} :\, d\boldsymbol{\varepsilon}^{p(i_q)}/3}, \quad d\widetilde{\tau}^{a(i_q)} \equiv \sqrt{1.5 \langle d\mathbf{s}^a \rangle^{(i_q)} : \langle d\mathbf{s}^a \rangle^{(i_q)}}. \tag{16.84}$$

where $H^{(i_q)}$ is the plastic tangent modulus, derived from the uniaxial stress-plastic strain curve of the matrix material. From the requirement of consistency of the plastic deformation process we come up with

$$\frac{\partial \widetilde{f}^{(i_q)}}{\partial \langle \boldsymbol{\sigma} \rangle^{(i_q)}} : d\langle \boldsymbol{\sigma} \rangle^{(i_q)} + \frac{\partial \widetilde{f}^{(i_q)}}{\partial \mathbf{a}^{p(i_q)}} :\ d\mathbf{a}^{p(i_q)} + \frac{\partial \widetilde{f}^{(i_q)}}{\partial \gamma^{(i_q)}} d\gamma^{(i_q)} = 0. \tag{16.85}$$

The partial derivatives in (16.85) are found under the following assumptions: $\boldsymbol{\varepsilon}^{p(i_q)}$, $\langle \boldsymbol{\sigma} \rangle^{p(i_q)} = \text{const}$ $(q = 1, \ldots, r)$. From (16.76), (16.82) and (16.83), (16.84) the relations for the differential of average stresses and hardening parameters inside the component $v^{(q)}$ can be derived as

$$d\langle \boldsymbol{\sigma} \rangle^{(i_q)} = \mathbf{B}^{*(i_q)}\, d\boldsymbol{\sigma}^0 - \sum_{k=1}^{r} \mathbf{C}^{*(i_q i_k)} \left[\lambda^{(i_k)} \frac{\partial \widetilde{f}^{(i_k)}}{\partial \langle \boldsymbol{\sigma} \rangle^{(i_k)}} - \lambda^{(0)} \frac{\partial \widetilde{f}^{(0)}}{\partial \langle \boldsymbol{\sigma} \rangle^{(0)}} \right], \tag{16.86}$$

$$d\gamma^{(i_q)} = d\lambda^{(i_q)}, \quad d\mathbf{a}^{p(i_q)} = d\lambda^{(i_q)} A(\gamma^{(i_q)}) \langle \mathbf{s}^a \rangle^{(i_q)}. \tag{16.87}$$

In Eq. (16.87) $d\lambda^{(0)} \equiv 0$ and $\partial\widetilde{f}^{(0)}/\partial\langle\boldsymbol{\sigma}\rangle^{(0)} \equiv 0$ if the conditions (16.80) are not met for the component $v^{(0)}$.

Substitution of the equations (16.86), (16.87) in (16.85) leads to the following relation for the proportionality factor:

$$d\lambda^{(i_q)} = \sum_{s=1}^{r} \beta_{qs} b_s, \qquad b_s = \frac{\partial \widetilde{f}^{(i_s)}}{\partial\langle\boldsymbol{\sigma}\rangle^{(i_s)}} : \mathbf{B}^{*(i_s)} : d\boldsymbol{\sigma}^0. \tag{16.88}$$

The elements of the inverse of the matrix $(\boldsymbol{\beta})$ are given by

$$\boldsymbol{\beta}^{-1}_{qs} = \frac{\partial \widetilde{f}^{(i_q)}}{\partial\langle\boldsymbol{\sigma}\rangle^{(i_q)}} \left[\mathbf{C}^{*(i_q i_s)}(\delta_{i_s 0}-1)+\delta_{i_s 0}\sum_{n=1}^{r} \mathbf{C}^{*(i_q i_n)}\right] \frac{\partial \widetilde{f}^{(i_s)}}{\partial\langle\boldsymbol{\sigma}\rangle^{(i_s)}} + 2\frac{\partial F^{(i_q)}}{\partial\gamma^{(i_q)}}\delta_{qs}. \tag{16.89}$$

Plastic deformations in the component $v^{(i_q)}$ take place if the condition $d\lambda^{(i_q)} > 0$ is satisfied together with the local yield criterion (16.80), (16.81_1). By this means for a prescribed loading path $\boldsymbol{\sigma}^0(t)$ (where t is a monotonically varying parameter) the system of equations (16.86) becomes a system of six ordinary differential equations for the determination of all components of the tensor $\langle\boldsymbol{\sigma}\rangle^{(q)}$. In so doing only five of these six equations are independent by virtue of the yield condition (16.80). After integration of the system (16.86) the averaged plastic strains inside each subdomain are found and the averaged strains are given by $\langle\boldsymbol{\varepsilon}\rangle^{(i_q)} = \mathbf{M}^{(0)}\langle\boldsymbol{\sigma}\rangle^{(i_q)} + \boldsymbol{\varepsilon}^{p(i_q)}$. The overall plastic strains $\boldsymbol{\varepsilon}^p_*$ and the total overall strains are defined by the relations $\boldsymbol{\varepsilon}^p_* = \sum_j (\mathbf{B}^{*(j)})^\top \boldsymbol{\varepsilon}^{p(j)} c^{(j)}$, $(j = 0,\ldots,n^c+1)$, $\langle\boldsymbol{\varepsilon}\rangle = \mathbf{M}^*\boldsymbol{\sigma}^0 + \boldsymbol{\varepsilon}^p_*$. The numerical integration of Eq. (16.86) under the restriction (16.81_1) can be carried out by different integration schemes analogous to the treatment of homogeneous materials with vertex yield surfaces as, for example, described in [806], [841], [857], [926]. The same methodology usually is applied in particular versions of mean field methods in the mechanics of composite materials [300], [493], [494], [640]. In this work a backward difference method with a variable integration step has been used as the integration scheme for Eq. (16.82). An iterative procedure is used, starting with the elastic predictor (16.88) which provides a first estimate of the stress increment in the component. This means that at each step an Euler–backward scheme is used in combination with a Newton method, with the initial guess being determined by the elastic predictor.

16.5.4 Numerical Results

We consider an isotropic composite consisting of a steel matrix and identical spherical carbidic inclusions with elastoplastic parameters:

Table 16.2. *Elastoplastic constants*

Phase	k [GPa]	ν	τ_0 [GPa]	h [GPa]	n
matrix	175	0.3	2.75	1.5	0.5
inclusions	300	0.25	∞	–	–

At first we consider the case of a dilute concentration of the inclusions $c^{(1)} \ll 1$. This leads to $\mathbf{Y} \equiv \mathbf{I}$, see (16.75), and we have the possibility to check the quality of our local plasticity model by the use of finite element analysis (FEA). Comparison of the accumulated effective plastic strains $\gamma(\mathbf{x})/c^{(1)}$ ($\mathbf{x} \in v_i^c$), calculated by the proposed analytical method with results obtained by finite element analysis (FEA), are represented in Fig. 16.11 for hydrostatic loading $\boldsymbol{\sigma}^0 = \sigma_0^0 \boldsymbol{\delta}$. FEA results are presented for two points in the coating v_i^c, near the boundaries with the core v_i^s ($|\mathbf{x}| = a + 0$) and with the matrix $v^{(0)}$ ($|\mathbf{x}| = a(1+\xi) - 0$), respectively; here ξ is the relative thickness of the coating, and two thicknesses are considered in Fig. 16.11: $\xi = 0.033$, and $\xi = 0.1$. In the local plasticity theory the subdivision of the coating v_i^c into one layer of individual subdomains v_{ij}^c $(j = 1, \ldots, n^c)$ is employed by the use of cross-sections $\theta = \pi/n^c$ in a spherical coordinate system (θ, ϕ, r), the origin of which coincides with the center of the inclusion. Hereafter we will restrict our problems to axisymmetric loading and, therefore, the plastic strains in the ribbon $a < r < a(1+\xi)$, $k\pi/n^c < \phi < (k+1)\pi/n^c$ $(k = 0, \ldots, n^c - 1)$ can be obtained by the rotation around the axis $\theta = 0$. As we see in Fig. 16.11 for the considered case of $n^c = 11$ and for a sufficiently thin coating ($\xi \le 0.1$) the proposed analytical model provides satisfactory exactness and, as a consequence, from here on we will consider the thickness $\xi = 0.1$ only.

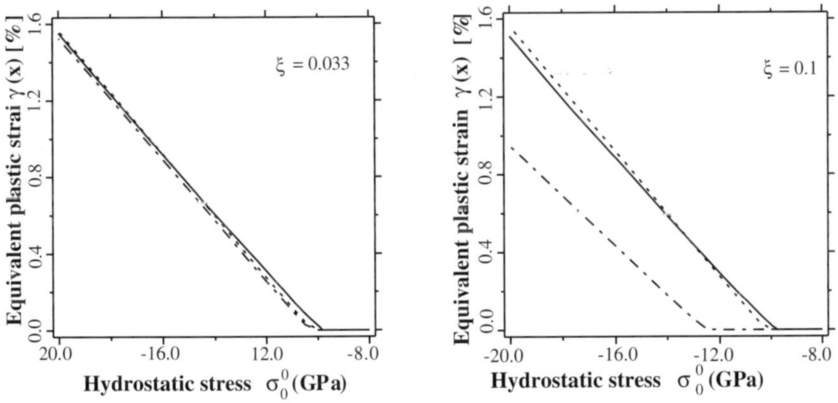

Fig. 16.11. Accumulated effective plastic strains $\gamma(\mathbf{x})/c^{(1)}$ as a function of hydrostatic loading σ_0^0 calculated via FEA (dashed curve for $|\mathbf{x}| = a + 0$, dot-dashed curve for $|\mathbf{x}| = a(1+\xi) - 0$) and by the proposed model (solid curve). a) — $\xi = 0.033$, b) — $\xi = 0.1$.

For uniaxial loading $\sigma_{ij}^0 = \sigma_{33}^0 \delta_{i3}\delta_{j3}$, $\sigma_{33}^0 \sim t$ the influence of the mesh width of the uniform subdivision (π/n^c) of the coating on the local values of the von Mises effective plastic strains $\gamma^{(i)}$ $(i = 2, \ldots, n^c + 1)$ (16.84) is studied; here t is the time, i.e. a monotonically increasing parameter. The results shown in Fig. 16.12 were calculated for $\xi = 0.1$ and for both $n^c = 11$ and for $n^c = 33$. In Fig. 16.12 the values $\gamma(\mathbf{x})$ as a function of the polar angle $0 < \theta < 90°$ are shown for $\sigma_{33}^0 = 2.8$ GPa, whereas onset of yielding takes place at $\sigma_{33}^0 = 2.2$ GPa. As can be seen for the considered case both subdivisions of the coating ($n^c = 11$ and $n^c = 33$) lead to similar results for the local values of the von Mises effective plastic strains $\gamma^{(i)}$ $(i = 2, \ldots, n^c + 1)$ (16.84). The increase of the degree of subdivision, $n^c = 33$, leads to an increase in the calculated values ε_*^{pn} which is, however, smaller than 0.5%.

Let us now study the finite concentration of carbidic inclusions ($c^{(1)} = 0.25$), for which the FEA–unit cell analysis is not able to capture real random arrangements of the inclusions. Whereas the solution of the corresponding linear problem can be found by any known method (see Chapters 8 and 9), in the current presentation we will use only the solution of the linear elastic problem (16.75), (16.76) obtained by the two–particle approximation of MEFM. Let us assume uniaxial tensional loading $\sigma_{ij}^0 = \sigma_{33}^0\delta_{i3}\delta_{j3}$, $\sigma_{33}^0 \sim t$. Figure 16.13 shows the comparison of the overall plastic strains ε_{*33}^p calculated once by the use of traditional mean field method assumptions (15.10) and once by the proposed assumptions (16.78). As can be seen, the concentration of plastic strains in the coating leads to a "softening" effect of the coated inclusions. The limiting case of the "softening" of the inclusions is the replacement of them by voids, resulting in a significant increase of the overall plastic strains [such a result is presented in Fig. 16.13 by the use of the traditional scheme (15.10), (16.80), (16.81)]. For the qualitative comparison in Fig. 16.13 the overall plastic strains are calculated also by a FEA-unit cell model for face–centered–cubic (FCC) packing of the inclusions, when the orientation of unit cell coincides with the global coordinate system. As can be seen in Fig. 16.13 the error of the proposed model (in comparison with FEA) is much smaller than that of the traditional mean field method (15.10).

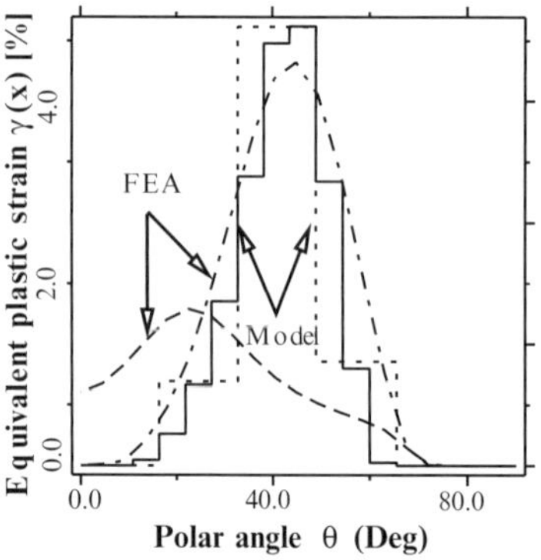

Fig. 16.12. $\gamma(\mathbf{x})$ vs θ, calculated via FEA (dot–dashed curve for $|\mathbf{x}| = a + 0$, dotted curve for $|\mathbf{x}| = a(1 + \xi) - 0$) and by the proposed model (dashed curve for $n^c = 11$, solid curve for $n^c = 33$).

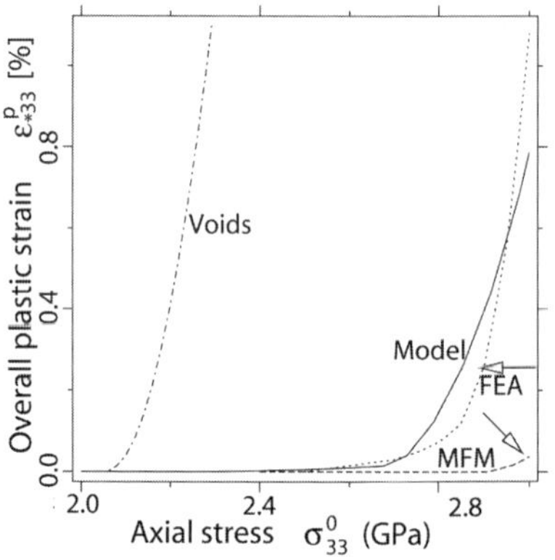

Fig. 16.13. ε_{*33}^p vs σ_{33}^0 calculated via the proposed model (16.78) (solid curve), by FEA (dotted curve) and by the traditional MFM (15.10) (dashed curve). ε_{*33}^p for model material with the voids (dot–dashed curve).

Let us now compare the estimations obtained by the proposed model and by other popular methods. There are different versions of mean field methods employing flow theory [298] and secant concept method (see, e.g., [910] and also Section 16.2). In Fig. 16.14 the curves 4 and 5 are calculated in the framework of the mean field method (15.10) (at $\xi = 0$) by the secant concept method and by the flow theory, respectively. Modified improved methods utilize the estimations of statistical averages in each component of either the stress potential or the second moment of stresses (15.11) that can be found by perturbation methods and by the method of integral equations (see Chapter 13) as well as by variational methods (for references see [897]. For our elastoplastic analysis in the framework of homogeneity of plastic strains in the matrix ($\xi = 0$) we will use the exact perturbation method in combination with the secant concept method [131] as well as the method of integral equations in combination with flow theory. The corresponding curves 2 and 3, respectively, are presented in Fig. 16.14. Finally, the proposed local plasticity model takes into account the inhomogeneity of plastic strains in some matrix layer around the inclusions ($\xi = 0.1$) as well as the inhomogeneity of the first–order moment of the stresses in the coating (curve 1). As can be seen from Fig. 16.14 the mean field method approach (curves 4 and 5) yields predictions that are stiffer than the predictions of the modified approach (curves 2 and 3). The secant concept model (curves 2 and 4) seems softer by comparison with flow theory (curves 3 and 5) both for classical and for modified methods, based on the first-order moments and second-order moments of the stresses in each individual phase, respectively. The known question regarding the correctness of the secant concept model is not discussed here. It is just mentioned that even for radial external loading the stress path at the local level is not a radial one. At the same time in the considered case ($\sigma^0_{33} = 3$ GPa) the modified secant approach (curve 2) predicts values of the overall plastic strain ε^p_{*33} which are considerably smaller than those calculated via the use of both FEA–unit cell model and the local model of plasticity for random packing of the inclusions (curve 1). As can be seen in Fig. 16.14 the better discretization of the stress–plastic strain state in the proposed model (curve 1) leads to significantly improved predictions of the overall plastic strains (with respect to the comparison with the FEA–unit cell model).

As a further example we consider a cyclic external loading, described $\boldsymbol{\sigma}^0 = \boldsymbol{\sigma}^{\text{fix}} + \boldsymbol{\sigma}^{\max} + (\boldsymbol{\sigma}^{\min} - \boldsymbol{\sigma}^{\max}))(t-1)[H(t-1) - H(1-t)]$, which is a combination of a constant hydrostatic loading $\sigma^{\text{fix}}_{ij} = \sigma^{\text{fix}}_0 \delta_{ij}$ and a uniaxial cyclic stress with zero mean-stress and an amplitude $\sigma^{am} = \sigma^{\max}_{33}$, $\boldsymbol{\sigma}^{\max} = -\boldsymbol{\sigma}^{\min}$, $\sigma^{\max}_{ij} = \sigma^{\max}_{33}\delta_{i3}\delta_{j3}$; here H is the Heaviside step function and t is the time. From Fig. 16.15 we see that the overall plastic strain component ϵ^p_{*33} after the first cycle (with $\sigma^{\max}_{33} = 2.8$ GPa) converges to a steady-state response (or steady cycle), i.e., a closed loop is obtained. The hydrostatic loading leads to a considerable variation of the closed loop of the overall plastic strains (plastic shakedown). It should be mentioned that the use of the criterion (15.10) tends to an overprediction of the initial yield stress of the composite, $\sigma^{0Y}_{33} = 2.92$ GPa instead of $\sigma^{0Y}_{33} = 2.34$ GPa obtained by the proposed method, and to independence of plastic deformations on hydrostatic loading $\boldsymbol{\sigma}^{\text{fix}} = \sigma^{\text{fix}}_0 \boldsymbol{\delta}$. Therefore, in the considered range of external loading $\sigma^{\max}_{33} < 2.92$ GPa by using (15.10) the composite material would deform elastically, and for $\sigma^{\max}_{33} > 2.92$ GPa after the first cycle of elastoplastic deformation the process would result in elastic shakedown.

The local model proposed can be applied to a wide class of nonlinear problems for which the local properties of the components become location-dependent: nonlinear elasticity and conductivity, viscosity and creeping, and viscoplasticity. It can be generalized easily to any number of thin coating layers of nonellipsoidal

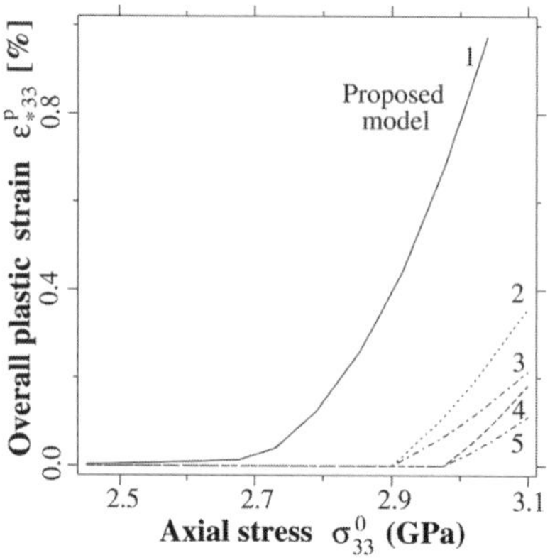

Fig. 16.14. ε^p_{*33} vs σ^0_{33} calculated via the proposed model (16.78) (1), via a modified approach (2 – secant concept method, 3 – flow theory), and via the MFM (4 – secant concept method, 5 – flow theory).

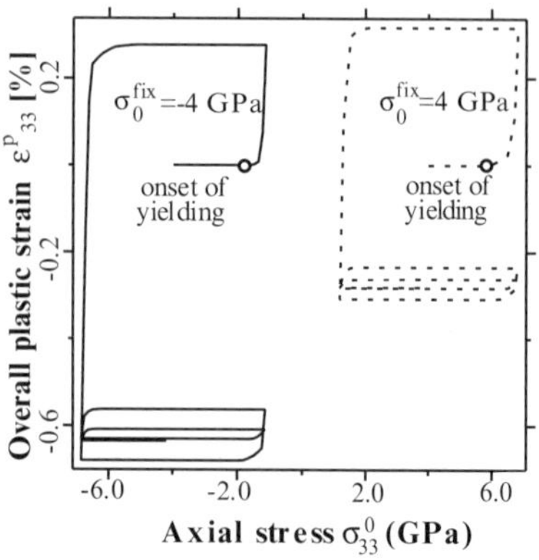

Fig. 16.15. ε^p_{*33} vs σ^0_{33} calculated for different constant hydrostatic contributions $\sigma_0^{\mathrm{fix}} = -4$ GPa (solid curve), 4 GPa (dotted curve); $\sigma_{33}^{\max} = 2.8$ GPa, (o)–onset of yielding.

inclusions (see [184]) with a distribution of size, shape, orientation and properties, to a consideration of statistical averages of the second moments of stresses in both the matrix and the coating. The possible constitutive relations are not limited to the v.Mises yield criterion assumed in this study, and modification of the present method to accommodate general yield criteria and general hardening laws can be performed. It is only important that response of a "coated" inclusion is defined by the Eqs. (16.72) and (16.74$_1$), notwithstanding the inclusion can be considered as some sort of a "black box". In particular, a coin-shaped cohesive micro-crack (BarenblattDugdale type) [681] can be analyzed as such a "black box". The principal limitation of this approach is due to the assumption of statistical homogeneity of the composite microstructure. Nevertheless, in the light of the researches on functionally graded composites performed by the MEFM (see Chapter 12), obviously, consideration of the local model of elastoplastic deformation of graded materials are also possible.

17

Some related problems

17.1 Conductivity

17.1.1 Basic Equations and General Analysis

Determining the relation between the macroscopic properties of a material and its microstructure is a very important problem not only in mechanics but also in physics of a transport phenomena. As it was mentioned in Chapter 3.8, the equations describing the steady-state conditions of such processes as heat and mass transfer, the electric conductivity and permittivity, and filtration of a Newtonian liquid in undeformable cracked-porous media are mathematically equivalent. For reasons of this mathematical analogy, we will consider the general results for estimation of effective conductivity coefficients (such as the coefficient of thermal conductivity and diffusion, electrical conductivity, dielectric constants, magnetic permeability, Darcy's constant). So, we have the basic equations of steady-state transfer process with no source terms

$$\nabla \mathbf{q} = 0, \quad \mathbf{q} = -\boldsymbol{\kappa} \nabla T, \tag{17.1}$$

where T is a potential (e.g., temperature), $\boldsymbol{\kappa}$ is a conductivity tensor, $\mathbf{q}$ is a flux vector and the first Eq. (17.1) is a transfer equation (Fourier, Ficks, Ohm, Darcy, etc.). In the case of the ideal (or perfect) contact at the interphase boundary $\mathbf{x} \in \partial v_{01}$ between the phases $v^{(0)}$ and $v^{(1)}$, the normal component of the flux $\mathbf{q_n}(\mathbf{x}) = \mathbf{q}(\mathbf{x})\mathbf{n}(\mathbf{x})$ and potential $T(\mathbf{x})$ are continuous across the interphase $[[\mathbf{q_n}(\mathbf{x})]] = 0$, $[[T(\mathbf{x})]] = 0$, where $[[\mathbf{f}]]$ is the jump of $\mathbf{f} = \mathbf{q_n}, T$ across $\mathbf{x} \in \partial v_{01}$ with the normal $\mathbf{n}(\mathbf{x}) \perp \partial v_{01}$. Interphase boundary conditions are accompanied by one of two homogeneous boundary conditions ($\mathbf{x} \in \partial w$):

$$T(\mathbf{x}) = \mathbf{G}^{\partial w} \cdot \mathbf{x}, \quad \text{or} \quad \mathbf{q_n}(\mathbf{x}) = \mathbf{Q}^{\partial w} \cdot \mathbf{n}^{\partial w}(\mathbf{x}), \tag{17.2}$$

at the boundary ∂w ($v^{(0)}, v^{(1)} \subset w$) of the domain w containing statistically large number of inhomogeneities; here $\mathbf{G}^{\partial w}$ and $\mathbf{Q}^{\partial w}(\mathbf{x})$ are the given constant symmetric matrix.

Equations (17.1) have the same structure as Eqs. (6.1) and (6.2) but the lower dimension which simplifies its analysis. Because of this, the analysis of average schemes of the conductivity equation apart from their intrinsic interest can be considered as an encouragement of the analysis of relevant elastic problems. It is sufficient to use the effective conductivity coefficients (such as the coefficients of thermal conductivity and diffusion, electrical conductivity, dielectric constant, permeance, Darcy's constant etc.) for the medium as a whole.

The solution found in Section 3.8 for single and pair-interacting inclusions in the infinite matrix enables one to realize any average scheme described in Chapter 8. Depending on the numerous ways of closing the integral equations involved (analogous to Eqs. (8.8)–(8.11), the methods in this case are classified in an analogy with Chapter 8. Four groups of methods for determining the effective coefficients are known. The first group is that of model treatments, replacing the real stochastic structure of composites by specific ones that have intuitive physical interpretation and lead to particular exact results. Probably the most famous special models are the fractal-type microinhomogeneous media composed from composite Hashin's spherical or cylindrical coated inclusions [430], [440] (see also [65], [775]) of all possible sizes, from finite down to the infinitesimally small, as well as Dykhne's [304] two-dimensional, two-phase composites with statistically equivalent phases. Another sort of a special structure are the regular ones ([285], [556], [753], [756]) or periodic structures containing randomly dispersed inhomogeneities in the unit cell ([414], [627], [964]); percolation models, in particular, belong to this group ([957], [1005]). The perturbation methods of the second group give correct results in the case of the availability of a small parameter such as either the small differences of the conduction coefficients ([976], [977], [995]) of the components or the small concentration of inhomogeneities as in the perturbation method of the second order (9.18) ([233], [498], [706], [707]). Variational methods providing two wide bounds of effective properties for the strong-contrast composites are presented by two-point bounds by Hashin and Shtrikman [440] (see also [430]), and by three-point bounds by Beran [67] (see also [771], [772], [1102], as well as Sections 6.5 and 6.6). The fourth group of self-consistent methods is based on expressing the solution of the steady-state transfer equation with random rapidly oscillating coefficients in terms of the Green's function of the analogous equation for a homogeneous medium. Depending on the numerous ways of closing the integral equations obtained, the methods in this case are classified as the method of effective medium ([120], [429], [579]), generalized self-consistent method [769], differential method ([774], [873], [1234]), Mori-Tanaka-Eshelby method of the average fields ([51], [212], [443], [444], [1166]), the singular approximation method [1007], the method of conditional moments [568], and the MEF method [165]. Below we briefly summarize mentioned results keeping in mind that they were usually obtained by the methods analogous to ones considered in Chapter 8 (see also [112], [644], [734], [775], [995], [1106]).

For statistically homogeneous media with homogeneous boundary conditions either (17.2_1) or (17.2_2), the analog of Hill's condition (6.15) $\langle \mathbf{q}\nabla T\rangle = \langle \mathbf{q}\rangle\langle\nabla T\rangle$ yields equivalence of two definitions of effective conductivity $\boldsymbol{\kappa}^*$:

$$\langle \mathbf{q}\rangle = -\boldsymbol{\kappa}^*\langle\nabla T\rangle, \quad \langle \mathbf{q}\nabla T\rangle = -\boldsymbol{\kappa}^*\langle\nabla T\rangle\langle\nabla T\rangle \tag{17.3}$$

or resistivity k^*

$$\langle \nabla T \rangle = -\mathbf{k}^* \langle \mathbf{q} \rangle, \quad \langle \mathbf{q} \nabla T \rangle = -\mathbf{k}^* \langle \mathbf{q} \rangle \langle \mathbf{q} \rangle, \tag{17.4}$$

where the local conductivity and resistivity are each other's inverse $\boldsymbol{\kappa}(\mathbf{x}) = \mathbf{k}^{-1}(\mathbf{x})$ by definition, while the consistency of the definitions of the effective conductivity and effective resistivity $\boldsymbol{\kappa}^* = (\mathbf{k}^*)^{-1}$ is followed from comparison of (17.3_2) and (17.4_2). The effective parameters $\boldsymbol{\kappa}^*$ and $\mathbf{k}^*$ can be represented by means of concentrator factors $\mathbf{A}^*(\mathbf{x})$ and $\mathbf{B}^*(\mathbf{x})$:

$$\boldsymbol{\kappa}^* = \langle \boldsymbol{\kappa} \mathbf{A}^* \rangle, \quad \mathbf{k}^* = \langle \mathbf{k} \mathbf{B}^* \rangle, \tag{17.5}$$

where the tensors $\mathbf{A}^*(\mathbf{x})$ and $\mathbf{B}^*(\mathbf{x})$ with the properties $\langle \mathbf{A}^* \rangle = \boldsymbol{\delta}$, $\langle \mathbf{B}^* \rangle = \boldsymbol{\delta}$ are defound by the equalities: $\nabla T(\mathbf{x}) = \mathbf{A}^*(\mathbf{x})\mathbf{G}^{\partial w}$, $\mathbf{q}(\mathbf{x}) = \mathbf{B}^*(\mathbf{x})\mathbf{Q}^{\partial w}$ for the boundary conditions (17.2_1) and (17.2_2), respectively. Introduction of the concentrator factors $\mathbf{A}^*(\mathbf{x})$ and $\mathbf{B}^*(\mathbf{x})$ is simply a notational convenience providing, however, a natural and useful guide for construction of different approximate schemes for the estimations of effective properties.

We will perform an estimation of effective properties basing on the general integral equation analogous to (7.19) and (7.20):

$$\nabla T(\mathbf{x}) = \langle \nabla T \rangle(\mathbf{x}) + \int \mathbf{U}(\mathbf{x}-\mathbf{y})[\boldsymbol{\kappa}_1(\mathbf{y})\nabla T(\mathbf{y}) - \langle \boldsymbol{\kappa}_1 \nabla T \rangle(\mathbf{y})]\, d\mathbf{y}, \tag{17.6}$$

$$\mathbf{q}(\mathbf{x}) = \langle \mathbf{q} \rangle(\mathbf{x}) + \int \boldsymbol{\Gamma}(\mathbf{x}-\mathbf{y})[\mathbf{k}_1(\mathbf{y})\mathbf{q}(\mathbf{y}) - \langle \mathbf{k}_1 \mathbf{q} \rangle(\mathbf{y})]\, d\mathbf{y}, \tag{17.7}$$

where $\mathbf{U}(\mathbf{x}) = \nabla\nabla \mathbf{G}^c(\mathbf{x})$ is a second derivative of a Green's function $\mathbf{G}^c$ (3.214) for an infinite homogeneous matrix with the conductivity $\boldsymbol{\kappa}^c(\mathbf{x}) \equiv$const, $\boldsymbol{\Gamma}(\mathbf{x}) = -\boldsymbol{\kappa}^c[\boldsymbol{\delta}\delta(\mathbf{x}) - \mathbf{U}(\mathbf{x})\boldsymbol{\kappa}^c]$, and $\mathbf{f}_1(\mathbf{y}) \equiv \mathbf{f}^{(1)}(\mathbf{y}) - \mathbf{f}^c$ ($\mathbf{f} = \boldsymbol{\kappa}, \mathbf{k}$), where $\mathbf{f}^c$ is the properties of comparison medium.

Justification of the term $\langle \boldsymbol{\kappa}_1 \nabla T \rangle(\mathbf{y})$ providing an absolute convergence of the integral (17.6) was considered at a different level of rigor via the method of an additional subtraction term [40], via the renormalization method ([465], [498], [499]), and by the method [823] estimated influence of a macroscopic boundary. Obviously, the more general Eqs. (17.6) and (17.7) proposed for statistically inhomogeneous media can be easily verified analogously to Eq. (7.19).

17.1.2 Perturbation Methods

Let us consider the weakly inhomogeneous macroscopically isotropic media with the local conductivity represented in the form

$$\boldsymbol{\kappa}(\mathbf{x}) = \bar{\boldsymbol{\kappa}} + \boldsymbol{\kappa}'(\mathbf{x}) \tag{17.8}$$

with a small fluctuation of $\boldsymbol{\kappa}'(\mathbf{x})$ about its mean value $\bar{\boldsymbol{\kappa}} = \langle \boldsymbol{\kappa} \rangle = \bar{\varkappa}\boldsymbol{\delta}$: $\max_{\mathbf{x}\in w} ||\boldsymbol{\kappa}'(\mathbf{x})\bar{\boldsymbol{\kappa}}^{-1}|| \ll 1$. Following the scheme of the correlation approximation method [995], substitution of Eq. (17.8) into Eq. (17.6) at $\boldsymbol{\kappa}^c \equiv \langle \boldsymbol{\kappa} \rangle$ and its first iteration of the successive approximation method lead to

$$\langle \boldsymbol{\kappa}'(\mathbf{x})\nabla T(\mathbf{x}) \rangle = -\frac{1}{3\bar{\varkappa}}\mathbf{B}^{\varkappa}(\mathbf{0})\mathbf{G}^{\partial w} \tag{17.9}$$

Here we used a Green's function (3.215) with the corresponding Eshelby tensor $\mathbf{S}(\mathbf{x}) = \boldsymbol{\delta}/3$ (see (3.225)) and a central symmetry of the binary correlation function of the conductivity tensor $\mathbf{B}^{\varkappa}(\mathbf{x}) = \langle \boldsymbol{\kappa}'(\mathbf{0}) \otimes \boldsymbol{\kappa}'(|\mathbf{x}|)\rangle$, related to the binary correlation function of the indicator function $\mathbf{B}^V_{ij}(\mathbf{x}) \equiv \langle V^{(i)\prime}(\mathbf{0})V^{(j)\prime}(\mathbf{x})\rangle$ $[V^{(i)\prime}(\mathbf{x}) \equiv V^{(i)}(\mathbf{x}) - c^{(i)}]$: $\mathbf{B}^{\varkappa}(\mathbf{x}) = \sum_i \sum_j \boldsymbol{\kappa}^{(i)\prime} \otimes \boldsymbol{\kappa}^{(j)\prime} \mathbf{B}^V_{ij}$ if the conductivity tensor $\boldsymbol{\kappa}(\mathbf{x}) = \sum_i \boldsymbol{\kappa}^{(i)} V^{(i)}(\mathbf{x})$ is described by a piecewise function: $\boldsymbol{\kappa}^{(i)} \equiv$ const., $i = 0, 1, \ldots$. In its turn, $\mathbf{B}^V_{ii}(\mathbf{x})$ is the autocovariance of phase $v^{(i)}$ (5.84). In particular, for the macroscopically isotropic composites $\boldsymbol{\kappa}^* = \varkappa^* \boldsymbol{\delta}$ with isotropic phases $\boldsymbol{\kappa}(\mathbf{x}) = \varkappa(\mathbf{x})\boldsymbol{\delta}$, substitution of Eq. (17.9) into (17.5$_1$) yields the following relation for effective conductivity

$$\frac{\varkappa^*}{\bar{\varkappa}} = 1 - \frac{1}{3}\frac{\langle \varkappa^2 \rangle - \bar{\varkappa}^2}{\bar{\varkappa}^2} \tag{17.10}$$

proposed in [113], and rederived by many authors (see e.g. [642]). In a similar manner, we can get a formula for the effective resistivity $\mathbf{k}^* = k^* \boldsymbol{\delta}$: $k^*/\bar{k} = 1 - 2(\langle k^2 \rangle - \bar{k}^2)/(3\bar{k}^2)$, where $\mathbf{k}^* \not\equiv (\boldsymbol{\kappa}^*)^{-1}$, $\mathbf{k}(\mathbf{x}) = k(\mathbf{x})\delta$.

Continuation of the described scheme of the successive approximation method and holding of subsequent terms allows the expansion of the effective conductivity to be found in powers of the difference in the phase conductivities. Hence Sen and Torquato [983] estimated the effective conductivity of anisotropic two-phase composite medium with the isotropic phases $\boldsymbol{\kappa}^{(p)} = \varkappa^{(p)} \boldsymbol{\delta}$ as

$$\boldsymbol{\kappa}^* = \boldsymbol{\kappa}^{(p)}\boldsymbol{\delta} + \sum_{n=1}^{m} \mathbf{a}_n^{(p)} \Big(\frac{\varkappa^{(p)} - \varkappa^{(q)}}{\varkappa^{(q)}}\Big)^n + o\Big(\Big(\frac{\varkappa^{(p)} - \varkappa^{(q)}}{\varkappa^{(q)}}\Big)^n\Big) \tag{17.11}$$

where $(p = 1 - q,\ q = 0, 1)$ and the anisotropic tensors are expressed through the n-point correlation functions (5.82) of the indicator function $V^{(p)}$. In particular, for macroscopically isotropic media in dD case $(d = 2, 3)$, the first parameters $\mathbf{a}_n^{(p)} = a_n^{(p)} \boldsymbol{\delta}$ are

$$a_1^{(p)} = c^{(p)}, \quad a_2^{(p)} = -d^{-1} c^{(p)} c^{(q)}, \quad a_3^{(p)} = d^{-2} c^{(p)} c^{(q)} [c^{(q)} + (d-1)\zeta_p], \tag{17.12}$$

where the so-called ζ_p $(p = 1,\ q = 0)$ parameter is defined by Eq. (6.252) at $d = 3$ and is presented in [1106]) at $d = 2$. Formulae (17.10) and (17.12) shows that to the order $o((\varkappa_1^{(1)})^2)$ the effective conductivity of macro-isotropic media is not sensitive to the specific details of the microstructure described by the function $\mathbf{B}^V_{pp}(|\mathbf{x}|)$ and depends only on $\mathbf{B}^V_{pp}(\mathbf{0})$. However, the higher orders $m \geq 3$ will depend on the total information about the n-point probability densities $S_m^{(p)}(\mathbf{y}^m)$ (5.82). In so doing $\mathbf{a}_2^{(p)}$ depends on $\mathbf{B}^V_{pp}(\mathbf{x})$ for the general case of macro-anisotropic media.

Let us consider another sort of perturbation methods when a concentration of inhomogeneities is a small parameter in the expansion of the effective conductivity of matrix composites

$$\boldsymbol{\kappa}^* = \boldsymbol{\kappa}^{(0)} + \sum_{n=1}^{m} \mathbf{b}_n c^n + o(c^m), \tag{17.13}$$

where the coefficients $\mathbf{b}_n$ are independent of the volume concentration of inclusions $c = c^{(1)}$ but depend on the conductivities of the phases.

At the dilute concentration of inhomogeneities when an interaction between the inclusions is negligible and an effective field acting on each inclusion coincides with an applied one (17.2$_1$) $\overline{\nabla T} = \mathbf{G}^{\partial w}$, we get

$$\boldsymbol{\kappa}^* = \boldsymbol{\kappa}^{(0)} + n\mathbf{R}^t + o(c), \tag{17.14}$$

where n is a numerical density of inhomogeneities: $n = c/\bar{v}_i$, and the tensor $\mathbf{R}^t$ is defined through the averaged polarization tensor $\bar{v}_i\boldsymbol{\kappa}_1\nabla T(\mathbf{x}) = \mathbf{R}_i^t(\mathbf{x})\mathbf{G}^{\partial w}$, $\mathbf{R}^t \equiv \langle\mathbf{R}_i^t\rangle$ of all inclusions $\mathbf{x} \in v_i$ analogously to the tensor $\mathbf{R}_i^\epsilon(\mathbf{x})$ (4.12). The polarizability tensor $\mathbf{R}_i^t(\mathbf{x})$ of the inclusion v_i is also called the Pólya-Szegó matrix [892]. For the homogeneous ellipsoidal inclusion, the tensor $\mathbf{R}_i^t(\mathbf{x}) \equiv \mathbf{R}_i^t = \text{const.}$ is analytically expressed through the *depolarization* tensor $\mathbf{S}_i$ (3.225) similar to the Eshelby tensor (3.135)

$$\mathbf{R}_i^t = \bar{v}_i\boldsymbol{\kappa}_1^{(i)}\mathbf{A}_i^t, \quad \mathbf{A}_i^t = (\boldsymbol{\delta} + \mathbf{S}_i\mathbf{k}^{(0)}\boldsymbol{\kappa}_1^{(i)})^{-1}. \tag{17.15}$$

In particular for the isotropic matrix and isotropic spheroidal inclusion with axis symmetry $\mathbf{n}$, we have (see (3.225))

$$\mathbf{S}_i = (1-2M)\boldsymbol{\eta}+M\boldsymbol{\nu}, \;\; \mathbf{R}_i^t = \frac{\varkappa_1^{(1)}\varkappa^{(0)}\bar{v}_i}{\varkappa^{(0)} + \varkappa_1^{(1)}(1-2M)}\boldsymbol{\eta}+\frac{\varkappa_1^{(1)}\varkappa^{(0)}\bar{v}_i}{\varkappa^{(0)} + \varkappa_1^{(1)}M}\boldsymbol{\nu}, \tag{17.16}$$

where the depolarization factor M is defined by Eq. (3.229) and a representation for $\mathbf{R}_i^t$ was obtained by the use of the operations with the basis tensors $\boldsymbol{\eta} = \mathbf{n}\otimes\mathbf{n}$ and $\boldsymbol{\nu} = \boldsymbol{\delta} - \boldsymbol{\eta}$ described in Appendix A.2.2. For a spherical anisotropic inclusion having an isotropic depolarization factor $\mathbf{S}_i = \boldsymbol{\delta}/3$, the polarizability tensor is presented by a simple formula $\mathbf{R}_i^t = \bar{v}_i\varkappa^{(0)}(\boldsymbol{\kappa}^{(1)} - \varkappa^{(0)}\boldsymbol{\delta})(\boldsymbol{\kappa}^{(1)} + 2\varkappa^{(0)}\boldsymbol{\delta})^{-1}$. If such the spheres have uniform distributed random orientation, their average polarizability tensor $\mathbf{R}^t = \langle\mathbf{R}_i^t\rangle$ is isotropic

$$\mathbf{R}^t = \bar{v}_i\boldsymbol{\delta}\varkappa^{(0)}\sum_{i=1}^{3}\frac{\lambda_i - \varkappa^{(0)}}{3(\lambda_i + 2\varkappa^{(0)})}, \tag{17.17}$$

where λ_i are the eigenvalues of $\boldsymbol{\kappa}^{(1)}$. In a similar manner, for the spheroidal inhomogeneities (17.16) with uniformly distributed random orientation, the tensor $\mathbf{R}^t$ is also isotropic

$$\mathbf{R}^t = R_0^t\boldsymbol{\delta}, \; R_0^t = \frac{\bar{v}_i}{3}\left[\frac{2\varkappa_1^{(1)}\varkappa^{(0)}}{\varkappa^{(0)} + \varkappa_1^{(1)}M} + \frac{\varkappa_1^{(1)}\varkappa^{(0)}}{\varkappa^{(0)} + \varkappa_1^{(1)}(1-2M)}\right], \tag{17.18}$$

where we used the averages $\langle\boldsymbol{\eta}\rangle = \boldsymbol{\delta}/3$, $\langle\boldsymbol{\nu}\rangle = 2\boldsymbol{\delta}/3$. In the limiting cases of spheroidal inclusions with $M = 1/2$ (cylinders) and $M = 0$ (disc-shape inclusions) the representation (17.18) becomes

$$R_0^t = \frac{\bar{v}_i\varkappa_1^{(1)}(5\varkappa^{(0)} + \varkappa^{(1)})}{3(\varkappa^{(0)} + \varkappa^{(1)})}, \quad R_0^t = \frac{\bar{v}_i\varkappa_1^{(1)}(\varkappa^{(0)} + 2\varkappa^{(1)})}{\varkappa^{(1)}}, \tag{17.19}$$

respectively.

Estimation of the higher order terms in the expansion (17.13), starting with $o((c^{(1)})^2)$, requires an analysis of m multiparticle interaction. So, Jeffrey [498] solved exactly the problem of two identical spheres in a uniform field $\mathbf{Q}$ prescribed at infinity by using bispherical coordinates and Legendre polynomials. He obtained an explicit representation for the c^2 coefficient of the expansion (17.13) in terms of the radial distribution function $g(r)$, which was generalized in [236] (see also [273], [467]) to the d-dimensional case. The problem for two spherical isotropic inclusions with different radii and properties was solved in [1043]. Pair interactions of coated or debonded spheres was estimated in [706] by using a twin spherical expansion technique. The same problem for two aligned spheroidal inclusions was solved in [707] via the boundary collocation method; Chen and Yang [224] used a generalization of the method [789] of polynomial representations of the effective fields and stress fields inside the inclusions; consequently, the case the matrix anisotropy was analyzed via the use of an appropriate coordinate transform reducing the initial anisotropic problem to to an analogous problem but with an isotropic matrix. The fast multipole method was used in [401] for the interaction analysis of numerous inclusions of arbitrary shape. Thorpe [1089] proposed conformal mapping to study a medium containing the non-conductive or superconducted inclusions of polygonal shape.

Buryachenko and Murov [165] used the solution [498] in the form of a slowly convergent series to demonstrate the high accuracy of the solution obtained in the framework of the hypothesis **H1)** analogous to Eq. (4.57). Namely, the exact expression of the order c^2 (10.99) can easily be recast in the form

$$\boldsymbol{\kappa}^* = \boldsymbol{\kappa}^{(0)} + n\mathbf{R}^t + n^2\mathbf{R}^t\mathbf{P}^0\mathbf{R}^t + n\mathbf{R}^t\mathbf{L}_{ij}, \tag{17.20}$$

where the term $\mathbf{L}_{ij}$ is defined by the solution of binary interacting inclusions v_i and v_j (10.28). In the case of its approximate estimation in the framework of the effective field hypothesis **H1**, Eq. (10.121) is reduced to

$$\mathbf{L}_{ij}=\int\mathbf{T}^t_{ij}(\mathbf{x}_i-\mathbf{x}_j)\Big\{[\boldsymbol{\delta}-\mathbf{R}^t\mathbf{T}^t_{ij}(\mathbf{x}_i-\mathbf{x}_j)]^{-1}\varphi(v_j,\mathbf{x}_j|;v_i,\mathbf{x}_j)-(1-V^0_i(\mathbf{x}_j)n\Big\}d\mathbf{x}_j\mathbf{R}^t; \tag{17.21}$$

where analogously to the tensor $\mathbf{T}_{ij}$ (4.56) for the elastic problem ($\mathbf{y}\in v_j,\ \mathbf{x}\in v_j$): $\mathbf{T}^t_i(\mathbf{x}-\mathbf{x}_j) = \langle\mathbf{U}(\mathbf{x}-\mathbf{y})\rangle_j$, $\mathbf{T}^t_{ij}(\mathbf{x}_i-\mathbf{x}_j) = \langle\mathbf{T}^t_j(\mathbf{x}-\mathbf{x}_j)\rangle_i$. For the d-dimensional ($d = 2,3$) spherical inclusions in the isotropic matrix the tensor $\mathbf{T}_j(\mathbf{r})$ ($\mathbf{r}=\mathbf{x}-\mathbf{x}_j$) is well known [1106]: $\mathbf{T}^t_j(\mathbf{r}) = (\,d\mathbf{n}\otimes\mathbf{n}-\boldsymbol{\delta})/[2(d-1)\pi\varkappa^{(0)}r^d]$, where $\mathbf{n} = \mathbf{r}/r,\ r = |\mathbf{r}|$. The tensor $\mathbf{L}_{ij}$ has a meaning of a perturbation of the effective field acting on the inclusion v_i introduced by the direct impact of surrounding inclusions v_j. Obviously, $\mathbf{L}_{ij}$ exactly vanishes in the case of exploiting of conductivity analog of the quasi-crystalline approximation by Lax [648] (see Eq. (8.65)). The approximation mentioned was implicitly used in [775] in the form of an assumption about well-separated inclusions. The formula (17.20) with $\mathbf{L}_{ij} = \mathbf{0}$ can also be obtained by expanding the Hashin-Shtrikman lower bound (see Section 6.5) through second the order in c. In parallel with an approximate estimation of the tensor $\mathbf{L}_{ij}$ (17.21), we will reproduce the accurate result $\mathbf{L}_{ij} = L^0_{ij}\boldsymbol{\delta}$ obtained in [498] for identical spheres in the form of a slowly convergent series, whose first terms are

$$L^0_{ij} = c\Big(\frac{\gamma^2}{4} + \frac{3}{16}\frac{\varkappa^{(1)}}{2\varkappa^{(0)} + 3\varkappa^{(1)}}\gamma^2 + \frac{1}{64}\gamma^3 + \ldots\Big), \quad \Big(\gamma = \frac{R^t_0}{\varkappa^{(0)}\overline{v}}\Big). \tag{17.22}$$

17.1.3 Self-Consistent Methods

If the volume fraction of inhomogeneities is high enough ($c \geq 0.1$), a correctness of the dilute approximation ignoring interaction between inclusions becomes questionable. A large number of approximations for estimating this interactions have been proposed with different levels of justification. All of them use as a basic element a solution for one inhomogeneity in the infinite matrix. We will consider the results of applicability to the conductivity problems of so-called self-consistent methods described in detail in Chapter 8 with respect to the elastic problems. These are the effective medium method, Mori-Tanaka method, MEFM, differentiation method, and their modifications.

A basic assumption of the classical self-consistent method is to consider each inhomogeneity as isolated, but incorporated into a matrix with the unknown effective properties of the composite. Then, the effective conductivity $\boldsymbol{\kappa}^*$ can be found from Eq. (17.14) where the concentrator factor $\mathbf{A}^t_i$ (17.15) of the phase $v^{(i)}$ ($i = 1, \ldots, N$) is estimated by the formula (17.15_2) by the use of replacement $\boldsymbol{\kappa}^{(0)} \to \boldsymbol{\kappa}^*$:

$$\boldsymbol{\kappa}^* = \boldsymbol{\kappa}^{(0)} + \sum_{i=1}^{N} c^{(i)}(\boldsymbol{\kappa}^{(i)} - \boldsymbol{\kappa}^{(0)})\mathbf{A}^t_i(\boldsymbol{\kappa}^*), \tag{17.23}$$

where $\mathbf{A}^t_i(\boldsymbol{\kappa}^*) = [\boldsymbol{\delta} + \mathbf{S}_i(\boldsymbol{\kappa}^*)(\boldsymbol{\kappa}^*)^{-1}(\boldsymbol{\kappa}^{(i)} - \boldsymbol{\kappa}^*)]^{-1}$. For the spheroidal isotropic inclusions with uniform random orientation in an isotropic matrix Eq. (17.23) can be simplified by the use of Eq. (17.18):

$$\varkappa^* = \varkappa^{(0)} + \sum_{i=1}^{N} c^{(i)}\Big[\frac{2(\varkappa^{(i)} - \varkappa^*)\varkappa^*}{\varkappa^*(1 - M_i) + \varkappa^{(i)}M_i} + \frac{(\varkappa^{(i)} - \varkappa^*)\varkappa^*}{2\varkappa^* M_i + \varkappa^{(i)}(1 - 2M_i)}\Big]. \tag{17.24}$$

For spherical inclusions ($M_i = 1/3$), Eq. (17.24) is reduced to the form generalized to dD ($d = 2, 3$) case

$$\sum_{i=0}^{N} \frac{c^{(i)}(\varkappa^{(i)} - \varkappa^*)}{\varkappa^{(i)} + (d-1)\varkappa^*} = 0 \tag{17.25}$$

proposed in [114] for two-phases composites (see also [643], [1232]), and in [1106] for dD case.

Since the obtained representation (17.25) is symmetric with respect to the simultaneous interchange $\varkappa^{(i)} \leftrightarrow \varkappa^{(j)}$, $c^{(i)} \leftrightarrow c^{(j)}$ ($i = 0, 1, \ldots, N$), it is often called a symmetric effective medium (EM) approximation. The mentioned symmetry of Eq. (17.25) implies that each phase is treated on the same footing, although we initially considered a matrix structure composite with one continuous phase $v^{(0)}$ containing the other dispersed phases. Notwithstanding the fact of this inherent inconsistency, the EM approach is a realizable at least for Eq.

(17.25) ($N = 1$) in the sense that there is a particular random structure composite whose effective conductivity exactly coincides with the EM estimation (17.25).

Such a fractal-like (or self-similar in all -scale) structure justified in [774] consists of two-sort granular $v^{(0)}$ and $v^{(1)}$ of dilute concentration randomly distributed in the matrix consisting in turn of much smaller similar grains, and so on from finite size of inclusions to infinitesimally small. Realizability of the EM approach (17.25) provides a location of the estimations (17.25) within Hashin-Shtrikman bounds. However, Torquato [1106] demonstrated that the EM approach can violate three-point bounds (6.263) taking more realistic geometrical model of composites into account. Another inconsistency is related with the assumption that each inclusion is incorporated into the homogeneous medium with effective properties of the composite that completely ignores the availability of some matrix gaps surrounding the test inclusion. Acrivos and Chang [5] eliminated this disadvantage by covering the test inclusion by a layer of the matrix with subsequent incorporation of the composed inhomogeneity into the effective medium. A popular choice justification of a thickness h of the matrix layer surrounding inclusion with a radius a as $a^3/(a+h)^3 = c^{(1)}$ cannot be recognized as a rigorous.

It should be mentioned that the EM approach (17.23) is based on Eq. (17.14) obtained by the averaging of the flux $\mathbf{q}(\mathbf{x}) = \boldsymbol{\kappa}(\mathbf{x})\nabla T(\mathbf{x})$ over the macrodomain w at the boundary conditions (17.2_1). The average field approximation [889] uses the averaging operations of the different fields such as, e.g., a temperature gradient $\nabla T(\mathbf{x}) = \mathbf{k}(\mathbf{x})\mathbf{q}(\mathbf{x})$ and a polarization tensor $\boldsymbol{\kappa}_1(\mathbf{x})\nabla T(\mathbf{x})$. For example, averaging of the field $\nabla T(\mathbf{x}) = \mathbf{k}(\mathbf{x})\mathbf{q}(\mathbf{x})$ at the boundary condition (17.2_2) and at the same assumption for estimation of a concentrator factor $\mathbf{B}^*$ (17.5) as in the EM approach leads to the effective resistance representation

$$\mathbf{k}^* = \mathbf{k}^{(0)} + \sum_{i=1}^{N} c^{(i)}(\mathbf{k}^{(i)} - \mathbf{k}^{(0)})\mathbf{B}_i^t(\mathbf{k}^*), \tag{17.26}$$

$\mathbf{B}_i^t(\mathbf{k}^*)$ is a concentrator factor for a single inclusion v_i in the medium with the effective properties $\varkappa^*$: $\langle \mathbf{q} \rangle_i = \mathbf{B}_i^t(\mathbf{k}^*)\mathbf{Q}^{\partial w}$. For an ellipsoidal homogeneous inclusion this tensor is well known $\mathbf{B}_i^t(\mathbf{k}^*) = [\boldsymbol{\delta} + \mathbf{Q}_i(\mathbf{k}^*)(\mathbf{k}^*)^{-1}(\mathbf{k}^{(i)} - \mathbf{k}^*)]^{-1}$, $\mathbf{Q} = \boldsymbol{\kappa}^* - \boldsymbol{\kappa}^*\mathbf{S}_i$. For the aligned ellipsoidal inclusions with the same shape, the formula coincides with the Hashin-Shtrikman lower (upper) bound (6.261) if all inclusions are more (less) conducting than the matrix phase.

It is interesting, that for the isotropic composites containing homogeneous isotropic spherical inclusions, Eq. (17.26) can be presented in a symmetric form

$$\sum_{i=0}^{N} \frac{c^{(i)}(k^{(i)} - k^*)}{(d-1)k^{(i)} + k^*} = 0 \tag{17.27}$$

exactly coinciding with Eq. (17.25), and that is why average field approximations are sometimes confused with the EM approach. However, the mentioned two self-consistent methods are based on different assumptions leading to the different representations of effective properties for a general case of homogeneous ellipsoidal inclusions: $\mathbf{k}^{*(17.26)} \neq (\boldsymbol{\kappa}^{*(17.23)})^{-1}$ (see also [814]).

The implicit Eq. (17.23) can in general be solved only numerically via, e.g., the iteration scheme:

$$\boldsymbol{\kappa}^{*[n+1]} = \boldsymbol{\kappa}^{(0)} + \sum_{i=1}^{N} c^{(i)}(\boldsymbol{\kappa}^{(i)} - \boldsymbol{\kappa}^{*[n]})\mathbf{A}_i^t(\boldsymbol{\kappa}^{*[n]}), \tag{17.28}$$

with an initial approximation $\boldsymbol{\kappa}^{*[0]} = \boldsymbol{\kappa}^{[0]}$. It is plausible rather than rigorous that $\boldsymbol{\kappa}^{*[n]} \to \boldsymbol{\kappa}^*$ and $\boldsymbol{\kappa}^{*[n]} \to (\mathbf{k}^{*[n]})^{-1}$ as $n \to \infty$ where $\mathbf{k}^{*[n]}$ is a resistivity obtained in a similar iteration scheme (17.28) recast in terms of resistivity $\mathbf{k}^{(i)}$. However, an explicit solution of Eq. (17.23) is possible to obtain in some particular cases. For example, for two-phase composites, Eq. (17.25) is a quadratic equation for the effective conductivity with explicit solution:

$$\varkappa^* = \frac{1}{4}[b + (b^2 + 8\varkappa^{(0)}\varkappa^{(1)})^{1/2}],\ b = (dc^{(0)} - 1)\varkappa^{(0)} + (dc^{(1)} - 1)\varkappa^{(1)} \tag{17.29}$$

simplifying in two limiting cases of either the perfectly conducting particles ($\varkappa^{(1)}/\varkappa^{(0)} = \infty$) or ideally insulating inclusions ($\varkappa^{(1)}/\varkappa^{(0)} = 0$):

$$\frac{\varkappa^*}{\varkappa^{(0)}} = \begin{cases} (1 - dc^{(1)})^{-1}, & \text{if } \varkappa^{(1)}/\varkappa^{(0)} = \infty, \\ (1 - dc^{(1)}/(d-1))^{-1}, & \text{if } \varkappa^{(1)}/\varkappa^{(0)} = 0. \end{cases} \tag{17.30}$$

In the considered limiting cases of $\varkappa^{(1)}/\varkappa^{(0)}$ (17.30), Eq. (17.24) becomes linear with the solutions

$$\varkappa^* = (1 - \gamma_1)^{-1}\varkappa^{(0)}, \quad \text{and} \quad \varkappa^* = (1 - \gamma_2)\varkappa^{(0)}, \tag{17.31}$$

for $\varkappa^{(i)}/\varkappa^{(0)} = \infty$ and $\varkappa^{(i)}/\varkappa^{(0)} = 0$ ($i = 1, \ldots, N$), respectively, where $\gamma_1 = \sum_{i=1}^{N} c^{(i)}(2 - 3M_i)/[3M_i(1 - 2M_i)]$ and $\gamma_2 = \sum_{i=1}^{N} c^{(i)}(1 + 3M_i)/[6M_i(1 - M_i)]$. In particular, for the perfectly insulating ($k^{(i)} = 0$) spheroids degenerating into coin-shaped cracks ($a_1 = a_2 = a,\ a_3 = a\xi,\ \xi \to 0$) we have, according to Eq. (3.227), $M = \pi\xi/4$ that leads Eq. (17.31_2) to

$$\varkappa^* = (1 - \frac{8}{9}\langle\epsilon\rangle)\varkappa^{(0)}, \tag{17.32}$$

where $\langle\epsilon\rangle$ is the average crack density parameter $\epsilon_i = n^{(i)}a_i^3$ of microcracks with the radius a_i.

It should be mentioned that the formulae (17.30)–(17.32) predict either infinite or zero effective conductivity if the volume fraction $c^{(1)}$ reaches so-called *percolation threshold* $c_p^{(1)}$ (see the elastic case in Chapter 9.2). For example, for the spherical inclusions with $\varkappa^{(i)}/\varkappa^{(0)} = \infty$ (17.30) the percolation threshold $c_p^{(1)} = 0.3$ which is not sensitive to the microstructure details (such as, e.g., a fractional composition of inclusions and their radial distribution function) and can be quite different from the real percolation threshold of a concrete material. The percolation theory establishes a fundamental law for effective conductivity of random percolation in networks and continua in the vicinity of $c_p^{(1)}$: $\varkappa^* \sim \varkappa^{(0)}(c^{(1)} - c_p^{(1)})^t$ at $\varkappa^{(1)}/\varkappa^{(0)} \to \infty$, where the symbol $\sim$ stands asymptotically proportional, as $c^{(1)} \to c_p^{(1)}$. The critical exponent t (unlike $c_p^{(1)}$) has a universal value (approximately 2.0 and 1.3 for d=3 and 2, respectively) for different lattices. A detailed discussion of this theory is beyond the scope of the current presentation and can be found, e.g., in [250], [576], [957].

Another extension of the EM approach is the differential effective-medium (DEM) method which is also implicit but not symmetric with respect to the interchange $\varkappa^{(i)} \leftrightarrow \varkappa^{(j)}$, $c^{(i)} \leftrightarrow c^{(j)}$ $(i = 0, 1, \ldots, N)$. The DEM approach was extensively described for the elastic problems in Chapter 9.2. A conductivity analog of one version of the DEM method (9.77) can be recast for a two-phase isotropic composites with spherical isotropic inclusions in the following form:

$$\frac{\partial \varkappa^*}{\partial c} = \frac{d\varkappa^*}{1-c} \frac{\varkappa^{(1)} - \varkappa^*}{\varkappa^{(1)} + (d-1)\varkappa^*}, \tag{17.33}$$

with the initial condition $\varkappa^*(c=0) = \varkappa^{(0)}$ and $d = 2, 3$. Analytical integrating gives the equation

$$\Big(\frac{\varkappa^{(1)} - \varkappa^*}{\varkappa^{(1)} - \varkappa^{(0)}}\Big)\Big(\frac{\varkappa^{(0)}}{\varkappa^*}\Big)^{1/d} = 1 - c, \tag{17.34}$$

which, like the EM approach (17.30), is reduced to the simple solutions

$$\varkappa^* = (1-c)^{-d}\varkappa^{(0)}, \quad \text{or} \quad \varkappa^* = (1-c)^{d/(d-1)}\varkappa^{(0)}, \tag{17.35}$$

for two limiting cases of perfectly conducting particles (17.35_1) or ideally insulating inclusions (17.35_2), respectively. As can be seen, the percolation thresholds are absent in the DEM approach (17.35_1) and (17.35_2).

It is interesting that Sen *et al.* [984] obtained got excellent agreement of the predicted conductivity $\varkappa^* = (1-c)^{3/2}\varkappa^{(0)}$ (17.35_2) with experimental data for an electrically insulating fused glass beads saturated by a conducting fluid. The DEM approach (17.33) was generalized in [818] to the composites with a variety of shape and compositions of inclusions. The DEM scheme was also considered for randomly oriented ellipsoidal inclusions [385] and for cylindrical fibers [754]. Avelaneda [22] considered realizability of the DEM scheme for multiphase composites containing anisotropic inclusions of arbitrary shape.

In another group of self-consistent methods either one or few inclusions are immersed into a homogeneous medium possessing the properties of the matrix. The tested inclusions are subjected to some effective fields fount from the appropriate self-consistent integral equations, and that is why these methods of effective fields belong to the class of self-consistent methods. In particular, the conductivity analog of *explicit* Mori-Tanaka formula (8.102) can be recast for $N+1$-phase composites with homogeneous ellipsoidal inclusions in the form

$$\sum_{i=0}^{N} c^{(i)}(\boldsymbol{\kappa}^* - \boldsymbol{\kappa}^{(i)})\mathbf{A}_i^t = \mathbf{0}, \tag{17.36}$$

where $\mathbf{A}_0^t \equiv \boldsymbol{\delta}$ and $\mathbf{A}_i^t$ $(i = 1, \ldots, N)$ are defined by Eq. (17.15). For the spherical inclusions in dD cases $(d = 2, 3)$, Eq. (17.36) is reduced to the famous equation

$$\frac{\varkappa^* - \varkappa^{(0)}}{\varkappa^* + (d-1)\varkappa^{(0)}} = \sum_{i=1}^{N} \frac{c^{(i)}(\varkappa^{(i)} - \varkappa^{(0)})}{\varkappa^{(i)} + (d-1)\varkappa^{(0)}} \tag{17.37}$$

attributed at $d = 3$, $N = 1$ to Mossotti (1850) [790] and Clausius (in the dielectric context), Lorenz (in the refractivity context), and Maxwell (in the conductivity context); Markov [734] and Scaife [968] presented comprehensive reviews of the 150-year history of this formula with extensive references. Mossotti [790] (especially Clausius) pioneered the introduction of the effective field concept as a local field acting on the inclusions and differing from the applied macroscopic one. Among a few hypotheses used in [790], the most important one was in fact the quasi-crystalline approximation proposed 100 years later by Lax [649] (8.65) in a modern concise form. However, for the spherical inclusions the Clausius–Mossoti formula (17.37) as well as a Mori-Tanaka one (17.36) (see also [51]) are equivalent to the effective field method (MEF) (8.85) which are different one from another in a general case of ellipsoidal inclusions (see for details Chapter 8). For the limiting cases of either superconducting or insulating inclusions, Eq. (17.37) reduces to the representations

$$\varkappa^* = \frac{1+(d-1)c^{(1)}}{1-c^{(1)}}\varkappa^{(0)}, \quad \text{or} \quad \varkappa^* = \frac{(d-1)(1-c^{(1)})}{d-1+c^{(1)}}\varkappa^{(0)}, \tag{17.38}$$

at $\varkappa^{(1)}/\varkappa^{(0)} = \infty$ or $\varkappa^{(1)}/\varkappa^{(0)} = 0$, respectively, with the trivial percolation thresholds $c_p^{(1)} = 1$.

At last two-particle approximation of the MEFM scheme (10.34) and (10.38) can be recast in the terms conductivity tensors for two-phase composites as

$$\boldsymbol{\kappa}^* = \boldsymbol{\kappa}^{(0)} + n^{(1)}\mathbf{R}_1^t[\boldsymbol{\delta} - n^{(1)}\mathbf{P}^0\mathbf{R}_1^t - \mathbf{L}_{ij}]^{-1} \tag{17.39}$$

where the analytical representation for the tensor $\mathbf{L}_{ij}$ (17.20) was obtained for the spherical identical inclusions by both the approximative effective field method (17.21) and exact series expansion (17.22) in [498]. The graphs of $L_{ij}^0/c \sim \log(\varkappa^{(1)}/\varkappa^{(0)})$ calculated from the approximate (17.21) and exact (17.22) formulae obeying the well-stirred approximation (5.57) are shown in Fig. 17.1. Formula (17.21) can be assumed to be satisfactorily accurate. In the limiting case of superconducting identical spherical inclusions with well-stirred approximation of the radial distribution function of inclusions (5.57) and the isotropic matrix, we can find from Eq. (17.39) the effective conductivity

$$\varkappa^*/\varkappa^{(0)} = 1 + 3c^{(1)}(1-1.304c^{(1)})^{-1}, \quad \varkappa^*/\varkappa^{(0)} = 1 + 3c^{(1)}(1-1.503c^{(1)})^{-1}, \tag{17.40}$$

for the tensors $\mathbf{L}_{ij}$ (17.21) and (17.22), respectively. The representations (17.40_1) and (17.40_2) have the percolation thresholds $c_p^{(1)} = 0.77$ and $c_p^{(1)} = 0.66$, respectively.

Following [165], we compare the experimental data [761] and various methods of calculation of $\boldsymbol{\kappa}^*$ considered above for the example of the electrical conducting of a composite with an isotropic matrix filled via ideally conducted identical spherical inclusions with well-stirred approximation of the radial distribution function (5.57). Estimations were also obtained by the method of conditional moments [560]: $\varkappa^*/\varkappa^{(0)} = 1 + 3c^{(1)}(1 - c^{(1)}/3)/[1 - c^{(1)}]$. We see in Fig. 17.2 that calculation from the proposed relations (17.40_1) and (17.40_2) agrees better with experimental data than calculations via the EM method (17.30_1), via the Clausius-Mossotti formula (17.38_1), and by via conditional moments.

17.1.4 Nonlinear and Nonlocal Properties

The nonlinear and nonlocal problems of transport phenomena are worthy of more attention. It is interesting that the scheme of the perturbation method of the estimation of the second moment of fields (13.20) was proposed precisely in thermal conductivity theory [74]. The variance of the fields within each phase were derived in [68] and [23]. In so doing Cheng and Torquato [226] detected a double-peak character of probability density functions of fields for two-dimensional dispersions of different shapes and, therefore, the second moment of the field is a crude characteristic of the field fluctuation in the composite; they used Monte Carlo simulation techniques for occupancy of 64 inclusions in a periodic unit cell with subsequent numerical solution of integral equations involved in the fast multipole method (for references see [400], [402]). Knowledge of field distribution is important in understanding many crucial materials properties such as the breakdown phenomenon and the nonlinear behavior of composites. So, Stroud and Hui [1053] obtained an exact expression for the effective nonlinear conductivity $\boldsymbol{\kappa}^*$ for composites with a dilute concentration of nonlinear inclusions in a linear host. Zeng *et al.* [1216] proposed an approximate method for the estimation of $\boldsymbol{\kappa}^*$ for composites with a weakly nonlinear constitutive relation when the linearization (13.15), (16.10) and (16.11) was used to accompany the initial approximation (16.7) of $\boldsymbol{\kappa}(\mathbf{x})$. The same approximation was used in [664], [1052] to accompany the effective medium method. The scheme mentioned was generalized more recently in [130] (see also Chapter 11) in the framework of the MEF method. Variational method of estimation of the effective energy and conductivity was used in [382] through the estimation of the effective conductivity for the linear comparison composite.

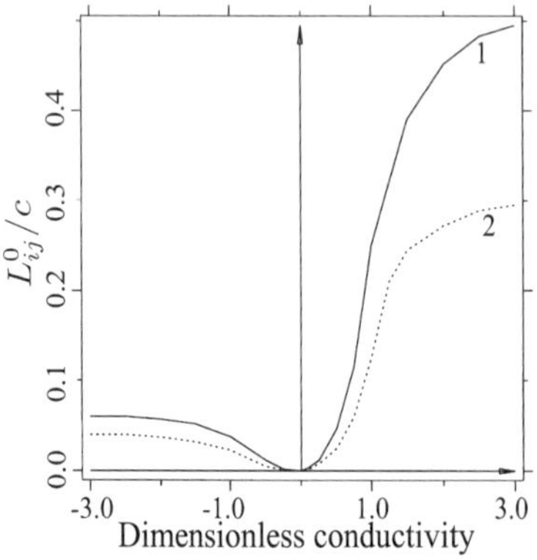

Fig. 17.1. The isotropic components L^0_{ij} vs dimensionless conductivity estimated via Eq. (17.22) (curve 1) and via Eq. (17.21) (curve 2).

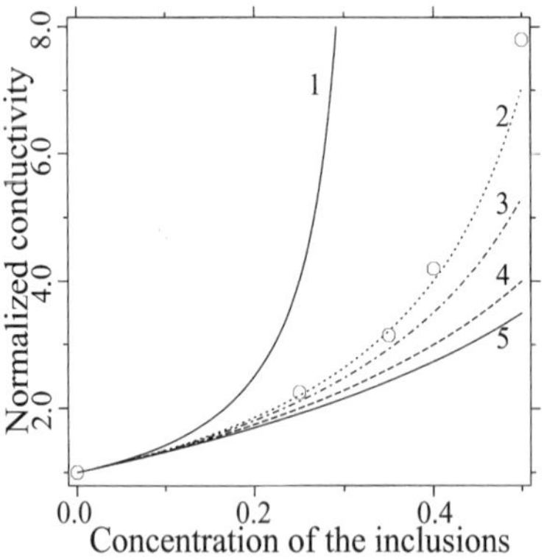

Fig. 17.2. $\varkappa^*/\varkappa^{(0)}$ vs $c^{(1)}$ estimated by Eqs. (17.30$_1$) (curve 1), (17.40$_2$) (curve 2), (17.40$_1$) (curve 3), (17.38$_1$) (curve 4), conditional moments (curve 5).

Progress in research of nonlocal effects is more moderate and was achieved under special assumptions. So, Beran and McCoy [70] used a perturbation method for the analysis of statistically homogeneous composite materials with small conductivity mismatch implying the application of assumptions for the flux $\mathbf{Q}$ analogous to Eq. (9.10); for the estimation of nonlocal effective properties they used Fourier transform method described in Subsection 12.3.3. Also by the use of the perturbation method [365], the variation of the local effective conductivity was detected near a wall having unsymmetric distribution of a constituent with respect to the boundary (see for comparison Fig. 14.3). Stagfen [1035] considered the case of very long and thin highly conducting fibers of so-called semidilute concentration; via the use of statistical averaged equations and slender body theory, the perturbation created by the surrounding fibers is demonstrated to be an integral of the temperature gradient about any point weighted by a function that is the appropriate nonlocal conductivity.

Among the papers dedicated to graded materials the following articles are noteworthy. In the analysis of the low-volume-fraction composites the pair-interacting inclusions with a special renormalization were considered [555] to avoid divergent integrals similar to [498] for statistically homogeneous suspensions that allow one to obtain concrete numerical results only for a step function $c(\mathbf{x}) = c^{\infty} H(x_1)$; a boundary layer effect similar to the elastic case depicted in Fig. 12.11 was detected. Buevich and Ustinov [122] presented a more heuristic analysis based on the effective medium method; they explicitly assumed that the mean temperatures of different components are the same and obtained effective constitutive equation containing only the first derivative of the potention [compare with Eqs. (12.72)–(12.74)]. A more general numerical model of a nonuniform composite with spherical inclusions was considered [732] taking into account a difference between the average temperatures in the phases. It is interesting that scaler effect analogous to the one detected in [181] (see also Chapter 12 and Fig. 12.11) for elastic graded composite was estimated in [874] (see also [1206] numerically for effective conductivity over the cross-section of the composite layer of finite thickness between two homogeneous half-spaces.

17.2 Thermoelectroelasticity of Composites

17.2.1 General Analysis

Piezoelectric materials have the property of converting mechanical energy into electrical energy (direct piezoelectric effect) and vice versa (inverse piezoelectric effect). In the case of composite constituents exhibiting electromechanical coupling the external electric field can produce mechanical deformations, which generate an internal stress field in the composite. These fields can influence the material's macroscopic response, so that the coupling effects must be taken into account when estimating the overall properties of composites with piezoelectric components. This electromechanical coupling behavior makes piezoelectric ceramics very attractive materials in sensor, actuator and resonator applications. Piezoelectric materials and especially piezoelectric composites open completely new possibilities in the design of adaptive or smart structure, and very interesting technological applications have been proposed, ranging from aeronautical and automotive structures to miniature positioning devises. From another aspect, geophysical investigations are usually based on the petrophysical analysis of deep drilling cores and on measured parameters of seismic wave beams such as group velocity, polarization, and divergence. In the long wavelength approach (as compared to rock crystallite sizes) these parameters are controlled by effective physical–mechanical characteristics of rocks. Pronounced anisotropy, texture, and piezoelectric activity are inherent in some polycrystal minerals ([811], [858], [1026], [1127]). The approach of macroscopic isotropy is not applicable to all minerals and rocks. Even a single component polycrystal may be textured, which causes a macroscopic anisotropy ([46], [1127], [1128], [1136], [1172]). Another problem with a multicomponent polycrystal is differences in connectivity of its various components. Two limiting cases are a connection of a single type, or the presence of connections of essentially different types, when one of the components is singly connected (matrix) and the other (or others) is multiply connected.

A natural property of piezoelectric polycrystal is their heterogeneity manifested at multiple length scales [680]. So, piezoelectric ceramics are usually prepared by synthesizing compressed powers of oxide, which produces macroscopically isotropic microporous ceramic ([228], [241]). This ceramic is initially nonpolled because of the effective polarization of ferroelectric crystallite is the vector sum of the polarization of all randomly oriented domains with permissible configurations dictated by the symmetry of the crystal. However, after a sufficient external electric field is applied, the individual domain switching generates the effective polarization of each crystal which is the vector sum of the polarization of each domain in individual crystal to be approximated to the applied electric field direction in this way to minimize the body energy. Moreover, not only does each grain itself have a domain structure, but the arrangement of the grains in the polycrystal also leads to heterogeneity produced by standard ceramic processing techniques. After polarizing, the piezoelectric ceramics can exhibit piezoelectric property suitable for functional materials. However, some monolithic piezoelectric ceramics have a low value of hydrostatic strain coefficient, limiting their applicability. Hence, composite materials such as piezoelectric ceramic fibers embedded in a soft pure elastic matrix are often a better technological solution for a large number of applications such as underwater and medical ultrasonic imaging [325] and transducers for underwater and sonar projector applications. They have superior electromechanical coupling characteristics (better acoustic impedance matching with water and (or) tissue) to conventional piezoelectric ceramic [1021]. Piezocomposites may provide material properties largely superior to conventional piezoelectric materials.

Let us consider a composite with elastic piezoelectric constituents. All random field under consideration are assumed to be statistically homogeneous and comply with the ergodicity hypothesis (see Subsection 5.2.2). Therefore averaging over the ensemble of realizations X may be replaced by that over the macrovolume w. For notational convenience the elastic and electric variable will be treated on equal footing, and with this in mind we recast the local linear constitutive relations of thermoelectroelasticity for this material (see, e.g., [747], [861]) in the notation introduced in [38]. Then the basic equation can be presented in the form analogous to Eqs. (6.1)–(6.3)

$$\boldsymbol{\mathcal{D}}\boldsymbol{\Sigma} = \mathbf{0}, \ \boldsymbol{\mathcal{E}} = \boldsymbol{\mathcal{D}}\boldsymbol{\mathcal{U}}, \tag{17.41}$$

$$\boldsymbol{\mathcal{E}} = \mathbb{M}\boldsymbol{\Sigma} + \boldsymbol{\Lambda}, \quad \boldsymbol{\Sigma} = \mathbb{L}(\boldsymbol{\mathcal{E}} - \boldsymbol{\Lambda}) \tag{17.42}$$

where

$$\boldsymbol{\mathcal{E}} = \left\| \begin{matrix} \varepsilon \\ \mathbf{E} \end{matrix} \right\|, \quad \boldsymbol{\Sigma} = \left\| \begin{matrix} \boldsymbol{\sigma} \\ \mathbf{D} \end{matrix} \right\|, \quad \mathbf{U} = \left\| \begin{matrix} \mathbf{u} \\ \phi \end{matrix} \right\|, \quad \boldsymbol{\mathcal{D}} = \left\| \begin{matrix} \text{def} & 0 \\ 0 & \text{grad} \end{matrix} \right\| \tag{17.43}$$

$$\mathbb{M} = \left\| \begin{matrix} \mathbf{M} & \mathbf{d}^\top \\ \mathbf{d} & -\mathbf{b} \end{matrix} \right\|, \quad \mathbb{L} = \left\| \begin{matrix} \mathbf{L} & \mathbf{e}^\top \\ \mathbf{e} & -\mathbf{k} \end{matrix} \right\|, \quad \boldsymbol{\Lambda} = \left\| \begin{matrix} \boldsymbol{\beta} \\ \mathbf{q}\theta \end{matrix} \right\|, \tag{17.44}$$

here $\mathbf{D}$ and $\mathbf{E}$ are the vectors of induction and electric field intensity, θ is a derivation of a stationary temperature field from a given value, $\mathbf{k}$ and $\mathbf{b}$ are the tensors of dielectric permeability and impermeability, $\mathbf{q}$ is the pyroelectric coefficients, $\mathbf{e}$ and $\mathbf{d}$ are the piezoelectric moduli, and ϕ is the electric potential. To obtain a symmetric matrix of coefficients we replaced the electric field $\mathbf{E}$ by $-\mathbf{E}$, and the tensors $\mathbf{k}$ and $\mathbf{b}$ by $-\mathbf{k}$ and $-\mathbf{b}$ on the right-hand sides of (17.42). Hereinafter, we assume summation over the repeated indexes. The Greek indexes range from 1 to 4, and the Latin ones range from 1 to 3.

Equations (17.41) and (17.42) should be complemented by the boundary conditions of the first or the second type. We suppose that the contact between components is ideal, and the normal components of the stress tensor and the electric induction vector are continuous, as well as the tangential components of the strain tensor and electrical field intensity. These conditions are fulfilled for both the matrix composites and multicomponent polycrystals. In the latter case, we should admit $\bar{v}^{(0)} = 0$. Except for notations, these equations coincide with the equations of linear thermoelasticity (6.1)-(6.3). Because of this the theory of PCM retraces at a particular instant the path of development of the theory of microinhomogeneous elastic media, exhibiting substantial progress (see Chapters 2–9). In light of the analogy mentioned in our brief survey we will not consider in detail the average scheme of PCM, one may refer instead to the appropriate scheme of Chapter 8.

So, early simplified PCM models, based on schemes of the parallel or successive connection of rheological elements and their combinations ([204], [724], [1016]) based on their three-dimensional connectivity characteristics for the analysis of the effective electroelastic moduli are approximate and can be used only for rough estimations. Banno [30], [31] generalized this approach by consideration of a discontinuous reinforcement effect through a cubes scheme. One [403] developed an extension of the well-known concentric cylinder Hashin's model of the cell to coupled electroelastic behaviors in order to predict the effective electroelastic moduli of continuous fibers reinforced composites. Coupled electroelastic field in piezoelectric ellipsoidal inhomogeneity embedded in an infinite matrix was expressed in terms of Fourier's transforms of electroelastic Green's functions in [1158] and [1156], which has enabled one to exploit this solution in a perturbation method for estimation of the effective properties (see also [552], [1029]). Based on the contour integral representation of Green's function derived in [267], Dunn and Taya [290] obtained explicit expressions for a set of four tensors corresponding to Eshelby's elastic tensors for ellipsoidal inclusion and incorporated this solution into several generalized micromechanical models such as Mori and Tanaka's approach, dilute concentration approach, the differential method of successive smoothing, and self-consistent model (see also [287], [288], [289], [291], [677], [1078]). It cannot be doubted that application of the most popular generalized Mori-Tanaka approach (see, e.g., [88], [219], [228], [480], [622], [1226]) for the investigation of PCMs with a general case of random orientation of piezoelectric fibers must lead the incorrect results (see, e.g., Section 8.3). The known micromechanical approaches considered in Chapter 8 were generalized in a straightforward manner for the analysis of PCM; these are the multi-inclusion model ([806], [678]), conditional moment method ([565], [567]), generalized singular approximation [996], effective medium method [679], one-particle approximation of the MEFM [670], and the MEFM [173], [176].

Any average scheme of effective properties estimations of the PCM uses a solution of a single inclusion in an infinite piezoelectric solid. This problem has been analyzed in ([54], [290], [481], [765], [1156]) as well as by Shermergor and Yakovlev [996] who considered the surface integral representation of the Green's function $\boldsymbol{\mathcal{G}}$ and proved the property of polynomial conservativity of an integral operator with the kernel $\boldsymbol{\mathcal{DGD}}$ for the ellipsoidal homogeneous inclusion. Shermergor and Yakovlev [996] suggested generalization of the interface operator by Hill [464] for piezoelectric contacted materials (at $\boldsymbol{\Lambda} \equiv \mathbf{0}$), while Buryachenko and Shermergor [194] analyzed a similar problem at $\boldsymbol{\Lambda} \not\equiv \mathbf{0}$ by other more compact method. This solution allows one to estimate the field distribution not only inside the inclusion, but also in the matrix in the vicinity of inclusion as well; moreover, recasting the solution of thermoelasticity for coated ellipsoidal inclusion [184] (see Chapter 4) in terms of generalized Green's function and interface operator suggested one can obtain easily the solution for the coated ellipsoidal inclusion for piezoelectric materials.

Of course, the absence of coupled field effect for the matrix essentially simplifies the micromechanical problems because in such a case the Green's function $\boldsymbol{\mathcal{G}}$ has a block-diagonal structure defined by the Green's functions of uncoupled fields. Then the generalized Eshelby tensor as well as the tensors (3.131) have the same structure and as it is well known, leads to nothing more than a slight modification of softwares supporting the approaches described in Chapter 8 for elastic problems; this simplification was used, e.g., in [176], [505]. Usually a comparison medium without coupled effects is used (see [567], [802]) whereas Buryachenko and Parton [173] introduced for the matrix the new field variables without coupling effects. The field coupling in the matrix complicates essentially a micromechanical analysis that extends the importance of a fundamental trend in micromechanics of coupled fields based on the diagonalization of response matrixes $\mathbb{M}$ by a matrix decomposition technique proposed and taking the beginning from the work by Straley [1048], who showed that by the use of a linear transform of the real fluxes, the problem of evaluation effective thermoelectric properties for a two-phase composite can be reduced to the conductivity problem of a composite. This method was generalized in [767] (see also [61], [218], [220]) where more references may be found) to the analysis of any number of coupled fields with scalar potentials; then for the field being considered any known results existing in the single-field effect can be transformed into coupled-field responses.

17.2.2 Generalized Hill's Conditions and Effective Properties

The general exact relations for the PCM should be mentioned. So, Benveniste [55], [56], [57] (see also [194]) obtained the effective eigenstress and stiffness for multiphase composites generalized the relations (6.20), (6.44), (6.77). We will reproduce these and other exact relations following [194]. We will obtain some exact expressions for volume-averaged values of the first and second moments of elastic electric and temperature fields. We will derive relationships, extending Hill's conditions to the electrothermoelastic fields. It will be shown that the total energy may be represented by a sum of two components, which are caused by electroelastic and thermal effects correspondingly. We will present an explicit expression for a tensor operator of the dependent fields when intersecting the component interface. Under certain assumption, we will calculate local fluctuations of the fields in a matrix domain being in the vicinity of inhomogeneous.

We extend the Hill's [464] condition known in the elasticity theory (see Chapter 6) to the piezoelectric composites. For precision, we consider homogeneous boundary conditions in the form

$$\Sigma_{\alpha\beta}n_\beta(\mathbf{x}) = \langle\Sigma_{\alpha\beta}\rangle n_\beta(\mathbf{x}) = t_\alpha(\mathbf{x}), \quad \mathbf{x}\in\partial w, \quad n_4 \equiv \mathbf{0} \tag{17.45}$$

where n_α are the components of an external normal to the domain boundary ∂w, t_i $(i = 1, 2, 3)$ and t_4 are the normal components of the mechanical stress tensor and vector of electric induction. In the case of ergodic fields, one may replace integrating over macrovolume by ensemble averaging. Hereinafter, the Latin indexes range from 1 to 3, and the Greek ones range from 1 to 4. Therefore in accordance with (17.41), the following equalities hold true:

$$\begin{aligned}\langle\Sigma_{\alpha\beta}\mathcal{E}_{\alpha\beta}\rangle &= \bar{w}^{-1}\int\langle\Sigma_{\alpha\beta}\mathcal{E}_{\alpha\beta}\rangle\, d\mathbf{x} = \bar{w}^{-1}\Big\langle\int\Sigma_{\alpha\beta}\mathcal{E}_{\alpha\beta}\, d\mathbf{x}\Big\rangle\\ &= \bar{w}^{-1}\Big\langle\int t_\alpha U_\alpha\, d\mathbf{s}\Big\rangle = \bar{w}^{-1}\int\langle t_\alpha U_\alpha\rangle\, d\mathbf{s},\end{aligned} \tag{17.46}$$

where integration in the surface integral is performed over the boundary ∂w, and the Dirac brackets denote ensemble averaging. Separating the displacement field fluctuations $\delta U_\alpha(\mathbf{x}) \equiv U_\alpha(\mathbf{x}) - \langle U_\alpha(\mathbf{x})\rangle$ in the surface integral, we have

$$\int\langle t_\alpha(\mathbf{x})U_\alpha(\mathbf{x})\rangle\, ds = \int\langle t_\alpha(\mathbf{x})\rangle\langle U_\alpha(\mathbf{x})\rangle\, ds + \int\langle t_\alpha(\mathbf{x})\delta U_\alpha(\mathbf{x})\rangle\, ds. \tag{17.47}$$

The last integral in (17.47) is zero because, in view of the boundary conditions (17.45) and the ergodicity, we may replace the averaging over the volume by that over ensemble. Then, we obtain from (17.46) and (17.47) $\langle\Sigma_{\alpha\beta}\mathcal{E}_{\alpha\beta}\rangle = \bar{w}^{-1}\int\langle t_\alpha(\mathbf{x})\rangle\langle U_\alpha(\mathbf{x})\rangle\, ds$. But, in accordance with (17.41) and (17.45),

$$\langle\boldsymbol{\mathcal{E}}\rangle = \langle\nabla_{(\alpha}U_{\beta)}\rangle, \quad \langle t_\alpha(\mathbf{x})\rangle = \langle\Sigma_{\alpha\beta}\rangle n_\beta(\mathbf{x}). \tag{17.48}$$

Therefore using the Gauss theorem, we find from (17.41), (17.47), and (17.48)

$$\langle\Sigma_{\alpha\beta}(\mathbf{x})\mathcal{E}_{\alpha\beta}(\mathbf{x})\rangle = \langle\Sigma_{\alpha\beta}(\mathbf{x})\rangle\langle\mathcal{E}_{\alpha\beta}(\mathbf{x})\rangle. \tag{17.49}$$

The equality (17.49) extends the Hill condition to piezoelectric materials and can be obtained, if, for the surface ∂w, it is possible to formulate the first boundary problem with given homogeneous fields of generalized strains $\boldsymbol{\mathcal{E}}$. Equation (17.49) was derived without any assumption on the correlation between fields the $\boldsymbol{\sigma}$ and $\boldsymbol{\mathcal{E}}$. The only requirement is that the fields be statistically and kinematically acceptable, i.e., they satisfy conditions (17.41). Formula (17.49) was obtained by Buryachenko and Shermergor [194] and independently in [680].

To derive the macroscopic equation of state, we assume that the local and macroscopic values are related as

$$\boldsymbol{\Sigma}(\mathbf{x}) = \mathbb{B}(\mathbf{x})\langle\boldsymbol{\Sigma}\rangle + \mathbb{F}(\mathbf{x}) \tag{17.50}$$

Due to the linearity of the problem, tensors $\mathbb{B}$ and $\mathbb{F}$ exist. Averaging tensor $\mathbb{B}$ over the volume of the ith component $\langle\mathbb{B}(\mathbf{x})\rangle_i \equiv \mathbb{B}_i$, we obtain the tensor of generalized stress concentration. The tensors $\mathbb{B}$ and $\mathbb{F}$ depend on the physical and geometric properties of the material components and may be found from the solution of the corresponding boundary problem for a microinhomogeneous medium. Averaging of (17.50) at $\theta \equiv 0$ gives an expression determining the effective tensors of generalized compliance $\mathbb{M}^*$ and expansion $\boldsymbol{\Lambda}^*$ in the macroscopic equation of state: $\langle\boldsymbol{\Sigma}\rangle = \mathbb{M}^*\langle\boldsymbol{\sigma}\rangle + \boldsymbol{\Lambda}^*$. At the thermodynamic equilibrium condition stated as $\theta = \langle\theta\rangle$, we find from (17.42) and (17.50): $\mathbb{M}^* = \langle\mathbb{M}\mathbb{B}\rangle$, $\boldsymbol{\Lambda}^* = \langle\boldsymbol{\Lambda}\rangle + \langle\mathbb{M}\mathbb{F}\rangle$, where $\langle\mathbb{B}\rangle = \mathrm{diag}(\mathbf{I}, \boldsymbol{\delta}) \equiv \mathbb{I}$, $\langle\mathbb{F}\rangle = \mathbf{0}$.

To find another representation of the tensor $\boldsymbol{\Lambda}^*$, we express the fields of generalized stresses, strains, and thermal expansion in the form (see (6.34), (6.35))

$$\boldsymbol{\Sigma}(\mathbf{x}) = \boldsymbol{\Sigma}^I(\mathbf{x}) + \boldsymbol{\Sigma}^{II}, \quad \boldsymbol{\mathcal{E}}(\mathbf{x}) = \boldsymbol{\mathcal{E}}^I(\mathbf{x}) + \boldsymbol{\mathcal{E}}^{II}(\mathbf{x}), \quad \boldsymbol{\Lambda}(\mathbf{x}) = \boldsymbol{\Lambda}^I(\mathbf{x}) + \boldsymbol{\Lambda}^{II}(\mathbf{x}), \tag{17.51}$$

where $\boldsymbol{\Sigma}^I$ and $\boldsymbol{\mathcal{E}}^I$ satisfy boundary conditions (17.45) and $\boldsymbol{\Lambda}^I \equiv \mathbf{0}$, and the fields $\boldsymbol{\Sigma}^{II}$ and $\boldsymbol{\mathcal{E}}^{II}$ satisfy zero boundary conditions $t_\alpha \equiv \mathbf{0}$ and $\boldsymbol{\Lambda}^{II} \equiv \boldsymbol{\Lambda}(\mathbf{x})$. In accordance with (17.50), we have $\boldsymbol{\Sigma}^{II}(\mathbf{x}) = \mathbb{F}(\mathbf{x})$. Then for the field under consideration, the local and macroscopic equation of state may be expressed in the form

$$\boldsymbol{\Sigma}^I(\mathbf{x}) = \mathbb{L}(\mathbf{x})\boldsymbol{\mathcal{E}}^I(\mathbf{x}), \quad \boldsymbol{\Sigma}^{II}(\mathbf{x}) = \mathbb{L}(\mathbf{x})[\boldsymbol{\mathcal{E}}^{II}(\mathbf{x}) - \boldsymbol{\Lambda}(\mathbf{x})], \tag{17.52}$$

$$\langle\boldsymbol{\Sigma}^I(\mathbf{x})\rangle = \mathbb{L}^*(\mathbf{x})\langle\boldsymbol{\mathcal{E}}^I(\mathbf{x})\rangle, \quad \langle\boldsymbol{\Sigma}^{II}(\mathbf{x})\rangle = \mathbb{L}^*(\mathbf{x})[\langle\boldsymbol{\mathcal{E}}^{II}(\mathbf{x})\rangle - \boldsymbol{\Lambda}^*(\mathbf{x})]. \tag{17.53}$$

Since $\boldsymbol{\Sigma}$, $\boldsymbol{\Sigma}^I$, $\boldsymbol{\Sigma}^{II}$, $\boldsymbol{\mathcal{E}}$, $\boldsymbol{\mathcal{E}}^I$, and $\boldsymbol{\mathcal{E}}^{II}$ are statically and kinematically admissible fields, then, taking into account Hill's condition (17.49), we have

$$\langle\boldsymbol{\Sigma}^{I\top}\boldsymbol{\mathcal{E}}\rangle = \langle\boldsymbol{\Sigma}^{I\top}\rangle\langle\boldsymbol{\mathcal{E}}\rangle, \quad \langle\boldsymbol{\mathcal{E}}^{I\top}\boldsymbol{\Sigma}\rangle = \langle\boldsymbol{\mathcal{E}}^{I\top}\rangle\langle\boldsymbol{\Sigma}\rangle, \tag{17.54}$$

$$\langle\boldsymbol{\Sigma}^{I\top}\boldsymbol{\mathcal{E}}^I\rangle = \langle\boldsymbol{\Sigma}^{I\top}\rangle\langle\boldsymbol{\mathcal{E}}^I\rangle, \quad \langle\boldsymbol{\Sigma}^{II\top}\boldsymbol{\mathcal{E}}^{II}\rangle = \langle\boldsymbol{\Sigma}^{II\top}\rangle\langle\boldsymbol{\mathcal{E}}^{II}\rangle = 0. \tag{17.55}$$

Subtracting equalities (17.54_1) and (17.54_2) from one another and using (17.52), (17.53), we find $\langle\boldsymbol{\Sigma}^{I\top}\boldsymbol{\Lambda}\rangle = \langle\boldsymbol{\Sigma}^{I\top}\rangle\langle\boldsymbol{\Lambda}^*\rangle$. Substituting (17.50) into the last equation, in view of arbitrariness $\langle\boldsymbol{\Sigma}\rangle$, we have

$$\boldsymbol{\Lambda}^* = \langle\mathbb{B}^\top\boldsymbol{\Lambda}\rangle. \tag{17.56}$$

We see that the matrix of effective generalized thermal expansion $\boldsymbol{\Lambda}^*$ is determined not only by its local value $\boldsymbol{\Lambda}(\mathbf{x})$ but also by the matrix $\mathbb{B}(\mathbf{x})$, which relates the local and averaged values of elastic and electric fields. Thus, analogously to the thermoelastic case (6.48), the effective generalized thermal expansion $\boldsymbol{\Lambda}^*$ is spatial and orientational averages of local magnetostriction strain in which the generalized stress concentrator factors $\mathbb{B}(\mathbf{x})$ serve as weighting factors. An analogous formula was obtained for the effective magnetostriction of composites and polycrystals with particulate and columnar microstructures in [222].

In some cases, another form of relationship (17.56) is more convenient. To derive it, we use the obvious equalities $\langle \boldsymbol{\mathcal{E}}^{II}\rangle = \langle V\boldsymbol{\mathcal{E}}^{II}\rangle + \langle (1-V)\boldsymbol{\mathcal{E}}^{II}\rangle$, $\langle V\boldsymbol{\Sigma}^{II}\rangle + \langle (1-V)\boldsymbol{\Sigma}^{II}\rangle = 0$. Then, assuming the homogeneity of matrix material, we obtain

$$\boldsymbol{\Lambda}^* = \langle \boldsymbol{\Lambda}\rangle + \langle \mathbb{M}_1\boldsymbol{\Sigma}^{II}V\rangle. \tag{17.57}$$

Formulae (17.56) and (17.57) generalize the similar results (6.44) and (6.49), respectively, of the thermoelastisity theory for piezoelectric composites. This results are commonly obtained by resolving the field $\boldsymbol{\sigma}$ and $\boldsymbol{\varepsilon}$ into components $\boldsymbol{\sigma}^I$ and $\boldsymbol{\varepsilon}^I$ and $\boldsymbol{\sigma}^{II}$ and $\boldsymbol{\varepsilon}^{II}$, similar to (17.51). The same approach was applied to evaluation of the elastic interaction energy of the $\boldsymbol{\sigma}^I$ and $\boldsymbol{\sigma}^{II}$, which will be used in the analysis of inhomogeneous piezoelectric composites.

17.2.3 Effective Energy Functions

Substituting (17.51) in the formula for the energy of generalized strains (compare with (6.61) $\langle W\rangle = \langle \sigma_{ij}(\varepsilon_{ji} - \beta_{ji})\rangle + \langle D_i(E_i - q_i\theta)\rangle \equiv \langle \boldsymbol{\Sigma}^\top(\boldsymbol{\mathcal{E}} - \boldsymbol{\Lambda})\rangle$ as a result, we obtain a decomposition, similar to the thermoelastic case $\langle W\rangle = \langle W^I\rangle + \langle W^{II}\rangle + \langle W^{\rm int}\rangle$. The terms on the right-hand side of the equality correspond to the energies of the fields $\boldsymbol{\Sigma}^I$ and $\boldsymbol{\Sigma}^{II}$, and to their interaction energy. Averages (17.54) and (17.55) give

$$2\langle W^I\rangle = \langle \boldsymbol{\Sigma}^{I\top}\rangle\langle \boldsymbol{\mathcal{E}}^I\rangle, \quad 2\langle W^{II}\rangle = -\langle \boldsymbol{\Sigma}^{II\top}\boldsymbol{\Lambda}\rangle, \tag{17.58}$$

$$\langle W^{\rm int}\rangle = \langle \boldsymbol{\Sigma}^{II\top}\rangle\langle \boldsymbol{\mathcal{E}}^I\rangle = \langle \boldsymbol{\Sigma}^{I\top}\rangle\Big(\langle \boldsymbol{\mathcal{E}}^{II}\rangle - \boldsymbol{\Lambda}^*\Big). \tag{17.59}$$

For boundary conditions (17.45) with $\boldsymbol{\Sigma}^{II}(\mathbf{x}) \equiv \mathbf{0}$ and $\mathbf{x} \in \partial w$, we obtain

$$\langle \boldsymbol{\mathcal{E}}^I\rangle = \mathbb{M}^*\langle \boldsymbol{\Sigma}\rangle, \quad \langle \boldsymbol{\Sigma}^{II}\rangle = \mathbf{0}, \quad \langle \boldsymbol{\mathcal{E}}^{II}\rangle = \boldsymbol{\Lambda}^*, \tag{17.60}$$

$$\langle \boldsymbol{\Sigma}^{I\top}\boldsymbol{\mathcal{E}}^I\rangle = \langle \boldsymbol{\Sigma}^\top\rangle\mathbb{M}^*\langle \boldsymbol{\Sigma}\rangle, \quad \langle \boldsymbol{\Sigma}^I\boldsymbol{\mathcal{E}}^{II}\rangle = \langle \boldsymbol{\Sigma}\rangle\boldsymbol{\Lambda}^*, \tag{17.61}$$

$$\langle \boldsymbol{\Sigma}^{II\top}\boldsymbol{\mathcal{E}}^I\rangle = 0, \quad \langle \boldsymbol{\Sigma}^{II\top}\boldsymbol{\mathcal{E}}^{II}\rangle = 0. \tag{17.62}$$

In view of (17.59) and (17.60), the field interaction energy is $\langle W^{\rm int}\rangle \equiv 0$. Therefore the total energy can be expresses in the form

$$\langle W\rangle = \langle W^I\rangle + \langle W^{II}\rangle = \frac{1}{2}\Big(\langle \boldsymbol{\Sigma}^\top\rangle\mathbb{M}^*\langle \boldsymbol{\Sigma}\rangle - \langle \boldsymbol{\Sigma}^{II\top}\boldsymbol{\Lambda}\rangle\Big). \tag{17.63}$$

Note that along with the effective tensors $\mathbb{M}^*$ and $\boldsymbol{\Lambda}^*$, energy $\langle W^{II}\rangle \equiv -\langle \mathbb{F}\boldsymbol{\Lambda}\rangle/2$ also is an effective function that is determined only by temperature, microstructure geometry, and the material parameters $\mathbb{M}(\mathbf{x})$ and $\boldsymbol{\Lambda}(\mathbf{x})$ and does not depend upon the applied external field $\boldsymbol{\Sigma}^I$. According to the second of formulae (17.58), we have $\langle W^{II}\rangle = 0$ at $\boldsymbol{\Lambda}(\mathbf{x}) \equiv$ const. ($\mathbf{x} \in w$), and hence $\langle W^{II}\rangle$ is a measure of the generalized residual microstresses in a microinhomogeneous medium. The value W^{II} can be found both by computation from the definition (17.58) and experimentally. Thus if the variations of the temperature θ are small as compared with the reference value T_0, then, based on thermodynamic considerations, we obtain (compare with (6.129$_2$)): $2\langle W^{II}\rangle = (C_\sigma^{E*} - \langle C_\sigma^E\rangle)\theta^2/T_0$, where C_σ^{E*} and $\langle C_\sigma^E\rangle$ are the effective and averaged specific heats at constant hydrostatic pressure and the electric field intensity, which can be easily measured experimentally.

17.2.4 Two-Phase Composites

For a two-phase medium with homogeneous components, the problem of calculating the effective generalized tensor of the thermal expansion $\boldsymbol{\Lambda}^*$ reduces to determination of the effective generalized compliance $\mathbb{M}^*$, i.e. $\boldsymbol{\Lambda}^*$ is uniquely expressed through $\mathbb{M}^*$. In fact, assume the homogeneity of the material properties within a component $\mathbf{f}(\mathbf{x}) \equiv$ const, ($\mathbf{x} \in v^{(i)}$, $i = 0, 1$; $\mathbf{f} = \mathbb{L}, \mathbb{M}, \boldsymbol{\Lambda}$) and substitute the expressions for stress concentrators of components

$$\langle \mathbb{B}_i\rangle = (c^{(i)})^{-1}(\mathbb{M}^{(i)} - \mathbb{M}^{(j)})^{-1}(\mathbb{M}^* - \mathbb{M}^{(j)}) \qquad (17.64)$$

($c^{(i)} = \langle V^{(i)}\rangle$, $i = 0, 1$, $j = 1 - i$) into the expression for the effective tensor $\boldsymbol{\Lambda}^*$ (17.56). Then, we find

$$\boldsymbol{\Lambda}^* = (\mathbb{M}^* - \mathbb{M}^{(0)})(\mathbb{M}^{(1)} - \mathbb{M}^{(0)})^{-1}\boldsymbol{\Lambda}^{(1)} + (\mathbb{M}^* - \mathbb{M}^{(1)})(\mathbb{M}^{(0)} - \mathbb{M}^{(1)})^{-1}\boldsymbol{\Lambda}^{(0)}. \qquad (17.65)$$

Analogous relationships were derived in [55], and [287], and [1226]. A similar exact link between effective magnetostriction and effective moduli is also established [223]. Formula (17.65) shows that if the electroelastic fields are not coupled within the components ($\mathbf{c} = \mathbf{d} \equiv \mathbf{0}$), then the effective pyroelastic $\mathbf{q}^*$ and thermal expansion $\boldsymbol{\beta}^*$ coefficients are independent of one another in a composite. In this case, any of the coefficients $\boldsymbol{\beta}^*$ or $\mathbf{q}^*$ may be zero. However, this is not the case when the piezoelectric moduli are nonzero at least for one of the components. Then, the tensor $\boldsymbol{\beta}^*$(or $\mathbf{q}^*$) is nonzero even though $\boldsymbol{\beta}(\mathbf{x}) \equiv \mathbf{0}$ (or $\mathbf{q}(\mathbf{x}) \equiv \mathbf{0}$) and $\mathbf{q}(\mathbf{x}) \not\equiv$ const. (or $\boldsymbol{\beta}(\mathbf{x}) \not\equiv$ const).

If the tensor $\boldsymbol{\Lambda}^*$ is known, then one can find the field $\boldsymbol{\Sigma}^{II}$ averaged over the volume of each component i:

$$\langle \boldsymbol{\Sigma}^{II}\rangle^{(i)} = (c^{(i)})^{-1}(\mathbb{M}^{(i)} - \mathbb{M}^{(j)})^{-1}(\boldsymbol{\Lambda}^* - \langle\boldsymbol{\Lambda}\rangle). \qquad (17.66)$$

From (17.58) and (17.66) we find the exact expression for the averaged stored energy $2\langle W^{II}\rangle = (\boldsymbol{\Lambda}^{(1)} - \boldsymbol{\Lambda}^{(0)})(\mathbb{M}^{(1)} - \mathbb{M}^{(0)})^{-1}(\boldsymbol{\Lambda}^* - \langle\boldsymbol{\Lambda}\rangle)$.

For two-phase composites with anisotropic components, one may obtain not only the averaged fields of generalized thermoelastic stresses, but also their exact local values if the local stress concentrators $\mathbb{B}(\mathbf{x})$ defined by (17.50) for the corresponding generalized elastic problem are known [57]

$$\mathbb{F}(\mathbf{x}) = (\mathbb{B}(\mathbf{x}) - \mathbb{I})(\mathbb{M}^{(1)} - \mathbb{M}^{(0)})(\boldsymbol{\Lambda}^{(1)} - \boldsymbol{\Lambda}^{(0)}). \qquad (17.67)$$

Since $\langle \boldsymbol{\sigma}^{II} \rangle^{(i)} = \langle \mathbb{F} \rangle^{(i)}$, averaging of (17.67) over the volume of a component $v^{(i)}$ $(i = 0, 1)$ yields (17.66).

We note that the results of this section hold true only for two-component composites of matrix or stochastic (mixture) structure. The first type includes, for example, a composite composed of a simply connected homogeneous ambient phase (matrix) and inclusions of similar shape, orientation, and size. In the general case, rocks are textured polycrystals [1127], equations (17.64)- (17.67) are applicable under certain assumptions. Namely, a rock can be modeled as a composite with inhomogeneities such as cracks or inclusions of a different mineral (for example, in the form of relatively rigid plates of mica in quartz [1127]). In so doing, the properties of a matrix (actually, a single-phase polycrystal) are estimated via one of the well-known method such as the generalized singular approximation [996] or the method of effective media (see Chapter 9). The smaller the fluctuations of material properties of a crystal composing a matrix yields the higher accuracy of (17.64)–(17.67).

Exact expressions (17.64)- (17.67) obtained in terms of the theory of two-phase composite can be applied to polycrystals (which comprise the majority of rocks) as follows. Computation of the averaged material fields utilize the discretization of possible orientations of the crystallographic axes, the procedure being traditional in mechanics of polycrystals. Crystallites may be assumed to have two possible orientations only. Then (17.64)–(17.67) may be applied to the two-component medium specified by these orientations.

17.2.5 Discontinuities of Generalized Fields at the Interface Between Components

To determine jumps of generalized stresses and strains at a boundary ∂v_{12} represented by an ideal contact between components $v^{(1)}$ and $v^{(2)}$, we will use projection operators of the second rank $\boldsymbol{\eta}$ and $\boldsymbol{\nu}$, and these of the fourth rank $\mathbf{F}$ and $\mathbf{E}$ introduced in [464] (see Eqs. (3.84)) and having "orthogonal" properties (3.85). We construct the following interface projection operators: $\mathbb{H}^{\eta} = \mathrm{diag}(\mathbf{F}, \boldsymbol{\eta})$, $\mathbb{H}^{\boldsymbol{\nu}} = \mathrm{diag}(\mathbf{E}, \boldsymbol{\nu})$ with the properties

$$\mathbb{H}^{\eta} + \mathbb{H}^{\boldsymbol{\nu}} = \mathrm{diag}(\mathbf{I}, \boldsymbol{\delta}) \equiv \mathbb{I}, \quad \mathbb{H}^{\eta}\mathbb{H}^{\eta} = \mathbb{H}^{\eta}, \quad \mathbb{H}^{\boldsymbol{\nu}}\mathbb{H}^{\boldsymbol{\nu}} = \mathbb{H}^{\boldsymbol{\nu}}, \quad \mathbb{H}^{\mathbf{n}}\mathbb{H}^{\boldsymbol{\nu}} = \mathbf{0}. \quad (17.68)$$

Hill [464], and Shermergor and Yakovlev [996] showed that tensors $\mathbf{F}$, $\mathbf{E}$ and $\mathbb{H}^{\mathbf{n}}$, $\mathbb{H}^{\boldsymbol{\nu}}$, respectively, resolve the tensor g_{ij} and matrix $[g_{ij}, f_k]^\top$ into the normal and tangential components, where g_{ij} and f_k are an arbitrary symmetrical second rank tensor and an arbitrary vector respectively, both specified on the surface ∂v_{12}. The meaning of the resolution is obvious if the unit normal is chosen to be $\mathbf{n} = (0, 0, 1)$. In this case, after the application of the operator $\mathbb{H}^{\eta}$, only the components g_{ij} and f_k that have the index 3 are nonzero.

The operators $\mathbb{H}^{\eta}$ and $\mathbb{H}^{\boldsymbol{\nu}}$ allow the ideal contact condition to be written in the form

$$\mathbb{H}^{\eta}\boldsymbol{\Sigma}^{(1)} = \mathbb{H}^{\eta}\boldsymbol{\Sigma}^{(2)}, \quad \mathbb{H}^{\boldsymbol{\nu}}\boldsymbol{\mathcal{E}}^{(1)} = \mathbb{H}^{\boldsymbol{\nu}}\boldsymbol{\mathcal{E}}^{(2)}. \quad (17.69)$$

Now, we obtain the expressions relating generalized stresses on both sides of the interface ∂v_{12}. According to (17.68) and (17.69_1), we have

$$\boldsymbol{\mathcal{E}}^{(2)} = \mathbb{M}^{(2)}\boldsymbol{\Sigma}^{(2)} + \boldsymbol{\Lambda}^{(2)} = \mathbb{M}^{(2)}(\mathbb{H}^{\eta}\boldsymbol{\Sigma}^{(1)} + \mathbb{H}^{\boldsymbol{\nu}}\boldsymbol{\Sigma}^{(2)}) + \boldsymbol{\Lambda}^{(2)}. \tag{17.70}$$

Multiplying both sides of (17.70) by the operator $\mathbb{H}^{\boldsymbol{\nu}}$ and taking into account condition (17.69_2) we obtain

$$\boldsymbol{\Sigma}^{(2)} = \mathbb{B}(\mathbf{n})\Big(\boldsymbol{\mathcal{E}}^{(1)} - \mathbb{M}^{(2)}\mathbb{H}^{\eta}\boldsymbol{\Sigma}^{(1)} - \boldsymbol{\Lambda}^{(2)}\Big), \tag{17.71}$$

where $\mathbb{B}(\mathbf{n}) = (\mathbb{H}^{\boldsymbol{\nu}}\mathbb{M}^{(2)}\mathbb{H}^{\boldsymbol{\nu}})^{-1}\mathbb{H}^{\boldsymbol{\nu}}$. Finally, substituting the equation of state (17.42) into the right-hand side of (17.71), we find

$$\boldsymbol{\Sigma}^{(2)} = \Big[\mathbb{I} + \mathbb{B}(\mathbf{n})(\mathbb{M}^{(1)} - \mathbb{M}^{(2)})\Big]\boldsymbol{\Sigma}^{(1)} + \mathbb{B}(\mathbf{n})(\boldsymbol{\Lambda}^{(1)} - \boldsymbol{\Lambda}^{(2)}) \tag{17.72}$$

Similarly, using the conjugation condition for the normal components of the generalized strain tensor of the first material component, we obtain

$$\boldsymbol{\mathcal{E}}^{(2)} = \Big[\mathbb{I} + \mathbb{A}(\mathbf{n})(\mathbb{L}^{(1)} - \mathbb{L}^{(2)})\Big]\boldsymbol{\mathcal{E}}^{(1)} + \mathbb{A}(\mathbf{n})\Big[\mathbb{L}^{(2)}\boldsymbol{\Lambda}^{(2)} - \mathbb{L}^{(1)}\boldsymbol{\Lambda}^{(1)}\Big], \tag{17.73}$$

where $\mathbb{A}(\mathbf{n}) \equiv (\mathbb{H}^{\eta}\mathbb{L}^{(2)}\mathbb{H}^{\eta})^{-1}\mathbb{H}^{\eta}$. Since relationships (17.72) and (17.73) should be equivalent at $\mathbb{M} = \mathbb{L}^{-1}$, the operators $\mathbb{A}(\mathbf{n})$ and $\mathbb{B}(\mathbf{n})$ are related as

$$\mathbb{L}^{(2)}\mathbb{A}(\mathbf{n}) + \mathbb{B}(\mathbf{n})\mathbb{M}^{(2)} = \mathbb{A}(\mathbf{n})\mathbb{L}^{(2)} + \mathbb{M}^{(2)}\mathbb{B}(\mathbf{n}) = \mathbb{I}. \tag{17.74}$$

where operators $\mathbb{B}(\mathbf{n})$ and $\mathbb{A}(\mathbf{n})$, each dependent on the normal vector $\mathbf{n}\perp\partial v_{12}$ and tensor of the properties of a corresponding phase, relate generalized stresses and strains on both sides of the contact. At $\theta = 0$ (17.73), (17.74) obtained in [194] coincide with similar equations obtained earlier in [996].

17.2.6 Phase-Averaged First and Second Moments of the Field $\boldsymbol{\Sigma}$

First, we obtain the averaged single-point second moment of the field $\boldsymbol{\Sigma}$ in terms of the analytical function $\mathbb{M}^* = \mathbb{M}^*(\mathbb{M}^{(i)})$, $(i = 0, 1, \ldots)$ which may be calculated with any desired accuracy or measured experimentally. Averaging is performed over the component $v^{(i)}$. For a given field $\mathbf{t}(\mathbf{x}) \equiv \langle\boldsymbol{\Sigma}\rangle\mathbf{n}(\mathbf{x})$, $\langle\boldsymbol{\Sigma}\rangle = \text{const.}$, $\mathbf{x} \in \partial w$ (see (17.50)), we will vary the local tensor of generalized compliance $\mathbb{M}(\mathbf{x}) \to \mathbb{M}(\mathbf{x}) + \delta\mathbb{M}(\mathbf{x})$, which leads to the variation of the field $\boldsymbol{\Sigma}^I \to \boldsymbol{\Sigma}^I + \delta\boldsymbol{\Sigma}^I$. This technique was applied in the case of a purely elastic field (see Chapter 13). Then, from the relation $2\langle W^I\rangle = \langle\boldsymbol{\Sigma}^{I\top}\mathbb{M}^*\boldsymbol{\Sigma}^I\rangle = \langle\boldsymbol{\Sigma}^{I\top}\rangle\mathbb{M}^*\langle\boldsymbol{\Sigma}^I\rangle$ at $\theta \equiv 0$, we obtain

$$\langle\boldsymbol{\Sigma}^{I\top}\rangle(\delta\mathbb{M}^*)\langle\boldsymbol{\Sigma}^I\rangle = \langle\boldsymbol{\Sigma}^{I\top}(\delta\mathbb{M})\boldsymbol{\Sigma}^I + 2(\delta\boldsymbol{\Sigma}^{I\top})\mathbb{M}\boldsymbol{\Sigma}^I\rangle. \tag{17.75}$$

we transform the second term on the right-hand side of (17.75) by partial integration

$$\langle\delta\boldsymbol{\Sigma}^{I\top}\mathbb{M}\boldsymbol{\Sigma}^I\rangle = (\bar{w})^{-1}\int \delta\boldsymbol{\Sigma}^{I\top}\boldsymbol{\mathcal{U}}\,ds - \langle(\nabla\delta\boldsymbol{\Sigma}^I)\boldsymbol{\mathcal{U}}^I\rangle \tag{17.76}$$

where $\boldsymbol{\mathcal{U}}$ is the generalized displacement matrix defined by (17.41_2). The surface integral taken over the boundary of domain w is zero at the constant boundary conditions (17.45). The second term on the right-hand side of (17.76) is zero due to static condition (17.41_1). Therefore we obtain from (17.75) and (17.76)

$$\langle \boldsymbol{\Sigma}^{I\top} \rangle (\delta \mathbb{M}^*) \langle \boldsymbol{\Sigma}^I \rangle = \langle \boldsymbol{\Sigma}^{I\top} (\delta \mathbb{M}) \boldsymbol{\Sigma}^I \rangle. \tag{17.77}$$

We assume that the components of the generalized compliance tensor and their variations are homogeneous within each phase, i.e.,

$$\mathbb{M}(\mathbf{x}) \equiv \mathbb{M}^{(i)} V^{(i)}(\mathbf{x}), \ \delta\mathbb{M}(\mathbf{x}) \equiv \delta\mathbb{M}^{(i)} V^{(i)}, \tag{17.78}$$

where $\mathbb{M}^{(i)}$, $\delta\mathbb{M}^{(i)}$ = const. Then, the value $\delta\mathbb{M}$ in (17.77) may be factored outside the phase averaging, and (17.45) may be written in the following symbolic form:

$$\langle \Sigma^I_{\alpha\beta} \otimes \Sigma^I_{\gamma\delta} \rangle^{(i)} = \frac{1}{c^{(i)}} \frac{\partial \mathbb{M}^*_{\lambda\mu\nu\eta}}{\partial \mathbb{M}^{(i)}_{\alpha\beta\gamma\delta}} \langle \Sigma_{\lambda\mu} \rangle \langle \Sigma_{\nu\eta} \rangle, \tag{17.79}$$

where we took into account that $\langle \boldsymbol{\Sigma}^{II} \rangle \equiv \mathbf{0}$ and hence $\langle \boldsymbol{\Sigma}^I \rangle = \langle \boldsymbol{\Sigma} \rangle$. Expressing (17.79) in terms of generalized strains, we obtain

$$\langle \mathcal{E}^I_{\alpha\beta} \otimes \mathcal{E}^I_{\gamma\delta} \rangle^{(i)} = \frac{1}{c^{(i)}} \frac{\partial \mathbb{L}^*_{\lambda\mu\nu\eta}}{\partial \mathbb{L}^{(i)}_{\alpha\beta\gamma\delta}} \langle \mathcal{E}_{\lambda\mu} \rangle \langle \mathcal{E}_{\nu\eta} \rangle, \tag{17.80}$$

Here the calculations are carried out at constant values of $\boldsymbol{\Lambda}$. Equations (17.79) and (17.80) were obtained Buryachenko and Shermergor [194]. Li and Dunn [680] independently obtained the intermediate Eq. (17.77) (without final representation (17.79)) at the condition (17.78). Analogous representations for the field fluctuations describing the transport phenomena (i.e., conductivity) were proposed in [23].

Now, we find the exact relations between the energy $\langle W^{II} \rangle$ and second single point moment of the residual generalized stresses $\langle \boldsymbol{\Sigma}^{II} \otimes \boldsymbol{\Sigma}^{II} \rangle^{(i)}$ $(i = 0, 1, \ldots, N)$ at $\mathbf{t}(\mathbf{x}) \equiv \mathbf{0}$ and $\mathbf{x} \in \partial w$. Let the generalized compliance be varied $\mathbb{M}(\mathbf{x}) \to \mathbb{M}(\mathbf{x}) + \delta\mathbb{M}(\mathbf{x})$ at a given tensor $\boldsymbol{\Lambda}(\mathbf{x})$ and $\mathbf{t}(\mathbf{x}) \equiv \mathbf{0}$, $\mathbf{x} \in \partial w$, as is done solving the thermoelastic problem (see Chapter 13). Then, we obtain

$$2\langle \delta W^{II} \rangle = \langle \boldsymbol{\Sigma}^{II\top} (\delta\mathbb{M}) \boldsymbol{\Sigma}^{II} \rangle - 2\langle (\delta \boldsymbol{\Sigma}^{II\top} \boldsymbol{\Lambda}^{II} \rangle, \quad 2\langle \delta W^{II} \rangle = -\langle (\delta \boldsymbol{\Sigma}^{II}) \boldsymbol{\Lambda}^{II} \rangle, \tag{17.81}$$

where we used the condition $\langle (\delta \boldsymbol{\Sigma}^\top \boldsymbol{\mathcal{E}} \rangle = 0$. Therefore

$$\langle \boldsymbol{\Sigma}^{II\top} (\delta\mathbb{M}) \boldsymbol{\Sigma}^{II} \rangle = -2\langle \delta W^{II} \rangle. \tag{17.82}$$

Here, we took into account that in view of Hill's condition, $\langle (\delta \boldsymbol{\Sigma}^\top) \boldsymbol{\mathcal{E}} \rangle = 0$. Using the expression for variation of generalized compliance in the form (17.78), we find the final expression

$$\langle \Sigma^{II} \otimes \Sigma^{II} \rangle^{(i)} = -\frac{2}{c^{(i)}} \frac{\partial \langle \, W^{II} \rangle}{\partial \mathbb{M}^{(i)}}, \quad \text{or} \quad \langle \Sigma^{II}_{\alpha\beta} \otimes \Sigma^{II}_{\gamma\delta} \rangle^{(i)} = -\frac{2}{c^{(i)}} \frac{\partial \langle W^{II} \rangle}{\partial \mathbb{M}^{(i)}_{\alpha\beta\gamma\delta}}. \tag{17.83}$$

Equation (17.83) was obtained in [194]. Li and Dunn [680] independently obtained the intermediate Eq. (17.82) (without final representation (17.83)) at the condition (17.78).

If a material is subject not only to temperature fields θ but also to generalized fields $\mathbf{t}$ defined by expression (17.45), then, we obtain resolving the value W into two W^I and W^{II} (17.63) and using (17.79) and (17.83):

$$\langle \Sigma \otimes \Sigma \rangle^{(i)} = \frac{1}{c^{(i)}} \frac{\partial \mathbb{M}^*}{\partial \mathbb{M}^{(i)}} \langle \boldsymbol{\Sigma} \rangle \otimes \langle \boldsymbol{\Sigma} \rangle - \frac{2}{c^{(i)}} \frac{\partial \langle W^{II} \rangle}{\partial \mathbb{M}^{(i)}} + \frac{2}{c^{(i)}} \frac{\partial \boldsymbol{\Lambda}^*}{\partial \mathbb{M}^{(i)}} \langle \boldsymbol{\Sigma} \rangle, \quad (17.84)$$

where all partial derivatives are calculated at fixed $\boldsymbol{\Lambda}(\mathbf{x})$. Thus the fourth rank tensor of the second moment of the generalized stress field $\langle \Sigma \otimes \Sigma \rangle^{(i)}$ averaged over the phase volume $v^{(i)}$, is uniquely determined from the functional relationships $\mathbb{M}^* = \mathbb{M}^*(\mathbb{M}^{(i)})$, $\langle W^{II} \rangle = \langle W^{II}(\mathbb{M}^{(i)}) \rangle$, $\boldsymbol{\Lambda}^* = \boldsymbol{\Lambda}(\mathbb{M}^{(i)})$ in four-dimensional space.

Note that the exact expression for the phase-averaged first moment $\langle \boldsymbol{\Sigma}^I \rangle^{(i)}$ $(i = 0, 1)$ can be found only for a two-phase composite (17.64), which is not the case for general multicomponent media. However, the exact solution for the first moment of the residual field $\langle \boldsymbol{\Sigma}^{II} \rangle^{(i)}$ $(i = 0, 1, \ldots, N)$ can be found if function $\langle W^{II} \rangle = \langle W^{II}(\boldsymbol{\Lambda}^{(i)}) \rangle$ is known. For this purpose, let $\boldsymbol{\Lambda}(\mathbf{x}) \to \boldsymbol{\Lambda}(\mathbf{x}) + \delta\boldsymbol{\Lambda}(\mathbf{x})$ be varied in the same manner as in the thermoelastic case, assuming $\mathbb{M}(\mathbf{x})$ to be constant ant $\mathbf{t}(\mathbf{x}) \equiv \mathbf{0}$, $\mathbf{x} \in \partial w$. We also assume that the generalized parameter $\boldsymbol{\Lambda}$ and its variation within each phase of the composite are homogeneous $\boldsymbol{\Lambda}(\mathbf{x}) \equiv \boldsymbol{\Lambda}^{(i)} V^{(i)}(\mathbf{x})$, $\delta\boldsymbol{\Lambda}(\mathbf{x}) \equiv \delta\boldsymbol{\Lambda}^{(i)} V^{(i)}(\mathbf{x})$, where $\boldsymbol{\Lambda}^{(i)}, \delta\boldsymbol{\Lambda}^{(i)} = \text{const}$. Hence we obtain the energy variation in the form

$$-2\langle \delta W^{II} \rangle = \langle \boldsymbol{\Sigma}^{II\top} \delta\boldsymbol{\Lambda} \rangle + \langle (\delta\boldsymbol{\Sigma}^{II\top}) \boldsymbol{\Lambda}^I \rangle. \quad (17.85)$$

Let us transform the last term with use of the equation of state (17.42):

$$\langle (\delta\boldsymbol{\Sigma}^{II\top}) \boldsymbol{\Lambda}^I \rangle = \langle (\delta\boldsymbol{\Sigma}^{II\top}) \boldsymbol{\mathcal{E}}^{II} \rangle \langle \boldsymbol{\Sigma}^{II\top} \delta\boldsymbol{\Lambda} \rangle - \langle \boldsymbol{\Sigma}^{II\top} \delta\boldsymbol{\mathcal{E}}^{II} \rangle. \quad (17.86)$$

Since the admissible fields $\delta\boldsymbol{\Sigma}^{II}$, $\boldsymbol{\mathcal{E}}^{II}$ and $\boldsymbol{\Sigma}^{II}$, $\delta\boldsymbol{\mathcal{E}}^{II}$ obey Hill's condition (17.49), we find substituting (17.86) in (17.85)

$$\langle \boldsymbol{\Sigma}^{II}_{\alpha\beta} \rangle^{(i)} = \frac{1}{c^{(i)}} \frac{\partial \langle W^{II} \rangle}{\partial \boldsymbol{\Lambda}^{(i)}_{\alpha\beta}} \quad (17.87)$$

$(i = 0, 1, \ldots, N;\ k, l = 1, \ldots, 4)$. In a manner like (13.23), we can get

$$\langle \boldsymbol{\Sigma}^I \rangle^{(i)} = \frac{1}{c^{(i)}} \langle \boldsymbol{\Sigma} \rangle \frac{\partial \boldsymbol{\Lambda}^*}{\partial \boldsymbol{\Lambda}^{(i)}}. \quad (17.88)$$

The partial derivatives in Eqs. (17.87) and (17.88) are carried out at fixed values of $\mathbb{M}(\mathbf{x})$.

Relationship (17.84) allows one to determine the fluctuations of the field $\boldsymbol{\Sigma}$ at the point $\mathbf{x} \in v^{(i)}$ $(i = 0, 1, \ldots, N)$. For this purpose it is sufficient to find $\mathbb{M}^*$ at the variation of the generalized compliance in the vicinity of the considered point $\mathbf{x}$. This problem was solved in [194] in the framework of hypothesis **H1**) only (see Chapter 8) which is the main hypothesis of many micromechanical methods in thermoelasticity. The same is true for the piezoelectric matrix composites. That implies the homogeneity of the stress field single-point moment within the inclusion. This fact enables determination of the value $\langle \boldsymbol{\Sigma} \otimes \boldsymbol{\Sigma} \rangle^{(0)}$ in the matrix material and the second moment of the generalized stress field for the matrix material in the vicinity of a fixed inclusion v_i. Actually, at a point $\mathbf{x} \in \partial v_i$ of the inclusion boundary $\boldsymbol{\Sigma}^{(0)}$ with the external normal $\mathbf{n}$, the outer limiting value of the tensor $\boldsymbol{\Sigma}$ is determined by (17.71). Then, the second moment $\langle \boldsymbol{\Sigma} \otimes \boldsymbol{\Sigma} \rangle_x$ of the field $\boldsymbol{\Sigma}$ for the matrix material in the neighborhood of point $\mathbf{x} \in \partial v_i$ can be found from the expression

$$\begin{aligned}\langle \boldsymbol{\Sigma} \otimes \boldsymbol{\Sigma} \rangle_x &= \langle [\mathbb{I} + \mathbb{B}(\mathbf{n})\mathbb{M}_1]\boldsymbol{\Sigma} \otimes [\mathbb{I} + \mathbb{B}(\mathbf{n})\mathbb{M}_1]\boldsymbol{\Sigma} \rangle^{(i)} \\ &+ [\mathbb{I} + \mathbb{B}(\mathbf{n})\mathbb{M}_1]\langle \boldsymbol{\Sigma} \rangle^{(i)} \otimes \mathbb{B}(\mathbf{n})\boldsymbol{\Lambda}_1^{(i)} + \mathbb{B}(\mathbf{n})\boldsymbol{\Lambda}_1^{(i)} \otimes [\mathbb{I} + \mathbb{B}(\mathbf{n})\mathbb{M}_1]\langle \boldsymbol{\Sigma} \rangle^{(i)} \\ &+ [\mathbb{B}(\mathbf{n})\boldsymbol{\Lambda}_1^{(i)} \otimes \mathbb{B}(\mathbf{n})\boldsymbol{\Lambda}_1^{(i)}]. \end{aligned} \tag{17.89}$$

Here, we assumed the field homogeneity within each inclusion v_i. Therefore, (17.89) is approximate, although the exact expressions (17.84) for $\langle \boldsymbol{\Sigma} \otimes \boldsymbol{\Sigma} \rangle^{(i)}$ and (17.87), (17.88) for $\langle \boldsymbol{\Sigma} \rangle^{(i)}$ are used on the right-hand side. Thus the relationships obtained in this Section 17.2 are exact (except (17.89)) and valid for any statistically homogeneous composite with ideal contact between its components and homogeneous boundary conditions. One [176] used the second moment estimations of coupled fields (17.84) and (17.89) for the prediction of the limiting surface (15.11) and (15.17).

17.3 Wave Propagation in a Composite Material

17.3.1 General Integral Equations and Effective Fields

The problem of wave propagation of monochromatic waves through a composite is reduced to the solution of the system of the integral equation presented in the form of Eq. (3.234)

$$\mathbf{u}(\mathbf{x}) = \mathbf{u}^{(0)}(\mathbf{x}) + \int \mathbf{g}(\mathbf{x} - \mathbf{y})\nabla\boldsymbol{\tau}(\mathbf{y})\, d\mathbf{y} + \int \mathbf{g}(\mathbf{x} - \mathbf{y})\boldsymbol{\pi}(\mathbf{y})\, d\mathbf{y}, \tag{17.90}$$

where the polarizations $\boldsymbol{\tau}$ and $\boldsymbol{\pi}$ are nonzero only over the phase of inclusions with the characteristic function $V(\mathbf{y}) = \sum_i V_i(\mathbf{y})$, and $\mathbf{u}^0$ is the displacement amplitude of the incident harmonic field with frequency ω, which would exist in a homogeneous medium with parameters $\mathbf{L}^{(0)}$, $\rho^{(0)}$ with prescribed boundary conditions at the boundary of a macrodomain w in the absent of inclusions. Transforming the first integral by parts taking the equality $\nabla_\mathbf{y} = -\nabla_\mathbf{x}$ into account, we obtain

$$\mathbf{u}(\mathbf{x}) = \mathbf{u}^{(0)}(\mathbf{x}) + \int \nabla\mathbf{g}(\mathbf{x} - \mathbf{y})\boldsymbol{\tau}(\mathbf{y})\, d\mathbf{y} + \int \mathbf{g}(\mathbf{x} - \mathbf{y})\boldsymbol{\pi}(\mathbf{y})\, d\mathbf{y} + \int \mathbf{g}(\mathbf{x} - \mathbf{s})\boldsymbol{\tau}(\mathbf{s})\, d\mathbf{s}, \tag{17.91}$$

where the last surface integral vanishes only for a finite number of inclusions in an unbounded matrix.

Centering and differentiating both sides of Eq. (17.91), we obtain

$$\mathbf{u}(\mathbf{x}) = \mathbf{U}(\mathbf{x}) + \int \nabla\mathbf{g}(\mathbf{x} - \mathbf{y})[\boldsymbol{\tau}(\mathbf{y}) - \langle \boldsymbol{\tau} \rangle(\mathbf{y})]\, d\mathbf{y} + \int \mathbf{g}(\mathbf{x} - \mathbf{y})[\boldsymbol{\pi}(\mathbf{y}) - \langle \boldsymbol{\pi} \rangle(\mathbf{y})]\, d\mathbf{y},$$

$$\begin{aligned}\boldsymbol{\varepsilon}(\mathbf{x})=\mathbf{E}(\mathbf{x})&+\int\nabla\nabla\mathbf{g}(\mathbf{x}-\mathbf{y})[\boldsymbol{\tau}(\mathbf{y})-\langle\boldsymbol{\tau}\rangle(\mathbf{y})]\,d\mathbf{y}\\&+\int\nabla\mathbf{g}(\mathbf{x}-\mathbf{y})[\boldsymbol{\pi}(\mathbf{y})-\langle\boldsymbol{\pi}\rangle(\mathbf{y})]\,d\mathbf{y},\end{aligned}\tag{17.92}$$

which describe the random fields $\mathbf{u}(\mathbf{x})$ and $\boldsymbol{\varepsilon}(\mathbf{x})$ of both the statistically homogeneous and inhomogeneous (so called functionally graded) random structure matrix composites; here $\mathbf{U}(\mathbf{x}) = \langle\mathbf{u}(\mathbf{x})\rangle$, $\mathbf{E}(\mathbf{x}) = \langle\boldsymbol{\varepsilon}(\mathbf{x})\rangle$, $\varepsilon_{ij}(\mathbf{x}) = \nabla_{(i}u_{j)}(\mathbf{x})$. Hereafter, the standard notations of Chapter 7 are used for the statistical averages, and it is assumed that the distance $\rho'(\mathbf{x})$ from $\mathbf{x}$ in (17.92) to the boundary $s=\partial w$ of w is much greater than the characteristic dimensions of the inclusions $a_i^1/\rho \ll 1$. The representations (17.92) differer by the terms $\langle\boldsymbol{\tau}\rangle(\mathbf{y})$ and $\langle\boldsymbol{\pi}\rangle(\mathbf{y})$ from the appropriate equations [1182], [542], [545] proposed for statistically homogeneous composites.

Before proceeding to the related approaches to the scattering of elastic waves by random structure matrix composites, we present the basic Eqs. (17.92) in the form of the static equation (8.1):

$$\boldsymbol{\mathcal{E}}(\mathbf{x}) = \langle\boldsymbol{\mathcal{E}}\rangle(\mathbf{x}) + \int \mathbf{U}^{\omega}(\mathbf{x}-\mathbf{y})[\boldsymbol{\mathcal{T}}(\mathbf{y}) - \langle\boldsymbol{\mathcal{T}}\rangle(\mathbf{y})]\,d\mathbf{y},\tag{17.93}$$

where

$$\boldsymbol{\mathcal{E}}=\begin{pmatrix}\mathbf{u}\\ \boldsymbol{\varepsilon}\end{pmatrix},\quad \boldsymbol{\mathcal{T}}=\begin{pmatrix}\boldsymbol{\pi}\\ \boldsymbol{\tau}\end{pmatrix},\quad \mathbf{U}^{\omega}=\begin{pmatrix}\mathbf{g} & \nabla\mathbf{g}\\ \nabla\mathbf{g} & \nabla\nabla\mathbf{g}\end{pmatrix},$$

After that the scheme of derivation of effective properties is quite similar to the one given in Chapters 8 and 9 for the static case. So, for the case in which the mechanical properties of the components differ slightly, the perturbation method of the first order of an accuracy with respect to the concentration of inclusions $\overline{\boldsymbol{\mathcal{E}}}(\mathbf{x}_i) \equiv \langle\boldsymbol{\mathcal{E}}\rangle(\mathbf{x}_i)$ has been used in [1199] to estimate the effective wave propagation velocities and the attenuation factors. One [995], discarding a density fluctuation ($\rho_1 \equiv 0$), used a similar method of the correlation approximation for obtaining of the long- and short-wave asymptotic solutions of the dispersion equation. The effective-medium approach widely employed (see, e.g, [260], [261], [262], [360], [824], [547], [856], [955]) is based on solution of the problem for one ellipsoidal inclusion (see, e.g., [361], [1181]) in the matrix with the effective properties of the medium and with specified load at infinity. Sato and Shindo [965] considered scattering of elastic waves combining a dilute approximation approach with a numerical solution obtained by the boundary element method for a single partially debonded elliptical inclusion. Variational principles for dynamic problems [1185] was used in [1066], [1067] for estimation of the dispersion and attenuation of elastic waves in random composites with the ellipsoidal inclusions. It should be mentioned that the T-matrix approximation method [730], [1132] and polarization approximation method [1182], as well as the one-particle variant of the effective-field method ([209], [542], [545], [546]), consider an isolated inclusion in the matrix with a self-consistent field acting at infinity, whose parameters are independent of the properties of the inclusion in question ("quasi-crystalline" approximation). Use this "quasi-crystalline" approximation reduces the problem to a single-particle one, thus entailing substantial error in estimating the coefficient in the second degree with respect to the concentration of inclusions c, in the relationship between the effective parameters of the medium and small c value. A general approach for setting up an effective wave operator proposed in [169] and reproduced in this section is based on the generalization of the static version of the MEFM.

We average (17.92_1) on sets $X(\cdot|v_1$, $X(\cdot|v_1, v_2)$, and so forth, for fixed sets v_1; $v_1, v_2; \ldots$ by means of different distribution densities $\varphi(v_m|v_1, \ldots, v_n)$ introduced in Chapter 7. For $\mathbf{x} \in v_1, \ldots, v_n$ we obtain an infinite system of integral equations for the displacement amplitude

$$\langle \mathbf{u}(\mathbf{x})|\mathbf{x}_1\rangle - \int \nabla \mathbf{g}(\mathbf{x}-\mathbf{y})\nabla \boldsymbol{\tau}(\mathbf{y})V_1(\mathbf{y})|\mathbf{x}_1\rangle \, d\mathbf{y} - \int \mathbf{g}(\mathbf{x}-\mathbf{y})\langle \boldsymbol{\pi}(\mathbf{y})V_1(\mathbf{y})|\mathbf{x}_1\rangle \, d\mathbf{y}$$
$$= \mathbf{U}(\mathbf{x}) + \int \nabla \mathbf{g}(\mathbf{x}-\mathbf{y})[\langle \boldsymbol{\tau}(\mathbf{y})|\mathbf{y};\mathbf{x}_1\rangle - \langle \boldsymbol{\tau}\rangle] \, d\mathbf{y}$$
$$+ \int \mathbf{g}(\mathbf{x}-\mathbf{y})[\langle \boldsymbol{\pi}(\mathbf{y})|\mathbf{y};\mathbf{x}_1\rangle - \langle \boldsymbol{\pi}\rangle] \, d\mathbf{y}, \tag{17.94}$$

$$\langle \mathbf{u}(\mathbf{x}) \mid \mathbf{x}_1, \ldots, \mathbf{x}_n\rangle - \sum_{i=1}^{n} \Big\{ \int \nabla \mathbf{g}(\mathbf{x}-\mathbf{y})\nabla \boldsymbol{\tau}(\mathbf{y})V_1(\mathbf{y})|\mathbf{x}_1, \ldots, \mathbf{x}_n\rangle \, d\mathbf{y}$$
$$- \int \mathbf{g}(\mathbf{x}-\mathbf{y})\langle \boldsymbol{\pi}(\mathbf{y})V_1(\mathbf{y})|\mathbf{x}_1, \ldots, \mathbf{x}_n\rangle \, d\mathbf{y}\Big\}$$
$$= \mathbf{U}(\mathbf{x}) + \int \nabla \mathbf{g}(\mathbf{x}-\mathbf{y})[\langle \boldsymbol{\tau}(\mathbf{y})|\mathbf{y};\mathbf{x}_1, \ldots, \mathbf{x}_n\rangle - \langle \boldsymbol{\tau}\rangle] \, d\mathbf{y}$$
$$+ \int \mathbf{g}(\mathbf{x}-\mathbf{y})[\langle \boldsymbol{\pi}(\mathbf{y})|\mathbf{y};\mathbf{x}_1, \ldots, \mathbf{x}_n\rangle - \langle \boldsymbol{\pi}\rangle] \, d\mathbf{y}. \tag{17.95}$$

Analogously to the definitions (8.8)–(8.11) we will introduce the effective fields $\overline{\mathbf{u}}(\mathbf{x})$, $\overline{\mathbf{E}}(\mathbf{x})$ and $\widetilde{\mathbf{u}}_{1,\ldots,n}(\mathbf{x})$, $\widetilde{\varepsilon}_{1,\ldots,n}(\mathbf{x})$ acting on the inclusion $\mathbf{x} \in v_i$ and n inclusions $v_1, \ldots, v_n$, respectively. For example, each inclusion $\mathbf{x} \in v_i$ from the n selected $v_1, \ldots, v_n$ is in a field

$$\overline{\mathbf{u}}(\mathbf{x}) = \widetilde{\mathbf{u}}_{1,\ldots,n}(\mathbf{x}) + \sum_{i\neq j} \int [\nabla \mathbf{g}(\mathbf{x}-\mathbf{y})\langle \boldsymbol{\tau}(\mathbf{y})V_i(\mathbf{y})|\mathbf{y};\mathbf{x}_1, \ldots, \mathbf{x}_n\rangle$$
$$+ \mathbf{g}(\mathbf{x}-\mathbf{y})\langle \boldsymbol{\pi}(\mathbf{y})V_i(\mathbf{y})|\mathbf{y};\mathbf{x}_1, \ldots, \mathbf{x}_n\rangle] \, d\mathbf{y}. \tag{17.96}$$

From (17.92_2) we obtain equations analogous to (17.94) and (17.95) for $\overline{\varepsilon}(\mathbf{x})$ and $\widetilde{\varepsilon}_{1,\ldots,n}(\mathbf{x})$. Thus , the fields of the displacement and strain amplitudes inside inclusion v_i depend on fields $\overline{\mathbf{u}}(\mathbf{x})$, $\overline{\varepsilon}(\mathbf{x})$, which are, generally speaking, nonuniform. In the long wave approximation the fields $\overline{\mathbf{u}}(\mathbf{x})$ and $\overline{\varepsilon}(\mathbf{x})$ are considered as constant in each domain v_i $(i = 1, \ldots, n)$; in what follows, to avoid taking account of the dependence on $\mathbf{x}$ of the later, we will assume that $\mathbf{x} \in v_i$ coincides with $\mathbf{x}_i$.

In order to close and then approximately solve system (17.94), (17.95), we employ the hypotheses of the effective field method (see Chapter 8)

H1). Each ellipsoidal inhomogeneity v_i is in the homogeneous fields $\overline{\mathbf{u}}(\mathbf{x})$ and $\overline{\varepsilon}(\mathbf{x})$; we have a point approximation of inclusions in the analyzing the outside fields scattered by the inclusion.

H2). $n+1$ inclusions are in fields $\widetilde{\mathbf{u}}_{1,\ldots,n+1}$, $\widetilde{\varepsilon}_{1,\ldots,n+1}$ and, for sufficiently large n, we have the close

$$\widetilde{\mathbf{u}}_{1,\ldots,j,\ldots,n+1}(\mathbf{x}) = \widetilde{\mathbf{u}}_{1,\ldots,n}(\mathbf{x}), \quad \widetilde{\varepsilon}_{1,\ldots,j,\ldots,n+1}(\mathbf{x}) = \widetilde{\varepsilon}_{1,\ldots,n}(\mathbf{x}), \tag{17.97}$$

where $i = 1, \ldots, n+1$ and some index $j \neq i$ is omitted on the right side of the equations.

Hypothesis **H1** enables us to estimate, on the basis of fields $\overline{\mathbf{u}}(\mathbf{x})$ and $\overline{\varepsilon}(\mathbf{x})$, the corresponding fields inside (3.242) and outside (3.243) inclusion v_i. The effective fields $\overline{\mathbf{u}}(\mathbf{x})$ and $\overline{\varepsilon}(\mathbf{x})$, in turn, can be expressed through the effective fields $\widetilde{\mathbf{u}}_{1,\ldots,n}(\mathbf{x})$ and $\widetilde{\varepsilon}_{1,\ldots,n}(\mathbf{x})$ acting on n fixed interacting inclusions (see Eq. (3.243))

$$\begin{aligned}
&\overline{\mathbf{u}}(\mathbf{x}_i) - \sum_{j\neq i} \nabla \mathbf{G}^{\omega}(\mathbf{r})\mathbf{R}_j^{\omega}\overline{\varepsilon}(\mathbf{x}_j) + \mathbf{G}^{\omega}(\mathbf{r})\mathbf{F}_j^{\omega}\overline{\mathbf{u}}(\mathbf{x}_j) = \widetilde{\mathbf{u}}_{1,\ldots,n}(\mathbf{x}_i),\\
&\overline{\varepsilon}(\mathbf{x}_i) - \sum_{j\neq i} \nabla\nabla \mathbf{G}^{\omega}(\mathbf{r})\mathbf{R}_j^{\omega\epsilon}\overline{\varepsilon}(\mathbf{x}_j) + \nabla\mathbf{G}^{\omega}(\mathbf{r})\mathbf{F}_j^{\omega}\overline{\mathbf{u}}(\mathbf{x}_j) = \widetilde{\varepsilon}_{1,\ldots,n}(\mathbf{x}_i),
\end{aligned} \tag{17.98}$$

where $\mathbf{R}_j^{\omega} = \bar{v}_j \mathbf{L}_1^{(j)} \boldsymbol{\Lambda}_j$, $\mathbf{F}_j^{\omega} = \bar{v}_j \omega^2 \rho_1^{(j)} \boldsymbol{\lambda}_j$, and $\mathbf{r} = \mathbf{x}_j - \mathbf{x}_i$. System (17.98) is a system of linear algebraic equations with respect to $\overline{\mathbf{u}}(\mathbf{x}_i)$ and $\overline{\varepsilon}(\mathbf{x}_i)$, and can be solved by standard methods of linear algebra. For this we present the tensors (17.98) in equivalent matrix form, and generate matrix $\mathbf{Z}^{-1}$ with elements $\mathbf{Z}_{mk}$ $(m, k = 1, \ldots, n)$ in the form of (12×12) submatrices:

$$\mathbf{Z}_{mk}^{-1} = \begin{pmatrix} \mathbf{I}\delta_{mk} - (1-\delta_{mk})\mathbf{G}^{\omega}(\mathbf{r})\mathbf{F}_k^{\omega} & -(1-\delta_{mk})\nabla\mathbf{G}^{\omega}(\mathbf{r})\mathbf{R}_k^{\omega} \\ -(1-\delta_{mk})\nabla\mathbf{G}^{\omega}(\mathbf{r})\mathbf{F}_k^{\omega} & \mathbf{I}\delta_{mk} - (1-\delta_{mk})\nabla\nabla\mathbf{G}^{\omega}(\mathbf{r})\mathbf{R}_k^{\omega} \end{pmatrix}, \tag{17.99}$$

$(\mathbf{r} = \mathbf{x}_k - \mathbf{x}_m)$ in which case the solution of (17.98) can be represent as follows

$$\begin{pmatrix} \overline{\mathbf{u}}(\mathbf{x}_i) \\ \overline{\varepsilon}(\mathbf{x}_i) \end{pmatrix} = \sum_{j=1}^{n} \mathbf{Z}_{ij} \begin{pmatrix} \widetilde{\mathbf{u}}_{1,\ldots,n}(\mathbf{x}_j) \\ \widetilde{\varepsilon}_{1,\ldots,n}(\mathbf{x}_j) \end{pmatrix}, \quad \mathbf{Z}_{ij} = \begin{pmatrix} \mathbf{Z}_{ij}^{11} \ \mathbf{Z}_{ij}^{12} \\ \mathbf{Z}_{ij}^{21} \ \mathbf{Z}_{ij}^{22} \end{pmatrix}, \tag{17.100}$$

The solution of (17.98) can also be developed by the successive-approximation method analogous to the static limit (4.67):

$$\mathbf{Z}_{mk} = \begin{pmatrix} \mathbf{I}\delta_{mk} + (1-\delta_{mk})\mathbf{G}^{\omega}(\mathbf{r})\mathbf{F}_k^{\omega} & (1-\delta_{mk})\nabla\mathbf{G}^{\omega}(\mathbf{r})\mathbf{R}_k^{\omega} \\ (1-\delta_{mk})\nabla\mathbf{G}^{\omega}(\mathbf{r})\mathbf{F}_k^{\omega} & \mathbf{I}\delta_{mk} + (1-\delta_{mk})\nabla\nabla\mathbf{G}^{\omega}(\mathbf{r})\mathbf{R}_k^{\omega} \end{pmatrix}. \tag{17.101}$$

It should be mentioned that the widely employed quasi-crystalline approximation (see, e.g., [545], [648], [1132], [1182]) is equivalent to the assumption that $\mathbf{Z}_{ij} = \mathrm{diag}(\mathbf{I}\delta_{ij}, \mathbf{I}\delta_{ij})$. In the subsequent obtaining of the effective wave operator we will follow [542], [545] as well as Buryachenko and Parton [169], whose results are reduced to the first ones if a direct binary interaction is neglected.

17.3.2 Fourier Transform of Effective Wave Operator

The above solutions for one inclusions (3.242) and for a finite number of inclusions (17.100) in effective fields $\overline{\mathbf{u}}(\mathbf{x}_i)$, $\overline{\varepsilon}(\mathbf{x}_i)$ and $\widetilde{\mathbf{u}}_{1,\ldots,n}(\mathbf{x}_i)$, $\widetilde{\varepsilon}_{1,\ldots,n}(\mathbf{x}_i)$, as well as the use of hypothesis **H2**, enables us to obtain from (17.94), (17.95) a closed system of linear equations governing the fields $\widetilde{\mathbf{u}}_{1,\ldots,j}(\mathbf{x}_i)$, $\widetilde{\boldsymbol{\varepsilon}}_{1,\ldots,j}(\mathbf{x}_i)$ $(j=1,\ldots,n;\ i=1,\ldots,j)$

$$\begin{aligned}\widetilde{\mathbf{u}}_{1,\ldots,j}(\mathbf{x}_i) &= \mathbf{U}(\mathbf{x}_i) + \int \nabla\mathbf{G}^{\omega}(\mathbf{x}_q-\mathbf{x}_i)\Big\{\mathbf{R}_q^{\omega}\sum_{l=1}^{j+1}\Big[\mathbf{Z}_{ql}^{21}\widetilde{\mathbf{u}}_{1,\ldots,j+1}(\mathbf{x}_l)\\ &\quad + \mathbf{Z}_{ql}^{22}\widetilde{\boldsymbol{\varepsilon}}_{1,\ldots,j+1}(\mathbf{x}_l)\Big]\varphi(v_q,\mathbf{x}_q|;v_1,\mathbf{x}_1,\ldots,v_j,\mathbf{x}_j)-\langle\boldsymbol{\tau}\rangle\Big\}\,d\mathbf{x}_q\\ &\quad + \int \mathbf{G}^{\omega}(\mathbf{x}_q-\mathbf{x}_i)\Big\{\mathbf{F}_q^{\omega}\sum_{l=1}^{j+1}\Big[\mathbf{Z}_{ql}^{11}\widetilde{\mathbf{u}}_{1,\ldots,j+1}(\mathbf{x}_l)+\mathbf{Z}_{ql}^{12}\widetilde{\boldsymbol{\varepsilon}}_{1,\ldots,j+1}(\mathbf{x}_l)\Big]\\ &\quad \times\varphi(v_q,\mathbf{x}_q|;v_1,\mathbf{x}_1,\ldots,v_j,\mathbf{x}_j)-\langle\boldsymbol{\pi}\rangle\Big\}\,d\mathbf{x}_q, \end{aligned}\tag{17.102}$$

$$\begin{aligned}\widetilde{\boldsymbol{\varepsilon}}_{1,\ldots,j}(\mathbf{x}_i) &= \mathbf{E}(\mathbf{x}_i) + \int \nabla\nabla\mathbf{G}^{\omega}(\mathbf{x}_q-\mathbf{x}_i)\Big\{\mathbf{R}_q^{\omega}\sum_{l=1}^{j+1}\Big[\mathbf{Z}_{ql}^{21}\widetilde{\mathbf{u}}_{1,\ldots,j+1}(\mathbf{x}_l)\\ &\quad + \mathbf{Z}_{ql}^{22}\widetilde{\boldsymbol{\varepsilon}}_{1,\ldots,j+1}(\mathbf{x}_l)\Big]\varphi(v_q,\mathbf{x}_q|;v_1,\mathbf{x}_1,\ldots,v_j,\mathbf{x}_j)-\langle\boldsymbol{\tau}\rangle\Big\}\,d\mathbf{x}_q\\ &\quad + \int \nabla\mathbf{G}^{\omega}(\mathbf{x}_q-\mathbf{x}_i)\Big\{\mathbf{F}_q^{\omega}\sum_{l=1}^{j+1}\Big[\mathbf{Z}_{ql}^{11}\widetilde{\mathbf{u}}_{1,\ldots,j+1}(\mathbf{x}_l)+\mathbf{Z}_{ql}^{12}\widetilde{\boldsymbol{\varepsilon}}_{1,\ldots,j+1}(\mathbf{x}_l)\Big]\\ &\quad \times\varphi(v_q,\mathbf{x}_q|;v_1,\mathbf{x}_1,\ldots,v_j,\mathbf{x}_j)-\langle\boldsymbol{\pi}\rangle\Big\}\,d\mathbf{x}_q, \end{aligned}\tag{17.103}$$

with obvious closing equation at $j=n$ and $q=1,\ldots,n$ in the left-hand side of Eqs. (17.102), (17.103) and in the right-hand side of ones, respectively.

To obtain a result in analytical form, we will confine ourselves to the case $n=2$ and to the solution of the problem of binary interaction of inclusions in the form (17.101). We also assume that

$$\widetilde{\mathbf{u}}_{1,2}(\mathbf{x}_i)=\overline{\mathbf{u}}(\mathbf{x}_i)=\text{const.},\quad \widetilde{\boldsymbol{\varepsilon}}_{1,2}(\mathbf{x}_i)=\overline{\boldsymbol{\varepsilon}}(\mathbf{x}_i)=\text{const.},\tag{17.104}$$

$(i=1,2)$. If the solution $\overline{\mathbf{u}}(\mathbf{x}_i)$, $\overline{\boldsymbol{\varepsilon}}(\mathbf{x}_i)$ is found, the statistical average fields inside the inclusions can be determined by the use of Eq. (3.242).

To simplify the average operations in (17.102), (17.103), we will consider inclusions of the same size and materials; instead of tensors $\mathbf{R}_i^{\omega}$, $\mathbf{F}_i^{\omega}$, we employ their averaging over the possible orientation $\langle\mathbf{R}_i^{\omega}\rangle$, $\langle\mathbf{F}_i^{\omega}\rangle$. For statistically isotropic homogeneous inclusion fields being considered, we assume $\varphi(v_q,\mathbf{x}_q|;v_i,\mathbf{x}_i)=\varphi(|\mathbf{x}_q-\mathbf{x}_i|)$ and $\varphi(\mathbf{x}_q-\mathbf{x}_i)=0$ at $|\mathbf{x}_q-\mathbf{x}_i|\le 2a$. Then, with the adopted degree of accuracy with respect to ω, Eq. (17.102) is reduced to

$$\begin{aligned}\overline{\mathbf{U}}(\mathbf{x}_i) &= \mathbf{U}(\mathbf{x}_i) - \int \nabla\mathbf{G}^{\omega}(\mathbf{x}_q-\mathbf{x}_i)\langle\mathbf{R}_q^{\omega}\rangle\overline{\mathbf{E}}(\mathbf{x}_q)\psi(\mathbf{x}_q-\mathbf{x}_i)\,d\mathbf{x}_q\\ &\quad - \int \mathbf{G}^{\omega}(\mathbf{x}_q-\mathbf{x}_i)\langle\mathbf{F}_q^{\omega}\rangle\overline{\mathbf{U}}(\mathbf{x}_q)\psi(\mathbf{x}_q-\mathbf{x}_i)\,d\mathbf{x}_q\\ &\quad + \int \nabla\mathbf{G}^{\omega}(\mathbf{x}_q-\mathbf{x}_i)\langle\mathbf{R}_q^{\omega}\rangle\nabla\nabla\mathbf{G}^{\omega}(\mathbf{x}_q-\mathbf{x}_i)\langle\mathbf{R}_q^{\omega}\rangle\overline{\mathbf{E}}(\mathbf{x}_q)\varphi(\mathbf{x}_q-\mathbf{x}_i)\,d\mathbf{x}_q\\ &\quad + \int \nabla\mathbf{G}^{\omega}(\mathbf{x}_q-\mathbf{x}_i)\langle\mathbf{R}_q^{\omega}\rangle\nabla\mathbf{G}^{\omega}(\mathbf{x}_q-\mathbf{x}_i)\langle\mathbf{F}_q^{\omega}\rangle\overline{\mathbf{U}}(\mathbf{x}_q)\varphi(\mathbf{x}_q-\mathbf{x}_i)\,d\mathbf{x}_q\end{aligned}$$

$$
+\int \mathbf{G}^\omega(\mathbf{x}_q-\mathbf{x}_i)\langle\mathbf{F}_q^\omega\rangle\nabla\mathbf{G}^\omega(\mathbf{x}_q-\mathbf{x}_i)\langle\mathbf{R}_q^\omega\rangle\overline{\mathbf{E}}(\mathbf{x}_q)\varphi(\mathbf{x}_q-\mathbf{x}_i)\,d\mathbf{x}_q
$$

$$
+\int \mathbf{G}^\omega(\mathbf{x}_q-\mathbf{x}_i)\langle\mathbf{F}_q^\omega\rangle\mathbf{G}^\omega(\mathbf{x}_q-\mathbf{x}_i)\langle\mathbf{F}_q^\omega\rangle\overline{\mathbf{U}}(\mathbf{x}_q)\varphi(\mathbf{x}_q-\mathbf{x}_i)\,d\mathbf{x}_q, \qquad (17.105)
$$

where $\psi(\mathbf{x}_q-\mathbf{x}_i)=n-\varphi(\mathbf{x}_q-\mathbf{x}_i)$, and $\overline{\mathbf{U}}(\mathbf{x})=\langle\overline{\mathbf{u}}(\mathbf{x})\rangle$, $\overline{\mathbf{E}}(\mathbf{x})=\langle\overline{\varepsilon}(\mathbf{x})\rangle$. To estimate $\overline{\mathbf{U}}(\mathbf{x}_i)$, $\overline{\mathbf{E}}(\mathbf{x}_i)$, we resort to the Fourier transformation in (17.105):

$$
\overline{\mathbf{U}}(\mathbf{k})=\mathbf{U}(\mathbf{k})+\mathbf{T}(\mathbf{k})\overline{\mathbf{E}}(\mathbf{k})+\mathbf{t}(\mathbf{k})\overline{\mathbf{U}}(\mathbf{k}), \qquad (17.106)
$$

where we used the same notations for the Fourier transformations of the functions, but with the argument $\mathbf{x}$ replaced by $\mathbf{k}$, and

$$
\begin{aligned}
\mathbf{T}(\mathbf{k}) &= -\int \nabla\mathbf{g}(\mathbf{x})\langle\mathbf{R}_q^\omega\rangle\psi(\mathbf{x})e^{-i\mathbf{k}\cdot\mathbf{x}}\,d\mathbf{x}_q \\
&+\int \nabla\mathbf{G}^\omega(\mathbf{x})\langle\mathbf{R}_q^\omega\rangle\nabla\nabla\mathbf{G}^\omega(\mathbf{x})\langle\mathbf{R}_q^\omega\rangle\varphi(\mathbf{x})e^{-i\mathbf{k}\cdot\mathbf{x}}\,d\mathbf{x} \\
&+\int \mathbf{G}^\omega(\mathbf{x})\langle\mathbf{F}_q^\omega\rangle\nabla\mathbf{G}^\omega(\mathbf{x})\langle\mathbf{R}_q^\omega\rangle\varphi(\mathbf{x})e^{-i\mathbf{k}\cdot\mathbf{x}}\,d\mathbf{x}, \qquad (17.107)\\
\mathbf{t}(\mathbf{k}) &= -\int \mathbf{g}(\mathbf{x})\langle\mathbf{F}_q^\omega\rangle\psi(\mathbf{x})e^{-i\mathbf{k}\cdot\mathbf{x}}\,d\mathbf{x}_q \\
&+\int \nabla\mathbf{G}^\omega(\mathbf{x})\langle\mathbf{R}_q^\omega\rangle\nabla\mathbf{G}^\omega(\mathbf{x})\langle\mathbf{F}_q^\omega\rangle\varphi(\mathbf{x})e^{-i\mathbf{k}\cdot\mathbf{x}}\,d\mathbf{x} \\
&+\int \mathbf{G}^\omega(\mathbf{x})\langle\mathbf{F}_q^\omega\rangle\mathbf{G}^\omega(\mathbf{x})\langle\mathbf{F}_q^\omega\rangle\varphi(\mathbf{x})e^{-i\mathbf{k}\cdot\mathbf{x}}\,d\mathbf{x}, \qquad (17.108)
\end{aligned}
$$

where the first integrals in (17.107), (17.108) were presented in the framework of the point approximation $\mathbf{G}^\omega(\mathbf{x})=\mathbf{g}(\mathbf{x})$ (see Subsection 3.3.2) of inclusions. A similar equation for the Fourier transform $\overline{\mathbf{E}}(\mathbf{k})$ is

$$
\overline{\mathbf{E}}(\mathbf{k})=\mathbf{E}(\mathbf{k})+\mathbf{S}(\mathbf{k})\overline{\mathbf{E}}(\mathbf{k})+\mathbf{s}(\mathbf{k})\overline{\mathbf{U}}(\mathbf{k}), \qquad (17.109)
$$

$$
\begin{aligned}
\mathbf{S}(\mathbf{k}) &= -\int\nabla\nabla\mathbf{g}(\mathbf{x})\langle\mathbf{R}_q^\omega\rangle\psi(\mathbf{x})e^{-i\mathbf{k}\cdot\mathbf{x}}\,d\mathbf{x}_q+\int\nabla\mathbf{G}^\omega(\mathbf{x})\langle\mathbf{F}_q^\omega\rangle\nabla\mathbf{G}^\omega(\mathbf{x})\langle\mathbf{R}_q^\omega\rangle \\
&\times\varphi(\mathbf{x})e^{-i\mathbf{k}\cdot\mathbf{x}}\,d\mathbf{x}+\int\nabla\nabla\mathbf{G}^\omega(\mathbf{x})\langle\mathbf{R}_q^\omega\rangle\nabla\nabla\mathbf{G}^\omega(\mathbf{x})\langle\mathbf{R}_q^\omega\rangle\varphi(\mathbf{x})e^{-i\mathbf{k}\cdot\mathbf{x}}d\mathbf{x}, \qquad (17.110)
\end{aligned}
$$

$$
\begin{aligned}
\mathbf{s}(\mathbf{k}) &= \int \nabla\mathbf{G}^\omega(\mathbf{x})\langle\mathbf{F}_q^\omega\rangle\mathbf{G}^\omega(\mathbf{x})\langle\mathbf{F}_q^\omega\rangle\varphi(\mathbf{x})e^{-i\mathbf{k}\times\mathbf{x}}\,d\mathbf{x}-\int\nabla\mathbf{g}(\mathbf{x})\langle\mathbf{F}_q^\omega\rangle \\
&\times\,\psi(\mathbf{x})e^{-i\mathbf{k}\cdot\mathbf{x}}\,d\mathbf{x}_q+\int\nabla\nabla\mathbf{G}^\omega(\mathbf{x})\langle\mathbf{R}_q^\omega\rangle\nabla\mathbf{G}^\omega(\mathbf{x})\langle\mathbf{F}_q^\omega\rangle\varphi(\mathbf{x})e^{-i\mathbf{k}\cdot\mathbf{x}}\,d\mathbf{x}. \qquad (17.111)
\end{aligned}
$$

To simplify the calculations, we employ the long-wave approximation $|\mathbf{k}l|\ll 1$, when $\mathbf{U}(\mathbf{x})$, $\mathbf{E}(\mathbf{x})$ are the smooth fields so that their supports are placed in the $\mathbf{k}$-space domain $|\mathbf{k}l|\ll 1$. Here l is the correlation radius of the inclusion field defined as $l^2=n^{-1}\int_0^\infty\psi(r)rdr$, $r=|\mathbf{x}|$. This enables us to represent $\exp(-i\mathbf{k}\cdot\mathbf{x})$ in region $|\mathbf{x}|<l$ in the form of a series $\exp(-i\mathbf{k}\cdot\mathbf{x})\approx 1-ik_ix_i-k_ik_jx_ix_j/2$ and to express the following integral by a few terms of a series with respect to $|\mathbf{k}l|$

$$\frac{1}{n}\int \nabla\nabla \mathbf{g}_0(\mathbf{x})\psi(\mathbf{x})\exp^{-i(\mathbf{k}\cdot\mathbf{x})}\ d\mathbf{x} = -\mathbf{P}_0 - l^2\mathbf{P}_1(\mathbf{k}\otimes\mathbf{k}),$$

$$\mathbf{P}_0 = -\frac{1}{n}\int \nabla\nabla \mathbf{g}_0(\mathbf{x})\psi(\mathbf{x})\ d\mathbf{x},\quad \mathbf{P}_1 = \frac{1}{2}\int_{\Omega_1} \nabla\nabla \mathbf{g}_0(\mathbf{n})(\mathbf{n}\otimes\mathbf{n})\ d\mathbf{s}_1, \quad (17.112)$$

where $\mathbf{n} = \mathbf{x}/|\mathbf{x}|$ and Ω_1 is the surface of the unit sphere and the tensor $\mathbf{P}_0$ was also defined by Eq. (3.110). Then, for a specified degree of accuracy with respect to ω we obtain from (17.106) and (17.109):

$$\overline{\mathbf{E}}(\mathbf{k}) = \mathbf{D}(\mathbf{k})\mathbf{E}(\mathbf{k}),\quad \overline{\mathbf{U}}(\mathbf{k}) = \mathbf{d}\mathbf{U}(\mathbf{k}),\quad \mathbf{d} = \boldsymbol{\delta} - \mathbf{i}\omega^{\mathbf{3}}\rho_{\mathbf{1}}^{(\mathbf{1})}[\mathbf{c}\mathbf{g}_{\mathbf{1}}\mathbf{J} - \bar{\mathbf{v}}\mathbf{d}_{\mathbf{3}}], \quad (17.113)$$

$$\mathbf{d}_3 = \int\Big[\nabla\mathbf{G}_1(\mathbf{x})\langle\mathbf{R}_0\rangle\nabla\mathbf{G}_0(\mathbf{x}) + \nabla\mathbf{G}_0(\mathbf{x})\langle\mathbf{R}_0\rangle\nabla\mathbf{G}_1(\mathbf{x})\Big]\langle\mathbf{R}_0\rangle\varphi(\mathbf{x})d\mathbf{x} \quad (17.114)$$

$$\mathbf{D} = \mathbf{D}_0\Big\{\mathbf{I} + i\omega^3\Big[n\mathbf{P}_0\mathbf{K}_3^R - nJ\mathbf{K}_3\langle\mathbf{R}_0\rangle + \mathbf{S}^\omega\Big]\mathbf{D}_0$$

$$+ nl^2\mathbf{P}_1(i\mathbf{k}\otimes i\mathbf{k})\langle\mathbf{R}_0\rangle\mathbf{D}_0\Big\}, \quad (17.115)$$

$$\mathbf{D}_0 = \Big(\mathbf{I} - n\mathbf{P}_0\langle\mathbf{R}_0\rangle - \int \nabla\nabla\mathbf{G}_0(\mathbf{x})\langle\mathbf{R}_0\rangle\nabla\nabla\mathbf{G}_0(\mathbf{x})\langle\mathbf{R}_0\rangle\varphi(\mathbf{x})\ d\mathbf{x}\Big)^{-1} \quad (17.116)$$

$$\mathbf{S}^\omega = \int \nabla\nabla\mathbf{G}_0(\mathbf{x})\langle\mathbf{R}_0\rangle\Big[\nabla\nabla\mathbf{G}_0(\mathbf{x})\mathbf{K}_3^R + \nabla\nabla\mathbf{G}_3(\mathbf{x})\langle\mathbf{R}_0\rangle\Big]\varphi(\mathbf{x})\ d\mathbf{x}$$

$$+ \int \Big[\nabla\nabla\mathbf{G}_0(\mathbf{x})\mathbf{K}_3^R + \nabla\nabla\mathbf{G}_3(\mathbf{x})\langle\mathbf{R}_0\rangle\Big]\nabla\nabla\mathbf{G}_0(\mathbf{x})\langle\mathbf{R}_0\rangle\varphi(\mathbf{x})\ d\mathbf{x}, \quad (17.117)$$

$$\mathbf{R}_0 = \bar{v}\mathbf{L}_1^{(1)}\mathbf{A},\quad J = \textstyle\int \psi(\mathbf{x})/n\ d\mathbf{x},\quad \mathbf{K}_3^R = \langle\mathbf{R}_0\mathbf{K}_3\mathbf{R}_0\rangle, \quad (17.118)$$

where $c = n\bar{v}$, $\mathbf{R}_0$ is the static limit of the tensors $\bar{v}\mathbf{L}_1^{(1)}\boldsymbol{\Lambda}(\omega,\mathbf{a})$ (3.243) at $\omega = 0$, and $\mathbf{G}^\omega = \mathbf{G}_0 + i\omega\mathbf{G}_1 + i\omega^3\mathbf{G}_3$.

We substitute the resultant expressions for $\overline{\mathbf{U}}(\mathbf{k})$, $\overline{\mathbf{E}}(\mathbf{k})$ in (17.113) into $\mathbf{k}$-representation of (17.90)

$$\mathbf{u}(\mathbf{k}) = \mathbf{u}_0(\mathbf{k}) + n\mathbf{g}(\mathbf{k})\langle\mathbf{R}^\omega\rangle((i\mathbf{k})\otimes\overline{\boldsymbol{\varepsilon}}(\mathbf{k})) + n\mathbf{g}(\mathbf{k})\langle\mathbf{F}^\omega\rangle\overline{\mathbf{u}}(\mathbf{k}) \quad (17.119)$$

leading to the following representation:

$$\mathbf{u}_0(\mathbf{k}) = \mathbf{U}(\mathbf{k}) - n\mathbf{g}(\mathbf{k})\langle\mathbf{R}^\omega\rangle[(i\mathbf{k})\otimes\mathbf{D}((i\mathbf{k})\otimes\mathbf{U}(\mathbf{k}))] - n\mathbf{g}(\mathbf{k})\langle\mathbf{F}^\omega\rangle\mathbf{d}\mathbf{U}(\mathbf{k}). \quad (17.120)$$

Convolving the latter equation with tensor $\boldsymbol{\mathcal{L}}^{(0)}(\mathbf{k}) = -\mathbf{L}^{(0)}(\mathbf{k}\otimes\mathbf{k}) + \rho^{(0)}\omega^2\boldsymbol{\delta}$ (3.62), and taking the equalities $\boldsymbol{\mathcal{L}}^{(0)}(\mathbf{k})\mathbf{u}_0(\mathbf{k}) = \mathbf{0}$, $\boldsymbol{\mathcal{L}}^{(0)}(\mathbf{k})\mathbf{g}(\mathbf{k}) = -\boldsymbol{\delta}$ into account, we obtain a Helmholtz equation for $\mathbf{u}(\mathbf{k})$ in $(\mathbf{k},\omega)$ space

$$\boldsymbol{\mathcal{L}}^*(\mathbf{k},\omega)\mathbf{U}(\mathbf{k}) = \mathbf{0},\quad \boldsymbol{\mathcal{L}}^*(\mathbf{k},\omega) = -\mathbf{L}^*(\mathbf{k},\omega)(\mathbf{k}\otimes\mathbf{k}) + \boldsymbol{\rho}^*(\omega), \quad (17.121)$$

$$\mathbf{L}^*(\mathbf{k},\omega) = \mathbf{L}_s + l^2\mathbf{L}^R[\mathbf{P}_1(i\mathbf{k}\otimes i\mathbf{k})]\mathbf{L}^R + i\omega^3\mathbf{L}_\omega, \quad (17.122)$$

$$\mathbf{L}_s = \mathbf{L}^{(0)} + \mathbf{L}^R,\quad \mathbf{L}^R = n\langle\mathbf{R}_0\rangle\mathbf{D}_0, \quad (17.123)$$

$$\boldsymbol{\rho}^*(\omega) = \rho_s\boldsymbol{\delta} + i\omega^3 c\rho^{(1)}[f\mathbf{g}_1 + \bar{v}\mathbf{d}_3],\ \rho_s = \rho^{(0)} + c\rho_1^{(1)}, \quad (17.124)$$

$$\mathbf{L}_\omega = \mathbf{L}^R\Big[n\mathbf{P}_0\mathbf{K}_3^R\mathbf{D}_0 - J\mathbf{K}_3\mathbf{L}^R + \mathbf{S}^\omega\mathbf{D}_0\Big], \quad (17.125)$$

where $f = \bar{v} - cJ$.

The expressions for $\mathbf{L}^*(\mathbf{k},\omega)$, $\boldsymbol{\rho}^*(\omega)$ depend on $\mathbf{k}$, ω, and hence a composite medium exhibits temporal and spatial dispersion. In particular, the effective elastic operator $\boldsymbol{\mathcal{L}}^*$ in the $\mathbf{x}$-space has the form $\mathcal{L}^*_{ij} = \nabla_k L^*_{ikjl}(\nabla)\nabla_l + \rho^*_{ij}\omega^2$, where $\mathbf{L}^*(\nabla)$ is also differential operator of the second order depending on ω:

$$\mathbf{L}^*(\nabla) = \mathbf{L}_s + i\omega^3\mathbf{L}_\omega + l^2\mathbf{L}^R\mathbf{P}_1(\nabla\otimes\nabla)\mathbf{L}^R. \tag{17.126}$$

For static problems ($\omega = 0$), the elastic-modulus operator coincides with the analogous operator of the moment theory of elasticity with constrained rotation (see, e.g., [618]). This operator, also estimated in Chapter 12 by different methods, is reduced to the effective moduli $\mathbf{L}^*$ considered in Chapter 8 under the assumption of a uniform external loading; that study also offered a comparison with experiment and with calculations based on other methods. Equations for $\boldsymbol{\mathcal{L}}^*$ and $\mathbf{L}^*(\nabla)$ (17.121)–(17.126) differ from the analogous expressions obtained in [542], [545] by virtue of terms that describe binary interaction of inclusions, which are of the first-order smallness in c. The obtained results can be reduced to the last ones if $\mathbf{d}_3$, $\mathbf{S}^\omega \equiv \mathbf{0}$ and the integral in the representation of $\mathbf{D}_0$ (17.116) is dropped out.

17.3.3 Effective Wave Operator for Composites with Spherical Isotropic Inclusions

To obtain analytical representations of the effective wave operator, we will consider a composite material withe isotropic constituents filled by identical spherical homogeneous inclusions. In such a case the Helmholtz operator $\boldsymbol{\mathcal{L}}^*(\mathbf{k},\omega)$ is also isotropic:

$$\mathcal{L}^*_{ij}(\mathbf{k},\omega) = \Big[(k^* - \frac{2}{3}\mu^*)n_in_j + \mu^*\delta_{ij}\Big]k^2 - \rho^*(\omega)\omega^2\delta_{ij}, \quad \mathbf{n} = \mathbf{k}/k, \ k = |\mathbf{k}|. \tag{17.127}$$

Using the projection operators $\boldsymbol{\eta} = \mathbf{n}\otimes\mathbf{n}$, $\boldsymbol{\nu} = \boldsymbol{\delta} - \boldsymbol{\eta}$ with the properties described in Appendix A.2.1, we can express $\boldsymbol{\mathcal{L}}^*(\mathbf{k},\omega)$ in terms of the scalar operators $\boldsymbol{\mathcal{L}}^*_{\mathrm{L}}(\mathbf{k},\omega)$ and $\boldsymbol{\mathcal{L}}^*_{\mathrm{T}}(\mathbf{k},\omega)$, that define the law of longitudinal and transverse wave propagation:

$$\begin{aligned}
\boldsymbol{\mathcal{L}}^*(\mathbf{k},\omega) &= \boldsymbol{\mathcal{L}}^*_{\mathrm{L}}(\mathbf{k},\omega)\boldsymbol{\eta} + \boldsymbol{\mathcal{L}}^*_{\mathrm{T}}(\mathbf{k},\omega)\boldsymbol{\nu}, \ \mathbf{L}^*(\mathbf{k},\omega) = (3k^*, 2\mu^*), && (17.128)\\
L^*_{\mathrm{L}}(\mathbf{k},\omega) &= -k^2(k^* + 4\mu^*/3) + \rho^*\omega^2, \quad L^*_{\mathrm{T}}(\mathbf{k},\omega) = -k^2\mu^* + \rho^*\omega^2, && (17.129)\\
\mathbf{L}_l &= (3k_l, 2\mu_l), \quad \mathbf{L}_s = (3k_s, 2\mu_s), \quad \mathbf{L}_\omega = (3k_\omega, 2\mu_\omega), && (17.130)\\
k^* &= k_s + c^2l^2k^2k_l - i\omega^3cfk_\omega, \ \mu^* = \mu_s + c^2l^2k^2\mu_l - i\omega^3cf\mu_\omega. && (17.131)
\end{aligned}$$

Here the static effective moduli

$$\begin{aligned}
3k_s &= 3k^{(0)} + 3R_13d_1c, \ 2\mu_s = 2\mu^{(0)} + 2R_22d_2c, \ \mathbf{L}^{(0)} = (3k^{(0)}, 2\mu^{(0)}),\\
3d_1 &= (1 - c3P^0_13R_1 - 3J_1c)^{-1}, \ 2d_2 = (1 - c2P^0_22R_2 - 2J_2c)^{-1},\\
3J_1 &= 2\beta^23R_12R_2\textstyle\int\varphi(r)|r|^{-4}dr, \quad \mathbf{R}_0 = \bar{v}(3R_1, 2R_2), \ \mathbf{D}_0 = (3d_1, 2d_2),\\
3J_2 &= [2\beta^23R_12R_2 + (2R_2)^2(7\gamma^2 - \alpha^2 + \beta^2)]\textstyle\int\varphi(r)|r|^{-4}\,dr,\\
\beta &= (3k^{(0)} + 4\mu^{(0)})^{-1}, \ \alpha = (3\mu^{(0)})^{-1}, \ \gamma = \beta - \alpha.
\end{aligned} \tag{17.132}$$

coincide with those obtained in Chapter 8. The imaginary part of the wave operator can be expressed by using the components of the isotropic tensors

$$\begin{aligned}
\rho^*_\omega &= \rho_s + i\omega^3 c\rho_\omega, \quad \rho_\omega = \frac{f\rho_1^{(1)}(2+\eta^3)}{12\pi\rho_1^{(0)}k_{\rm T}^3} + \bar{v}\rho^{(1)}d_3^0, \\
3k_\omega &= 9k_R^2 3k^k(\bar{v}3P_1^0 3R_1 - 3J_1) + 3k_R 3S_1 3d_1, \ \mathbf{S}^\omega = (3S_1, 2S_2), \\
2\mu_\omega &= 4\mu_R^2 2\mu^k(\bar{v}2P_2^0 2R_2 - 2J_2) + 2\mu_R 2S_2 2d_2, \ \mathbf{P}_0 = (3P_1^0, 2P_2^0), \\
\mathbf{K}_3 &= -(3k^k, 2\mu^k), \ 3k^k = \frac{\eta^5}{36\pi\rho^{(0)}k_{\rm T}^5}, \ 2\mu^k = \frac{3+2\eta^5}{60\pi\rho^{(0)}k_{\rm L}^5}, \ \eta = \frac{k_{\rm L}}{k_{\rm T}}, \\
3P_1^0 &= \beta, \ 2P_2^0 = \frac{3}{5}\beta(k^{(0)} + 2\mu^{(0)}), \ (3k_R, 2\mu_R) = c(3R_1 3d_1, 2R_2 2d_2),
\end{aligned} \tag{17.133}$$

where $k_{\rm L} = \omega\sqrt{\rho^{(0)}/(\lambda^{(0)}+2\mu^{(0)})}$ and $k_{\rm T} = \omega\sqrt{\rho^{(0)}/\mu^{(0)}}$ are the wave numbers in the matrix.

Finally, the nonlocal elastic effects are described by the tensor $\mathbf{L}_l = (3k_l, 2\mu_l)$ with the components

$$k_l = \frac{4\mu_R}{105\mu^0}\Big[14k_R + \frac{1}{3}\mu_R(3+\eta^2)\Big], \quad \mu_l = \frac{\mu_R^2}{105\mu^{(0)}}(3+4\eta^2). \tag{17.134}$$

The presented formulae differ from proposed in [545] by the presence of terms (such as $\mathbf{S}^\omega$, $\mathbf{d}_3$, J_1, J_2) taking into account direct binary interaction of inclusions. The influence of these terms can be significant; for example, the consideration of J_1, J_2 result in a twofold improvement of the estimates for $\mathbf{L}_s$ for the case of rigid inclusions in an incompressible matrix with $c = 0.4$. Due to awkwardness of analytical representations of $\mathbf{S}^\omega$, $\mathbf{d}_3$ (see [169]) their numerical calculation is preferable. Moreover, the initial nonlocal integral equations (17.103) and (17.104) could be solved by any one of three methods considered in Chapter 12 rather than by a particular case of the Fourier transform method exploited in this section. All three methods mentioned have a series of advantages and disadvantages, and it is critical for the analyst to be aware of their range of applications.

Thus, we can assume that effective wave operator (17.128) has been constructed. We will determine the Green tensor of this operator by using the following expression, analogous to (17.128) $\mathbf{g}^*(\mathbf{k},\omega) = \mathbf{g}^*_{\rm L}(\mathbf{k},\omega)\boldsymbol{\eta} + \mathbf{g}^*_{\rm T}(\mathbf{k},\omega)\boldsymbol{\nu}$, where $\mathbf{g}^*_{\rm L}(\mathbf{k},\omega) = -[\mathbf{L}^*_{\rm L}(\mathbf{k},\omega)]^{-1}$, $\mathbf{g}^*_{\rm T}(\mathbf{k},\omega) = -[\mathbf{L}^*_{\rm T}(\mathbf{k},\omega)]^{-1}$. Converting to the $\mathbf{x}$ representation, we obtain an expression for the effective Green tensor of the same form as for a homogeneous isotropic medium:

$$\mathbf{g}^*(\mathbf{x},\omega) = \frac{1}{4\pi\mu_s}\Big\{\frac{\exp(k^*_{\rm T}|\mathbf{x}|)}{|\mathbf{x}|}\boldsymbol{\delta} - \nabla\nabla\Big[\frac{\exp(ik^*_{\rm T}|\mathbf{x}|)}{|\mathbf{x}|} - \frac{\exp(ik^*_{\rm L}|\mathbf{x}|)}{|\mathbf{x}|}\Big]\Big\}, \tag{17.135}$$

where the effective wave numbers are written in complex form:

$$k^*_{\rm L} = \omega\sqrt{\frac{\rho_s}{\varkappa_s}}\Big[1 + \frac{1}{2}i\omega^3 cf\Big(\frac{\rho_\omega}{\rho_s} + \frac{\varkappa_\omega}{\varkappa_s}\Big)\Big], \quad k^*_{\rm T} = k_{\rm L}\sqrt{\frac{\varkappa_s}{\mu_s}}, \tag{17.136}$$

where $\varkappa_\omega = k_\omega + \frac{4}{3}\mu_\omega$, $\varkappa_s = k_s + \frac{4}{3}\mu_s$.

In the one-particle approximation of the MEFM [542], it has been demonstrated that the longitudinal and transversal components of the dynamic Green's tensor describe two wave types, diverging from a point source. Waves of the first type with wave number $k_{\rm L}^*$ and $k_{\rm T}^*$ have attenuation factors that are proportional to $(\omega l)^4$. Waves of the second type explained by the nonlocal effects, on the other hand, attenuate much more rapidly than the first one (over distances of order l). In what follows, therefore, we will disregard nonlocal effects, and will assume that the term $l^2\mathbf{L}^R[\mathbf{P}_1(i\mathbf{k}\otimes i\mathbf{k})]\mathbf{L}^R$ in (17.122) is zero. The wave numbers $k_{\rm L}^*$, $k_{\rm T}^*$ (17.136) are complex ones with the real parts defining the effective propagation velocities $c_{\rm L}^* = \sqrt{(k_s+4\mu_s/3)/\rho_s}$, $c_{\rm T}^* = \sqrt{\mu_s/\rho_s}$ in the longitudinal and transverse direction, respectively. As can be seen, discarding of the terms of the order ω^2 as compared to unity in the real parts of all analyzed quantities leads to independence of $k_{\rm L}^*$ and $k_{\rm T}^*$ on the frequency ω, and, therefore, to the absence of velocity dispersion which can be detected if we hold the terms of the order ω^2. The imaginary parts of the wave numbers $\gamma_{\rm L}^* = {\rm Im}k_{\rm L}^*$ and $\gamma_{\rm T} = {\rm Im}k_{\rm T}^*$ determining the attenuation factors, referred to unit length, can also be found from Eqs. (17.136). Since the attenuation factors $\gamma_{\rm L}$ and $\gamma_{\rm T}$ must be positive due to their physical meaning, their proportionality factor f in (17.136) is also positive $f = \bar v - c\int \psi(\mathbf{x})/n\, d\mathbf{x}$. When the binary correlation function φ is taken in the form of a step-function $\varphi(r) = ng(r)$ (9.91), this is possible only for $c < 1/8$, which is an imposed constraint on the volume concentration c providing a physical consistent of the resultant formulae. In the case of more accurate approximations for φ (see Chapter 5), the domain of c values for which formula for f is consistent is considerably extended (see [545], [1182]).

18

Multiscale Mechanics of Nanocomposites

Carbon fibers and plates are currently considered to be most promising. The recent discovery of carbon nanotubes (CNTs) and nanoplates have gained ever-broaded interest due to providing unique properties generated by their structural perfection, small size, low density, high strength, and excellent electronic properties (for references see Subsection 1.1.2). Polymeric composites reinforced with nanoelements of nanometer scale recently attracted tremendous attention in material engineering. The general comprehensive reviews of the various types of nanocomposites were given in [486], [582], [927], [1099]. Due to the high surface area and small interlayer distances, the nanoelements can, in principle, change the properties of polymer matrix due to the changing of polymer morphology and chain conformation. The design and fabrication of these materials are performed on the nanometer scale with the ultimate goal to obtain highly desirable macroscopic properties. Many mechanical, physical, and chemical factors could potentially form the properties of nanocomposites, and a better understanding of their relative contributions is needed. A question of a correct choice of a structural model applicable to nanocomposites presents many challenges for mechanics and will be considered in this chapter from the point of view of allowing for the random orientation and clustering effect of nanoelements in the nanocomposites.

18.1 Elements of Molecular Dynamic (MD) Simulation

18.1.1 General Analysis of MD Simulation of Nanocomposites

The mechanical and thermal properties of CNTs have not often been measured due to the difficulties of creating homogeneous and uniform samples of nanotubes. As a result, it is desirable to develop the model forming the mechanical properties of CNTs over their unique topology that may overcome the limitations of atomistic simulation concerning the wide range of length scale ($10^{-9} - 10^{-6}$m). Unlike many other fields in science and engineering, the evolution of CNTs to their current level significantly depends on the contributions from modeling and simulation (see, e.g., [357], [907], [991], [1093], [1092], [1219]). One of the fundamental issues that needs to be addressed in modeling macroscopic mechanical behavior of nanostructured materials based on molecular structure is the large difference in length scales. On the opposite ends of the length scale spectrum are computational chemistry and solid mechanics, each of which consists of highly developed and reliable modeling methods. If a hierarchical approach is used to model the macroscopic behavior of nanostructured materials, then a methodology must be developed to link the molecular structure and macroscopic properties. There are only a few such studies combining nano-and micro-modeling playing significant roles in the areas of characterizing CNT-based composites. Among these limiting investigations, CNTs were modeled as cylindrical shells, beams, and the system of trusses (for references see [645], [907], [1093], [1219]). In particular, an approach to pass atomistic information to continuum analysis was proposed in [357] for nanostructures for which one dimension is significantly large and demonstrated that, under certain conditions, continuum mechanics is applicable on the nanometer scale. Continuum analysis was linked with atomistic simulation [1219] by incorporating interatomic potential and atomic structures of CNTs directly into the constitutive law through the equalization of the strain energy function on the continuum level to the potential energy stored in the atomic bonds due to an imposed deformation; it was shown that a multi-body interatomic potential is essential to accurately describe the interatomic potential of the carbon atoms of a nanotube. Reich *et al.* [929] probed the elastic properties of carbon nanotubes with density functional theory and found excellent agreement with the elastic continuum approximation. Normal mode analysis was used [422] from an atomistic representation of various sizes of armchair SWCNTs mapped onto the coarse-grained wormlike chain model to show that the intrinsic bending stiffness of a tube increases with increasing radius of the tubes.

The models of computational chemistry predict molecular properties based on known quantum interactions, and computational solid mechanics models predict the macroscopic mechanical behavior of materials idealized as continuous media based on known bulk material properties. However, a corresponding model does not exist in the intermediate length scale range. Because the polymer molecules are on the same size scale as the nanotubes, the interaction at the polymer/nanotube interface is highly dependent on the local molecular structure and bonding. We will consider in detail the methodology [825], [826] developing the equivalent continuum modeling technique for construction of constitutive models SWNT-reinforced polymer composite. First, a model of the molecular structure of the nanotube and the adjacent polymer chains is established by using the atomic structure that has been determined from the MD simulations. Second, an equivalent continuum model is developed in which the mechanical properties are determined based on the force constants that describe the bonded and nonbonded interactions of the atoms in the molecular model and reflect the local polymer and nanotube structure. Finally, the equivalent continuum model is used in micromechanical analyses to determine the bulk constitutive properties of the SWNT/polymer composite with aligned and random nanotube orientations and with various nanotube lengths and volume fractions. The bonded and nonbonded interactions of the atoms in a molecular structure can be quantitatively described by using molecular mechanics. The forces that exist for each bond, as a result of the relative atomic positions, are described by the force field. The molecular potential energy for a nanostructured material will be described by the sum of the individual energies associated with bond stretching, angle variation, and with the non-bonded interactions, which includes van der Waals and electrostatic effects.

From the foregoing, it is evident that integration over scales is a subtle and challenging task, sometimes prompting re-appraisals of the conceptual basis. Since nanoscale variations are on the order of atomic length, quasicontinuum (see, e.g., [1062]) methods must be adopted. The limitations of this approach in the case of CNTs (where tube thickness equals one atom diameter) have been circumvented (see [18]) by using of the differential geometry concept of the exponential map. Space-frame and truss models [672], [825], lattice-dynamical (e.g., honeycomb structures), representative volumetric and nanoscale (i.e., molecular Lennard-Jones based) finite elements, as well as bending simulations, have been used to replicate the structural mechanical paradigm at molecular levels. Kwon [637] uses a coupled nonlinear finite-element and molecular dynamics approach with a smeared continuum model. Application of this method yields consistent results in three situations: (a) local dislocation behavior at a crack tip, of a square array of atoms, (b) tensile response of a zig-zag SWNT, and (c) a multiscale analysis of a multilayered woven fabric, the fibers being made of bundles of SWNTs. Another such multiscale study [473] of fixed-end simple shear (in FCC nickel) combines MD, crystal plasticity, and macroscopic internal state variable theory.

It is observed that on a sufficiently large coarse-graining scale in MD, linear elastic theory effectively approximates the constitutive relations. Reference [391] can be remarked in this connection as a landmark analysis of the relevance of MD in continuum mechanical calculations a task that has not been undertaken to this depth before. Essentially, Goldhirsch and Goldenberg [391] observed that as for the explored nanoscale disordered systems where the lattice strains are nonuniform, the microscopically based derivations of elasticity are not, in general, applicable. As a first step, they derive microscopically exact expressions for the displacement, strain and stress fields. Conditions under which linear elastic constitutive relations hold are studied theoretically and numerically. It turns out that standard continuum elasticity is not self-evident, and applies only above certain spatial scales, which depend on the details of the system and boundary conditions under investigation.

The critical question, viz., the equivalence of molecular system to continuum models (or the relevance and applicability of virial stress to continuum mechanics), has been addressed in another recent notable study in [1224], where the mass-transport contribution (i.e. the molecular kinetic energy) to momentum change, and the overall formulation of the virial stress, is proved to be counter to the continuum-mechanical ("Cauchy") stress. The inherent difference between the spatial (Eulerian) and material (Lagrangian) approaches to momentum-balance, is shown to be one reason for this discrepancy. It is evident from the foregoing that one must establish the basic equivalence of MD-based analyses with the desired continuum mechanics calculations before embarking on an integrated solution. Current literature is rife with unquestioned adoption of MD methods to diverse situations where the critical question of physical correctness has not been addressed.

Zhou [1225] further developed a thermomechanical continuum interpretation of atomistic deformation by decomposing atomic particle velocity into a structural deformation part and a thermal oscillation part. He obtained full dynamic equivalence between the discrete (molecular) and continuum systems. The coupling between the structural deformation and the thermal conduction processes results from their phenomenological basis in the momentum balance equation at the fully time-resolved atomic level, through an inertial force term. Adopting this modified (fluctuation-based) viewpoint, Gusev *et al.* [413] (see also [405]) proposed a fluctuation formula reported to improve [in convergence and efficiency] over the Parrinello-Rahman [859] formula. As the gradient morphology at the interface evolves in the MD simulation, one will capture its effect on the stress-strain behavior by estimating the local moduli of subvolumes over different scales in the composite.

Understanding the molecular response (at the nanometer scale) in the vicinity of the interface and then relating that with the bulk response is a preferred approach for composite material design, as compared to one based only on bulk (macroscopic) response. Thus, the objective of such a study is to analyze the interface behavior and correlate that with the macroscopic response of composites. By varying the domain of observation, one can obtain both area- and volume-averages for the local stresses and strains. This information is key to understanding the gradient morphology, i.e., the variation of important material properties such as cohesive energy, elastic constants, and thermoelectrical properties across interfaces. In so doing, local elasticity tensors have been calculated in [726] as well as in [1131] using subvolume averaging and cross-section (planar) averaging, respectively. In fact, the latter is valid for inhomogeneous systems with planar symmetry; it relates the local stress to a homogeneous strain. Such a (local) study has been recently reported in [206] in the case of single-wall carbon nanotubes (SWNTs) with topological (Stone-Wales) defects (this structure is a periodic structure with random imperfections considered in Chapter 11). A conclusion they have drawn is that the decrease in load-carrying capacity in the vicinity of the defects can be attributed to kinetic and kinematic changes in that subdomain.

18.1.2 Foundations of MD Simulation and Their Use in Estimation of Elastic Moduli

Just for completeness, we will reproduce the Hamiltonian formulation for the MD equations of motion (for details see, e.g., [12], [356], [417]) assuming description of the mechanical state of simulated system by means of the generalized coordinates and momentum: $\dot{\mathbf{x}}_a = \partial H/\partial \mathbf{p}_a$, $\dot{\mathbf{p}}_a = -\partial H/\partial \mathbf{x}_a$. Here the Hamiltonian H of an N-particle ensemble (called NVE) with coordinates $\mathbf{x}_a$ $(a = 1, \ldots)$, constant volume w and constant energy is given by

$$H = \sum_a \frac{\mathbf{p}_a^\top \mathbf{p}_a}{2m_a} + U(\mathbf{x}_1, \ldots, \mathbf{x}_N), \tag{18.1}$$

with moments $\mathbf{p}_a = m_a \dot{\mathbf{x}}_a$, masses m_a and a conservative potential $U(\mathbf{x}_1, \ldots, \mathbf{x}_N) = \sum_a V_a(\mathbf{x}_a) + \sum_{a,b>a} V_2(\mathbf{x}_a, \mathbf{x}_b) + \sum_{a,b>a,c>b} V_3(\mathbf{x}_a, \mathbf{x}_b, \mathbf{x}_c) + \ldots$, where the first term (usually ignored) represents the energy due an external force field, such as gravitational or electrostatic, while all the n-body effects described by the potentials V_n are usually incorporated into V_2 in order to reduce the computational expense of the simulation. MD methods rely on well-parameterized pairwise potentials to capture the bonded and nonbonded interactions at the molecular (coarse-grained) scale. These potentials are obtained semiempirically from *ab initio* (quantum chemistry) calculations at the electronic structure scale; such calculations are computationally intensive, and in fact, very difficult for any but the simplest molecules. We will use the Lennard-Jones (LJ) (6,12) interatomic potential taking into account the short-range repulsive as well as longer-range attractive dispersion forces: $V(\mathbf{x}_i, \mathbf{x}_j) = V^{\mathrm{LJ}}(r) = 4\epsilon^{\mathrm{LJ}}[(\sigma^{\mathrm{LJ}}/r)^{12} - (\sigma^{\mathrm{LJ}}/r)^6]$, $r < r^{\mathrm{cutoff}} = 2^{1/6} r_0$ (and $V^{\mathrm{LJ}}(r) \equiv 0$ at $r > r^{\mathrm{cutoff}}$), where $r = |\mathbf{r}_{ij}| = |\mathbf{x}_i - \mathbf{x}_j|$ is a distance between two atoms i and j located in the points $\mathbf{x}_i$ and $\mathbf{x}_j$, σ^{LJ} is a collision diameter, ϵ^{LJ} indicates the minimum of the potential function to occur for an atomic pair in equilibrium. Intermolecular forces being conservative, the resulting (interaction) force is just $F = -\nabla_r V = 24(\epsilon^{\mathrm{LJ}}/\sigma^{\mathrm{LJ}})[2(\sigma^{\mathrm{LJ}}/r)^{13} - (\sigma^{\mathrm{LJ}}/r)^7]$. The bonded interactions are modeled by the empirical FENE (Finite Extensible Nonlinear Elastic) potential, defined by the equation: $U^{\mathrm{FENE}} = -(k_0 r_0^2)/2 \ln[1 - (r/r_0)^2]$, where k_0 is a force constant.

One obtains (see e.g. [413]) from (18.1) an identity:

$$\langle \varepsilon_{ik} \frac{\partial H}{\partial \varepsilon_{lm}} \rangle = k_B T \delta_{il} \delta_{km}, \tag{18.2}$$

which is valid to $O(N^{-1})$ in any ensemble; here k_B is the Boltzmann constant, and T is the temperature. Here this relationship yields the constitutive law as follows. The microscopic stress tensor based on the virial (see the references above), is given by:

$$\sigma_{ij} = \frac{1}{w} \left\{ \sum_a \frac{(p_a)_i (p_a)_j}{m_a} + \sum_{a>b} \frac{\partial V}{\partial r_{ab}} \frac{(r_{ab})_i (r_{ab})_j}{r_{ab}} \right\}. \tag{18.3}$$

If two arbitrary points in the system be connected by vector $\mathbf{r}' = (x_1', x_2', x_3')$ in the reference state (specified by average scaling matrix $\langle h \rangle$), and after homogeneous deformation, by the vector $\mathbf{r} = (x_1, x_2, x_3)$ in the instantaneous frame (with scaling matrix h), then the strain tensor is defined via $r^2 = r'^2 + 2\varepsilon_{ik} x_i' x_k'$, where the Einstein summation convention is implied (terms with repeated indices are summed over those indices) and the equality $x_i' = \langle h_{ik} \rangle h_{kl}^{-1} x_l$ is taken into account. This yields the tensor equation $\boldsymbol{\varepsilon} = 0.5\left[\langle \mathbf{h}^\top \rangle^{-1} \mathbf{h}^\top \mathbf{h} \langle \mathbf{h} \rangle^{-1} - \boldsymbol{\delta} \right]$ for crystallographic strain ($\boldsymbol{\delta}$ is the identity matrix). The second rank transformation tensor $\mathbf{h}$ relates crystallographic and Cartesian coordinates for the unit cell: $\mathbf{h} = (\mathbf{a}, \mathbf{b}, \mathbf{c})$ with $\mathbf{a}, \mathbf{b}, \mathbf{c}$ being the vectors defining the unit cell as a parallelepiped. For the small strains, alternatively, the finite element formulation in terms of isoparametric elements yields a simple and effective way to calculate local strains through the displacements.

Finally, at low temperatures, one can neglect the difference between instantaneous scaling matrix $\mathbf{h}$ and the average scaling matrices $\langle \mathbf{h} \rangle$ (see [413]), obtaining the constitutive law: $L_{iklm} = \langle \varepsilon_{ik}\sigma_{nj} \rangle \langle \varepsilon_{nj}\varepsilon_{lm} \rangle^{-1}$. In any case, the stress-strain relationships $\boldsymbol{\sigma} \sim \boldsymbol{\varepsilon}$ obtained for the varying strain and measured stress in the framework of an NVE ensemble (see, e.g., [355]) provides the overall mechanical response $\mathbf{L}$. Alternatively, the work [404] employed an isothermal–isobaric ensemble approach to apply external stress and to measure the corresponding strain.

After the system has been defined and equilibrated (usually, for the constant volume or pressure ensembles) using any of several (e.g. Langevin, Nosé-Hoover, etc. which control temperature/pressure/volume/stress) a set of six independent strains are applied. From the response (using integrators such as velocity Verlet, r-RESPA), elastic properties are calculated using time-averaged kinetics (position, velocity, and forces) via fluctuation formulae [413] based on the virial (see, e.g., [458], [509], [988], [1087]). Strains can be estimated in different ways: the finite element scheme (e.g., isoparametric formulation), crystallographic strain method, etc. While the former is a general method and can be used for arbitrarily shaped subdomains, the latter restricts the possible configurations to general parallelepipeds, i.e., those obtained by homogeneous strains. The isothermal elasticity tensor can be expressed purely in terms of thermal strain fluctuations at a zero applied stress [413]. A comprehensive review of MD (including tight-binding, or TBMD) and *ab initio* approaches, as applied to nanotube mechanics, is found in [907].

18.1.3 Interface Modeling of NC

Since the specific interface area is very large at the nanoscale, it gives rise to substantial property changes in large volume fractions of both matrix and fiber constituents. Some estimates suggest that as much as 30% of the matrix volume may be formed by polymer chains aligned with particle or fiber walls. Fiber functionalization is considered necessary to improve mechanical properties in nanofiber-reinforced composites by increasing the stress transfer between the nanofiber and the matrix of a nanocomposite structure. Fiber–matrix adhesion is governed by the chemical and physical interactions at the interface. An extensive literature exists on surface treatment of conventional carbon fibers by methods such as oxidation in gas and liquid phases and anodic etching (for references see [190]). Poor fiber-matrix adhesion may result in composite failure at the interface, resulting in decreased longitudinal and transverse mechanical properties of the composite. Surface modification of carbon nanofibers changes the graphitization extent of the fiber and increases the surface area of the fiber.

The analysis considers both matrix and effective fiber to be homogeneous and elastic, and is thus applicable at any scale. The effective fiber modeled is referenced, but it should be additionally mentioned that the property changes inherent at interfaces can be adequately described by an effective fiber. Since the specific interface area is very large at the nanoscale, it gives rise to substantial property changes in large volume fractions of both matrix and fiber constituents. At this small length scale, the lattice structures of the nanotube and the resin cannot be considered continuous, and their interfacial properties cannot be determined through continuum mechanics. So, the interfacial bonding of SWNT reinforced epoxy composites was investigated [394] using molecular mechanics and molecular dynamics simulations based on a cured epoxy resin model. Other work [799] has predicted via a quantum mechanics analysis that covalent bonding between an alkyl radical and a nanotube is energetically favorable, and that the tubes of smaller diameters have higher binding energies, and, hence, a high-stress transfer can be realized in polyethylene-based carbon nanotube composites in the presence of free-radical generators. It was found [1167] that the polymer molecules form discrete adsorbtion layers as a function of radial distance from the axis of the nanotube with picks in the RDF. The molecules within the adsorption layers prefer to align parallel to the tube axis that leads to an enhancement of the mechanical modulus of CM. Wagner [1144] concluded that the interfacial bonding stress between the nanotubes and polymer could reach as high as 500 MPa. It should be mentioned that most polymers are hydrophobic and are not compatible with hydrophilic clays (for references see [369]). In this case, pretreatment of either the clays or the polymers is necessary to convert the clay surface from hydrophilic to organophilic. In the context of single lamella, recent work on Montmorillonite (see [731]), based on continuum theory of thin plates, yielded the membrane stiffness $Ed_s = 246$–258 N/m, besides estimating the thickness of the lamella to be $d_s = 0.678$ nm, which is comparable to the spacing between the outermost layers of atoms in the atomically thin sheet. The recent publications show a very well parameterized forcefield for treating Montmorillonite clay in an organic matrix. This forcefield has been used to study uncrosslinked epoxy monomers in the vicinity of a clay layer. Many have shown, from classical molecular dynamics calculations, the dynamics and density profiles of various intercalates confined between two layers of clay (see, e.g., [15], [450], [623], [1014]).

A theoretical model [822] shows that an interfacial layer of a polymer matrix composite whose matrix represents an isotropic medium is transverse-isotropic in the local coordinate system bound with the interface due to the change of macromolecule confirmation in this layer. Therefore, the modeled coating is cylindrically anisotropic in the global coordinate system bound with the fiber center. Among experimental research, it is noteworthy that Lin and Argon [686] produced a highly textured nylon 6 that mimics a single crystal with orthotropic symmetry and obtained a complete set of nine elastic constants. The transcrystalline material of the interphase modeled as transversely isotropic with the plane of isotropy parallel to the polymer/particle interface was considered as a coating of CM inclusions in the shape of either the spheres [1122] or intercalated stacks of nanoplates [992]; homogeneity of this coating was assumed along the direction normal to the interface. The assumption that the interface coating is isotropic was used for silicate NC in [1228], and for the nanoparticle/polymer composites in [826]. The case of an ellipsoidal inclusion coated by anisotropic thin layer with varying along the inclusion surface properties was analyzed in [184] (see also 2-D case considered in [947]). The creation of complete experimental data describing variation of all components of a stiffness tensor in the interphase is coupled with the large difficulties, although these data can, in principle, be obtained by MD simulation by the use of estimation of the effective properties evaluated in the observation windows moving along the cross section of a large sample containing a single nanotube (see, e.g., [501] [846], [877]). It was demonstrated [878] by means of atomistic simulation that the anisotropy of a polymer interface of a rigid spherical inclusion is rather weak. The theoretical background of analysis of composites reinforced by the coated inclusions with an arbitrary inhomogeneous coating and an arbitrary inclusion shape is proposed in [184].

A MD model may serve as a useful guide, but its relevance for a covalent-bonded system of only a few atoms in diameter is far from obvious. Because of this, the phenomenological multiple column model that considers the interlayer radial displacements coupled through the van der Waals forces is used. In so doing, the formation of a strong interface could be reached through functionalization of the nanoelements, which then can be chemically bonded to the polymer matrix chains. Without strong chemically bounding, load transfer between the CNTs and the polymer matrix mainly comes from weak electrostatic and van der Waals interactions (see, e.g., [394], [682]). Numerous researchers (for references see [14]) have attributed lower-than-predicted CNT-polymer composite properties to the availability of only a weak interfacial bounding. So Frankland *et al.* [355] demonstrated by MD simulation that the shear strength of a polymer/nanotube interface with only van der Waals interactions could be increased by over an order of magnitude at the occurrence of covalent bonding for only 1% of the nanotube's carbon atoms to the matrix.

The recent force-field-based molecular-mechanics calculations in [702] demonstrated that the binding energies and frictional forces play only a minor role in determining the strength of the interface. The key factor in forming a strong bond at the interface is having a helical conformation of the polymer around the nanotube; polymer wrapping around nanotube improves the polymer nanotube interfacial strength, although configurationally thermodynamic considerations do not necessarily support these architectures for all polymer chains (see, e.g., [1145]). Thus, the strength of the interface may result from molecular-level entanglement of the two phases and forced long-range ordering of the polymer. To ensure the robustness of data reduction schemes that are based on continuum mechanics, a careful analysis of continuum approximations used in macromolecular models and possible limitations of these approaches at the nanoscale are in addition required that can be done by the fitting of the results obtained by the use of the proposed phenomenological interface model with the experimental data of measurement of the stress distribution in the vicinity of a nanotube.

18.2 Bridging Nanomechanics to Micromechanics in Nanocomposites

Computational approaches, based on the MD simulation, are currently limited to nanoscale and cannot deal with the micro-length scales which should be analyzed by a continuum mechanics approach. These studies rely on fitting of atomistic simulation results to determine the important elastic parameters such as elastic moduli that implicitly assumes a local nature of constitutive law of continuum mechanics at the nanoscale based, in turn, on the assumption that a field scale (inhomogeneity) infinitely exceeds a material scale (molecular inhomogeneity) (see the references mentioned above). In so doing, despite the existence of other numerical approaches, the FEM are normally used in the region described in the framework of continuum mechanics which tacitly assumes that a strain energy density functional exists for materials. In turn, the locality of the strain energy density functional is a crucial assumption when attempting to construct coupled methods in the case of comparable scales of a molecular scale of analyzed structures and the scale of inhomogeneity of internal stresses (see [253]).

In the case of comparable scales of a molecular scale of analyzed structures and the scale of inhomogeneity of internal stresses, which of necessity yields nonlocal character of the constitutive law in the area of coupling of the mentioned scales, this popular basic assumption acting as a bridge between nano- and micromechanics can be considered as an approximation of the real nonlocal constitutive law (see, e.g., [185] and Chapter 12). The approximation mentioned is necessary owing to the fundamental incompatibility of the nonlocal atomistic description and the local continuum description. For mitigation of the sharp transition from the nonlocal atomistic region to a local continuum, some different coupling methods (for references see [253], [993]) use a nonlocal continuum formulation (see, e.g., [316], [891]) to describe the energy of the element in the region of compatible scalers. So, Picu [876] has estimated nonlocal elasticity kernels by the use of atomistic simulations based on the atomic scale structure of glasses. In the framework of linear isotropic nonlocal elasticity, it was determined that the kernel for a model material is tensorial, a different function weighting each entry of the stiffness tensor.

18.2.1 General Representations for the Local Effective Moduli

The most challenging issue of nanotechnology ([143], [339], [378]) is how mechanics can contribute to our understanding of the bridging mechanism between the coupled scales, which is described by the nonlocal constitutive equations involving the parameters of a relevant effective nonlocal operator with the nanostructure. MD simulation takes into account the discrete nature of the atomic interactions at the nanometer length scale and the interfacial characteristics of the nanotube and the surrounding polymer matrix. MD simulation uses only for estimation of both the strain $\mathbf{A}_\epsilon(\mathbf{x})$ and stress $\mathbf{B}_\epsilon(\mathbf{x})$ concentrator factors for molecular system with one nanoinclusion v in a large matrix sample w ($\mathrm{diam}\, v \ll \mathrm{diam}\, w$) subjected to homogeneous loading. These concentrator factors establish the link between the local stress–strain fields and the homogeneous displacement boundary conditions (6.4) $\mathbf{u}(\mathbf{x}) = \langle\boldsymbol{\varepsilon}\rangle \cdot \mathbf{x}$ ($\mathrm{diam} v \ll |\mathbf{x}-\mathbf{y}|,\ \mathbf{x} \in \partial w,\ \mathbf{y} \in v$) associated with a symmetric tensor $\langle\boldsymbol{\varepsilon}\rangle$: $\boldsymbol{\varepsilon}(\mathbf{x}) = \mathbf{A}_\epsilon(\mathbf{x})\langle\boldsymbol{\varepsilon}\rangle,\ \ \boldsymbol{\sigma}(\mathbf{x}) = \mathbf{B}_\epsilon(\mathbf{x})\langle\boldsymbol{\varepsilon}\rangle$. In a similar manner, another set of concentrator factors are introduced for the homogeneous traction boundary conditions (6.5) associated with a symmetric constant tensor $\mathbf{t}^{\partial w}(\mathbf{x}) = \langle\boldsymbol{\sigma}\rangle \cdot \mathbf{n}^{\partial w}(\mathbf{x})$ ($\mathbf{x} \in \partial w, \mathbf{n}^{\partial w}(\mathbf{x}) \perp w$): $\boldsymbol{\varepsilon}(\mathbf{x}) = \mathbf{A}_\sigma(\mathbf{x})\langle\boldsymbol{\sigma}\rangle,\ \ \boldsymbol{\sigma}(\mathbf{x}) = \mathbf{B}_\sigma(\mathbf{x})\langle\boldsymbol{\sigma}\rangle$. We don't need to know for constitutive phases a concrete constitutive law, local or nonlocal

$$\boldsymbol{\sigma}(\mathbf{x}) = \mathbf{L}(\mathbf{x})\boldsymbol{\varepsilon}(\mathbf{x}) \ \text{ or } \ \boldsymbol{\sigma}(\mathbf{x}) = \int \boldsymbol{\mathcal{L}}(\mathbf{x},\mathbf{y})\boldsymbol{\varepsilon}(\mathbf{y})\, d\mathbf{y}, \tag{18.4}$$

respectively, which is just a loss of information contained in the mentioned concentrator factors. Only in the sample area away from the nanoinclusion, where a field scale significantly exceeds a material scale, we can estimate (with some restrictions) $\mathbf{L}(\mathbf{x}) = \mathbf{B}_\epsilon(\mathbf{x})[\mathbf{A}_\epsilon(\mathbf{x})]^{-1}$ which is taken as a stiffness of the homogeneous matrix if $\mathbf{L}(\mathbf{x}) = \mathbf{L}^{(0)} = \text{const}$. Only for the previous exploitation of the MEFM the tensor $\mathbf{L}(\mathbf{x})$ estimated inside and in vicinity of the nanoinclusion will be used. The nature of this simplification is described in Chapter 12 where a fundamentally new effect of nonlocal dependence of local effective moduli in the theory of functionally graded materials was discovered.

A goal of micromechanics lying in the prediction of effective moduli $\mathbf{L}^*$ from the phase local constitutive laws $\boldsymbol{\sigma}(\mathbf{x}) = \mathbf{L}(\mathbf{x})\boldsymbol{\varepsilon}(\mathbf{x})$ and microtopology is based on the classic averaging equation $\mathbf{L}^* = \langle \mathbf{L}\mathbf{A}^*_\epsilon \rangle$ (6.20$_1$) involving the strain concentrator factor $\boldsymbol{\varepsilon}(\mathbf{x}) = \mathbf{A}^*_\epsilon(\mathbf{x})\langle\boldsymbol{\varepsilon}\rangle$ (6.18$_1$). However, in the case of nonlocal constitutive law (18.4$_2$) established at nanoscale, the mentioned classic equation used in a lot of publications must be generalized as $\mathbf{L}^* = \langle \mathbf{B}^*_\epsilon \rangle$, where $\mathbf{B}^*_\epsilon(\mathbf{x})$ is a new stress concentrator factor $\boldsymbol{\sigma}(\mathbf{x}) = \mathbf{B}^*_\epsilon(\mathbf{x})\langle\boldsymbol{\varepsilon}\rangle$ introduced analogously to the concentrator factor for one inhomogeneity $\mathbf{B}_\epsilon(\mathbf{x})$. In a similar manner, a representation of the effective compliance $\mathbf{M}^*$ (6.20$_2$) corresponding to the homogeneous traction boundary conditions (6.5) should be recast in the form $\mathbf{M}^* = \langle \mathbf{A}^*_\sigma \rangle$, involving a strain concentrator factor $\boldsymbol{\varepsilon}(\mathbf{x}) = \mathbf{A}^*_\sigma(\mathbf{x})\langle\boldsymbol{\sigma}\rangle$ analogous to one $\mathbf{A}_\sigma(\mathbf{x})$ for one inhomogeneity inside an infinite matrix. Moreover, all forthcoming averaging methods based on the use of local constitutive laws of constituent phases must be reconstructed in terms of both the strain $\mathbf{A}^*_\sigma(\mathbf{x})$ and stress $\mathbf{B}^*_\epsilon(\mathbf{x})$ concentrator factors. In the next subsection we will propose the concrete averaging schemes of effective moduli estimations used the concentrator factors $\mathbf{A}_\epsilon(\mathbf{x})$, $\mathbf{B}_\epsilon(\mathbf{x})$, $\mathbf{A}_\sigma(\mathbf{x})$, $\mathbf{B}_\sigma(\mathbf{x})$ for one inhomogeneity inside the infinite matrix.

18.2.2 Generalization of Popular Micromechanical Methods to the Estimations of Effective Moduli of NCs

We consider a linear composite medium with the local (18.4$_1$) and nonlocal (18.4$_2$) constitutive equations for a homogeneous matrix and inclusions, respectively. Then the general integral equation (7.19) and (7.20) for both the statistically homogeneous and statistically inhomogeneous (so-called graded) random structure composites can be generalized in a straightforward manner to the case of nonlocal elastic properties of inclusions as

$$\boldsymbol{\varepsilon}(\mathbf{x}) = \langle\boldsymbol{\varepsilon}\rangle(\mathbf{x}) + \int \mathbf{U}(\mathbf{x}-\mathbf{y})\Big\{\boldsymbol{\tau}(\mathbf{y}) - \langle\boldsymbol{\tau}\rangle(\mathbf{y})\Big\}\, d\mathbf{y}, \tag{18.5}$$

$$\boldsymbol{\sigma}(\mathbf{x}) = \langle\boldsymbol{\sigma}\rangle(\mathbf{x}) + \int \boldsymbol{\Gamma}(\mathbf{x}-\mathbf{y})\Big\{\boldsymbol{\eta}(\mathbf{y}) - \langle\boldsymbol{\eta}\rangle(\mathbf{y})\Big\}\, d\mathbf{y}. \tag{18.6}$$

where the tensors

$$\boldsymbol{\tau}(\mathbf{y}) = \boldsymbol{\sigma}(\mathbf{y}) - \mathbf{L}^{(0)}\boldsymbol{\varepsilon}(\mathbf{y}), \quad \boldsymbol{\eta}(\mathbf{y}) = \boldsymbol{\varepsilon}(\mathbf{y}) - \mathbf{M}^{(0)}\boldsymbol{\sigma}(\mathbf{y}) \tag{18.7}$$

are called the stress and strain polarization tensors, respectively. As can be seen, Eqs. (18.5) and (18.6) are new because they do not explicitly depend on the elastic moduli of inclusions, in contrast to the classical cases where the strain and stress (see, e.g., [1184], and Eqs. 7.9 and 7.12)

$$\boldsymbol{\tau}(\mathbf{y}) = \mathbf{L}_1(\mathbf{y})\boldsymbol{\varepsilon}(\mathbf{y}), \quad \boldsymbol{\eta}(\mathbf{y}) = \mathbf{M}_1(\mathbf{y})\boldsymbol{\sigma}(\mathbf{y}) \tag{18.8}$$

polarization tensors have such a dependence. It should be mentioned that the definitions (18.7) exactly coincide with the definitions (6.171) and (6.172) attributed to Hill [461]. However, use of the definitions (18.7) in Eqs. (18.5) and (18.6) implies a nonlocal nature of the constitutive law inside the inclusions while [461] primarily assumed a locality of the constitutive law in any point.

In the original version of the MEFM, one used the main hypothesis (called the hypothesis **H1** of many micromechanical methods, according to which each inclusion is located inside a homogeneous so-called effective field (see Chapter 8). Exploiting the effective field hypothesis allows for replacement of the real inclusions with, perhaps, an arbitrary shape and structure by the fictitious ellipsoidal homogeneous inclusions with the same average thermoelastic response depending on the concentrator factors $\mathbf{A}_\sigma(\mathbf{x})$ and $\mathbf{B}_\epsilon(\mathbf{x})$ rather than stiffness $\mathbf{L}$ and compliance $\mathbf{M}$ of the inclusion which are unknown due to the possible nonlocality of their elastic properties. In so doing, the term inclusion is just used as a convenience notation of the area where the constitutive law (local or nonlocal) differs from the elastic properties of the homogeneous matrix and which embeds a real nanoinclusion. No restrictions are imposed on the microtopology of the microstructure and the shape of inclusions, as well as on the inhomogeneity of the stress field inside the inclusions and, moreover, on the constitutive laws of these inclusions which elastic responses are described by the concentrator factors $\mathbf{A}_\sigma(\mathbf{x})$ and $\mathbf{B}_\epsilon(\mathbf{x})$. For example, a new analog of Mori-Tanaka approach for the composite materials with nonlocal linear elastic properties of inclusions

$$\mathbf{M}^* = \mathbf{M}^{(0)} + \langle(\mathbf{A}_\sigma - \mathbf{M}^{(0)}\mathbf{B}_\sigma)V\rangle(c^{(0)}\mathbf{I} + \langle\mathbf{B}_\sigma V\rangle)^{-1} \tag{18.9}$$

depends on the concentrator factors $\mathbf{A}_\sigma(\mathbf{x})$ and $\mathbf{B}_\sigma(\mathbf{x})$ but not on the elastic moduli of inclusions as in the original Mori-Tanaka approach (8.95) to which Eq. (18.9) is reduced only for composites reinforced by the inclusions with the local elastic properties providing the equality $\mathbf{A}_\sigma(\mathbf{x}) = \mathbf{M}(\mathbf{x})\mathbf{B}_\sigma(\mathbf{x})$.

The equation (18.9) is a new step in micromechanics when for prediction of effective moduli of composites we don't need to know anything about nonlocal nature of constitutive lows of inclusions. We only need to know that these laws are linear and we need to know the concentrator factors $\mathbf{A}_\sigma(\mathbf{x})$ and $\mathbf{B}_\sigma(\mathbf{x})$, which can be found directly from MD simulations. In a similar manner any known micromechanical model (see Chapters 8 and 9) can be reformulated in terms of the concentrator factors. For instance, one particle approximation of the MEFM (8.79) and (8.85) can be generalized as

$$\mathbf{M}^* = \mathbf{M}^{(0)} + \Big[\mathbf{I} - \langle(\mathbf{A}_\sigma - \mathbf{M}^{(0)}\mathbf{B}_\sigma)\mathbf{Q}^0 V\rangle\Big]^{-1}\langle(\mathbf{A}_\sigma - \mathbf{M}^{(0)}\mathbf{B}_\sigma)V\rangle, \tag{18.10}$$

$$\mathbf{L}^* = \mathbf{L}^{(0)} + \Big[\mathbf{I} - \langle(\mathbf{B}_\epsilon - \mathbf{L}^{(0)}\mathbf{A}_\epsilon)\mathbf{P}^0 V\rangle\Big]^{-1}\langle(\mathbf{B}_\epsilon - \mathbf{L}^{(0)}\mathbf{A}_\epsilon)V\rangle. \tag{18.11}$$

Creation of a bridging mechanism is a challenging issue of nanotechnology and a valid one for further investigation. In the subsequent presentation, only the continuum mechanics approach will be explored with known thermoelastic properties of constituent phases. It should be mentioned that the principle of continualization (see [415]) is widely used in various divisions of physics. A classical particular example is the continuum theory of dislocation describing discrete defects in discrete systems in the framework of a continuum theory.

18.3 Modeling of Nanofiber NCs in the Framework of Continuum Mechanics

The present analysis should make it clear that micromechanical modeling based on continuum mechanics must be used with caution for nanocomposites. Once the matrix and embedded nanoelements are modeled in the framework of continuum mechanics, the length scale of the problem is significantly larger than the material scale analyzed. In such a case, the term conventional inclusion is used interchangeably with the term nanoelement when the reinforcement of the composite material is considered. Our goal is to better understand the origin of the reinforcing efficiency of nanocomposites. Due to their relative simplicity and instantaneousity, the micromechanical model proposed provides the ability to evaluate the key factors controlling the effective elastic behavior, and to explore large design spaces.

18.3.1 Statistical Description of NCs with Prescribed Random Orientation of NTs

Experimental research (see, e.g., [1097]) and molecular dynamic simulation mentioned earlier indicated that nanofibers can be effectively considered in the framework of continuum mechanics as the homogeneous prolate spheroidal anisotropic homogeneous inclusions with a large aspect ratio ($\alpha \cong 1000$). Current research focuses on the need to elucidate the fundamental reinforcement mechanisms in nanotube-based composites and develop tools to relate the nanoscale structure to the properties of nanotube-based composites. A nanocomposite is modeled as a linearly elastic composite medium subjected to homogeneous remote loading, which consists of a homogeneous matrix containing a statistically homogeneous random field of nanofibers with prescribed random orientation described by the orientation distribution function (ODF).

In the present research, one particle approximation of the MEFM based on the "quasi-crystalline" approximation by Lax [649] (see Chapter 8) will be used, which is equivalent in particular cases of homogeneous ellipsoidal inclusions to the Hashin-Shtrikman type estimations proposed in [898]. Even in this simplified case, the effective elastic moduli depend on the spatial correlation of inclusion location, which take particular "ellipsoidal" shape adopted to the spatial correlation. However, to the author's knowledge, no models exist that satisfy all theoretical criteria mentioned in Chapter 8 for arbitrary phase anisotropy and fiber-orientation distributions.

Let stresses and strains be related to each other via the constitutive equation (6.2) $\boldsymbol{\varepsilon}(\mathbf{x}) = \mathbf{M}(\mathbf{x})\boldsymbol{\sigma}(\mathbf{x}) + \boldsymbol{\beta}(\mathbf{x})$. We will use the notations and assumptions of Chapter 8. In the mesodomain w containing a statistically large set $X = (V_i, \mathbf{x}_i, \omega_i)$, $(i = 1, 2, \ldots)$ of ellipsoids v_i with characteristic functions V_i, centers x_i, semiaxes a_i^j $(j = 1, 2, 3)$ and an aggregate of Euler angles ω_i, a characteristic function W is defined. For the description of the random structure of the composite let us introduce a conditional probability density $\varphi(v_m, \mathbf{x}_m | v_i, \mathbf{x}_i)$, which is the probability density to find the mth inclusion in the domain v_m with a center $\mathbf{x}_m$ given fixed inclusions in the domains v_i with the the centers $\mathbf{x}_i$ (see for details Chapters 5 and 8). The area where $\varphi(v_m, \mathbf{x}_m | v_i, \mathbf{x}_i)$ equals 0 is called an excluded volume (or correlation hole). In a simplest case the one point function $\varphi(v_m, \mathbf{x}_m) = \varphi(v_m)$ is reduced to the orientation distribution function (ODF) f specifying the volume fraction of fibers with an orientation placed between $\mathbf{g}$ and $\mathbf{g} + d\mathbf{g}$: $dV/V \equiv f(g)\ d\mathbf{g}$ (for details see Appendix A.1, as well as [373], [616], [786], [1031]). We will consider three different ODF for the Euler angles parametric representation (5.110)–(5.112) analyzed in Section 5.4:

$$f\ d\mathbf{g} = (8\pi^2)^{-1} \sin(\theta)\ d\theta\ d\phi_1\ d\phi_2, \tag{18.12}$$
$$f\ d\mathbf{g} = (2\pi)^{-1}\delta(\theta - \pi/2)\delta(\phi_1 - 0)\ d\theta\ d\phi_1\ d\phi_2, \tag{18.13}$$
$$f\ d\mathbf{g} = (2\pi)^{-2}\delta(\theta - \pi/2)\ d\theta\ d\phi_1\ d\phi_2, \tag{18.14}$$

The ODF (18.12), (18.14), and (18.12) will be termed as the 1-D, 2-D, and 3-D, respectively, uniformly random distributions.

18.3.2 One Nanofiber Inside an Infinite Matrix

This and the next sections attempt to quantitatively investigate the performance of the approach presented in Chapter 8 to the estimation of effective elastic properties $\mathbf{L}^*$ [(8.85) and (8.102)] of nanocomposites. The results are directly compared with solutions obtained from simplified assumptions (such as Mori-Tanaka approach) and they are presented in order to place the advantages and limitations of the refined approach in evidence. Let us consider a nanocomposite made from the Epox 862 reinforced by the prolate spheroidal nanofibers. The matrix is described by isotropic elastic properties with Young's modulus $E^{(0)} = 3.01$ GPa, Poisson's ratio $\nu^{(0)} = 0.41$, and coefficient of thermal expansion (CTE) $\boldsymbol{\beta}^{(0)} = \beta_0^{(0)}\boldsymbol{\delta}$, $\beta_0^{(0)} = 10^{-4}(K^{-1})$ (see [1072]). Experimental research and molecular dynamic simulation indicated that nanofibers can be effectively considered in the framework of continuum mechanics as the homogeneous prolate spheroidal anisotropic homogeneous inclusions (see [825]) with a large aspect ratio ($\alpha^{\mathrm{r}} \geq 1000$). In the parametric analysis we will consider both isotropic [$E^{(1)} = 240.06$ GPa, $\nu^{(1)} = 0.38$, (see [1097]) $\beta_0^{(1)} = -1.6 \cdot 10^{-6}(K^{-1})$ (see [950]) and transversally anisotropic fibers [$L_{1111}^{(1)} = 457.6$ GPa, $L_{1122}^{(1)} = 8.4$ GPa, $L_{2222}^{(1)} = 14.3$ GPa, $L_{2233}^{(1)} = 5.5$ GPa, $L_{2323}^{(1)} = 27.0$ GPa, see [825]].

It should be mentioned that even in the case of the straight form, the nanofibers are cylinders with the smooth ends rather than the prolate spheroids. Therefore, the modeling of the nanofiber shape by the shape of a prolate spheroid is only approximately accepted by the authors with the aim to use the analytical Eshelby solution for a single homogeneous ellipsoidal inclusion inside the homogeneous matrix subjected to the homogeneous remote loading. Following [154], we begin the modeling of nanocomposites from an elastic analysis of a cylinder fiber ∂v_i with the smooth ends described by the equation in the local coordinate system with the symmetry axes which, in turn, are parallel to the global coordinate system

$$\partial v_i = \begin{cases} (|x_1| - l)^2/a^2 + x_2^2/a^2 + x_3^2/a^2 = 1, & \text{if } l \le |x_1| \le l + a, \\ x_2^2/a^2 + x_3^2/a^2 = 1, & \text{if } |x_1| \le l, \end{cases} \tag{18.15}$$

where $l + a$ is the length of the fiber with aspect ratio $\alpha^r = a/(a + l)$. For the construction of a correlation hole, we inscribe into the domain v_i the spheroid $v_i^{\rm Ins}$ of maximum possible volume bounded by the surface $\partial v_i^{\rm Ins}$: $x_1^2/b^2 + x_2^2/a^2 + x_3^2/a^2 = 1$ with semiaxes a and $b = l + a$. After that we will consider circumscribed spheroid $v_i^{\rm circ}$: $x_1^2/b_v^2 + x_2^2/a_v^2 + x_3^2/a_v^2 = 1$ with the same aspect ratio α^v as the spheroid $v_i^{\rm Ins}$: $\alpha^v = a_v/b_v = a/b$. A simple geometrical analysis leads to the boundaries of the semiaxes of the spheroid $v^{\rm circ}$: $a\sqrt{a^2+l^2}/\rho_1 \le a_v \le a\sqrt{a^2+b^2}/\rho_2$, $b\sqrt{a^2+l^2}/\rho_1 \le b_v \le b\sqrt{a^2+b^2}/\rho_2$, where $\rho_k^2 = b^2/(1 - \epsilon^2 \cos^2 \phi_k)$ $(k = 1, 2)$ is a polar representation of a cross section $(\partial v_i^{\rm Ins} \cap \{x_2 = 0\})$ with an eccentricity $\epsilon = \sqrt{1 - \min(a,b)^2/\max(a,b)^2}$, and $\phi_1 = \mathrm{arctg}(l/a)$, $\phi_2 = \mathrm{arctg}(b/a)$. A final step is the construction of a correlation hole v_i^0 with semiaxes $2b_v$, $2a_v$, $2a_v$ homothetic to the spheroid $v_i^{\rm circ}$ which, in turn, is homothetic to $v_i^{\rm Ins}$ with the semiaxes $b = l + a$ and a coinciding with the characteristic sizes of the fiber v_i.

The analyses of the effect of fiber and matrix elastic properties on the stress distribution both inside and in the vicinity of a cylindrical fiber embedded in the infinite matrix were performed via the analytical model (see, e.g., [801], [917]) and by FEA ([199], [389]). The fiber axial ratio, α (varying from 200 until 1000), and the ratio of the Young's modulus of the fiber to that of the matrix was considered in [389] in the analysis of elastic load transfer inside the cylindrical, ellipsoidal, paraboloidal, and conical fibers embedded in the matrix subjected to the axial (only) loading. For example, for a cylindrical fiber, the axial stress distribution along the fiber axis estimated via FEA and approximate analytical models was greatest at the center, decreased steadily over most of the fiber length and fell rapidly to zero near the fiber end (see [389], [801], [917]). In a similar manner the nanocomposites reinforced with the bone-shaped nanofibers synthesized in [1201] can be investigated. In the present research, a single cylindrical fiber with the smooth ends embedded in a large matrix sample is analyzed for six different unit external loading that allows a strain polarization tensor averaged over the volume of the fiber.

To estimate the tensors $\mathbf{B}(\mathbf{x})$, $\mathbf{C}(\mathbf{x})$ and $\mathbf{Q}_i^0$, it is necessary to solve a series of 3-D elasticity theory problems with the only nonzero component of the stress tensor $\overline{\boldsymbol{\sigma}}$ or eigenstrain tensor $\boldsymbol{\beta}^{\rm f}$ (see Chapter 4). Doing so by the finite element method is a rather time-consuming procedure. It is possible, however, to reduce the overall computational effort greatly by taking the symmetry of geometry and linearity of the problem into account. Specifically, the problem of evaluation of the tensors $\mathbf{B}(\mathbf{x})$, $\mathbf{C}(\mathbf{x})$ and $\mathbf{Q}_i^0$ can be reformulated for the first octant only.

Moreover, we can replace an infinite matrix domain with the finite one, namely, a cube with side size l^* : as numerical study shows, already for $l^* = 20(l+a)$ an effect of the outer cube boundary is negligible.

Within the inclusion v_i and/or in the hole v_i^0 the eigenstrain $\beta_j^{\rm f} = 1$, $\beta_k^{\rm f} = 0$ $(j \neq k)$, $j, k = 1, 2, ..., 6$. is prescribed. In the case, when one of the normal strains acts, i.e. $j = 1$ or $j = 2$ or $j = 3$, the surfaces $x_1 = 0$, $x_2 = 0$ and $x_3 = 0$ are the surfaces of symmetry. In the cases when act one of the shear strains, i.e., $j = 4$ or $j = 5$ or $j = 6$, the surfaces $x_1 = 0$, $x_2 = 0$ and $x_3 = 0$ are the surfaces of symmetry or asymmetry. For example, for $\beta_4^{\rm f} = 1$, $\beta_k^{\rm f} = 0$ $(k \neq 4)\, j, k = 1, 2, ..., 6$, i.e., for $\beta_{23}^{\rm f} = 1$, the surface $x_1 = 0$ is the surface of symmetry, and surfaces $x_2 = 0$ and $x_3 = 0$ are the surfaces of antisymmetry; the stress and strain fields are symmetric relative to surface $x_1 = 0$ and others are antisymmetric (the same value with opposite sign) relative to surface $x_2 = 0$ and $x_3 = 0$.

To obtain a reliable solution of the described problem, the sufficiently fine FE mesh has to be used. Moreover, it must be refined in the vicinity of expected stress concentration points, i.e., around the ends of the inclusion or correlation hole domain, as well as on the inclusion surface or on the surface of the correlation hole. At the same time, the FE solution must be not very time-consuming: noteworthy, the higher the aspect ratio of inclusion is, the more difficulties arise in obtaining the convergent numerical solution. In our numerical study, the 3-D 10-node tetrahedral elements with quadratic approximation of displacements and linear stress approximation were utilized. Here, for the aspect ratio α^r=10 the required number of nodes of the FE mesh is about 26000, whereas for $\alpha^r = 100$ their number grows up to 70000.

At first, we will analyze a stress concentrator factors $\langle\sigma_{11}\rangle_i/\langle\sigma_{11}\rangle$ and $\langle\sigma_{22}\rangle_i/\langle\sigma_{22}\rangle$ at one isolated spheroidal fiber in the infinite matrix. When aspect ratios of the fibers are larger than 50 the difference between the stress concentrator factors obtained from the general expressions for the Eshelby tensor and expressions for infinitely long fibers differ by less than 0.5%. Now we will analyze stress concentration factors $\sigma_{11}(\mathbf{x})/\langle\sigma_{11}\rangle$ and $\sigma_{22}(\mathbf{x})\ /\langle\sigma_{22}\rangle$ at the axis $\mathbf{x} = (x_1, 0, 0)^\top$ $(x_1 \in [-l-a, l+a])$ in the relative coordinate $x_{r1} = x_1/(l+a) \in [0, 1]$ of one isolated anisotropic fiber v_i in the infinite matrix subjected to the homogeneous remote loading $\langle\sigma_{ij}\rangle = \delta_{1i}\delta_{1j}$ and $\langle\sigma_{ij}\rangle = \delta_{2i}\delta_{2j}$, respectively (see Fig. 18.1). For comparison, for the points removed from the fiber ends at the distances $0.2l$ and $0.02l$ for $\alpha = 10$ and $\alpha = 100$, respectively, the transverse stress concentration factors the stress distributions $\sigma_{22}(\mathbf{x})/\langle\sigma_{22}\rangle$ unessentially differ from the appropriate Eshelby solutions for spheroidal fiber with the same aspect ratios.

It is interesting to consider a radial variation of the axial and transverse stresses $\sigma_{ij}(\mathbf{x}^{\rm cs})\ /\sigma_{ij}(\mathbf{x}_0^{\rm cs})$ (no summations over i, j) along the lines $\mathbf{x}^{\rm cs} = (x_1^{\rm cs}, 0, x_3)^\top$ $(x_3 \in [0, a])$ in the cross sections $x_1 = x_1^{\rm cs}$; here $\mathbf{x}_0^{\rm cs} = (x_1^{\rm cs}, 0, 0)^\top$ are the points at the fiber axis. The maximum variation of the relative stress concentration factor $\sigma_{ij}(\mathbf{x}^{\rm cs})/\sigma_{ij}(\mathbf{x}_0^{\rm cs})$ is achieved for the axial stresses $i = j = 1$ (for details see [154]). In so doing, the transverse stresses $\sigma_{33}(\mathbf{x}^{\rm cs})/\sigma_{33}(\mathbf{x}_0^{\rm cs})$ vary only 1.5% in the radial direction.

18.3.3 Numerical Results for NCs Reinforced with Nanofibers

Let us consider the composites with aligned fibers parallel to the axis Ox_1 (1D uniform random orientation) (18.12) that can be experimentally achieved e.g. by mechanically stretching the composites above the glass transition temperature of the polymer and then releasing the road at 300 K [107] (see also [1090], [1099]). The MEF (8.85) and MTM (8.102) lead, in general, to the different estimation of effective elastic moduli even for the isotropic fibers if the fiber shape is not ellipsoidal. For the spheroidal fibers with the "correlation hole" v_i^0 homothetic to fiber v_i when $\mathbf{Q}_i^0 \equiv \mathbf{Q}_i$, the MEF (8.85) and MTM (8.102) lead to identical estimation of effective elastic moduli and effective CTE.

Figure 18.2 demonstrates the dependencies of the effective Young's moduli

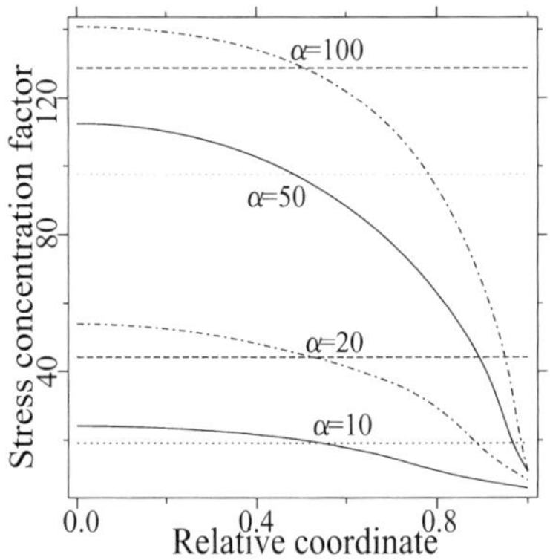

Fig. 18.1. A stress concentration factor vs x_{r1} for the spheroidal (straight lines) and cylindrical (curved lines) fibers for α=100, 50, 20, 10.

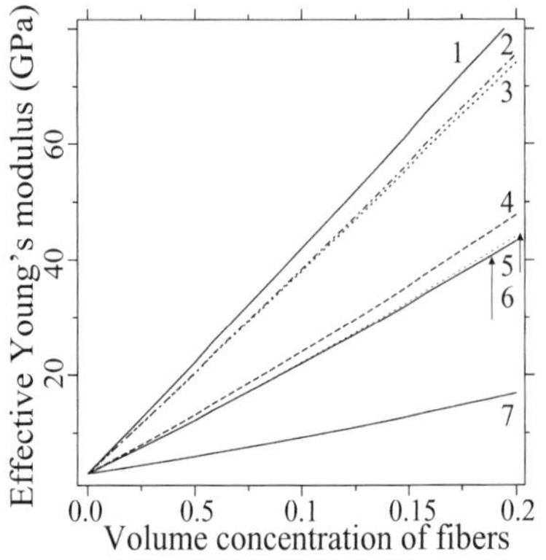

Fig. 18.2. Effective Young's modulus E_1^* (GPa) vs c for $\alpha = 100$ (1,2,3), $\alpha = 30$ (3,5,6), and $\alpha = 10$ (curve 3) of fibers with spheroidal shape (1,4,7) and cylindrical shape estimated by the MTM (2,5) and by MEF (3,6).

E_1^* as the functions of volume concentrations of anisotropic fibers estimated by the MEF and MTM approaches. It can be seen in Fig. 18.2 that the difference between estimations by the MEF and MTM methods is small and decreases with reduction of the aspect ratio from $\alpha = 100$ to $\alpha = 10$. Effective moduli E_1^* for composites reinforced by cylindrical fibers take the lower values than the composites with spheroidal fibers with the same aspect ratio. However, for $\alpha = 10$, the influence of the different fiber shape (cylindrical or spheroidal) and the estimation method (MEF or MTM) on E_1^* is negligible and less than 0.03%. In so doing, the mentioned influence of the fiber shape and estimation method on the effective transverse Young's modulus E_3^* does not exceed 0.14% for the aspect ratio varying in the range $10 \leq \alpha \leq 100$. One demonstrates (see [189]) that effective stiffness is defined by the axial stiffness of fibers rather than by their transversal stiffness.

Let us consider the 3-D uniform random orientation of both the anisotropic and isotropic fibers in the isotropic matrix. For isotropic inclusions, the estimations (8.85) and (8.102) coincide and give the symmetric representations of effective compliances. The effective moduli (8.130) were also analyzed for the case of 3-D uniform random orientation of anisotropic fibers. For the estimation of averages we used Simpson's quadrature formula with $\Delta\theta = 2\Delta\phi = \pi/m$ and evaluated the convolutions $k^* = L^*_{iikk}/9$, $\mu^* = (L^*_{ikik} - \frac{1}{3}L^*_{iikk})/10$ (see Eq. (2.199)). The components of the isotropic tensor $\mathbf{L}^{*\text{is}} \equiv (3k^*, 2\mu^*)$ differ from the relevant components of the numerical estimation of $\mathbf{L}^*$ by the values $6.0 \cdot 10^{-3}\%$ and $6.0 \cdot 10^{-5}\%$ at $m = 50, 200$, respectively, at $c = 0.2$. In so doing, the corresponding bulk and shear components of tensors $\mathbf{L}^*$ estimated by the MEF and MTM differ by values 0.9% and 1.6%, respectively. It gives confidence that the tensors $\mathbf{L}^*$ estimated by the MEF and MTM are really isotropic and very close to each other but do not coincide. This conclusion is also confirmed by concrete numerical analyses presented in Fig. 18.3 where a comparison with experimental data [639] is presented. Experimental data were obtained for nanocomposites produced from the polymer matrix Epox 862 reinforced by the nanofibers either PR-24-PS (□) or PR-24 HHT (○). The PR-24-PS nanofibers served as the baseline against which the PR-24 HHT nanofibers are compared. PR-24 HHT nanofibers are PR-24-PS fibers that have been heat-treated at 3000°C. After heat treating of pristine nanofibers to a temperature of 3000°C, graphene layers became straight, and minimum interlayer spacing was reached (PR-24 HHT). The difference between the effective Young's moduli estimated by the MEF and MTM for spheroidal fibers is no more than 1.5% at $0 \leq c \leq 0.1$, while the mentioned difference for cylindrical fibers increases to 1.9%. However, the estimated effective moduli are significantly more sensitive to the fiber shape rather than to the estimation method (either MEF or MTM). So, the difference of the values of E^* for the composites reinforced with the spheroidal and cylindrical fibers equal 9.2% and 7.8% for $\alpha = 100$ and $\alpha = 30$, respectively. It is interesting that experimental data are close to the numerical estimations obtained by the MEF for the anisotropic fibers with $\alpha = 100$ and the properties evaluated by an equivalent-continuum modeling method [825].

The difference of the MTM and MEF is more dramatically demonstrated at the analysis of stress concentrator factors $\mathbf{B}^*_i$ (8.107) estimated by these methods. For 3-D uniform random orientation of spheroidal fiber with $\alpha = 100$, the tensor $\mathbf{B}^*_i$ estimated by the MTM is isotropic ($B^*_{i|1111} = B^*_{i|2222} = B^*_{i|3333}$, $B^*_{i|1212} = B^*_{i|1313} = B^*_{i|2323}$) and does not depend on the orientation of the fiber v_i. For definiteness sake, we will consider the fixed fiber specified by the orientation $\mathbf{n} = \mathbf{e}_1$. As can be seen in Fig. 18.4, the components $B^*_{i|3333}$ and $B^*_{i|2323}$ evaluated by the MEF and MTM differ from one another at $c = 0.2$ no more than on 31% (for the component $B^*_{i|3333}$) and 13% (for the component $B^*_{i|2323}$), respectively. In so doing, the components $B^*_{i|1111}$ and $B^*_{i|1212} = B^*_{i|1313}$ estimated by the MEF vary in the range $0.996 \div 1.0$ that differ from the appropriate tensor components estimated by the MTM by the factors 4.5 and 6.6, respectively.

Now we will consider in more detail the case of 2-D uniformly random orientation (18.14) of anisotropic fibers. It was demonstrated ([188], [189], [190]) that if the "correlation hole" v^0_i is homothetic to spheroidal fiber v_i when $\mathbf{Q}^0_i \equiv \mathbf{Q}_i$ then the estimations by both the MTM and MEF coincide for E^*_1 and E^*_3 and differ from one another in the case of cylindrical fibers. In so doing the transversal Young's moduli E^*_3 estimated for both the anisotropic and isotropic fibers are very close. Replacement of spheroidal fibers with $\alpha = 100$ on cylindrical ones leads to lowering of predicted E_1^* on 13.4% and 9.5% using the MEF and MTM, respectively. The values E_3^* estimated by the MEF and MTM differ from one another on 0.014% and 0.028% at $\alpha = 100$ and $\alpha = 30$, respectively, for composites with volume concentration $c = 0.2$ of cylindrical fibers. However, we can observe a significantly more dramatic situation of the loss of symmetry of the matrix of effective elastic moduli estimated by both the MEF and MTM: $L_{1133} \neq L_{3311}^*$ for both spheroidal and cylindrical fibers. In that way, with estimated accuracy $L_{1133}^{*\text{MEF}} = L_{3311}^{*MTM}$ and $L_{3311}^{*\text{MEF}} = L_{1133}^{*MTM}$ for the spheroidal inclusions. These comparative estimations of nondiagonal elements of stiffness matrix L_{1133}^* and L_{3311}^* for spheroidal fibers are depicted in Fig. 18.5. Thus, the assumption of taking the distribution of fibers (described by the shape of v_i^0) around a given fiber to have a spheroidal shape with the same aspect ratio as the given fiber leads to well-known asymmetry of the MTM estimates (this claim was not known before for the MEF) for the effective stiffness in the case of 2-D uniformly random orientation of anisotropic fibers of both spheroidal and cylindrical shapes.

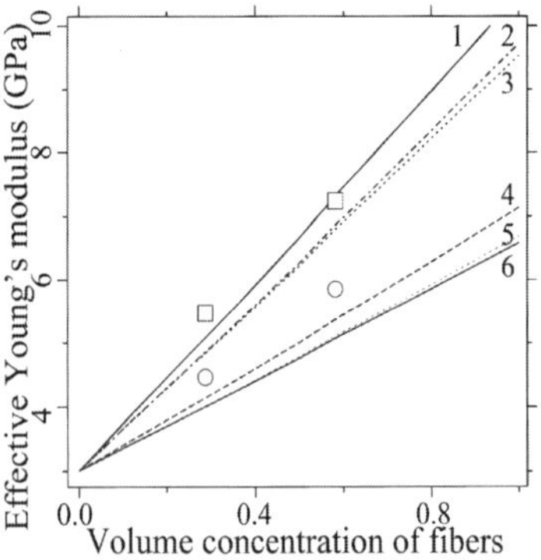

Fig. 18.3. Effective Young's modulus E_3^* vs c estimated by the MEF (1,3, 5, 6) and MTM (2, 5) for the spheroidal (1, 4) and cylindrical (2, 3, 5, 6) fibers with $\alpha = 100$ (1, 2, 3) and $\alpha = 30$ (4, 5, 6).

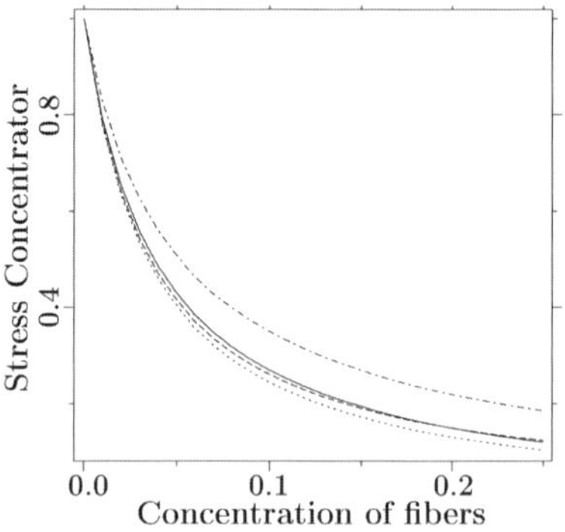

Fig. 18.4. $B_{i|3333}^*$ (solid and dot-dashed curves) and $B_{i|2323}^*$ (dotted and dashed curves) vs c estimated by the MEF (solid and dotted curves) and MTM (dot-dashed and dashed curves).

To avoid this contradiction with a physical requirement for 2-D uniform random distribution of fibers, we will choose another appropriate shape of a correlation hole v_i^0 enclosing the fiber v_i $(i = 1, \ldots)$ and prohibited for the centers of surrounding fibers. Indeed, let a representative fiber v_i be enclosed by the layer Ω_i^0 parallel to the plate Ox_1x_2 and the plate of symmetry of this layer contains the axis of the fiber symmetry. Then the surrounding fibers outside the layer Ω_i^0 can have any orientation and location that is compatible with previous choice of v_i^0 with $\mathbf{Q}_i^0 = \mathbf{Q}_i$. However, for the fibers lying inside the layer Ω_i^0 the choice of the mentioned v_i^0 leads to a significant reduction of a real "correlation hole" prohibited for surrounding fibers. Indeed, the centers of fibers v_j lying in the layer Ω_i^0 and perpendicular to the referred fiber v_i cannot be placed closer than a_1 to this referred fiber. Because of this, we can intuitively choose the domain v_i^0 as an oblate spheroid (with the semiaxes $a_1^0 = a_2^0 = a_1 + a_2$ parallel to the plane

Ox_1x_2, and $a_3^0 = 2a_3$) enclosing the prolate spheroid v_i. The arguments given are plausible rather than rigorous and would not apply for the general case of prescribed random orientation of fibers with a small aspect ratio. For example, for composites with the isotropic constituents and 3-D uniform random distribution of short fibers, Ponte Castaneda and Willis [898] used spherical symmetry of the domain v_i^0. In Fig. 18.6 the effective Young's moduli E_1^* of the composite material with 2-D uniform random orientation of anisotropic fibers are analyzed for the identical shape of oblate ellipsoidal spatial correlation of inclusion location $v_i^0 = \text{const.}$ ($a_1^0 = a_2^0 = a_1+a_2$, $a_3^0 = 2a_3$). The diagonal elements of the matrixes L_{ijkl} estimated by the MEF and by the MTM are closely allied at $c < 0.5$ (see e.g. Fig. 18.6), however, the nondiagonal elements can differ in a few instances. In so doing the matrix L_{ijkl}^* is not symmetric ($L_{1133}^{*MTM} \neq L_{3311}^{*MTM}$) (see Figs. 18.5) and symmetric ($L_{1133}^{*MEF} = L_{3311}^{*MEF} \approx L_{3311}^{*MTM}$) for the estimations obtained by the MTM and MEF, respectively.

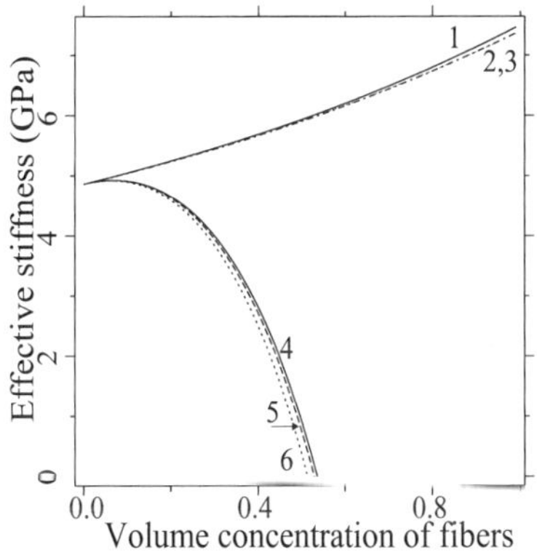

Fig. 18.5. Effective L_{1133}^* (GPa) (2, 4, 6) and L_{3311}^* (GPa) (1, 3, 5) vs c estimated by the MEF (2, 5) and MTM (1, 3, 4, 6) for the spheroidal (1, 6) and cylindrical (2, 3, 4, 5) fibers with α=100.

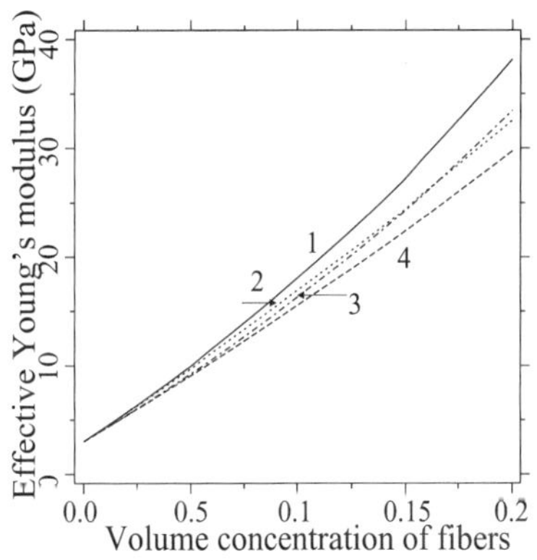

Fig. 18.6. Effective Young's modulus E_1^* (GPa) vs c estimated by the MEF (1, 3) and MTM (2, 4) for the spheroidal (1, 2) and cylindrical (3, 4) fibers with α=100.

It should be mentioned that in Fig. 18.5, such a wide range $0 \leq c \leq 0.8$ was deliberately used just for more distinct demonstration of incorrectness of the MTM's used in the analyzed case of a random orientation of anisotropic fibers. However, even for the small volume concentration $c \leq 0.15$ the advantage of the MEF over the MTM is also obvious. The measure of 'inconsistency' $\eta = 2|L_{1133}^* - L_{3311}^*|/(L_{1133}^* + L_{3311}^*)$ reflecting the effect of symmetry loss of $\mathbf{L}^*$ was estimated for both MEF and MTM method. It was found that the MTM and MEF for the spheroidal fibers with $\alpha = 100$ and $c = 0.05, 0.1, 0.15$ produce the parameters $\eta = 1.5\%, 4.2\%, 10.1\%$ and $\eta = 0.3\%, 0.4\%, 0.6\%$, respectively. In so doing, the values of the parameter η estimated by the MEF fall within the tolerance zone of the estimation error of $\mathbf{L}^*$ related with the numerical evaluation of appropriate averaging integrals.

The analysis presented can cherish hopes for the assumption that since the volume concentration of CNT is small, the dilute approximation (DA) formula to the first order in concentration $\mathbf{L}^{*\text{dil}} = \mathbf{L}^{(0)} + \langle \mathbf{R}^{\varepsilon v} V \rangle$ may be adequate for those who might be interested in such things; here $\mathbf{R}^{\epsilon v}$ (8.87) can be defined for the ellipsoidal fibers through the Eshelby tensor $\mathbf{S} = \mathbf{PL}^{(0)}$: $\mathbf{R}^{\epsilon v} = \mathbf{L}_1^{(1)}(\mathbf{I} + \mathbf{PL}_1^{(1)})^{-1}$. However, detailed analysis demonstrates that in the considered case of 2-D uniform random orientation of nanofibers, dilute approximation $\mathbf{L}^{*\mathrm{dil}}$ overestimate the effective Young modulus E_3^* evaluated by the MEF on 22.7%, 58.4%, 93.6% at $c = 0.05, 0.1, 0.15$, respectively. Therefore, the dilute approximation, meaning neglect of actions of surrounding fibers on the individual fiber being considered, cannot be accepted as an appropriate method for estimation of $\mathbf{L}^*$ even for the small fiber volume concentration $c = 0.05$. We have justified that for the $2D$ uniform random orientation of nanofibers exhibiting both the very large elastic mismatch with respect to the matrix (with the factor 150) and high aspect ratio ($100 < \alpha$), both simplified methods MTM (8.105) and DA cannot be selected (opposite to the MEM) for the estimation of $\mathbf{L}^*$ even for moderately small volume concentration of nanofiber $c \leq 0.15$.

18.4 Modeling of Clay NCs in the Framework of Continuum Mechanics

18.4.1 Existing Modeling of Clustered Materials and Clay NCs

The assumption of so-called *statistical homogeneity* in the case of clay NCs being considered is valid for completely exfoliated nanocomposites which were intensively analyzed in the framework of conventional micromechanics by the MTM (see [739], [1159], [1160]), by Halpin-Tsai (see [421]) equation [115], by the generalized Takayanagi's model method taking the interphase properties into account [504], as well as by the finite element analysis of multi-inclusion periodic model [412], [414]. Gusev [412] demonstrated a significant difference between the predictions obtained either by the numerical simulation or by the Halpin-Tsai equation for composites by the aligned fibers with the aspect ratio between 10 and 1000.

At the same time, it is known that the eventual abandonment of this so–called hypothesis of statistically homogeneous fields leads to a *nonlocal* coupling between statistical averages of the stress $\langle \boldsymbol{\sigma} \rangle(\mathbf{x})$ and strain $\langle \boldsymbol{\varepsilon} \rangle(\mathbf{x})$ tensors when the statistical average stress is given by an integral of the field quantity weighted by some tensorial function, i.e., the *nonlocal* effective elastic properties take place (see for references Chapter 12). A particular case of composite materials containing particle-rich and particle-poor regions is the clustered material where such regions have a concrete shape and random location that does not assumes the loss of statistical homogeneity of the composite structure. Usually a random location of points in the clusters is modeled by the sequential adsorption algorithm which was quantitatively compared with another simulation methods in Chapter 5. The concept of clusters is similar to that of "fractal structures" (see, e.g., [329], [1047]), which were successfully applied in [1077] (see also [263], [594], [1010]) for the statistical analyses of clustered materials by the use of such parameters of statistical descriptors as cluster size ζ, the fractal dimension D, and the radius of gyration R_g^2. Even though these and some other parameters are used for identification of clustered structures, it is not sufficient to predict the overall properties of composites for several reasons. The first one is that these parameters are not complete enough for the characterization of the micromorphology of fillers simply because one can present other morphology with the same descriptors. More informative characteristics of the random configurations use statistical second-order quantities that examine the association fillers relative to other particles in an immediate local neighborhood of the reference filler. Another one that is that the prediction of mechanical properties requires one or another micromechanical model. Sheng *et al.* [992] (see also [115], [588]) have emphasized that the notions of the matrix and inclusions widely used in conventional micromechanics can no longer be directly applied to nanoclay composites due to the hierarchical structure of clusters which, nevertheless, can be treated as a sort of equivalent (or effective) inhomogeneities embedded, in turn, in the bulk matrix material. A similar approach is well known in the engineering material science of conventional agglomerated composite materials (for references see, e.g., [310]) simply because in a polymeric matrix the filler (even if present in minimum quantities) is always more or less agglomerated. Microscopic studies have confirmed the existence of two kinds of primary structures in composites, i.e., aggregates with particles bound together firmly enough, and agglomerates—system of weakly interrelated aggregates. On the opposite hand, it was demonstrated [512] that at the accumulation of damage, the cavities tend to be gathered in clusters where the coalescence must be accelerated much earlier than it would be in a homogeneously voided material. In so doing, random porous microstructures without clusters and microstructures with a connected cluster are the hardest and the softest configurations, respectively, whereas microstructures with disconnected clusters lead to intermediate responses [85]. Nielsen [809] considered agglomerates as a plurality of contacting primary particles behaving like hard effective particles below a certain stress threshold where a high stress should cause the particles constituting an agglomerate to move relative to each other. Multiscale modeling strategy accounts for the hierarchical morphology of nanocomposite uses; therefore, a concept of "effective particle" in which the properties, size, and shape are not unique and depend on the concrete micromechanical model and the concrete condition of "equivalence" explored by a researcher.

A description of the composite cluster structure as well as micromechanical modeling are significantly simplified in the case of clusters of a deterministic structure. For the most part such the clusters (called then the clusters of periodic structure) can be described as "cut-out" from the infinite periodic media. A popular micromechanical modeling can be conditionally described by two limiting kinds of approaches, both of them based on the concept of effective properties of clusters. In the first case the effective properties of a cluster are estimated for the infinite periodic media (as well as for the infinite statistically homogeneous media) from which a cluster was mentally "cut-out." The effective properties of the layered clusters were estimated by the Halpin-Tsai equation (see [421], [115]) and by MTM (see [1160]). Among the related approaches concerned with the conventional clustered random structure composites, it might be well to point out the analysis of clustered fiber composites [82] by the MTM (see also analysis of nanofiber composites [997]), modeling of clustered particulate composite by the effective medium method [321] and [1177], as well as the investigation of cavity coalescence in inhomogeneously voided ductile solids [652] using the self-consistent method. In the second sort of approaches based on the "cut-out" of a cluster, the effective elastic properties of a cluster extracted from the composite are estimated through the equalization of the strain energy function at the set of inclusions combining the cluster to the potential energy stored in a homogeneous area with effective properties and the sizes of the cluster. It should be mentioned that the choice of the area referred to as a cluster and "cut-out" from the composite for the subsequent analysis by one or another micromechanical method is not unique and defined by the subjective partiality of a researcher. So, in both papers [721] and [992], the silicate aligned clusters in nanocomposites of random structure were modeled in the framework of laminated theory (see e.g. [237]) as the laminated structures containing n silicate plates with the thickness d_s separated by the matrix layer of the thickness d. However, Luo and Daniel [721] estimated the effective properties of a cluster bounded by two matrix layers $d/2$ in thick that with necessity leads to an invariantness of the cluster effective moduli on the layer number n. Sheng *et al.* [992] (see also [115] have proposed an alternative cluster bounded by two silicate plates that yields the dependence of the cluster effective moduli on n simply because the volume concentration of silica in such a cluster is not a constant $d/(d+d_s)$ as in the model [721] but varies from $2d/(2d+d_s)$ at $n=2$ till $d/(d+d_s)$ at $n=\infty$.

Both above mentioned groups of approaches implicitly assume a local nature of effective constitutive law for a fictitious area with effective properties of the cluster based in turn on the assumption that a field scale (internal stress inhomogeneity) infinitely exceeds a material scale (e.g., a distance between the inclusions inside a cluster). The correctness of this popular basic assumption acting as a bridge between two microscale (internal structure of an individual cluster and the structure of the composite) is questionable in the usual comparability of both the mentioned field and material scales inside the cluster where even the local effective moduli are varying over a cross section of the cluster of either periodic (Chapter 11 and [1061]) or statistically homogeneous structure (Chapter 12), while a volume concentration of inclusions in either the periodic or random structures is constant. It leads, in turn, to a new class of phenomena of the macroscopic scale based on the use of nonlocal continua (for references see Chapters 11, 12 and [166], [595], [718]), described by the constitutive laws relating the macroscopic stress and strain by either the differential or integral operators explicitly depending on the underlying microstructure without any additional assumptions on the macrolevel. One reason for the mentioned incorrectness is explicable on the basis of the fact that the boundary conditions prescribed in both model clusters (see cited papers in [721], [992], [1159], [1160]) are not compatible with the real contact conditions at the surface of the real considered cluster in the composite material. The role of the boundary conditions in the estimation of the effective moduli is drastically increased in importance if the scale of the cluster internal structure approaches the total cluster scale. It is spectacularly demonstrated in the series of papers (see [484], [506], [548], [843], [865]) concerned with analyses of effective elastic moduli of a representative volume element (RVE) of the periodic structure composites estimated at both the tractional and displacement boundary conditions. So, Jiang *et al.* [506] demonstrated that the effective moduli of composites with the RVE containing just one inclusion (stiffer than the matrix in 10 times) can be varied on 40% fold if the traction and displacement boundary conditions at the RVE surface are replaced by one or the other. Furthermore, increasing the RVE leads to convergence of the effective moduli $\mathbf{L}^*$ estimated at both mentioned boundary conditions to the $\mathbf{L}^*$ estimated at the periodic boundary condition. For example, for the RVE containing 16 inclusions, this variation decreases till 10%.

The modeling and simulation of random nano- and microstructures are becoming more and more ambitious due to the advances in modern computer software and hardware (see Chapter 4 and [24], [339], [537], [999], [1055], [1237]). The problem of obtaining accurate predictions from relatively small computer models was reached by exploring periodic boundary conditions at the unit cell containing just a few tens of inclusions generated by Monte Carlo simulation (see, e.g., Subsection 10.5.1 and [412], [537], [631], [982], [1237]). Numerical analysis of nanocomposites of periodic structures in the framework of continuum mechanics was performed by the FE (see, e.g., [93], [412], [414], [725]), and by the BIE [698]. The clustering effect of particulate composites on both the elastoplastic deformation and on the damage accumulation was intensively investigated by the FE analyses (for references see[1116]); a similar approach was applied in [992] in the 2-D case of nanosilicate composites modeling the clusters by three rectangular parallel inclusions. However, the combination of computational micromechanics with analytical micromechanics seems to be very promising because it allows for exploring the most powerful features of both mentioned groups of methods (see Chapters 8-10). In this light, the method of solution of a finite number of interacting spheroidal inclusions developed by Kushch [625], [628] and based on the multipole expansion technique is best suited for incorporation into the numerical MEFM. The basic idea of the method consists in expansion of the displacement vector into a series over a set of vectorial functions satisfying the governing equations of elastic equilibrium. The re-expansion formulae derived for these functions provide exact satisfaction of the interfacial boundary conditions. Incorporation of the multipole expansion technique into the MEFM in this chapter will make it possible to abandon the majority of simplified assumptions exploring at the analysis of random structure composites reinforced by the clusters of a deterministic structure.

18.4.2 Estimations of Effective Thermoelastic Properties and Stress Concentrator Factors of Clay NCs via the MEF

Let a linear elastic body occupy an open bounded domain $w \subset R^3$ with a smooth boundary Γ and with a characteristic function W. The domain w contains a homogeneous matrix $v^{(0)}$ and a statistically homogeneous set $X = (\Omega_i)$ of clusters Ω_i with characteristic functions V_i and bounded by the closed smooth surfaces Γ_i $(i = 1, 2, \ldots)$. It is assumed that the clusters can be grouped into components (phases) $\Omega^{(k)}$ $(k = 1, 2, \ldots, N)$ with identical mechanical and geometrical properties (such as the shape, size, orientation, and microstructure of clusters). Since the inclusions v_i introduced in Section 18.5 have no restrictions on the inhomogeneity of thermoelastic properties, they can be considered as the clusters

$\Omega_i \subset \Omega^{(k)}$ containing a deterministic field of inclusions $v_1, \ldots, v_m$ $(m = n^{\Omega(k)})$ with the thermoelastic properties $\mathbf{M}_j^{v(k)}(\mathbf{x})$ and $\boldsymbol{\beta}_j^{v(k)}$ $(j = 1, \ldots, n^{\Omega(k)})$. In a similar manner we can define the tensors $\mathbf{B}_i, \mathbf{C}_i, \mathbf{R}_i, \mathbf{F}_i$ for a cluster Ω_i analogous to Eqs. (3.11) and (3.12) for the inclusion v_i. It should be mentioned that for clustered materials being considered, $\mathbf{R}_k(\mathbf{x}), \mathbf{F}_k(\mathbf{x}) \equiv \mathbf{0}$ for the matrix's part of the cluster $\mathbf{x} \in \Omega^{(k)} \setminus \cup_j v_j$ $(j = 1, \ldots, n^{\Omega(k)};\ k = 1, \ldots, N)$. Therefore,

$$\mathbf{R}_k = \sum_{j=1}^{n^{\Omega(k)}} \int_{v_j} \mathbf{M}_{j1}^{v(k)}(\mathbf{x})\mathbf{B}_j^{v(k)}(\mathbf{x})\, d\mathbf{x}, \tag{18.16}$$

$$\mathbf{F}_k = \sum_{j=1}^{n^{\Omega(k)}} \int_{v_j} \left[\mathbf{M}_{j1}^{v(k)}(\mathbf{x})\mathbf{C}_j^{v(k)}(\mathbf{x}) + \boldsymbol{\beta}_{j1}^{v(k)}\right] d\mathbf{x}, \tag{18.17}$$

where $\mathbf{f}^{(k)}(\mathbf{x}) = \mathbf{f}_j^{v(k)}(\mathbf{x})$ at $\mathbf{x} \in v_i \subset \Omega^{(k)}$. As can be seen from Eq. (18.16) and (18.17), the concentrator tensors $\mathbf{R}_k$ and $\mathbf{F}_k$ do not depend explicitly on the shape and the size of the cluster Ω_k but only on the geometrical and mechanical properties of inclusions v_j constituent the cluster Ω_k. This dependence becomes apparent through the tensors $\mathbf{B}_k(\mathbf{x})$ and $\mathbf{C}_k(\mathbf{x})$ describing the thermoelastic solution for $n^{v(k)}$ interacting inclusions v_j $(j = 1, \ldots, n^{\Omega(k)})$ inside the infinite matrix subjected to the remote homogeneous loading $\overline{\boldsymbol{\sigma}}(\mathbf{x}_k)$. This solution was obtained by the method of multipole expansion (see for details [155]). Therefore the tensors $\mathbf{B}_j^{v(k)}(\mathbf{x})$ and $\mathbf{C}_j^{v(k)}(\mathbf{x})$ depend on the mechanical and geometrical parameters of all inclusions v_j $(j = 1, \ldots, n^{\Omega(k)})$ in the cluster Ω_k being considered. These tensors are estimated following the scheme considered in Subsection 4.2.2 and based on the solution obtained for a single cluster in [155] for both the arbitrary homogeneous far load tensor and thermal expansion allowing all components of the tensors $\mathbf{B}_j^{v(k)}(\mathbf{x})$ and $\mathbf{C}_j^{v(k)}(\mathbf{x})$.

It should be mentioned that in Eqs. (18.16)–(18.17) the shape of the cluster $\Omega_i \subset \Omega^{(k)}$ is not assumed to be ellipsoidal. It is only used in the hypothesis **H1** (8.14) and (8.15) for simplification of the subsequent calculations which can, in principle, be performed without assumption about the ellipsoidal shape of clusters. However, the ellipsoidal shape assumption is exploited more fundamentally in Eq. (8.67) through the choice of an ellipsoidal shape of the correlation hole Ω_i^0 that makes possible to accept the hypothesis (8.14). In the case of aligned identical clusters being considered, it is logical to assume that the shape of Ω_i^0 is homothetic to the shape of Ω_i with a scale factor 2.

After estimation of the tensors $\mathbf{R}_k$ and $\mathbf{F}_k$, we can accept the hypotheses **H1** and **H2** of Chapters 8 and 9 applied to the clusters Ω_i rather than to the inclusions v_i. It allows us to explore any of the versions of the MEFM and some related methods (such as e.g. the methods of the effective medium, differential, and Mori-Tanaka) considered in Chapters 8 and 9. Such a scheme was realized with respect to the clustered materials in [155] for the different average methods. It was proposed some sort of the classical relations by Levin [665] (see also Chapter 8) between the effective compliance and effective eigenstrain obtained just in the framework of the hypothesis **H1** for one sort of cluster. However, the numerical parametric analysis was carried out only via the one-particle approximation of the MEFM [describing by Eqs. (8.79) and (8.80) with the appropriate tensors $\mathbf{R}, \mathbf{F}, \mathbf{Q}^0$] for the identical aligned clusters with statistically homogeneous distribution in the space. The average thermoelastic response (i.e., the tensors $\mathbf{R}$, $\mathbf{F}$) of any cluster is the same as that of some fictitious ellipsoidal (see also Eq. (3.189) and (3.190)) homogeneous inclusion with thermoelastic parameters

$$\mathbf{M}_1^{f(k)} = \mathbf{R}_k(\mathbf{I}\overline{\Omega}_k - \mathbf{Q}_k\mathbf{R}_k)^{-1}, \quad \boldsymbol{\beta}_1^{f(k)} = \overline{\Omega}_k^{-1}(\mathbf{M}_1^{f(k)}\mathbf{Q}_k + \mathbf{I})\mathbf{F}_k. \tag{18.18}$$

The parameters (18.18) of fictitious ellipsoidal inclusions are simply a notational convenience and explicitly depend on the volume $\overline{\Omega}_k$ of the cluster. No restrictions are imposed on the microtopology of the cluster as well as on the inhomogeneity of the stress state in the cluster. They were introduced only as a tribute to tradition and can be used just for estimations of the tensors $\mathbf{R}_k$ and $\mathbf{F}_k$ by the use of the attached Eshelby' solution, although the tensors $\mathbf{R}_k$ and $\mathbf{F}_k$ used for introduction of the fictitious properties (18.18) must be previously evaluated by Eqs. (18.16) and (18.17) from the thermoelastic solution for a single cluster in the infinite matrix. As it will be demonstrated, the dependence of effective properties $\mathbf{M}^*$ (8.79) and $\boldsymbol{\beta}^*$ (8.80) on the internal structure and mechanical properties of clusters take place only through the tensors $\mathbf{R}_k$ and $\mathbf{F}_k$ describing the solution for one cluster inside the infinite matrix as well as on the tensor $\mathbf{Q}_i^0$ taking into account the shape of the correlation hole Ω_i^0 prohibited for the location of surrounding clusters. Construction of the domain Ω_i^0 is performed on the basis of a concrete geometrical structure of the clusters in the following manner.

For simplicity, the internal structure of an intercalated clay particle is idealized as a multilayer stack containing $n^{\Omega(1)} = n$ single silicate plate with uniform interlayer spacing d, as depicted in Fig. 18.7a. Exfoliated clay nanolayers will be considered as aligned oblate spheroids v_i with a small aspect ratio ($\alpha = b/a \ll 1$, typically on the order of 0.01) described by the equation $x_1^2/a^2+x_2^2/a^2+x_3^2/b^2 = 1$ in the local coordinate system connected with the spheroid semiaxes which in turn are parallel to the global coordinate system. Intercalated clusters can vary in thickness and interlayer spacing d depending on the degree of intercalation. Usually the plate thickness observed was roughly 1 nm, the interlayer spacing ranged from 2 to over 5 nm, and the number of plates per stack varied from 1 to 50. It is assumed that all stacks are randomly located, aligned and identical, i.e. have the same geometrical structures (a, b, d, n) and elastic properties of nanoplates (see Fig. 18.7a). The volume concentration of inclusions (oblate spheroids) v_i in the composite material is $c = \frac{4}{3}\pi a^2 b n^{(1)} n^{\Omega(1)}$, where $n^{(1)}$ is a cluster number density.

The stack containing n ellipsoids v_i is inscribed in the surface $\partial\Omega_i^{\text{stack}}$ described by the equations

$$\partial\Omega_i = \begin{cases} x_1^2/a^2 + x_2^2/a^2 + (x_3 - h)^2/b^2 = 1 & \text{if } h \le x_3 \le h + b\ , \\ x_1^2/a^2 + x_2^2/a^2 = 1 & \text{if } |x_3| \le h, \\ x_1^2/a^2 + x_2^2/a^2 + (x_3 + h)^2/b^2 = 1 & \text{if } -h - b \le x_3 \le -h, \end{cases} \tag{18.19}$$

where $h = (2b + d)(n - 1)/2$ (see Fig. 18.7b). Thus, the value $(h + b)/a$, called an aspect ratio of the cluster, explicitly depends on $b/a, d/a$ and n. We can inscribe into the domain Ω_i^{stack} the spheroid Ω_i^{ins} of maximum possible volume bounded by the surface $x_1^2/a^2+x_2^2/a^2+x_3^2/c^2=1$ with semiaxes a and $\bar{b}=h+b$. We will assume that the the cluster Ω_i is a spheroid of minimum possible volume $x_1^2/a_\Omega^2+x_2^2/a_\Omega^2+x_3^2/b_\Omega^2=1$ with the same aspect ratio as the spheroid Ω_i^{Ins}: $\alpha^\Omega=a_\Omega/b_\Omega=a/\bar{b}$. A simple geometrical analysis leads to the boundaries of the semiaxes of the spheroid Ω_i: $a\sqrt{a^2+h^2}/\rho_1\le a_\Omega\le a\sqrt{a^2+\bar{b}^2}/\rho_2$, $\bar{b}\sqrt{a^2+h^2}/\rho_1\le b_\Omega\le \bar{b}\sqrt{a^2+\bar{b}^2}/\rho_2$, where $\rho_k^2=\bar{b}^2/(1-\epsilon^2\cos^2\phi_k)$ $(k=1,2)$ is a polar representation of a cross-section $(\partial\Omega_i^{\text{Ins}}\cap\{x_2=0\})$ with an eccentricity $\epsilon=\sqrt{1-[\min(a,\bar{b})/\max(a,\bar{b})]^2}$, and $\phi_1=\text{arctg}(h/a)$, $\phi_2=\text{arctg}(\bar{b}/a)$.

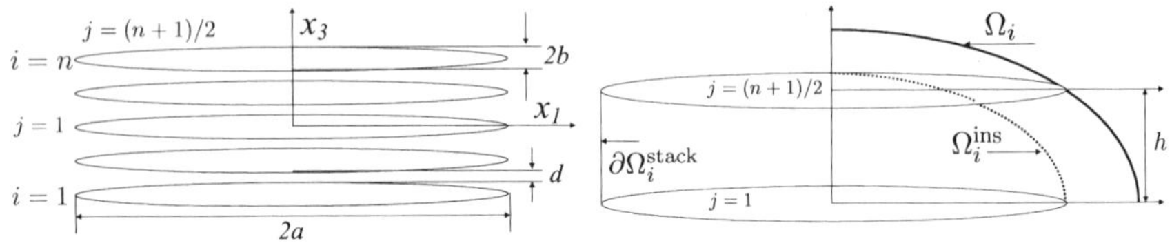

a) b)

Fig. 18.7. Schematic representation of a cluster. a) A stack of oblate spheroids. b) Definition of the cluster shape Ω_i.

It should be mentioned that in the case of exploiting of the one particle approximation of the MEFM being used for obtaining of numerical results, the estimations of the effective properties obtained depend only on the shape of the correlation hole Ω_i^0 (which coincides with the shape of the cluster Ω_i constructed above) rather than on its size. In so doing, the estimations of effective properties (8.79) and (8.80) were obtained for the multiply connected "inclusions" called clusters that generalize the approach proposed in Chapter 8 for 1-connected inclusions. The proposed method is logically close to the morphological-pattern approach [101] (see also [1213]) although the concrete estimation method of inclusion interactions inside the cluster (see the tensors $\mathbf{R}_j$, $\mathbf{F}_j$) and between the clusters [see the tensors $\mathbf{Y}_{ij}$ (8.66), (8.68)] are different.

18.4.3 Numerical Solution for a Single Cluster in an Infinite Medium

Let us consider an infinite elastic solid containing a finite cluster of n oblate spheroidal nanoplates. The matrix Epox 862 is described by isotropic elastic properties with Young's modulus $E^{(0)}=3.01GPa$, Poisson's ratio $\nu^{(0)}=0.41$ (see [1072]). In the parametric analysis we will consider a nanocomposite reinforced by the isotropic nanoinclusions with the properties $E^{(1)}=300$ GPa, $\nu^{(1)}=0.4$ (see [992]) arranged into the randomly located aligned clusters containing n inclusions with $a=1$ and $\alpha\le 0.01$, and with the axis of symmetry coinciding with the axis Ox_3 (see Fig. 18.7a). Numerical analysis of a finite number of inclusions (see Chapter 4) is performed in all cases for number of harmonics $n^p=11$ that provide a relevant accuracy of obtained solutions [155]. One of the advantages of the utilized method is its insensitivity to the aspect ratio of inclusions as well as the inclusion-matrix properties ratio. Estimation of the accuracy is presented in [625], [628], [630]. There, the inclusions of various aspect ratio including the extreme case of penny-shaped cracks were studied and no problems neither with convergence nor with accuracy were detected.

At first, we will analyze the average stress concentrator factors $\langle\sigma_{11}\rangle_{(i)}/\langle\sigma_{11}\rangle$, $(i = 1,\ldots,n,\ \alpha = 0.01$ and $d = 4b,\ a = 1)$ at the inclusions of one isolated cluster Ω_k in the infinite matrix subjected to the homogeneous loading $\langle\sigma_{ij}\rangle = \delta_{1i}\delta_{1j}$. In Fig. 18.8 it can be seen that the stress concentrator factor $\langle\sigma_{11}\rangle_{(i)}/\langle\sigma_{11}\rangle$ $(i = 1,\ldots,n)$ significantly (almost in three times) decrease from the outer spheroid to the central spheroid of the cluster. Hereafter one depicts the stress concentrator factors of inclusions v_i $(i = (n+1)/2,\ldots,n)$ numbered by the subscript $j = 1,\ldots,(n+1)/2$ $(j = i-(n-1)/2$, see Fig. 18.7a). This decreasing of stress concentrator factor growths with the extension of an inclusion number in the cluster. So, variation of $\langle\sigma_{11}\rangle_{(i)}/\langle\sigma_{11}\rangle$ inside the stack reaches 2.7 times at $n = 17$ while for $n = 5$ the stress concentrator factor varies on just 40%. If the number of the spheroids is much enough (more than 13) then the average stress concentrator factors in three external spheroids differ between the clusters with 13 and 17 spheroids on just 1% (see Fig. 18.8). Such a behavior is qualitatively confirmed by 2-D finite element analysis in [992].

The large stress in the outer inclusions counteracts to the stress concentrator at the internal inclusions of the cluster (shielding effect). Only a small amount of load is transferred to the middle silicate layer, due to the low shear modulus of the galleries. In so doing, a variation of stress concentrator factor $\langle\sigma_{33}\rangle_{(i)}/\langle\sigma_{33}\rangle$ (at $\langle\sigma_{ij}\rangle = \delta_{3i},\delta_{3j}$) in the different inclusions in a cluster is less than 20% and slightly increase from outer inclusions to the internal inclusions of the cluster only for a large number of inclusions in the stack. This effect is distinctive for both the functionally graded and clustered materials of random structure composites reinforced by the spherical inclusions when the local effective stiffness increase from the cluster boundary to the cluster center that with necessity yield the increasing of stress concentrator factor in the cluster center (for details see Chapter 12).

The influence of the interlayer spacing $d = 2,3,5,9b$ is analyzed in Fig. 18.9

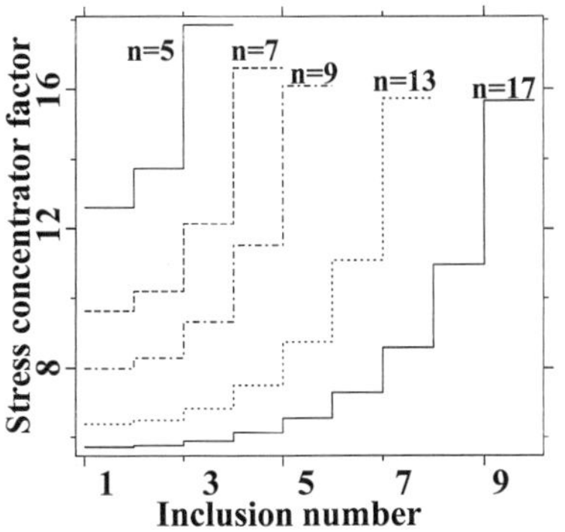

Fig. 18.8. Average stress concentrator factor $\langle\sigma_{11}\rangle_{(j)}/\langle\sigma_{11}\rangle$ in the different spheroids $j = 1,\ldots,(n+1)/2$ in the clusters containing n inclusions.

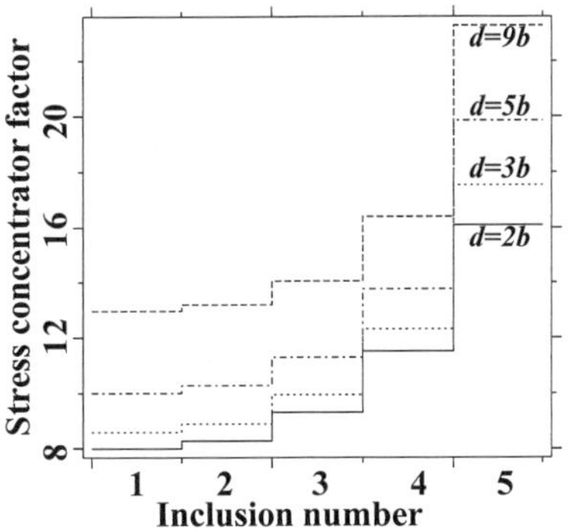

Fig. 18.9. Average stress concentrator factor $\langle\sigma_{11}\rangle_{(j)}/\langle\sigma_{11}\rangle$ in the different spheroids $j = 1,\ldots,(n+1)/2$ in the clusters with different inter-layer spacing of inclusions.

at the fixed values of all remaining ones ($n = 9$, $\alpha = 0.01$). The values $\langle\sigma_{11}\rangle_{(i)}/\langle\sigma_{11}\rangle$ increase, as it is expected, with extension of d. However, in the range of the inter-layer spacing of principal practical interest $d = 2 \div 9b$ the average stress concentrator factors at the spheroids in the cluster significantly below of the limiting stress concentrator factor $\langle\sigma_{11}\rangle_{(i)}/\langle\sigma_{11}\rangle = 38.9$ for an isolated inclusion in the infinite matrix at $d \to \infty$. It should be mentioned that analogous stress concentrator factor for the isolated spheroid with $\alpha = 0.001$ is $\langle\sigma_{11}\rangle_{(i)}/\langle\sigma_{11}\rangle = 85.4$ in excess of the value $\langle\sigma_{11}\rangle_{(i)}/\langle\sigma_{11}\rangle = 38.9$ ($\alpha = 0.01$) by a factor 2.2.

18.4.4 Numerical Estimations of Effective Properties of Clay NCs

We now turn our attention to the analysis of effective elastic moduli of composites reinforced by randomly located identical aligned ($\alpha = 0.01$, $d = 4b$, $a = 1$) with a different number of oblate spheroids in the clusters $n = 1, 2, 5, 13, 17$. As can be seen in Figs. 18.10 and 18.11, increasing the inclusion number in the clusters leads to decreasing the effective Young moduli E_3^* and E_1^* in both the transversal symmetry direction Ox_3 and the longitudinal direction Ox_1. The reason for such a behavior is explained by decreasing of the average stress concentrator factors $\mathbf{B}_i^{v(1)}$ (3.11) ($i = 1, \ldots, N$) at the individual inclusions in a single cluster in the infinite matrix with a rise in the inclusion number in a cluster (see Fig. 18.8). In so doing, influence of surrounding clusters on the chosen one Ω_k in the composite material described by the effective field $\langle\overline{\boldsymbol{\sigma}}\rangle_k$ does not culminate in elimination of the slackening effect of the stress concentrator factors in an individual cluster.

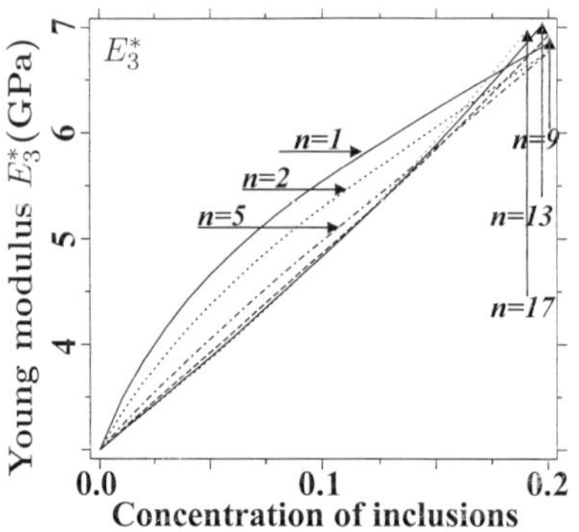

Fig. 18.10. Effective elastic modulus E_3^* vs volume concentration of inclusions for the different inclusion number $n^{(1)} = 1, 2, 5, 13, 17$ in the clusters.

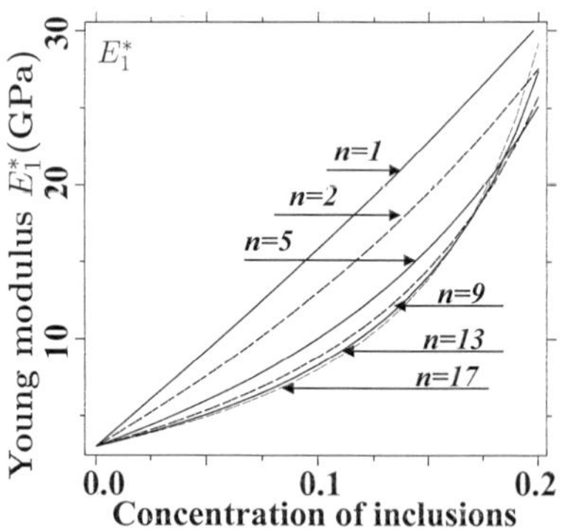

Fig. 18.11. Effective elastic modulus E_1^* vs volume concentration of inclusions for the different inclusion number $n = 1, 2, 5, 13, 17$ in the clusters.

It should be mentioned that the case $n = 1$ corresponds to the Mori-Tanaka approach (see [52], [787], [1073], [1159] as well as the comprehensive review in Chapter 8). This approach was applied in [1000] to the analyses of unidirectional nanocomposites with dispersed and parallel flake-like inclusions at the simplified assumption of complete exfoliation and full dispersion of nanoplates. In so doing a disagreement of the prediction of effective moduli and experimental data was attributed [1000] to debonding between clay nanolayers and the matrix. However, we can conclude now, from Figs. 18.10 and 18.11, that the reduction of effective moduli of nanocomposites with the clusterization does not uniquely depend on debonding and can be explained by a pure elastic shielding effect of delamination of stress concentrator factors at the internal nanoinclusions in the clusters (see Fig. 18.8). In a similar manner, it was mentioned [508] that the effective elastic moduli increase significantly only before the volume concentration of clay reaches a certain value (5%–10%), and no evident increase of effective moduli will be observed with the further input of clay. Feng [332] explained this effect at the intuitive level as a sequential conversion of clay arrangement: completely exfoliated → intercalated arrangement → tactoidal structure with the increase of clay concentration. Now, this explanation can be quantitatively confirmed.

As can be seen in Figs. 18.10 and 18.11, the effective elastic moduli are not sensitive to the inclusion number of spheroids inside the clusters if $n \geq 9$. It is interesting that it was also obtained [332] a weak sensitivity of effective elastic properties of clay nanocomposites by FEA if the number of plates in the stacks is more than 9. Because of this, we will investigate the inter-layer space $d = 2, 3, 5, 9b$ on the effective Young moduli E_1^* (see Fig. 18.12) and E_3^* (see Fig. 18.13) at the fixed $\alpha = 0.01$ and $n = 9$. As one would expect, increasing of the interlayer spacing slacking the interaction of the spheroids inside the cluster according to Fig. 18.9 leads to a rise of the effective Young moduli. In so doing, E_1^* and E_3^* rise 2.1 and 1.3 times, respectively, at $c = 0.1$. Moreover, an accelerated growth of E_1^* at $n = 9$ and $c \geq 0.05$ falls even more significantly than the increasing of E_1^* for a single spheroid in the clusters (see Fig. 18.11), which is explained by some sort of shielding effect in the composite material at $n = 1$.

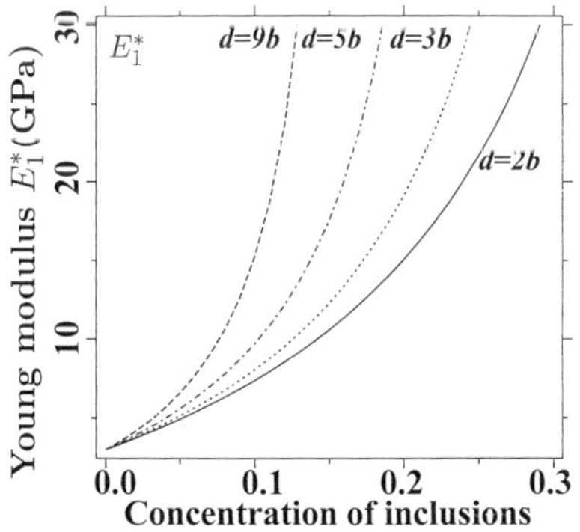

Fig. 18.12. Effective elastic modulus E_1^* vs volume concentration of inclusions for the different interlayer spacing $d = 2, 3, 5, 9b$ ($n = 9$).

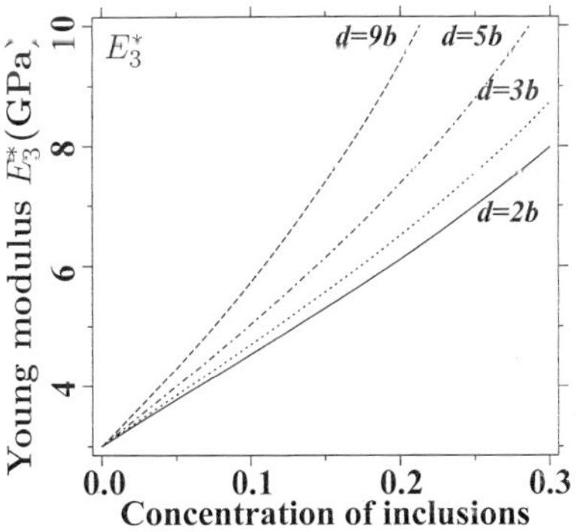

Fig. 18.13. Effective elastic modulus E_3^* vs volume concentration of inclusions for the different interlayer spacing $d = 2, 3, 5, 9b$ ($n = 9$).

Thus, we come to a practically important conclusion that the clustering of stiff oblate spheroidal inclusions leads to the weakening of the reinforcement effect of nanocomposites. This result has a qualitative confirmation by numerical simulation by Sheng *et al.* [992] (see also [1228]), who described this effect in 2-D finite element analysis of periodic structure composites with a quasi-random location of a number of clusters containing the aligned three rectangular inclusions in the large unit cell. Furthermore, well known experimental data (see, e.g., [589], [697], [1223]) also point to the decreasing of effective elastic properties (and fracture toughness) with the growth of clustering with the maximum effective stiffness of nanocomposites corresponding to the completely exfoliated structures. Kornmann *et al.* [589] explained this behavior at the intuitive level of justification: as the interlamellar spacing is increased, the effective particle volume fraction is also increased. The corresponding reduction of particle stiffness is a much weaker effect that increases the volume fraction. Now, this intuitive explanation can be replaced by rigorous justification confirmed by Figs. 18.8 and 18.11.

However, the mentioned behavior of decreasing of the effective stiffness with the increasing of the number of nanoplates in the clusters is drastically changing with the variation of the inclusion shape. So, one of the most critical technological problems is a significant growth of the effective viscosity of suspensions at the clustering of spherical inclusions. For a fixed filler content, the viscosity of a system with agglomerates of spherical inclusions is always higher than that of the "well-dispersed" sample (see for references, e.g., [310], [512], [671], [688], [809]). In Figs. 18.14 and 18.15 the functional dependences $E_1^* \sim c$ and $E_3^* \sim c$ are presented for the spherical inclusions in the clusters ($a = 1,\ \alpha = 1,\ d = 0.04a,\ n = 1, 2, 5, 9, 13, 17$). As can be seen, both the effective Young's E_1^* and E_3^* at the fixed c increase with the increasing of n. However, the dependence $E_1^* \sim c$ varies very slightly with the variation of n (just on 0.9% at $c = 0.2$) as opposite to the dependence $E_3^* \sim c$ varying in 1.9 times at $c = 0.2$)

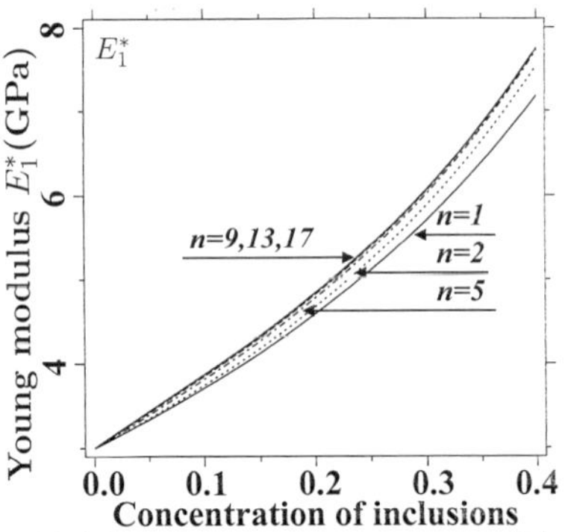

Fig. 18.14 Effective elastic modulus E_1^* vs volume concentration of inclusions c for the different inclusion number $n = 1, 2, 5, 13, 17$ in the clusters.

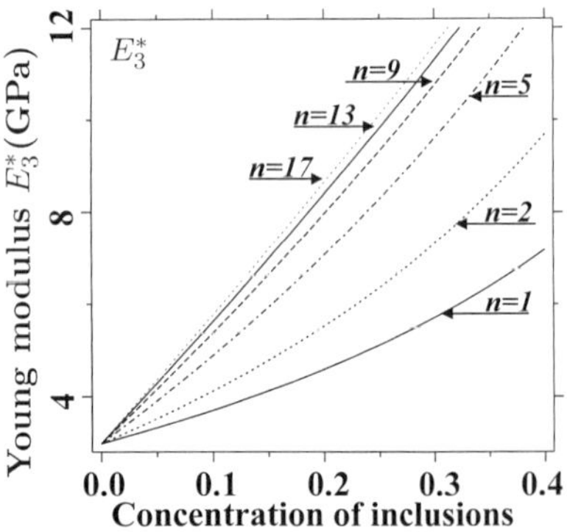

Fig. 18.15. Effective elastic modulus E_3^* vs volume concentration of inclusions c for the different inclusion number $n = 1, 2, 5, 13, 17$ in the clusters.

In light of this, we will analyze in more detail the influence of the inclusion shape in the composite materials reinforced by the clusters with $n = 9$ which are large enough for generating of maximum cluster effect. In Figs. 18.16 and 18.17 an influence of an aspect ratio $\alpha = b/a$ varying from 0.001 until 10 is investigated with the fixed $d = 0.04a,\quad a \equiv 1,\quad n = 9$. It is interesting that the behavior of the dependences $E_1^* \sim c$ for the fixed α and $E_1^* \sim \alpha$ at the fixed c are both monotonic. In so doing the dependence $E_3^* \sim c$ for the fixed α is monotonically growing as the functions $E_1^* \sim c$ for the fixed α while the variation of dependences $E_3^* \sim \alpha$ at the fixed c is not monotonic and achieves its minimum for the investigated values of α at $\alpha = 0.1$. Nonmonotonic behavior of functions $E_3^* \sim \alpha$ at the fixed c for the analyzed clustered material correlates well with the analogous dependencies $E_3^* \sim \alpha$ at the fixed c for completely exfoliated system corresponding to MTM estimations archiving its minimum for investigated values of α at $\alpha = 0.3$. Thus, we investigated the influence of the shape (α varies from 0.001 to 1) on the effective moduli of clustered materials which quantitatively confirmed intuitively expected decreasing and reverse of the "shielding" effects for the clusters with the less-oblate inclusions.

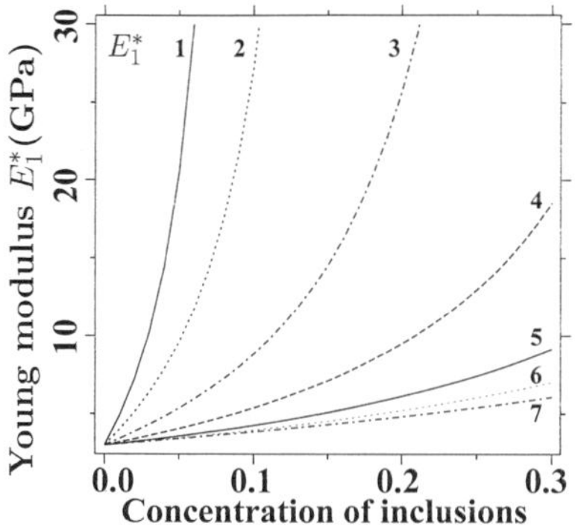

Fig. 18.16 Effective elastic modulus E_1^* vs volume concentration of inclusions for the different aspect ratio of inclusions α =0.001 (curve 1); 0.003 (2);0.01 (3); 0.03 (4), 0.1 (5), 0.3 (6), 1.0 (7).

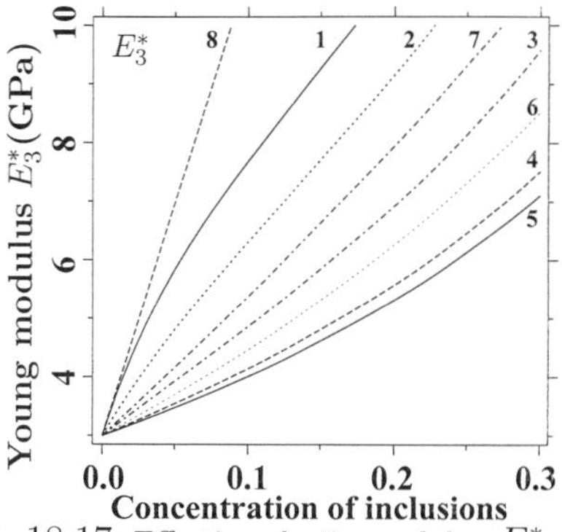

Fig. 18.17. Effective elastic modulus E_3^* vs volume concentration of inclusions for the different aspect ratio of inclusions α = 0.001 (curve 1); 0.003 (2);0.01 (3); 0.03 (4), 0.1 (5), 0.3 (6), 1.0 (7), 10.0 (8).

We are coming now to the estimation of an effective coefficient of thermal expansion (CTE) $\boldsymbol{\beta}^*$. The normalized CTE $(\beta_{11}^* - \beta_0^{(0)})/\beta_{01}^{(1)}$ and $(\beta_{33}^* - \beta_0^{(0)})/\beta_{01}^{(1)}$ are presented in Figs. 18.18 and 18.19, respectively, as the functions of the

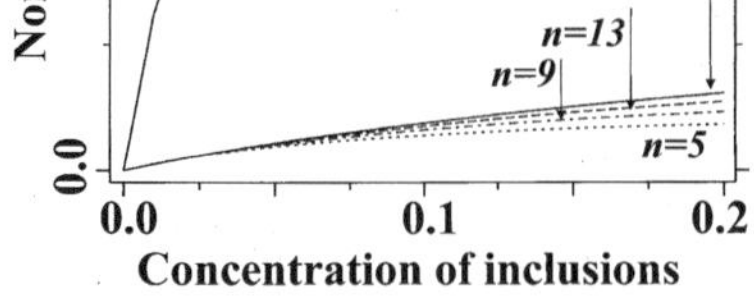

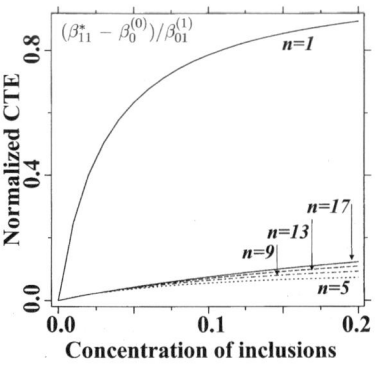

Fig. 18.18. Normalized CTE $(\beta_{11}^* - \beta_0^{(0)})/\beta_{01}^{(1)}$ vs volume concentration of inclusions $c^{(1)}$ for the different inclusion number $n = 1, 5, 9, 13, 17$ in the clusters.

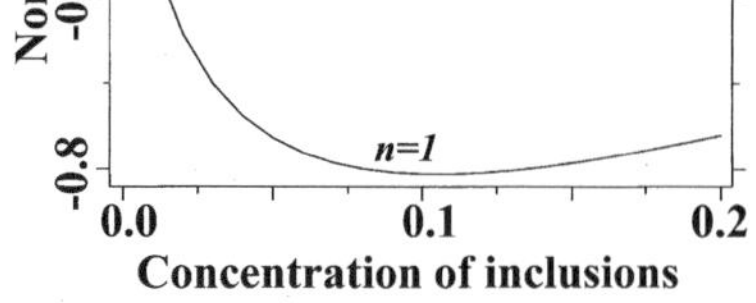

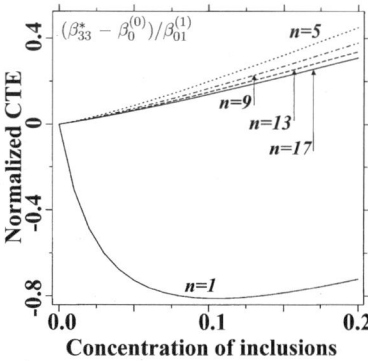

Fig. 18.19. Normalized CTE $(\beta_{33}^* - \beta_0^{(0)})/\beta_{01}^{(1)}$ vs volume concentration of inclusions $c^{(1)}$ for the different inclusion number $n = 1, 5, 9, 13, 17$ in the clusters.

volume concentration of nanoinclusions $c^{(1)}$ for an isotropic CTE of both constituents $\boldsymbol{\beta}^{(0)} = \beta_0^{(0)}\boldsymbol{\delta}$, $\boldsymbol{\beta}^{(1)} = \beta_0^{(1)}\boldsymbol{\delta}$, and $\boldsymbol{\beta}_1^{(1)} = \beta_{01}^{(1)}\boldsymbol{\delta}$. Nonmonotonical behavior of the value $(\beta^*_{33} - \beta^{(0)}_0)/\beta^{(1)}_{01} \sim c^{(1)}$ for the completely exfoliated nanocomposites ($n = 1$) is explained by the Poisson effect. Indeed, even for the small concentration of isotropic nanoplates with positive CTE, the residual strains $\langle \varepsilon_{33} \rangle$ at $\langle \boldsymbol{\sigma} \rangle \equiv \mathbf{0}$ are governed by the shrinkage of oblate stiff nanoelements in the Ox_1-direction leading to the negative deformation of the matrix in a perpendicular direction. With the growth of nanoelement concentration, the mentioned effect is compensated by the shrinkage of nanoelements in the transversal direction yielding the decrease of the effective CTE in this direction. For the clustered nanocomposites ($n > 1$), the weak dependencies of $\boldsymbol{\beta}^*$ vs $c^{(1)}$ are explained by the shielding effect analogous to the pure elastic case (see Fig. 18.8).

We analyzed a particular case of an arrangement of inclusions inside the cluster when the inclusion centers are placed on a straight line. In a similar manner, we can consider a practically more important case of spherical inclusion clusters with the inclusion centers occupying the ellipsoidal aligned domains. Another limiting case of aligned fibers ($\alpha \gg 1$) arranged into the clusters with the fiber centers placed in the elliptic domains can also be considered analogously. This case is correspondent to the primary product of current methods of SWCNT synthesis containing not individual, exfoliated SWCNTs, but rather bundles of nanotubes or nanoropes consisting of a few to a few hundred SWCNTs of uniform size arranged in hexagonal order (see [705], [907], [959]).

18.5 Some Related Problems in Modeling of NCs Reinforced with NFs and Nanoplates

A necessary step of NC investigation is a statistical analysis of their morphology. Progress in nanomechanics is based on methods of allowing for the statistical mechanics of a multiparticle system considering n-point correlation functions and direct multiparticle interaction of inclusions. Estimation of one-point statistical functions (such as volume filler concentration and their ODF, see, e.g., [245], [585], [701], [1195]) have been universally accepted as a necessary first step of a statistical description of NCs morphology. Improved materials processing allows us to manufacture the NCs with prescribed orientation of nanoelements. Although the aligned nanofibers and nanoplates, as was demonstrated in Sections 18.3 and 18.4 (see also [1090], [1099]), are desirable, the real orientation of fibers is usually described by a single parameter of a standard deviation of the Gaussian distribution with the center coinciding with the flow direction (see, e.g., [696], [1195]). In the investigation of clay orientation in NCs, Koo *et al.* [585] (see also [701]) have obtained Gaussian distribution with the center, which indicates that the clay platelets' normals are aligned in the direction perpendicular to the flow direction. They demonstrated quite different shear-induced orientation behaviors with the concentration of silicates. Vaia and Wagner [1126] mentioned that the critical concentration at which packing considerations dictate orientation correlation between nanoelements is 10^{-3} for a plate and 10^{-4} for a rod. These low volume fractions imply that NCs are not fundamentally isotropic, but have a tendency to exhibit anisotropic properties or a domain-like texture that contain nanoelements with a local preferred orientation even though macroscopically the global orientation is isotropic.

In connection with the mentioned practical interests, the necessity of estimation of effective elastic moduli of NCs with an arbitrary prescribed random orientation of nanoelements emerges. In so doing the randomly oriented nanofibers and the stacks of nanoplates demonstrate strongly anisotropic behavior at least in the sense of effective properties. The difficulties of this problem are strikingly illustrated in Section 18.3 which dictates the importance of a correct evaluation of also 2-point correlation functions $\varphi(v_m, \mathbf{x}_m|; v_i, \mathbf{x}_i)$ describing mutual location of the nanoplates and nanotubes. As a minimum, we need to know the shape of the correlation hole v_i^0. The construction of a radial distribution function $\varphi(v_m, \mathbf{x}_m|; v_i, \mathbf{x}_i)$ as well as v_i^0 is a well known problem for the particulate composites (for references see [168], [1106], [1107]). Significantly less research in the trend of statistical modeling has been performed for CM with random orientation of fibers (for references see [203]). Due to the absence of experimental estimation of both $\varphi(v_m, \mathbf{x}_m|; v_i, \mathbf{x}_i)$ and v_i^0, their numerical simulation is extremely desirable. The numerical MEFM presents a universally rigorous scheme of both analyses of the microstructures and prediction of macroscopic properties and the statistical distributions of the local stress fields.

We should also mention the special features of load transfer, in tension and in compression, in MWNT-epoxy composites studied in [969] where it was detected that load transfer in tension was poor in comparison with load transfer in compression, implying that during load transfer to MWNTs, only the outer layers are stressed in tension due to the telescopic inner-wall sliding (reaching at the shear stress 0.5 MPa, see [1212], [1211]), whereas all the layers respond in compression. In the framework of continuum mechanics, both the SWNTs [825] and MWNTs (see [1091], [1092]) modeled as the hollow cylinder and multilayer one, respectively, are sometimes replaced by the solid homogeneous cylinder with effective fictitious properties. Obviously, these approaches ignoring interaction of the NT with the surrounding matrix are associated with the second type of "cut-out" modeling of clustered materials considered in Subsection 18.4.1 accompanied with an explanation of their approximative nature. These issues deserve further investigation, for example, by explicitly including graded interface layers at least at the level of parametric analysis (see, e.g., [91], [1130]). In so doing, one usually uses the popular multilayer coating inclusion model which assumes the homogeneous elastic properties in each layer [875] (for references, see, also Chapter 4 and [184]). Nevertheless, Pipes and Hubert [881] investigated the effective mechanical properties of a large array of CNs assembled in helical geometries of circular cross-section without the matrix when the nanowall has a cylindrical anisotropy. A cylindrical anisotropy of an SWNT was also estimated [673] via the molecular structural mechanics method indicating that the radial modulus decreases with increasing tube diameter while the circumferential modulus is insensitive to tube diameter, and roughly equal to the axial modulus. In relation to this topic, the MD simulation in [404] (for references see also [34]) of CNT embedded in polyethylene is noteworthy. They performed MD simulation to derive average stress-strain curves of a periodic system at six different unit external loading with subsequent evaluation of effective elastic moduli of CM.

An alternative approach taking into account interfacial effect is based on the concept of surface stress and surface tension which can be clarified by the additional relation between interface/surface stress tensor, the deformation-dependent surface energy, and the deformation-independent surface/interfacial tension (see [488]). Sharma and Ganti [990] (see also [283]) have extended the Eshelby's original formalism for ellipsoidal inclusion to nanoinclusion and reached an interesting conclusion in that only inclusions with a constant curvature admit a uniform elastic state, thus restricting this remarkable property only to spherical and cylindrical inclusions. A similar problem for a spherical inhomogeneity inside an elastic half-space was considered in [762]. Another generalization of the Eshelby formalism proposed in [723] is based on the higher order continuum theories of elasticity and Cosserat's pseudocontinuum theory to construct the continuum theory with scale effects. A particular model is considered that may serve as a basis for investigations of the scale effects due to cohesion interactions.

One feature characteristic of fiber nanocomposites is that the reinforcing NTs are not straight but rather have curvature or waviness varying through the composites ([108], [1140], [905]); see also helical NT in [907], [997]. This is partially due to the low bending stiffness of nanotubes of the small diameter (~1 nm) and small elastic Young's modulus in the transversal direction. It is reasonable to assume that the waviness reduces the effectiveness of the NT reinforcement of the polymer at least in the fiber orientation direction. The estimation method of $\mathbf{L}^*$ was proposed in [108] by the use of Mori-Tanaka approach. In the case of isotropic fibers being assumed, the problem was reduced to the estimation of average strain concentrator factor for an individual waved fiber in the infinite matrix. This model is invariant with respect to the shape of a correlation hole v_i^0 (simply because $\mathbf{L}^*$ doesn't depend on v_i^0) and cannot be used for analysis of anisotropic fibers even for their similar orientation (because in such a case $\langle \boldsymbol{\eta} \rangle_i \neq \langle \mathbf{L}_1 V_i \rangle_{(i)} \langle \boldsymbol{\varepsilon} \rangle_i$). As shown in Chapter 8, the effective properties $\mathbf{L}^*$ are defined by an average concentrator of stress polarization factor in the fiber rather than average strain concentrator factor. The correct method of estimation of the fictitious properties (18.18) is based on the matching of a fictitious inclusion with the same elastic response describe by the tensor $\mathbf{R}^{(k)}$ as a real one. It fundamentally differs from the choice of $\mathbf{L}^{f(k)}$ proposed in [342] and based on the use of the mixture rule only for estimation of the fictitious Young modulus in the direction of the fiber orientation. Moreover, the effective properties estimated for isotropic waved fiber in [108] via Eq. (18.18) are anisotropic and equivalent. Therefore, a correctness of their exploiting for the estimation of effective properties $\mathbf{L}^*$ of the NCs with an arbitrary prescribed random orientation of waved nanofibers by the MTM is questionable.

Although the FEA approach is quite powerful for the mentioned problems, the analytical models are worthy of more attention. The elegant analytical models [108] and [997], recently proposed for the fibers of small curvature (both the waved and helical shapes), are based on an assumption that the average stress concentrator factor in each section of NT locally approximated by a straight infinite cylinder approaches that of an infinitely long straight NT oriented at the same angle and embedded in an infinite matrix. In this connection, the generalized iteration version of the MEFM is noteworthy. The generalized version takes into account inhomogeneity of the effective fields in the longitudinal fiber direction and its homogeneity over the transversal direction (this idea is initiated by both the numerical results of Subsection 18.3.2 and by the binary model [1204]); the last reduces the number of estimated variables. This approach is based on the numerical solution of an appropriate singular integral equation using a new subtraction technique (see for references on general description, e.g., [268] and Chapter 4) eliminating singularities by exploiting the solutions [108] and [997]. Substitution of nonlocal solutions for one and two fibers subjected to the inhomogeneous effective fields into the general integral equations for microinhomogeneous media will allow one to obtain implicit representation for the nonlocal integral operators of the effective elastic properties and stress concentrator factor.

It was demonstrated [1040] that the large amount of CNTs is concentrated in agglomerates if the average volume concentration of CNTs in a sample reaches 7.5%. Agglomeration means statistical inhomogeneity of local volume concentration of CNTs when some local regions with a higher concentration of CNTs than the average volume fraction in the material take place. These regions with concentrated CNTs called clusters are usually assumed to be an ellipsoidal domain and cut out from the infinite statistically homogeneous composite sample (see also Chapter 12). Shi *et al.* [997] used a known approach (see, e.g., analysis of fiber clustering in conventional composites in [553]) for replacement of these clusters by the fictitious homogeneous inclusions of the same size and the properties estimated as the effective properties for an appropriate infinite statistically homogeneous medium. Such an approach is correct in the limiting case of an infinite difference of scales of both the cluster sizes and the fiber lengths. The reason for an error introduced in the range of comparable mentioned scales was explained in Subsection 18.4.1. An isolated SWNT bundle without matrix was analyzed [694] via a hybrid atom/continuum model, in which the intratube interaction energy is calculated using the MD-based continuum approach while the intertube interaction energy is modeled by a usual Lennard-Jones type potential. Linearized bulk elastic properties of the SWNT bundle with respect to a stable configuration are transversely isotropic and characterized by five independent elastic moduli estimated in this chapter. The prescribed approach of cluster analysis is obviously associated with the second type of "cut-out" modeling of clustered materials considered in Subsection 18.4.1 accompanied with an explanation of their approximative nature. Deterministic cluster structure of nanoropes is best suited for the application of the cluster model considered in Subsection 18.4.2 with obvious replacement of both the arrangement and shape (from oblate spheroidal shape to prolate spheroidal one) of nanoelements in the cluster. An additional challenging issue of nanotechnology considered in [906] is that the nanorope contains not parallel NTs, but rather twisted or woven NTs leading to much better load transfer mechanism in tension than a straight bundle would have. It would be interesting to analyze this problem by the approximate method proposed in the previous paragraph for the investigation of individual waved fiber with incorporation of a solution for a single fiber into the rope problem with subsequent insertion of the solution for an individual rope into the MEFM for the estimation of effective properties of NCs.

A degenerate case of clustered materials is the functionally graded materials (FGMs) where the 1- and n-point correlation functions are the deterministic function of coordinates. Usually FGMs are associated with CM where the volume concentration of inclusions varies through the space. However, in the simple case, the fiber volume concentration is a constant $c(\mathbf{x}) \equiv$ const. but ODF $f(\mathbf{g}) \equiv f(\mathbf{g}, \mathbf{x}) \neq$ const. Even in such a case, the problem has the same level of sophisticated mathematics than the general case of statistically inhomogeneous materials with the ergodicity violation and the difference between the ensemble and volume averages. An involved nonlocal integral equation of the MEFM can be solved by an iteration method with the initial approximation coinciding with an appropriate solution for statistically homogeneous media. The generalized version of the MEFM takes into account inhomogeneity of the effective fields in the longitudinal fiber direction and its homogeneity over the transversal direction. The prescribed approach is a generalization of a known approach where one assumes a simplified microstructure characterized by groups of aligned fibers or clay flakes interacting with other aligned groups, although alignment between groups is not necessarily the same (see [335]). This two-step model consists of the first step of estimation of effective properties for a statistically homogeneous medium that can be used as an initial approximation in the solution of an appropriate nonlocal integral equation. Another limiting case of fiber FGMs is CM reinforced by the aligned fibers with the volume fiber concentration described by a step function that is a feature of CM reinforced by the aligned carbon nanotube forests (see [904]). This case is more simple because all correlation holes v_i^0 are parallel and effective fields $\overline{\boldsymbol{\sigma}}_i(\mathbf{y})$ $(i = 1, 2, \ldots)$ varying along axial fiber direction are identical for all fibers (at least in the case of Lax [649] approximation).

A high aspect ratio of CNT and extraordinary mechanical properties (strength and flexibility) make them ideal reinforcing fibers in nanocomposites, which can produce advanced nanocomposites with improved stiffness and strength. These extraordinary properties of nanoelements are specified by their periodic structure which is not, however, perfect forever and can contain some imperfections (such as defects, see e.g. [47], [206], [1034]). It was found [998] that the critical strain of defect nucleation of a CNT is sensitive to its chiral angle but not to its diameter, and that with the increase in the Young's modulus of matrix, the critical strain of CNTs decreases. Because of this, the development of the theory of homogenization of initially periodic structures with the introduced infinite number of imperfections of unit cells randomly distributed in space has an essentially practical meaning (see also Chapter 10 and [146]).

Thus the problems mentioned above have a technical rather than a fundamental character. The MEFM appears to be simple enough in both theoretical and computational aspects and provides the calculation with reasonable accuracy of the effective properties of composites reinforced by either the randomly oriented nanofibers or clustered nanoplates for a whole range of parameters.

19

Conclusion. Critical Analysis of Some Basic Concepts of Micromechanics

T is strange,- but true, for Truth is always strange-
Stranger than fiction ...

-Lord Byron. *Don Juan.*

If you can look into the seeds of time,
And say which grain will grow and which will not,
Speak then to me ...

-William Shakespeare. *Macbeth.*

Let us discuss the main scheme as well as the short sketch of limitations and ideas as well as possible generalizations of some basic concepts of the linear version of the MEFM and some related methods (see Chapters 7-14). This sketch does not pretend to be rigorous and may contain controversial statements, too personal or one-sided arguments which are deliberately presented with the aim to give free reins to our imagination.

Let us consider the basic equations of linear elasticity of composites in micropoint (see Chapter 8 for details):

$$\nabla\boldsymbol{\sigma} = \mathbf{0}, \qquad \boldsymbol{\sigma}(\mathbf{x}) = \mathbf{L}(\mathbf{x})\boldsymbol{\varepsilon}(\mathbf{x}), \qquad \boldsymbol{\varepsilon}(\mathbf{x}) = \text{Def}\mathbf{u}(\mathbf{x}). \tag{19.1}$$

Substituting (19.1_2) and (19.1_3) into the equilibrium equation (19.1_1) we obtain a differential equation:

$$\nabla\mathbf{L}^c\nabla\mathbf{u}(\mathbf{x}) = -\nabla\mathbf{L}_1(\mathbf{x})\nabla\mathbf{u}(\mathbf{x}) \tag{19.2}$$

with the constant coefficients in the left-hand side that allows the general integral equation (7.19)

$$\boldsymbol{\varepsilon}(\mathbf{x}) = \langle\boldsymbol{\varepsilon}\rangle(\mathbf{x}) + \int \mathbf{U}^c(\mathbf{x}-\mathbf{y})[\boldsymbol{\tau}(\mathbf{y}) - \langle\boldsymbol{\tau}\rangle]d\mathbf{y}. \tag{19.3}$$

where the tensor $\mathbf{U}^c = \nabla\nabla\mathbf{G}^c$ is defined by the Green function $\mathbf{G}^c$ (3.3) for the infinite homogeneous medium with an elastic modulus $\mathbf{L}^c$.

Equation (19.3) can be rearranged in the form of an infinite system of integral equations (8.8)–(8.11) that can be solved by the different approximate methods considered in Chapters 8 and 9. The main hypothesis of many micromechanical methods called effective field hypothesis **H1)** (8.14) and (8.15) allows the infinite system for integral equations with respect the smooth effective fields for any shape and structure of inclusions [see Eqs. (8.18) and (8.19)]. However, a generality of the last statements is an illusion because arbitrariness of the inclusion shape does not guarantee a protection from the necessity of using of Eshelby tensor in the MEFM. Indeed, the notion Eshelby tensor has a fundamental conceptual sense rather than only an analytical solution of some particular problem. Exploiting the Eshelby tensor concept in the MEFM is based on the ellipsoidal shape of the correlation hole v_{ij}^0 rather than on the inclusion shape v_i. An abandonment of the assumption of the v_{ij}^0's ellipsoidal shape leads of necessity to the inhomogeneity of the effective field $\overline{\boldsymbol{\sigma}}_i$ acting on the inclusion v_i that is prohibited for the classical version of the MEFM (see Sections 9.4).

For statistically homogeneous medium subjected to homogeneous boundary conditions, two-particle approximation of a closing hypothesis **H2** (8.39):

$$\langle \widetilde{\boldsymbol{\sigma}}(\mathbf{x})_{1,2}\rangle_i = \langle \overline{\boldsymbol{\sigma}}(\mathbf{x})\rangle_i = \text{const.} \quad (i = 1, 2). \tag{19.4}$$

reduces this system to the system of algebraic Eqs. (8.40). Further simplification is accomplished by "quasi-crystalline" approximation by Lax (8.65):

$$\langle \overline{\boldsymbol{\sigma}}_i(\mathbf{x})|v_i, \mathbf{x}_i; v_j, \mathbf{x}_j\rangle = \langle \overline{\boldsymbol{\sigma}}_i\rangle, \quad \mathbf{x} \in v_i \tag{19.5}$$

which is weaker than Mori-Tanaka assumption (8.90):

$$\overline{\boldsymbol{\sigma}}_i \equiv \langle \boldsymbol{\sigma}\rangle^{(0)}. \tag{19.6}$$

The Mori-Tanaka hypothesis (19.6) is probably the most popular hypothesis of micromechanics which at the same time exerted most deleterious impact on the development of this subject. Harmful effect is explained not only by the loss of the effective modulus symmetry (see Subsections 8.3.3 and Section 18.3) in some general cases but, that is crucial, by self-closing nature of the hypothesis (19.6). The seeming simplicity and illusory assuredness hide the ways of generalization of this hypothesis while an extension of "quasi-crystalline" approximation (19.5) to the closing hypothesis (19.4) is obvious.

Summarizing what we said above, we will present now the contractions of some concepts and assumptions erroneously recognized in micromechanics as basic ones: a) constitutive equation, b) homogeneous comparison medium, c) Green function, d) Eshelby tensor, e) effective field hypothesis **H1**.

The fallacy in these recognitions will be justified in an inverse order.

Indeed, we demonstrated in the book two methods of analyses of the inhomogeneous effective fields $\overline{\boldsymbol{\sigma}}_i$. In the first case of statistically homogeneous medium subjected to the homogeneous boundary conditions, the inhomogeneity of $\overline{\boldsymbol{\sigma}}_i$ takes place if the correlation hole v_{ij}^0 is nonellipsoidal that leads to the problem of the estimation of stress distribution in the domain v_{ij}^0 undergoing to the constant eigenstrains $\boldsymbol{\beta}_1(\mathbf{x}) \equiv \text{const}$, $\mathbf{L}_1(\mathbf{x}) \equiv \mathbf{0}$ ($\mathbf{x} \in v_{ij}^0$) (see Subsections 4.2.3 and Section 9.4). A more general second method applicable for the analysis of both statistically homogeneous and statistically inhomogeneous fields in both the infinite and bounded media, and for any shape of v_{ij}^0 is reduced to the estimation of perturbations produced by the inclusion v_j in both inside and outside in some vicinity of the inclusion v_j (see Eq. (14.63)). In so doing these perturbations evaluated by any either numerical or analytical method available for a researcher are reduced to the classical Eshelby internal and external tensors only in some particular simplest cases.

Thus the popular effective field hypothesis **H1** is not fundamental and can be easily avoided if it is known the numerical solution just for one inclusion obtained by any numerical method available for researcher. Known numerical methods such as FEA, VIE, BIE, and the complex potentials, which can be used for micromechanical analysis, have a series of advantages and disadvantages. It is crucial for the analyst to be aware of their range of applications. This solution only in simplest cases is expressed through the Green function and Eshelby tensor which are not, in general, necessary tools in micromechanics. In the case of taking of multiparticle interactions into account, a solution for a single inclusion should be complemented by a solution for n interacting inclusions in the comparison medium that also can be performed without the notions Green function and Eshelby tensor (see Chapters 10 and 14) by the use of operator representations of these solutions. In so doing, the existence of the mentioned operators was justified through the Green function's technique.

The challenge of modern micromechanics is a development of the general methodology incorporating the solution for multiple interacting inhomogeneities obtained by highly accurate numerical methods into the most general scheme of analytical micromechanics (see Chapters 10-12, and 14). A fundamental difference between the proposed methodology and the ones published earlier is a systematic analysis of statistical distributions $\langle\boldsymbol{\sigma}(\otimes\boldsymbol{\sigma})^n\rangle^{(k)}(\mathbf{x})$ ($n=0,1,\ldots;\ k=0,1,\ldots,N;\ \mathbf{x}\in v^{(k)}$) of local microfields rather than only effective properties based on the average fields inside the phases $\langle\boldsymbol{\sigma}\rangle^{(k)}$. Moreover, this approach allows us to estimate the conditional moments of stresses inside the fixed inclusions v_i: $\langle\boldsymbol{\sigma}(\otimes\boldsymbol{\sigma})^n\rangle_i(\mathbf{x}^l)$ ($\mathbf{x}^l\in v_i\subset v^{(k)}$), which are more informative than $\langle\boldsymbol{\sigma}(\otimes\boldsymbol{\sigma})^n\rangle^{(k)}(\mathbf{x})$ ($\mathbf{x}\in v^{(k)}$) (see Chapter 14). Obviously, the analysis of nonlinear phenomena such as plasticity, creep, and damage should use precisely $\langle\boldsymbol{\sigma}(\otimes\boldsymbol{\sigma})^n\rangle_i(\mathbf{x}^l)$ ($\mathbf{x}^l\in v_i\subset v^{(k)}$), rather than $\langle\boldsymbol{\sigma}(\otimes\boldsymbol{\sigma})^n\rangle^{(k)}(\mathbf{x})$ ($\mathbf{x}\in v^{(k)}$) (see Chapters 15 and 16), which is a too rough descriptor of the stress distribution especially in the case of statistically inhomogeneous media in bounded domains (see Chapters 10, 13, and 14).

Furthermore, popular assumptions about homogeneity of the comparison medium and their unboundedness (19.2) were in reality accepted exclusively for introduction of a much more powerful tool of micromechanical research such as Green's function; see Eq. (19.3). However, the eventual abandonment of the Green function concept (see Chapters 10, 11, and 14) removed any restrictions on the comparison medium in the case of linear elastic problems. A comparison medium can either coincide with the whole space or be bounded by any surface (either simple connected or multiply connected). Elastic properties and eigenstrains of the comparison medium can be either homogeneous or inhomogeneous. A single restriction on the inhomogeneity of a comparison medium is a deterministic nature of this inhomogeneity. So, a comparison medium can have either continuously varying elastic moduli or piecewise constant elastic properties such as e.g. periodic structure (see Chapter 11) or two joint elastic half spaces. The bounded or unbounded medium with one or a few fixed macroinclusions (e.g. macrocracks) can also be considered as a comparison medium. However, at any rate, we need only to know the numerical solutions for the mentioned comparison medium containing one and a few interacting micro inclusions with any their possible location in the comparison medium.

At last, must we really know a local constitutive law (19.1_2) if we are going to estimate the effective elastic moduli of composite materials? At first glance this question is beyond the scope of common sense. However, despite the apparent paradoxicality, the correct answer to this question is no, in the following sense. Indeed, it was demonstrated in Section 18.2 that we don't need to know anything about either local or nonlocal nature of constitutive laws of inclusions. We only need to know that these laws are linear and also need to know the concentrator factors $\mathbf{A}_\sigma(\mathbf{x})$ and $\mathbf{B}_\sigma(\mathbf{x})$ (18.10) (which can be found directly, e.g., from MD simulations) at least for one inclusion in a comparison medium mentioned above.

Shortening what we said above, we can conclude that for linear elastic problems for microinhomogeneous medium of random structure:

a) we do not need to know a constitutive law for the inclusions,
b) we do not need to assume the homogeneity and infiniteness of a comparison medium,
c) we do not need to know the Green tensor and Eshelby tensor, and
d) we do not need to assume homogeneity of the effective fields $\overline{\boldsymbol{\sigma}}(\mathbf{x})$.

What we really need is the numerical solutions for one and a few interacting inclusions with any their possible location in the comparison medium described above.

Thus, in the framework of a unique scheme of the proposed MEFM, we have undertaken in this book an attempt to analyze the wide class of statical and dynamical, local and nonlocal, linear and nonlinear multiscale problems of composite materials with deterministic (periodic and nonperiodic), random (statistically homogeneous and inhomogeneous, so-called graded) and mixed (periodic structures with random imperfections) structures in bounded and unbounded domains, containing coated or uncoated inclusions of any shape and orientation and subjected to coupled or uncoupled, homogeneous or inhomogeneous external fields of different physical natures.

A

Appendix

A.1 Parametric Representation of Rotation Matrix

Three parametric representations of the rotation tensor $\mathbf{g} \in SO(3)$ (2.6) have been widely used for the description of the local system. In the first case, let the matrix $\mathbf{g} = \|g_{ij}\|$, $(i, j = 1, \ldots, 3)$ be the rotation matrix for coordinates $\mathbf{x}'$ of an arbitrary point in the reference frame orthonormalized $\mathbf{e}'$ $(\mathbf{e}'_i \cdot \mathbf{e}'_j = \delta_{ij})$ expressed by means of the coordinates $\mathbf{x}$ in the crystal orthonormalized coordinate system $\mathbf{e}$ $(\mathbf{e}_i \cdot \mathbf{e}_j = \delta_{ij})$ with the same origin of coordinates:

$$\mathbf{x}' = \mathbf{g}\mathbf{x}. \tag{A.1.1}$$

We require that the orthonormalized basis is right-handedness $\mathbf{e}_1 \cdot \mathbf{e}_2 \otimes \mathbf{e}_3 = 1$ and $\mathbf{e}_i \cdot \mathbf{e}_i = \delta_{ij}$ $(\mathbf{e}_i \cdot \mathbf{e}_i = \delta_{ij})$, where $\mathbf{e}_i \cdot \mathbf{e}_j$ $(\mathbf{e}_i \cdot \mathbf{e}_j)$ denotes the projection of the ith basic vector on the jth one (or vice versa). The expansion of the new basis $\mathbf{e}_1, \mathbf{e}_2, \mathbf{e}_3$ into the vectors of the old one $\mathbf{e}'_{i'} = g_{i'j}\mathbf{e}_j$ is defined by coefficients $g_{i'j}$ called a *cosine matrix*, because each element $g_{i'j}$ is the cosine of the angle between two corresponding axes $g_{i'j} = \mathbf{e}'_i \cdot \mathbf{e}_j = \cos\alpha_{i'j}$. Since the rotation does not alter the length or angles, it leaves the scalar production of any two vectors invariant. That is, if $\mathbf{x}' = \mathbf{g}\mathbf{x}$ and $\mathbf{y}' = \mathbf{g}\mathbf{x}$, then the substitution of Eq. (A.1.1) into the left-hand side of the equality $(\mathbf{x}', \mathbf{y}') = (\mathbf{x}, \mathbf{y})$ leads to $\sum_{i,k,l} g_{ik}g_{il}x_k x_l = \sum_k x_k y_k$. Equating the coefficient of $x_k y_l$ yields the equality $\mathbf{g}^\top\mathbf{g} = \boldsymbol{\delta}$ meaning the *orthogonality* of the matrix $\mathbf{g}$: $\mathbf{g}^\top = \mathbf{g}^{-1}$. Thus, the matrix $\|g_{ji'}\|$ is simultaneously both the inverse of the initial matrix $\|g_{i'j}\|$ and its *transpose*. Relation $\mathbf{g}^\top\mathbf{g} = \boldsymbol{\delta}$ provides six independent constant conditions on the components of $\mathbf{g}$ such that only three components of $\mathbf{g}$ are independent. The set of all second-order orthogonal tensors form a grope under the inner product called the three-dimensional orthogonal group $O(3)$. The orthogonal transformation can be divided into proper rotation (if $\Delta \equiv \det \|g_{i'j}\| = 1$, which form the special orthogonal group $SO(3) \subset O(3)$) and improper rotation, or rotation with reflection (if $\Delta = -1$). The unity element of the group $SO(3)$ in the dyadic representation is $\boldsymbol{\delta} = \delta_{ij}\mathbf{e}_i \otimes \mathbf{e}_j$.

The second one is the axis-angle parameterization of the rotation tensor $\mathbf{g} \in SO(3)$ specified by an axis of rotation $\mathbf{n}$ $(|\mathbf{n}| = 1)$ and the angle of rotation,

ω, taken in the right-handed sense about $\mathbf{n}$ (the interested reader is referred to [1031]. Then the components of $\mathbf{g}$ are completely described by the vector $\omega\mathbf{n}$:

$$g_{ij}(\omega\mathbf{n}) = \delta_{ij}\cos\omega - \varepsilon_{ijk}n_k\sin\omega + (1-\cos\omega)n_in_j, \tag{A.1.2}$$

where ε_{ijk} are the components of a permutation tensor. A particularly useful class of such axis-angle parameterizations, the neo-Eulerian (see, e.g., [616]), are obtained by scaling the axis of rotation by a function of the angle $\omega\mathbf{n} \to f(\omega)\mathbf{n}$ which includes the Rodrigues' vector $\boldsymbol{\rho} = \tan(\omega/2)\mathbf{n}$ with the property that rotations about a fixed axis trace straight lines in the parametric space. In the case of all possible rotations, the direction of $\mathbf{n}$ is restricted to the surface of unit hemisphere centered on the origin of $\mathbf{n}$: $-\pi < \omega \leq \pi$. Points lying in this "ball" of parameters (called π sphere) specify all physically possible $\omega\mathbf{n}$ rotations. In so doing, the rotation vectors $\pm\pi\mathbf{n}$ have identical physical meaning. If θ and ψ are the ordinary spherical coordinates of the vector $\mathbf{n}$, then $\mathbf{n} = \sin\theta\cos\psi\mathbf{e}_1 + \sin\theta\sin\psi\mathbf{e}_2 + \cos\theta\mathbf{e}_3$, with the range of allowed variables $0 \leq \theta \leq \pi$ and $0 \leq \psi < 2\pi$, and

$$\mathbf{g} = \begin{pmatrix} mpr - ns & ms + npr & qr \\ -mps - nr & mr - nps & -qs \\ -mq & -nq & p \end{pmatrix}, \tag{A.1.3}$$

where $m = \cos\theta$, $n = \sin\theta$, $p = \cos\psi$, $q = \sin\psi$, $r = \cos\omega$, and $s = \sin\omega$. The orthogonal tensor $\mathbf{g}(\omega\mathbf{n})$ can also be represented by the formula (see, e.g., [411], [758]) $\mathbf{g}(\omega\mathbf{n}) = \boldsymbol{\delta} + \sin\omega\mathbf{N} + (1-\cos\omega)\mathbf{N}^2 = e^{\omega\mathbf{P}}$, where the three-dimensional skew-symmetric tensor $\mathbf{N}$ with components $N_{ij} = -\epsilon_{ijk}n_k$ is introduced to represent the unit vector $\mathbf{n}$, and the exponential representation $e^{\omega\mathbf{N}} \equiv \sum_{k=1}^{\infty}\mathbf{N}^k/k!$ is widely used in computational solid mechanics (see [758]).

However, we will usually use the third parameterization for the rotation matrix $\mathbf{g} \in SO(3)$ exploring the Euler angles ϕ_1, θ, ϕ_2 (see e.g. [124]) with the range of allowed parameters constituting a rectangular box with two of its bounded surfaces excluded $0 \leq \phi_1 < 2\pi$, $0 \leq \phi_2 < 2\pi$, $0 \leq \theta \leq \pi$. Let the coordinate axes (Ox, Oy, Oz) are rotated into the axes (Ox', Oy', Oz') (initially coinciding one with another) using the matrix $\mathbf{g}(\phi_1, \theta, \phi_2)$, which can be presented in the form of a product of three successive rotations: the rotation $\mathbf{g}(\phi_1, 0, 0)$ through the angle ϕ_1 about the axis Oz (the axes (Ox, Oy, Oz)transform into the axes $(O\bar{x}, O\bar{y}, Oz)$ the rotation $g(0, \theta, 0)$ to the angle θ around the axis $O\bar{x}$ (the axes $(O\bar{x}, O\bar{y}, Oz)$ transform into the axes $O\bar{x}, Oy'', Oz'$), and the rotation $\mathbf{g}(0, 0, \phi_2)$ to the angle ϕ_2 ones again about the axis Oz' (axes $(O\bar{x}, Oy'', Oz')$ transform into the axes (Ox', Oy', Oz')):

$$\mathbf{g}(\phi_1, 0, 0) = \begin{pmatrix} c_2 & s_2 & 0 \\ -s_2 & c_2 & 0 \\ 0 & 0 & 1 \end{pmatrix}, \quad \mathbf{g}(0, \theta, 0) = \begin{pmatrix} 1 & 0 & 0 \\ 0 & c_1 & s_1 \\ 0 & -s_1 & c_1 \end{pmatrix}, \quad \mathbf{g}(0, 0, \phi_2) = \begin{pmatrix} c_3 & s_3 & 0 \\ -s_3 & c_3 & 0 \\ 0 & 0 & 1 \end{pmatrix},$$

where $c_1 = \cos(\theta)$, $s_1 = \sin(\theta)$, $c_2 = \cos(\phi_1)$, $s_2 = \sin(\phi_1), c_3 = \cos(\phi_2)$, $s_3 = \sin(\phi_2)$. Then an arbitrary matrix $\mathbf{g}(\phi_1, \theta, \phi_2)$ can be express as a multiplication of the matrices corresponding to the Euler rotations with the angles ϕ_1, θ, ϕ_2: $\mathbf{g}(\phi_1, \theta, \phi_2) = \mathbf{g}(0, 0, \phi_2)\mathbf{g}(0, \theta, 0)\mathbf{g}(\phi_1, 0, 0)$, or, in more detail,

$$\mathbf{g}=\begin{pmatrix} c_2c_3-c_1s_2s_3 & s_2c_3+c_1c_2s_3 & s_1s_3 \\ -c_2s_3-c_1s_2c_3 & -s_2s_3+c_1c_2c_3 & s_1c_3 \\ s_1s_2 & -s_1c_2 & c_1 \end{pmatrix}. \tag{A.1.4}$$

The matrix $\mathbf{g}(\phi_1,\theta,\phi_2)$ can be also determined in the term of the matrices (A.1.2): $\mathbf{g}(\phi_1,\theta,\phi_2) = g(\phi_1\mathbf{e}_1)g(\theta\mathbf{e}_3)g(\phi_2\mathbf{e}_1)$ when the unit vector $\mathbf{e}_3 = \mathbf{e}_3(\theta,\psi)$, $\psi = \pi/2-\phi_1$, specifies points on the rotation unit sphere S^2. If we substitute $\phi_1 \to \pi-\phi_2$ and $\phi_2 \to \pi-\phi_1$ in the matrix $\mathbf{g}$ (A.1.4), we obtain its transpose, which coincides with the inverse matrix according to (A.1.4). Therefore, the rotation matrix $\mathbf{g}(\phi_1,\theta,\phi_2)$ has the inverse matrix $\mathbf{g}(\pi-\phi_2,\theta,\pi-\phi_1)$.

Thus, if the angle between the new and old axes is known, Eq. (A.1.1) is used, if the axis and the angle of rotation are known, Eq. (A.1.2) is exploited, and if the Eulerian angles are known, Eq. (A.1.4) is used.

A.2 Second and Fourth-Order Tensors of Special Structures

In a number of micromechanical problems, one uses the tensors of a special kind depending on the Kronecker delta and on a unit vector. The expansion of these tensors in a special basis allows us to simplify the algebraic operations with ones such as a multiplication and an inversion (see [544], [618], [986], [1152]).

A.2.1 E-basis

Let us consider the fourth-order tensors made up fro the unit vector $\mathbf{n}$ and the Kronecker deltas $\boldsymbol{\delta}$. These tensors are assumed to be symmetric in the first and the second pairs of indexes but, in general, nonsymmetric with respect to permutation of the pairs. Al tensors of this structure can be represented as a linear combination of six linear independent basis tensors

$$\begin{aligned} &E^{(1)}_{ijkl} = \delta_{i(k}\delta_{l)j}, \quad E^{(2)}_{ijkl} = \delta_{ij}\delta_{kl}, \quad E^{(3)}_{ijkl} = \delta_{ij}n_k n_l, \\ &E^{(4)}_{ijkl} = n_i n_j \delta_{kl}, \quad E^{(5)}_{ijkl} = n_{i)}n_{(l}\delta_{k)(j}, \quad E^{(6)}_{ijkl} = n_i n_j n_k n_l. \end{aligned} \tag{A.2.1}$$

The tensors $\mathbf{E}^{(k)}$ can be represented in matrix form

$$\mathbf{E}^{(1)} = \begin{pmatrix} 1&0&0&0&0&0 \\ 0&1&0&0&0&0 \\ 0&0&1&0&0&0 \\ 0&0&0&\frac{1}{2}&0&0 \\ 0&0&0&0&\frac{1}{2}&0 \\ 0&0&0&0&0&\frac{1}{2} \end{pmatrix}, \quad \mathbf{E}^{(2)} = \begin{pmatrix} 1&1&1&0&0&0 \\ 1&1&1&0&0&0 \\ 1&1&1&0&0&0 \\ 0&0&0&0&0&0 \\ 0&0&0&0&0&0 \\ 0&0&0&0&0&0 \end{pmatrix},$$

$$\mathbf{E}^{(3)} = \begin{pmatrix} n_1^2 & n_2^2 & n_3^2 & n_1n_2 & n_1n_3 & n_2n_3 \\ n_1^2 & n_2^2 & n_3^2 & n_1n_2 & n_1n_3 & n_2n_3 \\ n_1^2 & n_2^2 & n_3^2 & n_1n_2 & n_1n_3 & n_2n_3 \\ 0&0&0&0&0&0 \\ 0&0&0&0&0&0 \\ 0&0&0&0&0&0 \end{pmatrix}, \mathbf{E}^{(4)} = \begin{pmatrix} n_1^2 & n_2^2 & n_3^2 & 0&0&0 \\ n_1^2 & n_2^2 & n_3^2 & 0&0&0 \\ n_1^2 & n_2^2 & n_3^2 & 0&0&0 \\ n_1n_2 & n_1n_2 & n_1n_2 & 0&0&0 \\ n_1n_3 & n_1n_3 & n_1n_3 & 0&0&0 \\ n_2n_3 & n_2n_3 & n_2n_3 & 0&0&0 \end{pmatrix},$$

$$\mathbf{E}^{(5)} = \frac{1}{4}\begin{pmatrix} 4n_1^2 & 0 & 0 & 2n_1n_2 & 2n_1n_3 & 0 \\ 0 & 4n_2^2 & 0 & 2n_1n_2 & 0 & 2n_2n_3 \\ 0 & 0 & 4n_3^2 & 0 & 2n_1n_3 & 2n_2n_3 \\ 2n_1n_2 & 2n_1n_2 & 0 & n_1^2+n_2^2 & n_2n_3 & n_1n_3 \\ 2n_1n_3 & 0 & 2n_1n_3 & n_2n_3 & n_1^2+n_3^2 & n_1n_2 \\ 0 & 2n_2n_3 & 2n_2n_3 & n_1n_3 & n_1n_2 & n_2^2+n_2^3 \end{pmatrix},$$

$$\mathbf{E}^{(6)} = \begin{pmatrix} n_1^4 & n_1^2n_2^2 & n_1^2n_3^2 & n_1^3n_2 & n_1^3n_3 & n_1^2n_2n_3 \\ n_1^2n_2^2 & n_2^4 & n_2^2n_3^2 & n_1n_2^3 & n_1n_2^2n_3 & n_2^3n_3 \\ n_1^2n_3^2 & n_2^2n_3^2 & n_3^4 & n_1n_2n_3^2 & n_1n_3^3 & n_2n_3^3 \\ n_1^3n_2 & n_1n_3^3 & n_1n_2n_3^2 & n_1^2n_2^2 & n_1^2n_2n_3 & n_1n_2^2n_3 \\ n_1^3n_3 & n_1n_2^2n_3 & n_1^3n_3 & n_1^2n_2n_3 & n_1^2n_3^2 & n_1n_2n_3^2 \\ n_1^2n_2n_3 & n_2^2n_3 & n_2n_3^3 & n_1n_2^2n_3 & n_1n_2n_3^2 & n_2^2n_3^2 \end{pmatrix}, \qquad \text{(A.2.2)}$$

In particular, the linear space of the isotropic tensors symmetric with respect to permutation of the first and the second pairs is defined only by the two tensors $\mathbf{E}^{(1)}$ and $\mathbf{E}^{(2)}$. The basis tensors $\mathbf{E}^{(1)}$ and $\mathbf{E}^{(2)}$ forms the orthogonal basis $\mathbf{N}_1 = \mathbf{E}_2/3$, and $\mathbf{N}_2 = \mathbf{E}_1 - \mathbf{E}_2/3$ (2.17) in which the arbitrary isotropic tensor of fourth-order can be decomposed as in Eq. (2.197).

Let us define the operation of multiplication (contraction over two indices):

$$(\mathbf{E}^{(\alpha)} : \mathbf{E}^{(\beta)})_{ijkl} \equiv \mathbf{E}^{(\alpha)}_{ijpq} E^{(\beta)}_{pqkl}. \qquad \text{(A.2.3)}$$

The linear tensor space, spanned by the basis $\mathbf{E}^{\alpha}$ ($\alpha = 1, \ldots, 6$) is closed with respect to the operation of multiplication (A.2.3), and therefore forms a closed algebra where the inner product (A.2.3)

$$(\mathbf{E}^{(\alpha)} : \mathbf{E}^{(\beta)})_{ijkl} = a^{\alpha\beta}_{\gamma} E^{(\gamma)}_{ijkl} \qquad \text{(A.2.4)}$$

is defined by the matrix $a^{\alpha\beta}_{\gamma}$ (the left column represents the left factor in the lefthand side of Eq. (A.2.4))

Table A.2.1. "Multiplication" matrix $a^{\alpha\beta}_{\gamma}$ of the basic tensors $\mathbf{E}^{(k)}$ ($k = 1, \ldots, 6$)

	$\mathbf{E}^{(1)}$	$\mathbf{E}^{(2)}$	$\mathbf{E}^{(3)}$	$\mathbf{E}^{(4)}$	$\mathbf{E}^{(5)}$	$\mathbf{E}^{(6)}$
$\mathbf{E}^{(1)}$	$\mathbf{E}^{(1)}$	$\mathbf{E}^{(2)}$	$\mathbf{E}^{(3)}$	$\mathbf{E}^{(4)}$	$\mathbf{E}^{(5)}$	$\mathbf{E}^{(6)}$
$\mathbf{E}^{(2)}$	$\mathbf{E}^{(2)}$	$3\mathbf{E}^{(2)}$	$3\mathbf{E}^{(2)}$	$\mathbf{E}^{(2)}$	$\mathbf{E}^{(3)}$	$\mathbf{E}^{(3)}$
$\mathbf{E}^{(3)}$	$\mathbf{E}^{(3)}$	$\mathbf{E}^{(2)}$	$\mathbf{E}^{(3)}$	$\mathbf{E}^{(2)}$	$\mathbf{E}^{(3)}$	$\mathbf{E}^{(3)}$
$\mathbf{E}^{(4)}$	$\mathbf{E}^{(4)}$	$3\mathbf{E}^{(4)}$	$3\mathbf{E}^{(6)}$	$\mathbf{E}^{(4)}$	$\mathbf{E}^{(6)}$	$\mathbf{E}^{(6)}$
$\mathbf{E}^{(5)}$	$\mathbf{E}^{(5)}$	$\mathbf{E}^{(4)}$	$\mathbf{E}^{(6)}$	$\mathbf{E}^{(4)}$	$\frac{1}{2}(\mathbf{E}^{(5)} + \mathbf{E}^{(6)})$	$\mathbf{E}^{(6)}$
$\mathbf{E}^{(6)}$	$\mathbf{E}^{(6)}$	$\mathbf{E}^{(4)}$	$\mathbf{E}^{(6)}$	$\mathbf{E}^{(4)}$	$\mathbf{E}^{(6)}$	$\mathbf{E}^{(6)}$

A.2.2 P-basis

The analysis of the general transversally isotropic tensors for which all directions perpendicular to some unit vector $\mathbf{n}$ are equivalent, we will begin from the second-order tensors. We introduce the basis containing two tensors $\eta_{ij} = n_i n_j$ and $\nu_{ij} \equiv \delta_{ij} - \eta_{ij}$

$$\nu_{ik}\nu_{kj} = \nu_{ij}, \quad \eta_{ik}\eta_{kj} = \eta_{ik}, \quad \nu_{ik}\eta_{kj} = \nu_{jk}\eta_{ki} = 0 \tag{A.2.5}$$

decomposing the Kronecker delta as $\delta_{ij} = \nu_{ij} + \eta_{ij}$. The most general second-order tensor $\mathbf{A}$ reflecting the transverse isotropy is defined by the construction $A_{ij} = a_1\nu_{ij} + a_2\eta_{ij}$. By the use of a straitforward checking, we can confirm that the inverse $\mathbf{A}^{-1}$ of the tensor $\mathbf{A}$ ($\mathbf{A}^{-1}\mathbf{A} = \boldsymbol{\delta}$), as well as the product $\mathbf{AC}$ with the tensor $C_{ij} = c_1\nu_{ij} + c_2\eta_{ij}$ can be found in the closed form

$$(\mathbf{A}^{-1})_{ij} = a_1^{-1}\nu_{ij} + a_2^{-1}\eta_{ij}, \quad (\mathbf{AC})_{ij} = a_1c_1\nu_{ij} + a_2c_2\eta_{ij}. \tag{A.2.6}$$

Let us consider now the tensors of the fourth-order tensors constructed from Kronecker deltas and the unit vectors $\mathbf{n}$, which are symmetric in the first and the second pairs of indices, but not symmetric with respect to permutation of the pairs. The tensors with such a structure can be constructed as the various independent"outer" products of ν_{ij} and n_{kl} and can be represented as a linear combination of the following six linearly independent tensors, which, thus, play the role of a basis:

$$\begin{aligned} &P^{(1)}_{ijkl} = \nu_{ij}\nu_{kl}, \quad P^{(2)}_{ijkl} = \nu_{i(k}\nu_{l)j} - \nu_{ij}\nu_{kl}/2, \quad P^{(3)}_{ijkl} = \nu_{ij}n_kn_l, \\ &P^{(4)}_{ijkl} = n_in_j\nu_{kl}, \quad P^{(5)}_{ijkl} = n_{i)}n_{(k}\nu_{l)(j}, \quad P^{(6)}_{ijkl} = n_in_jn_kn_l. \end{aligned} \tag{A.2.7}$$

The basis tensors $\mathbf{E}^{(k)}$ and $\mathbf{P}^{(k)}$ are related one with another by the relations $\mathbf{P}^{(i)} = a^{ij}\mathbf{E}^{(j)}$, $\mathbf{E}^{(i)} = (a^{-1})^{ij}\mathbf{P}^{(j)}$, with the matrixes

$$a^{ij} = \begin{pmatrix} 0 & 1 & -1 & -1 & 0 & 1 \\ 1 & -\frac{1}{2} & \frac{1}{2} & \frac{1}{2} & -2 & \frac{1}{2} \\ 0 & 0 & 1 & 0 & 0 & -1 \\ 0 & 0 & 0 & 1 & 0 & -1 \\ 0 & 0 & 0 & 0 & 1 & -1 \\ 0 & 0 & 0 & 0 & 0 & 1 \end{pmatrix}, \quad (a^{-1})^{ij} = \begin{pmatrix} \frac{1}{2} & 1 & 0 & 0 & 2 & 1 \\ 1 & 0 & 1 & 1 & 0 & 1 \\ 0 & 0 & 1 & 0 & 0 & 1 \\ 0 & 0 & 0 & 1 & 0 & 1 \\ 0 & 0 & 0 & 0 & 1 & 1 \\ 0 & 0 & 0 & 0 & 0 & 1 \end{pmatrix}. \tag{A.2.8}$$

Six basis tensors (A.2.7) form also a closed algebra with respect to the operation of multiplication described by the matrix $a_\gamma^{\alpha\beta}$ (the left column represents the left factor in the lefthand side of Eq. (A.2.8))

Table A.2.2 "Multiplication" matrix $a_\gamma^{\alpha\beta}$ of the basic tensors $\mathbf{P}^{(k)}$ $(k = 1, \ldots, 6)$

	$\mathbf{P}^{(1)}$	$\mathbf{P}^{(2)}$	$\mathbf{P}^{(3)}$	$\mathbf{P}^{(4)}$	$\mathbf{P}^{(5)}$	$\mathbf{P}^{(6)}$
$\mathbf{P}^{(1)}$	$2\mathbf{P}^{(1)}$	0	$2\mathbf{P}^{(3)}$	0	0	0
$\mathbf{P}^{(2)}$	0	$\mathbf{P}^{(2)}$	0	0	0	0
$\mathbf{P}^{(3)}$	0	0	0	$\mathbf{P}^{(1)}$	0	$\mathbf{P}^{(3)}$
$\mathbf{P}^{(4)}$	$2\mathbf{P}^{(4)}$	0	$2\mathbf{P}^{(6)}$	0	0	0
$\mathbf{P}^{(5)}$	0	0	0	0	$0.5\mathbf{P}^5$	0
$\mathbf{P}^{(6)}$	0	0	0	$\mathbf{P}^{(4)}$	0	$\mathbf{P}^{(6)}$

It should be mentioned that this production operation is not commutative, so $\mathbf{P}^{(3)}\mathbf{P}^{(4)} = \mathbf{P}^{(1)}$ but $\mathbf{P}^{(4)}\mathbf{P}^{(3)} = 2\mathbf{P}^{(6)}$.

The general foruth-order tensor for a transverse isotropy may be constructed as an arbitrayi linear combination of the six elementary tensors

$$\mathbf{X} = \sum_{k=1}^{6} X_k \mathbf{P}^{(k)} \equiv (X_1,\ X_2,\ X_3,\ X_4,\ X_5,\ X_6), \qquad \text{(A.2.9)}$$

where the second equality uses the briefer symbolic notation. Any inner product $(\mathbf{XY})_{ijkl} = X_{ijpq}Y_{pqkl}$ of two tensors $\mathbf{X} = \sum_{k=1}^{6} X_k\mathbf{P}^{(k)}$ and $\mathbf{Y} = \sum_{k=1}^{6} Y_k\mathbf{P}^{(k)}$, which is generally noncommutative, can be calculated readily means of the multiplication Table A.2.1 for the elementary tensors:

$$\mathbf{XY} = (2X_1Y_1 + X_3Y_4, X_2Y_2, 2X_1Y_3 + X_3Y_6, 2X_4Y_1 + X_6Y_4, X_5Y_5/2, X_6Y_6 + 2X_4Y_3). \qquad \text{(A.2.10)}$$

By reducing this product to the unit tensor $\mathbf{XY} = \mathbf{I}$, we may calculate the inverse of $\mathbf{X}$ as

$$\mathbf{X}^{-1} = \Big(\frac{X_6}{2\Delta},\ \frac{1}{X_2},\ -\frac{X_3}{\Delta},\ -\frac{X_4}{\Delta},\ \frac{4}{X_5},\ \frac{2X_1}{\Delta}\Big), \qquad \text{(A.2.11)}$$

where $\Delta = 2(X_1X_6 - X_3X_4)$. In the case, when the tensor X_{ijkl} is symmetric with respect to permutation of the pairs ij and kl, one can show that the inverse tensor $(\mathbf{X}^{-1})_{ijkl}$ posses the same symmetry. However, this property is generally lost in an inner product of the tensors with such symmetry.

In particular, if Ox_3 is the axis of transverse symmetry and, therefore, $\mathbf{n} = (0,0,1)^\top$, the tensors $\mathbf{P}^{(k)}$ $(k = 1,\ldots,6)$ (A.2.7) have the following non-zero tensor components:

$$\begin{aligned}
&P^{(1)}_{1111} = P^{(2)}_{2222} = P^{(1)}_{1122} = P^{(1)}_{2211} = 1, \quad P^{(6)}_{3333} = 1,\\
&P^{(2)}_{1212} = P^{(2)}_{2121} = P^{(2)}_{1221} = P^{(2)}_{2112} = P^{(2)}_{1111} = P^{(2)}_{2222} = \frac{1}{2},\\
&P^{(2)}_{1122} = P^{(2)}_{2211} = -\frac{1}{2}, \quad P^{(3)}_{1133} = P^{(3)}_{2233} = 1, \quad P^{(4)}_{3311} = P^{(4)}_{3322} = 1,\\
&P^{(5)}_{1313} = P^{(5)}_{2323} = P^{(5)}_{1331} = P^{(5)}_{2332} = P^{(5)}_{3131} = P^{(5)}_{3232} = P^{(5)}_{3131} = P^{(5)}_{3232} = \frac{1}{4}.
\end{aligned} \qquad \text{(A.2.12)}$$

In particular, a general form of transversally isotropic fourth-order tensor $\mathbf{A} = (a_1,\ldots,a_6)$ has the following components

$$\mathbf{A} = ((A_{1111} + A_{1122})/2,\ 2A_{1212},\ A_{1133},\ A_{3311},\ 4A_{1313},\ A_{3333}). \qquad \text{(A.2.13)}$$

The isotropic tensors $\mathbf{L} = (L_1,\ldots,L_6)$ and $\mathbf{M} = (M_1,\ldots,M_6)$ of elastic moduli and stiffness have the components $\mathbf{L} = (\lambda+\mu, 2\mu, \lambda, \lambda, 4\mu, \lambda+2\mu)$, and

$$\mathbf{M} = \Big(\frac{(1-\nu)}{4\mu(1+\nu)}, \frac{1}{2\mu}, \frac{-\nu}{2\mu(1+\nu)}, \frac{-\nu}{2\mu(1+\nu)}, \frac{1}{2\mu}, \frac{1}{2\mu(1+\nu)}\Big), \qquad \text{(A.2.14)}$$

respectively, where $\lambda = 2\mu\nu/(1-2\nu)$. Unit fourth rank tensor has representation $\mathbf{I} = (1/2, 1, 0, 0, 2, 1)$.

If we have a uniform distribution of orientations (statistical isotropy) of inclusions with the properties represented through the tensors $\mathbf{P}^{(i)}$ (A.2.7) and $\boldsymbol{\eta}$, $\boldsymbol{\nu}$ (A.2.5), it can be useful the averaged tensors given by

$$\langle \mathbf{P}^{(1)} \rangle = \frac{2}{15}(10\mathbf{N}_1 + \mathbf{N}_2), \quad \langle \mathbf{P}^{(2)} \rangle = \frac{4}{15}(5\mathbf{N}_1 + 2\mathbf{N}_2), \quad \langle \mathbf{P}^{(5)} \rangle = \frac{1}{5}\mathbf{N}_2,$$
$$\langle \mathbf{P}^{(3)} \rangle = \langle \mathbf{P}^{(4)} \rangle = \frac{2}{15}(5\mathbf{N}_1 - \mathbf{N}_2), \quad \langle \mathbf{P}^{(6)} \rangle = \frac{1}{15}(5\mathbf{N}_1 + 2\mathbf{N}_2), \tag{A.2.15}$$
$$\langle \boldsymbol{\eta} \rangle = \frac{1}{3}\boldsymbol{\delta}, \quad \langle \boldsymbol{\nu} \rangle = \frac{2}{3}\boldsymbol{\delta}. \tag{A.2.16}$$

A.2.3 B-basis

It turns out that a symbolic representation by Walpole [1152] (see also [331], [462], [1150]) generalizing the form (2.197) for the isotropic tensors holds for any fourth-order tensors with transverse isotropy with respect to the axis $\mathbf{n}$. The basis (called Walpole's basis) in the space of fourth-order pair-symmetric tensors is given by the tensors

$$B^{(1)}_{ijkl} = \nu_{ij}\nu_{kl}/2, \quad B^{(2)}_{ijkl} = \eta_{ij}\eta_{kl}, \quad B^{(3)}_{ijkl} = \nu_{i(k}\nu_{l)j} - \nu_{ij}\nu_{kl}/2,$$
$$B^{(4)}_{ijkl} = 2\eta_{i)(k}\nu_{l)(j}, \quad B^{(5)}_{ijkl} = \eta_{ij}\nu_{kl}, \quad B^{(6)}_{ijkl} = \nu_{ij}\eta_{kl}, \tag{A.2.17}$$

related with another basis tensors $\mathbf{P}^{(i)}$ $(i = 1, \ldots, 6)$ (A.2.7) according to the following rule

$$\mathbf{B}^{(1)} = \mathbf{P}^{(1)}/2, \ \ \mathbf{B}^{(2)} = \mathbf{P}^{(6)}, \ \ \mathbf{B}^{(3)} = \mathbf{P}^{(2)},$$
$$\mathbf{B}^{(4)} = 2\mathbf{P}^{(5)}, \ \ \mathbf{B}^{(5)} = \mathbf{P}^{(4)}, \ \ \mathbf{B}^{(6)} = \mathbf{P}^{(3)}. \tag{A.2.18}$$

By testing we immediately obtain a multiplication-convolution of two tensors

$$\mathbf{B}^{(\alpha)} \cdot \mathbf{B}^{(\beta)} = a_{\gamma}^{\alpha\beta}\mathbf{B}^{(\gamma)}, \tag{A.2.19}$$

where the matrix $a_{\gamma}^{\alpha\beta}$ is prescribed by Table A.2.3 (the left column of the table represents the left factor)

Table A.2.3 "Multiplication" matrix $a_{\gamma}^{\alpha\beta}$ of the basic tensors $\mathbf{B}^{(k)}$ $(k = 1, \ldots, 6)$

	$\mathbf{B}^{(1)}$	$\mathbf{B}^{(2)}$	$\mathbf{B}^{(3)}$	$\mathbf{B}^{(4)}$	$\mathbf{B}^{(5)}$	$\mathbf{B}^{(6)}$
$\mathbf{B}^{(1)}$	$\mathbf{B}^{(1)}$	0	0	0	0	$\mathbf{B}^{(6)}$
$\mathbf{B}^{(2)}$	0	$\mathbf{B}^{(2)}$	0	0	$\mathbf{B}^{(5)}$	0
$\mathbf{B}^{(3)}$	0	0	$\mathbf{B}^{(3)}$	0	0	0
$\mathbf{B}^{(4)}$	0	0	0	$\mathbf{B}^{(4)}$	0	0
$\mathbf{B}^{(5)}$	$\mathbf{B}^{(5)}$	0	0	0	0	$2\mathbf{B}^{(2)}$
$\mathbf{B}^{(6)}$	0	$\mathbf{B}^{(6)}$	0	0	$2\mathbf{B}^{(1)}$	0

It should be mentioned that while $\mathbf{B}^{(1)}, \mathbf{B}^{(2)}, \mathbf{B}^{(3)}$ and $\mathbf{B}^{(4)}$ are full-symmetric and idempotent, $\mathbf{B}^{(5)}$ and $\mathbf{B}^{(6)}$ are only pair-symmetric and because of this the production operation (A.2.19) is not commutative, so $\mathbf{B}^{(5)}\mathbf{B}^{(6)} = 2\mathbf{B}^{(2)}$ but $\mathbf{B}^{(6)}\mathbf{B}^{(5)} = 2\mathbf{B}^{(1)}$. The general fourth-order tensor $\mathbf{X}$ for a transverse isotropy can be decomposed in the basis $\mathbf{B}^{(\alpha)}$ $(\alpha = 1, \ldots, 6)$

$$\mathbf{X} = \sum_{\alpha=1}^{6} X_\alpha \mathbf{B}^{(\alpha)} \equiv (X_1, X_2, X_3, X_4, X_5, X_6)^W, \tag{A.2.20}$$

where the second equality introduces Walpole's 6×1 vector of symbolic notation. The superscript ^{W} (A.2.20) is introduced for indication of dissimilarity from the analogous tensor decomposition in the tensor basis $\mathbf{P}^\alpha$ ($\alpha = 1, \ldots, 6$) (see Eq. (A.2.18)). Walpole's components X_α ($\alpha = 1, \ldots 6$) of the tensor $\mathbf{X}$ are found through a suitable contraction of the indices of $\mathbf{X}$:

$$\begin{aligned} X_1 &= \frac{1}{2} B^{(1)}_{mmij} X_{ijkl} B^{(1)}_{klnn}, \quad X_2 = B^{(2)}_{mmij} X_{ijkl} B^{(1)}_{klnn}, \\ X_3 &= \frac{1}{2} B^{(3)}_{mnij} X_{ijkl} B^{(3)}_{klmn}, \quad X_4 = \frac{1}{2} B^{(4)}_{mnij} X_{ijkl} B^{(4)}_{klmn}, \\ X_5 &= \frac{1}{2} B^{(2)}_{mmij} X_{ijkl} B^{(1)}_{klnn}, \quad X_1 = \frac{1}{2} B^{(1)}_{mmij} X_{ijkl} B^{(2)}_{klnn}. \end{aligned} \tag{A.2.21}$$

It should be mentioned that if the first four components of the decomposition (A.2.20) are positive, the tensor $\mathbf{X}$ is positively defined. If $X_5 = X_6$, then the tensor $\mathbf{X}$ is diagonally symmetric (i.e. symmetric with respect to permutation of the pairs ij and kl) and the Walpole's components (A.2.20) are connected with five familiar independent elastic moduli:

$$X_1 = 2k, \;\; X_2 - 2X_5^2/X_1 = E_3, \;\; X_5/X_1 = \nu_3, \;\; X_3 = 2\mu_{12}, \;\; X_4 = 2\mu_{13} \tag{A.2.22}$$

namely, with the plane-strain bulk modulus k, the transverse shear modulus μ_{12}, the axial shear modulus μ_{13}, and the axial Young modulus and Poisson's ratio E_3 and ν_3, respectively.

The space of pair-symmetric transversely isotropic tensors, spanned by the basis $\mathbf{B}^{(\alpha)}$ is closed with respect to sum $\mathbf{X} + \mathbf{Y}$ and product $\mathbf{XY}$ operations between two tensors, and hence forms an algebra. While the sum of Walpole's associated vectors $\mathbf{X} = (X_1, \ldots, X_6)^W$ and $\mathbf{Y} = (\mathbf{Y}_1, \ldots, Y_6)^W$ is defined as the sum of the homologous components

$$\mathbf{X} + \mathbf{Y} = (X_1 + Y_1, \ldots, X_6 + Y_6)^W \tag{A.2.23}$$

the inner product, which is generally noncommutative, can be calculated readily means of the multiplication Table A.2.3 for the elementary tensors

$$\mathbf{XY} = (X_1Y_1 + 2X_6Y_5, X_2Y_2 + 2X_5Y_6, X_3Y_3, X_4Y_4, X_5Y_1 + X_2Y_5, X_6Y_2 + X_1Y_6)^W. \tag{A.2.24}$$

By reducing this product to the unit tensor $\mathbf{XY} = \mathbf{I}$, we may calculate the inverse of $\mathbf{X}$ as

$$\mathbf{X}^{-1} = \Big(\frac{X_2}{\Delta}, \frac{X_1}{\Delta}, \frac{1}{X_3}, \frac{1}{X_4}, -\frac{X_5}{\Delta}, -\frac{X_6}{\Delta} \Big)^W, \tag{A.2.25}$$

where $\Delta = X_1X_2 - 2X_5X_6$ and the Walpole's associated vector of the identity tensor $\mathbf{I} = (1, 1, 1, 1, 0, 0)^W$. In the case, when the tensor X_{ijkl} is symmetric with respect to permutation of the pairs ij and kl, one can show that the inverse tensor $(\mathbf{X}^{-1})_{ijkl}$ posses the same symmetry. However, this property is generally lost in an inner product of the tensors with such symmetry.

A.3 Analytical Representation of Some Tensors

A.3.1 Exterior-Point Eshelby Tensor

Let us rewrite the tensor $\mathbf{T}_{ij}(\mathbf{x}_i - \mathbf{x}_j)$ (4.56) in the form $\mathbf{T}_{ij}(\mathbf{x}_i - \mathbf{x}_j) = \mathbf{L}^{(0)}\mathbf{T}_{ij}^S(\mathbf{x}_i - \mathbf{x}_j)$ where the dimensionless tensor $\mathbf{T}_{ij}^S(\mathbf{x}_i - \mathbf{x}_j)$ for isotropic medium $\mathbf{L}^{(0)}$ with two d-dimensional ($d = 2, 3$) spherical inclusions is known

$$\mathbf{T}_{ij}^S(\mathbf{x}_i - \mathbf{x}_j) = \sum_{p=1}^{6} b_p^S \mathbf{E}^{(p)}, \tag{A.3.1}$$

where the basis tensors $\mathbf{E}^{(p)}$ were introduced by Eq. (A.2.1) and the values b_p^S $(p = 1, \ldots, 6)$ are defined by the equations

$$\begin{aligned}
b_1^S &= \frac{1}{\omega_d(1-\nu)r^d}(1 - 2\nu^{(0)} + \frac{d}{d+2}\rho^2), \\
b_2^S &= \frac{1}{2\omega_d(1-\nu)r^d}(-1 + 2\nu^{(0)} + \frac{d}{d+2}\rho^2), \\
b_3^S &= \frac{d}{2\omega_d(1-\nu)r^d}(1 - \nu^{(0)}\rho^2), \quad b_4^S = \frac{d}{2\omega_d(1-\nu)r^d}(1 - \nu^{(0)} - \nu^{(0)}\rho^2), \\
b_5^S &= \frac{2d\nu^{(0)}}{\omega_d(1-\nu)r^d}(1 - \rho^2), \quad b_6^S = \frac{d+2}{2\omega_d(1-\nu)r^d}(-d + (d+4)\rho^2),
\end{aligned} \tag{A.3.2}$$

and $\nu^{(0)} \equiv (3k^{(0)} - 2\mu^{(0)})/(6k^{(0)} + 2\mu^{(0)})$ is Poisson's ratio of the matrix, $r \equiv |\mathbf{x}_i - \mathbf{x}_j|$, $\rho^2 \equiv [(a_i)^2 + (a_j)^2]/r^2$. For two identical inclusions the representations (A.3.1) and (A.3.2) are reduced to the equivalent relations obtained in [1189], and in [515] for the 3-D problem as well as in [525] for the 2-D problem. The case of 3-D spheres of different sizes is considered in [179].

We will reproduce an analytical representation for external Eshelby tensor $\mathbf{S}(\mathbf{x})$ ($\mathbf{x} \notin v_i$) following [519]. Let us consider an ellipsoidal inclusion v_i: $x_i x_i / a_I^2 \le 1$ with the semi-axes a_i and the center placed at the original of the Cartesian coordinate system $\mathbf{x} = (x_1, x_2, x_3)^\top$. Herein the tensorial notations by Mura [794] are used: repeated lowercase indices are summed up from 1 to 3, while uppercase indices which take on the same numbers as the lowercase ones are not summed up. For any point $\mathbf{x} = (x_1, x_2, x_3)^\top$ located outside the ellipsoid v_i, one can define an imaginary ellipsoidal v_i^{image} $x_i x_i (a_I^2 + \lambda)^{-1} = 1$ (λ =const.>0), with the unit outward normal vector $n_i = x_i\Big[(a_I^2 + \lambda)\sqrt{\theta(\lambda)}\Big]^{-1}$, at the point $\mathbf{x} \in v_i^{\text{image}}$ placed at the surface of the v_i^{image}, where $\theta(\lambda) = \theta_i(\lambda)\theta_i(\lambda)$, and $\theta_i(\lambda) = x_i/(a_I^2 + \lambda)$. Then the external-point Eshelby tensor was found as

$$\begin{aligned}
S_{ijkl} &= S_{IK}^{(1)}(\lambda)\delta_{ij}\delta_{kl} + S_{IJ}^{(2)}(\lambda)(\delta_{ik}\delta_{jl} + \delta_{il}\delta_{jk}) + S_I^{(3)}(\lambda)(\lambda)\delta_{ij}n_k n_l \\
&+ S_K^{(4)}\delta_{kl}n_i n_j + S_I^{(5)}(\lambda)(\delta_{ik}n_j n_l + \delta_{il}n_j n_k) \\
&+ S_J^{(6)}(\lambda)(\delta_{jk}n_i n_l + \delta_{jl}n_i n_k) + S_{IJKL}^{(7)}(\lambda)n_i n_j n_k n_l
\end{aligned} \tag{A.3.3}$$

with the explicit compact representation for the tensors

$$S_I^{(3)} = \frac{\rho^3(\lambda)}{2(1-\nu)}[1-\rho_I^2(\lambda)], \quad S_I^{(4)} = \frac{\rho^3(\lambda)}{2(1-\nu)}[1-2\nu-\rho_I^2(\lambda)],$$
$$S_I^{(5)} = \frac{\rho^3(\lambda)}{2(1-\nu)}[\nu-\rho_I^2(\lambda)], \quad S_I^{(6)} = \frac{\rho^3(\lambda)}{2(1-\nu)}[\nu-\rho_I^2(\lambda)],$$
$$S_{IJKL}^{(7)} = \frac{\rho^3(\lambda)}{2(1-\nu)}\Big\{2[\rho_I^2(\lambda)+\rho_J^2(\lambda)+\rho_K^2(\lambda)+\rho_L^2(\lambda)]$$
$$+\rho_m(\lambda)\rho_m(\lambda) - \frac{4\rho_M^2(\lambda)\theta_m(\lambda)\theta_m(\lambda)}{\theta(\lambda)} - 5\Big\}, \tag{A.3.4}$$

where $\rho_I(\lambda) = a_I(a_I^2+\lambda)^{-1/2}$, $\rho(\lambda) = [\rho_1(\lambda)\rho_2(\lambda)\rho_3(\lambda)]^{1/3}$. The tensors $\mathbf{S}^{(1)}$ and $\mathbf{S}^{(2)}$ are represented by the standard elliptic integrals, which can be integrated explicitly for the spheroidal inclusions ($a_1 \neq a_2 = a_3$) with aspect ratio $\alpha = a_1/a_2$, and the following representations take place

$$4(1-\nu)S_{11}^{(1)}(\lambda) = -\Big[4\nu+\frac{2}{\alpha^2-1}\Big]g(\lambda) - \frac{2\rho_1^3(\lambda)}{3(\alpha^2-1)}$$
$$+\Big[4\nu+\frac{2}{\alpha^2-1}\Big]\rho_1(\lambda)\rho_2^2(\lambda),$$
$$4(1-\nu)S_{12}^{(1)}(\lambda) = -\Big[4\nu-\frac{2\alpha^2+1}{\alpha^2-1}\Big]g(\lambda) + \Big[4\nu-\frac{2\alpha^2}{\alpha^2-1}\Big]\rho_1(\lambda)\rho_2^2(\lambda),$$
$$4(1-\nu)S_{21}^{(1)}(\lambda) = -\Big[2\nu+\frac{2\alpha^2+1}{\alpha^2-1}\Big]g(\lambda) - \frac{2\alpha^2}{\alpha^2-1}\rho_1(\lambda)\rho_2^2(\lambda),$$
$$4(1-\nu)S_{22}^{(1)}(\lambda) = -\Big[2\nu-\frac{4\alpha^2-1}{4(\alpha^2-1)}\Big]g(\lambda) + \frac{\alpha^2\rho_2^4(\lambda)}{2(\alpha^2-1)\rho_1(\lambda)},$$
$$4(1-\nu)S_{11}^{(2)}(\lambda) = -\Big[4\nu-\frac{4\alpha^2-2}{\alpha^2-1}\Big]g(\lambda) - \frac{2\rho_1^3(\lambda)}{3(\alpha^2-1)}$$
$$-\Big[4\nu-\frac{4\alpha^2-2}{\alpha^2-1}\Big]\rho_1(\lambda)\rho_2^2(\lambda),$$
$$4(1-\nu)S_{12}^{(2)}(\lambda) = -\Big[\nu+\frac{\alpha^2+2}{\alpha^2-1}\Big]g(\lambda) - \Big[2\nu+\frac{2}{\alpha^2-1}\Big]\rho_1(\lambda)\rho_2^2(\lambda),$$
$$4(1-\nu)S_{22}^{(2)}(\lambda) = \Big[2\nu-\frac{4\alpha^2-7}{4(\alpha^2-1)}\Big]g(\lambda) + \frac{\alpha^2\rho_2^4(\lambda)}{2(\alpha^2-1)\rho_1(\lambda)}, \tag{A.3.5}$$

where $S_{13}^{(1)} = S_{12}^{(1)}$, $S_{31}^{(1)} = S_{21}^{(1)}$, $S_{23}^{(1)} = S_{32}^{(1)} = S_{33}^{(1)} = S_{22}^{(1)}$, and $S_{13}^{(2)} = S_{21}^{(2)} = S_{31}^{(2)} = S_{12}^{(2)}$, $S_{23}^{(2)} = S_{32}^{(2)} = S_{33}^{(2)} = S_{22}^{(2)}$. Here

$$g(\lambda) = \begin{cases} -\frac{\alpha^2}{\alpha^2-1}\frac{\rho_2^2(\lambda)}{\rho_1(\lambda)} + \frac{\alpha}{(\alpha^2-1)^{3/2}}\ln\Big[(\alpha^2-1)^{1/2}\rho_2(\lambda)+\frac{\alpha\rho_2(\lambda)}{\rho_1(\lambda)}\Big], & \text{for } \alpha>1, \\ -\frac{\alpha^2}{\alpha^2-1}\frac{\rho_2^2(\lambda)}{\rho_1(\lambda)} + \frac{\alpha}{(1-\alpha^2)^{3/2}}\tan^{-1}\frac{\alpha}{(1-\alpha^2)^{1/2}\rho_1(\lambda)}, & \text{for } \alpha<1, \end{cases}$$
$$\lambda = 0.5\Big[r^2-a_1^2-a_2^2+\sqrt{(r^2+a_1^2-a_2^2)^2-4(a_1^2-a_2^2)x_1^2}\Big], \; r=|\mathbf{x}|. \tag{A.3.6}$$

In particular, we have for a circle cylinder ($\alpha\to\infty$, $x_1=0$, $a=a_2=a_3$, $j=2,3$) $g(\alpha) = -a_2^2r^{-2}$, $\lambda = r^2-a_2^2$, $\rho_1=1, \rho_j = a_j/r$, $\theta_1 = n_1 = 0$. $\theta_j = n_j = x_j/r, \theta=1$, and

$$
\begin{aligned}
&S^{(1)}_{11} = 8\nu\rho_2^2, \quad S^{(1)}_{12} = 2(2\nu-1)\rho_2^2, \quad S^{(1)}_{21} = 2\rho_2^2, \quad S^{(1)}_{22} = (2\nu-1)\rho_2^2 + \rho_2^4/2,\\
&S^{(2)}_{11} = 8\rho_2^2, \quad S^{(2)}_{12} = (1-\nu)\rho_2^2, \quad S^{(2)}_{22} = (1-2\nu)\rho_2^2 + \rho_2^4/2,\\
&S^{(7)}_{ijkl} = \frac{\rho_2^2}{(1-\nu)}\Big\{\rho_i^2(\lambda) + \rho_j^2(\lambda) + \rho_k^2(\lambda) + \rho_l^2(\lambda) - 2\rho_2^2 - 2\Big\}, \qquad \text{(A.3.7)}
\end{aligned}
$$

For a disc ($\alpha \to 0$)

$$
\begin{aligned}
&g(\alpha) = 0, \quad \lambda = 0.5\Big[r^2 - a_2^2 + \sqrt{(r^2-a_2^2)^2 + 4a_2^2x_1^2}\Big],\\
&\rho_1 = 0,\ \rho_j = 2a_j\Big[r^2 + a_j^2 + \sqrt{(r^2-a_j^2)^2 + 4a_j^2x_1^2}\Big]^{-1/2},\\
&\theta_1 = x_1/\lambda,\ \theta_j = 2x_j\Big[r^2 + a_j^2 + \sqrt{(r^2-a_j^2)^2 + 4a_j^2x_1^2}\Big]^{-1/2}, \qquad \text{(A.3.8)}
\end{aligned}
$$

and, in particular, for $r = x_1$: $\lambda = r^2$, $\rho_1 = 0$, $\rho_j = a_j(r^2 + a_j^2)^{-1/2}$, $\theta_1 = r^{-1}$, $\theta_j = 0$, $\theta = r^{-2}$, $n_1 = 1, n_j = 0$. For spherical inclusions ($\alpha = 1$):

$$
\begin{aligned}
S_{ijkl} = &\frac{\rho^3}{30(1-\nu)}\Big[(3\rho^2 + 10\nu - 5)\delta_{ij}\delta_{kl} + (3\rho^2 - 10\nu + 5)(\delta_{ik}\delta_{jl} + \delta_{il}\delta_{jk})\\
&+ 15(1-\rho^2)\delta_{ij}n_kn_l + 15(1-2\nu-\rho^2)\delta_{kl}n_in_j + 15(7\rho^2-5)n_in_jn_kn_l\\
&+ 15(\nu-\rho^2)(\delta_{ik}n_jn_l + \delta_{il}n_jn_k + \delta_{jk}n_in_l + \delta_{jl}n_in_k)\Big], \qquad \text{(A.3.9)}
\end{aligned}
$$

where $n_i = x_i/r$ and $\rho = \rho_i = a/r$ ($i = 1,2,3$). In the case of the spherical inclusions v_i and $\mathbf{x} \in v_j$, we have $\mathbf{T}^S_{ij}(\mathbf{x}_i - \mathbf{x}_j) \equiv \langle \mathbf{S}(\mathbf{x})\rangle_j$.

A.3.2 Some Tensors Describing Fluctuations of Residual Stresses

The notations of the Eqs. (13.45) and (13.46):

$$
\begin{aligned}
\mathbf{H}^1_{ip} &= \int [\mathbf{T}_{ip}(\mathbf{x}_i - \mathbf{x}_p)\mathbf{S}_{pi} \otimes \mathbf{T}_{ip}(\mathbf{x}_i - \mathbf{x}_p)\mathbf{S}_{pi}]\varphi(v_p, \mathbf{x}_p|; v_i, \mathbf{x}_i)\, d\mathbf{x}_p,\\
\mathbf{H}^2_{ip} &= \int [\mathbf{T}_{ip}(\mathbf{x}_i - \mathbf{x}_p)\mathbf{S}_{pp} \otimes \mathbf{T}_{ip}(\mathbf{x}_i - \mathbf{x}_p)\mathbf{S}_{pp}]\varphi(v_p, \mathbf{x}_p|; v_i, \mathbf{x}_i)\, d\mathbf{x}_p,\\
\mathbf{H}^3_{ip} &= \int [\mathbf{T}_{ip}(\mathbf{x}_i - \mathbf{x}_p)\mathbf{S}_{pp} \otimes \mathbf{T}_{ip}(\mathbf{x}_i - \mathbf{x}_p)\mathbf{S}_{pi}]\varphi(v_p, \mathbf{x}_p|; v_i, \mathbf{x}_i)\, d\mathbf{x}_p,\\
\mathbf{H}^4_{ip} &= \int [\mathbf{T}_{ip}(\mathbf{x}_i - \mathbf{x}_p)\mathbf{S}_{pi} \otimes \mathbf{T}_{ip}(\mathbf{x}_i - \mathbf{x}_p)\mathbf{S}_{pp}]\varphi(v_p, \mathbf{x}_p|; v_i, \mathbf{x}_i)\, d\mathbf{x}_p,\\
\mathbf{P}^1_{ip} &= \int [\mathbf{T}_{ip}(\mathbf{x}_i - \mathbf{x}_p)\mathbf{S}_{pp} \otimes \mathbf{T}_{ip}(\mathbf{x}_i - \mathbf{x}_p)\boldsymbol{\gamma}_p\\
&\quad + \mathbf{T}_{ip}(\mathbf{x}_i - \mathbf{x}_p)\boldsymbol{\gamma}_p \otimes \mathbf{T}_{ip}(\mathbf{x}_i - \mathbf{x}_p)\mathbf{S}_{pp}]\varphi(v_p, \mathbf{x}_p|; v_i, \mathbf{x}_i)\, d\mathbf{x}_p,\\
\mathbf{P}^2_{ip} &= \int [\mathbf{T}_{ip}(\mathbf{x}_i - \mathbf{x}_p)\mathbf{S}_{pi} \otimes \mathbf{T}_{ip}(\mathbf{x}_i - \mathbf{x}_p)\boldsymbol{\gamma}_p\\
&\quad + \mathbf{T}_{ip}(\mathbf{x}_i - \mathbf{x}_p)\boldsymbol{\gamma}_p \otimes \mathbf{T}_{ip}(\mathbf{x}_i - \mathbf{x}_p)\mathbf{S}_{pi}]\varphi(v_p, \mathbf{x}_p|; v_i, \mathbf{x}_i)\, d\mathbf{x}_p,\\
\mathbf{P}^3_{ip} &= \int [\mathbf{T}_{ip}(\mathbf{x}_i - \mathbf{x}_p)\boldsymbol{\gamma}_p \otimes \mathbf{T}_{ip}(\mathbf{x}_i - \mathbf{x}_p)\boldsymbol{\gamma}_p]\varphi(v_p, \mathbf{x}_p|; v_i, \mathbf{x}_i)\, d\mathbf{x}_p, \qquad \text{(A.3.10)}
\end{aligned}
$$

$$\mathbf{H}_p^{10} = \int [\mathbf{T}_p(\mathbf{x}_0 - \mathbf{x}_p)\mathbf{R}_p \otimes \mathbf{T}_p(\mathbf{x}_0 - \mathbf{x}_p)\mathbf{R}_p]\varphi(v_p, \mathbf{x}_p|; v_0, \mathbf{x}_0)\, d\mathbf{x}_p,$$

$$\mathbf{P}_p^{10} = \int [\mathbf{T}_p(\mathbf{x}_0 - \mathbf{x}_p)\mathbf{R}_p \otimes \mathbf{T}_p(\mathbf{x}_0 - \mathbf{x}_p)\mathbf{F}_p + \mathbf{T}_p(\mathbf{x}_0 - \mathbf{x}_p)\mathbf{F}_p \otimes \mathbf{T}_p(\mathbf{x}_0 - \mathbf{x}_p)\mathbf{R}_p]\varphi(v_p, \mathbf{x}_p|; v_0, \mathbf{x}_0)\, d\mathbf{x}_p,$$

$$\mathbf{P}_p^{30} = \int [\mathbf{T}_p(\mathbf{x}_0 - \mathbf{x}_p)\mathbf{F}_p \otimes \mathbf{T}_p(\mathbf{x}_0 - \mathbf{x}_p)\mathbf{F}_p]\varphi(v_p, \mathbf{x}_p|; v_0, \mathbf{x}_0)\, d\mathbf{x}_p, \quad \text{(A.3.11)}$$

respectively, where $\mathbf{S}_{kl} = \mathbf{Z}_{kl}\mathbf{R}_l$, $\gamma_k = \sum_l \mathbf{Z}_{kl}\mathbf{F}_l$, $(k, l = i, p)$.

A.3.3 Integral Representations for Stress Intensity Factors

The magnitudes of the modes I, II and III SIFs K_j $(j = 1, 2, 3)$ at the given point $\mathbf{z} = (0, R^c\cos\varphi, R^c\sin\varphi)^\top$ along the front of a crack of radius R^c due to an arbitrary distributed of the normal traction $p(\rho_0, \varphi_0)$ and the shear one $\tau(\rho_0, \varphi_0)$ (arbitrary inclined to the y, z axes: $\tau = \tau_x + i\tau_y$) are given by the formulae [324]

$$K_1(\mathbf{z}) = \frac{1}{\pi\sqrt{\pi R^c}} \int_0^{2\pi}\int_0^{R^c} \frac{\sqrt{R^{c2} - \rho_0^2}\, p(\rho_0, \varphi_0)\rho_0}{R^{c2} + \rho_0^2 - 2R^c\cos(\varphi - \varphi_0)}\, d\rho_0\, d\varphi_0,$$

$$K_2(\mathbf{z}) + iK_3(\mathbf{z}) = \frac{1}{\pi\sqrt{\pi R^c}} \int_0^{2\pi}\int_0^{R^c} \sqrt{R^{c2} - \rho_0^2} \left\{ \frac{e^{-i\varphi}\tau(\rho_0, \varphi_0)}{R^{c2} + \rho_0^2 - 2R^c\cos(\varphi - \varphi_0)} + \frac{\nu}{2-\nu} \frac{e^{i\varphi}\{3R^c - \rho_0 e^{i(\varphi-\varphi_0)}\}\bar\tau(\rho_0, \varphi_0)}{R^c\{R^c - \rho_0 e^{i(\varphi-\varphi_0)}\}^2} \right\} \rho_0\, d\rho_0\, d\varphi_0, \quad \text{(A.3.12)}$$

where an over-bar denotes a complex conjugate. According to [370], [530] we change from polar coordinate (ρ_0, φ_0) to the new coordinate (v, ζ)

$$v = \left[1 + d^{-1}\sqrt{R^{c2} - \rho_0^2}\right]^{-1}, \quad \zeta = \sin^{-1}\left\{(\rho_0/d)\sin(\varphi_0 - \varphi)\right\}, \quad \text{(A.3.13)}$$

where $d^2 = R^{c2} + \rho_0^2 - 2R^c\rho_0\cos(\varphi_0 - \varphi)$ is the square of the distance between the point (ρ_0, φ_0) and the point (R^c, φ) along the crack front. Then (A.3.12) is converted to the integral form

$$K_1(\mathbf{z}) = \frac{4\sqrt{\pi R^c}}{\pi^2} \int_{-\pi/2}^{\pi/2}\int_0^1 p(v, \zeta)\frac{(1-v)^2\cos\zeta}{[(1-v)^2 + v^2]^2}\, dv\, d\zeta,$$

$$K_2(\mathbf{z}) + iK_3(\mathbf{z}) = \frac{4\sqrt{\pi R^c}}{\pi^2} \int_{-\pi/2}^{\pi/2}\int_0^1 \frac{(1-v)^2\cos\zeta}{[(1-v)^2 + v^2]^2} \times \left\{\tau e^{-i\varphi} + \frac{2\nu}{(2-\nu)}\bar\tau e^{i(\varphi-\zeta)}\left[e^{-i\zeta} + \frac{v^2\cos\zeta}{(1-v)^2 + v^2}\right]\right\} dv\, d\zeta, \quad \text{(A.3.14)}$$

which are nonsingular. Here one corrects some misprints in [529], [530], in Eqs. (A.3.13) and (A.3.14).

References

1. ABAQUS (2001) *Theory Manual for Version 6.2-1.* Hibbitt, Karlsson, and Sorenson, Inc., Pawtucket, RI
2. Aboudi J, Pindera M-J, Arnold SM (1999) Higher-order theory for functionally graded materials. *Composites* **B30**:777–832
3. Accorsi ML, Nemat-Nasser S (1986) Bounds on the overall elastic and instantaneous elastoplastic moduli of periodic composites. *Mech Mater* **5**:209–220
4. Achenbach JD, Zhu H (1990) Effect of interfaces on micro and macromechanical behavior of hexagonal-array fiber-composites. *J Appl Mech,* **57**:956–963
5. Acrivos A, Chang E (1986) A model for estimating transport quantities in two-phase materials. *Phys Fluids,* **29**:3–4
6. Adams BL, Morris PR, Wang TT, Willden KS, Wright SI (1987) Description of orientation coherence in polycrystalline materials. *Acta Metall,* **35**:2935–2946
7. Adams BL, Olson T (1998) The metho structure – properties linkage in polycrystals. *Prog Mater Sci,* **43**:1–88
8. Agarwal GS, Dutta Gupta S (1988) T-matrix approach to the nonlinearb susceptibilities of heterogeneous media. *Phys Rev,* **A38**:5678–5687
9. Aifantis EC (1999) Gradient deformation models at nano, micro, and macro scales. *J Engng Mater Technol,* **121**:189–202
10. Alexandrov K (1965) Average values of tensorial variables. *DAN SSSR (Soviet Phys-Docl),* **164**:800–802 (In Russian)
11. Alexandrov K, Aisenberg L (1966) A method of calculating the physical constants of polycrystalline materials. *DAN SSSR,* **167**:1028 (In Russian. English translation: *Soviet Phys-Docl,* **11**:323–325)
12. Allen MP, Tildesley DJ (1987) *Computer Simulation of Liquids.* Clarendon Press, Oxford
13. Al-Ostaz A, Jasiuk, I, Lee, M (1998) Circular inclusion in half-plane: effect of boundary conditions *J Engng Mech,* **124**:293–300
14. Andrews R, Weisenberger MC (2004) Carbon nanotube polymer composites. *Curr Opin Solid State Mater Sci,* **8**:31–37
15. Anderson KL, Sinsawat A, Vaia RA, Farmer BL (2005) Control of silicate nanocomposite morphology in binary fluids: coarse-grained molecular dynamics simulation. *J Polym Sci Part B: Polym Phys,* **43**:1014–1024
16. Armington M (1991) Limit distributions of the states and inhomogenization in random media. *Acta Mechan,* **88**:27–59
17. Arnould JF (1982) Etude de l'endommagement des materiaux composites par decohesion fibre-matrice. In: Corvino A et al. (eds), *Comptes Rendus des Troisiemes Journees Nationales sur les Composites (JNC-3).* AMAS, Paris, 159–169

18. Arroyo M, Belytschko T (2002) An atomistic-based finite deformation membrane for single layer crystalline films. *J Mech Phys Solids,* **50**:1941–1977
19. Arsenault RJ, Taya M (1987) Thermal residual stress in metal matrix composite. *Acta Metall,* **35**:651–659
20. Asaro RJ, Barnett DM (1975) The non-uniform transformation strain problem for an anisotropic ellipsoidal inclusions. *J Mech Phys Solids,* **23**: 77–83
21. Atkin RJ, Fox N (1990) *An Introduction to the Theory of Elasticity.* Longman, London
22. Avellaneda M (1987) Iterated homogenization, differential effective medium theory and applications. *Commun Pure Appl Math,* **40**: 527–554
23. Axel F (1992) Bounds for field fluctuations in two-phase materials. *J Appl Phys,* **72**:1217–1220
24. Babuska I, Anderson B, Smith PJ, Levin K (1999) Damage analysis of fiber composites. Part I: Statistical analysis on fiber scale. *Comput Methods Appl Mech Engng,* **172**:27–77
25. Bakhvalov NG, Panasenko G (1989) *Homogenization: Averaging Processes in Periodic Media.* Kluwer, Dordrecht
26. Ball JM (1977) Convexity conditions and existence theorems in nonlinear elasticity. *Arch Rational Mech Anal,* **63**:337–403
27. Ballabh TK, Paul M, Middya TR, Basu AN (1992) Theoretical multiple-scattering calculation of nonlinear elastic constants of disordered solids. *Phys Rev,* **B45**:2761–2771
28. Ballas J, Sladek J, Sladek V (1989) *Stress Analysis by Boundary Element Methods.* Elsevier, Amsterdam
29. Banerjee PK (1994) *The Boundary Element Methods in Engineering.* McGraw-Hill, London, New York
30. Banno H (1983) Recent developments of piezoelectric ceramic products and composites of synthetic rubber and piezoelectric ceramic particles. *Ferroelectrics,* **50**:3–12
31. Banno H (1988) Recent development in piezoelectric composites in Japan. In: Saite S (ed), *Adv Ceramics.* Oxford University Press and Ohmsh, Ltd., New York and Tokyo, 8–26
32. Bao G, Sao Z (1992) Remarks on crack-bridging concepts. *Appl Mech Rev,* **45**:355–366
33. Bao G, Hutchinson JW, McMeeking RM (1991) Particle reinforcement of ductile matrices against plastic flow and creep. *Acta Metall,* **39**:1871–1880
34. Barbier D, Brown D, Grillet AC, Neyertz S (2004) Interface between end-functionalized PEO oligomers and a silica nanoparticle studied by molecular dynamics simulations *Macromoleculas,* **37**:4695–4710
35. Barker GC (1993) Computer simulation of granular materials. In: Mehta A (ed) *Granular Matter - An Interdisciplinary Approach.* Springer Verlag, Berlin
36. Barnett DM (1972) The precise evaluation of derivatives of the anisotropic elastic Green's functions. *Phys Stat Sol (b),* **49**:741–748
37. Barnett DM, Asaro RJ (1972) The fracture mechanics of slit like cracks in anisotropic elastic media. *J Mech Phys Solids,* **20**:353–366
38. Barnett DM, Lothe J (1975) Dislocation and line charges in anisotropic piezoelectric insulators. *Phys Stat Solids (b),* **67**:105–111
39. Bar-yoseph P, Avrashi J (1986) New variational-asymptotic formulations in elastic composite materials. *J Appl Math Phys,* **37**:305–321
40. Batchelor GK (1972) Sedimentation in a dilute dispersion of spheres. *J Fluid Mech,* **52**:245–268

41. Batchelor GK (1974) Transport properties of two-phase materials with random structure. *Ann-Rev Fluid Mech,* **6**:227–255
42. Batchelor GK, Green JT (1972) The determination of the bulk stress in a suspension of spherical particles to order c^2. *J Fluid Mech,* **56**:401–427
43. Bauer F, Böhm HJ, Fischer FD, Rammerstorfer FG (1987) Finite element analysis of drastic shape changes of cylinders due to thermal cycling. *Comput Struct,* **26**: 263–274
44. Baus M, Colot JL (1987) Thermodynamics and structure of a fluid of hard rods, disks, spheres, of hyperspheres from rescaled virial expansion. *Phys Rev,* **A36**:3912–3925
45. Bazant ZP, Jirasek M (2002) Nonlocal integral formulations of plasticity and damage: Survey of progress. *J Engng Mech,* **128**:1119–1149
46. Belikov VP, Aleksandrov KS, Ryzhova TV (1970) *Elastic Properties of Rock-Forming Minerals and Rocks.* Nauka, Moscow (In Russian)
47. Belytschko T, Xiao SP, Schatz GC, Ruoff RS (2002) Atomistic simulation of nanotube fracture. *Phys Rev,* **B65**:235430-1-8
48. Bensoussan A, Lions JL, Papanicolaou G (1978) *Asymptotic Analysis for Periodic Structures.* North-Holland, Amsterdam, New York
49. Bennet CH (1972) Serially deposited amorphous aggregates of hard spheres. *J Appl Phys,* **43**:2727–2734
50. Benveniste Y (1985) The effective mechanical behavior of composite materials with imperfect contact between the constituents. *Mech Mater,* **4**:197–208
51. Benveniste Y (1986) On the effective thermal conductivity of multiphase composites. *J Appl Math Phys (ZAMP),* **37**:696–713
52. Benveniste Y (1987a) A new approach to application of Mori-Tanaka's theory in composite materials. *Mech Mater,* **6**:147–157
53. Benveniste Y (1987b) A differential effective medium theory with a composite sphere embedding. *J Appl Mech,* **54**:466–468
54. Benveniste Y (1992) The determination of the elastic and electric fields in piezoelectric inhomogeneity. *J Appl Phys,* **72**:1086–1095
55. Benveniste Y (1993a) Universal relations in piezoelectric composites with eigenstress and polarization fields I, binary media: local fields and effective behavior. *J Appl Mech,* **60**:265–269
56. Benveniste Y (1993b) Universal relations in piezoelectric composites with eigenstress and polarization fields II, multiphase media: effective behavior. *J Appl Mech,* **60**:270–275
57. Benveniste Y (1994) Exact results concerning the local fields and effective properties in piezoelectric composites. *J Eng Mater Technol,* **116**:260–267
58. Benveniste Y (2006) A general interface model for a three-dimensional curved thin anisotropic interphase between two anisotropic media. *J Mech Phys Solids,* **54**:708–734
59. Benveniste Y, Dvorak GJ (1990a) On a correspondence between mechanical and thermal effects in two–phase composites. In: Weng GJ, Taya M, Abe H (eds) *Micromechanics and Inhomogeneity, The Toshio Mura 65th Anniversary Volume.* Springer-Verlag, New York, 65–81
60. Benveniste Y, Dvorak GJ (1990b) On a correspondence between mechanical and thermal effects in composites with slipping interfaces. In: Dvorak GJ (ed), *Inelastic Deformation of Composite Materials.* IUTAM Symposium, Springer-Verlag, Berlin, 77–98
61. Benveniste Y, Dvorak GJ (1992) On uniform fields and universal relations in piezoelectric composites *J Mech Phys Solids,* **40**:1295–1312

62. Benveniste Y, Dvorak GJ, Chen T (1989) Stress fields in composites with coated inclusions. *Mech Mater,* **7**:305–317
63. Benveniste Y, Dvorak GJ, Chen,T (1991) On diagonal and elastic symmetry of the approximate effective stiffness tensor of heterogeneous media. *J Mech Phys Solids,* **39**:929–946
64. Benveniste Y, Miloh T (1986) The effective conductivity of composites with imperfect contuct at constituent interfaces. *Int J Eng Sci,* **24**:1537–1552
65. Benveniste Y, Milton GW(2003) New exact results for the e'ective electric, elastic, piezoelectric and other properties of composite ellipsoid assemblages. *J Mech Phys Solids,* **51**, (2003) 1773–1813
66. Beran MJ (1968) *Statistical Continuum Theories.* John Wiley & Sons, New York
67. Beran M (1974) Application of statistical theories for the determination of thermal, electrical and magnetic properties of heterogeneous materials. In: Sendeckyj GP (ed), *Mechanics of Composite Materials.* Academic Press, New York, **2**:209–249
68. Beran M (1980) Field fluctuations in a two-phase random medium. *J Math Phys,* **21**:2583–2585
69. Beran MJ, McCoy JJ (1970a) Mean field variations in a statistical sample of heterogeneous linearly elastic solids. *Int J Solid Struct,* **6**:1035–1054
70. Beran MJ, McCoy JJ (1970b) Mean field variation in random media. *Quart Appl Math,* **28**:245–257
71. Beran M, Molyneux J (1966) Use of classical variational principles for the effective bulk modulus in heterogeneous media. *Quart Appl Math,* **24**:107–125
72. Berdichevsky VL (1983) *Variational Principle of Continuum Mechanics.* Nauka, Moskow (In Russian)
73. Bergander H (1995) Finite plastic constitutive laws for finite deformations. *Acta Mechan,* **109**:79–99
74. Bergman DJ (1978) The dielectric constant of a composite material — a problem of classical physics. *Phys Rep,* **43C**:377–407
75. Berryman JG (1983) Random close packing of hard spheres and disks. *Phys Rev,* **A 27**:1053–1061
76. Berryman JG (1985) Measurement of spatial correlation functions using image processing techniques. *J Appl Phys,* **57**:2374–2384
77. Berryman JG (2005) Bounds and self-consistent estimates for elastic constants of random polycrystals with hexagonal, trigonal, and tetragonal symmetries. *Int J Solids Struct,* **53**:2141–2173
78. Berryman JG, Berge PA (1996) Critique of two explicit schemes for estimating elastic properties of multiphase composites. *Mech Mater* **22**:149–164
79. Berveiller M, Fassi-Fenri O, Hihi A (1987) The problem of two plastic and heterogeneous inclusions in an anisotropic medium. *Int J Engng Sci,* **25**:691–709
80. Berveiller M, Zaoui A (1979) An extension of the self-consistent scheme to plastically flowing polycrystals. *J Mech Phys Solids,* **26**:325–344
81. Bharata S, Levinson M (1978) Signorini's perturbation scheme for a general reference configuration in finite electrostatics. *Arch Rat Mech Anal,* **67**:365–394
82. Bhattacharyya A, Lagoudas DC (2000) Effective elastic moduli of two-phase transversely isotropic composites with aligned clustered fibers. *Acta Mechan,* **145**:65–93
83. Bhattacharyya A, Weng GJ (1994) The elastoplastic behavior of a class of two–phase composites containing rigid inclusions. *Appl Mech Rev,* **47**(Part 1):S45–S65
84. Bildstein B, Kahl G (1994) Triplet correlation functions for hard-spheres: computer simulation results. *J Chem Phys,* **100**:5882–5893
85. Bilger N, Auslender F, Bornert M, Jean-Claude Michel J-C, Moulinec H, Suquet P, Zaoui A (2005) Effect of a nonuniform distribution of voids on the plastic response of voided materials: a computational and statistical analysis. *Int J Solids Struct,* **42**:517–538

86. Bimberg D, Grundmann M, Ledentsov NN (1998) *Quantum Dot Heterostructures.* John Wiley and Sons, New York
87. Binder K, Heerman DW (1997) *Monte Carlo Simulation in Statistical Physics: an Introduction.* Springer, Berlin NY
88. Bing J, Daining F, Kehchih H (1997) The effective properties of piezocomposites, Part II: The effective electroelastic moduli. *Acta Mechan Sinica,* **13**:347–354
89. Bobeth M, Diener G (1986) Field fluctuations in multicomponents mixtures. *J Mech Phys Solids,* **34**:1–17
90. Bobeth M, Diener G (1987) Static elastic and thermoelastic field fluctuations in multiphase composites. *J Mech Phys Solids* **35**:137–145
91. Bogetti TA, Wang T, VanLandingham MR, Gillespie JW (1999) Characterization of nanoscale property variations in polymer composite systems: 2. Numerical modeling. *Composites: Part A,* **30**:85–94
92. Böhlke T, Bertran A (2001) The evolution of Hooke's law due to texture development in FCC polycrystals. *Int J Solids Struct,* **38**:9437–9459
93. Böhm H (2004) Continuum models for the thermomechanical behavior of discontinuously reinforced materials. *Adv Engng Mater* **6**:626–633
94. Böhm HJ, Eckschlager A, Han W (2002) Multi-inclusion unit cell models for metal matrix composites with randomly oriented discontinuous reinforcements. *Comput Mater Sci,* **25**:42–53
95. Böhm HJ, Rammerstorfer FG, Weissenbek E (1993) Some simple models for micromechanical investigations of fiber arrangement effect in MMCs. *Comput Mater Sci,* **1**:177–194
96. Bolotin VV (1993) Random initial defects and fatigue life prediction. In: Sobczuk K (ed), *Stochastic Approach to Fatigue: Experiments, Modeling and Reliability Estimation.* Springer-Verlag, Vienna, 121–163
97. Bolotin VV, Moskalenko VN (1968) Determination of the elastic constants of a microinhomogeneous medium. *Zh Priklad Mekh Tekhn Fiz (J Appl Mech Tech Phys)* N1:66–72 (In Russian)
98. Bompard P, Dan W, Guennounl F, Francois D (1987) Mechanics and fracture behavior of porous materials. *Engng Fract Mech,* **28**:627–642
99. Borisenko AI, Tarapov IE (1968) *Vector and Tensor Analysis with Applications.* Prentice-Hall, Englewood Cliffs, NY
100. Bornert M, Hervé E, Stolz C, Zaoui A (1994) Self-consistent approaches and strain heterogeneities in two-phase elastoplastic materials. *Appl Mech Rev,* **47**(1, Part 2):S66–S76
101. Bornert M, Stolz C, Zaoui A (1996) Morphologically representative pattern-based bounding in elasticity. *J Mech Phys Solids,* **44**:307–331
102. Bostroem A, Olsson P, Datta SK (1992) Effective plane wave propagation through a medium with spheroidal inclusions surrounded by thin interface layers. *Mech Mater* **14**:59–66
103. Boudreaux DS, Gregor JM (1977) Structure simulation of transition-metal-metalloid glasses. *J Appl Phys,* **48**:152–158
104. Boutin C (1996) Microstructural effects in elastic composites. *Int J Solids Struct,* **33**:1023–1051
105. Bouyge F, Jasiuk I, Ostoja-Starzewski M (2001) A micromechanically based couple-stress model of an elastic two-phase composite. *Int J Solids Struct,* **38**:1721–1735
106. Bover AF, Ortiz W (1993) The influence of grain size on the toughness of monolithic ceramic. *J Engng Mater Technol,* **115**:228–236
107. Bower C, Rosen R, Jinb L, Han J, Zhoua O (1999) Deformation of carbon nanotubes in nanotube-polymer composites. *Appl Phys Lett,* **74**:3317–3319

108. Bradshaw RD, Fisher FT, Brinson LC (2003) Fiber waviness in nanotube-reinforced polymer composites: II. Modeling via numerical approximation of the dilute strain concentration tensor. *Compos Sci Technol,* **63**:1705–1722
109. Brebbia CA, Telles JCF, Wrobel LC (1984) *Boundary Element Techniques.* Springer-Verlag, Berlin
110. Brechet YJM (1994) Clusters, plasticity and damage: a missing link? *Mater Sci Engng,* **A175**:63–69
111. Bristow JR (1960) Microcracks, and the static and dynamic elastic constants of annealed heavily coldworked metals. *Br J Appl Phys,* **11**:81–85
112. Brosseau C (2006) Modelling and simulation of dielectric heterostructures: a physical survey from an historical perspective. *J Phys D: Appl Phys,* **39**:1277–1294
113. Brown WF (1955) Solid mixture permittivities. *J Phys Chem,* **23**:1514–1517
114. Bruggeman DAG (1935) Berechnung verschiedener physikalischer Konstante von heterogenete substanze I: Dielektrizitätskonstanten und leitfähigkeiten der misckörper aus isotropen substanzen. *Annal Physik,* **24**:636–679
115. Brune DA, Bicerano J (2002) Micromechanics of nanocomposites: comparison of tensile and compressive elastic moduli, and prediction of effects of incomplete exfoliation and imperfect alignment on modulus. *Polymer,* **43**:369–387
116. Bruno OP (1991) Taylor expansions of bounds for the effective conductivity and the effective elastic moduli of multicomponent composites and polycrystals. *Asympt Anal,* **4**:339–365
117. Budiansky Y (1965) On the elastic moduli of some heterogeneous material. *J Mech Phys Solids,* **13**:223–227
118. Budiansky B, Hutchinson JW, Lambropoulos JC (1983) Continuum theory of dilatant transformation toughening in ceramics. *Int J Solids Struct,* **19**:337–355
119. Budiansky B, O'Connel RJ (1976) Elastic moduli of cracked solids. *Int J Solids Struct,* **12**:81–91
120. Buevich YA (1992) Heat and mass transfer in disperse media. I. Average field equations. *Int J Heat Mass Transfer,* **35**:2445–2452
121. Buevich YuA, Shelchkova, IN (1978) Flow of dense suspensions. *Prog Aerospace Sci,* **18**:121–150
122. Buevich YA, Ustinov VA (1995) Effective conductivity of a macroscopically inhomogeneous dispersions. *Int J Heat Mass Transfer,* **38**:381–389
123. Bui HD (1978) Some remarks about the formulation of three-dimensional thermoelastic problems by integral equations. *Int J Solid Struct,* **14**:935–939
124. Bunge HJ (1982) *Texture Analysis in Mater Science.* Butterworths, Boston
125. Buryachenko VA (1987) Correlation function of stress field in matrix composites. *Mekhanika Tverdogo Tela* $N^0 3$, 69–76 (In Russian. Engl Transl. *Mech Solids,* **22**: 66–73)
126. Buryachenko VA (1990a) Effective viscoplasticity parameters of suspension. *Inzhenerno Fiz Zhurnal,* **58**:452–456 (In Russian. Engl Transl. *J Engng Phys,* **58(3)**:331–334)
127. Buryachenko VA (1990b) Prediction of macroproperties of composite medium by the elastoplastic effective field method. *Probl Prochnosti* (11), 57–76 (In Russian. Engl Transl. *Strength Mater,* **22**:1645–1650)
128. Buryachenko VA (1993a) *Effective Physicomechanical Properties of Random Structure Composites.* D. Sc. Thesis, 397p., SP Timoshenko Institute of Mechanics of NAS of Ukraine, Kiev (In Russian)
129. Buryachenko VA (1993b) Mechanics of random structure composites. In: Miravete A (ed) *Proc. of 9th Int. Conf. on Composites.* Univ. Zaragoza, Spain **3**:398–405

130. Buryachenko VA (1993c) Effective strength properties of elastic physically nonlinear composites. In: Marigo JJ, Rousselier G (eds), *Proc. of the MECAMAT Conf. Micromechanics of Materials.* Editions Eyrolles, Paris, 567–578
131. Buryachenko VA (1996) The overall elastoplastic behavior of multiphase materials. *Acta Mechan,* **119**:93–117
132. Buryachenko VA (1998) Some nonlocal effects in graded random structure matrix composites. *Mech Res Commun,* **25**:117–122
133. Buryachenko VA (1999a) Triply periodical particulate matrix composites in varying external stress fields. *Int J Solids Struct,* **36**:3837–3859
134. Buryachenko VA (1999b) Effective thermoelastic properties of graded doubly periodic particulate composites in varying external stress fields. *Int J Solids Struct,* **36**:3861–3885
135. Buryachenko VA (1999c) Thermo-elastoplastic deformation of elastically homogeneous materials with a random field of inclusions. *Int J Plast,* **15**:687–720
136. Buryachenko VA (2000a) Internal residual stresses in elastically homogeneous solids: I. Statistically homogeneous stress fluctuations. *Int J Solids Struct,* **37**:4185–4210
137. Buryachenko VA (2000b) Internal residual stresses in elastically homogeneous solids: II. Stress fluctuations near a crack tip and effective energy release rate. *Int J Solids Struct,* **37**:4211–4238
138. Buryachenko VA (2001a) Multiparticle effective field and related methods in micromechanics of composite materials. *Appl Mech Rev,* **54**:1–47
139. Buryachenko VA (2001b) Multiparticle effective field and related methods in micromechanics of random structure composites. *Math Mech Solids,* **6**:577–612
140. Buryachenko VA (2001c) A simple method of multiple inclusion interaction problem. *Int J Comput Civil Struct Engng,* **1**:7–25
141. Buryachenko VA (2001d) Locality principle and general integral equations of micromechanics of composite materials. *Math Mech Solids,* **6**:299–321
142. Buryachenko VA (2002) Stress fluctuations near a crack tip in ceramic materials and effective energy release rate. In: Brand RC, Sakai M, Munz D, Shevchenko VJ, White KW (eds) *Fracture Mechanics of Ceramics, V.13: Crack-Microstructure interection, R-Curve Behavior, Environmental Effects in Fracture, and Standardtization.* Kluver, Norwell MA, 47–62
143. Buryachenko VA (2004a) Foreword to Special Issue on Recent Advances in Micromechanics of Composite Materials. *Int J Multiscale Comput Engng,* **2**:vii–viii
144. Buryachenko VA (2004b) Multiscalar mechanics of nonlocal effects in heterogeneous materials. *Int J Multiscale Comput Engng,* **2**:1–14
145. Buryachenko VA (2005a) Effective elastic moduli and stress concentrator factors in random structure aligned fiber composites. *ZAMP,* **56**:1107–1115
146. Buryachenko VA (2005b) Effective elastic moduli of triply periodic particulate matrix composites with imperfect unit cells. *Int J Solids Struct,* **42**:4811–4832
147. Buryachenko VA (2007) Generalization of the multiparticle effective field method in static of random structure matrix composites. *Acta Mechan,* **188**:167–208
148. Buryachenko VA, Bechel VT (2000) A volume integral equation method for multiple inclusion interaction problems. *Compos Sci Technol,* **60**:2465–2469
149. Buryachenko VA, Böhm HJ, Rammerstorfer FG (1996) Modelling of the overall elastoplastic behavior of multiphase materials by the effective field method. In: Pineau A, Zaoui Z (eds) *IUTAM Symp. Micromech. of Plasticity and Damage of multiphase materials.* Kluwer, Dordrecht, 35–42
150. Buryachenko VA, Goikhman BD (1985) Integrodifferential criterion of the strength of aging polymers. *Problemy Prochnosti,* **17**(8):88–91 (In Russian. Engl Transl *Strength of Mater* **17**:1138–1141)

151. Buryachenko VA, Kreher WS (1995) Internal residual stresses in heterogeneous solids — a statistical theory for particulate composites. *J Mech Phys Solids,* **43**:1105–1125
152. Buryachenko VA, Kushch VI (2006) Effective transverse elastic moduli of composites at non-dilute concentration of a random field of aligned fibers. *ZAMP,* **57**:491–505
153. Buryachenko VA, Kushch VI (2007) Statistical properties of local residual micro stresses in elastically homogeneous half-space. *Int J Multiscale Comput Engng* (In press)
154. Buryachenko VA, Kushch VI, Dutka VA, Roy A (2007) Effective elastic properties of nanocomposites reinforced by cylindrical nanofibers. (Submitted)
155. Buryachenko VA, Kushch VI, Roy A (2007) Effective thermoelastic properties of random structure composites reinforced by the clusters of deterministic structure (application to clay nanocomposites). *Acta Mechan* (In press)
156. Buryachenko VA, Lipanov AM (1986a) Stress concentration ellipsoidal inclusions and effective thermoelastic properties of composite materials. *Priklad Mekh,* (11):105–111 (In Russian. Engl Transl. *Soviet Appl Mech,* **22**(11):1103–1109)
157. Buryachenko VA, Lipanov AM (1986b) Equations of mechanics for gas-saturated porous media. *Priklad Mekh Tekhn Fiz,* (4):106–109 (In Russian. Engl Transl. *J Appl Mech Tech Phys,* **27**:577–581)
158. Buryachenko VA, Lipanov AM (1988) Equation of isotropic deformation for gas-saturated materials with allowance for large strains of spherical pores. *Priklad Mekh Tekhn Fiz,* (4), 120–124 (In Russian. Engl Transl. *J Appl Mech Tech Phys,* **29**:565–569)
159. Buryachenko VA, Lipanov AM (1989a) Effective field method in the theory of perfect plasticity of composite materials. *Priklad Mekh Tekhn Fiz,* (3):149–155 (In Russian. Engl Transl. *J Appl Mech Tech Phys,* **30**:482–487)
160. Buryachenko VA, Lipanov AM (1989b) Predicting the parameters of a nonlinear flow of multicomponent mixtures. *Priklad Mekh Tekhn Fiz,* (4):53–57 (In Russian. Engl Transl. *J Appl Mech Tech Phys,* **30**:558–562)
161. Buryachenko VA, Lipanov AM (1990a) Effective moduli of elasticity of composites of third order. *Priklad Mekh Tekhn Fiz,* **31**(6):118–123 (In Russian. Engl Transl. *J Appl Mech Tech Phys,* **31**:909–913)
162. Buryachenko VA, Lipanov AM (1990b) Effective characteristics of elastic physically nonlinear composites. *Prikladnaya Mekhanika,* **26**(1):12–16 (In Russian. Engl Transl. *Sov Appl Mech,* **26**:9–13)
163. Buryachenko VA, Lipanov AM (1992) Thermoelastic stress concentration at ellipsoidal inclusions in matrix composites in the region of strongly varying external stress and temperature fields. In: Naimark OB, Evlampieva SE (eds) *Deformation and Fracture of Structural-Inhomogeneous Materials.* AN SSSR, Sverdlovsk, 12–19 (In Russian)
164. Buryachenko VA, Lipanov AM, Kozhevnikova YG (1991) Effective properties of elastoplastic porous materials. *Problemy Prochnosti,* (1):36–39 (In Russian. Engl Transl. *Strength Mater,* **23**:41–45)
165. Buryachenko VA, Murov V A (1991) Effective conductivity of matrix composites. *Inzhenerno Fiz Zhurnal,* **61**(2):305–312 (In Russian. Engl Transl. *J Engng Phys,* **61**:1041–1047)
166. Buryachenko VA, Pagano NJ (2003) Nonlocal models of stress concentrations and effective thermoelastic properties of random structure composites. *Math Mech of Solids,* **8**:403–433

167. Buryachenko VA, Pagano NJ (2005) Multiscale analysis of multiple interacting inclusions problem: finite number of interacting inclusions. *Math Mech Solids,* **10**:25–62
168. Buryachenko VA, Pagano NJ, Kim RY, Spowart JE (2003) Quantitative description of random microstructures of composites and their effective elastic moduli. *Int J Solids Struct,* **40**:47–72
169. Buryachenko VA, Parton VZ (1990a) Effective Helmholtz operator for matrix composites. *Izv AN SSSR, Mekh Tverd Tela* (3):55–63 (In Russian. Engl Transl. *Mech Solids,* **25**:60–69
170. Buryachenko VA, Parton VZ (1990b) One-particle approximation of the effective field method in the statics of composites. *Mekh Kompoz Mater,* (3):420–425 (In Russian. Engl Transl. *Mech Compos Mater,* **26**(3):304–309)
171. Buryachenko VA, Parton VZ (1990c) Effective parameters of statistically inhomogeneous matrix composites. *Izv AN SSSR, Mekh Tverd Tela,* (6):24–29 (In Russian. Engl Transl. *Mech Solids,* **25**:22–28)
172. Buryachenko VA, Parton VZ (1990d) Bounds of effective moduli of composite materials. *Mekh Kompoz Mater,* (3):420–425 (In Russian)
173. Buryachenko VA, Parton VZ (1991) Effective parameters of static conjugating physical-mechanical fields in matrix composites. *Fiziko-Khimichescaja Mech Mater,* **27**(4):105–111 (In Russian. Engl Transl. *Sov Mater Sci,* **27**:428–433)
174. Buryachenko VA, Parton VZ (1992a) Effective field method in the statics of composites. *Priklad Mekh Tekhn Fiz,* (5):129–140 (In Russian. Engl Transl. *J Appl Mech Tech Phys,* **33**:735–745)
175. Buryachenko VA, Parton VZ (1992b) Multi-particle differential methods in the statics of composites. *Priklad Mekh Tekhn Fiz,* (3):148–156 (In Russian. Engl Transl. *J Appl Mech Tech Phys,* **33**:455–462)
176. Buryachenko VA, Parton VZ (1992c) Effective strength parameters of composites in coupled physicomechanical fields. *Priklad Mekh Tekhn Fiz,* (4), 124–130 (In Russian. Engl Transl. *J Appl Mech Tech Phys,* **33**:589–593)
177. Buryachenko VA, Rammerstorfer FG (1996a) Thinly coated inclusion with stress free strains in an elastic medium. *Mech Res Commun,* **23**:505–509
178. Buryachenko VA, Rammerstorfer FG (1996b) Elastoplastic behavior of elastically homogeneous materials with a random field of inclusions. In: Markov K (ed), *Continuum Models and Discrete Systems.* World Scientific, Singapore, 140–147
179. Buryachenko VA, Rammerstorfer FG (1997a) Elastic stress fluctuations in random structure particulate composites. *Eur J Mech A/Solids,* **16**:79–102
180. Buryachenko VA, Rammerstorfer FG (1997b) A local theory of elastoplastic deformations of random structure composites. *Z Angew Math Mech,* **77**(S1):S61–S62
181. Buryachenko VA, Rammerstorfer FG (1998a) Micromechanics and nonlocal effects in graded random structure matrix composites. In: Bahei-El-Din YA, Dvorak GJ (eds) *IUTAM Symp. on Transformation Problems in Composite and Active Materials.* Kluwer, Dordrecht, 197–206
182. Buryachenko VA, Rammerstorfer FG (1998b) Thermoelastic stress fluctuations in random structure coated particulate composites. *Eur J Mechanics A/Solids,* **17**:763–788
183. Buryachenko VA, Rammerstorfer FG (1999) On the thermoelasticity of random structure particulate composites. *Z Angew Math Phys,* **50**:934–947
184. Buryachenko VA, Rammerstorfer FG (2000) On the thermostatics of composites with coated inclusions. *Int J Solids Struct,* **37**:3177–3200
185. Buryachenko VA, Rammerstorfer FG (2001) Local effective thermoelastic properties of graded random structure composites. *Arch Appl Mech,* **71**:249–272

186. Buryachenko VA, Rammerstorfer FG, Plankensteiner AF (1997) A local theory of elastoplastic deformations of random structure composites. *Z Angew Math Mech,* **7**(S1):S61–S62
187. Buryachenko VA, Rammerstorfer FG, Plankensteiner AF (2002) A local theory of elastoplastic deformation of two-phase metal matrix random structure composites. *J Appl Mech,* **69**:489–496
188. Buryachenko VA, Roy A (2005a) Effective elastic moduli of nanocomposites with prescribed random orientation of nanofibers. *Composites B,* **36**:405–416
189. Buryachenko VA, Roy A (2005b) Effective thermoelastic moduli and stress concentrator factors of nanocomposites. *Acta Mechan,* **177**:149–169
190. Buryachenko VA, Roy A, Lafdi K, Anderson KL, Chellapilla S (2005) Multi-scale mechanics of nanocomposites including interface: experimental and numerical investigation. *Comp Sci Technol,* **65**:2435–2465
191. Buryachenko VA, Schoeppner G (2004) Effective elastic and failure properties of fiber aligned composites. *Int J Solids Struct,* **41**:4827–4844
192. Buryachenko VA, Scorbov YS, Gunin SV (1991a) Effective parameters of the strength of matrix composites. *Probl Prochnosti* (12):47–51 (In Russian. Engl Transl. *Strength Mater,* **23**(12))
193. Buryachenko VA, Scorbov YS, Gunin SV, Lysov VV (1991b) Effective rheological parameters of suspensions with polyfractional filler. *Inzhenerno Fiz Zhurnal,* **61**:928–933 (In Russian. Engl Transl. *J Engng Phys,* **61**:1478–1482)
194. Buryachenko VA, Shermergor TD (1995) Material and field characteristics of piezoelectric rocks. Some exact results. *Fiz Zemli,* (8):32–42 (In Russian. Engl Transl. *Phys of the Solid Earth,* (1996) **31**:665–672)
195. Buryachenko VA, Tandon GP (2004) Estimation of effective elastic properties of random structure composites for arbitrary inclusion shape and anisotropy of components using finite element analysis. *Int J Multiscale Comput Engng,* **2**:29–45
196. Cargill III GS (1994) Random packing for amorphous binary alloys. *J Phys Chem Solids,* **55**:1375–1380
197. Carlson DE (1972) Linear thermoelasticity. In: Truesdell C (ed) *Flügge's Handbuch der Physik VI a/2.* Springer-Verlag, Berlin-Heidelberg-New York, 297–346
198. Carrado K (2003) Polymer-clay nanocomposites. In: Shonaike GO, Advani SA (eds), *Advance Polymer Materials.* CRC Press, Boca Raton, London. NY, 349–437
199. Carrara AS, McGarry FG (1968) Matrix and interface stresses in a discontinuous fibre composite model. *J Compos Mater,* **2**:222–243
200. Cemlins A (1988) Representation of two-phase flows by volume averaging. *Int J Multiphase Flow,* **14**:81–91
201. Cesarano III J, McEuen MJ, Swiler T (1995) Computer simulation of particle packing. *Intern SAMPE Technical Conf,* **27**:658–665
202. Chaban IA (1965) Self-consistent field approach to calculation of the effective parameters of microinhomogeneous media. *Akust Zhurn,* **10**:351–358 (In Russian. Engl Transl. *Soviet Physics-Acoustics,* **10**:298–302)
203. Chaikin PM, Donev A, Man W, Frank H. Stillinger FH, Torquato S (2006) Some observations on the random packing of hard ellipsoids. *Ind Eng Chem Res,* **45**:6960–6965
204. Chan HLW, Unsworth J (1989) Simple model for piezoelectric ceramic/polymer 1–3 composites used in ultrasonic transducer applications. *IEEE Trans. Ultrasonics Ferroelectr Frequen Control,* **36**:434–441

205. Chandra A, Huang Y, Hu RX (1997) Crack-size dependence of overall responses of fiber-reinforced composites with matrix cracking. *Int J Solids Struct,* **34**:3837–3857
206. Chandra N, Namilae S, Shet C (2004) Local elastic properties of carbon nanotubes in the presence of Stone-Wales defects. *Phys Rev,* **B69**:094101
207. Chang JS, Cheng CH (1992) Thermoelastic properties of composites with short coated fibers. *Int J Solids Struct,* **29**:2259–2279
208. Chang T, Huajian Gao H (2003) Size-dependent elastic properties of a single-walled carbon nanotube via a molecular mechanics model. *J Mech Phys Solids,* **51**:1059–1074
209. Chekin BC (1970) Effective parameters of elastic medium with randomly distributed cracks. *Izv AN SSSR, Fiz Zemli,* N10:13–21 (In Russian. Engl Transl. *Phys Solid Earth,* **5**)
210. Chen C, Lu TJ, Fleck NA (1999) Effect of imperfections on the yielding of two-dimensional foams. *J Mech Phys Solids,* **47**:2235–2272
211. Chen C, Fleck NA (2002) Size effects in the constrained deformation of metallic foams. *J Mech Phys Solids,* **50**:955–977
212. Chen CH, Wang YC (1996) Effective thermal conductivity of misoriented short-fiber reinforced composites. *Mech Mater,* **23**:217–228
213. Chen FG, Young K (1977) Inclusions of arbitrary shape in an elastic medium. *J Math Phys,* **18**:1412–1416
214. Chen G, Zhou J (1992) *Boundary Element Methods.* Academic Press, London
215. Chen HS, Acrivos A (1978a) The solution of the equations of linear elasticity for an infinite region containing two spherical inclusions. *Int J Solids and Struct,* **14**:331–348
216. Chen HS, Acrivos A (1978b) The effective elastic moduli of composite materials containing spherical inclusions at non-dilute concentrations. *Int J Solids Struct,* **14**:349–364
217. Chen T (1993a) Thermoelastic properties and conductivity of composites reinforced by spherically anisotropic inclusions, *Mechan of Mater* **14**:257–268
218. Chen T (1993b) Piezoelectric properties of multiphase fibrous composites: some theoretical results. *J Mech Phys Solids,* **41**:1781–1794
219. Chen TY (1994) Micromechanical estimates of the overall thermoelectroelastic moduli of multiphase fibrous composites. *Int J Solids Struct,* **31**:3099–3111
220. Chen T (1999) Exact moduli and bounds of two-phase composites with coupled multifield linear responses *J Mech Phys Solids,* **45**:385–398
221. Chen T, Dvorak GJ, Yu CC (2007) Size-dependent elastic properties of unidirectional nano-composites with interface stresses. *Acta Mechan,* **188**:39–54
222. Chen T, Nan C-W, Weng GJ, Chen G-X (2003a) Unified approach for the estimate of effective magnetostriction of composites and polycrystals with particulate and columnar microstructures. *Phys Rev,* **B68**:224406
223. Chen T, Nan C-W, Weng GJ (2003b) Exact connections between effective magnetostriction and effective elastic moduli of fibrous composites and polycrystals. J Appl Phys, **94**:491–495
224. Chen T, Yang S-H (1995) The problem of thermal conductivity for two ellipsoidal inhomogeneities in an anisotropic medium and its relevance to composite materials. *Acta Mechan,* **111**:41–58
225. Chen X, Papathanasiou TD (2004) Interface stress distributions in transversely loaded continuous fiber composites: parallel computation in multi-fiber RVEs using the boundary element method. *Comp Sci Technol,* **64**:1101–1114
226. Cheng H, Torquato S (1997) Electric-field fluctuations in random dielectric composites. *Phys Rev,* **B56**:8060–8068

227. Cheng J, Jordan EH, Walker KP (1997) Gauss integration applied to a Green's function formulation for cylindrical fiber composites. *Mechan Mater* **26**:247–267
228. Cheng J, Wang B, Du S (2002) A statistical model prediction of effective electroelastic properties of polycrystalline ferroelectric ceramics with randomly oriented defects. *Mechan Mater* **34**:643–655
229. Cheng YF, Guo SJ, Lay HY (2000) Dynamic simulation of random packing of spherical particles. *Powder Technol,* **107**:123–130
230. Cherkaoui M, Sabar H, Berveiller M (1995) Elastic composites with coated reinforcements: a micromechanical approach for non homothetic topology. *Int J Engng Sci,* **33**:829–843
231. Chernikh KF (1986) *Nonlinear Theory of Elasticity in Machine-Building Calculations.* Mashinostroenie, Leningrad (In Russian)
232. Chernikh KF, Shubina IM (1978) Allowance for the compressibility of rubber. Mekhanika Elastomerov. *Kuban. Univ, Krasnodar,* **263**:56–62 (In Russian)
233. Chiew Y-C, Glandt ED (1987) Effective conductivity of dispersion: the effect of resistance at the particle surfaces. *Chem Engng Sci,* **42**:2677–2685
234. Chow TS, Hermans JJ (1969) The elastic constants of fiber reinforced materials. *J Compos Mater* **3**:382–396
235. Chou T-W (1992) *Microstructural Design of Fiber Composites.* Cambridge Solid State Science Series. Cambridge University Press, Cambridge, UK, 99–168
236. Choy TA, Alexandropoulos A, Thorpe MF (1998) Dielectric function for a material containing hyperspherical inclusions in $O(c^2)$: I. Multipole expansions; II. Method of images. *Proc Roy Soc Lond Ser,* **A454**(1975):1973–1992, 1993–2013
237. Christensen RM (1979) *Mechan of Composite Materials.* Wiley Interscience, New York
238. Christensen RM (1990) A critical evaluation for a class of micromechanics models. *J Mech Phys Solids,* **38**:379–404
239. Christensen RM, Lo KH (1979) Solutions for effective shear properties in three phase sphere and cylinder models. *J Mech Phys Solids,* **27**:315–330
240. Chudnovsky A, Wu S (1993) Evaluation of energy release rate in the crack-microcrack interaction problem. *Int J Solids Structure,* **29**:1699–1709
241. Chueng HT, Kim HG (1987) Characteristics of domain in tetragonal phase PZT ceramics. *Ferroelectrics,* **76**:327–333
242. Clarke AS, Willey JD (1987) Numerical simulation of the dense random packing of a binary mixture of hard spheres: amorphous metals. *Phys Rev,* **B35**:7350–7356
243. Cleary MP, Chen LW, Lee SM (1980) Self-consistent techniques for heterogeneous solids. *J Engng Mech,* **106**(5):861–871
244. Conlon KT, Wilkinson DS (1996) Microstructural inhomogeneity and the strength of particulate metal matrix composites. In: Pineau A, Zaoui A (eds) *IUTAM Symp. on Micromechanics of Plasticity and Damage of Multiphase Materials,* Kluwer, Dordrecht, 347–354
245. Cooper CA, Ravich D, Lips D, Mayer J, Wagner HD (2002) Distribution and alignment of carbon nanotubes and nanofibrils in a polymer matrix. *Comp Sci Technol,* **62**:1105–1112
246. Corson PB (1974) Correlation function for predicting properties of heterogeneous materials. I. Experimental measurement of spatial correlation functions in multiphase solids. *J Appl Phys,* **45**:3159–3164
247. Corvasce F, Lipinski P, Berveiller M (1990) The effects of thermal, plastic and elastic stress concentration on the overall behavior of metal matrix composites. In: Dvorak GJ (ed), *Inelastic Deformation of Composite Materials.* Springer-Verlag, New York, 389–408

248. Cox BN, Dadkhah MS, James MR, Marshall DB, Morris WL, Shaw M (1990) On determining temperature dependent interfacial shear properties and bulk residual stresses in fibrous composites. *Acta Metall,* **38**:2425–2433
249. Cox DR, Isham V (1980) *Point Processes.* Chapman and Hall, London and NY
250. Creswik RJ, Farah HA, Poole CP (1998) *Introduction to Renormalization Group Methods Physics.* John Wiley & Sons, New York
251. Cristescu ND, Craciun E-M, Soós E (2003) *Mechanics of Elastic Composites.* Chapman & Hall/CRC, Boca Raton, FL
252. Curtin WA (1999) Stochastic damage evaluation and failure in fiber-reinforced composites. Hutchinson JW, Wu TJ (eds), *Adv Appl Mech,* **36**:163–253
253. Curtin WA, Miller RE (2003) Atomistic/continuum coupling in computational materials science. *Modelling Simul Mater Sci Engng,* **11**:R33–R68
254. Cutler RA, Vircar AV (1985) The effect of binder thickness and residual stresses on the fracture toughness of cement carbide. *J Mater Sci,* **20**:3557–3573
255. Dai LH, Huang GJ (2001) Incremental micromechanical scheme for nonlinear particulate composites *Int J Mech Sci,* **43**:1179–1193
256. Dao M, Gu P, Maewal A, Asaro RJ (1997) A micromechanical study of residual stresses in functionally graded materials. *Acta Mater,* **45**:3265–3276
257. Darinskii BM, Fokin AG, Shermergor TD (1967) The calculation of elastic moduli of polycrystalline agragates. *Zh Prikl Mekh Thekhn Fiziki,* **6**(5):123 (In Russian. Engl Transl. *J Appl Mech Tech Phys,* **8**:79–82)
258. Darinskii BM, Shermergor TD (1964) Temperature relaxation in polycrystals of cubic structure. *Fizika Metallov i Metallovedenie,* **18**:645 (In Russian)
259. Darinskii BM, Shermergor TD (1965) Elastic moduli of cubic polycrystals. *Zh Prikl Mekh Thekhn Fiziki,* **6**(4):79–82 (In Russian)
260. Datta SK (1976) Scattering of elastic waves by a distribution of inclusions. *Archives of Mech,* **28**:317–324
261. Datta SK (1977a) Diffraction of plane elastic waves by ellipsoidal inclusions. *J Acoust Soc Am,* **61**:1432–1437
262. Datta SK (1977b) A self-consistent approach to multiple scattering by elastic ellipsoidal inclusions. *J Appl Mech,* **44**:657–661
263. Dave RN, Rosato AD, Bhaswan K (1995) Characterization of clustering microstructure in highly elastic low density uniform granular shear flows. *Mech Res Commun,* **22**:335–342
264. Davies GJ (1973) *Solidification and Casting.* Appl Sci Publication, Oxford
265. Davis IL, Carter RG (1989) Random particle packing by reduced dimension algorithms. *J Appl Phys,* **67**:1022–1029
266. Dederichs P H, Zeller R (1973) Variational treatment of the elastic constants of disordered materials. *Z Physik,* **259**:103–116
267. Deeg WF (1980) *The Analysis of Dislocation, Crack and Inclusion Problems in Piezoelectric Solids.* PhD Thesis, Stanford University, Stanford, CA
268. Delves LM, Mohamed JL (1985) *Computational Methods for Integral Equations.* Cambridge University Press, Cambridge, UK
269. Desrumaux F, Meraghi F, Benzeggagh ML (2001) Generalized Mori-Tanaka scheme to model anisotropic damage using numerical Eshelby tensor. *J Compos Mater* **35**:603–624
270. Diener G, Hurrich A, Weissbarth J (1984) Bounds on the non–local effective elastic properties of composites. *J Mech Phys Solids,* **32**:21–39
271. Dienes JK, Zuo QH, Kershner JD (2006) Impact initiation of explosives and propellants via statistical crack mechanics. *J Mech Phys Solids*, **54**, 1237–1275
272. Diggle PJ (2003) *Statistical Analysis of Spatial Point Patterns* (2nd edn). Academic Press, New York

273. Djiordjevic BR, Hetherington JH, Thorpe, MF (1996) Spectral function for a conducting sheet containing circular inclusions. *Phys Rev,* **B35**:14862–14871
274. Döge G (2000) Grand canonical simulation of hard-disc systems by simulated tempering. In: Mecke KR, Stoyan D (eds) *Statistical Physics and Spatial Statistics: The Art of Analyzing and Modeling Spatial Structures and Pattern Formation.* Lecture Notes in Physics, Vol. 554, Berlin
275. Dong CY, Bonnet M (2002) An integral formulation for steady-state elastoplastic contact over a coated half-plane. *Comput Mech,* **28**:105–121
276. Dong CY, Lo, SH, Cheung YK (2002) Application of the boundary-domain integral equation in elastic inclusion problems. *Engng Anal Bound Elements,* **26**:471–477
277. Dong CY, Lo SH, Cheung YK (2004) Numerical solution for elastic half-plane inclusion problems by different integral equation approaches. *Engng Anal Bound Elements,* **28**:123–130
278. Drugan WJ (2000) Micromechanics-based variational estimations for a higher-order nonlocal constitutive equation and optimal choice of effective moduli for elastic composites. *J Mech Phys Solids,* **48**:1359–1387
279. Drugan WJ, Willis JR (1996) A micromechanics-based nonlocal constitutive equation and estimates of representative volume elements for elastic composites. *J Mech Phys Solids,* **44**:497–524
280. Du JK, Shen YP, Ye DY, Yue FR (2004) Scattering of anti-plane shear waves by a partially debonded magneto-electro-elastic circular cylindrical inhomogeneity. *Int J Engin Sci,* **42**:887–913
281. Du X, Ostoja-Starzewski M (2006) On the scaling from statistical to representative volume element in thermoelasticity of random materials. *Networks Heterogeneous Media,* **1**:259–274
282. Du ZZ, McMeeking R M, Schmauder S (1995) Transverse yielding and matrix flow past the fibers in metal matrix composites. *Mech Mater* **21**:159–167
283. Duan HL, Wang J, Huang ZP, Karihaloo BL (2005) Eshelby formalism for nano-inhomogeneities. *Proc Roy Soc Lond Ser,* **A461**:3335–3353
284. Duan ZP, Kienzler R, Herrmann G (1986) An integral equation method and its application to defect mechanics. *J Mechan Phys Solids,* **34**:539–561
285. Dul'nev GH, Malarev BI (1990) Percolation theory in the conductivity theory of inhomogeneous media. *Inzhenerno Fiz Zhurnal,* **39**:522–539 (In Russian)
286. Dumonted H (1986) Study of a boundary layer problem in elastic composite materials. *Math Model Numer Anal,* **20**:265–286
287. Dunn ML(1993) Exact relations between the thermoelectroelastic moduli of heterogeneous materials. *Proc Roy Soc Lond,* **A441**:549–557
288. Dunn ML (1994a) Electroelastic Green's functions for transversely isotropic piezoelectric media and their applications to the solutions of inclusion and inhomogeneity problems. *Int J Engrg Sci,* **32**:119–131
289. Dunn ML (1994b) Thermally induced Relds in electroelastic composite materials: average Relds and effective behavior. *J Engng Mat Technol,* **116**:200–207
290. Dunn ML, Taya M (1993a) Micromechanics predictions of the effective electroelastic moduli of piezoelectric composites. *Int J Solids Struct,* **30**:161–175
291. Dunn ML, Taya M (1993b) An analysis of piezoelectric composite materials containing ellipsoidal inhomogeneities. *Proc R Soc Lond: A* **443**:265–287
292. Duschlbauer D, Pettermann HE, Böhm HJ (2003) Mori Tanaka based evaluation of inclusion stresses in composites with nonaligned reinforcements. *Scripta Materialia,* **48**:223–228

293. Dutta A, Mashelkar RA (1989) Thermal conductivity of structured liquids. In: Hartnet JP, Irvin TF (eds), *Advances in Heat Transfer,* Academic Press, New York, **18**:161–239
294. Duva JM (1988) A constitutive description of nonlinear materials containing voids. *Mech Mater* **5**:137–144
295. Dvorak GJ (1990) On uniform fields in heterogeneous media. *Proc Roy Soc Lond,* **A431**:89–110
296. Dvorak GJ (1992a) On some exact results in thermoelasticity of composite materials. *J Thermal Stresses,* **15**:211–228
297. Dvorak GJ (1992b) Transformation field analysis of inelastic composite materials. *Proc R Soc Lond,* **A437**:311–327
298. Dvorak GJ (1993) Nadai lecture–micromechanics of inelastic composite materials: theory and experiment. *J Engng Mater Tech,* **115**:327–338
299. Dvorak GJ (2000) Composite materials: inelastic behavior, damage, fatigue, and fracture. *Int J Solids Struct,* **37**:155–170
300. Dvorak GJ, Bahei-El-Din YA, Wafa AM (1994) The modeling of inelastic composite materials with the transformation field analysis. *Modelling Simul Mater Sci Engng,* **2**:571–585
301. Dvorak GJ, Benveniste Y (1992a) On the thermomechanics of composites with imperfectly bonded interfaces and damage. *Int J Solids Struct,* **29**:2907–2919
302. Dvorak GJ, Benveniste Y (1992b) On transformation strains and uniform fields in multiphase elastic media. *Proc Roy Soc Lond,* **A437**:291–310
303. Dvorak GJ, Srinivas MV (1999) New estimations of overall properties of heterogeneous solids. *J Mech Phys Solids,* **47**:899–920
304. Dykhne AM (1970) Conductivity of a two-dimensional two-phase system. *J Experiment Theor Phys (JETP)* 59:110–116 (In Russian. Engl Trunsl. *Soviet Phys,* (1971) **32**:63–65)
305. Dyskin AV, Mühlhaus HB (1995) Equilibrium bifurcations in dipole asymptotics model of periodic crack arrays. In: Mühlhaus HBM (ed) *Continuum Models for Materwith Microstructure.* John Wiley & Sons, New York, 69–104
306. Ebbesen TW (1997) *Carbon Nanotubes: Preparation and Properties.* CRC Press, Boca Raton, FL
307. Eischen JW, Torquato S (1993) Determining elastic behavior of composites by the boundary element method. *J Appl Phys,* **74**:159–170
308. Ellaway SW, Fauxa DA (2002) Effective elastic stiffnesses of InAs under uniform strain. *J Appl Phys,* **92**:3027–3033
309. Entchev PB, Lagoudas DC (2002) Modeling porous shape memory alloys using micromechanical averaging techniques *Mechan Mater.* **34**:1–24
310. Enikolopyan NS, Fridman ML, Stalnova IO, Popov VL (1990) Filled polymers: mechanical properties and processability. *Adv Polym Sci,* **96**:1–67
311. Erdogan F (1995) Fracture mechanics of functionally graded materials. *Compos Engng,* **5**:753–770
312. Eringen AC (1968) Mechanics of micromorphic continua. In: Kröner E (ed), *Mechan of Generalized Continua.* Springer-Verlag, Berlin, 18–35
313. Eringen AC (1976) Nonlocal polar field theories. In: Eringen AC (ed), *Continuum Phys. Vol. IV. Polar and Nonlocal Field Theories.* Academic Press, New York, 205–267
314. Eringen AC (1978) Nonlocal continuum mechanics and some applications. In: Barut AO (ed) *Nonlinear Equations in Phys and Mathematics.* Reidel, Dordrecht, 271–318
315. Eringen AC (1999) *Micromedium Field Theories I. Foundations and Solids.* Springer-Verlag, Berlin

316. Eringen AC (2002) *Nonlocal Continuum Field Theories.* Springer-Verlag, New York
317. Eshelby JD (1956) The continuum theory of lattice defects. In: Seitz F, Turnbull D (eds), *Solid State Phys.* Academic Press, New York, **3**:79–144
318. Eshelby JD (1957) The determination of the elastic field of an ellipsoidal inclusion, and related problems. *Proc Roy Soc Lond,* **A241**:376–396
319. Eshelby JD (1961) Elastic inclusion and inhomogeneities. In: Sneddon IN, Hill R (eds), *Prog in Solid Mechan.* North-Holland, Amsterdam, **2**:89–140
320. Eskin GI (1981) *Boundary Value Problems for Elliptic Pseudodifferential Equations.* American Mathematical Society, Providence, RI
321. Estevez R, Maire E, Franciosi P, Wilkinson DS (1999) Effect of particle clustering on the strengthening versus damage rivalry in particulate reinforced elastic plastic materials: a 3-D analysis from a self-consistent modeling. *Eur J Mech A/Solids,* **18**:785–804
322. Evans AG (1987) Microfracture from thermal expansion anisotropy; I. Single phase systems. *Acta Metall,* **26**:1845–1853
323. Evans AG (1989) The new high-toughness ceramics. In: Wei RP, Gangloff RP (eds), *Fracture Mechanics: Perspectives and Directions (Twentieth Symposium)* Amer Soc Testing and Mater, 267–291
324. Fabrikant VI (1989) Complete solutions to some mixed boundary value problems in elasticity. In: Hutchinson JW, Wu TJ (eds) *Advances in Applied Mechanics* Academic Press, New York, 153–223
325. Fakri N, Azrar LL, Bakkali LE (2003) Electroelastic behavior modeling of piezoelectric composite materials containing spatially oriented reinforcements. *Int J Solids Struct,* **40**:361–384
326. Fan X-J, Phan-Thien N, Zheng R (1998) Complemented double layer boundary element method for periodic suspensions. *Z Angew Math Phys,* **49**:167–193
327. Fassi–Fehri O, Hihi A, Berveiller M (1989) Multiple site self consistent scheme. *Int J Engng Sci,* **27**:495–502
328. Feder J (1980) Random sequential adsorption *J Theor Biol,* **87**:237–254
329. Feder J (1988) *Fractals.* Plenum Press, New York
330. Fedoryuk MV (1988) Diffraction of acoustic waves on a triaxial ellipsoid. *J Acoustics,* **34**:160–164 (In Russian)
331. Federico S, Grillo A, Herzog W (2004) A transversally isotropic composite with a statistical distribution of spheroidal inclusions: a geometrical approach to overall properties. *J Mech Phys Solids,* **52**:2309–2327
332. Feng X (2001) Effective elastic moduli of polymer-layered silicate nanocomposites. *Chin Sci Bull,* **46**:1130–1133
333. Ferrante FJ, Arwade SR, Graham-Brady LL (2005) A translation model for non-stationary, non-Gaussian random processes. *Probabilistic Engin Mech,* **20**:215–228
334. Ferrari M (1991) Asymmetry and the high concentration limit of the Mori–Tanaka effective medium theory. *Mech Mater,* **11**:251–256
335. Fertig III RS, Garnich MR (2004) Influence of constituent properties and microstructural parameters on the tensile modulus of a polymer/clay nanocomposite *Compos Sci Technol,* **64**:2577–2588
336. Fett T, Munz D (1993) Influence of bridging interactions on the lifetime behavior of coarse-grained Al_2O_3. *J Eur Ceram Soc,* **12**:131–138
337. Filatov AN, Sharov LV (1979) *Integral Inequalities and the Theory of Nonlinear Oscillations.* Nauka, Moscow (In Russian)
338. Fish J (1992) The s-version of the finite element method. *Comput Struct,* **43**:539–547

339. Fish J (2006) Bridging the scales in nano engineering and science. *J Nanoparticle Res,* **8**:577–594
340. Fish J, Wagiman A (1993) Multiscale finite element method for a locally nonperiodic heterogeneous medium. *Comput Mech,* **12**:164–180
341. Fisher FD, Mayrtofer K (1997) The crack-closure effect for a penny-shaped crack bridged by an arbitrary located single fiber. *Composites,* **B28**:425–432
342. Fisher FT, Bradshaw RD, Brinson LC (2003) Fiber waviness in nanotube-reinforced polymer composites: I. Modulus predictions using effective nanotube properties. *Composites Sci and Technol,* **63**:1689–1703
343. Fleck NA, Hutchinson JW (1997) Strain gradient plasticity. In: Hutchinson JW, Wu TY (eds), *Advances in Appl Mech,* Academic Press, New York, **33**:295–361
344. Fokin AG (1984) A method of solving problems of the linear theory of elasticity. *Prikl Math Mech,* **48**:436–446 (In Russian. Engl Transl. *J Appl Math Mech,* **48**:315–323)
345. Fokin AG, Shermergor TD (1968a) The boundaries of the effective elastic moduli for inhomogeneous solids. *Zhurnal Prikl Mekh Tekhnic Fiziki,* **9**(4):39–46 (In Russian. Engl Transl. *J Appl Mech Tech Phys,* **9**:381–388)
346. Fokin AG, Shermergor TD (1968b) Calculation of elastic moduli of inhomogeneous materials. *Mekhanika Polymerov,* **4**:624–630 (In Russian. Engl Transl. *Mech Comp Mater* **4**:481–486)
347. Fokin AG, Shermergor TD (1969) Calculation of the effective elastic moduli of composite materials with multiphase interactions taken into consideration. *Zhurnal Prikl Mekh Tekhnic Fiziki,* **10**:51–57 (In Russian. Engl Transl. *J Appl Mech Tech Phys,* **10**:48–54)
348. Foldy LL (1945) The multiple scattering of waves. I. General theory of isotropic scattering by randomly distributed scatters. *Phys Rev,* **67**:107–117
349. Fond C, Riccardi A, Schirrer R, Montheillet F (2001) Mechanical interaction between spherical inhomogeneities: an assessment of a method based on the equivalent inclusion. *Eur J Mech A/Solids,* **20**:59–75
350. Forest S, Sab K (1998) Cosserat overall modeling of heterogeneous materials *Mech Res Commun,* **25**:449–454
351. Fornes TD, Paul DR (2003) Modeling properties of nylon 6/clay nanocomposites using composite theories *Polymers,* **44**:4993–5013
352. Fotiu P, Irschik H, Ziegler F (1991) Micromechanical foundations of dynamic plasticity with applications to damaging structures In: Brüller O, Mannl V, Najar J (eds), *Advances in Continuum Mechan.* Springer-Verlag, Berlin, 338–349
353. Franciosi P (2005) On the modified Green operator integral for polygonal, polyhedral and other non-ellipsoidal inclusions. *Int J Solids Struct,* **42**:3509–3531
354. Franciosi P, Lebail H (2004) Anisotropy features of phase and particle spatial pair distributions in various matrix/inclusions structures. *Acta Mater,* **52**:3161–3172
355. Frankland SJV, Calgar, A, Brener DW, Griebel M (2002) Molecular simulation of the interface of chemical crosslinks on the shear strength of carbon nanotube-polymer interface. *J Phys Chem,* **B106**:3046–3048
356. Frenkel D, Smit B (2002) *Understanding Molecular Simulation: From Algorithms to Application.* Academic Press, San Diego
357. Friesecke G, James RD (2000) A scheme for the passage from atomic to continuum theory for thin films, nanotubes and nanorods. *J Mech Phys Solids,* **48**:1519–1540
358. Frisch HL (1965) Statistics of random media. *Trans Soc Rheol,* **9**:293–312
359. Fu LS (1982) Mechanical aspects of NDE by sound and ultrasound. *Appl Mech Rev,* **35**:1047–1057

360. Fu LS (1987) Dynamic moduli and located damage in composites. In: Vary A (ed), *Material Analysis by Ultrasonics: Metals, Ceramics, Composites.* Noyes Data Corp, New York, 225–248
361. Fu LS, Mura T (1983) The determination of elastodynamic fields of an ellipsoidal inhomogeneity. *J Appl Mech,* **50**:390–396
362. Fu Y, Evans AG (1985) Some effects of microcracks on the mechanical properties of brittle solids; I. Stress-strain relations. *Acta Metall,* **33**:1515–1523
363. Fu Y, Klimkowski KJ, Rodin GJ, Berger E, Browne JC, Singer JK, Van De Geijn, RA, Vemaganti KS (1998) A fast solution method for three-dimensional many-particle problems of linear elasticity. *Int J Numer Methods Engng,* **42**:1215–1229
364. Fung YC (1965) *Foundations of Solid Mechanics.* Prentice-Hall, Englewood Cliffs, NJ
365. Furmañski P (1997) Head conduction in composites: homogenization and macroscopic behavior. *Appl Mech Rev,* **50**:327–356
366. Furukawa K, Imai K, Kurashige M (2000) Simulated effect of box size and wall on porosity of random packing of spherical particles. *Acta Mechan,* **140**:219–231
367. Gambin B, Kröner E (1989) High order terms in the homogenized stress-strain relation of periodic elastic media. *Phys Stat Sol,* **151**:513–519
368. Ganghoffer JF, de Borst R (2000) A new framework in nonlocal mechanics. *Int J Engng Sci,* **38**:453–486
369. Gao F (2004) Clay/polymer composites: the story. *Mater Today,* **7**:50–55
370. Gao H, (1988) Nearly circular shear mode cracks. *Int J Solids Struct,* **24**:177–193
371. Garboczi EJ, Day (1995) An algorithm for computing the effective linear properties of heterogeneous materials. *J Mechan Phys Solids,* **43**:1349–1362
372. Gavazzi AC, Lagoudas DC (1990) On the evaluation of Eshelby's tensor and its application to the elastoplastic fibrous composites. *Comput Mech,* **7**:13–19
373. Gel'fand IM, Milos RA, Shapiro SY (1963) *Representations of the Rotation and Lorentz Groups and Their Applications.* Pergamon, Oxford
374. Gel'fand IA, Shilov G (1964) *Generalized Functions.* Academic Press, **1**, New York
375. Giannelis EP, Krishnamoorti R, Manias E (1999) Polymer-silicate nanocomposites: model systems for confined polymers and polymer brushes. *Adv Polym Sci,* **138**:107–147
376. Gibson LJ, Ashby MF (1997) *Cellular Solids: Structure and Properties,* (2nd edn), Cambridge University Press, Cambridge, UK
377. Ghassemieh E, Nassehi V (2001) Prediction of failure and fracture mechanisms of polymeric composites using finite element analysis. Part 2: Fibre reinforced composites. *Polym Compos,* **22**:542–554
378. Ghoneim NM, Busso EP, Kioussis N, Huang H (2003) Multiscale modeling of nanomechanics and micromechanics: an overview. *Philos Mag,* **83**:3475–3528
379. Ghosh S, Moorthy S (2004) Three dimensional Voronoi cell finite element model for microstructures with ellipsoidal heterogeneties. *Comput Mech,* **34**:510–531
380. Ghosh S, Mukhopadhyay S N (1991) A two-dimensional automatic mesh generator for finite element analysis for random composites. *Compos & Struct,* **41**:245–256
381. Ghosh S, Nowak Z, Lee K (1997) Quantitative characterization and modeling of composite microstructures by Voronoi cells. *Acta Mater,* **45**:2215–2234
382. Gibiansky L, Torquato S (1998) New approximation for the effective energy of nonlinear conducting composites. *J Appl Phys,* **84**:301–305
383. Gibson L, Ashby M (1998) *Cellular Solids.* Cambridge University Press, Cambridge, UK

384. Gilormini P, Brechet Y (1999) Syntheses: mechanical properties of heterogeneous media: Which material for which model? Which model for which material? *Modelling Simul Mater Sci Engng,* **7**:805–816
385. Giordano S (2005) Order and disorder in heterogeneous material microstructure: electric and elastic characterisation of dispersion of pseudo-oriented spheroids. *Int J Engng Sci,* **43**:1033–1058
386. Glas F (1991) Coherent stress relaxation in a half space: modulated layers, inclusions, steps, and a general solution. *J Appl Phys,* **70**:3556–3571
387. Glas F 2002 Analytical calculation of the strain field of single and periodic misfitting polygonal wires in a half-space. *Philos Mag,* **A82**:2591–2608
388. Glas F (2003) Elastic relaxation of a truncated circular cylinder with uniform dilatational eigenstrain in a half space. *Phys Stat Sol,* **B237**:599–610
389. Goh KL, Aspden RM, Mathias KJ, Hukins DW (2004) Finite-element analysis of the effect of material properties and the fiber shape on stresses in an elastic fiber embedded in an elastic matrix in a fiber-composite material. *Pro Roy Soc London A,* **460**:2339–2352
390. Goikhman BD, Buryachenko VA (1980) One variant of the topokinetic model of fracture of solids. *Problemy Prochnosti,* (2):28–31 (In Russian. Engl Trans. *Strength Mater* **12**:152–157
391. Goldhirsch I, Goldenberg C (2002) On the microscopic foundations of elasticity. *Eur Phys J,* **E9**:245–251
392. Golovchan VT (1974) The solution of static boundary-value problems for the elastic body constrained by spherical surfaces. *Dokl AN Ukr SSSR* (1):61–64 (In Ukrainian)
393. Gonzalez A, Roman FL, White JA (1999) A test-particle method for the calculation of the three-particle distribution function of the hard-sphere fluid: density functional theory and simulation. *J Phys: Condens Matter,* **11**:3789–3998
394. Gou J, Minaie B, Wang B, Liang Z, Zhang C (2004) Computational and experimental study of interfacial bonding of single-walled nanotube reinforced composites. *Comput Mater Sci,* **31**:225–236
395. Graham-Brady LL, Siragy EF, Baxter SC (2003) Analysis of heterogeneous composites based on moving-window techniques. *J Engng Mech,* **129**:1054–1064
396. Gray LJ, Ghosh D, Kaplan T (1996) Evaluation of the anisotropic Green's function in three dimensional elasticity. *Comput Mech,* **17**:255-261
397. Green AE, Adkins JE (1970) *Large Elastic Deformations.* Clarendon Press, Oxford
398. Green AE, Zerna W (1968) *Theoretical Elasticity* (2nd edn). Oxford University Press, Oxford
399. Green PI, Sibson R (1977) Computing Dirichlet tesselations in the plane. *Computer J,* **21**:168–173
400. Greengard L, Helsing J (1998) On the numerical evaluation of elastostatic fields in locally isotropic two-dimensional composites. *J Mechan Phys of Solids,* **46**:1441–1462
401. Greengard L, Moura M (1994) On the numerical evaluation of electrostatic fields in composite materials. *Acta Numerica,* **3**:379–410
402. Greengard L, Rokhlin V (1997) A new version of the fast multipole method for the Laplace equation in three dimensions. *Acta Numerica,* **6**:229–270
403. Grekov AA, Karamarov SO, Kuprienko AA (1989) Effective properties of a transversely isotropic piezocomposite with cylindrical inclusions. *Ferroelectrics,* **99**:115–126

404. Griebel M, Hamaekers J (2004) Molecular dynamics simulations of the elastic moduli of polymer-carbon nanotube composites. *Computer Meth Appl Mech Engng,* **193**:1773–1788
405. Grigoras S, Gusev AA, Santos S, Suter UW (2002) Evaluation of elastic constants of nanoparticles from atomistic simulations. *Polymers,* **43**:489–494
406. Grujicic M, Zhag Y (1998) Determination of effective elastic properties of functionally graded materials using Voronoi cell finite element method. *Mater Sci Engng,* **A251**:64–76
407. Gubernatis IE (1979) Long-wave approximations for the scattering of elastic waves from flaws with applications to ellipsoidal voids and inclusions. *J Appl Phys,* **50**:4046–4058
408. Gubernatis IE, Domany E, Krymhansl IA, Huberman M (1977) The Born approximation in the theory of the scattering of elastic waves by flows. *J Appl Phys,* **48**:2812–2819
409. Gurson AL (1977) Continuum theory of ductile rupture by void nucleation and growth: Part I–Yield criteria and flow rules for porous ductile media. (Ed. C Truesdell), *J Eng Mater Technol,* **99**:2–15
410. Gurtin ME (1972) The linear theory of elasticity. In: Truesdell C (ed) *Flügge's Handbuch der Physik VI a/2.* Springer-Verlag, Berlin, 1–295
411. Gurtin ME (1976) *An Introduction to Continuum Mechanics.* Academic Press, New York
412. Gusev AA (2001) Numerical identification of the potential of whisker-filled polymers. *Macromolecules,* **34**:3081–3093
413. Gusev AA, Zehnder MM, Suter UW (1996) Fluctuation formula for elastic constants. *Phys Rev,* **B 52**:52–65
414. Guseva O, Lusti HR, Gusev AA (2004) Matching thermal expension of mica-polymer nanocomposites and metals. *Model Simul Mater Sci Engng,* **12**:S101–S105
415. Guz AN, Rushchitskii YY (2003) Nanomaterials: on the mechanics of nanomaterials. *Int Appl Mech,* **39**:1271–1293
416. Hahn UA, Micheletti R, Pohlink R, Stoyan D, Wendrock H (1999) Stereological analysis and modelling of gradient sructures. *J Microsc,* **195**:113–124
417. Haile JM (1992) *Molecular Dynamics Simulation: Elementary Methods.* Wiley-Interscience, New York
418. Hall P (1988) *Introduction to the Theory of Coverage Processes.* John Wiley & Sons, New York
419. Effective moduli of cellular materials. *J Reinf Plast Comp,* **12**:186–197
420. Halle DK (1976) The physical properties of composite materials *J Mater Sci,* **11**:2105–2141
421. Halpin JC, Kardos JL (1976) The Halpin-Tsai equations: a review. *Polym Engng Sci,* **16**:344–352
422. Hamm M, Elliott JA, Smithson HJ, Windle AH (2004) Multiscale modeling of carbon nanotubes. *Mat Res Soc Symp Proc,* **788**:L10.11.6
423. Han CD (1974) Rheological properties of calcium carbonate filled polypropylene melts. *J Appl Polym Sci,* **18**:821–829
424. Hansen JP, McDonald IR (1986) *Theory of Simple Liquids.* Academic Press, New York
425. Harlow DG, Phoenix SL (1978) The chain of bundles probability model for the strength of fibrous materials I: Analysis and conjectures. *J Composite Mater* **12**:195–214
426. Harrison P (2000) *Quantum Wells, Wires and Dots: Theoretical and Computational Phys.* John Wiley & Sons, New York

427. Hashin Z (1962) The elastic moduli of heterogeneous materials. *J Appl Mech,* **29**:143–150
428. Hashin Z (1965) On elastic behavior of fiber reinforced materials of arbitrary transverse phase geometry. *J Mech Phys Solids,* **13**:119–134
429. Hashin Z (1968) Assessment of the self consistent scheme approximation: conductivity of particulate composites. *J Compos Mater,* **2**:284–300
430. Hashin Z (1972) *Theory of Fiber Reinforced Materials.* NASA Contractor report CR-1974, NASA, Washington, DC
431. Hashin Z (1979) Analysis of properties of fiber composites with anisotropic constituents. *J Appl Mech,* **46**:543–550
432. Hashin Z (1983) Analysis of composite materials–a survey. *J Appl Mech,* **50**:481–505
433. Hashin Z (1984) Thermal expansion of polycrystalline aggregates: I. Exact analysis. *J Mech Phys,* **32**:149–157
434. Hashin Z (1985) Large isotropic elastic deformation of composites and porous media. *J Mech Phys Solids,* **21**:711–720
435. Hashin Z (1988) The differential scheme and its application to cracked materials. *J Mech Phys Sol,* **36**:719–733
436. Hashin Z (1991) Thermoelastic properties of particular composites with imperfect interface. *J Mech Phys Solids,* **39**:745–762
437. Hashin Z, Rosen BW (1964) The elastic moduli of fiber-reinforced materials. *J Appl Mech,* **31**:223–232
438. Hashin Z, Shtrikman S (1962a) On some variational principles in anisotropic and nonhomogeneous elasticity. *J Mech Phys Solids,* **10**:335–342
439. Hashin Z, Shtrikman S (1962b) A variational approach to the theory of the elastic behavior of polycrystals. *J Mech Phys Solids,* **10**:343–352
440. Hashin Z, Shtrikman S (1962c) A variational approach to the theory of the effective magnetic permeability of multiphase materials. *J Appl Phys,* **35**:3125–3131
441. Hashin Z, Shtrikman S (1963) A variational approach to the theory of the behavior of multiphase materials. *J Mech Phys Solids,* **11**:127–140
442. Hasimoto H (1959) On the periodic fundamental solutions of the Stokes equations and their application to the viscous flow past a cubic array of spheres. *J Fluid Mech,* **5**:317–328
443. Hatta H, Taya M (1985) Effective thermal conductivity of a misoriented short fiber composite. *J Appl Phys,* **58**:2478–2486
444. Hatta H, Taya M (1986) Equivalent inclusion method for steady state heat conduction in composites. *Int J Engng Sci,* **24**:1159–1172
445. Hatta Y, Taya M (1987) Thermal stress in a coated short fiber composite. *J Engng Mater,* **109**:59–63
446. Hazanov S (1999) On micromechanics of imperfect interfaces in heterogeneous bodies smaller than the representative volume. *Int J Engng Sci,* **37**:847–861
447. Hazegawa H, Lee VG, Mura T (1993) Hollow circular cylindrical inclusion at the surface of a half-space. *J Appl Mech,* **60**:33–40
448. He D, Ekere NN (2001) Structure simulation of concentrated suspensions of hard spherical particles *AIChE J,* **47**:53–59
449. He D, Ekere NN, Cai L (1999) Computer simulation of random packing of unequal particles. *Phys Rev,* **E60**:7098
450. Heinz H, Koerner H, Anderson KL, Vaia RA, Farmer BL (2005) A force field for mica-type silicates and dynamics of octadecylammonium chains grafted to montmorillonite. *Chem Mater,* **17**:5658–5669
451. Helsing J (1995) Estimating effective properties from cross-sectional photographs. *J Comput Phys,* **117**:281–288

452. Henderson D, Plischke M (1987) Sum rules for the pair-correlation function of inhomogeneous fluid: results for the hard-sphere-hard-wall system. *Proc Roy Soc Lond,* **A410**:409–420
453. Herczynski R, Pienkowska I (1980) Toward a statistical theory of suspension. *Annu Rev Fluid Mech,* **12**:137–269
454. Herrmann KP, Mihovsky IM (1994) On the modeling of the inelastic thermomechanical behavior and the failure of fibre-reinforced composites–a unified approach. In: Markov KZ (ed) *Advances in Math Modelling of Composite Materials.* World Scientific, Singapore, 141–191
455. Hershey V (1954) The elasticity of an isotropic aggregate of anisotropic cubic crystals. *J Appl Mech,* **21**:236–241
456. Hervé E (2002) Thermal and thermoelastic behaviour of multiply coated inclusion-reinforced composites. *Int J Solids Struct,* **39**:1041–1058
457. Hervé E, Zaoui A (1993) N-layered inclusion-based micromechanical modeling. *Int J Engng Sci,* **31**:1–10
458. Hess S, Kroger M, Voigt H (1998) Thermomechanical properties of the WCA-Lennard-Jones model system in its fluid and solid states. *Physica,* **A250**:58–82
459. Hill R (1952) The elastic behavior of a crystalline aggregate. *Proc Phys Soc,* **A65**:349–354
460. Hill R (1963a) Elastic properties of reinforced solids: some theoretical principles. *J Mech Phys Solids,* **11**:357–372
461. Hill R (1963b) New derivations of some elastic extremum principles. *Prog in Appl Mechanics.* The Prager Anniversary Volume. Macmillan, New York, 99–106
462. Hill R (1964) Theory of mechanical properties of fiber-strength materials: I. Elastic behavior. *J Mech Phys Solids,* **12**:199–212
463. Hill R (1965) A self-consistent mechanics of composite materials. *J Mech Phys Solids,* **13**:212–222
464. Hill R (1983) Interfacial operators in the mechanics of composite media. *J Mech Phys Solids,* **31**:347–357
465. Hinch EJ (1977) An averaged-equation approach to particle interactions in a fluid suspension. *J Fluid Mech,* **83**:695–720
466. Hinrichsen E L, Feder J, Jossang T (1986) Geometry of random sequential adsorption. *J Statist Phys,* **44**:793–827
467. Honein E, Honein T, Herrmann G (1990) On two circular inclusions in harmonic problems. *Quart Appl Math,* **50**:479–499
468. Honein T, Herrmann G (1990) On bounded inclusions with circular or straight boundaries in plane elastostatic *J Appl Phys,* **57**:850–856
469. Hori M, Kubo J (1998) Analysis of probabilistic distribution and range of average stress in each phase of heterogeneous materials. *J Mech Phys Solids,* **46**:537–556
470. Hori M, Nemat-Nasser S (1987) Interacting microcracks near the tip in the process zone of a macrocrack. *J Mech Phys Solids,* **35**:601–629
471. Hori M, Nemat-Nasser S (1993) Double-inclusion model and overall moduli of multi-phase composites. *Mech Mater,* **14**:189–206
472. Horii H, Nemat-Nasser S (1985) Elastic field of interacting inhomogeneities. *Int J Solids Struct,* **21**:731–745
473. Horstemeyer MF, Baskes MI, Prantil VC, Philliber J, Vanderheide S (2003) A multi-scale analysis of fixed-end simple shear using molecular dynamics, crystal plasticity, and a macroscopic internal state variable theory. *Model Simul Mater Sci Engng,* **11**:265–286
474. Hu G (1996) A method of plasticity for general aligned spheroidal void or fiber-reinforced composites. *Int J Plast,* **12**:439–449

475. Hu G (1997) Composite plasticity based on matrix average second order stress moment. *Int J Solids Struct,* **34**:1007–1015
476. Hu G, Tian F (1997) Micromechanical analysis of fatigue properties of metal-matrix composites. *Mech Res Commun,* **24**:65–68
477. Hu GK, Weng GJ (1998) Influence of thermal residual stresses on the composite macroscopic behavior. *Mech Mater* **27**:229–240
478. Hu SM (1989) Stress from a parallelepipedic thermal inclusion in a semispace. *J Appl Phys,* **66**:2741–2743
479. Huang JH, Furuhashi R, Mura T (1993) Frictional sliding inclusions. *J Mech Phys Solids,* **41**:247–265
480. Huang JH, Liu H-K, Dai W-L (2000) The optimized fiber volume fraction for magnetoelectric coupling effect in piezoelectric-piezomagnetic continuous fiber reinforced composites. *Int J Engng Sci,* **38**:1207–1217
481. Huang JH, Yu JS (1994) Electroelastic Eshelby tensors for an ellipsoidal piezoelectric inclusion. *Compos Engng,* **4**:1169–1182
482. Huang Y, Hu KX, Chandra AA (1994) A generalized self-consistent mechanics method for composite materials with multiphase inclusions. *J Mech Phys Solids,* **94**:491–502
483. Huang Y, Hu KX (1995) A generalized self-consistent mechanics method for solids containing elliptical inclusions. *J Appl Mech,* **62**:566–572
484. Huet C (1990) Application of variational concepts to size effects in elastic heterogeneous bodies. *J Mech Phys Solids,* **38**:813–841
485. Huet C (1997) An integrated micromechanics and statistical continuum thermodynamics approach for studying the fracture behavior of microcracked heterogeneous materials with delayed response. *Engng Fracture Mech,* **58**:459–556
486. Hussain F, Hojjati M, Okamoto M, Gorga RE (2006) Review article: Polymer-matrix nanocomposites, processing, manufacturing, and application: An overview. *J Compos Mater,* **40**:1511–1575
487. Hutchinson JW (1976) Bounds and self–consistent estimates for creep of polycrystalline materials. *Proc Roy Soc Lond Series* **A348**:101–127
488. Ibach H (1997) The role of surface stress in reconstruction, epitaxial growth and stabilization of mesoscopic structures. *Surf Sci Rep,* **29**(5-6):193–263
489. Ibrahim IA, Mohammed FA, Lavernia EJ (1991), Particulate reinforced metal matrix composites–a review. *J Mater Sci,* **26**:1137–1156
490. Iijima S (1991) Helical microtubes of graphitic carbon. *Nature,* **354**:56–58
491. Imam A, Johnson GC, Ferrari M (1995) Determination of the overall moduli in second order incompressible elasticity. *J Mech Phys Solids,* **43**:1087–1104
492. Ingber MS, Papathanasiou TD (1997) A parallel-supercomputing investigation of the stiffness of aligned, short-fiber-reinforced composites using the boundary element method. *Int J Numer Methods Engng,* **40**:3477–3491
493. Isupov LP (1996a) Constitutive equations of plastic anisotropic composite medium. In: Pineau A, Zaoui Z (eds), *IUTAM Symp Micromech of Plasticity and Damage of multiphase materials,* Kluwer, Dordrecht, 91–98
494. Isupov LP (1996b) A variant of the theory of plasticity of two-phase composite media. *Acta Mechan,* **119**:65–78
495. Jasiuk I, Kouider MW (1993) The effect of an inhomogeneous interphase on the elastic constants of transversely isotropic composites. *Mech Mater,* **15**:53–63
496. Jasiuk I, Mura T, Tsuchida E (1988) Thermal stresses and thermal expansion coefficient of short fiber composites with sliding interfaces. *J Engng Mater Tech,* **110**:96–100
497. Jasiuk I, Sheng PY, Tsuchida E (1997) A spherical inclusion in an elastic half-space under shear. *J Appl Mech,* **64**:471–479

498. Jeffrey DJ (1973) Conduction through a random suspension of spheres. *Proc Roy Soc Lond,* **A335**:355–367
499. Jeffrey DJ (1974) Group expansion for the bulk properties of a statistically homogeneous, random suspension. *Proc Roy Soc London,* **A338**:505–516
500. Jensen EBV (1998) *Local Stereology.* World Science, Singapore
501. Jensen LR, Pyrz R (2004) Molecular dynamics modeling of carbon nanotubes and their composites. In: Ghosh S, Lee JK (eds), *MaterProcessing and Design: Modelling, Simulation Application, NUMIFORM 2004.* American Institute of Physics, 1559–1564
502. Jeong H, Hsu DK, Shannon RE, Lian PK (1994) Characterization of anisotropic elastic constants of silicon-carbide particulate reinforced titanium metal matrix composites: Part II. Theory. *Metall Mater Trans,* **25A**:811–819
503. Jeulin D (2001) Random structure models for homogenization and fracture satistics. In: Jeulin D, Ostoja-Starzewski M (eds), *Mechanics of random and multiscale microstructures*, Springer, Wien-New York, 33–91
504. Ji XL, Jing JK, Jiang W, Jiang BZ (2002) Tensile modulus of polymer nanocomposites. *Polym Eng Sci,* **42**:983–993
505. Jiang B, Fang D-N, Hwang K-C (1999) A unified model for piezocomposites with non-piezoelectric matrix and piezoelectric inclusions. *Int J Solids Structure,* **37**:2707–2733
506. Jiang M, Jasiuk I, Ostoja-Starzewski M (2002) Apparent elastic and elastoplastic behavior of periodic composites. *Int J Solids and Struct,* **439**:199–212
507. Jikov VV, Kozlov SM, Oleinik OA (1994) *Homogenization of Differential Operators and Integral Functionals.* Springer-Verlag, Berlin
508. Jimenez G, Ogata N, Kawai H, Ogihara T (1997) Structure and thermal/ mechanical properties of poly (-caprolactone)-clay blend. *J Appl Polymer Sci,* **64**:2211–2220
509. Jin Y, Yuan FG (2003) Simulation of elastic properties of single-walled carbon nanotubes. *Composites Sci Technol,* **63**:1507–1515
510. Jodrey WS, Tory M (1985) Computer simulation of close random packing of equal spheres. *Phys Rev,* **A32**:2347
511. Johnson WC, Earmme YY, Lee JK (1980) Application of the strain field associated with an inhomogeneous precipitate. I: Theory. *J Appl Mech,* **47**:775–780
512. Joly P (1992) *Etude de la rupture d'aciers inoxydables austéno-ferritiques moulés, fragilisés par vieillissement à 400°C.* Thèse de Doctorat de l'Ecole de Paris, Paris, France
513. Ju JW, Chen TM (1992) Micromechanics and effective moduli of elastic composites with randomly dispersed inhomogeneities. *Macroscopic Behavior of Heterogeneous Materials from the Microstructure.* ASME, ADM 147 NY, 95–109
514. Ju JW, Chen TM (1994a) Effective elastic moduli of two-dimensional brittle solids with interacting microcracks, I: Basic formulations. *J Appl Mech,* **61**:349–357
515. Ju JW, Chen TM (1994b) Micromechanics and effective moduli of elastic composites containing randomly dispersed ellipsoidal inhomogeneities. *Acta Mechan,* **103**:103–121
516. Ju JW Chen TM (1994c) Effective elastic moduli of two-phase composites containing randomly dispersed spherical inhomogeneities. *Acta Mechan,* **103**:123–144
517. Ju JW, Chen TM (1994e) Micromechanics and effective elastoplastic behavior of two-phase metal matrix composites. *J Engng Mater Tech,* **116**:310–318
518. Ju JW, Lee HK (2001) A micromechanical damage model for effective elastoplastic behavior of partially debonded ductile matrix composites. *Int J Solids Struct,* **38**:6307–6332

519. Ju JW, Sun, LZ (1999) A novel formulation for the exterior point Eshelby's tensor of an ellipsoidal inclusion. *J Appl Mech,* **66**:570–572
520. Ju JW, Sun LZ (2001) Effective elastoplastic behavior of metal matrix composites containing randomly located aligned spheroidal inhomogeneities. Part I: micromechanics. *Int J Solids Struct,* **38**:183–201
521. Ju JW, Tseng KH (1992) A three-dimensional micromechenical theory for brittle solids with interacting microcracks. *Int J Damage Mech,* **1**:102–131
522. Ju JW, Tseng KN (1994) Effective elastoplastic behavior of two-phase metal matrix composites: micromechanics and computational algorithms. In: Voyiadjis GZ, Ju JW (eds), *Inelasticity and Micromechanics of Metal Matrix Composites.* Elsevier, Amsterdam, 121–141
523. Ju JW Tseng KH (1995) Improved two-dimensional micromechanical theory for brittle solids with randomly located interacting microcracks. *Int J Damage Mech,* **4**:23–57
524. Ju JW, Tseng KH (1996) Effective elastoplastic behavior of two-phase ductile matrix composites: a micromechanical framework. *Int J Solids Struct,* **33**:4327–4291
525. Ju JW, Zhang XD (1998) Micromechanics and effective transverse elastic moduli of composites with randomly located aligned circular fibers. *Int J Solids Struct,* **35**:941–960
526. Ju JW, Zhang XD (2001) Effective elastoplastic behaviour of ductile matrix composites containing randomly located aligned circular fibers. *Int J Solids Struct,* **38**:4045–4069
527. Kachanov LM (1971) *Foundation of Theory of Plasticity.* North-Holland, Amsterdam
528. Kachanov M (1987) Elastic solids with many cracks: a simple method of analysis *Int J Solids Struct,* **23**:23–43
529. Kachanov M (1993) Elastic solids with many cracks and related problems. In: Hutchinson JW, Wu TJ (eds), *Adv Appl Mechan.* Academic Press, New York, **30**:259–445
530. Kachanov M, Laures J-P (1989) Three-dimensional problem of strongly interacting arbitrarily located penny-shaped cracks. *Int J Fracture,* **41**:289–313
531. Kaganer VM, Jenichen B, Paris G, Ploog KH, Konovalov O, Mikulik P, Arai S (2002) Strain in buried quantum wires: analytical calculations and x-ray diffraction study. *Phys Rev,* **B66**:035310
532. Kalamkarov AL (1992) *Composite and Reinforced Elements of Construction.* John Wiley & Sons, New York
533. Kalamkarov AL, Kolpakov AG (1997) *Analysis, Design and Optimization of Composite Structures.* John Wiley & Sons, New York
534. Kalamkarov AL, Kudriavtsev BA, Parton VZ (1990) The boundary-layer method in the fracture mechanics of composites of periodic structure. *Prikl Matem Mech,* **54**:322–328 (In Russian, Engl Transl. *J Appl Math Mech,* **54**:266–271)
535. Kukarni R, Ochoa O (2006) Transverse and longitudinal CTE measurements of carbon fibers and their impact on interfacial residual stresses in composites. *J Compos Mater*, **40**, 734–754
536. Kaloerov SA, Goryanskaya ES (1995) Two-dimensional stress state of multiply-connected anisotropic solid with holes and cracks. *Theor Appl Mech,* **25**:45–56
537. Kaminski MM (2005) *Computational Mechanics of Composite Materials: Sensitivity, Randomness, and Multiscale Behaviour.* Springer-Verlag, London
538. Kanaun SK (1977) Self-consistent field approximation for an elastic composite medium. *Zhurnal Prikladnoi i Tehknich Fiziki,* **18**(2):160–169 (In Russian. Engl Transl. *J Appl Mech Techn Phys,* **18**:274–282)

539. Kanaun SK (1982) The effective field method in linear problems of statics of composite media. *Prikl. Matem. Mechanika,* **46**:655–665. (In Russian. Engl. Transl. *J Appl Math Mech,* **46**:520–528)
540. Kanaun SK (1983) Elastic medium with random fields of inhomogeneities. In: Kunin IA *Elastic Media with Microstructure.* Springer–Verlag, Berlin, **2**:165–228
541. Kanaun SK (1990) Self-consistent averaging schemes in the mechanics of matrix composite materials. *Mekhanika Kompozitnikh Materialov,* **26**:702–711 (In Russian. Engl Transl. *Mech Compos Mater,* **26**:984–992)
542. Kanaun SK, Levin VM (1984) Development of effective wave operator for medium with isolated inhomogeneities. *Mech Solids,* (5):67–76 (In Russian)
543. Kanaun SK, Levin VM (1986) Propagation of elastic waves through media with thin crack-like inclusions. *Prikladnaya Matematika i Mekhanika,* **50**:309-319 (In Russian. Engl Transl. *J Appl Math Mech,* **50**:231–239
544. Kanaun SK, Levin VM (1993) *Effective Field Method in Mechanics of Composite Materials.* University of Petrozavodsk, Petrozavodsk (In Russian)
545. Kanaun SK, Levin VM (1994) Effective field method on mechanics of matrix composite materials. In: Markov KZ (ed), *Advances in Math Modelling of Composite Materials.* World Scientific, Singapore, 1–58
546. Kanaun SK, Levin VM (2005) Propagation of shear elastic waves in composites with a random set of spherical inclusions (effective field approach) *Int J Solids Struct,* **42**:3971–3997
547. Kanaun SK, Levin VM, Sabina FJ (2004) Propagation of elastic waves in composites with random set of spherical inclusions (effective medium approach). *Wave Motion,* **40**:69–88
548. Kanit T, Forest S, Galliet I, Mounoury V, Jeulin D (2003) Determination of the size of the representative volume element for random composites: statistical and numerical approach. *Int J Solids Struct,* **40**:3647–3679
549. Kansal AR, Truskett TM, Torquato S (2000) Nonequilibrium hard-disk packing with controlled orientational order. *J Chem Phys,* **113**:4844–4851
550. Kantorovich LV, Krylov VI (1964) *Approximate Methods of Higher Analysis.* John Wiley & Sons, New York
551. Karlsson LM, Liljeborg A (1994) Second-order stereology for pores in translucent alumina studied by confocal scanning laser microscopy. *J Microsc,* **175**:186–194
552. Kashevskiy BE, Poroshin YuV (1988) Effective viscosity of a magnetic particle suspension in electrically conductive fluid. *Magnitnaya Gidrodinamika,* (2):78–82 (In Russian)
553. Kataoka Y, Taya M (2000) Analysis of mechanical behavior of a short fiber composite using macromechanics based model (effect of fiber clustering on composite stiffness and crack initiation). *JSME Int J,* **A43**:46–52
554. Keer LM, Lin W (1990) Analysis of cracks in transversally isotropic media. In: Weng GJ, Taya M, Abe H (eds), *Micromechanics and Inhomogeneity. The Toshio Mura 65the Anniversary Volume.* Springer-Verlag, New York, 187–195
555. Keiller RA, Feuillebous F (1993) Head conduction through an inhomogeneous suspension. *Proc Roy Soc Lond,* **A440**:717–726
556. Keller JB (1963) Conductivity of a medium containing a dense array of perfectly conducting spheres or cylinders or nonconducting cylinders. *J Appl Phys,* **34**:991–993
557. Keller JB (1987) Effective conductivity of reciprocal media. In: Papanicolau G (ed), *Random Media.* Springer-Verlag, New York, 183–188
558. Kerans RJ, Hay RS, Parthasaraty TA, Cinibulk MK (2002) Interface design for oxidation-resistant ceramic composites. *J Am Ceram Soc,* **85**:2599–2632

559. Khoroshun LP (1967) The theory of isotropic deformation of elastic bodies with random inhomogeneities. *Priklad Mech,* **3**:12–19 (In Russian)
560. Khoroshun LP (1977) About heat conductivity equations of composites. *Docladu Acad Nauk Ukraine.* **A**(7):630–634 (In Russian)
561. Khoroshun LP (1978) Random functions theory in problems on the macroscopic characteristics of microinhomogeneous media. *Priklad Mekh,* **14**(2):3–17 (In Russian. Engl Transl. *Soviet Appl Mech,* **14**:113–124)
562. Khoroshun LP (1987) Conditional-moment method in problems of the mechanics of composites. *Priklad Mekh,* **23**(10):100–108 (In Russian. Engl Transl. *Soviet Appl Mech,* **23**:989–998)
563. Khoroshun L (1996) On a mathematical model for inhomogeneous deformation of composites. *Priklad Mekh,* **32**(5):22–29 (In Russian. Engl Transl. *Int Appl Mech,* **32**:341–348)
564. Khoroshun LP (2000) Mathematical models and methods of the mechanics of stochastic composites. *Prikl Mekh,* **30**(10):30–62 (In Russian. Engl Transl. *Int Appl Mech,* **30**:1284–1316
565. Khoroshun LP, Dorodnykh TI (2004) The effective piezoelectric properties of polycrystals with the trigonal symmetry. *Acta Mechan,* **169**:203–219
566. Khoroshun LP, Maslov BP (1993) *Nonlinear Properties of Composite Materials of Stochastic Structure.* Naukova Dumka, Kiev (In Russian)
567. Khoroshun LP, Maslov BP, Leshchenko PV (1989) *Prediction of Effective Properties of Piezoactive Composites.* Naukova Dumka, Kiev (In Russian)
568. Khoroshun LP, Maslov BP, Shikula EN, Nazarenko LV (1993) *Statistical Mechanics and Effective Properties of Materials.* Naukova Dumka, Kiev (In Russian)
569. Khoroshun LP, Vesalo YA (1987) Theory of the effective properties of ideally plastic composites. *Priklad Mekh,* **23**(1):86–90 (In Russian. Engl Transl. *Int Appl Mech,* **23**:76–79)
570. Kim JC, Auh KH, Martin DM (2000) Multi-level particle packing model of ceramic agglomerates. *Model Simul Mater Sci Engng,* **8**:159–168
571. Kim RY, Pagano NJ (1991) Crack initiation in unidirectional brittle-material composite. *J Am Ceram Soc,* **74**:1082–1090
572. Kim S, Karrila SJ (1991) *Microhydrodynamics.* Butterworth–Heinemann, Oxford
573. Kim S, Mifflin RT (1985) The resistance and mobility functions of two equal spheres in low-Reynolds-number flow. *Phys Fluid,* **28**:2033–2045
574. Kinoshita N, Mura T (1971) Elastic fields of inclusions in anisotropic media. *Phys Stat Sol,* **(a)5**:759–768
575. Kiriakie K, Polyzos D, Valavanides M (1997) Low-frequency scattering of coated spherical obstacles. *J Engng Math,* **31**:379–395
576. Kirkpatrick S (1973) Percolation and cunductivity. *Rev Mod Phys,* **45**:574–588
577. Kirkwood JG (1935) Statistical mechanics of fluid mixtures. *J Chem Phys,* **3**:300–313
578. Knott GM, Jackson TL, Buckmaster J (2001) Random packing of heterogeneous propellants. *AIAA J,* **39**:678–686
579. Koelman JM, Kuijper A (1997) An effective medium model for the effective conductivity of N-component anisotropic and percolating mixture. *Physica,* **A247**:10–22
580. Koiter WT (1959) An infinite row of collinear cracks in an infinite elastic sheet. *Ing Arch,* **28**:168–172
581. Koiter WT (1960) General theorems for elastic-plastic solids. In: Sneddon IN, Hill IR (eds), *Progress in Solid Mechan.* North-Holland, Amsterdam, 165–221
582. Komarneni S (1992) Nanocomposites. *J Mater Chem,* **2**:1219–1230

583. Kondrachuk AV, Shapovalov GG, Kartuzov VV (1997) Simulation modeling of the randomly nonuniform structure of powders. Two-dimensional formulation of the problem. *Poroshkovaya Metallurgiya,* (1-2):111-118 (In Russian. Engl Translation. *Powder Metall Metal Ceram,* **36**:101–106)
584. König D, Carvajal-Gonzalz S, Downs AM, Vassy J, Rigaut JP (1991) Modelling and analysis of 3-D arrangements of particles by point process with examples of application to biological data obtained by confocal scanning light microscopy. *J Microscopy,* **161**:405–433
585. Koo CM, Kim SO, Chung IE (2003) Study on morphology evolution, orientational behavior, and anisotropic phase formation of highly filled polymer-layered silicate nanocomposites. *Macromolecules,* **36**:2748–2757
586. Koplik J, Needleman A (1988) Void growth and coalescence in porous plastic solids. *Int J Solids Struct,* **24**:835–853
587. Korneev VA, Petrashen GI (1987) Calculation of diffraction fields from an elastic cylinder. *Prob Dyn Theory Seismic Wave Propaga,* **27**:45–69
588. Kornmann X (2001) *Synthesis and Characterization of Thermoset Layered Silicate Nanocomposites.* PhD Thesis, Lulea University of Technol, Sweden
589. Kornmann X, Thomann R, Mülhaupt R, Finter J, Berglund L (2002) High performance epoxy-layered silicate nanocomposites. *Polym Engng Sci,* **42**:1815–1826
590. Korsunsky AM (1997) An axisymmetric inclusion in one of two perfectly bonded dissimilar elastic half-space. *J Appl Mech,* **64**:697–700
591. Kosheleva AA (1983) Method of multipolar expansion in the mechanics of matrix composites. *Mekhanika Kompozititnykh Materialov,* **19**(3):416–422 (In Russian. Engl. Transl. *Mech Compos Mater* **19**:301–307)
592. Kosmodamiansky AS (1972) *Stress Distribution in the Isotropic Multiply-Connected Solids.* University Publ, Donetsk (In Russian)
593. Kozaczek KJ, Sinharoy A, Ruud CO, Mcllree AR (1995) Micromechanical modeling of microstress fields around carbide precipitates in alloy 600. *Model Simul Mater Sci Engng,* **3**:829–843
594. Kozlov GV, Lipatov YS (2003) Description of the structure of the polymer matrix of particulate-filled polymer composites. *Mekh Kompoz Mater,* **39**(1):89–96 (In Russian. Engl Transl. *Mech Compos Mater,* **39**(1):65–70)
595. Kouznetsova VG, Geers MGD, Brekelmans WAM (2004) Multi-scale second-order computational homogenization of multi-phase materials: a nested finite element solution strategy. *Comput Methods Appl Mech Engng,* **193**:5525–5550
596. Krajcinovic D (1996) *Damage Mechan.* Elsevier, Amsterdam
597. Kreger IW (1972) Rheology of monodisperse lattices. *Adv Colloid and Interface Sci,* **3**:111–136
598. Kreher W (1988) Internal stresses and relations between effective thermoelastic properties of stochastic solids–some exact solutions. *Z Angew Math Mech,* **68**:147–154
599. Kreher W (1990) Residual stresses and stored elastic energy of composites and polycrystals. *J Mech Phys Solids,* **38**:115–128
600. Kreher W, Janssen R (1992) On microstructural residual stresses in particle reinforced ceramics. *J Eur Ceram Soc,* **10**:167–173
601. Kreher W, Molinari A (1993) Residual stresses in polycrystals as influenced by grain shape and texture. *J Mech Phys Solids,* **41**:1955–1977
602. Kreher W, Pompe W (1989) *Internal Stresses in Heterogeneous Solids.* Akademie-Verlag, Berlin
603. Kremesec VJ, Slattery JC (1977) The apparent stress-deformation of spheres in power-model fluid. *J Rheol,* **21**:459–461

604. Krivoglaz M, Cherevko A (1959) On the elastic moduli of a two-phase solid. *Phiz Metallov Metallovedenie,* **8**(2):161–168 (In Russian. Engl Transl. *Phys Metals Metall,* **8**:1–4)
605. Kröner E (1953) Das Fundamentalintegral der anisotropen elastischen Differentialgleichungen. *Z Physik,* **136**:402–410
606. Kröner E (1958) Berechnung der elastischen Konstanten des Vielkristalls aus den Konstanstanten des Einkristalls. *Z Physik,* **151**:504–518
607. Kröner E (1961) Zur plastischen Verformung des Vielkristalls, *Acta Metall,* **9**:155–161
608. Kröner E (1967a) Elastic moduli of perfectly disodered composite material. *J Mech Phys Solids,* **15**:319–329
609. Kröner E (1967b) Elasticity theory of materials with long range cohesive forces. *Int J Solid Struct,* **3**:731–742
610. Kröner E (1972) *Statistical Continuum Mechanics.* Springer-Verlag, Vienna–New York
611. Kröner E (1977) Bounds for effective moduli of disordered materials. *J Mech Phys Solids,* **25**:137–155
612. Kröner E (1986) Statistical modeling. In: Gittus J, Zarka J (eds), *Modeling Small Deformations of Polycrystals.* Elsevier, London/NY, 229–291
613. Kröner E (1990) Modified Green function in the theory of heterogeneous and/or anisotropic linearly elastic media. In: Weng GJ, Taya M, Abe H (eds) *Micromechanics and Inhomogeneity. The Toshio Mura 65th Anniversary Volume.* Springer–Verlag, New York, 197–211
614. Kröner E (1994) Nonlinear elastic properties of micro-heterogeneous media. *J Engng Mater Technol,* **116**:325–330
615. Kröner E, Datta BK (1970) Non-local theory of elasticity for a finite inhomogeneous medium–a derivation from lattice theory. In: Simmons J, de Wit R, Bullough R (eds), *Fundamental Aspects of Dislocation Theory.* Nat Bur Stand (US), Washington, 737–746
616. Kumar A, Dawson PR (1998) Modeling crystallographic texture evolution with finite elements over neo-Eulerian orientation spaces. *Comput Methods Appl Mech Engng,* **153**:259–302
617. Kunin IA (1963) Theory of dislocations. In: Shouten AY. (ed), *Tensorial Analysis for Physicists.* Nauka, Moskow, 373–450 (In Russian)
618. Kunin IA (1983) *Elastic Media with Microstructure.* Springer-Verlag, Berlin, **2**
619. Kunin IA (1984) On foundations of the theory of elastic media with microstructure *Int J Solids Struct,* **22**:969–978
620. Kunin IA, Sosnina EG (1971) Ellipsoidal inhomogeneity in the elastic medium. *Dokladi AN SSSR,* **37**:306–315 (In Russian. Engl Transl. *Sov Phys Dokl,* **16**:571–575)
621. Kunin IA, Vaisman AM (1970) On problems of the non-local theory of elasticity. In: Simmons J, de Wit R, Bullough R (eds) *Fundamental Aspects of Dislocation Theory.* Nat Bur Stand (US), Washington, 747–757
622. Kuo WS, Huang JH (1997) On the effective electroelastic properties of piezoelectric composites containing spatially oriented inclusions. *Int J Solids Struct,* **34**:2445–2461
623. Kuppa V, Menakanit S, Krishnamoorti R, Manias E (2003) Simulation insights on the structure of nanoscopically confined poly(ethylene oxide). *J Polymer Sci: Part B: Polym Phys,* **41**:3285–3298
624. Kushch VI (1987) Computation of the effective elastic moduli of a granular composite material of regular structure. *Priklad Mekh,* (4):57–61 (In Russian. Engl Transl. *Soviet Appl Mech,* **23**:362–364)

625. Kushch VI (1996) Elastic equilibrium of a medium containing finite number of aligned spheroidal inclusions. *Int J Solids Struct,* **33**:1175–1189
626. Kushch VI (1997a) Microstresses and effective elastic moduli of a solid reinforced by periodically distributed spheroidal particles. *Int J Solids Struct,* **34**:1353–1366
627. Kushch VI (1997b) Conductivity of a periodic particle composite with transversely isotropic phases. *Proc Roy Soc Lond,* **A453**:65–76
628. Kushch VI (1998a) Elastic equilibrium of a medium containing a finite number of arbitrarily oriented spheroidal inclusions. *Int J Solids Struct,* **35**:1187–1198
629. Kushch VI (1998b) Interacting cracks and inclusions in a solid by multiple expansion method. *Int J Solids and Struct,* **35**:1751–1762
630. Kushch VI (1998c) *The Stress State and Effective Thermoelastic Properties of Piece-Homogeneous Solids with Spheroidal Interfaces.* Dr. Sci. Thesis. Institute of Mechanics of the National Academy of Sciences, Kiev, Ukraine
631. Kushch VI, Sevostianov I (2004) Effective elastic properties of the particulate composite with transversely isotropic phases. *Int J Solids Struct,* **41**:885–906
632. Kushch VI, Shmegera SV, Buryachenko VA (2005) Interacting elliptic inclusions by the method of complex potentials. *Int J Solids Struct,* **42**:5491–5512
633. Kushch VI, Shmegera SV, Buryachenko VA (2006) Elastic equilibrium of a half plane containing a finite array of elliptic inclusions. *Int J Solids Struct,* **43**:3459–3483
634. Kuster GT, Toksöz MN (1974) Velocity and attenuation of seismic waves in two-phase media: I. Theoretical formulation. *Geophysics,* **39**:587–606
635. Kutlu Z, Chang FK (1995) Composite panels containing multiple through-the-width delaminations and subjected to compression. Part I: Analysis. *Compos Struct,* **31**:273–296
636. Kuznetsov SV (1991) Microstructural stress in porous media. *Priklad Mech,* **27**(11):23–28 (In Russian. Engl Transl. *Soviet Appl Mech,* **27**:750–755)
637. Kwon Y (2003) Discrete atomic and smeared continuum modeling for static analysis. *Engng Comput,* **20**:964–978
638. Kwon YW, Eren H (2000) Micromechanical study of interface stresses and failure in fibrous composites using boundary element method. *Polym Polym Compos,* **8**:369–386
639. Lafdi K, Matzek M (2003) Carbon nanofibers as a nano-reinforcement for polymeric nanocomposites. *The 35th Int SAMPE Technical Conference,* Dayton, Ohio
640. Lagoudas DS, Gavazzi AC, Nigam H (1991) Elastoplastic behavior of metal matrix composites based on incremental plasticity and the Mori–Tanaka averaging scheme. *Comput Mech,* **8**:193–203
641. Lambropolous JC (1986) Shear, shape and orientation effects in transformation toughening. *Int J Solids Struct,* **22**:1083–1106
642. Landau LD, Lifshitz EM (1960) *Electrodynamics of Continuum Media.* Pergamon Press, Oxford
643. Landauer R (1952) The electrical resistance of binary metallic mixtures. *J Appl Phys,* **23**:779–784
644. Landauer R (1978) Electric conductivity in inhomogeneous media. In: Garland JC, Tanner DB (eds) *Electric, Transport and Optical Properties of Inhomogeneous Media.* American Institute of Physics, New York, 2–43
645. Lau K-T, Gu C, Hui D (2006) A critical review on nanotube and nanotube/nanoclay related polymer composite materials *Composites,* **B37**:425–436
646. Laws N (1973) On the thermostatics of composite materials. *J Mech Phys Solids,* **21**:9–17
647. Laws N, Lee JC (1989) Microcracking in polycrystalline ceramics: elastic isotropy and thermal anisotropy. *J Mech Phys Solids,* **37**:603–618

648. Lax M (1951) Multiple scattering of waves. *Rev Modern Phys,* **23**:287–310
649. Lax M (1952) Multiple scattering of waves II. The effective fields dense systems. *Phys Rev,* **85**:621–629
650. Lay PT (1967) Potentiels élastiques, tenseurs de Green et de Neumann. *J Mećanique,* **6**:212–242
651. LeBaron PC, Wang Z, Pinnavaia TJ (1999) Polymer-layered silicate nanocomposites: an overview. *Appl Clay Sci,* **15**:11–29
652. Leblond JD, Perrin G (1999) A self-consistent approach to coalescence of cavities in inhomogeneously voided ductile solids. *J Mech Phys Solids,* **47**:1823–1841
653. Lee BJ, Mear ME (1991) Effect of inclusion shape on the stiffness of non-linear two-phase composites. *J Mech Phys Solids,* **39**:627–649
654. Lee HK, Simonovic S (2001) A damage constitutive model of progressive debonding in aligned discontinuous fiber composites. *Int J Solids Struct,* **38**:875–895
655. Lee HK, Simonovic S, Shin DK (2004) A computational approach for prediction of the damage evolution and crushing behavior of chopped random fiber composites. *Comput Mater Sci,* **29**:459–474
656. Lee J, Mal A (1997) A volume integral equation technique for multiple inclusion and crack interaction problems. *J Appl Mech,* **64**:23–31
657. Lee J, Mal A (1998) Characterization of matrix damage in metal matrix composites under transverse loads. *Comput Mech,* **21**:339–346
658. Lee J, Choi S, Mal A (2001) Stress analysis of an unbounded elastic solid with orthotropic inclusions and voids using a new integral equation technique. *Int J Solids Struct,* **38**:23–31
659. Lee JA, Mykkanen DL (1987) *Metal and Polymer Matrix Composites.* Noyes Data Corporation, New York
660. Lee JK, Johnson WC (1984) Elastic interaction and elastoplastic deformation of inhomogeneities. In: Bilby BA, Miller KJ, Willis JR (eds), *Fundamentals of deformation and Fracture. Eshelby Memorial Symposium* Cambridge University Press, Cambridge, UK, 145–162
661. Lee M, Jasiuk I, Tsuchida E (1991) The sliding circular inclusion in an elastic half-plane. *J Appl Mech,* **59**:S57–S64
662. Legrosa B, Mogilevskaya SG Crouch SL (2004) A boundary integral method for multiple circular inclusions in an elastic half-plane. *Engng Anal Boundary Elements,* **28**:1083–1098
663. Lekhnitskii AG (1963) *Theory of Elasticity of an Anisotropic Elastic Body.* Holder Day, San Francisco
664. Levi O, Bergman DJ (1994) Critical behavior of the weakly nonlinear conductivity and flicker noise of two-component composites. *Phys Rev,* **B50**:3652–3660
665. Levin VM (1967) Thermal expansion coefficient of heterogeneous materials. *Izv AN SSSR, Mekh Tverd Tela,* (2):88–94 (In Russian. Engl Transl. *Mech Solids,* **2**(2):58–61)
666. Levin VM (1971) The relation between mathematical expectations of stress and strain tensors in elastic microheterogeneous media. *Prikladnaya Matematika i Mekhanika* (In Russian. Engl Transl. *J Appl Math Mech,* **35**:694–701)
667. Levin VM (1975) Determination of effective elastic moduli of composite materials. *Docl Akad Nauk SSSR,* **220**:1042–1045 (In Russian. Engl Transl. *Sov Phys Docl,* **20**:147–148)
668. Levin VM (1976) Determination of composite material elastic and thermoelastic constants. *Izv AN SSSR, Mech Tverd Tela,* (6):137–145 (In Russian. Engl Transl. *Mech Solids,* **11**(6):119–126)

669. Levin VM (1977) On the stress concentration in inclusions in composite materials. *Prikl Matem Mekh,* **41**:735–743 (In Russian. Engl Transl. *J Appl Mathem Mech,* **41**:735–743)
670. Levin VM, Rakovskaja M. I, Kreher W. S (1999) The effective thermoelectroelastic properties of microinhomogeneous materials. *Int J Solids Struct,* **36**:2683–2705
671. Levis TB, Nielsen LE (1968) Viscosity of dispersed and aggregated suspensions of spheres. *J Rheol,* **12**:421–443
672. Li C, Chou TW (2003) A structural mechanics approach for the analysis of carbon nanotubes. *Int J Solids Struct,* **40**:2487–2499
673. Li C, Chou TW (2004) Elastic properties of single-walled carbon nanotubes in transverse directions *Phys Rev,* **B69**:073401
674. Li C, Ellyin F (1998) A mesomechanical approach to inhomogeneous particulate composites undergoing localized damage: part I–a mesodomain simulation. *Int J Solids Struct,* **36**:5529–5544
675. Li H, Zhong W-F, Li G-F (1985) On the method of equivalent inclusion in elastodynamics and scattering fields of ellipsoidal inhomogeneities. *Appl Math Mech,* **6**:489–498
676. Li JY (1999) On micromechanics approximation for the effective thermoelastic moduli of multi-phase composite materials. *Mech Mater* **31**:149–159
677. Li JY (2000a) The effective electroelastic moduli of textured piezoelectric polycrystalline aggregates. *J Mech Phys Solids,* **48**:529–552
678. Li JY (2000b) Magnetoelectroelastic multi-inclusion and inhomogeneity problems and their applications in composite materials. *Int J Engng Sci,* **38**:1993–2011
679. Li JY (2004) The effective pyroelectric and thermal expansion coefficients of ferroelectric ceramics. *Mechan Mater* **36**:949-958
680. Li JY, Dunn ML (1999) Analysis of microstructural fields in heterogeneous piezoelectric solids. *Int J Engng Sci,* **37**:665–685
681. Li S, Wang G (2004) On damage theory of a cohesive medium. *Int J Engng Sci,* **42**:861–885
682. Liao K, Li S (2001) Interfacial characteristics of a carbon nanotube polystyrene composite system. *Appl Phys Lett,* **79**:4225–4227
683. Lifshitz IM, Rozenzweig LN (1946) Theory of elastic properties of polycristals. *Zh Eksp Teor Fiz,* **16**:967–980 (In Russian)
684. Lifshitz IM, Rozenzweig LN (1947) On the construction of the Green's tensor for the basic equation of the theory of elasticity of an anisotropic infinite medium. *Zh Eksp Teor Fiz,* **17**:783–791 (In Russian)
685. Lifshitz IM, Rozenzweig LN (1951) Corrections of the paper "Properties of polycrystals". *Zh Eksp Teor Fiz,* **21**:1184 (In Russian)
686. Lin L, Argon AS (1992) Deformation resistance in oriented nylon 6. *Macromolecules,* **25**:4011–4024
687. Lin SC, Yang C, Mura T, Iwakuma T (1992) Average elastic-plastic behavior of composite materials. *Int J Solids Struct George Herrmann 70th Anniversary Issue,* **28**:1859–1872
688. Lipatov YS (1995) *Polymer Reinforcement. ChemTech Publishing.* Toronto, Ontario, Canada
689. Lipetzky P, Kreher W (1994) Statistical analysis of crack advance in ceramic composites. *Mech Mater,* **20**:225–240
690. Lipinski P, Barhdadi EH, Cherkaoui M (2006) Micromechanical modelling of an arbitrary ellipsoidal multi-coated inclusion. *Philos Mag,* **86**:1305–1326
691. Lipinski P, Berveiller M (1989) Elastoplasticity of micro-inhomogeneous metals at large strains. *Int J Plast,* **5**:149–172

692. Lipinski P, Berveiller, M, Reubrez E, Morreale J (1995) Transition theories of elastic-plastic deformation of metallic polycrystals. *Arch Appl Mech,* **65**:295–311
693. Lipton R (2003) Assessment of the local stress state through macroscopic variables. *Philos Trans Roy Soc Lond,* **361**:921–946
694. Liu JZ, Zheng Q-Z, Wang L-F, Jiang Q (2005) Mechanical properties of single-walled carbon nanotube bundles as bulk materials *J Mech Phys Solids,* **53**:123–142
695. Liu S, Wang Q (2005) Elastic field due to eigenstrains in a half-space. *J Appl Mech,* **72**:871–878
696. Liu T, Kumar S (2003) Quantitative characterization of SWNT orientation by polarized Raman spectroscopy. *Chem Phys Lett,* **378**:257–262
697. Liu W, Hoa SV, Pugh M (2005) Organoclay-modified high performance epoxy nanocomposites. *Compos Sci Technol,* **65**:307–316
698. Liu YJ, Nishimura N, Otani Y (2005) Large-scale modeling of carbon-nanotube composites by a fast multipole boundary element method. *Comput Mater Sci,* **34**:173–187
699. Lomakin VA (1970) *Statistical Problems of the Mechanics of Solid Deformable Bodies.* Nauka, Moscow (In Russian)
700. Lomakin VA, Sheinin VI (1974) Stress concentration at the boundary of a randomly inhomogeneous elastic body. *Mekh Tverdogo Tela,* **9**(2):65–70 (In Russian Engl Transl *Mech Solids,* **9**(2):58–63)
701. Loo LS, Gleason KK (2004) Investigation of polymer and nanoclay orientation distribution in nylon 6/montmorillonite nanocomposite. *Polymers,* **45**:5933–5939
702. Lordi V, Yao N (2000) Molecular mechanics of binding in carbon-nanotube-polystyrene composite system. *J Mater Res,* **15**:2770–2779
703. Lotwick HW (1982) Simulations on some spatial hard core models, and the complete packing problem *J Statist Comp Simul,* **15**:295–314
704. Lu GQ, Ti LB, Ishizaki K (1994) A new algorithm for simulating the random packing of monosized powder in CIP processes. *Mater Manufact Processes,* **9**:601–621
705. Lu JP (1997) Elastic properties of carbon nanotube and nanoropes. *Phys Rev Lett,* **79**:1297–1300
706. Lu SY, Lin HC (1996a) Effective conductivity of composites with spherical inclusions: effect of coating and detachment *J Appl Phys,* **79**:609–618
707. Lu SY, Lin HC (1996b) Effective conductivity of composites containing aligned spheroidal inclusions of finite conductivity *J Appl Phys,* **79**:6761–6769
708. Lubachevsky BD, Stillinger FH (1990) Geometric properties of random disk packing. *J Statist Phys,* **60**:561–583
709. Lubachevsky BD, Stillinger FH, Pinson EN (1991) Disks vs spheres: contrasting properties of random packing. *J Statist Phys,* **64**:501–524
710. Lubarda VA (1998) A note on the effective Lamé constants of polycrystalline aggregate of cubic crystals. *J Appl Mech,* **65**:769–770
711. Lubarda VA (2002) *Elastoplasticity Theory.* CRC Press, Boca Raton, FL
712. Lubarda VA, Markenscoff X (1998a) On the absence of Eshelby properties for non-ellipsoidal inclusions. *Int J Solids Struct,* **35**:3405–3411
713. Lubarda VA, Markenscoff X (1998b) On the stress field in sliding ellipsoidal inclusions with shear eigenstrain *J Appl Mech,* **65**:858–862
714. Lubarda VA, Richmond O (1999) Second-order analysis of dilute distribution of spherical inclusions. *Mech Mater* **31**:1–8
715. Luciano R, Willis JR (2000) Bounds of nonlocal effective relations for random composites loaded by configuration-dependent body force. *J Mech Phys Solids,* **48**:1827–1849

716. Luciano R, Willis JR (2001) Non-local effective relations for fibre-reinforced composites loaded by configuration-dependent body forces. *J Mech Phys Solids,* **49**:2705–2717
717. Luciano R, Willis JR (2003) Boundary-layer correlations for stress and strain field in randomly heterogeneous materials. *J Mech Phys Solids,* **51**:1075–1088
718. Luciano R, Willis JR (2004) Non-local constitutive equations for functionally graded materials. *Mech Mater,* **36**:1195–1206
719. Luciano R, Willis JR (2005) FE analysis of stress and strain fields in finite random composite bodies. *J Mech Phys Solids,* **53**:1505–1522
720. Luo HA, Weng GJ (1987) On Eshelby's inclusion problem in a three-phase spherically concentric solid, and a modification of Mori–Tanaka's method. *Mech Mater,* **6**:347–361
721. Luo J-J, Daniel IM (2003) Characterization and modeling of mechanical behavior of polymer/clay nanocomposites. *Compos Sci Technol,* **63**:1607–1616
722. Lurie AI (1990) *Nonlinear Theory of Elasticity.* North-Holland, Amsterdam, The Netherlands
723. Lurie S, Belov P, Volkov-Bogorodsky D, Tuchkova N (2003) Nanomechanical modeling of the nanostructures and dispersed composites. *Comput Mater Sci,* **28**:529–539
724. Lushcheykin GA (1987) Polymeric and composite materials. *Izv Acad Sci USSR Phys Solid Earth,* **51**:2273–2276 (In Russian)
725. Lusti HR, Gusev AA (2004) Finite element predictions for the thermoelastic properties of nanotube reinforced polymers. *Modelling Simul Mater Sci Engng,* **12**:107–119
726. Lutsko JP (1988) Stress and elastic constants in anisotropic solids: Molecular dynamics techniques. *J Appl Phys,* **64**:1152–1154
727. Ma Q, Clarke DR (1994) Piezospectroscopic determination of residual stresses in polycrystalline alumina. *J Am Ceram Soc,* **77**:298–302
728. Maekava ZI, Hamada H Yokoyama A (1989) Lamination theory of composite material with complex fiber orientation. *Proc ICCS,* **5**:701–714
729. Mal A (1999) Private communication
730. Mal AK, Knopoff L (1967) Elastic wave velocities in two-component systems. *I. Inst Math Appl,* **3**:376–387
731. Manevitch OL, Rutledge GC (2004) Elastic properties of a single lamella of montmorillonite by molecular dynamics simulation. *J Phys Chem,* **B108**:1428–1435
732. Marchioro M, Prosperitti A (1999) Heat conduction in a non-uniform composite with spherical inclusions. *Proc Roy Soc Lond,* **A455**:1483–1508
733. Markov KZ (1981) "One-particle" approximation in mechanics of composite materials. In: Brulin O, Hsieh RKT (eds), *Continuum Models of Discrete Systems.* North-Holland, Amsterdam, 441–448
734. Markov KZ (1999) Elementary micromechanics of heterogeneous media. In: Markov K, Preziosi L (eds), *Heterogeneous Media. Micromechanics, Modelling, Methods, and Simulations.* Birkhäuser, Boston, 1–162
735. Markov KZ, Willis JR (1998) On the two-point correlation function for dispersions of nonoverlapping spheres. *Math Models Methods Appl Sci,* **8**:359–377
736. Markworth AJ, Ramesh KS, Parks WP (1995) Review. Modeling studies applied to functionally graded materials. *J Mater Sci,* **30**:2183–2192
737. Martin JB (1975) *Plasticity: Fundamentals and General Results.* MIT Press, Cambridge, MA
738. Maruyama B, Alam K (2002) Carbon nanotubes and nanofibers in composite materials. *SAMPE J,* **38**:60–69

739. Masenelly-Varlot K, Reynaud E, Vigier G, Varlet J (2002) Mechanical properties of clay-reinforced polyamide. *J Polym Sci Part B: Polym Phys,* **40**:272–283
740. Maslov BP (1979) Macroscopic third-order elastic moduli. *Prikl Mekhanika,* **15**:57–61 (In Russian)
741. Maslov BP (1981) Effective constants of the theory of geometrically nonlinear solids. *Prikl Mekh,* **17**(5):45–50 (In Russian. Engl Transl. *Soviet Appl Mech,* **17**:439–444)
742. Maslov BP (2000) Stress concentration in incompressible multicomponent materials. *Prikl Mekh,* **36**(3):118-114 (In Russian. Engl Transl. *Int Appl Mech,* **36**:384–390)
743. Maslov BP, Khoroshun LP (1972) Effective characteristics of elastic, physically nonlinear inhomogeneous media. *Izvestia AN SSSR Mech Tverd Tela,* (2):149–153 (In Russian)
744. Masumura RA, Chou YT (1982) Antiplane eigenstrain problem of an elliptic inclusion in an anisotropic half space. *J Appl Mech,* **49**:52–54
745. Materon G (1989) *Random Set and Integral Geometry.* John Wiley & Sons, New York
746. Matthes S, Humbert M (1995) On the principle of a geometrical mean of even-rank tensors for textured polycrystals. *J Appl Cryst,* **28**:254–266
747. Maugin GA (1988) *Continuum Mechanics of Electromagnetic Solids.* North-Holland, Amsterdam
748. Mazilu P, Ju JW (1996) Comments and author's reply on "Micromechanics and effective moduli of elastic composites containing randomly dispersed ellipsoidal inclusions" by Ju JW, Chen TM, *Acta Mechan 1994* **103**:103–121. *Acta Mechan,* **114**:235–239
749. McCartney LN (1987) Mechanics of matrix cracking in brittle-matrix fiber-reinforced composites. *Proc Roy Soc Lond,* **A409**:329–350
750. McCoy JJ (1970) On the displacement field in an elastic medium with random variation of material properties. *Rec Adv Engng Sci,* **5**, Gordon and Breach, New York
751. McCoy JJ (1979) On the calculation of bulk properties of heterogeneous materials. *Q Appl Math,* **36**:137–149
752. McCoy JJ (1981) Macroscopic response of continue with random microstructure. In: Nemat-Nasser S (ed) *Mechanics Today.* Pergamon Press, Oxford, **6**:1–40
753. McKenzie DR, McPedran RC, Derrick GH (1978) The conductivity of lattices of spheres. II. The body centered and face centered cubic lattices. *Proc Roy Soc Lond,* **A362**:211–232
754. McLaughlin R (1977) A study of the differential scheme for composite materials. *Int J Engrg Sc,* **15**:237–244
755. McMeeking RM, Evans AC (1982) Mechanics of transformation toughening in brittle materials. *J Amer Ceram Soc,* **69**:242–246
756. McPedran RC, McKenzie DR (1978) The conductivity of lattices of spheres. I. The simple cubic lattice. *Proc Roy Soc Lond,* **A359**:45–63
757. McPedran RC, Movchan AB (1994) The Rayleigh multiple method for linear elasticity. *J Mechan Phys Solids,* **42**:711–727
758. Mehrabadi MA, Cowin SC, Jaric J (1995) Six-dimensional orthogonal tensor representation of the rotation about an axis in three dimensions. *Int J Solids Struct,* **32**:439–449
759. Meisner MJ, Kouris DA (1995) Interaction of two elliptic inclusions. *Int J Solids Struct,* **32**:451–466

760. Meraghni F, Blaman CJ, Benzeggagh ML (1996) Effect of interfacial decohesion on stiffness reduction in a random discontinuous-fibre composite containing matrix microcracks. *Compos Sci Technol,* **56**:541–555
761. Meridith RE, Tobias CW (1960) Resistance to potential flow through a cubical array of spheres. *J Appl Phys,* **31**:1270–1274
762. Mi C, Kouris DA (2006) Nanopaticles under the influence of surface/interface elasticity. *J Mech Mater Solids,* **1**:763–791
763. Micata Y, Taya M (1986) Thermal stress in a coated short fiber composite. *J Appl Mech,* **53**:681–689
764. Michel JC, Suquet P (1992) The constitutive law of nonlinear viscous and porous materials. *J Mech Phys Solids,* **40**:783–812
765. Mikata Y (2001) Explicit determination of piezoelectric Eshelby tensors for a spheroidal inclusion. *Int J Solids Struct,* **38**:7045–7063
766. Mikhlin SG, Proössdorf S (1980) *Singular Integral Operators.* Springer-Verlag, Berlin, New York
767. Milgrom M, Shtrikman S (1989) Linear response of two-phase composites with cross moduli: exact universal relations. *Phys Rev,* **A40**:1568–1575
768. Miller MJ (1969) Bounds for effective bulk modulus of heterogeneous materials. J *Mater Phys,* **10**:2005–2019
769. Miloh T, Benveniste Y (1988) A generalized self-consistent method for the effective conductivity of composites with ellipsoidal inclusions and cracked bodies *J Appl Phys,* **63**:789–7796
770. Miloh T, Benveniste Y (1999) On the effective conductivity of composites with ellipsoidal inhomogeneities and highly conducting interfaces. *Proc Roy Soc Lond,* **A455**:2687–2706
771. Milton GW (1981) Bounds on the transport and optical properties of a two-component composite material. *J Appl Phys,* **52**:5294–5304
772. Milton GW (1982) Bounds on the elastic and transport properties of two-component composites. *J Mech Phys Solids,* **30**:177–191
773. Milton GW (1984) Correlation of the electromagnetic and elastic properties of composites and microgeometries corresponding with effective medium approximations. In: Johnson DL, Sen PN (eds), *Physics and Chemistry of Porous Media.* American Institute of Physics, New York, 66–77
774. Milton GW (1985) The coherent potential approximation is a realizable effective medium scheme. *Commun Math Phys,* **99**:463–500
775. Milton GW (2002) *The Theory of Composites.* Cambridge University Press, Cambridge, UK
776. Milton GW, Phan-Tien N (1982) New bounds on effective elastic moduli of two-component materials. *Proc Roy Soc Lond,* **A380**:305–331
777. Mindlin RD (1948) Stress distribution around a hole near the edge of a plate under tension. *Proc Soc Exp Stress Anal,* **5**:56–68
778. Miracle DB (2005) Metal matrix composites–From science to technological significance. *Comp Sci Technol,* **65**:2526–2540
779. Mishchenko MI, Videen G, Babenko VA, Khlebtsov NG, Wriedt T (2004) T-matrix theory of electromagnetic scattering by particles and its applications: a comprehensive reference database *J Quant Spectrosc Radiat Transfer,* **88**:357–406
780. Mishnaevsky LL (2006) Functionally gradient metal matrix composites: Numerical analysis of the microstructure-strength relationships. *Comp Sci Technol,* **66**:1873–1887
781. Molchanov IS (1997) *Statistics of the Boolean Model Model foer Practitioners and Mathematicans.* John Wiley & Sons, New York

782. Molinari A, El Mouden M (1996) The problem of elastic inclusions at finite concentration. *Int J Solid Struct,* **33**:3131–3150
783. Monetto I, Drugan WJ (2004) A micromechanics-based nonlocal constitutive equation for elastic composites containing randomly oriented spheroidal heterogeneities *J Mech Phys Solids,* **52**:359–393
784. Morawiec A (1994) Review of deterministic methods of calculations of physical elastic constants. *Textures Microstruct,* **22**:139–167
785. Morawiec A (1996) The effective elastic constants of quasi-isotropic polycrystalline materials composed of cubic phase. *Phys Stat Sol,* **A155**:353–364
786. Morawiec A, Field DP (1996) Rodrigues parameterization for orientation and misorientation distributions. *Philos Mag,* **A73**:1111–1128
787. Mori T, Tanaka K (1973) Average stress in matrix and average elastic energy of materials with misfitting inclusions. *Acta Metall,* **21**:571–574
788. Morse PM, Feshbach H (1953) *Methods of Theoretical Physics.* Parts I and II. McGraw-Hill, Maidenhead
789. Moschovidis ZA, Mura T (1975) Two-ellipsoidal inhomogeneities by the equivalent inclusion method. *J Appl Mech,* **42**:847–852
790. Mossotti OF (1850) Discussione analitica sul'influenza che l'azione di un mezzo dielettrico ha sulla distribuzione dell'electricitá alla superficie di piú corpi elettrici disseminati in eso. *Mem Mat Fis della Soc Ital di Sci in Modena,* **24**:49–74
791. Movchan NV, Willis JR (1997) Influence of spatial correlations on crack bridging by frictional fibers. *Engng Fracture Mech,* **58**:571–579
792. Movchan NV, Willis JR (1998) Penny-shape crack bridged by fibres. *Quart Appl Math,* **56**:327–340
793. Müller WH, Gao H, Chiu C-H, Schmauder S (1996) A semi-infinite crack in front of a circular thermally mismatched heterogeneity. *Int J Solids Struct,* **33**:731–746
794. Mura T (1987) *Micromechanics of Defects in Solids.* Martinus Nijhoff, Dordrecht
795. Mura T, Shodia HM, Hirose Y (1996) Inclusions problems *Appl Mechan Rev,* **49**(10), Part 2):S118–127
796. Murakami Y (1987) *Stress Intensity Factors Handbook.* Pergamon Press, Oxford, **2**
797. Murata I, Mori T, Nakagawa M Continuous energy Monte Carlo calculations of randomly distributed spherical fuels in high-temperature gas-cooled reactor based on a statistical geometry model. *Nuclear Sci Engng,* **123**:96–109
798. Muskhelishvili NI (1953) *Some Basic Problems of the Mathematical Theory of Elasticity.* P. Noordhoff, Groningen, XXXI
799. Mylvaganam K, Zhang LC (2004) Chemical bonding in polyethylene-nanotube composites: a quantum mechanics prediction. *J Phys Chem,* **B108**:5217–5220
800. Nadeau JC, Ferrari M (1999) Microstructural optimization of a functionally graded transversely isotropic layer. *Mechan of Mater* **31**:637–651
801. Nairn JA (1997) On the use of shear-lag methods for analysis of stress transfer in unidirectional composites. *Mech Mater,* **26**:63–80
802. Nan C-W, Clarke DR (1997) Effective properties of ferroelectric and/or ferromagnetic composites: a unified approach and its application. *J Am Ceram Soc,* **60**:1333–1340
803. Nakamura T, Suresh S (1993) Effects of thermal residual stresses and fiber packing on deformation of metal-matrix composites *Acta Metall Mater,* **41**:1665–1681
804. Needleman A (2000) Computational mechanics at the mesoscale. *Acta mater,* **48**:105–124
805. Needleman A, Tvergaard V (1993) Comparison of crystal plasticity and isotropic hardening predictions for metal-matrix composites. *J Appl Mech,* **60**:70–76

806. Nemat-Nasser S, Hori M (1993) *Micromechanics: Overall Properties of Heterogeneous Materials.* Elsevier, North-Holland
807. Nemat–Nasser S, Iwakuma T, Hejazi M (1982), On composites with periodic structure. *Mech Mater* **1**:239–267
808. Nemat-Nasser S, Obata M (1986) Rate–dependent, finite deformation of polycrystals. *Proc Roy Soc Lond,* **A407**:377–404
809. Nielsen LE (1979) Dynamic mechanical properties of polymers filled with aggregated particles. *J Polym Sci: Polym Phys,* **17**:1897–1901
810. Nigmatulin RI (1979) Spatial averaging in the mechanics of heterogeneous and dispersed systems. *Int J Multiphase Flow,* **5**:353–385
811. Nikitin AN, Rusakova EI, Ivankina TI (1989) On the theory of origination of piezoelectric textures in rocks. *Izv Acad Sci USSR Phys Solid Earth,* (6):49–60 (In Russian)
812. Noda N-A, Matsumo T (1998) Singular integral equation method for interaction between elliptic inclusions. *J Appl Mech,* **65**:310–319
813. Noda N, Nakai S, Tsuji T (1998) Thermal stresses in functionally graded materials of particle-reinforced composite. *JSME Int J,* **A41**:178–184
814. Noh TW, Song PH, Sievers AJ (1991) Self-consistency conditions for the effective medium approximation in composite materials. *Phys Rev,* **B44**:5459–5464
815. Nolan GT, Kavanagh PE (1992) Computer simulation of random packing of hard spheres. *Powder Technol,* **72**:149–155
816. Norris AN (1985) A differential scheme for the effective moduli of composites. *Mech Mater,* **4**:1–16
817. Norris AN (1989) An examination of the Mori-Tanaka effective medium approximation for multiphase composites. *J Appl Mech,* **56**:83–88
818. Norris AN, Callegari AJ, Sheng PA (1985) A generalized differential effective medium theory. *J Mech Phys Solids,* **33**:525–543
819. Nozaki H, Shindo Y (1998) Effect of interface layers on elastic wave propagation in a fiber-reinforced metal-matrix composite. *Int J Engin Sci,* **36**:383–394
820. Nunan KC, Keller JB (1984) Effective elasticity tensor of a periodic composite. *J Mech Phys Sol,* **32**:259–280
821. Nye JF (1957) *Physical Properties of Crystals.* Oxford University Press, Oxford
822. Obraztsov IF, Yanovskii YG, Vlasov AN, Zgaevskii VE (2001) Poisson's ratios for interfacial layers of polymer matrix composites. *Doclady Academii Nauk,* **378**:336–338 (In Russian. Engl Transl. *Sov Doclady Phys,* **46**:366–368)
823. O'Brian RW (1979) A method for the calculation of the effective transport properties of suspensions of interacting particles. *J Fluid Mech,* **91**:17–39
824. O'Connel RJ, Budiansky B (1974) Seismic velocities in dry and saturated cracked solids. *J Geophys Res,* **79**:5412–5426
825. Odegard GM, Gates TS, Wise KE, Park C, Siochi EJ (2003) Constitutive modeling of nanotube-reinforced polymer composites. *Compos Sci Technol,* **63**:1671–1687
826. Odegard GM, Clancy TC, Gates TS (2005) Modeling of the mechanical properties of nanoparticle/polymer composites. *Polymers,* **46**:553–562
827. Oden JT, Zohdi TI (1997) Analysis and adaptive modeling of highly heterogeneous elastic structures. *Comput Methods Appl Mechan Engng,* **148**:367–391
828. Ogata Y, Tanemura M (1981) Estimation of interaction potentials of spatial point-patterns through the maximum-likelihood procedure. *Ann Inst Statist Math,* **33**:315–338
829. Ogata Y, Tanemura M (1984) Likelihood analysis of spatial point-patterns. *J Roy Statist Soc,* **B46**:496–518

830. Ogden RW (1972) Large deformation isotropic elasticity: on the correlation of theory and experiment for compressible rubberlike solids. *Proc Roy Soc Lond,* **A328**:567–583
831. Ogden RW (1974) On the overal moduli of non-linear elastic composite materials *J Mech Phys Solids,* **22**:541–554
832. Ogden RW (1997) *Non-linear Elastic Deformation.* Dover, Mineola, New York
833. Ogen L, Troadec JP, Gervois A, Medvedev N (1998) Computer simulation and tessellations of granular materials. In: Rivier N, Sadoc JF (eds), *Foams and Emulsions.* Kluwer, Dordrecht, 527–545
834. Ohno N, Wang J-D (1994) Kinematic hardening rules for simulation of ratchetting behavior. *Eur J Mech A/Solids,* **13**:519–531
835. Ohser J, Mücklich F (2000) *Statistical Analysis of Microstructures in Material Science.* John Wiley & Sons, Chichester
836. Okabe A, Boots B, Sugihara K (1992) *Spatial Tessellations.* John Wiley & Sons, New York
837. Olsson M, Giannakopoulos AE, Suresh S (1995) Elastoplastic analysis of thermal cycling: ceramic particles in a metallic matrix. *J Mech Phys Solids,* **43**:1639–1671
838. O'Rourke JP, Ingber MS, Weiser MW (1997) The effective elastic constants of solids containing spherical exclusions. *J Compos Mater* **31**:910–934
839. Ornstein LS, Zernike F (1914) Accidental deviation of density and opalescence at the critical points of a single substance. *Proc Acad Sci (Amsterdam),* **17**:793–806.
840. Ortiz M, Molinari A (1988) Microstructural thermal stresses in ceramic materials. *J Mech Phys Solids,* **36**:385–400
841. Ortiz M, Popov EP (1985) Accuracy and stability of integration algorithms for elastoplastic constitutive relations. *Int J Numer Methods Engng,* **21**:1561–1576
842. Ortiz M, Suresh S (1993) Statistical properties of residual stresses and intergranular fracture in ceramic materials. *J Appl Mech,* **60**:77-84
843. Ostoja-Starzewski M (2001) Mechanics of random materials: stochastic, scale effects and computation. In: Jeulin D, Ostoja-Starzewski M (eds), *Mechanics of random and multiscale microstructures*, Springer, Wien-New York, 93–161
844. Ostoja-Starzewski M (2002) Lattice models in micromechanics. *Appl Mech Rev* **55**:35–60
845. Ovid'ko IA, Sheinerman AG (2005) Elastic fields of inclusions in nanocomposite solids. *Rev Adv Mater Sci,* **9**:17–33
846. Ozmusul MS, Picu RC (2002) Elastic moduli of particulate composites with graded filler-matrix interfaces. *Polym Compos,* **23**:110–119
847. Pagano NJ (1974) Exact moduli of anisotropic laminates. In: Sendeckyj GP (ed), *Composite Materials.* Academic Press, New York, **2**:23–45
848. Pagano NJ (1987) Free-edge stress fields in composite laminates. *Int J Solids Struct,* **14**:401–406
849. Pagano NJ, Kim RY (1994) Progressive microcracking in unidirectional brittle matrix composites. *Mech Compos Mater Struct,* **1**:3–29
850. Pagano NJ, Rybicki EF (1974) On the significance of effective modulus solution for fibrous composites. *J Compos Mater* **8**:214–228
851. Pagano NJ, Yuan FG (2000) Significance of effective modulus theory (homogenization) in composite laminate mechanics. *Compos Sci Technol,* **60**:2471–2488
852. Pan HH, Weng GJ (1993) Thermal stress relief by plastic deformation in aligned two-phase composites. *Compos Engng,* **3**:219–234
853. Pan L, Adams DO, Rizzo FJ (1999) Boundary element analysis for composite materials and a library of Green's functions. *Composites & Struct,* **66**:685–693
854. Pan YC, Chou TW (1976) Point force solution for an infinite transversely isotropic solid. *J Appl Mech,* **43**:608–612

855. Panasenko GP (1983) Averaging processes in frame constructions with random properties. *Zhurnal Vychislitel'noi Matematiki i Matematicheskoi Fiziki,* **23**:1098–1109 (In Russian. Engl Transl. *USSR Comput Maths Math Phys,* **23**(5):48–55)
856. Pan'kov AA (2000) Generalized self-consistent method for determining effective dynamic elastic and diffraction properties of composites with random structures. *Int Confer Control Oscillat Chaos,* Proc N1:97–100
857. Papadopoulos P, Taylor RL (1994) On the application of multi-step integration methods to infinitesimal elastoplasticity. *Int J Numer Methods Engng,* **37**:3169–3184
858. Parkhomenko EI (1968) *Electrization in Rocks.* Nauka, Moscow (In Russian)
859. Parrinello M, Rahman AR (1982) Strain fluctuations and elastic constants. *J Chem Phys,* **76**:2662–2666
860. Parton VZ, Buryachenko VA (1990) Stress fluctuation in elastic composites. *Dokladi AN SSSR,* **310**:1075–1078 (In Russian. Engl Transl. *Sov Phys Docl,* **35**:191–193)
861. Parton VZ, Kudryavtsev BA (1988) *Electromagnetic Elasticity of Piezoelectric and Electrically Conductive Bodies.* Nauka, Moscow
862. Parton VZ, Perlin PI (1982) *Integral Equation Method in Elasticity.* MIR, Moscow
863. Paul B (1960) Prediction of the elastic constants of multiphase materials. *Trans Am Inst Min Metail Pet Engng,* **218**:36–41
864. Paxton JP, Mowles ED, Lukehart CM, Witzig AJ (2001) Polymer material property enhancement through nanocomposite technology. In: Hyer MW (ed), *Proc of the American Society for Composites. Sixteenth Technical Conference.* Loos AG, Blacksburg, VA
865. Pecullan S, Gibianski LV, Torquato S (1999) Scale effects on the elastic behavior of periodic behavior of periodic and hierarchical two-dimensional composites. *J Mech Phys Solids,* **47**:1509–1542
866. Peirce AP, Napier JAL (1995) A spectral multipole method for effeccient solution of large-scale boundary element models in elastostatics. *Int J Numer Methods Engng,* **38**:4009–4039
867. Percus JK, Yevick GJ (1958) Analysis of classical statistical mechanics by means of collective coordinates. *Phys Rev,* **110**:1–13
868. Perez MM, Wrobel LC (1996) An integral-equation formulation for anisotropic elastostatics. *J Appl Mech,* **63**:891–902
869. Perlin PI (1976) Application of the regular representation of singular integrals to the solution of the second fundamental problem of the theory of elasticity. *Prikl Metem Mekhan,* **40**:366–371 (In Russian. Engl Transl. *J Appl Math Mech,* **40**:342–347)
870. Peterson RE (1974) *Stress Concentration Factors.* John Wiley & Sons, New York
871. Pettermann HE, Bohm HJ, Rammerstorfer FG (1997) Some direction dependent properties of matrix–inclusion type composites with given reinforcement orientation distributions. *Composites,* **B28**:253–265
872. Phan-Thien N, Kim S (1994) *Microstructures in Elastic Media: Principles and Computational Methods.* Oxford University Press, New York
873. Phan-Thien N, Pham DC (2000) Differential multiphase models for polydispersed spheroidal inclusions: thermal conductivity and effective viscosity. *Int J Engng Sci,* **38**:73–88
874. Phelan PE, Niemann RC (1998) Effective thermal conductivity of a thin, randomly oriented composite material. *J Heat Transfer,* **120**:971–976
875. Piat R, Schnack E (2003) Hierarchical material modeling of carbon/carbon composites. *Carbon,* **41**:2121–2129

876. Picu RC (2002) Non-local elasticity kernels extracted from atomistic simulations. In: Zavaliangos A, Tikare V, Olevsky EA (eds), *Modelling and Numerical Simulation of Materials Behavior and Evolution.* Mater Res Soc, Warrendale, PA, **731**:71–76
877. Picu RC, Ozmusul MS (2003) Structure of linear polymeric chains confined between impenetrable spherical walls. *J Chem Phys,* **118**:11239–11248
878. Picu RC, Sarvestani A, Ozmusul MS (2004) Elastic moduli of polymer nanocomposites derived from their molecular structure. In: Harik VM (ed), *Trends in Nanoscale Mechanics: Analysis of Nanostructured Materials and Multiscale Modeling.* Kluwer, Dordrecht, 61–88
879. Pindera M-J, Aboudi J, Arnold SM (1995) Limitations of the uncoupled, RVE–based micromechanical approach in the analysis of functionally graded composites. *Mech Mater* **20**:77–94
880. Pindera M-J, Dunn P (1997) Evaluation of the higher-order theory for functionally graded materials via the finite-element method. *Compos Engng,* **7**:109–119
881. Pipes RB, Hubert P (2003) Helical carbon nanotube arrays: thermal expansion. *Compos Sci Technol,* **63**:1571–1579
882. Pipkin AC (1991) *A Course on Integral Equations.* Springer-Verlag, New York
883. Plankensteiner AF, Böhm HJ, Rammerstorfer FG, Buryachenko VA (1996) Hierarchical modeling of the mechanical behavior of high speed steels as layer–structured particulate MMCs. *J Physique IV,* **6**:C6-395–C6-402
884. Plankensteiner AF, Böhm HJ, Rammerstorfer FG, Buryachenko VA, Hackl G (1997) Modeling of layer–structured high speed tool steel. *Acta Metall Mater,* **45**:1875–1887
885. Pobedrya BY (1984) *Mechan of Composite Materials.* MGU, Moscow (In Russian)
886. Podalkov VV, Romanov VA (1984) Deformation of an elastic anisotropic micro-inhomogeneous half-space. *J Appl Math Mech,* **47**:383–388
887. Podil'chuk YuN (1984) *The Boundary Value Problems of Statics of Elastic Body.* Naukova Dumka, Kiev (In Russian)
888. Podil'chuk YuN (2001) Exact analytical solutions of three-dimensional static thermoelastic problems for a transversely isotropic body in curvilinear coordinate systems. *Int Appl Mech,* **37**:728–761
889. Polder D, Van Santen JH (1946) The effective permeability of mixtures of solids. *Physica,* **XII**:257–271
890. Polizzotto C (1994) On elastic plastic structure under cyclic load. *Eur J Mech A/Solids,* **13**(4-Suppl):149–173
891. Polizzotto C, Fuschi P, Pisano AA (2006) A nonhomogeneous nonlocal elasticity model. *Eur J Mech A/Solids,* **25**, 308–333
892. Pólya G, Szegó G (1951) *Isoperimetric Inequalities in Mathematical Physics.* Prinston University Press, Prinston, NJ
893. Ponte Castañeda P (1988), New variational principles in plasticity and their application to composite materials. *J Mech Phys Solids,* **40**: 1757–1788
894. Ponte Castañeda P (1991a) The effective mechanical properties of nonlinear isotropic composites. *J Mech Phys Solids,* **39**:45–71
895. Ponte Castañeda P (1991b) Effective properties in power-law creep. In: Cocks ALF, Ponter ARS (eds) *Micromechanics of Creep of Brittle Materials.* Elsevier, London, **2**:218–229
896. Ponte Castañeda P (1992) New variational principles in plasticity and their application to composite materials. *J Mech Phys Solids,* **40**:1757–1788
897. Ponte Castañeda P, Suquet P (1998) Nonlinear composites. In: Hutchinson JW, Wu TJ (eds), *Adv Appl Mech* **34**:171–302

898. Ponte Castañeda P, Willis JR (1995) The effect of spatial distribution on the effective behavior of composite materials and cracked media. *J Mech Phys Solids,* **43**:1919–1951
899. Praveen GN, Reddy JN (1998) Nonlinear transient thermoelastic analysis of functionally graded ceramics-metal plates. *Int J Solids Struct,* **35**:4437–4476
900. Prosperetti A (1998) Ensemble averaging techniques for disperse flows. In: Drew DA, Joseph DD, Passma SL (eds) *Particulate Flows Processing and Rheology.* Springer-Verlag, New York 99–136
901. Pyrz R (1994) Quantitative description of the microstructure of composites. Part I: Morphology of unidirectional composite systems. *Compos Sci Technol,* **50**:197–208
902. Pyrz R (2004) Microstructural description of composites–statistical methods. In: Böhm H (ed), *CISM Courses and Lectures.* Springer, Udine, **464**:173–233
903. Pyrz R, Bochenek B (1998) Topological disorder of microstructure and its relation to the stress field. *Int J Solids Struct,* **35**:2413–2427
904. Qi HJ, Teo KBK, Lau KKS, Boyce MC, Milne WI, Robertson J, Gleason KK (2003) Determination of mechanical properties of carbon nanotubes and vertically aligned carbon nanotube forests using nanoindentation. *J Mech Phys Solids,* **51**:2213–2237
905. Qian D, Dickey EC, Andrews R, Rantell T (2000) Load transfer and deformation mechanisms in carbon nanotube- polystyrene composites *Appl Phys Lett,* **76**: 2868–2870
906. Qian D, Liu WK, Ruoff RS (2003) Load transfer mechanism in carbon nanotube ropes. *Compos Sci Technol,* **63**:1561–1569
907. Qian D, Wagner GJ, Liu WK, Yu M-F, Ruoff RS (2002) Mechanics of carbon nanotubes. *Appl Mech Rev,* **55**:495–533
908. Qiu YP, Weng GJ (1990) On the application of Mori-Tanaka's theory involving transversely isotropic spheroidal inclusions. *J Engng Sci,* **28**:1121–1137
909. Qiu YP, Weng GJ (1991a) Elastic moduli of thickly coated particle and fiber–reinforced composites. *J Appl Mech,* **58**:388–398
910. Qiu YP, Weng GJ (1991b) The influence of estimation shape on the overall elastoplastic behavior of a two-phase isotropic composite. *Int J Solids Struct,* **27**:1537–1550
911. Qiu YP, Weng GJ (1992) A theory of plasticity for porous materials and particle–reinforced composites. *J Appl Mech,* **59**:261–268
912. Qiu YP, Weng GJ (1993) Plastic potential and yield function of porous materials with aligned and randomly oriented spheroidal voids. *Int J Plast,* **9**:271–290
913. Qiu YP, Weng GJ (1995) An energy approach to the plasticity of a two-phase composite containing aligned inclusions. *J Appl Mech,* **62**:1039–1046
914. Qu J (1993) The effect of slightly weakened interfaces on the overall elastic properties of composite materials. *Mech Mater,* **14**:269–281
915. Quintanilla J (1999) Microstructure and properties of random heterogeneous materials: a review of theoretical results. *Polym Engng Sci,* **39**:559–585
916. Quintanilla J, Torquato S (1997) Microstructure functions for a model of statistically inhomogeneous random media. *Phys Rev,* **E55**:1558–1565
917. Räisänen VI, Herrmann HJ (1999) Stress transfer in dilute short-fiber reinforced composites. *J Mater Sci,* **34**:897–904
918. Raghavan P, Moorthy S, Ghosh S, Pagano NJ (2001) Revisiting the composite laminate problem with an adaptive multi-level computational model. *Composite Sci Technol,* **61**:1017–1040
919. Rammerstorfer FG (1992) *Leichtbau–Repetitorium.* Oldenbourg-Verlag, Wien-München

920. Räisänen VI, Herrmann HJ (1997) Stress transfer in dilute short-fiber reinforced composites. *J Mater Sci,* **34**:897–904
921. Raghavan P, Moorthy S, Ghosh S, Pagano NJ (2001) Revisiting the composite laminate problem with an adaptive multi-level computational model. *Compos Sci Technol,* **61**:1017–1040
922. Rajapakse RKND, Shah AH (1988) Hybrid modeling of semi-infinite media. *Int J Solids Struct,* **24**:1205–1224
923. Rajendran AM, Grove DJ (1996) Modeling the shock response of silicon carbide, boron carbide, and titanium diboride. *Int J Impact Eng,* **18**:611–631
924. Rank E, Krause R (1997) A multiscale finite-element method. *Computers Struct,* **64**:139–144
925. Rankenburg IC, Zieve RJ (2001) Influence of shape on ordering of granular systems in two dimensions. *Phys Rev,* **E63**:61303-1–61303-9
926. Ray SK, Utku S (1989), A numerical model for the thermo-elasto-plastic behavior of a material. *Int J Numer Methods Engng,* **28**:1103–1114
927. Ray SS, Okamoto M (2003) Polymer/layered silicate nanocomposites: a review from preparation to processing. *Prog Polymer Sci,* **28**:1539–1641
928. Reddy JN (Ed) (1994) *Mechan of Composite Materials. Selected works of N. J. Pagano.* Kluwer, Dordrecht
929. Reich S, Thomsen C, Ordejon P (2002) Elastic properties of carbon nanotubes under hydrostatic pressure. *Phys Rev,* **B65**:153407
930. Reifsnider KL, Gao Z (1991) A micromechanics models for composites under fatigue loading. *Int J Fatigue,* **13**:149–156
931. Reiter T, Dvorak GJ (1998) Micromechanical models of functionally graded composite materials. In: Bahei-El-Din YA, Dvorak GJ (eds), *IUTAM Symp. on Transformation Problems in Composite and Active Materials.* Kluwer, Dordrecht, 173–184
932. Reiter T, Dvorak GJ, Tvergaard V (1997) Micromechanical models for graded composite materials. *J Mech Phys Solids,* **45**:1281–1302
933. Renton JD (1987) *Appl Elasticity.* John Wiley & Sons, Chichester, UK
934. Reuss A (1929) Berechnung der Fliessgrenze von Mischkristallen auf Grund der Plastizitatsbedingung fur Einkristalle. *Z Angew Math Mech,* **9**:49–58
935. Rice JR (1985) Three-dimensional elastic crack tip interaction with transformation strains and dislocations. *Int J Solids Struct,* **21**:781–791
936. Rice JR (1989) Weight function theory for three-dimensional elastic crack analysis. In: Wei RP, Gangloff RP (eds), *Fracture Mechanics: Perspectives and Directions (Twentieth Symp).* Amer Soc Test Mater, Philadelphia, 29–57
937. Rice RW, Pohanka RC (1979) The grain size dependence of spontaneous-cracking in ceramics. *J Am Ceram Soc,* **62**:559–563
938. Ripley BD (1977) Modeling spatial patterns. *J Roy Statist Soc,* **B39**:172–212
939. Ripley BD (1981) *Spatial Statistic.* John Wiley & Sons, New York
940. Rodin GJ (1993) The overall elastic response of materials containing spherical inhomogeneities *Int J Solids Struct,* **30**:1849–1863
941. Rodin GJ, Hwang YL (1991) On the problem of linear elasticity for an infinite region containing a finite number of spherical inclusions. *Int J Solids Struct,* **27**:145–159
942. Rogula D (1982) Nonlocal theory of material media. *CISM Courses and Lectures,* **268**. Springer-Verlag, Vienna, New York
943. Roscoe R (1952) The viscosity of a suspension of rigid spheres. *Br J Appl Phys,* **3**:267–268
944. Rosen BW (1970) Thermoelastic energy functions and minimum energy principles for composite materials. *Int J Engng Sci,* **8**:5–18

945. Rosen BW, Hashin Z (1970) Effective thermal expansion coefficients and specific heat of composite materials. *Int J Engng Sci,* **8**:157–173
946. Ru CQ (2003) Eshelby inclusion of arbitrary shape in an anisotropic plane or half-plane. *Acta Mechan,* **160**:219–234
947. Ru CQ, Schiavone P (1997) A circular inclusion with circumferentially inhomogeneous interface in antiplane shear. *Proc Roy Soc Lond,* **A453**:2551–2572
948. Ru CQ, Schiavone P, Mioduchowski A (2001) Elastic fields in two jointed half-planes with an inclusion of arbitrary shape. *Z Angew Math Phys,* **52**:18–32
949. Rubinstein AA (1987) Semi-infinite array of cracks in a uniform stress field. *Engng Fracture Mech,* **26**:15–21
950. Ruoff RS, Lorents DC (1995) Mechanical and thermal properties of carbon nanotubes. *Carbon,* **33**:925–930
951. Russ JC (2002) *The Image Processing Handbook.* CRC Press, Boca Raton, FL
952. Russel WB, Saville DA, Schowalter WR (1989) *Colloidal Dispersions.* Cambridge University Press, Cambridge, UK
953. Rybicki EF, Pagano NJ (1976) A study of the influence of microstructure on the modified effective modulus approach for composite laminates. *Proc 1975 Int Conf Composite Mater* **2**:198–207
954. Sab K, Nedjar B (2005) Periodization of random media and representative volume element size for linear composites. *C R Mecanique,* **333**:187–195
955. Sabina FJ, Smyshlyaev VP, Willis JR (1993) Self-consistent analysis of waves in a matrix-inclusion composite.–I. Aligned spheroidal inclusions. *J Mech Phys Solids,* **41**:1573–1588
956. Safadi AY (1996) *A Numerical Technique for Determining the Elastic Field of an Arbitrary Shaped Inhomogeneity.* PhD. Thesis, Northwestern University, Evanston
957. Sahimi M (1998) Non-linear and non-local transport processes in heterogeneous media: from long-range correlated percolation to fracture and materials breakdown. *Phys Rep,* **306**:213–395
958. Salerno GM, Watt JP (1986) Walpole bounds on the effective elastic moduli of isotropic multicomponent composites. *J Appl Phys,* **60**:1618–1624
959. Salvetat J-P, Briggs GA, Bonard J-Marc, Bacsa RR, Kulik AJ (1999) Elastic and shear moduli of single-walled carbon nanotube ropes. *Phys Rev Letter,* **82**:944–947
960. Sanchez-Palencia E (1980) *Homogenization Techniques and Vibration Theory.* Lecture Notes in Physics, No. 127, Springer-Verlag, Berlin
961. Sanchez-Palencia E (1987) Boundary layers and edge effects in composites. In: Sanchez-Palencia E, Zaoui A (eds), *Homogenization Techniques for Composite Media.* Lecture Notes in Physics, Springer-Verlag, Berlin **272**:121–147
962. Sangani A, Lu W (1987) Elastic coefficients of composites containing spherical inclusions in a periodic array. *J Mech Phys Sol,* **35**:1–21
963. Sangani AS, Mo G (1997) Elastic interactions in particulate composites with perfect as well as imperfect interfaces. *J Mech Phys Solids,* **45**:2001–2031
964. Sangani AS, Yao C (1997) Transport processes in random array of cylinders. I. Thermal conduction. *Phys Fluids,* **31**:2426–2434
965. Sato H, Shindo Y (2002) Influence of microstructure on scattering of plane elastic waves by a distribution of partially debonded elliptical inclusions. *Mech Mater* **34**:401–409
966. Savin GN (1961) *Stress Concentration Around Holes.* Pergamon Press, New York
967. Savin GN, Khoroshun LP (1972) Problem of elastic constants of randomly reinforced materials. *Mechanics of Composite Media and Related Problems of Analysis.* Nauka, Moscow, 437–444 (In Russian)
968. Scaife BKP (1989) *Principle of Dielectrics.* Oxford University Press, Oxford, UK

969. Schadler LS, Giannaris SC, Ajayan PM (1998) Load transfer in carbon nanotube epoxy composites. *Appl Phys Lett,* **73**:3842–3844
970. Schapery RA (1968) Thermal expansion coefficients of composite materials based on energy principles. *J Compos Mater,* **2**:380–404
971. Schapery RA (1986) A micromechanical model for non-linear viscoelastic behavior of particle-reinforced rubber with distributed damage. *Eng Fract Mech,* **25**:845–867
972. Schclar NA (1994) *Anisotropic Analysis Using Boundary Elements.* Comput Mech Publ, Southampton and Boston
973. Schmauder S (2002) Computational mechanics. *Annu Rev Mater Res,* **32**, 437–465
974. Schnack E, Szikrai S, Türke K (1998) Local effects in engineering with macro-elements. *Comput Methods Appl Mech Engng,* **157**:299–309
975. Schraad MW, Triantafyllidis N (1997) Scale effects in media with periodic and nearly periodic microstructures. I. Macroscopic properties. *J Appl Mech,* **64**:751–62
976. Schulgasser K (1976a) Relationship between single-crystal and polycrystal electrical conductivity. *J Appl Phys,* **47**:1880–1886
977. Schulgasser K (1976b) On the conductivity of fiber-reinforced materials. *J Math Phys,* **17**:382–387
978. Schulgasser K (1987) Thermal expansion of polycrystalline aggregates with texture. *J Mech Phys Solids,* **35**:35–42
979. Scorohod VV (1961) Calculation of the effective isotropic moduli of disperse solid systems. *Poroshkovaya Metallurgiya (Powder Metall),* (1):50–51 (In Russian)
980. Sedov LL (1966) *Foundations of the Non-Linear Mechanics of Continua.* Pergamon Press, Oxford
981. Segurado J, Gonzallez C, Llorca J (2003) A numerical investigation of the effect of particle clustering on the mechanical properties of composites. *Acta Mater,* **51**:2355–2369
982. Segurando J, Llorca J (2002) A numerical approximation to the elastic properties of sphere-reinforced composites. *J Mech Phys Solids,* **50**:2107–2121
983. Sen AK, Torquato S (1989) Effective conductivity of anisotropic two-phase composite medium. *Phys Rev,* **B39**:4504–4515
984. Sen P, Scala C, Cohen MH (1981) A self-similar model for sedimentary rocks with application to the dielectric constant of fused glass beards. *Geophysics,* **46**:781–795
985. Seo K, Mura T (1979) The elastic field in a half space due to ellipsoidal inclusions with uniform dilatation eigenstrains. *J Appl Mech,* **46**:568–572
986. Sevostianov I, Kachanov M (2002) Explicit cross-property correlations for anisotropic two-phase composite materials. *J Mech Phys Solids,* **30**:252–282
987. Sevostianov IB, Levin VM, Pompe W (1998) Evaluation of the mechanical properties of ceramics during drying. *Phys Stat Sol,* **a166**:817–828
988. Sewell TD, Menikoff R, Bedrov D, Smith GD (2003) A molecular simulation study of elastic properties of HMX. *J Chem Phys,* **119**:7417–7426
989. Shan Z Gokhale AM (2002) Representative volume element for non-uniform micro-structure. *Comput Mater Sci,* **24**:361–379
990. Sharma P, Ganti S (2004) Size-dependent Eshelbys tensor for embedded nano-inclusions incorporating surface/interface energies. *J Appl Mech,* **71**:663–671
991. Shen S, Atluri SN (2004) Computational nano-mechanics and multi-scale simulation. *Computers Mater Continua,* **1**:59–90
992. Sheng N, Boyce MC, Parks DM, Rutledge GC, Abes JI, Cohen RE (2004) Multi-scale micromechanical modeling of polymer/clay nanocomposites and the effective clay particle. *Polymers,* **45**:487–506

993. Shenoy VB, Miller R, Tadmor EB, Rodney D, Phillipsa R, Ortiz M (1999) An adaptive finite element approach to atomic-scale mechanics–the quasicontinuum method. *J Mech Phys Solids,* **36**:500–531
994. Sherman DI (1959) On the problem of plane strain in non-homogeneous media. In: Olszak W (ed), *Nonhomogeneity in Elasticity and Plasticity.* Pergamon Press, 3–20
995. Shermergor TD (1977) *The Theory of Elasticity of Microinhomogeneous Media.* Nauka, Moscow (In Russian)
996. Shermergor TD, Yakovlev VB (1993) Concentration of coupled electrical mechanical fields on a crystallite surface in textured quartz. *Izv Acad Sci Russ Phys Solid Earth,* **32**:89–94 (In Russian)
997. Shi D-L, Feng X-Q, Huang YY, Hwang K-C, Gao H (2004) The effect of nanotube waviness and agglomeration on the nanotube-reinforced composites. *J Engng Mater Technol,* **126**:250–257
998. Shi DL, Feng XQ, Jiang HQ, Huang Y, Huang KS (2005) Multiscale analysis of fracture of carbon nanotubes embedded in composites. *Int J Fracture,* **134**:369–386
999. Shi J, Ghanem R (2006) A stochastic nonlocal model for materials with multiscale behavior. *Int J Multiscale Comput Engng,* **4**:501–520
1000. Shia D, Hui CY, Burnside SD, Giannelis EP (1998) An interface model for the prediction of Young's modulus of layered silicate-elastomer nanocomposites. *Polym Compos,* **19**:608–617
1001. Shibata S, Jasiuk I, Mori T, Mura T (1990) Successive iteration method applied to composites containing sliding inclusions: effective modulus and elasticity. *Mech Mater* **9**:229–243
1002. Shidlovskii AK, Glushkov EH, Reztsov VF, Snarskaya GI (1989) Generalized form of Eshelby principle in electodynamic of inhomogeneous media and some its application. *Dokladi AN Ukraine SSR,* **A**(3):82–86
1003. Shilov GE (1968) it Generalized Functions and Partial Differential Equations. Gordon & Breach, New York
1004. Shindo Y, Nozaki H, Datta SK (1995) Effect of interface layers on elastic wave propagation in a metal matrix composite reinforced by particles. *J Appl Mech,* **62**:178–185
1005. Shklovskii BI, Efros AL (1979) *Electronic Properties of Doped Semiconductors.* Nauka, Moscow (In Russian)
1006. Shubin AB (1995) On maximum density of random packing of the identical solid spheres. *Rasplavy* (1):92–97 (In Russian)
1007. Shvidler MI (1985) *Statistical Hydrodynamics of Porous Media.* Nauka, Moscow. (In Russian)
1008. Siboni G, Benveniste Y (1991) A micromechanical model for the effective thermomechanical behaviour of multiphase composite media. *Mech Mater* **11**:107–122
1009. Sihn S, Roy AK (2004) Modeling and prediction of bulk properties of open-cell carbon foam. *J Mech Phys Solids,* **52**:167–191
1010. Sigrist S, Jullien R, Lahaye J (2001) Agglomeration of solid particles. *Cement Concrete Compos,* **23**:153–156
1011. Silva MJ, Hayes WC, Gibson LJ (1995) The effect of non-periodic microstructure on the elastic properties of two-dimensional cellular solids. *Int J Mech Sci,* **37**:1161–1177
1012. Simmons G, Wang H (1971) *Single Crystal Elastic Constants and Calculates Aggregate Properties.* MIT Press, Cambridge/London
1013. Sinelnikov NN, Mazo MA, Berlin AA (1997) Dense packing of random binary assemblies of disks. *J Phys I France,* **7**:247–254

1014. Sinsawat A, Anderson KL, Vaia RA, Farmer BL (2003) Influence of polymer matrix composition and architecture on polymer nanocomposite formation: coarse-grained molecular dynamics simulation. *J Polym Sci Part B: Polym Phys,* **41**:3272–3284
1015. Sirotin YuI, Shaskolskaya MP (1982) *Fundamentals of Crystal Physics.* Mir Publishers, Moscow
1016. Skinner BP, Newnham RE, Gross LE (1978) Flexible composite transducers. *Mater Res Bull,* **13**:599
1017. Sladek V, Sladek J (1998) Singular integrals and boundary elements. *Comput Meth Appl Mech Engng,* **157**:251–266
1018. Sloan CS, Cowell MD, Lehnhoff TF (1999) The effect of a large hole on the stress concentration factor of a satellite hole in a tension field. *J Pressure Vessel Piping,* **121**:252–256
1019. Smith JC (1974) Correction and extension of van der Poels method for calculating the shear modulus of a particulate composite. *J Res Natl Bur Stand Sect,* **A78**:355–361
1020. Smith JC (1976) Experimental values for the elastic constants of a particulate-filled glassy polymer. *J Res Nat Bur Stand US,* **80A**:45–49
1021. Smith WA (1989) The role of piezocomposites in ultrasonic transducers. *Proceedings of the IEEE 1989 Ultrasonic Symposium,* 755–766
1022. Smith P, Torquato S (1988) Computer simulation results for the two-point probability functions of composite media. *J Comput Phys,* **76**:176–191
1023. Smyshlyaev VP, Cherednichenko KD (2000) A rigorous derivation of strain gradient effects in the overall behavior of periodic heterogeneous media. *J Mech Phys Solids,* **48**:1325–1357
1024. Smyshlyaev VP, Fleck NA (1996) The role of strain gradients in the grain size effects for polycrystals. *J Mech Phys Solids,* **44**:465–495
1025. Sobczuk K, Spencer BF (1991) *Random Fatigue: From Data to Theory.* Academic Press, New York
1026. Sobolev GA, Demin VN (1980) *Mechanoelectric Phenomena in the Earth.* Nauka, Moscow (In Russian)
1027. Sokolkin YV, Tashkinov AA (1984) *Deformation and Fracture Mechanics of Structurally Inhomogeneous Bodies.* Nauka, Moscow (In Russian)
1028. Sokolnikov IS (1983) *Mathematical Theory of Elasticity* (2nd edn). Robert E. Krieger, Melbourne
1029. Sokolov AYu, Tvardovskiy VV (1988) On determination of the effective characteristics of piezoelectric composites. In: Leksovskiy AM (ed), *Physics of Strength of Heterogeneous Materials.* FTI AN SSSR, Leningrad, 227–236 (In Russian)
1030. Sotiropolous DA, Achenbach JD, Zhu H (1987) An inverse scattering method to characterize inhomogeneities in elastic solids. *J Appl Phys,* **62**:2771–2777
1031. Spenser AJM (1980) *Continuum Mechanics.* John Wiley & Sons, New York
1032. Spitzig WA, Kelly JF, Richmond O (1985) Quantitative characterization of second-phase populations. *Metallography,* **18**:235–261
1033. Spowart JE, Maruyama B, Miracle DB (2001) Multi-scale characterization of spatially heterogeneous systems: implications for discontinuously reinforced metal-matrix composite microstructures. *Mater Sci Engng,* **A307**:51–66
1034. Srivastava D, Wei C, Cho K (2003) Nanomechanics of carbon nanotubes and composites. *Appl Mech Rev,* **56**:215–230
1035. Stagfen ESG (1988) A nonlocal theory for the heat transport in composites containing highly conducting fibrous inclusions. *Phys Fluids,* **31**:2405–2425
1036. Stang H (1986) Strength of composite materials with small cracks in the matrix. *Int J Solids Struct,* **22**:1259–1277

1037. Stein E, Zhang G (1992) Theoretical and numerical shakedown analysis for kinematic hardening materials. In: Owen DRJ, Oñate E, Hinton E (eds), *Computational Plasticity. Fundamental and Applications.* Pineridge Press, Swansea 1:1–25
1038. Stell G (1991) Statistical mechanics applied to random-media problem. In: Owen DRJ, Oñate E, Hinton E (eds) *Mathematics of Random Media.* Lectures in Applied Mathematics **27**:109–127
1039. Stell G, Rirvold PA (1987) Polydispersity in fluid, dispersions, and composites: some theoretical results. *Chem Engng Commun,* **51**:233–260
1040. Stephan C, Nguen TP, Chapelle ML, Lefrant S (2000) Characterization of single-walled carbon nanotubes-PMMA composites. *Synth Methods,* **108**:139–149
1041. Sternberg E, Sadowsky MA (1952) On the axisymmetric problem of the elasticity for an infinite region containing two spherical inclusions. *J Appl Mech,* **19**:19–27
1042. Stigh U (2006) Continuum damage mechanics and the life-fraction rule. *J Appl Mech,* **73**:702–704
1043. Stoy RD (1989) Solution procedure for the Laplace equation in bispherical coordinates for two sphere in uniform external field: parallel orientation. *J Appl Phys,* **65**:2611–2615
1044. Stoyan D (1998) Random sets: models and statistics. *Int Statistical Rev,* **66**:1–27
1045. Stoyan D (2000) Basic ideas of spatial statistics. In: Mecke KR, Stoyan D (eds) *Statistical Physics and Spatial Statistics: The Art of Analyzing and Modeling Saptial Structures and Pattern Formation.* Lecture Notes in Physics, Springer, Berlin, 554
1046. Stoyan D, Kendall WS, Mecke J (1995) *Stochastic Geometry and Its Applications.* John Wiley & Sons, Chichester
1047. Stoyan D, Stoyan H (1994) *Fractals, Random Shapes and Point Fields. Methods of Geometric Statistics.* J Wiley & Sons, Chichester
1048. Straley JP (1981) Thermoelectric properties of inhomogeneous materials. *J Phys D: Appl Phys,* **14**:2101–2105
1049. Stratonovich RL (1963) *Topics in the Theory of Random Noise.* Gordon and Breach, New York
1050. Stratton JA (1941) *Electromagnetic Theory.* McGraw-Hill, New York
1051. Strouboulis T, Copps K, Babuska I (2001) The generalized finite element method. *Computer Meth Appl Mech Engng,* **190**:4081–4193
1052. Stroud D (1998) The effective medium approximation: some recent development. *Superlatt Microstruct,* **23**:567–573
1053. Stroud D, Hui PM (1988) Nonlinear succeptibilities of granular materials. *Phys Rev,* **B37**:8719–8724
1054. Stump DM, Budiansky B (1989) Crack-growth resistance in transformation-toughened ceramics. *Int J Solids Struct,* **25**:635–646
1055. Sukumar N, Chopp DL, Moes N, Belytschko T (2001) Modeling holes and inclusions by level sets in the extended finite-element method. *Computer Meth Appl Mech Engng,* **190**:6183–6200
1056. Sun CT, Zhou SG (1988) Failure of quasi-isotropic composite laminates with free edges. *J Reinf Plast Comp,* **7**:515
1057. Sun YF, Peng YZ (2003) Analytic solutions for the problems of an inclusion of arbitrary shape embedded in a half-plane. *Appl Math Comput,* **140**:105–113
1058. Suquet P (1995) Overall properties of nonlinear composites: A modified secant moduli theory and its link with Ponte Castañeda's nonlinear variational procedure. *C R Acad Sci Paris, Série IIb,* **320**:563–571
1059. Suresh S, Mortensen A (1998) *Fundamentals of Functionally Graded Materials: Processing and Thermomechanical Behaviour of Graded Metals and Metal-Ceramic Composites.* IOM Communications, London

1060. Suresh S, Mortensen A, Needlman A (1993) *Fundamentals of Metal-Matrix Composites.* Butterworth-Heinemann, Boston
1061. Svistkov AL, Evlampieva SE (2003) Using of smoothing avaraging operator to evaluate macroscopic parameters in structurally inhomogeneous materials. *Priklad Mekh Tekhn Fiz,* **44**(5):150–160 (In Russian. Engl Transl. *J Appl Mech Tech Phys,* **44**:727–735)
1062. Tadmor EB, Ortiz M, Phillips R (1996) Quasicontinuum analysis of defects in solids. *Philos Mag,* **A73**:1529–1563
1063. Taggart DG, Bassani JL (1991) Elastic–plastic behavior of particle reinforced composites—influence of residual stresses. *Mechan of Mater* **12**:63–80
1064. Takano N, Okuno Y (2004) Three-scale finite element analysis of heterogeneous media by asymptotic homogenisation and mesh superposition method. *Int J Solids Struct,* **41**:4121–4135
1065. Takao Y, Taya M (1985) Thermal expansion coefficients and thermal stresses in an aligned short fiber composite with application to a short carbon fiber/aluminum. *J Appl Mech,* **52**:806–810
1066. Talbot DR, Willis JR (1982a) Variational estimates for dispersion and attenuation of waves in random composites. I. General theory. *Int I Solids Struct,* **18**:673–683
1067. Talbot DRS, Willis JR (1982b) Variational estimates for dispersion and attenuation of waves in random composites. II. Isotropic composites. *Int I Solids Struct,* **18**:685–698
1068. Talpaert YR (2002) *Tensor Analysis and Continuum Mechanics.* Kluwer, Dordrecht
1069. Tanaka M, Sladek V, Sladek J (1994) Regularization techniques applied to boundary element methods. *Appl Mech Rev,* **47**:457–499
1070. Talbot DRS, Willis JR (1985) Variational principles for nonlinear inhomogeneous media. *IMA J Appl Math,* **35**:39–54
1071. Tandon GP, Kim RY, Bechel VT (2004) Construction of the fiber-matrix interfacial failure in a polymer matrix composites. *Int J Multiscale Comput Engng,* **2**:101–114
1072. Tandon GP, Kim RY, Rice BP (2002) Influence of vapor-grown carbon nanocomposites on thermomechanical properties of graphite-epoxy composites. *Proc. American Society for Composites 17th Technical Conference.* Purdue University, West Lafayette, IN, Paper 2039
1073. Tandon GP, Weng GJ (1984) The effect of aspect ratio of inclusions on the elastic properties of unidirectionally aligned composites. *Polym Compos,* **5**:327–333
1074. Tandon GP, Weng GJ (1986) Average stress in the matrix and effective moduli of randomly oriented composites. *Compos Sci Technol,* **27**:111–132
1075. Tandon GP, Weng GJ (1988) A theory of particle–reinforced plasticity. *J Appl Mech,* **55**:126–135
1076. Tanemura M (1979) On random complete packing by discs. *Ann Inst Statist Math,* **31**:351–365
1077. Taya M (1990) Some thoughts on inhomogeneous distribution of fillers in composites. In: Weng GJ, Taya M, Abe H (eds), *Micromechanics and Inhomogeneity, The Toshio Mura 65th Anniversary Volume.* Springer-Verlag, New York, 433–447
1078. Taya M (1995) Micromechanics modelling of electronic composites. *J Engng Mater Technol,* **117**:462–469
1079. Taya M, Arsenault RJ (1989) *Metal Matrix Composites. Thermomechanical Behavior.* Pergamon Press, Oxford

1080. Taya M, Hayashi S, Kobayashi AS, Yoon HS (1990) Toughening of a particulate-reinforced ceramic-matrix composite by thermal residual stress. *J Amer Ceram Soc,* **73**:1382–1391
1081. Telles JCF, Brebbia CA (1991) Boundary element solution for half-plane problems. *Int J Solids Struct,* **17**:1149–1158
1082. Teodosiu C (1982) *Elastic Models of Crystal Defects.* Academei, Bucuresty, Springer, Berlin
1083. Terrones M (2003) Science technology of the twenty-first century: synthesis, properties, and applications of carbon nanotubes. *Annu Rev Mater Res,* **33**:419–501
1084. Tewari A, Gokhale AM, Spowart JE, Miracle DB (2004) Quantitative characterization of spatial clustering in three-dimensional microstructures using two-point correlation functions. *Acta Mater,* **52**:307–319
1085. Theocaris PS (1987) *The Concept of Mesophase in Composites.* Springer-Verlag, Berlin
1086. Theocaris PS (1991) The elliptic paraboloidal failure criterion for cellular solids and brittle forms. *Acta Mechan,* **89**:93–121
1087. Theodorou DN, Suter UW (1986) Local structure and the mechanism of response to elastic deformation of a glassy polymer. *Macromolecules,* **19**:379–387
1088. Thomson R, Zhou SJ, Carlsson AE, Tewary VK (1992) Lattice imperfections studied by use of lattice Green's functions. *Phys Rev,* **B17**:10613–10622
1089. Thorpe MF (1992) The conductivity of a sheet containing a few polygonal holes and/or superconducting inclusions *Proc Roy Soc Lond,* **A437**:215–227
1090. Thostenson ET, Chou T-W (2002) Aligned multi-walled carbon nanotube-reinforced composites: processing and mechanical characterization. *J Phys D: Appl Phys,* **35**:L77–L80
1091. Thostenson ET, Chou T-W (2003) On the elastic properties of carbon nanotube-based composites: modeling and characterization. *J Phys D: Appl Phys,* **36**:573–582
1092. Thostenson ET, Li CY, Chou TW (2005) Nanocomposites in context. *Compos Sci Technol,* **65**:491–516
1093. Thostenson ET, Ren ZF, Chou T-W (2001) Advances in the science and technology of carbon nanotubes and their composites: a review. *Compos Sci and Technol,* **61**:1899–1912
1094. Thovert JE, Acrivos A (1989) The effective thermal conductivity of a random polydispersed suspension of spheres to order c^2. *Chem Engng Commun,* **82**:177–191
1095. Thovert JE, Kim IC, Torquato S, Acrivos A (1990) Bounds on the effective properties of polydispersed suspensions of spheres: an evaluation of two relevant morphological parameters. *J Appl Phys,* **67**:6088–6098
1096. Throop GJ, Bearman RJ (1965) Numerical solution of the Percus-Yervick equation for the hard-sphere potential. *J Chem Phys,* **42**:2408–2411
1097. Tibbetts GG, McHugh JJ (1999) Mechanical properties of vapor-grown carbon fiber composites with thermoplastic matrices. *J Mater Res,* **14**:2871–2880
1098. Ting TCT (1996) *Anisotropic Elasticity. Theory and Applications.* Oxford University Press, New York, Oxford
1099. Tjong SC (2006) Structural and mechanical properties of polymer nanocomposites. *Mater Sci Engng,* **R53**:73–197
1100. Tobochnik J, Chapin PM (1988) Monte Carlo simulation of hard spheres near random closest packing using sphrical boundary conditions. *J Chem Phys,* **88**:5824–5830
1101. Toebes ML, van Heeswijk JMP, Bitter JH, van Dillen AJ, de Jong KP (2004) The influence of oxidation on the texture and the number of oxygen-containing surface groups of carbon nanofibers. *Carbon,* **42**:307–315

1102. Torquato S (1980) *Microscopic Approach to Transport in Two-Phase Random Media.* PhD Thesis, State University of New York at Stony Brook
1103. Torquato S (1991) Random heterogeneous media: microstructure and improved bounds on effective properties. *Appl Mech Rev,* **44**(2):37–75
1104. Torquato S (1997) Effective stiffness tensor of composite media – I. Exact series expansion. *J Mech Phys Solids,* **45**:1421–1448
1105. Torquato S (1998) Effective stiffness tensor of composite media: II. Application to isotropic dispersions. *J Mech Phys Solids,* **45**:1421–1448
1106. Torquato S (2002a) *Random Heterogeneous Materials: Microstucture and Macroscopic Properties.* Springer-Verlag, New York, Berlin
1107. Torquato S (2002b) Statistical description of microstructures. *Annu Rev Mater Res,* **32**:77–111
1108. Torquato S, Lado F (1992) Improved bounds on the effective elastic moduli of random arrays of cylinders. *J Appl Mech,* **59**:1–6
1109. Torquato S, Rintoul MD (1995) Effect of the interface on the properties of composite media. *Phys Rev Lett,* **75**:4067–4070
1110. Torquato S, Stell G (1985) Microstructure of two-phase random media. *J Chem Phys,* **82**:980–987
1111. Torquato S, Truskett TM, Debenetti PG (2000) Is random close packing of spheres well defined? *Phys Rev Lett,* **84**:2064–2067
1112. Treves F (1980) *Introduction to Pseudodifferential and Fourier Integral Operators.* Plenum Press, New York, **1**
1113. Trias D, Costa J, Mayugo JA, J.E. Hurtado JE (2006) Random models versus periodic models for fibre reinforced composites. *Comput Mater Sci,* **38**:316–324
1114. Truesdell CA (1991) *A First Course in Rational Continuum Mechanics.* (2nd edn). Academic Press, New York, **1**
1115. Tsai SW, Wu EM (1971) A general theory of strength for anisotropic materials. *J Compos Mater,* **5**:58–80
1116. Tszeng TC (1998) The effect of particle clustering on the mechanical behavior of particle reinforced composites. *Composites* **29B**:299–308
1117. Turnbull D, Cormia RL (1960) A dynamic hard sphere model. *J Appl Phys,* **31**:674–678
1118. Turner PA, Signorelli JW, Bertinetti MA, Bolmaro RE (1999) Explicit method for calculating the effective properties and micromechanical stresses: an application to an alumina-SiC composites. *Philos Mag,* **79**:1379–1394
1119. Tvergaard V, Needleman A (1997) Nonlocal effects on localization in a void-sheet. *Int J Solids Struct,* **34**:2221–2238
1120. Tvergaard V, Hutchinson JW (1988) Microcracking in ceramics by thermal expansion or elastic anisotropy. *J Am Ceram Soc,* **71**:157–166
1121. Twersky V (1978) Acoustic bulk parameters in distribution of pair correlated scatterers. *J Acoust Soc Am,* **64**:1710–1719
1122. Tzika PA, Boyce MC, Parks DM (2000) Micromechanics of deformation in particle-toughened polyamides. *J Mech Phys Solids,* **48**:1893–1929
1123. Umetani S-I (1988) *Adaptive Boundary Element Methods in Elastostatics.* Computational Mech Publ, Southampton, Boston
1124. Usuki A, Kojima Y, Kawasumi M, Okada A, Fukushima Y, Kurauchi T et al (1993) Synthesis of nylon 6-clay hybrid. *J Mater Res,* **8**:1179–1184
1125. Vaia RA, Giannelis EP (2001) Polymer nanocomposites: status and opportunities. *MRS Bull,* 394–401
1126. Vaia RA, Wagner HD (2004) Framework for nanocomposites. *Mater Today,* **7**:32–37

1127. Valter K, Kurtasov SF, Nikitin AN, Torina EG (1993a) Modeling of deformation textures in high temperature quartz. *Izv Acad Sci Russ Phys Solid Earth,* (6):25–48 (In Russian)
1128. Valter K, Nikitin AN, Shermergor TD, Yakovlev VD (1993b) Determination of effective electroelastic constants of polycrystalline rocks. *Izv Acad Sci Russ Phys Solid Earth* (6):83–88 (In Russian)
1129. Vanin GA (1996) Plane strain gradient theory of multilevel media. *Mekh Tverdogo Tela,* (3):5–15 (In Russian. Engl Transl. *Mech Solids,* **31**(3):2–11)
1130. VanLandinghama MR, Dagastinea RR, Eduljeea RF, McCullougha RL, JW Gillespie JrJW (1999) Characterization of nanoscale property variations in polymer composite systems: 1. Experimental results. *Composites,* **A30**:75–83
1131. van Workum K, de Pablo JJ (2003) Computer simulation of the numerical properties of amorphous polymer nanostructures. *Nanoletters,* **3**:1405–1410
1132. Varadan VK, Ma Y, Varadan VV (1985) A multiple scattering theory for elastic wave propagation in discrete random media. *J Acoust Soc Amer,* **77**:375–385
1133. Varadan VK, Varadan VV (Editors) (1980) *Acoustic Electomagnetic and Elastic Wave-Scattering Focus on the T-Matrix Approach.* Pergamon, New York
1134. Varga RS (2000) *Matrix Iterative Analysis.* Springer, Berlin
1135. Vasiliev VV (1993) *Mechanics of Composites Structures.* Taylor & Francis, Washington
1136. Venk GP (1993) Development of deformation textures in rocks. *Izv Acad Sci Russ Phys Solid Earth,* (6):5–36 (In Russian)
1137. Verlet L, Weis JJ (1972) Perturbation theory for the thermodynamic properties of simple liquids. *Mol Phys,* **24**:1013–1024
1138. Verruijt A (1997) A complex variable solution for a deforming circular tunnel in an elastic half-plane. *Int J Numer Analyt Methods Geomechan,* **21**:77–89
1139. Verruijt A (1998) Deformations of an elastic half plane with a circular cavity. *Int J Solid Struct,* **35**:2795–2804
1140. Vigolo B, Penicaud AP, Couloun C, Sauder S, Pailler R, Journet C, Bernien P, Poilin P (2000) Macroscopic fibers and ribbons of oriented carbon nanotubes. *Science,* **290**:1331–1334
1141. Vinson JR, Sierakowski RL (2003) *The Behavior of Structures Composed of Composites Materials.* Kluwer, Dordrecht, Boston
1142. Voight W (1889) Uber die Beziehung zwischen den beiden Elastizitatskonstanten isotroper Korper. *Wied Ann,* **38**:573–587
1143. Voight W (1910) *Lehrbuch der Kristallphysik.* B.G. Teubner, Leipzig und Berlin
1144. Wagner HD (2002) Nanotube–polymer adhesion: a mechanics approach. *Chem Phys Lett,* **361**:57–61
1145. Wagner HD, Vaia RA (2004) Nanocomposites: issue at the interface. *Mater Today,* **7**:38–42
1146. Wakashima K, Tsukamoto H (1991) Mean-field micromechanics model and its application to the analysis of thermomechanical behavior of composite materials. *Mater Sci Engng,* **A32**:883–892
1147. Walker KP, Jordan EH, Freed AD (1990) Equivalence of Green's function and the Fourier series representation of composites with periodic microstructure. In: Weng GJ, Taya M, Abe H (eds) *Micromechanics and Inhomogeneity.* Springer-Verlag, New York, 535–558
1148. Walpole LJ (1966a) On the bounds for the overall elastic moduli of inhomogeneous system. I. *J Mech Phys Solids,* **14**:151–162
1149. Walpole LJ (1966b) On the bounds for the overall elastic moduli of inhomogeneous system. II. *J Mech Phys Solids,* **14**:289–301

1150. Walpole LJ (1969) On the overall elastic moduli of composite materials. *J Mech Phys Solids,* **17**:235–251
1151. Walpole LJ (1978) A coated inclusion in an elastic medium. *Math Proc Camb,* **83**:495–506
1152. Walpole LJ (1981) Elastic behavior of composite materials: theoretical foundations. *Adv Appl Mech,* **21**:169–242
1153. Walpole LJ (1997) An inclusion in one of two joined isotropic elastic half-spaces. *IMA J Appl Math,* **59**:193–209
1154. Walsh JB (1965) The effect of cracks on the compressibility of rocks. *J Geophys Res,* **70**:381–389
1155. Wang B (1990) A general theory on media with randomly distributed inclusions: Part I. The average field behaviors. *J Appl Mech,* **57**:857–862
1156. Wang B (1992) Three-dimentional analysis of an ellipsoidal inclusion in a piezoelectric material. *Int J Solids Struct,* **29**:293–308
1157. Wang B (1994) Effective behaviour of piezoelectric composites. In: Ostoja-Starzevski M, Jasiuk I (eds), *Micromechanics of Random Media. Appl Mech Rev,* **47**:112–121
1158. Wang B, Liu Y (1990) The average field in piezoelectric media with randomly distributed inclusions. In: Hsieh KTN (ed), *Mechanical Modeling of New electromagnetic Materials.* Elsevier, Amsterdam, 313–318
1159. Wang J, Pyrz R (2004a) Prediction of the overall moduli of layered silicate-reinforced nanocomposites-part I: basic theory and formulas. *Composites Sci Technol,* **64**:925–934
1160. Wang J, Pyrz R (2004b) Prediction of the overall moduli of layered silicate-reinforced nanocomposites-part II: analyses. *Compos Sci Technol,* **64**:935–944
1161. Warren WE, Kraynik AM (1997) Linear elastic behavior of a low-density Kelvin foam with open cells. *J Appl Mech,* **64**:787–794
1162. Watt JP (1976) The elastic properties of composite materials. *Rev Geophys Res,* **14**:541–563
1163. Watt JP (1979) Hashin-Shtrikman bounds of the effective elastic moduli of polycrystals with orthorhombic symmetry. *J Appl Phys,* **50**:6290–6295
1164. Watt JP (1980) Hashin-Shtrikman bounds of the effective elastic moduli of polycristals with monoclinic symmetry. *J Appl Phys,* **51**:1520–1524
1165. Watt JP, Peselnic L (1980) Clarification of the Hashin-Shtrikman bounds of the effective elastic moduli of polycristals with hexagonal, trigonal, and tetragonal symmetries. *J Appl Phys,* **51**:1525–1531
1166. Weber L, Fischer C, Mortensen A (2003) On the influence of the shape of randomly oriented, non-conducting inclusions in a conducting matrix on the effective electrical conductivity. *Acta Mater,* **51**:495–505
1167. Wei C, Srivastava D, Cho K (2004) Structural ordering in nanotube polymer composites. *Nano Lett,* **4**:1949–1952
1168. Weissenbek E, Böhm HJ, Rammerstorfer FG (1994) Micromechanical investigations of arrangement effects in particle reinforced metal matrix composites. *Comput Mater Sci,* **3**:263–278
1169. Weissenbek E, Pettermann HE, Suresh S (1997) Numerical simulation of plastic deformation in compositionally graded metal-ceramic structures. *Acta Mater,* **45**:3401–3417
1170. Weng GJ (1984) Some elastic properties of reinforced solids with special reference to isotropic ones containing spherical inclusions. *Int J Engng Sci,* **22**:845–856
1171. Weng GJ (1990) The theoretical connection between Mori–Tanaka's theory and the Hashin–Shtrikman–Walpole bounds. *Int J Engng Sci,* **28**:1111–1120

1172. Wenk HR, Van Houtte P (2004) Texture and anisotropy. *Rep Prog Phys,* **67**:1367–1428
1173. Werner E, Siegmund T, Fischer FD (1994) A computer study of the thermomechanical deformation behavior of a duplex steel. *Comput Mater Sci,* **3**:279–285
1174. Wertheim MS (1963) Exact solution of the Percus-Yevick integral equation for hard spheres. *Phys Rev Lett,* **10**:321–323
1175. Wiener O (1912) Die theorie des mischkörpers fr̈ das feld des stationar̈en strm̈ung. Erste abhandlung die mttelswertsẗze fr̈ kraft, polarisation und energie. *Abt Math-Physichen Klasse Königl Sächsh Gessel Wissen,* **36**(6):509–604
1176. Widom W (1966) Random sequential addition of hard spheres to a volume. *J Chem Phys,* **44**:3888–3894
1177. Wilkinson DS, Pompe W, Oeschener M (2001) Modeling the mechanical behavior of heterogeneous multi-phase materials. *Prog in Mater Sci,* **46**:379–405
1178. Willis JR (1965) The elastic interaction energy of dislocation loops in anisotropic media. *Q J Mech Appl Math,* **18**:419–433
1179. Willis JR (1977) Variational and related methods for the overall properties and self-consistent estimates for the overall properties. *J Mech Phys Solids,* **25**:185–202
1180. Willis JR (1978) Variational principles and bounds for the overall properties of composites. In: Provan JW (ed), *Continuum Models of Disordered Systems.* University of Waterloo Press, Waterloo 185–215
1181. Willis JR (1980a) A polarization approach to the scattering of elastic waves I. Scattering by a single inclusion. *J Mech Phys Solids,* **28**:287–305
1182. Willis JR (1980b) A polarization approach to the scattering of elastic waves. II: Multiple scattering from inclusions. *J Mech Phys Solids,* **28**:307–326
1183. Willis JR (1980c) Relationships between derivatives of the overall properties of composites by perturbation expansion and variational principles. In: Nemat-Nasser S (ed), *Variational Methods in Mechanics of Solids.* Pergamon Press, New York, 59–66
1184. Willis JR (1981a) Variational and related methods for the overall properties of composites. *Adv Appl Mech.* **21**:1–78
1185. Willis IR (1981b) Variational principles for dynamic problems for inhomogeneous elastic media. *Wave Motion,* **3**:1–11
1186. Willis JR (1982) Elasticity theory of composites. In: Hopkins HA, Sewell MI (eds), *Mechanics of Solids, The Rodney Hill 60th Anniversary Volume.* Pergamon Press, Oxford, 653–686
1187. Willis JR (1983) The overall elastic response of composite materials. *J Appl Mech,* **50**:1202–1209
1188. Willis JR (1985) The nonlocal influence of density variations in a composite. *Int J Solids Struct,* **21**:805–817
1189. Willis JR, Acton JR (1976) The overall elastic moduli of a dilute suspension of spheres. *Q J Mechan Appl Math,* **29**:163–177
1190. Willis JR, Talbot DRS (1990) Variational methods in the theory of random composite materials. In: Maugin GA (ed) *Continuum Models and Discrete systems.* Longman Scientific & Technical, Harlow, UK, **1**:113–131
1191. Wilson RB, Cruse TA (1978) Efficient implementation of anisotropic three dimensional boundary-integral equation stress analysis. *Int J Numer Meth Engng,* **12**:1283–1397
1192. Wilson WK, Yu LW (1979) The use of the J-integral in thermal stress crack problems. *Int J Fracture,* **15**:377–387
1193. Withers PJ (1989) The determination of the elastic field of an ellipsoidal inclusion in a transversally isotropic medium, and its relevance to composite materials. *Philos Magazine,* **A59**:750–781

1194. Wong SC, Mai YW (2003) Performance synergism in polymer-based hybrid materials. In: Shonaike GO, Advani SG (eds), *Adv Polymeric Materials. Structure Property Relationships.* CRC Press, Boca Raton, FL, 439–477
1195. Wood JR, Zhao Q, Wagner HD (2001) Orientation of carbon nanotubes in polymers and its detection by Raman spectroscopy. *Composites,* **A32**:391–399
1196. Wu LZ (2003) The elastic field induced by a hemispherical inclusion in the half-space. *Acta Mech Sinica,* **19**:253–262
1197. Wu LZ, Du SY (1996) The elastic field in a half-space with a circular cylindrical inclusion. *J Appl Mech,* **63**:925–932
1198. Wu MS (1993) Effective moduli of finite anisotropic media with cracks. *Mech Mater* **15**:139–158
1199. Wu RS, Aki K (1985) Elastic wave scattering by a random medium and small-scale inhomogeneities in the lithosphere. *J Geoph Res,* **B90**:10261–10273
1200. Xia M, Hamada H, Maekawa Z (1995) Flexural stiffness of injection molded glass fiber reinforced thermoplastics. *Int Polym Process,* **10**:74–81
1201. Xu TT, Fisher FT, Brinson LC, Ruoff RS (2003) Bone-shape nanomaterials for nanocomposite application. *Nano Lett,* **3**:1135–1139
1202. Yang B, Pan E (2003) Elastic fields of quantum dots in multilayered semiconductors: A novel Green's function approach. *J Appl Mech,* **70**:161–168
1203. Yang B, Tewary VK (2005) Green's function-based multiscale modeling of defects in a semi-infinite silicon substrate. *Int J Solids Struct,* **42**:4722–4737
1204. Yang Q, Cox B (2003) Spatially averaged local strains in textile composites via the binary model formulation. *J Engng Mater Technol,* **125**:418–425
1205. Yang S, Gokhale AM, Shan Z (2000) Utility of microstructure modeling for simulation of micro-mechanical response of composites containing non-uniformly distributed fibers. *Acta Mater,* **48**:2307–2322
1206. Yin HM, Paulino GH, Buttlar WG, Sun LZ (2005) Effective thermal conductivity of two-phase functionally graded particulate composites. *J Appl Phys,* **98**:063704
1207. Yin HM, Paulino GH, Buttlar WG, Sun LZ (2007) Micromechanics-based thermoelastic model for functionally graded particulate materials with particle interactions. *J Mech Phys Solids,* **55**:132–160
1208. Yoon BJ, Kim S (1987) Note on the direct calculation of mobility functions for two equal-size spheres in Stokes flow. *J Fluid Mech,* **185**:437–446
1209. Yu HY, Sanday SC (1990) Axisymmetric inclusion in a half space. *J Appl Mech,* **57**:74–77
1210. Yu HY, Sanday SC, Rath BB, Chang CI (1995) Elastic field due to defects in transversely isotropic bimaterials. *Proc Roy Soc Lond,* **A449**:11–30
1211. Yu M-F, Dyer MJ, Chen J, Qian D, Liu WK, Ruoff RS (2001) Locked twist in multiwalled carbon-nanotube ribbons. *Phys Rev,* **B64**:241407(R)
1212. Yu M-F, Yakobson BI, Ruo RS (2000) Controlled sliding and pullout of nested shells in individual multiwalled nanotubes. *J Phys Chem,* **B104**:8764–8767
1213. Zaoui A (2002) Continuum micromechanics: Survey. *J Engng Mech,* **128**:808–816
1214. Zapparov KI, Perlin PI (1976) Numerical solution of plane elasticity theory problems for regions of complicated configuration. *Prikl Mekhan,* **12**(5):103–108 (in Russian)
1215. Zeman J (2003) *Analysis of Composite Materials with Random Microstructures.* PhD Thesis. Czech TU in Prague
1216. Zeng X, Bergman DJ, Hui PM, Stroud D (1988) Effective medium theory for weakly nonlinear composites. *Phys Rev,* **B38**:10970–10973
1217. Zerda AS, Lesser AJ (2001) Intercalated clay nanocomposites: morphology, mechanics, and fracture behavior. *J Polym Sci Part B: Polym Phys,* **39**:1137–1146

1218. Zhang HT, Chou YT (1985) Antiplane eigenstrain problem of an elliptic inclusion in a two-phase anisotropic medium. *J Appl Mech,* **52**:87–90
1219. Zhang P, Huang Y, Geubelle PH, Klein PA, Hwang KC (2002) The elastic modulus of single-wall carbon nanotubes: a continuum analysis incorporating interatomic potential. *Int J Solids Struct,* **39**:3893–3906
1220. Zhao YH, Weng GJ (2002) The effect of debonding angle on the reduction of effective moduli of particle and fiber-reinforced composites. *J Appl Mech,* **69**, 292–302
1221. Zheng Q-S, Du D-X (2001) An explicit and universally applicable estimate for the effective properties of multiphase composites which accounts for inclusion distribution. *J Mech Phys Solids,* **49**:2765–2788
1222. Zhiging J, Tennysin RC (1989) Closure of cubic tensor polynomial failure surface. *J Comp Mech,* **23**:208–231
1223. Zhou G, Lee LJ, Castro J (2003) Nano-clay and long fiber reinforced composites based on epoxy and phenolic resins. *Annual Technical Conference–ANTEC Society of Plastic Engng.* Nashville, **2**:2094–2098
1224. Zhou M (2003) A new look at the atomic level virial stress: on continuum-molecular system equivalence. *Proc Roy Soc Lond,* **A459**:2347–2392
1225. Zhou M (2005) Thermomechanical continuum interpretation of atomistic deformation. *Int J Multiscale Comput Engng,* **3**:177–197
1226. Zhou SA (1990) Materials multiple mechanics of elastic dielectric composites. *Appl Math Mech (China),* **11**:215–237
1227. Zhou SA, Hsieh RK (1986) Statistical theory of elastic materials with micro-defects. *Int J Engng Sci,* **24**:1195–1206
1228. Zhu LJ, Narh KA (2004) Numerical simulation of the tensile modulus of nanoclay-filled polymer composites. *J Polym Sci Part B: Polym Phys,* **42**:2391–2406
1229. Zhu ZG, Weng GJ (1989) Creep deformation of particle-strengthened metal-matrix composites. *J Engng Mater Technol,* **111**:99–105
1230. Zhu ZG, Weng GJ (1990) A local theory for the calculation of overall creep strain of particle–reinforced composites. *Int J Plast,* **6**:449–469
1231. Ziegler F (1992) Developments in structural dynamic viscoplasticity including ductile damage. *Z Angew Math Mech,* **72**(4):T5–T15
1232. Ziman JM (1979) *Models of Disorder.* Cambridge University Press, New York
1233. Zimmerman RW (1991) Elastic moduli of a solid containing spherical inclusions. *Mech Mater,* **12**:17–24
1234. Zimmerman RW (1996) Effective conductivity of a two-dimensional medium containing elliptical inhomogeneities *Proc Roy Soc Lond,* **A452**:1713–1727
1235. Zinchenko AZ (1994) Algorithm for random close packing of spheres with periodic boundary conditions. *J Comput Phys,* **114**:298–307
1236. Zohdi T, Wriggers P (1999) A domain decomposition method for bodies with heterogeneous microstructure based on material regularization. *Int J Solids Struct,* **36**:2507–2525
1237. Zohdi TI, Wriggers P (2005) *Introduction to Computational Micromechanics.* Springer, Berlin
1238. Zuiker JR (1995) Functionally graded materials: choice of micromechanics model and limitations in property variation. *Compos Engng,* **5**:807–819
1239. Zuiker JR, Dvorak GJ (1994) The effective properties of functionally graded composites–I. Extension of the Mori–Tanaka method to linearly varying fields. *Compos Engng,* **4**:19–35
1240. Zuo QH, Addessio FL, Dienes JK, Lewis MW (2006) A rate-dependent damage model for brittle materials based on the dominant crack. *Int J Solids Struct,* **43**:3350–3380

Index

Printed in the United States of America